INTELLIGENT GENE EDITING IN PLANTS

INTELLIGENT GENE EDITING IN PLANTS

THE CONVERGENCE OF AI AND CRISPR

Edited by

AFTAB AHMAD

Natural and Medical Sciences Research Center, University of Nizwa, Nizwa, Oman

BAOHONG ZHANG

Department of Biology, East Carolina University, Greenville, NC, United States

KAMEL A. ABD-ELSALAM

Plant Pathology Research Institute, Agricultural Research Center, Giza, Egypt

AHMED AL-HARRASI

Natural and Medical Sciences Research Center, University of Nizwa, Nizwa, Oman

ELSEVIER

ACADEMIC PRESS

An imprint of Elsevier

Academic Press is an imprint of Elsevier
125 London Wall, London EC2Y 5AS, United Kingdom
50 Hampshire Street, 5th Floor, Cambridge, MA 02139, United States

ISBN: 978-0-443-44079-3

For information on all Academic Press publications visit our website at
https://www.elsevier.com/books-and-journals

Publisher: Andre Wolff
Acquisitions Editor: Chiara Giglio
Editorial Project Manager: Rahul Basak
Production Project Manager: Fahmida Sultana
Cover Designer: Mark Rogers

Typeset by TNQ Tech

Contents

Part I

An introduction to gene editing technologies

1. Gene editing technologies: A comprehensive overview

2. Revolution of CRISPR—Cas: Advancements in genome editing for agriculture

3. The CRISPR—Cas toolkit for gene editing in plants

13. AI and CRISPR: Studying gene functions and regulatory networks in plants

Kamel A. Abd-Elsalam, Rawan K. Hassan, Mohamed A. Aref, Adham M. AlDakrany, and Toka E. Abdelkhalek

14. AI-driven gene editing for disease resistance in plants

Rajrishi Awadhesh Kumar, Anshuman Mishra, and Siddhesh B. Ghag

15. Artificial intelligence tools for predicting off-targets in plant genome editing

Fredy Davi Albuquerque Silva, Eulálio Gutenberg Dias dos Santos, and Elizabeth Pacheco Batista Fontes

16. AI and multiplex gene editing: The future of gene editing in plants

Benoit Mermaz, Alessandra Alves de Souza, and Cintia H.D. Sagawa

17. Artificial intelligence and CRISPR technologies for enhancing food security

Nabarun Roy, Shalini Singh, and Hari Shankar Gaur

18. Intelligent gene editing for stress tolerance in plants

Jyotsana Tilgam, Shbana Begam, Deepanshu Jayaswal, Kuldip Jayaswall, Joy Das, and Rakesh Kumar

19. AI-guided CRISPR in crop yield enhancement

Elshan Musazade, Isack Ibrahim Mrisho, Diming Zhang, Jinshan Gao, Zhiyuan Cheng, and Xianzhong Feng

20. AI-assisted regulation of gene-edited plants: Opportunities and challenges

Muhammad Riaz, Hamad Khan, Muhammad Faheem, Ahmed Al-Harrasi, Muhammad Umer Masood, Nayla Munawar, Paulo A. Zaini, Muhammad Rasheed, Stanislaus Antony Ceasar, and Aftab Ahmad

21. AI and CRISPR: Biosafety, governance, and public perception of gene-edited organisms

Leire Escajedo San-Epifanio, Javier Uncetabarrenechea Larrabe, and Igor Filibi

26. Future landscape of CRISPR and AI in plant gene editing

Oseias R. Feitosa-Junior

Contributors

Kamel A. Abd-Elsalam Plant Pathology Research Institute, Agricultural Research Center, Giza, Egypt

Toka E. Abdelkhalek Biotechnology English Program, Faculty of Science, Cairo University, Giza, Egypt

Mughair Abdul Aziz Department of Integrative Agriculture, College of Agriculture and Veterinary Medicine, United Arab Emirates University, Al Ain, United Arab Emirates

Aftab Ahmad Natural and Medical Sciences Research Center, University of Nizwa, Nizwa, Oman

Mariam Akhtar National Institute for Biotechnology and Genetic Engineering, Faisalabad, Pakistan

Emre Aksoy Middle East Technical University, Ankara, Turkey

Ahmed Al-Harrasi Natural and Medical Sciences Research Center, University of Nizwa, Nizwa, Oman

Adham M. AlDakrany School of Biotechnology, Nile University, Giza, Egypt

Qasim Ali Punjab Institute of Cardiology, Lahore, Pakistan

Alessandra Alves de Souza Citrus Research Center "Sylvio Moreira", Agronomic Institute - IAC, Cordeirópolis, Brazil

Imran Amin National Institute for Biotechnology and Genetic Engineering, Faisalabad, Pakistan

Mohamed A. Aref School of Biotechnology, Nile University, Giza, Egypt

Sajjad Asaf Natural and Medical Sciences Research Center (NMSRC), University of Nizwa, Nizwa, Oman

Mohamed A.M. Atia Bioinformatics Laboratory, College of Computing, University Mohammed VI Polytechnic, Ben Guerir, Morocco

Shbana Begam ICAR-National Institute for Plant Biotechnology, New Delhi, India

Abdelghani Belgaid Bioinformatics Laboratory, College of Computing, University Mohammed VI Polytechnic, Ben Guerir, Morocco

Stanislaus Antony Ceasar Department of Biotechnology, Karunya Institute of Technology and Sciences, Coimbatore, Tamil Nadu, India

Monoswi Chakraborty Genetics, Genomics and Bioinformatics (GGB) Program, University of California, Riverside, CA, United States; Institute for Integrative Genome Biology (IIGB), University of California, Riverside, CA, United States; Department of Chemical and Environmental Engineering, Marlan and Rosemary Bourns College of Engineering, University of California, Riverside, CA, United States

Punam Chakraborty Molecular Biology and Biotechnology Division, Council of Scientific and Industrial Research (CSIR)-National Botanical Research Institute, Lucknow, Uttar Pradesh, India; Academy of Scientific and Innovative Research (AcSIR), Ghaziabad, Uttar Pradesh, India

Prosenjit Chakraborty ICAR-Central Institute for Cotton Research, Nagpur, Maharashtra, India

Ming Cheng School of Agriculture, Food and Wine, Adelaide University, Waite Campus, Urrbrae, SA, Australia; Plant Synthetic Biology Australia, Adelaide University, Waite Campus, Urrbrae, SA, Australia

Zhiyuan Cheng Key Laboratory of Soybean Molecular Design Breeding, Northeast Institute of Geography and Agroecology, Chinese Academy of Sciences, Changchun, PR China; University of Chinese Academy of Sciences, Beijing, PR China

Ansu Chetri Department of Chemistry & Biochemistry, Sharda School of Engineering & Sciences, Sharda University, Greater Noida, Uttar Pradesh, India

Srirupa Gupta Choudhury Départment de biologie cellulaire et moléculaire, Université du Québec à Trois-Rivières, Trois-Rivières, QC, Canada; CSIR-Institute of Genomics and Integrative Biology, New Delhi, India

Joy Das ICAR-Central Institute for Cotton Research, Nagpur, Maharashtra, India

Fredy Davi Albuquerque Silva Department of Biochemistry and Molecular Biology/Bioagro, National Institute of Science and Technology in Plant-Pest Interactions, Federal University of Viçosa, Viçosa, Brazil

Achraf El Allali Bioinformatics Laboratory, College of Computing, University Mohammed VI Polytechnic, Ben Guerir, Morocco

Muhammad Faheem Department of Biochemistry, University of Agriculture, Faisalabad, Pakistan

Oseias R. Feitosa-Junior Department of Plant Pathology, Microbiology and Biotechnology, North Dakota State University, Fargo, ND, United States

Xianzhong Feng Key Laboratory of Soybean Molecular Design Breeding, Northeast Institute of Geography and Agroecology, Chinese Academy of Sciences, Changchun, PR China; University of Chinese Academy of Sciences, Beijing, PR China

Sajid Fiaz Institute of Molecular Biology and Biotechnology, The University of Lahore, Lahore, Pakistan

Igor Filibi EU-Frames TransR. Department of Public Law, Historical-Legal Sciences and History of Political Thought, University of the Basque Country, Bilbao, Spain

Jinshan Gao Key Laboratory of Soybean Molecular Design Breeding, Northeast Institute of Geography and Agroecology, Chinese Academy of Sciences, Changchun, PR China; University of Chinese Academy of Sciences, Beijing, PR China

Mallikarjuna Garladinne Plantigen Agrisolutions, Pulivendula, Andhra Pradesh, India

Hari Shankar Gaur School of Agriculture, Galgotias University, Greater Noida, Uttar Pradesh, India

Siddhesh B. Ghag Department of Atomic Energy, Centre for Excellence in Basic Sciences, School of Biological Sciences, University of Mumbai, Mumbai, Maharashtra, India

Manish Ghumnani ICAR-National Institute for Plant Biotechnology, New Delhi, India

Zheng Gong School of Agriculture, Food and Wine, Adelaide University, Waite Campus, Urrbrae, SA, Australia; School of Agriculture and Food Sustainability, The University of Queensland, St Lucia, QLD, Australia

Vinita Gotmare ICAR-Central Institute for Cotton Research, Nagpur, Maharashtra, India

Eulálio Gutenberg Dias dos Santos, Department of Biochemistry and Molecular Biology/Bioagro, National Institute of Science and Technology in Plant-Pest Interactions, Federal University of Viçosa, Viçosa, Brazil

Rawan K. Hassan Smart Engineering Systems Research Centre (SESC), Nile University, Cairo, Egypt

Mustabshira Ikram Department of Land Resources and Environmental Sciences, Montana State University, Bozeman, MT, United States

Ifrah Imran National Institute for Biotechnology and Genetic Engineering, Faisalabad, Pakistan

Abdullah Jamil Plant Breeding and Genetics Program, Gulf Coast Research and Education Center (GCREC), University of Florida, Wimauma, FL, United States

Sohini Jash JMS School of Life Sciences, University of Sussex, Brighton, United Kingdom

Deepanshu Jayaswal ICAR-National Institute of Seed Science & Technology, Maunath Bhanjan, Uttar Pradesh, India

Kuldip Jayaswall ICAR-National Institute of Seed Science & Technology, Maunath Bhanjan, Uttar Pradesh, India

Hamad Khan Natural and Medical Sciences Research Center (NMSRC), University of Nizwa, Nizwa, Oman

Muhammad Danyal Khan Department of Entomology, Louisiana State University, Baton Rouge, LA, United States

Hira Khursheed Scholarious Scientific, Montreal, QC, Canada; Mojoki BioPartners, Gujrat, Pakistan; Clinical and Translational Sciences (CaTS) Lab, Douglas Research Center, McGill University, Montreal, QC, Canada

Rajrishi Awadhesh Kumar Department of Atomic Energy, Centre for Excellence in Basic Sciences, School of Biological Sciences, University of Mumbai, Mumbai, Maharashtra, India

Rakesh Kumar Division of Crop Improvement, ICAR-Central Institute for Cotton Research, Nagpur, Maharashtra, India; ICAR-Central Institute for Cotton Research, Nagpur, Maharashtra, India

Javier Uncetabarrenechea Larrabe EU-Frames TransR. Department of Public Law, Historical-Legal Sciences and History of Political Thought, University of the Basque Country, Bilbao, Spain

Yufeng Liu Yale University, New Haven, CT, United States

Sayanti Mandal Department of Chemistry & Biochemistry, Sharda School of Engineering & Sciences, Sharda University, Greater Noida, Uttar Pradesh, India

Khaled Masmoudi Department of Integrative Agriculture, College of Agriculture and Veterinary Medicine, United Arab Emirates University, Al Ain, United Arab Emirates

Muhammad Umer Masood Department of Plant Sciences and Plant Pathology, Montana State University, Bozeman, MT, United States

Kasi Rao Mediga Godrej Agrovet Limited, Khammam, Telangana, India

Benoit Mermaz Department of Molecular, Cellular, and Developmental Biology, Yale University, New Haven, CT, United States

Anshuman Mishra Department of Atomic Energy, Centre for Excellence in Basic Sciences, School of Biological Sciences, University of Mumbai, Mumbai, Maharashtra, India

Vishal Monga Department of Plant Sciences and Plant Pathology, Montana State University, Bozeman, MT, United States

Isack Ibrahim Mrisho College of Chinese Medicinal Material, Jilin Agricultural University, Changchun, PR China

Nayla Munawar Department of Chemistry, United Arab Emirates University, Al-Ain, United Arab Emirates

Ahmad Munir Department of Biochemistry, University of Agriculture, Faisalabad, Pakistan

Elshan Musazade Key Laboratory of Soybean Molecular Design Breeding, Northeast Institute of Geography and Agroecology, Chinese Academy of Sciences, Changchun, PR China

Sabavat Raju Naik International Crops Research Institute for Semi-Arid Tropics, Hyderabad, Telangana, India; Department of Genetics and Plant Breeding, Parbhani, Maharashtra, India

Rubab Zahra Naqvi National Institute for Biotechnology and Genetic Engineering, Faisalabad, Pakistan

Zahra Noreen Department of Botany, Division of Science and Technology, University of Education, Lahore, Pakistan

Elizabeth Pacheco Batista Fontes Department of Biochemistry and Molecular Biology/Bioagro, National Institute of Science and Technology in Plant-Pest Interactions, Federal University of Viçosa, Viçosa, Brazil

Supratik Palchoudhury ICAR-Central Institute for Cotton Research, Nagpur, Maharashtra, India

Shivali Pandita Molecular Biology and Biotechnology Division, Council of Scientific and Industrial Research (CSIR)-National Botanical Research Institute, Lucknow, Uttar Pradesh, India; Department of Zoology, University of Lucknow, Lucknow, Uttar Pradesh, India

Manisha Parashar ICAR-National Institute for Plant Biotechnology, New Delhi, India

Aqsa Perveen Department of Botany, Division of Science and Technology, University of Education, Lahore, Pakistan

Dominic Previtera School of Agriculture and Food Sustainability, The University of Queensland, St Lucia, QLD, Australia

K.P. Raghavendra ICAR-National Institute for Research on Commercial Agriculture, Rajahmundry, Andhra Pradesh, India; ICAR-Central Institute for Cotton Research, Nagpur, Maharashtra, India

Muhammad Rasheed Department of Physiology, School of Basic Medical Science, Chongqing Medical University, Chongqing, China

Muhammad Zawwad Raza Scholarious Scientific, Montreal, QC, Canada; Mojoki Bio-Partners, Gujrat, Pakistan

Muhammad Riaz Yale University, New Haven, CT, United States

Nabarun Roy School of Agriculture, Galgotias University, Greater Noida, Uttar Pradesh, India

Cintia H.D. Sagawa PreScouter Inc., Chicago, IL, United States

Muhammad Waseem Sajjad National Institute for Biotechnology and Genetic Engineering, Faisalabad, Pakistan

Leire Escajedo San-Epifanio EU-Frames TransR. Department of Public Law, Historical-Legal Sciences and History of Political Thought, University of the Basque Country, Bilbao, Spain

Kameshwaran Senthil ICAR-National Institute for Plant Biotechnology, New Delhi, India

Shalini Singh School of Agriculture, Galgotias University, Greater Noida, Uttar Pradesh, India

Rohini Sreevathsa ICAR-National Institute for Plant Biotechnology, New Delhi, India

Ramsha Syed Department of Chemistry & Biochemistry, Sharda School of Engineering & Sciences, Sharda University, Greater Noida, Uttar Pradesh, India

Ashwini Talakayala International Crops Research Institute for Semi-Arid Tropics, Hyderabad, Telangana, India

Jyotsana Tilgam ICAR-National Bureau of Agriculturally Important Micro-organisms, Maunath Bhanjan, Uttar Pradesh, India

Deepak Kumar Upadhyay Department of Chemistry & Biochemistry, Sharda School of Engineering & Sciences, Sharda University, Greater Noida, Uttar Pradesh, India

Praveen C. Verma Molecular Biology and Biotechnology Division, Council of Scientific and Industrial Research (CSIR)-National Botanical Research Institute, Lucknow, Uttar Pradesh, India; Academy of Scientific and Innovative Research (AcSIR), Ghaziabad, Uttar Pradesh, India

V.N. Waghmare ICAR-Central Institute for Cotton Research, Nagpur, Maharashtra, India

Afifa Younas Department of Botany, Baba Guru Nanak University, Nankana Sahib, Pakistan

Mehmooda Yousaf Department of Botany, Division of Science and Technology, University of Education, Lahore, Pakistan

Paulo A. Zaini Department of Plant Sciences, University of California, Davis, CA, United States

Diming Zhang Key Laboratory of Soybean Molecular Design Breeding, Northeast Institute of Geography and Agroecology, Chinese Academy of Sciences, Changchun, PR China; University of Chinese Academy of Sciences, Beijing, PR China

Yan Zhang School of Agriculture and Food Sustainability, The University of Queensland, St Lucia, QLD, Australia

About the editors

Aftab Ahmad

Aftab Ahmad earned his PhD and completed his postdoctoral research in Plant Molecular Biology at the University of Shizuoka, Japan, under the prestigious 21st Century Center of Excellence Program. He also served as a Visiting Scientist in the Department of Plant Sciences at the University of California, Davis, United States. With over a decade of teaching experience in biochemistry and molecular biology at the University of Agriculture Faisalabad, Pakistan, Dr. Aftab now continues to advance plant genome editing research aimed at improving crop resilience and productivity at the Natural and Medical Sciences Research Center, University of Nizwa, Oman. He has published in leading international peer-reviewed journals, such as *Frontiers in Plant Science*, *Journal of Biomedical Science*, *Plant Molecular Biology Reporter*, *Journal of Cereal Science*, and *Genetic Resources and Crop Evolution*, and many more.

Baohong Zhang

Baohong Zhang is a THCAS Distinguished Professor in the Department of Biology at East Carolina University and an internationally recognized leader in genomics and biotechnology. He earned his Bachelor's degree in Plant Genetics and Breeding from China Agricultural University and his PhD in Environmental Toxicology from Texas Tech University. For over 20 years, Dr. Zhang has conducted pioneering research in transgenic technologies, genome editing, and functional genomics, supported by major funding agencies, such as the NSF, NIH, DoD, and USDA, as well as industry partners, including the North Carolina Biotechnology Center and DuPont. He is a prolific Author, with several publications listed as Highly Cited Papers by the ISI Web of Science. He is the Editor-in-Chief, Associate Editor, or Guest Editor for multiple international journals, including *Functional & Integrative Genomics*, *Plants*, *Scientific Reports*, *Plant Biotechnology Journal*, *Molecular Biotechnology*, *Plant Molecular Biology*, and *Journal of Cotton Research*. A Fellow of the American Association for the Advancement of Science, Dr. Zhang has received numerous honors, including, more recently, the ECU Achievement in International Research and Creative Activity Award (2017), the Cotton Researcher of the Year (2018), the Lifetime Achievement Research Award (2019), and the Cotton Biotechnology Award (2020).

Kamel A. Abd-Elsalam

Kamel A. Abd-Elsalam, PhD, is currently a Research Professor at the Plant Pathology Research Institute, Agricultural Research Center, Giza. He earned his PhD in molecular plant pathology through a joint program between Christian Albrechts University of Kiel, Germany, and Suez Canal University, Egypt, followed by a postdoctoral fellowship at Christian Albrechts University. His research focuses on nanobiotechnology applications in agriculture and plant protection, and he has authored numerous peer-reviewed articles and edited books with major international publishers. His scientific contributions have been recognized nationally and regionally, including awards from the Federation of Arab Scientific Research Councils for research excellence in biotechnology and the Egyptian State Award for Agricultural Scientific Excellence from the Academy of Scientific Research and Technology. He has been recognized among the world's top 2% most influential scientists in nanobiotechnology for three consecutive years (2021–2023), according to global rankings developed by Stanford University. Since 2019, he has served as the Editor-in-Chief of two Elsevier book series, *Nanobiotechnology for Plant Protection* and *Genome Modified Plants and Microbes in Food and Agriculture*.

Ahmed Al-Harrasi

Ahmed Al-Harrasi is the Vice-Chancellor for Graduate Studies, Research, and External Relations and teaches Organic Chemistry at the University of Nizwa, Oman, where he also founded and chairs the Natural and Medical Sciences Research Center. He earned his BSc (1997) in Chemistry from Sultan Qaboos University and his MSc (2002) and PhD (2005) in Organic Chemistry from the Free University of Berlin as a DAAD Fellow under the supervision of Prof. Hans-Ulrich Reissig. He later pursued postdoctoral research in chemical biology as a Fulbright Scholar at Cornell University, United States. A Fellow of The Royal Society of Chemistry and a Member of UNESCO's International Basic Sciences Program, Professor Al-Harrasi has led over 40 funded research projects and is an accomplished and prolific Author. His contributions to research have been widely recognized through the Order of Royal Commendation from His Majesty. He is also a recipient of the National Research Award from the Research Council of Oman. From 2019 to 2024, he has been consistently listed among the World's Top 2% of Scientists by Stanford University.

Preface

Agriculture stands at a critical crossroads as the global food system faces escalating pressures from climate change, population growth, emerging pests and diseases, environmental degradation, and shrinking natural resources. Meeting these challenges requires transformative scientific approaches that move beyond conventional breeding and incremental biotechnological improvements. Among the most disruptive and promising innovations reshaping modern plant science are CRISPR-based gene-editing technologies and their accelerating convergence with artificial intelligence (AI).

Since its introduction, CRISPR–Cas technology has revolutionized plant gene editing by enabling precise, efficient, and versatile genetic modifications. From targeted gene knockouts and analytic replacements to multiplex gene editing and regulatory element modulation, CRISPR has dramatically expanded the scope of achievable crop improvements. These advances have facilitated the development of plants with enhanced yield, stress tolerance, disease resistance, and nutritional quality while also accelerating functional genomics research. However, as CRISPR applications expand in complexity and scale, several technical translational challenges have become increasingly evident. These include optimizing guide RNA design, minimizing off-target effects, predicting DNA repair outcomes, improving delivery systems, and ensuring safety and regulatory compliance.

In parallel, artificial intelligence has emerged as a transformative force across biological sciences. Advances in machine learning, deep learning, and data-driven modeling are redefining how large-scale biological datasets are analyzed and translated into actionable insights. In plant gene editing, AI has become instrumental in predicting gRNA efficiency and specificity, modeling on- and off-target activities, forecasting repair outcomes, deciphering gene regulatory networks, integrating multiomics datasets, and linking genotypes to complex phenotypes. The integration of AI with CRISPR technologies is thus enabling a new generation of intelligent gene-editing pipelines that are more predictive, efficient, scalable, and adaptable to real-world agricultural needs.

This edited volume, *Intelligent Gene Editing in Plants: The Convergence of AI and CRISPR*, is designed to provide a comprehensive, authoritative, and forward-looking overview of this rapidly evolving interdisciplinary landscape. By bringing together leading experts from plant biology, genome editing, artificial intelligence, bioinformatics, regulatory science, and policy, the book captures both the scientific foundations and the applied potential of AI-enabled CRISPR technologies in plants.

The book is organized into three interlinked parts that guide readers from fundamental concepts to advanced applications and emerging horizons. Part I establishes the biological and technological foundations of gene editing in plants. It covers the evolution of gene-editing technologies, the CRISPR–Cas toolkit, delivery systems, and applications across food and nonfood crops, while also introducing the conceptual basis for integrating AI into gene-editing workflows.

Part II focuses on the practical synergies between CRISPR and AI. The chapters in this section explore intelligent gRNA design, the prediction of on-target efficiency and off-target risks, the modeling of DNR repair outcomes, multiplex gene editing, and AI-driven approaches for improving disease resistance, stress tolerance, crop yield, and food security. Importantly, this part also addresses the biosafety, governance, regulatory framework, and public perception, emphasizing the growing role of AI in risk assessment, regulatory decision-making, and responsible innovations.

Part III looks toward the future of plant genome engineering. It highlights the emerging approaches, such as the editing of cis-regulatory elements, AI-integrated omics analyses, synthetic biology, next-generation gene-editing platforms, and the long-term vision for intelligent data-driven plant breeding systems. These chapters collectively illustrate how the convergence of AI and CRISPR will shape the next phase of sustainable and climate-resilient agriculture.

A defining strength of this volume is its global and interdisciplinary perspective. The contributions span diverse crops, environments, and regulatory contexts, reflecting the worldwide relevance of intelligent gene-editing technologies. The editors have sought to balance conceptual depth with applied relevance, ensuring that the book serves both as a reference for established researchers and as an entry point for readers new to the field.

We anticipate that this book will not only inform research and development efforts, but also stimulate new collaborations and ideas at the interface of plant biology, computational science, and policy. The convergence of AI and CRISPR represents more than a technological advancement; it offers a critical pathway toward resilient food systems, sustainable agriculture, and global food security in the decades ahead.

The Editors

PART I

An introduction to gene editing technologies

CHAPTER

1

Gene editing technologies: A comprehensive overview

Nayla Munawar[1], Ahmad Munir[2], Ahmed Al-Harrasi[3], Muhammad Riaz[4], Sajjad Asaf[3], and Aftab Ahmad[3]

[1]Department of Chemistry, United Arab Emirates University, Al-Ain, United Arab Emirates; [2]Department of Biochemistry, University of Agriculture, Faisalabad, Pakistan; [3]Natural and Medical Sciences Research Center, University of Nizwa, Oman; [4]Yale University, New Haven, CT, United States

1.1 Background

Ensuring global food security remains one of the most urgent challenges confronting humanity. Despite significant technological advances in agriculture, hunger and malnutrition persist in many regions, impeding progress toward the United Nations Sustainable Development Goals (SDGs), particularly the aim of eradicating hunger by 2030 (Ndhlovu and Mokone, 2025). Global food systems are under growing pressure from climate change, population growth, and limited natural resources, which are further intensified by socioeconomic disruptions caused by the COVID-19 pandemic (Citaristi, 2022). According to the World Food Program (WFP, www.wfp.org), over 135 million individuals currently experience acute food insecurity, and projections suggest this number could rise over 840 million by 2030 (Citaristi, 2022; Gatton and Gallegos, 2023). The situation is compounded by the rapid degradation of arable land, declining water availability, and the emergence of climate-resilient pathogens that continue to threaten agricultural sustainability and productivity.

Achieving sustainable food security requires coordinated action across multiple sectors, such as scientific, governmental, and industrial sectors. Translating cutting-edge agricultural research into practical, policy-driven applications is critical to addressing the root causes of hunger and poverty. Sustainable agriculture must not only increase productivity but also ensure environmental conservation and nutritional enhancement. Therefore, scientists are

Intelligent Gene Editing in Plants
https://doi.org/10.1016/B978-0-443-44079-3.00006-1

working to develop climate-resilient crops that can maintain yield, nutritional value, and stress tolerance under variable environmental conditions (Barrett, 2021).

The Green Revolution, which began in the mid-1940s under the leadership of Norman E. Borlaug (Nobel Peace Prize in 1970), marked a pivotal milestone in global agriculture by introducing high-yielding crop varieties that dramatically increased grain output (Pingali, 2022). However, the limitations of conventional breeding have become increasingly evident (Acquaah, 2024). Traditional selection methods rely on natural or induced variability and require long generational cycles, making them inadequate to meet the demands of a growing population under the stresses of climate change. As a result, the focus of modern plant science has shifted toward precision biotechnology, incorporating genetically modified (GM) and genome-editing (GE) technologies capable of accelerating crop improvement with unprecedented specificity and efficiency (Husaini and Sohail, 2024).

These New Breeding Techniques (NBTs) rely on a detailed understanding of plant genomes, enabling targeted gene modifications to introduce desired traits (Fernández Ríos et al., 2024). The convergence of OMICs (genomics, transcriptomics, proteomics, and metabolomics) has enabled scientists to link molecular data to phenotypic performance for genetic improvement of crops. The concept of the "pangenome", representing the complete set of genes across all strains of a species, provides essential insights into intraspecific genetic diversity (Secomandi et al., 2025). Within the pangenome, the "core genome" consists of genes shared among all individuals, while the "accessory genome" includes genes unique to specific strains that contribute to environmental adaptability and phenotypic plasticity. Expanding this approach to a "super-pangenome", that encompasses wild relatives, allows researchers to explore genetic diversity across entire genera (He et al., 2025). Such resources enhance the identification of loci associated with stress resistance, nutrient metabolism, and yield, thereby providing a genetic foundation for genomics-assisted breeding (GAB) and marker-based selection.

Developing superior crop varieties further depends on identifying functional genes, markers, and allelic variants linked to the specific traits. In this regard, genome editing (GE), introduced in the early 2000s, has become a central technology. Unlike conventional transgenic methods that often rely on introducing exogenous genes, GE enables the modification of native sequences within a species' own genome, making it both more precise and generally safer in terms of biosafety and regulatory evaluation (Pacesa et al., 2024). GE techniques have rapidly gained public and scientific attention as they allow targeted, small-scale alterations without foreign DNA integration. Because of its precision and minimal off-target effects, GE is expected to gradually supersede GM technologies in agricultural applications over the next decade (Gupta et al., 2021).

As its core, genomic editing operates through specialized enzymes collectively known as site-directed nucleases (SDNs), which introduce double-stranded breaks (DSBs) at specific genomic locations. These breaks activate the cell's intrinsic DNA repair mechanisms, such as homologous recombination (HR) and non-homologous end joining (NHEJ), facilitating the incorporation, deletion, or substitution of genetic material. SDNs can be guided to their target sequences through either engineered DNA-binding domains or complementary RNA molecules. The primary classes of SDNs include meganucleases (MegaNs), zinc-finger nucleases (ZFNs), transcription activator-like effector nucleases (TALENs), and clustered regularly interspaced short palindromic repeats (CRISPR)/CRISPR-associated proteins (Cas) systems. Among these technologies, CRISPR/Cas9 has emerged as the most powerful and widely adopted technology because of its simple design, low cost, modular nature, and broad applicability across organisms.

The voyage toward precise genome manipulation has deep historical roots in classical genetics. Traditional plant improvement strategies relied on "forward genetics", where breeders selected phenotypic variants generated by spontaneous or induced mutations, often through radiation or chemical mutagenesis (Aklilu, 2021). While these approaches were instrumental in early genetic improvement, they were time-consuming and laborious and lacked molecular level control. The emergence of "reverse genetics" fundamentally changed this paradigm by enabling researchers to start with a known gene sequence, modify it purposefully, and analyze the resulting phenotypic outcomes (Choudhary et al., 2023). Within this framework, genome engineering represents the broader discipline concerned with the intentional design and modification of genomic architecture, while genome editing and gene editing describe specific techniques for introducing targeted alterations using the cell's endogenous DNA repair pathways such as HDR and NHEJ as mentioned above and shown in Fig. 1.1. "Gene editing" generally targets individual genes, while "genome editing"

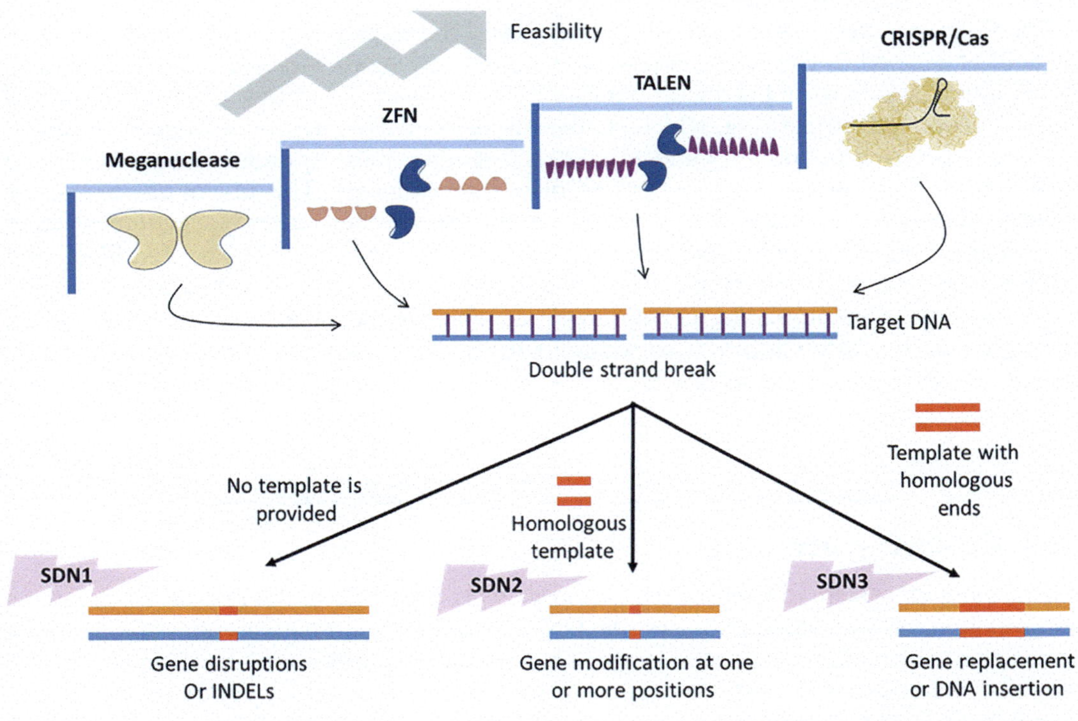

FIGURE 1.1 Feasibility and mechanistic overview of site-directed nuclease (SDN) genome editing technologies. The top panel illustrates the feasibility of genome editing to achieve complex biological outcomes. The middle and bottom panels detail the mechanisms of SDN-mediated genome modification, showing the induction of a double-strand break (DSB) and the subsequent repair via either NHEJ for small indels (SDN1) or HDR for precise small (SDN2) or large (SDN3) sequence insertions.

may involve simultaneous manipulation/modification of multiple loci or regulatory regions across the genome.

The conceptual foundation of genome editing emerged in the 1970s, when researchers demonstrated that a DNA fragment flanked by homologous sequences could integrate into a host genome via homologous recombination (HR) (Fatollahi Arani and Zeinoddini, 2023). However, HR alone proved inefficient due to random integration events and low targeting frequency. The breakthrough came with the realization that introducing a DSB at a specific genomic locus could greatly enhance recombination efficiency. This discovery led to the development of programmable nucleases capable of generating targeted DSBs, thereby revolutionizing gene manipulation by activating repair pathways such as HR and NHEJ (Ahmad et al., 2024a, 2024b, 2024c). Although the efficiency of DSB repair is influenced by factors such as cell type, replication status, and chromatin context, these mechanisms underpin all modern genome editing systems. When DSBs are repaired via NHEJ, small insertions or deletions (indels) may occur (SDN-1), often disrupting gene function. Alternatively, when repair occurs via HDR using a donor DNA template, specific and precise edits can be introduced at the desired site (SDN-2 & SDN-3). These features enable researchers to insert, delete, or replace DNA sequences to repair defective genes, introduce novel traits, or regulate gene expression (Fig. 1.1).

Building upon decades of molecular genetics and the sequencing of the human genome, the field of gene editing now encompasses two major strategies: targeted gene replacement, which introduces localized sequence modifications without disrupting gene function, and broader genomic restructuring, which enables more extensive alterations in gene organization and expression patterns (Li and Xia, 2020). Together, these approaches have redefined the boundaries of functional genomics and plant biotechnology. The advent of site-specific nucleases has been pivotal to this progress, paving the way for five major genome editing platforms, Cre-loxP system (Nakamura et al., 2025), MegaNs, ZFNs, TALENs, and CRISPR/Cas system (Aslam et al., 2022), each representing a step toward greater precision and versatility in manipulating genetic material. This chapter focuses on genome-editing techniques particularly CRISPR/Cas technology, as a key innovation shaping sustainable agriculture and global food security.

1.2 Genome editing techniques

1.2.1 Cre-loxP system

The Cre-loxP system, developed in the early 1990s, was among the first efficient genetic manipulation tools to achieve site-specific DNA recombination in eukaryotic cells. The system originated from the P1 bacteriophage, which naturally employs this mechanism to integrate its DNA into the host genome. It consists of two essential components: (i) Cre recombinase, a 38 kDa tyrosine recombinase enzyme, and (ii) loxP sites, which are 34-base pair DNA sequences specifically recognized by Cre. Each loxP site comprises two 13-bp inverted palindromic repeats separated by an 8-bp asymmetric spacer region with 6bp variable except for two in the middle that define a certain orientation of the loxP sequence (McLellan et al., 2017). The mechanism of the Cre-loxP system is shown in Fig. 1.2.

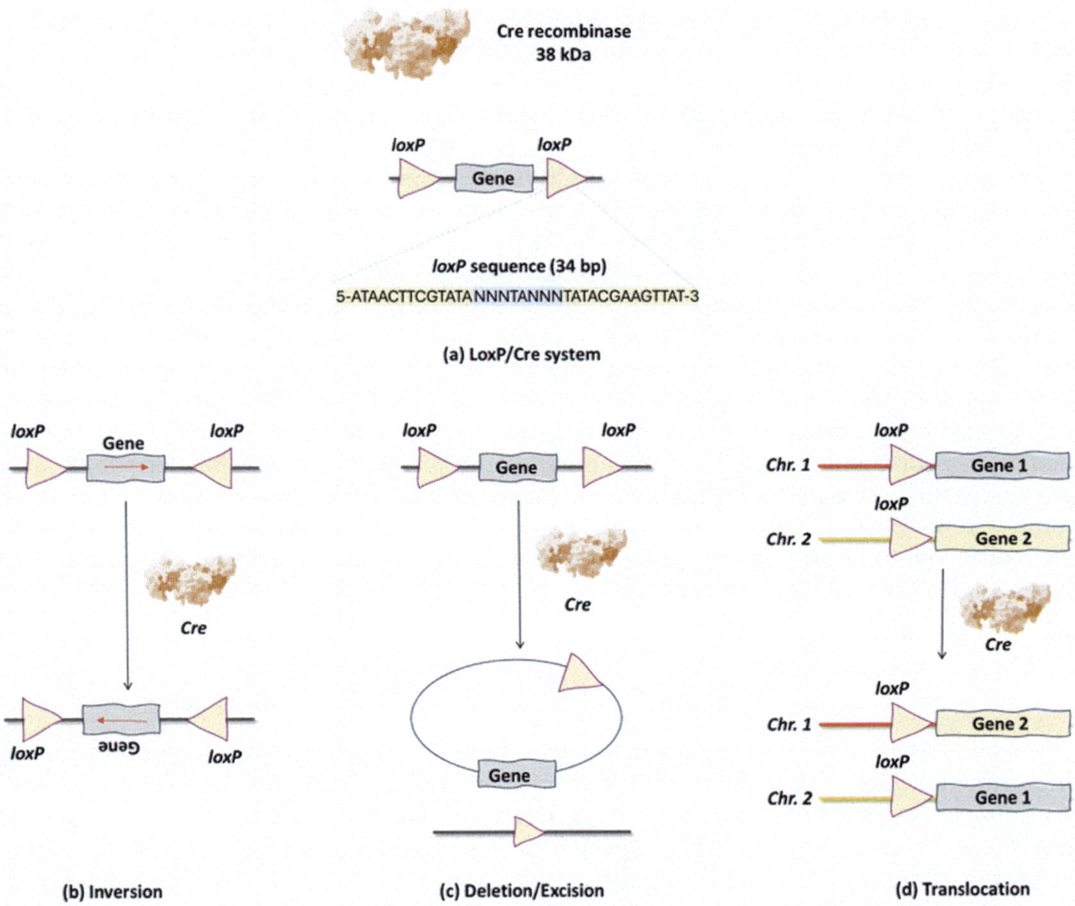

FIGURE 1.2 Mechanism of the Cre-loxP system. (a) This diagram illustrates the Cre recombinase enzyme recognizing the specific 34 bp loxP site to catalyze site-specific DNA recombination. Cre can perform three main genetic modifications. (b) inversion: Flanking a gene with loxP sites in opposite orientations flips the DNA segment. (c) Deletion/excision: Flanking a gene with loxP sites in the same orientation removes the intervening DNA (creating conditional knockouts). (d) Translocation/insertion: Recombination between loxP sites on separate DNA molecules allows for DNA segment exchange or insertion.

In this system, two Cre molecules bind independently to each 13-bp inverted repeat within a loxP site, forming a dimer. Two such dimers positioned at different loxP sites then associate to form a tetrameric complex that juxtaposes the two DNA segments. Cre subsequently introduces DSBs at both loxP sites, and the host's DNA ligase efficiently repairs the resulting nicks. This recombination event can mediate three distinct types of DNA rearrangements, depending on the relative orientation and chromosomal position of the loxP sites, i. Inversion: When two loxP sites are present on the same DNA molecule but in opposite orientations, the intervening sequence is inverted. This process is reversible, allowing the DNA segment to toggle between orientations; ii. Excision (Deletion): When loxP sites are oriented in the same direction on a single DNA molecule, Cre-mediated recombination excises the intervening segment, resulting in an irreversible deletion; iii. Translocation: When loxP sites are located on different DNA molecules, Cre recombination facilitates the exchange of DNA fragments, generating reciprocal translocations that can be reversible (Hoersten et al., 2024).

The simplicity and specificity made the Cre-loxP system a widely used platform for conditional gene editing in mammalian systems. Many scientists generated transgenic and knockout mouse models using this system (Ren et al., 2022). The lack of endogenous loxP sequences in eukaryotic genomes minimizes non-specific binding, while the relatively long 34-bp recognition sequence further reduces off-target recombination. Consequently, Cre-loxP has been particularly valuable for precise gene excision and conditional knockout studies, especially through the development of inducible systems such as the tetracycline-or tamoxifen-regulated CreERT models (Vizoso et al., 2022). Although the applications of Cre-loxP extend across disciplines from neurobiology to cancer research, the efficiency of Cre-mediated recombination can be constrained by the physical distance between loxP sites and chromatin accessibility. Furthermore, the labor-intensive and time-consuming construction of Cre-loxP-based transgenic systems and the lack of a centralized global database for efficiently collecting and sharing information with the research community restricted the wider applications of this technique (Ren et al., 2022).

1.3 Meganucleases (homing endonucleases)

MegaNs, also known as homing endonucleases (HEs), comprise a class of naturally occurring endonucleases found in archaea, bacteria, eukaryotes, and their respective viruses. These nucleases recognize and cleave long DNA sequences, typically 20–30 base pairs in length, making them among the most specific nucleases, as shown in Fig. 1.3a. Based on their conserved catalytic motifs and structural features, MegaNs are classified into five main families: LAGLIDADG, HNH, GIY-YIG, His-Cys box, and PD-(D/E)XK (Bhagtaney and Sundarrajan, 2023). Among these, the LAGLIDADG family (also referred to as LHEs) is prevalent and extensively characterized.

Members of the LHE family are distinguished by their extensive DNA-protein contacts and ability to bind long, unique DNA sequences with exceptional specificity compared to other HEs families. Members of the LHE family are characterized as both homodimeric and monomeric forms, and recognize unique DNA sequences through protein-DNA interactions. The B-sheet domains of LHEs extend deeply into the DNA major groove, where they establish multiple hydrogen bonds and hydrophobic interactions with both the nucleobases and the flanking phosphodiester backbone, ensuring high-affinity and sequence-specific binding (Silva et al., 2011). Owing to this sophisticated mode of recognition, considerable efforts have been devoted to re-engineering LAGLIDADG-DNA interactions through advanced protein design and directed evolution techniques to expand their target range. Despite these efforts, structural analyses have revealed that even closely related LHEs may employ distinct subsets of amino acid residues to recognize similar DNA motifs, underscoring the complexity and variability of their sequence recognition mechanisms. This

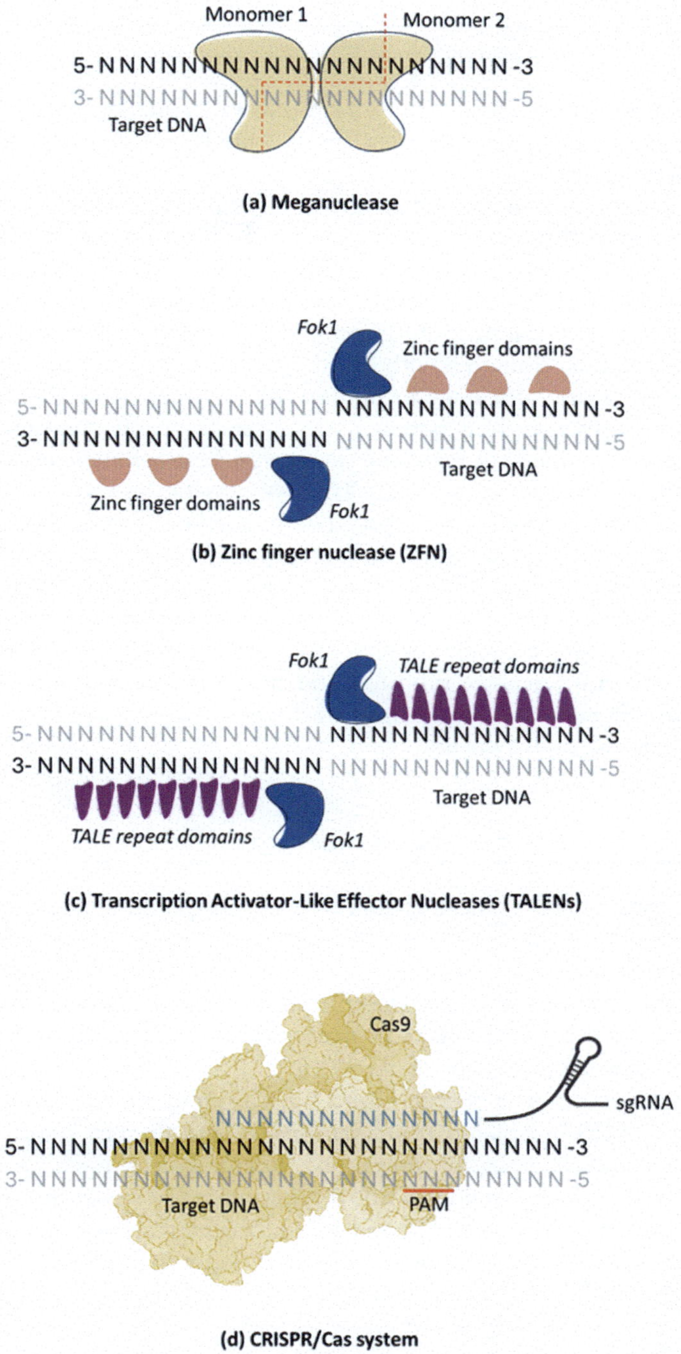

FIGURE 1.3 Comparison of major site-specific nuclease technologies for genome editing. This figure illustrates four primary classes of engineered nucleases. (a) Meganuclease: Naturally occurring dimeric enzymes (monomers 1 and 2) that recognize long (12–40 bp) and highly specific DNA sequences, causing a DSB. (b) ZFN: Engineered enzymes that use an array of modular ZF domains to specifically recognize and bind DNA triplets. The binding domains are fused with the non-specific cleavage domain of the *Fok*I endonuclease, which must dimerize to introduce a DSB at the target site. (c) TALENs: Engineered enzymes that utilize TALE repeat domains, where each domain recognizes a single base pair. Like ZFNs, binding domains of TALENs are fused with *Fok*I endonuclease, requiring dimerization for DNA cleavage. (d) CRISPR/Cas system: The modern and widely used system, employing the Cas9 endonuclease guided by a single guide RNA (sgRNA). The sgRNA uses base pairing to recognize the target DNA sequence, while Cas9 recognizes a protospacer-adjacent motif (PAM) sequence to introduce a DSB at the target site.

context-dependent interaction pattern poses a significant challenge to rationally redesigning LHEs for novel genomic targets while preserving catalytic activity and structural stability (Pâques and Duchateau, 2007).

Despite these challenges, engineered MegaNs have demonstrated targeted cleavage in both plant and mammalian genomes, including maize (Daboussi et al., 2015) and human genes such as XPC and RAG1 (Dupuy et al., 2013). Their relatively small molecular mass (approximately 40 kDa) facilitates cellular delivery and makes them attractive for applications where vector size is limited. In addition, hybrid systems such as MegaTALs, which fuse meganuclease catalytic domains with TALE DNA-binding motifs, have improved targeting flexibility and editing efficiency in human cells (Dupuy et al., 2013).

Nevertheless, the tight coupling between DNA-binding and cleavage domains, often involving more than 50 amino acid residues, poses significant challenges to reprogramming MegaNs for new targets (Danilo et al., 2022). Their enzymatic activity can also be influenced by DNA methylation and chromatin structure, thereby further limiting their versatility (Bhagtaney and Sundarrajan, 2023). Although industrial pipelines have been developed to engineer MegaNs with altered specificities, these approaches require advanced protein design expertise and remain costly. Additionally, the risk of off-target activity cannot be eliminated. While MegaNs paved the way for precision genome modification, their re-engineering constraints have led to the preference for more modular and user-friendly technologies such as ZFNs and TALENs, which offer easier customization, faster design cycles, and broader applicability across biological systems (Bhardwaj and Nain, 2021).

1.4 Zinc-finger nucleases (ZFNs)

ZFNs represent one of the earliest generations of programmable endonucleases developed for targeted genome modification. These synthetic enzymes consist of two discrete modules: (i) a programmable DNA-binding domain composed of ZF proteins (ZFPs) and (ii) a DNA-cleavage domain derived from a restriction endonuclease (Fig. 1.3b). The modular architecture enables ZFNs to recognize and cleave virtually any desired genomic sequence with high precision (Zubair et al., 2025).

Zinc-finger proteins were first characterized by Klug and colleagues in 1985 from transcription factor IIIA (TFIIIA) isolated from *Xenopus oocytes*, thereby marking the discovery of one of the most abundant DNA-binding motifs in eukaryotes. Each ZF domain typically comprises a conserved 30-amino-acid module folded into a ββα structure, stabilized by a single zinc ion coordinated by two cysteine and two histidine residues in a Cys_2His_2 configuration (Shamshirgaran et al., 2022). The α-helix of each finger fits into the DNA major groove, where it recognizes a triplet or quartet of base pairs (3-4 bp) through hydrogen bonding and hydrophobic interactions. Notably, six amino acid residues at positions −1, +1, +2, +3, +5, and +6 of the α-helix are responsible for direct DNA base recognition, while the remaining conserved residues provide structural stability. This modularity makes ZFPs reprogrammable-altering these residues enables the generation of new DNA-binding specificities that can be tailored to virtually any sequence. Consequently, tandem arrays of three to six zinc fingers can be engineered to bind extended targets ranging from 9 bp to 18 bp with high affinity (Cui et al., 2021).

To convert DNA recognition into site-specific cleavage, a catalytic domain capable of introducing DSBs must be fused to the ZFP array. DSBs activate endogenous repair pathways such as NHEJ and HDR, which can introduce small insertions, deletions, or sequence replacements at the target locus. Early studies demonstrated that classical type II restriction endonucleases, such as *EcoRI* and *BamHI,* were unsuitable for programmable editing because their catalytic and recognition functions are inseparable; altering their recognition sequence invariably disrupts catalytic activity. A major advance occurred with the characterization of the type IIS restriction enzyme *Fok*I from *Flavobacterium okeanokoites*, which possesses two independent domains: an N-terminal DNA binding region (~41 kDa) and a C-terminal catalytic domain (~25 kDa) (Shaltz and Jinks-Robertson, 2021). FokI recognizes the non-palindromic pentanucleotide 5′-GGATG-3′ and cleaves the DNA 9 and 13 nucleotides downstream of this site. Importantly, the cleavage domain can function independently of the recognition domain, enabling its use as a generic nuclease module. When linked to a sequence-specific ZFP array via a flexible peptide linker, FokI acts as a catalytic moiety that introduces DSBs adjacent to the chosen DNA sequence. The resulting chimeric enzyme, a zinc-finger nuclease, combines customizable recognition with targeted cleavage capability. Since FokI functions as a dimer, efficient DSB formation requires two ZFN monomers binding to opposite DNA strands in an inverted tail-to-tail orientation. Upon binding, the two FokI catalytic domains dimerize, generating a staggered DSB between the two recognition sites (Sobeh et al., 2025). Repair of this break by NHEJ often introduces small indels, leading to gene disruption, whereas the presence of a donor template allows HDR-mediated insertion or sequence correction.

Importantly, fusion of the ZFP array with the FokI cleavage domain does not impair the DNA-binding ability of the zinc fingers nor the catalytic activity of the nuclease. This property established ZFNs as one of the first customizable nucleases successfully applied to genome editing in both plant and animal systems. Furthermore, ZFPs have been adapted to create zinc-finger activators (ZFAs) and zinc-finger repressors (ZFRs) by fusing them with transcriptional activator or repressor domains, thereby enabling sequence-specific modulation of gene expression (Mukhametzyanova et al., 2024).

Despite their versatility, ZFNs present several practical limitations. Designing highly specific ZFP arrays is effort-intensive, prolonged, and resource-demanding, as the binding specificity of each finger is strongly influenced by its neighboring modules. For example, the presence of an aspartate residue at position +2 in the α-helix can extend recognition from 3 bp to 4 bp, thereby altering the specificity of adjacent fingers and complicating accurate target prediction. These inter-finger dependencies, combined with the need for extensive screening, make the construction of reliable ZFN arrays challenging. Another major drawback is the potential for off-target cleavage, which can induce cytotoxicity in mammalian cells. Off-target effects are concentration-dependent and can be minimized by regulating ZFN expression through inducible vectors or by extending the number of fingers in each array to increase recognition length and specificity. Moreover, rationally designed ZFPs with optimized binding affinity have been employed to mitigate unintended DNA breaks (Manikishore et al., 2025).

Although substantial improvements have enhanced ZFN performance, the complexity of design, context-dependent recognition, and off-target toxicity have limited their widespread application in therapeutic settings. These constraints ultimately led to the development of next-generation platforms such as TALENS and the CRISPR/Cas9 system, which combine modular design with greater efficiency, lower cytotoxicity, and simplified customization (Bhardwaj and Nain, 2021).

1.5 Transcription activator-like effector nucleases (TALENs)

TALENs represent the second major generation of programmable nucleases, developed to overcome the design complexity and target limitations associated with zinc-finger nucleases. TALENs are chimeric enzymes consisting of two functional domains: (i) a customizable DNA-binding domain derived from TALEs of *Xanthomonas* bacteria, and (ii) a DNA-cleavage domain from the *Fok*I endonuclease (Fig. 1.3c (Bhagtaney and Sundarrajan, 2023)).

TALEs are naturally occurring proteins secreted by plant-pathogenic *Xanthomonas* species that bind to promoter regions in host genomes to activate susceptibility (S) genes, facilitating infection. Each TALE contains a central domain composed of tandem, nearly identical 33–35 amino acid repeat units, each of which specifically recognizes one nucleotide in the DNA target sequence. Within each repeat, the amino acids at positions 12 and 13, known as repeat-variable diresidues (RVDs), determine nucleotide specificity through direct base recognition. The four most common RVDs, NI, HD, NG, and NN, preferentially bind to adenine (A), cytosine (C), thymine (T), and guanine (G), respectively. This one-to-one correspondence between repeat modules and DNA bases makes the TALE code inherently simple and predictable compared to the context-dependent recognition of zinc-finger domains. As a result, TALENs can be designed rapidly and with higher success rates for virtually any target sequence that begins with a thymine residue, a characteristic requirement for optimal binding (Jain et al., 2021).

To convert TALEs into programmable nucleases, their DNA-binding domains are fused to the catalytic domain of FokI, creating a hybrid protein capable of introducing DSBs at specific genomic locations (Fig. 1.3c). Like ZFNs, TALENs function as dimers; each monomer binds to opposite strands of the target DNA in a tail-to-tail configuration, allowing the FokI domains to dimerize and cleave the DNA within the spacer region between the binding sites. The DSBs are subsequently repaired through endogenous DNA repair mechanisms, either error-prone NHEJ, which introduces small insertions or deletions that disrupt gene function, or precise HDR, which allows precise sequence replacement or insertion by providing a repair template (Zaman et al., 2025).

TALENs offer several advantages over ZFNs. Their DNA-binding domains exhibit high modularity and predictability, enabling efficient targeting of longer and more diverse genomic sequences with reduced off-target activity. Each repeat in a TALE domain interacts independently with its corresponding base, and the absence of extensive inter-repeat context dependency simplifies the design process. Moreover, TALENs tolerate broader sequence contexts and demonstrate high editing efficiency across diverse organisms, including plants, animals, and human cells (Liu et al., 2025a, 2025b).

Applications of TALENs have been demonstrated with practical studies in diverse biological fields. In plants, TALENs have been used to generate targeted gene knockouts and to improve agronomic traits such as disease resistance and abiotic stress tolerance. In animal systems, they have been used to develop transgenic models and gene therapies, including the correction of genetic defects in human cells. Their versatility also extends to genomic modifications in non-model organisms, highlighting their adaptability and cross-species functionality (Sakuma and Yamamoto, 2023; Zaman et al., 2025).

Despite its superior specificity, the applications of TALEN technology face certain challenges. For example, the repetitive nature of TALE repeat sequences complicates cloning due to recombination events during vector assembly, requiring specialized cloning methods such as *Golden Gate Assembly. In addition, the large size of TALE-based constructs (~3 kb per monomer) presents challenges for in vivo delivery, particularly using viral vectors with limited packaging capacity. Nonetheless, these constraints have been mitigated by optimized repeat-variable di-residue libraries and improved assembly systems, which have significantly increased design efficiency and reproducibility (Sakuma and Yamamoto, 2017; Cermak et al., 2014).

Because each repeat module in TALENs recognizes a single nucleotide independently through its specific repeat-variable di-residue (RVD), target recognition is highly predictable and precise. This one-to-one base recognition mechanism renders TALENs inherently more target-specific than ZFNs, whose binding specificity is often influenced by interactions among adjacent zinc-finger motifs, leading to variable and context-dependent recognition patterns. Overall, TALENs represented a major advancement in the evolution of programmable nucleases, combining the modular predictability of DNA recognition with the catalytic reliability of FokI- mediated cleavage. Their introduction marked a transition toward more accessible and versatile genome-editing tools, paving the way for the emergence of the most specific, simple, and efficient CRISPR/Cas systems (Malzahn et al., 2017).

1.6 CRISPR–Cas systems

CRISPR/Cas represents a powerful genome-editing platform that has transformed modern molecular biology. Discovered serendipitously in the late 1980s during *Escherichia coli* genome sequencing, CRISPR loci were initially described as "short regularly spaced repeats." Subsequent comparative genomic studies revealed their widespread occurrence, present in nearly 40% of sequenced bacteria and approximately 90% of archaea, and their close association with *cas* genes, which encode proteins bearing nuclease, helicase, polymerase, or nucleotide-binding functions. It was later established that the CRISPR/Cas system operates as an adaptive immune mechanism, providing microorganisms with sequence-specific protection against bacteriophages and plasmids.

The natural CRISPR immune process involves three distinct phases (Boettcher and McManus, 2015). In the adaptation phase, short fragments of foreign DNA (spacers), such as those from bacteriophages, are incorporated into the CRISPR array as molecular memory of infection. During expression, the CRISPR locus is transcribed into a precursor CRISPR RNA (Pre-RNA), which is subsequently processed into mature CRISPR RNAs (crRNAs),

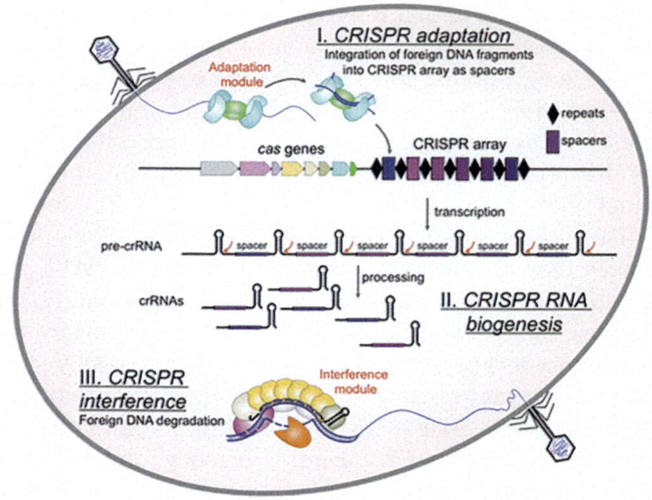

FIGURE 1.4 Mechanism of natural CRISPR/Cas immunity system. Three phases of the mechanism of action of natural CRISPR/Cas immunity system (I. CRISPR adaptation, II. CRISPR RNA biogenesis, III. CRISPR interference). *Adopted from Munawar and Ahmad (2022).*

each containing a spacer sequence derived from a previously encountered invading genetic element. In the final interference stage, crRNA makes a complex with one or more Cas nuclease to recognize and cleave the complementary sequence in the invading nucleic acid, thereby neutralizing the threat (Fig. 1.4).

CRISPR systems are broadly classified into two major classes, subdivided into six types and over 20 subtypes, based on their effector composition and gene architecture. Class 1 systems (Type I, III, and IV) employ multi-subunit protein complexes for target degradation, whereas Class 2 systems (Type II, V, and VI) depend on a single, multidomain Cas effector protein to execute DNA or RNA cleavage. Among these, the Type II CRISPR–Cas system from *Streptococcus pyogenes* has become the most widely harnessed for genome engineering, owing to its simplicity, high efficiency, and broad adaptability across diverse organisms (Sasnauskas and Siksnys, 2020).

In its simplest engineered form, the CRISPR/Cas9 complex comprises two essential components: the Cas9 nuclease and a single-guide RNA (sgRNA), as shown in Fig. 1.3d. Cas9 acts as a programmable endonuclease, functioning as the molecular "scissors" that creates targeted DNA double-stranded breaks. The protein contains two catalytic domains, HNH and RuvC, that cleave complementary and non-complementary DNA strands, respectively. The sgRNA is a synthetic chimeric RNA generated by fusing two natural RNA molecules: the CRISPR RNA (crRNA), which defines the target sequence, and the trans-activating CRISPR RNA (tracrRNA), which serves as a structural scaffold for Cas9 binding. Guided by the 20-nucleotide complementary sequence within the sgRNA, Cas9 identifies and binds to the target DNA adjacent to a short motif known as the protospacer adjacent motif (PAM). For *S. pyogenes* Cas9, the canonical PAM sequence is 5′-NGG-3′, where "N" can represent any nucleotide base. PAM recognition is essential for discriminating between self-and non-self DNA, as it prevents Cas9 from cleaving the bacterial CRISPR array itself. Upon sgRNA binding, Cas9 undergoes a conformational rearrangement that activates its catalytic domains, introducing a DSB approximately three base pairs upstream of the PAM (Jiang and Doudna, 2017). Cellular repair mechanisms are then engaged, primarily NHEJ and HDR. NHEJ, an error-prone process, often results in indels that disrupt gene function, whereas HDR uses a homologous DNA template to precisely repair or modify the locus, enabling targeted insertion or correction of genes (Amitai and Sorek, 2016).

Unlike earlier genome-editing platforms such as ZFNs and TALENs, which rely on engineered protein-DNA interactions, CRISPR/Cas9 employs an RNA-DNA base-pairing mechanism for sequence recognition. This fundamental difference provides several key advantages, such as i. simplified design (retargeting requires only modification of the sgRNA sequence, eliminating the need for protein redesign), ii. high scalability (multiple loci can be targeted simultaneously (multiplex editing) using several sgRNAs), iii. predictable specificity (off-target sites can be computationally identified due to well-defined RNA-DNA complementarity, iv. broad applicability (the system functions efficiently in a wide range of species and cell types). Furthermore, because Cas9 activity can be driven by RNA delivery rather than direct protein manipulation, the CRISPR approach is significantly faster and more cost-effective than ZFN or TALEN systems. These attributes have led to its rapid adoption across fields, including functional genomics, crop improvement, disease modeling, and gene therapy (Rath et al., 2015).

Despite these advantages, CRISPR/Cas9 is not without limitations. Its primary technical challenge is off-target cleavage, which may occur when sgRNA exhibits partial complementarity with unintended genomic sites. Advances such as high-fidelity Cas9 variants, improved gRNA design algorithms, and the use of truncated or chemically modified sgRNAs have substantially reduced off-target effects. Another constraint lies in the PAM dependency, which restricts the range of editable sequences. However, engineered Cas9 variants have extended targeting capabilities, like SpCas9-NG recognizes NG, NAG, or NGT PAMs, increasing the number of targetable sites by approximately 1.5-2 fold , while xCas9 recognizes an even broader range of PAMs, including NG, GAA, and GAT. Together, these variants have broadened PAM compatibility, enhanced targeting specificity, and reduced off-target effects (Modrzejewski et al., 2020). The system's flexibility has also led to the development of functional derivatives. Cas9 nickases (nCas9), created by mutating one of the two nuclease domains (H840A in HNH or D10A in RuvC), generate single-stranded breaks (SSBs) instead of DSBs, thereby reducing off-target effects. These have been further adapted into base editors and prime editors, enabling precise nucleotide substitutions or targeted insertions without inducing DSBs (Wolt et al., 2016).

Comparatively, while ZFNs and TALENs exhibit higher intrinsic accuracy per edit, their design and assembly are far more laborious and costly. CRISPR/Cas9, in contrast, offers unmatched programmability, scalability, and experimental throughput (Table 1.1). These characteristics have cemented its status as the most accessible and versatile genome editing platform developed to date.

A holistic understanding of the CRISPR/Cas9 system, its origin, molecular components, and operational principles forms the basis for assessing its transformative potential and ongoing challenges. A SWOT analysis (Strengths, Weaknesses, Opportunities, and Threats) of the system provides a structured evaluation of this technology (Munywoki and Munywoki, 2021). The "strengths" include simplicity, cost-efficiency, and the ability to target multiple genomic loci simultaneously, while the "weaknesses" encompass off-target effects,

TABLE 1.1 Comparison of major programmable nuclease systems: ZFNs, TALENs, and CRISPR/Cas9, highlighting differences in design complexity, specificity, and multiplexing capacity.

Features		Meganucleases 	ZFNs 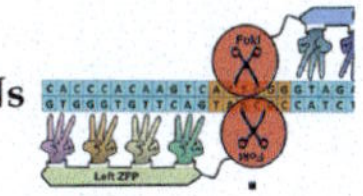	TALENs 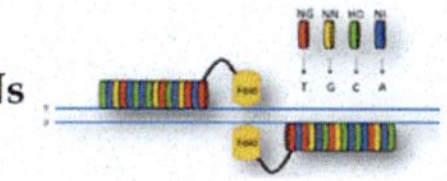	CRISPR/Cas9 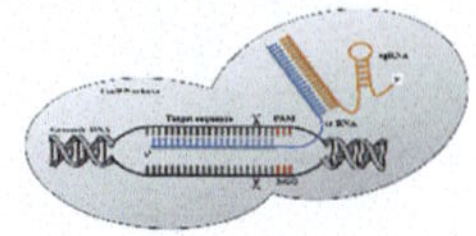
Origin & y ear of introduction		Microbial mobile genetic elements (1990)	*Xenopus* oocystes (1996)	*Xanthomonas* species (2010)	Immune system of bacteria and archaea (2012)
Natural function		Mobile genetic elements	DNA binding motifs in eukaryotes	Plant pathogenic protein	Prokaryotic defense protein
Mechanism of action	Target recognition	Protein-DNA	Protein-DNA (one to triplet)	Protein-DNA (one to one)	RNA-DNA (one to one)
	Target cleavage	Homing nucleases	Fok-1 nuclease	Fok-1 nuclease	Cas nucleases
Target specificity	Target site/ length	12–24 bp	~9–36 nt	~12–50 nt	~20–23 nt
	Target efficiency	Binds with DNA bases in the major groove and phosphodiester backbone	It binds to a triplet of DNA bases	Requires T nt at 5′	Requires PAM (5′NGG)
	Off-targets	Low	Low	Very few	High
Engineering constrains	Design feasibility	Engineering of protein required for novel targets	Engineering of the DNA-binding domain required for every target	Engineering of the DNA-binding domain required for every target DNA	RNA complementary to the target DNA is required
	Assembly	No assembly required	Difficult	Technical but easy	Easy
	Time	Time consuming	Highly time-consuming	Less time-consuming	Less time-consuming
	Cost	Expensive	Expensive	Not cost-effective	Cost effective
Commercialization tools		No independent database available	Available	Available	Easily available
Uses	Genome editing possibilities	Gene addition, gene deletion	Gene disruption, gene deletion, gene correction, gene addition,t ag ligation, ObLiGaRe	Gene activation, gene repression,g ene disruption, gene deletion,g ene correction, gene addition,t ag ligation, ObLiGaRe	Gene disruption, gene deletion,g ene correction, gene addition.

The table is adopted from *Munawar and Ahmad (2022)*.

PAM sequence limitations, and variable repair pathway efficiency. The "opportunities" lie in the expansion of Cas variants, therapeutic genome correction, crop bioengineering, and synthetic biology applications, while "threats" include ethical controversies, potential misuse, intellectual property disputes, and regulatory uncertainty surrounding human germline editing. This analysis underscores that while CRISPR/Cas9 has already reshaped biotechnology, its continued evolution will depend on responsible innovation and refined molecular precision (Fino et al., 2020).

Importantly, the CRISPR toolkit extends far beyond Cas9. Novel Cas effectors have been identified with distinct catalytic activities and targeting specificities, such as Cas12 (Cpfl) for staggered DNA cleavage, Cas13 for RNA editing, Cas6 for RNA processing, and Cas3 for processive DNA degradation. Recently characterized complexes like Cas7-11 and the compact CasX protein further diversify the CRISPR landscape, enabling finer control over gene regulation and epigenetic modification. These emerging systems, discussed in the following sections, exemplify the ongoing diversification and refinement of CRISPR-based technologies, thereby broadening their applications across molecular biology, medicine, and agriculture (Pyzocha and Chen, 2018).

1.7 Diversity of CRISPR–Cas variants: Cas12, Cas13, Cas6, Cas7-11, Cas3, and CasX

Since its adaptation for eukaryotic genome editing in 2013, the CRISPR/Cas9 system has become a cornerstone of molecular biology. Yet, continuous exploration of microbial genomes has revealed a remarkable diversity of CRISPR/Cas variants that differ in structure, function, and substrate specificity. These alternative effectors collectively broaden the CRISPR toolbox and have expanded genome manipulation beyond the DNA double-strand breaks characteristic of Cas9, enabling finer control of gene expression, epigenetic modification, and RNA editing (Koonin et al., 2017). The best-known CRISPR effectors, Cas9 (Type II), Cas12 (Type V), and Cas13 (Type VI) belong to Class 2, reflecting their structural simplicity and high suitability for genome editing (Dutta et al., 2025).

1.8 Cas12 (Cpfl) – type V DNA-targeting nucleases

Following Cas9, the Cas12 (Cpfl) enzyme, first identified in *Prevotella and Francisella* species, emerged as a distinct DNA-targeting nuclease belonging to Class 2, Type V CRISPR systems. Unlike Cas9, Cas12 requires only a crRNA (without a tracerRNA) to guide target recognition and introduce staggered DSBs, producing cohesive 5′ overhangs rather than blunt ends (Hillary and Ceasar, 2023). Cas12 recognizes a T-rich PAM sequence (typically 5′-TTTV-3′, where V = A, C, or G) located upstream of the target site, thereby expanding its targeting scope to AT-rich genomic regions that are less accessible to Cas9. Upon crRNA-guided binding, Cas12 cleaves both DNA strands using a single RuvC-like catalytic domain, enabling high-fidelity genome editing and efficient gene insertion. The structural simplicity and sticky-end cleavage pattern of Cas12 facilitate efficient homology-directed repair (HDR) and base editing, making it particularly suitable for plant genome engineering and diagnostics. Moreover, its unique collateral cleavage activity upon target recognition forms the basis for CRISPR-based biosensing platforms, such as DETECTR and HOLMES, enabling rapid pathogen and nucleic acid detection (Yan et al., 2019).

1.9 Cas13 – type VI RNA-targeting nucleases

In contrast to Cas9 and Cas12, which target DNA, Cas13 represents a family of RNA-guided RNA nucleases (Type VI) within Class 2 systems. First identified in 2016, Cas13 employs a crRNA to recognize single-stranded RNA (ssRNA) sequences without requiring a PAM. Instead, recognition depends on a short protospacer-flanking site (PFS) that prevents self-targeting (Yan et al., 2019). Cas13 contains two higher eukaryotes and prokaryotes nucleotide-binding (HEPN) domains responsible for RNA cleavage. Once Cas13 binds to its target RNA, it undergoes conformational activation, cleaving both the target as well as nearby non-specific RNAs, a phenomenon known as collateral cleavage. This property has been ingeniously harnessed in molecular diagnostic assays such as SHERLOCK, which enables ultrasensitive detection of RNA viruses, including SARS-CoV-2 (Cox et al., 2017). The general mechanism of Cas 13 is shown in Fig. 1.5 below.

The Cas13 family exhibits considerable phylogenetic and structural diversity. Based on sequence, domain organization, and functional properties, Cas13 proteins are grouped into several subtypes, including Cas13a, Cas13b, Cas13c, Cas13d, and more recently described compact orthologs such as Cas13X and Cas13Y. Despite these variations, all Cas13 enzymes share two key functional HEPN motifs responsible for catalytic activity, although their domain positions differ across subtypes. In Cas13a, Cas13c, and Cas13d, the HEPN domains occupy central and C-terminal locations, whereas in Cas13b, Cas13X, and Casl3Y, they are found at the N- and C-termini. The cleavage mechanism remains conserved, with one HEPN motif initiating the phosphodiester bond cleavage and the other stabilizing the reaction intermediate. The guide RNA (gRNA) spacer that directs Cas 13 typically ranges from 24 to 30 nucleotides, depending on the enzyme subtype. Among these, Cas13c remains the least functionally characterized due to limited structural and biochemical data (O'Connell, 2019).

Beyond its role in natural bacterial immunity, Cas13 has been harnessed for programmable RNA manipulation in eukaryotic systems. Catalytically inactive mutants, known as dead Cas13 (dCas13), have been engineered by inactivating one or both HEPN domains. These variants retain RNA-binding capacity without cleavage activity and serve as versatile scaffolds for RNA tracking, translational repression, and post-transcriptional regulation. The most impactful adaptation of this system is its application in RNA base editing (RBE). By fusing dCas13b with adenosine deaminases such as ADAR2, researchers achieved precise "A-to-I" conversions, where adenosine is deaminated to inosine, which is interpreted as guanosine during translation (Deng et al., 2023). Similarly, coupling engineered cytidine deaminase variants of ADAR2 to dCas13b enables "C-to-U″ transitions, in which cytidine is converted into uracil within an mRNA duplex. These modifications occur through targeted mismatches between the guide RNA and mRNA, producing controlled deamination events

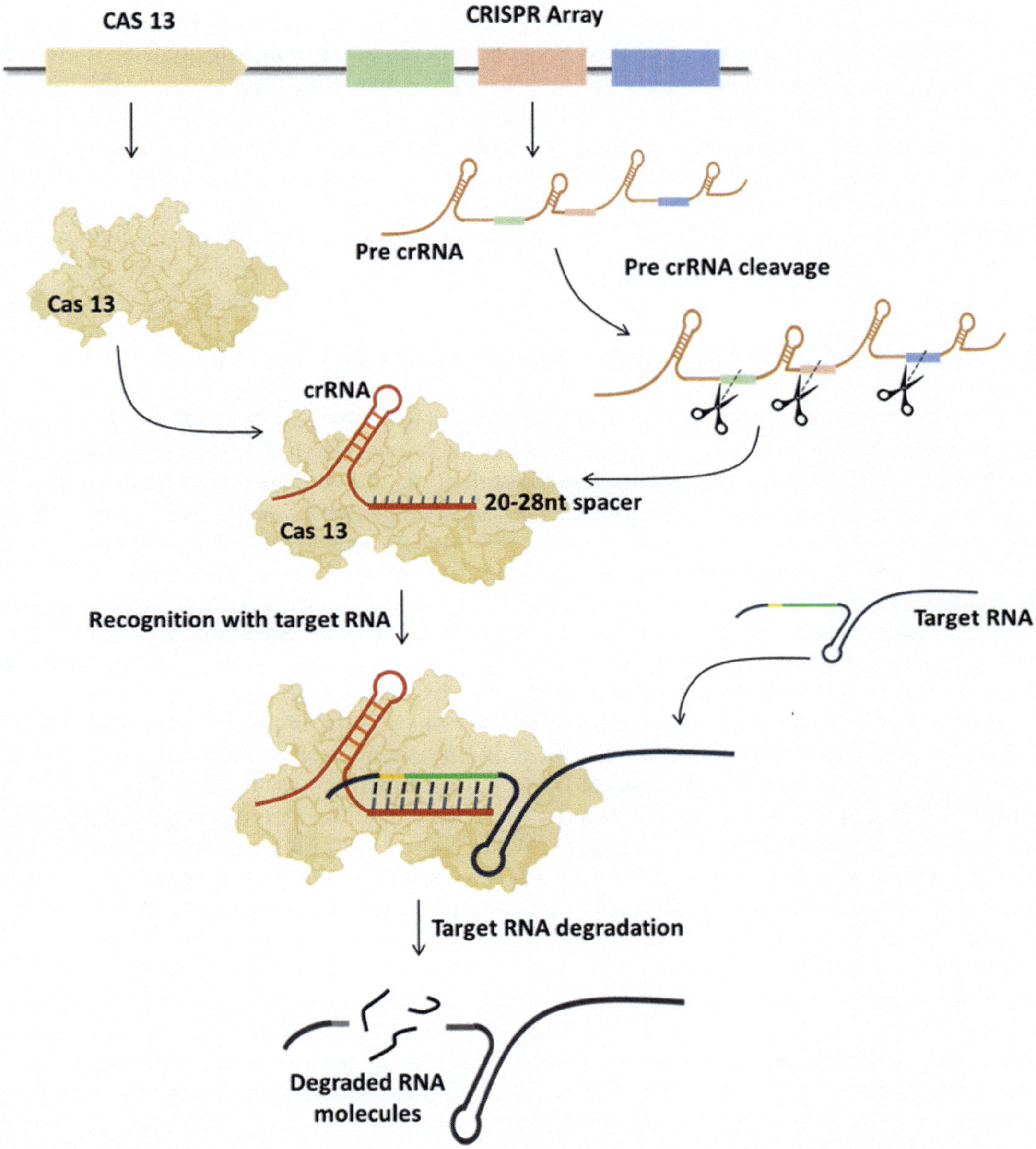

FIGURE 1.5 Mechanism of Cas13 for targeting and degrading RNA molecules: This figure details the function of the RNA-targeting CRISPR–Cas13 system. The Cas13 enzyme is guided by a specific crRNA (CRISPR RNA) molecule, derived from the CRISPR array, to form an active complex. The crRNA directs Cas13 to a complementary target RNA sequence. Upon binding, Cas13 becomes an activated RNase, resulting in the rapid degradation of the target RNA and other surrounding RNA molecules.

without altering the underlying genomic DNA (Yang and Patel, 2024). Cas13's versatility extends to both transient and therapeutic RNA applications. Because its activity is restricted to RNA molecules, edits are inherently reversible and less likely to introduce heritable mutations, providing an important safety advantage for clinical use. This property positions Cas13 as a valuable tool for correcting RNA-level defects, regulating gene expression, and controlling viral RNA replication in infectious diseases. In addition, Cas13's ability to process precursor crRNAs into mature, functional crRNAs enables autonomous guide RNA maturation, facilitating multiplexed RNA targeting within a single system (Chen et al., 2024).

1.10 Cas14 – Ultra-compact single-stranded DNA-targeting nucleases

The systematic microbial and archaeal genome analysis by Jennifer Doudna's research group revealed the presence of several small, uncharacterized *cas* loci, including *cas1*, *cas2*, *cas4*, and the novel Cas14 family (Harrington et al., 2018). Cas14 proteins are remarkably compact, with molecular weights ranging between 40 and 70 kDa, corresponding to approximately 400–700 amino acids, roughly half the size of canonical Cas nucleases such as Cas9 or Cas12. Despite their small size, Cas14 proteins retain a conserved RuvC-like endonuclease domain, that serves as the catalytic center for nucleic acid cleavage. Comparative genomic analyses indicate that Cas14 homologs are exclusively found in archaeal species, suggesting that they represent an ancient, possibly primordial, branch of the CRISPR lineage predating bacterial Cas systems (Hillary and Ceasar, 2023).

The *casl4* genes have been classified into at least 24 variants, organized into three major subgroups, Cas14a, Cas14b, and Cas14c, based on phylogenetic clustering and sequence homology. Functional assays demonstrated that Cas14 functions as a stand-alone effector, requiring no auxiliary proteins for activity, a feature that further distinguishes it from many multi-component CRISPR complexes. When cloned with its associated CRISPR array and expressed heterologously in *Escherichia coli*, Cas14 was shown to form a functional ribonucleoprotein complex capable of recognizing and cleaving single-stranded DNA (ssDNA) (Savage, 2019).

The discovery that Cas14 specifically targets ssDNA, in contrast to the double-stranded DNA substrates of Cas9 and Cas12 or the RNA substrates of Cas13, provided important insight into its evolutionary role. This activity suggests that Cas14 likely evolved as a defense mechanism against single-stranded DNA viruses or mobile genetic elements that replicate through single-stranded intermediates. Notably, Cas14-mediated cleavage is guided by a complementary gRNA, similar to other CRISPR systems, but it does not require a PAM to recognize the target site. The absence of PAM dependency significantly expands its potential targeting range and enables discrimination of sequences at single-nucleotide resolution. Experimental characterization revealed that Cas14's ssDNA cleavage is more specific than the activity observed for Cas12 or Casl3, as it strictly depends on the presence of a correctly paired gRNA target duplex (Zhou et al., 2023). This precision makes Cas14 particularly valuable for applications requiring single-nucleotide discrimination, such as single-nucleotide polymorphism (SNP) detection or high-fidelity diagnostic assays. Furthermore, its small molecular size enhances compatibility with delivery systems that face packaging constraints, such as adeno-associated viral (AAV) vectors, broadening its potential for therapeutic use (Munir et al., 2024a, 2024b).

1.11 Cas6 – RNA processing and crRNA maturation

Cas6, a small endoribonuclease found primarily in Class 1 systems, plays a critical role in crRNA biogenesis by processing precursor CRISPR transcripts (pre-crRNA) into mature functional crRNAs (Niewoehner et al., 2013). It recognizes and cleaves specific hairpin structures within CRISPR repeat sequences to generate guide RNAs required for interference. While Cas6 itself does not function as an effector nuclease for editing, its structural and catalytic mechanisms have been repurposed in engineered systems to enhance RNA processing efficiency and gRNA array construction for multiplex genome editing. Cas6's precise RNA recognition ability has also inspired the development of synthetic tools for RNA tracking and regulation, contributing to modular CRISPR architectures that combine Cas9 or Cas12 with engineered RNA-processing modules for multi-gene targeting (Sokolowski et al., 2014).

1.12 Cas7-11 – hybrid RNA-targeting enzymes

Recently, the Cas7-11 complex was discovered in Type III-E CRISPR systems, bridging the evolutionary gap between multi-protein Class 1 and single-protein Class 2 systems. This unusually large, single-polypeptide effector combines multiple Cas7-like domains with a Cas11-like C-terminal domain, forming a compact, RNA-guided RNA endonuclease. Cryo-electron microscopy studies of the active Cas7-11 enzyme from *Desulfonema ishimotonii* (DiCas7-11) revealed an extended, modular conformation resembling the backbone of canonical Class 1 effector complexes, composed of Cas7 and Cas11 subunits. However, in Cas7-11, these domains have diverged evolutionarily, resulting in a self-contained structure with distinct mechanistic properties (Liu et al., 2025a, 2025b). The first Cas7-like domain of DiCas7-11 is primarily responsible for processing and recognizing CRISPR RNA (crRNA), encapsulating approximately the final 15 nucleotides of the repeat sequence. This domain alone retains most of the RNA recognition and processing activity. The subsequent Cas7-like domains mediate metal-dependent RNA cleavage, each contributing to precise and programmable RNA degradation. Structural and mutagenesis analyses indicate that a variable insertion within the fourth Cas7 domain does not significantly affect the enzyme's activity, implying that this region could be modified or removed to generate more compact versions of the effector (Özcan et al., 2021).

The functional characterization of Cas7-11 represents a major advance over previous RNA-targeting CRISPR systems. Prior to its discovery, Cas13 was the only CRISPR nuclease known to target RNA; however, its collateral cleavage activity caused widespread degradation of non-target transcripts once activated. This property made Cas13 useful for diagnostics but unsuitable for therapeutic RNA editing, where precision is essential. In contrast,

Cas7-11 demonstrates high target specificity without collateral cleavage, allowing controlled RNA editing and interference analogous to how Cas9 mediates site-specific DNA cleavage. Structural studies capturing Cas7-11 in complex with its target RNA and guide RNA have provided deep mechanistic insights into its mode of action. The enzyme forms a stable ribonucleoprotein complex in which the guide RNA directs target recognition, while the catalytic domains execute precise RNA cleavage. Mapping of the active site revealed the specific residues and conformational changes essential for target binding and catalysis, paving the way for the rational engineering of the enzyme for tailored RNA manipulation in eukaryotic systems (Kato et al., 2022).

Despite its advantages, the full-length Cas7-11 protein is notably large, posing challenges for cellular delivery using conventional viral vectors such as adeno-associated viruses (AAVs).. To overcome this limitation, researchers led by Abudayyeh and Gootenberg analyzed the structural data and identified a dispensable region that contributed to the enzyme's bulk but not to its activity (Özcan et al., 2021). Through targeted deletion of this segment, they engineered a smaller and functionally equivalent variant named Cas7-11S. The compact Cas7-11S retained its RNA-targeting efficiency while fitting within a single AAV vector, enabling successful delivery and targeted RNA interference in mammalian cells. The development of Cas7-11 and its optimized derivative, Cas7-11S, provides a versatile platform for high-precision RNA editing and regulation. Its inherent specificity, absence of collateral activity, and programmable RNA-guided mechanism position it as a promising alternative to Cas13-based tools for therapeutic and research applications. The ongoing structural refinements and domain engineering of Cas7-11 are expected to enhance its potential further as a next-generation RNA-targeting system suitable for in vivo genetic regulation and disease intervention (Abudayyeh et al., 2017).

1.13 Cas3 – type I system with processive DNA degradation

Cas3, the hallmark effector of Type I Class 1 CRISPR systems, differs substantially from Cas9 and Cas12 in both structure and mechanism. Rather than producing a localized DNA cut, Cas3 functions as a helicase-nuclease enzyme that unwinds and degrades DNA processively along one strand after recruitment by the Cascade (CRISPR-associated complex for antiviral defense) complex. This processive degradation capability allows Cas3 to remove long stretches of DNA, enabling large-fragment deletions spanning several kilobases (Sinkunas et al., 2011). Its unique activity has been exploited for generating targeted chromosomal deletions and studying large-scale genome rearrangements. Cas3's ability to induce extensive excision offers potential for therapeutic gene removal or viral genome eradication, but it requires precise control to avoid excessive degradation (Csörgö et al., 2020).

1.14 CasX – compact RNA-guided DNA nucleases

CasX (also known as Cas12e) was identified through metagenomic mining of groundwater bacterial communities and represents one of the smallest RNA-guided nucleases characterized to date (–980 amino acids). Despite lacking sequence homology with Cas9 or Cas12, CasX performs similar DNA cleavage functions using an RNA-DNA recognition mechanism. CasX recognizes a 5′-TTCN PAM motif and induces staggered double-strand breaks, akin to Cas12 but with reduced protein size, facilitating delivery via viral and non-viral vectors where cargo capacity is limited. Structural analyses reveal that CasX forms a compact architecture with a single RuvC catalytic domain, optimized for minimal off-target activity and enhanced editing precision. Recently engineered variants, such as DpbCasX-R3-v2 and PlmCasX-R1-v2, demonstrate improved efficiency in mammalian systems, underscoring CasX's potential for in vivo therapeutic applications (Liu et al., 2019; Tsuchida et al., 2022).

1.15 Bridge RNA – A novel platform for programmable genome recombination

The growing diversity of genome-editing platforms, from MegaNs and ZFNs to the CRISPR–Cas family, has established a foundation for increasingly precise and programmable genetic manipulation. Yet, even with the efficiency of the Cas9, Cas12, and Cas13 systems, these methods predominantly rely on introducing DSBs to trigger repair by endogenous cellular mechanisms. While effective, DSB-dependent editing carries the risk of undesired mutations, chromosomal rearrangements, and genomic instability. In response to these challenges, researchers have explored novel, non-nuclease-based genome-manipulation systems capable of performing accurate DNA rearrangements without inducing DSBs. One such promising innovation is the Bridge RNA system, a newly characterized programmable RNA-guided mechanism for site-specific DNA recombination (Hiraizumi et al., 2024).

The concept of Bridge RNA originated from investigations into IS110 insertion sequences, a family of small autonomous mobile genetic elements widely distributed among prokaryotes. These elements mediate genomic rearrangements through site-specific recombination, facilitated by recombinase enzymes encoded within the IS110 family. Unlike CRISPR–Cas nucleases, which cut DNA directly, the IS110 recombination system functions through precise template alignment and recombination between homologous regions of DNA. Detailed analysis of these elements led to the identification of structured noncoding RNAs that bridge the interaction between target DNA and donor DNA, now referred to as Bridge RNAs. These RNAs act as molecular adaptors that align two DNA molecules into proximity, guiding recombination without cleaving the DNA strands (Durrant et al., 2024).

Structurally, a Bridge RNA contains two internal stem-loop domains, each with a distinct hybridization function. The target-binding loop recognizes and base-pairs with the genomic DNA sequence to be modified, while the donor-binding loop interacts with the donor DNA segment that carries the intended insertion or correction sequence. This dual recognition allows the Bridge RNA molecule to form a transient ternary complex, comprising the RNA, target DNA, and donor DNA, that promotes recombination mediated by an IS110-family recombinase. The recombinase introduces a single-strand nick or local structural distortion, allowing precise exchange between the target and donor regions. This mechanism supports insertions, deletions, and inversions with high sequence fidelity (Shapiro et al., 2007).

One of the defining characteristics of Bridge RNA technology is its modularity. Each binding loop can be independently reprogrammed to recognize a different sequence, allowing the same RNA scaffold to direct diverse genomic rearrangements at chosen loci, giving greater flexibility and control over specific editing (Kurth and Mochizuki, 2009). This property reflects the programmability shared with the CRISPR guide RNA system, but there is a crucial mechanistic difference. Whereas CRISPR–Cas uses RNA-DNA base pairing to direct nuclease-mediated cleavage, Bridge RNA guides recombination by aligning homologous sequences for recombinase-mediated DNA exchange. As a result, it enables precise genomic modifications without triggering cellular DSB repair pathways, thereby minimizing off-target alterations within the genome.

1.16 Applications of CRISPR-based genome editing in plants

The implementation of CRISPR-based genome editing has transformed modern plant biotechnology by enabling precise manipulation of genes associated with growth, development, and stress tolerance. Over the past decade, the number of studies applying CRISPR/Cas systems in plants has increased dramatically, contributing to substantial progress in crop improvement, stress management, yield enhancement, and nutritional quality. To date, genome editing has been successfully performed in more than 40 crop species, including major staples such as rice, maize, wheat, tomato, and potato (Liu et al., 2017; Zhu et al., 2020).

Abiotic stresses such as drought, salinity, and extreme temperatures remain among the most severe constraints on global agricultural productivity and are projected to intensify under ongoing climate change. CRISPR-mediated editing of genes responsible for stress perception and tolerance has proven effective in producing resilient crop varieties. For example, the knockout of OsRR22, which encodes a type-B response regulator in rice, enhanced salinity tolerance, while mutation of the Zinc-finger nuclease gene in *Oryza sativa* (indica cv. MTU1010) resulted in improved leaf water retention and reduced stomatal density (Santosh Kumar et al., 2020). The altered physiology of the *dst* mutant conferred greater tolerance to drought and salt, along with increased grain yield. Similarly, in tomato, targeted disruption of *SlARF4* promoted enhanced root development and the accumulation of soluble sugars and chlorophyll under osmotic stress, improving plant performance in saline conditions (Mustafa et al., 2025). The CRISPR/Cas9 system has also been instrumental in enhancing drought tolerance in several crops. In maize, the precise insertion of the native GOS2 promoter upstream of the *ARGOS8* gene increased yield stability under flowering stress, improving grain yield by approximately five bushels per acre without compromising performance under well-watered conditions (Shi et al., 2017). In tomato, knockout of the *SlMAPK3* gene led to differential expression of drought-responsive genes (*SlLOX, SlGST*, and *SlDREB*), conferring better protection against oxidative damage and improving drought adaptability (Vidya and Arun, 2023).

Temperature extremes have similarly been addressed through targeted genome modification. In tomato, the deletion of SlAGAMOUS-LIKE 6 (SlAGL6) improved fruit set and productivity under high-temperature stress (Klap et al., 2017), while in rice, the knockout of

OsMYB30, a β-amylase-related gene regulating starch degradation, resulted in enhanced cold tolerance through increased maltose accumulation and stabilization of starch metabolism (Zeng et al., 2020). CRISPR has also played a pivotal role in improving plant resistance to pathogens, including bacteria, fungi, and viruses. Modifying plant susceptibility (*Su*) genes has proven a successful strategy for disease resistance. In rice, the targeted knockout of *Xa13*, *OsSWEET13*, and *OsSWEET14* genes significantly enhanced resistance to bacterial blight, while disruption of *TaMLO* and *TaEDR1* genes in wheat conferred durable resistance to powdery mildew (Li et al., 2021). The *eIF4G* gene in rice, which acts as a susceptibility factor for the rice tungro virus, was also edited to produce resistant varieties. Furthermore, CRISPR-mediated modification of fungal virulence genes such as *ALB1* and *RSY1* in rice reduced the severity of rice blast disease, and disruption of *USTA* and *UvSLT2* minimized smut infection. Viral resistance has been achieved by directly targeting viral genomes; for instance, editing of replication-related regions (*MP, CP, Rep*, and *IR*) provided effective protection against Tomato Yellow Leaf Curl Virus (TYLCV) and Wheat Dwarf Virus (WDV) in tomato and barley, respectively (Kalinina et al., 2020).

In addition to improving stress and disease resistance, CRISPR technology has been used to increase yield potential and grain quality by manipulating key quantitative trait loci (QTLs). In rice, silencing of *OsGS3*, a negative regulator of grain size, resulted in larger grains, while knockout of *OsGW2* and *TaCKX2-D1* enhanced grain weight and number. Similarly, editing of the *SlCLV3* promoter and the *SlENO* gene in tomato led to increased fruit size and yield, demonstrating the potential of CRISPR for fine-tuning developmental traits that influence productivity. Beyond yield improvement, CRISPR has been employed to enhance nutritional content and food quality. For instance, targeted mutagenesis of *IbGBSSI* and *IbSBEII* in sweet potato improved starch composition, while suppression of *LCYe* in banana increased β-carotene levels, contributing to biofortification. Editing of *alpha-gliadin* genes in wheat successfully reduced gluten content, minimizing allergenicity. Similarly, disruption of *FAD2* in *Brassica napus* and peanut increased oleic acid accumulation, enhancing oil quality. In rice, knockout of *SBEIIb* shifted starch composition toward higher amylose content, while CRISPR-mediated disruption of amylose biosynthetic genes in cassava and potato modified starch quality. In strawberry, editing of *FvebZIPs1.1* altered sugar content, improving fruit sweetness and flavor (Munir et al., 2024a, 2024b).

CRISPR-based genome editing has also emerged as a valuable tool for manipulating breeding systems. The technology has been applied to induce male sterility, a key factor in hybrid seed production, by knocking out fertility-related genes such as *TMS5* in rice and *TaNP1* in wheat. These mutations produced genic male-sterile lines useful for controlled hybridization. Moreover, editing of the S-RNase locus in potato eliminated self-incompatibility, enabling self-fertilization and accelerating breeding cycles. CRISPR has further contributed to haploid induction by disrupting genes involved in pollen development, such as *OsMTL*, which, when knocked out, caused chromatin fragmentation in spermatids and resulted in the formation of haploid plants. These strategies collectively streamline the generation of homozygous lines and enhance genetic stability in breeding programs (Abd-Elsalam et al., 2024).

Overall, the versatility of the CRISPR/Cas system in plants has established it as an indispensable platform for advancing crop genetics. Its applications span from enhancing resilience to environmental stressors and pathogens to optimizing yield, nutritional value, and reproductive efficiency. As genome-editing technologies continue to evolve with the integration of next-generation variants such as Cas12 and Cas13, their deployment in plant biotechnology promises to accelerate the development of climate-resilient, nutritionally enriched, and high-yielding crop varieties essential for future global food security (Abd-Elsalam et al., 2024).

1.17 Ethical challenges of genome editing

Over the past decade, CRISPR–Cas genome editing has stood out as one of the most significant scientific breakthroughs, offering an unparalleled ability to modify the genetic structure of living organisms, including plants, animals, and humans (Caplan et al., 2015). While this technology has revolutionized biological research and agricultural science, it has simultaneously raised complex ethical, social, and regulatory questions regarding its applications and long-term consequences (van der Meer, 2025). The increasing capacity to rewrite genomes compels scientists, policymakers, and ethicists to deliberate on issues of safety, equity, and moral responsibility, emphasizing the need for comprehensive ethical and legal frameworks to guide the responsible use of genome editing (Kang et al., 2017). The first major international discussion on the ethical and biomedical implications of CRISPR took place at the Napa Valley meeting in 2015, where leading scientists, including the developers of CRISPR–Cas9, joined ethicists to address the moral boundaries of genome editing (Gregorowius et al., 2017). This meeting prompted global engagement from institutions such as the U.S. National Academies of Sciences, Engineering, and Medicine (NASEM), the Chinese Academy of Sciences, and the UK Royal Society, which convened the International Summit on Human Gene Editing to deliberate on responsible research practices (Mulvihill et al., 2017). Following these discussions, the Nuffield Council on Bioethics (2016) expanded the dialogue to include agriculture, industrial biotechnology, and environmental applications, recognizing that ethical considerations must extend beyond human medicine to encompass ecological and societal impacts (Gyngell et al., 2019).

In the agricultural context, CRISPR–Cas systems have emerged as a preferred technology for crop improvement due to their precision, efficiency, and adaptability. Yet, the rapid commercialization of genome-edited crops has reignited moral and biosafety debates concerning their ecological and societal implications (Mulvihill et al., 2017). Critics question whether the deliberate alteration of plant genomes might disrupt natural ecosystems or create unforeseen risks to biodiversity. Nonetheless, unlike traditional transgenic approaches, CRISPR often produces cisgenic or intragenic modifications, meaning that no foreign DNA is permanently integrated into the host genome (Doxzen and Halpern, 2020). From an ethical standpoint, such modifications are viewed as morally neutral; their acceptability depends on how the technology is applied rather than on the technique itself (Schultz-Bergin, 2018). A key point in the debate concerns whether CRISPR-edited crops should be treated the same as conventional genetically modified organisms (GMOs). Genome-edited plants typically undergo small, targeted nucleotide changes using the organism's own repair mechanisms, most commonly through NHEJ. These changes are indistinguishable from mutations that occur naturally or are induced by conventional chemical or radiation mutagenesis (Ahmad et al., 2024a, 2024b, 2024c). Moreover, transient delivery methods or ribonucleoprotein (RNP) complexes allow editing without foreign DNA integration, and any temporary expression cassettes introduced can be segregated out in subsequent generations. Consequently, genome-edited crops differ fundamentally from transgenic GMOs, posing lower potential risk to human health and the environment. From a utilitarian ethical perspective, CRISPR-based modification, especially when free from transgene insertion, can be considered less morally contentious and more acceptable than previous genetic engineering methods (Brokowski and Adli, 2019).

Despite this, broader ethical discussions in plant genome editing extend beyond biosafety. The technology intersects with long-standing critiques of industrial agriculture, raising questions of justice, ownership, and intergenerational responsibility. Concerns have been voiced regarding unequal access to advanced biotechnologies, potential monopolization of intellectual property, and the socio-economic divide between countries that can afford to adopt genome editing and those that cannot. Ethical reflections also consider ecological stewardship and the rights of future generations to inherit a sustainable environment. Therefore, societal dialogue must accompany scientific advancement to define appropriate boundaries for genome editing in agriculture (Zhang et al., 2020). In the agricultural sector, the proliferation of CRISPR technologies also highlights disparities in regulatory approaches among countries. Some jurisdictions, such as the United States and Argentina, evaluate genome-edited crops based on the final product rather than the process used, while others, including the European Union, apply stringent GMO-like regulations. This divergence complicates global trade, research collaboration, and policy harmonization. The rapid diffusion of CRISPR innovations has thus prompted scholars in the social sciences and humanities to explore its broader societal consequences (Courtier-Orgogozo, 2025). Their analyses examine questions of governance, public perception, and ethical communication, providing critical insights into how emerging technologies interact with socio-economic systems and cultural norms (Doxzen and Halpern, 2020).

The socio-ethical landscape of gene editing in agriculture demands evidence-based policymaking supported by interdisciplinary research. Comprehensive evaluations must integrate ethical reasoning with economic, ecological, and cultural dimensions to ensure that deployment of genome-editing promotes sustainability and equity. This is particularly relevant in developing nations, such as India, where food security pressures are high, yet awareness of gene-edited products remains limited. Although the country has not approved any GM food crops for cultivation, products derived from genetically modified sources, such as edible oils and Bt cotton derivatives, are already in circulation, often without clear labeling or consumer awareness. This situation highlights the urgent need for robust governance, transparent communication, and ethical literacy across the agricultural and regulatory sectors. In conclusion, CRISPR-based plant genome editing exemplifies the dual nature of technological innovation, a tool of immense promise that also demands moral scrutiny. The ongoing challenge lies not only in assessing its scientific risks and benefits but also in addressing the ethical responsibilities that accompany its application. Establishing inclusive dialogue among scientists, policymakers, and the public will be essential for aligning innovation with societal values and ensuring that genome editing contributes responsibly to sustainable agriculture and global food security.

Declaration

The authors declare that no generative artificial intelligence tools were used in the writing or preparation of this chapter. However, Professional Grammarly was used to improve the clarity and English editing of this chapter. The content reflects the authors' own scholarly work, and the authors assume full responsibility for all statements, interpretations, and conclusions presented.

References

Abd-Elsalam, K.A., Ahmed, A., Munir, A., 2024. CRISPRized fruit, vegetable, and ornamental crops: a note from. In: CRISPRized Horticulture Crops. Elsevier, pp. 3–20.

Abudayyeh, O.O., Gootenberg, J.S., Zhang, F., 2017. RNA editing with CRISPR-Cas13. Nature 558 (7711), 599–604.

Acquaah, G., 2024. Conventional plant breeding to modern plant breeding: evolution, achievements, and limitations. In: Plant Molecular Breeding in Genomics Era: Concepts and Tools, pp. 1–42.

Ahmad, A., Abd-Elsalam, K.A., Ali, M., Majeed, H.N., Arif, A., Munir, A., Munawar, N., 2024a. Regulatory triggers of CRISPR-edited crops. In: Global Regulatory Outlook for CRISPRized Plants. Elsevier, pp. 91–112.

Ahmad, A., Arif, A., Munir, A., 2024b. CRISPR technology commercialization and biosafety. In: Global Regulatory Outlook for CRISPRized Plants. Elsevier, pp. 461–514.

Ahmad, A., Munir, A., Zafar, H., Zahoor, M.K., Hassan, S., Khan, S.H., 2024c. Tracking footprints of CRISPR-based genome editing. In: Global Regulatory Outlook for CRISPRized Plants. Elsevier, pp. 113–145.

Aklilu, E., 2021. Review on forward and reverse genetics in plant breeding. Life 14 (1), 127–135. https://doi.org/10.1080/26895293.2021.1888810.

Amitai, G., Sorek, R., 2016. CRISPR–Cas adaptation: insights into the mechanism of action. Nat. Rev. Microbiol. 14 (2), 67–76. https://doi.org/10.1038/nrmicro.2015.14.

Aslam, S., Munir, A., Usman Aslam, H.M., Khan, S.H., Ahmad, A., 2022. Genome Editing Advances in Soybean Improvement against Biotic and Abiotic Stresses Soybean Improvement: Physiological, Molecular and Genetic Perspectives. Springer International Publishing, Pakistan, pp. 241–274. https://doi.org/10.1007/978-3-031-12232-3_13.

Barrett, C.B., 2021. Overcoming global food security challenges through science and solidarity. Am. J. Agric. Econ. 103 (2), 422–447. https://doi.org/10.1111/ajae.12160.

Bhagtaney, L., Sundarrajan, P., 2023. An Overview of Tools for Genome Editing: ZFNs, Mega Nucleases, and Talens CRISPR/Cas-Mediated Genome Editing in Plants. Apple Academic Press, India, pp. 37–64. https://doi.org/10.1201/9781003331759-3.

Bhardwaj, A., Nain, V., 2021. TALENs—an indispensable tool in the era of CRISPR: a mini review. J. Genet. Eng. Biotechnol. 19 (1), 125. https://doi.org/10.1186/s43141-021-00225-z.

Boettcher, M., McManus, M.T., 2015. Choosing the right tool for the job: RNAi, TALEN, or CRISPR. Mol. Cell 58 (4), 575–585. https://doi.org/10.1016/j.molcel.2015.04.028.

Brokowski, C., Adli, M., 2019. CRISPR ethics: moral considerations for applications of a powerful tool. J. Mol. Biol. 431 (1), 88–101. https://doi.org/10.1016/j.jmb.2018.05.044.

Caplan, A.L., Parent, B., Shen, M., Plunkett, C., 2015. No time to waste - the ethical challenges created by CRISPR: CRISPR/Cas, being an efficient, simple, and cheap technology to edit the genome of any organism, raises many ethical and regulatory issues beyond the use to manipulate human germ line cells. EMBO Rep. 16 (11), 1421–1426. https://doi.org/10.15252/embr.201541337.

Cermak, T., Starker, C.G., Voytas, D.F., 2014. Efficient design and assembly of custom TALENs using the Golden Gate platform. In: Chromosomal Mutagenesis. Springer, pp. 133–159.

Chen, F., Zhang, C., Xue, J., Wang, F., Li, Z., 2024. Molecular mechanism for target RNA recognition and cleavage of Cas13h. Nucleic Acids Res. 52 (12), 7279–7291. https://doi.org/10.1093/nar/gkae324.

Choudhary, J.R., Bhavyasree, R.K., Sheoran, S., Choudhary, M., Chandra, S., Kaswan, V., Wani, S.H., 2023. Forward and reverse genetics in crop breeding. Advanced Crop Improvement: Volume 1: Theory and Practice 1, 257–275. https://doi.org/10.1007/978-3-031-28146-4_10.

Citaristi, I., 2022. World food programme—WFP. In: The Europa Directory of International Organizations 2022. Routledge, pp. 302–306.

Courtier-Orgogozo, V., 2025. Risks associated with CRISPR homing gene drive. C. R. Biol. 348 (G1), 211–227. https://doi.org/10.5802/crbiol.182.

Cox, D.B., Gootenberg, J.S., Abudayyeh, O.O., Franklin, B., Kellner, M.J., Joung, J., Zhang, F., 2017. RNA editing with CRISPR-Cas13. Science 358 (6366), 1019–1027.

Csörgö, B., León, L.M., Chau-Ly, I.J., Vasquez-Rifo, A., Berry, J.D., Mahendra, C., Bondy-Denomy, J., 2020. A compact Cascade–Cas3 system for targeted genome engineering. Nat. Methods 17 (12), 1183–1190.

Cui, Z., Liu, H., Zhang, H., Huang, Z., Tian, R., Li, L., Fan, W., Chen, Y., Chen, L., Zhang, S., Das, B.C., Severinov, K., Hitzeroth, I.I., Debata, P.R., Jin, Z., Liu, J., Huang, Z., Xie, W., Xie, H., Lang, B., Ma, J., Weng, H., Tian, X., Hu, Z., 2021. The comparison of ZFNs, TALENs, and SpCas9 by GUIDE-seq in HPV-targeted gene therapy. Mol. Ther. Nucleic Acids 26, 1466–1478. https://doi.org/10.1016/j.omtn.2021.08.008.

Daboussi, F., Stoddard, T.J., Zhang, F., 2015. Engineering meganuclease for precise plant genome modification. In: Advances in New Technology for Targeted Modification of Plant Genomes. Springer, pp. 21–38.

Danilo, B., Montes, É., Archambeau, H., Lodé, M., Rousseau-Gueutin, M., Chèvre, A.M., Mazier, M., 2022. I-SceI and customized meganucleases-mediated genome editing in tomato and oilseed rape. Transgenic Res. 31 (1), 87–105. https://doi.org/10.1007/s11248-021-00287-2.

Deng, X., Osikpa, E., Yang, J., Oladeji, S.J., Smith, J., Gao, X., Gao, Y., 2023. Structural basis for the activation of a compact CRISPR-Cas13 nuclease. Nat. Commun. 14 (1). https://doi.org/10.1038/s41467-023-41501-5.

Doxzen, K., Halpern, J., 2020. Focusing on Human Rights: a framework for CRISPR germline genome editing ethics and regulation. Perspect. Biol. Med. 63 (1), 44–53. https://doi.org/10.1353/pbm.2020.0003.

Dupuy, A., Valton, J., Leduc, S., Armier, J., Galetto, R., Gouble, A., Lebuhotel, C., Stary, A., Pâques, F., Duchateau, P., Sarasin, A., Daboussi, F., Xu, S.-yong, 2013. Targeted gene therapy of xeroderma pigmentosum cells using meganuclease and TALEN™. PLoS One 8 (11), e78678. https://doi.org/10.1371/journal.pone.0078678.

Durrant, M.G., Perry, N.T., Pai, J.J., Jangid, A.R., Athukoralage, J.S., Hiraizumi, M., McSpedon, J.P., Pawluk, A., Nishimasu, H., Konermann, S., Hsu, P.D., 2024. Bridge RNAs direct programmable recombination of target and donor DNA. Nature 630 (8018), 984–993. https://doi.org/10.1038/s41586-024-07552-4.

Dutta, S., Mondal, D., Bhowmick, R., Bhattacharjee, S., 2025. Exploring the Diversity: Variants of the CRISPR-Cas System. In: Principles and Applications of Genome Editing in Field, Horticultural and Forest Crops. Iterative International Publishers, ISBN 978-93-7020-910-7.

Fatollahi Arani, S., Zeinoddini, M., 2023. Gene editing: biosecurity challenges and risks. Journal of Police Medicine 12 (1), 1–19.

Fernández Ríos, D., Benítez Candia, N., Soerensen, M.C., Goberna, M.F., Arrúa, A.A., 2024. Regulatory landscape for new breeding techniques (NBTs): insights from Paraguay. Front. Bioeng. Biotechnol. 12, 1332851.

Fino, E., Montagna, D.M., Nocera, G., Amendola, A., 2020. Spotlight on biotechnology: from a technological breakthrough to valuation challenges. A case study on crispr therapeutics and crispr/cas 9 revolution. SSRN. https://doi.org/10.2139/ssrn.3709239.

Gatton, M.L., Gallegos, D., 2023. A 5-year review of prevalence, temporal trends and characteristics of individuals experiencing moderate and severe food insecurity in 34 high income countries. BMC Public Health 23 (1). https://doi.org/10.1186/s12889-023-17139-9.

Gregorowius, D., Biller-Andorno, N., Deplazes-Zemp, A., 2017. The role of scientific self-regulation for the control of genome editing in the human germline: the lessons from the Asilomar and the Napa meetings show how self-regulation and public deliberation can lead to regulation of new biotechnologies. EMBO Rep. 18 (3), 355–358.

Gupta, S., Kumar, A., Patel, R., Kumar, V., 2021. Genetically modified crop regulations: scope and opportunity using the CRISPR-Cas9 genome editing approach. Mol. Biol. Rep. 48 (5), 4851–4863. https://doi.org/10.1007/s11033-021-06477-9.

Gyngell, C., Bowman-Smart, H., Savulescu, J., 2019. Moral reasons to edit the human genome: picking up from the Nuffield report. Journal of medical ethics 45 (8), 514–523.

Harrington, L.B., Burstein, D., Chen, J.S., Paez-Espino, D., Ma, E., Witte, I.P., Cofsky, J.C., Kyrpides, N.C., Banfield, J.F., Doudna, J.A., 2018. Programmed DNA destruction by miniature CRISPR-Cas14 enzymes. Science 362 (6416), 839–842. https://doi.org/10.1126/science.aav4294.

He, W., Li, X.X., Qian, Q., Shang, L., 2025. The developments and prospects of plant super-pangenomes: demands, approaches, and applications. Plant Commun. 6 (2). https://doi.org/10.1016/j.xplc.2024.101230.

Hillary, V.E., Ceasar, S.A., 2023. A review on the mechanism and applications of CRISPR/Cas9/Cas12/Cas13/Cas14 proteins utilized for genome engineering. Mol. Biotechnol. 65 (3), 311–325. https://doi.org/10.1007/s12033-022-00567-0.

Hiraizumi, M., Perry, N.T., Durrant, M.G., Soma, T., Nagahata, N., Okazaki, S., Athukoralage, J.S., Isayama, Y., Pai, J.J., Pawluk, A., Konermann, S., Yamashita, K., Hsu, P.D., Nishimasu, H., 2024. Structural mechanism of bridge RNA-guided recombination. Nature 630 (8018), 994–1002. https://doi.org/10.1038/s41586-024-07570-2.

Hoersten, J., Ruiz-Gómez, G., Paszkowski-Rogacz, M., Gilioli, G., Guillem-Gloria, P.M., Lansing, F., Pisabarro, M.T., Buchholz, F., 2024. Engineering spacer specificity of the Cre/loxP system. Nucleic Acids Res. 52 (13), 8017–8031. https://doi.org/10.1093/nar/gkae481.

Husaini, A.M., Sohail, M., 2024. Agrochemical-free genetically modified and genome-edited crops: towards achieving the United Nations sustainable development goals and a 'greener' green revolution. J. Biotechnol. 389, 68–77. https://doi.org/10.1016/j.jbiotec.2024.04.015.

Jain, S., Shukla, S., Yang, C., Zhang, M., Fatma, Z., Lingamaneni, M., Abesteh, S., Lane, S.T., Xiong, X., Wang, Y., Schroeder, C.M., Selvin, P.R., Zhao, H., 2021. TALEN outperforms Cas9 in editing heterochromatin target sites. Nat. Commun. 12 (1). https://doi.org/10.1038/s41467-020-20672-5.

Jiang, F., Doudna, J.A., 2017. CRISPR-Cas9 structures and mechanisms. Annu. Rev. Biophys. 46, 505–529. https://doi.org/10.1146/annurev-biophys-062215-010822.

Kalinina, N.O., Khromov, A., Love, A.J., Taliansky, M.E., 2020. CRISPR applications in plant virology: virus resistance and beyond. Federation Phytopathology 110 (1), 18–28. https://doi.org/10.1094/PHYTO-07-19-0267-IA.

Kang, X.J., Caparas, C.I.N., Soh, B.S., Fan, Y., 2017. Addressing challenges in the clinical applications associated with CRISPR/Cas9 technology and ethical questions to prevent its misuse. Protein Cell 8 (11), 791–795. https://doi.org/10.1007/s13238-017-0477-4.

Kato, K., Zhou, W., Okazaki, S., Isayama, Y., Nishizawa, T., Gootenberg, J.S., Abudayyeh, O.O., Nishimasu, H., 2022. Structure and engineering of the type III-E CRISPR-Cas7-11 effector complex. Cell 185 (13), 2324. https://doi.org/10.1016/j.cell.2022.05.003.

Klap, C., Yeshayahou, E., Bolger, A.M., Arazi, T., Gupta, S.K., Shabtai, S., Usadel, B., Salts, Y., Barg, R., 2017. Tomato facultative parthenocarpy results from SlAGAMOUS-LIKE 6 loss of function. Plant Biotechnol. J. 15 (5), 634–647. https://doi.org/10.1111/pbi.12662.

Koonin, E.V., Makarova, K.S., Zhang, F., 2017. Diversity, classification and evolution of CRISPR-Cas systems. Curr. Opin. Microbiol. 37, 67–78.

Kurth, H.M., Mochizuki, K., 2009. Non-coding RNA: a bridge between small RNA and DNA. RNA Biol. 6 (2). https://doi.org/10.4161/rna.6.2.7792.

Li, S., Xia, L., 2020. Precise gene replacement in plants through CRISPR/Cas genome editing technology: current status and future perspectives. aBIOTECH 1 (1), 58–73. https://doi.org/10.1007/s42994-019-00009-7.

Li, J., Li, Y., Ma, L., 2021. Recent advances in CRISPR/Cas9 and applications for wheat functional genomics and breeding. aBIOTECH 2 (4), 375–385.

Liu, X., Wu, S., Xu, J., Sui, C., Wei, J., 2017. Application of CRISPR/Cas9 in plant biology. Acta Pharm. Sin. B 7 (3), 292–302. https://doi.org/10.1016/j.apsb.2017.01.002.

Liu, J.J., Orlova, N., Oakes, B.L., Ma, E., Spinner, H.B., Baney, K.L.M., Chuck, J., Tan, D., Knott, G.J., Harrington, L. B., Al-Shayeb, B., Wagner, A., Brötzmann, J., Staahl, B.T., Taylor, K.L., Desmarais, J., Nogales, E., Doudna, J.A., 2019. CasX enzymes comprise a distinct family of RNA-guided genome editors. Nature 566 (7743), 218–223. https://doi.org/10.1038/s41586-019-0908-x.

Liu, Q., Lin, L., He, S., Yu, J., Xie, C., Gai, C., Han, Y., Liu, C., Huang, F., Chen, D., Song, Y., Qin, G., Zeng, R., 2025a. A simple and efficient TALEN system for genome editing in plants. Plant Mol. Biol. 115 (1). https://doi.org/10.1007/s11103-025-01551-3.

Liu, Y., Zhang, S., Hu, C., 2025b. Cas7 Meets Cas14: A Strategic Partnership in the Type VII CRISPR-Cas. Oxford University Press US.

Malzahn, A., Lowder, L., Qi, Y., 2017. Plant genome editing with TALEN and CRISPR. Cell Biosci. 7 (1), 21.

Manikishore, M., Maurya, S.K., Rathee, S., Patil, U.K., 2025. Genome editing approaches using zinc finger nucleases (ZFNs) for the treatment of motor neuron diseases. Curr. Pharm. Biotechnol. 26 (10), 1514–1531. https://doi.org/10.2174/0113892010307288240526071810.

McLellan, M.A., Rosenthal, N.A., Pinto, A.R., 2017. Cre-loxP-mediated recombination: general principles and experimental considerations. Current protocols in mouse biology 7 (1), 1–12.

Modrzejewski, D., Hartung, F., Lehnert, H., Sprink, T., Kohl, C., Keilwagen, J., Wilhelm, R., 2020. Which factors affect the occurrence of off-target effects caused by the use of CRISPR/cas: a systematic review in plants. Front. Plant Sci. 11. https://doi.org/10.3389/fpls.2020.574959.

Mukhametzyanova, L., Schmitt, L.T., Torres-Rivera, J., Rojo-Romanos, T., Lansing, F., Paszkowski-Rogacz, M., Hollak, H., Brux, M., Augsburg, M., Schneider, P.M., Buchholz, F., 2024. Activation of recombinases at specific DNA loci by zinc-finger domain insertions. Nat. Biotechnol. 42 (12), 1844–1854. https://doi.org/10.1038/s41587-023-02121-y.

Mulvihill, J.J., Capps, B., Joly, Y., Lysaght, T., Zwart, H.A., Chadwick, R., 2017. International human genome organisation (HUGO) committee of ethics, law, and society (CELS). Ethical issues of CRISPR technology and gene editing through the lens of solidarity. Br. Med. Bull. 122 (1), 17–29.

Munawar, N., Ahmad, A., 2022. An Introduction to Genome Editing Techniques the CRISPR/Cas Tool Kit for Genome Editing. Springer Nature, United Arab Emirates, pp. 1–28. https://doi.org/10.1007/978-981-16-6305-5_1.

Munir, A., Ali, M., Qari, S.H., Munawar, N., Saleem, M.S., Ahmad, A., 2024a. CRISPR workflow solutions: cargos and versatile delivery platforms in genome editing. In: CRISPRized Horticulture Crops. Elsevier, pp. 67–90.

Munir, A., Asma, N., Ashraf, S., Zaini, P., Munawar, N., Zhang, B., Ahmad, A., 2024b. Advancing metabolic engineering in plants using CRISPR/cas. In: Gene-Edited Crops. CRC Press, pp. 241–256.

Munywoki, G., Munywoki, J., 2021. A SWOT analysis of the implications of CRISPR/cas9 technology in crop production: a review. Agrotechnology 10, 236.

Mustafa, M.D., Mazhar, S., Syed, A., Ali, U., 2025. A mini review on genome editing via CRISPR-CAS system to enhance agronomic traits of wheat and tomato against environmental stresses. Pure and Applied Biology (PAB) 14 (2), 468–478.

Nakamura, K., Ito, S., Ohguchi, Y., Jimbo, T., Wada, Y., Nakajima, R., Kanou, M., Yamana, K., Ueda, H., Asakura, A., 2025. Establishment of Cre/LoxP-mediated multifunctional reporter knock-in rats with the CRISPR system. PLoS One 20 (6), e0325444. https://doi.org/10.1371/journal.pone.0325444.

Ndhlovu, E., Mokone, N., 2025. Leveraging sustainable food systems for achieving Sustainable Development Goals: an overview. International Journal of Research in Business and Social Science (2147- 4478) 14 (5), 265–274. https://doi.org/10.20525/ijrbs.v14i5.4206.

Niewoehner, O., Jinek, M., Doudna, J.A., 2013. Evolution of CRISPR RNA recognition and processing by Cas6 endonucleases. Nucleic Acids Res. 42 (2), 1341–1353.

O'Connell, M.R., 2019. Molecular mechanisms of RNA targeting by Cas13-containing type VI CRISPR–Cas systems. Journal of molecular biology 431 (1), 66–87.

Özcan, A., Krajeski, R., Ioannidi, E., Lee, B., Gardner, A., Makarova, K.S., Koonin, E.V., Abudayyeh, O.O., Gootenberg, J.S., 2021. Programmable RNA targeting with the single-protein CRISPR effector Cas7-11. Nature 597 (7878), 720–725. https://doi.org/10.1038/s41586-021-03886-5.

Pacesa, M., Pelea, O., Jinek, M., 2024. Past, present, and future of CRISPR genome editing technologies. Cell 187 (5), 1076–1100.

Pâques, F., Duchateau, P., 2007. Meganucleases and DNA double-strand break-induced recombination: perspectives for gene therapy. Curr. Gene Ther. 7 (1), 49–66. https://doi.org/10.2174/156652307779940216France.

Pingali, P., 2022. Are the Lessons from the Green Revolution Relevant for Agricultural Growth and Food Security in the Twenty-First Century? Springer Science and Business Media LLC, pp. 21–32. https://doi.org/10.1007/978-981-19-5542-6_2.

Pyzocha, N.K., Chen, S., 2018. Diverse class 2 CRISPR-cas effector proteins for genome engineering applications. ACS Chem. Biol. 13 (2), 347–356. https://doi.org/10.1021/acschembio.7b00800.

Rath, D., Amlinger, L., Rath, A., Lundgren, M., 2015. The CRISPR-Cas immune system: biology, mechanisms and applications. Biochimie 117, 119–128. https://doi.org/10.1016/j.biochi.2015.03.025.

Ren, H., Yin, A., Wu, P., Zhou, H., Zhou, J., Yu, Y., Lu, H., 2022. Establishment of a cre-loxP system based on a leaky LAC4 promoter and an unstable panARS element in kluyveromyces marxianus. Microorganisms 10 (6), 1240. https://doi.org/10.3390/microorganisms10061240.

Sakuma, T., Yamamoto, T., 2017. Current overview of TALEN construction systems. Genome Editing in Animals: Methods Protoc. 25–36.

Sakuma, T., Yamamoto, T., 2023. Updated overview of TALEN construction systems. Genome Editing in Animals: Methods Protoc. 27–39.

Santosh Kumar, V., Verma, R.K., Yadav, S.K., Yadav, P., Watts, A., Rao, M., Chinnusamy, V., 2020. CRISPR-Cas9 mediated genome editing of drought and salt tolerance (OsDST) gene in indica mega rice cultivar MTU1010. Physiol. Mol. Biol. Plants 26 (6), 1099–1110.

Sasnauskas, G., Siksnys, V., 2020. CRISPR adaptation from a structural perspective. Curr. Opin. Struct. Biol. 65, 17–25. https://doi.org/10.1016/j.sbi.2020.05.015.

Savage, D.F., 2019. Cas14: big advances from small CRISPR proteins. Biochemistry 58 (8), 1024–1025. https://doi.org/10.1021/acs.biochem.9b00035.

Schultz-Bergin, M., 2018. Is CRISPR an ethical game changer? J. Agric. Environ. Ethics 31 (2), 219–238. https://doi.org/10.1007/s10806-018-9721-z.

Secomandi, S., Gallo, G.R., Rossi, R., Rodríguez Fernandes, C., Jarvis, E.D., Bonisoli-Alquati, A., Gianfranceschi, L., Formenti, G., 2025. Pangenome graphs and their applications in biodiversity genomics. Nat. Genet. 57 (1), 13–26. https://doi.org/10.1038/s41588-024-02029-6.

Shaltz, S., Jinks-Robertson, S., 2021. Mutagenic repair of a ZFN-induced double-strand break in yeast: effects of cleavage site sequence and spacer size. DNA Repair 108, 103228. https://doi.org/10.1016/j.dnarep.2021.103228.

Shamshirgaran, Y., Liu, J., Sumer, H., Verma, P.J., Taheri-Ghahfarokhi, A., 2022. Tools for efficient genome editing; ZFN, TALEN, and CRISPR. Methods Mol. Biol. 2495, 29–46. https://doi.org/10.1007/978-1-0716-2301-5_2.

Shapiro, B.A., Yingling, Y.G., Kasprzak, W., Bindewald, E., 2007. Bridging the gap in RNA structure prediction. Curr. Opin. Struct. Biol. 17 (2), 157–165. https://doi.org/10.1016/j.sbi.2007.03.001.

Shi, J., Gao, H., Wang, H., Lafitte, H.R., Archibald, R.L., Yang, M., Hakimi, S.M., Mo, H., Habben, J.E., 2017. ARGOS8 variants generated by CRISPR-Cas9 improve maize grain yield under field drought stress conditions. Plant Biotechnol. J. 15 (2), 207–216. https://doi.org/10.1111/pbi.12603.

Silva, G., Poirot, L., Galetto, R., Smith, J., Montoya, G., Duchateau, P., Pâques, F., 2011. Meganucleases and other tools for targeted genome engineering: perspectives and challenges for gene therapy. Curr. Gene Ther. 11 (1), 11–27. https://doi.org/10.2174/156652311794520111.

Sinkunas, T., Gasiunas, G., Fremaux, C., Barrangou, R., Horvath, P., Siksnys, V., 2011. Cas3 is a single-stranded DNA nuclease and ATP-dependent helicase in the CRISPR/Cas immune system. EMBO J. 30 (7), 1335–1342. https://doi.org/10.1038/emboj.2011.41.

Sobeh, E.I., Elfeky, S.A., Korany, R.M.S., Saleh, H.M., Ahmed-Farid, O.A., El-ghannam, G., 2025. Exploring the radioprotective shield of ZnFe2O4/cysteine-folic acid nanocomposite in γ-irradiated male Wistar rats. J. Drug Deliv. Sci. Technol. 112. https://doi.org/10.1016/j.jddst.2025.107297.

Sokolowski, R.D., Graham, S., White, M.F., 2014. Cas6 specificity and CRISPR RNA loading in a complex CRISPR-Cas system. Nucleic Acids Res. 42 (10), 6532–6541.

Tsuchida, C.A., Zhang, S., Doost, M.S., Zhao, Y., Wang, J., O'Brien, E., Fang, H., Li, C.P., Li, D., Hai, Z.Y., Chuck, J., Brötzmann, J., Vartoumian, A., Burstein, D., Chen, X.W., Nogales, E., Doudna, J.A., Liu, J.J.G., 2022. Chimeric CRISPR-CasX enzymes and guide RNAs for improved genome editing activity. Mol. Cell 82 (6), 1199. https://doi.org/10.1016/j.molcel.2022.02.002.

van der Meer, H., 2025. Commercial implications of CRISPR technology: opportunities and ethical challenges. J. Commerc. Biotechnol. 30 (3), 292–302.

Vidya, N., Arun, M., 2023. Updates and applications of CRISPR/cas technology in plants. J. Plant Biol. 66 (6), 499–518. https://doi.org/10.1007/s12374-023-09383-8.

Vizoso, M., Ej Pritchard, C., Bombardelli, L., van den Broek, B., Krimpenfort, P., Beijersbergen, R.L., van Rheenen, J., 2022. A doxycycline-and light-inducible Cre recombinase mouse model for optogenetic genome editing. Nat. Commun. 13 (1), 6442.

Wolt, J.D., Wang, K., Sashital, D., Lawrence-Dill, C.J., 2016. Achieving plant CRISPR targeting that limits off-target effects. Plant Genome 9 (3). https://doi.org/10.3835/plantgenome2016.05.0047.

Yan, F., Wang, W., Zhang, J., 2019. CRISPR-Cas12 and Cas13: the lesser known siblings of CRISPR-Cas9. Cell biology and toxicology 35 (6), 489–492.

Yang, H., Patel, D.J., 2024. Structures, mechanisms and applications of RNA-centric CRISPR–Cas13. Nat. Chem. Biol. 20 (6), 673–688. https://doi.org/10.1038/s41589-024-01593-6.

Zaman, W., Khalil, A.A.K., Amin, A., 2025. TALEN-interceded genome editing in plants: unveiling new frontiers in secondary metabolite improvement and genetic diversity. Plants 14 (19), 3024.

Zeng, Y., Wen, J., Zhao, W., Wang, Q., Huang, W., 2020. Rational improvement of rice yield and cold tolerance by editing the three genes OsPIN5b, GS3, and OsMYB30 with the CRISPR–cas9 system. Front. Plant Sci. 10. https://doi.org/10.3389/fpls.2019.01663.

Zhang, D., Hussain, A., Manghwar, H., Xie, K., Xie, S., Zhao, S., Larkin, R.M., Qing, P., Jin, S., Ding, F., 2020. Genome editing with the CRISPR-Cas system: an art, ethics and global regulatory perspective. Plant Biotechnol. J. 18 (8), 1651–1669. https://doi.org/10.1111/pbi.13383.

Zhou, B., Yang, R., Sohail, M., Kong, X., Zhang, X., Fu, N., Li, B., 2023. CRISPR/Cas14 provides a promising platform in facile and versatile aptasensing with improved sensitivity. Talanta 254, 124120. https://doi.org/10.1016/j.talanta.2022.124120.

Zhu, H., Li, C., Gao, C., 2020. Applications of CRISPR–Cas in agriculture and plant biotechnology. Nat. Rev. Mol. Cell Biol. 21 (11), 661–677.

Zubair, A., Ali, M., Ahmad, F., Ayidh Althobaiti, S., 2025. Unlocking the role of transcription activator-like effector nuclease (TALENs) and zinc finger nuclease (ZFN) in the treatment of HIV. Mol. Biol. Rep. 52 (1). https://doi.org/10.1007/s11033-025-10993-3.

CHAPTER

2

Revolution of CRISPR–Cas: Advancements in genome editing for agriculture

Rakesh Kumar[1], *Joy Das*[3], *K.P. Raghavendra*[2, 3], *Supratik Palchoudhury*[3], *Prosenjit Chakraborty*[3], *Vinita Gotmare*[3], *and V.N. Waghmare*[3]

[1]Division of Crop Improvement, ICAR-Central Institute for Cotton Research, Nagpur, Maharashtra, India; [2]ICAR-National Institute for Research on Commercial Agriculture, Rajahmundry, Andhra Pradesh, India; [3]ICAR-Central Institute for Cotton Research, Nagpur, Maharashtra, India

2.1 Introduction

Food and environmental security are among the most serious challenges the world is facing today. The global population expansion, along with climate variation, evolving biocide resistant pests and pathogens, and diminishing natural resources, puts food security and agricultural sustainability at risk (Liu et al., 2021; Das et al., 2022; Hamdan et al., 2022; Chen et al., 2024). To overcome these challenges, improved agricultural strategies have changed and progressed through time, yet with each new generational technological advancement, new challenges have arisen (Liu et al., 2021; Hamdan et al., 2022; Aljabali et al., 2024; Kumar et al., 2024). The Clustered Regularly Interspaced Short Palindromic Repeats (CRISPR) and CRISPR-associated proteins (Cas), initially identified as essential components of bacterial defense mechanism, have been harnessed as a potent genome editing platform that addressed the fundamental constraints of the predecessor technologies with remarkable efficiency and simplicity (Jinek et al., 2012; Cong et al., 2013; Mali et al., 2013; Kumar et al., 2024). This technological breakthrough revolutionized the domain of genome editing, enabling precise genome modifications easy and accessible to researchers worldwide. Guided RNA (gRNA) based tagging system removed the need for the complex protein

Intelligent Gene Editing in Plants
https://doi.org/10.1016/B978-0-443-44079-3.00021-8

engineering that requires in Transcription Activator-Like Effector Nucleases (TALENs) and Zinc Finger Nucleases (ZFNs) (Christian et al., 2010; Liu et al., 2021; Kumar et al., 2024); thus higher efficiency, better accuracy, and least off-target effects were observed due to optimized gRNA and better Cas variants (Jinek et al., 2012; Mali et al., 2013; Knott and Doudna, 2018). Most importantly, CRISPR enables multiplex gene editing such as modification of several traits at once (Cong et al., 2013). Its newer derivatives such as base editing and prime editing have also increased the flexibility of the system whereby one or few nucleotide deletion, insertion, or substitutions are made and double-strand breaks (DSBs) are not generated (Gaudelli et al., 2017; Anzalone et al., 2019). The CRISPR–Cas technology has demonstrated remarkable success in agricultural applications, facilitating rapid development of crop genotypes with enhanced yields, enriched nutritional profile, increased stress tolerance and climate resilience (Arora and Narula, 2017; Zhu et al., 2020; Chen et al., 2024). It meets the dire need for sustenance by controlling the technical, ethical, or regulatory boundaries associated with older genetic engineering technologies (Turnbull et al., 2021). In this context, this review systematically explores the CRISPR–Cas technology from inception to milestones, underlying molecular mechanisms, and comparative assessment with its predecessor genome editing platforms. Furthermore, the review highlights its agricultural applications and future prospects while addressing prevailing challenges and ethical considerations.

2.2 CRISPR–Cas technology: Brief overview

2.2.1 Origin of CRISPR–Cas from bacterial adaptive immunity

CRISPR was first discovered in *Escherichia coli* genome in 1987 and later identified in archaea, *Haloferax mediterranei*, in 1993 (Ishino et al., 1987; Mojica et al., 1993). During early 2000s, researchers observed similarities between CRISPR spacer sequences and those of bacteriophages, plasmids, and archeal viruses. This pointed to its role in adaptive immunity (Mojica et al., 2000, 2005; Bolotin et al., 2005; Pourcel et al., 2005). These discoveries showed that CRISPR–Cas systems help prokaryotes defend against the invading hostile genetic elements by incorporating short DNA segments in the CRISPR assembly (Makarova et al., 2006; Barrangou et al., 2007). The term CRISPR was coined in 2002 to describe arrays made up of short, conserved repeats separated by variable spacers. These are located in intergenic regions and often have a common leader sequence (Jansen et al., 2002). Researchers identified associated *cas* genes as crucial parts of the system. Later studies in comparative genomics confirmed that CRISPR–Cas functions as an RNA-guided adaptive immune machinery in prokaryotes. This process shares functional analogy with RNA interference mechanism observed in eukaryotes (Mojica et al., 2005; Makarova et al., 2006).

2.2.2 CRISPR–Cas system: Classification and mechanism

CRISPR–Cas are categorized into two classes based on their number of effector protein molecules. Class 1 CRISPR–Cas functions through multiple protein complexes, whereas class 2 assembly operates via a single effector protein for genome targeting (Makarova et al., 2015; Shmakov et al., 2017). Each class of assembly contains three types, i.e., class 1 with I (Cas3 Cascade complex), III (Cas7-11 Cascade complex), IV (Cas7 type genes) types and class 2 with II (Cas9), V (Cas12), VI (Cas13) types. Due to its simpler structure, class 2 systems, especially Cas9, Cas12, and Cas13, have become popular tools for genome modification (Ishino et al., 2018; Pickar-Oliver and Gersbach, 2019). Among these, CRISPR–Cas9 is the most commonly used system, with SpCas9 (*Streptococcus pyogenes* Cas9) being the first variant deployed for genome editing in eukaryotes, including human and mouse cells (Jinek et al., 2012; Cong et al., 2013; Mali et al., 2013).

The CRISPR–Cas9 system includes three essential components, namely Cas9 nuclease, CRISPR RNA (crRNA) and trans-activating crRNA (tracrRNA) (Fig. 2.1), of which the two RNA molecules combine to form a single-guide RNA (sgRNA) that keeps CRISPR-Cas9 functional, allowing efficient genome editing (Jinek et al., 2012; Mali et al., 2013). The sgRNA directs Cas9 protein to a complementary DNA sequence next to a three nucleotide Protospacer Adjacent Motif (PAM), typically 5′-NGG-3′. This positioning enables Cas9 to unwind the DNA and start the cleavage process (Jinek et al., 2012; Cong et al., 2013; Sternberg et al., 2014). The Cas9 protein consists of two functional domains, namely the recognition (REC) lobe containing REC1 and REC2 that interact with sgRNA, and the nuclease

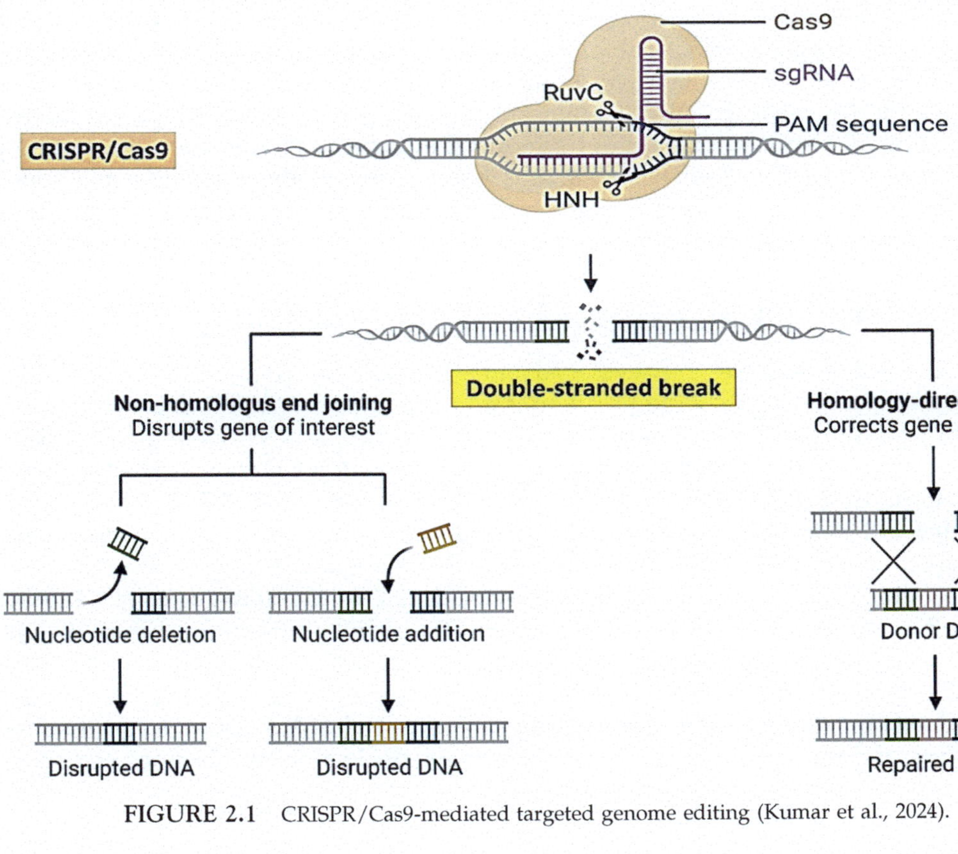

FIGURE 2.1 CRISPR/Cas9-mediated targeted genome editing (Kumar et al., 2024).

(NUC) lobe comprising HNH, RuvC nuclease domain, and PAM-interacting (PI) domain (Nishimasu et al., 2014; Ishino et al., 2018). The HNH domain nicks the DNA strand complementary to crRNA, while RuvC cleaves the non-target strand, causing double-strand break (DSB) (Mojica et al., 2009; Jinek et al., 2012; Cong et al., 2013; Mali et al., 2013).

These DSBs induced by Cas9 are repaired either through template-independent non-homologous end joining (NHEJ) or via homology-directed repair (HDR) in presence of donor template (Cong et al., 2013). NHEJ is a rapid but error-prone repair mechanism that reconnects broken DNA ends causing insertions or deletions (indels). These indels often results in frame-shift or nonsense mutations, making NHEJ particularly amenable for gene knockout applications (Cong et al., 2013; Mali et al., 2013). In contrast, HDR is a high-fidelity DNA repair process that employs a homologous donor template sequence to introduce precise gene correction or target sequence insertion at specific genomic loci (Fig. 2.1) (Sun et al., 2016; Knott and Doudna, 2018; Pickar-Oliver and Gersbach, 2019; Liu et al., 2019; Hamdan et al., 2022).

The CRISPR–Cas technology has seen substantial advancements with the development of various Cas variants, each providing different mechanisms and uses. xCas9, an engineered version of Cas9, has broader PAM compatibility and improved DNA specificity, allowing for more flexible and precise genome modification (Hu et al., 2018; Hossain et al., 2025). dCas9, a non-active form of Cas9, can still bind to DNA without cutting, making it useful for transcription regulation, epigenetic changes, and gene silencing (Cai et al., 2023). Cas12 (Cpf1) is another DNA-targeting nuclease that exhibits unique PAM specificity and ability to generate staggered cuts, enabling multiplex-genome editing and precise gene insertions (Zetsche et al., 2015; Kleinstiver et al., 2019). Conversely, Cas13 nuclease targets single-stranded RNA, offering versatile applications in RNA editing, transcriptome modulation, and viral RNA detection without altering the genome (Abudayyeh et al., 2016; Yang and Patel, 2024). Together, these variants have increased the versatility, precision, and functionality of CRISPR-based genome and transcriptome engineering (Hillary and Ceasar, 2023; Gupta et al., 2022; Cai et al., 2023).

2.2.3 Merits of CRISPR–Cas over other gene editing methods

CRISPR–Cas technology presents several advantages over other genome editing tools like ZFNs and TALENs, driving its widespread adoption and impact. CRISPR–Cas systems are much easier to design and use since specificity relies on simple RNA-DNA pairing. This removes the need for the complex protein design needed for ZFNs or TALENs (Jinek et al., 2012; Cong et al., 2013; Mali et al., 2013; Kumar et al., 2024). CRISPR–Cas also shows higher editing efficiency and accuracy, which results in more predictable outcomes and fewer off-target effects, especially with improved gRNA design and high-fidelity Cas variants (Jinek et al., 2012; Cong et al., 2013; Kumar et al., 2024). The lower cost of CRISPR-Cas makes it more accessible to a wider scientific community, supporting research and development in various applications (Mali et al., 2013; Aljabali et al., 2024). Researchers can modify several genes concurrently using multiple guide RNAs, allowing for changes to complex traits or whole biochemical pathways (Cong et al., 2013). Moreover, advancements like base and prime editing enable accurate single-nucleotide substitutions and small

TABLE 2.1 Comparative features of crop improvement technologies.

Parameter	Traditional breeding	Transgenic technologies (GMOs)	ZFNs & TALENs	CRISPR–Cas (incl. Base/Prime editing)
Precision/ Targeting	Low (random recombination)	Low (random insertion)	Moderate (site-specific, protein-based)	High (sequence-specific, RNA-guided)
Design complexity	Not applicable (natural process)	Moderate to high	High (complex protein engineering)	Low (simple RNA design)
Cost	Resource-intensive (land, labor)	High (development & regulatory)	High	Relatively low
Efficiency	Variable, unpredictable	Variable stable integration	Moderate	High
Speed to market/ Development time	Very long (multiple generations)	Long (stable lines, regulatory approval)	Long (design & validation)	Short (single generation, rapid modification)
Multiplexing	Limited (random)	Limited	Difficult	High (multiple gRNAs)
Off-target effects	Not applicable (natural variation)	Possible (random insertion)	Minimized (optimized design)	Possible (gRNA mismatches)
Regulatory status/ Public perception	High acceptance (natural)	Stringent regulation, public skepticism	Less defined, often treated as GMOs	Varied (often "non-GMO" for SDN1/SDN2), growing acceptance

indels, without causing DSBs. These innovations have expanded the application of CRISPR-Cas system beyond standard gene editing applications, leading to improved precision and fewer off-target effects (Gaudelli et al., 2017; Anzalone et al., 2019). Genome editing via CRISPR–Cas exhibits great versatility, making it adept to diverse organisms and cell types (Stella and Montoya, 2016; Zhu et al., 2020). Table 2.1 summarizes key differences among contemporary crop improvement approaches, ranging from conventional breeding to advanced genome editing tools.

2.3 Chronicles of genome editing (from DNA double helix to CRISPR–Cas)

The journey of genome editing reflects human curiosity and technological progress. It has evolved from basic discoveries in molecular biology to the precise tools we have today.

2.3.1 Elucidation of DNA structure (1953)

The Watson and Crick's model of DNA double-helix structure was fundamentally derived from X-ray diffraction data elucidated by Rosalind Franklin. This crucial moment in understanding DNA structure along with its role as genetic material set the stage for all future advancements in genetic engineering (Watson and Crick, 1953; Maddox, 2003).

2.3.2 Discovery of restriction enzymes (1960s–1970s)

Werner Arber's hypothesis about bacteria-controlled defense against invading phage DNA and subsequent isolation of restriction enzymes from *E. coli* between 1963 and 1974 were key advancements in molecular biology. These "molecular scissors" can recognize and cut foreign DNA at specific sites. Another modification enzymes protect the host genome by blocking self-DNA cleavage. This discovery was essential for developing recombinant DNA technology (Smith and Wilcox, 1970; Arber, 1974).

2.3.3 Genetic engineering and transgenics (1970s–1990s)

The first instances of recombinant DNA technology emerged in the 1970s when biologists constructed recombinant DNA molecules for the first time due to their ability to use plasmid as vectors to carry gene of interest (Cohen et al., 1973). Genetic engineering was first initiated when organisms were able to transfer genes cross the reproductive barriers system (Cohen et al., 1973; Berg et al., 1975). By the 1980s and 1990s, this advances led to the development of genetically modified (GM) crops, wherein exogenous genes (transgenes) were introduced to plant genomes to confer improved traits such as pest resistance and herbicide tolerance. In 1994, the "Flavr Savr" tomato became the first GM crop to be commercialized (Bruening and Lyons, 2000). GM crops especially cotton, soybean, canola, and maize have contributed more than 186 billion US Dollars to global farm incomes (Brookes and Barfoot, 2018). However, GM crops face problems, including expensive and long regulatory scrutiny of foreign DNA, public resistance in many countries (Turnbull et al., 2021). On the other hand, CRISPR–Cas based genome editing has considerably alleviated regulatory barriers and ethical constraints, thereby accelerating the development and commercialization of superior crop genotypes compared to lengthy transgenic development pipelines (Turnbull et al., 2021).

2.3.4 Development of early gene editing tools (ZFNs and TALENs - 2000s)

Transgenic technology made it possible to add new genes, but it often caused random incorporation of DNA segments in the host genome. This led to unpredictable expression patterns and possible off-target effects. Such challenges drove the creation of more precise gene editing tools. In the early 2000s, ZFNs represent the first generation of programmable nucleases capable of inducing DSBs at specific DNA sites. Structurally, ZFNs consist of a DNA-binding zinc finger domain and a *Fok*I nuclease (Kim et al., 1996). Around 2010, TALENs were developed and offered better specificity than ZFNs. TALENs used engineered DNA-binding domains from TAL effectors, for recognition of specific sequences, paired with the *Fok*I nuclease (Christian et al., 2010). Later, fragrant rice variety was developed by disrupting the *OsBADH2* gene using TALEN (Shan et al., 2015). However, both ZFNs and TALENs needs complex protein engineering for each target sites, limiting their widespread adoption (Liu et al., 2021; Aljabali et al., 2024; Kumar et al., 2024).

2.3.5 The revolutionary discovery of CRISPR–Cas (2012)

A landmark breakthrough in genome editing occurred in 2012 when Jennifer Doudna and Emmanuelle Charpentier repurposed the CRISPR–Cas system for precise, programmable genetic modifications (Jinek et al., 2012). This system, which originated as a bacterial immune mechanism, offered unmatched simplicity, efficiency, and economical sustainability compared to earlier techniques (Jinek et al., 2012; Cong et al., 2013; Das et al., 2022). Its design allows custom guide RNAs to be programmed to target nearly any DNA sequence. The advancement revolutionized genome editing, earning Doudna and Charpentier the 2020 Nobel Prize in Chemistry. Subsequently, CRISPR–Cas technology witnessed rapid expansion in applied research including therapeutics and agriculture (Stella and Montoya, 2016; Ishino et al., 2018). CRISPR–Cas has since emerged as a leading tool in genome modification platform, recognized for its operational simplicity and wide applicability across diverse biological systems (Knott and Doudna, 2018).

2.4 Applications in crop improvement

CRISPR–Cas technology has greatly influenced crop improvement. It provides precise tools to enhance various traits and tackle agricultural problems.

2.4.1 Enhancing yield and quality

CRISPR–Cas applications have shown significant potential in improving both yield and quality. This leads to more productive and appealing crops. Boosting photosynthetic efficiency is essential for crop yield. CRISPR–Cas technology can reprogram key enzymes involved in photosynthesis. For example, improving photosynthesis can occur by introducing elements that avoid photorespiration or by reprogramming RuBisCO through CRISPR-mediated DNA changes. These methods can enhance photosynthesis and biomass (Das et al., 2022; Yamori et al., 2025). Recent studies target genes like *OsHXK1* to improve photosynthetic efficiency and yield in rice through genome editing (Zheng et al., 2021). Efficient nutrient uptake is also crucial for crop growth and yield. The CRISPR–Cas can be applied to optimize genes involved in nutrient uptake from the soil. In rice, editing of cis-regulatory element of *OsNAS* gene has increased iron and zinc accumulation, leading reduced micronutrient deficiencies (Ludwig et al., 2024). Knocking out *ZmHMA3* gene deteriorated zinc tolerance in maize by enhancing permeability of cell membrane and accumulating metal ions (Lv et al., 2023). Inducing an indel mutation in the fatty acid desaturase (*GhFAD2-1*) gene created transgene-free high-oleic acid containing mutants in cotton (Chen et al., 2021).

Lodging, or the permanent bending of the stem, can reduce crop yield. While the snippets do not detail specific CRISPR methods for lodging resistance, genome editing can target genes related to stem strength and structure to help prevent this. CRISPR–Cas can modify genes affecting sensory qualities of crops. Base editing has been used to fine-tune the flavor profiles of peas and tomatoes by extending shelf life of perishable horticultural produce, thereby minimizing postharvest losses (Chen et al., 2024). In tomatoes, CRISPR–Cas9 induced knockout of *SlPL* gene led to healthier fruits and longer shelf life without compromising taste or nutrition (Uluisik et al., 2016; Wang et al., 2019). Similarly, targeted modification of *MaACO1* gene supressed ethylene production in bananas, resulting in delayed ripening, increased vitamin C and sugar content while extending shelf life (Hu et al., 2021).

2.4.2 Improving abiotic stress tolerance

Abiotic stresses like salinity, osmotic, cold, and hyperthermal stress critically lower agricultural productivity. CRISPR–Cas offers effective solutions to improve crop tolerance against abiotic stress. This technology allows for targeted genetic edits that increase drought tolerance by focusing on genes that control water absorption and transfer in plants. For example, modifying the *ZmHDT103* gene in maize has decreased water loss rate (Wang et al., 2024). Editing the *TaRPK1* gene can improve water uptake efficiency in wheat (Rahim et al., 2023). Additionally, manipulating the *Sal1* gene can improve drought tolerance by increasing osmoprotectant production, like proline (Mohr et al., 2022). Salinity is one of the major constraints for crop growth in coastal areas. CRISPR–Cas9 has been applied to modify *OsRR22* gene in rice, leading to enhanced salt tolerance (Xu et al., 2022). Extreme temperature conditions can disrupt plant development and often damages crops. For instance, application of CRISPR–Cas technology helps in modifying heat shock proteins and associated factors to improve heat tolerance, whereas, overexpression of *HsfA1b* increased heat resilience in wheat (Tian et al., 2019). Editing genes like *bHLH57* in rice has proved effective in combating against chilling stress and improved yield (Zhang et al., 2023).

2.4.3 Enhancing pest and disease resistance (biotic stress)

Biotic stresses from pests, viruses, bacteria, and fungi cause significant crop losses. These challenges can be effectively addressed through CRISPR–Cas mediated development of improved crop varieties. In particular, Cas13 effectively targets RNA genomes of viruses, such as sweet potato chlorotic stunt virus blocking their reproduction (Yu et al., 2022). On the other hand, CRISPR–Cas9 based modification has boosted viral resistance by targeting host susceptibility-related genes, like *TaPDIL5-1* in wheat, leading to virus resistance (Kan et al., 2022). Similarly, disrupting *ZmPDRP1/2* gene in maize has limited the replication and spread of the sugarcane mosaic virus (Xie et al., 2024). In rice, CRISPR–Cas9 driven editing of *OsSWEET11/13/14* or *OsPUB9* gene has enhanced resistance against bacterial blight in rice (Kim et al., 2024; Oliva et al., 2019). Multi-allelic mutations introduced in *MLO* gene via CRISPR–Cas9 conferred resistance to powdery mildew disease in wheat (Wang et al., 2014). Targeted changes to the *GmUGT* gene have strengthened soybean defenses against phytophagous insects (Zhang et al., 2022). In cotton, knocking out the *G. hirsutum14-3-3d* gene has improved resistance to *Verticillium dahliae* (Zhang et al., 2018b).

2.4.4 Biofortification

Biofortification aims to improve a crop's nutritional profile and application of CRISPR–Cas based gene modification might be a major avenue in combating malnutrition. In rice, CRISPR–Cas based gene insertion increased pro-vitamin A levels by boosting carotenoid accumulation, offering a promising way to tackle vitamin A deficiency (Dong et al., 2020). It has also been proven useful in improving essential amino acid profiles; for example, CRISPR–Cas9-based editing of the *LKR/SDH* genes significantly boosted lysine levels in rice (Rastogi et al., 2025). Since iron deficiency is a leading cause of anemia, biofortification efforts have targeted genes that regulate iron metabolism. Modification of *OsNRAMP2* gene in rice enhanced reallocation of iron (Chang et al., 2022). In wheat, targeted editing of *TaIPK1* gene lowered phytic acid levels while boosting iron and zinc in the grains (Ibrahim et al., 2022). Furthermore, low-gluten varieties of wheat with potential health benefits were developed through CRISPR–Cas9-based gene modification of *α-gliadin* gene (Sánchez-León et al., 2018).

2.5 Advantages of CRISPR–Cas over traditional breeding and transgene technologies

CRISPR–Cas has many benefits over traditional breeding and former transgenic technologies. It stands out in terms of specificity, swiftness, accuracy, and economic sustenance. CRISPR–Cas provides exceptional precision in modifying gene with the help of a complementary RNA molecule (Jinek et al., 2012; Cong et al., 2013; Mali et al., 2013). This approach reduces off-target mutations that occur in places beyond the intended target (Fu et al., 2013; Li et al., 2019; Han et al., 2020; Spasskaya et al., 2023).

Traditional breeding relies on random genetic mixing through sexual reproduction. This makes it arduous to incorporate desirable traits without the addition of undesirable one (Hamdan et al., 2022). Mutational breeding, which uses chemical or physical mutagens to induce random mutations in plants, often causes widespread off-target effects and requires extensive screening of large populations to identify individuals with desirable traits (Hamdan et al., 2022). Transgenic technologies, or GMOs, often integrate foreign DNA randomly into the host genome. This can result in unpredictable gene expressions, gene silencing, or interference with other genes (Hamdan and Tan, 2024; Chen et al., 2024). Although older sequence-specific nuclease (SSN)-based tools like ZFNs and TALENs provided more specificity than random insertion, they were not as precise, and were more costly and complex than CRISPR–Cas (Zhang et al., 2018a; Das et al., 2022; Aljabali et al., 2024).

CRISPR–Cas speeds up the breeding process significantly. Genetic modifications can be achieved within one or two generations, enabling rapid development of improved cultivars compared to traditional breeding (Chen et al., 2019; Das et al., 2022; Sampath et al., 2023). Traditional breeding is slow and labor-intensive, often requiring several generations of crossing and backcrossing to get the right trait combinations and eliminate unwanted DNA (Chen et al., 2019; Das et al., 2022; Hamdan et al., 2022).

Although transgenic technologies are quicker than traditional methods, developing stable transgenic lines and securing regulatory approval with consumer acceptance can still take a long time (Hamdan et al., 2022; Hamdan and Tan, 2024). CRISPR–Cas is quite cost-effective due to its straightforward design. This makes it accessible to more research institutions and possibly smaller breeding programs (Chen et al., 2019, 2024; Liu et al., 2021). Traditional breeding can use a lot of resources, requiring large areas of land and considerable labor over many generations. Transgenic technologies come with high development and regulatory costs, limiting their use to a few major crops and large companies (Hamdan et al., 2022; Hamdan and Tan, 2024).

CRISPR–Cas shows high editing efficiency, which results in a better success rate for achieving desired genetic modifications (Jinek et al., 2012; Cong et al., 2013; Mali et al., 2013; Knott and Doudna, 2018). New CRISPR variants like base and prime editing have enhanced editing efficiency and expanded the possibilities of gene editing (Gaudelli et al., 2017; Anzalone et al., 2019). The efficiency of traditional breeding depends on natural genetic variation and recombination, which can be unpredictable and inefficient for complex traits. Although transgenic technologies are effective at introducing new traits, the efficiency of stable integration and expression can vary, often needing extensive screening (Hamdan et al., 2022).

Many CRISPR-edited crops are viewed as "non-GMO" in some regulatory areas, especially when no foreign DNA is used. To some extent this can make it easier for the public to accept and simplify regulatory processes (Turnbull et al., 2021; Hamdan et al., 2022; Ahmad et al., 2023). Traditional breeding usually enjoys high public acceptance because it is seen as a natural process. In contrast, transgenic technologies often face public skepticism and strict regulatory challenges in many parts of the world. This is due to concerns about introducing foreign DNA and unforeseen long-term effects (Hamdan et al., 2022; Kuzma, 2023).

2.6 Challenges and ethical considerations

The rapid advancement of CRISPR–Cas9 technology has significantly expanded its applicability in plant science. Its preliminary applications established successful genome editing in model plants like *Arabidopsis* and *Nicotiana*, before extending to economically significant staple crops such as rice and wheat (Li et al., 2013; Shan et al., 2013). This success led to its widespread adoption across various crop systems (Chen et al., 2024). The first CRISPR–Cas9-edited crop to gain USDA regulatory approval was a non-browning white button mushroom in 2016, as it contained no foreign DNA (Globus and Qimron, 2018). That same year, CRISPR-edited waxy corn developed by DuPont Pioneer also avoided regulatory scrutiny for similar reasons.

Despite its great potential, applying CRISPR–Cas technology in agriculture comes with technical challenges and important ethical issues. While CRISPR–Cas is very specific, off-target mutations can still happen, especially when the guide RNA sequence closely resembles other genomic regions (Fu et al., 2013; Li et al., 2019; Han et al., 2020; Spasskaya et al., 2023). These unintentional edits can cause unwanted changes or affect gene function. Ongoing research focuses on creating high-precision Cas variants and adjusting guide RNA design to lessen off-target cleavage events (Spasskaya et al., 2023). Gene editing, even when precise, can sometimes cause pleiotropic effects. This means that modifying one gene can unintentionally influence multiple unrelated traits. This complicates the breeding process; a desired change in one trait might disrupt others negatively (Troadec et al., 2019). Careful phenotypic characterization of edited plants is essential to recognize and reduce such effects.

Delivering Cas protein and guiding RNA into plant cells safely and efficiently remains a big challenge. Common methods include *Agrobacterium*-mediated transformation, protoplast transformation followed by regeneration, and biolistics (gene gun). Each method has its limits regarding efficiency, host range, and potential for off-target integration or somaclonal variation, especially in recalcitrant crop species (Levengood et al., 2024). The rules governing gene-edited crops vary greatly by country, mainly depending on whether the final product has foreign DNA. Gene-edited products are usually categorized based on how the genetic alteration is made. Site-Directed Nuclease 1 (SDN1) includes small indels at the target site, resembling naturally occurring mutations, without adding foreign DNA. Many countries view SDN1 products as non-GMO, which leads to simpler regulatory pathways (Kuzma, 2023; Hamdan et al., 2022). Site-Directed Nuclease 2 (SDN2) includes precise base changes or small substitutions using a DNA template, but without stable integration of foreign DNA. These products are also often seen as non-GMO (Ahmad et al., 2023; Kuzma, 2023). Site-Directed Nuclease 3 (SDN3) involves inserting larger DNA sequences, including foreign genes, using HDR. These products generally fall under the traditional GMO classification and face stricter regulation. The lack of a global regulatory standard creates uncertainty for developers and can slow down the commercialization and international trade of gene-edited crops (Ahmad et al., 2023; Kuzma, 2023).

The ability to modify genomes precisely raises various ethical concerns. These include worries about the "naturalness" of gene-edited organisms, possible unintended ecological effects like gene flow to wild relatives, and issues regarding fair access to the technology and its advantages, especially for smallholder farmers in developing nations. There are also wider societal discussions around the responsible application of powerful genetic technologies and their lasting effects on food systems and the environment. To address public

TABLE 2.2 Regulatory categories of gene-edited crops (SDN classification).

Category	Description of genetic alteration	Presence of foreign DNA	Typical regulatory status	Implication for approval/Trade
SDN1	Small deletions or insertions (indels) at target site	No	Non-GMO in many countries	Streamlined regulatory pathways
SDN2	Precise base changes or small substitutions using DNA template	No	Non-GMO in many countries	Streamlined regulatory pathways
SDN3	Insertion of larger DNA fragments, including foreign genes	Yes	Analogous to traditional GMOs	Subject to more rigorous regulation

concerns and build trust, it is vital to communicate transparently about the science, applications, and oversight of CRISPR–Cas technology. Open dialog, public education, and engagement with stakeholders are necessary to ensure that ethical considerations and societal values are part of developing and using gene-edited crops (Hamdan et al., 2022; Kuzma, 2023). A detailed overview of these regulatory categories is presented in Table 2.2.

2.7 Future perspectives and outlook

New opportunities for innovation in plant breeding and agriculture are emerging as CRISPR–Cas technology continues to advance rapidly is opening. It holds strong potential to reshape strategies for ensuring food security worldwide and addressing climate-related challenges. Advanced technologies like base and prime editing move beyond the traditional CRISPR–Cas9 system that relies on DSBs, offering enhanced precision and versatility. Base editing employs a modified CRISPR system that couples a Cas9 nickase with a deaminase enzyme, enabling precise base substitution (e.g., A to G, C to T) without introducing DSBs. This minimizes off-target effects by precisely introducing point mutations (Gaudelli et al., 2017). Prime editing represents an advanced genome editing approach, wherein a Cas9 nickase fused to a reverse transcriptase coupled with a prime editing gRNA (pegRNA) introduces targeted DNA modification without generating DSBs. This provides unmatched flexibility and precision for a broader range of genetic manipulation (Anzalone et al., 2019). These advancements are likely to improve gene editing capabilities. They make it possible to introduce precise changes that mimic natural variations. CRISPR–Cas, especially with these emerging tools, is set to transform plant breeding significantly. It enables faster crop improvement with better attributes, including higher productivity, better dietary content, and greater resilience to adverse environment (Liu et al., 2021; Das et al., 2022; Chen et al., 2024). This precise breeding can substantially shorten the time span of crop development in comparison to traditional methods. It brings better varieties to farmers and consumers more quickly. As the global population grows and climate change worsens, the need for resilient and productive crops will only rise. CRISPR–Cas technology provides a powerful platform to develop crops that can thrive in hostile environments like drought-prone areas and saline soils, resist new pests and diseases, and deliver higher nutritional value (Liu et al., 2021; Das et al., 2022; Hamdan et al., 2022; Chen et al., 2024). This directly helps improve food security and build sustainable agricultural systems that can withstand future environmental challenges. Realizing the full potential of CRISPR–Cas technology requires sustained research efforts and innovation. This encompasses exploring new Cas enzymes with different PAM requirements or functions, improving delivery methods to various plant species, increasing editing efficiency and accuracy, and advancing computational algorithms for better gRNA design while minimizing off-target risks. Additionally, research into the complex genetic networks behind desirable traits will be vital for guiding effective gene editing strategies.

2.8 Conclusion

The emergence of CRISPR–Cas technology represents a milestone in genetic engineering, marking a revolution in genome editing. It builds on foundational discoveries like the DNA double helix and restriction enzymes, and it goes beyond the limitations of earlier gene editing tools viz., TALEN or ZFN. CRISPR–Cas has introduced unmatched precision, speed, efficiency, and cost-effectiveness to genetic manipulation. Current key achievements include CRISPR–Cas based crop improvement in yield and quality (e.g., enhanced photosynthesis, better nutrient uptake, longer shelf life), strengthening abiotic stresses tolerance and creating strong resistance to different diseases and pests. Its role in biofortification, by improving the nutritional value in staple food crops, also helps address global malnutrition. Despite these impressive advancements, challenges remain in reducing off-target and pleiotropic effects, refining delivery systems, navigating complex regulatory frameworks, and addressing ethical concerns. However, the ongoing evolution of CRISPR–Cas, especially with base and prime editing, promises even greater precision and versatility, further establishing its importance. CRISPR–Cas is a vital tool in addressing global agricultural issues. It highlights its transformative potential in this field. It provides a powerful and sustainable path to boost food production, improve nutrition, and create climate-resilient crops. This significantly contributes to meeting global food demands as the population grows and environmental conditions shift. The vast potential of future CRISPR–Cas advancements calls for ongoing research, ethical innovation, and open public dialog to fully realize its possibilities in creating an environmentally resilient future with reliable food access.

Acknowledgments

The authors gratefully acknowledges the financial support from the Indian Council of Agricultural Research, Department of Agricultural Research and Education, Government of India, under the ICAR EFC on the genome editing scheme: ‘Crop Science for Food and Nutritional Security’ (sub-Scheme 10: ‘Enhancing climate resilience and ensuring food security with genome editing tools’).

Data availability statement

Not applicable.

Declaration

"The authors declare that no generative artificial intelligence tools were used in the writing or preparation of this chapter. The content reflects the authors’ own scholarly work, and the authors assume full responsibility for all statements, interpretations, and conclusions presented."

Ethical approval

Not applicable.

Conflicts of interest

The authors declare that they have no conflicting or competing interests.

References

Abudayyeh, O.O., Gootenberg, J.S., Konermann, S., Joung, J., Slaymaker, I.M., Cox, D.B., Shmakov, S., Makarova, K. S., Semenova, E., Minakhin, L., Severinov, K., Regev, A., Lander, E.S., Koonin, E.V., Zhang, F., 2016. C2c2 is a single-component programmable RNA-guided RNA-targeting CRISPR effector. Science (New York) 353 (6299), aaf5573. https://doi.org/10.1126/science.aaf5573.

Ahmad, A., Jamil, A., Munawar, N., 2023. GMOs or non-GMOs? The CRISPR Conundrum. Front. Plant Sci. 14, 1232938. https://doi.org/10.3389/fpls.2023.1232938.

Aljabali, A.A.A., El-Tanani, M., Tambuwala, M.M., 2024. Principles of CRISPR-Cas9 technology: advancements in genome editing and emerging trends in drug delivery. J. Drug Deliv. Sci. Technol. 92, 105338. https://doi.org/10.1016/j.jddst.2023.105338.

Anzalone, A.V., Randolph, P.B., Davis, J.R., Sousa, A.A., Koblan, L.W., Levy, D.R., Chen, P.J., Wilson, C., Newby, G. A., Raguram, V., Liu, D.R., 2019. Search-and-replace genome editing without double-strand breaks or donor DNA. Nature 576 (7785), 149–157. https://doi.org/10.1038/s41586-019-1711-4.

Arber, W., 1974. DNA restriction and modification mechanisms. Annu. Rev. Biochem. 43, 301–338. https://doi.org/10.1146/annurev.bi.43.070174.001505.

Arora, L., Narula, A.M., 2017. Gene editing and crop improvement using CRISPR-Cas9 system. Front. Plant Sci. 8, 1–21. https://doi.org/10.3389/fpls.2017.01932.

Barrangou, R., Fremaux, C., Deveau, H., Richards, M., Boyaval, P., Moineau, S., Romero, D.A., Horvath, P., 2007. CRISPR provides acquired resistance against viruses in prokaryotes. Science (New York) 315 (5819), 1709–1712. https://doi.org/10.1126/science.1138140.

Berg, P., Baltimore, D., Boyer, H.W., Cohen, S.N., Davis, R.W., Hogness, D.S., Nathans, D., Roblin, R., Watson, J.D., Weissman, S.M., Zinder, N.D., 1975. Summary statement of the asilomar conference on recombinant DNA molecules. Proc. Natl. Acad. Sci. 72 (6), 1981–1984. https://doi.org/10.1073/pnas.72.6.1981.

Bolotin, A., Quinquis, B., Sorokin, A., Ehrlich, S.D., 2005. Clustered regularly interspaced short palindrome repeats (CRISPRs) have spacers of extrachromosomal origin. Microbiology (Reading, England) 151 (Pt 8), 2551–2561. https://doi.org/10.1099/mic.0.28048-0.

Brookes, G., Barfoot, P., 2018. Farm income and production impacts of using GM crop technology 1996-2016. GM Crops Food 9 (2), 59–89. https://doi.org/10.1080/21645698.2018.1464866.

Bruening, G., Lyons, J.M., 2000. The case of the FLAVR SAVR tomato. Calif. Agric. 54 (4), 6–7. https://doi.org/10.3733/ca.v054n04p6.

Cai, R., Lv, R., Shi, X., Yang, G., Jin, J., 2023. CRISPR/dCas9 tools: epigenetic mechanism and application in gene transcriptional regulation. Int. J. Mol. Sci. 24 (19), 14865. https://doi.org/10.3390/ijms241914865.

Chang, J.D., Gao, W., Wang, P., Zhao, F.J., 2022. OsNRAMP5 is a major transporter for lead uptake in rice. Environmental Science & Technology 56 (23), 17481–17490. https://doi.org/10.1021/acs.est.2c06384.

Chen, F., Chen, L., Yan, Z., Xu, J., Feng, L., He, N., Guo, M., Zhao, J., Chen, Z., Chen, H., Yao, G., Liu, C., 2024. Recent advances of CRISPR-based genome editing for enhancing staple crops. Front. Plant Sci. 15, 1478398. https://doi.org/10.3389/fpls.2024.1478398.

Chen, K., Wang, Y., Zhang, R., Zhang, H., Gao, C., 2019. CRISPR/Cas genome editing and precision plant breeding in agriculture. Annu. Rev. Plant Biol. 70, 667–697. https://doi.org/10.1146/annurev-arplant-050718-100049.

Chen, Y., Fu, M., Li, H., Wang, L., Liu, R., Liu, Z., Zhang, X., Jin, S., 2021. High-oleic acid content, nontransgenic allotetraploid cotton (*Gossypium hirsutum* L.) generated by knockout of GhFAD2 genes with CRISPR/Cas9 system. Plant Biotechnol. J. 19 (3), 424–426. https://doi.org/10.1111/pbi.13507.

Christian, M., Cermak, T., Doyle, E.L., Schmidt, C., Zhang, F., Hummel, A., et al., 2010. Targeting DNA double-strand breaks with TAL effector nucleases. Genetics 186 (2), 757–761. https://doi.org/10.1534/genetics.110.120717.

Cohen, S.N., Chang, A.C.Y., Boyer, H.W., Helling, R.B., 1973. Construction of biologically functional bacterial plasmids in vitro. Proc. Natl. Acad. Sci. 70 (11), 3240–3244. https://doi.org/10.1073/pnas.70.11.3240.

Cong, L., Ran, F.A., Cox, D., Lin, S., Barretto, R., Habib, N., et al., 2013. Multiplex genome engineering using CRISPR/Cas systems. Science 339 (6121), 819–823. https://doi.org/10.1126/science.1231143.

Das, D., Singha, D.L., Paswan, R.R., Chowdhury, N., Sharma, M., Reddy, P.S., Chikkaputtaiah, C., 2022. Recent advancements in CRISPR/Cas technology for accelerated crop improvement. Planta 255, 109. https://doi.org/10.1007/s00425-022-03893-y.

Dong, O.X., Yu, S., Jain, R., Zhang, N., Duong, P.Q., Butler, C., Li, Y., Lipzen, A., Martin, J.A., Barry, K.W., Schmutz, J., Tian, L., Ronald, P.C., 2020. Marker-free carotenoid-enriched rice generated through targeted gene insertion using CRISPR-Cas9. Nat. Commun. 11 (1), 1178. https://doi.org/10.1038/s41467-020-14981-y.

Fu, Y., Foden, J.A., Khayter, C., Maeder, M.L., Reyon, D., Joung, J.K., Sander, J.D., 2013. High-frequency off-target mutagenesis induced by CRISPR-Cas nucleases in human cells. Nat. Biotechnol. 31 (9), 822–826. https://doi.org/10.1038/nbt.2623.

Gaudelli, N.M., Gehrke, A.C., Kuo, S.C., Kimura, S., Kim, Y.B., Lo, J.M., Roth, D.B., Liu, D.R., 2017. Programmable base editing of A•T to G•C in genomic DNA without DNA cleavage. Nature 551 (7681), 464–471. https://doi.org/10.1038/nature24644.

Globus, R., Qimron, U., 2018. A technological and regulatory outlook on CRISPR crop editing. J. Cell. Biochem. 119 (2), 1291–1298. https://doi.org/10.1002/jcb.26303.

Gupta, R., Ghosh, A., Chakravarti, R., Singh, R., Ravichandiran, V., Swarnakar, S., Ghosh, D., 2022. Cas13d: a new molecular scissor for transcriptome engineering. Front. Cell Dev. Biol. 10, 866800. https://doi.org/10.3389/fcell.2022.866800.

Hamdan, M.F., Tan, B.C., 2024. Genetic modification techniques in plant breeding: a comparative review of CRISPR/Cas and GM technologies. Hortic. Plant J. 11 (5), 1807–1829. https://doi.org/10.1016/j.hpj.2024.02.012.

Hamdan, M.F., Mohd Noor, S.N., Abd-Aziz, N., Pua, T.-L., Tan, B.C., 2022. Green revolution to gene revolution: technological advances in agriculture to feed the world. Plants 11 (10), 1297. https://doi.org/10.3390/plants11101297.

Han, H.A., Pang, J.K.S., Soh, B.S., 2020. Mitigating off-target effects in CRISPR/Cas9-mediated in vivo gene editing. J. Mol. Med. 98 (4), 615–632. https://doi.org/10.1007/s00109-020-01893-6.

Hillary, V.E., Ceasar, S.A., 2023. A review on the mechanism and applications of CRISPR/Cas9/Cas12/Cas13/Cas14 proteins utilized for genome engineering. Mol. Biotechnol. 65 (3), 311–325. https://doi.org/10.1007/s12033-022-00567-0.

Hossain, K.A., Nierzwicki, L., Orozco, M., Czub, J., Palermo, G., 2025. Flexibility in PAM recognition expands DNA targeting in xCas9. bioRxiv : The Preprint Server for Biology. https://doi.org/10.1101/2024.08.26.609653, 2024.08.26.609653.

Hu, C., Sheng, O., Deng, G., He, W., Dong, T., Yang, Q., Dou, T., Li, C., Gao, H., Liu, S., Yi, G., Bi, F., 2021. CRISPR/Cas9-mediated genome editing of MaACO1 (aminocyclopropane-1-carboxylate oxidase 1) promotes the shelf life of banana fruit. Plant Biotechnol. J. 19 (4), 654–656. https://doi.org/10.1111/pbi.13534.

Hu, J.H., Miller, S.M., Geurts, M.H., Tang, W., Chen, L., Sun, N., Zeina, C.M., Gao, X., Rees, H.A., Lin, Z., Liu, D.R., 2018. Evolved Cas9 variants with broad PAM compatibility and high DNA specificity. Nature 556 (7699), 57–63. https://doi.org/10.1038/nature26155.

Ibrahim, S., Saleem, B., Rehman, N., Zafar, S.A., Naeem, M.K., Khan, M.R., 2022. CRISPR/Cas9 mediated disruption of Inositol Pentakisphosphate 2-Kinase 1 (TaIPK1) reduces phytic acid and improves iron and zinc accumulation in wheat grains. J. Adv. Res. 37, 1–10. https://doi.org/10.1016/j.jare.2021.08.013.

Ishino, Y., Krupovic, M., Forterre, P., 2018. History of CRISPR-Cas from encounter with a mysterious repeated sequence to genome editing technology. J. Bacteriol. 200 (7), e00580-17. https://doi.org/10.1128/JB.00580-17.

Ishino, Y., Shinagawa, H., Makino, K., Amemura, M., Nakata, A., 1987. Nucleotide sequence of the iap gene, responsible for alkaline phosphatase isozyme conversion in *Escherichia coli*, and identification of the gene product. J. Bacteriol. 169 (12), 5429–5433. https://doi.org/10.1128/jb.169.12.5429-5433.1987.

Jansen, R., Embden, J.D., Gaastra, W., Schouls, L.M., 2002. Identification of genes that are associated with DNA repeats in prokaryotes. Mol. Microbiol. 43 (6), 1565–1575. https://doi.org/10.1046/j.1365-2958.2002.02839.x.

Jinek, M., Chylinski, K., Fonfara, I., Hauer, A., Doudna, J.A., Charpentier, E., 2012. A programmable dual-RNA–guided DNA endonuclease in adaptive bacterial immunity. Science 337 (6096), 816–821. https://doi.org/10.1126/science.1225829.

Kan, J., Cai, Y., Cheng, C., Jiang, C., Jin, Y., Yang, P., 2022. Simultaneous editing of host factor gene TaPDIL5-1 homoeoalleles confers wheat yellow mosaic virus resistance in hexaploid wheat. New Phytol. 234 (2), 340–344. https://doi.org/10.1111/nph.18002.

Kim, M.-S., Le, V.T., Jung, Y.J., Kang, K.-K., Cho, Y.-G., 2024. *OsPUB9* gene edited by CRISPR/Cas9 enhanced resistance to bacterial leaf blight in rice (*Oryza sativa* L.). Int. J. Mol. Sci. 25 (13), 7145. https://doi.org/10.3390/ijms25137145.

Kim, Y.G., Cha, J., Chandrasegaran, S., 1996. Hybrid restriction enzymes: zinc finger fusions to FokI cleavage domain. Proc. Natl. Acad. Sci. 93 (3), 1156–1160. https://doi.org/10.1073/pnas.93.3.1156.

Kleinstiver, B.P., Sousa, A.A., Walton, R.T., Tak, Y.E., Hsu, J.Y., Clement, K., Welch, M.M., Horng, J.E., Malagon-Lopez, J., Scarfò, I., Maus, M.V., Pinello, L., Aryee, M.J., Joung, J.K., 2019. Engineered CRISPR-Cas12a variants with increased activities and improved targeting ranges for gene, epigenetic and base editing. Nat. Biotechnol. 37 (3), 276–282. https://doi.org/10.1038/s41587-018-0011-0.

Knott, G.J., Doudna, J.A., 2018. CRISPR-Cas guides the future of genetic engineering. Science 361 (6405), 866–869. https://doi.org/10.1126/science.aas9120.

Kumar, R., Das, J., Puttaswamy, R.K., et al., 2024. Targeted genome editing for cotton improvement: prospects and challenges. Nucleus 67, 181–203. https://doi.org/10.1007/s13237-024-00479-1.

Kuzma, J., 2023. Governance of gene-edited plants: insights from the history of biotechnology oversight and policy process theory. Sci. Technol. Hum. Val. 48 (6), 1260–1291. https://doi.org/10.1177/01622439221150495.

Levengood, H., Zhou, Y., Zhang, C., 2024. Advancements in plant transformation: from traditional methods to cutting-edge techniques and emerging model species. Plant Cell Rep. 43 (11), 273. https://doi.org/10.1007/s00299-024-03359-9.

Li, J.F., Norville, J.E., Aach, J., McCormack, M., Zhang, D., Bush, J., Church, G.M., Sheen, J., 2013. Multiplex and homologous recombination-mediated genome editing in Arabidopsis and Nicotiana benthamiana using guide RNA and Cas9. Nat. Biotechnol. 31 (8), 688–691. https://doi.org/10.1038/nbt.2654.

Li, J., Manghwar, H., Sun, L., Wang, P., Wang, G., Sheng, H., Zhang, J., Liu, H., Qin, L., Rui, H., et al., 2019. Whole genome sequencing reveals rare off-target mutations and considerable inherent genetic or/and somaclonal variations in CRISPR/Cas9-edited cotton plants. Plant Biotechnol. J. 17 (5), 858–868. https://doi.org/10.1111/pbi.13019.

Liu, M., Rehman, S., Tang, X., Gu, K., Fan, Q., Chen, D., Ma, W., 2019. Methodologies for improving HDR efficiency. Front. Genet. 9, 691. https://doi.org/10.3389/fgene.2018.00691.

Liu, Q., Yang, F., Zhang, J., Liu, H., Rahman, S., Islam, S., Ma, W., She, M., 2021. Application of CRISPR/Cas9 in crop quality improvement. Int. J. Mol. Sci. 22 (8), 4206. https://doi.org/10.3390/ijms22084206.

Ludwig, Y., Dueñas Jr., C., Arcillas, E., Macalalad-Cabral, R.J., Kohli, A., Reinke, R., Slamet-Loedin, I.H., 2024. CRISPR-mediated promoter editing of a *cis*-regulatory element of *OsNAS2* increases Zn uptake/translocation and plant yield in rice. Front. Genom. Ed. 5. https://doi.org/10.3389/fgeed.2023.1308228.

Lv, G., Li, Y., Chen, J., Wu, Z., Wu, W., Wu, X., Lin, H., Wang, T., 2023. ZmHMA3 enhances heavy mental Zn stress tolerance and mediates Zn transport in Maize. Res. Square Pre-print. https://doi.org/10.21203/rs.3.rs-2970282/v1.

Maddox, B., 2003. Rosalind Franklin: The Dark Lady of DNA. Harper Perennial.

Makarova, K.S., Grishin, N.V., Shabalina, S.A., Wolf, Y.I., Koonin, E.V., 2006. A putative RNA-interference-based immune system in prokaryotes: computational analysis of the predicted enzymatic machinery, functional analogies with eukaryotic RNAi, and hypothetical mechanisms of action. Biol. Direct 1, 7. https://doi.org/10.1186/1745-6150-1-7.

Makarova, K.S., Wolf, Y.I., Alkhnbashi, O.S., Costa, F., Shah, S.A., Saunders, S.J., Barrangou, R., Brouns, S.J., Charpentier, E., Haft, D.H., Horvath, P., Moineau, S., Mojica, F.J., Terns, R.M., Terns, M.P., White, M.F., Yakunin, A.F., Garrett, R.A., van der Oost, J., Backofen, R., Koonin, E.V., 2015. An updated evolutionary classification of CRISPR-Cas systems. Nat. Rev. Microbiol. 13 (11), 722–736. https://doi.org/10.1038/nrmicro3569.

Mali, P., Yang, L., Esvelt, K.M., Aach, J., Guell, M., DiCarlo, J.E., et al., 2013. RNA-guided human genome engineering via Cas9. Science 339 (6121), 823–826. https://doi.org/10.1126/science.1232033.

Mohr, T., Horstman, J., Gu, Y.Q., Elarabi, N.I., Abdallah, N.A., Thilmony, R., 2022. CRISPR-CAS9 gene editing of the SAL1 gene family in wheat. Plants 11 (17), 2259. https://doi.org/10.3390/plants11172259.

Mojica, F.J.M., Díez-Villaseñor, C., García-Martínez, J., Almendros, C., 2009. Short motif sequences determine the targets of the prokaryotic CRISPR defence system. Microbiology (Reading, England) 155 (Pt 3), 733–740. https://doi.org/10.1099/mic.0.023960-0.

Mojica, F.J., Díez-Villaseñor, C., García-Martínez, J., Soria, E., 2005. Intervening sequences of regularly spaced prokaryotic repeats derive from foreign genetic elements. J. Mol. Evol. 60 (2), 174–182. https://doi.org/10.1007/s00239-004-0046-3.

Mojica, F.J., Díez-Villaseñor, C., Soria, E., Juez, G., 2000. Biological significance of a family of regularly spaced repeats in the genomes of Archaea, Bacteria and mitochondria. Mol. Microbiol. 36 (1), 244–246. https://doi.org/10.1046/j.1365-2958.2000.01838.x.

Mojica, F.J., Juez, G., Rodríguez-Valera, F., 1993. Transcription at different salinities of Haloferax mediterranei sequences adjacent to partially modified PstI sites. Mol. Microbiol. 9 (3), 613–621. https://doi.org/10.1111/j.1365-2958.1993.tb01721.x Top of Form.

Nishimasu, H., Ran, F.A., Hsu, P.D., Konermann, S., Shehata, S.I., Dohmae, N., Ishitani, R., Zhang, F., Nureki, O., 2014. Crystal structure of Cas9 in complex with guide RNA and target DNA. Cell 156 (5), 935–949. https://doi.org/10.1016/j.cell.2014.02.001.

Oliva, R., Ji, C., Atienza-Grande, G., Huguet-Tapia, J.C., Perez-Quintero, A., Li, T., Eom, J.S., Li, C., Nguyen, H., Liu, B., Auguy, F., Sciallano, C., Luu, V.T., Dossa, G.S., Cunnac, S., Schmidt, S.M., Slamet-Loedin, I.H., Vera Cruz, C., Szurek, B., Frommer, W.B., White, F.F., Yang, B., 2019. Broad-spectrum resistance to bacterial blight in rice using genome editing. Nat. Biotechnol. 37 (11), 1344–1350. https://doi.org/10.1038/s41587-019-0267-z.

Pickar-Oliver, A., Gersbach, C.A., 2019. The next generation of CRISPR–Cas technologies and applications. Nat. Rev. Mol. Cell Biol. 20 (8), 490–507. https://doi.org/10.1038/s41580-019-0131-5.

Pourcel, C., Salvignol, G., Vergnaud, G., 2005. CRISPR elements in Yersinia pestis acquire new repeats by preferential uptake of bacteriophage DNA, and provide additional tools for evolutionary studies. Microbiology (Reading, England) 151 (Pt 3), 653–663. https://doi.org/10.1099/mic.0.27437-0.

Rahim, A.A., Uzair, M., Rehman, N., Fiaz, S., Attia, K.A., Abushady, A.M., Yang, S.H., Khan, M.R., 2023. CRISPR/Cas9 mediated TaRPK1 root architecture gene mutagenesis confers enhanced wheat yield. J. King Saud Univ. Sci. 36 (2), 103063. https://doi.org/10.1016/j.jksus.2023.103063.

Rastogi, K., Mankar, S.P., Ibarra, O., Molina-Risco, M., Faion-Molina, M., Gracia, M.B., Rathore, K.S., Thomson, M.J., Septiningsih, E.M., 2025. CRISPR/Cas9 multiplex genome editing to enhance grain lysine concentration in a US rice cultivar. Curr. Plant Biol. 42, 100490. https://doi.org/10.1016/j.cpb.2025.100490.

Sampath, V., Rangarajan, N., Sharanappa, C.H., Deori, M., Veeraragavan, M., Ghodake, B.D., Kaushal, 2023. Advancing crop improvement through CRISPR technology in precision agriculture trends-a review. International Journal of Environment and Climate Change 13 (11), 4683–4694. https://doi.org/10.9734/ijecc/2023/v13i113437.

Sánchez-León, S., Gil-Humanes, J., Ozuna, C.V., Giménez, M.J., Sousa, C., Voytas, D.F., Barro, F., 2018. Low-gluten, nontransgenic wheat engineered with CRISPR/Cas9. Plant Biotechnol. J. 16 (4), 902–910. https://doi.org/10.1111/pbi.12837.

Shan, Q., Wang, Y., Li, J., Zhang, Y., Chen, K., Liang, Z., Zhang, K., Liu, J., Xi, J.J., Qiu, J.L., Gao, C., 2013. Targeted genome modification of crop plants using a CRISPR-Cas system. Nat. Biotechnol. 31 (8), 686–688. https://doi.org/10.1038/nbt.2650.

Shan, Q., Zhang, Y., Chen, K., Zhang, K., Gao, C., 2015. Creation of fragrant rice by targeted knockout of the OsBADH2 gene using TALEN technology. Plant Biotechnol. J. 13 (6), 791–800. https://doi.org/10.1111/pbi.12292.

Shmakov, S., Smargon, A., Scott, D., Cox, D., Pyzocha, N., Yan, W., Abudayyeh, O.O., Gootenberg, J.S., Makarova, K.S., Wolf, Y.I., Severinov, K., Zhang, F., Koonin, E.V., 2017. Diversity and evolution of class 2 CRISPR-Cas systems. Nat. Rev. Microbiol. 15 (3), 169–182. https://doi.org/10.1038/nrmicro.2016.184.

Smith, H.O., Wilcox, K.W., 1970. A restriction enzyme from Hemophilus influenzae. J. Mol. Biol. 51 (2), 379–391. https://doi.org/10.1016/0022-2836(70)90137-9.

Spasskaya, D.S., Davletshin, A.I., Bachurin, S.S., Tutyaeva, V.V., Garbuz, D.G., Karpov, D.S., 2023. Improving the on-target activity of high-fidelity Cas9 editors by combining rational design and random mutagenesis. Appl. Microbiol. Biotechnol. 107 (7), 2385–2401. https://doi.org/10.1007/s00253-023-12440-6.

Stella, S., Montoya, G., 2016. The genome editing revolution: a CRISPR–Cas TALE off–target story. Bioessays 38 (10), 939–950. https://doi.org/10.1002/bies.201600072.

Sternberg, S.H., Redding, S., Jinek, M., Greene, E.C., Doudna, J.A., 2014. DNA interrogation by the CRISPR RNA-guided endonuclease Cas9. Nature 507 (7490), 62–67. https://doi.org/10.1038/nature13011.

Sun, Y., Zhang, X., Wu, C., He, Y., Ma, Y., Hou, H., Guo, X., Du, W., Zhao, Y., Xia, L., 2016. Engineering herbicide-resistant rice plants through CRISPR/Cas9-Mediated homologous recombination of acetolactate synthase. Mol. Plant 9 (4), 628–631. https://doi.org/10.1016/j.molp.2016.01.001.

Tian, X., Wang, F., Zhao, Y., Lan, T., Yu, K., Zhang, L., Qin, Z., Hu, Z., Yao, Y., Ni, Z., Sun, Q., Rossi, V., Peng, H., Xin, M., 2019. Heat shock transcription factor A1b regulates heat tolerance in wheat and Arabidopsis through OPR3 and jasmonate signalling pathway. Plant Biotechnol. J. 18 (5), 1109–1111. https://doi.org/10.1111/pbi.13268.

Troadec, M.B., Pagès, J.C., the Scientific Committee of the High Council for biotechnology, France, 2019. Where are we with unintended effects in genome editing applications from DNA to phenotype: focus on plant applications. Transgenic Res. 28 (Suppl. 2), 125–133. https://doi.org/10.1007/s11248-019-00146-1.

Turnbull, C., Lillemo, M., Hvoslef-Eide, T.A.K., 2021. Global regulation of genetically modified crops amid the gene edited crop boom - a review. Frontiers in plant Science 12, 630396. https://doi.org/10.3389/fpls.2021.630396.

Uluisik, S., Chapman, N.H., Smith, R., Poole, M., Adams, G., Gillis, R.B., Besong, T.M., Sheldon, J., Stiegelmeyer, S., Perez, L., Samsulrizal, N., Wang, D., Fisk, I.D., Yang, N., Baxter, C., Rickett, D., Fray, R., Blanco-Ulate, B., Powell, A.L., Harding, S.E., Seymour, G.B., 2016. Genetic improvement of tomato by targeted control of fruit softening. Nat. Biotechnol. 34 (9), 950–952. https://doi.org/10.1038/nbt.3602.

Wang, D., Samsulrizal, N.H., Yan, C., Allcock, N.S., Craigon, J., Blanco-Ulate, B., Ortega-Salazar, I., Marcus, S.E., Bagheri, H.M., Perez Fons, L., Fraser, P.D., Foster, T., Fray, R., Knox, J.P., Seymour, G.B., 2019. Characterization of CRISPR mutants targeting genes modulating pectin degradation in ripening tomato. Plant Physiol. 179 (2), 544–557. https://doi.org/10.1104/pp.18.01187.

Wang, J., Wang, J., Zhang, X., Zhang, H., Wu, X., Li, X., et al., 2014. Targeted mutagenesis of TaMLO genes using CRISPR/Cas9 system in wheat. Mol. Plant 7 (10), 1547–1550. https://doi.org/10.1093/mp/ssu091.

Wang, X., Guo, Y., Wang, Y., Peng, Y., Zhang, H., Zheng, J., 2024. ZmHDT103 negatively regulates drought stress tolerance in maize seedlings. Agronomy 14 (1), 134. https://doi.org/10.3390/agronomy14010134.

Watson, J.D., Crick, F.H.C., 1953. Molecular structure of nucleic acids: a structure for deoxyribose nucleic acid. Nature 171 (4356), 737–738. https://doi.org/10.1038/171737a0.

Xie, J., Fei, X., Yan, Q., Jiang, T., Li, Z., Chen, H., Wang, B., Chao, Q., He, Y., Fan, Z., Wang, L., Wang, M., Shi, L., Zhou, T., 2024. The C4 photosynthesis bifunctional enzymes, PDRPs, of maize are co-opted to cytoplasmic viral replication complexes to promote infection of a prevalent potyvirus sugarcane mosaic virus. Plant Biotechnol. J. 22 (7), 1812–1832. https://doi.org/10.1111/pbi.14304.

Xu, R., Li, Z., Li, Y., Wang, L., Zhao, Q., Luan, X., et al., 2022. CRISPR/Cas9-mediated editing of OsRR22 enhances salt tolerance in rice. Plant Biotechnol. J. 20 (10), 1957–1968. https://doi.org/10.1111/pbi.13876.

Yamori, W., Nakazato, I., Qu, Y., Sanga, Y., Miyata, T., Uehara, R., Noto, Y., Namba, K., Fukayama, H., Matsumura, H., Arimura, S., 2025. Chloroplast genome editing of Rubisco boosts photosynthesis and plant growth. bioRxiv. https://doi.org/10.1101/2025.01.02.631008.

Yang, H., Patel, D.J., 2024. Structures, mechanisms and applications of RNA-centric CRISPR-Cas13. Nat. Chem. Biol. 20 (6), 673–688. https://doi.org/10.1038/s41589-024-01593-6.

Yu, Y., Pan, Z., Wang, X., Bian, X., Wang, W., Liang, Q., Kou, M., Ji, H., Li, Y., Ma, D., Li, Z., Sun, J., 2022. Targeting of SPCSV-RNase3 via CRISPR-Cas13 confers resistance against sweet potato virus disease. Mol. Plant Pathol. 23 (1), 104–117. https://doi.org/10.1111/mpp.13146.

Zetsche, B., Gootenberg, J.S., Abudayyeh, O.O., Slaymaker, I.M., Makarova, K.S., Essletzbichler, P., Volz, S.E., Joung, J., van der Oost, J., Regev, A., Koonin, E.V., Zhang, F., 2015. Cpf1 is a single RNA-guided endonuclease of a class 2 CRISPR-Cas system. Cell 163 (3), 759–771. https://doi.org/10.1016/j.cell.2015.09.038.

Zhang, L., Xiang, Z., Li, J., Wang, S., Chen, Y., Liu, Y., Mao, D., Luan, S., Chen, L., 2023. bHLH57 confers chilling tolerance and grain yield improvement in rice. Plant Cell Environ. 46 (4), 1402–1418. https://doi.org/10.1111/pce.14513.

Zhang, Y., Guo, W., Chen, L., Shen, X., Yang, H., Fang, Y., Ouyang, W., Mai, S., Chen, H., Chen, S., Hao, Q., Yuan, S., Zhang, C., Huang, Y., Shan, Z., Yang, Z., Qiu, D., Zhou, X., Cao, D., Jiao, Y., 2022. CRISPR/Cas9-Mediated targeted mutagenesis of *GmUGT* enhanced soybean resistance against leaf-chewing insects through flavonoids biosynthesis. Front. Plant Sci. 13, 802716. https://doi.org/10.3389/fpls.2022.802716.

Zhang, Y., Massel, K., Godwin, I.D., et al., 2018a. Applications and potential of genome editing in crop improvement. Genome Biol. 19, 210. https://doi.org/10.1186/s13059-018-1586-y.

Zhang, Z., Ge, X., Luo, X., Wang, P., Fan, Q., Hu, G., Xiao, J., Li, F., Wu, J., 2018b. Simultaneous editing of two copies of *Gh14-3-3d* confers enhanced transgene-clean plant defense against *Verticillium dahliae* in allotetraploid upland cotton. Front. Plant Sci. 9, 842. https://doi.org/10.3389/fpls.2018.00842.

Zheng, S., Ye, C., Lu, J., Liufu, J., Lin, L., Dong, Z., et al., 2021. Improving the rice photosynthetic efficiency and yield by editing OsHXK1 via CRISPR/Cas9 system. Int. J. Mol. Sci. 22 (17), 9594. https://doi.org/10.3390/ijms22179594.

Zhu, H., Li, C., Gao, C., 2020. Applications of CRISPR-Cas in agriculture and plant biotechnology. Nat. Rev. Mol. Cell Biol. 21 (11), 661–677. https://doi.org/10.1038/s41580-020-00288-9.

Further reading

Boch, J., Scholze, H., Schornack, S., Landgraf, A., Hahn, S., Kay, S., Lahaye, T., Nickstadt, A., Bonas, U., 2009. Breaking the code of DNA binding specificity of TAL-type III effectors. Science (New York) 326 (5959), 1509–1512. https://doi.org/10.1126/science.1178811.

Cardi, T., Murovec, J., Bakhsh, A., d'Ambrosio, M., Perrotta, G., D'Alessandro, R., Ferrante, P., 2023. CRISPR/Cas-mediated plant genome editing: outstanding challenges a decade after implementation. Trends Plant Sci. 28 (10), 1144–1165. https://doi.org/10.1016/j.tplants.2023.08.006.

Cathomen, T., Keith Joung, J., 2008. Zinc-finger nucleases: the next generation emerges. Mol. Ther. : The Journal of the American Society of Gene Therapy 16 (7), 1200–1207. https://doi.org/10.1038/mt.2008.114.

Chen, Y., Yu, D., Li, J., Liu, H., Wang, Y., Zhang, Y., et al., 2022. CRISPR/Cas9-mediated editing of bHLH57 improves cold tolerance and grain yield in rice. Plant Biotechnol. J. 20 (12), 2327–2339. https://doi.org/10.1111/pbi.13904.

Garneau, J.E., Dupuis, M.È., Villion, M., Romero, D.A., Barrangou, R., Boyaval, P., Fremaux, C., Horvath, P., Magadán, A.H., Moineau, S., 2010. The CRISPR/Cas bacterial immune system cleaves bacteriophage and plasmid DNA. Nature 468 (7320), 67–71. https://doi.org/10.1038/nature09523.

Gasiunas, G., Barrangou, R., Horvath, P., Siksnys, V., 2012. Cas9-crRNA ribonucleoprotein complex mediates specific DNA cleavage for adaptive immunity in bacteria. Proc. Natl. Acad. Sci. 109 (39), E2579–E2586. https://doi.org/10.1073/pnas.1208507109.

Jiang, W., Zhou, H., Bi, H., Fromm, M., Yang, B., Weeks, D.P., 2013. Demonstration of CRISPR/Cas9/sgRNA-mediated targeted gene modification in Arabidopsis, tobacco, sorghum and rice. Nucleic Acids Res. 41 (20), e188. https://doi.org/10.1093/nar/gkt780.

Khan, M.A., Khan, M.A., Javed, M.A., 2020. CRISPR/Cas9: a new era in plant genome editing. Journal of Agriculture and Food 5 (1), 23–30.

Kumar, A., Singh, R., Singh, A., 2020. CRISPR/Cas9-mediated genome editing for enhancing abiotic stress tolerance in plants. J. Genet. Eng. Biotechnol. 18 (1), 1–10. https://doi.org/10.1186/s43141-020-00057-z.

Kumar, R., Singh, A., Singh, R., 2023. CRISPR-Cas9-mediated genome editing for improving disease resistance in plants. Plant Cell Rep. 42 (10), 1775–1786. https://doi.org/10.1007/s00299-023-03058-2.

Kumlehn, J., Pietralla, J., Hensel, G., Pacher, M., Puchta, H., 2018. The CRISPR/Cas revolution continues: from efficient gene editing for crop breeding to plant synthetic biology. J. Integr. Plant Biol. 60 (9), 834–846. https://doi.org/10.1111/jipb.12658.

Liberal, Â., Pinela, J., Vívar-Quintana, A.M., Ferreira, I.C.F.R., Barros, L., 2020. Fighting iron-deficiency anemia: innovations in food fortificants and biofortification strategies. Foods 9 (12), 1871. https://doi.org/10.3390/foods9121871.

Mohanta, M., Panda, M.K., Mohapatra, S., Jena, S.K., 2020. CRISPR/Cas9: a revolutionary tool for crop improvement. In: Recent Trends in Biotechnology and Biodiversity. Springer, Singapore. https://doi.org/10.1007/978-981-15-5813-8_13.

Mohd Noor, S.N., Hamdan, M.F., Abd-Aziz, N., Pua, T.-L., Tan, B.C., 2023. CRISPR-Cas9 system: a new approach for crop improvement. Plant Biotechnology Reports 17 (5), 1–12. https://doi.org/10.1007/s11816-023-00099-2.

Moscou, M.J., Bogdanove, A.J., 2009. A simple cipher governs DNA recognition by TAL effectors. Science (New York, N.Y.) 326 (5959), 1501. https://doi.org/10.1126/science.1178817.

Nidhi, S., Anand, U., Oleksák, P., Tripathi, P., Lal, J.A., Thomas, G., Kuča, K., Tripathi, V., 2021. Novel CRISPR–cas systems: an updated review of the current achievements, applications, and future research perspectives. Int. J. Mol. Sci. 22 (17), 9594. https://doi.org/10.3390/ijms22179594.

Sharma, A., Kumar, R., Singh, R., 2024. Advances in CRISPR-Cas9 technology for crop improvement. J. Plant Biotechnol. 12 (1), 1–10.

CHAPTER

3

The CRISPR–Cas toolkit for gene editing in plants

Muhammad Umer Masood[1], *Abdullah Jamil*[2], *Mustabshira Ikram*[3], *Vishal Monga*[1], *Monoswi Chakraborty*[4], *Ahmed Al-Harrasi*[5], *Muhammad Danyal Khan*[6], *and Aftab Ahmad*[5]

[1]Department of Plant Sciences and Plant Pathology, Montana State University, Bozeman, MT, United States; [2]Plant Breeding and Genetics Program, Gulf Coast Research and Education Center (GCREC), University of Florida, Wimauma, FL, United States; [3]Department of Land Resources and Environmental Sciences, Montana State University, Bozeman, MT, United States; [4]Institute for Integrative Genome Biology (IIGB), University of California, Riverside, CA, United States; [5]Natural and Medical Sciences Research Center (NMSRC), University of Nizwa, Nizwa, Oman; [6]Department of Entomology, Louisiana State University, Baton Rouge, LA, United States e-mail address: mkhan@agcenter.lsu.edu

3.1 Introduction

CRISPR (Clustered Regularly Interspaced Short Palindromic Repeats) technology has transformed the field of plant science in the last 10 years. Building on the older and more complex genome editing tools such as zinc finger nucleases and TALENs, the advent of CRISPR/Cas9 has enabled plant genome editing that has achieved breakthroughs never imagined using traditional breeding methods (Erdogan et al., 2023). Compared to other genomic modification tools that are tedious and time-consuming, CRISPR/Cas is remarkably simple, efficient, and low-cost, allowing target genetic changes to be made with unmatched ease (Mainkar et al., 2023). Because of the simplicity and versatility of CRISPR, agriculture is currently undergoing what many have referred to as a new era in crop improvement, essentially a genetic revolution (Azameti and Dauda, 2021). Not only is CRISPR/Cas9 an RNA-guided endonuclease system adapted from the bacterial adaptive

Intelligent Gene Editing in Plants
https://doi.org/10.1016/B978-0-443-44079-3.00008-5

immunity system, but it is also composed of the programmable single guide RNA (sgRNA) and the Cas9 nuclease. It can be targeted to any genomic sequence of interest by designing a 20-nucleotide guide; however, once a sequence is designed, it is highly flexible and scalable for plant research (Erdogan et al., 2023). The first successful attempts of CRISPR-mediated gene editing in plants were reported in 2013, and the use of CRISPR has been explored since then (Arora and Narula 2017). Today, CRISPR has become one of the most powerful tools of plant biotechnology, allowing for advanced research with functional genomics programs and breeding programs (Razzaq et al., 2021). Its versatility in producing heritable, transgene-free mutations has made it an essential technology for applied plant biotechnology for both basic research and crop improvement (Zhu et al., 2020). As can be seen, the rise of CRISPR/Cas technology has allowed plant biologists to precisely rewrite DNA and open up new avenues for addressing the pressing agricultural questions and challenges, such as food security and climate resilience (Mallick et al., 2025).

3.1.1 CRISPR-based applications

CRISPR-based genome modification has quickly moved from a laboratory tool to widespread use in a variety of plant species for both basic and applied research purposes. Since its original use in model plant systems such as *Arabidopsis thaliana* and tobacco, CRISPR technology has been used in nearly all major crops, especially cereals (rice, wheat, maize), legumes, and fiber crops. CRISPR technology has also been applied in a variety of horticultural species, including tomato, potato, citrus, and grapes (Miladinovic et al., 2021; Rukavtsova et al., 2022). Because of the universal nature of the CRISPR system, integrating a guide RNA (gRNA) instead of specific interactions between proteins and DNA of specific species, CRISPR can potentially edit nearly every plant genome with proper transformation methods. This potential has led to the broad genetic improvements in crops in a relatively short amount of time (Scheben et al., 2017). For example, CRISPR/Cas9 has enabled researchers to produce disease-resistant varieties by knocking out susceptibility genes (for instance, eIF4E for virus resistance in cucumber) (Atarashi et al., 2020), to improve tolerance to abiotic stresses by modifying stress-responsive regulators in rice and wheat (Zafar et al., 2020). CRISPR has been applied to a range of plant types to improve agronomic traits like yield, plant architecture, as well as quality traits that are important to consumer groups (Sedeek et al., 2019). Plants can be engineered with traits to improve resistance to pathogens and insects, enhanced tolerance to drought, salinity, and other stresses, changed flowering or ripening times to extend their growing period, and higher nutritional content and flavor profiles of their fruits (Tyagi et al., 2021; Mathiazhagan et al., 2021). For example, in fruits, CRISPR edits have delayed fruit softening in tomato, leading to improved shelf life (Adaskaveg and BlancoUlate, 2023) and developed seedless types of citrus (Vardi et al., 2008). CRISPR has been used to increase grain protein concentration and develop herbicide tolerance (Zhang et al., 2021). All these applications highlight the tremendous potential of CRISPR and its associated techniques, such as CRISPRi, CRISPRa, base editing, and prime editing for the precise genetic improvement of plants.

The diversity of applications argues that CRISPR is an all-purpose plant breeding tool. So far, dozens of endogenous plant genes have been targeted, and the new phenotypes obtained from CRISPR genetic edits have not been as easy to attain or unattainable using traditional plant breeding alone. Furthermore, CRISPR-edited plants can be produced in a transgene-free format (by segregating out the CRISPR transgene, or using RNP delivery), so some of these enhancements may avoid regulatory constraints imposed on transgenic organisms and therefore may expedite their advancement to the field. Overall, CRISPR's impact spans across all major categories of plants from annual grain crops to perennial fruit trees and is transforming how we breed plants for agriculture, horticulture, and sustainability. This chapter will review representative examples of CRISPR applications on different plant species and the important achievements and limitations to date.

3.1.2 Scope of the chapter

This chapter provides a thorough overview of the CRISPR–Cas toolkit for plant genome editing. We first follow the story of CRISPR from its origins as a bacterial immune system through to its development as a genome editing platform in plants. We then introduce the main CRISPR–Cas systems used in plant research, such as Cas9, Cas12, Cas13, and other emerging variants, and discuss the essential molecular components, including guide RNAs, PAM sites, Cas effectors, and donor templates. Next, we clarify the molecular mechanisms of CRISPR-mediated gene editing, including the key processes of target recognition, DNA cleavage, and DNA repair. A separate section is provided to note the variety of delivery systems that can be used to deliver CRISPR components to plant cells, such as Agrobacterium-mediated transformation, particle bombardment, viral vectors, and nanotechnology. We then review some new developments and possibilities in editing paradigms, including base editing, prime editing, and multiplex genome engineering. Finally, we provide an overview of applications in plant biotechnology for enhancing disease resistance, crop quality, abiotic stress tolerance, and metabolic traits. The chapter provides examples of representative case studies in major crops to illustrate real-world occurrences. The chapter also attempts to address challenges facing researchers today (e.g., editing efficiency, delivery obstacles, regulatory issues), and clarifies that off-target effects are less critical in plants than in human health applications. To conclude, we address the future of plant gene editing with CRISPR 3.0, artificial intelligence, synthetic biology, and RNA-guided alternative platforms like bridge-RNA and seek-RNA. The chapter is intended to be a foundational reference and look-ahead for CRISPR technology in plant science.

3.2 Early discoveries and bacterial immune function

The CRISPR–Cas story began with the discovery of unusual repetitive DNA sequences in bacteria. In 1987, Ishino et al. serendipitously identified a series of short repeats interspaced with unique sequences in *E. coli* while cloning an unrelated gene (Ishino et al., 2018). Throughout the 1990s and early 2000s, similar repeat arrays were found in many bacteria and archaea. Mojica and colleagues in Spain recognized that these "clustered regularly interspaced short palindromic repeats" (Sunwasiya and Mondal, 2024), later coined CRISPR by Jansen et al. (2002), shared common features across species. Jansen's team not only named the repeats but also noted they were consistently flanked by a set of conserved genes (now called Cas genes, for "CRISPR-associated") that had been previously misannotated as DNA repair proteins (Jansen et al., 2002). The functions of the mysterious CRISPR sequences remained elusive until 2005, when comparative genomics provided a crucial clue: multiple groups independently reported that the intervening spacer sequences matched fragments of bacteriophage and plasmid genomes (Guy and Bolt, 2005). In 2005, scientists hypothesized that CRISPR arrays might record past viral infections, implying a role in adaptive immunity (Mojica and Montoliu, 2016). Indeed, another team observed the same phage-spacer correlation in Yersinia, strengthening the idea that CRISPR could provide immunity against phages (Deveau et al., 2010). Around the same time, Bolotin et al. (2005) discovered a CRISPR locus in *Streptococcus thermophilus* that lacked some known Cas genes but instead carried a novel large cas gene (now known as Cas9) and an adjacent short motif in the phage-targeted spacers. This motif, termed the protospacer adjacent motif (PAM), was found to be critical for target recognition, presaging our understanding of how CRISPR targets invaders. These early discoveries raised the possibility that CRISPR–Cas was not a repair system at all, but rather part of a microbial adaptive immune system (Bolotin et al., 2005). In 2006, Koonin's group formally proposed that CRISPR–Cas functions like a prokaryotic RNA interference mechanism defending against viruses (Makarova et al., 2006). This hypothesis was decisively confirmed the next year: in 2007, Barrangou, Horvath, and colleagues at Danisco experimentally demonstrated that bacteria acquire phage resistance via CRISPR. They showed that *S. thermophilus* cells challenged with bacteriophages integrated new phage-derived spacers into their CRISPR locus and became phage-resistant, whereas deleting spacers or cas genes abolished this immunity. This landmark study established that CRISPR–Cas is an adaptive immune system in bacteria, storing snippets of invader DNA as memories and using them to fend off future infections (Ishino et al., 2018).

Following the breakthrough of 2007, researchers rapidly unraveled the molecular mechanism of CRISPR–Cas immunity. In 2008, van der Oost and collaborators found that the spacer sequences are transcribed into small RNAs (later named CRISPR RNAs or crRNAs) that guide Cas protein complexes to complementary viral DNA targets. Later that same year, Marraffini and Sontheimer showed that CRISPR interference acts on DNA rather than RNA, in contrast to eukaryotic RNAi, by demonstrating that phage DNA (not mRNA) is the target of the interference complex. This was a pivotal insight, as it made clear that CRISPR immunity involves a DNA-cleaving mechanism and hinted that it could be repurposed beyond bacteria (Lander, 2016). By 2010, the role of the Cas proteins in effecting cleavage was illuminated: Moineau's group identified that the Cas9 enzyme from *S. thermophilus* introduces site-specific double-strand breaks (DSBs) in target DNA exactly three nucleotides upstream of the PAM sequence. Notably, Cas9 appeared to be the sole protein required to execute the DNA cut in this Type II CRISPR system, unlike Type I and III systems, which use multiprotein complexes (Lander, 2016). The final pieces of the CRISPR–Cas puzzle fell into place in 2011. Emmanuelle Charpentier's lab discovered a second RNA, the transactivating crRNA (tracrRNA), in *Streptococcus pyogenes*, which forms a duplex with the crRNA and is essential for guiding Cas9 to its DNA target. Together with an RNase III enzyme, tracrRNA enables the maturation of crRNA and loading of Cas9, explaining how the DNA-cutting complex is assembled. By reconstituting the *S. thermophilus* CRISPR–Cas9 locus in *E. coli*, Siksnys and colleagues further confirmed in late 2011 that all necessary components for immunity were contained within the locus and could function in a nonnative host (Lander, 2016). In sum, by 2011, researchers had decoded how bacteria capture foreign DNA snippets as spacers, transcribe them into gRNAs, and employ Cas nucleases (like Cas9) to recognize and cleave invading genetic elements. This understanding of CRISPR–Cas as an RNA-guided adaptive immune response set the stage for its transformation into a revolutionary tool for genetics (Ishino et al., 2018).

3.2.1 Classification of CRISPR–Cas systems

Comparative genomics has revealed significant diversity among CRISPR–Cas systems, characterized by distinct Cas gene sets and varying architectures of CRISPR loci across species. In 2011, Makarova, Koonin, and colleagues provided the first unified classification, identifying three major CRISPR–Cas types (I, II, III), each defined by signature Cas genes and unique mechanisms. Type I and III systems, prevalent in bacteria and archaea, utilize multiprotein complexes (e.g., Cascade in Type I), while Type II systems feature the single multidomain effector Cas9. Type I systems rely on Cas3 nuclease-helicase for interference, Type III systems include Cas10, and Type II systems employ Cas9 alone. The original classification subdivided these types further (e.g., IA to IF) based on Cas gene variations and CRISPR repeats (Makarova et al., 2015).

By 2015, additional CRISPR types necessitated an updated framework, introducing two overarching classes: Class 1 (Types I, III, IV) with multiprotein complexes, and Class 2 (Types II, V, VI) utilizing single effector proteins. This expansion added Type V systems exemplified by Cas12a (Cpf1), an enzyme distinct from Cas9. Shortly thereafter, in 2016, RNA-targeting Type VI systems featuring Cas13a (C2c2) emerged, broadening CRISPR functionality to RNA cleavage (Makarova et al., 2015).

Today, CRISPR classification encompasses two classes, six primary types (VI), and numerous subtypes. Class 2's enzymes Cas9, Cas12, and Cas13 have expanded molecular toolkits with unique targeting and biotechnological applications. Despite ongoing discoveries and refinements, the current classification provides an essential phylogenetic and functional foundation for understanding CRISPR–Cas diversity and exploiting these systems for biotechnological advancements (Makarova et al., 2020).

3.2.2 Repurposing CRISPR for genome editing

As the biological mechanisms of CRISPR–Cas immunity became clearer, its potential as a versatile gene editing tool emerged. Early indications appeared in 2008, when Sontheimer and Marraffini proposed that transferring CRISPR systems into other organisms could enable programmable genetic silencing (Marraffini and Sontheimer, 2008). The pivotal breakthrough occurred in 2012 when researchers successfully reconstituted and demonstrated the bacterial CRISPR–Cas9 system's programmable DNA-cutting capability (Gasiunas et al., 2012). Virginijus Šikšnys' team first characterized the Cas9 complex from *S. thermophilus*, establishing that its DNA targeting specificity was directed by a 20-nucleotide guide sequence within the crRNA (Gasiunas et al., 2012). Shortly thereafter, Jennifer Doudna reported similar findings with *Streptococcus pyogenes* Cas9 (Jinek et al., 2012).

Critically, Jinek et al. (2012) engineered a simplified single-guide RNA (sgRNA), merging crRNA and tracrRNA into one molecule, vastly enhancing ease of use.

By early 2013, independent groups applied CRISPR–Cas9 in living mammalian cells. Feng Zhang's lab at the Broad Institute demonstrated precise genome editing in human and mouse cells, optimizing Cas9 for eukaryotic systems and establishing both nonhomologous end-joining and homology-directed repair as feasible editing pathways. Concurrently, George Church's group independently validated these findings in human cells. Compared to earlier technologies (zinc-finger nucleases and TALENs), CRISPR's simplicity and versatility, requiring only redesign of the ~20 nt RNA sequence, rapidly accelerated its global adoption (Wang and Qi, 2016).

Expanding beyond Cas9, in 2015, researchers identified Cas12a (Cpf1), a Type V enzyme offering distinct PAM recognition and staggered DNA cleavage, complementing Cas9's functionalities. Subsequently, Cas13a (Type VI), an RNA-targeting nuclease, broadened applications to include viral diagnostics and RNA editing (Chen and Doudna, 2017). The contemporary CRISPR toolbox now encompasses various specialized editors (base editors, prime editors, epigenome editors, RNA editors), greatly diversifying genome and transcriptome manipulation capabilities. This evolution from bacterial immune defense to sophisticated genome engineering highlights the transformative impact of fundamental scientific research, making CRISPR–Cas among modern biology's most significant technological advances (Chen and Doudna, 2017).

3.3 Major types of CRISPR–Cas systems used in plants (Cas9, Cas12, Cas13)

CRISPR–Cas systems are categorized into Class 1 (multiprotein effector complexes) and Class 2 (single-protein effectors). The Class 2 systems have been most widely adopted for genome editing due to their simplicity, especially Cas9 (Type II), Cas12 (Type V), and Cas13 (Type VI). Each of these nucleases has distinct characteristics, including different target molecules (DNA vs. RNA), guide RNA requirements, and cleavage patterns. Below, we outline the major types of CRISPR–Cas systems used in plants, along with notable variants (Hillary and Ceasar, 2023).

3.3.1 Cas9 (Type II)

Cas9 is the hallmark of Type II CRISPR systems and the first CRISPR effector harnessed for gene editing in plants. SpyCas9 (*Streptococcus pyogenes* Cas9) introduced the CRISPR revolution and has been applied successfully in many organisms, including numerous plant species. Cas9 is an RNA-guided DNA endonuclease that forms a complex with a dual-guide RNA (crRNA: tracrRNA, often combined as a single-guide RNA) to recognize a target DNA sequence, followed by a short protospacer-adjacent motif (PAM, typically 5′NGG3′ for SyCas9). Upon target binding, Cas9's two catalytic domains (RuvC and HNH) each cut one strand of the DNA, creating a double-strand break (DSB) with blunt ends. In plant cells, these DSBs are predominantly repaired by nonhomologous end joining (NHEJ), leading to small insertions or deletions that can knock out gene function. Cas9's robust activity and reliability have made it the most widely used CRISPR tool in plant gene editing (Fonfara et al., 2014). However, Cas9 has some limitations, such as its strict PAM requirement and off-target cleavage potential. Ongoing protein engineering and discovery of Cas9 orthologs aim to address these issues. For example, engineered variants like SpCas9NG and SpCas9SpRY can recognize broader or relaxed PAM sequences, and high-fidelity versions (e.g., SpCas9HF1, eSpCas9) reduce off-target effects. Cas9's large size (~1368 amino acids for SpyCas9) can pose delivery challenges, but smaller orthologs (e.g., SaCas9 from *Staphylococcus aureus*, ~1053 aa) and truncated forms have been explored for delivery in plants when using size-limited vectors. Overall, Cas9 remains a cornerstone of plant genome editing for its efficiency in creating targeted DNA mutations (Fonfara et al., 2014; Pandita, 2022; Hillary and Ceasar, 2023).

3.3.2 Cas12/Cpf1 (Type V)

Cas12 enzymes (formerly known as Cpf1 for the first discovered subtype Cas12a) belong to the Class 2 Type V systems and have broadened the plant genome editing toolbox. Like Cas9, Cas12 nucleases are RNA-guided endonucleases that target DNA, but they exhibit several key differences. Cas12a (Cpf1), the best-characterized Type VA effector, recognizes a T-rich PAM (e.g., 5′TTTN3′ for FnCas12a from *Francisella*) instead of the G-rich PAMs of Cas9. Cas12a binds its target as a single crRNA-protein complex, without requiring a tracrRNA. Upon binding, Cas12a induces a DSB but with staggered cuts, leaving 5′ overhangs of typically four to five nucleotides (a "sticky end") rather than Cas9's blunt ends. This staggered cleavage is a result of Cas12's single RuvC domain cutting both DNA strands at different positions. Another unique feature of Cas12a is its ability to process precursor CRISPR arrays autonomously: after cleaving one target, the activated Cas12a can cleave tandemly arrayed crRNA transcripts to release additional guides, which have been exploited for multiplex gene editing in plants. Cas12 enzymes are typically similar in size to Cas9 (about 1100–1300 amino acids). Several Cas12 variants have been applied in plants, including LbCas12a (from *Lachnospiraceae bacterium*) and engineered versions thereof with broadened PAM compatibilities (e.g., LbCas12aRVR recognizes 5′TYCV3′). Beyond Cas12a, other Cas12 family effectors such as Cas12b (C2c1) have been adapted for editing in eukaryotes for instance, an *Alicyclobacillus* Cas12b variant was engineered to function at ~37°C and successfully edit human cells, suggesting potential for plant applications. A remarkable property of Cas12 (especially Cas12a and Cas12b) is their collateral cleavage activity: once activated by a target DNA, the Cas12 enzyme can nonspecifically degrade single-stranded DNA in solution. While in gene editing, one aims to avoid collateral activity; this feature has been repurposed for sensitive DNA detection assays in plant diagnostics (analogous to how Cas13 is used for RNA detection). In summary, Cas12/Cpf1 provides an alternative to Cas9 for plant gene editing, with advantages in certain contexts such as targeting AT-rich genomic regions, easier multiplexing, and novel applications in diagnostics (Pandita, 2022; Hillary and Ceasar, 2023).

3.3.3 Cas13 (Type VI)

Cas13 effectors (Type VI CRISPR systems) are unique in that they target RNA instead of DNA. This property makes Cas13 an RNA-guided RNA endonuclease, well-suited for applications that require transient gene knockdown or antiviral defense in plants without permanently altering the genome. Cas13 enzymes (including subtypes Cas13a, Cas13b, Cas13c, and the compact Cas13d) use a single crRNA to recognize a single-stranded RNA sequence, and they contain two Higher Eukaryotes and Prokaryotes nucleotide-binding (HEPN) domains that catalyze RNA cleavage. Notably, Cas13 does not require a PAM, but some Cas13 orthologs recognize a protospacer flanking site (PFS), a short motif adjacent to the RNA target that can influence target recognition (for example, Cas13a from *Leptotrichia* prefers a non-G nucleotide immediately 3′ of the target). Upon binding to its RNA target, Cas13 performs site-specific cleavage of that target RNA. In addition, like Cas12, Cas13 exhibits collateral RNase activity, indiscriminately cutting other nontarget RNAs in the vicinity. In a cellular context, this can lead to cell toxicity if uncontrolled, but it has been harnessed for diagnostic platforms (e.g., the SHERLOCK system uses Cas13a collateral activity to detect viral RNAs via fluorescent reporters). Cas13 proteins are generally in the 900–1300 aa size range, comparable to Cas9/Cas12. In plants, Cas13 has been deployed to combat RNA viruses and to knock down endogenous transcripts. For instance, a codon-optimized LshCas13a (from *Leptotrichia shahii*) was used to confer viral resistance by targeting Turnip Mosaic Virus (TMV) RNA in *Nicotiana* species. Because Cas13 does not create DNA breaks, it avoids the risk of off-target genomic mutations; its effects are temporary (lasting as long as the RNA or Cas13 is expressed). This makes Cas13 a valuable tool for cases where permanent genome edits are undesirable, and for novel RNA-focused techniques (such as REPAIR, which couples a deactivated Cas13 with an RNA-editing enzyme to achieve site-specific base changes in RNA). Overall, Cas13 expands the CRISPR toolbox into the realm of transcriptome engineering and viral diagnostics in plant biology (Gupta et al., 2022; Hillary and Ceasar, 2023).

3.3.4 Other Cas variants (Cas3, CasX, Cas14, Cas7-11)

Beyond Cas9, Cas12, and Cas13, researchers have discovered and repurposed additional CRISPR–Cas systems that expand the genome editing toolkit for plants. These include both Class 1 systems (which use multiprotein complexes) and novel Class 2 effectors.

- **Cas3 (Type I);** In Type I CRISPR systems, Cas3 is the signature nuclease/helicase that works with a Cascade complex of multiple proteins. Unlike the precise cuts of Cas9/12, Cas3 provides a "DNA shredder" activity: once recruited to the target by the Cascade, Cas3 degrades the DNA processively, often resulting in large deletions spanning kilobases. This has been harnessed for making large genomic deletions or rearrangements. For example, Type IE CRISPR–Cas (from *Thermobifida fusca*) with Cas3 was shown to induce deletions up to 100 kb in human cells, and a Cas3-based approach was tested in maize (*Zea mays*) protoplasts to demonstrate feasibility in plants. As of now, Cas3 systems have not been widely used in live plants, partly due to delivery complexity (multiple Cascade subunits must be expressed), but they hold promise for large-scale genome engineering in plants (Kim et al., 2024).

- **CasX (Cas12e, Type VE)**; CasX is a recently discovered compact DNA nuclease from the Cas12 family (Type VE) characterized by a relatively small size (~980 amino acids). CasX (now Cas12e) was shown to cleave DNA in an RNA-guided manner in bacteria and human cells. Its smaller size makes it attractive for plant gene editing because it could be easier to deliver (for instance, via viral vectors or as RNP complexes). Cas12e uses a dual RNA guide (crRNA plus tracrRNA) similar to Cas9, and it recognizes a short PAM (reported as 5′TA3′ in one study). While plant applications of CasX/Cas12e are still in early stages, its efficient activity in other systems suggests it could be adapted to plants, expanding options for targeting when smaller Cas effectors are needed (Yang and Patel, 2019).
- **Cas14 (Cas12f, Type VF);** Cas14 refers to a group of ultrasmall CRISPR–Cas effectors (~400–700 amino acids) identified in archeal genomes, now classified as Cas12f. Cas14/Cas12f is unique for targeting single-stranded DNA and is capable of distinguishing single-nucleotide differences in target sequences. Because of their small size and single-strand specificity, Cas14 enzymes have been engineered as sensitive diagnostic tools. In plants, a Cas14 system was leveraged to detect single stranded DNA plant viruses (such as geminiviruses) with high specificity. For example, CRISPR–Cas14 diagnostics can be designed to cleave a reporter molecule only in the presence of virus-derived ssDNA, enabling early and accurate detection of plant viral infections. The exceptionally small size of Cas12f also raises the possibility of easier delivery in plant cells (e.g., via nanoparticle or viral delivery), although their use for genome editing (cutting double-stranded genomic DNA) is still being optimized due to the difficulty of these tiny nucleases in unwinding and cleaving stable DNA duplexes (Savage, 2019).
- **Cas7-11 (Type IIIE);** Cas7-11 is a novel effector that blurs the line between Class 1 and Class 2 systems. It originates from a Type III CRISPR (Class 1) but functions as a single large protein, essentially a fusion of multiple Cas7 and Cas11 subunits that can independently process crRNA and cleave RNA targets. Cas7-11 is remarkable for making precise, guide-directed cuts in RNA without collateral damage to other RNAs. This contrasts with Cas13, which has collateral activity; Cas7-11's specific cleavage is thus viewed as a potentially safer RNA editing tool. The drawback is Cas7-11's size, which was initially too large to fit in standard delivery vectors. However, researchers have recently engineered a more compact version called Cas7-11S by removing nonessential regions, retaining its RNA targeting function while reducing size. While use of Cas7-11 in plants has yet to be reported widely, its ability to precisely knock down transcripts without the bystander effect could be valuable for plant biotechnology, for example, in fine-tuning gene expression or conferring virus resistance with minimal off-target effects at the RNA level. As delivery methods improve, Cas7-11 and similar RNA-targeting effectors may become part of the plant gene-editing toolkit alongside Cas13 (Kato et al., 2022).

In summary, the continual discovery of new CRISPR–Cas variants (from the multi-subunit Cas3 and Cas10 systems to compact Cas12f and chimeric Cas7-11) is rapidly expanding the possibilities for plant genome engineering. Each new system brings unique capabilities such as large deletions, smaller delivery size, or precise RNA editing, allowing researchers to choose the best tool for a given application in plant research and breeding (Hillary and Ceasar, 2023).

3.4 Components of the CRISPR toolkit

A functional CRISPR-mediated genome editing system in plants typically comprises a few core components:

(1) a guide RNA (gRNA) that specifies the target sequence
(2) the Cas effector protein that introduces the cut or edit, and
(3) a donor template (if precise editing via homology-directed repair (HDR) is desired) (Kondrateva et al., 2021).

Additional elements like promoters for expression and delivery vectors are also part of the toolkit, but here we focus on the primary CRISPR components and their variants. All these components can be tailored for plant systems. For example, using plant codon-optimized Cas genes, U-rich promoters for gRNA expression in plant nuclei, and appropriate delivery strategies (Agrobacterium, gene gun, etc.). Below, we discuss each component in a general plant biology context (Aksoy et al., 2022).

3.4.1 Guide RNA (gRNA)

Guide RNAs (gRNAs) direct Cas nucleases to target DNA or RNA via base pairing. In bacteria, they comprise a crRNA (spacer + repeat) and a tracrRNA, but plant gene editing typically uses a single guide RNA (sgRNA), combining both. Cas9 sgRNAs include a ~20 nt spacer and an ~80 nt scaffold; Cas12a and Cas13 use only crRNAs with hairpin-forming repeats. gRNAs are driven by U-rich Pol III promoters (e.g., U6/U3), requiring a precise start nucleotide. Spacer design is critical: it must be genome unique and adjacent to a compatible PAM. Tools like CRISPR-P or CHOPCHOP aid in designing gRNA for different organisms. For multiplexing, multiple gRNAs can be expressed via self-cleaving tRNA or ribozyme systems. The gRNA defines Cas targeting specificity (Moon et al., 2019).

3.4.2 Protospacer adjacent motif (PAM)

PAMs are short sequences flanking the target site, essential for Cas binding and activity. SpCas9 requires 5′NGG3′; variants like SaCas9, SpCas9NG, and SpRY broaden PAM compatibility. Cas12 enzymes prefer T-rich PAMs (e.g., LbCas12a uses 5′TTTV3′); engineered versions expand target range (e.g., TYCV, TATV). Cas12 PAMs are 5′-proximal, unlike Cas9′s 3′side PAM. Cas13, targeting RNA, lacks a PAM but may require a protospacer flanking site (PFS). Selecting appropriate PAMs is key to choosing the target site. Mismatches in PAMs may be tolerated but reduce efficiency (Cencic et al., 2014).

3.4.3 Cas proteins and variants

Cas proteins perform the cleavage. SpCas9 (~4.2 kb) is widely used but PAM-limited; SaCas9 (~3.2 kb) and others (e.g., StCas9, CjCas9) offer alternatives with different PAMs. Engineered variants (SpCas9NG, SpRY, HF1, xCas9) expand PAMs or improve fidelity. Cas9 nickase and dCas9 are used for base editing or gene regulation. Cas12a/b/j/f variants vary in activity, PAM specificity, and size. LbCas12a and FnCas12a are commonly used in plants. CasΦ (Cas12j) is compact and has been successfully demonstrated in *Nicotiana*. Cas13 variants (a-d) target RNA and enable viral defense or RNA editing. CasRx (Cas13d) is small and effective. Variant choice depends on size, PAM, target type, and delivery method (CebrianSerrano and Davies, 2017).

3.4.4 Donor templates for HDR

HDR enables precise edits but is inefficient in plants due to NHEJ dominance. Donor templates (ssODNs or dsDNA) provide the intended sequence flanked by homology arms. ssODNs suit small edits; dsDNA is used for insertions, requiring longer arms (0.51 kb). Strategies to enhance HDR include cell cycle synchronization, NHEJ inhibition, geminivirus replicons, and template tethering (e.g., Cas9VirD2 fusions). Donor design must include PAM disruption to prevent recutting. Though HDR frequencies are low (<1%), successful trait insertions (e.g., herbicide resistance) highlight the donor template's value in precise plant genome editing (Song and Stieger, 2017).

3.5 Mechanisms of CRISPR-mediated gene editing

CRISPR-mediated gene editing operates in three general steps: target recognition and binding, DNA cleavage, and DNA repair. An engineered single-guide RNA (sgRNA), which includes a CRISPR RNA (crRNA) sequence complementary to the target DNA, first directs the Cas protein to the intended target site in the genome. A short protospacer adjacent motif (PAM), present adjacent to the target site (for example, 5′NGG3′ for *Streptococcus pyogenes* Cas9), is required for initial binding; the Cas9sgRNA complex scans the genome for PAM sites and, upon finding a matching target adjacent to a PAM, the sgRNA forms base pairs with the DNA target strand. This RNA-DNA hybrid formation triggers conformational changes that activate the nuclease function of Cas9 for cleavage (Asmamaw and Zawdie, 2021).

3.5.1 Target recognition and binding

Target recognition is accomplished by the RNA-guided Cas9 protein searching for a PAM sequence and then testing adjacent DNA for complementarity to the 20-nucleotide guide sequence of the sgRNA. Upon PAM recognition, local DNA unwinding allows the sgRNA to hybridize with the target DNA strand if fully complementary (Asmamaw and Zawdie, 2021). Complete base pairing between the guide RNA and target strand induces a stable R-loop structure (RNA-DNA duplex), effectively "locking" the complex onto the target site. Notably, Cas9 remains catalytically inactive until the correct sgRNA-target pairing is achieved. This multistep surveillance mechanism (PAM scanning followed by sequence-specific binding) underlies the high targeting specificity of CRISPR/Cas9 (Lim et al., 2016).

3.5.2 DNA cleavage

Once the target DNA is recognized and the Cas9sgRNA complex is stably bound, Cas9's nuclease domains cleave the DNA. The Cas9 protein contains two endonuclease domains: HNH and RuvC, each responsible for cutting one strand of the DNA double helix. The HNH domain cuts the DNA strand that is complementary to the guide RNA, while the RuvC domain cuts the noncomplementary strand. In the case of SpCas9, these cuts occur three base pairs upstream of the PAM site, generating a site-specific double-strand break (DSB), typically with blunt ends (Lim et al., 2016). Other CRISPR nucleases (e.g., Cas12a/Cpf1) introduce staggered cuts and have different PAM requirements, but the fundamental principle is the same: a targeted DSB is induced at the genomic locus defined by the guide RNA. This DSB is the critical lesion that triggers the cell's DNA repair processes (Bandyopadhyay et al., 2020).

3.5.3 DNA repair pathways

The cell's response to the CRISPR-induced DSB ultimately produces the genome editing outcome. Eukaryotic cells predominantly rely on two DSB repair pathways: nonhomologous end joining (NHEJ) and homology directed repair (HDR). NHEJ is an error-prone repair mechanism that directly ligates broken DNA ends without the need for a homologous template (Fig. 3.1). Because NHEJ can insert or delete a few base pairs at the break site, it often yields small insertions or deletions (indels) at the cut site. These indels can disrupt reading frames or introduce premature stop codons, effectively knocking out gene function (Xue and Greene, 2021).

3.6 Delivery systems for CRISPR components in plants

Effective delivery of CRISPR components (typically the Cas nuclease and the guide RNA) into plant cells is a critical step for successful gene editing. Plant cells present unique challenges for delivery because of their rigid cell walls and often genotype-dependent transformability. A variety of delivery systems have been developed in plant biology, ranging from biological vectors to physical and chemical methods (Cardi et al., 2023). The goal is to introduce the CRISPR/Cas machinery into either somatic or germline cells of the plant and, ideally, to do so without permanently integrating foreign DNA into the plant's genome. Below, we outline the major delivery methods: *Agrobacterium* mediated transformation, biolistic particle bombardment, viral vector delivery, and nanoparticle/protoplast-based methods (Zulfiqar et al., 2019).

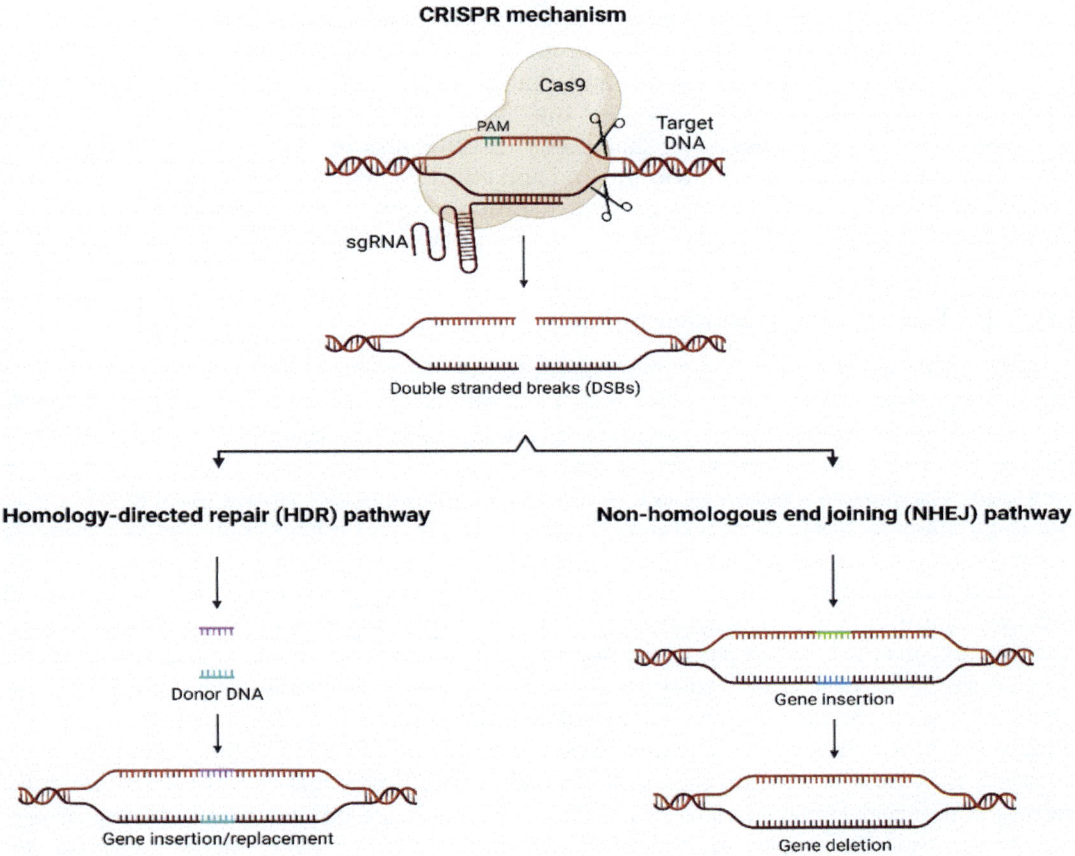

FIGURE 3.1 Cas9 is guided by an sgRNA to a DNA sequence adjacent to a PAM, where it introduces a double-strand break. The break can be repaired by NHEJ, which generates indels and disrupts genes, or by HDR, which uses a donor template for precise insertion or replacement. This balance between repair pathways determines the editing outcome.

3.6.1 *Agrobacterium* mediated transformation

The soil bacterium *Agrobacterium tumefaciens* has long been used as a tool to stably transform plants, and it remains the principal vehicle for delivering CRISPR gene editing cassettes into many plant species. In this method, the Cas gene and sgRNA expression cassette are cloned into a T-DNA region of a binary plasmid, which *Agrobacterium* then transfers into plant cells during infection. The T-DNA (carrying the CRISPR/Cas construct) typically integrates into the plant genome, leading to stable expression of Cas9 and sgRNA in transformed cells. Using *Agrobacterium*, researchers can regenerate whole plants from transformed cells (via tissue culture), obtaining transgenic plants that carry and express the CRISPR components. These plants can pass gene edits to the next generation, especially if germline cells were edited. However, *Agrobacterium* transformation has limitations: many plant species and elite cultivars are recalcitrant to *Agrobacterium*, meaning they do not infect or transform readily (AziziDargahlou and Pouresmaeil, 2024). A notable workaround in certain plants (e.g., *Arabidopsis thaliana*) is the "floral dip" procedure, where *Agrobacterium* is applied to flowering plants so that seeds directly acquire transgenes (Zhang et al., 2006). Overall, *Agrobacterium* is a widely used and effective delivery system for CRISPR components in plants, but its host range and regeneration requirements motivate the use of alternative methods in many cases (AziziDargahlou and Pouresmaeil, 2024).

3.6.2 Biolistic (particle bombardment)

Biolistic delivery, or particle bombardment, is a physical method of introducing DNA or ribonucleoproteins (RNPs) into plant cells by high-velocity bombardment of microscopic particles. In this method, gold or tungsten microprojectiles coated with plasmid DNA, mRNA, or even Cas9 proteins. gRNA-RNP complexes are shot into plant tissue using a gene gun. The particles penetrate cell walls and membranes, delivering the CRISPR cargo internally. Biolistic has become a common alternative for crop species that are less amenable to *Agrobacterium* mediated transformation, such as many cereals and other monocots (Liu et al., 2019). An important advantage of biolistic is that it imposes essentially no biological host range limitation: it can be applied to virtually any plant species or genotype and is even effective on cells that *Agrobacterium* cannot infect. It is also well-suited for codelivery multiple pieces of DNA or RNPs simultaneously, which is useful for multiplex genome editing or for providing a repair template to stimulate HDR (Li et al., 2022). However, particle bombardment has drawbacks. The integration of delivered DNA is random; when plasmids are used, they may insert into the plant genome in multiple copies and at off-target sites, leading to complex transgene insertions. This can complicate downstream breeding or regulatory approval if transgene-free plants are desired. Despite these challenges, biolistic delivery has been successfully used to create a number of genome-edited crops and is especially valuable for plants that lack efficient *Agrobacterium* transformation protocols (Zhu et al., 2020).

3.6.3 Viral vectors

Plant viruses have been engineered as vectors to deliver CRISPR components, offering a means of transient delivery that can circumvent some tissue culture steps. Because many plant viruses can systemically infect cells of an intact plant, they provide a way to edit plant genomes in planta (within the living plant) and potentially reach reproductive cells for heritable edits. Viral delivery strategies in plants generally involve using DNA viruses (like geminiviruses) or RNA viruses to carry parts of the CRISPR machinery. Viral vector systems thus represent a powerful toolkit for CRISPR delivery, especially for achieving transient editing (where no transgene integration is desired) and for tackling the challenge of editing multiple cells or the germline in a whole plant (Zaidi and Mansoor, 2017).

3.6.4 Nanoparticles and other methods

In recent years, researchers have begun exploring nanoparticle-mediated delivery and other innovative techniques to overcome the barriers of plant cell transformation. Nanoparticles, such as mesoporous silica nanoparticles, carbon nanotubes, and various polymers, can serve as carriers to ferry CRISPR reagents into plant cells. These nanomaterials can be functionalized to bind DNA or protein and are small enough to pass through cell walls or be taken up by plant cells (for instance, via leaf infiltration or high-pressure spraying). One major advantage of nanoparticle-mediated delivery is that it can enable gene editing without stable DNA integration by delivering preassembled Cas9–sgRNA ribonucleoproteins (RNPs) or RNA transcripts, both of which function transiently and are degraded after their activity is complete. In plants, however, truly efficient nanoparticle-based CRISPR delivery remains challenging, largely because a very high proportion of cells must be transformed to regenerate an edited plant (Demirer et al., 2021).

3.7 Recent innovations

The CRISPR–Cas toolkit is continually evolving, and several innovative genome editing strategies have emerged in recent years to address limitations of classic CRISPR and to expand its capabilities. Three notable advances are base editing, prime editing, and multiplex genome editing. Base and prime editing are often called "CRISPR 2.0" technologies because they allow more precise nucleotide changes without making double-strand breaks, an especially important feature for applications where creating random indels is not sufficient (Liu et al., 2017) (Fig. 3.2). Below, we describe these innovations and their relevance in the context of plant genome editing.

3.7.1 Base editing

Base editing is a method that uses a modified, catalytically impaired Cas enzyme (often Cas9 nickase or dead Cas9) fused to a DNA-modifying enzyme to directly convert one base pair into another at a target site without introducing a DNA break. The most common base editors are cytosine base editors (CBEs) and adenine base editors (ABEs). CBEs typically consist of Cas9 nickase tethered to a cytidine deaminase enzyme; when guided to a target site, this complex converts a C·G base pair into a T·A base pair (through deamination of cytosine to uracil, which is then resolved to thymine after DNA replication or repair). ABEs similarly convert A·T to G·C by deaminating adenine to inosine (which is read as guanine) in the DNA (Anzalone et al., 2020). In plant systems, base editing has quickly become a powerful technique (Molla et al., 2021). Base editors leverage endogenous base excision repair pathways to install targeted nucleotide changes. Because no double-strand break is made, the frequency of undesired indels is greatly reduced, and cellular stress responses to DNA damage are minimized (Molla et al., 2021).

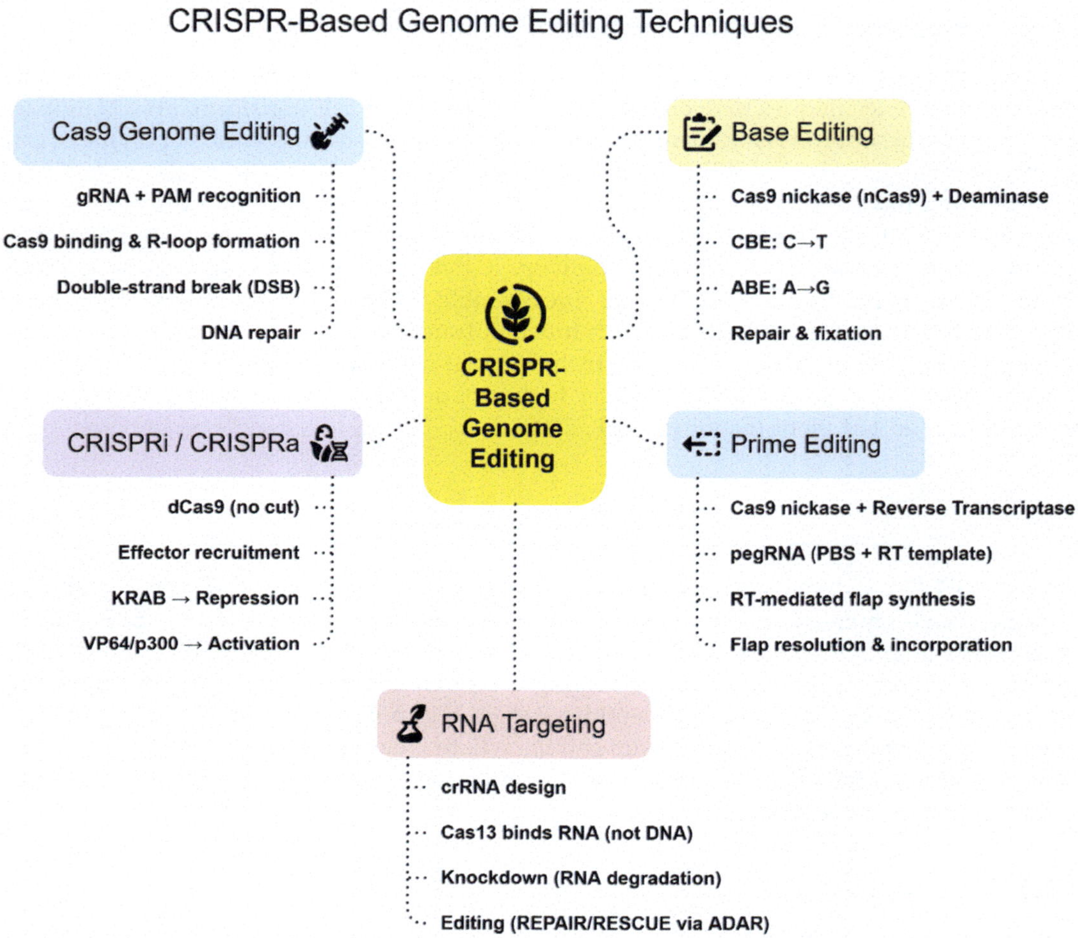

FIGURE 3.2 CRISPR tools have expanded beyond Cas9 cutting into multiple modalities. Base editing introduces precise single-nucleotide changes without DSBs, while prime editing enables diverse small edits using a reverse transcriptase and pegRNA. CRISPRi/a modulates transcription through dCas9–effector fusions, and Cas13 systems target RNA for knockdown or editing. Together, these approaches broaden the scope of genetic and epigenetic control.

3.7.2 Prime editing

Prime editing is a newer CRISPR-derived technology that can be thought of as a "search and replace" gene editing system. It expands the scope of possible edits beyond what base editors or standard CRISPR/Cas9 can achieve, allowing insertions, deletions, and all types of base substitutions in a targeted manner, all without requiring donor DNA templates or inducing DSBs. The prime editing process works as follows: the Cas9 nickase-RT complex binds to the target DNA via the pegRNA, makes a single-strand nick in the target strand, and then the reverse transcriptase synthesizes new DNA at the nick using the pegRNA's template sequence. This installs a modified sequence at the break site. The cell's repair processes then incorporate the edited strand into the genome, fixing the desired change (Mikhaylova et al., 2024). Initial demonstrations in rice, wheat, and tomato showed that prime editors could, for example, introduce precise point mutations conferring herbicide resistance or correct disease-susceptibility alleles, with fewer off-target effects than classical HDR-based editing (Lin et al., 2020; Lu et al., 2020). Researchers are actively improving prime editing efficiency in various crops, and it holds promise for introducing complex traits like allele swaps or gene corrections in crop improvement programs.

3.7.3 Multiplex gene editing

Multiplex gene editing refers to the simultaneous editing of multiple target sites in the same organism. With CRISPR technology, this is typically achieved by expressing multiple sgRNAs along with a Cas nuclease, allowing the nuclease to introduce mutations at several target sites at once. In plants, multiplex editing is particularly valuable because many agronomic traits are controlled by gene families or multiple pathways, and knocking out or modifying several genes in parallel can create a more pronounced phenotype or confer a desired trait that single-gene edits cannot (McCarty et al., 2020). Using such strategies, researchers have been able to edit two, three, or even more target genes simultaneously in plants. Multiplex CRISPR tools have revolutionized plant biology and breeding by making it feasible to induce complex genetic changes that used to require years of crossing different mutants (Najera et al., 2019). The approach accelerates research as well: it enables dissection of redundant gene functions by mutating all redundant family members together and allows engineering of traits like heterosis or polygenic stress tolerances by stacking mutations (McCarty et al., 2020).

3.8 Applications in plant biotechnology

CRISPR/Cas technology has rapidly become a central tool in plant biotechnology, livestock, and bacterial gene editing allowing precise and efficient genome modifications. Its applications span disease resistance, yield and quality improvement, abiotic stress tolerance, and metabolic engineering (Asmamaw and Zawdie, 2021) (Fig. 3.3). In these areas, CRISPR has been employed to knock out genes, introduce precise allele edits, or modulate gene expression to enhance plant performance and product value. A growing number of experimental and field-based studies in major crops validate its potential for agricultural innovation (Liu et al., 2021).

This section highlights peer-reviewed, experimentally validated CRISPR applications in three major crops, rice, maize, and tomato, chosen for their global significance. For each crop, we examine how CRISPR has been applied to address biotic and abiotic stress, boost productivity, and improve nutritional or metabolic profiles. Each case study includes the crop's importance, CRISPR targets and strategies, and the resulting phenotypic or agronomic outcomes, underscoring the transformative impact of CRISPR-driven genome editing in plant science.

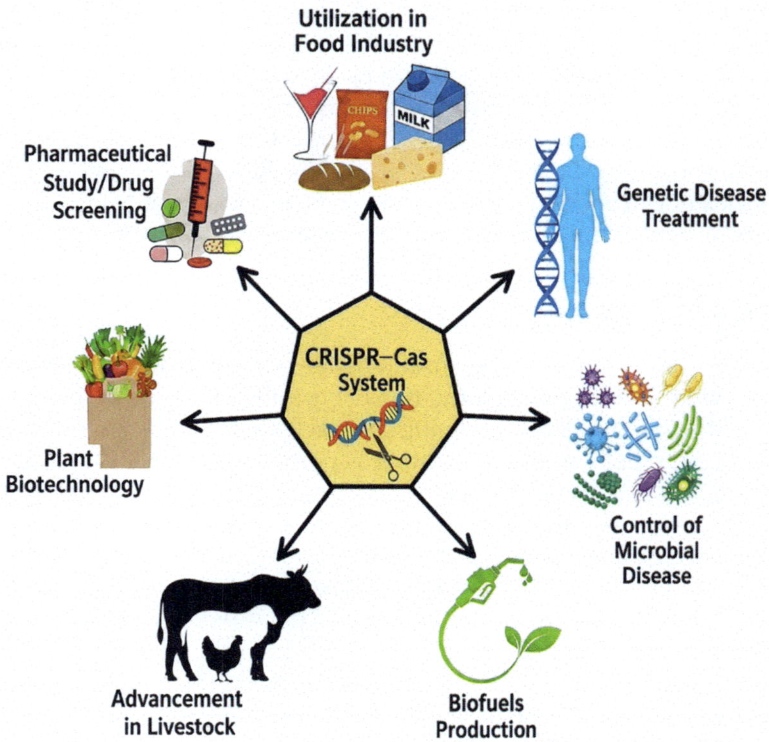

FIGURE 3.3 CRISPR systems are applied widely in medicine, agriculture, and industry. They enable gene therapy and disease modeling, improve crop traits and livestock, and drive innovations in microbial engineering, food production, and biofuels. These applications demonstrate CRISPR's role as a transformative technology across biological sciences.

3.8.1 Rice (*Oryza sativa*)

Rice is a staple cereal for over half of the world's population, making improvements in its resilience and nutritional quality highly impactful. CRISPR–Cas technology has been widely adopted in rice due to efficient transformation and regeneration, enabling precise edits for trait improvement. Below, we highlight key CRISPR applications in rice across different trait categories.

3.8.1.1 Disease resistance

CRISPR-mediated gene editing has conferred resistance to major rice diseases by targeting susceptibility (S) genes. A prominent example is bacterial blight (BB) caused by *Xanthomonas oryzae*. Researchers used CRISPR–Cas9 to introduce small deletions in the promoter of the *OsSWEET14* gene (a known S gene activated by pathogen TAL effectors). In an elite rice cultivar, an engineered 6-bp promoter deletion prevented pathogen-induced *OsSWEET14* expression, resulting in dramatically shorter lesion lengths (~17 cm) compared to wild-type plants (1829 cm lesions) under *Xanthomonas* inoculation. The *OsSWEET14* edited lines exhibited high BB resistance without any yield penalty, demonstrating that knocking out or altering S gene promoters via CRISPR can provide effective disease resistance (Zafar et al., 2020). Similar strategies have been applied to other rice diseases; for instance, CRISPR knockouts of *OsERF922* confer resistance to rice blast fungus, and *OsSEC3A* edits provide blast tolerance, highlighting the broad utility of genome editing for enhancing rice disease resistance (Wang et al., 2016).

3.8.1.2 *Yield enhancement*

Yield-related traits in rice have been improved by targeting negative regulators of grain size, number, and plant architecture. Multiple studies report that CRISPR–Cas9 knockouts of genes such as *Gn1a* (Grain number 1a), *DEP1* (Dense and Erect Panicle 1), *GS3* (Grain Size 3), *GW2*, and *TGW6* lead to increased grain number, larger grain size, or heavier 1000 grains weight (Li et al., 2016; Achary and Reddy, 2021). For example, CRISPR was used to create novel alleles of *Gn1a* and *DEP1*. Field evaluations showed that one edited *Gn1a* allele and three *DEP1 alleles* each conferred significantly higher grain yield than even the best-performing natural alleles. These yield gains were achieved by elevating panicle grain number (*Gn1a* knockout increases inflorescence meristem activity) and by producing a compact, erect panicle architecture (*DEP1* knockout) that supports more grain without lodging (Huang et al., 2018). Likewise, editing grain size genes like *GS3* (a negative regulator of grain length) results in larger grains and higher yield per plant (Sun et al., 2018). Overall, CRISPR enabled mutagenesis of key yield genes in rice has proven to be a rapid breeding strategy to improve grain productivity, validating phenotypes previously observed in traditional mutants.

3.8.1.3 *Abiotic stress tolerance*

Enhancing stress tolerance in rice via CRISPR has focused on genes that modulate drought and salinity responses. One case study is the *OsDST gene* ("Drought and Salt Tolerance" regulator): Santosh Kumar et al. (2020) used CRISPR–Cas9 to knockout *OsDST* in an indica rice cultivar, which produced mutants with broader leaves, reduced stomatal density, and superior leaf water retention under drought stress. These physiological changes translated into improved drought tolerance in edited plants. Another example involves the ABA signaling pathway gene *OsPYL9* (an ABA receptor). Precise CRISPR edits of *OsPYL9* yielded rice lines with enhanced drought resilience and even increased grain yield under water stress. The *OsPYL9* mutants had altered circadian and stress-responsive gene expression, improving their performance during drought (Usman et al., 2020). For salinity tolerance, knocking out *OsRR22* (a negative regulator in the cytokinin pathway) via CRISPR has been shown to confer higher salt tolerance in rice seedlings (Zhang et al., 2019). Similarly, CRISPR disruption of *OsERA1* (ABA signaling repressor) imparted greater drought tolerance and deeper rooting in edited plants (Ogata et al., 2020). These cases illustrate that CRISPR can create alleles for climate resilience, enabling rice plants to maintain yield and survive under adverse environmental conditions.

3.8.1.4 *Metabolic and nutritional trait engineering*

CRISPR–Cas is also accelerating improvements in rice grain quality and nutrition. A notable achievement is the creation of highly resistant starch (high amylose) rice lines through multiplex gene editing. Using CRISPR, researchers simultaneously knocked out four starch branching enzyme (SBE) genes in the rice cultivar 'Presidio'. Homozygous quadruple mutants were obtained, and some lines showed up to a 15% increase in resistant starch content in the grain. This higher amylose content is nutritionally beneficial, as resistant starch has a lower glycemic index and promotes gut health (Biswas et al., 2023). Another metabolic trait engineered by CRISPR is grain aroma: disrupting the *BADH2 gene* (which normally suppresses synthesis of 2-acetyl-1-pyrroline, the compound conferring fragrance) via CRISPR has led to new aromatic rice varieties (Imran et al., 2023). Additionally, gene editing has been used to improve food safety by reducing toxic element accumulation. For instance, CRISPR knockouts of *OsNRAMP5* and *OsARM1* in rice have been shown to significantly reduce cadmium and arsenic levels in rice grains, addressing health concerns (Zhang and Huang, 2019). These examples underscore how CRISPR-mediated modifications in metabolic pathways can enhance the nutritional profile of rice (e.g., provitamin A, mineral content, amylose) and reduce harmful components, all while maintaining agronomic performance.

3.8.2 Maize (*Zea mays*)

Maize is the second most widely cultivated cereal in 2020. It serves as a staple food, animal feed, and industrial crop (Erenstein et al., 2022). Modern maize breeding emphasizes high yield, stress resilience, and grain quality. CRISPR–Cas genome editing has emerged as a valuable tool to introduce targeted variations into maize's often elite but genetically narrow germplasm, enabling trait improvements that are otherwise difficult to achieve by conventional breeding. Below, we review CRISPR case studies in maize for disease resistance, yield, stress tolerance, and metabolic traits.

3.8.2.1 Disease resistance

CRISPR has been deployed in maize to develop resistance against viral and fungal pathogens that significantly impact yield. A recent example involves maize rough dwarf disease, caused by a virus. Liu et al. (2022) used CRISPR–Cas9 to target and knock out the maize gene *ZmGDIα*, which encodes a factor exploited by Rough Dwarf Virus (RDV). The edited *zmGDIα* mutants showed strongly increased resistance to maize rough dwarf virus, with no observable agronomic penalties in growth or yield. In another study, knocking out the maize *LOX3* gene (lipoxygenase 3) via CRISPR conferred enhanced resistance to fungal infection by *Ustilago maydis* (corn smut). The lox3 mutants exhibited an amplified reactive oxygen species burst upon infection, which is linked to durable immunity against the pathogen. These findings demonstrate that editing host susceptibility genes can create maize lines with improved disease resistance (Wang et al., 2022). Additionally, CRISPR editing has been applied to combat Gibberella stalk rot: for instance, deleting the *ZmCOI1a* gene (a regulator of jasmonate signaling) resulted in heightened resistance to *Fusarium graminearum* in stalks (Wang et al., 2022). Collectively, such case studies highlight CRISPR's versatility in engineering disease-resistant maize by either knocking out susceptibility factors or enhancing innate defense pathways.

3.8.2.2 *Yield enhancement*

Improving grain yield is a primary goal in maize, and CRISPR has enabled precise modifications of genes controlling ear development and plant architecture. One compelling example is the manipulation of kernel row number (KRN), a key determinant of ear size and yield. An et al. (2022) identified a gene (referenced as Zm00001d016075) that acts as a negative regulator of kernel row number. Using CRISPR–Cas9, they created loss-of-function mutations in this gene, which led to a significant increase in the number of kernel rows per ear and a corresponding rise in grain yield per plant. This targeted approach effectively enhanced yield potential by altering inflorescence meristem development (An et al., 2022). Another yield-related application is the generation of shorter, sturdier maize plants for high-density planting. For instance, editing the *ZmGA20ox3* gene (involved in gibberellin biosynthesis) produced semidwarf maize lines with improved lodging resistance. Such semidwarf CRISPR mutants can sustain higher planting densities and fertilizer inputs, indirectly boosting overall field yield (Liu et al., 2024). Moreover, a landmark field demonstration by Shi et al. (2017) replaced the native promoter of the *ARGOS8* gene with a strong constitutive promoter using CRISPR, achieving elevated *ARGOS8* expression. The edited maize showed a five bushel/acre (~5%10%) yield increase under drought stress conditions in multilocation field trials (Shi et al., 2017). This result illustrates how gene editing can create "superior alleles" that outperform natural variants in yield under challenging environments. Overall, CRISPR has opened new avenues to fine-tune maize yield components (kernel number, size, plant stature) with greater speed and precision than traditional breeding.

3.8.2.3 *Abiotic stress tolerance*

Given maize's sensitivity to drought and other abiotic stresses, CRISPR has been employed to fortify stress tolerance mechanisms. A notable success is the *ARGOS8* promoter editing mentioned above, which not only raised yield but specifically improved drought tolerance. *ARGOS8* is a negative regulator of ethylene responses; by boosting its expression via CRISPR promoter replacement, maize plants had reduced ethylene sensitivity and thus sustained higher yields under water-limited conditions. In essence, this created a drought tolerant maize line that performs better in dry field environments (Shi et al., 2017). Another example is improving resistance to stalk lodging (often caused by wind or rain storms): CRISPR edits of the *stiff1 gene* (which negatively regulates stalk strength) introduced a frameshift that resulted in thicker, stronger stalks. The edited plants are less prone to lodging, maintaining yield in high-wind conditions or high-density stands (Zhang et al., 2020). Salt tolerance has also been addressed; for instance, knocking out *ZmCLCg* (a chloride channel gene) was found to adjust ionic homeostasis under salinity stress, although further validation in field environments is ongoing (Wang et al., 2022). Additionally, discoveries from CRISPR functional genomics are identifying new targets: *ZmSRL5* was shown to be essential for cuticular wax and drought tolerance, suggesting that CRISPR knockout of *ZmSRL5* can make leaves more drought-hardy (Pan et al., 2020). In summary, CRISPR-based edits in maize have yielded traits like drought tolerance and lodging resistance by modulating stress-response genes, thereby stabilizing production in the face of climate stressors.

3.8.2.4 *Metabolic and nutritional traits*

Maize is not only a caloric staple but also an industrial crop, so metabolic trait engineering via CRISPR has significant value. A prime example is the development of waxy maize through gene editing. Waxy maize (high in amylopectin starch) is desired for food and bioindustry uses. Traditionally, waxy hybrids were made by introgression a natural loss-of-function allele of the *Wx1 gene* (which encodes granule-bound starch synthase). Using CRISPR–Cas9, researchers were able to directly knock out *Wx1* in multiple elite inbred lines, creating isogenic waxy versions of those lines. The resulting CRISPR-derived waxy hybrids showed the expected starch composition (almost 100% amylopectin) while maintaining the high yield and agronomic performance of the original hybrids. In fact, the CRISPR-waxy corn performed better than conventionally bred waxy lines, which often carry yield drag due to linked genes. This case, culminating in a limited commercial release of CRISPR waxy corn in the US Midwest, exemplifies how genome editing can fast-track specialty trait development without sacrificing productivity (Miladinovic et al., 2021; Wang et al., 2024). Another metabolic trait edited in maize is aroma/flavor: by simultaneously knocking out two homologs (*ZmBADH2A* and *ZmBADH2B*), researchers created maize kernels that accumulate aromatic 2-acetyl-1-pyrroline (analogous to fragrant rice). Quality protein improvement is also being pursued; while classic Quality Protein Maize was achieved via natural mutations (Opaque2), CRISPR is now enabling more precise edits in storage protein or amino acid biosynthesis genes to enhance lysine and tryptophan content in maize endosperm (Wang et al., 2021). Furthermore, ongoing studies use CRISPR to modify oil biosynthesis genes for higher oleic acid corn oil and to increase provitamin A carotenoids by targeting carotenoid hydroxylases. Though many of these nutritional edits are in research stages, the wax.

3.8.3 Tomato (*Solanum lycopersicum)*

Tomato is one of the most important vegetable crops worldwide, grown for its fruits, which are a key source of vitamins and antioxidants in the human diet. It is the second most consumed vegetable after potato, with annual production on the order of 100–180 million tons globally (Calıskan et al., 2022). Tomato is also a model system for fleshy fruit biology. Enhancing disease resistance, yield, stress tolerance, and nutritional quality in tomato has broad implications for both food security and nutritional health. CRISPR–Cas genome editing has been avidly adopted in tomato research, given the ease of transformation and the availability of wild genetic diversity that can be harnessed by “de novo domestication” strategies. Below, we outline key CRISPR applications in tomato across the major improvement categories.

3.8.3.1 *Disease resistance*

Several studies have leveraged CRISPR to impart resistance to tomato diseases by knocking out host susceptibility genes. A striking example is the broad-spectrum disease resistance achieved by editing the *SlDMR61 gene* in tomato. *DMR6* encodes an enzyme that is a negative regulator of plant defense; Thomazella et al. (2021) created CRISPR knockouts of *SlDMR61* and found that the mutants were resistant to multiple pathogens, including bacterial speck (*Pseudomonas syringae*), oomycete late blight (*Phytophthora infestans*), and fungal wilt, without visible fitness costs. This single-gene edit conferring Multi disease resistance is especially significant for a crop like tomato, which suffers from diverse pathogens (Thomazella et al., 2021). Another well-known case is resistance to powdery mildew. Nekrasov et al. (2017) generated an *SlMLO1 gene* knockout in tomato using CRISPR; *MLO1* absence is known to prevent powdery mildew fungal infection. The edited tomato plants showed complete resistance to *Oidium neolycopersici* (powdery mildew), with no mildew lesions under challenge, confirming the expected resistant phenotype. Similarly, CRISPR knockout of *SlPMR4*, another powdery mildew-related susceptibility gene, significantly reduced mildew disease severity (Koseoglou, 2017). Beyond fungal diseases, CRISPR has been used to engineer virus resistance in tomato. Wang et al. (2018) knocked out tomato *DCL2 genes* (DICERlike 2, involved in antiviral RNA interference); double mutants in *SlDCL2a/2b* exhibited enhanced resistance to tomato mosaic virus, as the absence of certain *DCL2* isoforms can trigger a compensatory antiviral defense pathway. Additionally, CRISPR has been harnessed in a more direct antiviral approach: researchers designed sgRNAs targeting conserved motifs of tomato yellow leaf curl virus (TYLCV) and, by expressing Cas9 in tomato, achieved significant reductions in TYLCV replication in infected plants (Tashkandi et al., 2018). These studies collectively demonstrate that CRISPR–Cas editing can fortify tomato against a spectrum of diseases by either removing susceptibility factors or by directly targeting viral genomes, thereby reducing reliance on chemical controls and increasing crop durability.

3.8.3.2 Yield enhancement

Enhancing tomato yield through CRISPR has involved editing genes that control plant architecture, flowering time, and fruit size/number traits collectively determining productivity. One pioneering demonstration was the de novo domestication of a wild tomato relative *(Solanum pimpinellifolium)* using multiplex CRISPR editing. Zsögön et al. (2018) simultaneously edited six loci in the wild tomato (including genes for growth habit, flowering, fruit size, and nutrient content) to create a new, high-yielding cultivated-type tomato in a single generation. The resulting lines had dramatically improved yield traits: fruit size was increased ~3 fold and fruit number ~10 fold compared to the wild progenitor, owing to edits such as *SELF PRUNING (SP gene)* (for determinate growth and concentrated fruit set) and FW2.2/CLV3 (for larger fruit). Such CRISPR-derived plants effectively transformed a wild, low-yield tomato into a productive cultivar while also boosting nutritional lycopene (Zsögön et al., 2018). In more targeted studies on cultivated tomato, CRISPR has been used to tweak flowering genes to accelerate and increase fruit yield. For example, knocking out the *SP5G gene* (a florigen suppressor) causes early flowering; when combined with a partial loss of function in *SELF PRUNING* (for determinacy), it produces ultra-early bearing, compact tomato plants ideal for high-density cultivation and even urban farming. These edits lead to tomato varieties that flower and set fruit weeks earlier than normal, potentially allowing additional harvest cycles per season (Soyk et al., 2017). Another yield-related trait is inflorescence branching: the gene *J2 (Jointless2)* and *IAA9* have been edited to create highly branched trusses with more flowers and fruits, thereby increasing fruit count per plant in certain tomato lines (Baranov et al., 2024). While some architecture edits can trade off fruit size for number, the overall yield (total fruit weight) can be optimized. Importantly, many of these yield-oriented edits can be stacked; for instance, combining *SP/SP5G* (earliness and determinacy) with *CLV3* (fruit size) and multiflora (branching) yields a tomato ideotype with high yield potential per unit area (Larriba et al., 2024). In summary, CRISPR is enabling precise assembly of tomato yield traits achieving in a few editing steps what took decades in traditional breeding thus accelerating the development of tomato varieties with improved productivity and suitability to modern cultivation systems.

3.8.3.3 *Abiotic stress tolerance*

Tomatoes are often challenged by drought, salinity, and heat in open field and greenhouse production. CRISPR-based gene editing has offered new strategies to improve tomato's abiotic stress resilience. One example is enhancing drought tolerance by altering hormonal signaling. IllouzEliaz et al. (2020) used CRISPR to modify the *SlGID1* gene, which encodes a gibberellin receptor involved in growth and stress responses. The edited tomatoes had a partial loss of function GID1 *allele*, resulting in plants with reduced GA signaling. Under drought conditions, these *SlGID1* edited plants maintained higher leaf water content and exhibited improved survival and yield compared to wildtype, indicating enhanced drought tolerance. This improvement is attributed to GID1 mutants having smaller stomatal apertures and more robust roots, thus conserving water (IllouzEliaz et al., 2020). In another study, CRISPR was used to knocout *SlLBD40* (a LATERAL ORGAN BOUNDARIES DOMAIN transcription factor) in tomato. *SlLBD40* knockout mutants showed significantly greater drought resistance, likely by modulating root development and stress response genes (Liu et al., 2020). Tran et al. (2021) edited the *SlHyPRP1 gene*, which encodes a proline-rich protein. Precise deletion of a repressor domain in *SlHyPRP1* led to tomato lines that germinated better and grew more vigorously under high salinity, due to enhanced osmo-protection and salt stress signaling. Similarly, Bouzroud et al. (2020) knocked out *SlARF4* (an auxin response factor) and observed that the mutants were more salt-tolerant, presumably by altering ion transport and leaf cuticle composition. For heat stress, targeting regulatory genes like *SlMAPK3* have been explored: CRISPR-edited *SlMAPK3* mutants maintained higher antioxidant enzyme activities and showed better fruit set under heat, indicating improved thermotolerance (Wang et al., 2017). These cases illustrate that CRISPR edits can harden tomato plants against abiotic stresses by recalibrating their growth-regulatory networks. The engineered stress-tolerant tomatoes can better withstand episodes of drought or salinity, thereby stabilize yield and extending the range of environments for tomato cultivation.

3.8.3.4 *Metabolic and nutritional trait engineering*

Tomato fruits are renowned for health-promoting compounds (vitamins, antioxidants), and CRISPR has enabled the enhancement of nutritional traits as well as other metabolic features like flavor. A landmark example is the development of a high-GABA tomato in Japan, the first CRISPR-edited food to be commercialized. Scientists edited the tomato gene *SlGAD3* (which encodes glutamate decarboxylase, an enzyme that limits GABA accumulation). By knocking out an autoinhibitory domain in *SlGAD3,* the fruits of the edited "Sicilian Rouge High GABA" tomato accumulate 45 times more γ-aminobutyric acid (GABA) than the fruits of a conventional tomato. GABA is a bioactive compound believed to confer health benefits such as lowering blood pressure (Nonaka et al., 2017). Another trait improved by

CRISPR is fruit carotenoid content. In the aforementioned de novo domestication study, one of the six edited loci was *CRTISO*, a carotenoid isomerase gene. The edited wild tomatoes showed a 500% increase in lycopene content (a red pigment and antioxidant) compared to standard cultivated tomatoes. The crtiso knockout disrupted upstream carotenoid flux, effectively biofortifying the fruit with lycopene (resulting in a deep red color and higher nutritional value) (Zsögön et al., 2018). CRISPR has also been applied to improve tomato flavor and postharvest qualities: for instance, deleting genes *INV* and *LIN5* (involved in sugar metabolism) can increase fruit sugar content and sweetness, and knocking out PG or PL (polygalacturonase and pectate lyase) leads to fruits with enhanced firmness and shelf life by reducing cell wall breakdown (Wang et al., 2021). CRISPR's ability to multiplex, or edit several genes at once, is particularly useful in this context; for example, simultaneous editing of carotenoid-related genes can generate tomato fruits with different colors, such as yellow, orange, and pink, each enriched with specific carotenoids. In conclusion, CRISPR–Cas technology has proven to be a powerful tool in tomato for enhancing nutritional quality (from amino acids like GABA to vitamins and antioxidants) and modifying metabolic pathways to meet consumer preferences, all while maintaining or even improving yield and agronomic performance.

3.9 Challenges and limitations

Even as CRISPR-based genome editing tools transform biology, several challenges and limitations temper their applications. These range from molecular issues (like unintended mutations and variable efficiency) to practical obstacles in delivery, organism development, and socio-legal acceptance. Below, we examine key challenges and how they impact different contexts, particularly contrasting plant genome editing with animal and medical applications (Park et al., 2025).

3.9.1 Off-target effects

Off-target editing, where CRISPR–Cas induces unintended mutations at non-target genomic sites, is a major concern, particularly in medical and animal systems. Such off-target cuts can cause unpredictable insertions/deletions, disrupt gene regulation, or activate oncogenes, posing serious safety risks. Accordingly, several strategies have been developed to improve targeting precision, including high-fidelity Cas9 variants, paired nickases, truncated guide RNAs, and comprehensive off-target screening via sequencing and computational tools (Manghwar et al., 2020; Kalter et al., 2025).

In contrast, off-target effects in plants are generally less critical due to the ability to remove undesired mutations through backcrossing in sexually reproducing crops. This natural filtering mechanism allows selection of individuals with the intended edit but free from off-target changes, an advantage not available in animal systems. Still, plant researchers remain cautious, using in-silico and in-vitro prediction tools to assess gRNA specificity and minimize off-target potential. Thus, while off-target effects are a key safety concern in therapeutic applications, they are a manageable challenge in plant genome editing, addressed through both molecular design and genetic segregation (Hajiahmadi et al., 2019).

3.9.2 Editing efficiency and mosaicism

Editing efficiency, the percentage of cells or organisms incorporating the intended modification, varies widely depending on the cell type, DNA repair pathway, and nature of the edit. Precise insertions via HDR are typically much less efficient than simple NHEJ-mediated knockouts. This often necessitates screening large numbers of individuals to identify those carrying the desired edits, particularly for knock-ins or multiplexed targets (Wang and Han, 2024).

Mosaicism arises when CRISPR edits occur after the first cell division, resulting in organisms with a mix of edited and unedited cells. In animals, especially when using zygotic CRISPR injections, mosaicism is common and complicates both phenotypic analysis and inheritance, as not all tissues or offspring may carry the edit. Strategies to reduce mosaicism include early-stage delivery of CRISPR components or using pre-edited donor cells via somatic cell nuclear transfer, though these approaches are technically demanding (Mehravar et al., 2019).

In plants, mosaicism can occur during tissue regeneration from edited calli, where not all cells are uniformly modified. This is typically addressed by generating multiple independent events and selecting non-mosaic, true-breeding lines in subsequent generations. Still, both low editing efficiency and mosaicism require considerable validation and screening effort (Soyars et al., 2018).

To address these challenges, researchers are optimizing delivery systems and employing alternatives like base and prime editors that bypass HDR, as well as refining expression timing and Cas/gRNA constructs to boost efficiency. Despite progress, mosaicism and variable editing rates remain significant technical hurdles in both plant and animal systems (Wang and Han, 2024).

3.9.3 Regulatory and public acceptance

Regulation and public perception shape CRISPR's real-world deployment. Globally, regulatory frameworks for gene-edited organisms vary: the European Union has historically regulated CRISPR-edited crops as GMOs, though recent proposals aim to exempt edits without foreign DNA. In contrast, the US, Brazil, Japan, and others take more lenient views, often excluding small, transgene-free edits from strict oversight. These differences affect innovation timelines and commercialization potential (Ahmad et al., 2021).

Public acceptance is mixed but improving—surveys suggest gene-edited crops are viewed more favorably than traditional GMOs, especially when edits mimic natural mutations. Transparency and education are key to fostering trust. In human health, the controversial 2018 germline editing case spurred international consensus against heritable genome modification. Most countries now permit somatic gene editing (e.g., for disease therapy) under strict regulation, while germline editing remains largely banned. Ethical engagement and clear communication about CRISPR's benefits (e.g., disease resistance, reduced pesticide use) are critical for public support (Nguyen et al., 2023).

3.9.4 Intellectual property and access

CRISPR's legal landscape is complex, with competing patents from the Broad Institute and the CVC group (UC Berkeley, Charpentier, and Vienna) creating a fragmented IP environment. The Broad holds patents on CRISPR use in eukaryotic cells, while UC controls foundational bacterial/plant applications. Globally, parallel disputes are ongoing. Over 11,000 CRISPR-related patent families exist, often requiring users to navigate multiple licenses for a single application (Sanner, 2016; Holman, 2022).

This poses challenges for startups, public researchers, and developing countries, especially in agriculture and therapeutics. While academic use is generally unimpeded via non-profit licenses (e.g., through Addgene), commercial development requires formal licensing, raising cost and complexity. Equity concerns also arise for CRISPR-based therapies, which may be prohibitively expensive. Open-access initiatives and humanitarian licensing could help ensure broader global access to this transformative technology (Meyer, 2024).

3.10 Future prospects

Looking ahead, the CRISPR field is rapidly evolving into new technologies and interdisciplinary integrations often dubbed "CRISPR 3.0 and beyond". Future gene editing tools will be more precise, versatile, and intelligent, frequently aided by computational power and novel engineering. Below we explore emerging prospects: next-generation CRISPR systems, the convergence of CRISPR with artificial intelligence, the fusion of gene editing with synthetic biology circuits, and other cutting-edge directions (such as RNA-guided transposases). These developments promise to expand the genome editing "toolkit" from basic cut-and-repair toward an era of programmable gene writing, smart therapeutics, and complex genetic programming. The recent development of DNA-guided Cas12a shows that synthetic CRISPR DNA can reprogram Cas12a for sequence-specific RNA recognition and cleavage, offering a new route for programmable transcript regulation (Wu et al., 2026). In plants, such systems could complement genome editing by enabling transient knockdown of target genes, detection of plant RNA viruses, and rapid functional analysis of stress- or disease-related transcripts without permanently altering the genome. Similarly, plant-optimised artificial intelligence designed editors (PAiD) can further expand the potential of CRISPR to develop gene-edited plants based on CRISPR–Cas. Artificial nucleases like Open-CRISPR-1 will enable researchers to improve editing outcomes and efficiencies of various editing techniques such as base editing and prime editing (Das et al., 2026).

3.10.1 "CRISPR" and beyond

CRISPR technology has evolved from its early form (CRISPR 1.0) based on wild-type Cas9, creating double-strand breaks (DSBs) for knockouts, to CRISPR 2.0 systems like base editors and dCas9-fusion tools enabling precise, non-DSB modifications (Knutsen, 2024). Now, Next-generation tools like prime editing, which uses a Cas9 nickase fused to a reverse transcriptase and guided by a pegRNA to perform precise "search-and-replace" edits, including all 12 base conversions, insertions, and deletions without requiring donor DNA or DSBs (Mikhaylova et al., 2024). Despite its versatility, prime editing faces hurdles like bulky machinery and pegRNA optimization.

Beyond prime editing, emerging genome editors such as RNA-guided transposases are gaining traction. These systems, dubbed "CRISPR 4.0″ by some, enable large DNA insertions (~5 kb) at precise loci via bridge RNA/seekRNA without DSBs. If adapted to eukaryotes, they could enable synthetic genome construction ("genome writing") (de la Gándara et al., 2024).

3.10.2 Integration with artificial intelligence

As CRISPR tools grow more sophisticated, AI and machine learning are playing a transformative role across the editing workflow. AI assists in sgRNA design, optimizing sequences for high on-target activity and minimal off-target risks using models trained on large datasets. Deep learning frameworks also help predict off-target cleavage sites, making genome editing more precise and efficient (Khoshandam et al., 2024).

In protein engineering, AI helps discover and optimize new CRISPR enzymes, predicting 3D structures and suggesting beneficial mutations. Notably, Profluent's OpenCRISPR-1, an AI-designed Cas9 variant, demonstrates the power of generative models in creating novel, functional proteins. AI is also being used to tailor CRISPR-based therapies, guiding multiplex editing combinations, and automating experimental design. From delivery vector design to circuit optimization, AI is accelerating CRISPR development, transforming it into a data-driven discipline with the potential for predictive, personalized editing solutions (Kim et al., 2025).

3.10.3 Emerging directions

Several promising avenues are expanding CRISPR's reach. RNA editing using dCas13 fused to ADAR enzymes enables precise, reversible transcript modification (e.g., A→I), useful for temporary corrections without altering the genome. This opens possibilities for safer therapeutic applications or dynamic gene regulation (Zhu et al., 2024).

Epigenome editing using tools like CRISPRoff/CRISPRon enables heritable gene regulation without DNA sequence changes, offering a reversible approach to gene control. Convergence with nanotechnology and computational protein design may yield smarter delivery systems and novel editors (Nuñez et al., 2021).

3.11 Conclusion

CRISPR-based genome editing has rapidly grown from a fundamental discovery into a multi-faceted toolkit transforming biology. In this report, we examined the components of this toolkit – from the classic Cas9 nuclease that launched the genome editing revolution, to newer base editors, prime editors, and RNA-guided transposases that extend our capabilities. We reviewed how these tools are being applied across domains: in agriculture (to create disease-resistant or high-yield crops), in biomedical research and therapy (to model diseases, knock out viral genes, and attempt cures for genetic disorders), in synthetic biology (to construct genetic circuits and cell-based devices), and beyond. Alongside the excitement, we discussed the challenges and limitations that have become apparent: off-target effects that demand careful control, variable editing outcomes like mosaicism, difficulties in delivering editors to the right cells (or regenerating whole organisms from edited cells), and external hurdles in regulation, public acceptance, and intellectual property. Each of these challenges is being met with innovative strategies, whether technical improvements (e.g., high-fidelity Cas variants, better delivery vehicles) or policy and community engagement efforts. Finally, we cast an eye to the future prospects of genome editing an era we termed "CRISPR 3.0 and beyond", characterized by smarter and more powerful systems. These include prime editing and other next-gen enzymes that can write large DNA segments, integration of AI to design and optimize editors, incorporation of CRISPR into complex genetic circuits for programming cell behavior, and novel modalities like RNA editing and epigenome editing that expand what "gene editing" means. In conclusion, the CRISPR toolkit is steadily maturing into a comprehensive "genetic surgery" suite. Its components (nucleases, base editors, prime editors, etc.) each serve specific roles, and in combination they allow researchers to edit genomes with increasing finesse. The applications range from curing diseases to engineering ecosystems, underscoring CRISPR's broad impact as platform technology. Challenges remain, but they are surmountable with interdisciplinary effort biologists, engineers, computer scientists, ethicists, and policymakers working together to ensure the technology develops responsibly and reaches its full potential. CRISPR has often been likened to a word processor for DNA; if so, we are now upgrading it with better find-and-replace functions, spell-check (error reduction), and even predictive text (AI guidance). The trajectory of CRISPR points toward a future where we can correct genetic typos in cells as easily as fixing a document and even write new biological functions into genomes. Such capabilities carry profound promises for improving agriculture, environmental sustainability, and human health. Realizing that promise will require continued innovation and responsible stewardship, but the progress to date gives confidence that the "CRISPR-enabled" future once purely speculative is now on the horizon.

Declaration

"The authors declare that no generative artificial intelligence tools were used in the writing or preparation of this chapter. The content reflects the authors' own scholarly work, and the authors assume full responsibility for all statements, interpretations, and conclusions presented."

References

Achary, V.M.M., Reddy, M.K., 2021. CRISPR-Cas9 mediated mutation in GRAIN WIDTH and WEIGHT2 (GW2) locus improves aleurone layer and grain nutritional quality in rice. Sci. Rep. 11 (1), 21941.

Adaskaveg, J.A., Blanco-Ulate, B., 2023. Targeting ripening regulators to develop fruit with high quality and extended shelf life. Curr. Opin. Biotechnol. 79, 102872.

Ahmad, A., Ghouri, M.Z., Munawar, N., Ismail, M., Ashraf, S., Aftab, S.O., 2021. Regulatory, ethical, and social aspects of CRISPR crops. In: CRISPR Crops: The Future of Food Security. Springer Singapore, Singapore, pp. 261–287.

Aksoy, E., Yildirim, K., Kavas, M., Kayihan, C., Yerlikaya, B.A., Çalik, I., Demirel, U., 2022. General guidelines for CRISPR/Cas-based genome editing in plants. Mol. Biol. Rep. 49 (12), 12151–12164.

An, Y., Chen, L., Li, Y.X., Li, C., Shi, Y., Zhang, D., Wang, T., 2022. Fine mapping qKRN5. 04 provides a functional gene negatively regulating maize kernel row number. Theoretical and Applied Genetics 135 (6), 1997–2007.

Anzalone, A.V., Koblan, L.W., Liu, D.R., 2020. Genome editing with CRISPR–Cas nucleases, base editors, transposases and prime editors. Nat. Biotechnol. 38 (7), 824–844.

Arora, L., Narula, A., 2017. Gene editing and crop improvement using CRISPR-Cas9 system. Frontiers in plant science 8, 1932.

Asmamaw, M., Zawdie, B., 2021. Mechanism and applications of CRISPR/Cas-9-mediated genome editing. Biol. Targets & Ther. 353–361.

Atarashi, H., Jayasinghe, W.H., Kwon, J., Kim, H., Taninaka, Y., Igarashi, M., Nakahara, K.S., 2020. Artificially edited alleles of the eukaryotic translation initiation factor 4E1 gene differentially reduce susceptibility to cucumber mosaic virus and potato virus Y in tomato. Front. Microbiol. 11, 564310.

Azameti, M.K., Dauda, W.P., 2021. Base editing in plants: applications, challenges, and future prospects. Front. Plant Sci. 12, 664997.

Azizi-Dargahlou, S., Pouresmaeil, M., 2024. Agrobacterium tumefaciens-mediated plant transformation: a review. Mol. Biotechnol. 66 (7), 1563–1580.

Bandyopadhyay, A., Kancharla, N., Javalkote, V.S., Dasgupta, S., Brutnell, T.P., 2020. CRISPR-Cas12a (Cpf1): a versatile tool in the plant genome editing tool box for agricultural advancement. Frontiers in plant science 11, 584151.

Baranov, D., Dolgov, S., Timerbaev, V., 2024. New advances in the study of regulation of tomato flowering-related genes using biotechnological approaches. Plants 13 (3), 359.

Biswas, S., Ibarra, O., Shaphek, M., Molina-Risco, M., Faion-Molina, M., Bellinatti-Della Gracia, M., Septiningsih, E. M., 2023. Increasing the level of resistant starch in 'Presidio'rice through multiplex CRISPR–Cas9 gene editing of starch branching enzyme genes. Plant Genome 16 (2), e20225.

Bolotin, A., Quinquis, B., Sorokin, A., Ehrlich, S.D., 2005. Clustered regularly interspaced short palindrome repeats (CRISPRs) have spacers of extrachromosomal origin. Microbiology 151 (8), 2551–2561.

Bouzroud, S., Gasparini, K., Hu, G., Barbosa, M.A.M., Rosa, B.L., Fahr, M., Zouine, M., 2020. Down regulation and loss of auxin response factor 4 function using CRISPR/Cas9 alters plant growth, stomatal function and improves tomato tolerance to salinity and osmotic stress. Genes 11 (3), 272.

Calıskan, M.E., Bakhsh, A., Jabran, K. (Eds.), 2022. Potato Production Worldwide. Academic Press.

Cardi, T., Murovec, J., Bakhsh, A., Boniecka, J., Bruegmann, T., Bull, S.E., Van Laere, K., 2023. CRISPR/Cas-mediated plant genome editing: outstanding challenges a decade after implementation. Trends Plant Sci. 28 (10), 1144–1165.

Cebrian-Serrano, A., Davies, B., 2017. CRISPR-Cas orthologues and variants: optimizing the repertoire, specificity and delivery of genome engineering tools. Mamm. Genome 28 (7), 247–261.

Cencic, R., Miura, H., Malina, A., Robert, F., Ethier, S., Schmeing, T.M., Pelletier, J., 2014. Protospacer adjacent motif (PAM)-distal sequences engage CRISPR Cas9 DNA target cleavage. PLoS One 9 (10), e109213.

Chen, J.S., Doudna, J.A., 2017. The chemistry of Cas9 and its CRISPR colleagues. Nat. Rev. Chem 1 (10), 0078.

Das, P., Saha, R., Panda, D., Ghosh, C., Avinash, S.P., Panda, S., Baig, M.J., Molla, K.A., 2026. AI-designed nuclease performs robust knock out, base editing and prime editing in plants. bioRxiv, 2026-01. https://doi.org/10.64898/2026.01.20.700739.

de la Gándara, Á., Spínola-Amilibia, M., Araújo-Bazán, L., Núñez-Ramírez, R., Berger, J.M., Arias-Palomo, E., 2024. Molecular basis for transposase activation by a dedicated AAA+ ATPase. Nature 630 (8018), 1003–1011.

Demirer, G.S., Silva, T.N., Jackson, C.T., Thomas, J.B., W. Ehrhardt, D., Rhee, S.Y., Landry, M.P., 2021. Nanotechnology to advance CRISPR–Cas genetic engineering of plants. Nat. Nanotechnol. 16 (3), 243–250.

Deveau, H., Garneau, J.E., Moineau, S., 2010. CRISPR/Cas system and its role in phage-bacteria interactions. Annu. Rev. Microbiol. 64, 475–493.

Erdoğan, İ., Cevher-Keskin, B., Bilir, Ö., Hong, Y., Tör, M., 2023. Recent developments in CRISPR/Cas9 genome-editing technology related to plant disease resistance and abiotic stress tolerance. Biology 12 (7), 1037.

Erenstein, O., Jaleta, M., Sonder, K., Mottaleb, K., Prasanna, B.M., 2022. Global maize production, consumption and trade: trends and R&D implications. Food Secur. 14 (5), 1295–1319.

Fonfara, I., Le Rhun, A., Chylinski, K., Makarova, K.S., Lecrivain, A.L., Bzdrenga, J., Charpentier, E., 2014. Phylogeny of Cas9 determines functional exchangeability of dual-RNA and Cas9 among orthologous type II CRISPR-Cas systems. Nucleic acids research 42 (4), 2577–2590.

Gasiunas, G., Barrangou, R., Horvath, P., Siksnys, V., 2012. Cas9–crRNA ribonucleoprotein complex mediates specific DNA cleavage for adaptive immunity in bacteria. Proc. Natl. Acad. Sci. 109 (39), E2579–E2586.

Gupta, R., Ghosh, A., Chakravarti, R., Singh, R., Ravichandiran, V., Swarnakar, S., Ghosh, D., 2022. Cas13d: a new molecular scissor for transcriptome engineering. Front. Cell Dev. Biol. 10, 866800.

Guy, C.P., Bolt, E.L., 2005. Archaeal Hel308 helicase targets replication forks in vivo and in vitro and unwinds lagging strands. Nucleic acids research 33 (11), 3678–3690.

Hajiahmadi, Z., Movahedi, A., Wei, H., Li, D., Orooji, Y., Ruan, H., Zhuge, Q., 2019. Strategies to increase on-target and reduce off-target effects of the CRISPR/Cas9 system in plants. Int. J. Mol. Sci. 20 (15), 3719.

Hillary, V.E., Ceasar, S.A., 2023. A review on the mechanism and applications of CRISPR/Cas9/Cas12/Cas13/Cas14 proteins utilized for genome engineering. Mol. Biotechnol. 65 (3), 311–325.

Holman, C.M., 2022. The broad Institute scores another victory in its battle with the university of California over the patenting of CRIPSR. Biotechnol. Law Rep. 41 (3), 122–137.

Huang, L., Zhang, R., Huang, G., Li, Y., Melaku, G., Zhang, S., Hu, F., 2018. Developing superior alleles of yield genes in rice by artificial mutagenesis using the CRISPR/Cas9 system. The Crop Journal 6 (5), 475–481.

Illouz-Eliaz, N., Nissan, I., Nir, I., Ramon, U., Shohat, H., Weiss, D., 2020. Mutations in the tomato gibberellin receptors suppress xylem proliferation and reduce water loss under water-deficit conditions. J. Exp. Bot. 71 (12), 3603–3612.

Imran, M., Shafiq, S., Tang, X., 2023. CRISPR-Cas9-mediated editing of BADH2 gene triggered fragrance revolution in rice. Physiol. Plantarum 175 (1), e13871.

Ishino, Y., Krupovic, M., Forterre, P., 2018. History of CRISPR-Cas from encounter with a mysterious repeated sequence to genome editing technology. J. Bacteriol. 200 (7), 10–1128.

Jansen, R., Embden, J.D.V., Gaastra, W., Schouls, L.M., 2002. Identification of genes that are associated with DNA repeats in prokaryotes. Mol. Microbiol. 43 (6), 1565–1575.

Jinek, M., Chylinski, K., Fonfara, I., Hauer, M., Doudna, J.A., Charpentier, E., 2012. A programmable dual-RNA–guided DNA endonuclease in adaptive bacterial immunity. Science 337 (6096), 816–821.

Kalter, N., Fuster-García, C., Silva, A., Ronco-Díaz, V., Roncelli, S., Turchiano, G., Hendel, A., 2025. Off-target effects in CRISPR-cas genome editing for human therapeutics: progress and challenges. Mol. Ther. Nucleic Acids 36 (3), 102636.

Kato, K., Zhou, W., Okazaki, S., Isayama, Y., Nishizawa, T., Gootenberg, J.S., Nishimasu, H., 2022. Structure and engineering of the type III-E CRISPR-Cas7-11 effector complex. Cell 185 (13), 2324–2337.

Khoshandam, M., Soltaninejad, H., Hosseinkhani, S., Saltanatpour, Z., Goudarzi, M.T.H., Hamidieh, A.A., 2024. CRISPR/Cas and artificial intelligence to improve precision medicine: future perspectives and potential limitations. Human Gene 42, 201356.

Kim, D.Y., Lee, S.Y., Ha, H.J., Park, H.H., 2024. Structural basis of Cas3 activation in type IC CRISPR-Cas system. Nucleic Acids Res. 52 (17), 10563–10574.

Kim, M.G., Go, M.J., Kang, S.H., Jeong, S.H., Lim, K., 2025. Revolutionizing CRISPR technology with artificial intelligence. Exp. Mol. Med. 1–13.

Knutsen, A., 2024. CRISPR 2.0 Kinder, Gentler, More Powerful: unlike CRISPR 1.0, which hacks through DNA, CRISPR 2.0 is capable of surgical incisions, making it suitable for a broader range of gene therapy applications. Genetic Engineering & Biotechnology News 44 (6), 44–46.

Kondrateva, E., Demchenko, A., Lavrov, A., Smirnikhina, S., 2021. An overview of currently available molecular Cas-tools for precise genome modification. Gene 769, 145225.

Koseoglou, E., 2017. The Study of SlPMR4 CRISPR/Cas9-mediated Tomato Allelic Series for Resistance against Powdery Mildew. Wageningen University, Wageningen, The Netherlands.

Lander, E.S., 2016. The heroes of CRISPR. Cell 164 (1), 18–28.

Larriba, E., Yaroshko, O., Pérez-Pérez, J.M., 2024. Recent advances in tomato gene editing. Int. J. Mol. Sci. 25 (5), 2606.

Li, M., Li, X., Zhou, Z., Wu, P., Fang, M., Pan, X., Li, H., 2016. Reassessment of the four yield-related genes Gn1a, DEP1, GS3, and IPA1 in rice using a CRISPR/Cas9 system. Frontiers in plant science 7, 377.

Li, X., Sun, B., Qian, H., Ma, J., Paolino, M., Zhang, Z., 2022. A high-efficiency and versatile CRISPR/Cas9-mediated HDR-based biallelic editing system. J. Zhejiang Univ. - Sci. B 23 (2), 141–152.

Lim, Y., Bak, S.Y., Sung, K., Jeong, E., Lee, S.H., Kim, J.S., Kim, S.K., 2016. Structural roles of guide RNAs in the nuclease activity of Cas9 endonuclease. Nat. Commun. 7 (1), 13350.

Lin, Q., Zong, Y., Xue, C., Wang, S., Jin, S., Zhu, Z., Gao, C., 2020. Prime genome editing in rice and wheat. Nat. Biotechnol. 38 (5), 582–585.

Liu, C., Kong, M., Yang, F., Zhu, J., Qi, X., Weng, J., Xie, C., 2022. Targeted generation of Null Mutants in ZmGDIα confers resistance against maize rough dwarf disease without agronomic penalty. Plant Biotechnol. J. 20 (5), 803.

Liu, G., Li, J., Godwin, I.D., 2019. Genome editing by CRISPR/Cas9 in sorghum through biolistic bombardment. In: Sorghum: Methods and Protocols. Springer New York, New York, NY, pp. 169–183.

Liu, H., Ding, Y., Zhou, Y., Jin, W., Xie, K., Chen, L.L., 2017. CRISPR-P 2.0: an improved CRISPR-Cas9 tool for genome editing in plants. Mol. Plant 10 (3), 530–532.

Liu, L., Zhang, J., Xu, J., Li, Y., Guo, L., Wang, Z., Zhang, N., 2020. CRISPR/Cas9 targeted mutagenesis of SlLBD40, a lateral organ boundaries domain transcription factor, enhances drought tolerance in tomato. Plant Sci. 301, 110683.

Liu, Q., Yang, F., Zhang, J., Liu, H., Rahman, S., Islam, S., She, M., 2021. Application of CRISPR/Cas9 in crop quality improvement. Int. J. Mol. Sci. 22 (8), 4206.

Liu, Y., Chen, Z., Zhang, C., Guo, J., Liu, Q., Yin, Y., Liu, X., 2024. Gene editing of ZmGA20ox3 improves plant architecture and drought tolerance in maize. Plant Cell Rep. 43 (1), 18.

Lu, Y., Tian, Y., Shen, R., Yao, Q., Zhong, D., Zhang, X., Zhu, J.K., 2020. Precise genome modification in tomato using an improved prime editing system. Plant Biotechnol. J. 19 (3), 415.

Mainkar, P., Manape, T.K., Satheesh, V., Anandhan, S., 2023. CRISPR/Cas9-mediated editing of PHYTOENE DESATURASE gene in onion (Allium cepa L.). Front. Plant Sci. 14, 1226911.

Makarova, K.S., Grishin, N.V., Shabalina, S.A., Wolf, Y.I., Koonin, E.V., 2006. A putative RNA-interference-based immune system in prokaryotes: computational analysis of the predicted enzymatic machinery, functional analogies with eukaryotic RNAi, and hypothetical mechanisms of action. Biol. Direct 1 (1), 7.

Makarova, K.S., Wolf, Y.I., Alkhnbashi, O.S., Costa, F., Shah, S.A., Saunders, S.J., Koonin, E.V., 2015. An updated evolutionary classification of CRISPR–Cas systems. Nat. Rev. Microbiol. 13 (11), 722–736.

Makarova, K.S., Wolf, Y.I., Iranzo, J., Shmakov, S.A., Alkhnbashi, O.S., Brouns, S.J., Koonin, E.V., 2020. Evolutionary classification of CRISPR–Cas systems: a burst of class 2 and derived variants. Nat. Rev. Microbiol. 18 (2), 67–83.

Mallick, A., Pramanik, B., Debnath, S., 2025. CRISPR Technologies for Climate-Smart and Sustainable Agriculture: An Updated Overview. CRISPR for Climate-Smart and Sustainable Agriculture, pp. 1–25.

Manghwar, H., Li, B., Ding, X., Hussain, A., Lindsey, K., Zhang, X., Jin, S., 2020. CRISPR/Cas systems in genome editing: methodologies and tools for sgRNA design, off-target evaluation, and strategies to mitigate off-target effects. Adv. Sci. 7 (6), 1902312.

Marraffini, L.A., Sontheimer, E.J., 2008. CRISPR interference limits horizontal gene transfer in staphylococci by targeting DNA. Science 322 (5909), 1843–1845.

Mathiazhagan, M., Chidambara, B., Hunashikatti, L.R., Ravishankar, K.V., 2021. Genomic approaches for improvement of tropical fruits: fruit quality, shelf life and nutrient content. Genes 12 (12), 1881.

McCarty, N.S., Graham, A.E., Studená, L., Ledesma-Amaro, R., 2020. Multiplexed CRISPR technologies for gene editing and transcriptional regulation. Nat. Commun. 11 (1), 1281.

Mehravar, M., Shirazi, A., Nazari, M., Banan, M., 2019. Mosaicism in CRISPR/Cas9-mediated genome editing. Dev. Biol. 445 (2), 156–162.

Meyer, L., 2024. Intellectual property challenges in synthetic biology: strategies for successful commercialization. J. Commerc. Biotechnol. 29 (4), 165–174.

Mikhaylova, E.V., Kuluev, B.R., Gerashchenkov, G.A., Chemeris, D.A., Garafutdinov, R.R., Kuluev, A.R., Chemeris, A.V., 2024. Prime-editing methods and pegRNA design programs. Mol. Biol. 58 (1), 17–32.

Miladinovic, D., Antunes, D., Yildirim, K., Bakhsh, A., Cvejić, S., Kondić-Špika, A., Hilioti, Z., 2021. Targeted plant improvement through genome editing: from laboratory to field. Plant Cell Rep. 40 (6), 935–951.

Mojica, F.J., Montoliu, L., 2016. On the origin of CRISPR-Cas technology: from prokaryotes to mammals. Trends Microbiol. 24 (10), 811–820.

Molla, K.A., Sretenovic, S., Bansal, K.C., Qi, Y., 2021. Precise plant genome editing using base editors and prime editors. Nat. Plants 7 (9), 1166–1187.

Moon, S.B., Ko, J.H., Kim, J.S., Kim, Y.S., 2019. Improving CRISPR genome editing by engineering guide RNAs. Trends Biotechnol. 37 (8), 870–881.

Najera, V.A., Twyman, R.M., Christou, P., Zhu, C., 2019. Applications of multiplex genome editing in higher plants. Curr. Opin. Biotechnol. 59, 93–102.

Nekrasov, V., Wang, C., Win, J., Lanz, C., Weigel, D., Kamoun, S., 2017. Rapid generation of a transgene-free powdery mildew resistant tomato by genome deletion. Sci. Rep. 7 (1), 482.

Nguyen, T.H., Ben Taieb, S., Moritaka, M., Ran, L., Fukuda, S., 2023. Public Acceptance of foods derived from genome editing technology: a review of the technical, social and regulatory aspects. J. Int. Food Agribus. Mark. 35 (4), 397–427.

Nonaka, S., Arai, C., Takayama, M., Matsukura, C., Ezura, H., 2017. Efficient increase of γ-aminobutyric acid (GABA) content in tomato fruits by targeted mutagenesis. Sci. Rep. 7 (1), 7057.

Nuñez, J.K., Chen, J., Pommier, G.C., Cogan, J.Z., Replogle, J.M., Adriaens, C., Weissman, J.S., 2021. Genome-wide programmable transcriptional memory by CRISPR-based epigenome editing. Cell 184 (9), 2503–2519.

Ogata, T., Ishizaki, T., Fujita, M., Fujita, Y., 2020. CRISPR/Cas9-targeted mutagenesis of OsERA1 confers enhanced responses to abscisic acid and drought stress and increased primary root growth under nonstressed conditions in rice. PLoS One 15 (12), e0243376.

Pan, Z., Liu, M., Zhao, H., Tan, Z., Liang, K., Sun, Q., Qiu, F., 2020. ZmSRL5 is involved in drought tolerance by maintaining cuticular wax structure in maize. J. Integr. Plant Biol. 62 (12), 1895–1909.

Pandita, D., 2022. CRISPR/Cas-mediated genome editing technologies in plants. In: Plant Abiotic Stress Physiology. Apple Academic Press, pp. 293–330.

Park, S.J., Lee, G.E., Cho, S.M., Choi, E.H., 2025. Recent applications, future perspectives, and limitations of the CRISPR-Cas system. Mol. Ther. Nucleic Acids.

Razzaq, M.K., Aleem, M., Mansoor, S., Khan, M.A., Rauf, S., Iqbal, S., Siddique, K.H., 2021. Omics and CRISPR-Cas9 approaches for molecular insight, functional gene analysis, and stress tolerance development in crops. Int. J. Mol. Sci. 22 (3), 1292.

Rukavtsova, E.B., Zakharchenko, N.S., Lebedev, V.G., Shestibratov, K.A., 2022. CRISPR-Cas genome editing for horticultural crops improvement: advantages and prospects. Horticulturae 9 (1), 38.

Sanner, J.W., 2016. The struggle for CRISPR: a billion dollar question in intellectual property. U. Ill. JL Tech. & Pol'y 431.

Santosh Kumar, V.V., Verma, R.K., Yadav, S.K., Yadav, P., Watts, A., Rao, M.V., Chinnusamy, V., 2020. CRISPR-Cas9 mediated genome editing of drought and salt tolerance (OsDST) gene in indica mega rice cultivar MTU1010. Physiol. Mol. Biol. Plants 26 (6), 1099–1110.

Savage, D.F., 2019. Cas14: big advances from small CRISPR proteins. Biochemistry 58 (8), 1024–1025.

Scheben, A., Wolter, F., Batley, J., Puchta, H., Edwards, D., 2017. Towards CRISPR/Cas crops–bringing together genomics and genome editing. New Phytol. 216 (3), 682–698.

Sedeek, K.E., Mahas, A., Mahfouz, M., 2019. Plant genome engineering for targeted improvement of crop traits. Frontiers in plant science 10, 114.

Shi, J., Gao, H., Wang, H., Lafitte, H.R., Archibald, R.L., Yang, M., Habben, J.E., 2017. ARGOS 8 variants generated by CRISPR-Cas9 improve maize grain yield under field drought stress conditions. Plant Biotechnol. J. 15 (2), 207–216.

Song, F., Stieger, K., 2017. Optimizing the DNA donor template for homology-directed repair of double-strand breaks. Molecular Therapy-Nucleic Acids 7, 53–60.

Soyars, C.L., Peterson, B.A., Burr, C.A., Nimchuk, Z.L., 2018. Cutting edge genetics: CRISPR/Cas9 editing of plant genomes. Plant Cell Physiol. 59 (8), 1608–1620.

Soyk, S., Müller, N.A., Park, S.J., Schmalenbach, I., Jiang, K., Hayama, R., Lippman, Z.B., 2017. Variation in the flowering gene SELF PRUNING 5G promotes day-neutrality and early yield in tomato. Nat. Genet. 49 (1), 162–168.

Sun, S., Wang, L., Mao, H., Shao, L., Li, X., Xiao, J., Zhang, Q., 2018. A G-protein pathway determines grain size in rice. Nat. Commun. 9 (1), 851.

Sunwasiya, D.K., Mondal, S., 2024. Applications of CRISPR/Cas9 guided genome editing in livestock: an update. EC Clinical and Medical Case Reports 7, 01–10.

Tashkandi, M., Ali, Z., Aljedaani, F., Shami, A., Mahfouz, M.M., 2018. Engineering resistance against Tomato yellow leaf curl virus via the CRISPR/Cas9 system in tomato. Plant Signal. Behav. 13 (10), e1525996.

Thomazella, D.P.D.T., Seong, K., Mackelprang, R., Dahlbeck, D., Geng, Y., Gill, U.S., Staskawicz, B., 2021. Loss of function of a DMR6 ortholog in tomato confers broad-spectrum disease resistance. Proc. Natl. Acad. Sci. 118 (27), e2026152118.

Tran, M.T., Doan, D.T.H., Kim, J., Song, Y.J., Sung, Y.W., Das, S., Kim, J.Y., 2021. CRISPR/Cas9-based precise excision of SlHyPRP1 domain (s) to obtain salt stress-tolerant tomato. Plant Cell Rep. 40 (6), 999–1011.

Tyagi, S., Kumar, R., Kumar, V., Won, S.Y., Shukla, P., 2021. Engineering disease resistant plants through CRISPR-Cas9 technology. GM Crops Food 12 (1), 125–144.

Usman, B., Nawaz, G., Zhao, N., Liao, S., Liu, Y., Li, R., 2020. Precise editing of the OsPYL9 gene by RNA-guided Cas9 nuclease confers enhanced drought tolerance and grain yield in rice (Oryza sativa L.) by regulating circadian rhythm and abiotic stress responsive proteins. Int. J. Mol. Sci. 21 (21), 7854.

Vardi, A., Levin, I., Carmi, N., 2008. Induction of seedlessness in citrus: from classical techniques to emerging biotechnological approaches. J. Am. Soc. Hortic. Sci. 133 (1), 117–126.

Wang, B., Li, N., Huang, S., Hu, J., Wang, Q., Tang, Y., Yu, Q., 2021. Enhanced soluble sugar content in tomato fruit using CRISPR/Cas9-mediated SlINVINH1 and SlVPE5 gene editing. PeerJ 9, e12478.

Wang, F., Qi, L.S., 2016. Applications of CRISPR genome engineering in cell biology. Trends Cell Biol. 26 (11), 875–888.

Wang, F., Wang, C., Liu, P., Lei, C., Hao, W., Gao, Y., Zhao, K., 2016. Enhanced rice blast resistance by CRISPR/Cas9-targeted mutagenesis of the ERF transcription factor gene OsERF922. PLoS One 11 (4), e0154027.

Wang, L., Han, H., 2024. Strategies for improving the genome-editing efficiency of class 2 CRISPR/Cas system. Heliyon 10 (19).

Wang, L., Chen, L., Li, R., Zhao, R., Yang, M., Sheng, J., Shen, L., 2017. Reduced drought tolerance by CRISPR/Cas9-mediated SlMAPK3 mutagenesis in tomato plants. J. Agric. Food Chem. 65 (39), 8674–8682.

Wang, Q., Gao, H., Liu, K., Wang, H., Zhang, F., Wei, L., Yuan, H., 2024. CRISPR/Cas9-mediated enhancement of semi-dwarf glutinous traits in elite Xiangdaowan rice (Oryza sativa L.): targeting SD1 and Wx genes for yield and quality improvement. Front. Plant Sci. 15, 1333191.

Wang, Y., Liu, X., Zheng, X., Wang, W., Yin, X., Liu, H., Wang, F., 2021. Creation of aromatic maize by CRISPR/Cas. J. Integr. Plant Biol. 63 (9), 1664–1670.

Wang, Y., Tang, Q., Pu, L., Zhang, H., Li, X., 2022. CRISPR-Cas technology opens a new era for the creation of novel maize germplasms. Front. Plant Sci. 13, 1049803.

Wang, Z., Hardcastle, T.J., Pastor, A.C., Yip, W.H., Tang, S., Baulcombe, D.C., 2018. A novel DCL2-dependent miRNA pathway in tomato affects susceptibility to RNA viruses. Genes Dev. 32 (17–18), 1155–1160.

Wu, X., Lam, W.H., Zhao, Z., Cao, Y., Lin, H., Feng, X., Zhai, Y., Hsing, I.M., 2026. DNA-guided CRISPR–Cas12a effectors for programmable RNA recognition and cleavage. Nat. Biotechnol. 1–1.

Xue, C., Greene, E.C., 2021. DNA repair pathway choices in CRISPR-Cas9-mediated genome editing. Trends Genet. 37 (7), 639–656.

Yang, H., Patel, D.J., 2019. CasX: a new and small CRISPR gene-editing protein. Cell Res. 29 (5), 345–346.

Zafar, K., Khan, M.Z., Amin, I., Mukhtar, Z., Yasmin, S., Arif, M., Mansoor, S., 2020. Precise CRISPR-Cas9 mediated genome editing in super basmati rice for resistance against bacterial blight by targeting the major susceptibility gene. Frontiers in plant science 11, 575.

Zafar, S.A., Zaidi, S.S.E.A., Gaba, Y., Singla-Pareek, S.L., Dhankher, O.P., Li, X., Pareek, A., 2020. Engineering abiotic stress tolerance via CRISPR/Cas-mediated genome editing. J. Exp. Bot. 71 (2), 470–479.

Zaidi, S.S.E.A., Mansoor, S., 2017. Viral vectors for plant genome engineering. Front. Plant Sci. 8, 539.

Zhang, A., Liu, Y., Wang, F., Li, T., Chen, Z., Kong, D., Luo, L., 2019. Enhanced rice salinity tolerance via CRISPR/Cas9-targeted mutagenesis of the OsRR22 gene. Mol. Breed. 39 (3), 47.

Zhang, S., Zhang, R., Gao, J., Song, G., Li, J., Li, W., Li, G., 2021. CRISPR/Cas9-mediated genome editing for wheat grain quality improvement. Plant Biotechnol. J. 19 (9), 1684.

Zhang, X., Henriques, R., Lin, S.S., Niu, Q.W., Chua, N.H., 2006. Agrobacterium-mediated transformation of Arabidopsis thaliana using the floral dip method. Nat. Protoc. 1 (2), 641–646.

Zhang, Y., Huang, C.F., 2019. Reduction in cadmium accumulation in japonica rice grains by CRISPR/Cas9-mediated editing of OsNRAMP5. J. Integr. Agric. 18 (3), 688–697.

Zhang, Z., Zhang, X., Lin, Z., Wang, J., Liu, H., Zhou, L., Lin, Z., 2020. A large transposon insertion in the stiff1 promoter increases stalk strength in maize. Plant Cell 32 (1), 152–165.

Zhu, G., Zhou, X., Wen, M., Qiao, J., Li, G., Yao, Y., 2024. CRISPR–Cas13: pioneering RNA editing for nucleic acid therapeutics. BioDesign Research 6, 0041.

Zhu, H., Li, C., Gao, C., 2020. Applications of CRISPR–Cas in agriculture and plant biotechnology. Nat. Rev. Mol. Cell Biol. 21 (11), 661–677.

Zsögön, A., Čermák, T., Naves, E.R., Notini, M.M., Edel, K.H., Weinl, S., Peres, L.E.P., 2018. De novo domestication of wild tomato using genome editing. Nat. Biotechnol. 36 (12), 1211–1216.

Zulfiqar, M., Qamar, I., Murtaza, M., Haider, Z., Javed, A., Abbas, A., Iftikhar, A., 2019. A novel plant genome editing CRISPR/Cas9 System: to modify stress tolerance responses in plants. Am. J. Biotechnol. Biosci 3, 16.

CHAPTER

4

CRISPR cargo delivery systems for intelligent plant genome engineering

Monoswi Chakraborty[1,2,3], *Srirupa Gupta Choudhury*[4,5], *and Muhammad Umer Masood*[6]

[1]Genetics, Genomics and Bioinformatics (GGB) Program, University of California, Riverside, CA, United States; [2]Institute for Integrative Genome Biology (IIGB), University of California, Riverside, CA, United States; [3]Department of Chemical and Environmental Engineering, Marlan and Rosemary Bourns College of Engineering, University of California, Riverside, CA, United States; [4]Département de biologie cellulaire et moléculaire, Université du Québec à Trois-Rivières, Trois-Rivières, QC, Canada; [5]CSIR-Institute of Genomics and Integrative Biology, New Delhi, India; [6]Department of Plant Sciences and Plant Pathology, Montana State University, Bozeman, MT, United States

4.1 Introduction

The onset of CRISPR–Cas technology has set the pace and radically transformed plant genetics. CRISPR/Cas enzymes (originally bacterial adaptive immunity components) and their programmable guide RNAs enable precise, efficient genome editing, far surpassing earlier tools (ZFNs, TALENs) in versatility and ease of use (Nadakuduti and Enciso-Rodríguez, 2021; Laforest and Nadakuduti, 2022). The first CRISPR-edited plants appeared in 2013 (e.g., *Arabidopsis,* rice), and since then, CRISPR has been used in dozens of plant genera across diverse taxa (Nadakuduti and Enciso-Rodríguez, 2021). In plants, CRISPR has rapidly become a powerful functional genomics and breeding tool: it accelerates the development of improved cultivars by creating targeted knockouts or changes in genes controlling yield, stress resistance, domestication, and quality (Zhu et al., 2020; Laforest and Nadakuduti,

Intelligent Gene Editing in Plants
https://doi.org/10.1016/B978-0-443-44079-3.00024-3

2022; Chen et al., 2024). Recent innovations (new Cas proteins and variants, base and prime editors) have expanded CRISPR's impeccable capabilities for precise nucleotide changes and trait stacking in crops (Laforest and Nadakuduti, 2022; Chen et al., 2024). Advancements as such facilitate global food security by building resilience, nutrition, and high yield in staples like rice, wheat, maize, and others (Chen et al., 2024).

Despite remarkable progress in the field, plant genome editing faces unique challenges. Plant cells are walled, polyploid, and mostly recalcitrant to transformation and regeneration (Nadakuduti and Enciso-Rodríguez, 2021; Laforest and Nadakuduti, 2022; Cardi et al., 2023). The rigid cell wall acts as a barrier to the entry point of editing reagents, and most delivery vectors (*Agrobacterium* or biolistics) work especially only in particular species or genotypes (Nadakuduti and Enciso-Rodríguez, 2021; Laforest and Nadakuduti, 2022). Moreover, many crops are polyploid or have complex genomes, complicating efficient editing of all gene copies (Laforest and Nadakuduti, 2022). Traditional workflows rely on tissue culture and regeneration, which are slow and often species-specific. For example, even advanced methods often require weeks of callus culture and crossing to segregate away transgenes, making genome editing laborious (Weiss et al., 2025). In summary, a rigid cell wall, lack of broadly effective transformation protocols, and difficult regeneration have been major bottlenecks for plant gene editing (Nadakuduti and Enciso-Rodríguez, 2021; Laforest and Nadakuduti, 2022).

Another key factor is the form of the CRISPR cargoes and their delivery method. The CRISPR cargoes includes the Cas nuclease (or nuclease variant) and guide RNA(s), delivered as DNA constructs, RNA, or ribonucleoprotein (RNP) complexes (Hassan et al., 2021). Beyond the original SpCas9, a variety of Cas effectors are in use: other Cas9 orthologs (SaCas9, StCas9, Cas9-NG, xCas9), Cas12a/b enzymes, and even miniature nucleases like CasΦ and TnpB (Nadakuduti and Enciso-Rodríguez, 2021). Many Cas variants have been engineered for smaller size, expanded PAM recognition or higher fidelity. In addition to nucleases, specialized "editors" such as cytosine or adenine base editors and the search-and-replace prime editors are now routinely used in plants. These fusion proteins (Cas nickases or dead Cas linked to deaminases or reverse transcriptase) enable precise single-base conversions or small insertions without inducing double-strand breaks (Laforest and Nadakuduti, 2022; Yang et al., 2024). Altogether, modern CRISPR toolkits for plants include knock-out nucleases, base editors, prime editors, and multiplex gRNA arrays, allowing knockouts, knock-ins, and all combinations of base substitutions (Liu et al., 2022; Laforest and Nadakuduti, 2022; Yang et al., 2024).

To deliver these cargoes, multiple strategies exist. The conventional method is *Agrobacterium*-mediated transformation (AMT), where *A. tumefaciens* transfers T-DNA encoding Cas and gRNAs into plant cells (often via leaf infiltration or explant infection). Biolistic (particle bombardment) delivery can shoot DNA, RNA, or RNPs into tissues, bypassing host-range limits of *Agrobacterium*. Protoplast transfection (PEG or electroporation) is another route to introduce DNA, RNA, or RNP into plant cells freed of their walls (Laforest and Nadakuduti, 2022). Each method has trade-offs: *Agrobacterium* often yields stable transgenics, biolistics works in recalcitrant species (but can cause tissue damage), and RNP delivery (DNA-free biolistics or protoplasts) enables transgene-free editing (Laforest and Nadakuduti, 2022). Beyond these, emerging approaches leverage viral vectors and nanomaterials.

Plant viruses (e.g., geminiviruses, tobacco rattle virus) have been engineered to carry gRNA and even compact Cas proteins systemically; for example (Weiss et al., 2025), used an engineered TRV (tobacco rattle virus) to deliver a small TnpB nuclease and guide RNA into *Arabidopsis*, achieving transgene-free germline edits in one step (Weiss et al., 2025). Nanoparticles (carbon nanotubes (CNTs), clay nanosheets, etc.) can adsorb Cas/gRNA ribonucleo-complexes and penetrate cell walls, offering another DNA-free delivery route. Recent reviews highlight that micro- and nano-carriers can "surpass the defensive primary cell wall barrier" and protect CRISPR cargo, potentially improving transformation efficiency across species (Shivashakarappa et al., 2025). In practice, scientists also use auxin/meristem induction, floral dip (in *Arabidopsis*), or grafting techniques to extend delivery. Overall, the choice of Cas variant (size, PAM, editing function) and cargo form (DNA/RNA/RNP) must be coordinated with an appropriate delivery system (*Agrobacterium*, biolistics, nanoparticle, or viral vector) for each plant species.

In this chapter, we survey the evolving landscape of CRISPR "cargoes" and their delivery platforms in plants. We cover the spectrum from model species (e.g., *Arabidopsis*, *Nicotiana*) to major crops (rice, maize, wheat, tomato, etc.) and emerging or non-model species (wild relatives, orphan crops, forest trees). We highlight recent high-impact studies on new Cas effectors, base/prime editors and multiplex gRNA designs, and on novel delivery methods that bypass tissue culture. By contrasting different cargoes and delivery strategies, this chapter aims to discuss the state-of-the-art in plant genome editing and to identify gaps and future directions for achieving efficient, genotype-independent, and transgene-free editing in diverse plant taxa. Altogether, this chapter highlights the artificial intelligence (AI) and machine learning (ML)-driven strategies for method prediction, protocol optimization, and rational cargo design. The emerging framework of "intelligent CRISPR delivery" acts as a decisive factor, thereby shedding light into this growing technology of achieving precise, scalable, genotype-independent, and transgene-free editing across diverse plant taxa.

4.2 CRISPR cargoes in plants

CRISPR cargoes consist of the molecular components essential for precise genome and transcriptome engineering in plants. Among these, the primary ones are Cas nucleases (e.g., SpCas9, SaCas9, Cas12a/b, CasΦ, CasMINI) that introduce site-specific DNA breaks or edits (Zhu et al., 2020). The nucleases are guided by gRNAs, typically expressed from U6/U3 promoters, which can be multiplexed to edit multiple loci simultaneously (Hassan et al., 2021). For error-free, precise alterations, donor templates enable homology-directed repair (HDR), though efficiency is relatively low in plants (Chen et al., 2019). Additionally, advanced editors such as prime editors and base editors (CBEs, ABEs) broaden the scope of CRISPR by enabling targeted nucleotide substitutions and small indels without double-strand breaks (Anzalone et al., 2020). Collectively, these cargoes provide an all-round, versatile toolkit for functional genomics, crop improvement, and plant virus resistance.

4.2.1 Cas nucleases

The choice of Cas nuclease has a remarkable effect on the efficiency, specificity, and delivery of genome editing module in plants. The standard SpCas9 (from *Streptococcus pyogenes*) nuclease is ~1368 amino acids (~160 kDa) and needs an NGG PAM (Qin et al., 2019). While SpCas9 is robust and well-characterized, its lengthy size and strict PAM requirements limit editing efficiencies in certain genomic contexts. To address this, alternative Class 2 nucleases have been introduced.

4.2.1.1 *SaCas9* (Staphylococcus aureus *Cas9)*

Shorter in length (~1053 aa) with activity comparable to SpCas9 (Qin et al., 2019). SaCas9 identifies a longer PAM (NNGRRT) than SpCas9 (Qin et al., 2019), and engineered variants (e.g., SaCas9-KKH) can accept relaxed NNNRRT PAMs (Qin et al., 2019). The compact size of SaCas9 enhances delivery (e.g., via viruses or RNPs) in plant systems (Hassan et al., 2021).

4.2.1.2 *Cas12a (Cpf1) and Cas12b (type V–B)*

These are distinct monomeric nucleases (~1200–1500 aa) that recognize T-rich PAMs (e.g., Cas12a requires 5′-TTTV-3′ (Safari et al., 2019). Unlike Cas9, Cas12a uses a single crRNA (no tracrRNA), and cleaves with staggered (5′ overhang) cuts (Safari et al., 2019). The T-rich PAMs complement Cas9's G-rich PAMs, expanding targetable sites. For example, LbCas12a and AsCas12a have been widely used in plants for multiplexed editing and insertion of sequences (staggered DSBs can favor directional HDR) (Safari et al., 2019). Cas12b (C2c1) is another type V nuclease recently shown to edit crops(Strecker et al., 2019). *Alicyclobacillus acidiphilus* Cas12b was demonstrated to efficiently edit multiple targets in rice, establishing Cas12b as another promising genome Cas editor after Cas9 and Cas12a (Ming et al., 2020).

4.2.1.3 *Miniature Cas variants (type V–F and V-J)*

Recently discovered ultracompact nucleases such as Cas12f and Cas12j (also known as CasΦ) are <600 aa (roughly half the size of Cas9) (Gong et al., 2024). These "mini-Cas" enzymes function in human cells and have been adapted to plants as well (Gong et al., 2024; Wu et al., 2024). For example, in a plant geminivirus replicon system, Cas12j2 and SpCas12f1 were shown to edit *Nicotiana benthamiana* genomes (Gong et al., 2024). Engineered derivatives (e.g., CasMINI, an optimized Cas12f variant) have similarly been co-expressed with guide RNAs in plants (Gong et al., 2024). The reduced size of these systems makes them attractive for delivery by size-limited vectors (like plant viruses) (Gong et al., 2024).

4.2.1.4 *Cas13 (type VI)*

Unlike the above DNA-targeting nucleases, Cas13 enzymes cleave RNA. In plants, Cas13 has been deployed for virus resistance and transcript knockdown (Robertson et al., 2022). Aman et al. (2018) showed that LshCas13a expressed in *Nicotiana benthamiana* and *Arabidopsis* could specifically degrade Turnip Mosaic Virus RNA (targeting HC-Pro and CP genes) (Kavuri et al., 2022). Mahas et al. (2019) further demonstrated that Cas13d (also called CasRx) confers "robust" RNA virus interference, achieving higher knockdown of multiple viruses (e.g., TMV, PVX) than Cas13a/b (Kavuri et al., 2022). This RNA-focused editing is particularly useful for crop protection against RNA viruses without modifying the plant DNA.

While *Streptococcus pyogenes* Cas9 (SpCas9) remains, the predominant nuclease employed in plant genome editing, the expanding repertoire of Cas orthologs and engineered variants has markedly broadened the CRISPR toolkit. These include nucleases that differ in PAM requirements, molecular size, and cleavage mechanisms, collectively enabling more flexible and precise editing across diverse plant genomes. However, this diversification introduces an invaluable factor: the choice of an appropriate delivery vehicle tailored to the needs of biophysical and biochemical properties of each Cas variant. Larger nucleases may exceed the packaging limits of certain viral or nanocarrier systems, whereas smaller effectors such as Cas12f or CasΦ are more amenable to compact delivery platforms. Consequently, the co-optimization of nuclease architecture and delivery modality (biological, physical, or nano-formulated) has become a central determinant of editing efficiency, tissue specificity, and transgene-free outcomes in plants.

4.2.2 Guide RNAs (gRNAs)

CRISPR editing relies on short guide RNAs to program the Cas nuclease to its target. In plants, gRNAs (single-guide RNAs for Cas9/Cas12 systems or crRNAs for Cas13) are typically transcribed from Pol III promoters, most commonly *Arabidopsis* U6 or U3 promoters (Hsieh-Feng and Yang, 2020). U6/U3 promoters are constitutively active in all tissues, yielding high levels of gRNA expression (Hsieh-Feng and Yang, 2020).

The remarkable feature of plant CRISPR is multiplexing: implying the expression of several gRNAs in one construct so as to edit multiple loci simultaneously. Various strategies (e.g., tRNA-processing or ribozyme linkers) enable a single polycistronic transcript to yield multiple gRNAs. For example, the polycistronic tRNA–gRNA (PTG) strategy produces tandem gRNAs separated by tRNA sequences; endogenous RNase P and Z enzymes then cut the transcript into individual gRNAs. This approach allows precise processing and efficient production of multiple gRNAs and has enabled simultaneous editing of several genes (Xie et al., 2015; Hsieh-Feng and Yang, 2020). In fact, PTG-based constructs have been shown to enable Cas9 to target multiple sites at once in plants (Hsieh-Feng and Yang, 2020). Such multiplex editing has been applied to knock out entire gene families or stack traits: for instance, PTG arrays have been used in rice, wheat, and maize to modify quantitative trait loci, metabolic pathways, and stress-resistance gene sets. Because multiplex constructs need only one promoter, they can be kept small, which is advantageous for viral delivery or transformation (Hsieh-Feng and Yang, 2020). In summary, plant genome engineering routinely uses gRNAs under U6/U3 promoters for strong ubiquitous expression (Dong et al., 2017) and leverages multiplex gRNA arrays to target multiple loci or genes in parallel (Hsieh-Feng and Yang, 2020; McCarty et al., 2020)

4.2.3 Donor templates

When a precise change (gene insertion or allele replacement) is needed, the repair of a CRISPR-induced break must be directed by Homology-Directed Repair (HDR) rather than error-prone non-homologous end joining (NHEJ). Unfortunately, HDR is inherently inefficient in plants. Plant cells overwhelmingly favor NHEJ, and HDR occurs only in S/G2 phases and typically at low frequency (Zhou et al., 2025). As a result, most plant genome edits (knockouts/indels) rely on NHEJ. Enhancing HDR in plants has been a significant research goal.

One key strategy is to increase the availability of donor template DNA at the break site. For example, deconstructed geminivirus replicons (e.g., Bean Yellow Dwarf Virus) have been used as high-copy "amplifiers" of donor templates in plant cells (Gong et al., 2024). These viral vectors replicate by rolling-circle replication, producing many copies of the donor repair template. Early work by Baltes et al. (2014) showed that geminiviral replicons carrying both CRISPR reagents and donor DNA could dramatically raise HDR rates in tobacco cells (Gong et al., 2024). Such replicons have since been used in rice, wheat, and other crops to boost precise gene targeting. Additionally, coupling the donor template directly to the editing machinery has proven effective. A recent innovation fused Cas9 to a viral Rep protein, creating a Cas9–Rep fusion that "tethers" the donor DNA (which contains a viral replication origin) to the Cas9 complex (Zhou et al., 2025) demonstrated that this tethering enriched donor DNA at the DSB and raised HDR insertion rates ~4–7-fold in rice (Zhou et al., 2025). The Cas9–Rep fusion bound 66× more donor template in vivo than Cas9 alone (Zhou et al., 2025).

Other approaches aim to bias the cell toward HDR: synchronizing transformation or editing to the S/G2 cell cycle stages or using small molecules that inhibit NHEJ factors. However, despite these advances, robust HDR in plants remains difficult. Most precise editing still yields low efficiencies and often a mixture of HDR and alternative end-joining outcomes (Zhou et al., 2025). In practice, gene knock-ins and allele replacements are far less routine than simple knockouts. The challenge of efficient HDR limits some applications (e.g., inserting large transgenes or making precise base changes) in crop improvement.

4.2.4 Advanced editors

Beyond simple nucleases, the emergence of "second-generation" CRISPR editors has significantly broadened the range and precision of achievable genetic modifications. These systems transcend the double-strand break–dependent repair paradigm of classical Cas9, enabling direct and programmable nucleotide conversions or small sequence insertions without requiring donor templates or extensive regeneration workflows. By coupling catalytically impaired or nickase variants of Cas enzymes with deaminases, reverse transcriptases, or other functional domains, these editors—such as cytosine and adenine base editors, prime editors, and their subsequent optimized derivatives—offer unprecedented flexibility for fine-tuning gene function, rectifying point mutations, and introducing targeted allelic diversity across diverse plant genomes.

Base editors (BEs) fuse a Cas9 nickase (Cas9n) to a nucleotide deaminase, allowing direct conversion of one base to another without creating a DSB (Jeong et al., 2025). Two main classes exist: cytosine base editors (CBEs) that convert C•G→T•A, and adenine base editors (ABEs) that convert A•T→G•C (Jeong et al., 2025). Since no DSB or donor is required, BEs can operate more efficiently, resulting in fewer collateral edits. In plants, BEs have already been used successfully. For example, optimized CBEs and ABEs have produced targeted point mutations in rice, wheat, maize, tomato, and other crops for herbicide resistance, disease resistance, yield traits, and more (Yang et al., 2024). One study in *Arabidopsis* and soybean protoplasts identified particular CBE and ABE variants that gave high editing at many sites (Jeong et al., 2025). Overall, base editors enable precise "single base" trait engineering in plants without introducing foreign DNA (Jeong et al., 2025).

Prime editors (PEs) are even more versatile: a Cas9 nickase is fused to a reverse transcriptase and programmed by a "prime editing guide RNA" (pegRNA) that both directs Cas9 and encodes the desired edit (substitution, insertion, or deletion) (Nelson et al., 2022). In principle, PE can install virtually any small change at a target site in a single step (Nelson et al., 2022). Prime editing was first demonstrated in human cells (Anzalone et al., 2020) and has been applied to plants. Proof-of-concept reports have achieved base substitutions, small insertions and deletions in rice, maize, and *Arabidopsis* using PE (Nelson et al., 2022). However, plant prime editing efficiencies have so far been modest and variable. Editing rates are often much lower than with BEs or nucleases, and success depends critically on pegRNA design, expression, and stability. Recent advances (e.g., engineered pegRNAs with added motifs) can boost PE in mammalian cells (Nelson et al., 2022); similar optimizations are being tested in plants. In summary, while prime editing offers a "search-and-replace" approach without DSBs, it remains a cutting-edge technique in plant biotechnology. Ongoing work is focused on optimizing the prime editor protein, pegRNA scaffolds, and delivery methods to improve efficiency in crops.

4.3 CRISPR cargo delivery systems in plants

4.3.1 DNA-based delivery methods

4.3.1.1 Agrobacterium-mediated transformation (AMT)

AMT has long been the mainstay for plant genetic engineering. *Agrobacterium tumefaciens* naturally transfers a T-DNA into plant genomes; CRISPR cassettes (Cas9/gRNA) can be carried on T-DNA to generate stable edits. AMT is efficient in many dicots (e.g., tomato, tobacco, soybean) and some monocots (rice, wheat) (Cunningham et al., 2018; Niedbała et al., 2021; Ebrahimi and Hashemi, 2024). For instance, Cas9 delivered via AMT has been used to edit wheat, tomato and tobacco, improving grain weight and disease resistance (Ebrahimi and Hashemi, 2024). However, AMT's host range is limited – many monocots or cultivar lines are "recalcitrant" to Agrobacterium infection (Cunningham et al., 2018). Moreover, AMT typically leads to random integration of the CRISPR transgene into the genome (Cunningham et al., 2018). While effective for stable transformation, this genomic integration means regulatory hurdles, and it requires lengthy tissue culture steps.

4.3.1.2 Biolistics (particle bombardment)

This method is an alternative physical method: DNA (or RNA/RNP) is coated on microprojectiles (gold/tungsten) and shot into tissues (usually embryogenic callus or meristems) (Liang et al., 2018). Biolistics is species-independent and can deliver large DNA fragments or RNPs (Cunningham et al., 2018; Ozyigit and Yucebilgili Kurtoglu, 2020; Rustgi et al., 2022; Ebrahimi and Hashemi, 2024). It has been successfully used in cereals and other crops (maize, wheat, rice, barley) (Ebrahimi and Hashemi, 2024). Unlike *Agrobacterium*, biolistics can also transform chloroplast or mitochondrial genomes (Cunningham et al., 2018). However, it causes tissue damage and often requires regeneration from callus (Cunningham et al., 2018). It also tends to yield low and uneven transgene expression. Complex insertion patterns (multiple copies, rearrangements) frequently occur (Cunningham et al., 2018; Laforest and Nadakuduti, 2022), complicating breeding of the edited plants. Nonetheless, biolistics has the advantage of co-delivering multiple cargoes (DNA and RNP) simultaneously across all species.

4.3.2 DNA-free delivery methods

4.3.2.1 Direct delivery of Cas9 mRNA or RNP complexes

Protoplast transfection (PEG or electroporation) offers a DNA-free route in vitro. Enzymatically stripped cells (protoplasts) can take up DNA, RNA or RNPs by PEG treatment or electroporation (Laforest and Nadakuduti, 2022). This allows transient expression of CRISPR reagents without genomic integration. Many species (even recalcitrant ones) yield protoplasts, making this broadly applicable (Laforest and Nadakuduti, 2022). Reports show Cas9 RNPs delivered by PEG into rice, maize, potato, and other protoplasts yield high editing rates without off-target plasmid DNA (Banakar et al., 2021; Laforest and Nadakuduti, 2022). Editing can be multiplexed in protoplasts by co-transfecting several gRNAs (Laforest and Nadakuduti, 2022). However, one notable weakness in *plant regeneration*: only a few crops (e.g., tobacco, *Arabidopsis*, potato) have robust protocols to regenerate whole plants from protoplasts (Laforest and Nadakuduti, 2022). In many crops, protoplasts either do not redivide or create aberrant plants (somaclonal variation) (Laforest and Nadakuduti, 2022; Liu et al., 2022). Until regeneration systems are available, protoplast editing remains mainly a screening/validation step rather than a field-useable method.

4.3.2.2 In planta *methods*

Methods such as pollen tube or flower dip transformation bypass tissue culture. Agrobacterium or biolistics can target pollen or ovules directly. For example, pollen magnetofection (attaching DNA to magnetic nanoparticles) has enabled CRISPR RNP delivery to lily (Laforest and Nadakuduti, 2022; Aydin, 2024; Alam, 2025). However, these approaches often produce few viable edited plants; pollen viability and chimerism are issues (Laforest and Nadakuduti, 2022). The classic floral dip (*Agrobacterium* infiltration of flowering plants) is a simple, DNA-free method that mostly works in Arabidopsis and related species. In larger crops, methods like vacuum infiltration (into buds) or shoot meristem injection are being explored to introduce RNPs in situ. These avoid transgenesis but remain technically challenging and low efficiency today (Hu and Liu, 2025; Sasaki et al., 2025).

4.3.3 Nanoparticle-based delivery methods

Nanotechnology has emerged as a disruptive approach for plant transformation (Cunningham et al., 2018; Demirer et al., 2021). Nanocarriers (1–100 nm particles) can traverse plant cell walls without the brute force of biolistics. Many types have been tested: CNTs, carbon dots, mesoporous silica nanoparticles (MSNs), layered double hydroxides (LDH), DNA nanostructures, and lipid/polymer nanoparticles. These can adsorb or conjugate DNA, RNA, RNPs, or proteins and ferry them into cells. Nanoparticle carriers offer several advantages. They can be designed to target cell types or organelles, and do not require genomic integration. For example, CNTs decorated with DNA/RNA spontaneously enter diverse plant cells, including chloroplasts (Demirer et al., 2021; Laforest and Nadakuduti, 2022). Demirer et al. showed that positively charged nanotubes deliver genes into Arabidopsis leaves without DNA integration (Demirer et al., 2021). MSNs were used to deliver *Cre* recombinase protein into maize embryos to excise a loxP-flanked marker, demonstrating that proteins (even enzymes) can be imported (Demirer et al., 2021). In general, nanoparticle methods are genotype-independent: the same particles can transfect monocots or dicots equivalently, as entry is mechanical/chemical rather than biological (Cunningham et al., 2018; Demirer et al., 2021). Notably, nanoparticles can even enter intact leaves, pollen, or seeds if applied correctly, raising the possibility of field-level spraying of CRISPR cargo (Mujtaba et al., 2021; Squire et al., 2023).

However, major challenges remain. Reported editing efficiencies with nanocarriers in plants are still low. Unlike mammalian cells (where RNP editing via nanoparticles has been demonstrated (Laforest and Nadakuduti, 2022), no fully DNA-free nanoparticle CRISPR editing of a multicellular plant has yet been published. Laforest et al. noted that RNP delivery by nanoparticles "has yet to be achieved in plants" largely due to the extremely high delivery needed for whole-plant editing(Laforest and Nadakuduti, 2022). In plants, the payload often needs to get to meristem cells, and the cell wall/cuticle barriers are more complex. Stability of CRISPR RNPs on particles, controlling release inside cells, and avoiding toxicity are active issues. Many early studies used DNA plasmids (which may integrate permanently in the genome) rather than purified RNPs.

4.3.4 Virus-mediated delivery

4.3.4.1 Plant viral vectors (TRV, PVX, geminiviruses, foxtail mosaic virus)

Plant viruses are natural delivery agents, able to infect most tissues and even spread systemically (Li et al., 2024). Several viruses have been repurposed as CRISPR carriers (X(Wu et al., 2024). RNA viruses like tobacco rattle virus (TRV), barley stripe mosaic virus (BSMV), foxtail mosaic virus (FoMV), and sonchus yellow net rhabdovirus can carry gRNAs into plants expressing Cas9 (Ebrahimi and Hashemi, 2024; Steinberger and Voytas, 2025; Weiss et al., 2025). For example, TRV can be agro-infiltrated into Cas9-transgenic tomato or tobacco to drive systemic sgRNA delivery. Sonchus yellow net virus (SYNV, a rhabdovirus) has been engineered to deliver both SpCas9 and sgRNA for heritable editing in pepper (Laforest and Nadakuduti, 2022; Ebrahimi and Hashemi, 2024). A recent breakthrough by Weiss et al. (2025) used TRV to carry a *compact* RNA-guided nuclease (TnpB/ISYmu1) and its gRNA, achieving *transgene-free, germline* edits in *Arabidopsis* in one infection (Luo et al., 2024; Weiss et al., 2025). This work overcame the usual viral cargo limitation by using a smaller nuclease; the TRV vector alone inserted the functional editing machinery without needing to transform a Cas9 plant (Luo et al., 2024).

Despite these successes, viral methods have trade-offs. Most plant viruses have small genomes (4-8 kb), so they cannot accommodate large payloads like the standard Cas9 (~4.2 kb). This is why many studies use plants already expressing Cas9 and have the virus deliver just the gRNA(s). Alternatively, they use smaller nucleases (Cas12f, CasΦ, TnpB) that fit (Achiam et al., 2023). Even then, virus-induced editing is often mosaic and not fully heritable, because viruses usually do not infect germ cells efficiently (Laforest and Nadakuduti, 2022; Weiss et al., 2025). TRV, for instance, typically does not transmit to seeds, though its infection of meristem cells can induce heritable mutations (Weiss et al., 2025). By contrast, DNA viruses (geminiviruses) offer another route: they can replicate to high copy number in nuclei and carry larger inserts.

4.3.4.2 *Replicon-based systems*

Geminivirus replicons (derived from Geminiviridae) are circular ssDNA viruses (e.g., bean yellow dwarf virus) that are engineered to amplify a DNA fragment in planta (Mahmood et al., 2025). Modern strategies use deconstructed geminivirus vectors delivered via Agrobacterium T-DNA (Mahmood et al., 2025). A geminivirus replicon contains an intergenic region (IR) flanking the gene of interest, and when Rep protein is expressed, it self-replicates to thousands of copies. This was exploited to boost HDR: Cas9 and a repair template on a replicon means the template is present in high copy, greatly increasing insertional editing efficiency (Mahmood et al., 2025). Indeed, targeted gene insertions have been achieved in wheat and rice using geminivirus replicon systems (Yu et al., 2020). Geminivirus vectors can also carry multiple gRNAs or a donor array. Previous studies highlight that geminivirus vectors are poised to revolutionize plant editing by easing nucleic acid deployment and reducing tissue culture needs (Mahmood et al., 2025). Key challenges for virus methods include containment (virus escape), off-target virus effects, and public perception. However, they offer one of the most efficient in vivo delivery platforms for CRISPR cargoes (especially for woody or large crops where regeneration is complex).

4.3.5 Emerging non-traditional approaches

Beyond the discussed modes of cargo delivery systems, several other innovative methods are still in development. Some of the key approaches are.

4.3.5.1 *Cell-penetrating peptides (CPPs)*

CPPs are short peptides (often arginine-rich) that can ferry cargo across cell membranes (Wu et al., 2022). rCPPs are derived from protein transduction domains that can bind and carry DNA, RNA, or proteins into plant cells (Wu et al., 2022). For example, a CPP fused to Cas9 RNP can be incubated with leaf tissue, resulting in uptake and targeted mutations without any DNA vectors. CPPs can also transport developmental regulators to induce cell division. These approaches yield sophisticated delivery and have shown success in rice, maize, and other crops. CPPs are minimally toxic and avoid transgenesis concerns (Wu et al., 2022). The limitation is that efficiency is modest and often requires wound sites or surfactants to penetrate the cuticle.

4.3.5.2 *Electroporation and microinjection*

Lab-on-chip devices and microinjection are capable of delivering RNPs into specific cells or tissues (Lino et al., 2018). For instance, cathodic electroporation of flower apical cells or shoot meristems was attempted in certain species. Ultrasound and biocompatible surfactants can transiently open stomatal pores to admit nanoparticles. These are essentially experimental. Pollen-based methods (pollen magnetofection, pollen tube injection) aim to introduce editing agents into gametes (Zhao et al., 2017). Vejlupkova et al. (2020) attempted a similar approach in maize. They reported some transgene-free edits, but only in dicots (Laforest and Nadakuduti, 2022). In general, exploiting reproductive cells is appealing to bypass regeneration, but success in important cereals remains elusive.

4.3.5.3 *Developmental regulator fusions*

Another trend is fusing CRISPR components or repair enzymes with plant developmental regulators. For example, VirD2 relaxase fusions help tether repair templates to DSB sites (Pramanik et al., 2021). Overexpression of *WUSCHEL* or *BBM* genes in transformed tissues can induce somatic embryogenesis of edited cells (making transformation genotype-independent). These methods improve the editing pipeline, often combined with Agrobacterium. They are not standalone delivery methods, but rather enhancers.

In each case, a key distinction is in vitro (protoplast, tissue culture) versus in vivo (whole plant, non-lab). Protoplast and explant methods require lab regeneration; pollen- or viral-based methods can operate in intact plants. The choice depends on the crop and application. For high-throughput screening or trait discovery, in vitro methods suffice; for field-ready crops, in vivo methods that avoid tissue culture are preferable (Table 4.1).

4.3.6 Bioinformatics and AI for intelligent design of CRISPR cargoes in plants

Bioinformatics and artificial intelligence (AI) have been in the forefront for revolutionizing how CRISPR cargoes are being delivered in plants. Intelligent computational pipelines now assist researchers in discovering, selecting, and even designing delivery systems (viral vectors, nanoparticles, RNP complexes, etc.) tailored to specific plant species, tissues, and cellular contexts. Integrating CRISPR technologies with machine learning and data-driven tools is increasingly seen as key to overcoming delivery bottlenecks (Chen et al., 2024). These approaches enable predictive modeling of which delivery method will be most compatible and efficient for a given plant system, based on genomic and physiological data. Below, we highlight several emerging AI-powered and bioinformatics platforms driving more efficient, targeted CRISPR delivery as part of next-generation genome editing workflows in agriculture.

4.3.7 LLM Co-pilots (e.g., CRISPR-GPT)

Large language model agents are now being used to automate experiment planning for plant genome editing (Acharya et al., 2025). Notably, CRISPR-GPT is an AI system that assists in end-to-end design of gene editing experiments, including automated selection of

TABLE 4.1 Comparative analysis of major CRISPR delivery platforms in plants.

Delivery method (type)	CRISPR cargo compatibility	Expression & integration	Target tissues (culture)	Species scope & efficiency	Key considerations (limitations)
Agrobacterium-mediated transformation (biological)	DNA (T-DNA vectors)	Primarily stable expression via T-DNA integration (transgene present; not transgene-free)	Tissue culture explants (mostly dicots; limited efficacy in monocots) – requires regeneration in vitro	High efficiency in many dicots; very limited success in most monocots (genotype-dependent)	Widely used standard; limited host range (many monocots are recalcitrant) and integrates foreign DNA, necessitating removal of transgenes in progeny.
Biolistic (gene gun) (physical)	DNA, RNA, RNPs	Can be transient (with RNPs), but DNA bombardment often causes stable, random genomic insertions (transgene integration)	Embryogenic callus or meristems – requires tissue culture steps for plant regeneration	Broadly applicable to both monocots and dicots; moderate-to-high editing frequencies (varies by species and genotype)	Species-independent delivery of various cargos; however, high-velocity particles can cause tissue damage and fragmented, multi-copy insertions in the genome.
Protoplast transfection (PEG/ Electroporation) (Physical/ Chemical)	DNA, RNA, RNPs	Transient expression (especially with RNPs); enables DNA-free editing with no integration into genome	Isolated protoplast cells – requires plant regeneration from protoplast (often challenging)	In vitro applicable to diverse species (dicots and monocots); very high editing rates in protoplasts, but low overall efficiency in recovering whole gene-edited plants	Bypasses agrobacterium and vectors by direct RNP delivery (avoids integration); major bottleneck is that protoplast regeneration is not established for many crops.
Nanoparticle-mediated delivery (Chemical/ Nanotechnological)	DNA, RNA, RNPs, proteins	Typically transient, non-integrative delivery (especially with RNA or RNP cargo; low risk of stable integration)	Various tissues (e.g., leaves, seeds) – often can be applied without traditional tissue culture	Potentially broad species applicability; currently low-to-moderate editing efficiency (still being improved in research)	Promising DNA-free approach (no agrobacterium or viral vector needed); early-stage – efficiency and delivery consistency remain low, requiring further optimization.

TABLE 4.1 Comparative analysis of major CRISPR delivery platforms in plants.—cont'd

Delivery method (type)	CRISPR cargo compatibility	Expression & integration	Target tissues (culture)	Species scope & efficiency	Key considerations (limitations)
Viral vectors (biological)	DNA or RNA vectors (can carry gRNAs; some systems deliver small Cas variants or RNP)	Both stable and transient possible; often used as replicating episomes for transient expression (usually no genome integration)	Whole plants (meristems, somatic or reproductive tissues) – no tissue culture needed for infection in susceptible hosts	Species-dependent on virus host range (often effective in dicots with compatible virus); moderate editing rates, with systemic spread through the plant	Enables systemic, heritable editing without transgene integration; key limits are small cargo capacity (large Cas9 won't fit some viruses) and narrow host range – also requires biosafety containment for plant virus use.
Cell-penetrating peptides (CPPs) (Biological/ Chemical)	RNPs (Cas9 protein + gRNA)	Transient delivery of Cas9–gRNA RNP complexes (direct protein uptake; no DNA, thus no integration)	Leaf tissues, protoplasts (and potentially pollen in experiments) – currently limited to lab-scale applications (not yet a field method)	Demonstrated mainly in model dicots so far; monocot delivery is still experimental; low editing efficiency in current protocols	Innovative protein-delivery method that achieves DNA-free genome edits; however, it is an early-stage technology with very limited efficiency and scope to date (significant optimization needed).

appropriate delivery methods and protocol optimization (Qu et al., 2025). CRISPR-GPT's "delivery agent" parses the user's target species/tissue, retrieves relevant literature, and then matches the scenario to the best delivery strategy – for example, recommending *Agrobacterium*-mediated T-DNA transfer versus a nanoparticle or viral vector based on what has the highest success in similar contexts (Qu et al., 2025). Impressively, in benchmarking tests spanning 50 diverse gene-editing scenarios, CRISPR-GPT's AI-driven recommendations of primary and secondary delivery methods aligned closely with expert consensus, outperforming baseline GPT-4 models, especially for challenging or recalcitrant cell types (Qu et al., 2025). This LLM co-pilot takes it a step further by generating step-by-step experimental protocols once a delivery mode is chosen. For instance, CRISPR-GPT autonomously provided a complete protocol for lentiviral vector production and infection when that method was identified as optimal (Qu et al., 2025). Such agentic automation of delivery system selection and experimental setup exemplifies how AI can act as a "lab assistant", streamlining the design-build process for plant genome editing.

Here are the steps that can be adopted to use LLMs for deciding CRISPR cargo choice.

1. **Input Phase**
   User Inputs: Target plant species, genome type, transformation history, tissue type, desired edit (knockout, base edit, prime edit), regulatory constraints.
   System Query: CRISPR-GPT prompts the user or extracts metadata to define project context.
2. **Knowledge Integration**
   Database Access: Integrates published datasets on Cas orthologs, delivery efficiencies, and plant transformation outcomes.
   Model Library: Includes curated knowledge of delivery vehicles (*Agrobacterium*, viral vectors, biolistics, nanocarriers, etc.) and compatibility with CRISPR cargoes (Cas9, Cas12, CasΦ, Cas13, base/prime editors).
3. **AI Reasoning Layer**
   LLM Analysis: CRISPR-GPT analyzes user inputs and database features using natural language reasoning and structured data mapping.
   AI/ML Models: Predict delivery success probabilities based on parameters such as cargo size, PAM requirement, tissue tropism, and prior species performance.
4. **Decision and Recommendation**
   Cargo-Delivery Matching: Suggests optimal cargo–delivery pairs (e.g., CasΦ + viral replicon for small genomes; Cas9 RNP + nanoparticle for transgene-free edits).
   Output Visualization: Displays ranked recommendations with rationale (efficiency, scalability, heritability, risk factors).
5. **Experimental Optimization**
   Protocol Generation: Auto-generates draft protocols, reagent lists, and construct maps.
   Feedback Loop: Integrates user or experimental results to retrain AI models, refining future recommendations.
6. **Outcome**
   Result: AI-guided, context-specific CRISPR delivery recommendation tailored to plant species, desired edit type, and lab resources (Fig. 4.1).

4.3.8 Predictive delivery modeling

Beyond LLMs, other AI models leverage plant genomic and physiological data to predict the compatibility and efficiency of different delivery methods (Chen et al., 2019). Machine learning enables analysis of features namely, cell wall composition, tissue type, or species-specific transformation records to suggest which delivery strategy (e.g., PEG-mediated RNP transfection vs. biolistic DNA delivery) would likely yield the highest editing efficiency. By analogy, in plants, these models can recommend species-optimized delivery methods to facilitate error-free and efficient CRISPR reagent uptake. For example, an AI tool may predict certain monocot crops responding better to electroporation-based delivery or nanoparticle systems (due to Agrobacterium host range limitations) (Achiam et al., 2023; Chen et al., 2024). However, in a succulent or woody tissue, a viral replicon might systemically penetrate more. Previous efforts depict that AI could assist in restructuring vectors

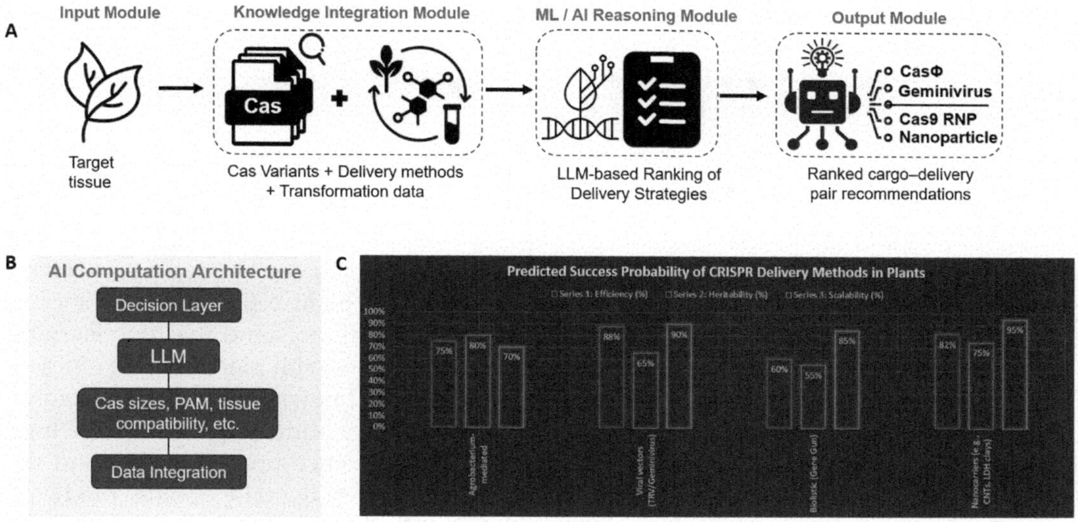

FIGURE 4.1 LLM-guided conceptual framework for intelligent cargo–delivery selection in plants: Schematic representation of an intelligent decision framework for selecting optimal CRISPR cargo–delivery systems in plants. The model integrates biological inputs, predictive learning, and AI reasoning to propose tailored, efficient, and transgene-free genome editing strategies. A) Input-to-decision pipeline, B) AI reasoning architecture, C) output visualization.

and standardizing protocols to the target context (Dixit et al., 2023). This in turn would effectively determine which delivery approach would be most compatible before committing to lab trials. This data-driven optimization helps researchers avoid costly trial-and-error and adapt strategies to recalcitrant species (Achiam et al., 2023).

4.3.9 AI-driven vector and nanocarrier design

Bioinformatic tools are also advancing the design of the delivery vehicles themselves. One emerging frontier is using AI for in silico engineering of viral vectors and nanoparticles with improved targeting and stability for plant cells. In human gene therapy, for instance, deep learning is being used to engineer viral capsids (e.g., AAV) with novel tropisms and higher delivery efficiency (Li et al., 2025). Similar approaches are being explored to modify plant viral vectors' host specificity or cargo capacity. Likewise, machine learning guides the design of polymeric and lipid nanoparticles: by mining large datasets of delivery experiments, algorithms can predict which nanoparticle formulations will efficiently traverse cell barriers (Gong et al., 2022; Ortiz-Perez et al., 2024). Egorov et al. (2021) demonstrated that combining robotics, microfluidics, and AI can vastly accelerate the screening of liposomes and polymer carriers for drug/gene delivery (Dixit et al., 2023). Adapting this to agriculture, researchers can computationally screen nanomaterials (e.g., CNTs, silica, or clay nanosheets) to identify surface chemistries that bind and protect CRISPR RNPs while maximizing uptake in plant tissues (Kocsisova and Coneva, 2023). AI models even help design supporting elements – *promoters, enhancers,* etc. – to ensure the delivered CRISPR components are expressed at the right level and tissue specificity (Dixit et al., 2023). Overall, a new generation of AI-assisted delivery platforms is emerging to optimize how CRISPR cargos are packaged and delivered, from smarter viral vectors to custom nanoparticles engineered for plant systems (Ortiz-Perez et al., 2024).

4.3.10 Peptide-based delivery innovations

Another bioinformatics-driven advance is the discovery of novel carrier peptides that can ferry CRISPR cargoes into plant cells (Squire et al., 2023). CPPs are short amino acid sequences capable of traversing cell membranes and even the plant cell wall, and they can be conjugated to Cas proteins or nucleic acids. Computational tools and machine learning are proving invaluable in identifying or designing such peptides with plant-specific efficacy. For example, the *CPP site* database now catalogs hundreds of known CPPs, and a machine-learning web server (KELM-CPPpred) has predicted new peptide sequences with high membrane-translocating potential (Liu et al., 2023). These in silico predictions expand the toolkit of delivery peptides beyond those discovered experimentally. Some predicted CPPs have been validated to form complexes with CRISPR ribonucleoproteins and facilitate their uptake in cells (Liu et al., 2023). Additionally, bioinformatics can help tailor peptides for organelle-specific targeting: analyses of transit peptide motifs have guided the design of fusions that direct CRISPR cargoes to chloroplasts or mitochondria in planta. *Sprayable* peptide-nucleotide formulations are even being explored, aiming to deliver CRISPR reagents by simple foliar application (Chen et al., 2019). While these approaches are still maturing, AI-driven peptide design illustrates how non-traditional, DNA-free delivery methods can be rationally optimized for plant genome editing (Chen et al., 2019).

In summary, the convergence of AI, bioinformatics, and genome editing is yielding intelligent delivery systems that complement the physical innovations in CRISPR cargo delivery. By computationally modeling and optimizing each protocol step – from choosing vector system to designing construct and protocol parameters – researchers can achieve more stabilized and targeted delivery of CRISPR reagents in plants. These advances not only accelerate the pace of crop genome engineering but also improve reproducibility and success rates. The integration of an expert knowledge base, predictive algorithms, and automated planning tools heralds a new era of agentic genome editing framework, where bioinformatics and AI guide the selection of the right tool for the job – ensuring that the CRISPR cargo reaches its plant target effectively and accelerates precise genetic improvements in crops.

4.4 Challenges and future directions in CRISPR delivery for plants

4.4.1 Overcoming tissue culture bottlenecks and genotype dependency

A persistent challenge in plant genome editing is the untimely dependence on in vitro tissue culture for transformation and regeneration, which is consistently time-consuming, labor-intensive, and highly genotype-dependent (Yuan et al., 2023; Li et al., 2024; Han et al., 2025). Plenty of crop species require tons of months of culturing to obtain edited plants, and certain cultivars suffer from somaclonal variation during regeneration or remain recalcitrant (Sebiani-Calvo et al., 2024). To bypass these bottlenecks, researchers are developing tissue culture-independent (*in planta*) transformation techniques that introduce CRISPR reagents directly into plants with minimal or no callus phase (Sebiani-Calvo et al., 2024). Classic examples include floral dip in *Arabidopsis* and newer protocols that infect meristems or ovules, enabling edited seeds without extensive culturing (Sebiani-Calvo et al., 2024). Another innovative approach is *haploid inducer-mediated editing (HI-Edit)*, which utilizes specially bred pollen donors to deliver CRISPR components to zygotes, producing edited embryos without tissue culture (Delzer et al., 2024; Han et al., 2025). Collectively, these strategies – from *in planta* techniques to haploid-based methods – are converging on solutions to make plant transformation faster and genotype-flexible, thereby unleashing the full potential of CRISPR in diverse crops.

4.4.2 Improving homology-directed repair efficiency in plants

Precise genome edits via HDR remain challenging in plants, as DNA breaks are predominantly resolved by non-homologous end joining (NHEJ) (Chen et al., 2022). The native frequency of HDR in plant somatic cells is extremely low, which has been a major barrier to gene targeting for trait improvement (Chen et al., 2022). Recent research has focused on biasing repair pathways and increasing donor availability to enhance HDR outcomes. One approach is to suppress or bypass NHEJ transiently – for instance, using mutations or inhibitors against key NHEJ enzymes (like DNA ligase IV) – thereby channeling more breaks into the HDR pathway (Borphukan et al., 2025; Haider and Mussolino, 2025). At the same time, methods to increase the concentration and persistence of repairtemplates at the cut site have shown success (Haider and Mussolino, 2025). Geminivirus-derived replicons exemplify this strategy: these small DNA viruses can autonomously replicate a donor template inside plant cells, providing a high-copy template for HDR. Using a deconstructed wheat dwarf virus replicon, researchers achieved ~10% HDR-mediated precise editing in rice, a striking improvement over baseline frequencies (Molla et al., 2022; Borphukan et al., 2025). Such replicon systems avoid the immediate integration of the donor into the genome, keeping it available for recombination (Borphukan et al., 2025). Other innovations include engineering the CRISPR nuclease itself to favor HDR – for example, fusing Cas9 to peptide domains from recombination proteins (like VirD2 or RAD52) to tether donor DNA or recruit repair factors to the break (Li et al., 2021; Chen et al., 2022). Alternative Cas nucleases with staggered-cutting activity (e.g., Cas12a/Cpf1) have also been explored for their potential to yield overhangs more amenable to HDR (Chen et al., 2022; Vu et al., 2020). While true gene targeting in crops is still in its infancy, these combined approaches have begun to elevate HDR efficiencies to practical levels. The progress suggests that precise allele replacements and gene insertions – once considered elusive in plants – could become routine as delivery of donor templates and manipulation of repair pathways continue to improve (Borphukan et al., 2025).

4.4.3 Expanding viral vector cargo capacities for CRISPR delivery

Plant viruses are attractive CRISPR delivery vehicles due to their innate ability to infect cells systemically and express transgenes; however, limited cargo capacity has constrained their use (Mikhaylova, 2025). Many commonly used viral vectors cannot accommodate the large ~4.2 kb sequence encoding *Streptococcus pyogenes* Cas9 (SpCas9) along with regulatory elements. For example, Tobacco rattle virus (TRV), a popular RNA virus vector, has too small a genome size to carry a Cas9 expression cassette, so early virus-induced genome editing systems delivered only sgRNAs into transgenic Cas9-expressing plants (Mikhaylova, 2025). To overcome this, researchers are engineering viruseswith larger or more flexible genomes and deploying multi-component delivery schemes. Some filamentous RNA viruses (e.g., certain Potexvirus and Bunyavirus vectors) can tolerate large inserts due to their extensible particle structure or segmented genomes (Mikhaylova, 2025). Notably, a foxtail mosaic virus (FoMV) vector (Potexvirus) with a wide host range was shown to deliver Cas9 and sgRNA, demonstrating virus-mediated editing without a transgenic nuclease in solanaceous plants (Baysal et al., 2025; Mikhaylova, 2025). Likewise, bullet-shaped rhabdoviruses such as Sonchus yellow net virus (SYNV) have been used to carry the entire CRISPR/Cas machinery (~6 kb) in *Nicotiana* species (Mikhaylova, 2025). These successes are tempered by trade-offs: viruses with removed genes (e.g., coat or movement proteins) can accept bigger inserts but may lose the ability to spread through the plant, sometimes forcing a return to tissue culture for infection (Mikhaylova, 2025). DNA viral replicons (such as geminivirus-derived vectors) offer a different solution: by replicating within nuclei without encapsidation, they effectively lift the size constraint and have been used to express large cargos like Cas9 or HDR templates, albeit confined to the initially transformed cells (Mikhaylova, 2025). Beyond retooling vectors, shrinking the payload is another complementary strategy. Researchers are actively searching for smaller Cas proteins ("mini-Cas" nucleases) that could fit into standard virus backbones (Mikhaylova, 2025). For instance, Cas12a is somewhat smaller than SpCas9, and even more compact Cas variants (e.g., engineered Cas9 truncations or new orthologs) are under investigation (Tuncel et al., 2025). In summary, expanding the carrying capacity of plant viral vectors – through clever virology and protein engineering – is a key frontier. Recent advances suggest that transient viral delivery of CRISPR (even including the Cas enzyme) to whole plants is becoming feasible, opening the door to high-efficiency genome editing via simple inoculation of crops (Uranga and Daròs, 2023).

4.4.4 Enhancing nanoparticle delivery systems for field use and targeting

Nanoparticle-mediated delivery is emerging as a versatile platform to deploy CRISPR reagents in plants, with an eye toward direct field applications. Nanomaterials can be engineered to cross the plant cell wall, a barrier that often complicates traditional transformation (Wang et al., 2019; Dunbar et al., 2022). Recent studies have shown that functionalized CNTs can bind and ferry plasmid DNA or RNA into intact plant tissues. For example, CNTs loaded with plasmids were found to diffuse through rice leaf cells and embryos, driving transient expression of fluorescent reporters without any wounding or transgene integration (Dunbar et al., 2022). Researchers extended this to deliver CRISPR–Cas constructs (targeting a reporter gene) into rice embryos *in planta*, with the goal of inducing heritable edits directly in the germline (Dunbar et al., 2022). Other nanocarriers, such as MSNs, DNA nanostructures, liposomes, and layered double hydroxide clay nanosheets, have likewise been explored as vehicles to introduce CRISPR components (DNA, RNA, or ribonucleoproteins) into crop seedlings and even mature plants (Zhi et al., 2022; Li et al., 2025). These approaches are typically *species-independent* – e.g., the same CNT formulation can potentially be used on different crops – highlighting their promise for broad agricultural use without bespoke protocols.

A significant advantage of nanoparticle delivery is the prospect of field deployment: one can imagine sprayable or seed-coating formulations that farmers could apply to edit crops on-site (Lew et al., 2020). Achieving this requires that nanoparticles are biocompatible, targeted, and effective under outdoor conditions (Sharma and Lew, 2022; Yan et al., 2022). Rational design of nanocarriers is therefore focused on stability and specificity. In the field, leaves are the usual entry point for nanomaterials, so particles must withstand sunlight, temperature fluctuations, rain, and microbial enzymes long enough to release their CRISPR cargo (Sharma and Lew, 2022). Surface coatings (polymers, peptides, etc.) can be tuned to protect the cargo (e.g., shielding an RNA guide from nucleases) and to facilitate entry through stomata or even delivery to specific cell types (Sharma and Lew, 2022) For instance, *magnetic nanoparticles* have been used in a targeted manner through pollen magnetofection: DNA- or RNP-coated magnetic iron oxide particles are bound to pollen grains, and a magnetic field is used to drive this pollen into pistils. This technique has yielded genetically modified or edited seeds in crops like tomato without any tissue culture, essentially editing the next generation via reproduction (Zhao et al., 2017; Sandhya et al., 2020; Park et al., 2024). Such targeting to reproductive cells or meristems is crucial for producing heritable edits in a single treatment. Just as importantly, modern nanomaterials are being designed to be biodegradable and non-toxic, ensuring that after delivering their payload they do not persist in the plant or soil. Early evaluations indicate that certain DNA nanostructures and polymer-coated nanoparticles can break down in plant tissues or the environment over time, mitigating biosafety concerns (Sharma and Lew, 2022). Although nanoparticle-based plant genome editing is still nascent, continued progress in materials science and plant nanobiotechnology is expected to yield safe, high-efficiency delivery tools. These could enable farmers to implement genome editing much like a chemical treatment – a transformative step toward on-demand, in-field crop trait improvement (Demirer et al., 2021) (Bhattacharya and Sharma, 2020).

4.4.5 Toward non-transgenic, heritable, field-deployable genome editing

Across all these advances lies a unifying goal: to achieve genome edits in crops without introducing permanent foreign DNA, in a manner that is heritable and easily deployable outside the lab. Non-transgenic (DNA-free) genome editing is highly desirable because regulatory agencies in many countries treat such edits as conventional mutations rather than GMOs (Tuncel et al., 2025). In fact, plants carrying only small edits (indels or base changes) with no transgene integration are already exempt from GMO regulations in the US, Japan, Australia, and several other nations, expediting their path to market (Mikhaylova, 2025). Realizing a workflow for "transgene-free" editing typically involves delivering CRISPR–Cas reagents in a transient form – as RNA or protein or as a short-lived DNA vector – so that the editing tool is active only briefly and does not integrate into the plant's chromosomes (Mikhaylova, 2025). Multiple proof-of-concept studies have validated this approach. For example, delivering pre-assembled Cas9–sgRNA ribonucleoproteins (RNPs) via protoplast transfection or biolistic bombardment has produced edited plants in species like rice, wheat, and lettuce with no transgene present; the only genomic change is the intended mutation itself (Mikhaylova, 2025). Although RNP delivery can be less efficient than DNA-based methods, it bypasses the need for subsequent segregation of transgenes and has already yielded fertile, stably edited crop lines in a single generation (Mikhaylova, 2025). Likewise, the virus-induced and nanoparticle-based techniques described above are designed to leave no trace after editing: the virus or nanoparticle degrades or is diluted out, and only the edited allele remains in the plant's progeny (Han et al., 2025; Mikhaylova, 2025). Achieving consistent germline edits (so that a high proportion of progeny carry the edit) is an area of active development, often tackled by targeting meristematic or reproductive cells for editing.

Encouragingly, the convergence of novel delivery methods with genotype-independent regeneration techniques is bringing truly field-deployable genome editing closer to reality (Han et al., 2025). We are beginning to see the outlines of an era in which a farmer or breeder might apply a formulation (a viral inoculum, a nanoparticle spray, or a pollen coat) to a plant and recover an edited crop variety without any lab-based transformation pipeline. This one-step, in-field editing paradigm would dramatically accelerate crop improvement and could democratize the technology beyond large laboratories or companies (Mikhaylova, 2025). While further work is essential to ascertain efficiency and safety in open environments, the prospect of non-transgenic, heritable genome editing in the field is on the horizon (Han et al., 2025).

4.5 Conclusion

Efficient delivery of CRISPR components remains one of the most significant technical challenges in applying genome editing across diverse plant taxa. Traditional transformation approaches are often limited by species specificity, tissue culture dependency, and low scalability. However, recent advances in nanotechnology, plant viral vectors, and direct RNP delivery offer promising transformative possibilities, thereby paving the way for transgene-free editing and tissue culture–independent methods which could be readily applied in both laboratory and field settings.

Alongside, AI and bioinformatics are now playing an innovative and pivotal role in optimizing CRISPR delivery. Machine learning models are being employed to design more efficient guide RNAs, predict editing outcomes, and identify genotype-specific transformation bottlenecks. Tools such as CRISPR-GPT have taken over the market and automated experimental design, accelerated discovery manifold and reduced trial-and-error in delivery optimization. These in silico advancements ensure more targeted, efficient, and adaptable gene-editing workflows.

Looking forward, the future of plant biotechnology will depend on scalable, precise, and regulation-conscious CRISPR delivery systems. As transgene-free technologies mature and are aligned with evolving regulatory frameworks, genome-edited crops may see relatively rapid commercialization and broader public acceptance. With the convergence of molecular biology, materials science, and intelligent automation, CRISPR-based crop improvement is poised to become an integral part of sustainable and precision agriculture.

Declaration

"The authors declare that no generative artificial intelligence tools were used in the writing or preparation of this chapter. The content reflects the authors' own scholarly work, and the authors assume full responsibility for all statements, interpretations, and conclusions presented".

References

Acharya, D.B., Kuppan, K., Divya, B., Agentic, A.I., 2025. Autonomous intelligence for complex goals - a comprehensive survey. IEEE Access 13, 18912–18936. https://doi.org/10.1109/ACCESS.2025.3532853.

Achiam, J., Adler, S., Agarwal, S., Ahmad, L., Akkaya, I., Aleman, F.L., Almeida, D., Altenschmidt, J., Altman, S., Anadkat, S., Avila, R., Babuschkin, I., Balaji, S., Balcom, V., Baltescu, P., Bao, H., Bavarian, M., Belgum, J., Bello, I., Berdine, J., Bernadett-Shapiro, G., Berner, C., Bogdonoff, L., Boiko, O., Boyd, M., Brakman, A.L., Brockman, G., Brooks, T., Brundage, M., Button, K., Cai, T., Campbell, R., Cann, A., Carey, B., Carlson, C., Carmichael, R., Chan, B., Chang, C., Chantzis, F., Chen, D., Chen, S., Chen, R., Chen, J., Chen, M., Chess, B., Cho, C., Chu, C., Chung, H.W., Cummings, D., Currier, J., Dai, Y., Decareaux, C., Degry, T., Deutsch, N., Deville, D., Dhar, A., Dohan, D., Dowling, S., Dunning, S., Ecoffet, A., Eleti, A., Eloundou, T., Farhi, D., Fedus, L., Felix, N., Fishman, S.P., Forte, J., Fulford, I., Gao, L., Georges, E., Gibson, C., Goel, V., Gogineni, T., Goh, G., Gontijo-Lopes, R., Gordon, J., Grafstein, M., Gray, S., Greene, R., Gross, J., Gu, S.S., Guo, Y., Hallacy, C., Han, J., Harris, J., He, Y., Heaton, M., Heidecke, J., Hesse, C., Hickey, A., Hickey, W., Hoeschele, P., Houghton, B., Hsu, K., Hu, S., Hu, X., Huizinga, J., Jain, S., Jain, S., Jang, J., Jiang, A., Jiang, R., Jin, H., Jin, D., Jomoto, S., Jonn, B., Jun, H., Kaftan, T., Kaiser, Ł., Kamali, A., Kanitscheider, I., Keskar, N.S., Khan, T., Kilpatrick, L., Kim, J.W., Kim, C., Kim, Y., Kirchner, J.H., Kiros, J., Knight, M., Kokotajlo, D., Kondraciuk, Ł., Kondrich, A., Konstantinidis, A., Kosic, K., Krueger, G., Kuo, V., Lampe, M., Lan, I., Lee, T., Leike, J., Leung, J., Levy, D., Li, C.M., Lim, R., Lin, M., Lin, S., Litwin, M., Lopez, T., Lowe, R., Lue, P., Makanju, A., Malfacini, K., Manning, S., Markov, T., Markovski, Y., Martin, B., Mayer, K., Mayne, A., McGrew, B., McKinney, S.M., McLeavey, C., McMillan, P., McNeil, J., Medina, D., Mehta, A., Menick, J., Metz, L., Mishchenko, A., Mishkin, P., Monaco, V., Morikawa, E., Mossing, D., Mu, T., Murati, M., Murk, O., Mély, D., Nair, A., Nakano, R., Nayak, R., Neelakantan, A., Ngo, R., Noh, H., Ouyang, L., O'Keefe, C., Pachocki, J., Paino, A., Palermo, J., Pantuliano, A., Parascandolo, G., Parish, J., Parparita, E., Passos, A., Pavlov, M., Peng, A., Perelman, A., de Avila Belbute Peres, F., Petrov, M., de Oliveira Pinto, H.P., Pokorny, M., Pokrass, M., Pong, V.H., Powell, T., Power, A., Power, B., Proehl, E., Puri, R., Radford, A., Rae, J., Ramesh, A., Raymond, C., Real, F., Rimbach, K., Ross, C., Rotsted, B., Roussez, H., Ryder, N., Saltarelli, M., Sanders, T., Santurkar, S., Sastry, G., Schmidt, H., Schnurr, D., Schulman, J., Selsam, D., Sheppard, K., Sherbakov, T., Shieh, J., Shoker, S., Shyam, P., Sidor, S., Sigler, E., Simens, M., Sitkin, J., Slama, K., Sohl, I., Sokolowsky, B., Song, Y., Staudacher, N., Such, F.P., Summers, N., Sutskever, I., Tang, J., Tezak, N., Thompson, M.B., Tillet, P., Tootoonchian, A., Tseng, E., Tuggle, P., Turley, N., Tworek, J., Uribe, J. F.C., Vallone, A., Vijayvergiya, A., Voss, C., Wainwright, C., Wang, J.J., Wang, A., Wang, B., Ward, J., Wei, J., Weinmann, C.J., Welihinda, A., Welinder, P., Weng, J., Weng, L., Wiethoff, M., Willner, D., Winter, C., Wolrich, S., Wong, H., Workman, L., Wu, S., Wu, J., Wu, M., Xiao, K., Xu, T., Yoo, S., Yu, K., Yuan, Q., Zaremba, W., Zellers, R., Zhang, C., Zhang, M., Zhao, S., Zheng, T., Zhuang, J., Zhuk, W., Zoph, B., 2023. GPT-4 technical report. arXiv. https://doi.org/10.48550/arXiv.2303.08774.

Alam, A., 2025. Tissue culture-free genetic engineering and genome editing of peanut. Mol. Biotechnol. https://doi.org/10.1007/S12033-025-01476-8.

Aman, R., Ali, Z., Butt, H., Mahas, A., Aljedaani, F., Khan, M.Z., Ding, S., Mahfouz, M., 2018. RNA virus interference via CRISPR/Cas13a system in plants. Genom. Biol. 19 (1), 1. https://doi.org/10.1186/s13059-017-1381-1.

Anzalone, A.V., Koblan, L.W., Liu, D.R., 2020. Genome editing with CRISPR–Cas nucleases, base editors, transposases and prime. Nat. Biotechnol. 38 (7), 824–844. https://doi.org/10.1038/s41587-020-0561-9.

Aydin, M., 2024. Role of Nanotechnology in the Advancement in Genome Editing in Plants Applications of Genome Engineering in Plants. wiley, Turkey, pp. 78–90. https://doi.org/10.1002/9781394183913.ch5.

Baltes, N.J., Gil-Humanes, J., Cermak, T., Atkins, P.A., Voytas, D.F., 2014. DNA replicons for plant genome engineering. Plant Cell 26 (1), 151–163. https://doi.org/10.1105/tpc.113.119792.

Banakar, R., Schubert, M., Kurgan, G., Rai, K.M., Beaudoin, S.F., Collingwood, M.A., Vakulskas, C.A., Wang, K., Zhang, F., 2021. Efficiency, specificity and temperature sensitivity of Cas9 and Cas12a RNPs for DNA-free genome editing in plants. Frontiers in Genome Editing 3. https://doi.org/10.3389/fgeed.2021.760820.

Baysal, C., Kausch, A.P., Cody, J.P., Altpeter, F., Voytas, D.F., 2025. Rapid and efficient in planta genome editing in sorghum using foxtail mosaic virus-mediated sgRNA delivery. Plant J. 121 (2). https://doi.org/10.1111/tpj.17196.

Borphukan, B., Khatun, M., Fartyal, D., James, D., Reddy, M.K., 2025. A gemini virus-derived autonomously replicating system for HDR-mediated genome editing of the EPSP synthase gene in indica rice. Plants 14 (3), 477. https://doi.org/10.3390/plants14030477.

Cardi, T., Murovec, J., Bakhsh, A., Boniecka, J., Bruegmann, T., Bull, S.E., Eeckhaut, T., Fladung, M., Galovic, V., Linkiewicz, A., Lukan, T., Mafra, I., Michalski, K., Kavas, M., Nicolia, A., Nowakowska, J., Sági, L., Sarmiento, C., Yıldırım, K., Zlatković, M., Hensel, G., Van Laere, K., 2023. CRISPR/Cas-mediated plant genome editing: outstanding challenges a decade after implementation. Trends Plant Sci. 28 (10), 1144–1165. https://doi.org/10.1016/j.tplants.2023.05.012.

Chen, K., Wang, Y., Zhang, R., Zhang, H., Gao, C., 2019. CRISPR/Cas genome editing and precision plant breeding in agriculture. Annu. Rev. Plant Biol. 70 (1), 667–697. https://doi.org/10.1146/annurev-arplant-050718-100049.

Chen, H., Neubauer, M., Wang, J.P., 2022. Enhancing HR frequency for precise genome editing in plants. Front. Plant Sci. 13. https://doi.org/10.3389/fpls.2022.883421.

Chen, F., Chen, L., Yan, Z., Xu, J., Feng, L., He, N., Guo, M., Zhao, J., Chen, Z., Chen, H., Yao, G., Liu, C., 2024. Recent advances of CRISPR-based genome editing for enhancing staple crops. Front. Plant Sci. 15. https://doi.org/10.3389/fpls.2024.1478398.

Cunningham, F.J., Goh, N.S., Demirer, G.S., Matos, J.L., Landry, M.P., 2018. Nanoparticle-mediated delivery towards advancing plant genetic engineering. Trends Biotechnol. 36 (9), 882–897. https://doi.org/10.1016/j.tibtech.2018.03.009.

Delzer, B., Liang, D., Szwerdszarf, D., Rodriguez, I., Mardones, G., Elumalai, S., Johnson, F., Nalapalli, S., Egger, R., Burch, E., Meier, K., Wei, J., Zhang, X., Gui, H., Jin, H., Guo, H., Yu, K., Liu, Y., Breitinger, B., Poets, A., Nichols, J., Shi, W., Skibbe, D., Que, Q., Kelliher, T., 2024. Elite, transformable haploid inducers in maize. The Crop Journal 12 (1), 314–319. https://doi.org/10.1016/j.cj.2023.10.016.

Demirer, G.S., Silva, T.N., Jackson, C.T., Thomas, J.B., Ehrhardt, D.W., Rhee, S.Y., Mortimer, J.C., Landry, M.P., 2021. Nanotechnology to advance CRISPR–Cas genetic engineering of plants. Nat. Nanotechnol. 16 (3), 243–250. https://doi.org/10.1038/s41565-021-00854-y.

Dixit, S., Kumar, A., Srinivasan, K., Vincent, P.M.D.R., Ramu Krishnan, N., 2023. Advancing genome editing with artificial intelligence: opportunities, challenges, and future directions. Front. Bioeng. Biotechnol. 11. https://doi.org/10.3389/fbioe.2023.1335901.

Dong, F., Xie, K., Chen, Y., Yang, Y., Mao, Y., 2017. Polycistronic tRNA and CRISPR guide-RNA enables highly efficient multiplexed genome engineering in human cells. Biochem. Biophys. Res. Commun. 482 (4), 889–895. https://doi.org/10.1016/j.bbrc.2016.11.129.

Dunbar, T., Tsakirpaloglou, N., Septiningsih, E.M., Thomson, M.J., 2022. Carbon nanotube-mediated plasmid DNA delivery in rice leaves and seeds. Int. J. Mol. Sci. 23 (8), 4081. https://doi.org/10.3390/ijms23084081.

Ebrahimi, V., Hashemi, A., 2024. CRISPR-based gene editing in plants: focus on reagents and their delivery tools. Bioimpacts 15. https://doi.org/10.34172/bi.30019.

Gong, D., Ben-Akiva, E., Singh, A., Yamagata, H., Est-Witte, S., Shade, J.K., Trayanova, N.A., Green, J.J., 2022. Machine learning guided structure function predictions enable in silico nanoparticle screening for polymeric gene delivery. Acta Biomater. 154, 349–358. https://doi.org/10.1016/j.actbio.2022.09.072.

Egorov, E., Pieters, C., Korach-Rechtman, H., Shklover, J., Schroeder, A., 2021. Robotics, microfluidics, nanotechnology and AI in the synthesis and evaluation of liposomes and polymeric drug delivery systems. Drug Deliv. Transl. Res. 11 (2), 345–352. https://doi.org/10.1007/s13346-021-00929-2.

Gong, Z., Andrew Previtera, D., Wang, Y., Ramón Botella, J., 2024. Geminiviral-induced genome editing using miniature CRISPR/Cas12j (CasΦ) and Cas12f variants in plants. Plant Cell Rep. 43 (3). https://doi.org/10.1007/s00299-023-03092-9.

Haider, S., Mussolino, C., 2025. Fine-tuning homology-directed repair (HDR) for precision genome editing: current strategies and future directions. Int. J. Mol. Sci. 26 (9). https://doi.org/10.3390/ijms26094067.

Han, X., Deng, Z., Liu, H., Xiang, J., 2025. Current advancement and future prospects in simplified transformation-based plant genome editing. Plants 14 (6), 889. https://doi.org/10.3390/plants14060889.

Hassan, M.M., Zhang, Y., Yuan, G., De, K., Chen, J.G., Muchero, W., Tuskan, G.A., Qi, Y., Yang, X., 2021. Construct design for CRISPR/Cas-based genome editing in plants. Trends Plant Sci. 26 (11), 1133–1152. https://doi.org/10.1016/j.tplants.2021.06.015.

Hsieh-Feng, V., Yang, Y., 2020. Efficient expression of multiple guide RNAs for CRISPR/Cas genome editing. aBIOTECH 1 (2), 123–134. https://doi.org/10.1007/s42994-019-00014-w.

Hu, M., Liu, D., 2025. Unlocking the potential of genome editing in agriculture with tissue culture-free techniques. New Phytol. 246 (4), 1478–1484. https://doi.org/10.1111/nph.70046.

Jeong, Y.Y., Han, J.H., Yu, J., Bae, S., Seo, P.J., 2025. Identification of optimal adenine and cytosine base editors for genome editing in Arabidopsis and soybean. BMB Rep. 58 (7), 288–292. https://doi.org/10.5483/BMBRep.2024-0102.

Kavuri, N.R., Ramasamy, M., Qi, Y., Mandadi, K., 2022. Applications of CRISPR/Cas13-Based RNA editing in plants. Cells 11 (17), 2665. https://doi.org/10.3390/cells11172665.

Kocsisova, Z., Coneva, V., 2023. Strategies for delivery of CRISPR/Cas-mediated genome editing to obtain edited plants directly without transgene integration. Frontiers in Genome Editing 5. https://doi.org/10.3389/fgeed.2023.1209586.

Laforest, L.C., Nadakuduti, S.S., 2022. Advances in delivery mechanisms of CRISPR gene-editing reagents in plants. Frontiers in Genome Editing 4. https://doi.org/10.3389/fgeed.2022.830178.

Lew, T.T.S., Sarojam, R., Jang, I.C., Park, B.S., Naqvi, N.I., Wong, M.H., Singh, G.P., Ram, R.J., Shoseyov, O., Saito, K., Chua, N.H., Strano, M.S., 2020. Species-independent analytical tools for next-generation agriculture. Nat. Plants 6 (12), 1408–1417. https://doi.org/10.1038/s41477-020-00808-7.

Li, G., Wang, H., Zhang, X., Wu, Z., Yang, H., 2021. A Cas9–transcription factor fusion protein enhances homology-directed repair efficiency. J. Biol. Chem. 296, 100525. https://doi.org/10.1016/j.jbc.2021.100525.

Li, J., Zhang, Q., Wang, Z., Liu, Q., 2024. The roles of epigenetic regulators in plant regeneration: exploring patterns amidst complex conditions. Plant Physiol. 194 (4), 2022–2038. https://doi.org/10.1093/plphys/kiae042.

Li, Z., Khan, W.U., Bai, G., Dong, C., Wang, J., Zhang, Y., Wang, C., Zhang, H., Wang, W., Luo, M., Chen, F., 2025. From code to life: the AI-driven revolution in genome editing. Adv. Sci. 12 (30). https://doi.org/10.1002/advs.202417029.

Liang, Z., Chen, K., Zhang, Y., Liu, J., Yin, K., Qiu, J.-L., Gao, C., 2018. Genome editing of bread wheat using biolistic delivery of CRISPR/Cas9 in vitro transcripts or ribonucleoproteins. Nat. Protoc. 13 (3), 413–430. https://doi.org/10.1038/nprot.2017.145.

Lino, C.A., Harper, J.C., Carney, J.P., Timlin, J.A., 2018. Delivering crispr: a review of the challenges and approaches. Drug Deliv. 25 (1), 1234–1257. https://doi.org/10.1080/10717544.2018.1474964.

Liu, B.R., et al., 2023. Cell-penetrating peptides for use in development of transgenic plants. Molecules 28 (8), 3367. https://doi.org/10.3390/MOLECULES28083367.

Liu, G., Lin, Q., Jin, S., Gao, C., 2022. The CRISPR-Cas toolbox and gene editing technologies. Mol. Cell 82 (2), 333–347. https://doi.org/10.1016/j.molcel.2021.12.002.

Luo, Y., Tang, X., Yu, C., Qiu, Y., Nowak, J.S., Lu, Q.S., Tian, L., Na, R., 2024. Development of a csy4-processing TRV-based CRISPR/Cas9 genome editing system in Nicotiana benthamiana. Plant Mol. Biol. Rep. 42 (4), 774–783. https://doi.org/10.1007/s11105-024-01459-0.

Mahas, A., Aman, R., Mahfouz, M., 2019. CRISPR-Cas13d mediates robust RNA virus interference in plants. Genom. Biol. 20 (1), 263. https://doi.org/10.1186/s13059-019-1881-2.

Mahmood, M.A., Sajjad, M.W., Imran, I., Naqvi, R.Z., Amin, I., Shafiq, M., Aslam, M.Q., Mansoor, S., 2025. Attaining the promise of geminivirus-based vectors in plant genome editing. Viruses 17 (5), 631. https://doi.org/10.3390/v17050631.

McCarty, N.S., Graham, A.E., Studená, L., Ledesma-Amaro, R., 2020. Multiplexed CRISPR technologies for gene editing and transcriptional regulation. Nat. Commun. 11 (1). https://doi.org/10.1038/s41467-020-15053-x.

Mikhaylova, E., 2025. Virus-induced genome editing (VIGE): one step away from an agricultural revolution. Int. J. Mol. Sci. 26 (10), 4599. https://doi.org/10.3390/ijms26104599.

Ming, M., Ren, Q., Pan, C., He, Y., Zhang, Y., Liu, S., Zhong, Z., Wang, J., Malzahn, A.A., Wu, J., Zheng, X., Zhang, Y., Qi, Y., 2020. CRISPR–Cas12b enables efficient plant genome engineering. Nat. Plants 6 (3), 202–208. https://doi.org/10.1038/s41477-020-0614-6.

Molla, K.A., Shih, J., Wheatley, M.S., Yang, Y., 2022. Predictable NHEJ insertion and assessment of HDR editing strategies in plants. Frontiers in Genome Editing 4. https://doi.org/10.3389/fgeed.2022.825236.

Mujtaba, M., Wang, D., Carvalho, L.B., Oliveira, J.L., Pereira, A.do E.S., Sharif, R., Jogaiah, S., Krishna Paidi, M., Wang, L., Ali, Q., Fernandes Fraceto, L., 2021. Nanocarrier-mediated delivery of miRNA, RNAi, and CRISPR-cas for plant protection: current trends and future directions. ACS Agricultural Science & Technology 1 (5), 417–435. https://doi.org/10.1021/acsagscitech.1c00146.

Nadakuduti, S.S., Enciso-Rodríguez, F., 2021. Advances in genome editing with CRISPR systems and transformation technologies for plant DNA manipulation. Front. Plant Sci. 11. https://doi.org/10.3389/fpls.2020.637159.

Nelson, J.W., Randolph, P.B., Shen, S.P., Everette, K.A., Chen, P.J., Anzalone, A.V., An, M., Newby, G.A., Chen, J.C., Hsu, A., Liu, D.R., 2022. Engineered pegRNAs improve prime editing efficiency. Nat. Biotechnol. 40 (3), 402–410. https://doi.org/10.1038/s41587-021-01039-7.

Niedbała, G., Niazian, M., Sabbatini, P., 2021. Modeling agrobacterium-mediated gene transformation of tobacco (Nicotiana tabacum)—a model plant for gene transformation studies. Front. Plant Sci. 12. https://doi.org/10.3389/fpls.2021.695110.

Ortiz-Perez, A., van Tilborg, D., van der Meel, R., Grisoni, F., Albertazzi, L., 2024. Machine learning-guided high throughput nanoparticle design. Digital Discovery 3 (7), 1280–1291. https://doi.org/10.1039/d4dd00104d.

Ozyigit, I.I., Yucebilgili Kurtoglu, K., 2020. Particle bombardment technology and its applications in plants. Mol. Biol. Rep. 47 (12), 9831–9847. https://doi.org/10.1007/s11033-020-06001-5.

Park, C.W., Choi, J.Y., Son, Y.J., Kim, D.H., Li, H., Liang, W., Lee, C., Jung, K.H., Kim, Y.J., 2024. Magnetofected pollen gene delivery system could generate genetically modified Cucumis sativus. Hortic. Res. 11 (8). https://doi.org/10.1093/hr/uhae179.

Pramanik, D., Shelake, R.M., Kim, M.J., Kim, J.-Y., 2021. CRISPR-mediated engineering across the central dogma in plant biology for basic research and crop improvement. Mol. Plant 14 (1), 127–150. https://doi.org/10.1016/j.molp.2020.11.002.

Qin, R., Li, J., Li, H., Zhang, Y., Liu, X., Miao, Y., Zhang, X., Wei, P., 2019. Developing a highly efficient and wildly adaptive CRISPR - Sa Cas9 toolset for plant genome editing. Plant Biotechnol. J. 17 (4), 706–708. https://doi.org/10.1111/pbi.13047.

Qu, Y., Huang, K., Yin, M., Zhan, K., Liu, D., Yin, Di, Cousins, H.C., Johnson, W.A., Wang, X., Shah, M., Altman, R. B., Zhou, D., Wang, M., Cong, Le, 2025. CRISPR-GPT for agentic automation of gene-editing experiments. Nat. Biomed. Eng. https://doi.org/10.1038/s41551-025-01463-z.

Robertson, G., Burger, J., Campa, M., 2022. CRISPR/Cas-based tools for the targeted control of plant viruses. Mol. Plant Pathol. 23 (11), 1701–1718. https://doi.org/10.1111/mpp.13252.

Rustgi, S., Naveed, S., Windham, J., Zhang, H., Demirer, G.S., 2022. Plant biomacromolecule delivery methods in the 21st century. Frontiers in Genome Editing 4. https://doi.org/10.3389/fgeed.2022.1011934.

Safari, F., Zare, K., Negahdaripour, M., Barekati-Mowahed, M., Ghasemi, Y., 2019. CRISPR Cpf1 proteins: structure, function and implications for genome editing. Cell Biosci. 9 (1). https://doi.org/10.1186/s13578-019-0298-7.

Sandhya, D., Jogam, P., Rao Allini, V., Abbagani, S., Alok, A., 2020. The present and potential future methods for delivering CRISPR/Cas9 components in plants. J. Genet. Eng. Biotechnol. 18 (1), 25. https://doi.org/10.1186/s43141-020-00036-8.

Sasaki, K., Urano, K., Mimida, N., Nonaka, S., Ezura, H., Imai, R., 2025. A long shelf-life melon created via CRISPR/Cas9 RNP-based in planta genome editing. Frontiers in Genome Editing 7. https://doi.org/10.3389/fgeed.2025.1623097.

Sebiani-Calvo, A., Hernández-Soto, A., Hensel, G., Gatica-Arias, A., 2024. Crop genome editing through tissue-culture-independent transformation methods. Frontiers in Genome Editing 6. https://doi.org/10.3389/fgeed.2024.1490295.

Sharma, P., Lew, T.T.S., 2022. Principles of nanoparticle design for genome editing in plants. Frontiers in Genome Editing 4. https://doi.org/10.3389/fgeed.2022.846624.

Shivashakarappa, K., Marriboina, S., Dumenyo, K., Taheri, A., Yadegari, Z., 2025. Nanoparticle-mediated gene delivery techniques in plant systems. Front. Nanotechnol. 7. https://doi.org/10.3389/fnano.2025.1516180.

Squire, H.J., Tomatz, S., Voke, E., González-Grandío, E., Landry, M., 2023. The emerging role of nanotechnology in plant genetic engineering. Nat. Rev. Bioeng. 1 (5), 314–328. https://doi.org/10.1038/s44222-023-00037-5.

Steinberger, A.R., Voytas, D.F., 2025. Virus-induced gene editing free from tissue culture. Nat. Plants 11 (7), 1241–1251. https://doi.org/10.1038/s41477-025-02025-6.

Strecker, J., Jones, S., Koopal, B., Schmid-Burgk, J., Zetsche, B., Gao, L., Makarova, K.S., Koonin, E.V., Zhang, F., 2019. Engineering of CRISPR-Cas12b for human genome editing. Nat. Commun. 10 (1). https://doi.org/10.1038/s41467-018-08224-4.

Tuncel, A., Pan, C., Clem, J.S., Liu, D., Qi, Y., 2025. CRISPR–Cas applications in agriculture and plant research. Nat. Rev. Mol. Cell Biol. 26 (6), 419–441. https://doi.org/10.1038/s41580-025-00834-3.

Uranga, M., Daròs, J.A., 2023. Tools and targets: the dual role of plant viruses in CRISPR–Cas genome editing. Plant Genome 16 (2). https://doi.org/10.1002/tpg2.20220.

Vu, T.V., et al., 2020. Highly efficient homology-directed repair using CRISPR/Cpf1-geminiviral replicon in tomato. Plant Biotechnol. J. 18 (10), 2133. https://doi.org/10.1111/PBI.13373.

Wang, J.W., Grandio, E.G., Newkirk, G.M., Demirer, G.S., Butrus, S., Giraldo, J.P., Landry, M.P., 2019. Nanoparticle-mediated genetic engineering of plants. Mol. Plant 12 (8), 1037–1040. https://doi.org/10.1016/j.molp.2019.06.010.

Weiss, T., Kamalu, M., Shi, H., Zheng, L., Amerasekera, J., Zhong, Z., Adler, B.A., Song, M.M., Vohra, K., Wirnowski, G., Chitkara, S., Ambrose, C., Steinmetz, N., Sridharan, A., Sahagun, D., Banfield, J.F., Doudna, J. A., Jacobsen, S.E., 2025. Viral delivery of an RNA-guided genome editor for transgene-free germline editing in Arabidopsis. Nat. Plants 11 (5), 967–976. https://doi.org/10.1038/s41477-025-01989-9.

Wu, H., Zhang, K., Zhang, Z., Wang, J., Jia, P., Cong, L., Jia, L., Duan, Y., Ke, F., Zhang, F., Liu, Z., Lu, F., Wang, Y., Li, Z., Chang, M., Zou, J., Zhu, K., 2022. Cell-penetrating peptide: a powerful delivery tool for DNA-free crop genome editing. Plant Sci. 324, 111436. https://doi.org/10.1016/j.plantsci.2022.111436.

Wu, H., Sun, Y., Wang, Y., Luo, L., Song, Y., 2024. Advances in miniature CRISPR-Cas proteins and their applications in gene editing. Arch. Microbiol. 206 (5). https://doi.org/10.1007/s00203-024-03962-0.

Xie, K., Minkenberg, B., Yang, Y., 2015. Boosting CRISPR/Cas9 multiplex editing capability with the endogenous tRNA-processing system. Proc. Natl. Acad. Sci. 112 (11), 3570–3575. https://doi.org/10.1073/pnas.1420294112.

Yan, Y., Zhu, X., Yu, Y., Li, C., Zhang, Z., Wang, F., 2022. Nanotechnology strategies for plant genetic engineering. Adv. Mater. 34 (7). https://doi.org/10.1002/adma.202106945.

Yang, X., Zhu, P., Gui, J., 2024. Advancements of CRISPR-mediated base editing in crops and potential applications in populus. Int. J. Mol. Sci. 25 (15), 8314. https://doi.org/10.3390/ijms25158314.

Yu, Y., Wang, X., Sun, H., Liang, Q., Wang, W., Zhang, C., Bian, X., Cao, Q., Li, Q., Xie, Y., Ma, D., Li, Z., Sun, J., 2020. Improving CRISPR-Cas-mediated RNA targeting and gene editing using SPLCV replicon-based expression vectors in Nicotiana benthamiana. Plant Biotechnol. J. 18 (10), 1993–1995. https://doi.org/10.1111/pbi.13384.

Yuan, H.Y., Kagale, S., Ferrie, A.M.R., 2023. Multifaceted roles of transcription factors during plant embryogenesis. Front. Plant Sci. 14. https://doi.org/10.3389/fpls.2023.1322728.

Zhao, X., Meng, Z., Wang, Y., Chen, W., Sun, C., Cui, B., Cui, J., Yu, M., Zeng, Z., Guo, S., Luo, D., Cheng, J.Q., Zhang, R., Cui, H., 2017. Pollen magnetofection for genetic modification with magnetic nanoparticles as gene carriers. Nat. Plants 3 (12), 956–964. https://doi.org/10.1038/s41477-017-0063-z.

Zhi, H., Zhou, S., Pan, W., Shang, Y., Zeng, Z., Zhang, H., 2022. The promising nanovectors for gene delivery in plant genome engineering. Int. J. Mol. Sci. 23 (15), 8501. https://doi.org/10.3390/ijms23158501.

Zhou, Z., Xiao, J., Yin, S., Chen, Y., Yuan, Y., Zhang, J., Xiong, L., Xie, K., 2025. Cas9-Rep fusion tethers donor DNA in vivo and boosts the efficiency of HDR -mediated genome editing. Plant Biotechnol. J. 23 (6), 2006–2017. https://doi.org/10.1111/pbi.70036.

Zhu, H., Li, C., Gao, C., 2020. Applications of CRISPR–Cas in agriculture and plant biotechnology. Nat. Rev. Mol. Cell Biol. 21 (11), 661–677. https://doi.org/10.1038/s41580-020-00288-9.

Further reading

Jat, S.K., Bhattacharya, J., Sharma, M.K., 2020. Nanomaterial based gene delivery: a promising method for plant genome engineering. J. Mater. Chem. B 8 (19), 4165–4175. https://doi.org/10.1039/D0TB00217H.

Warman, C., Panda, K., Vejlupkova, Z., Hokin, S., Unger-Wallace, E., Cole, R.A., et al., 2020. High expression in maize pollen correlates with genetic contributions to pollen fitness as well as with coordinated transcription from neighboring transposable elements. PLoS Genet. 16 (4), e1008462. https://doi.org/10.1371/journal.pgen.1008462.

CHAPTER

5

Evolution of CRISPR technology, progress, and its future implications for crop improvement

Afifa Younas[1], *Sajid Fiaz*[2], *Zahra Noreen*[3], *Aqsa Perveen*[3] and *Mehmooda Yousaf*[3]

[1]Department of Botany, Baba Guru Nanak University, Nankana Sahib, Pakistan; [2]Institute of Molecular Biology and Biotechnology, The University of Lahore, Lahore, Pakistan; [3]Department of Botany, Division of Science and Technology, University of Education, Lahore, Pakistan

5.1 Introduction

Site-specific nuclease-based genome editing technologies (GETs) enable transgene integration, genome engineering, and reverse genetics, producing accurate and effective outcomes. There are two types of conventional GETs: spontaneous mutations through hybridization and artificially induced mutations using various chemicals. These fall into two categories: physical and chemical mutagens. Physical mutagens include UV light, X-rays, and particle radiation, viz., thermal and fast neutrons (Chaudhary and Kumar, 2023). In the meantime, the majority of chemical mutagens are azides and alkylating agents, such as diepoxybutane (Usharani et al., 2016), sodium trinitride NaN3 (Szarejko et al., 2017), and ethylmethanesulfonate. These chemicals cause a range of mutations, such as chromosomal aberration, insertion, reversion, deletion, translocation, and point mutation. According to (Appels and Honeycutt, 2019), GETs available before 20 years ago were complex in character and could only diversify the DNA sequence.

Intelligent Gene Editing in Plants
https://doi.org/10.1016/B978-0-443-44079-3.00017-6

5.2 Outline of genome editing technologies

Genome editing tools, such as zinc finger nucleases (ZFNs), transcription activator-like effector nucleases (TALENs), and CRISPR-associated protein 9 (Cas9), are fundamental to genome editing . According to (Shamshirgaran et al., 2022), TALENs and ZFNs were the best technologies before 2013. Both rely on the *Fok1* endonuclease's nuclease domain to cleave double-stranded DNA. The most difficult task is selecting the appropriate target sequences, since ZFNs comprise multiple Cys2-His2 ZF domains, each of which binds to a distinct PAM. The intricate interactions among ZFNs make ZFN design challenging. A functional ZFN can be created using public DNA libraries to cover every 100 bp of DNA sequence; however, its efficiency is lower. Researchers need to purchase commercially available ZFNs, which are unaffordable, to boost efficiency.

To find a distinct 18–24 bp DNA sequence, two ZFNs form a dimer. Conversely, Barrangou et al. observed that TALENs generated by *Xanthomonas* plant pathogens are transported to the target via the injectisome during infection. To improve in vivo homologous recombination, strategies based on DNA repeats have been proposed. TALENs in vivo efficacy is diminished by their requirement for thymidine, and thymidine remains at the first place for targeted mutagenesis (Wang et al., 2017a, 2017b), which highlights the need for investigational confirmation of TALENs (Raigani et al., 2025). The use of ZFNs and TALENs is currently limited because of off-target risks, complex DBPs, and context-specific attachment requirements (Bhagtaney and Sundarrajan, 2023).

5.2.1 Cataloging of CRISPR/Cas systems

CRISPR/Cas system is categorized into two classes based on variations in effector module configurations. The first category comprises multi-subunit crRNA-effector complexes and includes types I, III, and IV. Class 2 is distinguished by an effector complex composed of a single large protein, Cas, and comprises types II, V, and VI. Type II CRISPR–Cas9 exhibits high on-target efficiency and RNA-guided DNA cleavage. Recent major developments in class 2, enabled by the RNA-guided DNA cleavage mechanism, include CRISPR from *Prevotella* and *Francisella 1* (CRISPR-Cpf1), a type V-A protein, and class 2 candidate 1 (CRISPR-C2c1), a type V–B protein, both of which have substantial potential for single-stranded RNA targeting and editing (Jin et al., 2025).

5.2.2 Evolution of the CRISPR/Cas9 system

Due to ease of handling and implementation, usefulness, and adaptability, CRISPR–Cas9 is currently a commonly used GET. CRISPR arrays with unknown biological characteristics were first identified in 1987 in the E. coli genome. Because several spacers share sequence homology with viral and plasmid DNA, experiments conducted in 2005 convincingly demonstrated that CRISPR arrays are involved in adaptive immunity. A highly efficient, target-specific RNA-guided DNA cleavage method was initially described by (Jinek et al., 2012). While the defense mechanism relies on RNA-mediated DNA targeting, Deltcheva et al. discovered that CRISPR arrays provide defense when foreign DNA and Cas9 protein are combined. CRISPR/Cas methods provide adaptive protection in archaea and bacteria, protecting them from attacks by nucleic acids such as viral DNA. They achieve this immunity by recognizing and cleaving foreign genetic material in a sequence-specific manner, using spacer sequences derived from previous exposure.

At the attachment point of a CRISPR locus, spacers between two neighboring repeats are integrated to confer immunity. Through repeated exposure to foreign DNA, the spacers are transcribed into 40-nt crRNAs. These spacers then associate with transactivating CRISPR RNA (tracrRNA) to activate the Cas9 nuclease. In foreign DNA, the tracrRNA guides Cas9 to cleave the homologous dsDNA sequence (protospacer). According to Hsu et al., the PAM located downstream on the target DNA is essential for the effective separation of 5′-NGG-3′ and, rarely, 5′-NAG-3. For RNA and target DNA to couple, a seed sequence located about 12 bp upstream of the PAM is essential (Zhong et al., 2018).

A distinctive feature of this system is its ability to modify a large number of genes simultaneously by introducing double-strand breaks (DSBs) at different loci (Lee et al., 2016). Because the Watson-Crick model recognizes specific regions, sequence analysis identifies non-specific sites, and careful gRNA design can correct changes at these sites, the CRISPR–Cas9 method is among the most dependable for target specificity. Conversely, the target specificity of TALENs and ZFNs depends on both unexpected and function-specific features of protein-DNA interactions. Furthermore, unlike ZFN's proprietary nature, the scientific community with CRISPR expertise has made significant contributions to disseminating new information to newcomers. Researchers can now access a number of online resources to help with any CRISPR-related issues.

5.2.3 Application of CRISPR/Cas system for crop advancement

5.2.3.1 Mechanism of CRISPR/Cas system and plant adaptations

In biotechnology, CRISPR system has quickly become a game-changing tool that enables precise and effective genome editing. Its widespread applications, adaptability, and versatility have transformed plant functional genomics and improved crops.

5.2.3.2 Components of CRISPR/Cas

CRISPR is built around three core elements: Cas nuclease, gRNA (Guide RNA), and the PAM (Protospacer adjacent motif) sequence. When these components are coupled, they enable extremely precise, targeted changes in plant genomes. The Cas9 protein derived from Streptococcus pyogenes is a DNA endonuclease that produces double-strand breaks at specific sites. A short synthetic RNA molecule known as gRNA comprises two parts: the trans-activating tracrRNA (CRISPR RNA), which binds to Cas9 and accelerates cleavage, and the crRNA (CRISPR RNA), which detects the target DNA site. A PAM sequence adjacent to the target site is required for Cas9; this is often the motif 5′-NGG-3′. According to (Jinek et al., 2012) and Hsu et al. (2014), this PAM requirement guarantees target specificity and inhibits self-targeting within the genome.

5.2.4 Types of CRISPR/Cas systems used in plants

Many modifications to CRISPR/Cas systems are being developed for plant genome editing. These systems vary in their targeting strategies and biochemical properties. Cas9-based systems remain the most widely used due to their efficiency and reliability. The genomic regions being altered have increased thanks to the tuning of two variants, SpCas9 and SaCas9, to recognize distinct PAM sequences (NGG and NNGRRT, respectively) (Makarova et al., 2020).

5.2.4.1 CRISPR/Cas12a

Cas12a (Cpf1) is a Class 2 system that recognizes T-rich PAM sequences (5′-TTTV-3′) and produces staggered cuts rather than blunt ends. This feature facilitates targeted gene insertions and enhances the efficiency of homology-directed repair (HDR). Additionally, because it requires only one crRNA, Cas12a streamlines multiplex editing methods (Tang et al., 2017).

5.2.4.2 Base editors and prime editors

Recent innovations that extend the CRISPR toolkit beyond conventional DSB-dependent procedures include base editors and prime editors. Cytidine and adenine base editors do not involve breaking DNA. They are particularly useful for correcting point mutations and investigating single-nucleotide polymorphisms (SNPs) that impact agronomic traits (Komor et al., 2016). Meanwhile, prime editors, which combine a Cas9 nickase, allow the insertion, deletion, or exchange of particular DNA sequences without the need for donor templates (Anzalone et al., 2019).

5.2.5 Delivery methods in plants

5.2.5.1 PEG-mediated protoplast transformation

Polyethylene glycol (PEG) is used in this important genetic transformation method. Protoplasts from a variety of species (maize, soybeans, *A. thaliana*, tobacco, rice, and wheat) respond well to this method. Plasmids carrying Cas9 and gRNA are delivered to protoplasts in the presence of PEG. The CaMV35S promoter was used for Cas9 and gRNA, respectively, and U3 and maize were first reported to use PEG-mediated CRISPR construct transformation (Liang et al., 2014). In a few studies, Cas9 was shown to target important genes, be expressed under specific promoters, and be tailored for particular plants (Lowder et al., 2016). After being dissolved in water and filtered to ensure sterility, plasmid DNA is combined with the protoplast suspension. Protoplasts are then gradually mixed with the required amount of PEG a few minutes later. Protoplasts are regenerated in appropriate regeneration media (Lowder et al., 2015). Transgene-free mutant plants are produced by applying Cas9/gRNA ribonucleoproteins with PEG-mediated delivery in various plants, including *Arabidopsis*, lettuce, rice, and tobacco. In lettuce mutants, editing occurred up to 46% of the time (Woo et al., 2015). Later, without using a vector or DNA, Kim et al. modified the genomes of tobacco and soybean plants using Cpf1/CrRNA ribonucleoproteins.

5.2.6 Biolistic method

The final outcome of this transformation approach is the development of a "gene gun" or "biolistic gun," in which microscopic Au, Ag, or W particles are coated with DNA vectors or Cas9/gRNA ribonucleoproteins. These coated microprojectiles are then propelled into plant explants under high pressure to facilitate efficient delivery of CRISPR/Cas9 components. For this process, ideal parameters such as target distance, particle size, and explant type are required. The explants after transformation are regenerated in regeneration media under suitable pressure. Cas9/gRNA ribonucleoproteins have been successfully delivered to Brassica (Murovec et al., 2018), potatoes (Andersson et al., 2018), and maize (Svitashev et al., 2015), leading to the regeneration of mutants. Because it involves less vector/DNA editing, the bombardment technique for Cas9/gRNA ribonucleoprotein delivery is in demand. However, a very low editing efficiency of 2.4%–9.7% is achieved because of the labor-intensive generation of altered tissues and selection pressure (Svitashev et al., 2015).

5.2.7 *Agrobacterium*-mediated delivery method

Agrobacterium-mediated delivery is the most widely used delivery strategy for various plants. The binary vector containing the gRNA expression cassette and Cas9 is introduced into an *Agrobacterium* strain. Additionally, CRISPR constructs are genetically transformed into the required plant explants, such as callus, leaf, and reproductive tissues, by *Agrobacterium*. More than 20 plant species have been effectively altered to date using *Agrobacterium*-mediated delivery of CRISPR/Cas9. This system is more effective and has demonstrated higher editing efficiency than the particle bombardment system. *Agrobacterium*-based genetic transformation is also commonly employed for monocot crops with lower regeneration and transformation capacities. *Agrobacterium*-compatible CRISPR/Cas9 binary vectors were developed due to their high editing efficiency in both monocots and dicots (Xing et al., 2014). The banana cultivar Rasthal demonstrated a 59% mutation rate in the phytoene desaturase gene when its genome was edited by Agrobacterium (Kaur et al., 2018).

5.2.8 Applications for crop improvement

The CRISPR/Cas method has rapidly drawn the attention of researchers and businesses and has become a sophisticated tool for precise genome and gene editing because of its promise to accelerate crop development and deliver major agronomic benefits. CRISPR/Cas-based genome editing has been successfully applied to the majority of important crops, including cotton, wheat, and rice.

5.2.8.1 Yield enhancement

Enhancing crop yield through conventional breeding as well as transgenic technology is a long-term objective. However, scientists have not yet achieved significant improvements using transgenic approaches because yield traits are complex and controlled by multiple genes and gene networks. Numerous studies demonstrate that specific genes, when expressed, can have negative effects on agricultural yields, even though it is practically not possible to identify a single gene that significantly controls production. Crop yield can be directly enhanced by suppressing or even deleting these genes. The rapidly evolving CRISPR/Cas system offers an ideal tool for deleting genes that adversely affect plant yield. According to a study carried out by Li et al. , larger grain size, number, and weight, as well as erect and densely packed panicles, are among the features associated with rice production that were considerably enhanced by deleting the genes that act as negative regulators of crop yield, namely the *dep1*, *gn1a*, or *gs3* gene (Li et al., 2016). In another study, Xu et al. (2016) used CRISPR/Cas9 technology to delete the negative regulators *gw2*, *gw5*, and *tgw6* of grain weight in rice. As a result, the genome editing mutants exhibit bigger grain size and weight (R Xu et al., 2016). The *vacuolar invertase2* (*vin2*) gene also regulates the size and weight of rice seeds; CRISPR/Cas9 knockout mutants exhibit smaller seeds and grain weights, mostly as a result of regulating the movement of sucrose toward developing seed grains (Lee et al., 2019).

5.2.8.2 Biotic stress resistance (pathogen and pest resistance)

Diseases caused by various pathogens are a serious problem for many plants, particularly important crops such as wheat and cotton. In addition to affecting plant development and growth, biotic factors also reduce crop yield and quality, and in some cases, they can cause total yield loss. Plants respond to different diseases through a complex process and mechanism that involves numerous genes. Transgenic strategies have enabled scientists to increase the activity of targeted genes and develop improved plants with considerable resistance to various pathogens. Currently, progress is limited because there is no single gene that plays a major, dominant role in the plant response to diseases. However, eliminating an unwanted gene can confer resistance to specific pathogens in plants. As a result, CRISPR/Cas9 is widely used to create mutants that provide tolerance to diseases caused by microorganisms in plants.

Additionally, this method was used to grow genome-edited plants that are highly resistant to viruses. In natural environments, plant viruses are common, infect plants, and have major effects on plant growth and development. In 2016, Pyott and coworkers successfully knocked out the *eIF(iso)4E* gene in *Arabidopsis* using CRISPR/Cas9 technology. Their findings showed that turnip mosaic virus, a significant pathogen in field-grown vegetable crops, could not infect *eIF(iso)4E* knockout mutants without compromising other characteristics such as biomass and flowering time (Pyott et al., 2016).

5.2.8.3 Abiotic stress tolerance (drought, salinity, heat, Cold)

Plant responses to these stresses involve a large number of functional and structural genes, as well as non-coding RNAs (Shi et al., 2017a, 2017b). Genes obtained using both conventional transgenic technology and the most recent sophisticated gene-editing technologies can be used to increase crop resistance to these environmental stresses. However, thousands of genes are interlinked to overcome a single stress, as a single gene is unable to play a dominant role under stress conditions. Therefore, advances in CRISPR/Cas technology are being used to create genome-editing mutants to enhance stress resistance in plants. The *AGROS8* gene is a negative regulator of the ethylene response. Recently, this gene was modified using CRISPR/Cas9 to enhance drought tolerance in maize . This shows how CRISPR/Cas technology may help plants become more resilient to abiotic stresses. Tomato plants can grow and produce fruit even under heat stress after deletion of the R gene *SlAGAMOUS-Like 6* (*SlAGL6*) using CRISPR/Cas9 technology (Klap et al., 2017). In wild-type tomato, this deletion greatly reduces fertilization-dependent fruit set. Genome-edited tomato plants showed enhanced fruiting conditions, and their capacity to bear fruit also increased under high-temperature treatment. Furthermore, heat stress does not affect tomato sexual reproduction potential (Klap et al., 2017). In a recent study, *Arabidopsis* mutants dpa4-sod7-aitr256, edited through CRISPR/Cas, showed enhanced growth under water stress (Chen et al., 2019). Suppression and removal of the ARF4 transcription factor gene improved water efficiency in tomato, and the plants were more resilient to salinity and osmotic stress (Bouzroud et al., 2020). G protein gene mutants (*gs3* and *dep1*), produced by CRISPR/Cas, improved the ability of rice to grow under various stress conditions, particularly salinity stress (Cui et al., 2020). The ability of rice to withstand alkaline stress was improved by ppa6 gene knockout (Wang et al., 2018a, 2018b).

5.2.8.4 Nutritional quality improvement

CRISPR/Cas technology is being used to modify the secondary metabolism of plants to enhance crop quality. Lycopene biosynthesis in tomato was improved through multiplex genome editing of multiple genes in the carotenoid metabolic pathway (Li et al., 2018a, 2018b). The main protein matrix is determined by protein structure. Using the CRISPR/Cas9 genome editing method, Sanchez-Leon and coworkers produced low-gluten, transgene-free wheat by targeting a specific type of a-gliadin gene (Sanchez-Leon et al., 2018). Knocking out the d-hordein gene using CRISPR/Cas increased the barley protein matrix around starch granules (Yang et al., 2020). In *Camelina sativa* seeds, the protein portfolio of the seed storage protein gene *CRUCIFERIN C* (*CRUC*) was altered by CRISPR/Cas9 technology (Lyzenga et al., 2019).

Carbohydrate production and composition in plants can also be altered through genome editing using CRISPR/Cas. Grains with starch tolerance and high amylose content are beneficial for human health. In allotetraploid rapeseed, mature seeds accumulated more starch after knockout of the *pat2/5* genes using CRISPR/Cas9 genome editing (Zhang et al., 2019). Sun et al. (2017) successfully suppressed the starch branching enzyme (SBE) gene *SBEIIb* using the CRISPR/Cas technique, resulting in high-amylose, genome-edited rice. A higher proportion of long-chain debranched amylopectin in rice seeds produced up to 25% amylose (Sun et al., 2017). Similar results were achieved in sweet potatoes, where the *SBEII* gene was deleted using CRISPR/Cas9 technology, although the genome-edited plants showed no change in the total amount of starch (Wang et al., 2018a, 2018b). CRISPR/Cas9 modification of the granule-bound starch synthase (*GBSS*) gene in potatoes also altered starch quality (Andersson et al., 2017).

Researchers hope that the CRISPR/Cas9 genome editing system can be applied to enhance oil content in plants. Karunarathna et al. (2020) achieved oil-rich seeds in rapeseed through immediate CRISPR knockdown of multiple genes, i.e., *BnSFAR4* and *BnSFAR5*, without negatively affecting other agronomic parameters such as seed germination. In carrot, deletion of the *tt8* gene increased oil and protein content by altering the fatty acid composition in the seed, and the genome-edited plant did not affect any yield attributes (Zhai et al., 2020).

5.2.8.5 Delay ripening and shelf life extension

The banana is a fruit with high nutritional value and significant commercial importance. It is considered a major staple food in underdeveloped nations. After exogenous ethylene induction, bananas ripen and decompose within a week. The short shelf life of bananas substantially restricts its marketing, storage, and transportation, resulting in significant postharvest loss (Krishnakumar and Thirupathi, 2014). The first factor considered when developing postharvest preservation technology is ethylene generation, which is intimately linked to the shelf life of bananas. The ripening process may be effectively delayed by genetic alteration that reduces endogenous ethylene production or impairs ethylene signal transduction (Elitzur et al., 2016).

CRISPR/Cas technology is used to produce edited banana plants with variations in the editing sequences (Chunhua et al., 2021). Under normal conditions, edited fruits showed delayed ripening and improved shelf life. Disruption of the *MaACO1* gene in mutant fruit restored responsiveness to ethephon, and the fruit ripened normally. The primary task of CRISPR/Cas9 was to target the *MaACO1* gene and develop long shelf life. These changes improved post-harvest germplasm and increased economic value.

5.2.9 CRISPR/Cas9 for hybrid breeding

The most important Cas-mediated gene deletion was observed in male-sterile lines. In previous studies, heat-sensitive male-sterile gene *TMS5* in maize, tm5 lines in rice (Li et al., 2017a, 2017b; Zhou et al., 2016), photosensitive male-sterile rice (Bouzroud et al., 2020), and *ms45* in wheat (Singh et al., 2018; Seki et al., 2011) were produced. Using CRISPR/Cas9, the *OsMATL* gene was knocked out to produce haploid rice (Yao et al., 2018). The harmful gene was removed from the rice genome, and fertility in japonica-indica hybrids was improved (Yu et al., 2018).

5.2.10 Functional genomics and traits discovery

5.2.10.1 Gene knockout studies

Gene knockdown is a key technique for discovering plant traits and advancing functional genomics. CRISPR/Cas9 can perform several tasks, e.g., introducing heritable mutations and modifying agronomic characteristics such as disease resistance, yield, and stress tolerance. By examining the phenotypic results of gene disruption, researchers can deduce gene function. Rice resistance to blast was increased by knocking down the gene *OsERF922*. In another study, scientists worked on powdery mildew and knocked out the gene *S1M1o1* (Nekrasov et al., 2017). These case studies show that CRISPR knockouts validate genetic trait corrections that were required in past traditional mutagenesis. Gene knockouts also clarify tolerance to biotic and abiotic stress (Huang et al., 2020).

The primary pathway for DSB (double-stranded break) repair in plants is non-homologous end-joining, which can result in small deletions or insertions (indels), usually less than 100 bp (Puchta, 2005). Most often, frameshift mutations caused by deletions in the coding site of a gene can lead to loss of gene function. The most commonly used method to knock out a gene is CRISPR/Cas9 because it is highly efficient and easy to use.

In plants, double-strand breaks are repaired by joining non-homologous regions, which can result in additions and deletions (indels) of fewer than 100 bps (Puchta, 2005). These additions and deletions occur in coding sequences, and the resulting frameshift mutations lead to loss of gene function. Nowadays, due to its precision and efficiency, CRISPR/Cas9 is widely used for targeted gene knockouts. Genome editing has been successfully used in *Arabidopsis*, maize, rice, tobacco, and soybean (Wang et al., 2017a, 2017b; Xie et al., 2015; Zhang et al., 2015; Char et al., 2016; Minkenberg et al., 2016; Chilcoat et al., 2017).

In previous studies, multiplex gene editing using a single vector carrying an expression cassette for six sgRNAs was used to produce the PYL/PYR sextuple hybrid. By the T3 generation, the homozygous sextuple mutant displayed the expected phenotype (Zhang et al., 2015). Multiplex gene editing is important for crop improvement because it allows stacking multiple desirable traits. It is also powerful for functional genomics and for analyzing redundant gene families and gene pathways. Using CRISPR/Cas9, knocking out the *GW2*, *GW5*, and *TGW6* genes (negative regulators) increased grain size and weight (R. Xu et al., 2016).

5.2.10.2 *Gene function validation*

Functional validation is the most critical stage for determining gene roles with specificity. After initial confirmation, transcriptomics, proteomics, and genome-wide association studies are used. Furthermore, CRISPR/Cas9 has greatly accelerated the generation of mutant variants in candidate genes. Functional validation uses gain- and loss-of-function approaches. *TaGW2*, a negative regulator of grain weight, was verified by CRISPR/Cas9 gene knockout in wheat. The edited mutants showed larger grain weight and higher yield in wheat (Wang et al., 2016a, 2016b). Activation of stress-responsive genes by CRISPR was used to verify positive effects on drought and salinity stress (Park et al., 2021). The combination of CRISPR and transcriptome analysis provides insight into regulatory pathways. These integrated strategies improve crop trait identification and provide more gene-trait relationships.

5.2.10.3 *Regulatory element dissection*

Protein-coding genes are important, including the enhancers, promoters, untranslated regions, and phenotypic expression. In recent technologies, CRISPR-mediated methods are used to analyze regulatory elements. In default or disturbed gene sequences, the cis-regulatory region can be replaced or deleted. For rice diseases, effector-binding elements in the promoter *OsSWEET14* affect pathogen-induced activation (Xu et al., 2019). In another example, altering the regulatory regions of *GmFT2a* in soybeans can change flowering time (Cai et al., 2020). Examining DNA methylation and histone acetylation states at particular loci is also made possible by the growing application of dCas9-based epigenetic modifiers (Papikian et al., 2019).

5.2.10.4 *Multiplex gene editing*

Features such as yield, nutrition, stress resistance, and adaptability are linked to a complex multiplex genome and interacting, functionally redundant genes. Multiplex gene editing with CRISPR/Cas9 is the most efficient approach, enabling the identification of expression and characteristics of several genes. Previous studies showed that *ARGOS8* homologs in maize enhance drought tolerance through ethylene signaling pathways (Shi et al., 2017a, 2017b). Another example is the Rice *OsGS1*, *OsGS2*, and *OsGLU* genes, which control nitrogen absorption and are subject to multiplex genome editing (Li et al., 2021).

5.2.11 Synthetic biology and metabolic engineering

Plants and microorganisms naturally produce a vast range of large molecules with significant activity. Therefore, these compounds are challenging to produce synthetically. These enzymes follow strict regulatory pathways that ensure controlled metabolite production in the vector. Earlier, metabolic engineering was restricted to deleting or overexpressing a gene in the native organism or reconstructing pathways in a specific host cell (Cravens et al., 2019).

Synthetic biology, also known as the engineering of biological systems, was introduced in the 1960s (Singh, 2014; Cameron et al., 2014; Wang et al., 2013). This is a multidisciplinary research area that integrates biology, chemistry, mathematics, and computer science (Auslander et al., 2017; Shapira et al., 2017). This field involves structuring biological components such as enzymes, circuits, modules, and systems, or redesigning biological systems by programming genetic information to achieve desired outcomes (Cameron et al., 2014). Moreover, this technology has shown new pathways for the regulated and desired development of species (Li et al., 2020; Andrianantoandro et al., 2006). It was introduced in the 1960s but has become prominent in the last decade.

5.2.11.1 Engineering novel biosynthetic pathways

This approach involves incorporating potential enzymes, and in some instances, entire pathways are transferred into a new host. In recent years, advanced metabolic engineering strategies have increasingly focused on constructing fully artificial biosynthetic pathways. These pathways are assembled using a combination of natural enzymes, engineered enzyme variants, and de novo-designed enzymes, enabling the production of metabolic routes that do not occur naturally (Siegel et al., 2015).

Through engineered microorganisms, researchers can refine GA biosynthetic pathways to achieve higher production efficiency, lower operational costs, and greater environmental sustainability (Wasser, 2004). In its native metabolic system, Escherichia coli utilizes the methylerythritol phosphate (MEP) pathway (Frank and Groll, 2017). Using synthetic biology approaches, scientists have established a GA biosynthetic pathway in Escherichia coli by introducing and expressing critical pathway enzymes, resulting in improved heterologous production efficiency (Wang et al., 2018a, 2018b).

Metabolic engineering provides an essential framework for reconstructing metabolic networks to maximize the synthesis of target compounds. The production of desired metabolites can be enhanced by redesigning and modifying metabolic pathways within the host organism. By applying metabolic flux analysis to identify metabolic constraints and then adjusting those steps, the biosynthetic efficiency of GA can be enhanced (Zhao et al., 2025). In plants, metabolic pathway engineering holds significant potential for transferring specialized metabolic traits from one species to another (Owen et al., 2017). Beyond simply transferring entire pathways, there is also the potential to redesign them to produce novel compounds tailored for specific applications (Reed et al., 2017).

5.2.11.2 *Production of high-value compounds (e.g., pharmaceuticals, nutraceuticals)*

Synthetic biology enables a new generation of microbial engineering for the biotechnological synthesis of medicines and other valuable molecules. The production of artemisinin, a crucial medication in the treatment of malaria, is the first and most well-known application of synthetic biology. Originally derived from plants, this antimalarial can now be produced in large quantities in heterologous hosts, such as Saccharomyces cerevisiae (Paddon et al., 2013). Synthetic biology is also used to generate vaccines using attenuated pathogens. For example, genome-editing modifications can reduce viral replication in poliovirus, converting it into a safer form (Coleman et al., 2008).

Nutraceuticals consist of various bioactive compounds that offer sustained medicinal benefits beyond basic nutrition or direct pharmaceutical effects, distinguishing them from functional foods and con drugs (Jain and Ramawat, 2013). Microbial-based metabolic engineering has recently achieved considerable improvements in the development and production of high-value nutraceuticals. Genetically modified enzymes have facilitated the engineering of microbial hosts to produce a range of nutraceutical compounds (Wang et al., 2016a, 2016b).

Expanded host systems (chassis) tailored for the production of medicines and other secondary metabolites have also advanced significantly. Researchers have successfully produced genome-minimized strains of the industrial bacterium Streptomyces avermitilis that grow more quickly than the wild type and exhibit significant heterologous expression of numerous secondary metabolite gene clusters (Komatsu et al., 2013). Many microorganisms, including Escherichia coli and Saccharomyces cerevisiae, are engineered as ideal production platforms, functioning as cell factories to generate key monoterpenoids for industrial applications (Keasling, 2012; Kim et al., 2016).

5.2.11.3 *Regulation of flowering time and developmental processes*

Flowering is an important stage in a plant's life and represents a sensitive developmental transition from the vegetative to the reproductive stage (Maple et al., 2024). Plants modulate their growth and development schedules according to their optimal flowering time for successful production and propagation (Johansson and Staiger, 2015). The establishment of floral meristem identity occurs through a precise genetic regulatory network that controls the expression of key identity genes. *LFY* directly activates *AP1*, which alleviates the strong repression of *SHORT VEGETATIVE PHASE, SUPPRESSOR OF CONSTANS1*, and *AGAMOUS-LIKE24* on *SEP3*. Then *SEP3*, together with *LFY*, activates genes associated with floral organ development (Liu et al., 2009).

The splicing process is catalyzed by 3′ splicing factors and their associated proteins that bind to the 3′ ss of pre-mRNAs. In Arabidopsis thaliana, several 3′ splicing factors are linked to the regulation of flowering time and development (Lee et al., 2023a, 2023b). Flowering time is known to be regulated by the AS of key player genes, including the circadian clock genes *PSEUDO-RESPONSE REGULATOR 7* (*PRR7*) and *PRR9*, temperature-dependent AS of *FLOWERING LOCUS M* (*FLM*) (Lee et al., 2020), *FLOWERING CONTROL LOCUS A* (*FCA*) (Wang and Brendel, 2006), and transcriptional control of a MADS-box transcription factor (Lee et al., 2023a, 2023b; Qi et al., 2019).

The *PHOTOPERIOD1* (*PPD1*) gene controls flowering time in wheat and barley when plants are exposed to long-day conditions (Turner et al., 2005; Seki et al., 2011). In *Arabidopsis*, FT protein moves via the phloem from the leaves to the shoot (Turck et al., 2008). To transition from vegetative to reproductive growth in the SAM, FT binds to the promoters of *APETALA1* (*AP1*) and *FRUITFULL* (*FUL*) and interacts with the bZIP transcription factor (TF) *FLOWERING LOCUS D* (*FD*) (Abe et al., 2005; Wigge et al., 2005).

5.2.12 Regulation, ethics, and public perception

Genome editing is used to achieve targeted improvements in plant breeding (Bearth et al., 2022; Busch et al., 2022). The ethical aspects of genome editing in plants address issues such as germline gene editing testing, implementation in plants, regulatory frameworks, and implications for market access and trade. Off-target mutations pose significant risks by altering non-target genes, resulting in unwanted consequences. This inherent risk correlates with the number and type of genetic modifications made, especially permanent changes that can persist for generations (Chen et al., 2023).

5.2.12.1 Regulatory landscape across Countries

In the U.S., rules are set by the Department of Defense, the FDA, and the EPA. Public opinion has also influenced scientists to adopt self-regulatory practices, as exemplified by the successful Recombinant DNA Guidelines. Due to the regulation of new gene technology applications, the 1998 de facto moratorium revoked authorizations for new transgenic editing of organisms across the EU, including the UK (Frewer et al., 2011).

5.2.12.2 Genome-edited versus transgenic crops (GMOs)

Mutation breeding, crossbreeding, and transgenic breeding were the primary crop enhancement technologies at the start of the 21st century. In 1993, the first tobacco gene-targeting experiment and the identification of DNA mutations improved the efficacy of gene-targeting. Through the use of artificial nucleases, genome-editing technologies (gene knockout, single-base editing, and guided editing) enable changes at specific genomic locations (Guo et al., 2023). Another equally popular technology is the production and application of transgenic plants. According to Zhang et al., the cultivation of transgenic plants lowers chemical use in agriculture and has enabled the transition to intensive agricultural technologies.

With efficient genome editing, specific modifications in the genome can be achieved in a shorter period of time. The Green Revolution, which boosted agricultural crop production by creating high-yielding cereal cultivars, especially dwarf wheat and rice, was the most significant phase in plant breeding (Lee et al., 2015). Among horticultural species, ornamental plants, particularly roses and chrysanthemums, saw the most success with induced mutagenesis (Guo et al., 2023; Das et al., 2023).

CRISPR/Cas9 is the most popular and sophisticated gene-editing technique for enhancing crop plants (Brookes, 2022). The CRISPR/Cas9 approach can improve plant resilience to biotic stress caused by bacteria, viruses, fungi, and nematodes. Numerous instances exist in which resistance to bacterial and fungal infections has been effectively improved.

Cisgenic genetic modification has led to the production of blight-resistant potatoes, as compared with synthetic or organic fungicides.

Genome editing technology has effectively improved the economic and nutritional aspects of horticultural plants and other crops, raising their productivity (Kayalvizhi et al., 2020; Hernandez-Munoz et al., 2019; Bhatta and Malla, 2020). Increased yields in cereal crops are due to genes encoding traits such as larger grain size, higher grain weight, and greater panicle number (Brookes, 2022).

5.2.12.3 *Consumer acceptance and ethical issues*

Genome editing is a more precise and promising target delivery technology compared with traditional breeding approaches (Brossard, 2019). Recent studies have highlighted consumer acceptance and trust in plant genome editing technology over transgenic modifications (Brossard, 2019; Beghin and Gustafson, 2021) (Busch et al., 2022).

However, when species are crossed during breeding, the public is more concerned than when genes from similar species are added (Kronberger et al., 2014). Furthermore, the insertion of foreign genes into a host, which is linked to contamination and "Frankenstein food," has more negative associations than gene editing within a host (Delwaide et al., 2015; Edenbrandt et al., 2018). Oppositional non-governmental organizations have not been very active in opposing the technology, since the media portrays genome editing as more beneficial than genetic modification (Brossard, 2019) (Zhong et al., 2018).

5.2.13 Delivery barriers and regeneration issues of CRISPR

The major delivery barriers for CRISPR include the large size of CRISPR components, limited biological stability due to degradation by nucleases, and difficulty crossing cell membranes because of their negative charge and hydrophilic nature. Once inside cells, CRISPR components risk degradation in lysosomes and endosomes, requiring delivery systems that provide protection, stability, efficient internalization, and targeted release. Commonly used viral vectors (such as AAV and lentiviruses) are limited by loading capacity, random integration, and immune response risks. However, non-viral carriers like lipid nanoparticles and polymers offer promising targeted delivery but face challenges crossing cell membrane barriers and in regeneration and therapeutic efficacy. Regeneration problems caused by CRISPR base gene editing are due to insufficient delivery approaches, low cell immunity, and potential safety concerns from off-target effects. Innovative base and prime editing tools reduce these risks, but progress in delivery to tissue regeneration remains the main focus.

5.2.14 Challenges and limitations of CRISPR

5.2.14.1 *Gene-editing errors and off-target effects*

Off-target effects arise from accidental mutations in the genome during gene editing. These mutations may disrupt gene function, leading to biological effects such as tumor development. To address this issue, scientists are developing gene sequencing methods and computational models to identify and mitigate the risk of these off-target mutations (Lino et al., 2018).

5.2.14.2 *Unintended mutations and genetic consequences*

Unintended mutations caused by CRISPR editing tools can lead to large-scale DNA deletions, rearrangements, or unintended insertions. Germline mutations can be passed from one generation to the next and raise concerns for future generations. In fact, editing a specific gene might disrupt other biological pathways in unpredictable ways (Tang et al., 2025).

5.2.15 Future prospects of CRISPR

CRISPR gene editing for plants provides a transformative and promising platform for sustainable agriculture, climate resilience, higher yields, enhanced nutrition, insect and pest resistance, and resistance to biotic and abiotic stresses. In the literature, there are many examples of CRISPR editing technology for crop improvement, such as increased viral disease resistance in cassava, improved nitrogen use efficiency in legumes and cereals, enhanced pest and insect resistance, and longer shelf life. These innovations, from laboratory to commercial applications, are benefiting digital agritech platforms (Jin et al., 2025).

Moreover, CRISPR's precision and efficiency in editing plant genomes democratize access for breeders to develop sustainable varieties aligned with agricultural challenges and preferences. The progressive incorporation of multi-omics technologies further enhances the identification and manipulation of key genes controlling desirable traits, leading to precise plant breeding. Overall, CRISPR is set to revolutionize plant breeding by making it faster, cheaper, and more effective at addressing global agricultural and environmental challenges (Dubey and Mostafavi, 2023).

5.3 Conclusion

CRISPR technology has emerged as a revolutionary technique in genome editing, particularly for crop improvement. This technology has rapidly advanced simple gene-editing tools into advanced platforms such as Cas9, Cas12a/Cpf1, base and prime editors, improving precision and expanding capabilities in plant genome modifications. These tools enable innovative approaches to hybrid breeding and the domestication of wild crop relatives, and promise targeted delivery by increasing resistance to biotic and abiotic stresses, as well as yield, quality, and shelf life. PEG-mediated transformation, biolistic methods, and Agrobacterium-assisted delivery support CRISPR techniques, but certain regeneration challenges persist. However, functional genomics using CRISPR, along with the incorporation of base and prime editing tools, facilitates gene function validation, gene regulation, and precise trait discovery. In short, this technology has improved plant breeding approaches to overcome future food insecurity and demands continuous research in a sustainable manner.

Declaration

"The authors declare that no generative artificial intelligence tools were used in the writing or preparation of this chapter. Professional Grammarly was used to improve the clarity and English language of this chapter. The content reflects the authors' own scholarly work, and the authors assume full responsibility for all statements, interpretations, and conclusions presented".

References

Abe, M., Kobayashi, Y., Yamamoto, S., Daimon, Y., Yamaguchi, A., Ikeda, Y., Ichinoki, H., Notaguchi, M., Goto, K., Araki, T., 2005. FD, a bZIP protein mediating signals from the floral pathway integrator FT at the shoot apex. Science 309 (5737), 1052–1056. https://doi.org/10.1126/science.1115983.

Andersson, M., Turesson, H., Nicolia, A., Fält, A.-S., Samuelsson, M., Hofvander, P., 2017. Efficient targeted multi-allelic mutagenesis in tetraploid potato (Solanum tuberosum) by transient CRISPR-Cas9 expression in protoplasts. Plant Cell Rep. 36 (1), 117–128. https://doi.org/10.1007/s00299-016-2062-3.

Andersson, M., Turesson, H., Olsson, N., Fält, A.S., Ohlsson, P., Gonzalez, M.N., Samuelsson, M., Hofvander, P., 2018. Genome editing in potato via CRISPR-Cas9 ribonucleoprotein delivery. Physiol. Plantarum 164 (4), 378–384. https://doi.org/10.1111/ppl.12731.

Andrianantoandro, E., Basu, S., Karig, D.K., Weiss, R., 2006. Synthetic biology: new engineering rules for an emerging discipline. Mol. Syst. Biol. 2. https://doi.org/10.1038/msb4100073, 2006.0028.

Anzalone, A.V., Randolph, P.B., Davis, J.R., Sousa, A.A., Koblan, L.W., Levy, J.M., Chen, P.J., Wilson, C., Newby, G. A., Raguram, A., Liu, D.R., 2019. Search-and-replace genome editing without double-strand breaks or donor DNA. Nature 576 (7785), 149–157. https://doi.org/10.1038/s41586-019-1711-4.

Appels, R., Honeycutt, R.L., 2019. rDNA: Evolution over a Billion Years. Informa UK Limited, pp. 81–136. https://doi.org/10.1201/9780429286285-5.

Auslander, S., Auslander, D., Fussenegger, M., 2017. Synthetic biology-the synthesis of biology. Angew. Chem. Int. Ed. 56 (23), 6396–6419.

Bearth, A., Kaptan, G., Kessler, S.H., 2022. Genome-edited versus genetically-modified tomatoes: an experiment on people's perceptions and acceptance of food biotechnology in the UK and Switzerland. Agric. Hum. Val. 39 (3), 1117–1131.

Beghin, J.C., Gustafson, C.R., 2021. Consumer valuation of and attitudes towards novel foods produced with new plant engineering techniques: a review. Sustainability 13 (20), 11348. https://doi.org/10.3390/su132011348.

Bhagtaney, L., Sundarrajan, P., 2023. An Overview of Tools for Genome Editing: ZFNs, Mega Nucleases, and Talens CRISPR/Cas-Mediated Genome Editing in Plants. Apple Academic Press, India, pp. 37–64. https://doi.org/10.1201/9781003331759-3.

Bhatta, B.P., Malla, S., 2020. Improving horticultural crops via crispr/cas9: current successes and prospects. Plants 9 (10), 1360.

Bouzroud, S., Gasparini, K., Hu, G., Barbosa, M.A.M., Rosa, B.L., Fahr, M., Bendaou, N., Bouzayen, M., Zsögön, A., Smouni, A., Zouine, M., 2020. Down regulation and loss of auxin response factor 4 function using CRISPR/Cas9 alters plant growth, stomatal function and improves tomato tolerance to salinity and osmotic stress. Genes 11 (3), 272. https://doi.org/10.3390/genes11030272.

Brookes, G., Genetically Modified (GM), 2022. Crop use 1996–2020: environmental impacts associated with pesticide use change. GM Crops Food 13 (1), 262–289. https://doi.org/10.1080/21645698.2022.2118497.

Brossard, D., 2019. Biotechnology, communication and the public: keys to delve into the social perception of science. Metode Science Studies Journal 2019 (9), 39–45. https://doi.org/10.7203/metode.9.11347.

Busch, G., Ryan, E., von Keyserlingk, M.A., Weary, D.M., 2022. Citizen views on genome editing: effects of species and purpose. Agric. Hum. Val. 39 (1), 151–164.

Cai, Y., et al., 2020. Editing of regulatory regions of GmFT2a modulates flowering time in soybean. Plant Biotechnol. J. 18 (8), 1854–1863.

Cameron, D.E., Bashor, C.J., Collins, J.J., 2014. A brief history of synthetic biology. Nat. Rev. Microbiol. 12 (5), 381–390.

Char, S.N., Neelakandan, A.K., Nahampun, H., Frame, B., Main, M., Spalding, M.H., Becraft, P.W., Meyers, B.C., Walbot, V., Wang, K., Yang, B., 2016. An Agrobacter ium delivered CRISPR/Cas9 system for high frequency targeted mutagenesis in maize. Plant Biotechnol. J. 15, 257–268.

Chaudhary, N., Kumar, G., 2023. Mutagenic radiations: X-rays, ionizing particles, and ultraviolet. Biotechnologies and Genetics in Plant Mutation Breeding 1, 45–67.

Chen, S., Zhang, N., Zhang, Q., Zhou, G., Tian, H., Hussain, S., Ahmed, S., Wang, T., Wang, S., 2019. Genome editing to integrate seed size and abiotic stress tolerance traits in Arabidopsis reveals a role for DPA4 and SOD7 in the regulation of inflorescence architecture. Int. J. Mol. Sci. 20 (11), 2695. https://doi.org/10.3390/ijms20112695.

Chen, J., Sheng, C.W., Peng, Y., Wang, K., Jiao, Y., Palli, S.R., Cao, H., 2023. Transcript level and sequence matching are key determinants of off-target effects in RNAi. J. Agric. Food Chem. 72 (1), 577–589.

Chilcoat, D., Liu, Z.B., Sander, J., 2017. Use of CRISPR/Cas9 for crop improvement in maize and soybean. Prog. Mol. Biol. Transl. Sci. 149, 27–46.

Chunhua, H., Ou, S., Guiming, D., Weidi, H., Tao, D., Qiaosong, Y., Tongxin, D., Chunyu, L., Huijun, G., Siwen, L., Ganjun, Y., Fangcheng, B., 2021. CRISPR/Cas9-mediated genome editing of MaACO1 (aminocyclopropane-1-carboxylate oxidase 1) promotes the shelf life of banana fruit. Plant Biotechnol. J. 19, 654–665.

Coleman, J.R., Papamichail, D., Skiena, S., Futcher, B., Wimmer, E., Mueller, S., 2008. Virus attenuation by genome-scale changes in codon pair bias. Science 320 (5884), 1784–1787. https://doi.org/10.1126/science.1155761.

Cravens, A., Payne, J., Smolke, C.D., 2019. Synthetic biology strategies for microbial biosynthesis of plant natural products. Nat. Commun. 10 (1), 2142.

Cui, Y., Jiang, N., Xu, Z., Xu, Q., 2020. Heterotrimeric G protein are involved in the regulation of multiple agronomic traits and stress tolerance in rice. BMC Plant Biol. 20 (1). https://doi.org/10.1186/s12870-020-2289-6.

Das, T., Anand, U., Pal, T., Mandal, S., Kumar, M., Radha, Gopalakrishnan, A.V., Pérez de la Lastra, J.M., Dey, A., 2023. Exploring the potential of CRISPR/Cas genome editing for vegetable crop improvement: an overview of challenges and approaches. Biotechnol. Bioeng. 120 (5), 1215–1228. https://doi.org/10.1002/bit.28344.

Delwaide, A.C., Nalley, L.L., Dixon, B.L., Danforth, D.M., Nayga, R.M., Van Loo, E.J., Verbeke, W., 2015. Revisiting GMOs: are there differences in European consumers' acceptance and valuation for cisgenically vs transgenically bred rice? PLoS One 10 (5). https://doi.org/10.1371/journal.pone.0126060.

Dubey, A.K., Mostafavi, E., 2023. Biomaterials-mediated CRISPR/Cas9 delivery: recent challenges and opportunities in gene therapy. Front. Chem. 11, 1259435. https://doi.org/10.3389/fchem.2023.1259435.

Edenbrandt, A.K., House, L.A., Gao, Z., Olmstead, M., Gray, D., 2018. Consumer acceptance of cisgenic food and the impact of information and status quo. Food Qual. Prefer. 69, 44–52. https://doi.org/10.1016/j.foodqual.2018.04.007.

Elitzur, T., Yakir, E., Quansah, L., Zhangjun, F., Vrebalov, J., Khayat, E., Giovannoni, J.J., Friedman, H., 2016. Banana MaMADS transcription factors are necessary for fruit ripening and molecular tools to promote shelf-life and food security. Plant Physiol. 171 (1), 380–391. https://doi.org/10.1104/pp.15.01866.

Frank, A., Groll, M., 2017. The methylerythritol phosphate pathway to isoprenoids. Chem. Rev. 117 (8), 5675–5703. https://doi.org/10.1021/acs.chemrev.6b00537.

Frewer, L.J., Bergmann, K., Brennan, M., Lion, R., Meertens, R., Rowe, G., Siegrist, M., Vereijken, C., 2011. Consumer response to novel agri-food technologies: implications for predicting consumer acceptance of emerging food technologies. Trends Food Sci. Technol. 22 (8), 442–456. https://doi.org/10.1016/j.tifs.2011.05.005.

Guo, Y., Zhao, G., Gao, X., Zhang, L., Zhang, Y., Cai, X., Yuan, X., Guo, X., 2023. CRISPR/Cas9 gene editing technology: a precise and efficient tool for crop quality improvement. Planta 258 (2). https://doi.org/10.1007/s00425-023-04187-z.

Hernandez-Munoz, S., Pedraza-Santos, M.E., López, P.A., Gomez-Sanabria, J.M., Morales-Garcia, J.L., 2019. Mutagenesis in the improvement of ornamental plants. Rev. Chapingo Ser. Hortic. 25 (3), 151–167.

Hsu, P.D., Lander, E.S., Zhang, F., 2014. Development and applications of CRISPR-Cas9 for genome engineering. Cell 157 (6), 1262–1278.

Huang, L., et al., 2020. The OsDST gene regulates drought and salt tolerance in rice. Plant Physiol. 182 (1), 447–460.

Jain, N., Ramawat, K.G., 2013. Nutraceuticals and Antioxidants in Prevention of Diseases Natural Products: Phytochemistry, Botany and Metabolism of Alkaloids, Phenolics and Terpenes. Springer Berlin Heidelberg, India, pp. 2559–2580. https://doi.org/10.1007/978-3-642-22144-6_70.

Jin, Y.Y., Zhang, P., Liu, D.P., 2025. Optimizing homology-directed repair for gene editing: the potential of single-stranded DNA donors. Trends Genet. 41 (9), 788–803. https://doi.org/10.1016/j.tig.2025.04.014.

Jinek, M., Chylinski, K., Fonfara, I., Hauer, M., Doudna, J.A., Charpentier, E., 2012. A programmable dual-RNA–guided DNA endonuclease in adaptive bacterial immunity. Science 337 (6096), 816–821. https://doi.org/10.1126/science.1225829.

Johansson, M., Staiger, D., 2015. Time to flower: interplay between photoperiod and the circadian clock. J. Exp. Bot. 66 (3), 719–730. https://doi.org/10.1093/jxb/eru441.

Karunarathna, N.L., et al., 2020. Elevating seed oil content in a polyploid crop by induced mutations in SEED FATTY ACID REDUCER genes. Plant Biotechnology Journal n/a n/a), (in press).

Kaur, N., Alok, A., Shivani, Kaur, N., Pandey, P., Awasthi, P., Tiwari, S., 2018. CRISPR/Cas9-mediated efficient editing in phytoene desaturase (PDS) demonstrates precise manipulation in banana cv. Rasthali genome. Funct. Integr. Genom. 18 (1), 89–99. https://doi.org/10.1007/s10142-017-0577-5.

Kayalvizhi, K., Ramesh Kumar, A., Sankari, A., Anand, M., 2020. Induction of mutation in flower crops-A review. Int. J. Curr. Microbiol. Appl. Sci. 9 (6), 1320–1329. https://doi.org/10.20546/ijcmas.2020.906.164.

Keasling, J.D., 2012. Synthetic biology and the development of tools for metabolic engineering. Metab. Eng. 14 (3), 189–195. https://doi.org/10.1016/j.ymben.2012.01.004.

Kim, J., Salvador, M., Saunders, E., González, J., Avignone-Rossa, C., Jiménez, J.I., 2016. Properties of alternative microbial hosts used in synthetic biology: towards the design of a modular chassis. Essays Biochem. 60 (4), 303–313. https://doi.org/10.1042/EBC20160015.

Klap, C., Yeshayahou, E., Bolger, A.M., Arazi, T., Gupta, S.K., Shabtai, S., Usadel, B., Salts, Y., Barg, R., 2017. Tomato facultative parthenocarpy results from Sl AGAMOUS - LIKE 6 loss of function. Plant Biotechnol. J. 15 (5), 634–647. https://doi.org/10.1111/pbi.12662.

Komatsu, M., Komatsu, K., Koiwai, H., Yamada, Y., Kozone, I., Izumikawa, M., Hashimoto, J., Takagi, M., Omura, S., Shin-ya, K., Cane, D.E., Ikeda, H., 2013. Engineered Streptomyces avermitilis host for heterologous expression of biosynthetic gene cluster for secondary metabolites. ACS Synth. Biol. 2 (7), 384–396. https://doi.org/10.1021/sb3001003.

Komor, A.C., Kim, Y.B., Packer, M.S., Zuris, J.A., Liu, D.R., 2016. Programmable editing of a target base in genomic DNA without double-stranded DNA cleavage. Nature 533 (7603), 420–424. https://doi.org/10.1038/nature17946.

Krishnakumar, T., Thirupathi, V., 2014. Shelf life extension of ethylene treated bananas at different storage temperature with the ethylene action inhibitor, 1-Methylcyclopropene. Trends Biosci 22, 3673–3679.

Kronberger, N., Wagner, W., Nagata, M., 2014. How natural is "more natural"? The role of method, type of transfer, and familiarity for public perceptions of cisgenic and transgenic modification. Sci. Commun. 36 (1), 106–130.

Lee, J., Chin, J.H., Ahn, S.N., Koh, H.J., 2015. Brief History and Perspectives on Plant Breeding Current Technologies in Plant Molecular Breeding: A Guide Book of Plant Molecular Breeding for Researchers. Springer, Netherlands, pp. 1–14. https://doi.org/10.1007/978-94-017-9996-6_1.

Lee, C.S., Wang, R.W., Chang, H.H., Capurso, D., Segal, M.R., Haber, J.E., 2016. Chromosome position determines the success of double-strand break repair. Proc. Natl. Acad. Sci. U. S. A 113 (2), E146. https://doi.org/10.1073/pnas.1523660113.

Lee, D.W., Lee, S.K., Rahman, M.M., Kim, Y.J., Zhang, D., Jeon, J.S., 2019. The role of rice vacuolar Invertase2 in seed size control. Mol. Cells 42 (10), 711–720. https://doi.org/10.14348/molcells.2019.0109.

Lee, K.C., Chung, K.S., Lee, H.T., Park, J.H., Lee, J.H., Kim, J.K., 2020. Role of Arabidopsis splicing factor SF1 in temperature-responsive alternative splicing of FLM pre-mRNA. Front. Plant Sci. 11. https://doi.org/10.3389/fpls.2020.596354.

Lee, K.C., Kim, Y.C., Kim, J.K., Lee, H., Lee, J.H., 2023a. Regulation of flowering time and other developmental plasticities by 3'Splicing factor-mediated alternative splicing in Arabidopsis thaliana. Plants 12 (19), 3508.

Lee, H.T., Park, H.Y., Lee, K.C., Lee, J.H., Kim, J.K., 2023b. Two Arabidopsis splicing factors, U2AF65a and U2AF65b, differentially control flowering time by modulating the expression or alternative splicing of a subset of FLC upstream regulators. Plants 12 (8). https://doi.org/10.3390/plants12081655.

Li, M., Li, X., Zhou, Z., Wu, P., Fang, M., Pan, X., Lin, Q., Luo, W., Wu, G., Li, H., 2016. Reassessment of the four yield-related genes Gn1a, DEP1, GS3, and IPA1 in rice using a CRISPR/Cas9 system. Front. Plant Sci. 7 (2016). https://doi.org/10.3389/fpls.2016.00377.

Li, C., et al., 2017a. A high-efficiency CRISPR/Cas9 system for targeted mutagenesis in Cotton (Gossypium hirsutum L.). Sci. Rep. 7, 43902.

Li, J., Zhang, H., Si, X., Tian, Y., Chen, K., Liu, J., Chen, H., Gao, C., 2017b. Generation of thermosensitive male-sterile maize by targeted knockout of the ZmTMS5 gene. J. Genet. Genomics 44 (9), 465–468. https://doi.org/10.1016/j.jgg.2017.02.002.

Li, C., Zong, Y., Wang, Y., Jin, S., Zhang, D., Song, Q., Gao, C., 2018a. Expanded base editing in rice and wheat using a Cas9-adenosine deaminase fusion. Genome Biol. 19 (1), 59.

Li, X., et al., 2018b. Lycopene is enriched in tomato fruit by CRISPR/Cas9-Mediated multiplex genome editing. Front. Plant Sci. 9 (559).

Li, H., Yang, Y., Hong, W., Huang, M., Wu, M., Zhao, X., 2020. Applications of genome editing technology in the targeted therapy of human diseases: mechanisms, advances and prospects. Signal Transduct. Targeted Ther. 5 (1). https://doi.org/10.1038/s41392-019-0089-y.

Li, S., et al., 2021. Multiplex genome editing of nitrogen metabolism genes enhances nitrogen-use efficiency in rice. Nat. Commun. 12, 7072.

Liang, Z., Zhang, K., Chen, K., Gao, C., 2014. Targeted mutagenesis in Zea mays using TALENs and the CRISPR/cas system. J. Genet. Genomics 41 (2), 63–68. https://doi.org/10.1016/j.jgg.2013.12.001.

Lino, C.A., Harper, J.C., Carney, J.P., Timlin, J.A., 2018. Delivering crispr: a review of the challenges and approaches. Drug Deliv. 25 (1), 1234–1257. https://doi.org/10.1080/10717544.2018.1474964.

Liu, C., Xi, W., Shen, L., Tan, C., Yu, H., 2009. Regulation of floral patterning by flowering time genes. Dev. Cell 16 (5), 711–722. https://doi.org/10.1016/j.devcel.2009.03.011.

Lowder, L.G., Zhang, D., Baltes, N.J., Paul, J.W., Tang, X., Zheng, X., Voytas, D.F., Hsieh, T.F., Zhang, Y., Qi, Y., 2015. A CRISPR/Cas9 toolbox for multiplexed plant genome editing and transcriptional regulation. Plant Physiol. 169 (2), 971–985. https://doi.org/10.1104/pp.15.00636.

Lowder, L., Malzahn, A., Qi, Y., 2016. Rapid evolution of manifold CRISPR systems for plant genome editing. Front. Plant Sci. 7, 1683.

Lyzenga, W.J., Harrington, M., Bekkaoui, D., Wigness, M., Hegedus, D.D., Rozwadowski, K.L., 2019. CRISPR/Cas9 editing of three CRUCIFERIN C homoeologues alters the seed protein profile in Camelina sativa. BMC Plant Biol. 19 (1). https://doi.org/10.1186/s12870-019-1873-0.

Makarova, K.S., Wolf, Y.I., Iranzo, J., Shmakov, S.A., Alkhnbashi, O.S., Brouns, S.J.J., Charpentier, E., Cheng, D., Haft, D.H., Horvath, P., Moineau, S., Mojica, F.J.M., Scott, D., Shah, S.A., Siksnys, V., Terns, M.P., Venclovas, Č., White, M.F., Yakunin, A.F., Yan, W., Zhang, F., Garrett, R.A., Backofen, R., van der Oost, J., Barrangou, R., Koonin, E.V., 2020. Evolutionary classification of CRISPR–Cas systems: a burst of class 2 and derived variants. Nat. Rev. Microbiol. 18 (2), 67–83. https://doi.org/10.1038/s41579-019-0299-x.

Maple, R., Zhu, P., Hepworth, J., Wang, J.W., Dean, C., 2024. Flowering time: from physiology, through genetics to mechanism. Plant Physiol. 195 (1), 190–212. https://doi.org/10.1093/plphys/kiae109.

Minkenberg, B., Xie, K., Yang, Y., 2016. Discovery of rice essential genes by characterizing CRISPR-edited mutation of closely related rice MAP kinase genes. Plant J. 89, 636–648.

Murovec, J., Guček, K., Bohanec, B., Avbelj, M., Jerala, R., 2018. DNA-free genome editing of Brassica oleracea and B. Rapa protoplasts using CRISPR-cas9 ribonucleoprotein complexes. Front. Plant Sci. 9. https://doi.org/10.3389/fpls.2018.01594.

Nekrasov, V., et al., 2017. Targeted mutagenesis in the model plant tomato using the CRISPR/Cas system. Nat. Biotechnol. 35 (4), 441–443.

Owen, C., Patron, N.J., Huang, A., Osbourn, A., 2017. Harnessing plant metabolic diversity. Curr. Opin. Chem. Biol. 40, 24–30. https://doi.org/10.1016/j.cbpa.2017.04.015.

Paddon, C.J., Westfall, P.J., Pitera, D.J., Benjamin, K., Fisher, K., McPhee, D., Leavell, M.D., Tai, A., Main, A., Eng, D., Polichuk, D.R., Teoh, K.H., Reed, D.W., Treynor, T., Lenihan, J., Jiang, H., Fleck, M., Bajad, S., Dang, G., Dengrove, D., Diola, D., Dorin, G., Ellens, K.W., Fickes, S., Galazzo, J., Gaucher, S.P., Geistlinger, T., Henry, R., Hepp, M., Horning, T., Iqbal, T., Kizer, L., Lieu, B., Melis, D., Moss, N., Regentin, R., Secrest, S., Tsuruta, H., Vazquez, R., Westblade, L.F., Xu, L., Yu, M., Zhang, Y., Zhao, L., Lievense, J., Covello, P.S., Keasling, J.D., Reiling, K.K., Renninger, N.S., Newman, J.D., 2013. High-level semi-synthetic production of the potent antimalarial artemisinin. Nature 496 (7446), 528–532. https://doi.org/10.1038/nature12051.

Papikian, A., et al., 2019. Epigenome editing in plants through dCas9-based targeted DNA methylation. Nat. Commun. 10, 4273.

Park, J., et al., 2021. CRISPR activation of drought-responsive genes enhances stress tolerance in rice. Front. Plant Sci. 12, 678950.

Puchta, H., 2005. The repair of double-strand breaks in plants: mechanisms and consequences for genome evolution. J. Exp. Bot. 56, 1–14.

Pyott, D.E., et al., 2016. Engineering of CRISPR/Cas9-mediated potyvirus resistance in transgene-free Arabidopsis plants. Mol. Plant Pathol. 17 (8), 1276–1288.

Qi, H.D., Lin, Y., Ren, Q.P., Wang, Y.Y., Xiong, F., Wang, X.L., 2019. RNA splicing of FLC modulates the transition to flowering. Front. Plant Sci. 10. https://doi.org/10.3389/fpls.2019.01625.

Raigani, M., Eftekhari, Z., Adeli, A., Kazemi-Lomedasht, F., 2025. Advancing gene editing therapeutics: clinical trials and innovative delivery systems across diverse diseases. Mol. Ther. Nucleic Acids 36 (3), 102666. https://doi.org/10.1016/j.omtn.2025.102666.

Reed, J., Stephenson, M.J., Miettinen, K., Brouwer, B., Leveau, A., Brett, P., Goss, R.J.M., Goossens, A., O'Connell, M. A., Osbourn, A., 2017. A translational synthetic biology platform for rapid access to gram-scale quantities of novel drug-like molecules. Metab. Eng. 42, 185–193. https://doi.org/10.1016/j.ymben.2017.06.012.

Sanchez-Leon, S., et al., 2018. Low-gluten, nontransgenic wheat engineered with CRISPR/Cas9. Plant Biotechnol. J. 16 (4), 902–910.

Seki, M., Chono, M., Matsunaka, H., Fujita, M., Oda, S., Kubo, K., Kato, K., 2011. Distribution of photoperiod-insensitive alleles Ppd-B1a and Ppd-D1a and their effect on heading time in Japanese wheat cultivars. Breed. Sci. 61 (4), 405–412.

Shamshirgaran, Y., Liu, J., Sumer, H., Verma, P.J., Taheri-Ghahfarokhi, A., 2022. Tools for efficient genome editing; ZFN, TALEN, and CRISPR. Methods Mol. Biol. 2495, 29–46. https://doi.org/10.1007/978-1-0716-2301-5_2.

Shapira, P., Kwon, S., Youtie, J., 2017. Tracking the emergence of synthetic biology. Scientometrics 112 (3), 1439–1469.

Shi, J., et al., 2017a. ARGOS8 variants generated by CRISPR/Cas9 improve maize grain yield under drought. Nat. Commun. 8, 14751.

Shi, J., Gao, H., Wang, H., Renee Lafitte, H., Archibald, R.L., Yang, M., Hakimi, S.M., Mo, H., Habben, J.E., 2017b. ARGOS 8 variants generated by CRISPR -Cas9 improve maize grain yield under field drought stress conditions. Plant Biotechnol. J. 15 (2), 207–216. https://doi.org/10.1111/pbi.12603.

Siegel, J.B., Smith, A.L., Poust, S., Wargacki, A.J., Bar-Even, A., Louw, C., Shen, B.W., Eiben, C.B., Tran, H.M., Noor, E., Gallaher, J.L., Jacob, B., Yoshikuni, Y., Gelb, M.H., Keasling, J.D., Stoddard, B.L., Lidstrom, M.E., Baker, D., 2015. Computational protein design enables a novel one-carbon assimilation pathway. Proc. Natl. Acad. Sci. 112 (12), 3704–3709. https://doi.org/10.1073/pnas.1500545112.

Singh, V., 2014. Recent advancements in synthetic biology: current status and challenges. Gene 535 (1), 1–11. https://doi.org/10.1016/j.gene.2013.11.025.

Singh, M., Kumar, M., Albertsen, M.C., Young, J.K., Cigan, A.M., 2018. Concurrent modifications in the three homeologs of Ms45 gene with CRISPR-Cas9 lead to rapid generation of male sterile bread wheat (Triticum aestivum L.). Plant Mol. Biol. 97 (4–5), 371–383.

Sun, Y., Jiao, G., Liu, Z., Zhang, X., Li, J., Guo, X., Du, W., Du, J., Francis, F., Zhao, Y., Xia, L., 2017. Generation of high-amylose rice through CRISPR/Cas9-Mediated targeted mutagenesis of starch branching enzymes. Front. Plant Sci. 8. https://doi.org/10.3389/fpls.2017.00298.

Svitashev, S., Young, J.K., Schwartz, C., Gao, H., Falco, S.C., Cigan, A.M., 2015. Targeted mutagenesis, precise gene editing, and site-specific gene insertion in maize using Cas9 and guide RNA. Plant Physiol. 169 (2), 931–945. https://doi.org/10.1104/pp.15.00793.

Szarejko, I., Szurman-Zubrzycka, M., Nawrot, M., Marzec, M., Gruszka, D., Kurowska, M., Maluszynski, M., 2017. Creation of a TILLING population in barley after chemical mutagenesis with sodium azide and MNU. Biotechnologies for plant mutation breeding 91–111.

Tang, X., Lowder, L.G., Zhang, T., Malzahn, A.A., Zheng, X., Voytas, D.F., Zhong, Z., Chen, Y., Ren, Q., Li, Q., Kirkland, E.R., Zhang, Y., Qi, Y., 2017. A CRISPR–Cpf1 system for efficient genome editing and transcriptional repression in plants. Nat. Plants 3 (3). https://doi.org/10.1038/nplants.2017.18.

Tang, S., Chen, X., Tong, X., Zhu, L., 2025. Overcoming the delivery challenges in CRISPR/cas9 gene editing for effective cancer treatment: a review of delivery systems. Int. J. Med. Sci. 22 (14), 3625–3649. https://doi.org/10.7150/ijms.112724.

Turck, F., Fornara, F., Coupland, G., 2008. Regulation and identity of florigen: FLOWERING LOCUS T moves center stage. Annu. Rev. Plant Biol. 59, 573–594.

Turner, A., Beales, J., Faure, S., Dunford, R.P., Laurie, D.A., 2005. Botany: the pseudo-response regulator Ppd-H1 provides adaptation to photoperiod in barley. Science 310 (5750), 1031–1034. https://doi.org/10.1126/science.1117619.

Usharani, K.S., Kumar, C.A., Vanniarajan, C., 2016. EMS-chemical mutagen for induction of mutations. Chem Sci Rev Lett 5, 286–289.

Wang, B.B., Brendel, V., 2006. Molecular characterization and phylogeny of U2AF35 homologs in plants. Plant Physiol. 140 (2), 624–636. https://doi.org/10.1104/pp.105.073858.

Wang, Y.H., Wei, K.Y., Smolke, C.D., 2013. Synthetic biology: advancing the design of diverse genetic systems. Annu. Rev. Chem. Biomol. Eng. 4 (1), 69–102.

Wang, F., Wang, F., Wang, F., et al., 2016a. Enhanced rice blast resistance by CRISPR/Cas9-mediated editing of OsERF922. Plant Biotechnol. J. 14 (8), 1647–1655.

Wang, J., Guleria, S., Koffas, M.A.G., Yan, Y., 2016b. Microbial production of value-added nutraceuticals. Curr. Opin. Biotechnol. 37, 97–104. https://doi.org/10.1016/j.copbio.2015.11.003.

Wang, Y., Zong, Y., Gao, C., 2017a. Targeted mutagenesis in hexaploid bread wheat using the TALEN and CRISPR/Cas systems. In: Wheat Biotechnology. Humana Press, New York, NY, pp. 169–185.

Wang, M., Mao, Y., Lu, Y., Tao, X., Zhu, J.k., 2017b. Multiplex gene editing in rice using the CRISPR-Cpf1 system. Mol. Plant 10 (7), 1011–1013. https://doi.org/10.1016/j.molp.2017.03.001.

Wang, C., Liwei, M., Park, J.B., Jeong, S.H., Wei, G., Wang, Y., Kim, S.W., 2018a. Microbial platform for terpenoid production: Escherichia coli and Yeast. Front. Microbiol. 9. https://doi.org/10.3389/fmicb.2018.02460.

Wang, Y., et al., 2018b. Simultaneous editing of three homoeologs of TaGW2 increases grain size and weight in wheat. Plant Biotechnol. J. 16 (11), 1810–1819.

Wasser, S.P., 2004. Reishi or Ling Zhi (Ganoderma Lucidum) Encyclopedia of Dietary Supplements. CRC Press, Israel, pp. 603–622. https://doi.org/10.1081/E-EDS-120022119.

Wigge, P.A., Kim, M.C., Jaeger, K.E., Busch, W., Schmid, M., Lohmann, J.U., Weigel, D., 2005. Integration of spatial and temporal information during floral induction in Arabidopsis. Science 309 (5737), 1056–1059. https://doi.org/10.1126/science.1114358.

Woo, J.W., Kim, J., Kwon, S.I., Corvalán, C., Cho, S.W., Kim, H., Kim, S.G., Kim, S.T., Choe, S., Kim, J.S., 2015. DNA-free genome editing in plants with preassembled CRISPR-Cas9 ribonucleoproteins. Nat. Biotechnol. 33 (11), 1162–1164. https://doi.org/10.1038/nbt.3389.

Xie, K., Minkenberg, B., Yang, Y., 2015. Boosting CRISPR/Cas9 multiplex editing capability with the endogenous tRNA processing system. Proc. Natl. Acad. Sci. U.S.A. 112, 3570–3575.

Xing, H.L., Dong, L., Wang, Z.P., Zhang, H.Y., Han, C.Y., Liu, B., Wang, X.C., Chen, Q.J., 2014. A CRISPR/Cas9 toolkit for multiplex genome editing in plants. BMC Plant Biol. 14 (1). https://doi.org/10.1186/s12870-014-0327-y.

Xu, R., et al., 2016. Rapid improvement of grain weight via highly efficient CRISPR/Cas9-mediated multiplex genome editing in rice. J. Genet. Genomics 43, 529–532.

Xu, R., Yang, Y., Qin, R., Hao, L., Qiu, C., Li, L., Wei, P., Yang, J., 2016. Rapid improvement of grain weight via highly efficient CRISPR/Cas9-mediated multiplex genome editing in rice. J. Genet. Genomics 43, 529–532.

Xu, Z., et al., 2019. Editing of effector-binding elements in the promoter of OsSWEET14 confers resistance to bacterial blight. Nat. Biotechnol. 37 (12), 1364–1370.

Yang, Q., et al., 2020. Mutation of the d-hordein gene by RNA-guided Cas9 targeted editing reducing the grain size and changing grain compositions in barley. Food Chem. 311, 125892.

Yao, L., Zhang, Y., Liu, C., Liu, Y., Wang, Y., Liang, D., Liu, J., Sahoo, G., Kelliher, T., 2018. OsMATL mutation induces haploid seed formation in indica rice. Nat. Plants 4 (8), 530–533. https://doi.org/10.1038/s41477-018-0193-y.

Yu, X., Zhao, Z., Zheng, X., Zhou, J., Kong, W., Wang, P., Bai, W., Zheng, H., Zhang, H., Li, J., Liu, J., Wang, Q., Zhang, L., Liu, K., Yu, Y., Guo, X., Wang, J., Lin, Q., Wu, F., Ren, Y., Zhu, S., Zhang, X., Cheng, Z., Lei, C., Liu, S., Liu, X., Tian, Y., Jiang, L., Ge, S., Wu, C., Tao, D., Wang, H., Wan, J., 2018. A selfish genetic element confers non-Mendelian inheritance in rice. Science 360 (6393), 1130–1132. https://doi.org/10.1126/science.aar4279.

Zhai, Y., Yu, K., Cai, S., Hu, L., Amoo, O., Xu, L., Yang, Y., Ma, B., Jiao, Y., Zhang, C., Khan, M.H.U., Khan, S.U., Fan, C., Zhou, Y., 2020. Targeted mutagenesis of BnTT8 homologs controls yellow seed coat development for effective oil production in Brassica napus L. Plant Biotechnol. J. 18 (5), 1153–1168. https://doi.org/10.1111/pbi.13281.

Zhang, Z., Mao, Y., Ha, S., Liu, W., Botella, J.R., Zhu, J.K., 2015. A multiplex CRISPR/Cas9 platform for fast and effi cient editing of multiple genes in Arabidopsis. Plant Cell Rep. 35, 1519–1533.

Zhang, K., Nie, L., Cheng, Q., Yin, Y., Chen, K., Qi, F., Zou, D., Liu, H., Zhao, W., Wang, B., Li, M., 2019. Effective editing for lysophosphatidic acid acyltransferase 2/5 in allotetraploid rapeseed (Brassica napus L.) using CRISPR-Cas9 system. Biotechnol. Biofuels 12 (1). https://doi.org/10.1186/s13068-019-1567-8.

Zhao, S., Su, Q., Liang, L., Gao, X., Wei, J., Lin, M., Su, Y., 2025. Biological functions, synthetic biology strategies, and application prospects of ganoderic acid in the food industry—a review. Journal of Future Foods. https://doi.org/10.1016/j.jfutfo.2024.11.024.

Zhong, Z., Zhang, Y., Qi, Y., Tang, X., Ren, Q., Liu, S., Yang, L., Wang, Y., Liu, X., Liu, B., Zhang, T., Zheng, X., Le, Y., Zhang, Y., Qi, Y., 2018. Plant genome editing using FnCpf1 and LbCpf1 nucleases at redefined and altered PAM sites. Mol. Plant 11 (7), 999–1002. https://doi.org/10.1016/j.molp.2018.03.008.

Zhou, H., He, M., Li, J., Chen, L., Huang, Z., Zheng, S., Zhu, L., Ni, E., Jiang, D., Zhao, B., Zhuang, C., 2016. Development of commercial thermo-sensitive genic male sterile rice accelerates hybrid rice breeding using the CRISPR/Cas9-mediated TMS5 editing system. Sci. Rep. 6 (1). https://doi.org/10.1038/srep37395.

CHAPTER

6

Gene editing beyond food crops

Sohini Jash[1], Deepak Kumar Upadhyay[2], Ansu Chetri[2], Ramsha Syed[2], and Sayanti Mandal[2]

[1]JMS School of Life Sciences, University of Sussex, Brighton, United Kingdom; [2]Department of Chemistry & Biochemistry, Sharda School of Engineering & Sciences, Sharda University, Greater Noida, Uttar Pradesh, India

6.1 Introduction

As the worldwide population is predicted to rise to approximately 9–10 billion by the decade 2050, world agriculture will be demanded to produce close to 70% more food, which is virtually doubling the current levels of food production (Raza et al., 2025). Currently, there are over 345 million people living under severe food insecurity with an estimated two billion people being causative of micronutrient deficiencies. This scenario reveals the duality of the necessity to increase the availability of food resources and the value of nutrition to provide comprehensive food security (Hussain et al., 2025).

The increased world food demands are requiring serious changes in crop production to be made without adding any more land, water, or energy to support the global food needs. These attempts are however, now becoming compromised by the escalating pace of global warming, which is worsening the current problems and posing a threat to the food industry. Vulnerable crop varieties are usually the affected ones by the consequences of the changing climate patterns, especially those who have contributed less on the current environmental crisis (Alotaibi, 2023; Atanasov et al., 2021; Li et al., 2023; Pinna and Dentali, 2023). It is estimated that global temperatures rise by about 2.0–6.4°C by the close of the 21st century, which will exert more heat waves, droughts, floods, and irregular rainfall affecting agricultural systems and food production networks due to the increasing temperature issues (Alotaibi, 2023).

In this regard, creating weather-resistant crops has taken a universal interest, and new genomic technologies are appeared as effective resolutions. CRISPR–Cas technology is among them; it has transformed modern farming because of its ability to aid in the exact, efficient, and accurate editing of the plant genes. Scientists have been effective in using CRISPR to solve abiotic stress in economically important crops such as rice, wheat, and

Intelligent Gene Editing in Plants
https://doi.org/10.1016/B978-0-443-44079-3.00012-7

soybean by making them more resilient to drought, salinity, and even weather changes (Bhat et al., 2021; Bhatt et al., 2025). The greatest advantages of the technology are its high accuracy, ability to multiplex gene-editing processes, and ability to decipher more complicated stress responses that are key to plant adaptation (Haque et al., 2018). With the discovery of CRISPR is still in its preliminary stages, but it has the potential to be a revolutionary solution to rising global food security concerns in a shifting climate environment despite its restrictions in the growing challenge of genomic breeding methods. In the recent past, Artificial Intelligence (AI) has been integrated in genome engineering alongside CRISPR that has significantly expanded the capabilities and accuracy of gene-editing. The capability to analyze highly diverse and large volumes of data surpassing that of any other method of bioinformatics is what makes AI see patterns and determine relationships as well as make predictions that no one previously had access to. AI now helps to predict the best version of guide RNA (gRNA) sequences, evaluate genomic contexts, and predict off-target activity with an impressive level of accuracy with the assistance of deep learning and machine learning models (Pandey et al., 2025; Kim et al., 2025; Majid Hurrah et al., 2026). Such predictive potentials enable researchers to refine genome editing to a desired high specificity and efficiency, reducing the trial-and-error period that is mostly linked with the use of CRISPR.

Moreover, AI-based systems can also be used to design new single guide RNAs (sgRNAs), to identify novel Cas protein isoforms, and can be more flexible and educate more versatile in genome editing (Li et al., 2025). Advanced computational analysis is used to screen large genomic and protein database collections, detects any unknown CRISPR-like system and predicts it. This potent combination of AI and CRISPR does not only simplify the design of the experimental process, it also reduces the volume of laboratory assays, minimizes possibilities of failure, and shortens the time of discovery. Altogether, the combination of CRISPR and AI can be viewed as a groundbreaking shift in the field of biological research-introducing a genome editing tool as a highly specific and carefully designed chemical instrument to a predictive and intelligent system. This chapter provides the background of this integration, discussing the ways AI can be used to empower CRISPR-based technologies to provide a greater degree of accuracy, speed, and reliability in the search of sustainable and climate-resistant agriculture.

6.2 Working mechanism of CRISPR–Cas9

A cutting-edge gene-editing tool that allows for precise DNA alterations is the CRISPR-Cas9 system. Guide RNA (gRNA) and CRISPR associated (Cas-9) proteins are two crucial components of the CRISPR/Cas9 system. The CRISPR RNA (crRNA) and the trans-activating CRISPR RNA (tracrRNA) are two RNA molecules that are combined to form the single-guide RNA (sgRNA) used in the system. This sgRNA is designed to align with a particular DNA sequence found in the genome. Therefore, by matching the nucleotides with the corresponding DNA sequence, it directs the Cas9 endonuclease to its target. For effective binding and cleavage, Cas9 needs a short-conserved region called the Protospacer Adjacent Motif (PAM), usually "NGG," next to the target DNA sequence.

Cas9 proteins are multi-domain DNA endonucleases which cleaves the target DNA and makes a DSB. Therefore, the mechanism involved in the CRISPR–Cas9 genome editing works by first recognizing the site at which the cut is to be made, secondly by cleaving the specific site and lastly by repairing the cleaved site. After the target sequence is identified by the nuclease, the Cas9 protein identifies the PAM sequence at 5′-NGG-3′ (where N can be any base in the form of a nucleotide). This appreciation makes the nuclease initiate local melting of the DNA and eventually a DNA is produced. The Cas9 protein then initiates DNA cleavage and produces DSB with blunt ends, ultimately, the DSBs are repaired by the host cellular mechanism (Mengstie and Wondimu, 2021)

After binding, Cas9 induces DSB in the DNA. The two nuclease domains that it uses in its effort to accomplish its task are the HNH domain, which breaks the DNA strand that is complementary to the sgRNA, and the RuvC-like domain, which will break that strand that is incompatible (forming a blunt-ended double-strand break).

The sgRNA guides the Cas9 in the presence of HNH and RuvC to the site where the DSB forms flat ends, resulting in the activation of DNA repair mechanism, i.e., Non-Homologous End Joining (NHEJ) and Homology Direct Repair (HDR). Moreover, NHEJ is an error-prone repair mechanism where the broken DNA ends are directly ligated without the need for a homologous template. NHEJ lacks precision and may lead to base mutations; the accurately repaired DNA strand is not readily identifiable from the target sequence and will be cleaved repeatedly by Cas9 unless the sgRNA becomes unrecognizable. Meanwhile HDR is a complex and precise process, it allows for precise repair or insertion of new genetic material. However, the probability of HDR occurrence in a mature is cell is much less than that of NHEJ (Li et al., 2023)

6.3 DNA interference by CRISPR–Cas9 in bacterial adaptive immunity

There are three phases of CRISPR mechanism i.e., Adaptation phase, crRNA Biogenesis, and Interference phase (Fig. 6.1).

6.3.1 Adaptation phase

In this phase foreign plasmid or viral DNA, invading the bacteria get integrated in the bacterial genome as a spacer sequence in the CRISPR locus or array. And these spacer sequence get isolated from each one with the help of palindromic repeats sequence, hence making each spacer sequence an individual unit. And these spacer sequence serve as a memory for the future immune response against invading foreign plasmid or viral DNA. There are three primary parts of the CRISPR locus. The gene that codes for trans-activating CRISPR RNA (tracrRNA) is located at the top, followed by the Cas operon, and finally, a CRISPR array.

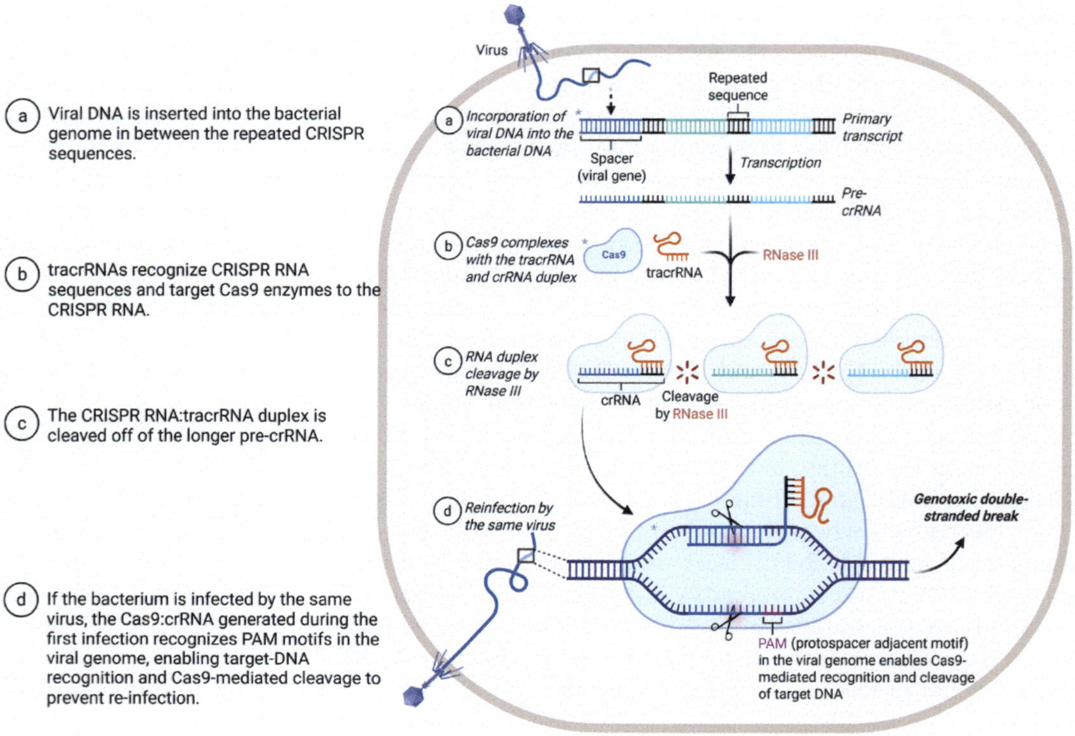

FIGURE 6.1 CRISPR–Cas9 working mechanism. *Images created with BioRender.*

6.3.2 CRISPR RNA (crRNA) biogenesis

CRISPR RNA (crRNA) consists of target sequence and hairpin loop. After transcription, each spacer sequence present in CRISPR array get transcribed into an individual precrRNA. Moreover, each precrRNA base pair with a tracrRNA, and form an RNA duplex. Hairpin loop is always present in synthetic gRNA, but not necessary in natural gRNA. Gene present in Cas operon encode for the Cas9 and other Cas proteins. PrecrRNA is further treated for becoming a mature crRNA. *RNaseIII* process the 5′ part of crRNA and 3′ tail is further processed and trimmed. Now RNA duplex of mature crRNA and tracrRNA present is known as sgRNA.

6.4 Interference phase

The interference phase is the final stage in the CRISPR immune defense system used by many archaea and bacteria to protect against invading genetic material, such as viruses.

During this phase, the CRISPR system uses previously stored genetic information to recognize and eliminate foreign DNA. Once a piece of viral DNA has been captured and stored in the bacterial genome as a "spacer" (part of the CRISPR array), the cell can later use this information if the virus invades again. The CRISPR array is translated into long RNA molecule, which is then processed into smaller pieces called crRNAs. Moreover, each crRNA contains a sequence that matches a segment of the viral DNA.

These crRNAs are then incorporated into a protein complex known as Cas. In the interference phase, the crRNA guides the Cas protein to any incoming DNA that matches the crRNA sequence. If a match is found, the Cas protein binds to the foreign DNA and cuts it, effectively destroying the invader before it can cause harm. This process is highly specific and it relies on the base-pair complementarity of the crRNA with the target DNA, and ensures that only the foreign DNA is targeted, not the own genetic material of the organism. Cas9 act as a genetical scissor. Cas9 protein has two regions, NUC lobe and REC lobe

i. **REC lobe** – it is also known as recognition lobe. REC lobe is consisting of REC1, REC2, and REC3 helical domain. REC lobe helps in binding of sgRNA with the viral or plasmid DNA. REC lobe is also known as the helical lobe.
ii. **NUC lobe**-it is also known as the nuclease lobe because it plays crucial role as a nuclease by cutting foreign DNA. It consists of three domain PI, Ruvc and HNH domain.
   1. PI (PAM interactive) domain or CT (C terminal) domain - It play crucial role in scanning and identifying the PAM sequence of invading DNA. For example, ribonucleoprotein complex of sgRNA and Cas9 protein does not blindly bind to the foreign DNA, it gets scanned by PI domain.
   2. HNH domain-It helps in cleaving the complementary strand at three base pair upstream from the PAM sequence. So, it is responsible for cleaving the target DNA strand that is complementary to the guide RNA.
   3. Ruvc domain-It also acts as a nuclease. Helps in cleavage of the noncomplementary strand of the target DNA.

6.4.1 Advances in CRISPR: Base and prime editing

Over the past few years, the approach based on CRISPR has made significant breakthroughs in our abilities to insert specific alterations and control different genomes (Li et al., 2021). Advancement of base editing (BE) technology has facilitated the further introduction of point mutations with high precision, without introducing DSBs, donor DNA templates, and independence on homology-directed repair (Gaudelli et al., 2017; Molla and Yang, 2019; Nishida et al., 2016). To date, two principal classes of base editors have been established: cytosine base editors (CBEs), which facilitate the conversion of cytosine–guanine (C–G) base pairs into thymine–adenine (T–A) pairs, and adenine base editors (ABEs), which catalyze adenine–thymine (A–T) to guanine–cytosine (G–C) transitions (Gaudelli et al., 2017). Collectively, CBEs and ABEs enable all four transition mutations—C→T, A→G, T→C, and G→A—which together constitute nearly 30% of known human pathogenic single-nucleotide variants. To either induce or repair transition mutations, these technologies have been demonstrated to effectively work in a host of cell types and species including diverse animal models of human genetic diseases (Molla and Yang, 2019). Successful implementations of CBEs (Liu et al., 2018b; Ren et al., 2019) and ABEs (Liu et al., 2018a; Hua et al., 2019) were initially pioneered in David Liu's laboratory at the Broad Institute of MIT and Harvard (Ryan Cross, 2019).

Modern base editors typically comprise a catalytically modified CRISPR–Cas enzyme, often a Cas9 nickase (nCas9) incapable of creating double-strand breaks, fused with a single-stranded DNA deaminase that directs the editing process to specific genomic targets. This fusion protein may also include DNA repair–modulating elements to enhance editing precision (Anzalone et al., 2020). The hybridization between the gRNA and complementary DNA strand on the interaction of Cas results in the creation of a ssDN R-loop with the PAM-containing strand on display (Jiang and Doudna, 2017). While nucleotides located close to the PAM site (typically within 6 ± 2 nucleotides) remain inaccessible due to steric hindrance from the Cas protein, those positioned farther from the PAM are exposed as ssDNA and become available for the deaminase enzyme to act upon (Jiang and Doudna, 2017).

CBEs work with the help of cytidine deaminases that convert cytosine base residues at R-loop into uracils which are later read by the DNA polymerases as thymines (Komor et al., 2017)b). Equally, ABEs use engineered versions of the TadA + deoxyadenosine deaminase to change adenosines to inosines that polymerases will treat as guanines (Gaudelli et al., 2017). Regardless of its accuracy, recent studies have cast doubt on the idea that in some cases, base editing can cause unexpected mutations as was the case in mouse embryos and rice models (Ryan Cross, 2019).

To target the delivery of therapeutics, adeno-associated viruses (AAVs) have become the chosen vectors thanks to their extensive tissue tropism and the relatively low immunogenicity potential (Kantor et al., 2020). Also, analytical software such as EditR can perform a precise quantification of the efficiency of base editing as based on a Sanger sequencing result (Kluesner et al., 2018).

Though CBE and ABE have a truly promising potential in terms of precise DNA editing, what used to be one of the key limitations of these (only recently) base-editing platforms, until recently, has been the impossibility to induce correct base conversion beyond the four canonical transition mutations (Kantor et al., 2020). To overcome these intrinsic constraints, the authors of this journal article created another type of genome editing procedure known as prime editing and was introduced by (Anzalone et al., 2019). Much like classical CRISPR-based BE systems, PE is not associated with the formation of any kind of a double-stranded DNA break (DSB), thus reducing the chances of unwanted insertions, deletions, or chromosomal unbalancing.

Prime editors are engineered molecular complexes that combine a nCas9 enzyme fused with a reverse transcriptase, directed by a particular prime-editing guide RNA (pegRNA). Also, it is noteworthy that the considered pegRNA is very different among the regular sgRNA typically employed by traditional models of CRISPR and constitutes the most important part of the whole prime editing process. The pegRNA contains two key parts, (a) the part having the complementary sequence to the target DNA sequence and matching the nCas9 with its specific locus in the genome, and (b) the part that encodes the desired genetic alteration (Anzalone et al., 2019).

During editing, the 5′ end of the pegRNA attaches to the five ends of the target DNA primer binding site (PBS) which then replaces the complementary strand. The case of PCR is then cut by the PAM-containing strand of DNA, leaving a single-stranded break, which acts as a primer to be used by reverse transcriptase enzyme attached to nCas9. Then, the RT stretches this nicked PAM strand by synthesizing a novel DNA sequence depending on the internal template of the pegRNA thus incorporating the modification of desired sequences in a highly programmable and controlled accessory.

It results in two temporary structures of DNA flaps: an edited three pieces flap, which is recently synthesized based on pegRNA template, and the unedited five pieces flap, which can be the initial sequence. The engagement between these flaps and the complementary DNA strand is achieved on a dynamic equilibrium. Although the five end bases may be ideally complementary, the unedited five surprisingly long base is normally broken down by cellular endonucleases that then functions during lagging-strand replication DNA synthesis (Hosfield et al., 1998).

Subsequently, the cell repair and replication complex acts on the resulting heteroduplex, which consists of one edited and one unedited strand, eventually correcting it and permanently establishing into the host genome the desired sequence. The technology makes prime editing possible to attain high accuracy in generating exact and location-specific genetic changes without destroying DSB or the use of donor DNA templates, which is a significant advance toward genome engineering accuracy and scalability.

6.5 Utilization of CRISPR in achieving abiotic stress resilience

Plant stress is any external environmental or physiological factors that cause adverse effect in the growth, development, or productivity of plants. Abiotic stress is one of the most important limitations to agricultural productivity around the world, and it includes undesirable environmental factors, including drought, salinity, temperature variations, being exposed to heavy metals, and radiation, among others, which have serious effects on crop productivity and quality (Kopecká et al., 2023; Gull et al., 2019). Stressors are further increased by accelerating effects of climate change, especially in developing countries, of which increased temperature and altered rainfall patterns completely hinder agricultural yield (Pereira, 2016).

Plants have developed highly coordinated defense mechanisms to endure such circumstances. They include complicated signal transduction pathways which converts the perception of stress into transcriptional responses, eventually leading to the activation or downregulation of genes necessary to guarantee stress tolerance (Kopecká et al., 2023; Gull et al., 2019). Given the growing uncertainty of the climatic conditions, future work in agriculture is seen to focus on creating climate-responsive crops that can withstand multiple, naturally occurring stresses. Moreover, CRISPR genome editing tool has become a paradigm shift to uncover genetic processes that govern tolerance to various abiotic stresses such as salinity, drought, temperature, and nutrient deficiency, in many crop varieties (Hurrah et al., 2024; Jash et al., 2026). Various allelic variations have been presented using this technology in fundamental genes behind the responses of the plants to climate induce stress. As an example, an instance of the use of the tool CRISPR is involved the alteration of the *OPEN STOMATA 2* (*OST2*) gene encoding the plasma membrane H^+-ATPase that regulatable, in *Arabidopsis* (Osakabe et al., 2016).

The formation of proton gradients which trigger stomatal opening is mediated by plasma membrane proton (PM) H(+)-ATPase (Merlot et al., 2007). When plants are under stress of dehydration or drought, the abscisic acid (ABA) itself attaches itself to the C-terminal portion of such ATPases and inhibits their activity, leading to stomatal closure. Interestingly the ost2 locus has been found to have two prevailing mutations that interfere with ABA-mediated inhibition resulting in constitutive activity of the proton pump and eventual necrotic lesions (Merlot et al., 2007).

With an adapted CRISPR–Cas9 system based on a truncated sgRNA (tru-sgRNA) more than 32% of mutagenesis of transgenic plants via Cas9 was obtained with no evidence of off-target effects. Examination of these mutants during ABA induced situation showed that the heightened level of stomatal closing and minimized water loss were profoundly experienced in the *OST2* (*Open Stomata 2*) locus CRISPR vegetation than in the wild type. These experiments showed that CRISPR induced mutations in the *OST2* locus increased drought resistance by vasoconstricting stomatal control (Joshi et al., 2020). Likewise, salt stress in plants is a complex physiological problem, which has three main elements; (i) sodium pressure, (ii) chloride accumulation, and (iii) osmotic imbalance. To elucidate the salt tolerance mechanisms, CRISPR was used in rice which helped the scientists to pinpoint the role of only the *OsRAV2* gene, an AP2/ERF family TF, as the only RAV gene that exhibited significant induction when salt was high (Farhat et al., 2019). Through the development of a synthetic gRNA, that targets the GT-1 element in the *OsRAV2* promoter region, the regulatory relationship between the GT-1 motif and *OsRAV2* under salt stress could be explained (Duan et al., 2016).

Besides, CRISPR-related studies were extended to the *OsPRX2* gene, which is also rather significant in the maintenance of the intracellular hydrogen peroxide concentrations. Comparative studies reported that the overexpressed *OsPRX2* rice plants and *OsPRX2* knockout mutants through CRISPR–Cas9 established strong antioxidant responses, were further regulated in ROS and were able to control stomata under the potassium-deficient conditions. These discoveries are pointers to the wide applications of CRISPR to understanding not only the role of a gene but also in designing better abiotic resistance against stress in staple crops (Ma et al., 2018).

In a significant breakthrough in the use of the gene-editing technology in the field of agriculture is the successful commercialization of a CRISPR–Cas9-edited tomato variety in Japan that contains much more γ-aminobutyric acid (GABA) than non-edited varieties (Ezura, 2022). Afterward, the CRISPR is an extensive tool used in a variety of plant species, both model plants and economically important crops. It has so far been used in model plants to include *Nicotiana benthamiana*. CRISPR technology has been applicable to facilitate multiplex (multiple gene) and homologous recombination (HR) genome editing in *Arabidopsis*, *Nicotiana benthamiana*, and *Nicotiana tabacum* by targeting gRNAs to specific sites of DNA to create DSBs to be repaired via HDR or NHEJ using a donor template to target precise genome localizations and gene knockouts (Donovan et al., 2020; Kumar et al., 2022). All these illustrate the wide versatility, effectiveness, and the transformative capabilities of the CRISPR platform in simple plant biology and in the agribusiness field.

6.6 Utilization of CRISPR in achieving biotic stress resilience

Plants are facing a great diversity of biotic stresses that are induced by living organisms which include fungus, viruses, bacteria, nematodes, and insects. These stress causing agents cause many diseases, infections and types of physical damages that have adverse effects on the crop health and eventually cause the degradation of agricultural productivity. Scientific research studies have over the years come up with diverse mechanisms and molecular approaches that alleviate these harmful effects (Gull et al., 2019). Bacterial, viral, and fungal infections and herbivores incursion are called biotic stresses and cause significant losses in agriculture, approaching 11%–30% of the total crop worldwide each year (van Esse et al., 2020).

Plant defense to pathogens is essential in supporting crop development and productivity which is even more critical in consideration of the increase in world population and the related food security demands. The use of gene-editing technologies and especially the CRISPR allows to create disease-resistant types of crops, which will contribute greatly to the decreased yield losses because of the strengthening of native plant defense systems. To date, CRISPR toolkit has been most implemented to develop resistance to viral, bacterial, and fungal pathogens. It has depicted outstanding accomplishments in creating crop plants that are more resistant against a wide range of pathogens (Bisht et al., 2019). The CLIPR-based systems have demonstrated exemplary capabilities in solving large-scale issues in the agricultural sector, such as loss of yield and the vulnerability to abiotic stress and biotic (Manghwar et al., 2020; Bortesi and Fischer, 2015; Ma et al., 2016; Wan et al., 2020). To illustrate, the genetics of providing resistance against various pathogens have been discovered in lettuce in the form of mutations that give the plant long-lasting resistance against diseases (Christopoulou et al., 2015). Also, in barley, directed mutagenesis of genome through a resistance locus, o (*MLO*) has been used to provide strong resistance against powdery mildew (Miklis et al., 2007). Specifically, the MLO mutation is especially important because the gene knockout mechanism has provided stable resistance to resist an allelic breakdown over 20 years (Li et al., 2022a; Panstruga and Schulze-Lefert, 2002).

Rust, mildew, rot, and smut amongst others as some of the fungal pathogens are continuing to draw massive losses annually not only by mitigating the crop yields of the crop, but also by compromising the quality of the products. CRISPR–Cas system has played a significant role in increasing the resistance to fungal diseases by utilizing genomic understanding of host-pathogen interaction (Wang et al., 2022). RNA-gCas9 endonucleases have been successfully deployed for the precise mutagenesis of *MLO* (Mildew Resistance Locus O) loci across diverse agricultural crops. This targeted genome-editing strategy has been effectively demonstrated in solanaceous species like tomato (*Solanum lycopersicum*), perennial fruit crops including grapevine (*Vitis vinifera*), and complex polyploid cereals such as wheat (*Triticum aestivum*) (Wang et al., 2014; Pramanik et al., 2021; Nekrasov et al., 2017; Malnoy et al., 2016). In addition, they have succeeded in producing transgene-free plants by applying visual marker-specific selection systems, e.g., Cas9-PF, to accelerate the screening of gene-edited lines (Liu et al., 2019).

Specific loss-of-function *mlo* alleles associated with other species such as *Arabidopsis thaliana*, barley and tomato confer also durable and broad-spectrum resistance against powdery mildew (Piffanelli et al., 2004; Consonni et al., 2006; Bai et al., 2009). It is important to note that the deletion site at the *MLO-B1* locus of the wheat genome produced through CRISPR/Cas9 deletion, Tamlo-R32 mutant, conserved normal growth and yield of the plants, and retained resistance to powdery mildew (Li et al., 2022a). Among the three *MLO* homeoalleles found in hexaploid wheat, *TaMLO-A1* mutated with CRISPR–Cas9 gave a significant resistance to infection with *Blumeria graminis* f. sp. tritici (Wang et al., 2014). On the same note, a transgene free tomato line with 48 bp deletion in *SlMlo1* gene reported a consistent resistance to *Oidium neolycopersici* which results in powdery mildew (Nekrasov et al., 2017).

Knockouts of *OsSEC3A* and *OsERF922* gene, which have been effectively employed in rice to impart resistance to rice blast disease, using CRISPR–Cas9 (Ma et al., 2018; Wang et al., 2016), have been applied in other plant species, such as wheat (Gan and Ling, 2022). Outside fungi, CRISPR/Cas gene editing technique has been used to combat plant viral infections. Researchers have expressed CRNAs to interfere with Tomato yellow leaf curl virus (TYLCV) genomic areas, including both noncoding and coding sequences, the intergenic region (IR), viral capsid protein (CP), and the RCRII motif of the replication protein (Rep) by using virus-specific sgRNA. Of them, the conserved IR stem-loop was the most effective in attaching the viral replication and accumulation (Ali et al., 2015). The same approach was used against the beet curly top virus (BCTV) and Merremia mosaic virus (MerMV)-two geminiviruses with conserved IR sequences, which resulted in the attenuated disease symptoms and indicated that the mutations are capable of self-cure at the same sgRNA (Ali et al., 2015).

The tobacco rattle virus (TRV) vector was further used to insert sgRNAs into a stably expressing Cas9 in *Nicotiana benthamiana* to knock out the TYLCV genome. CRISPR proved effective in providing antiviral resistance to plants with targeted genome interference, with transgenic lines expressing sgRNAs to express both cucumber mosaic virus (CMV) and tobacco mosaic virus (TMV) showing a reduction in viral accumulation of 40%–80% (Zhang et al., 2018).

6.7 Evolution of AI and its subsets

Artificial intelligence (AI) has its history in the 1950s when the first AI systems were developed, and it was based chiefly on rule-based reasoning to replicate elements of human thought (Nilsson, 1982). This period of development provided the conceptual basis to subsequent innovations leading to the field development. With time, machine learning (ML) became a vital and disruptive component of AI, and it includes a range of different paradigms, including unsupervised, reinforcement, and supervised learning (Simeone, 2018; Barreto and Armenteras, 2020; Hastie et al., 2001). With these methods machines have been able to learn by learning on their own, to syringe out trends, recognize correlation, and to make choices without any programming guidance.

The emergence and use of neural networks, and, in specific, recurrent neural networks (RNNs) and convolutional neural networks (CNNs) resulted in the massive elevation of deep learning (DL), a specialized division of ML that improved computational possibilities many times more (Oruh and Viriri, 2022; Janiesch et al., 2021). The artificially intelligent systems of deep learning transformed AI with its outstanding results in the complex and data-intensive fields of natural language processing (NLP), computer vision, and image recognition. The following technological advances were transformative in the world of AI, broadly expanding its possible uses in various fields, such as genomics and precision medicine, where ML and DL models are being used to process and analyze large and complex biological data (Quazi, 2022).

Reinforcement learning, supervised learning, unsupervised learning, and semi-supervised learning, are the four basic kinds of machine learning models (Chen et al., 2024). Every category is used in different purposes in terms of data availability and type, and the issue under consideration. In supervised learning, labeled datasets are used to train the models; that is, an input and a corresponding output are known a priori. In this way, the model will be trained to generalize the input features back to the labels in the output, and will produce accurate predictions when presented with other, previously unknown data. A supervised learning (Tarca et al., 2007) involves training a model with labeled data, i.e., the inputs of the model are associated with known labels. This strategy allows the model to estimate a mapping function which can accurately predict computations with new, unobserved information; thus, it is directly effective in the cases of the existence of labeled datasets. There are classification and regression methods of supervised learning algorithms which are widely separated. Classification algorithms are aimed at classifying data in discrete categories-e.g., identification of plant or crop diseases based on image data (Jiang et al., 2020). Moreover, on the other hand, regression algorithms are also used to comprehend and predict continuous numerical results, including for example forecasting agricultural output or crop productivity (Pantazi et al., 2016).

Unsupervised learning, on the other hand, involves poorly labelled data, meaning that it is not known what the correct outputs are. The objective of this technique is to reveal hidden structures, obscure patterns, and inherent relationships in data without any prior information. It frequently uses clustering technologies, which cluster the data points on common attributes, and dimensionality reduction technologies, which are used to reduce the volume of large datasets and preserve the most important attributes. In unsupervised learning, the two main techniques are dimensionality reduction and clustering. Clustering deals with sorting data points of similar features thus aiding in pattern discovery whereas dimensionality reduction deals with reducing the information in a large population dataset to a smaller number of variables without much loss of valuable information (Eckhardt et al., 2023). However, semi-supervised learning combines the advantage of the supervised and the unsupervised approaches. It relies on a small number of labelled samples and a bulk of unlabelled data to enhance the inference and extrapolation abilities of the model in particular cases where labelled samples are either costly or time-intensive to acquire (Wang et al., 2022).

Reinforcement learning (Barto, 2019) is a dynamic and interactive process that is unlike the supervised and unsupervised learning. Instead of running with fixed collections of data, RL algorithms do not pause but remain at work in the environment where they make a row of decisions and see the effect of their choices. It's a trial-and-error method, meaning that the algorithm obtains feedbacks in the form of rewarded/penalized. By continuously repeating the process the model tries to perfect its strategy which will yield maximum cumulative rewards by rewarding the behaviours which bring positive consequences and punishing the behaviours which favour the optimum behaviour. By this iterative optimization, the reinforcement learning evolves the capability to carry out context-sensitive, sequential decision-making to respond to changing environments and long-term goals. These features predispose the use of RL in situations that involve solving complex problems and involve adaptability and strategic decision-making, as well as improvement with time (Chen et al., 2024).

6.8 Incorporating AI and CRISPR

Artificial intelligence (AI) can be described as the capability of the computer systems to execute activities that would otherwise be performed by their cognitive abilities such as reasoning, learning, and problem-solving. ML is a major branch of AI, because it pays attention to algorithms able to learn a predictive or decision-making pattern, based on data (Sarker, 2021) (Table 6.1) (Fig. 6.2). ML consists of numerous different methods, such as reinforcement learning, supervised learning and unsupervised learning, among which they are used in analytical goals that differ. In this field, an artificially neural network-based form of ML as DL (LeCun et al., 2015) is a specialized subfield capable of processing and interpretation of complex, high-dimensional data to support various paradigms of learning and provide a breakthrough in the computational intelligence domain.

In contrast, to the traditional ML methods, DL is mathematically applicable in all three learning paradigms, including supervised, unsupervised, and reinforcement; which is why it can be considered extremely versatile to process large and complexness data sets. It is based on the principles of DL that this is a recent change, as generative AI has become a game changer in the realization of the creation of new data, content, or molecular structure based on the existing datasets through learning. Such generative models, such as large language models (LLMs), have transformed areas of language processing, molecular biology, and computational chemistry, through new outputs that reflect the complexity of the real world.

Other recent advances in DL also include prediction of protein structure directly based on the amino acid sequence, which is a significant landmark in the field of computational biology (Baek and Baker, 2022). In contrast to previous physics-based models that were based on biophysical simulation, AI-driven systems learn structural information based on large-scale patterns of data on known proteins. In 2020, AlphaFold1 was launched by the Google DeepMind company (Jumper et al., 2021), marking a new phase of a DL-based model of protein structures. In 2021, this was succeeded by AlphaFold2 (Bryant et al., 2022) that had almost experimental accuracy and transformed structural biology. Meanwhile, the Baker laboratory created another DL-based system of accurate protein predication named RoseTTAFold (Baek et al., 2021). The technology subsequently became RoseTTAFoldNA (Baek et al., 2024), a model to predict the structure of nucleic acids, and RoseTTAFold All-Atom (Krishna et al., 2024), which increased the resolution to the atomic level and allowed such modeling. In the most recent advancement, the AlphaFold3 (Abramson et al., 2024) was created, which combines generative AI models with predicting not only protein-protein but also with nucleic acids and other biomolecules. The 2024 Nobel Prize in Chemistry celebrated a monumental breakthrough in structural biology, specifically honouring

TABLE 6.1 Application of AI in plant breeding in genome editing.

Application of AI	Model used	Crop	Outcome	References
Prediction of off-target editing	CRISPRon, elevation	*Arabidopsis*, wheat	Prediction and minimization of effects by off-target genome editing	Listgarten et al. (2018); Kim et al. (2019)
Prediction of gene functionality	Support vector machines, random forest	Soybean, maize	Recognizes the function of genes	Liakos et al. (2018) Azodi et al. (2019)
Enhancement of g-RNA design	CRISPR-P, DeepCRISPR	Tomato, rice	Prediction of specific and efficient gRNA	Lei et al. (2014); Chuai et al. (2018)
Association of gene and trait	Genome-wide association study (GWAS) + machine learning, DeepCas9, OmicsAI, DeepSpCas9	Wheat having tolerance against wheat	Linking genes with complicated genes using various data	Zingaretti et al. (2020) Sandhu et al. (2021)
Prediction of genomic editing outcome	DeepSpCas9, DeepCas9	Rice, *Arabidopsis*	Foresee the efficiency of genomic editing by CRISPR-Cas systems	Wang et al. (2023)
Integrating data	AI fusion models, deep Omics	Brassica, tomato	Improvement in the interpretation of edited gene networks	Zingaretti et al. (2020)
Phenotype analysis of edited lines	PlantCV, Convulational neural networks (CNNs), UAV-DL	Rice edited by CRISPR	Discover traits that are edited using image-based machine learning tools	Singh et al. (2016) Ubbens and Stavness (2017)
Designing of genome editing strategy	AI simulations, bayesian networks	Stress-tolerant cotton	Designs network of genes for efficient and simultaneous genome editing	Kamilaris and Prenafeta-Boldú (2018)

the AI programs that solved the 50-year-old "protein folding problem." Together, these breakthroughs highlight the broad possibilities of AI-based technology, ranging from natural language processing to more complex biological simulation.

The incorporation of AI technologies within genome editing has also overstated most of the drawbacks of the traditional CRISPR solutions, hence creating improved precision, efficiency, and adaptability. Models based on AI are now central to the development of the optimum gRNA, and can predict off-target activities in large genomic datasets, which greatly

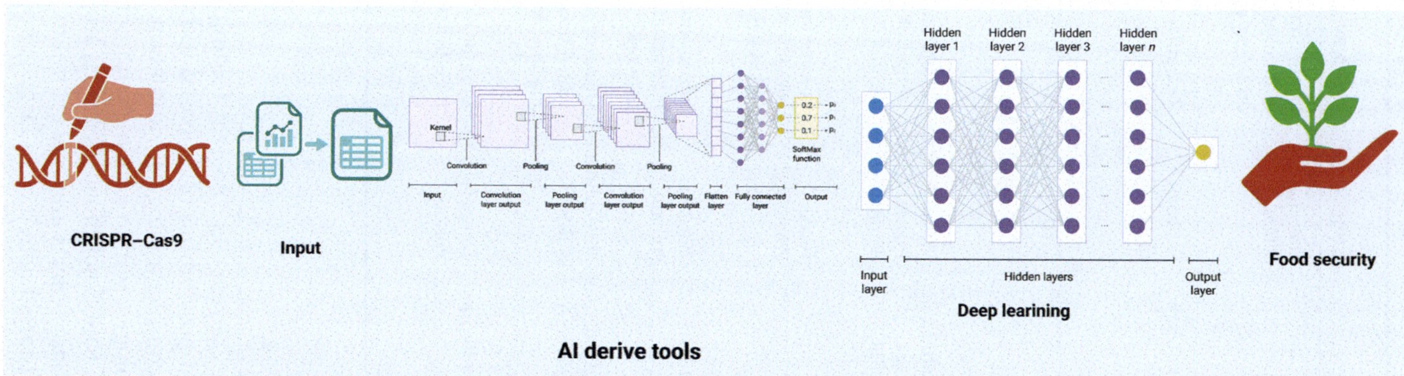

FIGURE 6.2 The architecture of deep learning using CRISPR/Cas9. An architecture of a deep learning model, under the artificial intelligence paradigm, illustrating the numerous steps that the predictor of on/off targets in CRISPR/Cas9 system takes. *Images created with BioRender.*

optimizes the performance and reliability of CRISPR. The automation and optimization of processes that also relied on trial-and-error before allow researchers to concentrate more on how to resolve complex genetic engineering problems with AI. In addition to data interpretation, AI-driven generative models are rapidly becoming able to generate new RNA, DNA, and amino acid sequences (Madani et al., 2023; Callaway, 2024), leading to new innovations in genome editing and synthetic biology that could not be achieved using the traditional experimental approach.

ML (machine learning) and CRISPRML, a subset of AI, improves the performance of computer systems by mimicking the way humans learn. By using algorithms and large datasets, machine learning models can automatically identify patterns and derive rules from input data, enabling them to make predictions or decisions even when encountering previously unseen information (Table 6.2).

TABLE 6.2 Databases related with studies on genome editing for the purpose of developing artificial intelligence models.

S. No.	Editing type	Machine learning model	Description	Reference
	Prime editing (PED)	DeePE	It comprises 54,856 pairs of pegRNAs and their corresponding target sequences	Kim et al. (2021)
	Base editing (BED)	XGBoost	Contains 1134 target sequences	Li et al. (2022a), (2022b)
	Base editing (BED)	GBT	The dataset including 10,638 sgRNA-target pairings was randomly partitioned into subsets	Pallaseni et al. (2022)
	Base editing (BED)	SVM, LR (L1 and L2 regularized), AdaBoost and random forest	Contains Cas9 variants	Kim et al. (2023)
	Base editing (BED)	Be-hive	The dataset including 10,638 sgRNA-target pairings was randomly partitioned into subsets	Arbab et al. (2020)
	CRISPRCas9 Genome editing	GBT	There are 201,934 off target sites which are dispersed throughout The human genome	Lazzarotto et al. (2020)
	CRISPRCas9 GED	Recurrent neural Network	It comprises of 50,000 gRNAs for each nuclease which collectively targets approximately 20,000 genes	Wang et al. (2019)
	CRISPRCas9 GED	Random forest	It comprises 33 sets of sgRNAs, which associated with its specified targets	Abadi et al. (2017)
	CRISPRCas9 GED	Convolutional neutral network	It comprises 400,000 sgRNA sequences taken from the genome CRISPR project	Rauscher et al. (2017)
	CRISPRCas9Genome editing	CRISTA	Nucleases from 2 human cell lines guided by RNA	Tsai et al. (2015)

Adapted from Dixit, S., Kumar, A., Srinivasan, K., Vincent, P.M.D.R., Ramu Krishnan, N., 2024. Advancing genome editing with artificial intelligence: opportunities, challenges, and future directions. Front. Bioeng. Biotechnol. 11. https://doi.org/10.3389/fbioe.2023.1335901

6.9 Generating Cas proteins and discovering novel proteins

Zhao et al. (2023) introduced an interaction, dynamics and conservation (IDC) strategy designed to minimize Cas protein size through the use of structural insights. This approach incorporates key features such as nucleoprotein interactions, conformational dynamics, and ortholog conservation to determine which regions of the protein are necessary for function. By applying AlphaFold2 to predict the structure of Cas13, the researchers generated five miniaturized variants using the IDC framework. These engineered proteins maintained functional enzymatic architecture while removing nonessential regions, effectively reducing protein size without compromising activity.

In another study (Yoon et al., 2024), a structure homology-based search workflow was developed that utilized a Foldseek-clustered AlphaFold database (Barrio-Hernandez et al., 2023; van Kempen et al., 2024). Using the two representative HEPN domains of Cas13 (HEPN1 and HEPN2) as query structures, they identified a novel ancestral clade of Cas13, termed Cas13an. Remarkably, Cas13an is only one-third the size of typical Cas13 proteins, yet it demonstrated strong RNA depletion capabilities and effective antiphage defense in *E. coli*, indicating that smaller Cas proteins can retain potent functional activity.

In addition to rational protein design, generative AI models have also been developed to create entirely new biological sequences. One such model, Evo (Nguyen et al., 2024), was designed to predict and generate sequences at scales ranging from individual molecules to whole genomes. Evo was trained on more than 80,000 prokaryotic and phage genome and metagenome datasets, enabling it to accurately predict functions across proteins, DNA, and RNA, as well as determine gene essentiality at single-nucleotide resolution. Impressively, Evo successfully generated innovative CRISPR complexes together with their corresponding transposable elements, gRNAs, and genome-scale coding sequences of over 650 kb.

In a related effort, Ruffolo et al. (2024) fine-tuned a language model similar to ProGen2 (Nijkamp et al., 2023) to generate large numbers of new proteins. Moreover, they leveraged the CRISPR–Cas Atlas—a dataset containing 1,246,163 CRISPR operons—to produce millions of novel CRISPR–Cas sequences. From this dataset, they selected 238,917 Cas9 sequences to train the model and generate Cas9-like variants. Out of the generated candidates, 209 were tested for activity in human cells, and the most promising candidate, OpenCRISPR-1, was identified. Notwithstanding its reduced sequence similarities to SpCas9, OpenCRISPR-1 exhibited superior activity. This research is the first successful demonstration of human genome editing via a wholly AI-designed CRISPR complex, highlighting the transformational potential of AI-driven methodologies for advanced gene-editing technologies.

6.10 Predicting genome editing efficiency with various gRNAs

The efficacy of gene editing using the CRISPR system can differ greatly depending on the chosen target site, and extensive research has been carried out to determine the factors that influence this variability. Numerous previous studies have demonstrated that various characteristics of the gRNA and its target site—including nucleotide composition, GC content, RNA secondary structure, and the epigenetic state of the DNA—are associated with editing performance (Doench et al., 2014, 2016; Szczelkun et al., 2014). However, these factors operate in complex and often interdependent ways, and purely experimental methods have not yet fully revealed the underlying rules. As a result, designing highly effective sgRNAs for specific genome editing applications continues to be a significant challenge. Therefore, to address this issue, ML-based strategies have emerged as powerful solutions, enabling the development of predictive models that leverage these features to improve the accuracy of efficiency prediction and facilitate the selection of high-performance sgRNAs (Chen et al., 2024).

Because different genome editing systems have distinct mechanisms and capabilities, their editing efficiencies can vary widely. In response, several specialized computational tools have been developed for specific systems such as BE and PE. Examples include FORECasT-BE (Pallaseni et al., 2022) for base editors and DeepPE (Kim et al., 2021) for prime editors. These models rely on intrinsic features of each editing system and incorporate efficiency data generated from high-throughput experimental assessments. Despite the availability of numerous models and tools designed to predict editing efficiency and guide RNA performance, their outcomes often differ substantially when tested on the same experimental dataset. These discrepancies are likely due to variations in feature selection, dataset composition, and the algorithmic methods employed (Haeussler et al., 2016).

6.11 Improving the specificity of editing

Various computing environments and programs are developed to make genome editing more specific, by anticipating the off-target effects. These tools are commonly divided into three main categories, which include: alignment based, hypothesis based and learning based. Allele-based systems such as Cas-OFFinder and CasOT work on the premise that genomic sequences that are highly like the gRNA are identified (Xiao et al., 2014; Bae et al., 2014). Therefore, hypothesis-driven tools such as CRISPOR and CHOPCHOP rely on empirical rules and known biological principles to estimate off-target risks (Labun et al., 2016; Concordet and Haeussler, 2018). Meanwhile, learning-based methods—including Elevation (Listgarten et al., 2018), DeepCRISPR (Chuai et al., 2018), and CrnnCrispr (Liu et al., 2021)—apply machine learning and deep learning approaches to build more complex predictive models. Compared to alignment- or hypothesis-based strategies, learning-based models are often more accurate because they integrate multiple features and uncover nonlinear relationships that may influence guide RNA specificity.

Learning-based approaches for gRNA specificity prediction generally follow a three-step process. First, all potential off-target sites across the entire genome are identified by scanning for sequences such as the gRNA. Second, machine learning models are constructed to assign scores to each potential off-target site using features such as sequence similarity, chromatin accessibility, and other genomic properties, since not every near-match will result in actual off-target cleavage. Third, these individual scores are combined into a single numerical specificity score using a predictive algorithm or model. As an illustration, prediction of off-target activities to design CRISPR gRNAs end-to-end (Listgarten et al., 2018) proposed a two-layer regression model, Elevation, which uses a seed-and-extension algorithm to identify close-matching Cas9 target motifs. The results are then aggregated with the use of Elevation-aggregate a gradient boosted regression tree model which combines individual scores to give an overall specificity predictor. Extensive experimental studies have also enabled the development of high-throughput methods to detect off-target cleavage sites, such as GUIDE-seq, CIRCLE-seq, and Discover-seq (Tsai et al., 2015; Wienert et al., 2019). Moreover, these methods generate large-scale datasets of validated off-target events, providing valuable training data for machine learning models. DeepCRISPR, a hybrid neural network model, takes advantage of genome-wide unlabeled epigenetic information and sgRNA sequences to pre-train its network. Subsequently, experimentally verified off-target data are used to fine-tune the model, and a CNN classifier is incorporated to predict off-target sites (Chuai et al., 2018). In contrast to the old ML models, the deep learning algorithms could scale the size of large amounts of data and form complex features without being engineered by hand, leading to improved predictive performance in terms of accuracy and reliability.

6.12 AI in CRISPR–Cas-specific gRNA design

Gene editing software largely also hinges on the right choice of gRNA sequence to use, since editing specificity and efficiency is directly proportional to the design of the gRNA used. There are gRNAs, which are capable of triggering almost all the target alleles in a particular cell group and others that are characterized by low or no activity at all (Lee et al., 2018). To overcome such variability, extensive collections of tools to design gRNA are already available, many of which rely on ML and DL algorithms. These computational models were designed to rival gRNA efficiency by combining complicated biological and sequence-based parameters.

In earlier experiments, it was stressed that the predictive methodology is influential in the creation of predictive gRNA efficiency and off-targeted and on-targeted binding activities (Doench et al., 2016). Although older models such as Linear Regression (LR) have demonstrated a moderate predictive performance (Moreno-Mateos et al., 2015), newer models have proven to be better, i.e., Gradient Boosted Trees (GBTs) (Doench et al., 2016) and Support Vector Machines (SVMs) (Chari et al., 2015; Wong et al., 2015) that are capable of uncovering complex and nonlinear associations among genomic features (McKinney et al., 2006)

To increase the predictive accuracy, several ML/DL-based gRNA prediction tools have been presented. The model created by (Wessels et al., 2024) is called TIGER (Targeted Inhibition of Gene Expression via gRNA design): a convolutional neural network (CNN) modeling that predicts the gRNA efficacy not only due to sequence composition but also due to external information in the context of the genome. In this study, they have created and tested close to 200,000 RfxCas13d gRNA containing systematic insertions, and deletions (indels), designed mismatches, at key genes in human cells. Findings showed that the integrated model of TIGER worked successfully at distinguishing between essential and nonessential genes and was more successful than the available AI-based prediction systems in predicting on-target and off-target activities.

Right selection of gRNA sequence that should be used is one of the key factors in the use of genome editing software, as the specificity and efficiency of a given editing relies directly on the design of the gRNA being used. One of the sets of RNA is gRNAs with the potential to activate nearly all the target alleles within a specific cell population and others with low or zero activity (Lee et al., 2018). To eliminate such variability, large sets of tools to design gRNA are already present, most of them based on ML, and DL algorithms. These computational models were made to compete with gRNA efficiency by integrating complex biological and sequence-based parameters.

In the previous experiments, it was emphasized that the mechanism of the predictive methodology is assertive in the establishment of the predictive gRNA activity and off-targeted as well as on-targeted binding functions (Doench et al., 2016). Even though the more ancient models like Linear Regression (LR) have already shown an intermediate level of predictive capabilities (Moreno-Mateos et al., 2015), more recent models have shown to be superior i.e., Gradient Boosted Trees (GBTs) (Doench et al., 2016) and Support Vector Machines (SVMs) (Chari et al., 2015; Wong et al., 2015), which can identify complex and nonlinear relationships between genomic features (McKinney et al., 2006)

To enhance the predictor accuracy, several tools of ML/DL-based gRNA prediction have been introduced. The model developed by (Wessels et al., 2024) is referred to as TIGER: a convolutional neural network (CNN) modeling that anticipates the gRNA efficacy not just in the arrangements of the sequence, but likewise through external data in the genomic setting. They have also designed and validated in this paper approximately 200,000 RfxCas13d guide RNA with systematic designed indels, deletions, and insertions at essential genes in human cells. However, the results indicated that the integrated model of TIGER was successful in separating between essential and non-essential genes and was more effective than the existing AI-based prediction systems used in predicting activities on-target and off-target.

A recent important advancement is SkipGuide (Louie et al., 2021), which is destined to be used in exon omissions as a genetic disease therapy. This works by taking advantage of CRISPR–Cas9 editing to both target splice donor and acceptor sites, which then allows the efficient selection of exon skipping to inhibit this exon expression of functional protein expression. SkipGuide was found to have high predictive reliability, as its predictive accuracy was observed in experiments which have used over 1000 gRNAs across 791 mouse cell splice sequences. With its application, efficient gRNAs to exon-skipping therapy are quickly identified and time and resources are saved.

All in all, the accurate choice of gRNA sequences is one of the foundations of a successful genome editing. Some of the most complex computational models consisting of SVMs and GBTs and CNN-based models such as TIGER, CRISPRon, and piCRISPR have proven to be remarkably predictive of on-target and off-target activities. All these innovations form an important step forward in utilizing the full power of CRISPR/Cas technologies to realize focused, effective, and safe genome editing and have far-reaching consequences in relation to the field of biomedical research as well as therapeutic development.

6.13 Utilization of artificial intelligence and base editing

BE is a very useful gene editing method, which allows precise modification of a single nucleotide in the genome to be done with high efficiency without creating any DSB (Rees and Liu, 2018)b). Nonetheless, the off-targets can still occur even with its accuracy. Besides the off-target effects that include the use of guide RNA and Cas proteins, the enzymes themselves could also cause unintended edits without depending on the sgRNA (Sretenovic et al., 2023; Wu et al., 2022). It is also found that in rice, CBEs have the capability of inducing extensive off-target mutation throughout the genome (Jin et al., 2019). This is mainly because the deaminases involved like those in CBEs like rAPOBEC1 could bind to ssDNA without the requirement of Cas9 and sgRNA and thus cause off-target single nucleotide variants (SNVs). Machine learning methods can help overcome these shortcomings with a quick approach to finding deaminases with mitigated off-target activities. In the article discovery of deaminase functions by structure-based protein clustering Jiaying Huang (Huang et al., 2023), all proteins with deaminase domains were selected and AlphaFold2 was applied to predict the structure of such proteins only with the help of amino acid sequences. A similarity of structure was then used to formulate these proteins into diverse clusters that possessed a given functional feature. After experimental validation, SCP1.201 clade was discovered to possess several new proteins with a ssDNA deaminating capacity which means that they can also be utilized in the process of base editing. Notably, other deaminases that were identified exhibited reduced off-target action than rAPOBEC1. The latter findings point to the rapidity of protein functions and properties discovery through machine learning, which, in the end, will aid the creation of deaminases that are more optimized in base editing. This plan opens the door to the development of more efficient and precise genome editing technique, which will further reduce the footprint and use of gene editing technique to improve the crop improvement programs (Chen et al., 2024). However (Evanoff and Komor, 2019), offer a detailed description of base editors in their study and focus on the modularity of this type of system and the abundance of customizable elements that can be used in every part of system. A range of ML and DL models have since been developed to promote the effectiveness of base editors, although the main aim was to predict and demonstrate efficient editing outcomes. BE-Hive was introduced (Arbab et al., 2020), created on a deep conditional autoregressive architecture that predicts the editing patterns and BE efficiency, whereas this model demonstrated great predictability of the genotypic outcomes and editing success in base editing. BE-Hive was subsequently used to calculate plans to fix many types of SNVs associated with disease, with an accuracy of 90 or higher, even between bystander bases. Subsequently (Pallaseni et al., 2022) came up with another ML model that helps to predict the base editing efficiency and the pattern of the outcome according to the sequence features and location information to approximate the editing profiles of the various kinds of base editors. The model was found to generalize well in a range of datasets with precision scores ranging between 0.49 and 0.72 based on the editor. A highlight of this study was the significant impact of sequence context on the process of base editing, especially with the nucleotides surrounding the target base, and this observation makes it rather difficult to predict how effectively a given site will be base edited or how many bystander edits will be formed. To deal with this problem (Song et al., 2020), trained a deep learning model that fused sequence-based features, as well as the positional nature of target nucleotides in the genome. Higher performance was realized by this model than current methods in the prediction of the efficiency of editing ABEs and CBEs with Pearson correlation having values of between 0.50 and 0.95. base editing outcome prediction has now become a significant concern in this direction, and most widely used is deep learning. BE-DICT, which was created by (Marquart et al., 2021), is another model implemented as a deep learning that forecasted the outcome of adenine- and cytosine-based editing with the help of an attention-based algorithm. BE-DICT uses a protospacer sequence as input and determines what is the probability a nucleotide at each position will be edited. The model was trained using several datasets and provided various AUC results of different editors: CBE4max: 0.94, ABEmax: 0.86, ABE8e: 0.66, and TargetAID: 0.97. Subsequently (Park and Kim, 2023), also proposed and made available two DL models, DeepABE and DeepCBE, which can be used as web-based tools to forecast the efficacy and results of ABEs and CBEs, respectively. Another study (Kim et al., 2024), tested nine variations of the Cas9 engineered to target various sequences of Protospacer Adjacent Motifs (PAM). They built a DL model named DeepCas9variants that was able to predict the most appropriate variant of Cas9 to target a particular genomic site according to desirable editing patterns. As a continuation, a computational model was also developed called DeepBE to validate the editing efficiency and productivity of 63 different BEs with Pearson and Spearman correlation values reported as 0.82 down to 0.95 and 0.80 down to 0.94, respectively (Li et al., 2022b), developed CAELM which is an ML model that uses chromatin accessibility and sequence context to predict the efficiency of CBEs. CAELM (Chromatin Accessibility Enabled Learning Model) first and second showed successful prediction of in situ base editing results. They had previously used a convolutional neural network (Feng et al., 2019) to predict GBE base editing efficiency, but in the current study chose the XGB Regressor because the dataset size containing editing data only contained 1134 target patterns, and XGB Regressor is frequently more effective than deep neural networks on each of those (Fernoaga et al., 2020). Pearson correlation was used to test the accuracy of the model, which gave a value of 0.64 between predicted and measured results. Besides this (Chen et al., 2022), created CGBEs (C•G• to G•C base editors) that have varied editing profiles. They tested these in mammalian cells in 10, 638, genomically integrated, target sites and discovered 10 promising CGBEs. They have developed ML models based on this dataset (Koblan et al., 2021) and were able to predict the quality and the yield of the obtained editing results ($R = 0.90$). These CGBEs made more than 90% accuracy in repairing and up to 70% effectiveness in correcting 546 transversion single-nucleotide mutations related to diseases of WT coding sequences.

Structural prediction tools such as diffusion models or AlphaFold2 which are powered by AI can also be used in the production of better versions of base editors. The article by (Huang et al., 2023) aimed to design tailored cytosine base editors by utilizing AlphaFold2 to act in a specific way. AlphaFold2 is an impressive predictor that can be applied to determine protein structures incredibly well; it has, however, some limitations when predicting proteins that bear a very similar sequence. In cases where the similarities are extremely high, AlphaFold2 will have difficulties distinguishing them, and will have problems to accurately detect fine structural or functional differences caused by Single Nucleotide Polymorphisms (SNPS). Altogether, the contribution of AI to the base editing technologies is a transformative one. The release of prediction models of AI have enabled the construction of better and more efficient base editors, which has generated huge progress in genome editing and even possible therapies. In addition, the influence of AI is not limited to sequence-based predictions but also structural innovation is incorporated. The adoption of AI, represented using such instruments as AlphaFold2, allows exploring and designing new base editors with new properties, which is a brilliant example of how AI can change the future of genetic medicine and genome editing.

6.14 Applications of AI and prime editing

PE is a more recent approach to genome editing, which includes the introduction of, programmed sequence alterations into DNA with the use of reverse transcription (Yan et al., 2020). The method is very flexible and can produce a wide range of genetic changes including base changes, insertions, and deletions. Nevertheless, to attain good editing specificity and purity of the product, the prime editing guide RNA (pegRNA) should be optimally developed by conducting experimental trials (Mathis et al., 2023). Prime editing is based on three major components such as RT, a pegRNA, and a nickage of Cas9. Moreover, the pegRNA contains the sequence of recognition of the target as well as the desired edit, which co-purify the specific modification in the DNA. The formulation of the functional pegRNAs is far harder and more intricate in comparison to the design of guides designed to be used in the conventional methods of CRISPR-based genome editing. Fortunately, there are computational software such as Easy-Prime (Li et al., 2021) and PrimeDesign (Hsu et al., 2021) that can facilitate and enable such a strenuous design process (Li et al., 2021). Easy-Prime, which was trained on an already created prime editing dataset (Li et al., 2021). This tool measures and ranks pegRNA candidates using already known and recently discovered sequence properties, and prediction-based RNA folding and secondary structure. With the help of their work, prime editing guides were successfully optimized to correct mutation in 89.5% of the 152,351 variants they discovered in Genome-Wide Association Studies (GWAS). Besides this, Easy-Prime can also produce optimal pegRNAs to a broad spectrum of genetic variations related to diseases. The other useful design tool is PrimeDesign that was created by (Hsu et al., 2021) and aids in the construction of pegRNA to perform various editing operations such as single nucleotide replacements and insertions, deletions, and inversions. To solve the problem of harmful genetic alleles, they created potential pegRNAs and ngRNAs that were based on pathogenic human variants in ClinVar8 ($n = 69,481$). They determined that 91.7% of these variants were all efficiently targeted with a single pegRNA spacer with a maximum of 34 nucleotides. Then they experimentally tested the pegRNAs and ngRNAs that were produced by PrimeDesign in various editing situations. They discovered, however, that not every one of the designs proposed resulted in the desired modifications at high frequencies. Thus, the users of PrimeDesign can still be required to refine and readjust their pegRNA choices to what they reviewed as their first-pass predictions.

In order to enhance the precision of predicting the efficiency and outcomes of prime editing, several ML and DL methods have been created(Koeppel et al., 2023). tested the variables which are the most important to determine the efficiency of prime editing insertions.

Through the observations, they were able to come up with an ML model which could predict efficiencies of insertion basing on the following characteristics; the structure and length of the nucleic acid, the 2 degrees structure of the insertion sequence, and the levels of expression of TREX1 and TREX2. However, the exonucleases have an important function in breaking down the 3′ flap of DNA which is a crucial intermediation in prime editing inserts. There was also another ML model that was created to estimate the likelihood to achieve insertion using prime editing by varying the length and composition of the insert sequence, the flanking DNA, and the presence of DNA repair proteins using sequence-based inputs. This model was developed and tested on many sequences, locus in genomes and human cell lines, and it was found that the success of insertions highly relies on sequence length, composition, and structural attributes.

Mathis et al. (2023), created a deep learning system PREDICT that is aimed at predicting the outcomes and efficiencies of prime editing events. PREDICT is trained with a recurrent neural network (RNN) of over 90,000 experiments, which have been performed so far on prime editing. The model recorded a high predictive focus, with the Spearman correlation coefficients of 0.85 and 0.78 of planned and unintended edits, respectively. This shows that the PREDICT can predict editing rates of a broad spectrum of small genomic changes. PE thus is an efficient and versatile gene editing platform that can correctly carry out base substitutions, insertions as well as deletions with a high potential of healing human disease-causing mutations. Its general effectiveness is determined by various factors such as the location of target, the design of the edit sequences and the repair pathway of DNA mismatch. Recent studies are paying more attention to making AI-based models more accurate in terms of prediction by adding new biological factors; one example of such a breakthrough was made related to TREX1 and TREX2 (Dixit et al., 2024).

6.15 Conclusion

CRISPR system is an innovative breakthrough in the world of gene editing that has turned everything upside down in terms of genome editing. It is an originally prokaryotic natural adaptive immunity system that allows bacteria and archaea to protect themselves against virus attackers. Moreover, he potential of this molecular tool was anticipated by the work of one of the pioneers of this field, Emmanuelle Charpentier and Jennifer Doudna, who became the 2020 Nobel Prize awardees in Chemistry due to their contributions to this field. Through the evolution of this natural defense system, scientists have created an extremely efficient and programmable gene-editing platform that can be used to create targeted modifications in the genome of eukaryotes.

CRISPR has developed to play a vital role in developing abiotic and biotic resistance traits of plants among various applications due to their serious threat to food security and agricultural yield in the world. Also, the emergence of viral infections in crops may result in significant losses of the crop and cause economic harm thus strategies based on CRISPR can be promising as a solution to the development of virus-resistant crops that will provide the agricultural system with resilience. Although having an impressive range of uses, CRISPR–Cas has some drawbacks. One of the biggest problems with CRISPR is that the

Cas protein can accidentally attach to and cut the wrong DNA paths if they look similar to the intended target. With such off-target effects, they might cause undesired mutations and even interfere with vital genes, and this is an ethical and safety concern (Naeem et al., 2020). To overcome these challenges, scientists have put their efforts on making the editing more specific and precise with the use of CRISPR. Major bottlenecks have been satisfied by the advancement of high fidelity Cas9 variants that have a better ability to discriminate between targets and sophisticated gRNA design algorithms that could predict and minimize potential off-target sites (Zafar et al., 2020).

AI combined with CRISPR technology has enhanced the development of genome engineering by enhancing the execution as well as the specificity. The AI will work best in the large-scale analysis of experimental data to detect the complex patterns and predictive characteristics connected with gene-editing efficiency. Through computational modeling, AI systems can quickly carry out complex analyses, such as the discovery of CRISPR-like systems among large genomic datasets using amino acid sequences or structural homology of proteins. In addition, AI-based protein design systems are being used to design new proteins that perform new genome engineering capabilities not existing in nature. AI predictive-optimization can predict studies in a much less laborious way, minimize the likelihood of failure, and provide researchers with the most effective and accurate methods to edit genomes, to reduce experimental workloads and predict outcomes.

Declaration

"The authors declare that no generative artificial intelligence tools were used in the writing or preparation of this chapter. The content reflects the authors' own scholarly work, and the authors assume full responsibility for all statements, interpretations, and conclusions presented".

References

Abadi, S., Yan, W.X., Amar, D., Mayrose, I., Han, X., 2017. A machine learning approach for predicting CRISPR-Cas9 cleavage efficiencies and patterns underlying its mechanism of action. PLoS Comput. Biol. 13 (10), e1005807. https://doi.org/10.1371/journal.pcbi.1005807.

Abramson, J., Adler, J., Dunger, J., Evans, R., Green, T., Pritzel, A., Ronneberger, O., Willmore, L., Ballard, A.J., Bambrick, J., Bodenstein, S.W., Evans, D.A., Hung, C.C., O'Neill, M., Reiman, D., Tunyasuvunakool, K., Wu, Z., Žemgulytė, A., Arvaniti, E., Beattie, C., Bertolli, O., Bridgland, A., Cherepanov, A., Congreve, M., Cowen-Rivers, A.I., Cowie, A., Figurnov, M., Fuchs, F.B., Gladman, H., Jain, R., Khan, Y.A., Low, C.M.R., Perlin, K., Potapenko, A., Savy, P., Singh, S., Stecula, A., Thillaisundaram, A., Tong, C., Yakneen, S., Zhong, E.D., Zielinski, M., Žídek, A., Bapst, V., Kohli, P., Jaderberg, M., Hassabis, D., Jumper, J.M., 2024. Accurate structure prediction of biomolecular interactions with AlphaFold 3. Nature 630 (8016), 493–500. https://doi.org/10.1038/s41586-024-07487-w.

Ali, Z., Abulfaraj, A., Idris, A., Ali, S., Tashkandi, M., Mahfouz, M.M., 2015. CRISPR/Cas9-mediated viral interference in plants. Genome Biol. 16 (1). https://doi.org/10.1186/s13059-015-0799-6.

Alotaibi, M., 2023. Climate change, its impact on crop production, challenges, and possible solutions. Not. Bot. Horti Agrobot. Cluj-Napoca 51 (1), 13020. https://doi.org/10.15835/nbha51113020.

Anzalone, A.V., Randolph, P.B., Davis, J.R., Sousa, A.A., Koblan, L.W., Levy, J.M., Chen, P.J., Wilson, C., Newby, G. A., Raguram, A., Liu, D.R., 2019. Search-and-replace genome editing without double-strand breaks or donor DNA. Nature 576 (7785), 149–157. https://doi.org/10.1038/s41586-019-1711-4.

Anzalone, A.V., Koblan, L.W., Liu, D.R., 2020. Genome editing with CRISPR–Cas nucleases, base editors, transposases and prime editors. Nat. Biotechnol. 38, 824–844. https://doi.org/10.1038/s41587-020-0561-9.

Arbab, M., Shen, M.W., Mok, B., Wilson, C., Matuszek, Ż., Cassa, C.A., Liu, D.R., 2020. Determinants of base editing outcomes from target library analysis and machine learning. Cell 182 (2), 463. https://doi.org/10.1016/j.cell.2020.05.037.

Atanasov, A.G., Zotchev, S.B., Dirsch, V.M., et al., 2021. Natural products in drug discovery: advances and opportunities. Nat. Rev. Drug Discov. 20 (3), 200–216. https://doi.org/10.1038/s41573-020-00114-z.

Azodi, C.B., Bolger, E., McCarren, A., Roantree, M., de Los Campos, G., Shiu, S.H., 2019. Benchmarking parametric and machine learning models for genomic prediction of complex traits. G3 (Bethesda, Md.) 9 (11), 3691–3702. https://doi.org/10.1534/g3.119.400498.

Bae, S., Park, J., Kim, J.-S., 2014. Cas-OFFinder: a fast and versatile algorithm that searches for potential off-target sites of Cas9 RNA-guided endonucleases. Bioinformatics 30, 1473–1475. https://doi.org/10.1093/bioinformatics/btu048.

Baek, M., Baker, D., 2022. Deep learning and protein structure modeling. Nat. Methods 19 (1), 13–14. https://doi.org/10.1038/s41592-021-01360-8.

Baek, M., DiMaio, F., Anishchenko, I., Dauparas, J., Ovchinnikov, S., Lee, G.R., Wang, J., Cong, Q., Kinch, L.N., Dustin Schaeffer, R., Millán, C., Park, H., Adams, C., Glassman, C.R., DeGiovanni, A., Pereira, J.H., Rodrigues, A.V., Van Dijk, A.A., Ebrecht, A.C., Opperman, D.J., Sagmeister, T., Buhlheller, C., Pavkov-Keller, T., Rathinaswamy, M.K., Dalwadi, U., Yip, C.K., Burke, J.E., Christopher Garcia, K., Grishin, N.V., Adams, P.D., Read, R.J., Baker, D., 2021. Accurate prediction of protein structures and interactions using a three-track neural network. Science 373 (6557), 871–876. https://doi.org/10.1126/science.abj8754.

Baek, M., McHugh, R., Anishchenko, I., Jiang, H., Baker, D., DiMaio, F., 2024. Accurate prediction of protein–nucleic acid complexes using RoseTTAFoldNA. Nat. Methods 21 (1), 117–121. https://doi.org/10.1038/s41592-023-02086-5.

Bai, X., Correa, V.R., Toruño, T.Y., Ammar, E.D., Kamoun, S., Hogenhout, S.A., 2009. AY-WB phytoplasma secretes a protein that targets plant cell nuclei. Mol. Plant Microbe Interact. 22 (1), 18–30. https://doi.org/10.1094/MPMI-22-1-0018.

Barreto, J.S., Armenteras, D., 2020. Open data and machine learning to model the occurrence of fire in the ecoregion of "Llanos Colombo–Venezolanos". Remote Sens. 12 (23), 3921. https://doi.org/10.3390/rs12233921.

Barrio-Hernandez, I., Yeo, J., Jänes, J., Mirdita, M., Gilchrist, C.L.M., Wein, T., Varadi, M., Velankar, S., Beltrao, P., Steinegger, M., 2023. Clustering predicted structures at the scale of the known protein universe. Nature 622 (7983), 637–645. https://doi.org/10.1038/s41586-023-06510-w.

Barto, A.G., 2019. Reinforcement learning: connections, surprises, challenges. AI Mag. 40 (1), 3–15. https://doi.org/10.1609/aimag.v40i1.2844.

Bhat, M.A., Mir, R.A., Kumar, V., Shah, A.A., Zargar, S.M., Rahman, S., Jan, A.T., 2021. Mechanistic insights of CRISPR/Cas-mediated genome editing towards enhancing abiotic stress tolerance in plants. Physiol. Plantarum 172 (2), 1255–1268. https://doi.org/10.1111/ppl.13359.

Bhatt, R., Kunal, Moulick, D., Bárek, V., Brestic, M., Ahmed, G., Skalicky, M., Hossain, A., 2025. Sustainable strategies to limit nitrogen loss in agriculture through improving its use efficiency—aiming to reduce environmental pollution. J. Agric. Food Res. 22, 101957. https://doi.org/10.1016/j.jafr.2025.101957.

Bisht, D.S., Bhatia, V., Bhattacharya, R., 2019. Improving plant-resistance to insect-pests and pathogens: the new opportunities through targeted genome editing. Seminars in Cell & Developmental Biology, CRISPR/Cas9 96, 65–76. https://doi.org/10.1016/j.semcdb.2019.04.008.

Bortesi, L., Fischer, R., 2015. The CRISPR/Cas9 system for plant genome editing and beyond. Biotechnol. Adv. 33 (1), 41–52. https://doi.org/10.1016/j.biotechadv.2014.12.006.

Bryant, P., Pozzati, G., Elofsson, A., 2022. Improved prediction of protein-protein interactions using AlphaFold2. Nat. Commun. 13, 1265. https://doi.org/10.1038/s41467-022-28865-w.

Callaway, E., 2024. ChatGPT for CRISPR' creates new gene-editing tools. Nature 629 (8011), 272. https://doi.org/10.1038/d41586-024-01243-w.

Chari, R., Mali, P., Moosburner, M., Church, G.M., 2015. Unraveling CRISPR-Cas9 genome engineering parameters via a library-on-library approach. Nat. Methods 12 (9), 823–826. https://doi.org/10.1038/nmeth.3473.

Chen, S., Liu, Z., Lai, L., Li, Z., 2022. Efficient C-to-G base editing with improved target compatibility using engineered deaminase–nCas9 fusions. CRISPR J. 5 (3), 389–396. https://doi.org/10.1089/crispr.2021.0124.

Chen, L., Liu, G., Zhang, T., 2024. Integrating machine learning and genome editing for crop improvement. aBIOTECH 5, 262–277. https://doi.org/10.1007/s42994-023-00133-5.

Christopoulou, M., Wo, S.R.C., Kozik, A., McHale, L.K., Truco, M.J., Wroblewski, T., Michelmore, R.W., 2015. Genome-wide architecture of disease resistance genes in lettuce. G3: Genes, Genomes, Genet. 5 (12), 2655–2669. https://doi.org/10.1534/g3.115.020818.

Chuai, G., Ma, H., Yan, J., et al., 2018. *DeepCRISPR*: optimized CRISPR guide RNA design by deep learning. Genome Biol. 19, 80. https://doi.org/10.1186/s13059-018-1459-4.

Concordet, J.P., Haeussler, M., 2018. CRISPOR: intuitive guide selection for CRISPR/Cas9 genome editing experiments and screens. Nucleic Acids Res. 46 (1), W242. https://doi.org/10.1093/nar/gky354.

Consonni, C., Humphry, M.E., Hartmann, H.A., Livaja, M., Durner, J., Westphal, L., Vogel, J., Lipka, V., Kemmerling, B., Schulze-Lefert, P., Somerville, S.C., Panstruga, R., 2006. Conserved requirement for a plant host cell protein in powdery mildew pathogenesis. Nat. Genet. 38 (6), 716–720. https://doi.org/10.1038/ng1806.

Dixit, S., Kumar, A., Srinivasan, K., Vincent, P.M.D.R., Ramu Krishnan, N., 2024. Advancing genome editing with artificial intelligence: opportunities, challenges, and future directions. Front. Bioeng. Biotechnol. 11. https://doi.org/10.3389/fbioe.2023.1335901.

Doench, J.G., Hartenian, E., Graham, D.B., Tothova, Z., Hegde, M., Smith, I., Sullender, M., Ebert, B.L., Xavier, R.J., Root, D.E., 2014. Rational design of highly active sgRNAs for CRISPR-Cas9-mediated gene inactivation. Nat. Biotechnol. 32 (12), 1262–1267. https://doi.org/10.1038/nbt.3026.

Doench, J.G., Fusi, N., Sullender, M., Hegde, M., Vaimberg, E.W., Donovan, K.F., Smith, I., Tothova, Z., Wilen, C., Orchard, R., Virgin, H.W., Listgarten, J., Root, D.E., 2016. Optimized sgRNA design to maximize activity and minimize off-target effects of CRISPR-Cas9. Nat. Biotechnol. 34 (2), 184–191. https://doi.org/10.1038/nbt.3437.

Donovan, S., Mao, Y., Orr, D.J., Carmo-Silva, E., McCormick, A.J., 2020. CRISPR-Cas9-Mediated mutagenesis of the rubisco small subunit family in Nicotiana tabacum. Frontiers in Genome Editing 2. https://doi.org/10.3389/fgeed.2020.605614.

Duan, Y.B., Li, J., Qin, R.Y., Xu, R.F., Li, H., Yang, Y.C., Ma, H., Li, L., Wei, P.C., Yang, J.B., 2016. Identification of a regulatory element responsible for salt induction of rice OsRAV2 through ex situ and in situ promoter analysis. Plant Mol. Biol. 90 (1–2), 49–62. https://doi.org/10.1007/s11103-015-0393-z.

Eckhardt, C.M., Madjarova, S.J., Williams, R.J., Ollivier, M., Karlsson, J., Pareek, A., Nwachukwu, B.U., 2023. Unsupervised machine learning methods and emerging applications in healthcare. Knee Surg. Sports Traumatol. Arthrosc. 31 (2), 376–381. https://doi.org/10.1007/s00167-022-07233-7.

Evanoff, M., Komor, A.C., 2019. Base editors: modular tools for the introduction of point mutations in living cells. Emerg. Top. Life Sci. 3 (5), 483–491. https://doi.org/10.1042/ETLS20190088.

Ezura, H., 2022. Letter to the editor: the world's first CRISPR tomato launched to a Japanese market: the social-economic impact of its implementation on crop genome editing. Plant Cell Physiol. 63 (6), 731–733. https://doi.org/10.1093/pcp/pcac048.

Farhat, S., Jain, N., Singh, N., Sreevathsa, R., Dash, P.K., Rai, R., Yadav, S., Kumar, P., Sarkar, A.K., Jain, A., Singh, N.K., Rai, V., 2019. CRISPR-Cas9 directed genome engineering for enhancing salt stress tolerance in rice. Semin. Cell Dev. Biol. 96, 91–99. https://doi.org/10.1016/j.semcdb.2019.05.003.

Feng, S., Zhou, H., Dong, H., 2019. Using deep neural network with small dataset to predict material defects. Mater. Des. 162, 300–310. https://doi.org/10.1016/j.matdes.2018.11.060.

Fernoaga, V., Sandu, V., Balan, T., 2020. Artificial intelligence for the prediction of exhaust back pressure effect on the performance of diesel engines. Appl. Sci. 10, 7370. https://doi.org/10.3390/app10207370.

Gan, W.C., Ling, A.P.K., 2022. CRISPR/Cas9 in plant biotechnology: applications and challenges. Biotechnologia 103 (1), 81–93. https://doi.org/10.5114/bta.2022.113919.

Gaudelli, N.M., Komor, A.C., Rees, H.A., Packer, M.S., Badran, A.H., Bryson, D.I., Liu, D.R., 2017. Programmable base editing of T to G C in genomic DNA without DNA cleavage. Nature 551 (7681), 464–471. https://doi.org/10.1038/nature24644.

Gull, A., Lone, A.A., Wani, N.U.I., Gull, A., Lone, A.A., Wani, N.U.I., 2019. Biotic and abiotic stresses in plants. In: Abiotic and Biotic Stress in Plants. IntechOpen. https://doi.org/10.5772/intechopen.85832.

Haeussler, M., Schönig, K., Eckert, H., Eschstruth, A., Mianné, J., Renaud, J.B., Schneider-Maunoury, S., Shkumatava, A., Teboul, L., Kent, J., Joly, J.S., Concordet, J.P., 2016. Evaluation of off-target and on-target scoring algorithms and integration into the guide RNA selection tool CRISPOR. Genome Biol. 17 (1). https://doi.org/10.1186/s13059-016-1012-2.

Haque, E., Taniguchi, H., Hassan, M.M., Bhowmik, P., Karim, M.R., Śmiech, M., Zhao, K., Rahman, M., Islam, T., 2018. Application of CRISPR/Cas9 genome editing technology for the improvement of crops cultivated in tropical climates: recent progress, prospects, and challenges. Front. Plant Sci. 9. https://doi.org/10.3389/fpls.2018.00617.

Hastie, T., Friedman, J., Tibshirani, R., 2001. The Elements of Statistical Learning. Springer Series in Statistics. Springer, New York, NY. https://doi.org/10.1007/978-0-387-21606-5.

Hosfield, D.J., Mol, C.D., Shen, B., Tainer, J.A., 1998. Structure of the DNA repair and replication endonuclease and exonuclease FEN-1: coupling DNA and PCNA binding to FEN-1 activity. Cell 95 (1), 135–146. https://doi.org/10.1016/S0092-8674(00)81789-4.

Hsu, J.Y., Grünewald, J., Szalay, R., Shih, J., Anzalone, A.V., Lam, K.C., Shen, M.W., Petri, K., Liu, D.R., Keith Joung, J., Pinello, L., 2021. PrimeDesign software for rapid and simplified design of prime editing guide RNAs. Nat. Commun. 12 (1). https://doi.org/10.1038/s41467-021-21337-7.

Hua, K., Tao, X., Zhu, J.-K., 2019. Expanding the base editing scope in rice by using Cas9 variants. Plant Biotechnol. J. 17, 499–504. https://doi.org/10.1111/pbi.12993.

Huang, J., Lin, Q., Fei, H., He, Z., Xu, H., Li, Y., Qu, K., Han, P., Gao, Q., Li, B., Liu, G., Zhang, L., Hu, J., Zhang, R., Zuo, E., Luo, Y., Ran, Y., Qiu, J.-L., Zhao, K.T., Gao, C., 2023. Discovery of deaminase functions by structure-based protein clustering. Cell 186, 3182–3195.e14. https://doi.org/10.1016/j.cell.2023.05.041.

Hurrah, I.M., Mohiuddin, T., Mandal, S., Ghorai, M., Bhattacharya, S., Nongdam, P., Al-Tawaha, A.R., Bursal, E., Shekhawat, M.S., Pandey, D.K., Dey, A., 2024. Strategies and Applications of Genomic Editing in Plants with CRISPR/Cas9 CRISPR and Plant Functional Genomics. CRC Press, India, pp. 59–80. https://doi.org/10.1201/9781003387060-4.

Hussain, M.A., Li, L., Kalu, A., Wu, X., Naumovski, N., 2025. Sustainable food security and nutritional challenges. Sustainability 17 (3), 874. https://doi.org/10.3390/su17030874.

Janiesch, C., Zschech, P., Heinrich, K., 2021. Machine learning and deep learning. Electron. Mark. 31, 685–695. https://doi.org/10.1007/s12525-021-00475-2.

Jash, S., Das, S., Hurrah, I.M., Kumari, A., Mandal, S., 2026. CRISPR-edited plants for resistance to viral diseases: methods and applications. CRISPR Technology for Combating Plant Pests and Pathogens 129–160. https://doi.org/10.1016/B978-0-443-43920-9.00007-5.

Jiang, F., Doudna, J.A., 2017. CRISPR-Cas9 structures and mechanisms. Annu. Rev. Biophys. 46, 505–529. https://doi.org/10.1146/annurev-biophys-062215-010822.

Jiang, F., Lu, Y., Chen, Y., Cai, Di, Li, G., 2020. Image recognition of four rice leaf diseases based on deep learning and support vector machine. Comput. Electron. Agric. 179, 105824. https://doi.org/10.1016/j.compag.2020.105824.

Jin, S., Zong, Y., Gao, Q., Zhu, Z., Wang, Y., Qin, P., Liang, C., Wang, D., Qiu, J.L., Zhang, F., Gao, C., 2019. Cytosine, but not adenine, base editors induce genome-wide off-target mutations in rice. Science 364 (6437), 292–295. https://doi.org/10.1126/science.aaw7166.

Joshi, R.K., Bharat, S.S., Mishra, R., 2020. Engineering drought tolerance in plants through CRISPR/Cas genome editing. 3 Biotech 10, 400. https://doi.org/10.1007/s13205-020-02390-3.

Jumper, J., Evans, R., Pritzel, A., Green, T., Figurnov, M., Ronneberger, O., Tunyasuvunakool, K., Bates, R., Žídek, A., Potapenko, A., Bridgland, A., Meyer, C., Kohl, S.A.A., Ballard, A.J., Cowie, A., Romera-Paredes, B., Nikolov, S., Jain, R., Adler, J., Back, T., Petersen, S., Reiman, D., Clancy, E., Zielinski, M., Steinegger, M., Pacholska, M., Berghammer, T., Bodenstein, S., Silver, D., Vinyals, O., Senior, A.W., Kavukcuoglu, K., Kohli, P., Hassabis, D., 2021. Highly accurate protein structure prediction with AlphaFold. Nature 596 (7873), 583–589. https://doi.org/10.1038/s41586-021-03819-2.

Kamilaris, A., Prenafeta-Boldú, F.X., 2018. Deep learning in agriculture: a survey. Comput. Electron. Agric. 147, 70–90. https://doi.org/10.1016/j.compag.2018.02.016.

Kantor, A., McClements, M.E., MacLaren, R.E., 2020. CRISPR-Cas9 DNA base-editing and prime-editing. Int. J. Mol. Sci. 21, 6240. https://doi.org/10.3390/ijms21176240.

Kim, D., Kim, D.E., Lee, G., Cho, S.I., Kim, J.S., 2019. Genome-wide target specificity of CRISPR RNA-guided adenine base editors. Nat. Biotechnol. 37 (4), 430–435. https://doi.org/10.1038/s41587-019-0050-1.

Kim, H.K., Yu, G., Park, J., Min, S., Lee, S., Yoon, S., Kim, H.H., 2021. Predicting the efficiency of prime editing guide RNAs in human cells. Nat. Biotechnol. 39 (2), 198–206. https://doi.org/10.1038/s41587-020-0677-y.

Kim, N., Choi, S., Kim, S., Song, M., Seo, J.H., Min, S., et al., 2023. Deep learning models to predict the editing efficiencies and outcomes of diverse base. Nat. Biotechnol. 1–4. https://doi.org/10.1038/s41587-023-01792-x.

Kim, N., Choi, S., Kim, S., Song, M., Seo, J.H., Min, S., Park, J., Cho, S.R., Kim, H.H., 2024. Deep learning models to predict the editing efficiencies and outcomes of diverse base editors. Nat. Biotechnol. 42 (3), 484–497. https://doi.org/10.1038/s41587-023-01792-x.

Kim, M.G., Go, M.J., Kang, S.H., Jeong, S.H., Lim, K., 2025. Revolutionizing CRISPR technology with artificial intelligence. Exp. Mol. Med. 57 (7), 1419–1431. https://doi.org/10.1038/s12276-025-01462-9.

Kluesner, M.G., Nedveck, D.A., Lahr, W.S., Garbe, J.R., Abrahante, J.E., Webber, B.R., Moriarity, B.S., 2018. EditR: a method to quantify base editing from sanger sequencing. CRISPR J. 1 (3), 239–250. https://doi.org/10.1089/crispr.2018.0014.

Koblan, L.W., Arbab, M., Shen, M.W., Hussmann, J.A., Anzalone, A.V., Doman, J.L., Newby, G.A., Yang, D., Mok, B., Replogle, J.M., Xu, A., Sisley, T.A., Weissman, J.S., Adamson, B., Liu, D.R., 2021. Efficient C•G-to-G•C base editors developed using CRISPRi screens, target-library analysis, and machine learning. Nat. Biotechnol. 39 (11), 1414–1425. https://doi.org/10.1038/s41587-021-00938-z.

Koeppel, J., Weller, J., Peets, E.M., Pallaseni, A., Kuzmin, I., Raudvere, U., Peterson, H., Liberante, F.G., Parts, L., 2023. Prediction of prime editing insertion efficiencies using sequence features and DNA repair determinants. Nat. Biotechnol. 41 (10), 1446–1456. https://doi.org/10.1038/s41587-023-01678-y.

Komor, A.C., Zhao, K.T., Packer, M.S., Gaudelli, N.M., Waterbury, A.L., Koblan, L.W., Kim, Y.B., Badran, A.H., Liu, D.R., 2017. Improved base excision repair inhibition and bacteriophage Mu Gam protein yields C:G-to-T:A base editors with higher efficiency and product purity. Sci. Adv. 3 (8). https://doi.org/10.1126/sciadv.aao4774.

Kopecká, R., Kameniarová, M., Černý, M., Brzobohatý, B., Novák, J., 2023. Abiotic stress in crop production. Int. J. Mol. Sci. 24 (7), 6603. https://doi.org/10.3390/ijms24076603.

Krishna, R., Wang, J., Ahern, W., Sturmfels, P., Venkatesh, P., Kalvet, I., Lee, G.R., Morey-Burrows, F.S., Anishchenko, I., Humphreys, I.R., McHugh, R., Vafeados, D., Li, X., Sutherland, G.A., Hitchcock, A., Neil Hunter, C., Kang, A., Brackenbrough, E., Bera, A.K., Baek, M., DiMaio, F., Baker, D., 2024. Generalized biomolecular modeling and design with RoseTTAFold All-Atom. Science 384 (6693). https://doi.org/10.1126/science.adl2528.

Kumar, M., Ayzenshtat, D., Marko, A., Bocobza, S., 2022. Optimization of T-DNA configuration with UBIQUITIN10 promoters and tRNA–sgRNA complexes promotes highly efficient genome editing in allotetraploid tobacco. Plant Cell Rep. 41 (1), 175–194. https://doi.org/10.1007/s00299-021-02796-0.

Labun, K., Montague, T.G., Gagnon, J.A., Thyme, S.B., Valen, E., 2016. CHOPCHOP v2: a web tool for the next generation of CRISPR genome engineering. Nucleic Acids Res. 44 (1), W272. https://doi.org/10.1093/NAR/GKW398.

Lazzarotto, C.R., Malinin, N.L., Li, Y., Zhang, R., Yang, Y., Lee, G.H., Cowley, E., He, Y., Lan, X., Jividen, K., Katta, V., Kolmakova, N.G., Petersen, C.T., Qi, Q., Strelcov, E., Maragh, S., Krenciute, G., Ma, J., Cheng, Y., Tsai, S.Q., 2020. CHANGE-seq reveals genetic and epigenetic effects on CRISPR–Cas9 genome-wide activity. Nat. Biotechnol. 38 (11), 1317–1327. https://doi.org/10.1038/s41587-020-0555-7.

LeCun, Y., Bengio, Y., Hinton, G., 2015. Deep learning. Nature 521, 436–444. https://doi.org/10.1038/nature14539.

Lee, C.M., Davis, T.H., Bao, G., 2018. Examination of CRISPR/Cas9 design tools and the effect of target site accessibility on Cas9 activity. Exp. Physiol. 103, 456–460. https://doi.org/10.1113/EP086043.

Lei, Y., Lu, L., Liu, H.Y., Li, S., Xing, F., Chen, L.L., 2014. CRISPR-P: a web tool for synthetic single-guide RNA design of CRISPR-system in plants. Mol. Plant 7 (9), 1494–1496. https://doi.org/10.1093/mp/ssu044.

Li, Y., Chen, J., Tsai, S.Q., Cheng, Y., 2021. Easy-Prime: a machine learning–based prime editor design tool. Genome Biol. 22 (1). https://doi.org/10.1186/s13059-021-02458-0.

Li, S., Lin, D., Zhang, Y., Deng, M., Chen, Y., Lv, B., Li, B., Lei, Y., Wang, Y., Zhao, L., Liang, Y., Liu, J., Chen, K., Liu, Z., Xiao, J., Qiu, J.-L., Gao, C., 2022a. Genome-edited powdery mildew resistance in wheat without growth penalties. Nature 602 (7897), 455–460. https://doi.org/10.1038/s41586-022-04395-9.

Li, S., An, J., Li, Y., Zhu, X., Zhao, D., Wang, L., Sun, Y., Yang, Y., Bi, C., Zhang, X., Wang, M., 2022b. Automated high-throughput genome editing platform with an AI learning in situ prediction model. Nat. Commun. 13 (1). https://doi.org/10.1038/s41467-022-35056-0.

Li, T., Yang, Y., Qi, H., Cui, W., Zhang, L., Fu, X., He, X., Liu, M., Li, P.-feng, Yu, T., 2023. CRISPR/Cas9 therapeutics: progress and prospects. Signal Transduct. Targeted Ther. 8 (1). https://doi.org/10.1038/s41392-023-01309-7.

Li, Z., Khan, W.U., Bai, G., Dong, C., Wang, J., Zhang, Y., Wang, C., Zhang, H., Wang, W., Luo, M., Chen, F., 2025. From code to life: the AI-driven revolution in genome editing. Adv. Sci. 12 (30). https://doi.org/10.1002/advs.202417029.

Liakos, K., Busato, P., Moshou, D., Pearson, S., Bochtis, D., 2018. Machine learning in agriculture: a review. Sensors 18 (8), 2674. https://doi.org/10.3390/s18082674.

Listgarten, J., Weinstein, M., Kleinstiver, B.P., Sousa, A.A., Joung, J.K., Crawford, J., Gao, K., Hoang, L., Elibol, M., Doench, J.G., Fusi, N., 2018. Prediction of off-target activities for the end-to-end design of CRISPR guide RNAs. Nat. Biomed. Eng. 2 (1), 38–47. https://doi.org/10.1038/s41551-017-0178-6.

Liu, Z., Lu, Z., Yang, G., Huang, S., Li, G., Feng, S., Liu, Y., Li, J., Yu, W., Zhang, Y., Chen, J., Sun, Q., Huang, X., 2018a. Efficient generation of mouse models of human diseases via ABE- and BE-mediated base editing. Nat. Commun. 9 (1). https://doi.org/10.1038/s41467-018-04768-7.

Liu, Z., Chen, M., Chen, S., Deng, J., Song, Y., Lai, L., Li, Z., 2018b. Highly efficient RNA-guided base editing in rabbit. Nat. Commun. 9 (1). https://doi.org/10.1038/s41467-018-05232-2.

Liu, Y., Zeng, J., Cheng, Y., Guo, Y., Yu, H., Li, Y., Huang, C., 2019. Cas9- PF , an early flowering and visual selection marker system, enhances the frequency of editing event occurrence and expedites the isolation of genome-edited and transgene-free plants. Plant Biotechnol. J. 17 (7), 1191–1193. https://doi.org/10.1111/pbi.13118.

Liu, N., Zhou, L., Qu, J., Yao, S., 2021. Recent advances in CRISPR/Cas9 directed base editing. Curr. Gene Ther. 21, 327–337. https://doi.org/10.2174/1566523221666210322090638.

Louie, W., Shen, M.W., Tahiry, Z., Zhang, S., Worstell, D., Cassa, C.A., Sherwood, R.I., Gifford, D.K., 2021. Machine learning based CRISPR gRNA design for therapeutic exon skipping. PLoS Comput. Biol. 17 (1). https://doi.org/10.1371/JOURNAL.PCBI.1008605.

Ma, X., Zhu, Q., Chen, Y., Liu, Y.G., 2016. CRISPR/Cas9 platforms for genome editing in plants: developments and applications. Mol. Plant 9 (7), 961–974. https://doi.org/10.1016/j.molp.2016.04.009.

Ma, J., Chen, J., Wang, M., Ren, Y., Wang, S., Lei, C., Cheng, Sodmergen, Z., 2018. Disruption of OsSEC3A increases the content of salicylic acid and induces plant defense responses in rice. J. Exp. Bot. 69 (5), 1051–1064. https://doi.org/10.1093/jxb/erx458.

Madani, A., Krause, B., Greene, E.R., Subramanian, S., Mohr, B.P., Holton, J.M., Olmos, J.L., Xiong, C., Sun, Z.Z., Socher, R., Fraser, J.S., Naik, N., 2023. Large language models generate functional protein sequences across diverse families. Nat. Biotechnol. 41 (8), 1099–1106. https://doi.org/10.1038/s41587-022-01618-2.

Majid Hurrah, I., Banday, L., Mohi, T., Din, U., Rasool, I., Mandal, S., 2026. In: Chen, J.-T. (Ed.), Machine Learning for Data Mining in Plant Research. AI-Enhanced Plant Omics, pp. 236–258. https://doi.org/10.1079/9781800628649.0011.

Malnoy, M., Viola, R., Jung, M.H., Koo, O.J., Kim, S., Kim, J.S., Velasco, R., Kanchiswamy, C.N., 2016. DNA-free genetically edited grapevine and apple protoplast using CRISPR/Cas9 ribonucleoproteins. Front. Plant Sci. 7 (2016). https://doi.org/10.3389/fpls.2016.01904.

Manghwar, H., Li, B., Ding, X., Hussain, A., Lindsey, K., Zhang, X., Jin, S., 2020. CRISPR/Cas systems in genome editing: methodologies and tools for sgRNA design, off-target evaluation, and strategies to mitigate off-target effects. Adv. Sci. 7 (6). https://doi.org/10.1002/advs.201902312.

Marquart, K.F., Allam, A., Janjuha, S., Sintsova, A., Villiger, L., Frey, N., Krauthammer, M., Schwank, G., 2021. Predicting base editing outcomes with an attention-based deep learning algorithm trained on high-throughput target library screens. Nat. Commun. 12 (1). https://doi.org/10.1038/s41467-021-25375-z.

Mathis, N., Allam, A., Kissling, L., Marquart, K.F., Schmidheini, L., Solari, C., Balázs, Z., Krauthammer, M., Schwank, G., 2023. Predicting prime editing efficiency and product purity by deep learning. Nat. Biotechnol. 41 (8), 1151–1159. https://doi.org/10.1038/s41587-022-01613-7.

McKinney, B.A., Reif, D.M., Ritchie, M.D., Moore, J.H., 2006. Machine learning for detecting gene-gene interactions: a review. Appl. Bioinf. 5 (2), 77–88. https://doi.org/10.2165/00822942-200605020-00002.

Mengstie, M.A., Wondimu, B.Z., 2021. Mechanism and applications of crispr/cas-9-mediated genome editing. Biol. Targets & Ther. 15, 353–361. https://doi.org/10.2147/BTT.S326422.

Merlot, S., Leonhardt, N., Fenzi, F., Valon, C., Costa, M., Piette, L., Vavasseur, A., Genty, B., Boivin, K., Müller, A., Giraudat, J., Leung, J., 2007. Constitutive activation of a plasma membrane H+-ATPase prevents abscisic acid-mediated stomatal closure. EMBO J. 26 (13), 3216–3226. https://doi.org/10.1038/sj.emboj.7601750.

Miklis, M., Consonni, C., Bhat, R.A., Lipka, V., Schulze-Lefert, P., Panstruga, R., 2007. Barley MLO modulates actin-dependent and actin-independent antifungal defense pathways at the cell periphery. Plant Physiol. 144 (2), 1132–1143. https://doi.org/10.1104/pp.107.098897.

Molla, K.A., Yang, Y., 2019. CRISPR/Cas-Mediated base editing: technical considerations and practical applications. Trends Biotechnol. 37 (10), 1121–1142. https://doi.org/10.1016/j.tibtech.2019.03.008.

Moreno-Mateos, M.A., Vejnar, C.E., Beaudoin, J.D., Fernandez, J.P., Mis, E.K., Khokha, M.K., Giraldez, A.J., 2015. CRISPRscan: designing highly efficient sgRNAs for CRISPR-Cas9 targeting in vivo. Nat. Methods 12 (10), 982–988. https://doi.org/10.1038/nmeth.3543.

Naeem, M., Majeed, S., Hoque, M.Z., Ahmad, I., 2020. Latest developed strategies to minimize the off-target effects in CRISPR-cas-mediated genome editing. Cells 9 (7), 1608. https://doi.org/10.3390/cells9071608.

Nekrasov, V., Wang, C., Win, J., Lanz, C., Weigel, D., Kamoun, S., 2017. Rapid generation of a transgene-free powdery mildew resistant tomato by genome deletion. Sci. Rep. 7 (1). https://doi.org/10.1038/s41598-017-00578-x.

Nguyen, E., Poli, M., Durrant, M.G., Kang, B., Katrekar, D., Li, D.B., Bartie, L.J., Thomas, A.W., King, S.H., Brixi, G., Sullivan, J., Ng, M.Y., Lewis, A., Lou, A., Ermon, S., Baccus, S.A., Hernandez-Boussard, T., Re, C., Hsu, P.D., Hie, B.L., 2024. Sequence modeling and design from molecular to genome scale with Evo. Science 386 (6723). https://doi.org/10.1126/science.ado9336.

Nijkamp, E., Ruffolo, J.A., Weinstein, E.N., Naik, N., Madani, A., 2023. ProGen2: exploring the boundaries of protein language models. Cell Syst. 14 (11), 968. https://doi.org/10.1016/j.cels.2023.10.002.

Nilsson, N.J., 1982. Artificial intelligence: engineering, science, or slogan? AI Mag. 3, 2. https://doi.org/10.1609/aimag.v3i1.359.

Nishida, K., Arazoe, T., Yachie, N., Banno, S., Kakimoto, M., Tabata, M., Mochizuki, M., Miyabe, A., Araki, M., Hara, K.Y., Shimatani, Z., Kondo, A., 2016. Targeted nucleotide editing using hybrid prokaryotic and vertebrate adaptive immune systems. Science (New York, N.Y.) 353 (6305), aaf8729. https://doi.org/10.1126/science.aaf8729.

Oruh, J., Viriri, S., 2022. Long short-term memory recurrent neural network for automatic recognition of spoken English digits. In: Chbeir, R., Manolopoulos, Y., Prasath, R. (Eds.), Mining Intelligence and Knowledge Exploration. Springer International Publishing, Cham, pp. 188–198. https://doi.org/10.1007/978-3-031-21517-9_19.

Osakabe, Y., Watanabe, T., Sugano, S.S., Ueta, R., Ishihara, R., Shinozaki, K., Osakabe, K., 2016. Optimization of CRISPR/Cas9 genome editing to modify abiotic stress responses in plants. Sci. Rep. 6. https://doi.org/10.1038/srep26685.

Pallaseni, A., Peets, E.M., Koeppel, J., Weller, J., Vanderstichele, T., Ho, U.L., Crepaldi, L., van Leeuwen, J., Allen, F., Parts, L., 2022. Predicting base editing outcomes using position-specific sequence determinants. Nucleic Acids Res. 50, 3551–3564. https://doi.org/10.1093/nar/gkac161.

Pandey, S., Choudhari, J.K., Tripathi, A., Singh, A., Antony, A., Chouhan, U., 2025. Artificial intelligence-based genome editing in CRISPR/Cas9. In: Mandal, S. (Ed.), Artificial Intelligence (AI) in Cell and Genetic Engineering. Springer US, New York, NY, pp. 273–282. https://doi.org/10.1007/978-1-0716-4690-8_16.

Panstruga, R., Schulze-Lefert, P., 2002. Live and let live: insights into powdery mildew disease and resistance. Mol. Plant Pathol. 3, 495–502. https://doi.org/10.1046/j.1364-3703.2002.00145.x.

Pantazi, X.E., Moshou, D., Alexandridis, T., Whetton, R.L., Mouazen, A.M., 2016. Wheat yield prediction using machine learning and advanced sensing techniques. Comput. Electron. Agric. 121, 57–65. https://doi.org/10.1016/j.compag.2015.11.018.

Park, J., Kim, H.K., 2023. Prediction of base editing efficiencies and outcomes using DeepABE and DeepCBE. In: Bae, S., Song, B. (Eds.), Base Editors: Methods and Protocols. Springer US, New York, NY, pp. 23–32. https://doi.org/10.1007/978-1-0716-2879-9_3.

Pereira, A., 2016. Plant abiotic stress challenges from the changing environment. Front. Plant Sci. 7. https://doi.org/10.3389/fpls.2016.01123.

Piffanelli, P., Ramsay, L., Waugh, R., Benabdelmouna, A., D'Hont, A., Hollricher, K., Jørgensen, J.H., Schulze-Lefert, P., Panstruga, R., 2004. A barley cultivation-associated polymorphism conveys resistance to powdery mildew. Nature 430 (7002), 887–891. https://doi.org/10.1038/nature02781.

Pinna, G., Dentali, M., 2023. The impact of climate change on field crop productivity. Int. J. Geogr. Geol. Environ. 5, 203–205. https://doi.org/10.22271/27067483.2023.v5.i2b.232.

Pramanik, D., Shelake, R.M., Park, J., Kim, M.J., Hwang, I., Park, Y., Kim, J.Y., 2021. CRISPR/Cas9-mediated generation of pathogen-resistant tomato against tomato yellow leaf curl virus and powdery mildew. Int. J. Mol. Sci. 22 (4), 1–18. https://doi.org/10.3390/ijms22041878.

Quazi, S., 2022. Retracted article: artificial intelligence and machine learning in precision and genomic medicine. Med. Oncol. 39 (8). https://doi.org/10.1007/s12032-022-01711-1.

Rauscher, B., Heigwer, F., Breinig, M., Winter, J., Boutros, M., 2017. GenomeCRISPR - a database for high-throughput CRISPR/Cas9 screens. Nucleic Acids Res. 45 (D1), D679. https://doi.org/10.1093/nar/gkw997.

Raza, A., Khare, T., Zhang, X., Rahman, M.M., Hussain, M., Gill, S.S., Chen, Z.H., Zhou, M., Hu, Z., Varshney, R.K., 2025. Novel strategies for designing climate-smart crops to ensure sustainable agriculture and future food security. J. Sustain. Agric. Environ. 4 (2). https://doi.org/10.1002/sae2.70048.

Rees, H.A., Liu, D.R., 2018. Base editing: precision chemistry on the genome and transcriptome of living cells. Nat. Rev. Genet. 19 (12), 770–788. https://doi.org/10.1038/s41576-018-0059-1.

Ren, B., Liu, L., Li, S., Kuang, Y., Wang, J., Zhang, D., Zhou, X., Lin, H., Zhou, H., 2019. Cas9-NG greatly expands the targeting scope of the genome-editing toolkit by recognizing NG and other atypical PAMs in rice. Mol. Plant 12 (7), 1015–1026. https://doi.org/10.1016/j.molp.2019.03.010.

Ruffolo, J.A., Nayfach, S., Gallagher, J., Bhatnagar, A., Beazer, J., Hussain, R., Russ, J., Yip, J., Hill, E., Pacesa, M., Meeske, A.J., Cameron, P., Madani, A., 2024. Design of highly functional genome editors by modeling the universe of CRISPR-Cas sequences. bioRxiv. https://doi.org/10.1101/2024.04.22.590591.

Ryan Cross, 2019. CRISPR base editors make mistakes. C&EN Global Enterp 97, 3. https://doi.org/10.1021/cen-09709-leadcon.

Sandhu, K.S., Lozada, D.N., Zhang, Z., Pumphrey, M.O., Carter, A.H., 2021. Deep learning for predicting complex traits in spring wheat breeding program. Front. Plant Sci. 11. https://doi.org/10.3389/fpls.2020.613325.

Sarker, I.H., 2021. Machine learning: algorithms, real-world applications and research directions. SN Comput. Sci. 2 (3). https://doi.org/10.1007/s42979-021-00592-x.

Simeone, O., 2018. A very brief introduction to machine learning with applications to communication systems. IEEE Trans. Cogn. Commun. Netw. 4 (4), 648–664. https://doi.org/10.1109/tccn.2018.2881442.

Singh, A., Ganapathysubramanian, B., Singh, A.K., Sarkar, S., 2016. Machine learning for high-throughput stress phenotyping in plants. Trends Plant Sci. 21 (2), 110–124. https://doi.org/10.1016/j.tplants.2015.10.015.

Song, M., Kim, H.K., Lee, S., Kim, Y., Seo, S.Y., Park, J., Choi, J.W., Jang, H., Shin, J.H., Min, S., Quan, Z., Kim, J.H., Kang, H.C., Yoon, S., Kim, H.H., 2020. Sequence-specific prediction of the efficiencies of adenine and cytosine base editors. Nat. Biotechnol. 38 (9), 1037–1043. https://doi.org/10.1038/s41587-020-0573-5.

Sretenovic, S., Green, Y., Wu, Y., Cheng, Y., Zhang, T., Van Eck, J., Qi, Y., 2023. Genome- and transcriptome-wide off-target analyses of a high-efficiency adenine base editor in tomato. Plant Physiol. 193 (1), 291–303. https://doi.org/10.1093/plphys/kiad347.

Szczelkun, M.D., Tikhomirova, M.S., Sinkunas, T., Gasiunas, G., Karvelis, T., Pschera, P., Siksnys, V., Seidel, R., 2014. Direct observation of R-loop formation by single RNA-guided Cas9 and Cascade effector complexes. Proc. Natl. Acad. Sci. 111 (27), 9798–9803. https://doi.org/10.1073/pnas.1402597111.

Tarca, A.L., Carey, V.J., Chen, X.-wen, Romero, R., Drăghici, S., Lewitter, F., 2007. Machine learning and its applications to biology. PLoS Comput. Biol. 3 (6), e116. https://doi.org/10.1371/journal.pcbi.0030116.

Tsai, S.Q., Zheng, Z., Nguyen, N.T., Liebers, M., Topkar, V.V., Thapar, V., Wyvekens, N., Khayter, C., Iafrate, A.J., Le, L.P., Aryee, M.J., Joung, J.K., 2015. GUIDE-seq enables genome-wide profiling of off-target cleavage by CRISPR-Cas nucleases. Nat. Biotechnol. 33 (2), 187–198. https://doi.org/10.1038/nbt.3117.

Ubbens, J.R., Stavness, I., 2017. Deep plant phenomics: a deep learning platform for complex plant phenotyping tasks. Front. Plant Sci. 8. https://doi.org/10.3389/fpls.2017.01190.

van Esse, H.P., Reuber, T.L., van der Does, D., 2020. Genetic modification to improve disease resistance in crops. New Phytol. 225, 70–86. https://doi.org/10.1111/nph.15967.

van Kempen, M., Kim, S.S., Tumescheit, C., Mirdita, M., Lee, J., Gilchrist, C.L.M., Söding, J., Steinegger, M., 2024. Fast and accurate protein structure search with Foldseek. Nat. Biotechnol. 42 (2), 243–246. https://doi.org/10.1038/s41587-023-01773-0.

Wan, D.Y., Guo, Y., Cheng, Y., Hu, Y., Xiao, S., Wang, Y., Wen, Y.Q., 2020. CRISPR/Cas9-mediated mutagenesis of VvMLO3 results in enhanced resistance to powdery mildew in grapevine (Vitis vinifera). Hortic. Res. 7 (1). https://doi.org/10.1038/s41438-020-0339-8.

Wang, Y., Cheng, X., Shan, Q., Zhang, Y., Liu, J., Gao, C., Qiu, J.L., 2014. Simultaneous editing of three homoeoalleles in hexaploid bread wheat confers heritable resistance to powdery mildew. Nat. Biotechnol. 32 (9), 947–951. https://doi.org/10.1038/nbt.2969.

Wang, F., Wang, C., Liu, P., Lei, C., Hao, W., Gao, Y., Liu, Y.-G., Zhao, K., Wilson, R.A., 2016. Enhanced rice blast resistance by CRISPR/Cas9-Targeted mutagenesis of the ERF transcription factor gene OsERF922. PLoS One 11 (4), e0154027. https://doi.org/10.1371/journal.pone.0154027.

Wang, D., Zhang, C., Wang, B., Li, B., Wang, Q., Liu, D., Wang, H., Zhou, Y., Shi, L., Lan, F., Wang, Y., 2019. Optimized CRISPR guide RNA design for two high-fidelity Cas9 variants by deep learning. Nat. Commun. 10 (1). https://doi.org/10.1038/s41467-019-12281-8.

Wang, Y., Zafar, N., Ali, Q., Manghwar, H., Wang, G., Lu, Y., Ding, X., Ding, F., Hong, N., Wang, G., Jin, S., 2022. CRISPR/Cas genome editing technologies for plant improvement against biotic and abiotic stresses: advances, limitations, and future perspectives. Cells 11 (23), 3928. https://doi.org/10.3390/cells11233928.

Wang, X., Zeng, H., Lin, L., Huang, Y., Lin, H., Que, Y., 2023. Deep learning-empowered crop breeding: intelligent, efficient and promising. Front. Plant Sci. 14. https://doi.org/10.3389/fpls.2023.1260089.

Wessels, H.H., Stirn, A., Méndez-Mancilla, A., Kim, E.J., Hart, S.K., Knowles, D.A., Sanjana, N.E., 2024. Prediction of on-target and off-target activity of CRISPR–Cas13d guide RNAs using deep learning. Nat. Biotechnol. 42 (4), 628–637. https://doi.org/10.1038/s41587-023-01830-8.

Wienert, B., Wyman, S.K., Richardson, C.D., Yeh, C.D., Akcakaya, P., Porritt, M.J., Morlock, M., Vu, J.T., Kazane, K. R., Watry, H.L., Judge, L.M., Conklin, B.R., Maresca, M., Corn, J.E., 2019. Unbiased detection of CRISPR off-targets in vivo using DISCOVER-Seq. Science 364 (6437), 286–289. https://doi.org/10.1126/science.aav9023.

Wong, N., Liu, W., Wang, X., 2015. Wu-CRISPR: characteristics of functional guide RNAs for the CRISPR/Cas9 system. Genome Biol. 16, 218. https://doi.org/10.1186/s13059-015-0784-0.

Wu, Y., Ren, Q., Zhong, Z., Liu, G., Han, Y., Yu, B., Liu, L., Xiang, S., Liu, S., Tang, X., Zhou, J., Zheng, X., Simon, S., Zhang, T., Qi, Y., Zhang, Y., 2022. Genome-wide analyses of PAM-relaxed Cas9 genome editors reveal substantial off-target effects by ABE8e in rice. Plant Biotechnol. J. 20 (9), 1670–1682. https://doi.org/10.1111/pbi.13838.

Xiao, A., Cheng, Z., Kong, L., Zhu, Z., Lin, S., Gao, G., Zhang, B., 2014. CasOT: a genome-wide Cas9/gRNA off-target searching tool. Bioinformatics 30 (8), 1180–1182. https://doi.org/10.1093/bioinformatics/btt764.

Yan, J., Cirincione, A., Adamson, B., 2020. Prime editing: precision genome editing by reverse transcription. Mol. Cell 77, 210–212. https://doi.org/10.1016/j.molcel.2019.12.016.

Yoon, P.H., Zhang, Z., Loi, K.J., Adler, B.A., Lahiri, A., Vohra, K., Shi, H., Rabelo, D.B., Trinidad, M., Boger, R.S., Al-Shimary, M.J., Doudna, J.A., 2024. Structure-guided discovery of ancestral CRISPR-Cas13 ribonucleases. Science 385 (6708), 538–543. https://doi.org/10.1126/science.adq0553.

Zafar, M.M., Razzaq, A., Farooq, M.A., Rehman, A., Firdous, H., Shakeel, A., Mo, H., Ren, M., 2020. Insect resistance management in Bacillus thuringiensis cotton by MGPS (multiple genes pyramiding and silencing). Journal of Cotton Research 3 (1). https://doi.org/10.1186/s42397-020-00074-0.

Zhang, T., Zheng, Q., Yi, X., An, H., Zhao, Y., Ma, S., Zhou, G., 2018. Establishing RNA virus resistance in plants by harnessing CRISPR immune system. Plant Biotechnol. J. 16 (8), 1415–1423. https://doi.org/10.1111/pbi.12881.

Zhao, F., Zhang, T., Sun, X., Zhang, X., Chen, L., Wang, H., Li, J., Fan, P., Lai, L., Sui, T., Li, Z., 2023. A strategy for Cas13 miniaturization based on the structure and AlphaFold. Nat. Commun. 14, 5545. https://doi.org/10.1038/s41467-023-41320-8.

Zingaretti, L.M., Gezan, S.A., Ferrão, L.F.V., Osorio, L.F., Monfort, A., Muñoz, P.R., Whitaker, V.M., Pérez-Enciso, M., 2020. Exploring deep learning for complex trait genomic prediction in polyploid outcrossing species. Front. Plant Sci. 11. https://doi.org/10.3389/fpls.2020.00025.

CHAPTER

7

Convergence of gene editing and AI: The future of gene editing in plants

Mughair Abdul Aziz, and Khaled Masmoudi

Department of Integrative Agriculture, College of Agriculture and Veterinary Medicine, United Arab Emirates University, Al Ain, United Arab Emirates

7.1 Introduction

Agriculture experiences challenges of different forms both biotic and abiotic. Two of the important tools in agriculture at present are artificial intelligence (AI) and genome editing. As we know, conventional breeding can be slow and unpredictable, but not like them, AI and genome editing allow the scientists to modify genes precisely. Many have thought that this new combined approach has brought new ways for improving traits (Yıldırım et al., 2023; Luo et al., 2024). Stress tolerance, disease resistance, and nutrient enhancement of crops are likely traits to be improved by these combined approaches.

But many still prospect that challenges are remaining in spite of these advances are made. Managing large genomic datasets is often needed by the researchers often. They have to select and optimize editing enzymes, not only that, but need to identify suitable gene targets, and have to be very careful in predicting and validating the editing outcomes, which can be wrong and affect huge amount of efforts (Venkataraman et al., 2005). In our viewpoint, genetic diversity is a major culprit for raising problems, because the traits controlled by many genes, and the danger of off-target effects add complexity that requires everyone to be both careful in experiments and computational approaches.

What you are observing nowadays is the use of AI used in every field, and it can increasingly help to solve these challenges. This is reported and also, we know that machine learning (ML) and AI techniques can help predict the efficiency of genome editing. Everyone has seen, in early development of the 1950s, the use of AI to explore complex biological datasets. Now you can guess that today researchers will require assistance for developing their well-informed genome editing strategies (Rashid and Kausik, 2024; Dublino and Ercolano,

Intelligent Gene Editing in Plants
https://doi.org/10.1016/B978-0-443-44079-3.00010-3

2025). This help will come in providing the choice of optimal gene targets by analyzing genomic data (Gandomi et al., 2023).

Here, the AI provides information on targeted genome editing interventions based on their determined strategies. The thing that will happen is that it makes genome editing more data-informed and efficient for the development of enhanced crop varieties (Kassem, 2025). You found that the entire editing system is reorganized of the entire editing system, from target selection to outcome optimization. For the world, it will definitely be a significant progress in modern agricultural biotechnology; a tool to overcome the food security and climate-related agricultural challenges.

Our chapter discusses the growing combination between genome editing technologies and AI in plant science. It describes the development of genome editing tools, indicating the shift from early protein-based nucleases to modern genome editing–based systems, while depicting the growing complexity and scale of plant genomic data. The chapter overviews the advancements in AI, emphasizing the ML and deep learning (DL) models emergence as important enablers for interpreting complex biological information and guiding genome editing decisions. We focus on the convergence of genome editing and AI for crop trait improvement, illustrating data-driven predictions that enhance the precision and effectiveness of targeted genetic modifications. Practical applications of AI-driven genome editing in agriculture are also reviewed. In this chapter, we report the prospective impact of this convergence on the future of agricultural processes.

7.2 Genome editing technologies: Engineered nucleases to CRISPR–Cas systems

Genome editing includes molecular techniques that enable the deliberate modification of an organism's DNA sequence to alter specific genetic traits. These modifications may include insertions, deletions, replacements, or targeted mutations that result in observable changes in biological characteristics (Graham et al., 2020). Over the years, several genome editing tools have been developed to precisely target and manipulate specific loci within the genome (Fig. 7.1). Early methods include meganucleases, zinc finger nucleases (ZFNs), and transcription activator-like effector nucleases (TALENs), each relying on engineered proteins to recognize and cut DNA at particular sequences (González Castro et al., 2021). More recently, the clustered regularly interspaced short palindromic repeats and associated proteins (CRISPR–Cas) system has been developed as an efficient genome editing tool (Abdul Aziz et al., 2022).

Meganucleases are naturally derived enzymes that function as endodeoxyribonucleases, capable of identifying and cutting DNA at specific sequences typically 20 to 30 base pairs in length. Everyone uses meganucleases because of their extraordinary specificity, which enables precise genome modifications. It does this by inducing double-strand breaks at selected genomic sites. As described by Ashworth et al. (2006), their potential can be broadened by remaking the DNA-binding domains again to recognize new DNA sequences. Yet, this process is far from being direct; it demands detailed structural adjustments and extensive protein engineering. Due to this, our finding is that because of their technical

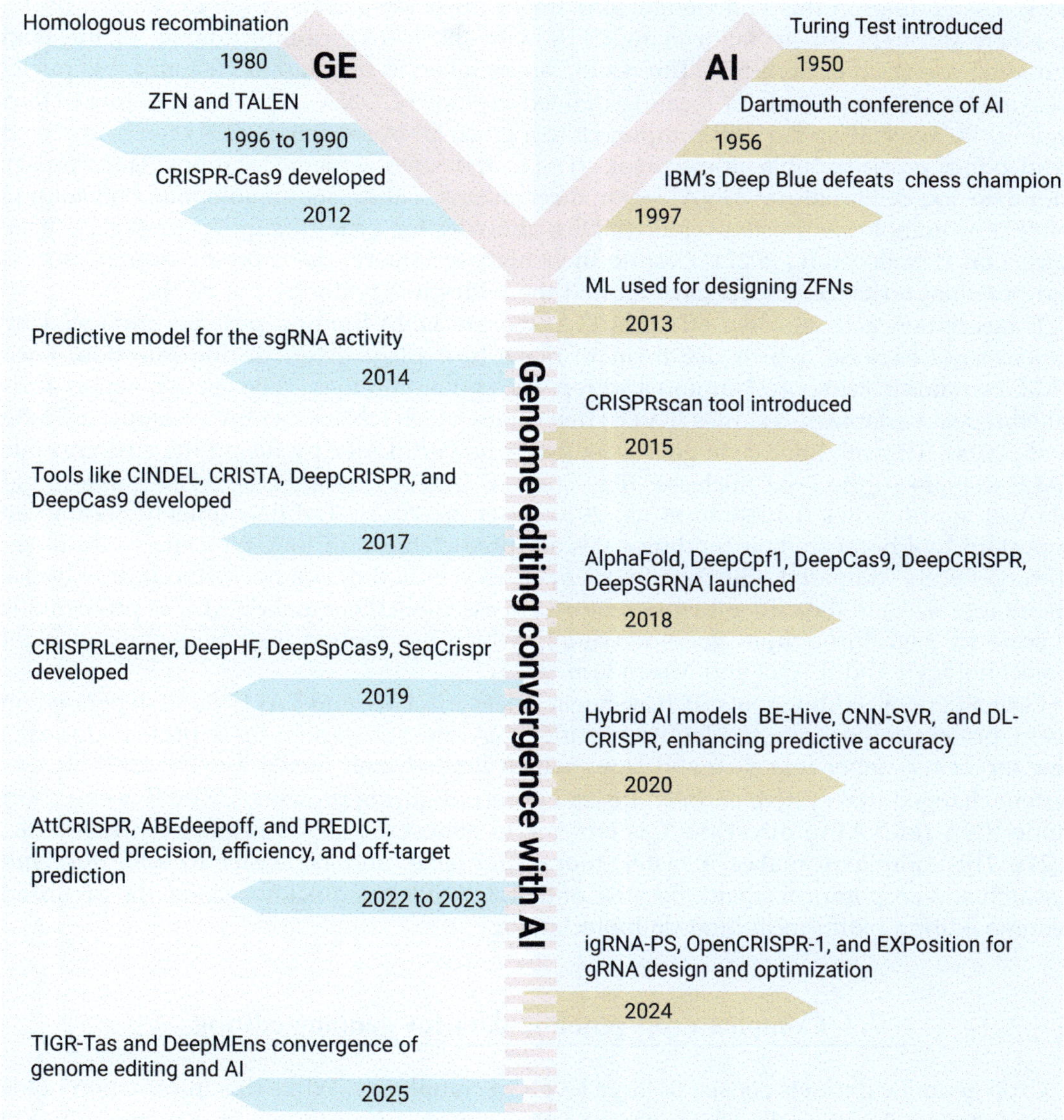

FIGURE 7.1 **Significant milestones in genome editing (GE) and AI.** Genome editing (GE) and AI convergence milestones. Initially developed along separate paths, both technologies begin to converge around 2013, leading to the emergence of integrated approaches that use AI to enhance the precision and efficiency of genome editing.

complexity and time investment, meganucleases are not commonly used in most current genome editing systems. We clarified how zinc finger domains function in recognizing specific DNA motifs through using the earlier studies, which showed that it is possible to engineer zinc finger proteins (ZFPs) to bind new genomic targets (Desjarlais and Berg,

1993). Depending on this knowledge, zinc finger nucleases (ZFNs) were developed by researchers through linking tailor-made ZFPs with the DNA-cleaving domain of the FokI endonuclease (Kim et al., 1996). The result can be taken as a group of customizable nucleases, and these are capable of locating defined regions of DNA and cutting defined DNA regions. By associating the ZFP component to a given DNA sequence, ZFNs can be directed to introduce precise double-strand breaks (DSBs) at desired genomic locations. These breaks stimulate the cell's natural DNA repair mechanisms, either nonhomologous end joining (NHEJ) or homology-directed repair (HDR), allowing for gene disruption or precise gene correction (Danilo et al., 2022). Despite their high specificity, the complex design process and potential off-target effects have limited the widespread adoption of ZFNs.

Transcription activator-like effectors (TALEs) are DNA-binding proteins produced by *Xanthomonas* bacteria, which use them to make host plant genes during infection. Each TALE contains a string of 34–amino acid repeats, and within these repeats, two amino acids called repeat variable di-residues (RVDs) determine which DNA bases are recognized (Deng et al., 2014). This modular code gave scientists a powerful idea by fusing the customizable TALE domain to the FokI nuclease; they created TALENs, synthetic enzymes able to cut DNA at specific sites (Christian et al., 2010; Li et al., 2011). TALENs quickly became an important tool because their binding rules are predictable, and they typically cause fewer off-target edits compared with zinc finger nucleases. Still, they are not without drawbacks. Their large size and the difficult work required to assemble them make TALENs problematic to scale for high-throughput genome editing, which has limited their wider use in plant biotechnology (Abdul Aziz and Masmoudi, 2025).

Earlier genome editing tools such as meganucleases, ZFNs, and TALENs, all depended on heavy protein engineering to identify specific DNA sites. Designing these proteins for each new target was labor-intensive and slow, which limited their flexibility. The CRISPR–Cas system changed that system entirely. Instead of reconfiguring proteins, CRISPR uses a short guide RNA (gRNA) to direct the Cas enzyme to almost any DNA sequence (Chen et al., 2024). This simplicity makes it faster, more affordable, and far easier to scale for both research and crop improvement. Because of that, CRISPR has quickly become the preferred genome editing platform in modern biotechnology.

7.3 Complex crop genomic data for genome editing

Crop genomic datasets consist of large levels of complexity. While this information holds great promise, it also makes data interpretation and its practical use in genome editing a demanding task. One of the primary hurdles lies in the vast diversity in genome size and structure across crop species (Abdul Aziz et al., 2023). For instance, the genome of *Oryza sativa* ranges approximately 430 million base pairs, whereas the *Triticum aestivum* genome surpasses 16 billion base pairs, making it more than five times larger than the human genome (IWGSC, 2018). These genomes not only contain tens of thousands of protein-coding genes but also vast expanses of noncoding regions, repetitive elements, and a wide array of genetic variants. It complicates the task of identifying genes associated with crucial agronomic traits, such as yield potential, disease resistance, drought tolerance, and nutrient efficiency

(Yıldırım et al., 2023). A number of genetic elements have been used for crop modification through the genome editing strategies (Table 7.1). However, identifying meaningful genetic components from this complex data remains a fundamental challenge in plant genome editing.

In addition to their complexity and scale, the heterogeneity and diversity of crop genomic data present further analytical challenges. Gene expression patterns can vary significantly not only between different crop species and cultivars but also across varied environmental conditions, growth stages, and tissue types (Luo et al., 2024). This variability complicates data integration, normalization, and comparative analysis, requiring efficient bioinformatic

TABLE 7.1 Genome editing strategies modification of different plant characteristics using the genomic data.

Genome editing	Plant	Genes	Plant characteristics	References
TALEN mutagenesis	Soybean	*FAD2-1A, FAD2-1B, FAD3A*	Oleic acid content enhancement	Demorest et al. (2016)
CRISPR–Cas9 mutagenesis	Tomato	*SlWUS, SlCLV3, SlWOX9, SlTFL1*	Inflorescence branching, fruit size, and plant architecture changes	Rodríguez-Leal et al. (2017)
CRISPR–Cas9 mutagenesis	Strawberry	*FaPG1*	Fruit firmness enhancement	López-Casado et al. (2023)
CRISPR–Cas9 mutagenesis	Wheat	*TaGW2*	Yield improvement	Wang et al. (2018)
CRISPR–Cas9 mutagenesis	Maize	*ARGOS8*	Drought tolerance	Shi et al. (2017)
CRISPR-based editing	Maize	*ZmALS1, ZmALS2*	Resistance to herbicides	Li et al. (2020)
Epigenome editing	Cassava	*MeSWEET10α*	Improved resistance to bacterial blight	Veley et al. (2023)
Editing of chloroplast genome	Lettuce	*psaA, psbA, rrn16*	Resistance to herbicides	Mok et al. (2022)
Prime editing	Rice	*Xa5, Xa23*	Improved resistance to bacterial blight	Gupta et al. (2023)
CRISPR-based editing	Strawberry	*FvebZIPs1.1*	Tuned sugar content	Xing et al. (2020)
CRISPR–Cas9 mutagenesis	Soybean	*AIP2a, AIP2b*	Improved protein content	Shen et al. (2022)
CRISPR-based editing	Wheat	*ALS*	Resistance to herbicides	Zhang et al. (2019)
ZFNs mutagenesis	Maize	*IPK1*	Reduction in phytate level and resistance to herbicide	Shukla et al. (2009)

pipelines to ensure accuracy and consistency. As genome sequencing technologies continue to evolve, the volume of data generated has grown exponentially, encompassing not just DNA sequences but also transcriptomics, epigenomics, proteomics, and phenotypic datasets (Shamim et al., 2024)). This data overflow places tremendous pressure on storage infrastructure, data management systems, and computational capacity. Traditional computational models and statistical approaches are increasingly inadequate for handling such high-throughput and multidimensional data.

The growing complexity and volume of crop genomic data limits the use of traditional analytical approaches, such as genome-wide association studies (GWAS) and variant calling. These methods often struggle with scalability, computational efficiency, and the ability to capture nonlinear relationships within high-dimensional datasets (Abdul Aziz and Masmoudi, 2023a,b). In response, AI, particularly ML and DL, is emerging as an essential alternative for managing and extracting big genomic data.

7.4 Development of AI from data interpretation to biological design

The concept of AI was developed before its application in genome editing. In 1950, mathematician Alan Turing proposed the Turing Test, an idea to assess whether machines could exhibit human-like intelligence, placing the philosophical and technical basis of AI (Fig. 7.1) (Schmidhuber, 2022). This was followed by a major turning point in 1956 at the Dartmouth Summer Research Project on AI, where the term "artificial intelligence" was formally introduced, predicting machines that could simulate aspects of human learning and reasoning (McCarthy et al., 2006). The early decades of AI focused on symbolic logic, rule-based systems, and problem-solving, with limited applications due to insufficient computational power and data. However, this changed dramatically in the 2010s. In 2012, Geoffrey Hinton achieved an advancement in the ImageNet competition by applying deep convolutional neural networks (CNNs) to image recognition tasks, significantly performing than previous models and sparking DL technology (Krizhevsky et al., 2012).

The release of Google's Transformer model in 2017 is thought to be the turning point in AI, bringing new possibilities for the way machines process and understand complex information. From that point came models such as BERT, GPT, and T5, each setting new standards for language understanding, translation, and text generation (Vaswani et al., 2017). Therefore, what began as a revolution in natural language processing has now expanded further. Recent developments in AI, particularly in diffusion models and large language models (LLMs), are now retrained for interpreting the languages of biology such as DNA, RNA, and proteins (Dublino and Ercolano, 2025). This shift has been driven by the explosion of publicly available biological data, including massive libraries of genomic, transcriptomic, and proteomic information. For now, these datasets are determining the next generation of generative AI models designed to predict, design, and optimize genome editing outcomes.

Unlike traditional deep learning models that focus mainly on prediction, generative AI goes a step further, which is able to describe, imagine, and create. These systems not only analyze vast biological datasets to bring the unknown patterns, predict gene functions, or link genotypes with traits (Zhu et al., 2025), but they can also generate entirely new

TABLE 7.2 Categories of AI methods and their applications for genome editing.

Method	Applications	Algorithm used	References
CnnCrispr	Prediction of off-target potential for sgRNA	CNN and biLSTM	Liu et al. (2020)
CNN-SVR (hybrid)	Predicting gRNA cutting efficiency at intended sites	CNN and SVR	Zhang et al. (2020)
CROTON	Target site editing precision	CNN and NAS	Li et al. (2021)
ABEdeepoff and CBEdeepoff	Detection of nonspecific edits by base editors	biLSTM	Zhang et al. (2023)
DeepCRISTL	Predicting the on-target editing efficiency of CRISPR/Cas9	LSTM	Elkayam et al. (2024)
DeepMEns	Predicts sgRNA-target binding efficiency for more precise and efficient CRISPR–Cas9 genome editing	CNN, transformers, LSTM, LSTM, LSTM, LSTM, and attention mechanism	Ding et al. (2025)
EXPosition	Improving CRISPR sgRNA knockout accuracy by combining gene expression data with established prediction tools	CNN, LSTM, and SVM	Cohen et al. (2024)
DeepIndel	Predictive analysis of CRISPR/Cas9 editing efficiency	BERT-based deep learning model deep SHAP	Zhang et al. (2024)
igRNA-PS	Predicts and enhances the editing efficiency and accuracy of near-PAMless ABEs, minimizing unintended bystander edits and increasing overall precision	CNN and transfer learning	Li et al. (2024)
DeepAcr	Predicting anti-CRISPR	CNN and DNN	Li et al. (2022)

biological sequences. From synthetic DNA and RNA to protein designs that have never existed in nature, generative models are changing the way to explore biological possibility (Abdul Aziz and Masmoudi, 2026). This capability opens new edges in plant science, enabling the creation of new-to-nature biomolecules with desired properties for improved crop performance, stress tolerance, or nutritional content (Xu and Jiang, 2025). As these models continue to evolve, they are composed to become major tools in genome editing.

Depending on the biological language used for training, several categories of AI models have emerged, each providing unique applications in genome editing (Table 7.2). Protein models have rapidly gained importance due to their ability to analyze protein structures, simulate molecular interactions, and redesign functional proteins (Abdul Aziz et al., 2021). An achievement in this area is Google's AlphaFold, which accurately predicts three-dimensional protein structures and has significantly advanced molecular biology

(Jumper et al., 2021). Similarly, DNA language models have evolved since their initial introduction in 2021. Among the four models trained specifically on plant DNA, AgroNT, released in late 2023 by Google and InstaDeep (Mendoza-Revilla et al., 2024). It was trained on a massive dataset of 10 million genome sequences from 48 plant species, enabling high-precision gene prediction, trait association, and editing target design (Mendoza-Revilla et al., 2024). RNA-based models, while more established in human genomics, are beginning to gain momentum in plant science. Tools like scGPT, trained on single-cell RNA sequencing (scRNA-seq) data, show great promise for revealing the regulatory changes in gene expression during various developmental and stress-response stages in crops (Sun et al., 2024). The latest development involves multimodal AI models, which are designed to integrate and process diverse biological data types, DNA, RNA, and proteins. In 2024, InstaDeep and BioNTech introduced the first AI system capable of simultaneously analyzing genomic, transcriptomic, and proteomic datasets (de Almeida et al., 2025).

AI is rapidly becoming a core enabler in fields such as genome editing, precision agriculture, and climate-tolerant crop design. Its ability to handle large-scale, multidimensional datasets makes it an ideal partner for genome editing, allowing for smarter target selection, gRNA design, and off-target prediction (Jumper et al., 2021). As computational power, data availability, and algorithmic complexity continue to grow, AI is expected to play an even more essential role in accelerating discoveries and solving some of the most complex challenges of genome editing.

7.5 AI for precision genome editing and predictive crop enhancement

The integration of AI technologies into genomic research will enhance the entire analytical pipeline, from data preprocessing and characteristics selection to predictive modeling and biological interpretation, with significantly improved speed and efficiency. We found that and it is known by most that traditional statistical methods often struggle to get the tangled network of interactions that define most plant traits. In contrast, AI tends to perform well in this kind of complexity (Freda et al., 2024). In our viewpoint, one of its main advantages is its ability to detect delicate, nonlinear relationships among genes that govern yield, stress tolerance, and disease resistance traits, usually controlled by several interacting loci (Abdul Aziz and Masmoudi, 2023a,b). Nowadays, many of the research teams are using DL and ensemble models to examine how interaction happens between the plant genes, and we found that the results of these research teams are proving to be surprisingly accurate (Gerakari et al., 2025). What comes out is that these tools not only give and predict to us the outcomes, but they inform us the relationships among genes that our older approaches were missing. This gives information that will really matter for the plant performance.

We have seen presently that AI has started to change how many genes scientists look at in their studies, not only for predicting but also confirming their roles. In our understanding, these approaches are supporting and now providing a quicker way than the usual trial-based studies, which takes a lot of time. When ML and DL tools are applied to very large genomic datasets, what they do is that they can go through the same patterns that would otherwise take months to detect them by using manual analysis (Dublino and Ercolano,

2025). What we find interesting is that these models, if trained on existing data, can even inform functions for genes that have never been tested before.

For most of the individuals, this kind of progress is particularly important in crops because the identification of genes that contribute to yield, stress tolerance, or disease resistance is still a slow process. In this case, AI systems are now helping us to know which genes may be more relevant for our further work by taking information from sequence motifs, gene expression data, and known gene–interaction networks (Wu et al., 2024). In our view, this makes genome editing more targeted and possible, as it gives the researchers the most meaningful genetic regions. What Caudai et al. (2021) reported was that this is turning functional genomics into a more predictive and evidence-based science.

When you are engaged in making effective gRNAs, many of you can now make AI a tool for your help in CRISPR–Cas9 research. In our experience, rather than relying on the process of trial and error, we can make use of AI models because it can examine DNA sequence patterns and the surrounding genomic context. This will give us a sense of understanding how to efficiently edit a target site and how possibly off-target effects can occur in the process (Sherkatghanad et al., 2023). What we find useful is that these tools can suggest guides with higher precision and confidence, helping researchers to make better-informed choices. In addition, before doing any lab works, AI can simulate different editing outcomes, providing a clearer idea of which approaches are most promising and to proceed with. In doing this, it often helps us to save time, reduce material waste, and increase the success rate of experiments of genome-editing.

It is well known now that when an individual is editing multiple genes at once, in that situations also AI-based approaches play an essential part for designing and evaluating CRISPR strategies. Many of you have seen that these models are able to examine different combinations of gRNAs, trying to reduce the cross-interference between different targets and helping in maintaining the editing efficiency across several loci. We found in one of the recent studies that applied ML and DL to CRISPR design, suggesting that these tools are able to handle complicated situations of genome editing. This showed us the usefulness of selecting data-driven gRNA for capturing polygenic traits (Sherkatghanad et al., 2023). For most of you, it will be important because some traits such as yield, stress tolerance, and nutritional quality of plants are usually in control of a network that consists of interacting genes.

What is interesting is that DL models are making the gRNA design to go further away from single-gene prediction and moving it toward genome-wide applications. In our viewpoint, by joining sequence information with epigenetic data, these models can sometimes find delicate biological signals that are missed by the simpler tools. We provide an example of that in DeepCRISPR (Chuai et al., 2018). You can see that this tool automatically takes out main features and order than in their potential for acting gRNA candidates across large genomic regions. Many of us have observed that rather than relying only on sequence alignment, this toll also looks into the factors that may determine the efficiency and accuracy of editing, helping researchers to pick reliable and effective guides.

In the present conditions, we can find that global off-target prediction tools are increasingly using neural networks and hybrid models to search the entire genomes for viewing some possible unintended edits. What we have seen is that by taking into consideration the context of sequence and their genomic features, these systems are helping to reduce the number of false positives that often show up in simpler mismatch-based searches. Several deep learning models, such as ABEdeepoff and CBEdeepoff, are now available as web-based platforms, offering more reliable ways to assess and minimize off-target risks during genome editing (Zhang et al., 2023).

Predictive models trained on large repair-outcome datasets can estimate the distribution of indels and base-editing genotypes at a given cut site. Tools such as inDelphi, Apindel, and ML base-editing predictors (models behind BE-prediction resources) enable users to predict predominant alleles and to bias design toward desired edits (to avoid frameshifts or favor specific amino-acid substitutions) (Shen et al., 2018; Liu et al., 2022; Marquart et al., 2021).

In our experience, optimizing gene editing often benefits from using AI-based phenotype prediction models. For many of you, the way of predicting plant phenotypes often involves looking at their traits like you will consider yield. We do this by focusing into ways their genotypic information interacts with environmental conditions. It gives us important clues about which genes are determining its yield (Danilevicz et al., 2022; Ahmad et al., 2024). But now most of you know that the advanced ML and DL algorithms are picking up complex patterns and nonlinear relationships within large genomic and phenotypic datasets. What we view is that these models make it possible to build predictive strategies that help in performing gene editing with greater confidence. For instance, Kaya (2025) has provided an idea of how AI-based phenotype prediction tools can sometimes simulate specific genetic changes. This might manifest effects on the plant growth (Kaya, 2025). This kind of information can help different individuals to make more informed decisions that will be useful to them before performing edits (Khader et al., 2024).

AI is now becoming one of the aids from our viewpoint in terms of organizing the complicated logistics that happens behind the gene editing experiments. You know that most of these experiments need costly reagents, instruments, and trained individuals. It can be informed that these requirements make genome editing both expensive and demanding to run. But we can say that by using on data from earlier trials that AI programs can help build more efficient workflows and cut down on unnecessary waste (Aijaz et al., 2025). In many labs, you can expect that this type of predictive modeling will begin to guide how materials and time are allocated. What this will do is that it will help researchers to manage their budgets and daily tasks more effectively (Chavhan et al., 2025).

We have found out that AI-based tools are making genome editing a lot more accurate. Nowadays, many of you are using these tools since they make it easier to find out which genes really matter and can be chosen as editing targets (Zhu et al., 2025). For many, the tools such as CropGPT and the Intelligent Precision Design Breeding (IPDB) platform bring together different kinds of data such as genomic, transcriptomic, and even phenotypic to get a clearer idea of how certain genes behave and which ones are worth to be edited first. What makes this helpful is that it can lessen the need for complicated experiments, for instance, those that deal with plant–microbe interactions, when looking for genes involved in stress responses (Sabeem et al., 2022).

7.6 Integrating AI and genome editing: A convergent approach to modern crop trait engineering

We are now seeing that AI is changing the process of genome editing. You may know that what was used to be a manual effort is now led by data through which we make careful prediction. AI tools now play a role for identifying the most important genomic targets, designing effective gRNA sequences, and selecting the optimal CRISPR-associated nucleases for each application (Gan and Ling, 2022). We have observed that AI is not only improving the usual gene knockout methods, but it is also making ways for modifying genomes (Fig. 7.2). From what recent studies suggest, the newer models can design edits that fine-tune gene expression instead of just switching a gene off completely. From what we have seen in recent reports, this way of working sometimes called quantitative trait engineering (Wójcik-Gront et al., 2024) gives researchers a chance to handle complex traits with much more control. In our understanding, this is the direction that will likely move crop improvement closer to what many now refer to as true precision breeding.

We have noticed that one of the ongoing needs in plant genome engineering is to make genome editors that are both compact and consistent. You know this is not easy, since these proteins must still work well inside living cells and remain compatible with the plant's own immune responses. As several groups have informed, this kind of work usually needs large training datasets, strong computing systems, and collaboration across many fields such as molecular biology, bioengineering, and plant science (Abdul Aziz et al., 2025). In our viewpoint, AI is making a difference here. For many individuals, it allows them to model, test, and sometimes also predict from them the behavior of editing proteins before it is built. With the ongoing mix of AI, synthetic biology, and genome editing, researchers are moving closer to a time when designing plants with chosen traits may no longer feel like a distant goal (Huang et al., 2022). When looking for examples, we found that Huang et al. (2023) were using AlphaFold2 to look at the structural details of deaminase enzymes. Their analysis provides to us the regions that had not been recognized earlier, and what this information does is that it helps in building more efficient base editors. More recently, the creation of OpenCRISPR-1—a de novo-designed gene editor built with a large language model trained on over a million CRISPR operons—has shown how AI-driven design can expand the possibilities for biological innovation (Ruffolo et al., 2024).

An example of the way AI and gene editing are converging in agriculture comes from recent work on rice variety classification (Islam et al., 2025). Using computer vision and DL approaches, particularly convolutional neural networks (CNNs), researchers achieved highly accurate identification of rice varieties from image data. By automating some steps such as image capture, feature extraction, and classification, these AI systems were able to differentiate delicate phenotypic differences with high precision. More than variety identification, such AI-driven tools have strong potential for supporting genome editing. They are able to link visual phenotypes to underlying genetic signatures and can guide the selection of target varieties and traits for editing. When integrated with genomic datasets, this approach can strengthen trait–marker associations and improve the precision breeding pipelines, which will be accelerating the development of improved rice cultivars. Another example of the way AI and genome editing are found to change crop improvement comes

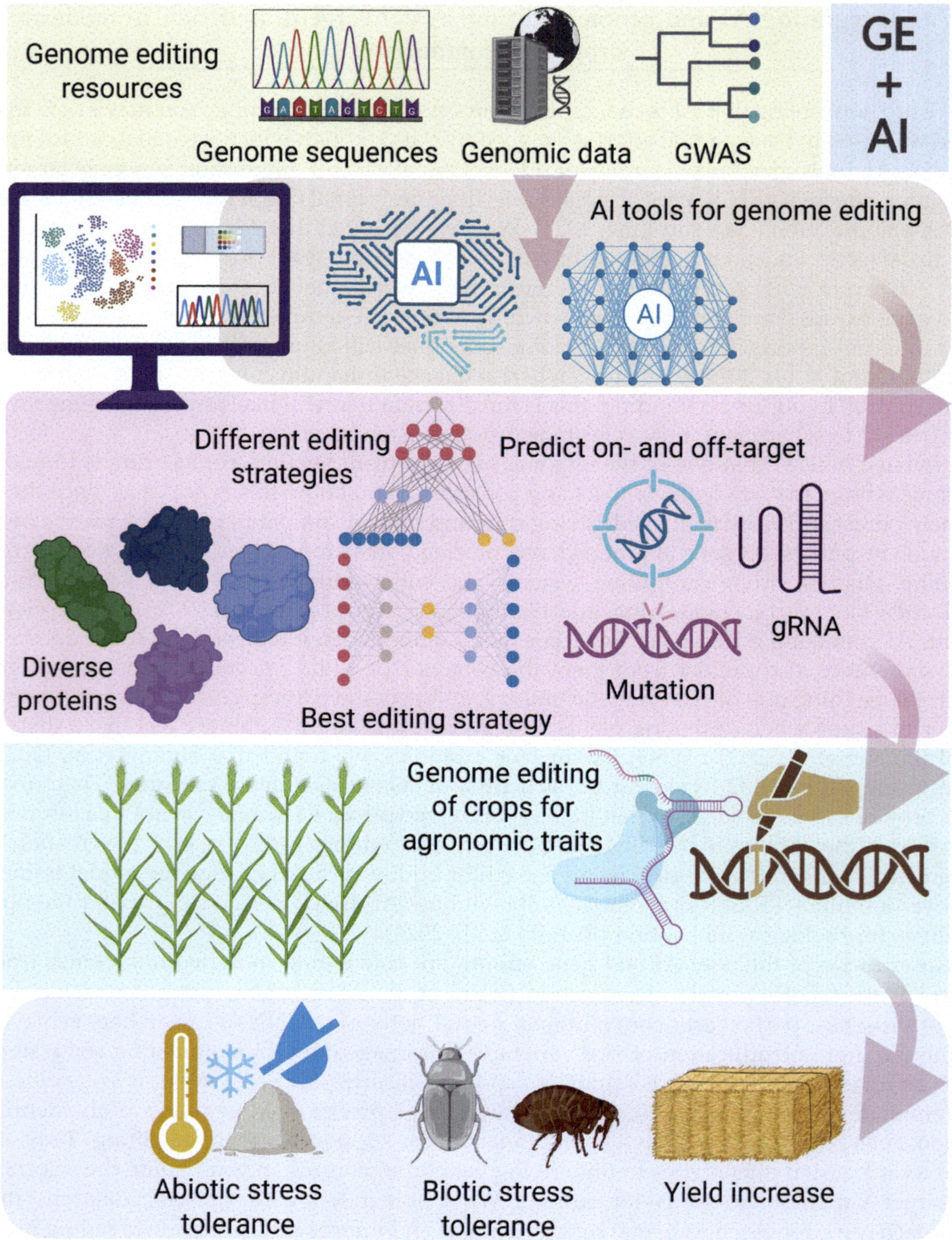

FIGURE 7.2 **Integration of genome editing (GE) and AI for improving the crops.** Genome editing (GE) and AI use in combination for improving the crops. AI analyzes genomic and phenotypic datasets for identifying optimal editing targets, predict on- and off-target effects, and to make highly efficient gRNAs. This integration brings precise genome modifications that are aimed at improving traits such as stress tolerance and yield of crops.

from recent work on wheat (Mushtaq et al., 2024). There has been challenges for developing varieties that will be tolerant to drought, heat, and salinity. For this, AI was used to examine through large genomic datasets. You may be knowing that ML models have been applied for detecting of patterns by researchers in gene expression and sequence variation across diverse wheat accessions. This will reveal for them major loci that are associated to stress tolerance. And then this information will direct their process to design CRISPR–Cas9 editing approach, where they target genes involved in osmotic regulation and antioxidant defense. Why these mechanisms, because they are the two phenomenon that allow plants to retain water and prevent cellular damage when under stress. After the tests were carried in the field, the wheat lines that were edited had got higher tolerance. The plants did it by maintaining their photosynthetic activity and water balance during drought, and they prevented oxidative damage by increasing their antioxidant systems activity. Yields also improved under various stress conditions. Now this shows the capability of targeted editing.

A disease-resistant rice was developed by using the CRISPR–Cas9 genome editing with AI (Sha et al., 2023). In this study, the focus was on a gene involved in cytidine diphosphate–diacylglycerol (CDP-DAG) synthesis, which is known to be an important lipid metabolic pathway that contributes to plant immune defense. When AI-based analysis of omics datasets was combined with high-throughput mutant screening, the team identified the most promising gene targets for precise editing. The resulting rice lines had resistance to three major pathogens, *Magnaporthe oryzae* (rice blast fungus), *Xanthomonas oryzae* pv. oryzae (bacterial blight), and *X. oryzae* pv. oryzicola (bacterial leaf streak).

In a recent study, the researchers were made able to apply in maize genome editing to aid crop improvement. They particularly used for knowing and predicting what will happen in the crossover events between maternal and paternal genomes (Farooq et al., 2024). This allowed them to identify genomic regions with higher recombination and mutation rates, which in turn increases the accuracy of genetic mapping and they will be able to selection their target genes. You can find it interesting that DNA methylation patterns are now being analyzed using AI tools (Sartor et al., 2019). This will enable these researchers to interpret the epigenetic signatures and differentiate functional genes from pseudogenes. This is going to help for identifying genomic regions that play key roles in stress adaptation. These regions are also identified through microbial interactions with plants under stress conditions (Masmoudi et al., 2021).

7.7 Implementing AI-enabled plant genome editing technologies in agricultural practices

Recent developments with AI's convergence into plant genome editing provide highly targeted and efficient trait modifications. In the United States, the biotechnology company TreeCo is working to engineer poplar trees with reduced lignin content to improve the paper production process (Service, 2023). Using an AI-driven platform, TreeCo can predict the effect of editing 21 lignin-related genes on wood composition, growth rate, and plant performance. The tool generated over 69,000 potential editing strategies, which were reduced to the top seven combinations using in silico simulations. These selected strategies are now being tested in the poplar genome.

Another example is Inari, a US-based company developing maize varieties with shorter stature and greater leaf biomass (Morrison, 2025). Their AI system analyzes mutations in promoter regions that influence specific plant traits, enabling precision edits to fine-tune gene expression. Inari is currently evaluating the performance of its genetically engineered maize in field trials in Belgium. Furthermore, the AlphaFold protein structure model has been used to redesign patatin, a native potato protein (Luo et al., 2024). Using deep learning–based structural predictions, patatin variants were engineered with enhanced thermostability and pH tolerance. These modified proteins were shown to improve the viscosity and elasticity of potato-based dough products. While this work focused on in vitro expression and characterization, the long-term aim included integrating these optimized protein traits into potato cultivars through genome editing or transgenic methods to enhance food processing qualities and nutritional value.

Large seed companies including BASF, Corteva, Bayer, and Syngenta are also turning to AI to enhance their genome editing programs. To strengthen their capabilities, many are collaborating with specialized biotech firms. For instance, BASF and Corteva have partnered with Tropic Biosciences, which provides proprietary AI platforms (Gershon, 2023). Bayer is also supporting startups such as Ukko (Weinreb, 2021), which integrate AI and genome editing technologies to develop innovative plant varieties.

7.8 Future perspectives

The convergence of genome editing and AI is composed to develop the future of crop improvement. The AI systems are expected to evolve from static prediction tools into autonomous, self-optimizing platforms that continuously learn from experimental and environmental databases. These systems will not only predict editing efficiency but also adapt editing strategies to the required conditions of plants, enhancing precision and reducing development cycles.

The multimodal AI models integrating genomic, epigenomic, transcriptomic, and environmental data will provide a broader understanding of trait expression. This will enable context-aware genome editing, where modifications are based on specific climates, stress conditions, or soil types, improving crop performance. The development of AI-designed regulatory elements, such as custom promoters and enhancers, will improve the control of gene expression, allowing for modified and tissue-specific inducible traits. Additionally, cross-species AI transfer learning will reduce the data dependency for understudied crops, increasing the adoption of advanced editing technologies across a broader range of species.

Future efforts will also emphasize scalability and field integration, enabling genome editing solutions to move beyond experimental settings. Integration with smart farming platforms and IoT-based phenotyping systems will make it possible to monitor, validate, and optimize genome edits directly under field conditions. To support these advancements, the establishment of standardized genomic databases, apparent AI frameworks, and globally consistent regulatory guidelines will be essential. Such system will ensure ethical deployment, facilitate collaboration, and enhance public trust in AI-based biotechnology. The integration of AI and genome editing is expected to form an agriculture system which will be based on data-driven, climate-challenges, and highly focused on food security needs.

7.9 Conclusion

The convergence of genome editing and AI is accelerating the potential to improve agricultural productivity through more precise genetic interventions. AI facilitates the interpretation of complex genomic datasets, guiding the identification of functional genomic elements and optimizing target site selection for highly accurate gene edits. AI and genome editing are reforming the future of crop improvement through enhancing yield, nutritional value, stress tolerance, and resource efficiency. While this is still in its early stages, initial outcomes are highly encouraging. As both technologies continue to develop, their combined application will enable new possibilities for agriculture and food security.

Declaration

"The authors declare that no generative artificial intelligence tools were used in the writing or preparation of this chapter. The content reflects the authors' own scholarly work, and the authors assume full responsibility for all statements, interpretations, and conclusions presented".

References

Abdul Aziz, M., Masmoudi, K., 2023a. Multifaceted roles of versatile LEA-II proteins in plants. In: Gupta, N.K., Shavrukov, Y., Singhal, R.K., Borisjuk, N. (Eds.), Multiple Abiotic Stress Tolerances in Higher Plants. CRC Press, Boca Raton, pp. 143–161.

Abdul Aziz, M., Masmoudi, K., 2023b. Insights into the transcriptomics of crop wild relatives to unravel the salinity stress adaptive mechanisms. Int. J. Mol. Sci. 24, 9813.

Abdul Aziz, M., Masmoudi, K., 2025. Molecular breakthroughs in modern plant breeding techniques. Hortic. Plant J. 11, 15–41.

Abdul Aziz, M., Sabeem, M., Mullath, S.K., Brini, F., Masmoudi, K., 2021. Plant group II LEA proteins: intrinsically disordered structure for multiple functions in response to environmental stresses. Biomolecules 11, 1662.

Abdul Aziz, M., Brini, F., Rouached, H., Masmoudi, K., 2022. Genetically engineered crops for sustainably enhanced food production systems. Front. Plant Sci. 13, 1027828.

Abdul Aziz, M., Masmoudi, K., 2026. Epigenetic stress memory for predictive and heritable crop epibreeding. Plant Stress 21, 101363. https://doi.org/10.1016/j.stress.2026.101363.

Abdul Aziz, M., Sabeem, M., Kutty, M.S., Rahman, S., Alneyadi, M.K., Alkaabi, A.B., Almeqbali, E.S., Brini, F., Vijayan, R., Masmoudi, K., 2023. Enzyme stabilization and thermotolerance function of the intrinsically disordered LEA2 proteins from date palm. Sci. Rep. 13, 11878.

Abdul Aziz, M., Shamim, A., Masmoudi, K., 2025. Pattern recognition receptors in plant immunity. In: Kishore, U., George, A.J.T. (Eds.), Innate Immunity: Pattern Recognition and Effector Mechanisms, Advances in Experimental Medicine and Biology, 1476. Springer, Cham, pp. 425–451.

Ahmad, M., Abdul Aziz, M., Sabeem, M., Kutty, M.S., Sivasankaran, S.K., Brini, F., Xiao, T.T., Blilou, I., Masmoudi, K., 2024. Date palm transcriptome analysis provides new insights on changes in response to high salt stress of colonized roots with the endophytic fungus *Piriformospora indica*. Front. Plant Sci. 15, 1400215.

Aijaz, N., Lan, H., Raza, T., Yaqub, M., Iqbal, R., Pathan, M.S., 2025. Artificial intelligence in agriculture: advancing crop productivity and sustainability. J. Agric. Food Res. 20, 101762.

Ashworth, J., Havranek, J.J., Duarte, C.M., Sussman, D., Monnat, R.J., Stoddard, B.L., Baker, D., 2006. Computational redesign of endonuclease DNA binding and cleavage specificity. Nature 441, 656–659.

Caudai, C., Galizia, A., Geraci, F., Le Pera, L., Morea, V., Salerno, E., Via, A., Colombo, T., 2021. AI applications in functional genomics. Comput. Struct. Biotechnol. J. 19, 5762–5790.

Chavhan, R.L., Jaybhaye, S.G., Hinge, V.R., Deshmukh, A.S., Shaikh, U.S., Jadhav, P.K., Kadam, U.S., Hong, J.C., 2025. Emerging applications of gene editing technologies for the development of climate-resilient crops. Front. Genome Ed. 7, 1524767.

Chen, F., Chen, L., Yan, Z., Xu, J., Feng, L., He, N., Guo, M., Zhao, J., Chen, Z., Chen, H., Yao, G., Liu, C., 2024. Recent advances of CRISPR-based genome editing for enhancing staple crops. Front. Plant Sci. 15, 1478398.

Christian, M., Cermak, T., Doyle, E.L., Schmidt, C., Zhang, F., Hummel, A., Bogdanove, A.J., Voytas, D.F., 2010. Targeting DNA double-strand breaks with TAL effector nucleases. Genetics 186 (2), 757–761.

Chuai, G., Ma, H., Yan, J., Chen, M., Hong, N., Xue, D., Zhou, C., Zhu, C., Chen, K., Duan, B., Gu, F., Qu, S., Huang, D., Wei, J., Liu, Q., 2018. DeepCRISPR: optimized CRISPR guide RNA design by deep learning. Genome Biol. 19 (1), 80. https://doi.org/10.1186/s13059-018-1459-4.

Cohen, S., Bergman, S., Lynn, N., Tuller, T., 2024. A tool for CRISPR-Cas9 sgRNA evaluation based on computational models of gene expression. Genome Med. 16, 152.

Danilevicz, M.F., Gill, M., Anderson, R., Batley, J., Bennamoun, M., Bayer, P.E., Edwards, D., 2022. Plant genotype to phenotype prediction using machine learning. Front. Genet. 13, 822173.

Danilo, B., Montes, É., Archambeau, H., Lodé, M., Rousseau-Gueutin, M., Chèvre, A.M., Mazier, M., 2022. I-SceI and customized meganucleases-mediated genome editing in tomato and oilseed rape. Transgenic Res. 31, 87–105.

de Almeida, B.P., Richard, G., Dalla-Torre, H., Blum, C., Hexemer, L., Pandey, P., Laurent, S., Rajesh, C., Lopez, M., Laterre, A., Lang, M., Şahin, U., Beguir, K., Pierrot, T., 2025. A multimodal conversational agent for DNA, RNA and protein tasks. Nat. Mach. Intell. 7, 928–941.

Demorest, Z.L., Coffman, A., Baltes, N.J., Stoddard, T.J., Clasen, B.M., Luo, S., Retterath, A., Yabandith, A., Gamo, M.E., Bissen, J., Mathis, L., Voytas, D.F., Zhang, F., 2016. Direct stacking of sequence-specific nuclease-induced mutations to produce high oleic and low linolenic soybean oil. BMC Plant Biol. 16, 225.

Deng, D., Yan, C., Wu, J., Pan, X., Yan, N., 2014. Revisiting the TALE repeat. Protein Cell 5 (4), 297–306. https://doi.org/10.1007/s13238-014-0035-2.

Desjarlais, J.R., Berg, J.M., 1993. Use of a zinc-finger consensus sequence framework and specificity rules to design specific DNA binding proteins. Proc. Natl. Acad. Sci. U.S.A. 90 (6), 2256–2260.

Ding, S., Zheng, J., Jia, C., 2025. DeepMEns: an ensemble model for predicting sgRNA on-target activity based on multiple features. Brief. Funct. Genomics 24, elae043.

Dublino, R., Ercolano, M., 2025. Artificial Intelligence Redefines Agricultural Genetics by Unlocking the Enigma of Genomic Complexity. Crop J., (in press).

Elkayam, S., Tziony, I., Orenstein, Y., 2024. DeepCRISTL: deep transfer learning to predict CRISPR/Cas9 on-target editing efficiency in specific cellular contexts. Bioinformatics 40, btae481.

Farooq, M.A., Gao, S., Hassan, M.A., Huang, Z., Rasheed, A., Hearne, S., Prasanna, B., Li, X., Li, H., 2024. Artificial intelligence in plant breeding. Trends Genet. 40, 891–908.

Freda, P.J., Ye, S., Zhang, R., Moore, J.H., Urbanowicz, R.J., 2024. Assessing the limitations of relief-based algorithms in detecting higher-order interactions. BioData Min. 17, 37.

Gan, W.C., Ling, A.P.K., 2022. CRISPR/Cas9 in plant biotechnology: applications and challenges. Biotechnologia 103, 81–93.

Gandomi, A.H., Chen, F., Abualigah, L., 2023. Big data analytics using artificial intelligence. Electronics 12, 957.

Gerakari, M., Katsileros, A., Kleftogianni, K., Tani, E., Bebeli, P.J., Papasotiropoulos, V., 2025. Breeding of solanaceous crops using AI: machine learning and deep learning approaches—a critical review. Agronomy 15, 757.

Gershon, G., September 26, 2023. Tropic Announces Collaboration with Corteva Agriscience Utilizing Tropic's Groundbreaking GEiGS® Technology to Develop Robust Disease Resistance Traits in Corn and Soybean. Press Release. Retrieved from. https://tropic.bio/tropic-announces-collaboration-with-corteva-agriscience-utilizing-tropics-groundbreaking-geigs-technology-to-develop-robust-disease-resistance-traits-in-corn-and-soybean/.

González Castro, N., Bjelic, J., Malhotra, G., Huang, C., Alsaffar, S.H., 2021. Comparison of the feasibility, efficiency, and safety of genome editing technologies. Int. J. Mol. Sci. 22, 10355.

Graham, N., Patil, G.B., Bubeck, D.M., Dobert, R.C., Glenn, K.C., Gutsche, A.T., Kumar, S., Lindbo, J.A., Maas, L., May, G.D., Vega-Sanchez, M.E., Stupar, R.M., Morrell, P.L., 2020. Plant genome editing and the relevance of off-target changes. Plant Physiol 183, 1453–1471.

Gupta, A., Liu, B., Chen, Q.J., Yang, B., 2023. High-efficiency prime editing enables new strategies for broad-spectrum resistance to bacterial blight of rice. Plant Biotechnol. J. 21, 1454–1464.

Huang, X., Huang, S., Han, B., Li, J., 2022. The integrated genomics of crop domestication and breeding. Cell 185, 2828–2839.

Huang, J., Lin, Q., Fei, H., He, Z., Xu, H., Li, Y., Qu, K., Han, P., Gao, Q., Li, B., Liu, G., Zhang, L., Hu, J., Zhang, R., Zuo, E., Luo, Y., Ran, Y., Qiu, J.L., Zhao, K.T., Gao, C., 2023. Discovery of deaminase functions by structure-based protein clustering. Cell 186, 3182–3195.e14.

International Wheat Genome Sequencing Consortium (IWGSC), 2018. Shifting the limits in wheat research and breeding using a fully annotated reference genome. Science 361, eaar7191.

Islam, M.M., Himel, G.M.S., Moazzam, M.G., Uddin, M.S., 2025. Artificial intelligence-based rice variety classification: a state-of-the-art review and future directions. Smart Agric. Technol. 10, 100788.

Jumper, J., Evans, R., Pritzel, A., Green, T., Figurnov, M., Ronneberger, O., Tunyasuvunakool, K., Bates, R., Žídek, A., Potapenko, A., Bridgland, A., Meyer, C., Kohl, S.A.A., Ballard, A.J., Cowie, A., Romera-Paredes, B., Nikolov, S., Jain, R., Adler, J., Back, T., Petersen, S., Reiman, D., Clancy, E., Zielinski, M., Steinegger, M., Pacholska, M., Berghammer, T., Bodenstein, S., Silver, D., Vinyals, O., Senior, A.W., Kavukcuoglu, K., Kohli, P., Hassabis, D., 2021. Highly accurate protein structure prediction with AlphaFold. Nature 596 (7873), 583–589.

Kassem, M.A., 2025. Harnessing artificial intelligence and machine learning for identifying quantitative trait loci (QTL) associated with seed quality traits in crops. Plants 14, 1727.

Kaya, C., 2025. Through high-throughput phenotyping and AI in controlled environments. Food Energy Secur. 14, e70050.

Khader, W., Kayali, M.K., Taj Eddin, T.M., Alameri, A.S., Abdul Aziz, M., Kizhakkayil, J., Senthilkumar, A., Al Dhaheri, A.S., Jaleel, A., 2024. Anticancer potential of *Calligonum comosum* L. Her. and *Calligonum crinitum* Boiss plants from United Arab Emirates. Adv. Tradit. Med. 1, 1.

Kim, Y.G., Cha, J., Chandrasegaran, S., 1996. Hybrid restriction enzymes: zinc finger fusions to Fok I cleavage domain. Proc. Natl. Acad. Sci. U.S.A. 93 (3), 1156–1160.

Krizhevsky, A., Sutskever, I., Hinton, G.E., 2012. ImageNet classification with deep convolutional neural networks. Adv. Neural Inf. Process. Syst. 25, 1097–1105.

Li, T., Huang, S., Jiang, W.Z., Wright, D., Spalding, M.H., Weeks, D.P., Yang, B., 2011. TAL nucleases (TALNs): hybrid proteins composed of TAL effectors and FokI DNA-cleavage domain. Nucleic Acids Res. 39 (1), 359–372.

Li, Y., Zhu, J., Wu, H., Liu, C., Huang, C., Lan, J., Zhao, Y., Xie, C., 2020. Precise base editing of non-allelic acetolactate synthase genes confers sulfonylurea herbicide resistance in maize. Crop J. 8, 449–456.

Li, V.R., Zhang, Z., Troyanskaya, O.G., 2021. CROTON: an automated and variant-aware deep learning framework for predicting CRISPR/Cas9 editing outcomes. Bioinformatics 37, i342–i348.

Li, Y., Wei, Y., Tan, Q., Zong, L., Wang, Y., Chen, J., Li, Y., 2022. DeepAcr: predicting anti-CRISPR with deep learning. bioRxiv. https://doi.org/10.1101/2022.04.02.486820, 2022.04.02.486820.

Li, B., Zhu, X., Zhao, D., Li, Y., Yang, Y., Li, J., Bi, C., Zhang, X., 2024. igRNA prediction and selection AI models (igRNA-PS) for bystander-less ABE base editing. J. Mol. Biol. 436, 168714.

Liu, Q., Cheng, X., Liu, G., Li, B., Liu, X., 2020. Deep learning improves the ability of sgRNA off-target propensity prediction. BMC Bioinf. 21, 51.

Liu, X., Wang, S., Ai, D., 2022. Predicting CRISPR/Cas9 repair outcomes by attention-based deep learning framework. Cells 11 (11), 1847. https://doi.org/10.3390/cells11111847.

López-Casado, G., Sánchez-Raya, C., Ric-Varas, P.D., Paniagua, C., Blanco-Portales, R., Muñoz-Blanco, J., Pose, S., Matas, A.J., Mercado, J.A., 2023. CRISPR/Cas9 editing of the polygalacturonase FaPG1 gene improves strawberry fruit firmness. Hortic. Res. 10, uhad011.

Luo, W., Sun, M., Zhang, A., Lin, C., Jin, Y., Wang, X., Huang, L., 2024. Comprehensive analysis of housekeeping genes, tissue-specific genes, and dynamic regulation across developmental stages in pearl millet. BMC Genom. 25, 1199.

Luo, X., Cao, L., Yu, L., Gao, M., Ai, J., Gao, D., Zhang, X., Lucas, W.J., Huang, S., Xu, J., Shang, Y., 2024. Deep learning-based characterization and redesign of major potato tuber storage protein. Food Chem. 443, 138556.

Marquart, K.F., Allam, A., Janjuha, S., Sintsova, A., Villiger, L., Frey, N., Krauthammer, M., Schwank, G., 2021. Predicting base editing outcomes with an attention-based deep learning algorithm trained on high-throughput target library screens. Nat. Commun. 12 (1), 5114. https://doi.org/10.1038/s41467-021-25375-z.

Masmoudi, K., Abdul Aziz, A., Shamim, A., Sabeem, M., Hazzouri, K., Amiri, K.M.A., 2021. Metagenomics of beneficial microbes in abiotic stress tolerance of date palm. In: Al-Khayri, J.M., Jain, S.M., Johnson, D.V. (Eds.), The Date Palm Genome. Springer, Cham, pp. 203–214.

McCarthy, J., Minsky, M., Rochester, N., Shannon, C.E., 2006. A proposal for the Dartmouth summer research Project on artificial intelligence, August 31, 1955. AI Mag. 27, 12–14.

Mendoza-Revilla, J., Trop, E., Gonzalez, L., Roller, M., Dalla-Torre, H., de Almeida, B.P., Richard, G., Caton, J., Lopez Carranza, N., Skwark, M., Laterre, A., Beguir, K., Pierrot, T., Lopez, M., 2024. A foundational large language model for edible plant genomes. Commun. Biol. 7, 835.

Mok, Y.G., Hong, S., Bae, S.-J., Cho, S.-I., Kim, J.-S., 2022. Targeted A-to-G base editing of chloroplast DNA in plants. Nat. Plants 8, 1378–1384.

Morrison, O., 2025. AI and gene editing poised to boost 'stagnating' corn yields: Inari. AgTechNavigator, 25 Mar 2025. Retrieved from. https://www.agtechnavigator.com/Article/2025/03/25/combining-ai-with-gene-editing-set-to-boost-stagnating-corn-yields-says-inari/.

Mushtaq, M.A., Ahmed, H.G.M.-D., Zeng, Y., 2024. Applications of artificial intelligence in wheat breeding for sustainable food security. Sustainability 16, 5688.

Rashid, A.B., Kausik, M.A.K., 2024. AI Revolutionizing Industries Worldwide: A Comprehensive Overview of Its Diverse Applications. Hybrid Advances 7, 100277. https://doi.org/10.1016/j.hybadv.2024.100277.

Rodríguez-Leal, D., Lemmon, Z.H., Man, J., Bartlett, M.E., Lippman, Z.B., 2017. Engineering quantitative trait variation for crop improvement by genome editing. Cell 171, 470–480.e8.

Ruffolo, J.A., Nayfach, S., Gallagher, J., Bhatnagar, A., Beazer, J., Hussain, R., Russ, J., Yip, J., Hill, E., Pacesa, M., Meeske, A.J., Cameron, P., Madani, A., 2024. Design of highly functional genome editors by modeling the universe of CRISPR-Cas sequences. bioRxiv 2024 04 (22), 590591.

Sabeem, M., Abdul Aziz, M., Mullath, S.K., Brini, F., Rouached, H., Masmoudi, K., 2022. Enhancing growth and salinity stress tolerance of date palm using *Piriformospora indica*. Front. Plant Sci. 13, 1037273.

Sartor, R.C., Noshay, J., Springer, N.M., Briggs, S.P., 2019. Identification of the expressome by machine learning on omics data. Proc. Natl. Acad. Sci. U.S.A. 116, 18119–18125.

Schmidhuber, J., 2022. Scientific integrity and the history of deep learning: the 2021 turing lecture, and the 2018 turing award. Tech. Rep. IDSIA-77-21 (v3) (IDSIA, Lugano, Switzerland).

Service, R.F., 2023. Genetically edited wood could make paper more sustainable. Science. https://doi.org/10.1126/science.adj7461.

Sha, G., Sun, P., Kong, X., Han, X., Sun, Q., Fouillen, L., Zhao, J., Li, Y., Yang, L., Wang, Y., Gong, Q., Zhou, Y., Zhou, W., Jain, R., Gao, J., Huang, R., Chen, X., Zheng, L., Zhang, W., Qin, Z., Zhou, Q., Zeng, Q., Xie, K., Xu, J., Chiu, T.Y., Guo, L., Mortimer, J.C., Boutté, Y., Li, Q., Kang, Z., Ronald, P.C., Li, G., 2023. Genome editing of a rice CDP-DAG synthase confers multipathogen resistance. Nature 618, 1017–1023.

Shamim, A., Abdul Aziz, M., Saeed, F., Kumari, R., Mary Joseph, A., Ponnachan, P., Kishore, U., Masmoudi, K., 2024. Revisiting surfactant protein D: an immune surveillance molecule bridging innate and adaptive immunity. Front. Immunol. 15, 1491175.

Shen, M.W., Arbab, M., Hsu, J.Y., Worstell, D., Culbertson, S.J., Krabbe, O., Cassa, C.A., Liu, D.R., Gifford, D.K., Sherwood, R.I., 2018. Predictable and precise template-free CRISPR editing of pathogenic variants. Nature 563 (7733), 646–651. https://doi.org/10.1038/s41586-018-0686-x.

Shen, B., Schmidt, M.A., Collet, K.H., Liu, Z.B., Coy, M., Abbitt, S., Molloy, L., Frank, M., Everard, J.D., Booth, R., Samadar, P.P., He, Y., Kinney, A., Herman, E.M., 2022. RNAi and CRISPR-Cas silencing E3-RING ubiquitin ligase AIP2 enhances soybean seed protein content. J. Exp. Bot. 73, 7285–7297.

Sherkatghanad, Z., Abdar, M., Charlier, J., Makarenkov, V., 2023. Using traditional machine learning and deep learning methods for on- and off-target prediction in CRISPR/Cas9: a review. Brief. Bioinform. 24, bbad131.

Shi, J., Gao, H., Wang, H., Lafitte, H.R., Archibald, R.L., Yang, M., Hakimi, S.M., Mo, H., Habben, J.E., 2017. ARGOS8 variants generated by CRISPR-Cas9 improve maize grain yield under field drought stress conditions. Plant Biotechnol. J. 15, 207–216.

Shukla, V.K., Doyon, Y., Miller, J.C., DeKelver, R.C., Moehle, E.A., Worden, S.E., Mitchell, J.C., Arnold, N.L., Gopalan, S., Meng, X., Choi, V.M., Rock, J.M., Wu, Y.Y., Katibah, G.E., Gu, Z., McCaskill, D., Simpson, M.A.,

Blakeslee, B., Greenwalt, S.A., Butler, H.J., Hinkley, S.J., Zhang, L., Rebar, E.J., Gregory, P.D., Urnov, F.D., 2009. Precise genome modification in the crop species Zea mays using zinc-finger nucleases. Nature 459, 437–441.

Sun, Y., Sun, J., Lin, C., Zhang, J., Yan, H., Guan, Z., Zhang, C., 2024. Single-cell transcriptomics applied in plants. Cells 13, 1561.

Vaswani, A., Shazeer, N., Parmar, N., Uszkoreit, J., Jones, L., Gomez, A.N., Kaiser, Ł., Polosukhin, I., 2017. Attention is all you need. Adv. Neural Inform. Proc. Syst. 30, 1.

Veley, K.M., Elliott, K., Jensen, G., Zhong, Z., Feng, S., Yoder, M., Gilbert, K.B., Berry, J.C., Lin, Z.-J.D., Ghoshal, B., Gallego-Bartolomé, J., Norton, J., Motomura-Wages, S., Carrington, J.C., Jacobsen, S.E., Bart, R.S., 2023. Improving cassava bacterial blight resistance by editing the epigenome. Nat. Commun. 14, 85.

Venkataraman, S., Zaruk, D., Hefferon, K., 2025. Recent developments, challenges and opportunities in genome editing for crop science from a societal perspective. Front. Genome 7, 1568072.

Wang, W., Simmonds, J., Pan, Q., Davidson, D., He, F., Battal, A., Akhunova, A., Trick, H.N., Uauy, C., Akhunov, E., 2018. Gene editing and mutagenesis reveal inter-cultivar differences and additivity in the contribution of TaGW2 homoeologues to grain size and weight in wheat. Theor. Appl. Genet. 131, 2463–2475.

Weinreb, G., 2021. Bayer leads $40m investment in Israeli biotech co Ukko. Globes, 27 January. Available at: https://en.globes.co.il/en/article-bayer-leads-40m-investment-in-israeli-biotech-co-ukko-1001358705.

Wójcik-Gront, E., Zieniuk, B., Pawełkowicz, M., 2024. Harnessing AI-powered genomic research for sustainable crop improvement. Agriculture 14, 2299.

Wu, Z., Sun, Y., Zhao, X., Liu, Z., Zhou, W., Niu, Y., 2024. Phenotype prediction in plants is improved by integrating large-scale transcriptomic datasets. NAR Genom. Bioinform 6, lqae184.

Xing, S., Chen, K., Zhu, H., Zhang, R., Zhang, H., Li, B., Gao, C., 2020. Fine-tuning sugar content in strawberry. Genome Biol. 21, 230.

Xu, Z., Jiang, D., 2025. AI-powered plant science: transforming forestry monitoring, disease prediction, and climate adaptation. Plants 14, 1626.

Yıldırım, K., Miladinović, D., Sweet, J., Akin, M., Galović, V., Kavas, M., Zlatković, M., de Andrade, E., 2023. Genome editing for healthy crops: traits, tools and impacts. Front. Plant Sci. 14, 1231013.

Zhang, R., Liu, J., Chai, Z., Chen, S., Bai, Y., Zong, Y., Chen, K., Li, J., Jiang, L., Gao, C., 2019. Generation of herbicide tolerance traits and a new selectable marker in wheat using base editing. Nat. Plants 5, 480–485.

Zhang, G., Dai, Z., Dai, X., 2020. A novel hybrid CNN-SVR for CRISPR/Cas9 guide RNA activity prediction. Front. Genet. 10, 1303.

Zhang, C., Yang, Y., Qi, T., Zhang, Y., Hou, L., Wei, J., Yang, J., Shi, L., Ong, S.G., Wang, H., Wang, H., Yu, B., Wang, Y., 2023. Prediction of base editor off-targets by deep learning. Nat. Commun. 14 (1), 5358. https://doi.org/10.1038/s41467-023-41004-3.

Zhang, G., Xie, H., Dai, X., 2024. DeepIndel: an interpretable deep learning approach for predicting CRISPR/Cas9-mediated editing outcomes. Int. J. Mol. Sci. 25, 10928.

Zhu, W., Li, W., Zhang, H., Li, L., 2025. Big data and artificial intelligence-aided crop breeding: progress and prospects. J. Integr. Plant Biol. 67, 722–739.

CHAPTER

8

From data to discovery: The role of AI in advancing gene editing

Emre Aksoy

Middle East Technical University, Ankara, Turkey

8.1 Introduction

Plant gene editing has changed a lot in the last two decades. Traditional breeding methods, which relied on selecting plants with desirable traits and crossbreeding them over many generations, took a long time and had a narrow focus. Genetic engineering made it possible to move specific genes from one organism to another. This led to the creation of genetically modified (GM) crops with traits like better stress resistance, herbicide resistance, and better nutritional content. However, these methods often faced regulatory and public acceptance challenges due to the introduction of foreign DNA into the plant genome.

The breakthrough happened in 2012 with the creation of CRISPR–Cas9 technology, which made it possible to edit the plant genome very precisely without the need for foreign DNA integration. This method enabled the targeted modification of specific genes, which made crops more resistant to disease, drought, and other problems. Since it was first developed, CRISPR technology has been used successfully on many crops, including rice, wheat, maize, and tomatoes.

Parallel to advancements in gene editing, the field of artificial intelligence (AI) has seen significant growth, particularly in its application to genomics. AI is made up of many different areas, such as machine learning (ML) and deep learning (DL), which have proven effective in analyzing complex biological data. In plant genomics, the combination of AI and CRISPR–Cas systems makes gene editing more useful outside of the lab by allowing the analysis of large-scale and multi-omics datasets, such as genomic, transcriptomic, and proteomic data. AI has been used to make sense of these complicated datasets, guess what genes do, find possible targets for gene editing, and speed up the process of making crops with good traits like disease resistance, drought tolerance, and better nutritional value. This combination of AI and gene editing technologies not only makes genome modifications more

Intelligent Gene Editing in Plants
https://doi.org/10.1016/B978-0-443-44079-3.00003-6

efficient and accurate, but it also has a lot of potential for solving global challenges like food security, climate change, and sustainable agriculture by making it easier to make precise changes and improve crop performance in certain environmental conditions.

Combining AI with gene editing tools, especially CRISPR–Cas systems, is changing the field of plant genomics by making crop improvement much more accurate and efficient. Traditional plant breeding methods are important, but they take a lot of time and do not cover a lot of ground. CRISPR–Cas systems reach their full potential when they are combined with AI. AI improves the design, prediction, and analysis of gene editing. AI makes gene editing more accurate and efficient by creating better CRISPR systems, predicting how well gRNA will work and how specific it will be to avoid off-target effects, and looking at the results of genetic changes. AI-powered models also predict how repairs will turn out after editing, which helps scientists figure out how gene modifications affect cells and makes gene editing results more reliable.

As the field grows, AI experts, plant biotechnologists, and regulatory bodies will need to work together across disciplines to deal with the challenges of AI-driven gene editing. Setting up strong ethical frameworks and rules will make sure that these technologies are used responsibly, getting the most out of their benefits while lowering the risks.

As we go through the chapter, it examines the role of AI in several aspects of the gene editing process, from the development of CRISPR systems to the analysis of editing outcomes. By highlighting current applications and future directions, this chapter aims to elucidate the current applications and future trajectories of AI in enhancing the accuracy, efficacy, and scope of gene editing in plants, thereby fostering more sustainable and resilient agricultural practices.

8.2 Foundations of AI in plant genomics: Understanding AI, ML, and DL

AI has changed the way researchers look at complicated biological data and sped up the process of editing genes. ML and DL are at the heart of this revolution because they make it possible to find useful patterns in huge, complex datasets. AI encompasses a spectrum of computational techniques that try to mimic human intelligence so that computers can do things like recognize patterns, make decisions, and solve problems. In plant genomics, AI systems are employed to analyze genomic sequences, predict gene functions, and model plant growth under varying environmental conditions. These abilities are very helpful for finding possible gene editing targets and improving breeding methods.

8.2.1 Learning from data via machine learning

Machine learning is a part of AI that uses algorithms to learn from data and make decisions or predictions without being told to do so. In plant genomics, ML algorithms are used to predict phenotypic traits from genotypic data. This makes it possible to find desirable traits in crops (Bose et al., 2024). Linear and logistic regression are two common ML methods used in plant genomics. Linear regression is used to predict continuous outcomes, while logistic regression is used to predict categorical outcomes. Support vector machines (SVMs) are another machine learning method used in plant genomics. They work well for classification tasks, like telling the difference between resistant plant varieties and those that are sensitive. Another way to choose features and work with big, complicated datasets is to use random forests and gradient boosting machines.

These ML techniques have shown efficacy in multiple applications (He et al., 2024). For example, they help map quantitative trait loci (QTLs) linked to complex traits, which helps us understand genetic architecture and look at gene expression patterns, which give us clues about regulatory mechanisms and gene function (Farooq et al., 2024). In genomic selection (GS), ML models use high-throughput genotyping technologies to collect genetic data from many markers. The predictive efficacy of these models arises from their capacity to formulate mathematical correlations between genotypic and phenotypic data obtained from training populations (Bose et al., 2024). This makes it easier to choose better genotypes. For instance, ML has helped make GS predictions more accurate for different crops, which has made breeding programs more effective (Montesinos-López et al., 2021). The integration of ML in GS not only simplifies the breeding process but also allows researchers to pinpoint and choose intricate traits that are challenging to evaluate via conventional techniques.

ML algorithms are increasingly used to find and predict protein variants that fit with certain engineering purposes. This is an important step in directed evolution in synthetic biology (Zhang et al., 2025). This ability is especially useful in plant protein engineering, where it used to take a lot of time and effort to figure out how changing one amino acid would change the properties of a protein. ML models can quickly screen and choose engineered proteins or enzymes. This speeds up the process of developing crops with desirable traits like higher yields, resistance to disease, and adaptability to different climates (Zhang et al., 2025).

Moreover, ML algorithms facilitate accurate phenotyping, essential for the progression of crop breeding (Gill et al., 2022). ML can make predictive models that help us understand and measure complex plant traits by looking at large phenotypic datasets (Bose et al., 2024). This is especially pertinent for complex traits affected by multiple genes × environmental interactions. The application of ML in phenotyping includes several aspects of agricultural research, such as identifying drought-resistant genes and predicting seed quality traits, as evidenced by studies on *Arabidopsis thaliana*, soybean, and maize (Sheikh et al., 2024). ML is a crucial instrument for contemporary plant genomics due to its capacity to analyze and comprehend vast quantities of data. This accelerates and enhances the precision of judgments about gene editing and agricultural enhancement.

8.2.2 Unveiling complex patterns by deep learning

Deep Learning (DL) is a subfield of ML that uses neural networks with multiple layers to model complicated relationships in data. This method is especially useful in plant genomics for combining different types of omics data (e.g., genomics, transcriptomics, and metabolomics) to get a complete picture of plant biology. It also improves genomic selection (GS) by finding non-linear relationships between genetic markers and phenotypic traits, and it automates the creation of gene editing tools like guide RNAs (gRNAs) for CRISPR systems by predicting off-target effects and improving targeting efficiency (Abbasi et al., 2025). This is a very important step because the success of gene editing depends a lot on how well gRNAs are designed. Deep learning (DL) algorithms can look at huge amounts of data on known gRNA sequences and their editing results to find complex patterns that can help them figure out the best gRNA designs. This is something that would be very hard to do with traditional methods (Abbasi et al., 2025).

Some important DL architectures are convolutional neural networks (CNNs), which are used to look at spatial data like imaging-based phenotyping, recurrent neural networks (RNNs), which are good for sequential data like time-series gene expression profiles, and Graph Neural Networks (GNNs), which are used to model relationships in graph-structured data like gene regulatory networks (Mienye and Swart, 2024). DL has made a lot of progress in plant genomics, such as making GS predictions more accurate and creating frameworks like Cropformer and CropARNet for crop phenotype prediction that can be understood (Wang et al., 2025; Zhou et al., 2025).

In addition, DL has changed the way protein engineering works in the context of gene editing. ProteinMPNN and other models like it quickly create candidate amino acid sequences for protein modeling. These models are more versatile and accurate in real life than traditional physics-based methods (Dauparas et al., 2022). This ability is important for creating new proteins or enzymes with specific functions that can then be added to plants to give them new traits. DL's ability to predict protein structures without needing to look at similar structures, as shown by AlphaFold, opens up even more possibilities for directed evolution in plant biotechnology (Zhang et al., 2025). AlphaFold is a neural network-based model that can reliably and accurately predict the three-dimensional shapes of proteins. This is important because it affects how stable and functional they are (Jumper et al., 2021). This is especially important for the heterologous expression of engineered proteins in plants, which makes sure they stay stable and work well.

In the broader scope of crop research, DL models have shown immense potential in precise phenotyping, which is essential for advancing crop breeding (Bose et al., 2024). By analyzing large-scale phenotypic datasets, DL can generate predictive models that help in understanding and quantifying complex plant traits. This includes the analysis of imaging-based phenotyping data using CNNs to identify and quantify traits like plant growth, disease symptoms, and yield components with high accuracy (Gill et al., 2022). The incorporation of multi-omics data using DL offers a comprehensive perspective on plant biology, enabling researchers to elucidate complex interactions among genes, environment, and phenotype, hence facilitating better educated gene editing methodologies (Maraveas, 2024).

8.2.3 Integrating AI in plant phenomics, genomics, and synthetic biology

Combining ML and DL techniques with traditional biological knowledge is how AI is being used in plant genomics to improve research results. This approach to different fields makes high-throughput phenotyping easier by automating the analysis of phenotypic data to speed up the discovery of traits. It also makes predictive breeding possible by using AI models to predict how well different plant varieties will do in different environments. Finally, it makes synthetic biology possible by creating new genetic constructs and pathways for improving crops (Bose et al., 2024; Sheikh et al., 2024). AI-powered tools also help scientists manage and analyze large genomic datasets, letting them find new insights that they could not before.

AI has made a big difference in high-throughput phenotyping, which is a major problem in plant breeding. Conventional phenotyping techniques frequently require significant labor and time and are susceptible to human error. AI, especially computer vision and deep learning, makes it possible to automatically and accurately measure different plant traits from images and other sensor data (He et al., 2025). For instance, CNNs can look at pictures taken by drones or ground-based platforms to see how plants are growing, find diseases, and guess how much they will produce. This gives quick and accurate phenotypic data that can be combined with genomic data to speed up breeding (He et al., 2025; Yang et al., 2025). This feature makes it possible to screen large groups of plants and find desirable traits faster and more accurately than ever before.

AI models are changing the way plant varieties are chosen and developed in predictive breeding. AI algorithms can predict how well new plant varieties will do in a variety of situations with great accuracy by using large sets of genomic, phenotypic, and environmental data (Xu et al., 2022; Pérez-Enciso et al., 2025). This includes making predictions about traits like how well a plant can handle drought, how well it uses nutrients, and how well it can fight off diseases. For example, advanced ML models are being used to guess the genetic value of people in breeding populations. This helps breeders make better choices about which plants to cross and choose, which greatly shortens the time it takes to breed new plants (Xu et al., 2022). This ability to predict is especially useful when it comes to climate change, where it is important to grow crops that can withstand changing weather.

AI is significantly impacting crop enhancement via synthetic biology, which encompasses the design and creation of innovative biological components, technologies, and systems. AI algorithms can facilitate the development of novel genetic constructions and pathways to enhance particular crops (Zhang et al., 2025). This encompasses identifying optimal gene pairings for certain qualities, forecasting the functionality of synthetic genetic systems, and enhancing gene circuits. Methods that are driven by AI have the ability to develop CRISPR guide RNAs that exhibit improved specificity and efficacy, which in turn reduces the number of off-target effects and increases the overall success rate of gene editing studies (Abbasi et al., 2025). The capacity to rapidly build and evaluate synthetic biological components in silico accelerates the improvement of crops with increased nutritional value, improved stress tolerance, or enhanced yields.

AI is also making a big difference in crop improvement through synthetic biology, which is the design and construction of new biological parts, devices, and systems. AI algorithms can help create new genetic constructs and pathways to improve specific crops (Zhang et al., 2025). This includes finding the best gene combinations for desired traits, predicting how synthetic genetic systems will work, and optimizing gene circuits. For instance, AI-powered tools can make CRISPR guide RNAs that are more specific and efficient, which lowers the number of off-target effects and raises the overall success rate of gene editing experiments (Abbasi et al., 2025). The ability to quickly design and test synthetic biological parts in silico speeds up the creation of crops with better nutritional value, better stress resistance, or higher yields.

Furthermore, AI is essential for organizing and evaluating the continuously growing number of genomic and multi-omics data produced in plant research. AI-driven systems may assimilate and evaluate varied datasets, such as genomics, transcriptomics, proteomics, and metabolomics, to reveal intricate biological linkages and pinpoint possible gene targets for editing (Khan et al., 2022; Xu et al., 2022). This holistic approach enables researchers to attain a profound comprehension of plant biology and to discover novel genes or pathways that may be modified for crop enhancement. Insights derived from AI-driven data analysis are essential for formulating accurate and successful gene editing techniques, hence enhancing sustainable and resilient agriculture practices worldwide (Bose et al., 2024).

8.3 AI-driven gRNA design and optimization for plants

The design of guide RNAs (gRNAs) is very important for the accuracy and speed of CRISPR–Cas systems when editing plant genes. These little RNA molecules tell the Cas enzyme where to go in the DNA to make changes. It is still very hard to make gRNAs that are both very good at targeting the right site (on-target activity) and very specific, so that they don't change similar sequences in other parts of the genome (off-target effects). AI, especially ML and DL, has become a powerful tool for solving this problem and has changed the way gRNA is designed and optimized for use in plants.

8.3.1 Predictive models for gRNA efficiency and specificity

AI-powered predictive models are changing how gRNAs are made, going from trial-and-error to data-driven optimization. These models look at huge amounts of data from gene editing experiments that worked and didn't work. They learn about the complex sequence features and structural characteristics that affect how well gRNA works. AI can use these insights to guess how well and specifically potential gRNA sequences will work before they are tested in the lab. This saves a lot of time and money on gene editing projects (Kim et al., 2025; Abbasi et al., 2025).

Plant genomes are very complicated because they have polyploidy, large genomes, and repeated sequences. This means that gRNA design tools and algorithms for plant genomes need to be made just for plants. Basic sequence alignment algorithms were often used in early gRNA design tools to find possible off-target sites. But modern AI-powered tools use advanced machine learning algorithms like support vector machines (SVMs), random forests, and deep neural networks to make better predictive models (Li et al., 2025). These models learn from experimental data from different plant species, which lets them find small sequence features and epigenetic changes that affect gRNA activity and specificity in plants (Jiang et al., 2025).

There are now a number of bioinformatics tools that are made just for designing plant gRNAs. CRISPR-PLANT is an online tool that helps scientists make very specific gRNAs for CRISPR–Cas9-mediated genome editing in different plant species. It does this by looking at possible off-target sites across the genome (Minkenberg et al., 2018). But most of these older tools use rule-based or heuristic algorithms instead of AI, which makes it harder for them to find the complex factors that affect gRNA activity. In recent times, tools that use DL have done a better job of predicting gRNA efficiency (Xie et al., 2014). These models can learn complicated, non-linear connections between the features of gRNA sequences and the results of their editing, which makes predictions more accurate. These kinds of improvements are very important for speeding up the creation of new crop varieties with better traits because they help researchers quickly find the best gRNAs for their target genes.

8.3.2 Minimizing off-target effects in plant gene editing

Off-target effects, where the CRISPR–Cas system cleaves DNA at unintended sites, pose a significant concern in gene editing, particularly for agricultural applications where precise and predictable outcomes are essential for regulatory approval and consumer acceptance. AI approaches are playing a pivotal role in minimizing these off-target effects by enhancing on-target activity and reducing unintended edits (Dutton, 2025).

AI models are employed to identify and prioritize gRNA sequences that exhibit high on-target activity while simultaneously minimizing off-target binding. This involves analyzing factors such as gRNA-DNA mismatch tolerance, chromatin accessibility, and the presence of single-nucleotide polymorphisms (SNPs) that might influence off-target cleavage. By combining these different data sources, AI algorithms can forcast the likelihood of off-target events for a given gRNA sequence, allowing researchers to select gRNAs with the most favorable specificity profiles (Dunka, 2022).

Recent advancements in AI-driven strategies for minimizing off-target effects encompass the creation of advanced DL models capable of predicting and quantifying off-target activity with unparalleled precision. For example, models utilizing CNNs and recurrent neural networks (RNNs) have been trained on extensive datasets of CRISPR–Cas9 cleavage sites and off-target occurrences, enabling the identification of nuanced sequence patterns that lead to undesired modifications (Zhang et al., 2023). These models can be utilized to create gRNAs with enhanced selectivity by circumventing sequences susceptible to off-target binding. A further intriguing strategy is utilizing AI to develop customized Cas enzymes with improved fidelity. Through the examination of the structural drivers of Cas specificity, AI algorithms can propose modifications that diminish off-target cleavage while preserving elevated on-target efficiency (Hajiahmadi et al., 2019). This has resulted in the creation of 'high-fidelity' Cas variants that are especially beneficial for applications necessitating exceptional accuracy, such as therapeutic gene editing in plants (Liu et al., 2022a). Additionally, AI is employed to enhance delivery mechanisms for CRISPR components, guaranteeing that the gene editing apparatus efficiently and accurately reaches target cells, thereby minimizing exposure to non-target cells and mitigating the risk of off-target effects (Modrzejewski et al., 2020).

In recent years, in addition to conventional and well-established gRNA design platforms, an increasing number of AI-based tools have been designed to increase Cas targeting efficiency while minimizing unintended off-target impacts (Table 8.1). In practice,

TABLE 8.1 Comparison of conventional and AI-driven CRISPR gRNA design platforms across species and algorithmic approaches.

Tool/ Platform	Species focus	Algorithm type	Off-target emphasis	Pros	Cons	Reference
CRISPR-P 2.0	Plants (various)	Rule-based with plant-specific heuristics	Basic genome-wide off-target search	Designed for plants; simple interface; integrates plant genome features (e.g., promoters, PAM sites)	Limited learning capacity; less accurate for complex efficiency prediction	Liu et al. (2017)
CROPSR	Crops (including polyploids)	ML scoring + heuristics	Integrated off-target scanning with genome complexity handling	Handles large and complex crop genomes; genome-scale design; database integration	Less customizable; limited deep learning predictive power	Müller Paul et al. (2022)
Benchling design tool	Broad (mammalian, some plant genomes supported)	Rule-based + Azimuth scoring	Yes (Doench et al., 2016) (algorithms)	User-friendly; integrates cloning and LIMS; team workflows	Proprietary backend; limited transparency of models	Uniyal et al. (2019)
Synthego design tool	Broad	Rule-based + Azimuth scoring	Yes (on/off-target)	Integrated synthesis; good accuracy; easy to use	Commercial; limited flexibility for academic customization	Roginsky (2018)
Deep CRISPR	Human (generalizable)	DL (unsupervised pre-training + CNN)	Joint on/off-target prediction	High predictive accuracy; learns complex features; public code	Requires computational resources; less plant-focused	Chuai et al. (2018)
DeepSp Cas9	Human/ SpCas9	DL (CNN trained on large datasets)	On-target only	Excellent generalization; benchmarked on multiple datasets	Limited to SpCas9; no off-target module	Kim et al. (2019)
CRISPRon	Broad	DL with data integration	On-target only	Uses large integrated datasets; accurate efficiency prediction; user-friendly web server	No explicit off-target prediction	Xiang et al. (2021)
CRISPR-DIPOFF	Human	Interpretable DL	Yes (interpretability emphasized)	Balances precision/recall; interpretable feature contributions	Not plant-specific; more computationally intensive	Toufikuzzaman et al. (2024)

these tools sit on a spectrum from purpose-built plant rule/heuristic systems to genome-scale and deep-learning models. For example, CRISPR-P 2.0 was developed specifically for plants and provides plant-aware sgRNA selection and scoring, improving on basic rule-based designs by incorporating plant genome context for more reliable candidate selection (Liu et al., 2017). CROPSR targets crop genomes (including polyploid and AT-rich genomes) with a genome-wide workflow and implements a new ML scoring model plus integrated primer design and database storage for whole-genome guide catalogs (Müller Paul et al., 2022). Its strength is genome-scale automation and explicit handling of crop genome complexity. In contrast, commercial lab platforms such as Benchling and Synthego's Design Tool provide user-friendly end-to-end design workflows (visualization, on/off-target scoring, cloning/plasmid assembly support, and integration with lab records). Benchling integrates widely used scoring algorithms (e.g., Doench rules) for on- and off-target ranking within a team-oriented Laboratory Information Management System (LIMS), while Synthego's tool bundles design with synthesis/validation pipelines and uses established scoring models (Azimuth/Doench) in its backend (Uniyal et al., 2019; Roginsky, 2018; Doench et al., 2016). On the DL side, DeepCRISPR introduced an unsupervised pre-training + CNN framework to concurrently forecast on-target effectiveness and comprehensive whole-genome off-target profiles, showing improved generalization versus older methods by learning representations from massive unlabeled sgRNA sets (Chuai et al., 2018). DeepSpCas9 is another high-performing deep model trained on large empirical datasets of SpCas9 indel frequencies and is notable for its strong generalization across independent datasets (Kim et al., 2019). CRISPRon (a later data-integration + DL model) further improved on-target efficiency predictions by training on a large, integrated dataset and providing an interactive web server for design (Xiang et al., 2021).

As a general guideline, rule-based tools rely on manually defined sequence heuristics (e.g., PAM motifs, GC content, Doench scoring), whereas ML–based tools employ statistical models trained on labeled datasets but lack the hierarchical feature extraction of deep architectures. In contrast, DL approaches (e.g., convolutional neural networks) can model complex, non-linear relationships and capture subtle sequence–context interactions. Plant-specific tools, on the other hand, emphasize genome availability and species-specific promoter or PAM characteristics rather than advanced learning algorithms. For off-target prediction with emphasis on model interpretability, CRISPR-DIPOFF implements interpretable deep models and integrated-gradient style explanations to reveal sequence subregions linked to off-target risks while attempting to balance precision/recall tradeoffs (Toufikuzzaman et al., 2024). Broad reviews and benchmarks show that learning-based and DL models (e.g., DeepCRISPR, DeepSpCas9, CRISPRon) generally outperform simple heuristic/rule-based scores for on-target ranking and can better model complex sequence/context interactions, while tools aimed at crops (CRISPR-P 2.0, CROPSR) emphasize species/genome-specific constraints and experimental validation workflows (Lee, 2023).

Case studeis illustrate the efficacy of AI in this domain. Researchers have created machine learning models to forecast the effectiveness and specificity of gRNA for various Cas9 variants, facilitating the selection of appropriate gRNAs for accurate gene editing in crops such as rice and wheat (Li et al., 2025). These models have demonstrated a substantial decrease in off-target mutations relative to conventional design methodologies. Another example pertains to the application of deep learning to examine extensive CRISPR screening data, discerning sequence patterns and structural attributes that influence off-target activity. This data is subsequently utilized to enhance gRNA design methods, resulting in the creation of more precise gRNAs (Liu et al., 2022a).

Additionally, AI is employed to create innovative Cas enzymes with enhanced selectivity. Through the examination of the structural and functional characteristics of diverse Cas proteins, AI can facilitate the development of novel Cas variants that demonstrate diminished off-target activity while maintaining on-target efficacy. This AI-driven iterative design method aims to provide next-generation gene editing tools that are safer and more accurate for plant applications. The use of AI into gRNA design and optimization processes is boosting the efficiency of gene editing while maintaining the safety and dependability of genetically modified crops for sustainable agriculture.

8.4 Explainable AI in plant bioinformatics

As AI models become increasingly complex and are adopted in essential applications such as gene editing and crop improvement, the need for transparency and interpretability, collectively known as Explainable AI (XAI), becomes paramount. XAI seeks to address the unclear characteristics of several sophisticated AI models, especially DNN, by elucidating the processes behind their predictions and choices (Karim et al., 2023). In plant bioinformatics, XAI is essential for fostering confidence among researchers, enabling model debugging, and guaranteeing that AI-generated suggestions are biologically valid and practical.

One of the primary motivations for XAI in plant bioinformatics is to enhance the understanding of complex biological processes. While AI models can accurately predict phenotypic traits from genomic data or optimize gene editing strategies, understanding why a particular prediction was made or how an optimization was achieved is often as important as the prediction itself. For instance, in genotype-to-phenotype prediction, XAI algorithms can identify the specific genetic polymorphisms or gene interactions that contribute most significantly to a predicted trait, providing valuable biological insights that can guide further research and breeding efforts (Novielli et al., 2024). This interpretability enables plant scientists to progress beyond simple correlation to comprehend causal linkages, hence expediting the identification of novel genes and pathways.

The benefit of this approach is clearly demonstrated in crop breeding. For example, when predicting complex agronomic traits, XAI models can pinpoint which known stress-responsive transcription factors or metabolic genes are driving a prediction for drought tolerance, thereby validating the model's decision against well-known biological information and building breeder trust (Danilevicz et al., 2025). Furthermore, XAI moves beyond validation to novel discovery. In an application to wheat genomics, an explainable deep learning model was used to analyze SNP data for yield prediction. The model not only achieved high accuracy but, through its interpretability layers, successfully identified and prioritized non-obvious SNPs in promoter regions and genes with previously uncharacterized functions in yield formation (Ruggeri and Vidács, 2025). This capability to highlight potentially causal genetic variants from vast genomic datasets transforms black-box predictors into powerful tools for hypothesis generation, directly guiding subsequent functional validation experiments with CRISPR–Cas9 to confirm the role of these candidate genes.

Diverse XAI methodologies are being modified and created for bioinformatics applications. These may be roughly classified as model-agnostic approaches, applicable to any AI model, and model-specific methods, designed for particular model architectures. Model-agnostic approaches such as SHAP (SHapley Additive exPlanations) and LIME (Local Interpretable Model-agnostic Explanations) elucidate the contribution of particular characteristics to a model's output (Karim et al., 2023). In plant genomics, these methodologies can elucidate predictions concerning disease resistance, yield, or stress tolerance by emphasizing the most significant genes or environmental variables. A study may employ SHAP to ascertain the genes most pivotal for a plant's drought tolerance, despite the complexity of the underlying AI model, as illustrated in a maize study where SHAP analysis of an ensemble model pinpointed essential transcription factors and osmotic stress genes influencing drought resilience predictions (Wu et al., 2025). Model-specific explainable artificial intelligence approaches are frequently utilized for deep learning models, especially in image-based phenotyping. Grad-CAM (Gradient-weighted Class Activation Mapping) can illustrate the areas of an image that a Convolutional Neural Network (CNN) concentrates on during prediction, such as detecting illness signs on a leaf (Mostafa et al., 2023). This visual elucidation aids researchers in determining if the model is acquiring physiologically pertinent properties or only depending on false connections. These insights are essential for troubleshooting models and guaranteeing their resilience in practical agricultural environments.

Beyond interpretability, XAI also addresses concerns related to bias and fairness in AI applications. In plant breeding, biased datasets could lead to AI models that favor certain genotypes or environmental conditions, potentially exacerbating existing inequalities or limiting genetic diversity. XAI can help identify and mitigate such biases by revealing the underlying factors influencing an AI model's decisions, ensuring that the technology is applied equitably and responsibly (Karim et al., 2023). This is particularly important as AI-driven tools become more integrated into decision-making processes for global food security.

While XAI offers significant advantages, its application in plant bioinformatics is still evolving. Challenges include the inherent complexity of biological data, the need for domain-specific adaptations of XAI methods, and the computational resources required for generating explanations for large-scale datasets. However, ongoing research is continuously developing more sophisticated and efficient XAI tools that are tailored to the unique demands of plant genomics and gene editing. The future of AI in plant bioinformatics will undoubtedly involve a deeper integration of XAI, fostering greater transparency, trust, and ultimately, more impactful discoveries in the quest for sustainable agriculture.

8.5 Enhancing plant genome editing outcomes with AI

The application of CRISPR–Cas systems in plants has progressed well beyond conventional gene knockouts. The current frontier focuses on achieving precise genetic modifications, including single-nucleotide substitutions, targeted insertions, and complex gene replacements, which demand a detailed understanding and precise control of cellular DNA repair mechanisms. AI is crucial in this transformation by expediting the advancement and refinement of next-generation genome editing technologies, including base editors and prime editors, thereby creating new opportunities for precise and effective crop enhancement.

8.5.1 Predicting repair outcomes post-editing

The outcome of a CRISPR-induced DSB is dictated by the intricate and competing cellular DNA repair mechanisms, chiefly non-homologous end joining (NHEJ) and homology-directed repair (HDR). The result, whether a minor insertion or deletion (InDel), a specific point mutation, or a smooth gene insertion, has traditionally been challenging to anticipate, frequently necessitating comprehensive experimental screening. Artificial intelligence is now utilized to simulate these repair mechanisms, predicting the most likely results for a specific target sequence and so facilitating the development of more efficient editing procedures. Predictive models, especially those utilizing deep learning, are developed using extensive databases of previous editing results. These models acquire the ability to correlate certain sequence contexts, chromatin accessibility characteristics, and epigenetic modifications with the resultant repair products. For example, a model may ascertain that a given nucleotide composition at the cut site is more likely to yield a +1 insertion instead of a deletion, or that the existence of microhomology regions significantly favors a particular deletion pattern (Özden and Minary, 2024).

The principles established in mammalian systems are now being adapted for plant genomics. Tools like Pythia, an AI-based system that learns DNA repair patterns to predict how cells will mend CRISPR-induced DSBs, demonstrate the potential for high-accuracy forecasting (Naert et al., 2025). While initially developed for human cells, the underlying computational framework is transferable. Pythia's AI-generated repair templates act as "molecular glue" to guide precise outcomes, a concept that could be revolutionary for plant gene targeting, where HDR efficiency is notoriously low. This transition from wet-lab-heavy experimentation to in silico prediction is a central theme in modern CRISPR research, as highlighted by systematic reviews that catalog the growing ecosystem of AI predictors designed to navigate the complexity of gene editing outcomes (Abbasi et al., 2025). Beyond Pythia, other DL architectures offer complementary capabilities. For instance, DeepIndel offers an interpretable deep learning framework tailored for predicting CRISPR/Cas9-mediated insertion and deletion (InDel) profiles, elucidating the most likely editing outcomes, essential for formulating effective knock-out strategies in plants (Zhang et al., 2024). Similarly, DL approaches like Apindel, which uses BiLSTM and attention mechanisms to forecast hundreds of potential mutation labels from sequence data alone, provide a blueprint for plant-specific model development (Liu et al., 2022c). However, predicting on-target outcomes is only one part of the challenge; ensuring specificity is equally critical. DL models are also being advanced to predict CRISPR off-target effects with high sensitivity, a vital step for improving the safety and regulatory acceptance of edited crops (Kumar, 2024). As comprehensively reviewed by Alipanahi et al. (2025), the synergy between diverse deep learning models, from CNNs to transformers, and curated genomic databases is creating an integrated knowledge base. This powerful combination is poised to accelerate the adaptation of these mammalian-tested AI tools, enabling more predictable, efficient, and safe precision genome editing in plants.

The reproducibility of repair distributions at a given target site suggests that well-trained predictive models could generalize across biological systems (Allen et al., 2019). For plant scientists, this means the future ability to input a target DNA sequence and receive a probabilistic report of the expected indels, informing the choice of target site or the need for specialized editors like base editors to achieve the desired change. This capability is paramount for applications like gene knock-in, where predicting the efficiency of HDR versus NHEJ is crucial for successfully integrating large DNA fragments, such as those encoding entire metabolic pathways for biofortification or stress resistance.

8.5.2 Advancements in base and prime editing for plants

Base editing and prime editing are new ways to edit genomes with great accuracy. They let you make specific changes to nucleotides without using DSBs or donor DNA templates. While powerful, their efficiency and product purity can vary significantly depending on the sequence context and the specific editor used. AI is becoming more and more important in making these new editing methods work better for improving crops.

Base editors (BEs) fuse a catalytically impaired Cas protein to a deaminase enzyme, allowing exact chemical conversion of one base pair into another (e.g., C•G to T•A). However, challenges such as off-target deamination, sequence context dependency, and predictable product purity (avoiding undesirable bystander edits) remain. AI-driven protein engineering is now playing a transformative role in overcoming these limitations. One important method is to utilize ML to look at high-throughput screening data from large, varied libraries of deaminase variations. For instance, by training on data from a single-round diversification library, ML models can predict mutations that simultaneously enhance editing efficiency, narrow the editing window, and alter sequence context preference, leading to the development of more specific and compact base editors (Lelanskyi et al., 2025). This approach effectively co-optimizes multiple fitness parameters, such as activity and specificity, which are often in tension during traditional engineering efforts (Ding et al., 2024).

Furthermore, the process of directed evolution, supercharged by machine learning, is accelerating the discovery of novel base editing enzymes. ML models can predict the functional outcomes of countless unsynthesized deaminase variants by learning from a fraction of experimentally tested sequences, dramatically reducing the experimental burden and guiding the survey of the massive protein sequences toward optimized editors with reduced off-target activity (Perrotta et al., 2024). This culminates in a powerful, integrated workflow where AI not only designs the editors themselves but also guides their application. Advanced ML models can now predict the optimal base editor for a given target sequence, factoring in the local sequence context to maximize on-target efficiency while minimizing bystander edits, a critical consideration for developing clean, predictable edits in crops (Fong and Wong, 2023). This end-to-end AI-driven pipeline, from enzyme creation to tool selection, represents a paradigm shift in the advancement and deployment of precision gene editing technologies for crop improvement.

Prime editing (PE) provides enhanced flexibility by employing a prime editing guide RNA (pegRNA) to drive a Cas9-reverse transcriptase fusion to a target locus, where it directly inscribes new genetic material into the genome. The creation of highly effective pegRNAs is intricate, necessitating the optimization of the primer binding site (PBS) and reverse transcriptase template (RTT) sequences. The intricacy of pegRNA design renders it an optimal challenge for AI, as evidenced by the creation of machine learning models such as DeepPE, which utilize extensive editing outcome data to forecast the efficacy of pegRNA designs and inform the selection of ideal parameters for maximal editing success (Kim et al., 2018; Mathis et al., 2023). Machine learning models are being trained on extensive datasets of prime editing outcomes to comprehend the principles that dictate prime editing efficiency. These models can forecast the success rate of a prospective pegRNA design, thereby obviating the necessity for expensive and time-intensive experimental screening of various designs (Mathis et al., 2023). In plants, where transformation and regeneration are protracted processes, this in silico pre-screening represents a significant enhancement in efficiency. Artificial intelligence can assist in addressing the specific obstacles associated with implementing precision editing in plants, including the management of the editing machinery's size for delivery and the optimization of its expression throughout various plant tissues.

This synergistic integration of AI with base and prime editing is fundamentally advancing the ambition of predictive genome design in plants. We are transitioning from an era of editing characterized by empirical testing to one defined by computational foresight. By modeling the complex cellular decision-making processes that follow the introduction of an editor, AI closes the critical loop between intervention and outcome. This capability enables a new paradigm of "smart" editing. Researchers can now in silico pre-screen thousands of potential edits, select the most precise molecular tool for the intended change, and forecast its phenotypic consequence with growing confidence. This empowers strategies previously deemed too complex, such as the fine-tuning of promoter elements for optimized gene expression, the intelligent introduction of allelic series from crop wild relatives, and the systematic correction of deleterious mutations underlying key agronomic traits. Eventually, the union of AI and precision editing is forging a direct, predictable pathway from genetic code to desired crop phenotype, enabling the programmable creation of novel genetic diversity and accompanying in a new age of design-driven crop enhancement.

8.6 Integration of AI in plant breeding and trait development

The true measure of success for any technological advancement in agriculture is its ability to deliver improved crops to the field. The integration of AI into the plant breeding pipeline, supercharged by the precision of CRISPR gene editing, is creating a new standard of accelerated crop improvement. This synergy is moving breeding from a largely phenotypic selection process to a predictive, data-driven science, enabling the direct and efficient development of crops with highly desirable traits. Traditional breeding is a slow process, often taking a decade or more to develop a new cultivar. AI is dramatically compressing this timeline by enhancing every stage, from gene discovery to the selection of elite breeding lines. This acceleration is achieved by tailoring gene edits with a precision that was previously impossible (Table 8.2).

The principal contribution of AI is its capacity to elucidate the intricate link between genotype and phenotype. By analyzing massive, multi-omics datasets (genomics, transcriptomics, proteomics, phenomics), AI models can identify the key genes and allelic variants that control agriculturally important traits. Once these genetic determinants are identified, CRISPR–Cas systems, optimized by AI for gRNA design and specificity, are used to introduce the precise genetic changes that will confer the desired trait. This process, often called "Breeding by Design" or "Predictive Breeding," allows scientists to move beyond random mutation and selection (Peleman and Van Der Voort, 2003). Instead, they can now design the ideal genotype on a computer, using AI to simulate the effects of different gene edits and their interactions (epistasis). AI models can forecast the impact of a particular modification on the plant's performance across diverse environmental circumstances, like drought,

TABLE 8.2 Applications of AI in different stages of plant gene editing and breeding.

Stage/Application area	AI Technique(s) involved	Biological data used	Outcome/Impact
Target discovery	ML, DL, GNN	Genomic, transcriptomic, proteomic	Identification of key genes, regulatory hubs, and allelic variants
gRNA design and optimization	ML, DL (CNN, RNN), interpretable models	gRNA sequence, chromatin, off-target datasets	Increased on-target efficiency, minimized off-target effects
Predicting repair outcomes	DL (BiLSTM, CNN)	Sequence context, chromatin, historical edits	Accurate prediction of indel patterns; guided base/prime editing
Trait prediction & genomic selection	ML, DL (multi-modal)	Genomic, phenomic, environmental	Accelerated selection of elite genotypes
Phenotyping & stress diagnostics	Computer vision, CNN, transformer models	Imaging, sensor, omics	Automated stress detection, high-throughput phenotyping
Synthetic biology & pathway engineering	ML, DL, protein language models	Protein sequences, regulatory elements	Design of novel proteins, regulatory circuits, synthetic promoters

heat, or nutritional stress (Xu et al., 2022). This facilitates the methodical design of climate-resilient crops prior to any laboratory modifications.

Moreover, AI-driven genomic selection (GS) is transforming the selection process. GS employs machine learning algorithms to forecast the breeding value of a plant only based on its genetic marker profile, eliminating the necessity for extensive field experiments in the initial generations (Farooq et al., 2024). The integration of gene editing to swiftly correct these advantageous traits in superior lineages results in an unparalleled acceleration of the breeding cycle. This enables breeders to promptly address emerging challenges, like novel disease strains or changing climatic trends, therefore safeguarding global food security (Crossa et al., 2017; Feng et al., 2024).

The theoretical potential of AI-enhanced gene editing is swiftly being actualized through tangible applications across a diverse array of crops and features. AI-driven methodologies are already being included into breeding processes to expedite gene identification, enhance gRNA design for intricate characteristics, and select changes with the highest anticipated agronomic significance. These improvements extend beyond primary crops such as rice, wheat, maize, and soybean; they are progressively being used to orphan and regionally significant crops, presenting new options to tackle local food security issues and adapt agriculture to varied agroecological environments. The integration of AI, high-throughput phenotyping, and multiplexed gene editing is set to facilitate the creation of crops exhibiting complex, polygenic characteristics, such as resistance to many concurrent stressors, which were previously unattainable using traditional approaches. Achieving this ambition necessitates strong data infrastructures, transparent AI models, and internationally coordinated regulatory frameworks to guarantee that the emerging breakthroughs are scientifically valid, egalitarian, safe, and broadly accessible. This new paradigm signifies a dramatic transition from observational breeding to predictive, design-oriented agriculture, radically altering the development of future crops.

8.6.1 Drought tolerance

Drought tolerance is a complicated polygenic characteristic, regulated by several genes and their intricate connections, rendering it exceedingly challenging to manipulate using traditional breeding methods. AI mitigates this complexity by synthesizing multi-omics data to elucidate the fundamental genetic architecture. Machine learning models were developed on genomic data and transcriptome profiles from plants experiencing water stress, facilitating the identification of essential regulatory genes and pivotal hub nodes within responsive molecular networks (Gupta et al., 2021; Kumar et al., 2023).

The implementation of machine learning in this field is diverse. A principal application is the discovery of drought-responsive genes using meta-analysis of extensive transcriptome datasets. This methodology has been effectively implemented across various species, including tomato (Chowdhury et al., 2023), barley (Panahi and Golkari, 2024), poplar (Tahmasebi et al., 2023), rice (Thanmalagan et al., 2022), wheat (He et al., 2025; Azimi et al., 2024), foxtail millet (Zhu et al., 2024), and the model plant Arabidopsis (Nazari et al., 2023). By aggregating data from multiple independent studies, ML algorithms can distinguish robust, conserved drought signatures from study-specific noise, revealing high-confidence candidate genes for functional validation. Beyond identification, ML models are being developed to predict transcriptional responses directly from genomic sequences or other features. For example, a model in rice was able to predict gene expression changes under heat and drought stress based on cis-regulatory elements and other genomic characteristics (Smet et al., 2023). Furthermore, network-based machine learning has proven powerful for pinpointing key regulators, such as predicting the transcription factors that act as master switches in drought resistance networks (Gupta et al., 2021).

Critically, these computational predictions are increasingly being followed by functional validation. In a notable example, ML and bioinformatics analyses identified the transcription factors *OsRAV11* and *OsRAV12* in rice while subsequent CRISPR–Cas9 experiments confirmed their role as negative regulators of drought tolerance, where knockout lines exhibited enhanced survival under water deficit (You et al., 2025). Finally, the scope of AI is expanding to include phenotypic diagnostics, with studies using ML and DL on agricultural imagery to directly identify drought stress in crops in field-like conditions (Ali et al., 2024).

8.6.2 Yield enhancement

Yield is the ultimate composite trait, influenced by numerous component traits like grain number, size, plant architecture, and photosynthetic efficiency. AI is essential for deconvoluting this complexity by integrating and analyzing the high-dimensional, multi-modal data that governs this final agricultural output. The application of AI in yield prediction and optimization spans from large-scale field monitoring to molecular-level analysis.

At the field level, deep learning and remote sensing have revolutionized yield forecasting by analyzing satellite and drone imagery to model the interactions between crop phenotypes and environmental conditions on a massive scale (Muruganantham et al., 2022). These techniques are a cornerstone of precision agriculture, enabling predictions based on in-season growth patterns (Jhajharia et al., 2023). The predictive power is significantly enhanced through data fusion, where DL integrates genetic, environmental, and management (GxExM) data into a unified model, moving beyond simple correlations to capture the complex interactions that ultimately determine crop performance and yield (Kick et al., 2023).

This integrative approach is exceptionally powerful in breeding. For instance, in maize, DL analytics have been used to enhance climate and genetic datasets, dramatically improving the accuracy of yield predictability and allowing breeders to select optimal genotypes for target environments (Sarzaeim et al., 2022). Similarly, a multi-modal DL framework in wheat breeding has demonstrated superior grain yield prediction by fusing genomic information with high-throughput phenomics data, effectively bridging the gap between the genetic potential of a plant and its expressed traits (Muruganantham et al., 2022).

The principles of AI-driven yield optimization are now being extended down to the molecular scale. The concept of "from sequence to yield" is being realized, where DL models are applied to forecast the output of biological systems, such as optimizing protein production, a foundational approach that can be adapted to engineer metabolic pathways for improved seed composition or biomass (Nikolados et al., 2021) AI synthesizes data from satellites, agricultural fields, genomes, and cellular processes, offering an unparalleled, comprehensive toolbox to elucidate yield factors and expedite the creation of enhanced crop varieties.

8.6.3 Nutritional quality (biofortification)

Enhancing the nutritional quality of crops, referred to as biofortification, is a primary objective in the fight against malnutrition. AI facilitates the mapping of intricate metabolic pathways that generate vitamins and minerals by synthesizing and analyzing multi-omics data to identify critical genetic and enzymatic regulatory nodes. This methodology transcends single-gene enrichment, aiming for a systems-level optimization of nutritional profiles. The initial crucial stage is the swift and precise assessment of nutritional characteristics, a procedure transformed by high-throughput phenotyping technologies. For example, in pulse crops, these platforms employ sophisticated sensors and machine learning to swiftly evaluate seed concentrations of iron, zinc, and protein, significantly expediting the identification of superior breeding lines (Madurapperumage et al., 2025). Hyperspectral imaging, in conjunction with machine learning, has been employed in barley to non-destructively forecast leaf nutrient concentrations, including nitrogen and phosphorus, enabling researchers to dynamically model the genetic regulation of nutrient uptake and allocation throughout the plant's life cycle (Grieco et al., 2022).

At the molecular level, AI is crucial in elucidating the genetic foundations of nutrition accumulation. An OMICs-based knowledge framework, integrating genomes, transcriptomics, and metabolomics, significantly depends on machine learning to elucidate the intricate network of genes and regulators governing micronutrient synthesis, therefore offering a blueprint for tailored biofortification (Khan et al., 2025). This is vividly demonstrated in rice, where machine learning approaches elucidated subspecies-specific ionomic variation, pinpointing genomic areas linked to the preferred accumulation of minerals such as iron, zinc, and selenium. This enables breeders to selectively merge alleles from several rice subspecies to develop new kinds with enhanced mineral content (Tiozon et al., 2024). Moreover, the purview of AI-enhanced enhancement transcends micronutrients to include comprehensive crop quality. Machine learning applications are currently employed to associate genetic markers with sensory and nutritional profiles in horticultural crops, facilitating the concurrent improvement of flavor compounds, vitamins, and shelf-life, thereby ensuring that biofortified crops are both nutritious and appealing to consumers (Ferrão et al., 2023). Utilizing these advanced computational methods, researchers may methodically develop crops with improved nutritional content, therefore establishing a more resilient and health-enhancing food system.

This integrated pipeline, spanning from the AI-driven discovery of key genetic determinants in drought tolerance and nutritional pathways to the CRISPR-mediated precision editing of these loci, and further to the AI-powered prediction of complex yield outcomes, demonstrates a paradigm shift from observational breeding to a predictive, design-based agriculture. The synergy between AI, which identifies the most impactful genetic edits and predicts their outcomes, and CRISPR–Cas, which executes them with molecular precision, is fundamentally reshaping how future crops will be developed. We are moving from a reality where breeders react to field performance over seasons to one where they can proactively design resilient, high-yielding, and nutrient-dense genotypes in silico. Realizing the full potential of this transformative synergy will require robust data infrastructures, transparent AI models, and globally coordinated regulatory frameworks to ensure these powerful innovations lead to equitable, safe, and widely accessible solutions for global food security.

8.7 AI in multi-omics data integration for plant research

The emergence of high-throughput technologies has produced extensive biological data across several levels, including genomic, transcriptomic, proteomic, metabolomic, and phenomic. Although each "omics" layer offers a significant insight into biological status, their full potential is realized through integration, uncovering the intricate, interrelated networks that regulate plant growth, development, and environmental response. The vast quantity and diversity of this data render conventional analytical methods insufficient. Artificial intelligence, especially deep learning, has become the crucial instrument for synthesizing these varied datasets to create a comprehensive understanding of plant biology and to propel the next phase of advancements in gene editing.

8.7.1 Combining genomic, transcriptomic, and proteomic data

The central challenge in multi-omics integration is the "curse of dimensionality", based on the fact that each dataset can contain millions of data points, and the complex, non-linear relationships between them. AI methodologies, including dimensionality reduction, feature learning, GNNs, and multi-modal DL, are uniquely suited to address this challenge (Kang et al., 2022). Techniques such as autoencoders (a type of neural network) compress high-dimensional omics data into a lower-dimensional, latent space representation. These latent features capture the most biologically relevant variations across datasets, effectively filtering out noise and identifying the core signals that link genotype to phenotype. For example, an autoencoder can integrate genomic variant data (SNPs) with transcriptomic (RNA-seq) data to identify genetic variants most predictive of gene expression (eQTLs) under specific conditions (Ballard et al., 2024).

Biological systems are fundamentally organized as graphs or networks, exemplified as gene regulatory networks and protein-protein interaction networks. GNNs, a category of deep learning models, are engineered to function on graph-structured data, amalgamating several omics layers by representing genes or proteins as nodes and their connections (e.g., co-expression, physical interaction) as edges. Through the dissemination of information across these edges, GNNs may forecast novel gene functions, pinpoint essential regulatory hubs, and deduce the potential repercussions of a perturbation (such as a gene edit) throughout the whole biological system (Xiao et al., 2023). For example, a GNN named EPGAT can amalgamate a protein-protein interaction network with transcriptome data from a drought-stressed plant to forecast which uncharacterized transcription factors are most pivotal to the stress response network (Schapke et al., 2021). Despite being in its nascent stages, GNN-based models have been utilized to examine protein-protein interaction networks in conjunction with transcriptomic data from drought-stressed plants, revealing critical uncharacterized transcription factors implicated in the stress response (Gupta et al., 2021).

The creation of tools such as DeepAProt illustrates the use of deep learning into functional genomics (Ahmed et al., 2023). DeepAProt is a web-based program intended to categorize protein sequences that respond to abiotic stressors, including drought, heat, cold, and salinity, in cereal crops of the Poaceae family. The tool employs an innovative activation function, Gaussian Error Linear Unit with Sigmoid (SIELU), which has exhibited enhanced performance compared to conventional activation functions like GeLU, as well as machine learning techniques such as Support Vector Machines (SVM) and Random Forest (RF). DeepAProt attained classification accuracies of 95.11%, 80.78%, 94.97%, and 81.69% for proteins subjected to cold, drought, heat, and salt stress, respectively. This technique offers researchers a quick and economical method to discover stress-responsive proteins, aiding in the creation of climate-resilient crop types. Furthermore, DeepAProt's intuitive interface and mobile application improve accessibility, advancing the overarching objective of food security and sustainable agriculture.

Multi-modal DL facilitates the concurrent processing of several data kinds. A model may include a CNN to extract features from leaf imagery (phenomics), an RNN to assess time-series gene expression data (transcriptomics), and fully connected networks to evaluate genotypic data, therefore developing joint representations that tie visual symptoms to molecular causes. This methodology has been utilized to forecast disease progression and resistance, such as rust and powdery mildew in cereals, by correlating early molecular signals with phenotypic results (Rubini and Kavitha, 2021; Sharma et al., 2023) underscored the need of combining many omics layers to provide a comprehensive understanding of plant responses to water stress. AI methods, namely deep learning models, were utilized to examine these high-dimensional information, pinpointing essential biomarkers and regulatory networks linked to drought tolerance. ML approaches were employed to forecast intricate features such as water consumption efficiency through the analysis of genotype-phenotype correlations. These models can analyze extensive phenotypic and genotypic data, facilitating the discovery of genetic loci associated with drought resistance. Predictive models enhance marker-assisted selection in breeding operations, expediting the creation of drought-resistant agricultural types.

This holistic, AI-driven integration moves research beyond single-gene analysis to a systems-level perspective. It allows scientists to understand not just which genes are important, but how they work together in pathways and networks to control complex traits. This is crucial for identifying the most effective and resilient targets for gene editing, as editing a central hub in a network is likely to have a more significant and stable effect than editing a peripheral gene. These progresses highlight the capacity of AI in improving our comprehension and manipulation of plant genomes for improved resilience and productivity.

8.7.2 Applications in stress tolerance and disease resistance

The integration of AI with multi-omics data is especially revolutionary for comprehending and enhancing plant resistance. Plant-pathogen interactions include a dynamic flow of molecular signals between the genomes of the host and the pathogen. AI models can integrate the transcriptome of the infected host plant with the genome of the invading pathogen to identify effector-target pairs, bottleneck genes, resistance network hubs, and master regulators. These models are now sophisticated enough to reverse-engineer complex interactions across multiple data types.

AI models can predict which pathogen effector proteins interact with specific host proteins to suppress immunity, and identify essential pathogen genes highly expressed during infection that could serve as targets for control strategies. Graph Convolutional Networks (GCNs) can be used to study the protein-protein interaction (PPI) network alongside sequence data of pathogen effectors in a trained model to predict physical interactions between pathogen effector proteins and host proteins (Liu et al., 2024). These models can learn to recognize complex sequences and structural features that dictate these interactions, which are crucial for pathogen-mediated suppression of immunity. The models can successfully predict several novel effector-target pairs, and CRISPR-mediated editing of the host target genes (e.g., silent mutations preventing effector binding) conferred enhanced resistance to pathogens in crops, providing a promising strategy for durable disease control.

Comparable AI-driven methodologies have likewise been utilized for abiotic stress responses. For example, deep learning models have been trained on transcriptomic and genomic datasets to forecast regulatory networks associated with drought and salinity tolerance in rice and maize, facilitating the identification of pivotal hub genes and transcription factors that govern stress adaptation (Li et al., 2023; Zeinalzadeh-Tabrizi and Nazari, 2025; Pradhan et al., 2025; Xie and Zhang, 2025; Sarijaloo et al., 2021). Additionally, SaGP, a machine learning framework, has been established to systematically identify genes associated with saline-alkali tolerance in plants by integrating genomic features, gene expression patterns, and functional annotations, thereby offering valuable candidates for breeding programs focused on enhancing stress tolerance (Qiao et al., 2025). A graph attention network (GAT) combined with environmental and omics data effectively predicted drought-responsive protein-protein interactions in Arabidopsis, resulting in the identification of new regulatory nodes in abscisic acid signaling (Bhati et al., 2025; Chang et al., 2024) devised a graph neural network model that integrates multi-source similarity network fusion and graph autoencoders to forecast abiotic stress-responsive miRNAs in plants, attaining an AUC of approximately 97.43%. This study illustrates that heterogeneous graph models can proficiently encapsulate miRNA-stress correlations derived from sequence, functional, and topological attributes (Chang et al., 2024; Li et al., 2025) introduced IRGL-RRI, an interpretable framework for graph representation learning aimed at predicting plant RNA–RNA interactions. The model enhanced noise robustness, interpretability, and cross-species generalizability in predicting regulatory RNA interactions under stress by masking segments of the RNA-RNA graph and integrating multi-scale feature embeddings using a KAN (Kolmogorov–Arnold Network) decoder (Liao et al., 2025). Moreover, transformer-based deep learning frameworks have been utilized to identify cis-regulatory motifs and promoter elements linked to abiotic stress responses in plants, facilitating the characterization of stress-responsive regulatory regions that can be targeted for crop enhancement (Peleke et al., 2024; Huang et al., 2025; Bhati et al., 2025). These examples underscore the extensive application of AI-driven network prediction approaches beyond biotic stress, providing potent tools for the creation of crops with enhanced abiotic stress tolerance.

AI can also be leveraged to identify central nodes within plant immune signaling networks whose targeted modulation through gene editing could amplify defense responses without inducing autoimmunity. While such applications have not yet been widely reported in crop science, parallel examples from other biological systems highlight this potential. For instance, NetBID2 identified MYC as an unknown driver in *KRAS*-mutant lung adenocarcinoma (LUAD), even though *MYC* expression was not significantly different (Dong et al., 2023). It reverse-engineers context-specific interaction networks using transcriptomics, proteomics, and phosphoproteomics, and infers the activity of candidate drivers. It identifies "hidden drivers" whose expression may not change but whose regulatory activity does.

Similarly, integration of multi-omic datasets has proven effective for identifying central regulatory nodes underlying stress tolerance in plants. For instance, integrating proteomic (interactome) and transcriptomic data in rice revealed a key negative regulator of immunity, and CRISPR-mediated editing of this gene conferred broad-spectrum resistance to multiple fungal pathogens without a yield penalty (Ponnurangan et al., 2025). In a parallel approach, combining chromatin accessibility data (ATAC-seq) with transcriptomics (Forestan et al., 2016) or metabolomics with transcriptomics (Liu et al., 2022b) allowed the identification of master transcription factors that coordinate the expression of hundreds of downstream stress-response genes. These integrative strategies provide a framework for applying AI-driven network inference in crops to pinpoint regulatory nodes whose precise manipulation could optimize defense responses while minimizing fitness costs.

ML models trained on multi-omics data from hundreds of plant genotypes grown under controlled stress conditions can learn the molecular "signatures" of resilience, then predict the stress tolerance of previously untested genotypes based on their molecular profiles, dramatically accelerating breeding and editing pipelines. One compelling example is DeepSTARR, a CNN adapted to link promoter sequences to quantitative gene expression responses under stress (e.g., salinity, oxidative stress) (de Almeida et al., 2022). The model was used to predict enhancer activity from sequence and was used to design synthetic regulatory elements experimentally. It can be used to generate thousands of in silico promoter variants for a stress-response gene from a wild crop relative. Although ML models were not utilized in designing synthetic stress-inducible promoters in plants(Bhadouriya et al., 2024), designed a synthetic salt-inducible promoter (cis-engineering), tested it *in planta,* and showed strong salt/drought induction versus *CaMV35S*(Bhadouriya et al., 2024). This supports that synthetic regulatory sequences can give high stress-specific expression. Moreover, CRISPR-mediated promoter engineering/replacement has produced crop variants with improved stress performance without detectable yield penalties in field tests (Shi et al., 2017). Collectively, these studies suggest that AI-assisted design coupled with accurate genome editing may provide promoter variations that provide robust, stress-inducible expression and improved tolerance while reducing trade-offs.

By harnessing AI to integrate information across diverse biological layers, researchers can move beyond correlative patterns toward a truly mechanistic and predictive understanding of plant resilience. These attempts enable the logical discovery of key regulatory nodes, effector–target interactions, and stress-responsive elements that can be strategically engineered through gene editing and synthetic biology. As AI models enhance their proficiency in deciphering the intricacies of plant systems, they will expedite target selection and facilitate the strategic creation of regulatory frameworks suited to particular environmental conditions. The convergence of AI-driven network inference, multi-omics integration, and precision genome engineering is poised to transform crop improvement from an empirical process into a predictive, design-driven discipline. Ultimately, this paradigm shift holds the promise of creating resilient crop varieties capable of thriving under the intertwined pressures of biotic and abiotic stress, thereby supporting sustainable agriculture in an era of global change.

8.8 Ethical, legal, and social Implications (ELSI) in plant gene editing

The integration of AI with gene editing in agriculture has significant potential, although it simultaneously prompts critical ethical, legal, and societal inquiries that require proactive consideration. Successfully traversing this intricate terrain is crucial for guaranteeing that these potent innovations are cultivated and used with responsibility, equity, and in service of the public interest (Eriksson et al., 2020).

8.8.1 Data privacy and security concerns

The foundation of AI in plant genomics lies in data, vast volumes of genomic, environmental, and phenotypic information. Managing this data presents substantial challenges. While plant data may seem less sensitive than human data, it constitutes a critical economic and strategic asset. The genomic sequence of a high-value, elite crop cultivar is commercially sensitive intellectual property, and AI models trained on such data could potentially be reverse-engineered or queried to reveal proprietary genetic information, raising risks of intellectual property theft or biopiracy (Bayern, 2022).

A related international debate centers on "Digital Sequence Information" (DSI), which refers to digital genomic data derived from genetic resources. Countries abundant in biodiversity, typically located in the Global South, contend that digital sequence information (DSI) derived from their indigenous flora should be governed by the same benefit-sharing requirements as tangible genetic resources in accordance with the Nagoya Protocol, a fundamental component of the Convention on Biological Diversity (Buck and Hamilton, 2011; Sirakaya, 2022). It was designed for a pre-digital era, regulating the physical transfer of genetic material; therefore, it currently lacks a clear, operational definition for DSI, creating a fundamental governance gap. The use of DSI in AI systems for crop improvement complicates this issue, as it is difficult to track data provenance and ensure fair and equitable sharing of benefits (Aubry, 2019; Rabone et al., 2019). Modern AI models, particularly deep learning architectures, do not merely access discrete sequences; they internalize patterns from vast, aggregated DSI datasets to make predictions or generate novel insights. This process of model training effectively "incorporates" the genetic information, blurring the lines between data access and genetic resource utilization. Consequently, when an AI model trained on global DSI repositories contributes to developing a new commercial crop, it becomes nearly impossible to attribute the contribution to specific genetic resources or to determine the corresponding obligations for benefit-sharing. This not only weakens the core ethical principle of the Nagoya Protocol but also creates significant legal and reputational risks for both public and private entities leveraging AI in plant biotechnology (Ambler et al., 2021). Without an international framework to resolve the status of DSI, the very data that fuels AI-driven innovation risks becoming a source of contention, potentially locking away valuable genetic information from biodiverse countries unwilling to see their resources used without reciprocity, thereby stifling global collaborative research essential for addressing food security challenges.

Additionally, AI cloud platforms and genomic databases are potential targets for cyberattacks. Data breaches could result in the loss of irreplaceable genetic datasets or the manipulation of existing data, which in turn could corrupt AI models and lead to flawed conclusions or ineffective gene edits. Ensuring robust cybersecurity measures and transparent data governance policies is therefore essential (Singh and Cheema, 2024). Taken together, safeguarding plant genomic data requires a multifaceted approach that combines legal frameworks, ethical benefit-sharing, and technological protections to preserve intellectual property, maintain scientific integrity, and support equitable innovation in agricultural AI.

8.8.2 Equity and accessibility in AI-driven plant gene editing

The AI-driven revolution in plant gene editing carries the risk of exacerbating existing inequalities in global agriculture. Developing and deploying advanced AI tools demands substantial computational resources, specialized expertise, and significant financial investment. This high barrier to entry tends to concentrate capabilities in the hands of a few large multinational corporations and well-funded institutions in developed countries, leaving public sector researchers and small- and medium-sized enterprises (SMEs), particularly in developing countries, at a disadvantage (Hackfort, 2021). If only a small number of entities control the most powerful AI platforms for gRNA design and trait discovery, they could dominate seed development, potentially marginalizing traditional breeding programs, limiting farmer choice, and increasing seed costs. These trends would disproportionately impact smallholder farmers, who contribute significantly to global food production (Montenegro de Wit, 2022).

To mitigate these risks, initiatives must focus on promoting open-source AI tools, expanding access to public cloud computing resources for agricultural research, and investing in bioinformatics and data science capacity-building for researchers in developing countries. Strategic partnerships among public institutions, CGIAR centers, and the private sector are essential to ensure that AI-driven improvements benefit staple foods and neglected crops critical for regional food security, rather than prioritizing only high-value cash crops (Gowri Shankar et al., 2024). Ultimately, fostering equitable access to AI technologies in plant gene editing is key to ensuring that the benefits of innovation are shared broadly, empowering diverse stakeholders and supporting a more inclusive and resilient global food system.

8.8.3 Regulatory frameworks and governance

The regulatory landscape for AI-assisted gene-edited plants is complex, fragmented, and rapidly evolving, with different jurisdictions adopting markedly different approaches. This

TABLE 8.3 Emerging governance and regulatory frameworks for AI-driven gene editing.

Region/ Country	Regulatory approach	Key features	Relevance to AI integration
USA	Product-based	Focus on end product; minimal restrictions for non-transgenic edits	Fast innovation, but limited AI-specific oversight
EU	Process-based (GMO directive)	Strict regulation of gene-edited organisms	Slow adoption; lacks specific AI guidance
UK	Precision breeding Act 2023	New category (PBOs), streamlined regulation	Encourages innovation; open to AI evidence in evaluation
India	Guidelines for SDN-1/2	Exemption from GMO biosafety; focus on the absence of foreign DNA	Emerging flexibility; AI use not yet addressed
China	Streamlined approval for natural-like edits	Clear commercialization pathway	Rapid deployment; regulatory recognition of new techniques
Brazil	Case-by-case	Balances innovation with environmental concerns	Potential to integrate AI-based evidence
Global initiatives	Nagoya protocol/DSI debates/CGIAR	Focus on benefit sharing, data sovereignty, and capacity building	Central for federated AI training and data governance

diversity creates significant uncertainty for developers and the public, particularly when deploying AI-driven technologies across international markets (Table 8.3).

The regulatory structure in the United States is centered on products. The Coordinated Framework for the Regulation of Biotechnology, overseen by the USDA, FDA, and EPA, governs gene-edited plants based on the final product rather than the methodology employed in their creation. This methodology has resulted in negligible constraints for genome-edited crops devoid of foreign DNA, facilitating innovation yet eliciting apprehensions over regulatory control (USDA, 2024; FDA, 2024). In contrast, the European Union has one of the most rigorous regulatory frameworks. In 2018, the European Court of Justice determined that gene-edited organisms are categorized under the GMO Directive (2001/18/EC), emphasizing the methodology above the outcome. In 2024, the European Commission suggested a new framework distinguishing between plants created by specific new genomic methods (NGTs) and transgenic GMOs, possibly alleviating requirements for particular gene-edited crops (EUROPEAN COMMISSION. 2024).

The United Kingdom, following Brexit, has moved in a different direction by establishing a more innovation-friendly environment. The Precision Breeding Act 2023 introduced a separate regulatory category for precision-bred organisms (PBOs). By focusing on the safety of the final product rather than the method of creation, the UK aims to stimulate domestic innovation and streamline the regulatory pathway for gene-edited crops (UK PARLIAMENT. 2023).

Regulatory advancements in Asia similarly illustrate this variance. India has initiated a move toward regulatory flexibility. In 2022, the Department of Biotechnology released rules indicating that SDN-1 and SDN-2 genome-edited goods, which lack foreign DNA, would be free from the stringent biosafety restrictions imposed on transgenic GMOs. These goods are evaluated by an Institutional Biosafety Committee to verify the absence of exogenous DNA (*DBT INDIA. 2022*). Nevertheless, the integration of AI into plant research and development is so young that particular regulatory frameworks for assessing AI-generated evidence are still inadequately created, creating a disparity between technical progress and regulatory control. Concurrently, China has just revised its regulatory framework for gene-edited crops. The Ministry of Agriculture and Rural Affairs released directives for the regulatory authorization of gene-edited crops, facilitating their commercialization. The recommendations indicate that gene-edited crops, especially those with modifications that may naturally occur, may experience an expedited approval procedure, therefore decreasing the time and expenses related to regulatory approval (Mallapaty, 2022).

Brazil's regulatory framework for gene-edited crops in Latin America is undergoing evolution. The government possesses particular rules regarding genetically modified organisms but employs a case-by-case methodology to ascertain if a gene-edited organism falls under GMO restrictions. This methodology seeks to harmonize innovation with safety and environmental factors (Zarate et al., 2023).

Taken together, these examples illustrate the highly heterogeneous nature of global regulatory frameworks for gene-edited crops. While some jurisdictions are moving toward streamlined, innovation-driven approaches, others continue to apply more cautious, process-based oversight. For AI-assisted gene editing, this regulatory patchwork presents both significant opportunities and formidable challenges. Flexible frameworks can accelerate innovation, deployment, and commercialization, but the lack of international harmonization complicates research collaborations, technology transfer, and global market approvals. A critical question arises: how can regulators confidently evaluate AI-generated outputs, such as the specificity of a predicted gRNA or the reliability of a model-predicted complex trait under field conditions? Developing clear international standards for auditing, validating, and interpreting AI models, including the adoption of Explainable AI (XAI) approaches in bioinformatics, will be essential for building regulatory trust and ensuring that AI-driven edits are precise, predictable, and safe (Nannini et al., 2023; Balasubramaniam et al., 2023).

Navigating this evolving landscape will require more than national regulatory adjustments. It will demand coordinated global governance. Harmonization efforts and international dialog, grounded in robust ethical principles and a commitment to equity, are crucial to ensure that the transformative potential of AI-powered gene editing is realized safely, fairly, and effectively across diverse agricultural systems (Gkikas et al., 2023).

8.9 Future perspectives

The integration of AI with gene editing will irrevocably transform plant biotechnology, enabling the development of predictive and autonomous systems to address issues such as climate change, food security, and sustainability (Fig. 8.1). Foundation models for plant biology will lead the next wave of innovation. These models will be trained on huge

FIGURE 8.1 Conceptual illustration summarizing the integration of AI and gene editing in plant biotechnology. The figure depicts the convergence of computational models, robotics, genomic technologies, and plant science to enable predictive breeding, precision trait engineering, and future autonomous crop improvement systems. This visual represents the thematic elements discussed throughout the chapter and highlights emerging technological directions for sustainable and resilient agriculture. *Illustration created using OpenAI DALL·E.*

multi-omics and environmental datasets to give general biological insights that can be adjusted for specific traits or crops (Kajrolkar, 2025; Xu et al., 2025). Simultaneously, self-driving laboratories that combine AI design, robotic automation, and high-throughput phenotyping will cut breeding cycles from years to months (Martin et al., 2023). Another big change will be predictive digital twins, which are multi-scale computational models that are connected to real-time field data. These models will let scientists test thousands of gene edits in a virtual setting, which will lower costs and risks (Dawn et al., 2024).

Future AI models will prioritize causal inference and explainability, pinpointing genetic determinants of intricate traits such as nitrogen-use efficiency rather than superficial correlations (Zhu et al., 2024). Gene editing will progress toward programmable systems, such as RNA-guided recombinases and spatiotemporally regulated CRISPR platforms, facilitating the integration of complete pathways and meticulous regulation of trait expression (Gao, 2021). AI will be very important for making these complicated changes and making them better. To make these breakthroughs happen, biologists, data scientists, ethicists, and policymakers need to work together closely. International collaboration is crucial for regulatory harmonization, open data sharing, and capacity building, especially in the Global South, to guarantee equitable access to innovations (Zafar et al., 2024; Turnbull et al., 2021).

Gene editing powered by AI will be a key part of regenerative agriculture, making crops that can survive in any climate, fertilize themselves, and have the best nutrition (Zaidi et al., 2020; Navarro et al., 2025; D'Imperio et al., 2024). For example, we will create crops with "weather-proof" traits, like cereals that can photosynthesize better when it's hot (C4 rice), roots that can store more carbon in the soil, and perennials that give stable yields while improving soil health. We will also engineer nitrogen fixation into major cereals, which will greatly reduce the need for synthetic fertilizers and their negative effects on the environment. Finally, we will design crops that are tailored to specific nutritional and health outcomes, like fruits with the right balance of phytonutrients or grains with better mineral bioavailability, which will help with personalized nutrition. These new ideas make it possible to grow more food with fewer resources and less harm to the environment. The combination of AI and gene editing is one of the most powerful tools for creating a safe, fair, and sustainable food system for the 21st century.

Declaration

"The authors declare that no generative artificial intelligence tools were used in the writing or preparation of this chapter. The content reflects the authors' own scholarly work, and the authors assume full responsibility for all statements, interpretations, and conclusions presented".

References

Abbasi, A.F., Asim, M.N., Dengel, A., 2025. Transitioning from wet lab to artificial intelligence: a systematic review of AI predictors in CRISPR. J. Transl. Med. 23, 153.

Ahmed, B., Haque, M.A., Iquebal, M.A., Jaiswal, S., Angadi, U.B., Kumar, D., Rai, A., 2023. DeepAProt: deep learning based abiotic stress protein sequence classification and identification tool in cereals. Front. Plant Sci. 13. https://doi.org/10.3389/fpls.2022.1008756.

Ali, T., Rehman, S.U., Ali, S., Mahmood, K., Obregon, S.A., Iglesias, R.C., Khurshaid, T., Ashraf, I., 2024. Smart agriculture: utilizing machine learning and deep learning for drought stress identification in crops. Sci. Rep. 14 (1). https://doi.org/10.1038/s41598-024-74127-8.

Alipanahi, R., Safari, L., Khanteymoori, A., 2025. Advancing CRISPR with deep learning: a comprehensive review of models and databases. Mol. Ther. Nucleic Acids.

Allen, F., Crepaldi, L., Alsinet, C., Strong, A.J., Kleshchevnikov, V., De Angeli, P., Páleníková, P., Khodak, A., Kiselev, V., Kosicki, M., Bassett, A.R., Harding, H., Galanty, Y., Muñoz-Martínez, F., Metzakopian, E., Jackson, S.P., Parts, L., 2019. Predicting the mutations generated by repair of Cas9-induced double-strand breaks. Nat. Biotechnol. 37 (1), 64–72. https://doi.org/10.1038/nbt.4317.

Ambler, J., Diallo, A.A., Dearden, P.K., Wilcox, P., Hudson, M., Tiffin, N., 2021. Including digital sequence data in the Nagoya Protocol can promote data sharing. Trends Biotechnol. 39 (2), 116–125. https://doi.org/10.1016/j.tibtech.2020.06.009.

Aubry, S., 2019. The future of digital sequence information for plant genetic resources for food and agriculture. Front. Plant Sci. 10. https://doi.org/10.3389/fpls.2019.01046.

Azimi, N., Ravash, R., Zinati, Z., 2024. Microarray meta-analysis and supervised machine learning to explore drought-tolerance-associated genes in wheat (Triticum aestivum). Genet. Resour. Crop Evol. 71, 3815–3831.

Balasubramaniam, N., Kauppinen, M., Rannisto, A., Hiekkanen, K., Kujala, S., 2023. Transparency and explainability of AI systems: from ethical guidelines to requirements. Inf. Software Technol. 159, 107197. https://doi.org/10.1016/j.infsof.2023.107197.

Ballard, J., Wang, Z., Li, W., Shen, L., Long, Q., 2024. Deep learning-based approaches for multi-omics data integration and analysis. BioData Min. 17 (1). https://doi.org/10.1186/s13040-024-00391-z.

Bayern, S., 2022. Reverse engineering (by) artificial intelligence. In: Research Handbook on Intellectual Property and Artificial Intelligence. Edward Elgar Publishing Ltd., United States, pp. 391–404. https://doi.org/10.4337/9781800881907.00029.

Bhadouriya, S.L., Karamchandani, A.N., Nayak, N., Mehrotra, S., Mehrotra, R., 2024. Artificially designed synthetic promoter for a high level of salt induction using a cis-engineering approach. Sci. Rep. 14 (1). https://doi.org/10.1038/s41598-024-64537-z.

Bhati, U., Sharma, A., Gupta, S., Kumar, A., Pradhan, U.K., Shankar, R., 2025. Decoding stress specific transcriptional regulation by causality aware Graph-Transformer deep learning. bioRxiv. https://doi.org/10.1101/2025.03.11.642297.

Bose, S., Banerjee, S., Kumar, S., Saha, A., Nandy, D., Hazra, S., 2024. Review of applications of artificial intelligence (AI) methods in crop research. J. Appl. Genet. 65 (2), 225–240. https://doi.org/10.1007/s13353-023-00826-z.

Buck, M., Hamilton, C., 2011. The Nagoya Protocol on access to genetic resources and the fair and equitable sharing of benefits arising from their utilization to the convention on biological diversity. Rev. Eur. Community Int. Environ. Law 20 (1), 47–61. https://doi.org/10.1111/j.1467-9388.2011.00703.x.

Chang, L., Jin, X., Rao, Y., Zhang, X., 2024. Predicting abiotic stress-responsive miRNA in plants based on multi-source features fusion and graph neural network. Plant Methods 20 (1). https://doi.org/10.1186/s13007-024-01158-7.

Chowdhury, R.H., Eti, F.S., Ahmed, R., Gupta, S.D., Jhan, P.K., Islam, T., Bhuiyan, M.A.R., Khayer, A., Rubel, M.H., 2023. Drought-responsive genes in tomato: meta-analysis of gene expression using machine learning. Research Square. https://doi.org/10.21203/rs.3.rs-3215460.

Chuai, G., Ma, H., Yan, J., Chen, M., Hong, N., Xue, D., Zhou, C., Zhu, C., Chen, K., Duan, B., Gu, F., Qu, S., Huang, D., Jia, W., Qi, L., 2018. DeepCRISPR: optimized CRISPR guide RNA design by deep learning. Genome Biol. 19 (1). https://doi.org/10.1186/s13059-018-1459-4.

Crossa, J., Pérez-Rodríguez, P., Cuevas, J., Montesinos-López, O., Jarquín, D., de los Campos, G., Burgueño, J., González-Camacho, J.M., Pérez-Elizalde, S., Beyene, Y., Dreisigacker, S., Singh, R., Zhang, X., Gowda, M., Roorkiwal, M., Rutkoski, J., Varshney, R.K., 2017. Genomic selection in plant breeding: methods, models, and perspectives. Trends Plant Sci. 22 (11), 961–975. https://doi.org/10.1016/j.tplants.2017.08.011.

Danilevicz, M.F., Upadhyaya, S.R., Batley, J., Bennamoun, M., Bayer, P.E., Edwards, D., 2025. Understanding plant phenotypes in crop breeding through explainable AI. Plant Biotechnol. J. 23 (10), 4200–4213. https://doi.org/10.1111/pbi.70208.

Dauparas, J., Anishchenko, I., Bennett, N., Bai, H., Ragotte, R.J., Milles, L.F., Wicky, B.I.M., Courbet, A., de Haas, R. J., Bethel, N., Leung, P.J.Y., Huddy, T.F., Pellock, S., Tischer, D., Chan, F., Koepnick, B., Nguyen, H., Kang, A., Sankaran, B., Bera, A.K., King, N.P., Baker, D., 2022. Robust deep learning–based protein sequence design using ProteinMPNN. Science 378 (6615), 49–56. https://doi.org/10.1126/science.add2187.

Dawn, N., Ghosh, S., Saha, A., Chatterjee, S., Ghosh, T., Guha, S., Sarkar, S., Mukherjee, P., Sanyal, T., 2024. A review on digital twins technology: a new frontier in agriculture. Artificial Intelligence and Applications 2 (4), 250–262. https://doi.org/10.47852/bonviewaia3202919.

DBT INDIA, 2022. Guidelines for the Regulation of Genome-Edited Plants. Department of Biotechnology, Government of India. https://dbtindia.gov.in/sites/default/files/DBT_Guidelines_Genome_Edited_Plants_2022.pdf. (Accessed 1 October 2025).

de Almeida, B.P., Reiter, F., Pagani, M., Stark, A., 2022. DeepSTARR predicts enhancer activity from DNA sequence and enables the de novo design of synthetic enhancers. Nat. Genet. 54 (5), 613–624. https://doi.org/10.1038/s41588-022-01048-5.

Ding, K., Chin, M., Zhao, Y., Huang, W., Mai, B.K., Wang, H., Liu, P., Yang, Y., Luo, Y., 2024. Machine learning-guided co-optimization of fitness and diversity facilitates combinatorial library design in enzyme engineering. Nat. Commun. 15 (1). https://doi.org/10.1038/s41467-024-50698-y.

Doench, J.G., Fusi, N., Sullender, M., Hegde, M., Vaimberg, E.W., Donovan, K.F., Smith, I., Tothova, Z., Wilen, C., Orchard, R., Virgin, H.W., Listgarten, J., Root, D.E., 2016. Optimized sgRNA design to maximize activity and minimize off-target effects of CRISPR-Cas9. Nat. Biotechnol. 34 (2), 184–191. https://doi.org/10.1038/nbt.3437.

Dong, X., Ding, L., Thrasher, A., Wang, X., Liu, J., Pan, Q., Rash, J., Dhungana, Y., Yang, X., Risch, I., Li, Y., Yan, L., Rusch, M., McLeod, C., Yan, K.-K., Peng, J., Chi, H., Zhang, J., Yu, J., 2023. NetBID2 provides comprehensive hidden driver analysis. Nat. Commun. 14 (1). https://doi.org/10.1038/s41467-023-38335-6.

Dunka, V., 2022. AI-powered computational biology for genome editing: developing deep learning models for CRISPR-Cas9 target site prediction, off-target effect minimization, and therapeutic efficacy enhancement. Artificial intelligence. Machine Learning, and Autonomous Systems 6, 1–40.

Dutton, G., 2025. AI-Enabled Gene Editing Produces Fewer Off-Target Outcomes: by creating gene editors not found in nature, or optimizing existing editors, AI can improve the accuracy, effectiveness, and accessibility of gene editing. Genetic Engineering and Biotechnology News 45 (2), 24–27. https://doi.org/10.1089/gen.45.02.09.

D'imperio, M., Palmitessa, D.O., Serio, F., 2024. Optimized plant nutrition for precision biofortification to meet nutrition custom human. In: Silicon Advances for Sustainable Agriculture and Human Health: Increased Nutrition and Disease Prevention. Springer.

Eriksson, D., Custers, R., Edvardsson Björnberg, K., Hansson, S.O., Purnhagen, K., Qaim, M., Romeis, J., Schiemann, J., Schleissing, S., Tosun, J., Visser, R.G.F., 2020. Options to reform the European union legislation on GMOs: scope and definitions. Trends Biotechnol. 38 (3), 231–234. https://doi.org/10.1016/j.tibtech.2019.12.002.

EUROPEAN COMMISSION, 2024. Proposal for a regulation of the European Parliament and of the Council on novel genomic techniques. https://ec.europa.eu/info/law/better-regulation/have-your-say/initiatives/13222-Novel-genomic-techniques_en. (Accessed 1 October 2025).

Farooq, M.A., Gao, S., Hassan, M.A., Huang, Z., Rasheed, A., Hearne, S., Prasanna, B., Li, X., Li, H., 2024. Artificial intelligence in plant breeding. Trends Genet. 40 (10), 891–908. https://doi.org/10.1016/j.tig.2024.07.001.

FDA, 2024. Foods Derived from Plants Produced Using Genome Editing: Guidance for Industry. U.S. Food and Drug Administration. https://www.fda.gov/regulatory-information/search-fda-guidance-documents/guidance-industry-foods-derived-plants-produced-using-genome-editing. (Accessed 1 October 2025).

Feng, W., Gao, P., Wang, X., 2024. AI breeder: genomic predictions for crop breeding. New Crops 1, 100010.

Ferrão, L.F.V., Dhakal, R., Dias, R., Tieman, D., Whitaker, V., Gore, M.A., Messina, C., Resende, M.F.R., 2023. Machine learning applications to improve flavor and nutritional content of horticultural crops through breeding and genetics. Curr. Opin. Biotechnol. 83. https://doi.org/10.1016/j.copbio.2023.102968.

Fong, J.H.C., Wong, A.S.L., 2023. Advancing CRISPR/Cas gene editing with machine learning. Current Opinion in Biomedical Engineering 28, 100477. https://doi.org/10.1016/j.cobme.2023.100477.

Forestan, C., Aiese Cigliano, R., Farinati, S., Lunardon, A., Sanseverino, W., Varotto, S., 2016. Stress-induced and epigenetic-mediated maize transcriptome regulation study by means of transcriptome reannotation and differential expression analysis. Sci. Rep. 6 (1). https://doi.org/10.1038/srep30446.

Gao, C., 2021. Genome engineering for crop improvement and future agriculture. Cell 184 (6), 1621–1635. https://doi.org/10.1016/j.cell.2021.01.005.

Gill, T., Gill, S.K., Saini, D.K., Chopra, Y., de Koff, J.P., Sandhu, K.S., 2022. A comprehensive review of high throughput phenotyping and machine learning for plant stress phenotyping. Phenomics 2 (3), 156–183. https://doi.org/10.1007/s43657-022-00048-z.

Gkikas, D.C., Theodoridis, P.K., Gkikas, M.C., 2023. Artificial intelligence (AI) use for e-governance in agriculture: exploring the bioeconomy landscape. In: Recent Advances in Data and Algorithms for E-Government. Springer.

Gowri Shankar, D.V., Kumari, P., Neelambari, D.B., 2024. Revolution agri-food systems: leveraging digital innovations for equitable sustainability and resilience. African Journal of Biological Sciences (South Africa) 6, 520–530.

Grieco, M., Schmidt, M., Warnemünde, S., Backhaus, A., Klück, H.C., Garibay, A., Tandrón Moya, Y.A., Jozefowicz, A.M., Mock, H.P., Seiffert, U., Maurer, A., Pillen, K., 2022. Dynamics and genetic regulation of leaf nutrient concentration in barley based on hyperspectral imaging and machine learning. Plant Sci. 315. https://doi.org/10.1016/j.plantsci.2021.111123.

Gupta, C., Ramegowda, V., Basu, S., Pereira, A., 2021. Using network-based machine learning to predict transcription factors involved in drought resistance. Front. Genet. 12. https://doi.org/10.3389/fgene.2021.652189.

Hackfort, S., 2021. Patterns of inequalities in digital agriculture: a systematic literature review. Sustainability 13 (22), 12345. https://doi.org/10.3390/su132212345.

Hajiahmadi, Z., Ali, M., Wei, H., Li, D., Orooji, Y., Ruan, H., Zhuge, Q., 2019. Strategies to increase on-target and reduce off-target effects of the CRISPR/Cas9 system in plants. Int. J. Mol. Sci. 20 (15), 3719. https://doi.org/10.3390/ijms20153719.

He, S., Li, X., Chen, M., Xu, X., Tang, F., Gong, T., Xu, M., Yang, W., Liu, W., 2024. Crop HTP technologies: applications and prospects. Agriculture 14 (5), 723. https://doi.org/10.3390/agriculture14050723.

He, J., Cui, B., Liu, P., Meng, X., Yan, J., 2025. Utilizing machine learning and bioinformatics analysis to identify drought stress responsive genes in wheat (Triticum aestivum L.). Front. Sustain. Food Syst. 9. https://doi.org/10.3389/fsufs.2025.1612009.

Huang, K., Tian, J., Sun, L., Hu, H., Huang, X., Zhou, S., Deng, A., Zhou, Z., Jiang, M., Li, G., Xie, P., Wang, Y., Jiang, X., 2025. TransGeneSelector: using a transformer approach to mine key genes from small transcriptomic datasets in plant responses to various environments. BMC Genom. 26 (1). https://doi.org/10.1186/s12864-025-11434-y.

Jhajharia, K., Mathur, P., Jain, S., Nijhawan, S., 2023. Crop yield prediction using machine learning and deep learning techniques. Procedia Comput. Sci. 218, 406–417. https://doi.org/10.1016/j.procs.2023.01.023.

Jiang, B., An, Z., Niu, L., Qin, D., 2025. Precise genome editing process and its applications in plants driven by AI. Funct. Integr. Genom. 25 (1). https://doi.org/10.1007/s10142-025-01619-9.

Jumper, J., Evans, R., Pritzel, A., Green, T., Figurnov, M., Ronneberger, O., Tunyasuvunakool, K., Bates, R., Žídek, A., Potapenko, A., Bridgland, A., Meyer, C., Kohl, S.A.A., Ballard, A.J., Cowie, A., Romera-Paredes, B., Nikolov, S., Jain, R., Adler, J., Back, T., Petersen, S., Reiman, D., Clancy, E., Zielinski, M., Steinegger, M., Pacholska, M., Berghammer, T., Bodenstein, S., Silver, D., Vinyals, O., Senior, A.W., Kavukcuoglu, K., Kohli, P., Hassabis, D., 2021. Highly accurate protein structure prediction with AlphaFold. Nature 596 (7873), 583–589. https://doi.org/10.1038/s41586-021-03819-2.

Kajrolkar, A., 2025. Deep learning applications in plant biology: bridging genotype and phenotype. Premier Journal of Plant Biology. https://doi.org/10.70389/pjpb.100011.

Kang, M., Ko, E., Mersha, T.B., 2022. A roadmap for multi-omics data integration using deep learning. Briefings Bioinf. 23, bbab454.

Karim, M.R., Islam, T., Shajalal, M., Beyan, O., Lange, C., Cochez, M., Rebholz-Schuhmann, D., Decker, S., 2023. Explainable AI for bioinformatics: methods, tools and applications. Briefings Bioinf. 24 (5). https://doi.org/10.1093/bib/bbad236.

Khan, M.H.U., Wang, S., Wang, J., Ahmar, S., Saeed, S., Khan, S.U., Xu, X., Chen, H., Bhat, J.A., Feng, X., 2022. Applications of artificial intelligence in climate-resilient smart-crop breeding. Int. J. Mol. Sci. 23 (19). https://doi.org/10.3390/ijms231911156.

Khan, I., Arif, U., Baloch, M.S., Khan, A.A., Shahzad, M.F., Khan, Q.U., Nadim, M.A., Saddozai, U.K., Fiaz, S., Shah, A.N., 2025. Crops biofortification through OMICs based knowledge. In: Crop Biofortification: Biotechnological Approaches for Achieving Nutritional Security under Changing Climate. wiley, Pakistan, pp. 51–60. https://doi.org/10.1002/9781394273270.ch04.

Kick, D.R., Wallace, J.G., Schnable, J.C., Kolkman, J.M., Alaca, B., Beissinger, T.M., Edwards, J., Ertl, D., Flint-Garcia, S., Gage, J.L., Hirsch, C.N., Knoll, J.E., de Leon, N., Lima, D.C., Moreta, D.E., Singh, M.P., Thompson, A., Weldekidan, T., Washburn, J.D., 2023. Yield prediction through integration of genetic, environment, and management data through deep learning. G3: Genes, Genomes, Genet. 13 (4). https://doi.org/10.1093/g3journal/jkad006.

Kim, H.K., Min, S., Song, M., Jung, S., Choi, J.W., Kim, Y., Lee, S., Yoon, S., Kim, H., 2018. Deep learning improves prediction of CRISPR-Cpf1 guide RNA activity. Nat. Biotechnol. 36 (3), 239–241. https://doi.org/10.1038/nbt.4061.

Kim, H.K., Kim, Y., Lee, S., Min, S., Bae, J.Y., Choi, J.W., Park, J., Jung, D., Yoon, S., Kim, H.H., 2019. SpCas9 activity prediction by DeepSpCas9, a deep learning–based model with high generalization performance. Sci. Adv. 5 (11). https://doi.org/10.1126/sciadv.aax9249.

Kim, M.G., Go, M.J., Kang, S.H., Jeong, S.H., Lim, K., 2025. Revolutionizing CRISPR technology with artificial intelligence. Exp. Mol. Med. 57 (7), 1419–1431. https://doi.org/10.1038/s12276-025-01462-9.

Kumar, A., 2024. Deep learning approaches for off-targets prediction in CRISPR-Cas9 genome editing to improve resistant in plants. In: Plant Elicitor Peptides: New Tool for Sustainable Agriculture. Springer Nature, India, pp. 319–334. https://doi.org/10.1007/978-981-97-6374-0_14.

Kumar, N., Mishra, B.K., Liu, J., Mohan, B., Thingujam, D., Pajerowska-Mukhtar, K.M., Mukhtar, M.S., 2023. Network biology analyses and dynamic modeling of gene regulatory networks under drought stress reveal major transcriptional regulators in Arabidopsis. Int. J. Mol. Sci. 24 (8). https://doi.org/10.3390/ijms24087349.

Lee, M., 2023. Deep learning in CRISPR-Cas systems: a review of recent studies. Front. Bioeng. Biotechnol. 11. https://doi.org/10.3389/fbioe.2023.1226182.

Lelanskyi, M., Wang, M., Scott, L., Rieber, L., Merrett, S., Schimunek, J., Mayr, A., Mcdowell, I., Klambauer, G., Bowen, T., 2025. Machine learning-driven optimization of specific, compact, and efficient base editors via single-round diversification. bioRxiv 2025 (07), 25.666807.

Li, J., Li, G., Wang, L., Li, D., Li, H., Gao, C., Zhuang, M., Zhuang, J., Zhou, H., Xu, S., Hu, Z., Wang, E., 2023. Predicting maize yield in Northeast China by a hybrid approach combining biophysical modelling and machine learning. Field Crops Res. 302, 109102. https://doi.org/10.1016/j.fcr.2023.109102.

Li, Z., Khan, W.U., Bai, G., Dong, C., Wang, J., Zhang, Y., Wang, C., Zhang, H., Wang, W., Luo, M., Chen, F., 2025. From code to life: the AI-driven revolution in genome editing. Adv. Sci. 12 (30). https://doi.org/10.1002/advs.202417029.

Liao, Q., Liu, X., Zhao, W., Tong, Y., Xu, F., Liu, X., Chen, Y., 2025. IRGL-RRI: interpretable graph representation learning for plant RNA–RNA interaction discovery. Front. Plant Sci. 16. https://doi.org/10.3389/fpls.2025.1617495.

Liu, H., Ding, Y., Zhou, Y., Jin, W., Xie, K., Chen, L.L., 2017. CRISPR-P 2.0: an improved CRISPR-Cas9 tool for genome editing in plants. Mol. Plant 10 (3), 530–532. https://doi.org/10.1016/j.molp.2017.01.003.

Liu, G., Lin, Q., Jin, S., Gao, C., 2022a. The CRISPR-Cas toolbox and gene editing technologies. Molecular cell 82, 333–347.

Liu, J., Zhang, L., Huang, L., Yang, T., Ma, J., Yu, T., Zhu, W., Zhang, Z., Tang, J., 2022b. Uncovering the gene regulatory network of maize hybrid ZD309 under heat stress by transcriptomic and metabolomic analysis. Plants 11, 677.

Liu, X., Wang, S., Ai, D., 2022c. Predicting CRISPR/Cas9 repair outcomes by attention-based deep learning framework. Cells 11, 1847.

Liu, Z., Wang, X., Cao, S., Lei, T., Chenzhu, Y., Zhang, M., Liu, Z., Lu, J., Ma, W., Su, B., Wang, Y., Zhou, Y., 2024. Deep learning facilitates precise identification of disease-resistance genes in plants. bioRxiv. https://doi.org/10.1101/2024.09.26.615248.

Madurapperumage, A., Naser, M.Z., Boatwright, L., Bridges, W., George, V., Thavarajah, D., 2025. High-throughput phenotyping platforms for pulse crop biofortification. Plants, People, PLANET 7 (1), 49–61. https://doi.org/10.1002/ppp3.10568.

Mallapaty, S., 2022. China's approval of gene-edited crops energizes researchers. Nature 602 (7898), 559–560. https://doi.org/10.1038/d41586-022-00395-x.

Maraveas, C., 2024. Image analysis artificial intelligence technologies for plant phenotyping: current state of the Art. AgriEngineering 6 (3), 3375–3407. https://doi.org/10.3390/agriengineering6030193.

Martin, H.G., Radivojevic, T., Zucker, J., Bouchard, K., Sustarich, J., Peisert, S., Arnold, D., Hillson, N., Babnigg, G., Marti, J.M., Mungall, C.J., Beckham, G.T., Waldburger, L., Carothers, J., Sundaram, S.S., Agarwal, D., Simmons, B.A., Backman, T., Banerjee, D., Tanjore, D., Ramakrishnan, L., Singh, A., 2023. Perspectives for self-driving labs in synthetic biology. Curr. Opin. Biotechnol. 79. https://doi.org/10.1016/j.copbio.2022.102881.

Mathis, N., Allam, A., Kissling, L., Marquart, K.F., Schmidheini, L., Solari, C., Balázs, Z., Krauthammer, M., Schwank, G., 2023. Predicting prime editing efficiency and product purity by deep learning. Nat. Biotechnol. 41 (8), 1151–1159. https://doi.org/10.1038/s41587-022-01613-7.

Mienye, I.D., Swart, T.G., 2024. A comprehensive review of deep learning: architectures, recent advances, and applications. Information 15 (12), 755. https://doi.org/10.3390/info15120755.

Minkenberg, B., Zhang, J., Xie, K., Yang, Y., 2018. CRISPR-PLANT v2: an online resource for highly specific guide RNA spacers based on improved off-target analysis. Plant Biotechnol. J. 17, 5.

Modrzejewski, D., Hartung, F., Lehnert, H., Sprink, T., Kohl, C., Keilwagen, J., Wilhelm, R., 2020. Which factors affect the occurrence of off-target effects caused by the use of CRISPR/Cas: a systematic review in plants. Front. Plant Sci. 11. https://doi.org/10.3389/fpls.2020.574959.

Montenegro de Wit, M., 2022. Can agroecology and CRISPR mix? The politics of complementarity and moving toward technology sovereignty. Agric. Hum. Val. 39 (2), 733–755. https://doi.org/10.1007/s10460-021-10284-0.

Montesinos-López, O.A., Montesinos-López, A., Hernandez-Suarez, C.M., Barrón-López, J.A., Crossa, J., 2021. Deep-learning power and perspectives for genomic selection. Plant Genome 14 (3). https://doi.org/10.1002/tpg2.20122.

Mostafa, S., Mondal, D., Karim, P., Kochian, L., Stavness, I., 2023. Explainable deep learning in plant phenotyping. Front. Artif. Intell. 6. https://doi.org/10.3389/frai.2023.1203546.

Müller Paul, H., Istanto, D.D., Heldenbrand, J., Hudson, M.E., 2022. CROPSR: an automated platform for complex genome-wide CRISPR gRNA design and validation. BMC Bioinf. 23 (1). https://doi.org/10.1186/s12859-022-04593-2.

Muruganantham, P., Wibowo, S., Grandhi, S., Samrat, N.H., Islam, N., 2022. A systematic literature review on crop yield prediction with deep learning and remote sensing. Remote Sens. 14 (9). https://doi.org/10.3390/rs14091990.

Naert, T., Yamamoto, T., Han, S., Röck, R., Horn, M., Bethge, P., Vladimirov, N., Voigt, F.F., Figueiro-Silva, J., Bachmann-Gagescu, R., Vleminckx, K., Helmchen, F., Lienkamp, S.S., 2025. Precise, predictable genome integrations by deep-learning-assisted design of microhomology-based templates. Nat. Biotechnol. https://doi.org/10.1038/s41587-025-02771-0.

Nannini, L., Balayn, A., Smith, A.L., 2023. Explainability in AI policies: a critical review of communications, reports, regulations, and standards in the Eu, Us, and UK. In: Proceedings of the 2023 ACM Conference on Fairness, Accountability, and Transparency.

Navarro, B.B., Machado, M.J., Figueira, A., 2025. Nitrogen Use efficiency in agriculture: integrating biotechnology, Microbiology, and novel delivery systems for sustainable agriculture. Plants 14, 2974.

Nazari, L., Ghotbi, V., Nadimi, M., Paliwal, J., 2023. A novel machine-learning approach to predict stress-responsive genes in Arabidopsis. Algorithms 16 (9), 407. https://doi.org/10.3390/a16090407.

Nikolados, E.-M., Aodha, O.M., Cambray, G., Oyarzún, D.A., 2021. From sequence to yield: deep learning for protein production systems. bioRxiv 1–10.

Novielli, P., Romano, D., Pavan, S., Losciale, P., Stellacci, A.M., Diacono, D., Bellotti, R., Tangaro, S., 2024. Explainable artificial intelligence for genotype-to-phenotype prediction in plant breeding: a case study with a dataset from an almond germplasm collection. Front. Plant Sci. 15, 1434229.

Özden, F., Minary, P., 2024. Learning to quantify uncertainty in off-target activity for CRISPR guide RNAs. Nucleic Acids Res. 52 (18), e87. https://doi.org/10.1093/nar/gkae759.

Panahi, B., Golkari, S., 2024. Identification of gene expression signature for drought stress response in barley (Hordeum vulgare L.) using machine learning approach. Curr. Plant Biol. 39, 100370. https://doi.org/10.1016/j.cpb.2024.100370.

Peleke, F.F., Zumkeller, S.M., Gültas, M., Schmitt, A., Szymański, J., 2024. Deep learning the cis-regulatory code for gene expression in selected model plants. Nat. Commun. 15 (1). https://doi.org/10.1038/s41467-024-47744-0.

Peleman, J.D., Van Der Voort, J.R., 2003. Breeding by design. Trends Plant Sci. 8 (7), 330–334. https://doi.org/10.1016/S1360-1385(03)00134-1.

Pérez-Enciso, M., Zingaretti, L., De Los Campos, G., 2025. Generative AI for predictive breeding: hopes and caveats. Theor. Appl. Genet. 138, 147.

Perrotta, R.M., Vinke, S., Ferreira, R., Moret, M., Mahas, A., Chiappino-Pepe, A., Riedmayr, L.M., Mehra, A.T., Lehmann, L.S., Church, G.M., 2024. Machine learning and directed evolution of base editing enzymes. bioRxiv. https://doi.org/10.1101/2024.05.17.594556.

Ponnurangan, V., Namachivayam, R., Pradeep, R.K.M., Jesudoss, D., Eswaran, K., Loganathan, A., Kumar, K.K., Vaikuntavasan, P., Maduraimuthu, D., Shanmugam, V., 2025. Biotechnological breakthroughs in rice disease management: an overview from transgenics to CRISPR. Mol. Biol. Rep. 52 (1). https://doi.org/10.1007/s11033-025-10701-1.

Pradhan, A.K., Gandham, P., Rajasekaran, K., Baisakh, N., 2025. Predictive prioritization of genes significantly associated with biotic and abiotic stresses in maize using machine learning algorithms. Front. Plant Sci. 16. https://doi.org/10.3389/fpls.2025.1611784.

Qiao, B., Gao, W., Zhang, X., Du, M., Wang, S., Liu, X., Pang, S., Yang, C., Wang, J., Zhao, Y., Xie, L., 2025. SaGP: identifying plant saline-alkali tolerance genes based on machine learning techniques. Front. Plant Sci. 16. https://doi.org/10.3389/fpls.2025.1629794.

Rabone, M., Harden-Davies, H., Collins, J.E., Zajderman, S., Appeltans, W., Droege, G., Brandt, A., Pardo-Lopez, L., Dahlgren, T.G., Glover, A.G., Horton, T., 2019. Access to marine genetic resources (Mgr): raising awareness of best-practice through a new agreement for biodiversity beyond national jurisdiction (bbnj). Front. Mar. Sci. 6. https://doi.org/10.3389/fmars.2019.00520.

Roginsky, J., 2018. Analyzing cRiSPR editing results: Synthego developed a tool called ice to be more efficient than other methods. Genetic Engineering & Biotechnology News 38, S24–S26.

Rubini, P., Kavitha, P., 2021. Deep learning model for early prediction of plant disease. In: 2021 Third International Conference on Intelligent Communication Technologies and Virtual Mobile Networks (ICICV). IEEE, pp. 1104–1107.

Ruggeri, D., Vidács, L., 2025. Advancing wheat single-nucleotide polymorphism data analysis with explainable deep learning models. Appl. Artif. Intell. 39 (1). https://doi.org/10.1080/08839514.2025.2565169.

Sarijaloo, F.B., Porta, M., Taslimi, B., Pardalos, P.M., 2021. Yield performance estimation of corn hybrids using machine learning algorithms. Artificial Intelligence in Agriculture 5, 82–89.

Sarzaeim, P., Muñoz-Arriola, F., Jarquín, D., Pieruschka, R., 2022. Climate and genetic data enhancement using deep learning analytics to improve maize yield predictability. J. Exp. Bot. 73 (15), 5336–5354. https://doi.org/10.1093/jxb/erac146.

Schapke, J., Tavares, A., Recamonde-Mendoza, M., 2021. EPGAT: gene essentiality prediction with graph attention networks. IEEE ACM Trans. Comput. Biol. Bioinf 19, 1615–1626.

Sharma, N., Raman, H., Wheeler, D., Kalenahalli, Y., Sharma, R., 2023. Data-driven approaches to improve water-use efficiency and drought resistance in crop plants. Plant Sci. 336, 111852. https://doi.org/10.1016/j.plantsci.2023.111852.

Sheikh, M., Iqra, F., Ambreen, H., Pravin, K.A., Ikra, M., Chung, Y.S., 2024. Integrating artificial intelligence and high-throughput phenotyping for crop improvement. J. Integr. Agric. 23 (6), 1787–1802. https://doi.org/10.1016/j.jia.2023.10.019.

Shi, J., Gao, H., Wang, H., Lafitte, H.R., Archibald, R.L., Yang, M., Hakimi, S.M., Mo, H., Habben, J.E., 2017. ARGOS8 variants generated by CRISPR-Cas9 improve maize grain yield under field drought stress conditions. Plant Biotechnol. J. 15 (2), 207–216. https://doi.org/10.1111/pbi.12603.

Singh, B., Cheema, S.S., 2024. Cybersecurity threats and mitigation strategies in agriculture 4.0 and 5.0: challenges and solutions in the digital transformation of agriculture. Journal Punjab Academy of Sciences 24, 15–32.

Sirakaya, A., 2022. Is the Nagoya Protocol designed to conserve biodiversity? Plants, People, PLANET 4 (1), 68–75. https://doi.org/10.1002/ppp3.10221.

Smet, D., Opdebeeck, H., Vandepoele, K., 2023. Predicting transcriptional responses to heat and drought stress from genomic features using a machine learning approach in rice. Front. Plant Sci. 14, 1212073.

Tahmasebi, A., Niazi, A., Akrami, S., 2023. Integration of meta-analysis, machine learning and systems biology approach for investigating the transcriptomic response to drought stress in Populus species. Sci. Rep. 13, 847.

Thanmalagan, R.R., Roy, A., Jayaprakash, A., Lakshmi, P.T.V., 2022. Comprehensive meta-analysis and machine learning approaches identified the role of novel drought specific genes in Oryza sativa. Plant Gene 32, 100382. https://doi.org/10.1016/j.plgene.2022.100382.

Tiozon, R.N., Buenafe, R.J.Q., Jain, V., Sen, P., Molina, L.R., Anacleto, R., Sreenivasulu, N., 2024. Machine learning technique unraveled subspecies-specific ionomic variation with the preferential mineral enrichment in rice. Cereal Chem. 101 (2), 367–381. https://doi.org/10.1002/cche.10706.

Toufikuzzaman, M., Hassan Samee, M.A., Sohel Rahman, M., 2024. CRISPR-DIPOFF: an interpretable deep learning approach for CRISPR Cas-9 off-target prediction. Briefings Bioinf. 25, bbad530.

Turnbull, C., Lillemo, M., Hvoslef-Eide, T.A., 2021. Global regulation of genetically modified crops amid the gene edited crop boom–a review. Frontiers in plant science 12, 630396.

UK PARLIAMENT. 2023. Precision Breeding Act 2023. https://www.legislation.gov.uk/ukpga/2023/12/contents/enacted (Accessed 1 October 2025).

Uniyal, A.P., Mansotra, K., Yadav, S.K., Kumar, V., 2019. An overview of designing and selection of sgRNAs for precise genome editing by the CRISPR-Cas9 system in plants. 3 Biotech 9 (6). https://doi.org/10.1007/s13205-019-1760-2.

USDA, 2024. Genome Editing in Agricultural Biotechnology. U.S. Department of Agriculture. https://www.fda.gov/food/agricultural-biotechnology/genome-editing-agricultural-biotechnology.

Wang, H., Yan, S., Wang, W., Chen, Y., Hong, J., He, Q., Diao, X., Lin, Y., Chen, Y., Cao, Y., Guo, W., Fang, W., 2025. Cropformer: an interpretable deep learning framework for crop genomic prediction. Plant Commun. 6 (3), 101223. https://doi.org/10.1016/j.xplc.2024.101223.

Wu, J., Xu, H., Li, Q., 2025. Integrating genomic selection and machine learning for predicting maize yield under drought. Maize Genomics Genet. 16.

Xiang, X., Corsi, G.I., Anthon, C., Qu, K., Pan, X., Liang, X., Han, P., Dong, Z., Liu, L., Zhong, J., Ma, T., Wang, J., Zhang, X., Jiang, H., Xu, F., Liu, X., Xu, X., Wang, J., Yang, H., Bolund, L., Church, G.M., Lin, L., Gorodkin, J.,

Luo, Y., 2021. Enhancing CRISPR-Cas9 gRNA efficiency prediction by data integration and deep learning. Nat. Commun. 12 (1). https://doi.org/10.1038/s41467-021-23576-0.

Xiao, S., Lin, H., Wang, C., Wang, S., Rajapakse, J.C., 2023. Graph neural networks with multiple prior knowledge for multi-omics data analysis. IEEE J. Biomed. Health Inform. 27 (9), 4591–4600. https://doi.org/10.1109/JBHI.2023.3284794.

Xie, T., Zhang, Z., 2025. Construction of large-scale maize drought resistance gene networks based on multi-omics integration and identification of key regulatory factors. Advances in Resources Research 5, 203–234.

Xie, K., Zhang, J., Yang, Y., 2014. Genome-wide prediction of highly specific guide RNA spacers for CRISPR–Cas9-mediated genome editing in model plants and major crops. Mol. Plant 7, 923–926.

Xu, Y., Zhang, X., Li, H., Zheng, H., Zhang, J., Olsen, M.S., Varshney, R.K., Prasanna, B.M., Qian, Q., 2022. Smart breeding driven by big data, artificial intelligence, and integrated genomic-enviromic prediction. Mol. Plant 15 (11), 1664–1695. https://doi.org/10.1016/j.molp.2022.09.001.

Xu, F., Wu, T., Cheng, Q., Wang, X., Yan, J., 2025. Foundation models in plant molecular biology: advances, challenges, and future directions. Front. Plant Sci. 16. https://doi.org/10.3389/fpls.2025.1611992.

Yang, H., Xu, P., Zhang, S., Kim, H., Shin, I., 2025. Construction of an intelligent analysis system for crop health status based on drone remote sensing data and CNN. IEEE Access 13, 31643–31657. https://doi.org/10.1109/access.2025.3540279.

You, Z., Ikram, M., Khan, N., Zhao, N., Derakhshani, B., Usman, B., Wang, H.-F., 2025. Machine learning and functional validation identify OsRAV11/12 as negative regulators of drought tolerance and early flowering in rice. Int. J. Biol. Macromol., 147709

Zafar, O., González, R.S., Wilkins, G., Morales, A., Ayday, E., 2024. Privacy-preserving data linkage across private and public datasets for collaborative agriculture research. arXiv. https://doi.org/10.48550/arXiv.2409.06069.

Zaidi, S.S.e.A., Mahas, A., Vanderschuren, H., Mahfouz, M.M., 2020. Engineering crops of the future: CRISPR approaches to develop climate-resilient and disease-resistant plants. Genome Biol. 21 (1). https://doi.org/10.1186/s13059-020-02204-y.

Zarate, S., Cimadori, I., Jones, M.S., Roca, M.M., Barnhill-Dilling, S.K., 2023. Assessing agricultural gene editing regulation in Latin America: an analysis of how policy windows and policy entrepreneurs shape agricultural gene editing regulatory regimes. Front. Bioeng. Biotechnol. 11. https://doi.org/10.3389/fbioe.2023.1209308.

Zeinalzadeh-Tabrizi, H., Nazari, L., 2025. Machine learning-based transcriptome mining to discover key genes for density stress in sweet corn. Ecological Genetics and Genomics 35, 100349. https://doi.org/10.1016/j.egg.2025.100349.

Zhang, G., Luo, Y., Dai, X., Dai, Z., 2023. Benchmarking deep learning methods for predicting CRISPR/Cas9 sgRNA on- and off-target activities. Briefings Bioinf. 24 (6). https://doi.org/10.1093/bib/bbad333.

Zhang, G., Xie, H., Dai, X., 2024. DeepIndel: an interpretable deep learning approach for predicting CRISPR/Cas9-Mediated editing outcomes. Int. J. Mol. Sci. 25, 10928.

Zhang, D., Xu, F., Wang, F., Le, L., Pu, L., 2025. Synthetic biology and artificial intelligence in crop improvement. Plant Commun. 6.

Zhou, S., Cheng, K., Lv, L., Jiang, J., Zhou, S., Zhou, Y., Xu, Z., Huang, Q., Yang, H., Chen, L., Xu, Y., Yao, Z., Zhao, T., 2025. CropARNet: a deep learning framework for crop genomic prediction with attention and residual modules. Crop Design 4 (4), 100118. https://doi.org/10.1016/j.cropd.2025.100118.

Zhu, C., Zhao, L., Zhao, S., Niu, X., Lin, L., Gao, H., Liu, J., Wang, L., Zhang, T., Cheng, R., Shi, Z., Zhang, H., Wang, G., 2024. Utilizing machine learning and bioinformatics analysis to identify drought-responsive genes affecting yield in foxtail millet. Int. J. Biol. Macromol. 277 (Pt 2), 134288. https://doi.org/10.1016/j.ijbiomac.2024.134288.

Further reading

Chen, X., Cui, X., Ma, H., 2025. The application of artificial intelligence and gene editing technologies in high-yield crop breeding: innovations and prospects for future agriculture. Advances in Resources Research 5, 146–166.

CHAPTER

9

Machine learning approaches in gene editing

Mohamed A.M. Atia[1], Abdelghani Belgaid[1], Kamel A. Abd-Elsalam[2], and Achraf El Allali[1]

[1]Bioinformatics Laboratory, College of Computing, University Mohammed VI Polytechnic, Ben Guerir, Morocco; [2]Plant Pathology Research Institute, Agricultural Research Center, Giza, Egypt

9.1 Introduction: Why machine learning for plant gene editing?

Machine learning (ML) has emerged as a problem-solving, and transformative domain, shifting from conventional rule-based to model-based programming, where algorithms learn directly from datasets. The integration of ML with bioengineering tools, particularly CRISPR-based gene editing, is reshaping the future of plant biotechnology by driving the selection of guide RNA, editor choice, and outcome prediction increasingly via data-driven approaches. Over a decade of dramatic increase in developing editing toolkits—in particular, the arrival of cytosine base editors (CBEs) and adenine base editors (ABEs) and, more recently, prime editors (PE)—has advanced editing chemistry beyond double-strand break (DSB)–dependent pathways, thereby creating more enriched solutions to determine more target design spaces that ML can navigate and optimize efficiently (Komor et al., 2016; Gaudelli, 2017; Anzalone et al., 2019). Concurrently, the availability of comprehensive, accurately labeled datasets from both high-throughput genome-wide off-target profiling as well as amplicon sequencing data has permitted the construction and validation of robust predictive models, paving the way for their operational deployment in plant research and bioengineering (Tsai et al., 2015; Lazzarotto et al., 2020; Clement et al., 2019).

The prediction activity of on-target events displays the revolution within this field from heuristic practices to modern statistical visualization and deep learning model development. Earlier, broadly adopted scoring functions explained sequence-level determinants of Cas9 activity and delivered optimized libraries (Doench et al., 2016). The next generation of deep models has been trained on tens to thousands of features to generalize rules/actions

Intelligent Gene Editing in Plants
https://doi.org/10.1016/B978-0-443-44079-3.00022-X

across numerous loci and cell contexts, capturing attributes of higher sequence levels when chromatin information is available (Kim et al., 2019). These advancements are directly applicable in plant gene/genome editing, where locus-, tissue-, and genotype-specific effects can be marked. All the same, estimation for off-target risk has benefited from orthogonal, genome-wide assays such as GUIDE-seq analysis in cells and CHANGE-seq analysis in vitro, which jointly disclose the contributions of mismatch patterns, chromatin state, and even donor genetic variation to Cas9 specificity-factors that ML models can integrate as attributes or domain-adapt via transfer learning (Tsai et al., 2015; Lazzarotto et al., 2020).

Crucially, precise quantification workflows have established the solid link between experimental data and model training. Recently developed tools, such as CRISPResso2, standardize the analysis, visualization, and comparison of deep sequencing data from GE experiments. In addition, CRISPResso2 offers accurate identification of edited bases as well as indel events generated from various nucleases and editors, allowing for invariant curation of training features and reproducible benchmarking of predictive algorithms (Clement et al., 2019). Beyond the developed task-specific models, the integration of foundation models (FMs) in genomics and sequence-to-regulatory-function models, which have the power to learn from long-range cis-regulatory logic, offers a wide range of effective transfer-learning foundations for gene editing tasks in plants, starting from guide prioritization in chromatinized contexts to editor-choice recommendations close to regulatory regions (Avsec et al., 2021).

Plant biology presents a rich domain containing distinctive constraints and opportunities that can shape the future of ML applications. Among these constraints is that many crops contain large, repeat-rich, and often polyploid genomes, which separately or together complicate the target selection process and increase the value of allele-aware design; better long-read assemblies and current pan-genome resources alleviate these issues and permit graph-based targetability analyses (Michael and VanBuren, 2020; Schreiber et al., 2024). Also, the practical limitations of both transformation and regeneration techniques, which often lead to limitations in experimental outcomes due to their genotype-dependent nature, underscore the importance of in silico approaches to concentrate wet-lab effort on the most promising guides, editors, and delivery strategies (Altpeter et al., 2016). Finally, while the CRISPR toolboxes were rapidly adapted to plant applications (Xie and Yang, 2013), cross-kingdom domain shifting means that models well-trained in mammalian contexts usually require retraining or recalibration on plant-derived measurements/records to maintain higher levels of accuracy; here, the dramatic expansion of plant-specific datasets, combined with the inherent portability of modern machine learning models, represents a significant competitive advantage (Doench et al., 2016; Kim et al., 2019) (Fig. 9.1).

This chapter explores ML models and approaches utilized across the plant editing workflow: (1) datasets and resources curation; (2) predictive modeling selection, editing outcomes, and assessment of off-target risk, model calibration, and uncertainty; (3) integration and optimization of multiobjective guide/editor with consideration of plant-specific constraints (such as polyploidy, pangenomes, regulatory context, and delivery/regeneration limits); and (3) case studies that exhibit recognizable improvements in either editing precision or throughput levels. Finally, we outline a roadmap toward closed-loop, artificial intelligence (AI)–assisted bioengineering—combining high-fidelity editors (Komor et al., 2016; Gaudelli, 2017; Anzalone et al., 2019), allele-aware target graphs (Schreiber et al., 2024), and reproducible/transferable sequence models (Avsec et al., 2021)—to speed up trait

FIGURE 9.1 Comprehensive plant genome editing workflow.

improvement while retaining exacting standards for specificity and reproducibility in crop species (Altpeter et al., 2016; Michael and VanBuren, 2020).

9.2 Data foundations for ML in plant gene editing

Building powerful ML approaches for plant gene editing begins with the data foundation procedure that explains if the plants are efficiently edited, regenerated, and phenotyped in reality, not just datasets simplified as they appear in benchmarks. Compared to the mammalian systems, plant gene editing holds unusual complexities, including polyploidy genomes having numerous homoeologous targets, cultivar-specific assemblies, organ/tissue-specific transformation or regeneration protocols, and ecological effects on editing outcomes as well as plant traits. Therefore, the substantial foundation deals with data as a connected skeleton containing hierarchical records covering the whole design, editing events, regeneration system, and phenotyping layout capturing, not just raw data about "what was edited," but also "how, where, and under which conditions" (Guo and Li, 2023).

9.2.1 Label types

Outcome tags/labels should be defined at considerable granularity. At the edit level, constant on-target efficacy (such as percentage of edited reads) must be coupled with allele or homeolog gene resolution levels so that ML models can learn efficiently from the subgenome-specific behavior in polyploid genomes. For DSB-nucleases, recording the genome-wide indel events (single-base insertion preferences, deletions binned by size, and frameshift probability) is more advantageous than documenting a single efficient editing event, as it supports downstream objectives, such as loss-of-function design. On the other side, for editor-specific labels, the base editing process requires identifying the conversion rates per-position within a certain activity window and a precise accounting of observer edits and product purity. Meanwhile, prime editing requires a precise calculation of the intended edit efficiency and the rates of indel events linked to prime editing guide RNA (pegRNA) design and the nicking strategy. For the off-target labels, they should discriminate between validated negative events, untested sites, and assay detection limits (for assays such as WGS, CIRCLE-seq, SITE-seq, and amplicon-seq), which is essential for honest ML supervision. Ultimately, the different classes of phenotype labels, including binary, ordinal, and continuous (e.g., herbicide tolerance class, days-to-flowering, biomass), must be linked to environment, replicates, and generation (T0/T1/T2) so that ML models can learn how to relate genotype-to-phenotype without confounding (Li et al., 2024).

9.2.2 Sources and plant design servers

Trustworthy editing outcomes mainly depend on the use of reliable resources and inputs. Therefore, utilizing reference genomes and annotations from resources or portals such as Phytozome and Ensembl Plants, which provide reference/ortholog gene sequences and repeat catalogs for model organisms, enables consistent mapping of targets across various cultivars/species and underpins transfer-learning experiments. Besides, the availability of regulatory/cis-regulatory resources (e.g., PlantTFDB/PlantRegMap) and transcription-factor context can explain locus-dependent variability (Tian et al., 2020).

Additionally, online design servers such as CRISPR-PLANT, CRISPR-P, CHOPCHOP, CRISPOR, and others furnish high-quality computations toward designing precise candidate gRNAs with the flexibility to set the specificity baselines and convenient negative sets (e.g., off-target candidates with mismatch/bulge profiles). Such equipped servers could be treated as powerful generators for labels and features, but not as basic ground, or snapshot versions, or record parameters, to avoid any slight drift across reruns. Therefore, when assembling cross-species outputs, aligning targets through synteny and/or orthogroups rather than gene symbols alone is recommended (Saini et al., 2025).

9.2.3 Features and context

The DNA sequence stays the primary signal or input for spacer-PAM encodings, k-mer determination, GC content calculation, seed-region weights, gRNA secondary structure prediction, and microhomology tract identification that collectively shape the indel repair precision levels. Recent studies reported that plant ML models improved markedly when the sequence data is enriched with other genomic and experimental contexts. The useful genomic features comprise a wide range of labels, including intron/exon/UTR class, coding frame, gene family size (as a sign of the functional redundancy level), isoform coverage, subgenome status (polyploids), repeat density, and distance to regulatory elements. For the editor- and construct-level attributes, many features, such as nuclease/editor variant (SpCas9, Cas12a, SpRY, PE/ABE/CBE), gRNA scaffold/tracrRNA, vector/promoter backbone, and multiplexing strategy, can modulate activity. Also, the experimental conditions, such as delivery technique (Agrobacterium, biolistic gene gun, viral), tissue type (meristem, callus, protoplast), growth regimen, regeneration staging, and even temperature level (especially for the prime editing), mostly explain large fractions of variance, which should be encoded as continuous or categorical features. Deeply defined inputs (such as read depth and replicate count) help ML models calibrate predicted probabilities. Eventually, consider learned representations of sequence language models, motif/scaffold embeddings, or latent cross-species spaces capable of capturing regularities that crafted features miss, especially when labeled data are scarce (LeBlanc and Charles, 2022).

9.2.4 Curation and leakage

The most disappointing cases of model failures are often due to data leakage rather than architectural choices. In plants, leakage predominantly occurs due to sequence similarity, which is a common phenomenon, especially when near-duplicate spacers that target homeologs/paralogs or allelic variants slip across the training and test datasets and yield inflated performance or low-specificity outcomes. Also, the leakage in context appears when the same locus/target is assessed in different tissues/constructs and appears in both splits. Assay leakage also occurs when postedit information (e.g., variant calls) is contaminated with input features. On the other side, cross-species leakage, especially in orthologous sites, can create an illusion of generalization. Among the other pitfalls is the tool-induced bias, which occurs when negatives are defined solely by one list, or a single assembly release does not reflect the true search space. To prevent or eliminate these, split the data based on the locus or gene-family clusters or the subgenome, not by reads or plant lines; group by the lab, batch, tissue, and generation (Sherkatghanad et al., 2023).

Furthermore, deduplication of sequences by strict identity thresholds and k-mer chopping can successfully separate the labels of "assay not performed" from other true negatives, in addition to freezing all upstream dependencies by registering assembly releases, as well as other parameter files. Besides, integrate the reports of ranking metrics and calibration and employ the bootstrap process to quantify predictive uncertainty, especially when deploying across multiple environments. Among the prospective challenges, new cultivars, tissues, or temperature regimes are invaluable determinants for detecting silent failure modes before model's impact design decisions. Ultimately, many practical attributes should be considered to eliminate as well as overcome the curation and leakage challenges, including standardizing a minimal, versioned architecture that stacks with the data of species/cultivars; the genome structure/polyploidy; the guide sequence and PAM; the editor type, scaffold, promoter, and delivery; targeted-tissue, environment, and technique details; deep sequencing read counts and depths; full outcome spectra; off-target proof coupled with detection limits; batch and generation; phenotype measurements; and well-defined references to the design servers. When the utilized datasets possess such a deeply defined structure, ML becomes a reliable and powerful tool, rather than an optimistic guess, sustaining plant gene editing that can be scalable across labs, cultivars, and species (Sherkatghanad et al., 2023) (Fig. 9.2).

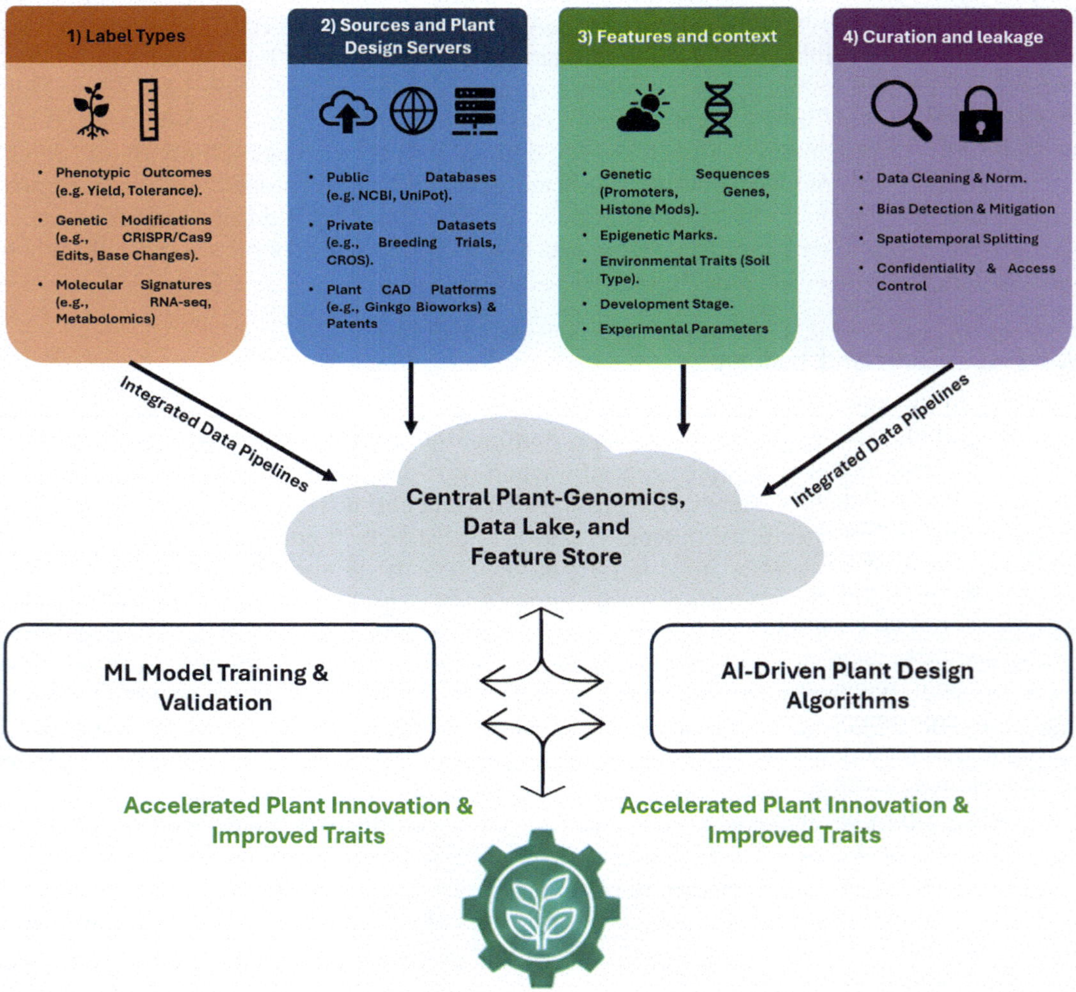

FIGURE 9.2 Machine learning data foundation framework for gene editing in plants.

9.3 Model landscape: From classical ML to foundation models

The predictive modeling landscape in gene editing has expanded from early feature-based algorithms to advanced ML algorithms, reflecting both the growing scale of genomic datasets and the complexity of biological-based algorithms. This family of models presents distinct advantages and limitations that contour their suitability for diverse gene editing applications.

The conventional or classical ML algorithms, such as logistic regression, support vector machines, and gradient-boosted trees (Chen and Guestrin, 2016; Ke et al., 2017), have proven their initial dominance in the field. Their reliance on engineered features, such as GC content, position-specific motifs, or microhomology scores, produces interpretable predictions with modest accuracy. While there are limited in representing higher-order dependencies, they remain strong baselines, especially when datasets are small, and their explainability is valuable for regulatory or agronomic contexts where interpretability is prioritized (Samek and Müller, 2019).

Among the modern models, deep sequence models introduced a qualitative leap. Recurrent architecture algorithms such as recurrent neural networks (RNNs) and long short-term memory (LSTM) (Hochreiter and Schmidhuber, 1997), as well as convolutional neural networks (CNNs) (Lecun et al., 1998), were used to learn motifs directly from one-hot encoded DNA sequences, thus bypassing handcrafted features. Furthermore, extended use of attention mechanisms could enhance interpretability by highlighting seed sequences or editing windows (Vaswani et al., 2017; Bahdanau et al., 2015). Although graph neural networks (Kipf and Welling, 2017; Scarselli et al., 2008) further integrate secondary structure features in addition to integrating sequence and structure context, plant applications remain preliminary.

The mixed or hybrid approaches incorporate advanced feature extraction methods with other traditional ML algorithms, for example, CNN coupled with gradient support vector regression (Smola and Schölkopf, 2004). Such approaches leverage the rendering power of deep neural networks while maintaining the calibration and stability of tree-based models, thereby offering accumulative consistent gains in predictive performance style. Meanwhile, ensemble approaches, including stacking and majority-vote architectures, have been mostly applied in CRISPR ML applications to stabilize predictions and decrease variance.

Transfer learning and domain adaptation address the bottleneck of limited plant-specific datasets. Among the leading strategies in computer vision is natural language processing (NLP), which is increasingly used and implemented in genome editing applications (Yosinski et al., 2014; Pan & Yang, 2009; Bengio et al., 2013).

Recently, many foundation models, such as pretrained transformer architectures on large corpora of DNA or RNA sequences, presented a new paradigm. The integrations with large language models in NLP established a framework that can capture long-range dependencies and universal sequence semantics and can adapt to plant-specific methods, editors, and species (Vaswani et al., 2017; Devlin et al., 2019; Brown et al., 2020). On the other hand, when coupled with transfer learning models, they guarantee a scalable adaptation to plant genomes and enable multitask frameworks that together predict on-target efficacy, repair spectra, off-target risk, and editor-specific outcomes.

Reinforcement learning (RL) provides another paradigm, where models iteratively optimize genome editing strategies through trial-and-error interactions with a simulated environment. Following the early policy optimization reshaping, RL frameworks were developed in a way that balances between the exploitation of known high-efficacy guides and exploration of novel sequence candidates. Although these frameworks are still emerging in genome editing applications, combined approaches with deep learning are promising for efficient navigation of extensive combinatorial search spaces and adaptive experimental design.

Eventually, recent quantum machine learning (QML) models are expected to represent a nascent and intriguing frontier. Leveraging quantum algorithms to process high-dimensional genomic feature spaces or combinatorial search problems may eventually complement classical and advanced ML in genome editing, though current applications in plants remain entirely prospective (Biamonte et al., 2017). Taken together, this continuum from interpretable linear baselines to multimodal foundation models illustrates the ongoing progression in algorithm complexity as well as the growing emphasis on generalizability, interpretability, and plant-specific calibration. The subsequent sections of these chapter illustrate how each family has been customized and operationalized for specific prediction tasks in nuclease, base, and prime editing.

9.3.1 On-target efficiency prediction

Accurate prediction of guide RNA (gRNA) activity at the locus of interest is essential for successful crop genome editing. However, considerable challenges remain, particularly in transformation and regeneration approaches, as well as the complexities of polyploidy, which frequently require multiallelic edits. These challenges present critical bottlenecks that must be overcome to advance the field of crop improvement. Sequence features including PAM-proximal motifs, GC%, guide/target structure, locus accessibility influence activity; cultivar SNPs and homeolog copy number add more plant-specific complexity (Doench et al., 2016; Niu et al., 2021).

In terms of current progress, human-trained baselines (Broad "Rule Sets") established feature-based prediction using gradient boosting/ensembles (Rule Set 2; later updates incorporating tracrRNA sequence and expanded training) and remain widely used as proxies in plants. Although the performance of those models is moderate (rank correlations on human data are between 0.4 and 0.6), the chromatin and sequence composition of different species make it problematic to transfer to plants. During the past few years, many specialized plant models have been growing: sgRNACNN (ensemble CNN) outperformed simpler classifiers with ~78–87% accuracy across crops after being trained on *Glycine max*, *Zea mays*, *Sorghum bicolor*, and *Triticum aestivum* to create discriminated high-efficiency guides (Niu et al., 2021). Deep CNN and RNN models from mammalian work (e.g., DeepSpCas9) motivate the development of new architectures that learn motifs and longer-range context (Kim et al., 2019; Niu et al., 2021); plant-specific epigenomic inputs (ATAC-seq) further improve AUC in rice/maize case studies (Niu et al., 2021Niu et al., 2021). However, many factors remain important limitations, as the available plant data sets are smaller and heterogenous (tissue, delivery, genotype), complicating generalization; polyploidy requires modeling multitarget success; cultivar variation implies pangenome-aware scoring. Evaluation practices (correlation vs. accuracy, per-gene vs. pooled) remain inconsistent, limiting cross-study comparability.

Practically, researchers can mitigate these issues by using consensus across tools (e.g., Rule Set 2/3 + a plant-trained model); prefer species/clade-matched training where available; design three to four gRNAs per gene to hedge uncertainty; check targets across cultivars or pangenomes. Incorporate simple rules (extreme GC, promoter-proximal sites) alongside ML and iteratively feed back empirical results to refine local predictions (Doench et al., 2016). For front-end design/enumeration in plants, CRISPR-P 2.0 (specific for plant genomes), CHOPCHOP v3, and CRISPOR are widely used portals that can feed ML scorers and pangenome checks; plant regulatory datasets such as PlantDHS help inject epigenomic context (Labun et al., 2019; Concordet and Haeussler, 2018; Liu et al., 2017; Zhang et al., 2016). Looking forward, larger plant screens and integration of plant epigenomes (CG/CHG/CHH methylation) should lift accuracy; multispecies transfer learning and active learning loops will reduce experimental burden for legumes and cereals (Lin et al., 2020).

9.3.2 Off-target prediction and specificity modeling

Large-sized, repetitive-enriched, and polyploid plant genomes create a real challenge in genome editing as they include many near-matches to a 20-nt spacer; only a subset is cleaved. ML aims to prioritize risky sites or off-targets and deliver aggregate specificity scores. Currently, two-step pipelines dominate (1) enumerate candidate sites (e.g., use of Cas-OFFinder to prioritize the genome-wide off-targets): and (2) score cleavage probability by ML (Bae et al., 2014). The Elevation model (gradient-boosted trees) and deep learning models used in such as DeepCRISPR improved precision–recall over heuristic scoring, leveraging mismatch position, local context and chromatin features (Listgarten et al., 2018; Chuai et al., 2018; Haeussler et al., 2016). Although most available machine learning models still come from humans, such as GUIDE-seq/CIRCLE-seq data, many plant-tailored platforms (like CRISPR-P 2.0) incorporate plant genomes (Liu et al., 2017; Tsai et al., 2015, 2017). Empirically, the full genome sequencing of rice revealed a few heritable off-targets under standard conditions, which supports a stringent prefilter plus a targeted validation approach (Tang et al., 2018).

Despite this progress, the lack of plant off-target ground, the high computational cost of analyzing the polyploid or rich-repetitive genomes, and the context matter (genic vs. intergenic, duplicated genes) remain challenging barriers. Cultivar-specific off-targets across pangenomes and variability across nucleases (SpCas9 variants, Cas12a) and editors (CBEs/ABEs) add more complexity levels. A practical workflow including the identification of all potential genome-wide off-target spots with $\leq$3–4 mismatches, prioritizing them using a calibrated ML model, and finally validating the top-scoring candidates via targeted-amplicon sequencing. Additionally, feasibly employing high-fidelity Cas variants or Cas12a nucleases to minimize off-target activity further. For crop improvement, gRNAs should be designed to target unique exonic regions and also cross-checked against considerable cultivar genomes or pangenome to ensure their specificity and transferability (Doench et al., 2016). In plants, pairing genome-wide enumerators (e.g., CRISPR-P 2.0) with learned scorers (Elevation/DeepCRISPR) works well as a two-step screen, especially across cultivar assemblies (Liu et al., 2017).

In the future, plant-trained specificity models (including bulges and repeat context), graph-indexed genome search for speed, enzyme-aware scoring, and editors' off-target models (esp. CBEs) will narrow candidate lists and lower false positives in crops (Wienert et al., 2019).

9.3.3 Predicting DSB repair outcomes (indel spectra)

Cas-induced breaks are predominantly repaired by NHEJ/MMEJ in plants, producing nonrandom indel spectra shaped by local sequence and microhomology, anticipating these distributions aids screening and allele design. To date, some developed tools (ML models), such as inDelphi (tree ensembles), FORECasT (random forests), and SPROUT, present predicted outcome distributions with good fidelity on human data. Meanwhile, newer deep models (e.g., CROTON, multitask CNN) added improved calibration, particularly for microhomology-mediated deletions (Allen et al., 2019; Leenay et al., 2019; Li et al., 2021). Despite being trained in mammalian datasets, sequence-deterministic rules are mostly applicable to plant sequences, where +1 insertions and short deletions are common (Leenay et al., 2019).

Nevertheless, plant-specific modifiers (cell type, culture, repair-pathway usage), pooled T0 mosaics, polyploid haplotypes, and limited quantitative plant outcome datasets complicate training and evaluation. In practice, outcome predictors select cut sites with high frameshift probability for knockouts; consider nuclease choice (e.g., Cas12a staggered cuts) to bias deletions; plan assays based on top predicted indels; in polyploids (e.g., wheat, potato), anticipate mixed outcomes across homeologs in polyploids and plan deeper screening, and adjust screening depth. Web tools such as inDelphi and FORECasT are directly usable today for site triage; newer interpretable models (e.g., X-CRISP, 2025) are emerging but not yet plant-trained (Shen et al., 2018; Allen et al., 2019) (Seale and Gonçalves, 2025). Looking ahead, plant-trained outcome models (thousands of plant edits), incorporation of epigenetic features, and coupling genotype outcome → function (frameshift probability) will make knockout design in cereals/legumes more deterministic (Allen et al., 2019).

9.3.4 Base editing: Efficiency and bystander outcome prediction

CBEs (C → T) and ABEs (A → G) install point mutations without DSBs but exhibit editor- and position-specific "windows" that can yield bystanders; ML predicts both efficiency and positional outcome profiles to guide design in crops (e.g., herbicide-resistance alleles in rice/wheat) (Pallaseni et al., 2022). At present, deep models such as BE-Hive and position-specific frameworks quantify within-window product distributions, achieving correlations up to ~0.7-0.8 in human benchmarking (Arbab et al., 2020; Pallaseni et al., 2022). Plant work corroborates motif and window dependencies; position-specific models (e.g., per-C probabilities) report Pearson's $R \approx 0.49$-0.72 across editors, offering actionable forecasts for bystanders (Pallaseni et al., 2022). Notably, WGS in rice revealed widespread gRNA-independent off-target deamination for early CBEs, spurring higher-fidelity deaminases, an area where ML-guided protein engineering is increasingly used (Jin et al., 2019). Still, data fragmentation across editor variants, scarcity of plant-native, quantitative by-position measurements, unmodeled RNA/DNA structure and expression/time effects, and mosaicism in T0 plants complicates mapping bulk predictions to whole-plant genotypes.

Practically, BE-Hive (interactive web app) and BE-DICT (attention model + web tool) provide per-position outcome distributions usable across editors; both are plant-usable for target triage (Marquart et al., 2021; Arbab et al., 2020). Researchers should use editor-aware predictors to choose targets with high on-base and low bystander probability; prefer narrow/shifted-window editors when a single base must be changed; in crops, compare alternative codons/sites yielding the same amino-acid change and prioritize the higher-scoring option; validate predicted bystanders by amplicon-seq in T0 and select lines accordingly. Future directions include unified multieditor models with explicit editor tokens, integration of DNA/RNA structure features, time-aware predictions, and ML-assisted deaminase design promise cleaner edits in cereals and legumes (Arbab et al., 2020).

9.4 Prime editing: pegRNA design and efficiency/purity prediction

Prime editing (PE) installs small substitutions/indels without DSBs via a Cas9-RT fusion and a pegRNA encoding PBS/RTT; it is attractive for precise crop trait engineering but shows variable efficiency in plants (Lin et al., 2020). So far, early plant demonstrations in rice and wheat validated feasibility but with modest efficiencies; design now benefits from data-driven models (Kim et al., 2021): (multi-model "DeepPE") linked pegRNA/PBS/RTT features to efficiency; OPED (transfer-learned nucleotide language model) generalizes across edit types and PE architectures, improving pegRNA ranking but is trained largely in mammalian data (Kim et al., 2021; Liu et al., 2023). These models, trained in mammalian cells, improve pegRNA ranking and are increasingly used to triage designs in crops pending plant-specific training (Lin et al., 2020).

Yet, low baseline activity in many plant contexts, multicomponent design (pegRNA + nicking sgRNA), and secondary structure effects, paucity of large, quantitative plant PE datasets, chromatin, and repair differences relative to mammalian systems constrain the adoption.

Practically, generating multiple pegRNAs varying PBS (≈10–13 nt) and RTT length (≈10–20 nt), selecting top-ranked designs by an ML scorer, and pairing with a strong nicking sgRNA chosen with established on-target models; in rice/maize protoplasts, efficiencies can be improved with optimized vector systems and thermal/expressional tweaks, empirical iteration guided by model saliency (e.g., penalizing extreme PBS GC) is effective (Lin et al., 2020; Kim et al., 2021). Practitioner tools now include PE-Designer and PE-Analyzer (web), DeepPrime (web + package), and OPED (transfer-learned nucleotide LM; paper) for pegRNA ranking; although trained mostly in mammalian data, they are plant-usable and complement empirical PBS/RTT scans (Liu et al., 2023; Yu et al., 2023).

In the future, plant-trained prime-editing models, detailed RNA folding features, multi-objective or generative design, and enzyme innovation guided by ML should raise in-planta rates for inducing precise edits in cereal/legume genomes (Liu et al., 2023). Table 9.1.

9.5 Conclusions and future perspectives

ML has emerged as a foundation of modern gene editing through reshaping the framework for analyzing, predicting, and engineering biological systems. Within the dramatically growing CRISPR landscape, ML has shifted genome engineering from an experimental concept (trial-and-error) toward modern computational rationality, depending on data-driven design. By incorporating multidimensional biological features such as sequence, chromatin accessibility, repair pathways, and experimental metadata, ML models promote the precise prediction of off-target editing, on-target activities, and editing outcomes across various editing modalities. In plant genomes, where polyploidy, highly repetitive DNA, and gene duplications present complex constraints, modern ML algorithms have proven their power to extract reliable signals from heterogenous datasets, assuring that in silico predictions could be translated into reproducible in vivo performance.

The progress recapped in this chapter displays how advanced ML models, starting from feature-based learner models to sequence-based deep learning and foundation models, are bridging gaps in data set scarcity, cross-species transferability, and interpretability. The resilience, robustness, and transparency of ML-guided frameworks proved to be enhanced through proper data curation, leakage-resistant splits, and standardization of plant-specific data sets. Altogether, these evolutions reflect a paradigm shift, as genome editing is no longer a totally experimental pursuit but a computationally guided domain where AI models drive every decision, from selecting the best suitable guide RNA to predicting the phenotypic outcomes.

Forthcoming advances modern genome editing will rely on coordination between three major frontiers: (1) availability of foundation and multitasking ML models trained on multiple plant species to permit generalization and pangenome-aware prediction across different plant families; (2) integration of various closed-loop active-learning approaches to improve the impact of prediction on experimental outcomes, and enabling adaptive optimization strategies can minimize laboratory costs; (3) establishing an accuracy-measuring framework of ML outcomes to enhance the user adoption and trust through quantifying the prediction reliability in realistic biological contexts. These frontiers, in combination with inventions of novel high-fidelity and programmable nucleases, as well as improving the delivery techniques and availability of comprehensive omics datasets, will help in developing intelligent genome editing platforms.

Ultimately, the convergence of AI and ML in plant biotechnology holds a promising signal for a new era of sustainability and precision-driven agriculture. Toward advancing food security, the integration of mature ML frameworks with holistic genomic understanding will empower the design of resilient and climate-adaptive crops. This vision requires persistent collaboration between molecular biologists, bioinformaticians, computational scientists, and breeders to guarantee that AI-driven gene editing will remain precise, ethical, and accessible worldwide.

Acknowledgments

The authors would like to express their gratitude to the College of Computing at Mohamed VI Polytechnic University for providing continued support.

TABLE 9.1 Machine learning–based tools for plant genomic and gene editing applications. This table summarizes validated plant-specific CRISPR-based machine learning (ML) tools, their ML model, primary aim, availability status (web service or codebase), supported crop species, and corresponding reference.

Tool	ML model	Aim/uses	Repository/platform URL	Status	Crops	Reference
Plant-specific tools (validated)						
QTG-Finder	Random forest	QTL causal gene prioritization	https://github.com/carnegie/QTG_Finder	Codebase	Arabidopsis, rice	Lin et al. (2019)
QTG-Finder2	Random forest	QTL causal gene prioritization (generalized)	https://github.com/carnegie/QTG_Finder	Codebase	Arabidopsis, rice, sorghum, setaria viridis	Lin et al. (2020)
sgRNACNN	Ensemble CNN	On-target efficacy (soybean/maize/sorghum/wheat)	https://github.com/nmt315320/sgRNACNN	Codebase	Soybean, maize, sorghum, wheat	Niu et al. (2021)
GRAiN (rice)	GRN + SVM	Rank drought TFs	http://rrn.uark.edu/shiny/apps/GRAiN/	Deployed	Rice	Gupta et al. (2021)
iCREPCP	Deep CNN	Plant core-promoter cis-element prediction	https://github.com/kaixuanDeng95/iCREPCP	Deployed	Arabidopsis, maize, sorghum	Deng et al. (2023)
CRISPR-P 2.0	Rule-based with custom scoring	Plant sgRNA design	http://rice.hzau.edu.cn/cgi-bin/rice_rs3/CRISPR_rice	Deployed	>20 plant species including rice, Arabidopsis, maize, wheat	Liu et al. (2017)
Plant-usable tools (not validated)						
CRISPRscan	Linear/regularized regression on in vivo data	On-target sgRNA efficiency (SpCas9)	https://www.crisprscan.org	Deployed	NA	Moreno-Mateos, M. A. et al. (2015)
SSC (sequence scan for CRISPR)	Linear regression, Elastic-Net	On-target efficiency (CRISPRko/CRISPRi/a)	http://crispr.dfci.harvard.edu/SSC/	Deployed	NA	Xu, H. et al. (2015)
inDelphi	Mixture model + ML	DSB repair outcome prediction	https://github.com/maxwshen/inDelphi-model	Deployed	NA	Shen et al. (2018)

(*Continued*)

TABLE 9.1 Machine learning–based tools for plant genomic and gene editing applications. This table summarizes validated plant-specific CRISPR-based machine learning (ML) tools, their ML model, primary aim, availability status (web service or codebase), supported crop species, and corresponding reference.—cont'd

Tool	ML model	Aim/uses	Repository/platform URL	Status	Crops	Reference
Plant-specific tools (validated)						
FORECasT	Logistic regression	DSB repair outcome prediction	https://github.com/felicityallen/SelfTarget	Deployed	NA	Allen et al. (2019)
BE-Hive	Deep learning ensemble	Base editing outcome distribution	https://www.crisprbehive.design/	Deployed	NA	Arbab et al. (2020)
BE-DICT	Attention-based deep learning model	Base editing outcome prediction	https://github.com/uzh-dqbm-cmi/crispr	Deployed	NA	Marquart et al. (2021)
OPED	Transfer-learned nucleotide LM	Prime editing pegRNA ranking	https://github.com/wenjiegroup/OPED	Deployed	NA	Liu et al. (2023)
DeepPrime	Deep learning	Prime editing efficiency prediction	https://deepcrispr.info/DeepPrime/	Deployed	NA	Yu et al. (2023)
DeepCRISPR	Deep learning (hybrid CNN + autoencoder)	On-target + genome-wide off-target	https://github.com/bm2-lab/DeepCRISPR	Deployed (may not work)	NA	Chuai et al. (2018)
PRIDICT/PRIDICT2	Deep learning/ensembles	Prime editing efficiency and by-products	https://github.com/maurerbot/PRIDICT; https://github.com/maurerbot/PRIDICT2	Deployed	NA	Mathis et al. (2025)
BEDICT 2.0	Attention-based DL (updated)	Base editing outcome prediction	https://github.com/uzh-dqbm-cmi/BEDICT-V2	Deployed	NA	Kissling et al. (2025)
X–CRISP	Interpretable neural network	DSB repair outcome prediction	https://github.com/joanagoncalveslab/xcrisp	Codebase	NA	Seale and Gonçalves (2025)
CRISPOR	CFD + ensemble heuristics	sgRNA design + off-target ranking	https://crispor.gi.ucsc.edu/	Deployed	NA	Concordet and Haeussler (2018)

Declaration

"The authors declare that no generative artificial intelligence tools were used in the writing or preparation of this chapter. The content reflects the authors' own scholarly work, and the authors assume full responsibility for all statements, interpretations, and conclusions presented".

AI Disclosure

During the preparation of this work, the author(s) used AI-assisted language tool in order to Declaration of Generative AI Use: During the preparation of this manuscript, the authors used an AI-assisted language tool exclusively for language editing and proofreading. Following this assistance, the authors carefully reviewed and revised the text as necessary and assume full responsibility for the content, interpretations, and conclusions presented in the published article. After using this tool/service, the author(s) reviewed and edited the content as needed and take(s) full responsibility for the content of the publication.

References

Allen, F., Crepaldi, L., Alsinet, C., Strong, A.J., Kleshchevnikov, V., De Angeli, P., Páleníková, P., Khodak, A., Kiselev, V., Kosicki, M., Bassett, A.R., Harding, H., Galanty, Y., Muñoz-Martínez, F., Metzakopian, E., Jackson, S.P., Parts, L., 2019. Predicting the mutations generated by repair of Cas9-induced double-strand breaks. Nat. Biotechnol. 37 (1), 64–72. https://doi.org/10.1038/nbt.4317.

Altpeter, F., Springer, N.M., Bartley, L.E., Blechl, A.E., Brutnell, T.P., Citovsky, V., Conrad, L.J., Gelvin, S.B., Jackson, D.P., Kausch, A.P., Lemaux, P.G., Medford, J.I., Orozco-Cárdenas, M.L., Tricoli, D.M., Van Eck, J., Voytas, D.F., Walbot, V., Wang, K., Zhang, Z.J., Neal Stewart, C., 2016. Advancing crop transformation in the era of genome editing. Plant Cell 28 (7), 1510–1520. https://doi.org/10.1105/tpc.16.00196.

Anzalone, A.V., Randolph, P.B., Davis, J.R., Sousa, A.A., Koblan, L.W., Levy, J.M., et al., 2019. Search-and-replace genome editing without double-strand breaks or donor DNA. Nature 576, 149–157. https://doi.org/10.1038/s41586-019-1711-4.

Arbab, M., Shen, M.W., Mok, B., Wilson, C., Matuszek, Ż., Cassa, C.A., Liu, D.R., 2020. Determinants of base editing outcomes from target library analysis and machine learning. Cell 182 (2), 463. https://doi.org/10.1016/j.cell.2020.05.037.

Avsec, Ž., Agarwal, V., Visentin, D., Ledsam, J.R., Grabska-Barwinska, A., Taylor, K.R., Kelley, D.R., 2021. Effective gene expression prediction from sequence by integrating long-range interactions. Nat. Methods 18 (10), 1196–1203.

Bae, S., Kweon, J., Kim, H.S., Kim, J.S., 2014. Microhomology-based choice of Cas9 nuclease target sites. Nat. Methods 11 (7), 705–706. https://doi.org/10.1038/nmeth.3015.

Bahdanau, D., Cho, K., Bengio, Y., 2015. Neural machine translation by jointly learning to align and translate. ICLR 2015.

Bengio, Y., Courville, A., Vincent, P., 2013. Representation learning: a review and new perspectives. IEEE Trans. Pattern Anal. Mach. Intell. 35 (8), 1798–1828.

Biamonte, J., Wittek, P., Pancotti, N., Rebentrost, P., Wiebe, N., Lloyd, S., 2017. Quantum machine learning. Nature 549 (7671), 195–202. https://doi.org/10.1038/nature23474.

Brown, T.B., Mann, B., Ryder, N., Subbiah, M., Kaplan, J., Dhariwal, P., Neelakantan, A., Shyam, P., Sastry, G., Askell, A., Agarwal, S., Herbert-Voss, A., Krueger, G., Henighan, T., Child, R., Ramesh, A., Ziegler, D.M., Wu, J., Winter, C., Hesse, C., Chen, M., Sigler, E., Litwin, M., Gray, S., Chess, B., Clark, J., Berner, C., McCandlish, S., Radford, A., Sutskever, I., Amodei, D., 2020. Language models are few-shot learners. Adv. Neural Inf. Process. Syst. 2020.

Chen, T., Guestrin, C., 2016. XGBoost: a scalable tree boosting system. In: Proceedings of the ACM SIGKDD International Conference on Knowledge Discovery and Data Mining, vols. 13–17. Association for Computing Machinery, United States, pp. 785–794. https://doi.org/10.1145/2939672.2939785.

Chuai, G., Ma, H., Yan, J., Chen, M., Hong, N., Xue, D., Liu, Q., 2018. DeepCRISPR: optimized CRISPR guide RNA design by deep learning. Genome biology 19 (1), 80.

Clement, K., Rees, H., Canver, M.C., Gehrke, J.M., Farouni, R., Hsu, J.Y., Cole, M.A., Liu, D.R., Keith Joung, J., Bauer, D.E., Pinello, L., 2019. CRISPResso2 provides accurate and rapid genome editing sequence analysis. Nat. Biotechnol. 37 (3), 224–226. https://doi.org/10.1038/s41587-019-0032-3.

Concordet, J.P., Haeussler, M., 2018. CRISPOR: intuitive guide selection for CRISPR/Cas9 genome editing experiments and screens. Nucleic Acids Res. 46 (1), W242. https://doi.org/10.1093/nar/gky354.

Deng, K., Zhang, Q., Hong, Y., Yan, J., Hu, X., 2023. iCREPCP: a deep learning-based web server for identifying base-resolution cis-regulatory elements within plant core promoters. Plant Commun. 4 (1), 100455. https://doi.org/10.1016/j.xplc.2022.100455.

Devlin, J., Chang, M.W., Lee, K., Toutanova, K., 2019. BERT: pre-training of deep bidirectional transformers for language understanding. In: NAACL HLT 2019 - 2019 Conference of the North American Chapter of the Association for Computational Linguistics: Human Language Technologies - Proceedings of the Conference, vol. 1. Association for Computational Linguistics (ACL), undefined, pp. 4171–4186.

Doench, J.G., Fusi, N., Sullender, M., Hegde, M., Vaimberg, E.W., Donovan, K.F., Smith, I., Tothova, Z., Wilen, C., Orchard, R., Virgin, H.W., Listgarten, J., Root, D.E., 2016. Optimized sgRNA design to maximize activity and minimize off-target effects of CRISPR-Cas9. Nat. Biotechnol. 34 (2), 184–191. https://doi.org/10.1038/nbt.3437.

Gaudelli, N.M., 2017. Programmable base editing of A●T to G●C in genomic DNA without DNA cleavage. Nature 551.

Guo, T., Li, X., 2023. Machine learning for predicting phenotype from genotype and environment. Curr. Opin. Biotechnol. 79, 102853.

Gupta, C., Ramegowda, V., Basu, S., Pereira, A., 2021. Using network-based machine learning to predict transcription factors involved in drought resistance. Front. Genet. 12. https://doi.org/10.3389/fgene.2021.652189.

Haeussler, M., Schönig, K., Eckert, H., Eschstruth, A., Mianné, J., Renaud, J.B., Schneider-Maunoury, S., Shkumatava, A., Teboul, L., Kent, J., Joly, J.S., Concordet, J.P., 2016. Evaluation of off-target and on-target scoring algorithms and integration into the guide RNA selection tool CRISPOR. Genome Biol. 17 (1). https://doi.org/10.1186/s13059-016-1012-2.

Hochreiter, S., Schmidhuber, J., 1997. Long short-term memory. Neural Comput. 9 (8), 1735–1780. https://doi.org/10.1162/neco.1997.9.8.1735.

Jin, S., Zong, Y., Gao, Q., Zhu, Z., Wang, Y., Qin, P., Gao, C., 2019. Cytosine, but not adenine, base editors induce genome-wide off-target mutations in rice. Science 364 (6437), 292–295.

Ke, G., Meng, Q., Finley, T., Wang, T., Chen, W., Ma, W., Liu, T.Y., 2017. Lightgbm: a highly efficient gradient boosting decision tree. Advances in Neural Information Processing Systems 30.

Kim, H.K., Kim, Y., Lee, S., Min, S., Bae, J.Y., Choi, J.W., Kim, H.H., 2019. SpCas9 activity prediction by DeepSpCas9, a deep learning–based model with high generalization performance. Sci. Adv. 5 (11), eaax9249.

Kim, H.K., Yu, G., Park, J., Min, S., Lee, S., Yoon, S., Kim, H.H., 2021. Predicting the efficiency of prime editing guide RNAs in human cells. Nat. Biotechnol. 39 (2), 198–206. https://doi.org/10.1038/s41587-020-0677-y.

Kipf T.N. Welling M. 2017 Semi-supervised classification with graph convolutional networks 5th International Conference on Learning Representations, ICLR 2017 - Conference Track Proceedings International Conference on Learning Representations, ICLR Netherlands https://dblp.org/db/conf/iclr/iclr2017.html.

Kissling, L., Mollaysa, A., Janjuha, S., Mathis, N., Marquart, K.F., Weber, Y., Moon, W.J., Lin, P.J.C., Fan, S.H.Y., Muramatsu, H., Vadovics, M., Allam, A., Pardi, N., Tam, Y.K., Krauthammer, M., Schwank, G., 2025. Predicting adenine base editing efficiencies in different cellular contexts by deep learning. Genome Biol. 26 (1). https://doi.org/10.1186/s13059-025-03586-7.

Komor, A.C., Kim, Y.B., Packer, M.S., Zuris, J.A., Liu, D.R., 2016. Programmable editing of a target base in genomic DNA without doublestranded DNA cleavage. Nature 533, 420–424. https://doi.org/10.1038/nature17946.

Labun, K., Montague, T.G., Krause, M., Torres Cleuren, Y.N., Tjeldnes, H., Valen, E., 2019. CHOPCHOP v3: expanding the CRISPR web toolbox beyond genome editing. Nucleic Acids Res. 47 (1), W171. https://doi.org/10.1093/nar/gkz365.

Lazzarotto, C.R., Malinin, N.L., Li, Y., Zhang, R., Yang, Y., Lee, G., Tsai, S.Q., 2020. CHANGE-seq reveals genetic and epigenetic effects on CRISPR–Cas9 genome-wide activity. Nat. Biotechnol. 38 (11), 1317–1327.

LeBlanc, N., Charles, T.C., 2022. Bacterial genome reductions: tools, applications, and challenges. Frontiers in Genome Editing 4, 957289.

Lecun, Y., Bottou, L., Bengio, Y., Haffner, P., 1998. Gradient-based learning applied to document recognition. Proc. IEEE 86 (11), 2278–2324. https://doi.org/10.1109/5.726791.

Leenay, R.T., Aghazadeh, A., Hiatt, J., Tse, D., Roth, T.L., Apathy, R., Shifrut, E., Hultquist, J.F., Krogan, N., Wu, Z., Cirolia, G., Canaj, H., Leonetti, M.D., Marson, A., May, A.P., Zou, J., 2019. Large dataset enables prediction of repair after CRISPR–Cas9 editing in primary T cells. Nat. Biotechnol. 37 (9), 1034–1037. https://doi.org/10.1038/s41587-019-0203-2.

Li, V.R., Zhang, Z., Troyanskaya, O.G., 2021. CROTON: an automated and variant-aware deep learning framework for predicting CRISPR/Cas9 editing outcomes. Bioinformatics 37 (Suppl. ment_1), i342–i348.

Li, J., Wu, P., Cao, Z., Huang, G., Lu, Z., Yan, J., Luo, M., 2024. Machine learning-based prediction models to guide the selection of Cas9 variants for efficient gene editing. Cell Rep. 43 (2).

Lin, F., Fan, J., Rhee, S.Y., 2019. QTG-Finder: a machine-learning based algorithm to prioritize causal genes of quantitative trait loci in Arabidopsis and rice. G3: Genes Genomes Genet. 9 (10), 3129–3138.

Lin, Q., Zong, Y., Xue, C., Wang, S., Jin, S., Zhu, Z., Wang, Y., Anzalone, A.V., Raguram, A., Doman, J.L., Liu, D.R., Gao, C., 2020. Prime genome editing in rice and wheat. Nat. Biotechnol. 38 (5), 582–585. https://doi.org/10.1038/s41587-020-0455-x.

Listgarten, J., Weinstein, M., Kleinstiver, B.P., Sousa, A.A., Joung, J.K., Crawford, J., Gao, K., Hoang, L., Elibol, M., Doench, J.G., Fusi, N., 2018. Prediction of off-target activities for the end-to-end design of CRISPR guide RNAs. Nat. Biomed. Eng. 2 (1), 38–47. https://doi.org/10.1038/s41551-017-0178-6.

Liu, H., Ding, Y., Zhou, Y., Jin, W., Xie, K., Chen, L.L., 2017. CRISPR-P 2.0: an improved CRISPR-cas9 tool for genome editing in plants. Mol. Plant 10 (3), 530–532. https://doi.org/10.1016/j.molp.2017.01.003.

Liu, F., Huang, S., Hu, J., Chen, X., Song, Z., Dong, J., Liu, Y., Huang, X., Wang, S., Wang, X., Shu, W., 2023. Design of prime-editing guide RNAs with deep transfer learning. Nat. Mach. Intell. 5 (11), 1261–1274. https://doi.org/10.1038/s42256-023-00739-w.

Marquart, K.F., Allam, A., Janjuha, S., Sintsova, A., Villiger, L., Frey, N., Krauthammer, M., Schwank, G., 2021. Predicting base editing outcomes with an attention-based deep learning algorithm trained on high-throughput target library screens. Nat. Commun. 12 (1). https://doi.org/10.1038/s41467-021-25375-z.

Mathis, N., Marquart, K.F., Allam, A., Krauthammer, M., Schwank, G., 2025. Systematic pegRNA design with PRIDICT2.0 and ePRIDICT for efficient prime editing. Nat. Protoc. https://doi.org/10.1038/s41596-025-01244-7.

Michael, T.P., VanBuren, R., 2020. Building near-complete plant genomes. Curr. Opin. Plant Biol. 54, 26–33.

Moreno-Mateos, M.A., Vejnar, C.E., Beaudoin, J.D., Fernandez, J.P., Mis, E.K., Khokha, M.K., Giraldez, A.J., 2015. CRISPRscan: designing highly efficient sgRNAs for CRISPR-Cas9 targeting in vivo. Nat. Methods 12 (10), 982–988.

Niu, M., Lin, Y., Zou, Q., 2021. sgRNACNN: identifying sgRNA on-target activity in four crops using ensembles of convolutional neural networks. Plant Mol. Biol. 105 (4), 483–495.

Pallaseni, A., Peets, E.M., Koeppel, J., Weller, J., Vanderstichele, T., Ho, U.L., Crepaldi, L., Van Leeuwen, J., Allen, F., Parts, L., 2022. Predicting base editing outcomes using position-specific sequence determinants. Nucleic Acids Res. 50 (6), 3551–3564. https://doi.org/10.1093/nar/gkac161.

Pan, S.J., Yang, Q., 2009. A survey on transfer learning. IEEE Transactions on Knowledge and Data Engineering 22 (10), 1345–1359.

Saini, H., Johar, V., Mondal, R., Vyas, S., Bisht, V., 2025. Integration of artificial intelligence with genome editing: a new era for molecular plant breeding. Physiol. Mol. Plant Pathol., 102952

Samek, W., Müller, K.R., 2019. Towards explainable artificial intelligence. Lect. Notes Comput. Sci. 11700, 5–22. https://doi.org/10.1007/978-3-030-28954-6_1.

Scarselli, F., Gori, M., Tsoi, A.C., Hagenbuchner, M., Monfardini, G., 2008. The graph neural network model. IEEE Trans. Neural Network. 20 (1), 61–80.

Schreiber, T., Prange, A., Schäfer, P., Iwen, T., Grützner, R., Marillonnet, S., et al., 2024. Efficient scar-free knock-ins of several kilobases in plants by engineered CRISPR-Cas endonucleases. Mol. Plant 17, 824–837. https://doi.org/10.1016/j.molp.2024.03.013.

Seale, C., Gonçalves, J.P., 2025. X-CRISP: domain-adaptable and interpretable CRISPR Repair outcome prediction. Bioinform. Adv., vbaf157

Shen, M.W., Arbab, M., Hsu, J.Y., Worstell, D., Culbertson, S.J., Krabbe, O., Cassa, C.A., Liu, D.R., Gifford, D.K., Sherwood, R.I., 2018. Predictable and precise template-free CRISPR editing of pathogenic variants. Nature 563 (7733), 646–651. https://doi.org/10.1038/s41586-018-0686-x.

Sherkatghanad, Z., Abdar, M., Charlier, J., Makarenkov, V., 2023. Using traditional machine learning and deep learning methods for on-and off-target prediction in CRISPR/Cas9: a review. Briefings Bioinf. 24 (3), bbad131.

Smola, A.J., Schölkopf, B., 2004. A tutorial on support vector regression. Stat. Comput. 14 (3), 199–222. https://doi.org/10.1023/B:STCO.0000035301.49549.88.

Tang, X., Liu, G., Zhou, J., Ren, Q., You, Q., Tian, L., Zhang, Y., 2018. A large-scale whole-genome sequencing analysis reveals highly specific genome editing by both Cas9 and Cpf1 (Cas12a) nucleases in rice. Genome Biol. 19 (1), 84.

Tian, F., Yang, D.C., Meng, Y.Q., Jin, J., Gao, G., 2020. PlantRegMap: charting functional regulatory maps in plants. Nucleic Acids Res. 48 (D1), D1104–D1113.

Tsai, S.Q., Zheng, Z., Nguyen, N.T., Liebers, M., Topkar, V.V., Thapar, V., Wyvekens, N., Khayter, C., Iafrate, A.J., Le, L.P., Aryee, M.J., Joung, J.K., 2015. GUIDE-seq enables genome-wide profiling of off-target cleavage by CRISPR-Cas nucleases. Nat. Biotechnol. 33 (2), 187–198. https://doi.org/10.1038/nbt.3117.

Tsai, S.Q., Nguyen, N.T., Malagon-Lopez, J., Topkar, V.V., Aryee, M.J., Joung, J.K., 2017. CIRCLE-seq: a highly sensitive in vitro screen for genome-wide CRISPR-Cas9 nuclease off-targets. Nat. Methods 14 (6), 607–614. https://doi.org/10.1038/nmeth.4278.

Vaswani, A., Shazeer, N., Parmar, N., Uszkoreit, J., Jones, L., Gomez, A., Kaiser Ł, N., Polosukhin, I., 2017. Attention is all you need. Adv. Neural Inf. Process. Syst. 2017, 5999–6009.

Wienert, B., Wyman, S.K., Richardson, C.D., Yeh, C.D., Akcakaya, P., Porritt, M.J., Morlock, M., Vu, J.T., Kazane, K. R., Watry, H.L., Judge, L.M., Conklin, B.R., Maresca, M., Corn, J.E., 2019. Unbiased detection of CRISPR off-targets in vivo using DISCOVER-Seq. Science 364 (6437), 286–289. https://doi.org/10.1126/science.aav9023.

Xie, K., Yang, Y., 2013. RNA-guided genome editing in plants using a CRISPR–Cas system. Mol. Plant 6 (6), 1975–1983.

Xu, H., Xiao, T., Chen, C.H., Li, W., Meyer, C.A., Wu, Q., Liu, X.S., 2015. Sequence determinants of improved CRISPR sgRNA design. Genome Res. 25 (8), 1147–1157.

Yosinski, J., Clune, J., Bengio, Y., Lipson, H., 2014. How transferable are features in deep neural networks? Adv. Neural Inf. Process. Syst. 4, 3320–3328.

Yu, G., Kim, H.K., Park, J., Kwak, H., Cheong, Y., Kim, D., Kim, H.H., 2023. Prediction of efficiencies for diverse prime editing systems in multiple cell types. Cell 186 (10), 2256–2272.

Zhang, T., Marand, A.P., Jiang, J., 2016. PlantDHS: a database for DNase I hypersensitive sites in plants. Nucleic Acids Res. 44 (D1), D1148–D1153.

PART II

Applications and synergies of CRISPR and AI for plant gene editing

CHAPTER

10

Practical design of gRNA: AI combined with CRISPR

Paulo A. Zaini

Department of Plant Sciences, University of California, Davis, CA, United States

10.1 A brief historical perspective

As is the case with most, if not all, applied molecular biology tools, CRISPR (clustered regularly interspaced short palindromic repeats) was initially discovered by studies on fundamental biological principles. As a natural adaptive immunity mechanism against invading viruses, bacteria and archaea use these "molecular scissors" to recognize and cut foreign nucleic acids. In the late 1980s, when CRISPR sequences were initially found (Ishino et al., 1987; Mojica et al., 1993), researchers still did not understand their function and widespread occurrence, as DNA sequencing was much more laborious and expensive. Subsequently, similar DNA repeat patterns were described in other bacteria and archaea, and it wasn't until the early 2000's when scientists finally started hinting at their functional role (Jansen et al., 2002; Mojica et al., 2005; Barrangou et al., 2007). During this period, researchers noted that spacers in the tandem repeats, which have homology to viral genes, share a common sequence at one end and are required for target recognition (Bolotin et al., 2005). This sequence was named the protospacer adjacent motif (PAM).

A major leap was the understanding that CRISPR systems relied on RNA molecules to direct specificity (Brouns et al., 2008). These RNAs, called guide RNAs (gRNAs), are transcribed from the CRISPR loci and recognize complementary viral or foreign DNA sequences through base pairing (Garneau et al., 2010). The single guide RNA (sgRNA), a fusion of two RNAs (crRNA and tracrRNA), directs the Cas9 nuclease to the precise target site in the genome, where Cas9 Ind.uces a targeted double-stranded break (Deltcheva et al., 2011; Jinek et al., 2012).

During the 2010's all the CRISPR components were understood at a high enough level to allow their use as DNA and RNA editing tools, similar to what had happened with DNA restriction enzymes decades earlier. With the fast reduction of DNA sequencing and synthesis costs occurring simultaneously, researchers could now design and "program" cells to

Intelligent Gene Editing in Plants
https://doi.org/10.1016/B978-0-443-44079-3.00013-9

express a nuclease such as Cas9 and sgRNAs to target specific sites in a given genome, as long as that genome could be transformed to express the transgenes encoding both CRISPR components. Many DNA vectors used in seminal publications became available in repositories such as Addgene, rapidly democratizing CRISPR tools to most molecular biology laboratories in the world. Indeed, its use exploded exponentially, as did modifications and new applications derived from the original works, rendering Emmanuelle Charpentier and Jennifer Doudna from the University of California, Berkeley, the Nobel prize in Chemistry in 2020 for their contributions to the development of CRISPR–Cas9 gene-editing technology. This abbreviated historical perspective is not exhaustive, and many details were left out, which can be found in dedicated reviews (Lander 2016; Ishino et al., 2018).

10.2 Early lessons on gRNA design

As mentioned above, before single guide RNAs were proposed, the CRISPR–Cas system was described as having crRNA (CRISPR RNA) and tracrRNA (trans-activating CRISPR RNA) components. While crRNAs are transcribed into long precursor RNAs and then further processed into small 39–44 nucleotide RNAs, tracrRNA forms a duplex with pre-crRNA, permitting RNase III-dependent processing, enabling the mature Cas9-crRNA-tracrRNA complex to target and cleave foreign DNA. Interestingly, only 20 nucleotides (depending on the Cas endonuclease) of the crRNA serve the guide function, pairing directly with the DNA target site (Fig. 10.1). Modifications in the crRNA sequence region within the sgRNA can alter the binding location, allowing for precise targeting of different DNA regions, effectively making it a programmable system for genome editing.

Although target specificity is achieved with the initial 20 nt region of the gRNA, shorter and longer sequences ranging from 17 to 24 nt have been successfully demonstrated (Hsu

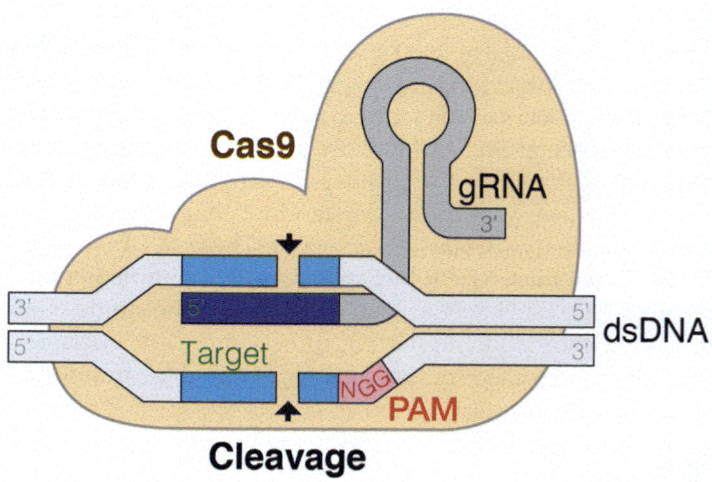

FIGURE 10.1 The simplified CRISPR–Cas system. A diagram of the Cas9 enzyme, gRNA, PAM and the double-stranded break induced in the target DNA are shown. Note the PAM sequence is not part of the gRNA. *https://en.wikipedia.org/ wiki/ CRISPR_gene_editing, by user Mariuswalter. License CC BY-SA 4.0.*

et al., 2013; Doench et al., 2014). These two studies used hundreds of different sgRNAs to build predictive models of specificity and activity, showing that longer sequences reduce the occurrence of off-target mutations, that is, the cutting of similar DNA sequences differing in one or two bases occurring at other sites in the genome. The authors also established the suggested minimum GC content of 50% (a range of 20%–80% is acceptable), as these bases form more stable interactions with the target DNA site, improving the cut efficiency.

The diagram above shows the double DNA strand cleavage once the target site is paired with the ribonucleoprotein (RNP) complex formed by Cas9 and the guide RNA. Natural DNA repair mechanisms can mend the cleavage, and in this process a mutation (deletion, insertion, or substitution) is commonly introduced at the target site. Depending on the mutation a frameshift will be created, compromising functionality of the encoded protein (a functional knock-out). The diagram also reveals an important aspect of this RNP, in which the cleavage occurs upstream of the PAM site, more specifically at or near three bases upstream of the PAM sequence. This was repeatedly observed in early works with Cas9, becoming one of the "working rules" of this specific nuclease. In the case of Cas9 we have observed deletions up to nine bases. Subsequent work with other nucleases later identified other PAM sequences and their working rules in terms cutting efficiency, PAM requirements and mutation window. It also revealed that gRNA requirements may vary, especially in length of the pairing region and scaffold sequence interacting with the nuclease.

As more Cas-like endonucleases were described and differences in the PAM sequence were noted, the CRISPR toolbox expanded the diversity of genomic sites that could be effectively targeted, alleviating some of the constraints imposed by the NGG sequence recognized by Cas9. Table 10.1 below lists most of the currently used Cas-like nucleases, and their validated PAM sequence requirements. These nucleases are available from vendors such as Addgene (for vectors encoding the nuclease gene) or New England Biolabs (for recombinant proteins). New and engineered variants are continually being discovered and characterized (see Adler et al., 2024; Cai et al., 2024; Wang et al., 2024 for recent comprehensive reviews).

NGG means any nucleotide followed by 2 Gs. NNGRRT means any two nucleotides, then G, R (A or G), R (A or G), and T. NNAGAAW where W is A or T. TTTV where V is A, C, or G. "None" indicates no strict PAM required per the cited source.

A closer look at the gRNA structure reveals in detail the target sequence region (also known as "spacer" in the original CRISPR description) pairing with the denatured DNA strand at its 5′ terminus and the scaffold region at the 3′ terminus (Fig. 10.2). This arrangement will vary from nuclease to nuclease, such as in Cas12 for example, but the general principle is maintained in terms of a target sequence region and a scaffold region. Note the gRNA does not contain the PAM sequence, recognized by the nuclease (Cas9 in the example shown). This is important when designing a DNA vector or ordering synthetic gRNAs from a vendor such as Sigma or GenScript.

An important limitation to consider is that despite the PAM and gRNA characteristics being well characterized for many nucleases, in practice, however, simple target recognition does not guarantee efficient cleavage. To increase the chances of success, researchers commonly design two gRNAs to target different sites of the gene to be knocked out.

TABLE 10.1 Commonly used Cas-like nucleases and their protospacer adjacent sequence and guide RNA characteristics.

Nuclease	PAM sequence	gRNA length (target) (nt)	gRNA scaffold length (nt)	Reference
SpCas9	NGG	20	~80	Jinek et al. (2012)
SaCas9	NNGRRT	21	42–45	Ran et al. (2015)
StCas9	NNAGAAW	22	40–50	Müller et al. (2016)
NmCas9	NNNNGATT	22	~50	Hou et al. (2013)
CjCas9	NNNNACA	22	40–50	Kim et al. (2017)
TdCas9	NAAAAC	20	45–60	Addgene CRISPR guide
SpCas9-VQR	NGA	20	~80	Kleinstiver et al. (2015)
SpCas9-VRER	NGCG	20	~80	Kleinstiver et al. (2015)
SpCas9-EQR	NGAG	20	~80	Kleinstiver et al. (2015)
SpCas9-NG	NG	20	~80	Nishimasu et al. (2018)
xCas9	NG, GAA, GAT	20	~80	Hu et al. (2018)
AsCas12a (Cpf1)	TTTV	20–24	19–20	Zetsche et al. (2020)
LbCas12a (Cpf1)	TTTV	20–24	21	Zetsche et al. (2020)
FnCas12a (Cpf1)	TTV	20–24	19	Zetsche et al. (2020)
Cascade + Cas3	Variable	27	Same 27	Westra et al. (2012)

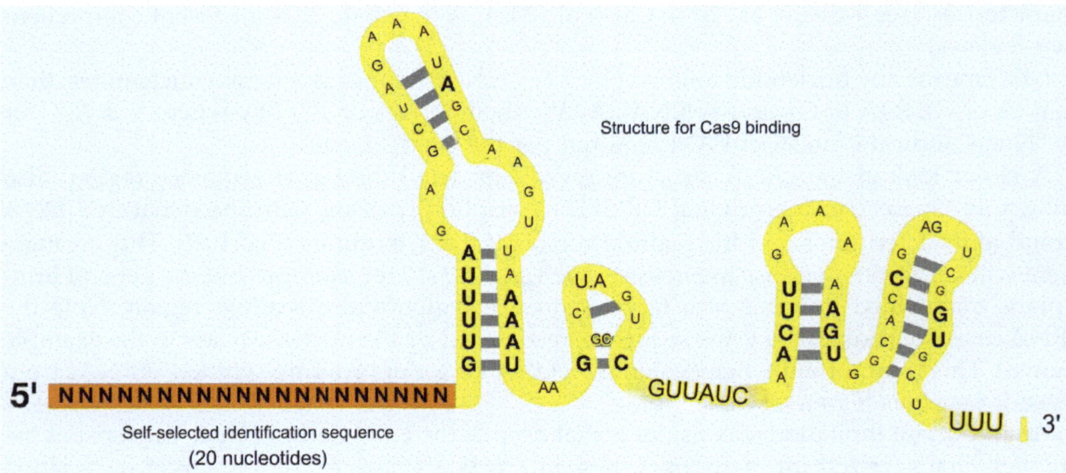

FIGURE 10.2 **The structure of sgRNA.** The variable (spacer or target) region is colored in orange at the 5′ terminal and the constant scaffold region used by Cas9 to bind the gRNA at the 3′ end shown in yellow. Note the PAM sequence is not present in the guide RNA sequence. *Modified from https://www.biotechacademy.dk*

Then by sequencing the target region encompassing both gRNA target sites, the frequency and types of mutations can be verified.

Another strategy now widely adopted is the in vitro testing of different gRNAs before larger investments are done for the development of vectors and knock-out cell lines or organisms. Recent years have witnessed tremendous advancements in RNA synthesis and modification technologies. Synthesis of complete gRNA (~100 nt) are now offered at an accessible cost, and the modifications increase their stability and functionality such as fluorescence labeling. The user simply informs the 20 nt sequence of the target region and the vendor will produce it attached to the constant ~80 nt scaffold sequence composing the complete sgRNA. To complement, the user can order recombinant Cas9 from multiple vendors as if it were a conventional DNA restriction enzyme. Combined, these two components can be used to assemble RNPs in vitro. To test functionality, the genomic region containing the target site(s) can be amplified by PCR and after clean-up be incubated with the assembled RNPs for DNA digestion. The product is then analyzed by electrophoresis, revealing which gRNA performs better. Large-scale screening of chemically synthesized gRNA libraries can be very informative but comes at a cost, which incentivizes the development of prediction models as will be discussed later.

10.3 General principles for gRNA selection

Once a target gene or genomic locus has been chosen, care should be taken if other similar sequences (that would be recognized by the same gRNA) are present in the same genome, as this would likely result in off-target mutations. The availability of high-quality genome sequences greatly helps to minimize this risk, especially those with phased haplotypes, covering all chromosome sets in a given genome. Similar sequences can easily be aligned, and decisions regarding the CRISPR target site within a gene or genomic region. This can be considered an initial screening of whether a specific site is even available. Target uniqueness might not be essential in all cases. For example, a guide RNA can be designed to target multiple similar genes (such as paralogs). Still, regarding the target site selection, consider the functional relevance of the site for the overall regulatory or encoded region. For knockout experiments, choose gRNAs targeting exons crucial for protein function, preferably early exons to truncate the mature encode protein early on its sequence.

To maximize cutting efficiency, it is crucial to design gRNAs that anneal properly to their target site. Using Cas9 as an example, the target or spacer region is 20 nt long, starting at the 5′ of the gRNA (Fig. 10.2). Ideally, this region would anneal perfectly to the target sequence to maximize the chance of cutting, but mismatches are allowed, especially on the first 10 nt. The later 10 nt of the spacer region, also known as the seed region, must pair with the target with no mismatches, otherwise cutting efficiency will drop significantly. Luckly researchers can rely on many online tools for gRNA selection, as explained in the next section.

10.4 Tools for gRNA selection

Once general principles of gRNA function were established, prediction tools were developed to allow selection based on relevant parameters such as GC content and predicted off-target sites. Effective off-target prediction requires knowledge of the complete genome sequence of the target organism, which is becoming increasingly accessible even for non-model species. Completeness of a genome sequence has layers such as prediction of all protein coding genes (still in thousands of fragments not assembled into chromosomes), complete sequence encompassing "telomere to telomere" chromosomal assemblies and finally phased genomes including all haplotypes in the given genome, with chromosomes fully assembled. Of course, the increasing detail comes at a cost and provides greater certainty for the prediction of off-target effects.

Already at the intermediate level of genome sequence completeness (assembled chromosomes), very good predictions regarding off-target effects can be made, as the different alleles are likely to have similar target sites. However, it is still recommended to investigate target regions in detail by resequencing, as even different cell lines (or cultivars of the same plant species for example) can have slightly different sequences previously unknown. The ideal situation involves completely sequencing the genome of the target organism and not relying on "reference" sequences.

Many of the tools developed for gRNA prediction offer a friendly user interface and are freely available online, democratizing access without requiring coding expertise. Many allow choosing the reference genome, the nuclease (Cas9, Cas12, Cas13 etc.) and will ask for a DNA region, usually of about 1000 bp within the desired genomic region containing the desired target site. The output will score and rank gRNA sequences (always 5′ to 3′) showing which DNA strand it targets (sense or antisense), GC% content, and predicted off-target sites. The user can then focus on a few for a detailed analysis and choice. Table 10.2 shows some notable examples of freely available online tools for gRNA prediction.

Several groups have used CRISPR screens (libraries of chemically synthesized gRNAs) to establish working rules for training prediction algorithms (Wang et al., 2019; Kim et al., 2020; Xiang et al., 2021; Riesenberg et al., 2025). Most of the modeling effort so far has centered on Cas9, but similar approaches should become available for all widely used similar programmable nucleases. Here we can begin to appreciate the value of computational tools to model different aspects of the CRISPR system, such as PAM sequence compatibility and gRNA-mediated targeting and cutting efficiency. In the next section we'll explore these in greater detail.

10.5 Use of AI to improve CRISPR systems

Artificial intelligence (AI) has undergone immense development in the last decade, with applications in many aspects of data science. For genome editing specifically, it is being used in several ways, such as predicting and enhancing precision DNA repair (Naert et al., 2025), designing safer (less off-target effects) more efficient nucleases (Silverstein et al., 2025),

TABLE 10.2 Popular validated tools for designing guide RNAs for CRISPR experiments.

Tool name	URL	Brief description
CRISPick	https://www.portals.broadinstitute.org/gppx/crispick/public	Tool by broad institute for efficient gRNA design; shows on/off-target scores and localization info
CHOPCHOP	https://www.chopchop.cbu.uib.no	Handles multiple CRISPR enzymes, visualizes off-target sites, batch processing options.
CRISPOR	https://www.crispor.tefor.net/	Advanced off-target scoring with multiple metrics, broad genome support, and cloning tools
GenScript sgRNA design tool	www.genscript.com/tools/gRNA-design-tool	Intuitive platform with updated on/off-target scoring and HDR design; supports many organisms
CRISPRedict	www.crispredict.org/	Interpretable predictions using a linear model, suitable for large datasets
DeepCRISPR	www.deepcrispr.net/	Deep learning-based tool for genome-wide gRNA efficiency and specificity prediction
Synthego CRISPR design tool	www.synthego.com/products/bioinformatics/crispr-design-tool	Fast, easy design for thousands of genomes and species; strong visualization tools
IDT CRISPR design checker	www.idtdna.com/site/order/designtool/index/CRISPR_SEQUENCE	Analyzes and scores guide sequences, helps with direct ordering and oligo selection
CRISPRon	https://www.rth.dk/resources/crispr/crispron/	High performance assessment of guide RNAs for CRISPR-spCas9 applications
CRISPR-GE	www.skl.scau.edu.cn/home/	Integrated platform focused on genome editing in plants; predicts gRNA efficiency and offers design tools for a range of plant genomes
CRISPR-P	http://www.cbi.hzau.edu.cn/CRISPR2/	Online tool specialized for plant gRNA design, supports multiple plant species and editing platforms
Cas-designer (RGEN tools)	www.rgenome.net/cas-designer/	Web tool allowing gRNA design for multiple cas nucleases and PAM types; supports customized target input and PAM choices

expanding the CRISPR toolkit (Yoon et al., 2024), and accelerating clinical and therapeutic translation (Agboola et al., 2025).

Perhaps the most relevant application of integrating AI and CRISPR for the topic of this chapter is the optimization of gRNA design and targeting. Machine learning models like DeepCRISPR (Chuai et al., 2018), CRISTA (Abadi et al., 2017), DeepHF (Wang et al., 2019), and CRISPR-M (Dixit et al., 2024) analyze genomic sequences to design optimal guide RNAs (gRNAs) that enhance targeting accuracy. These AI-driven tools reduce unintended edits, thereby increasing the safety and efficacy of CRISPR-based interventions. Collectively, these efforts have used tens of thousands of gRNAs to train the predictive models, reaching an optimization level of gRNA prediction that can be applied even to specific cell types for clinical applications. A current challenge is to validate these tools on non-model organisms, expanding their application to more areas such as industrial biotechnology and agriculture. Below are the main parameters in which AI can help to improve gRNA design for CRISPR experiments:

Efficiency: AI models, including deep learning and machine learning frameworks, can analyze vast numbers of gRNA sequences and their experimental outcomes to learn features associated with high on-target activity. This high-throughput capability can predict with high degree of certainty which gRNA sequences mediate successful cleavage and editing at the intended DNA target site.

Efficacy: Connecting a specific editing with a desired phenotype outcome is not always straightforward but is becoming increasingly predictable with the aid of AI, which can scan and summarize thousands of published CRISPR experiments and their outcomes. A gRNA can have high cleavage efficiency but variable efficacy if subsequent cellular processes do not reliably produce the desired phenotype.

Reduce off-targets: Just as AI can be used to predict on-target sites, it can also be used to predict off-target sites across the whole genome. This can be accomplished by ab initio methods trained in DNA sequences and also on published literature reporting off-targets for commonly targeted genes.

GC content: The percentage of guanine and cytosine bases in gRNA sequences is a crucial factor in CRISPR experiments. Values too high or too low outside the optimal ~40–80% range can destabilize the interaction with the nuclease and/or target DNA. Here again the capacity of AI to analyze thousands of gRNA sequences and their editing outcomes in published literature can greatly increase the predictability of success for each sequence, removing much of the guessing from traditional experiments.

Accessibility to target sites: Although structural predictions for nucleic acids are not as advanced as for proteins, yet, by integrating sequence data with epigenomic and chromatin structure information, AI can help researchers to avoid densely packed chromatin regions that is less accessible to the editing machinery. Models such as piCRISPR and other convolutional neural network approaches analyze both guide sequence and the local chromatin state, providing a more holistic assessment of whether a gRNA can effectively bind and mediate editing at its target site (Dixit et al., 2024). By learning from high-throughput datasets and experimental measurements like DNAse-seq or ATAC-seq (which map open chromatin), AI predicts editing success based on real accessibility, not only sequence matches.

Optimization of gRNA and nuclease concentrations: Many CRISPR applications and therapies rely on delivering RNPs (assembled gRNA + nuclease) directly inside target cells or tissues. Machine learning models can integrate multi-modal data—such as cell type, delivery method, and previous outcomes—to recommend specific concentrations and ratios for CRISPR components, tailoring protocols for particular experimental contexts. Here again it can scan thousands of published experiments to search for optimal parameters, which tend to become increasingly better as the training set enlarges.

Sequence context of the target site: Beyond the core target site, flanking DNA sequences impact activity by influencing DNA accessibility, chromatin structure, and the local DNA repair environment. This broader sequence context can modulate Cas9 cleavage efficiency and the type of DNA repair following cleavage. Understanding sequence context helps to explain why gRNAs that look similar in target sequences can have very different efficiencies in vivo, aiding in the selection of more effective guides and reducing experimental failures. Considering sequence context supports improved HDR (homology-directed repair) efficiency by choosing targets with favorable local sequence and epigenetic states for precise genome editing (Pallaseni et al., 2024).

Locating and appropriate PAM: As showed in Figs. 10.1 and 10.2, the protospacer adjacent motif (PAM) is a 2–6-base pair DNA sequence immediately following the DNA sequence targeted by the nuclease. AI supports precise and efficient identification of PAM sequences by combining bioinformatics, protein modeling, and data-driven prediction, thereby advancing both fundamental CRISPR research and applied genome editing technologies. Protein language models such as Protein2PAM predict PAM specificity directly from Cas protein sequences without relying on large experimental datasets. This method enables rapid, accurate PAM predictions and can even suggest reprogramming mutations to alter PAM specificity (Nayfach et al., 2025).

Predict repair outcomes: AI leverages large-scale experimental data and sophisticated neural network architectures to forecast the outcomes of CRISPR-induced DNA repair, enhancing the design and precision of genome editing experiments for research and therapeutic purposes. Tools such as "Pythia" simulate millions of genome editing outcomes by learning DNA repair patterns, enabling tailored design of gene edits with improved precision and reduced unwanted mutations (Naert et al., 2025).

Going beyond the design of gRNAs, AI can also be leveraged to automate the complete workflow of CRISPR experiment design and analysis. CRISPR-GPT (Qu et al., 2025) for example is a large language model (LLM) agent system that automates and enhances CRISPR-based gene-editing design and data analysis. It leverages the reasoning capabilities of LLMs for complex task decomposition, decision-making and interactive human–artificial intelligence (AI) collaboration. This system incorporates domain expertise, retrieval techniques, external tools and a specialized LLM that is continuously fine-tuned with open-forum discussions among scientists. It truly works as a complete solution platform, assisting users in selecting CRISPR systems, experiment planning, designing guide RNAs, choosing delivery methods, drafting protocols, designing assays and analyzing data. The authors demonstrated its potential by knocking out four genes with CRISPR–Cas12a in a human lung adenocarcinoma cell line and epigenetically activated two genes using CRISPR-dCas9 in a human melanoma cell line.

Alphabet Inc. is also emerging as a significant player in the use of AI to improve CRISPR applications. It has multiple subsidiaries contributing to different fronts. Google DeepMind is using its advanced AI models in genomics, and AlphaGenome is a tool introduced in 2025 capable of analyzing DNA sequences to predict how mutations affect gene expression and splicing patterns (Avsec et al., 2025). This aids the process of creating synthetic DNA for novel applications as well as understanding disease and designing new therapies. DeepMind also developed the AI Co-Scientist which helps researchers design experiments including gene editing, accelerating CRISPR democratization and applications (Gottweis et al., 2025). By leveraging the computational power of AI to genomic sciences, these Alphabet Inc. subsidiaries are expected to become important tools in the new wave of CRISPR research and applications. The speed and framework enabled by AI will allow us to transcend from the current state of "manual craft" in which researchers perform every step from choosing a target, designing gRNAs, performing mutations and analyzing results, in a workflow that can take up to months, to organized and streamlined pipelines that will increase safety and accelerate the learning curve and delivery of benefits to society. It is imperative, however, that concerns with data privacy and misuse of technology are addressed by scientists, regulators and society as a whole.

10.6 Challenges and perspectives

The development of "AI co-pilots" in genome engineering is certainly groundbreaking and will change the whole landscape of biotechnology, both for human health and bio-product development. A persistent challenge, especially for plant and fungal cells, is the delivery of CRISPR components, more so for plants that are aimed at being released in the environment. It is desirable (and even prohibitive in some countries) that nucleases and gRNAs are used to target specific sites without integrating any transgenes into the host genome, to prevent unintended consequences of spreading these genes in nature. The challenge of delivering CRISPR components to specific cell types is also true for animals, as they can easily be transformed and studied as cell cultures, but for clinical applications it is desired the CRISPR system be delivered to whole organisms. To this end partnerships like Mammoth Biosciences with Regeneron Pharmaceuticals are using AI to combine compact CRISPR systems with targeted delivery technologies, advancing precise gene editing in diverse tissues.

An area poised to witness fast developments in the coming years is the synergy of AI with structural biology. Developments such as AlphaFold have demonstrated its potential, and recent versions allow for protein modeling while interacting with RNA, DNA, small molecules and metal ions. Although nucleic acid structural modeling is still not as developed as protein (given the absence of important training sets such as the protein data bank), it is a matter of time until it catches up. This advancement will enable development of pipelines for designing synthetic nucleases and nucleic acids ab initio, potentially reaching similar or better functionality than current ones while potentially bypassing intellectual property that tends to stiffen new product and application development. It has been demonstrated in a CRISPR application known as prime editing that introducing structural features such as RNA G-quadruplex structures into sgRNA design (in this case called prime editing gRNA, or pegRNA) can increase successful mutation rates (Li et al., 2022). Another structural study showed that changes in the SL1 and SL2 regions of the scaffold region of the pegRNA improves prime editing efficiency without affecting the stability of the Cas9-pegRNA complex (Huszár et al., 2023). As the field progresses, current challenges and limitations of using LLMs for genome editing should be resolved, such as relying on statistical association, often leading to predictions that do not necessarily reflect the underlying biological processes, as reviewed in (Menon et al., 2025).

Personalized and precision medicine is another area in which AI is poised to significantly impact. As large amounts of data from imaging and omics methods can be quickly analyzed, pipelines tailored for clinical services can finally become a reality. As more patients will also have their genome completely sequenced, a continuous feedback learning environment will be established, and custom therapies that utilize the CRISPR toolbox will maximize the chance of a successful therapy for each patient, as in the case of cancer treatment for example. All the groundwork being laid with current machine learning approaches will pave the way for these personalized therapies (Abbasi et al., 2025; Pandey et al., 2025). For successful clinical applications beyond diagnostics and into therapeutics, improvements to mitigate or eliminate off-target effects to avoid unintended mutations must be achieved (Aussel et al., 2025; Park et al., 2025).

As a final comment, I want to emphasize the enormous potential of the synergy between AI and CRISPR. We are entering a phase of unlimited capabilities of genome engineering, in which ethical and safety matters must be considered, not simply technical possibilities. The tremendous power enabled by genome engineering can greatly benefit humanity, however it can also be misused to create new bioweapons that can be rapidly optimized and customized using CRISPR technologies. Society must grapple with this duality and avoiding it will not solve any threats, on the contrary, would leave us defenseless to new threats. I believe exhaustive discussion with different sectors of society will be essential to deal with these new technologies, and training a new generation of technical experts that can effectively identify and address new threats is a safer path to follow. A better-informed society will be able to address this more appropriately, both from technical and regulatory point of view. Prohibitions and obstacles such as declaring moratoriums or defunding research might hinder our development but will not stop all interested players outside these regulatory constraints. As with any powerful technology, it is here to stay. We can either surf the waves or be swallowed by them.

Declaration

The authors declare that no generative artificial intelligence tools were used in the writing or preparation of this chapter. The content reflects the authors' own scholarly work, and the authors assume full responsibility for all statements, interpretations, and conclusions presented."

References

Abadi, S., Yan, W.X., Amar, D., Mayrose, I., 2017. A machine learning approach for predicting CRISPR-cas9 cleavage efficiencies and patterns underlying its mechanism of action. PLoS Comput. Biol. 13 (10), e1005807. https://doi.org/10.1371/journal.pcbi.1005807.

Abbasi, A.F., Asim, M.N., Dengel, A., 2025. Transitioning from wet lab to artificial intelligence: a systematic review of AI predictors in CRISPR. J. Transl. Med. 23 (1), 153. https://doi.org/10.1186/s12967-024-06013-w.

Adler, B.A., Trinidad, M.I., Bellieny-Rabelo, D., et al., 2024. CasPEDIA database: a functional classification system for class 2 CRISPR-cas enzymes. Nucleic Acids Res. 52 (D1), D590–D596. https://doi.org/10.1093/nar/gkad890.

Agboola, O.E., Agboola, S.S., Odeghe, O.B., et al., 2025. Computational genome engineering through AI-CRISPR-precision medicine integration in modern therapeutics. Ann. Pharm. Fr. https://doi.org/10.1016/j.pharma.2025.08.001.

Aussel, C., Cathomen, T., Fuster-García, C., 2025. The hidden risks of CRISPR/Cas: structural variations and genome integrity. Nat. Commun. 16 (1), 7208. https://doi.org/10.1038/s41467-025-62606-z.

Avsec, Ž., Latysheva, N., Cheng, J., et al., 2025. AlphaGenome: Advancing regulatory variant effect prediction with a unified DNA sequence model. bioRxiv 2025–2026. https://doi.org/10.1101/2025.06.25.661532.

Barrangou, R., Fremaux, C., Deveau, H., et al., 2007. CRISPR provides acquired resistance against viruses in prokaryotes. Science 315 (5819), 1709–1712. https://doi.org/10.1126/science.1138140.

Bolotin, A., Quinquis, B., Sorokin, A., Dusko Ehrlich, S., 2005. Clustered regularly interspaced short palindrome repeats (CRISPRs) have spacers of extrachromosomal origin. Microbiology (Reading, England) 151 (Pt 8), 2551–2561. https://doi.org/10.1099/mic.0.28048-0.

Brouns, S.J.J., Jore, M.M., Lundgren, M., et al., 2008. Small CRISPR RNAs guide antiviral defense in prokaryotes. Science 321 (5891), 960–964. https://doi.org/10.1126/science.1159689.

Cai, Z., Xie, W., Bao, Z., 2024. Broadening the targetable space: engineering and discovery of PAM-flexible Cas proteins. Trends Microbiol. 32 (8), 728–731. https://doi.org/10.1016/j.tim.2024.05.004.

Chuai, G., Ma, H., Yan, J., et al., 2018. DeepCRISPR: optimized CRISPR guide RNA design by deep learning. Genome Biol. 19 (1), 80. https://doi.org/10.1186/s13059-018-1459-4.

Deltcheva, E., Chylinski, K., Sharma, C.M., et al., 2011. CRISPR RNA maturation by trans-encoded small RNA and host factor RNase III. Nature 471 (7340), 602–607. https://doi.org/10.1038/nature09886.

Dixit, S., Kumar, A., Srinivasan, K., Vincent, P.M.D.R., Krishnan, N.R., 2024. Advancing genome editing with artificial intelligence: opportunities, challenges, and future directions. Front. Bioeng. Biotechnol. 11 (January). https://doi.org/10.3389/fbioe.2023.1335901.

Doench, J.G., Hartenian, E., Graham, D.B., et al., 2014. Rational design of highly active sgRNAs for CRISPR-cas9-mediated gene inactivation. Nat. Biotechnol. 32 (12), 1262–1267. https://doi.org/10.1038/nbt.3026.

Garneau, J.E., Dupuis, M.-È., Villion, M., et al., 2010. The CRISPR/Cas bacterial immune system cleaves bacteriophage and plasmid DNA. Nature 468 (7320), 67–71. https://doi.org/10.1038/nature09523.

Gottweis, J., Weng, W.-H., Daryin, A., et al., 2025. Towards an AI co-scientist. arXiv. https://doi.org/10.48550/arXiv.2502.18864.

Hou, Z., Zhang, Y., Propson, N.E., et al., 2013. Efficient genome engineering in human pluripotent stem cells using Cas9 from Neisseria meningitidis. Proc. Natl. Acad. Sci. U. S. A 110 (39), 15644–15649. https://doi.org/10.1073/pnas.1313587110.

Hsu, P.D., Scott, D.A., Weinstein, J.A., et al., 2013. DNA targeting specificity of RNA-guided Cas9 nucleases. Nat. Biotechnol. 31 (9), 827–832. https://doi.org/10.1038/nbt.2647.

Hu, J.H., Miller, S.M., Geurts, M.H., et al., 2018. Evolved Cas9 variants with broad PAM compatibility and high DNA specificity. Nature 556 (7699), 57–63. https://doi.org/10.1038/nature26155.

Ishino, Y., Shinagawa, H., Makino, K., Amemura, M., Nakata, A., 1987. Nucleotide sequence of the iap gene, responsible for alkaline phosphatase isozyme conversion in Escherichia coli, and identification of the gene product. J. Bacteriol. 169 (12), 5429–5433. https://doi.org/10.1128/jb.169.12.5429-5433.1987.

Huszár, K., Welker, Z., Györgypál, Z., et al., 2023. Position-dependent sequence motif preferences of SpCas9 are largely determined by scaffold-complementary spacer motifs. Nucleic Acids Res. 51 (11), 5847–5863. https://doi.org/10.1093/nar/gkad323.

Ishino, Y., Krupovic, M., Forterre, P., 2018. History of CRISPR-cas from encounter with a mysterious repeated sequence to genome editing technology. J. Bacteriol. 200 (7), e00580-17. https://doi.org/10.1128/JB.00580-17.

Jansen, R., Jan, van Embden, D.A., Gaastra, W., Schouls, L.M., 2002. Identification of genes that are associated with DNA repeats in prokaryotes. Mol. Microbiol. 43 (6), 1565–1575. https://doi.org/10.1046/j.1365-2958.2002.02839.x.

Jinek, M., Chylinski, K., Fonfara, I., Hauer, M., Doudna, J.A., Charpentier, E., 2012. A programmable dual-RNA-guided DNA endonuclease in adaptive bacterial immunity. Science (New York, N.Y.) 337 (6096), 816–821. https://doi.org/10.1126/science.1225829.

Kim, E., Koo, T., Park, S.W., et al., 2017. In vivo genome editing with a small Cas9 orthologue derived from Campylobacter jejuni. Nat. Commun. 8 (February), 14500. https://doi.org/10.1038/ncomms14500.

Kim, N., Kim, H.K., Lee, S., et al., 2020. Prediction of the sequence-specific cleavage activity of Cas9 variants. Nat. Biotechnol. 38 (11), 1328–1336. https://doi.org/10.1038/s41587-020-0537-9.

Kleinstiver, B.P., Prew, M.S., Tsai, S.Q., et al., 2015. Engineered CRISPR-cas9 nucleases with altered PAM specificities. Nature 523 (7561), 481–485. https://doi.org/10.1038/nature14592.

Lander, E.S., 2016. The heroes of CRISPR. Cell 164 (1–2), 18–28. https://doi.org/10.1016/j.cell.2015.12.041.

Mojica, F.J., Juez, G., Rodríguez-Valera, F., 1993. Transcription at different salinities of haloferax mediterranei sequences adjacent to partially modified PstI sites. Mol. Microbiol. 9 (3), 613–621. https://doi.org/10.1111/j.1365-2958.1993.tb01721.x.

Li, X., Wang, X., Sun, W., et al., 2022. Enhancing prime editing efficiency by modified pegRNA with RNA G-quadruplexes. Mol. Cell Biol. 14 (4), mjac022. https://doi.org/10.1093/jmcb/mjac022.

Menon, A.V., Song, B., Chao, L., et al., 2025. Unraveling the future of genomics: CRISPR, single-cell omics, and the applications in cancer and immunology. Front. Genome Edit. 7, 1565387. https://doi.org/10.3389/fgeed.2025.1565387.

Mojica, F.J.M., Díez-Villaseñor, C., García-Martínez, J., Soria, E., 2005. Intervening sequences of regularly spaced prokaryotic repeats derive from foreign genetic elements. J. Mol. Evol. 60 (2), 174–182. https://doi.org/10.1007/s00239-004-0046-3.

Müller, M., Lee, C.M., Gasiunas, G., et al., 2016. Streptococcus thermophilus CRISPR-cas9 systems enable specific editing of the human genome. Mol. Ther. 24 (3), 636–644. https://doi.org/10.1038/mt.2015.218.

Naert, T., Yamamoto, T., Han, S., et al., 2025. Precise, predictable genome integrations by deep-learning-assisted design of microhomology-based templates. Nat. Biotechnol. 1–14. https://doi.org/10.1038/s41587-025-02771-0. August 12.

Nayfach, S., Bhatnagar, A., Novichkov, A., et al., 2025. Engineering of CRISPR-Cas PAM recognition using deep learning of vast evolutionary data. bioRxiv 2025.01.06, 631536. https://doi.org/10.1101/2025.01.06.631536.

Nishimasu, H., Shi, X., Ishiguro, S., et al., 2018. Engineered CRISPR-Cas9 nuclease with expanded targeting space. Science (New York, N.Y.) 361 (6408), 1259–1262. https://doi.org/10.1126/science.aas9129.

Pallaseni, A., Peets, E.M., Girling, G., et al., 2024. The interplay of DNA repair context with target sequence predictably biases Cas9-generated mutations. Nature Communications 15 (1), 10271. https://doi.org/10.1038/s41467-024-54566-7.

Pandey, S., Choudhari, J.K., Tripathi, A., Singh, A., Antony, A., Chouhan, U., 2025. Artificial intelligence-based genome editing in CRISPR/cas9. Methods Mol. Biol. 2952, 273–282. https://doi.org/10.1007/978-1-0716-4690-8_16.

Park, S.-J., Lee, Ga E., Cho, S.M., Choi, E.-H., 2025. Recent applications, future perspectives, and limitations of the CRISPR-cas system. Mol. Ther. Nucleic Acids 36 (3), 102634. https://doi.org/10.1016/j.omtn.2025.102634.

Qu, Y., Huang, K., Yin, M., et al., 2025. CRISPR-GPT for agentic automation of gene-editing experiments. Nat. Biomed. Eng. 1–14. https://doi.org/10.1038/s41551-025-01463-z. July 30.

Ran, F.A., Cong, Le, Yan, W.X., et al., 2015. In vivo genome editing using Staphylococcus aureus Cas9. Nature 520 (7546), 186–191. https://doi.org/10.1038/nature14299.

Riesenberg, S., Kanis, P., Rosa, K., Maricic, T., 2025. Robust prediction of synthetic gRNA activity and cryptic DNA repair by disentangling cellular CRISPR cleavage outcomes. Nat. Commun. 16 (May), 4717. https://doi.org/10.1038/s41467-025-59947-0.

Silverstein, R.A., Kim, N., Kroell, A.-S., et al., 2025. Custom CRISPR–Cas9 PAM variants via scalable engineering and machine learning. Nature 643 (8071), 539–550. https://doi.org/10.1038/s41586-025-09021-y.

Wang, D., Zhang, C., Wang, B., et al., 2019. Optimized CRISPR guide RNA design for two high-fidelity Cas9 variants by deep learning. Nat. Commun. 10 (1), 4284. https://doi.org/10.1038/s41467-019-12281-8.

Wang, M., Rieber, L., van Baaren, J., et al., 2024. Diverse class 2 CRISPR effectors as active nucleases with expanded targeting capabilities. CRISPR J. 7 (2), 120–130. https://doi.org/10.1089/crispr.2023.0058.

Westra, E.R., Paul, B.G. van E., Künne, T., et al., 2012. CRISPR immunity relies on the consecutive binding and degradation of negatively supercoiled invader DNA by cascade and Cas3. Mol. Cell 46 (5), 595–605. https://doi.org/10.1016/j.molcel.2012.03.018.

Xiang, X., Corsi, G.I., Anthon, C., et al., 2021. Enhancing CRISPR-Cas9 gRNA efficiency prediction by data integration and deep learning. Nat. Commun. 12 (1), 3238. https://doi.org/10.1038/s41467-021-23576-0.

Yoon, P.H., Zhang, Z., Loi, K.J., et al., 2024. Structure-guided discovery of ancestral CRISPR-Cas13 Ribonucleases. Science 385 (6708), 538–543. https://doi.org/10.1126/science.adq0553.

Zetsche, B., Strecker, J., Abudayyeh, O.O., Gootenberg, J.S., Scott, D.A., Zhang, F., 2020. A survey of genome editing activity for 16 Cas12a orthologs. Keio J. Med. 69 (3), 59–65. https://doi.org/10.2302/kjm.2019-0009-OA.

CHAPTER

11

AI for accurate prediction of on-target efficiency and minimization of off-targets in plants

Ming Cheng[1,2], *Dominic Previtera*[3], *Yan Zhang*[3], *and Zheng Gong*[1,3]

[1]School of Agriculture, Food and Wine, Adelaide University, Waite Campus, Urrbrae, SA, Australia; [2]Plant Synthetic Biology Australia, Adelaide University, Waite Campus, Urrbrae, SA, Australia; [3]School of Agriculture and Food Sustainability, The University of Queensland, St Lucia, QLD, Australia

11.1 Introduction

The development and continuous expansion of the CRISPR-based toolbox have provided an unprecedented capacity to manipulate plant genomes and epigenomes, facilitating rapid advancements in plant synthetic biology, biotechnology, genetic engineering, and the study of basic biology (Fig. 11.1) (Li et al., 2024; Zhan et al., 2021; Atia et al., 2024; Cardi et al., 2023; Demirer et al., 2021). The class II CRISPR/Cas is a two-component genome editing system that consists of a guide RNA (gRNA) that directs a universal Cas protein to the target sequence (Jinek et al., 2012; Cong et al., 2013). The fundamental and unmodified form of Cas is an endonuclease that mediates double-strand DNA breaks (DSB) at the site, which relies on endogenous DNA repair mechanisms to introduce genome edits (Zhang et al., 2019). Error-prone repair introduces mostly small insertion–deletion (InDel) mutations, while a supplied donor sequence could also be inserted (knock-in) at the break site through homology-directed repair (HDR) (Nambiar et al., 2022; Huang and Puchta, 2019). Developments of a nickase Cas variant have been utilized in base editors capable of nucleotide substitution (Gaudelli et al., 2017) and in prime editors for precise sequence manipulation (Anzalone et al., 2019). Catalytically inactive Cas ("dead" Cas) variants have been developed for targeted recruitment of transcriptional activators (Maeder et al., 2013;

Intelligent Gene Editing in Plants
https://doi.org/10.1016/B978-0-443-44079-3.00002-4

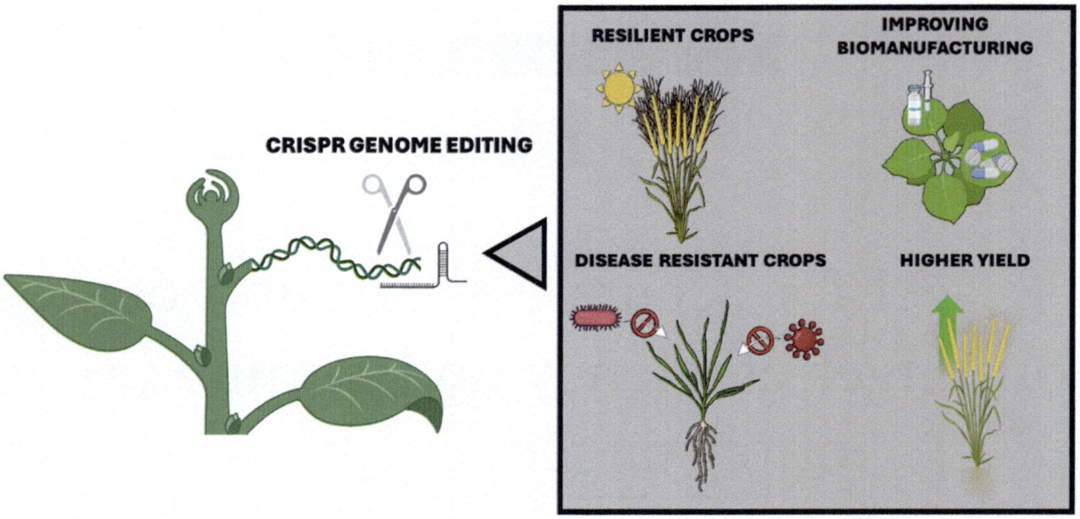

FIGURE 11.1 Schematic of the application of CRISPR genome editing in plants. CRISPR genome editing has revolutionized our capacity to modify genomes, with some example applications such as the development of climate- or environmental-resilient and disease-resistant crops with higher yield or improving plants as a chassis for biomanufacturing.

Lowder et al., 2018), repressors (Gilbert et al., 2014), and epigenome modifiers (Nakamura et al., 2021; Xu et al., 2025).

Using the extensive collection of CRISPR-based tools for genome modification is broadly dependent on targeting Cas and fused proteins to a genomic site (Konermann et al., 2015; Gaudelli et al., 2017; Gilbert et al., 2014; Nakamura et al., 2021). The CRISPR/Cas system is programmable to different DNA targets through Watson–Crick base pairing complementarity with the gRNA spacer sequence (Nakamura et al., 2021; Ma et al., 2013; Li et al., 2013; Jinek et al., 2012; Cong et al., 2013). This makes CRISPR-based technologies much more versatile than previous generations of genome editors that bind sequences through amino acid–DNA interactions (Atia et al., 2024). As such, the gRNA is a crucial determinant of genome editing efficiency and the overall specificity of CRISPR/Cas systems (Doench et al., 2014, 2016; Hiranniramol et al., 2020; Chuai et al., 2017; Liang et al., 2016). Optimizing gRNA design is, therefore, of particular interest in plant applications because of the long and resource-intensive process of tissue culture required to generate genome-edited plants (Cardi et al., 2023; Higa et al., 2025; Wang et al., 2023). Minimizing unspecific genome editing at off-target sites and maximizing efficiency at the intended target (hereon referred to as "on-target") are absolutely critical for widespread application of CRISPR tools in crop improvements and agriculture. However, the complexity and convolution of influential factors have made manual design of effective gRNAs impossible. The employment of machine learning (ML) and deep learning (DL) provides a unique opportunity to study large sets of gRNAs and their performance in cells. Integrating experimental datasets with ML, expanding scientific knowledge, and genomic resources enables the development of in silico tools for predicting gRNA on-target genome editing efficiency and pinpointing off-targets.

In this chapter, we discuss the application of CRISPR genome editing in plants, the importance of gRNA design, efforts into understanding factors that contribute to its specificity and on-target efficiency and, finally, the exciting progress in the development of cutting-edge computational tools for the design of superior gRNAs with a focus on plant applications.

11.2 CRISPR for plant genome editing and the gRNA design dilemma

CRISPR/Cas systems are originally known to form the adaptive immune mechanisms in prokaryotes against invading viruses and mobile genetic elements (Jinek et al., 2012; Mao et al., 2019). CRISPR systems are largely classified into two classes, with multiple types within each class (Marraffini, 2015; Koonin and Makarova, 2013). In the most widely studied, class II type II *Streptococcus pyogenes* CRISPR/Cas9 systems, the CRISPR array is transcribed to form a pre-CRISPR RNA (pre-crRNA). The repeat sequences in the pre-crRNAs are bound by the complementary region of a trans-activating CRISPR RNA (tracrRNA) (Jinek et al., 2012; Zhan et al., 2021; Doudna and Charpentier, 2014). tracrRNA binding facilitates crRNA maturation through processing by endogenous RNase III to release individual guide RNA (gRNA) containing one spacer and CRISPR repeat sequence (crRNA) bound by the tracrRNA (Jinek et al., 2012, 2014; Deltcheva et al., 2011). The tracrRNA forms a secondary structure that facilitates the complexation of a Cas effector with the gRNA, and to simplify the adoption of CRISPR/SpCas9 as a tool for genome editing, the tracrRNA was fused with crRNA to form a single-guide RNA (Jinek et al., 2012; Marraffini, 2015). The gRNA spacer sequence guides the Cas endonuclease to the target DNA, that is, to the invading phage sequence in bacteria or genomic targets for genome editing (Jinek et al., 2012; Doudna and Charpentier, 2014). Upon identifying a protospacer adjacent motif (PAM), the adjacent sequence of the target DNA is interrogated for complementarity to the gRNA spacer (Sternberg et al., 2014). In the presence of both a PAM and the target, the Cas9 effector mediates DNA cleavage with the formation of stable gRNA–DNA heteroduplexes (Sternberg et al., 2014). Unlike Cas9, the representative type V CRISPR/Cas12a system utilizes a single crRNA as the gRNA for DNA targeting (Zetsche et al., 2015; Bircheneder and Parniske, 2025). Cas12a cuts the DNA at a position farther from the PAM site, producing staggered ends rather than blunt cuts (Banakar et al., 2020; Wang and Han, 2024). Because it recognizes a 5′ T-rich PAM, Cas12a is extremely useful for efficiently targeting AT-rich genomic regions such as many plant promoters (Liu et al., 2025; Bircheneder and Parniske, 2025), which are often inaccessible to Cas9 due to its requirement for a G-rich PAM. The overall mechanism of DNA targeting using an RNA guide is core to enabling the high programmability and versatility of class II CRISPR systems. Furthermore, because each gRNA with its spacer sequence defines one target, multiple gRNAs could be coexpressed for multiplex genome editing (Xie et al., 2015). This allows simultaneous genome editing of multiple loci, which is crucial for crop improvement and engineering (Massel et al., 2021).

Over the last decade, both CRISPR/Cas9 and Cas12a systems have been extensively utilized for targeted genome modifications in not only research model plants such as *Arabidopsis thaliana* and *Nicotiana benthamiana* (Nekrasov et al., 2013; Li et al., 2013; Ma et al., 2013) but

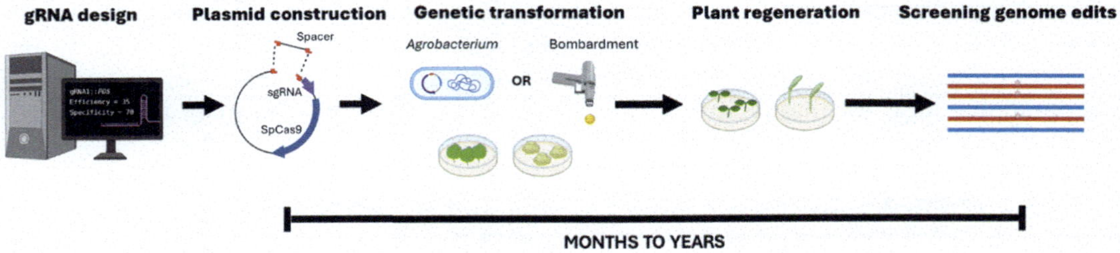

FIGURE 11.2 Schematic representation of a generalized workflow for CRISPR genome editing in plants. Typically, gene target-specific gRNA spacers are first designed and cloned into CRISPR genetic cassettes with the *SpCas9* protein gene. The genetic constructs are transformed into regenerative plant tissue through *Agrobacterium tumefaciens* or particle bombardment. Transformants are selected using antibiotics or herbicides. After the long duration of tissue culture, the regenerants are characterized and screened for genome edits. *(Created with BioRender.com)*

also in a broad number of agriculturally important crops such as tomato, maize, wheat, barley, rice, and sorghum (Jiang et al., 2013b; Feng et al., 2013; Brooks et al., 2014; Shan et al., 2013; Lawrenson et al., 2015; Liang et al., 2014; Zhang et al., 2014, 2025). CRISPR genome editing provides a more efficient and precise approach to study plant biology and to breed crops for greater food security (Cardi et al., 2023; Li et al., 2024; Zhan et al., 2021; Massel et al., 2021). To employ CRISPR-based tools for genome editing in plants, genetic constructs containing its components need to be delivered into regenerative tissues, which are intensively cultured under selection for transformants (Fig. 11.2) (Zhan et al., 2021; Shan et al., 2020). Plant transformation and regeneration are highly technical, resource-intensive, and time-consuming, taking months to years depending on the species (Oh et al., 2021; Kujur et al., 2021; Shan et al., 2020; Mao et al., 2019; Gong et al., 2021; Cheng et al., 2025). The regenerants of tissue culture could then be screened for mutants with genome edits, and at this stage, the effectiveness of the initially designed gRNAs is assessed (Mao et al., 2019; Cardi et al., 2023). Problematically, different CRISPR gRNAs have been shown to mediate extremely variable genome editing efficiencies at their intended target (hereon referred to as "on-target"). Multiple past studies in dicot model plants, tomato and *Nicotiana benthamiana*, have transiently expressed CRISPR/SpCas9 and different gRNAs and found largely distinct genome editing efficiencies between gRNAs, even when targeting the same gene (Gong et al., 2025; Naim et al., 2020; Slaman et al., 2023; Zhang et al., 2020b; Liang et al., 2019). Similarly, this variation in efficiency between gRNAs was also observed with CRISPR/Cas12a systems in plants (Slaman et al., 2024; Bernabé-Orts et al., 2019), reinstating the gRNA as a critical determinant of CRISPR functionality. Being able to confidently ensure successful and effective on-target genome editing when using user-customized gRNAs saves valuable scientific resources. Hence, it has become routine to experimentally validate gRNAs through in vitro digestion assays (Bente et al., 2020; Sagarbarria et al., 2023) or transient expression of CRISPR/Cas and the gRNA in plant cells (Wang et al., 2023; Elias et al., 2023; Higa et al., 2025; Nadakuduti et al., 2019; Shao et al., 2024; Nekrasov et al., 2013; Liang et al., 2019). However, in vitro digestion assays have been shown to be unreflective of genome editing efficiency in cells (Sagarbarria et al., 2023; Uusi-Mäkelä et al., 2018). Transient expression of CRISPR components in plant cells through agroinfiltration is generally limited to model plants, while protoplast systems are laborious and underdeveloped in many plant species.

Improving the specificity of CRISPR genome editors has also been a continuous area of interest. Genome editing at unintended genomic sites with sequence similarity to the gRNA spacer are known as "off-target effects" (Graham et al., 2020; Minkenberg et al., 2019). Although of slightly less concern in plant applications as compared to biomedicine, off-target edits may lead to unexpected phenotypic changes in plants that need to be screened for. It also poses regulatory concerns as well as public scrutiny on the safety of genome-edited plants as foods (Zhao and Wolt, 2017). Preventing off-target editing by CRISPR/Cas systems reduces these unforeseeable outcomes and the resources that need to be dedicated to identifying any off-target genome edits, particularly because an unbiased survey of off-target editing would require expensive techniques such as whole genome sequencing. Overall, these issues call for further research into understanding the underlying contributors to on-target efficiency and the occurrence of off-target genome editing. This expanding knowledge is important in enabling the development of valuable computational-based tools to assist in designing and selecting gRNAs that are highly active while mitigating off-target effects prior to plant transformation.

11.3 Spotlight on the gRNA: Critical determinants of CRISPR on-target genome editing efficiency

The gRNA is the only customizable part of a conventional CRISPR/Cas genome editing system, forming the basis of on-target specificity and undermining overall efficiency (Yan et al., 2017; Liang et al., 2019). Understanding what affects the on-target performance of a particular gRNA target forms the basis of developing tools for predicting and designing effective gRNAs, a crucial limiting step in the plant genome editing workflow. Effective gRNA design must consider both target binding capabilities and chromatin accessibility (Horlbeck et al., 2016). The binding capabilities are largely determined by the PAM and spacer sequence of gRNA, which not only dictates complementarity with the target but also influences the search time that Cas9 needs to discriminate against nontarget sequences across the genome (Moreb et al., 2020; Moreb and Lynch, 2021). Multiple studies have analyzed large datasets of CRISPR/SpCas9 gRNAs with experimentally characterized cleavage activity in search for factors enriched in effective ones. The overall nucleotide composition of the spacer target sequence showed a slight preference for adenine in functional gRNAs as compared to nonfunctional ones (Wong et al., 2015). More importantly, several independent studies have also found positional nucleotide preferences at certain locations along the spacer in effective gRNAs. The PAM proximal sequences, referring to nucleotides toward the end of the gRNA spacer, part of the seed, were shown to be a critical influence of on-target genome editing efficiency in animal models (Gagnon et al., 2014; Wang et al., 2014; Doench et al., 2014, 2016; Xiang et al., 2021; Xu et al., 2015; Chari et al., 2015; Konstantakos et al., 2022a). In particular, some have reported a preference for purines, with some disagreements between studies. Despite this, the single nucleotide at the last position of the gRNA spacer and directly upstream to the PAM has been repeatedly shown to be enriched in purines over pyrimidines (Konstantakos et al., 2022a). Several other reported nucleotide preferences for effective gRNA spacers include a preference for cytosine (C) at the 17th nucleotide, guanine (G) toward the end, and adenine (A) toward the middle of the spacer (Xu et al., 2015), generally disfavoring thymine (T) in the spacer (Wang et al., 2019; Moreno-Mateos et al., 2015; Corsi et al., 2022) and stretches of G or T nucleotides (Wong et al., 2015), among other rules.

Since Cas enzymes initiate cleavage at the PAM and proceed directionally toward the distal end of the target, nucleotides at the 3′ end of the gRNA spacer, where the seed region is located (Fig. 11.3a), are generally more critical for functionality than those at the 5′ end (Jiang et al., 2013a; Singh et al., 2016; Sternberg et al., 2014). Consistent with this, functional gRNAs often show higher accessibility at the 3′ end of the gRNA spacer, especially at bases 18–20, compared with nonfunctional ones (Wong et al., 2015). Studies in mammalian cells suggest that thymine (T) or cytosine (C) at the 20th nucleotide is disfavored, as these bases can pair with positions of the gRNA scaffold (crRNA repeat fused with tracrRNA) to form an extended stem loop that impairs activity (Doench et al., 2014; Wong et al., 2015). GC content is another critical factor to consider in gRNA design and is often used as a proxy for RNA structural stability. In rice, functional gRNAs were shown to have a wide range of GC contents (between 30% and 80%) (Liang et al., 2016). Nevertheless, moderate GC content is generally preferred, and extreme GC content is a feature of nonfunctional gRNAs due to excessive duplex stability with the target (Gerashchenkov et al., 2020; Liu et al., 2016; Wong et al., 2015). Studies in grapes revealed that gRNAs with 65% GC achieved the highest efficiencies (Ren et al., 2019). Comparable efficiencies were obtained at 50% GC, whereas low-GC gRNAs (20%–30%) showed at least a two-fold reduction, with the lowest efficiency (22.2 ± 3.7%) at 25% GC in "Char." In contrast, Malik et al. (2021) reported that overall GC content in gRNA spacers showed no significant differences among three groups (0%–40%, 41%–55%, and 56%–100%) in *Arabidopsis*. GC content in the PAM-proximal seed region strongly influenced cleavage efficiency, with higher GC levels significantly reducing gRNA activity (Malik et al., 2021). Similar findings in mammalian systems revealed that high GC content in the PAM-distal region favors editing efficiency (Labuhn et al., 2018).

In addition, the tendency for gRNAs to form secondary internal structures has been shown to negatively influence on-target performance (Thyme et al., 2016; Corsi et al., 2022; Riesenberg et al., 2022; Jensen et al., 2017; Moreb and Lynch, 2022). Self-folding and formation of hairpin structures within the spacer are critical to gRNA activity (Moreb and Lynch, 2022). Potential base pairing between the gRNA spacer and the rest of the gRNA scaffold should also be avoided (Thyme et al., 2016). Researchers have reported that misfolding in the seed region of gRNA spacer can reduce editing efficiency (Liu et al., 2016). Besides the abovementioned importance of positional nucleotide accessibility in the gRNA, the thermodynamic properties and energy changes of gRNA–DNA binding were shown to be another contributor to gRNA on-target performance, in which the binding free energy is essentially affected by the sequence and structural features of the gRNA (Corsi et al., 2022; Xiang et al., 2021). Modeling the binding energetics of gRNA–DNA was quite useful for differentiating efficient and inefficient gRNAs, as it would likely be influenced by many gRNA sequences and structural features (Corsi et al., 2022). Rather than modifying gRNA spacers to avoid unwanted secondary structures, researchers have also found that

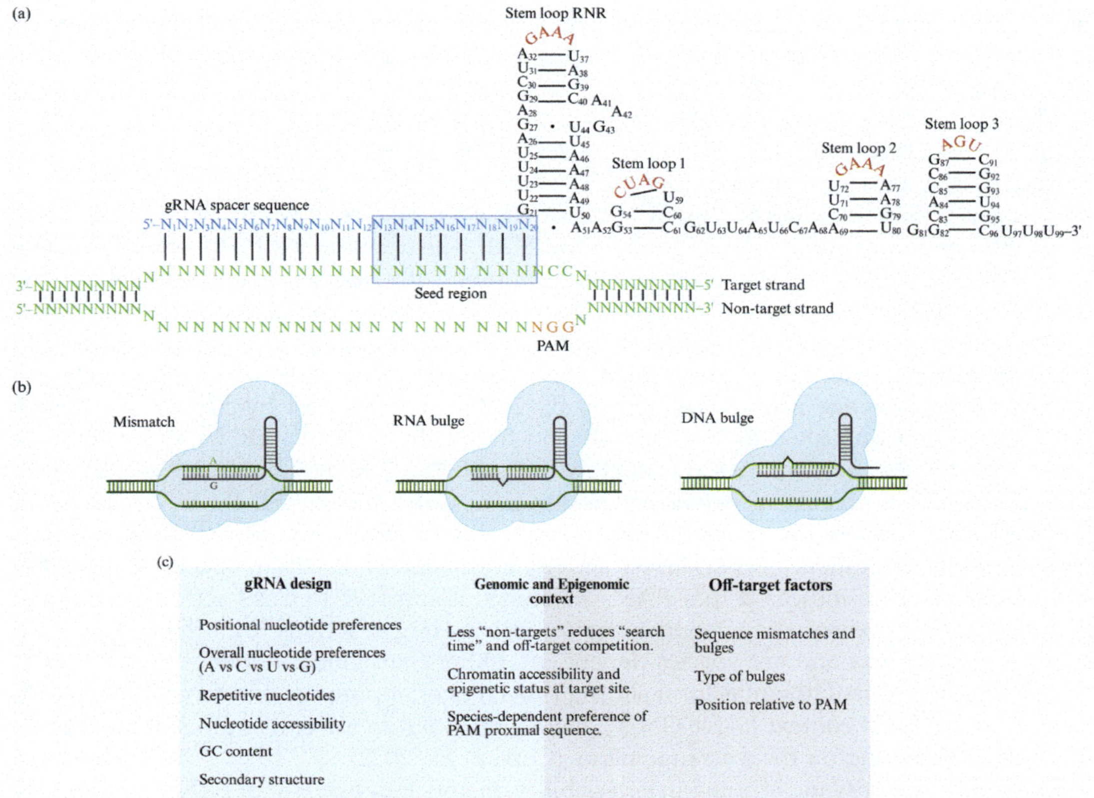

FIGURE 11.3 Importance of gRNA design in effective CRISPR genome editing. (A) Schematic of a single guide RNA (sgRNA or gRNA for coherency) in the CRISPR/Cas9 genome editing system. The 20-nt spacer sequence (blue) is located at the 5′ end of the gRNA and pairs with the target DNA strand. Genomic double-stranded DNA is shown in green, with the PAM sequence on the nontarget strand. The seed region (blue box) lies at the 3′ end of the spacer, and higher accessibility of the seed region is critical for gRNA function. Base pairing of N_{18}–N_{20} with A_{51}–A_{53} can lead to extended secondary structures, which are unfavorable for functional gRNAs. The gRNA scaffold contains four stem loop structures (red): RNR (GAAA), stem loop 1 (CUAG), stem loop 2 (GAAA), and stem loop 3 (AGU). Weak bonds are indicated by black dots, and complementary base pairing by short lines. (B) Three types of off-targets induced by CRISPR/Cas systems. Mismatches occur when the gRNA spacer sequence is not fully complementary to the target DNA. An RNA bulge refers to an insertion event, where one or more extra nucleotides are present in the gRNA relative to the DNA target. In contrast, a DNA bulge refers to a deletion event, where one or more nucleotides are missing in the gRNA compared to the DNA target. Multiple bulges can occur for off-target sites, and each bulge can be 1–5 bp long. (C) The key contributors affecting CRISPR/Cas on-target genome editing efficiency and the occurrence of off-target effects. Both gRNA design and off-target predictions need to consider chromatin accessibility and epigenetic context. *(Created with BioRender.com)*

altering the tracrRNA can improve gRNA stability and enhance Cas9 activity. For example, Scott et al. (2019) reported that substituting U→A at the 34th nucleotide of the tracrRNA (or U_{66} in Fig. 11.3a) or replacing a U–A bond with a G–C bond in stem loop 1 (CUAG) increased Cas9 activity (Scott et al., 2019). In rice, stem-loop elements such as the RAR (GAAA), second (GAAA), and third (AGU) loops were shown to influence gRNA on-target activity: the RAR is essential for pre-crRNA maturation and Cas9 cleavage, while stem loops 2 and 3 stabilize the DNA–RNA duplex (Jinek et al., 2012; Liang et al., 2016; Nishimasu et al., 2014).

The genomic context is not only involved in the off-target effect but also plays an influential role in CRISPR genome editing efficiency at the intended target. The number of nontargeted PAM sites in a genome that the Cas9-gRNA searches through before finding the correct target was previously shown to affect cleavage efficiency (Cong et al., 2013; Jinek et al., 2012; Sternberg et al., 2014; Moreb et al., 2020). The nontargeted PAM sites are known as the "search space," which when at a high number in a specific genome, increases the search time for the Cas9-gRNA complex. This can be broadly categorized as the genomic context (Moreb and Lynch, 2021; Moreb et al., 2020). When using Cas9 variants with tight PAM preferences, the immediate "search space" and time decrease, which effectively improves target cleavage activity (Moreb et al., 2020). Later studies by Moreb and Lynch (2021) examined published gRNA datasets and found that effective gRNA targets show a preference for the two variable nucleotides in PAM (the 2 N in 5′-NGGN-3′ for SpCas9). This preference is consistent between datasets from different species. On the other hand, the nucleotide preference of the PAM proximal dinucleotides in the gRNA spacer (N_{19} and N_{20}) and its influence on on-target activity correlated well within species but not between species. This indicated that PAM preference is intrinsic to Cas9 activity, while the gRNA's PAM proximal dinucleotides may affect on-target activity differently based on the genomic context and are species-dependent. Other than global nontargets and PAMs, adjacent local PAMs (known as local sliding PAMs) overlapping the on-target PAM, which constitutes its PAM context (5′-NGGN-3′), were also found to affect overall genome editing efficiency depending on their arrangements (Corsi et al., 2022).

Epigenetic features and chromatin accessibility, factors independent of gRNA sequence or structure, are another underlying contributor to variations in gRNA efficiency across cellular and experimental contexts (Jensen et al., 2017; Nakamura et al., 2021; Ito et al., 2024). Target site accessibility, chromatin status, and DNA methylation have been shown to impact CRISPR on-target activity across multiple species (Jensen et al., 2017; Wu et al., 2014; Horlbeck et al., 2016; Daer et al., 2017), including in several model plants (Přibylová et al., 2022; Weiss et al., 2022). In *Arabidopsis thaliana*, despite the same gRNA spacer sequence, heterochromatin sites showed lower genome editing efficiency compared to euchromatin sites, and similar results were observed with methylated and unmethylated sites (Weiss et al., 2022). However, one potential downside of this study is that preferences at the fourth nucleotide (N of 5′-NGGN-3′) of the PAM, as reported to influence on-target efficiency (Moreb and Lynch, 2021), were not considered when filtering potential targets using the 5′-NGG-3′ PAM (Weiss et al., 2022). Similarly, another study in *Nicotiana benthamiana* plants found that certain DNA methylation modifications affected CRISPR genome editing at the transgene target (Přibylová et al., 2022). Although the genomic and epigenomic context was shown as critical, the effects of these contexts can be hard to study and even harder to produce concrete, consistent relationships or rules due to high variability between organisms or different cell types (Ito et al., 2024; Feng et al., 2010). The combination of so many different complex and understudied factors that span these contextual, gRNA sequence, and structural features has raised the need to innovate computational tools to simplify and improve the design of gRNAs and the prediction of their performance (Fig. 11.3c).

11.4 From the lens of biologists: Integrating knowledge and AI for effective prediction of gRNA genome editing efficiency

Initial developments in bioinformatic tools were focused on streamlining adoption and utilized sequence alignment to facilitate the design of gRNAs for user-defined target sequences (Konstantakos et al., 2022a; Wang et al., 2020; Yuan et al., 2025). These tools searched sequences for canonical and noncanonical PAMs. Potential gRNAs adjacent to PAMs were labeled and outputted with some tools also capable of identifying possible off-target sites in the genome (Xie et al., 2014; Ma et al., 2013; Heigwer et al., 2014). Despite the rapidness and simplicity of alignment-based tools for gRNA design, they are unable to infer and predict the activity of gRNA in vivo. Further research into understanding the underlying factors that influence gRNA performance in mediating on-target genome editing (as reviewed previously) was necessary to enable the development of tools to predict it.

The second category of gRNA design tools utilizes empirically determined and experimentally supported hypotheses of known contributors toward gRNA on-target activity to devise rules, focusing mostly on factors in the spacer sequence. These rules are used in matrices to evaluate and score gRNA spacers, providing a prediction of gRNA performance and genome editing efficiency in cells. Representative web-based tools such as *E-CRISP* (Heigwer et al., 2014) and the plant-specific designer *CRISPR-P 2.0* (Liu et al., 2017) have incorporated gRNA GC content and some positional nucleotide preferences into their scoring matrix. These tools design gRNAs based on input sequence or particular genome sites and output an on-target or efficiency score for each gRNA. Several other alignment-based tools have not implemented scoring systems but provide additional supporting information, such as GC content, in order to aid gRNA selection (Park et al., 2015; Montague et al., 2014). However, overall, these hypothesis-driven selections of gRNA and rule-based efficacy scoring methods only take into account one or a few known contributing factors that are largely focused on the spacer sequence. How scores are adjusted for each gRNA spacer and how different contributing factors are weighed in the scoring matrix is largely limited to current knowledge toward what is believed to be important factors (Konstantakos et al., 2022a). Additional factors external to the spacer sequence itself, such as the surrounding sequence context, genomic context, the PAM distal sequence, gRNA secondary structure, and target accessibility (chromatin status and DNA methylation), were also shown to influence gRNA performance (as discussed previously) (Liu et al., 2020). This results in a complex and intertwined network that is difficult to unravel. Therefore, many hypothesis-driven efficiency scoring methods were shown to fall short of tools based on ML and DL algorithms (Konstantakos et al., 2022a; Wang et al., 2020; Yan et al., 2017). In this section, from the lens of biologists, we discuss the application of ML in the development of effective in silico tools for designing and predicting gRNA activity in biological systems.

Genomic resources combined with sequence alignment for gRNA identification and ML-based models for gRNA efficiency scoring have been a widely implemented workflow for gRNA design, utilized in several web-based tools. These gRNA design tools form the third category and offer superior capabilities compared to previous categories of tools. The generalized pipeline for developing ML algorithms requires an input training dataset consisting of hundreds to thousands of different gRNA spacers with quantified on-target activity in cells. The training dataset for supervised ML modeling is labeled based on measured efficiency, often by categorizing gRNAs into quantiles or percentiles. A subset of the initial experimental dataset is set aside for independent validation and evaluation of the final model. A multitude of predefined features of the gRNA, such as GC content, the gRNA sequence, motifs, PAM preferences, thermodynamic properties, and structural information, are used to train ML algorithms with the labeled dataset. In this way, ML overcomes issues with hypothesis-driven, rule-based scoring as it models the relationship between features and gRNA performance in a comprehensive training dataset and determines the weight of different features (Zhang et al., 2023a; O'Brien et al., 2020). With expanding knowledge, ML algorithms can also be improved by retraining using newly identified feature sets and compilation of larger training datasets from different studies (Wang et al., 2020). We have seen tremendous progress in the construction of many ML-based models and continuous efforts in improvements over past tools, particularly for gRNA efficiency prediction of the CRISPR/SpCas9 system. For clarity, we refer to different algorithms as: [first author name]' [year of publication], especially when no other names exist.

Wang'2014 (Wang et al., 2014) and *Rule Set 1* (Doench et al., 2014) are early implementations of ML for CRISPR/SpCas9 gRNA efficiency prediction. Pools of thousands of gRNAs were screened in mammalian cell lines using lentivirus transduction, and targeted gene knockout efficiency was quantified as a measure of genome editing efficiency. Both studies used the gRNA sequence as features with different labeling, but the latter further incorporated the surrounding sequence context of each gRNA into a 30 bp target (5′–NNNN–20 bp target-NGG–NNN–3′) as well as the gRNA GC content as features. To further accumulate a larger training dataset, Xu et al. (2015) combined the datasets from two different studies that relied on screening for gene knockouts to produce *Xu'2015* which was shown by the authors to outperform *Rule Set 1*. All 3 ML algorithms relied on training datasets where gRNA effectiveness was measured through phenotypic screens and gene knockout efficiency (Doench et al., 2014; Wang et al., 2014; Xu et al., 2015). *Rule Set 1* specifically excluded gRNAs that were less effective at generating a knockout due to the position of their target sites in the gene structure (such as exons toward the end), which is crucial in reducing bias (Doench et al., 2016). However, gene knockout still requires the formation of a particular type of mutation and is, therefore, not a direct assessment of gRNA activity and CRISPR cleavage efficiency (Hiranniramol et al., 2020). Another study overcame this issue by using next-generation targeted amplicon sequencing to assay and capture all types of mutations in a lentiviral transduced pooled gRNA screen and developed the ML-based efficiency scoring algorithm (Chari et al., 2015; Chari et al., 2015). Similar approaches for assessing gRNA activity were also adopted by Moreno-Mateos et al. (2015), which screened in vitro transcribed, pooled gRNA libraries in injected zebrafish to train a model and were implemented in the *CRISPRscan* web-based tool. Interestingly, comparative analysis showed that gRNA efficiency scores from Moreno-Mateos et al. (2015) were less correlated with experimental activity in datasets where gRNAs were expressed using U6 promoters as opposed to in vitro transcription (Haeussler et al., 2016; Moreno-Mateos et al., 2015). However, it is unclear if this is because of differences in species-specific preferences, as the in vitro transcription datasets were all derived from zebrafish systems (Moreb and Lynch, 2022; Haeussler et al., 2016). For experimental datasets with gRNAs expressed from U6 promoters (Haeussler et al., 2016), it was found that gRNA predictions scored by Doench et al. (2016) showed the highest correlation with activity (Doench et al., 2016; Doench et al., 2016), or *Rule Set 2* is the second iteration of *Rule Set 1* and was developed using a larger pool of gRNAs from phenotypic screens used to train an optimized ML model with the incorporation of more features such as the spacer melting temperature, positional-independent nucleotide counts, and identity of the variable dinucleotide in the SpCas9 PAM (5′-NGGN-3′) (Doench et al., 2016). Instead of classifying gRNA into effective and ineffective, like in Wang et al. (2014), or into the top 20% versus the remaining 80%, as with Doench et al. (2014, 2016), calculated a normalized rank for each gRNA that is scaled between 0 and 1, which allowed gRNAs to be further distinguished from each other (Doench et al., 2014, 2016; Wang et al., 2014). Owing to its substantiative improvement (Doench et al., 2016), it has been adopted in many web-based, all-in-one gRNA design tools such as *CRISPOR*, *CHOPCHOP*, *CRISPick*, and *crisprVerse* and is broadly used by many researchers to streamline the selection of effective gRNAs across biological systems (Labun et al., 2019; Concordet and Haeussler, 2018; Hoberecht et al., 2022; Doench et al., 2016).

Many gRNA efficiency prediction algorithms have focused on sequence features rather than the gRNA structure. Retaining the structure of the gRNA scaffold is important for Cas recognition, whereas the spacer region needs to remain accessible for target DNA binding (Wong et al., 2015; Riesenberg et al., 2022; Corsi et al., 2022). Wong et al. (2015) found that base pairing between certain nucleotides of the gRNA spacer, particularly at the end of the seed region, and nucleotides in the scaffold was shown to be more prevalent in less functional gRNAs. Combined with other sequence features, the authors developed the *WU-CRISPR* algorithm, which was shown to effectively distinguish functional and unfunctional gRNAs (Wong et al., 2015). However, likely because it was trained on extreme ends of gRNA functionality, the *WU-CRISPR* scores did not perform well and showed low correlation with activity in the comparative analysis by Haeussler et al. (2016), who used gRNA performance data from multiple studies. Another particularly interesting category of studies has also dived into parameterization of the energetic properties, kinetics, and physical properties of gRNA–target binding rather than deciphering spacer sequence preferences (Corsi et al., 2022). This approach was utilized in the development of *uCRISPR* and was explored in several studies for the prediction of gRNA on-target activity (Zhang et al., 2024b). Recently, Jin et al. (2024) used ML to study the kinetic and thermodynamic features of the gRNA–DNA R-loop, which enabled the development of a prediction tool that outperformed (Doench et al., 2016; Chari et al., 2015) when tested with public experimental gRNA datasets (Jin et al., 2024).

Additional developments using different feature sets and ML algorithms have seen continuous improvements in the prediction performance of gRNA activity compared to the benchmark. This illustrates one major advantage of learning-based tools in that there are opportunities to learn from their differences and integrate new mechanistic knowledge of CRISPR to allow further optimization and improvements. The third iteration, *Rule Set 3*, provided distinctions in the gRNA scaffold variants used by different datasets in addition to previous features in its predecessor (Doench et al., 2016; DeWeirdt et al., 2022). *Rule Set 3* was trained on nearly 50,000 gRNAs and their contextual sequences, showing greater predictive performance compared to Doench et al. (2016), which was trained on around one-tenth the number of gRNAs. Surprisingly, *Rule Set 3* also outperformed many DL-based tools, which are known to be more effective compared to supervised ML (DeWeirdt et al., 2022). Rather than relying on gRNAs designed to target genomic loci, Labuhn et al. (2018) used a lentivirus-delivered surrogate target where CRISPR/SpCas9-mediated genome editing would disrupt the frame of a fused fluorescent reporter. This provided the rapidness of gene knockout assays for assessing gRNA activity but avoids positional bias seen when targeting different locations in a gene (Labuhn et al., 2018). Combined with an analysis of gRNA datasets to derive features, the authors developed and tested a supervised ML model called *CRISPRater*. *CRISPRater* was tested and showed the capacity to effectively predict gRNA efficiency in much smaller independent datasets (20 or less gRNAs), emphasizing its reproducibility. Further work by the authors of *WU-CRISPR* led to the development of *sgDesigner*, which was derived from stacking models (two layers of modeling), shown to be a more effective approach than individual ML algorithms as used in many previous studies (Hiranniramol et al., 2020). *sgDesigner* showed a higher Spearman correlation to gRNA activity across multiple independent datasets compared to benchmarks (Hiranniramol et al., 2020). Concurrently, Peng et al. (2018) integrated and trained 2 ML algorithms to develop *TSAM*, which averages the gRNA efficiency predictions to boost performance (Peng et al., 2018). The first ML algorithm was trained on standard gRNA features and extracted those that were important. This was used as a feature to train a second ML algorithm with the addition of a gRNA sequence profile hidden Markov model as a feature, with the reasoning that there may be overall sequence similarities in more active gRNAs. Indeed, *TSAM* saw slightly better performance compared to Doench et al. (2016).

The use of different ML algorithms, how efficiency is quantified to enable the labeling of gRNA spacers, how gRNAs are labeled for activity, and the particular feature sets used may result in distinct predictive performance (Konstantakos et al., 2022a; Yuan et al., 2025; Haeussler et al., 2016; Chen and Wang, 2022; O'Brien et al., 2020; Corsi et al., 2023). The increasing number of gRNA design and efficiency prediction tools necessitates studies that independently evaluate and cross-validate their performance for different use cases. For example, Chen and Wang (2022) and Haeussler et al. (2016) found variable performance across multiple datasets from different species, with generally better performance when evaluated on datasets generated within the same study. Different gRNA features used in training ML algorithms are also an underlying factor for variations (Konstantakos et al., 2022a; O'Brien et al., 2020). The genomic context and epigenetic features are also core influencers of gRNA on-target activity, but they require large sets of genomic and epigenomic data for each species and are especially difficult with the heterogeneous nature of epigenetic state in different tissues. Another problem is that ML algorithms are trained with datasets of different experimental conditions. For example, the CRISPR/SpCas9 and gRNA could be delivered using different means, or the gRNA experimental on-target activity could also be measured using different methods. Recent works have emphasized the importance of techniques used to accurately quantify genome editing efficiency, particularly with methods such as high-throughput sequencing over gene knockout assays, as sequencing allows for direct assessment of mutation frequencies (Xiang et al., 2021; Corsi et al., 2023; Labuhn et al., 2018; Wilson et al., 2018). Tools such as *TUSCAN* were developed with this in mind, where the authors trained a new ML model using gRNA datasets where high-throughput amplicon sequencing (HTS) was used to measure activity, similar to what was used in curating the training dataset for developing (Chari et al., 2015; Wilson et al., 2018; Chari et al., 2015). We collated a table of ML-based gRNA efficiency prediction tools and the source of their training datasets (Table 11.1). Although some of these have been independently assessed and compared, there is a lack of effort in evaluating efficiency predictions across different use cases and species, with mammalian cells being the main focus in current studies (Wang et al., 2020). Hence, when designing gRNAs, referencing efficiency prediction scores across multiple tools may be necessary for greater confidence.

Other than classical ML-based models, DL-based approaches have also been explored for gRNA efficiency prediction. There have been less developments in DL-based algorithms, possibly because training DL-based models requires larger input datasets and more computational power (DeWeirdt et al., 2022; Zhang et al., 2023a; Ching et al., 2018). Regardless, in further benchmarking experiments, DL-based gRNA efficiency prediction has shown an overall stronger predictive performance over conventional ML (Chen and Wang, 2022; Labuhn et al., 2018). Therefore, we categorized DL-based tools as the fourth category of tools, which form the current cutting-edge. DL models are advantageous as they are capable of extracting features from labeled gRNA training datasets, potentially encompassing previously overlooked features or patterns in gRNAs (Wang et al., 2020). Predefined features could also be introduced to improve performance (Zhang et al., 2023a). In order to obtain the larger training dataset needed and avoid positional bias, many studies have utilized lentivirus to deliver barcoded surrogate targets and gRNA libraries into human cell lines, combined with mutation quantification using HTS for higher throughput screening over designing genome-targeting gRNAs (Labuhn et al., 2018). *DeepCpf1* and *DeepSpCas9* are DL-based gRNA efficiency prediction tools for Cas12a and Cas9, respectively, shown to outperform conventional ML algorithms (Kim et al., 2018, 2019). Both models were trained on large datasets generated using the method described with activity data in more than 10,000 gRNAs. Interestingly, the authors found that incorporating target site chromatin accessibility information improved the prediction performance of *DeepCpf1* for Cas12a but not *DeepSpCas9* for Cas9. Similar results were reported by Wang et al. (2019). The authors used a similar surrogate target system, measured the efficiency in human cell lines, and created a gRNA dataset for SpCas9 and high-specificity Cas9 variants, which was used to train a DL model (Wang et al., 2019). The DL model was fine-tuned for both U6 promoter-expressed and T7 in vitro transcribed gRNAs, outperforming many benchmarks of ML-based gRNA efficiency scoring, such as *TSAM* (Moreno-Mateos et al., 2015; Doench et al., 2016). Likewise, DL was implemented to produce *DeepSpCas9* variants, which is a comprehensive tool for predicting gRNA efficiency across different SpCas9 variants with higher specificity or diverse PAM recognition, bringing further versatility to the table (Kim et al., 2020a, 2020b).

In the work by Xiang et al. (2021), they found that the gRNA–DNA free energy was an important feature in their gRNA dataset targeting surrogates delivered using lentivirus. The binding energy was used to enhance the training of a DL model for gRNA efficiency prediction, *CRISPRon* (Xiang et al., 2021). Surprisingly, *CRISPRon* scores showed a higher

TABLE 11.1 Collection of machine learning-based algorithms for gRNA efficiency prediction in classic cleavage-based CRISPR/Cas systems.

Prediction algorithm/tool (alternative names)	Origin of training dataset	CRISPR system	Reference
Rule set 1 (Doench et al., 2014)	Human and mouse cells	Cas9	Doench et al. (2014)
Wang et al. (2014)	Human cells	Cas9	Wang et al. (2014)
(Xu et al., 2015) (SSC)	Human and mouse cells	Cas9	Xu et al. (2015)
(Chari et al., 2015) (sgRNAScorer)	Human cells	Cas9	Chari et al. (2015)
(Moreno-Mateos et al., 2015) (CRISPRscan)	Zebrafish	Cas9	Moreno-Mateos et al. (2015)
WU-CRISPR	Human and mouse cells	Cas9	Wong et al. (2015)
Rule set 2 (Doench et al., 2016)	Human and mouse cells	Cas9	Doench et al. (2016)
Ge-CRISPR	Zebrafish, human, and mouse cells	Cas9	Kaur et al. (2016)
CRISPRpred	Human and mouse cells	Cas9	Rahman and Rahman (2017)
predictSGRNA	Human cells	Cas9	Kuan et al. (2017)
TSAM	Human and mouse cells	Cas9	Peng et al. (2018)
TUSCAN	Human and mouse cells	Cas9	Wilson et al. (2018)
CRISPRater	Human cells	Cas9	Labuhn et al. (2018)
sgDesigner	Human cells	Cas9	Hiranniramol et al. (2020)
Cas13 design	Human cells	Cas13	Wessels et al. (2020)
Rule set 3	Human and mouse cells	Cas9	DeWeirdt et al. (2022)
CRISPRedict	Zebrafish and human cells	Cas9	Konstantakos et al. (2022b)
Deep learning-based tools			
DeepCRISPR	Human cells	Cas9	Chuai et al. (2018)
DeepCpf1	Human cells	Cas12a	Kim et al. (2018)
DeepSpCas9	Human cells	Cas9	Kim et al. (2019)
DeepHF	Human cells	Cas9	Wang et al. (2019)
DeepSpCas9 variants	Human cells	Cas9 variants	Kim et al. (2020b)
DeepxCas9 and DeepSpCas9-NG	Human cells	Cas9 variants	Kim et al. (2020a)
C-RNNCrispr	Human cells	Cas9	Zhang et al. (2020a)
CRISPRon	Human cells	Cas9	Xiang et al. (2021)
DeepGuide	*Yarrowia lipolytica*	Cas9 and Cas12a	Baisya et al. (2022)

TABLE 11.1 Collection of machine learning-based algorithms for gRNA efficiency prediction in classic cleavage-based CRISPR/Cas systems.—cont'd

Prediction algorithm/tool (alternative names)	Origin of training dataset	CRISPR system	Reference
CRISPRDB	Ensemble model	Cas9	Chen and Wang (2022)
AIdit-CRISPR	Human cells	Cas9	Zhang et al. (2023b)
DeepCas13	Human cells	Cas13	Chen et al. (2023)

correlation with activity across several independent datasets compared to *DeepSpCas9*. Others have also employed different methods and datasets to develop DL-based models and efficiency prediction tools such as *DeepCRISPR* (Chuai et al., 2018) and, notably, *AIdit-CRISPR* (Zhang et al., 2023b). *AIdit-CRISPR*'s on-target efficiency prediction model, *AIdit_ON* was trained with a staggering dataset consisting of 740,000 gRNAs, and the authors found that by adding sequences flanking the gRNA as input, they saw an increase in prediction performance. As expected, *AIdit_ON* outperformed other DL models such as *DeepSpCas9*, *DeepHF*, and *DeepCRISPR* among other conventional ML-based efficiency prediction models (Doench et al., 2016; Zhang et al., 2023b). Other than genome editing relying on mediating DNA DSBs, DL has also been applied in efficiency prediction for base editing (Kim et al., 2024), prime editing (Kim et al., 2021; Koeppel et al., 2023), and epigenome editing (Yang et al., 2023), important technologies in the CRISPR-based toolbox.

Recent years saw the advent of vast numbers of ML- and DL-based gRNA efficiency prediction tools. Chen and Wang (2022) used multiple SpCas9 gRNA datasets and determined the correlation between measured efficiency and prediction scores from several learning-based prediction tools. *CRISPRon*, *DeepHF*, and *DeepSpCas9*, tools based on DL models, were ranked in the top three, followed by conventional ML-based ones such as *TSAM* (Doench et al., 2016; Doench et al., 2016). Interestingly, the authors then employed model stacking similar to that used to develop *sgDesigner*. Knowing how different algorithms compare in efficiency prediction, the top-performing ones were combined in an ensemble model. The final model showed better gRNA efficiency prediction compared to individual models, including DL-based ones, and was implemented in a web-based gRNA design and efficiency prediction tool called *CRISPRDB* (Chen and Wang, 2022). The development of *CRISPRDB* illustrates the importance of cross-comparative studies. Continuous, independent efforts into evaluating the performances of different gRNA efficiency prediction tools using distinct experimental datasets are crucial to inform and understand the effectiveness of these tools in different use cases, as well as in directing future tool developments. As we described in this section, cutting-edge developments in learning-based gRNA efficiency prediction tools have, nevertheless, revolutionized and enabled effective design of gRNAs. However, many of these underlying models were trained with datasets in mammalian cell lines or animal models (Wang et al., 2020). How they perform in other biological systems needs to be evaluated, and retraining models may be necessary to develop gRNA efficiency predictions tailored for particular species.

11.5 Predicting gRNA on-target efficiency in plants

The integration of ML and DL with large-scale experimental gRNA evaluation has led to major advancements in not only our understanding of impactful gRNA features but has also simplified effective gRNA design through the development of computational tools capable of performance prediction. Despite this, however, these advancements were largely achieved in animal models, and their applicability in plants remains controversial (Liang et al., 2016). studied SpCas9 gRNAs that were efficient in plants, revealing GC content and secondary structures as main determinants of activity. Interestingly, positional nucleotide preferences that were shown to affect gRNA efficiency in animal cells do not apply to plant systems (Liang et al., 2016). This raises the question of whether hypothesis-driven or ML-based prediction tools developed and trained on datasets from animal systems remain applicable in plant genome editing. To our knowledge, there are two representative gRNA design tools tailored to plants, *CRISPR-PLANT* (Minkenberg et al., 2018) and *CRISPR-P* (Liu et al., 2017). Both of which have incorporated many nonmodel plant genomes; however, the former focuses on improving the prediction of potential off-target sites rather than on-target efficiency. Apart from providing a microhomology score shown to influence editing outcome and information of gRNA GC content and secondary structures formed, the updated *CRISPR-P 2.0* also integrated a hypothesis-driven scoring system for gRNA on-target efficiency (Liu et al., 2017). Comprehensive web-based gRNA design tools, *CRISPOR* and *CHOPCHOP,* have included plant genomes and are capable of alignment-based design of gRNAs while providing efficiency prediction scores from ML-based algorithms (Doench et al., 2016; Moreno-Mateos et al., 2015; Concordet and Haeussler, 2018; Labun et al., 2019). However, the adequacy of these for the prediction of gRNA activities in plants needs to be validated and further tested.

Benchmarking preexisting gRNA efficiency prediction algorithms is necessary to determine their applicability in plant genome editing and whether one tailored to a plant system needs to be developed. Using a transiently expressed, dual construct reporter for CRISPR-mediated on-target frameshift mutations in *N. benthamiana* plants (Liang et al., 2019), we first assessed the predictive performance of *CRISPR-P 2.0* and *CHOPCHOP*. However, the authors were inconclusive, as they found inconsistencies between predictions from both tools and with the experimental performance for the gRNAs tested (Liang et al., 2019). In a more systematic approach, one later study designed and assessed gRNAs targeting the transiently expressed *LUCIFERASE* reporter gene in *N. benthamiana* leaves (Naim et al., 2020). *In planta,* gRNA activity was conveniently measured as the frequency of reporter knockout, which allowed the authors to evaluate the predictive performance of several web-based computational gRNA design tools. Unfortunately, the study found that CRISPOR-integrated prediction models *Doench'2016* (Doench et al., 2016) (also used in Benchling) and *Moreno-Mateos'2015* (Moreno-Mateos et al., 2015), among several other popular tools, were all noncorrelated with the reduction of luminescence, concluding that the gRNA efficiency scoring tools assessed were unpredictable of their performance in plants (Naim et al., 2020). However, one potential downfall of both studies is the overreliance on gene knockouts over directly measuring the frequency of genome editing, as discussed earlier. In particular, Corsi et al. (2023), among others, have argued that it is unclear whether assaying reporter knockout efficiency actually captures subtle variations in genome editing frequencies between gRNAs. Therefore, the use of accurate techniques for quantifying genome edits, such as HTS, may be an overlooked but critical consideration when generating experimental datasets for evaluating and training gRNA on-target prediction algorithms (Corsi et al., 2023; Xiang et al., 2021). Indeed, a later study that quantified gRNA on-target genome editing efficiency using HTS found that groups of gRNAs with higher predicted (Doench et al., 2016) scores also mediated more efficient genome editing in tomato protoplasts (Slaman et al., 2023) and may be a useful indicator of gRNA activity *in planta*. In contrast to these results, Arndell et al. (2019) found that both the *Doench'2016* (Doench et al., 2016) and *WU-CRISPR* scores had little to no correlation with experimental efficiency of seven gRNAs, even when measured using HTS, in hexaploid wheat protoplasts. Notably, however, the latter study used much less gRNAs (seven compared to over 80 in tomato protoplasts) with their primary goal of establishing an efficient method of testing gRNA efficiencies in wheat (Arndell et al., 2019).

While the previous examples have discussed the use and comparison of these computational tools, there remains one primary limitation. Generating sufficient data to train an AI-based tool to predict plant gene editing efficiencies is far more difficult compared to bacterial, animal, or mammalian systems. High-throughput screening in plants remains underdeveloped, and past benchmarking efforts in plants have generally used small datasets (Annese et al., 2025). Plant genomes also show elevated levels of poly- and auto-polyploidy (He et al., 2024), which complicates gRNA design and raises the question of whether developed prediction tools are applicable in a plant genome context. Efforts to systematically benchmark gRNA efficiency prediction tools in plant genome editing have been limited and have mostly been a companion of larger studies. Further research is necessary to generate gRNA sets and accurately determine their experimental genome editing efficiency for a comprehensive evaluation of cutting-edge efficiency prediction models and computational tools in diverse plant species. The outcomes of these studies are critical persuasions of whether independent gRNA efficiency prediction models that are tailored for plant systems need to be developed.

11.6 Minimizing off-target effects through AI-based predictions

Off-target effects in CRISPR/Cas9 editing remain one of the most challenging barriers to achieving precise genome modification. These unwanted cleavage events arise when Cas9 binds and cuts DNA sequences that partially resemble the intended target. It is now well established that such off-target activity can result from mismatches or small InDels, which are often referred to as bulges (Fig. 11.3b) between the gRNA and target DNA. However, these are not the only factors at play. Cas9 specificity is influenced by a complex interplay of sequence features, thermodynamic stability, chromatin structure, and even the broader cellular environments. Advances in artificial intelligence, particularly in DL, have made it possible to uncover previously hidden determinants of off-target activity. These models can learn intricate relationships between the gRNA and DNA, revealing how variations in base pairing, nucleotide context, and chromatin accessibility shape Cas9 behavior.

In terms of sequence context, the number and placement of mismatches have a major impact on the risk of off-targeting. Lin et al. (2020) demonstrated with their *CRISPR-Net* model that Cas9 tolerates up to five or six mismatches when they are located far from the PAM, but cleavage activity drops sharply when mismatches occur within 10 base pairs of it. Similar trends were observed in later DL frameworks such as *CRISPR-SGRU* and *CRISPR-BERT*, which both revealed position-specific sensitivities within the guide region (Zhang et al., 2024a; Luo et al., 2024). The developers of *CRISPR-BERT*, for instance, analyzed their model for contributing sequence features. They identified several "hotspot" positions; mismatches at sites 1, 2, and 8 were often tolerated, whereas mismatches at position 14, a critical seed region, strongly inhibited cleavage (Luo et al., 2024).

The type of base substitution also matters. Transition mismatches, which substitute one purine for another or one pyrimidine for another, tend to be more tolerated than transversions that switch between purine and pyrimidine bases. This pattern reflects how chemical similarity between bases affects Cas9's ability to form a stable RNA–DNA hybrid. Models such as *CRISPR-Net* and *CRISPR-BERT* captured these effects clearly: G–A mismatches often retained activity, while C–G mismatches were strongly penalized (Luo et al., 2024; Lin et al., 2020). Other work, including *CRISPR-SGRU*, found that specific base combinations like G–C or C–G in PAM-proximal regions contribute disproportionately to off-target cleavage (Zhang et al., 2024a).

Local sequence composition adds another layer of complexity. Regions rich in GC content tend to promote stronger RNA–DNA binding, increasing the chance of cleavage. Yet when GC content becomes too high, excessive stability can prevent proper R-loop formation (Lin et al., 2020). The *CRISPR-M* model, which integrates thermodynamic features, further showed that higher melting temperature (T_m) correlates with increased off-target activity. This suggested that duplex stability is directly linked to Cas9 efficiency (Sun et al., 2024). In contrast, irregular GC patterns or secondary structures such as hairpins may cause guide misfolding, reducing hybridization and lowering cleavage probability. Importantly, mismatches rarely act independently. The effect of one mismatch often depends on neighboring bases or the presence of additional mismatches nearby. *CRISPR-Net*'s recurrent layers were able to learn this nonlinear dependency, confirming that mismatch effects are context-dependent rather than fixed for any single position.

While early prediction tools such as *Elevation* and *DeepCRISPR* could handle mismatches, they largely ignored InDels. Listgarten et al. (2018), Chuai et al. (2018), and Lin et al. (2014) introduced RNA or DNA bulges (Fig. 11.3b) to better describe InDel events, which are common in experimentally confirmed off-target sites and can substantially influence Cas9 activity. RNA bulges occur when the gRNA carries an extra nucleotide relative to the DNA, whereas DNA bulges arise when the DNA contains an extra base. However, not until 2020, *CRISPR-Net* became one of the first DL frameworks to explicitly encode and learn from bulge-containing sequence pairs (Lin et al., 2020). Incorporating bulge data improved the model's predictive accuracy: in CIRCLE-seq experiments, the area under the precision–recall curve (AUPRC) increased from 0.402 to 0.477, while GUIDE-seq data showed improvement from 0.212 to 0.254. These gains highlighted that bulge effects are essential for realistic modeling of Cas9 specificity. Moreover, the types of bulges were reported to influence binding differently. RNA bulges tend to be more tolerated, especially when they appear away from the PAM, since the gRNA structure can flexibly adjust to the distortion. DNA bulges, by contrast, disrupt the DNA backbone and hinder R-loop propagation, leading to a sharp drop in cleavage probability, particularly near the PAM-proximal region. Both *CRISPR-Net* and later studies, such as those by Yaish and Orenstein (2024), confirmed that bulge tolerance depends strongly on its position along the protospacer: PAM-proximal bulges almost never support cleavage, while mid-sequence or distal bulges can occasionally permit reduced activity.

Beyond sequence features, chromatin and epigenetic factors play an equally critical role in determining where Cas9 cuts (Lin et al., 2020). showed that integrating epigenetic data with *CRISPR-Net* significantly improved predictive power, raising the Area under the Receiver Operating Characteristic Curve (AUROC) from 0.803 to 0.981 in K562 cells. Open chromatin regions, such as those marked by CCCTC-binding factor, DNase hypersensitivity, and histone modification H3K4me3, are more accessible to Cas9 and therefore more prone to cleavage (Wang et al., 2025). More than just chromatin accessibility, cellular off-targets were reported to be enriched in chromatin states with active promoters, enhancers, and strong transcription (Lazzarotto et al., 2020). These results confirm that chromatin accessibility and histone modifications modulate Cas9 activity beyond what can be explained by sequence complementarity alone. Since epigenetic profiles show significant variation among cell types and under different experimental conditions, more studies in plants are needed to gain a deeper understanding of these patterns.

A final layer of complexity emerges when comparing in vitro and cellular systems (Yaish and Orenstein, 2024). analyzed large GUIDE-seq (in vivo) (Tsai et al., 2015) and CHANGE-seq (in vitro) (Lazzarotto et al., 2020) datasets and found that off-target sensitivity differs markedly between these environments. Cellular data revealed greater intolerance to multiple mismatches and stronger PAM-proximal sensitivity than in vitro assays. They also found that nucleotides around positions 15–17 had a greater influence on cleavage in cells than neighboring positions. Interestingly, DNA bulges near the PAM were even less tolerated in cells, while RNA bulges around position 5 showed unexpected sensitivity, likely reflecting the influence of chromatin structure or DNA repair mechanisms rather than Cas9's intrinsic biophysics. We collated a table of ML-based algorithms for analyzing the off-target effect (Table 11.2).

Collectively, these studies illustrate that off-target effects in CRISPR/Cas9 editing arise from a finely balanced combination of sequence mismatches, bulge formation, thermodynamic stability, chromatin accessibility, and cellular context. Advances in AI-driven modeling have made it possible to capture this multidimensional complexity, revealing that Cas9 specificity is not determined by sequence complementarity alone but by the broader molecular environment in which editing takes place. Understanding these factors continues to be crucial for developing safer and more precise genome-editing systems. So far, there has been limited research on AI-based off-target prediction in plants. However, such studies would be particularly valuable, as cellular context and epigenomic status can differ significantly between plants and mammals, while most existing research has been conducted in the latter.

TABLE 11.2 Collection of tools and algorithms for the prediction of off-targets.

Prediction algorithm/tool (alternative names)	Origin of training dataset	CRISPR system	Reference
CNN_std	Human cells	Cas9	Lin and Wong (2018)
Elevation	Human cells	Cas9	Listgarten et al. (2018)
AttnToMismatch_CNN	Human cells	Cas9 and Cas12a	Liu et al. (2019)
CRISPRoff	Human cells	Cas9	Alkan et al. (2018)
CRISPR-Net	Human cells	Cas9	Lin et al. (2020)
CRISPRitz	Human cells	Cas9	Cancellieri et al. (2020)
CRISPR-IP	Human cells	Cas9	Zhang and Jiang (2022)
CRISOT	Human cells	Cas9	Chen et al. (2023)
CRISPR-BERT	Human cells	Cas9	Luo et al. (2024)
CRISPR-DIPOFF	Human cells	Cas9	Toufikuzzaman et al. (2024)
CRISPR-M	Human cells	Cas9	Sun et al. (2024)
Crispr-SGRU	Human cells	Cas9	Zhang et al. (2024a)

11.7 Conclusion and future remarks

CRISPR/Cas has revolutionized the capacity for targeted genome editing due to its simplicity and programmability through the gRNA (Zhan et al., 2021). Concurrently, the importance of the gRNA and its design has been intensively studied to maximize on-target genome editing efficiency while limiting the off-target effects (Chuai et al., 2017). Cutting-edge advancements have integrated current knowledge on known contributors to the design of an effective and specific gRNA, large sets of experimental data, and ML approaches to understand and predict the performance of gRNAs (both on-target and off-target). There have been some efforts in combining ML-based on-target and off-target predictions into accessible, all-in-one tools for gRNA design, but these have mostly been limited. However, the improved accessibility of such developments is crucial in enabling an optimized and streamlined workflow for the design of gRNAs, particularly with the expansion of genomic resources. Tools for predicting gRNA performance have also been mostly focused on animal systems, and, with only a few exceptions, these tools have not been experimentally validated and evaluated systematically for their applicability in plants. There has also been very limited research into gRNA prediction and design tools that have been trained on plant-derived datasets and tailored for plant systems and genomes. Yet, this is an important avenue to improve the efficiency of the CRISPR genome editing workflow in agricultural crops, which is already restricted by the time-consuming process of tissue culture. We believe that this is critical for the widespread adoption and application of CRISPR technologies in plant biotechnology and crop genetic engineering.

Declaration

The authors declare that no generative artificial intelligence tools were used in the writing or preparation of this chapter. The content reflects the authors' own scholarly work, and the authors assume full responsibility for all statements, interpretations, and conclusions presented.

References

Alkan, F., Wenzel, A., Anthon, C., Havgaard, J.H., Gorodkin, J., 2018. CRISPR-Cas9 off-targeting assessment with nucleic acid duplex energy parameters. Genome Biol. 19 (1). https://doi.org/10.1186/s13059-018-1534-x.

Annese, D., Romani, F., Grandellis, C., Ives, L., Frangedakis, E., Buson, F.X., Molloy, J.C., Haseloff, J., 2025. Semi-automated workflow for high-throughput Agrobacterium-mediated plant transformation. Plant J. 122 (1). https://doi.org/10.1111/tpj.70118.

Anzalone, A.V., Randolph, P.B., Davis, J.R., Sousa, A.A., Koblan, L.W., Levy, J.M., Chen, P.J., Wilson, C., Newby, G. A., Raguram, A., Liu, D.R., 2019. Search-and-replace genome editing without double-strand breaks or donor DNA. Nature 576 (7785), 149–157. https://doi.org/10.1038/s41586-019-1711-4.

Arndell, T., Sharma, N., Langridge, P., Baumann, U., Watson-Haigh, N.S., Whitford, R., 2019. GRNA validation for wheat genome editing with the CRISPR-Cas9 system. BMC Biotechnol. 19 (1). https://doi.org/10.1186/s12896-019-0565-z.

Atia, M., Jiang, W., Sedeek, K., Butt, H., Mahfouz, M., 2024. Crop bioengineering via gene editing: reshaping the future of agriculture. Plant Cell Rep. 43 (4). https://doi.org/10.1007/s00299-024-03183-1.

Baisya, D., Ramesh, A., Schwartz, C., Lonardi, S., Wheeldon, I., 2022. Genome-wide functional screens enable the prediction of high activity CRISPR-Cas9 and -Cas12a guides in Yarrowia lipolytica. Nat. Commun. 13 (1). https://doi.org/10.1038/s41467-022-28540-0.

Banakar, R., Schubert, M., Collingwood, M., Vakulskas, C., Eggenberger, A.L., Wang, K., 2020. Comparison of CRISPR-Cas9/cas12a Ribonucleoprotein Complexes for genome editing efficiency in the rice Phytoene Desaturase (OsPDS) gene. Rice 13 (1). https://doi.org/10.1186/s12284-019-0365-z.

Bente, H., Mittelsten Scheid, O., Donà, M., 2020. Versatile in vitro assay to recognize Cas9-induced mutations. Plant Direct 4, e00269.

Bernabé-Orts, J.M., Casas-Rodrigo, I., Minguet, E.G., Landolfi, V., Garcia-Carpintero, V., Gianoglio, S., Vázquez-Vilar, M., Granell, A., Orzaez, D., 2019. Assessment of Cas12a-mediated gene editing efficiency in plants. Plant Biotechnol. J. 17 (10), 1971–1984. https://doi.org/10.1111/pbi.13113.

Bircheneder, M., Parniske, M., 2025. Engineering and comparison of Cas12a-based genome editing systems in plants. Plant J. 123 (5). https://doi.org/10.1111/tpj.70410.

Brooks, C., Nekrasov, V., Lippman, Z.B., Van Eck, J., 2014. Efficient gene editing in tomato in the first generation using the clustered regularly interspaced short palindromic repeats/CRISPR-associated9 system. Plant Physiol. 166 (3), 1292–1297. https://doi.org/10.1104/pp.114.247577.

Cancellieri, S., Canver, M.C., Bombieri, N., Giugno, R., Pinello, L., 2020. CRISPRitz: rapid, high-throughput and variant-aware in silico off-target site identification for CRISPR genome editing. Bioinformatics 36 (7), 2001–2008. https://doi.org/10.1093/bioinformatics/btz867.

Cardi, T., Murovec, J., Bakhsh, A., Boniecka, J., Bruegmann, T., Bull, S.E., Eeckhaut, T., Fladung, M., Galovic, V., Linkiewicz, A., Lukan, T., Mafra, I., Michalski, K., Kavas, M., Nicolia, A., Nowakowska, J., Sági, L., Sarmiento, C., Yıldırım, K., Zlatković, M., Hensel, G., Van Laere, K., 2023. CRISPR/Cas-mediated plant genome editing: outstanding challenges a decade after implementation. Trends Plant Sci. 28 (10), 1144–1165. https://doi.org/10.1016/j.tplants.2023.05.012.

Chari, R., Mali, P., Moosburner, M., Church, G.M., 2015. Unraveling CRISPR-Cas9 genome engineering parameters via a library-on-library approach. Nat. Methods 12 (9), 823–826. https://doi.org/10.1038/nmeth.3473.

Chen, Y., Wang, X., 2022. Evaluation of efficiency prediction algorithms and development of ensemble model for CRISPR/Cas9 gRNA selection. Bioinformatics 38, 5175–5181.

Chen, Q., Chuai, G., Zhang, H., Tang, J., Duan, L., Guan, H., Li, W., Li, W., Wen, J., Zuo, E., Zhang, Q., Qi, L., 2023. Genome-wide CRISPR off-target prediction and optimization using RNA-DNA interaction fingerprints. Nat. Commun. 14 (1). https://doi.org/10.1038/s41467-023-42695-4.

Cheng, M., Trusov, Y., Liu, G., Mao, Y., Botella, J.R., 2025. Establishing embryogenic tissue culture workflow for pineapple cultivar 73–50. Genes 16 (5). https://doi.org/10.3390/genes16050549.

Ching, T., Himmelstein, D.S., Beaulieu-Jones, B.K., Kalinin, A.A., Do, B.T., Way, G.P., Ferrero, E., Agapow, P.M., Zietz, M., Hoffman, M.M., Xie, W., Rosen, G.L., Lengerich, B.J., Israeli, J., Lanchantin, J., Woloszynek, S., Carpenter, A.E., Shrikumar, A., Xu, J., Cofer, E.M., Lavender, C.A., Turaga, S.C., Alexandari, A.M., Lu, Z., Harris, D.J., Decaprio, D., Qi, Y., Kundaje, A., Peng, Y., Wiley, L.K., Segler, M.H.S., Boca, S.M., Swamidass, S. J., Huang, A., Gitter, A., Greene, C.S., 2018. Opportunities and obstacles for deep learning in biology and medicine. J. R. Soc. Interface 15 (141). https://doi.org/10.1098/rsif.2017.0387.

Chuai, G.-H., Wang, Q.-L., Liu, Q., 2017. In Silico Meets In Vivo: towards Computational CRISPR-Based sgRNA Design. Trends Biotechnol. 35, 12–21.

Chuai, G., Ma, H., Yan, J., Chen, M., Hong, N., Xue, D., Zhou, C., Zhu, C., Chen, K., Duan, B., Gu, F., Qu, S., Huang, D., Jia, W., Qi, L., 2018. DeepCRISPR: optimized CRISPR guide RNA design by deep learning. Genome Biol. 19 (1). https://doi.org/10.1186/s13059-018-1459-4.

Concordet, J.P., Haeussler, M., 2018. CRISPOR: intuitive guide selection for CRISPR/Cas9 genome editing experiments and screens. Nucleic Acids Res. 46 (1), W242. https://doi.org/10.1093/nar/gky354.

Cong, L., Ran, F.A., Cox, D., Lin, S., Barretto, R., Habib, N., Hsu, P.D., Wu, X., Jiang, W., Marraffini, L.A., Zhang, F., 2013. Multiplex genome engineering using CRISPR/Cas systems. Science 339 (6121), 819–823. https://doi.org/10.1126/science.1231143.

Corsi, G.I., Qu, K., Alkan, F., Pan, X., Luo, Y., Gorodkin, J., 2022. CRISPR/Cas9 gRNA activity depends on free energy changes and on the target PAM context. Nat. Commun. 13 (1). https://doi.org/10.1038/s41467-022-30515-0.

Corsi, G.I., Anthon, C., Gorodkin, J., 2023. Testing on external independent datasets is necessary to corroborate machine learning model improvement. Bioinformatics 39, btad327.

Daer, R.M., Cutts, J.P., Brafman, D.A., Haynes, K.A., 2017. The impact of chromatin dynamics on Cas9-mediated genome editing in human cells. ACS Synth. Biol. 6 (3), 428–438. https://doi.org/10.1021/acssynbio.5b00299.

Deltcheva, E., Chylinski, K., Sharma, C.M., Gonzales, K., Chao, Y., Pirzada, Z.A., Eckert, M.R., Vogel, J., Charpentier, E., 2011. CRISPR RNA maturation by trans-encoded small RNA and host factor RNase III. Nature 471 (7340), 602–607. https://doi.org/10.1038/nature09886.

Demirer, G.S., Silva, T.N., Jackson, C.T., Thomas, J.B., Ehrhardt, D.W., Rhee, S.Y., Mortimer, J.C., Landry, M.P., 2021. Nanotechnology to advance CRISPR–Cas genetic engineering of plants. Nat. Nanotechnol. 16 (3), 243–250. https://doi.org/10.1038/s41565-021-00854-y.

DeWeirdt, P.C., McGee, A.V., Zheng, F., Nwolah, I., Hegde, M., Doench, J.G., 2022. Accounting for small variations in the tracrRNA sequence improves sgRNA activity predictions for CRISPR screening. Nat. Commun. 13 (1). https://doi.org/10.1038/s41467-022-33024-2.

Doench, J.G., Hartenian, E., Graham, D.B., Tothova, Z., Hegde, M., Smith, I., Sullender, M., Ebert, B.L., Xavier, R.J., Root, D.E., 2014. Rational design of highly active sgRNAs for CRISPR-Cas9-mediated gene inactivation. Nat. Biotechnol. 32 (12), 1262–1267. https://doi.org/10.1038/nbt.3026.

Doench, J.G., Fusi, N., Sullender, M., Hegde, M., Vaimberg, E.W., Donovan, K.F., Smith, I., Tothova, Z., Wilen, C., Orchard, R., Virgin, H.W., Listgarten, J., Root, D.E., 2016. Optimized sgRNA design to maximize activity and minimize off-target effects of CRISPR-Cas9. Nat. Biotechnol. 34 (2), 184–191. https://doi.org/10.1038/nbt.3437.

Doudna, J.A., Charpentier, E., 2014. The new frontier of genome engineering with CRISPR-Cas9. Science 346 (6213). https://doi.org/10.1126/science.1258096.

Elias, M.J., Hasley, J., Tian, M., Christopher, D.A., 2023. Development of a mesophyll protoplast-based system for gene editing of papaya. Vitro Cell Dev. Biol. Plant 59 (5), 517–535. https://doi.org/10.1007/s11627-023-10373-1.

Feng, S., Cokus, S.J., Zhang, X., Chen, P.-Y., Bostick, M., Goll, M.G., Hetzel, J., Jain, J., Strauss, S.H., Halpern, M.E., Ukomadu, C., Sadler, K.C., Pradhan, S., Pellegrini, M., Jacobsen, S.E., 2010. Conservation and divergence of methylation patterning in plants and animals. Proc. Natl. Acad. Sci. 107 (19), 8689–8694. https://doi.org/10.1073/pnas.1002720107.

Feng, Z., Zhang, B., Ding, W., Liu, X., Yang, D.L., Wei, P., Cao, F., Zhu, S., Zhang, F., Mao, Y., Zhu, J.K., 2013. Efficient genome editing in plants using a CRISPR/Cas system. Cell Res. 23 (10), 1229–1232. https://doi.org/10.1038/cr.2013.114.

Gagnon, J.A., Valen, E., Thyme, S.B., Huang, P., Ahkmetova, L., Pauli, A., Montague, T.G., Zimmerman, S., Richter, C., Schier, A.F., 2014. Efficient mutagenesis by Cas9 protein-mediated oligonucleotide insertion and large-scale assessment of single-guide RNAs. PLoS One 9 (5). https://doi.org/10.1371/journal.pone.0098186.

Gaudelli, N.M., Komor, A.C., Rees, H.A., Packer, M.S., Badran, A.H., Bryson, D.I., Liu, D.R., 2017. Programmable base editing of T to G C in genomic DNA without DNA cleavage. Nature 551 (7681), 464–471. https://doi.org/10.1038/nature24644.

Gerashchenkov, G.A., Rozhnova, N.A., Kuluev, B.R., Kiryanova, O.Y., Gumerova, G.R., Knyazev, A.V., Vershinina, Z.R., Mikhailova, E.V., Chemeris, D.A., Matniyazov, R.T., Baimiev, A.K., Gubaidullin, I.M., Baimiev, A.K., Chemeris, A.V., 2020. Design of guide RNA for CRISPR/Cas plant genome editing. Mol. Biol. 54 (1), 24–42. https://doi.org/10.1134/S0026893320010069.

Gilbert, L.A., Horlbeck, M.A., Adamson, B., Villalta, J.E., Chen, Y., Whitehead, E.H., Guimaraes, C., Panning, B., Ploegh, H.L., Bassik, M.C., Qi, L.S., Kampmann, M., Weissman, J.S., 2014. Genome-scale CRISPR-mediated control of gene repression and activation. Cell 159 (3), 647–661. https://doi.org/10.1016/j.cell.2014.09.029.

Gong, Z., Cheng, M., Botella, J.R., 2021. Non-GM genome editing approaches in crops. Frontiers in Genome Editing 3.

Gong, Z., Zhang, Y., Xia, D., Yoon, S., Crisp, P.A., Botella, J.R., 2025. Comprehensive benchmarking of genome editing quantification methods for plant applications. iScience 28 (6). https://doi.org/10.1016/j.isci.2025.112350.

Graham, N., Patil, G.B., Bubeck, D.M., Dobert, R.C., Glenn, K.C., Gutsche, A.T., Kumar, S., Lindbo, J.A., Maas, L., May, G.D., Vega-Sanchez, M.E., Stupar, R.M., Morrell, P.L., 2020. Plant genome editing and the relevance of off-target changes1[OPEN]. Plant Physiol. 183 (4), 1453–1471. https://doi.org/10.1104/pp.19.01194.

Haeussler, M., Schönig, K., Eckert, H., Eschstruth, A., Mianné, J., Renaud, J.B., Schneider-Maunoury, S., Shkumatava, A., Teboul, L., Kent, J., Joly, J.S., Concordet, J.P., 2016. Evaluation of off-target and on-target scoring algorithms and integration into the guide RNA selection tool CRISPOR. Genome Biol. 17 (1). https://doi.org/10.1186/s13059-016-1012-2.

He, B., Liu, W., Li, J., Xiong, S., Jia, J., Lin, Q., Liu, H., Cui, P., Wang, X., 2024. Evolution of plant genome size and composition. Genom. Proteom. Bioinform. 22 (5). https://doi.org/10.1093/gpbjnl/qzae078.

Heigwer, F., Kerr, G., Boutros, M., 2014. E-CRISP: fast CRISPR target site identification. Nat. Methods 11, 122–123.

Higa, L., Blank, M., Hampson, E., Matsuyama, J., Wilkes, K., Uehara, A., Bouwman, T., Lee, K., Wang, K., Muszynski, M., Du, Z.Y., 2025. Rapid assessment of CRISPR gRNAs with optimized protoplast transformation in Maize. Plant Cell Rep. 44 (10). https://doi.org/10.1007/s00299-025-03599-3.

Hiranniramol, K., Chen, Y., Liu, W., Wang, X., Luigi Martelli, P., 2020. Generalizable sgRNA design for improved CRISPR/Cas9 editing efficiency. Bioinformatics 36 (9), 2684–2689. https://doi.org/10.1093/bioinformatics/btaa041.

Hoberecht, L., Perampalam, P., Lun, A., Fortin, J.P., 2022. A comprehensive Bioconductor ecosystem for the design of CRISPR guide RNAs across nucleases and technologies. Nat. Commun. 13 (1). https://doi.org/10.1038/s41467-022-34320-7.

Horlbeck, M.A., Gilbert, L.A., Villalta, J.E., Adamson, B., Pak, R.A., Chen, Y., Fields, A.P., Park, C.Y., Corn, J.E., Kampmann, M., Weissman, J.S., 2016. Compact and highly active next-generation libraries for CRISPR-mediated gene repression and activation. eLife 5 (2016). https://doi.org/10.7554/eLife.19760.

Huang, T.K., Puchta, H., 2019. CRISPR/Cas-mediated gene targeting in plants: finally a turn for the better for homologous recombination. Plant Cell Rep. 38 (4), 443–453. https://doi.org/10.1007/s00299-019-02379-0.

Ito, Y., Inoue, S., Nakashima, T., Zhang, H., Yang, L., Kasuya, H., Matsukawa, T., Wu, Z., Yoshikawa, T., Kataoka, M., Ishikawa, T., Kagoya, Y., 2024. Epigenetic profiles guide improved CRISPR/Cas9-mediated gene knockout in human T cells. Nucleic Acids Res. 52 (1), 141–153. https://doi.org/10.1093/nar/gkad1076.

Jensen, K.T., Fløe, L., Petersen, T.S., Huang, J., Xu, F., Bolund, L., Luo, Y., Lin, L., 2017. Chromatin accessibility and guide sequence secondary structure affect CRISPR-Cas9 gene editing efficiency. FEBS (Fed. Eur. Biochem. Soc.) Lett. 591 (13), 1892–1901. https://doi.org/10.1002/1873-3468.12707.

Jiang, W., Bikard, D., Cox, D., Zhang, F., Marraffini, L.A., 2013a. RNA-guided editing of bacterial genomes using CRISPR-Cas systems. Nat. Biotechnol. 31 (3), 233–239. https://doi.org/10.1038/nbt.2508.

Jiang, W., Zhou, H., Bi, H., Fromm, M., Yang, B., Weeks, D.P., 2013b. Demonstration of CRISPR/Cas9/sgRNA-mediated targeted gene modification in Arabidopsis, tobacco, sorghum and rice. Nucleic Acids Res. 41 e188-e188.

Jin, L., Liyanage, R., Duan, D., Chen, S.J., 2024. Machine learning-inferred and energy landscape-guided analyses reveal kinetic determinants of CRISPR/Cas9 gene editing. bioRxiv. https://doi.org/10.1101/2024.04.30.591525.

Jinek, M., Chylinski, K., Fonfara, I., Hauer, M., Doudna, J.A., Charpentier, E., 2012. A programmable dual-RNA-guided DNA endonuclease in adaptive bacterial immunity. Science 337 (6096), 816–821. https://doi.org/10.1126/science.1225829.

Jinek, M., Jiang, F., Taylor, D.W., Sternberg, S.H., Kaya, E., Ma, E., Anders, C., Hauer, M., Zhou, K., Lin, S., Kaplan, M., Iavarone, A.T., Charpentier, E., Nogales, E., Doudna, J.A., 2014. Structures of Cas9 endonucleases reveal RNA-mediated conformational activation. Science 343 (6176). https://doi.org/10.1126/science.1247997.

Kaur, K., Gupta, A.K., Rajput, A., Kumar, M., 2016. Ge-CRISPR - an integrated pipeline for the prediction and analysis of sgRNAs genome editing efficiency for CRISPR/Cas system. Sci. Rep. 6. https://doi.org/10.1038/srep30870.

Kim, H.K., Min, S., Song, M., Jung, S., Choi, J.W., Kim, Y., Lee, S., Yoon, S., Kim, H., 2018. Deep learning improves prediction of CRISPR-Cpf1 guide RNA activity. Nat. Biotechnol. 36 (3), 239–241. https://doi.org/10.1038/nbt.4061.

Kim, H.K., Kim, Y., Lee, S., Min, S., Bae, J.Y., Choi, J.W., Park, J., Jung, D., Yoon, S., Kim, H.H., 2019. SpCas9 activity prediction by DeepSpCas9, a deep learning–based model with high generalization performance. Sci. Adv. 5 (11). https://doi.org/10.1126/sciadv.aax9249.

Kim, H.K., Lee, S., Kim, Y., Park, J., Min, S., Choi, J.W., Huang, T.P., Yoon, S., Liu, D.R., Kim, H.H., 2020a. High-throughput analysis of the activities of xCas9, SpCas9-NG and SpCas9 at matched and mismatched target sequences in human cells. Nat. Biomed. Eng. 4 (1), 111–124. https://doi.org/10.1038/s41551-019-0505-1.

Kim, N., Kim, H.K., Lee, S., Seo, J.H., Choi, J.W., Park, J., Min, S., Yoon, S., Cho, S.-R., Kim, H.H., 2020b. Prediction of the sequence-specific cleavage activity of Cas9 variants. Nat. Biotechnol. 38, 1328–1336.

Kim, H.K., Yu, G., Park, J., Min, S., Lee, S., Yoon, S., Kim, H.H., 2021. Predicting the efficiency of prime editing guide RNAs in human cells. Nat. Biotechnol. 39 (2), 198–206. https://doi.org/10.1038/s41587-020-0677-y.

Kim, N., Choi, S., Kim, S., Song, M., Seo, J.H., Min, S., Park, J., Cho, S.R., Kim, H.H., 2024. Deep learning models to predict the editing efficiencies and outcomes of diverse base editors. Nat. Biotechnol. 42 (3), 484–497. https://doi.org/10.1038/s41587-023-01792-x.

Koeppel, J., Weller, J., Peets, E.M., Pallaseni, A., Kuzmin, I., Raudvere, U., Peterson, H., Liberante, F.G., Parts, L., 2023. Prediction of prime editing insertion efficiencies using sequence features and DNA repair determinants. Nat. Biotechnol. 41 (10), 1446–1456. https://doi.org/10.1038/s41587-023-01678-y.

Konermann, S., Brigham, M.D., Trevino, A.E., Joung, J., Abudayyeh, O.O., Barcena, C., Hsu, P.D., Habib, N., Gootenberg, J.S., Nishimasu, H., Nureki, O., Zhang, F., 2015. Genome-scale transcriptional activation by an engineered CRISPR-Cas9 complex. Nature 517 (7536), 583–588. https://doi.org/10.1038/nature14136.

Konstantakos, V., Nentidis, A., Krithara, A., Paliouras, G., 2022a. CRISPR–Cas9 gRNA efficiency prediction: an overview of predictive tools and the role of deep learning. Nucleic Acids Res. 50, 3616–3637.

Konstantakos, V., Nentidis, A., Krithara, A., Paliouras, G., 2022b. CRISPRedict: a CRISPR-Cas9 web tool for interpretable efficiency predictions. Nucleic Acids Res. 50, W191–W198.

Koonin, E.V., Makarova, K.S., 2013. CRISPR-Cas: evolution of an RNA-based adaptive immunity system in prokaryotes. RNA Biol. 10 (5), 679–686. https://doi.org/10.4161/rna.24022.

Kuan, P.F., Powers, S., He, S., Li, K., Zhao, X., Huang, B., 2017. A systematic evaluation of nucleotide properties for CRISPR sgRNA design. BMC Bioinf. 18 (1). https://doi.org/10.1186/s12859-017-1697-6.

Kujur, S., Senthil-Kumar, M., Kumar, R., 2021. Plant viral vectors: expanding the possibilities of precise gene editing in plant genomes. Plant Cell Rep. 40, 931–934.

Labuhn, M., Adams, F.F., Ng, M., Knoess, S., Schambach, A., Charpentier, E.M., Schwarzer, A., Mateo, J.L., Klusmann, J.H., Heckl, D., 2018. Refined sgRNA efficacy prediction improves large and small-scale CRISPR-Cas9 applications. Nucleic Acids Res. 46 (3), 1375–1385. https://doi.org/10.1093/nar/gkx1268.

Labun, K., Montague, T.G., Krause, M., Torres Cleuren, Y.N., Tjeldnes, H., Valen, E., 2019. CHOPCHOP v3: expanding the CRISPR web toolbox beyond genome editing. Nucleic Acids Res. 47 (1), W171. https://doi.org/10.1093/nar/gkz365.

Lawrenson, T., Shorinola, O., Stacey, N., Li, C., Østergaard, L., Patron, N., Uauy, C., Harwood, W., 2015. Induction of targeted, heritable mutations in barley and Brassica oleracea using RNA-guided Cas9 nuclease. Genome Biol. 16 (1). https://doi.org/10.1186/s13059-015-0826-7.

Lazzarotto, C.R., Malinin, N.L., Li, Y., Zhang, R., Yang, Y., Lee, G.H., Cowley, E., He, Y., Lan, X., Jividen, K., Katta, V., Kolmakova, N.G., Petersen, C.T., Qi, Q., Strelcov, E., Maragh, S., Krenciute, G., Ma, J., Cheng, Y., Tsai, S.Q., 2020. CHANGE-seq reveals genetic and epigenetic effects on CRISPR–Cas9 genome-wide activity. Nat. Biotechnol. 38 (11), 1317–1327. https://doi.org/10.1038/s41587-020-0555-7.

Li, J.F., Norville, J.E., Aach, J., McCormack, M., Zhang, D., Bush, J., Church, G.M., Sheen, J., 2013. Multiplex and homologous recombination-mediated genome editing in Arabidopsis and Nicotiana benthamiana using guide RNA and Cas9. Nat. Biotechnol. 31 (8), 688–691. https://doi.org/10.1038/nbt.2654.

Li, B., Sun, C., Li, J., Gao, C., 2024. Targeted genome-modification tools and their advanced applications in crop breeding. Nat. Rev. Genet. 25 (9), 603–622. https://doi.org/10.1038/s41576-024-00720-2.

Liang, Z., Zhang, K., Chen, K., Gao, C., 2014. Targeted mutagenesis in Zea mays using TALENs and the CRISPR/cas system. J. Genet. Genomics 41 (2), 63–68. https://doi.org/10.1016/j.jgg.2013.12.001.

Liang, G., Zhang, H., Lou, D., Yu, D., 2016. Selection of highly efficient sgRNAs for CRISPR/Cas9-based plant genome editing. Sci. Rep. 6 (1). https://doi.org/10.1038/srep21451.

Liang, Y., Eudes, A., Yogiswara, S., Jing, B., Benites, V.T., Yamanaka, R., Cheng-Yue, C., Baidoo, E.E., Mortimer, J.C., Scheller, H.V., Loqué, D., 2019. A screening method to identify efficient sgRNAs in Arabidopsis, used in conjunction with cell-specific lignin reduction. Biotechnol. Biofuels 12 (1). https://doi.org/10.1186/s13068-019-1467-y.

Lin, J., Wong, K.C., 2018. Off-target predictions in CRISPR-Cas9 gene editing using deep learning. Bioinformatics 34 (17), i656–i663. https://doi.org/10.1093/bioinformatics/bty554.

Lin, Y., Cradick, T.J., Brown, M.T., Deshmukh, H., Ranjan, P., Sarode, N., Wile, B.M., Vertino, P.M., Stewart, F.J., Bao, G., 2014. CRISPR/Cas9 systems have off-target activity with insertions or deletions between target DNA and guide RNA sequences. Nucleic Acids Res. 42 (11), 7473–7485. https://doi.org/10.1093/nar/gku402.

Lin, J., Zhang, Z., Zhang, S., Chen, J., Wong, K.-C., 2020. CRISPR-net: a recurrent convolutional network quantifies CRISPR off-target activities with mismatches and indels. Adv. Sci. 7 (13). https://doi.org/10.1002/advs.201903562.

Listgarten, J., Weinstein, M., Kleinstiver, B.P., Sousa, A.A., Joung, J.K., Crawford, J., Gao, K., Hoang, L., Elibol, M., Doench, J.G., Fusi, N., 2018. Prediction of off-target activities for the end-to-end design of CRISPR guide RNAs. Nat. Biomed. Eng. 2 (1), 38–47. https://doi.org/10.1038/s41551-017-0178-6.

Liu, X., Homma, A., Sayadi, J., Yang, S., Ohashi, J., Takumi, T., 2016. Sequence features associated with the cleavage efficiency of CRISPR/Cas9 system. Sci. Rep. 6 (1). https://doi.org/10.1038/srep19675.

Liu, H., Ding, Y., Zhou, Y., Jin, W., Xie, K., Chen, L.L., 2017. CRISPR-P 2.0: an improved CRISPR-Cas9 tool for genome editing in plants. Mol. Plant 10 (3), 530–532. https://doi.org/10.1016/j.molp.2017.01.003.

Liu, Q., He, Di, Xie, L., Segata, N., 2019. Prediction of off-target specificity and cell-specific fitness of CRISPR-Cas System using attention boosted deep learning and network-based gene feature. PLoS Comput. Biol. 15 (10), e1007480. https://doi.org/10.1371/journal.pcbi.1007480.

Liu, G., Zhang, Y., Zhang, T., 2020. Computational approaches for effective CRISPR guide RNA design and evaluation. Comput. Struct. Biotechnol. J. 18, 35–44.

Liu, S., He, Y., Fan, T., Zhu, M., Qi, C., Ma, Y., Yang, M., Yang, L., Tang, X., Zhou, J., Zhong, Z., An, X., Qi, Y., Zhang, Y., 2025. PAM-relaxed and temperature-tolerant CRISPR-Mb3Cas12a single transcript unit systems for efficient singular and multiplexed genome editing in rice, maize, and tomato. Plant Biotechnol. J. 23 (1), 156–173. https://doi.org/10.1111/pbi.14486.

Lowder, L.G., Zhou, J., Zhang, Y., Malzahn, A., Zhong, Z., Hsieh, T.F., Voytas, D.F., Zhang, Y., Qi, Y., 2018. Robust transcriptional activation in plants using multiplexed CRISPR-Act2.0 and mTALE-act systems. Mol. Plant 11 (2), 245–256. https://doi.org/10.1016/j.molp.2017.11.010.

Luo, Y., Chen, Y., Xie, H.Z., Zhu, W., Zhang, G., 2024. Interpretable CRISPR/Cas9 off-target activities with mismatches and indels prediction using BERT. Comput. Biol. Med. 169, 107932. https://doi.org/10.1016/j.compbiomed.2024.107932.

Ma, M., Ye, A.Y., Zheng, W., Kong, L., 2013. A Guide RNA Sequence Design Platform for the CRISPR/Cas9 system for model organism genomes. BioMed Res. Int. 2013. https://doi.org/10.1155/2013/270805.

Maeder, M.L., Linder, S.J., Cascio, V.M., Fu, Y., Ho, Q.H., Joung, J.K., 2013. CRISPR RNA-guided activation of endogenous human genes. Nat. Methods 10 (10), 977–979. https://doi.org/10.1038/nmeth.2598.

Malik, A., Gul, A., Munir, F., Amir, R., Alipour, H., Babar, M.M., Bakhtiar, S.M., Paracha, R.Z., Khalid, Z., Hayat, M. Q., 2021. Evaluating the cleavage efficacy of CRISPR-Cas9 sgRNAs targeting ineffective regions of Arabidopsis thaliana genome. PeerJ 9. https://doi.org/10.7717/peerj.11409.

Mao, Y., Botella, J.R., Liu, Y., Zhu, J.K., 2019. Gene editing in plants: progress and challenges. Natl. Sci. Rev. 6 (3), 421–437. https://doi.org/10.1093/nsr/nwz005.

Marraffini, L.A., 2015. CRISPR-Cas immunity in prokaryotes. Nature 526 (7571), 55–61. https://doi.org/10.1038/nature15386.

Massel, K., Lam, Y., Wong, A.C.S., Hickey, L.T., Borrell, A.K., Godwin, I.D., 2021. Hotter, drier, CRISPR: the latest edit on climate change. Theor. Appl. Genet. 134 (6), 1691–1709. https://doi.org/10.1007/s00122-020-03764-0.

Minkenberg, B., Zhang, J., Xie, K., Yang, Y., 2018. CRISPR-PLANT v2: an online resource for highly specific guide RNA spacers based on improved off-target analysis. Plant Biotechnol. J. 17, 5.

Minkenberg, B., Zhang, J., Xie, K., Yang, Y., 2019. CRISPR - PLANT v2: an online resource for highly specific guide RNA spacers based on improved off-target analysis. Plant Biotechnol. J. 17 (1), 5–8. https://doi.org/10.1111/pbi.13025.

Montague, T.G., Cruz, J.M., Gagnon, J.A., Church, G.M., Valen, E., 2014. CHOPCHOP: a CRISPR/Cas9 and TALEN web tool for genome editing. Nucleic Acids Res. 42 (1), W401. https://doi.org/10.1093/nar/gku410.

Moreb, E.A., Lynch, M.D., 2021. Genome dependent Cas9/gRNA search time underlies sequence dependent gRNA activity. Nat. Commun. 12 (1). https://doi.org/10.1038/s41467-021-25339-3.

Moreb, E.A., Lynch, M.D., 2022. A meta-analysis of gRNA library screens enables an improved understanding of the impact of gRNA folding and structural stability on CRISPR-Cas9 activity. CRISPR J. 5 (1), 146–154. https://doi.org/10.1089/crispr.2021.0084.

Moreb, E.A., Hutmacher, M., Lynch, M.D., 2020. CRISPR-cas "non-target" sites inhibit on-target cutting rates. CRISPR J. 3, 550–561.

Moreno-Mateos, M.A., Vejnar, C.E., Beaudoin, J.D., Fernandez, J.P., Mis, E.K., Khokha, M.K., Giraldez, A.J., 2015. CRISPRscan: designing highly efficient sgRNAs for CRISPR-Cas9 targeting in vivo. Nat. Methods 12 (10), 982–988. https://doi.org/10.1038/nmeth.3543.

Nadakuduti, S.S., Starker, C.G., Ko, D.K., Jayakody, T.B., Buell, C.R., Voytas, D.F., Douches, D.S., 2019. Evaluation of methods to assess in vivo activity of engineered genome-editing nucleases in protoplasts. Front. Plant Sci. 10. https://doi.org/10.3389/fpls.2019.00110.

Naim, F., Shand, K., Hayashi, S., O'Brien, M., McGree, J., Johnson, A.A.T., Dugdale, B., Waterhouse, P.M., He, Z.-Y., 2020. Are the current gRNA ranking prediction algorithms useful for genome editing in plants? PLoS One 15 (1), e0227994. https://doi.org/10.1371/journal.pone.0227994.

Nakamura, M., Gao, Y., Dominguez, A.A., Qi, L.S., 2021. CRISPR technologies for precise epigenome editing. Nat. Cell Biol. 23 (1), 11–22. https://doi.org/10.1038/s41556-020-00620-7.

Nambiar, T.S., Baudrier, L., Billon, P., Ciccia, A., 2022. CRISPR-based genome editing through the lens of DNA repair. Mol. Cell 82 (2), 348–388. https://doi.org/10.1016/j.molcel.2021.12.026.

Nekrasov, V., Staskawicz, B., Weigel, D., Jones, J.D.G., Kamoun, S., 2013. Targeted mutagenesis in the model plant Nicotiana benthamiana using Cas9 RNA-guided endonuclease. Nat. Biotechnol. 31 (8), 691–693. https://doi.org/10.1038/nbt.2655.

Nishimasu, H., Ran, F.A., Hsu, P.D., Konermann, S., Shehata, S.I., Dohmae, N., Ishitani, R., Zhang, F., Nureki, O., 2014. Crystal structure of Cas9 in complex with guide RNA and target DNA. Cell 156 (5), 935–949. https://doi.org/10.1016/j.cell.2014.02.001.

Oh, Y., Kim, H., Kim, S.-G., 2021. Virus-induced plant genome editing. Curr. Opin. Plant Biol. 60, 101992.

O'brien, A.R., Burgio, G., Bauer, D.C., 2020. Domain-specific introduction to machine learning terminology, pitfalls and opportunities in CRISPR-based gene editing. Briefings Bioinf. 22, 308–314.

Park, J., Bae, S., Kim, J.-S., 2015. Cas-Designer: a web-based tool for choice of CRISPR-Cas9 target sites. Bioinformatics 31, 4014–4016.

Peng, H., Yi, Z., Blumenstein, M., Tao, D., Li, J., Hancock, J., 2018. CRISPR/Cas9 cleavage efficiency regression through boosting algorithms and Markov sequence profiling. Bioinformatics 34 (18), 3069–3077. https://doi.org/10.1093/bioinformatics/bty298.

Přibylová, A., Fischer, L., Pyott, D.E., Bassett, A., Molnar, A., 2022. DNA methylation can alter CRISPR/Cas9 editing frequency and DNA repair outcome in a target-specific manner. New Phytol. 235 (6), 2285–2299. https://doi.org/10.1111/nph.18212.

Rahman, M.K., Rahman, M.S., 2017. CRISPRpred: a flexible and efficient tool for sgRNAs on-target activity prediction in CRISPR/Cas9 systems. PLoS One 12 (8). https://doi.org/10.1371/journal.pone.0181943.

Ren, F., Ren, C., Zhang, Z., Duan, W., Lecourieux, D., Li, S., Liang, Z., 2019. Efficiency optimization of CRISPR/CAS9-mediated targeted mutagenesis in grape. Front. Plant Sci. 10. https://doi.org/10.3389/fpls.2019.00612.

Riesenberg, S., Helmbrecht, N., Kanis, P., Maricic, T., Pääbo, S., 2022. Improved gRNA secondary structures allow editing of target sites resistant to CRISPR-Cas9 cleavage. Nat. Commun. 13 (1). https://doi.org/10.1038/s41467-022-28137-7.

Sagarbarria, M.G.S., Caraan, J.A.M., Layos, A.J.G., 2023. Usefulness of current sgRNA design guidelines and in vitro cleavage assays for plant CRISPR/Cas genome editing: a case targeting the polyphenol oxidase gene family in eggplant (Solanum melongena L.). Transgenic Res. 32, 561–573.

Scott, T., Urak, R., Soemardy, C., Morris, K.V., 2019. Improved Cas9 activity by specific modifications of the tracrRNA. Sci. Rep. 9 (1). https://doi.org/10.1038/s41598-019-52616-5.

Shan, Q., Wang, Y., Li, J., Zhang, Y., Chen, K., Liang, Z., Zhang, K., Liu, J., Xi, J.J., Qiu, J.L., Gao, C., 2013. Targeted genome modification of crop plants using a CRISPR-Cas system. Nat. Biotechnol. 31 (8), 686–688. https://doi.org/10.1038/nbt.2650.

Shan, S., Soltis, P.S., Soltis, D.E., Yang, B., 2020. Considerations in adapting CRISPR/Cas9 in nongenetic model plant systems. Appl. Plant Sci. 8 (1). https://doi.org/10.1002/aps3.11314.

Shao, J., Peng, B., Zhang, Y., Yan, X., Yao, X., Hu, X., Li, L., Fu, X., Han, Z., Tang, K., 2024. A high-efficient protoplast transient system for screening gene editing elements in Salvia miltiorrhiza. Plant Cell Rep. 43 (2). https://doi.org/10.1007/s00299-023-03134-2.

Singh, D., Sternberg, S.H., Fei, J., Doudna, J.A., Ha, T., 2016. Real-time observation of DNA recognition and rejection by the RNA-guided endonuclease Cas9. Nat. Commun. 7. https://doi.org/10.1038/ncomms12778.

Slaman, E., Lammers, M., Angenent, G.C., de Maagd, R.A., 2023. High-throughput sgRNA testing reveals rules for Cas9 specificity and DNA repair in tomato cells. Frontiers in Genome Editing 5. https://doi.org/10.3389/fgeed.2023.1196763.

Slaman, E., Kottenhagen, L., de Martines, W., Angenent, G.C., de Maagd, R.A., 2024. Comparison of Cas12a and Cas9-mediated mutagenesis in tomato cells. Sci. Rep. 14 (1). https://doi.org/10.1038/s41598-024-55088-4.

Sternberg, S.H., Redding, S., Jinek, M., Greene, E.C., Doudna, J.A., 2014. DNA interrogation by the CRISPR RNA-guided endonuclease Cas9. Nature 507 (7490), 62–67. https://doi.org/10.1038/nature13011.

Sun, J., Guo, J., Liu, J., Sinha, S., Crispr, M., 2024. Predicting sgRNA off-target effect using a multi-view deep learning network. PLoS Comput. Biol. 20 (3), e1011972. https://doi.org/10.1371/journal.pcbi.1011972.

Thyme, S.B., Akhmetova, L., Montague, T.G., Valen, E., Schier, A.F., 2016. Internal guide RNA interactions interfere with Cas9-mediated cleavage. Nat. Commun. 7. https://doi.org/10.1038/ncomms11750.

Toufikuzzaman, M., Hassan Samee, M.A., Sohel Rahman, M., 2024. CRISPR-DIPOFF: an interpretable deep learning approach for CRISPR Cas-9 off-target prediction. Briefings Bioinf. 25, bbad530.

Tsai, S.Q., Zheng, Z., Nguyen, N.T., Liebers, M., Topkar, V.V., Thapar, V., Wyvekens, N., Khayter, C., Iafrate, A.J., Le, L.P., Aryee, M.J., Joung, J.K., 2015. GUIDE-seq enables genome-wide profiling of off-target cleavage by CRISPR-Cas nucleases. Nat. Biotechnol. 33 (2), 187–198. https://doi.org/10.1038/nbt.3117.

Uusi-Mäkelä, M.I.E., Barker, H.R., Bäuerlein, C.A., Häkkinen, T., Nykter, M., Rämet, M., 2018. Chromatin accessibility is associated with CRISPR-Cas9 efficiency in the zebrafish (Danio rerio). PLoS One 13 (4). https://doi.org/10.1371/journal.pone.0196238.

Wang, L., Han, H., 2024. Strategies for improving the genome-editing efficiency of class 2 CRISPR/Cas system. Heliyon 10 (19), e38588. https://doi.org/10.1016/j.heliyon.2024.e38588.

Wang, T., Wei, J.J., Sabatini, D.M., Lander, E.S., 2014. Genetic screens in human cells using the CRISPR-Cas9 system. Science 343 (6166), 80–84. https://doi.org/10.1126/science.1246981.

Wang, D., Zhang, C., Wang, B., Li, B., Wang, Q., Liu, D., Wang, H., Zhou, Y., Shi, L., Lan, F., Wang, Y., 2019. Optimized CRISPR guide RNA design for two high-fidelity Cas9 variants by deep learning. Nat. Commun. 10 (1). https://doi.org/10.1038/s41467-019-12281-8.

Wang, J., Zhang, X., Cheng, L., Luo, Y., 2020. An overview and metanalysis of machine and deep learning-based CRISPR gRNA design tools. RNA Biol. 17 (1), 13–22. https://doi.org/10.1080/15476286.2019.1669406.

Wang, Z., Shea, Z., Qi, L., Wang, K., Mills, K., Zhang, B., Zhao, B., 2023. Evaluate the guide RNA effectiveness via Agrobacterium-mediated transient assays in Nicotiana benthamiana. Front. Plant Sci. 14. https://doi.org/10.3389/fpls.2023.1111683.

Wang, G., Liu, X., Wang, A., Wen, J., Kim, P., Song, Q., Liu, X., Zhou, X., 2025. CRISPRoffT: comprehensive database of CRISPR/Cas off-targets. Nucleic Acids Res. 53 (D1), D914. https://doi.org/10.1093/nar/gkae1025.

Weiss, T., Crisp, P.A., Rai, K.M., Song, M., Springer, N.M., Zhang, F., 2022. Epigenetic features drastically impact CRISPR–Cas9 efficacy in plants. Plant Physiol. 190 (2), 1153–1164. https://doi.org/10.1093/plphys/kiac285.

Wessels, H.H., Méndez-Mancilla, A., Guo, X., Legut, M., Daniloski, Z., Sanjana, N.E., 2020. Massively parallel Cas13 screens reveal principles for guide RNA design. Nat. Biotechnol. 38 (6), 722–727. https://doi.org/10.1038/s41587-020-0456-9.

Wilson, L.O.W., Reti, D., O'Brien, A.R., Dunne, R.A., Bauer, D.C., 2018. High activity target-site identification using phenotypic independent CRISPR-Cas9 core functionality. CRISPR J. 1 (2), 182–190. https://doi.org/10.1089/crispr.2017.0021.

Wong, N., Liu, W., Wang, X., 2015. Wu-CRISPR: characteristics of functional guide RNAs for the CRISPR/Cas9 system. Genome Biol. 16, 218.

Wu, X., Scott, D.A., Kriz, A.J., Chiu, A.C., Hsu, P.D., Dadon, D.B., Cheng, A.W., Trevino, A.E., Konermann, S., Chen, S., Jaenisch, R., Zhang, F., Sharp, P.A., 2014. Genome-wide binding of the CRISPR endonuclease Cas9 in mammalian cells. Nat. Biotechnol. 32 (7), 670–676. https://doi.org/10.1038/nbt.2889.

Xiang, X., Corsi, G.I., Anthon, C., Qu, K., Pan, X., Liang, X., Han, P., Dong, Z., Liu, L., Zhong, J., Ma, T., Wang, J., Zhang, X., Jiang, H., Xu, F., Liu, X., Xu, X., Wang, J., Yang, H., Bolund, L., Church, G.M., Lin, L., Gorodkin, J., Luo, Y., 2021. Enhancing CRISPR-Cas9 gRNA efficiency prediction by data integration and deep learning. Nat. Commun. 12 (1). https://doi.org/10.1038/s41467-021-23576-0.

Xie, S., Shen, B., Zhang, C., Huang, X., Zhang, Y., Khodursky, A.B., 2014. sgRNAcas9: a software package for designing CRISPR sgRNA and evaluating potential off-target cleavage sites. PLoS One 9 (6), e100448. https://doi.org/10.1371/journal.pone.0100448.

Xie, K., Minkenberg, B., Yang, Y., 2015. Boosting CRISPR/Cas9 multiplex editing capability with the endogenous tRNA-processing system. Proc. Natl. Acad. Sci. 112, 3570–3575.

Xu, H., Xiao, T., Chen, C.H., Li, W., Meyer, C.A., Wu, Q., Wu, D., Cong, L., Zhang, F., Liu, J.S., Brown, M., Liu, X.S., 2015. Sequence determinants of improved CRISPR sgRNA design. Genome Res. 25 (8), 1147–1157. https://doi.org/10.1101/gr.191452.115.

Xu, D., Besselink, S., Ramadoss, G.N., Dierks, P.H., Lubin, J.P., Pattali, R.K., Brim, J.I., Christenson, A.E., Colias, P.J., Ornelas, I.J., Nguyen, C.D., Chasins, S.E., Conklin, B.R., Nuñez, J.K., 2025. Programmable epigenome editing by transient delivery of CRISPR epigenome editor ribonucleoproteins. Nat. Commun. 16 (1). https://doi.org/10.1038/s41467-025-63167-x.

Yaish, O., Orenstein, Y., 2024. Generating, modeling and evaluating a large-scale set of CRISPR/Cas9 off-target sites with bulges. Nucleic Acids Res. 52 (12), 6777–6790. https://doi.org/10.1093/nar/gkae428.

Yan, J., Chuai, G., Zhou, C., Zhu, C., Yang, J., Zhang, C., Gu, F., Xu, H., Wei, J., Liu, Q., 2017. Benchmarking CRISPR on-target sgRNA design. Briefings Bioinf. 19, 721–724.

Yang, Q., Wu, L., Meng, J., Ma, L., Zuo, E., Sun, Y., EpiCas, D.L., 2023. Predicting sgRNA activity for CRISPR-mediated epigenome editing by deep learning. Comput. Struct. Biotechnol. J. 21, 202–211. https://doi.org/10.1016/j.csbj.2022.11.034.

Yuan, H., Song, C., Xu, H., Sun, Y., Anthon, C., Bolund, L., Lin, L., Benabdellah, K., Lee, C., Hou, Y., Gorodkin, J., Luo, Y., 2025. An overview and comparative analysis of CRISPR-SpCas9 gRNA activity prediction tools. CRISPR J. 8 (2), 89–104. https://doi.org/10.1089/crispr.2024.0058.

Zetsche, B., Gootenberg, J.S., Abudayyeh, O.O., Slaymaker, I.M., Makarova, K.S., Essletzbichler, P., Volz, S.E., Joung, J., van der Oost, J., Regev, A., Koonin, E.V., Zhang, F., 2015. Cpf1 is a single RNA-guided endonuclease of a class 2 CRISPR-Cas system. Cell 163 (3), 759–771. https://doi.org/10.1016/j.cell.2015.09.038.

Zhan, X., Lu, Y., Zhu, J.K., Botella, J.R., 2021. Genome editing for plant research and crop improvement. J. Integr. Plant Biol. 63 (1), 3–33. https://doi.org/10.1111/jipb.13063.

Zhang, Z.R., Jiang, Z.R., 2022. Effective use of sequence information to predict CRISPR-Cas9 off-target. Comput. Struct. Biotechnol. J. 20, 650–661. https://doi.org/10.1016/j.csbj.2022.01.006.

Zhang, H., Zhang, J., Wei, P., Zhang, B., Gou, F., Feng, Z., Mao, Y., Yang, L., Zhang, H., Xu, N., Zhu, J.K., 2014. The CRISPR/Cas9 system produces specific and homozygous targeted gene editing in rice in one generation. Plant Biotechnol. J. 12 (6), 797–807. https://doi.org/10.1111/pbi.12200.

Zhang, Y., Malzahn, A.A., Sretenovic, S., Qi, Y., 2019. The emerging and uncultivated potential of CRISPR technology in plant science. Nat. Plants 5 (8), 778–794. https://doi.org/10.1038/s41477-019-0461-5.

Zhang, G., Dai, Z., Dai, X., 2020a. C-RNNCrispr: prediction of CRISPR/Cas9 sgRNA activity using convolutional and recurrent neural networks. Comput. Struct. Biotechnol. J. 18, 344–354.

Zhang, N., Roberts, H.M., Van Eck, J., Martin, G.B., 2020b. Generation and molecular characterization of CRISPR/Cas9-Induced mutations in 63 immunity-associated genes in tomato reveals specificity and a range of gene modifications. Front. Plant Sci. 11. https://doi.org/10.3389/fpls.2020.00010.

Zhang, G., Luo, Y., Dai, X., Dai, Z., 2023a. Benchmarking deep learning methods for predicting CRISPR/Cas9 sgRNA on- and off-target activities. Briefings Bioinf. 24, bbad333.

Zhang, H., Yan, J., Lu, Z., Zhou, Y., Zhang, Q., Cui, T., Li, Y., Chen, H., Ma, L., 2023b. Deep sampling of gRNA in the human genome and deep-learning-informed prediction of gRNA activities. Cell Discov. 9, 48.

Zhang, G., Luo, Y., Xie, H., Dai, Z., 2024a. CRISPR-SGRU: prediction of CRISPR/Cas9 off-target activities with mismatches and indels using stacked BiGRU. Int. J. Mol. Sci. 25, 10945.

Zhang, J., Guan, X., Moon, J., Zhang, S., Jia, Z., Yang, R., Hou, C., Guo, C., Pei, M., Liu, C., 2024b. Interpreting CRISPR-Cas12a enzyme kinetics through free energy change of nucleic acids. Nucleic Acids Res. 52, 14077–14092.

Zhang, Y., Cheng, M., Massel, K., Godwin, I.D., Liu, G., 2025. An accelerated transgene-free genome editing system using microparticle bombardment of sorghum immature embryos. aBIOTECH 6 (2), 202–214. https://doi.org/10.1007/s42994-025-00204-9.

Zhao, H., Wolt, J.D., 2017. Risk associated with off-target plant genome editing and methods for its limitation. Emerg. Top. Life Sci. 1 (2), 231–240. https://doi.org/10.1042/ETLS20170037.

CHAPTER

12

AI for improved prediction of CRISPR repair outcomes

Kameshwaran Senthil, Manisha Parashar, Manish Ghumnani, and Rohini Sreevathsa
ICAR-National Institute for Plant Biotechnology, New Delhi, India

12.1 Introduction to gene editing and the CRISPR–Cas system

Plant gene editing exhibits a transformative approach in the field of agricultural biotechnology, enabling precise modifications to enhance and enrich crop traits for high yield and stress resilience. In past centuries, chemical and irradiation-based techniques were introduced, which produced random mutagenesis in the genome (Ma et al., 2021). However, these methods have many limitations, like generation of nonspecific mutants, mutation in a large number of nucleotides, and sometimes even deletions, duplications, and relocation of genomic fragments. In addition, identification of mutations of interest is a lengthy and laborious process. Still, random mutagenesis methods are less effective for trait improvement in polyploid crops due to their extensive genetic redundancy (May et al., 2023). Hence, sequence-specific mutagenesis techniques emerged a few decades ago, such as the meganucleases, zinc finger nucleases (ZFNs), transcription activator-like effector nucleases (TALENs), and type II clustered regularly interspaced short palindromic repeat (CRISPR)/CRISPR-associated protein 9 (Cas9) (Bhagtaney and Sundarrajan, 2023), which has paved the way for targeted gene editing in plant genomes. The concept of gene editing, particularly through the CRISPR–Cas9 system, has completely revolutionized the world of plant biology since its adaptation for the eukaryotic genome. CRISPR–Cas9 is the modern-era genome editing technology, which became more well-liked than other genome editing approaches because of its simplicity, specificity, and versatility. CRISPR refers to sequence repeats separated by short stretches of variable regions in the genome that were first identified in prokaryotes (Sayyed and Mujumdar, 2025). Nearly all the archaea and most bacterial systems carry CRISPR as an adaptive immunity against invading extranucleic acid or host–virus interaction (Laub and Typas, 2024). Cas9 is a CRISPR-associated endonuclease protein that induces double-strand cleavage in the targeted genome by two protein domains,

Intelligent Gene Editing in Plants
https://doi.org/10.1016/B978-0-443-44079-3.00014-0

namely HNH (His-Asn-His) and RuvC RNaseH-like endonuclease, that bind to both strands of target DNA, respectively (Sternberg et al., 2015). Among all the types of CRISPR reported, the CRISPR–Cas9 (SpCas9) system has gained more attention. This was derived from *Streptococcus pyogenes*, which consists of a multidomain single protein, Cas9, and two small RNA molecules, crRNA (CRISPR RNA) and tracrRNA (trans-activating crRNA), to guide the Cas9 protein and interfere with the machinery through its protospacer targets (Zhang et al., 2023).

12.1.1 Mechanisms of gene editing and CRISPR

The basic mechanism of how gene editing works involves the recognition of the target-specific DNA sequences; precise cleavage at sites with nucleases, enabling the natural DNA repair mechanism to fix the breaks; and leading to a localized modification of genetic material. The molecularly manipulated CRISPR/Cas9 systems consist of a short 20-nucleotide DNA-based RNA template molecule called single guide RNA (sgRNA), a fusion of crRNA and tracrRNA. Cas9 is directed to the specific genomic site by sgRNA through its targeted complementary site, where it immediately harbors the protospacer adjacent motif (PAM), with the sequence of 5′-NGG-3′. From 3 bp upstream of PAM, Cas9 introduces a double-stranded break by HNH and RuvC binding domains (Mengstie and Wondimu, 2021). The double-strand break (DSB) is repaired by the cellular machinery through nonhomologous end joining (NHEJ), homologous-directed repair (HDR), or microhomology-mediated end joining (MMEJ). The deletion or insertion mutation at DSB is attainable by activating the error-prone NHEJ pathway. HDR requires a homologous DNA template near the DSB, which comes under the insertion-based genome editing method. Well-known as an alternative NHEJ, the MMEJ is an error-prone DSB repair mechanism that utilizes short homologous sequences to align and join the broken ends, often resulting in deletions or insertions at the junctions (Sung and Klein, 2006; Chang et al., 2017; Sfeir and Symington, 2015).

12.1.2 Importance of tracking footprints of Cas9 protein

"Cas9 footprints tracking system" refers to the pattern of checking the functional efficiency of Cas9, used in CRISPR gene editing, in its role of binding and cleaving the DNA at the target site. This ensures the editing is performed with high accuracy and safety on the part of both expected and unexpected outcomes. The importance of tracking Cas9 footprints relies on its critical role in successful gene editing. The suggestions for tracking these footprints are most mentioned by researchers for several reasons, such as verification of intended edits, detection of off-target edits, improvement of CRISPR–Cas9 specificity, and understanding Cas9 behavior. The process of tracking ensures that the desired genetic modification has been made at the intended target-specific site. Off-target effects, where Cas9 might cleave DNA at unplotted sites, induce a significant risk to safety, potentially leading to mutations. The tracking system helps identify and knock off these off-targets, enabling the researchers to make a safety edit (Lin and Luo, 2019). Refining the design of sgRNA to enhance its specificity, resulting in the reduction in the likelihood of off-target effects, is achieved by analyzing the footprints of Cas9 protein (Manghwar et al., 2020). The most important factor that is to be noted is how these tracking systems provide insights into the interaction between Cas9 and the plant genome, involving its binding and cleavage activities. This involves the study or knowledge of DNA-binding footprints, where the protein binds without cleavage, and cleavage footprints, where indels are established (Boyle et al., 2021). There are various methods to track the protein, like T7 endonuclease I assay, surveyor assay, Sanger sequencing, next-generation sequencing, chromatin immunoprecipitation, bisulfite sequencing, and CRISPR-specific off-target detection methods.

12.1.3 Challenges in predicting repair outcomes (off-targets, binding of Cas at off-target sites, recognition of noncanonical PAM sites)

Predicting the repair outcomes in CRISPR gene editing requires a basic understanding of where the Cas9 binds and cleaves the DNA, and what changes occur after this entire process. Challenges arise from off-target effects, binding of Cas at off-targets, and recognizing noncanonical PAM sites, which definitely lead the way to unexpected mutations. There are cases where Cas9 can bind to DNA without cutting, particularly using catalytically inactive Cas9 and with the notable mismatches in the gRNA. This can even influence epigenetic changes, greatly affecting gene expression, adding another layer to the prediction of repair outcomes. An unusual recognition of PAM sites, well known as "noncanonical PAMs," is also observed in a few reports, where "NAG" is considered instead of "NGG." This increases the risk of high off-target effects, making it way harder to ensure the precision of gene edits. The applications for predictions are vast and becoming more technologically advanced. Being a multifaceted challenge, the current science requires well-refined prediction models and experimental methods, ensuring the safety and efficacy of CRISPR applications in sustainable agriculture (Zhang et al., 2015).

12.2 Mechanisms of DSB repair

DSB repair mechanism occurs via two main pathways: homologous recombination (HR) and NHEJ. HR, an error-free repair process, uses a sister chromatid as a template to precisely fix the break, making it functional in the S/G2 phase of the cell cycle. NHEJ, a faster but error-prone method, directly joins the broken DNA ends without any template, still functioning throughout the cell cycle. Alternative pathways like MMEJ also contribute effectively to genome stability (Xue and Greene, 2021) (Fig. 12.1).

12.2.1 Nonhomologous end joining

This is a fundamental DNA repair pathway, which plays a critical role in maintaining genome integrity by fixing DSBs in eukaryotic cells. First coined as a term in 1996 by Moore and Haber, NHEJ is a mending mechanism that directly ligates the ends of the DSBs without the need for a homologous template, in contrast to HDR, which completely relies on a homologous sequence for guidance. This mechanism is active throughout the cell cycle, making it the more predominant DSB repair pathway in nondividing cells. Its error-prone nature often leads to insertions or deletions, which makes it a cornerstone in genome editing, where

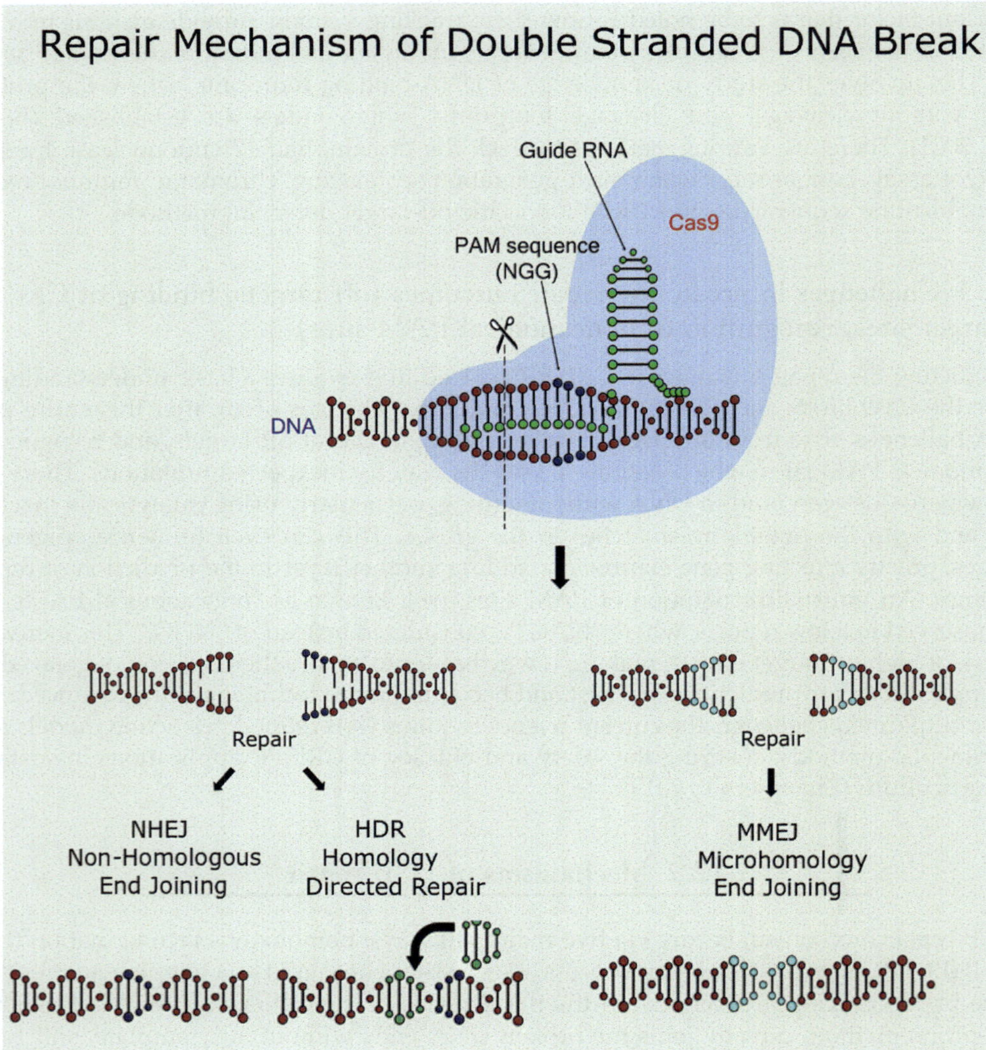

FIGURE 12.1 Various mechanisms of double-stranded DNA break repair during CRISPR/Cas9 gene editing.

it is used to introduce targeted mutations for gene knockouts. The process begins with the recognition of DSBs by the Ku70–Ku80 heterodimer, which binds to the broken DNA ends to prevent degradation and facilitate alignment. This is followed by the recruitment of the DNA-dependent protein kinase catalytic subunit in some organisms, though its role in plants is still less clear.

After the ends are prepared for ligation, the end processing starts, involving the resection and filling of gaps by specific nucleases and DNA polymerases. This is the stage that can often lead to a loss or a gain of nucleotides, contributing to the error-prone nature of the repair. The final step involves DNA ligase IV, in complex with XRCC4, joining the processed ends (Vogt and He, 2023).

12.2.2 Homology-directed repair

HDR in plants is a vital mechanism for precise genome editing, enabling targeted changes like gene replacements or insertions. However, it's less significant compared to NHEJ, presenting challenges for practical applications. HDR uses a homologous donor template to repair DSBs, involving steps like DSB recognition, strand invasion, DNA synthesis, and resolution. Submechanisms include synthesis-dependent strand annealing (SDSA) for small edits and DSB repair for more complex outcomes, active mainly in the S and G2 phases of the cell cycle. The process of repair starts upon the creation of a DSB; the DNA ends are recognized and processed by nucleases to generate 3′ single-stranded overhangs. These overhangs invade the homologous donor template, forming a D-loop structure. This step requires the presence of a sufficient donor repair template (DRT) for initiating HDR and is influenced by the cell's DNA repair machinery. The invading strand uses the DRT to synthesize new DNA, extending from the break site. This synthesis is error-free, ensuring fidelity, and is particularly important for precise edits like single nucleotide polymorphisms (SNPs) or insertions. The repair structure is resolved through various submechanisms, each with unique outcomes, such as single-strand annealing (SSA), SDSA, and DSB repair (Liao et al., 2024).

12.2.3 Microhomology-mediated end joining

MMEJ is a way plants repair breaks in their DNA using short matching sequences, often leading to deletions. Also known as alternative NHEJ, this is an error-prone DNA repair pathway that utilizes short sequences of homology to repair DSBs in plant cells. An interesting aspect is that MMEJ in plants often involves the addition of larger segments of DNA, called filler DNA, which can come from the plant's own genome or foreign DNA, making repair outcomes more complex than in other organisms. This process begins with the recognition of DSBs, followed by end resection mediated by nucleases like MRE11, creating single-stranded overhangs. This resection is crucial for exposing microhomologous sequences and is often triggered when classical NHEJ is inhibited, such as in Ku70 or Ku80 mutants. The single-stranded overhangs anneal at regions of microhomology, which are short complementary sequences. Following annealing, overhanging bases (flaps) are removed by nucleases like *Fen1*, and gaps are filled in by DNA polymerase, such as DNA polymerase θ in polymerase theta-mediated end joining (TMEJ), a specialized form of MMEJ. The ends are then ligated, often leading to deletions of the intervening sequence (Van Vu et al., 2021). In plants, NHEJ is the dominant DSB repair pathway, but MMEJ plays a significant role in specific contexts, particularly when classical NHEJ is compromised. The prevalence of MMEJ in plants is indeed less explored compared to HDR and NHEJ, but it is well defined that it can be significantly considered as a good backup mechanism.

12.3 Factors influencing DSB repair

DSBs are critical forms of DNA damage in plants, induced by environmental stresses, oxidative damage, or genome editing tools. Plants have evolved multiple repair pathways, including NHEJ, HDR, and MMEJ, each influenced by various intrinsic and extrinsic factors, which are relevant given the unique challenges of plant systems, such as large genomes and cell wall barriers.

Gene editing in plants, predominantly using CRISPR/Cas9, involves creating targeted DSBs, which are then repaired by the plant's DNA repair machinery. The choice and efficiency of these pathways are influenced by various factors, impacting the success of gene editing applications like crop improvement and stress resilience. The major factors that influence DSB repair are cell cycle stage, chromatin structure, genetic background, tissue type and developmental stage, presence of homologous templates, environmental conditions and plant hormones, genome architecture, the type of break, and in some cases, the polyploidy impact. During the cell cycle, S and G2 phases favor the HDR, whereas the rest of the phases support the NHEJ mechanism (Gutschner et al., 2016). The more tightly packed DNA, the more difficult it is to repair—this depicts the influence of chromatin structure (Gutschner et al., 2016). If a plant has mutations in its repair genes, then it faces a struggle to choose the right repair method for itself. The younger the stage, the better the DNA repair is, which leads to precise edits. The presence of a homologous template most likely results in the favor of HDR (Zhang et al., 2017). The stresses and plant hormones act antagonistically to maintain and enhance the repair mechanism, like any biotic or abiotic stress disturbs the repair, whereas plant hormones come in handy for a huge help and support (Garm et al., 2013). The complexity and gene duplications in the plant genome make it more vulnerable to mistakes during repair. Even the way the DNA has been cut changes the preference of choice in the repair mechanism, as with HDR or NHEJ. The most unexpected and interesting aspect is that plants with multiple sets of chromosomes, like wheat, might have more templates for HDR, but this can also lead to more repair errors, adding complexity to gene editing outcomes, especially to polyploid crops (Puchta, 2005).

12.4 CRISPR-based DNA modifications

12.4.1 Gene knockouts

CRISPR–Cas9 technology allows scientists to edit plant DNA by knocking out specific genes, which means disabling them to study their function or improve traits like disease resistance. This method has become a game-changer for plant breeding, where it helps create plants better suited to challenges like pests and climate stress. There are significant advancements made in CRISPR-based gene knockout across various plant species, particularly in staple and perennial crops. For example, in rice, knocking out the *OsHPP04* gene has reduced yield losses from root-knot nematodes (Huang et al., 2023), while in wheat, the *MLO* gene knockout has made plants more resistant to powdery mildew (Li et al., 2022). Unexpectedly, researchers have also used CRISPR to create seedless blackberries and non-browning avocados, expanding its applications beyond traditional crops. The future of CRISPR-based gene knockout in plants lies in addressing the challenges through improved delivery systems, enhanced specificity, and streamlined regulatory processes.

12.4.2 DNA insertions

CRISPR-based DNA insertions in plants are a promising biotechnological tool for improving crop traits. DNA insertions involve adding specific DNA sequences to achieve desired traits such as improved yield, disease resistance, and altered growth characteristics. This is particularly crucial, given the global food security challenges driven by population growth and climate change. The application of CRISPR in plants is now indispensable, with early studies focusing on important global crops. However, the process involves technical challenges, regulatory considerations, and ethical debates to be pondered. Two primary methods have emerged for achieving CRISPR-based DNA insertions in plants, each with distinct mechanisms and efficiencies, such as traditional CRISPR/Cas9 with HDR and prime editing (Schenke, 2020). The success of CRISPR-based DNA insertions depends heavily on delivery methods, which can influence whether the resulting plants are classified as transgenic (genetically modified organisms [GMOs]) or genome-edited. This technology, while transformative, requires careful consideration of biosafety, ethics, and global cooperation to realize its full benefits (Ahmad et al., 2023).

12.4.3 Gene replacement

CRISPR-based gene replacement in plants involves using advanced editing tools to swap out an existing gene with a new one, helping improve crops for better yields or resistance to diseases and herbicides. The main methods that are involved for gene replacement in plants are CRISPR/Cas9 with HDR, prime editing, and base editing (Li et al., 2017; Huang and Liu, 2023). The use of site-specific recombinases, combined with CRISPR, can improve targeted gene replacement efficiency. Despite its potential, CRISPR-based gene replacement faces several challenges: low HDR efficiency, off-target mutations, transgene integration, and immune responses. Gene replacement using CRISPR holds immense potential for plant improvement, but technical challenges need further optimization. Future research in prime editing and recombinase-based systems could enhance gene replacement strategies, making crop enhancement more precise and efficient.

12.5 Beyond gene editing

12.5.1 Gene regulation

CRISPR/Cas9, originally known for its gene-editing capabilities, has evolved to include gene regulation techniques that modulate gene expression without introducing permanent DNA changes. This is particularly important in plant biotechnology, where fine-tuning gene activity can enhance traits like yield, stress tolerance, and disease resistance without the regulatory hurdles associated with GMOs.

Gene regulation involves methods such as CRISPR interference (CRISPRi), CRISPR activation (CRISPRa), and novel approaches like guide-induced gene silencing (GIGS), each offering unique ways to control when, where, and how much a gene is expressed. These techniques are crucial for addressing global food security challenges, especially given the predicted population growth to 10 billion by 2050 and the need for sustainable agriculture amidst climate change. An unexpected detail is the discovery of GIGS, where guide RNAs alone can induce gene silencing, offering a compact and potentially nontransgenic approach. This could revolutionize plant regulation by reducing reliance on Cas proteins, but its regulatory status remains unclear, adding to global debates (Dong and Fan, 2024).

12.5.2 Genome imaging

Genome imaging in plants is an emerging and critical field within plant biology, focusing on the visualization and analysis of the functional aspects of the genome, particularly gene expression and protein dynamics, within living or fixed plant cells. Unlike genome sequencing, which provides the static sequence of DNA, genome imaging offers a dynamic view of how genes are expressed and how their products (proteins) function, which is essential for functional genomics. This is particularly important for understanding plant responses to environmental stresses and for applications in crop improvement. The field has gained attention due to the need to bridge the gap between genomic data and phenotypic outcomes, especially as plant genomic technologies have advanced rapidly. Imaging techniques are used to study complex traits related to growth, yield, and adaptation to biotic or abiotic stresses, such as disease, insects, drought, and salinity. Key techniques that are central to genome imaging are confocal microscopy and SNAP-tag technology. While challenges like autofluorescence and cell wall barriers exist, ongoing research continues to push the boundaries, promising enhanced crop resilience and a deeper understanding of plant genomes (Sadoine et al., 2021).

12.5.3 Base editing

Base editing is a genome editing technology that facilitates the targeted conversion of a single DNA base (adenine [A], thymine [T], cytosine [C], or guanine [G]) to another without inducing double-stranded DNA breaks (Fig. 12.2). This is typically achieved using CRISPR/Cas9-based systems, where a deactivated Cas9 (dCas9) or Cas9 nickases (nCas9) is fused with a base-modifying enzyme, such as cytosine deaminase for cytidine base editors (CBEs) or adenosine deaminase for adenine base editors (ABEs). These systems are guided by RNA to specific genomic loci, enabling precise single-nucleotide edits without the need for a DRT, which distinguishes them from HDR approaches (Komor et al., 2016). This method has gained traction due to its precision, simplicity, and ability to perform multiplex editing, making it a powerful tool for both functional genomics and crop improvement. This approach offers a promising avenue for precise genetic modification, with applications ranging from stress resistance to phenotypic alterations, addressing critical agricultural needs. While challenges such as PAM limitations, off-target effects, and delivery issues persist, the research community is actively developing solutions with significant advancements.

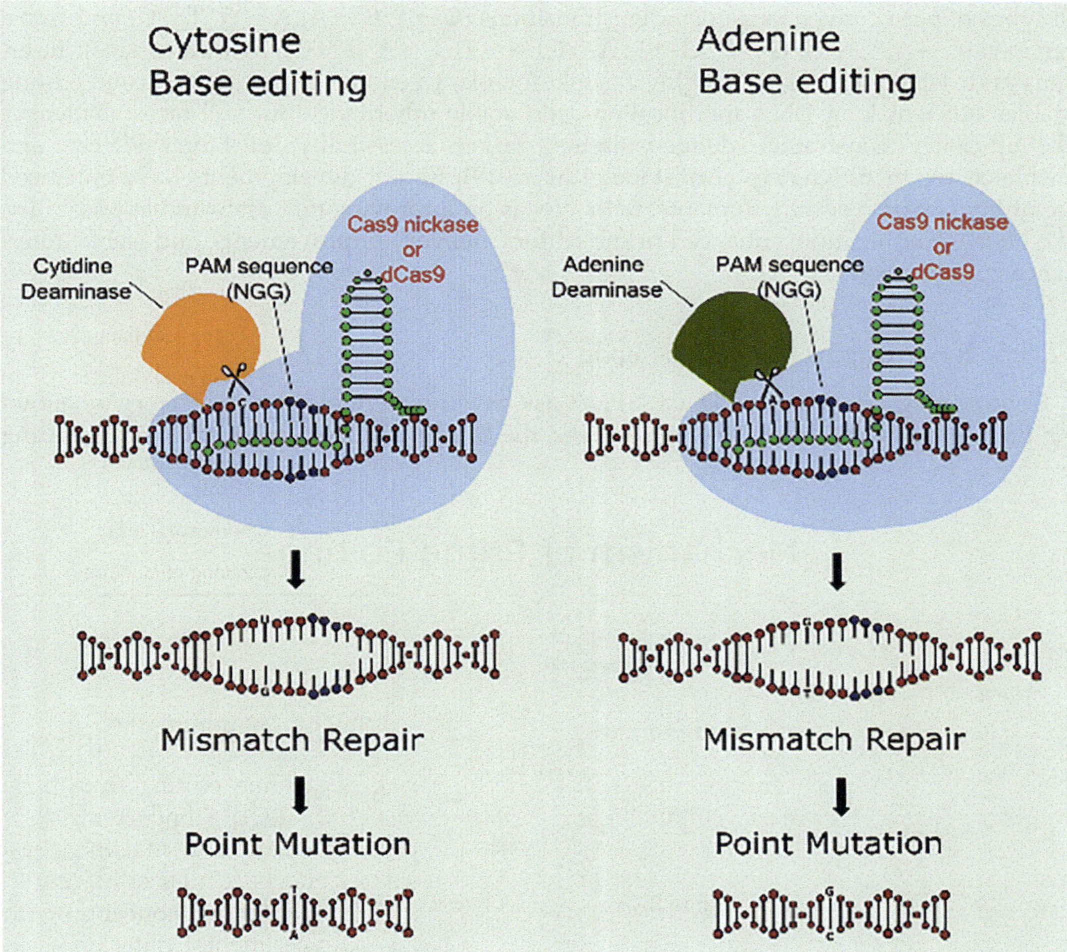

FIGURE 12.2 Deciphering the mechanisms of base editing: Cytosine and adenine base editing using cytidine and adenine deaminases, respectively.

12.5.4 Prime editing

Prime editing represents a significant advancement in plant genome editing, offering a versatile and precise method for modifying DNA without inducing double-stranded breaks or requiring donor DNA templates. This technology, an evolution of CRISPR/Cas9 and base editing, utilizes a prime editing guide RNA (pegRNA) and a fusion protein comprising a Cas9 nickase and a reverse transcriptase, enabling all 12 possible base-to-base conversions, insertions, and deletions. This completely operates by directing the fusion protein to a specific genomic location via pegRNA, which encodes the desired edit and facilitates reverse transcription to incorporate changes. This process avoids the need for DSBs, reducing the risk of unintended mutations compared to traditional CRISPR/Cas9. Its ability to perform all types of base conversions, including transitions (C→T, G→A, A→G, T→C) and transversions (C→A, C→G, G→C, G→T, A→C, A→T, T→A, T→G), as well as small insertions and deletions, makes it highly flexible for plant genome engineering. Prime editing enables efficient long DNA modifications and stable inheritance but still faces challenges like efficiency constraints, delivery issues, target accessibility, off-target effects, and mismatch repair differences (Anzalone et al., 2019). Recent developments have enhanced the utility of prime editing in plants, with a focus on improving efficiency and applicability, like gRNA optimization, enhanced prime editors, delivery improvements, and stable inheritance confirmation (Fig. 12.3).

12.5.5 Rewriting of the genetic code

Gene editing technologies like CRISPR/Cas9 have revolutionized plant science by allowing precise modifications to DNA. However, the frontier of plant biotechnology is shifting

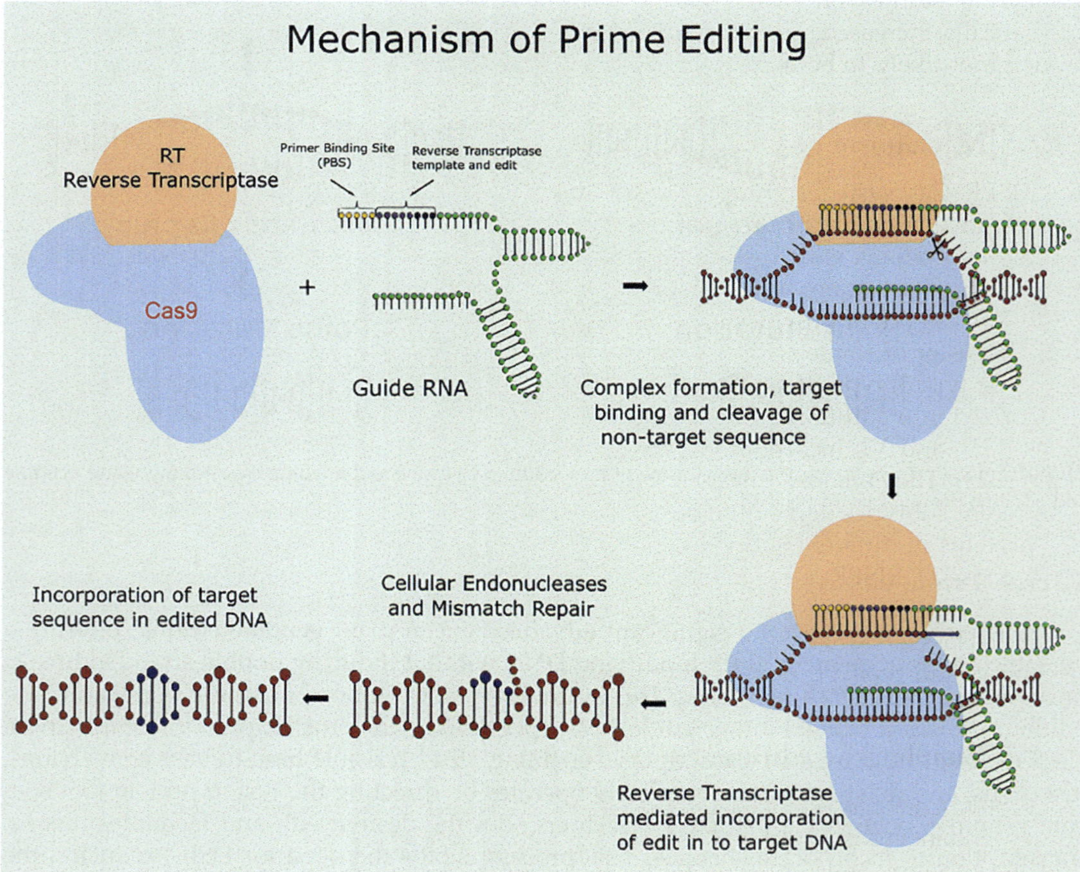

FIGURE 12.3 Mechanism of prime editing in plants.

beyond simple edits toward a complete rewriting of the genetic code. This advancement holds the potential to create crops with unprecedented traits, resilience, and productivity, paving the way for a new era of synthetic biology in agriculture. While CRISPR and other genome editing tools allow for targeted modifications, they primarily function by adding, deleting, or modifying specific genes. These methods, though powerful, are still limited by the existing genetic framework of an organism. Rewriting the genetic code moves beyond simple edits by altering the fundamental DNA translation mechanisms, effectively changing how genetic instructions are read and implemented by the plant's cellular machinery (Mukai et al., 2017). The potential applications are climate-resilient crops, stress resistance, enhanced nutritional content, and sustainable bioengineering. As with any breakthrough technology, rewriting the genetic code in plants presents ethical and regulatory challenges to be taken into concern.

12.6 Classification of gene-edited organisms based on the repair outcomes

The classification of gene-edited organisms is based on the repair mechanisms employed after DNA cleavage, influencing their regulatory status and public acceptance. Depending on whether the edits involve small mutations, precise modifications, or foreign DNA insertions, these organisms fall into distinct categories as SDN1 (site-directed nuclease 1), SDN2 (site-directed nuclease 2), and SDN3 (site-directed nuclease 3).

12.6.1 SDN1

SDN1 involves the use of genome editing tools to create targeted modifications in a plant's genome. This approach leads to plants with desired traits, which is achieved either through Agrobacterium-mediated transformation or by introducing a ribonucleoprotein (RNP) complex. This is a system where a nuclease creates a DSB at a specific location, followed by the cell's DNA repair mechanism repairing the break. In the case of SDN1, the repair occurs without the introduction of any foreign DNA, which is typically achieved by NHEJ, leading to small insertions or deletions, disrupting the gene function. The primary outcome of SDN1 is a targeted gene knockout without the incorporation of any foreign DNA. Several countries, including India, consider SDN1-based genome-edited plants under less or no regulation. Transgene-free gene-edited crops, obtained after removing the gene-editing tool through genetic segregation, are similar to conventional crops, as they harbor mutations that can occur naturally. These crops do not present added food safety or environmental risks compared to crops bred by traditional mutagenesis breeding and cannot be distinguished from conventional crops, unlike transgenic crops (GMOs). In many countries, such as the USA and Japan, transgene-free gene-edited crops are treated as conventional crops and are exempted from restrictive policies applied to GMOs. In India, genome-edited plants that use the SDN1 strategy are also exempted from the purview of regulations by current regulatory bodies (Roth et al., 2012; Waterworth et al., 2011).

12.6.2 SDN2

SDN2 refers to a genome-editing approach where a predefined genetic change is introduced at a specific site using CRISPR or other nuclease-based technologies, without integrating the foreign DNA. Unlike SDN1, which relies on random mutations via NHEJ, SDN2 uses HDR to introduce precise point mutations or small sequence modifications with the help of a short donor DNA template. SDN2 modifications are often indistinguishable from natural mutations or those created through traditional breeding, which has led some regulatory authorities to classify them separately from GMOs. Many countries, including Japan, Argentina, Brazil, and Australia, have relaxed regulations for SDN2 crops since they do not contain foreign DNA. The EU and some other regions still classify SDN2 as a form of genetic modification, subject to stricter regulations. In India, gene-edited crops generated using SDN1 and SDN2 approaches are exempted from GMO regulations.

12.6.3 SDN3

SDN3 is a genome editing technique that allows the integration of gene(s) into a predefined insertion site in the species' genome. This minimizes hazards associated with disrupting genes or regulatory elements. This technique facilitates targeted DNA insertion, distinguishing it from transgenesis, where the insertion site is not predefined. While SDN3 can induce off-target changes, these are generally fewer than those resulting from most mutagenesis techniques and are similar to those from conventional breeding. SDN3 works by inducing a DSB at a specific genomic location. This triggers precise repair processes, allowing for the seamless integration of exogenous genetic material at the targeted site.

When foreign DNA is integrated into the genome, these organisms may be classified and regulated as GMOs. This classification applies under frameworks such as the EU's GMO Directive. China has strict GMO regulations, especially for commercial crops. SDN3 organisms require biosafety approval and field trials before commercialization, making their acceptance more complex. In India, SDN3 remains classified as a GMO and requires extensive biosafety assessments. Public resistance to GMOs remains strong, affecting SDN3 acceptance (Podevin et al., 2013).

12.7 Need for precise prediction of repair outcomes

12.7.1 Why predict repair outcomes? To minimize off-targets

Predicting repair outcomes is essential to minimize off-target effects and ensure precise genome modifications. CRISPR-based genome editing relies on cellular repair pathways such as NHEJ and HDR. Since NHEJ is error-prone, it can introduce random insertions or deletions (indels), leading to unintended mutations. Accurate prediction of these repair events helps researchers design better gRNAs, optimize editing strategies, and reduce unwanted mutations, making genome editing more reliable and safer for agricultural applications (Movahedi et al., 2022).

Precise prediction of Cas footprints is necessary to reduce off-targets and increase public acceptance of gene-edited crops. The Cas footprint refers to the pattern of genetic changes introduced by Cas nucleases at the target site. Precise prediction of these footprints is essential for reducing off-target effects, which can cause unintended mutations in functionally important genes. By improving predictability, researchers can enhance the specificity of gene edits, reduce unintended consequences, and address regulatory and public concerns regarding genome-edited crops. Increased accuracy and transparency in genome editing processes foster greater public acceptance and regulatory approval, particularly in countries with strict biosafety laws (Lin and Luo, 2019).

12.7.2 Implications of genome-wide precise prediction of repair outcomes

The ability to predict genome editing repair outcomes with high precision on a genome-wide scale represents a transformative advancement in plant biotechnology, functional genomics, and molecular breeding. Accurate forecasting of both on-target and potential off-target indel profiles allows researchers to design gene editing strategies with unprecedented control and predictability. In the context of crop improvement, this predictive capability enables the generation of desired genetic variants with minimal risk of deleterious mutations, facilitating the development of elite cultivars with enhanced traits such as increased yield, improved stress tolerance, or disease resistance (Li, 2023).

From a functional genomics perspective, genome-wide prediction models empower systematic interrogation of gene function through targeted mutagenesis while minimizing confounding variables introduced by unintended repair outcomes. Moreover, in biotechnological applications, such models serve as critical tools for reducing genetic mosaicism and improving the stability and reproducibility of edited lines, particularly in clonally propagated crops or polyploid species. Crucially, precise prediction of repair profiles can also streamline regulatory frameworks by providing robust evidence for the intended nature and genomic integrity of edited plants. This level of precision distinguishes gene-edited organisms from transgenic GMOs, as it allows developers to demonstrate the absence of foreign DNA and off-target effects with high confidence. As a result, predictive models enhance the transparency, safety, and public acceptance of gene-edited crops, aligning with emerging regulatory paradigms focused on product-based rather than process-based assessments.

12.7.3 Limitations of conventional methods for assessing repair outcomes

Conventional approaches for analyzing genome editing outcomes—such as Sanger sequencing, PCR-based genotyping, and gel electrophoresis—present significant methodological and analytical limitations in the context of modern gene editing applications. While historically foundational, these techniques are inherently low-throughput, labor-intensive, and limited in their resolution and scalability. Their reliance on bulk population analysis often masks the heterogeneity of repair outcomes, particularly low-frequency mutations that may have significant phenotypic or regulatory consequences (Nadeem et al., 2018).

Importantly, traditional assays are ill-equipped to accurately detect complex genomic alterations such as large insertions or deletions, chromosomal rearrangements, and structural variants that can arise from DSB repair (Mehta and Nandha, 2018). Furthermore, these methods provide little to no predictive insight into the molecular mechanisms governing repair outcomes, nor do they offer the resolution required for genome-wide off-target effect profiling. As a result, they fall short in ensuring the comprehensive evaluation of genome editing events, particularly in highly repetitive or polyploid plant genomes.

To address these challenges, the field is increasingly turning to advanced computational frameworks and high-throughput sequencing technologies. Deep sequencing allows for base-pair resolution quantification of repair outcomes across large sample sets, while machine learning (ML) models trained on empirical repair data enable robust prediction of editing patterns based on sequence and chromatin features (Yuan et al., 2024). These innovations are critical for enhancing the accuracy, reproducibility, and safety assessments of gene-edited crops, thereby overcoming the limitations of legacy methodologies and meeting the stringent demands of regulatory and commercial applications.

12.8 AI tools for precise prediction of repair outcomes

12.8.1 Current tools to predict repair outcomes and limitations

The capability to sufficiently predict DNA repair outcomes is important for the optimization of genome editing technologies, especially CRISPR technologies. Conventional bioinformatics tools and experimental methods have given valuable insights into mechanisms of DNA repair, yet they tend to be limited by constraints of inaccuracy, unscalability, and nongeneralizability to various genomic settings (Abdi et al., 2024). In most cases, current tools depend on statistical models and heuristic algorithms, which are not as flexible and predictive as AI-based strategies. Traditional prediction methods are not able to grasp intricate sequence dependencies, repair pathway bias, and contextual genomic effects, thus constraining their precision and use for various experimental conditions. Moreover, they tend to be heavily reliant upon manual curation, thus becoming less practical for large-scale uses.

12.8.2 Artificial intelligence (AI) algorithms for predicting genome-wide repair outcome

The field of genome editing has shifted its gear with the advent of AI by providing robust prediction models that incorporate large sets of data and deep learning architectures. These AI-based models and algorithms have shown exemplary capability for genome sequence analysis, identification of complex sequence patterns, and accurate repair prediction with high precision. Various ML techniques, including supervised and unsupervised learning, probabilistic modeling, and reinforcement learning, are employed in these algorithms.

12.8.3 Deep learning tools for precise prediction of DSB repair outcomes

Deep learning can play a pivotal role in enhancing the precision of predictions regarding repair outcomes through the utilization of neural networks that analyze extensive genomic datasets. For instance, convolutional neural networks (CNNs) help spot specific sequence patterns related to repairs, whereas long short-term memory (LSTM) networks are utilized to investigate the dynamics of repair across temporal intervals (Valieris et al., 2020). Additionally, transformer-based architectures improve predictive accuracy by effectively capturing long-range dependencies present within genomic sequences.

12.8.4 Big data analysis and data integration approaches for repair outcomes

Big data analysis can be critical to improving model predictive ability with regard to DNA repair outcomes. The combination of multiomics data, covering chromatin accessibility, epigenetic landscapes, and transcriptomic maps, provides an integrated view of repair processes. The combination of diverse experimental data sources, such as CRISPR screens and high-throughput sequencing, ensures AI models gain information about a wide range of genomic environments. In addition, the implementation of feature selection and dimension reduction techniques optimizes model performance, while federated learning techniques allow access to distributed genomic sets while preserving data anonymity. These technological breakthroughs make AI-driven predictions ever more accurate, scalable, and widely applicable across different biological systems.

12.9 Key data sources for AI training to predict repair outcomes

12.9.1 CRISPR databases for repair outcomes

The success of AI-driven predictions for repair outcomes relies upon access to high-quality repair outcome training sets. Large-scale CRISPR databases serve as valuable tools for AI model training because they cover comprehensive repair outcomes documentation in various genomic contexts. Major databases like the CRISPR Repair Outcome Database (CROD) and GUIDE-seq systemically document repair activities and frequencies of indels and thus create a large dataset amenable to training models (Wang et al., 2025). In addition to those, experimental high-throughput sets like DISCOVER-seq add further to the AI model's ability to transfer predictions over different cell and genomic locations (Chia et al., 2025).

12.9.2 Sequence features influencing repair outcomes

Apart from CRISPR databases, sequence attributes significantly impact repair outcomes. Some of the reliable and important explanatory features for accurate predictions using AI models are several sequence determinants, GC content, repeat motifs, and DNA secondary structures. Elucidation of such sequence attributes enables improvement and fine-tuning of AI models to enhance repair outcome prediction for various genomic landscapes. Moreover, the PAM sequence plays an equally vital role in defining CRISPR–Cas9 binding efficiency (Cencic et al., 2014). PAM preference-driven AI-based analysis has led to enhanced selection of target sites, diminished off-target activities, and increased genome editing specificity.

12.9.3 Functional genomic database

Functional genomics databases, e.g., the ENCODE project and the Roadmap Epigenomics, contain large datasets that are relevant for training AI models (Fishman et al., 2023). These tools contain valuable data regarding gene expression patterns, chromatin structures, and transcription factor binding sites, all of which influence DNA repair mechanisms. Incorporation of such datasets for AI systems enables an integrated predictive model, thereby increasing the precision and reproducibility of the predictions regarding outcomes of repair.

12.9.4 Simulation of genomic data and synthetic biology to predict repair outcomes

Computational simulations and synthetic biology approaches contribute to training AI models by generating high-quality synthetic genomic data. In silico simulations of CRISPR–Cas9 editing outcomes, along with synthetic gene constructions, confirm AI-predicted patterns of repair and calibrate model performance. These methods fill the gap between experimental availability and the demands of an AI model for robust and generalizable prediction.

12.10 Applications of AI in predicting precise repair outcomes of DSB

The use of AI in predicting DSB repair outcomes can have transformative implications for genome editing.

12.10.1 Precise prediction of indel profile

AI models predict insertion and deletion (in/del) profiles with a great level of precision, allowing researchers to anticipate repair patterns with much accuracy. By examining many repair datasets, AI tools may help in adjusting repair templates for HDR, which provides more exact genetic changes (Jin et al., 2025). Such adjustments are really useful in therapeutic genome editing, where precise repair outcomes are imperative for applications like gene therapies.

12.10.2 Minimizing off-target effects

A big use of AI in genome editing is to reduce off-target effects. An improved genome editing safety can now be achieved by making editing more specific and finding possible unintended edits using AI prediction tools for off-target sites. These tools examine sequence similarity and chromatin access to predict likely off-target sites, allowing researchers to plan more exact CRISPR guides.

12.10.3 Bias prediction in the repair mechanism

AI models can predict biases or differences in repair mechanisms, as they show the contrast between NHEJ and HDR pathways (Smirnov and Yunusova, 2025). A knowledge of those differences allows for better experimental plans and gene correction strategies, which finally improve the usefulness and safety of genome editing. Apart from laboratory applications, AI-powered precision in genome editing helps increase public acceptance of gene-edited crops in society. By providing predictable trait changes and reducing unintended genomic changes, AI makes CRISPR-related agricultural advances more dependable. This increased precision helps with not only the regulatory approval process but also the risk assessment of gene-edited crops, as it provides transparency in safety checks and evaluations. Taking benefit from these AI-based predictions, regulatory bodies streamline their approval workflows, making genome editing more accessible for biotechnological and agricultural innovations. In summary, the AI-based prediction of accurate DSB repair outcomes offers enormous potential toward improving medicine, agriculture, and biotechnology through increasingly accurate, efficient, and safe genome editing technologies.

12.11 Current AI models

AI-driven models for predicting repair outcomes have significantly advanced the precision and efficiency of genome editing. Various interesting AI models have been proposed with their preferred abilities in predicting repair events of DNA.

12.11.1 DeepCRISPR

DeepCRISPR is an advanced and powerful deep learning-based tool to predict CRISPR–Cas9 genome editing outcomes with high accuracy (Table 12.1). This accounts for multilayers of genomic and epigenomic features comprising local nucleotide sequence context in addition to chromatin accessibility, as well as DSB repair pathway biases. The model utilizes CNNs to capture sequence motifs and associated features hierarchically without prior knowledge. DeepCRISPR significantly outperforms existing state-of-the-art models by benchmarking prediction accuracy on a variety of cell-specific and genome-wide experimental validations. Thus, DeepCRISPR attains better correlation coefficients with lower mean square errors than other methods in predicting on-target cleavage efficiency and editing specificity (Rafiq et al., 2025).

12.11.2 InDelphi

InDelphi is a state-of-the-art ML-based prediction tool for calculating the probability distribution of indel outcomes after CRISPR–Cas9-mediated DSBs in the plant genome (Table 12.1). Using large-scale datasets of CRISPR-induced repair outcomes in multiple plant

TABLE 12.1 Leading AI tools for predicting CRISPR repair outcomes.

Tool	Full name	AI/ML approach	Input features	Output predictions	Key applications	Limitations	Reference/ Source
SPROUT	Sequence and chromatin-based prediction of repair outcomes using transformers	Deep learning (transformer-based)	DNA sequence context, chromatin state, nucleosome positioning	Full spectrum of indel types with position-specific probabilities	Epigenome-aware CRISPR design, plant and animal genome editing	Requires chromatin data availability; model complexity	Abbasi et al. (2025)
FORECasT	Frame outcome repair estimator for CRISPR–Cas-induced tails	Random forest regression	Local nucleotide sequence around cut site	Likelihood of in-frame versus out-of-frame mutations	Therapeutic genome editing; maximizing gene knockout or functional restoration	Does not account for chromatin or cell-specific context	Dara et al. (2024)
inDelphi	Insertion and deletion prediction via machine learning	Deep neural network	Sequence window (35–50 bp) flanking CRISPR cut site	Probabilities of top-ranking repair outcomes	Guide RNA selection, high-fidelity CRISPR editing	Biased toward mammalian datasets; limited applicability in plant systems	Chen et al. (2019)
DeepCRISPR	—	CNN-based deep learning	sgRNA sequence, genomic context, epigenetic markers	sgRNA efficiency and off-target likelihood	Guide RNA efficiency optimization; CRISPR design scoring	Lacks direct modeling of repair outcome spectra	Zhou et al. (2020)
CRISPR-Net	—	Recurrent neural network (RNN)	Sequence context, including PAM site	Indel types and frameshifts	Predicting cell-type-specific editing results	Focused on small editing windows; limited plant genome data	Rafiq et al. (2025)
PE-Analyzer	Prime editing outcome analyzer	Pattern matching + probabilistic modeling	NGS reads of prime-edited sites	Frequency of correct edit, byproducts, indels, and RT insertions	Analysis of prime editing efficiency and fidelity	Limited to post-edit analysis; not predictive for untested sequences	Umar Ibrahim et al. (2023)

species, InDelphi annotates the mechanistic biases of DNA repair pathways and predicts site-specific indel patterns with high precision. This methodology provides a versatile platform for genome editing in plants and allows directed targeting of sites with high-yield repair-mediated products that enhance both the efficiency of editing and functional validation in plant biotechnology (Hamerlinck, 2022).

12.11.3 CRISPRpred

CRISPRpred is a computer program based on ML methods for predicting CRISPR–Cas9 genome editing efficacy and likely DNA repair responses in plants. It combines data from the DNA sequence in the vicinity of the target site and information regarding chromatin features, such as DNA accessibility within plant genomes. Using algorithms like random forests and gradient boosting, CRISPRpred learns complex patterns from these features to make accurate predictions. Testing on datasets from various plant species shows that CRISPRpred outperforms earlier models, providing stronger correlations between predicted and actual editing efficiencies and reducing prediction errors. It is used in designing efficient guide RNAs, thus increasing the accuracy and rate of plant genome editing (Muhammad Rafid, 2020).

12.11.4 SPROUT

SPROUT (Sequence and chromatin-based Prediction of Repair Outcomes Using Transformers) is a highly advanced AI model that was carefully crafted and trained for precise prediction of the outcomes of genome editing (Table 12.1). By drawing on a large collection of experimental dataset resources, SPROUT combines sequence-specific features and elements of the epigenetic landscape—namely, chromatin accessibility, histone modifications, and nucleosome positioning—to capture the complex interrelationships underlying DNA DSB repair processes. By considering localized DNA sequence environment and higher-order chromatin structure, respectively, SPROUT overcomes limitations of traditional sequence-based predictive frameworks and thus delivers a mechanistically guided and data-driven approach for predicting the diversity and prevalence of the outcomes of CRISPR–Cas9-mediated DSBs. By bringing these features together, SPROUT achieves state-of-the-art accuracy for predicting insertion–deletion (indel) profiles over a diverse range of cell types and of genomic loci and thus becomes a valuable tool for precision and therapeutic genome editing initiatives (Molla and Yang, 2020; Li et al., 2021).

12.11.5 FORECasT

FORECasT (Frame Outcome Repair Estimator for CRISPR–Cas-induced Tails) is an AI-powered model that predicts the repair outcomes of CRISPR–Cas9-induced DSBs (Table 12.1). It mainly focuses on distinguishing between in-frame and out-of-frame changes. Using ML strategies guided by deep empirical data sets, FORECasT embodies the sequence-dependent tendencies underlying the cellular DNA machinery of repair—mainly NHEJ and MMEJ. It carries out a comprehensive analysis of neighborhood sequence attributes around the site of the DSBs in order to forecast the distribution of frequencies of resulting insertions and deletions (indels) and thus estimate, accurately, probabilities of coding-frame preservation (Naert et al., 2020).

This function is particularly important in therapeutic genome editing, in which efficient restoration or disruption of a target gene relies on precise control of reading frame integrity. By modeling the likelihood of CRISPR-mediated changes maintaining or losing the open reading frame, FORECasT enables deliberate design of guide RNAs, specifically engineered for given therapeutic endpoints, including selective gene knockout and allele correction. It was extensively tested and proved on a series of different genomic contexts and cell types, and proved itself an important tool in aiding the development of gene editing technologies in preclinical and clinical applications.

In the years ahead, we can expect that genome editing technologies based on AI will employ more and more advanced ML approaches. For instance, reinforcement learning models may be employed to optimize repair pathway choices, and then hybrid AI models that combine symbolic reasoning and deep learning could further refine predictions. In addition, quantum computing approaches may be used for large-scale genomic simulations, thus enhancing the accuracy of predictions of complicated repair outcomes. These developments can act as a stepping-stone for future advancements in the field of plant biotechnology and eventually increase the precision, efficiency, and safety of applications of genome editing.

12.12 Challenges and future directions

The integration of CRISPR and AI could become a rare technology to change agriculture. However, there exist some challenges that need to be addressed in order to realize this technology's full potential. Scientific or technical hurdles, computational and infrastructural problems, and ethical, regulatory, and social impacts are part of those challenges. On the technical side, many crops have large, polyploid, and poorly annotated genomes. This makes CRISPR targeting and AI predictions less accurate and difficult. Moreover, many important agricultural traits are polygenic, influenced by several biotic and abiotic factors, which further limits the editing efficiency and prediction. While predictive AI models require large and high-quality data, plant phenotyping and multiomics data are often not consistent. When it comes to the computational part, combining omics data into usable AI models is still a big problem that slows down or halts this process. Even if AI models learn on one species or data set, they may not work well for different crop types. The most important developing countries still do not have the computational infrastructure to apply AI-driven CRISPR methods. Transcending scientific and computational challenges is just the beginning. Governing policies for gene-edited crops must be clear, overcoming the concerns about GMOs. Furthermore, questions about which sectors control large-scale agricultural genomic and phenotypic data sets must be answered, as this can significantly impede collaboration and its smooth application in society. Essentially, science is indeed a strong instrument; however, to fully realize its potency, the ideal data, sound models, and regulations, along with public trust, are required.

12.12.1 Future Directions

CRISPR–Cas systems, powered by AI, have revolutionized the potential for analysis of plant genomes and crop improvement. However, the journey of these systems to becoming dependable tools in agriculture is still quite a distance. Future progress will entail optimizing the editing machinery, designing superior computational models, refining technology practices, and assembling regulatory and societal acceptability.

12.12.2 Improving editing accuracy

One of the key goals of the future is increasing the accuracy of genome editing. Although conventional Cas9 endonucleases are highly efficient, there always exists a risk of them cutting off-target parts of the genome. Addressing this challenge, scientists are thus working on the construction of Cas9 mutants with higher stringency in sequence recognition and other nucleases like Cas12a that could perform with higher precision. The introduction of base editors and prime editors opens up the possibility of rewriting DNA at a single-nucleotide level or through minor insertions and deletions, without creating harmful double-stranded breaks. Such tools lower the dependency on error-prone repair pathways and offer a larger degree of predictability for the results. Additionally, gaining more understanding about how plants make the choice among the different repair mechanisms will allow scientists to steer editing outcomes toward more favorable modifications.

12.12.3 Building better plant datasets

AI models utilize extensive, high-quality training data sets in order to predict accurately. The majority of the training data available currently stems from mammalian systems, which are not directly transferable to plants. Thus, to bridge this gap, scientists need to create plant-specific data sets that reflect the typical attributes of crop genomes, particularly in species of polyploidy such as wheat and sugarcane. Compilation of data pertaining to the end outcomes of DNA repair, levels of gene expression, and structures of chromatin will enhance the reliability of AI instruments in plant systems. One of the necessary steps in achieving this will be to develop and maintain global collaborations and shared databases, thus making these datasets widely distributed and accessible, and consequently, less reliant on mammalian models.

12.12.4 Making AI more transparent

While current AI platforms show outstanding capability in predicting editing success, many of these platforms lack the ability to elucidate the processes behind these predictions. Future developments in AI will probably concentrate on explainable AI, where algorithms do not only present the answer but also indicate the particular sequence features or chromatin states that led to the prediction. Such explanatory ability is projected to facilitate scientists in designing better experiments and gain trust among the public and regulatory authorities. Integrating AI systems with multilayered biological information like transcriptomes and epigenomes will further provide an in-depth understanding of the interaction of edits with the larger biological framework of the plant. Methods of transfer learning and federated learning will increase the strength of predictions by enabling models to calibrate across different species without the need for large new data sets each time.

12.12.5 Improving delivery methods

Even highly designed editing tools and AI models are only effective if the CRISPR system can be efficiently delivered into the plant cells. Present techniques, such as *Agrobacterium*-mediated delivery and particle bombardment, are not always reliable, and their results often differ between species. Future development is anticipated with nanoparticle-based carriers, customized viral vectors, and direct RNP delivery systems that might enhance the effectiveness and lower the demand for foreign DNA integration from other organisms. Moreover, improvement in plant tissue culture and the regeneration process will be indispensable for generating complete plants from transformed tissues. It will be important to overcome these challenges in order to facilitate the reliable realization of practical breeding programs.

12.12.6 Strengthening regulation and public trust

Scientific progress by itself does not guarantee large-scale adoption; instead, the contribution of regulation and public trust is equally important. Existing regulations of gene-edited crops remain diverse among different countries, creating uncertainty for international commerce and commercialization initiatives. A key future task will be the development of harmonized international standards, which place more value on the qualities of the end product than the method of production used. Alongside formal regulation, a good and engaging public dialog is important. Concerns over food safety, ethical issues, and the environment can be addressed through open, evidence-based dialog. The appropriate and equitable use of the CRISPR-AI system improves sustainability, promotes climatic resilience, and achieves food security, contributing toward gaining wider public acceptability. Linking these technologies to broader initiatives, e.g., the United Nations Sustainable Development Goals, serves additionally to raise their importance at the global level.

12.12.7 The road ahead

Within the next decade, the experimental trials of CRISPR-AI are expected to move to a real agricultural impact. Plant-specific data-based predictive models, together with improvements in editing accuracy and delivery systems, will speed up breeding procedures. It is conceivable that farmers will, in the near future, have access to crop varieties that will give higher yields, will be able to withstand drought and heat, resist pests, and will have enhanced nutritional value. Combined with strong regulations and open disclosure, AI-improved CRISPR could be a major force for positive reform in the agricultural sector, ensuring food security for future generations.

Declaration

The authors declare that no generative artificial intelligence tools were used in the writing or preparation of this chapter. The content reflects the authors' own scholarly work, and the authors assume full responsibility for all statements, interpretations, and conclusions presented.

References

Abbasi, A.F., Asim, M.N., Dengel, A., 2025. Transitioning from wet lab to artificial intelligence: a systematic review of AI predictors in CRISPR. J. Transl. Med. 23 (1). https://doi.org/10.1186/s12967-024-06013-w.

Abdi, G., Jain, M., Barwant, M., Tendulkar, R., Tendulkar, M., Tariq, M., Amir, A., 2024. Unveiling the Dynamic Role of Bioinformatics in Automation for Efficient and Accurate Data Processing and Interpretation. Springer Science and Business Media LLC, pp. 279–319. https://doi.org/10.1007/978-981-99-8401-5_15.

Ahmad, A., Jamil, A., Munawar, N., 2023. GMOs or non-GMOs? The CRISPR Conundrum. Front. Plant Sci. 14. https://doi.org/10.3389/fpls.2023.1232938.

Anzalone, A.V., Randolph, P.B., Davis, J.R., Sousa, A.A., Koblan, L.W., Levy, J.M., Chen, P.J., Wilson, C., Newby, G. A., Raguram, A., Liu, D.R., 2019. Search-and-replace genome editing without double-strand breaks or donor DNA. Nature 576 (7785), 149–157. https://doi.org/10.1038/s41586-019-1711-4.

Bhagtaney, L., Sundarrajan, P., 2023. An Overview of Tools for Genome Editing: ZFNs, Mega Nucleases, and Talens CRISPR/Cas-Mediated Genome Editing in Plants. Apple Academic Press, India, pp. 37–64. https://doi.org/10.1201/9781003331759-3.

Boyle, E.A., Becker, W.R., Bai, H.B., Chen, J.S., Doudna, J.A., Greenleaf, W.J., 2021. Quantification of Cas9 binding and cleavage across diverse guide sequences maps landscapes of target engagement. Sci. Adv. 7 (8). https://doi.org/10.1126/sciadv.abe5496.

Cencic, R., Miura, H., Malina, A., Robert, F., Ethier, S., Martin Schmeing, T., Dostie, J., Pelletier, J., Fujii, H., 2014. Protospacer adjacent motif (PAM)-Distal sequences engage CRISPR Cas9 DNA target cleavage. PLoS One 9 (10), e109213. https://doi.org/10.1371/journal.pone.0109213.

Chang, H.H.Y., Pannunzio, N.R., Adachi, N., Lieber, M.R., 2017. Non-homologous DNA end joining and alternative pathways to double-strand break repair. Nat. Rev. Mol. Cell Biol. 18 (8), 495–506. https://doi.org/10.1038/nrm.2017.48.

Chen, W., McKenna, A., Schreiber, J., Haeussler, M., Yin, Y., Agarwal, V., Noble, W.S., Shendure, J., 2019. Massively parallel profiling and predictive modeling of the outcomes of CRISPR/Cas9-mediated double-strand break repair. Nucleic Acids Res. 47 (15), 7989–8003. https://doi.org/10.1093/nar/gkz487.

Chia, B.S., Seah, Y.F.S., Wang, B., Shen, K., Srivastava, D., Chew, W.L., 2025. Engineering a new generation of gene editors: integrating synthetic biology and AI innovations. ACS Synth. Biol. 14 (3), 636–647. https://doi.org/10.1021/acssynbio.4c00686.

Dara, M., Dianatpour, M., Azarpira, N., Omidifar, N., 2024. Convergence of CRISPR and artificial intelligence: a paradigm shift in biotechnology. Hum. Genet. 41, 201297. https://doi.org/10.1016/j.humgen.2024.201297.

Dong, G., Fan, Z., 2024. CRISPR/Cas-mediated germplasm improvement and new strategies for crop protection. Crop Health 2 (1). https://doi.org/10.1007/s44297-023-00020-x.

Fishman, V., Sindeeva, M., Chekanov, N., Shashkova, T., Ivanisenko, N., Kardymon, O., 2023. AI in Genomics and Epigenomics. Springer Science and Business Media LLC, pp. 217–243. https://doi.org/10.1007/978-3-031-35176-1_11.

Garm, C., Moreno-Villanueva, M., Bürkle, A., Larsen, L.A., Bohr, V.A., Christensen, K., Stevnsner, T., 2013. Genetic and environmental influence on DNA strand break repair: a twin study. Environ. Mol. Mutagen. 54 (6), 414–420. https://doi.org/10.1002/em.21791.

Gutschner, T., Haemmerle, M., Genovese, G., Draetta, G.F., Chin, L., 2016. Post-translational regulation of Cas9 during G1 enhances homology-directed repair. Cell Rep. 14 (6), 1555–1566. https://doi.org/10.1016/j.celrep.2016.01.019.

Hamerlinck, l., An optimized workflow for CRISPR/Cas9-mediated generation of indels and large deletions in induced pluripotent stem cells and neural stem cells. RARE-MED Symposium 2022. (2022).

Huang, Q., Lin, B., Cao, Y., Zhang, Y., Song, H., Huang, C., Sun, T., Long, C., Liao, J., Zhuo, Kan, 2023. CRISPR/Cas9-mediated mutagenesis of the susceptibility gene OsHPP04 in rice confers enhanced resistance to rice root-knot nematode. Front. Plant Sci. 14. https://doi.org/10.3389/fpls.2023.1134653.

Huang, Z., Liu, G., 2023. Current advancement in the application of prime editing. Front. Bioeng. Biotechnol. 11. https://doi.org/10.3389/fbioe.2023.1039315.

Jin, Y.Y., Zhang, P., Liu, D.P., 2025. Optimizing homology-directed repair for gene editing: the potential of single-stranded DNA donors. Trends Genet. 41 (9), 788–803. https://doi.org/10.1016/j.tig.2025.04.014.

Komor, A.C., Kim, Y.B., Packer, M.S., Zuris, J.A., Liu, D.R., 2016. Programmable editing of a target base in genomic DNA without double-stranded DNA cleavage. Nature 533 (7603), 420–424. https://doi.org/10.1038/nature17946.

Laub, M.T., Typas, A., 2024. Principles of bacterial innate immunity against viruses. Curr. Opin. Immunol. 89, 102445. https://doi.org/10.1016/j.coi.2024.102445.

Li, J., Sun, Y., Du, J., Zhao, Y., Xia, L., 2017. Generation of targeted point mutations in rice by a modified CRISPR/Cas9 system. Mol. Plant 10 (3), 526–529. https://doi.org/10.1016/j.molp.2016.12.001.

Li, S., Lin, D., Zhang, Y., Deng, M., Chen, Y., Lv, B., Li, B., Lei, Y., Wang, Y., Zhao, L., Liang, Y., Liu, J., Chen, K., Liu, Z., Xiao, J., Qiu, J.-L., Gao, C., 2022. Genome-edited powdery mildew resistance in wheat without growth penalties. Nature 602 (7897), 455–460. https://doi.org/10.1038/s41586-022-04395-9.

Li, V., 2023. CROTONdb: Genome-wide Variant Effect Prediction Database for CRISPR/Cas9 Editing Outcomes. bioRxiv.

Li, V.R., Zhang, Z., Troyanskaya, O.G., 2021. CROTON: an automated and variant-aware deep learning framework for predicting CRISPR/Cas9 editing outcomes. Bioinformatics 37 (Suppl. ment_1), i342. https://doi.org/10.1093/bioinformatics/btab268.

Liao, H., Wu, J., VanDusen, N.J., Li, Y., Zheng, Y., 2024. CRISPR-Cas9-mediated homology-directed repair for precise gene editing. Mol. Ther. Nucleic Acids 35 (4), 102344. https://doi.org/10.1016/j.omtn.2024.102344.

Lin, L., Luo, Y., 2019. Tracking CRISPR's footprints. Methods Mol. Biol. 1961, 13–28. https://doi.org/10.1007/978-1-4939-9170-9_2.

Ma, L., Kong, F., Sun, K., Wang, T., Guo, T., 2021. From classical radiation to modern radiation: past, present, and future of radiation mutation breeding. Front. Public Health 9. https://doi.org/10.3389/fpubh.2021.768071.

Manghwar, H., Li, B., Ding, X., Hussain, A., Lindsey, K., Zhang, X., Jin, S., 2020. CRISPR/Cas systems in genome editing: methodologies and tools for sgRNA design, off-target evaluation, and strategies to mitigate off-target effects. Adv. Sci. 7 (6). https://doi.org/10.1002/advs.201902312.

May, D., Paldi, K., Altpeter, F., 2023. Targeted mutagenesis with sequence-specific nucleases for accelerated improvement of polyploid crops: progress, challenges, and prospects. Plant Genome 16 (2). https://doi.org/10.1002/tpg2.20298.

Mehta, D.R., Nandha, A.K., 2018. Genome Editing in Plant Breeding. Informa UK Limited, pp. 605–644. https://doi.org/10.1201/b22473-18.

Mengstie, M.A., Wondimu, B.Z., 2021. Mechanism and applications of CRISPR/Cas-9-Mediated genome editing. Biol. Targets & Ther. 15, 353–361. https://doi.org/10.2147/btt.s326422.

Molla, K.A., Yang, Y., 2020. Predicting CRISPR/Cas9-Induced mutations for precise genome editing. Trends Biotechnol. 38 (2), 136–141. https://doi.org/10.1016/j.tibtech.2019.08.002.

Movahedi, A., Barati, B., Wang, S., Wei, H., Ruan, H., Zhuge, Q., 2022. Off-Target Effects of Crop Genome Editing and its Minimization Genome Editing Technologies for Crop Improvement. Springer Nature, China, pp. 185–208. https://doi.org/10.1007/978-981-19-0600-8_10.

Muhammad Rafid, A.H., 2020. CRISPRpred (SEQ): a sequence-based method for sgRNA on target activity prediction using traditional machine learning. BMC Bioinf. 21 (1), 223.

Mukai, T., Lajoie, M.J., Englert, M., Söll, D., 2017. Rewriting the genetic code. Annu. Rev. Microbiol. 71 (1), 557–577. https://doi.org/10.1146/annurev-micro-090816-093247.

Nadeem, M.A., Nawaz, M.A., Shahid, M.Q., Doğan, Y., Comertpay, G., Yıldız, M., Hatipoğlu, R., Ahmad, F., Alsaleh, A., Labhane, N., Özkan, H., Chung, G., Baloch, F.S., 2018. DNA molecular markers in plant breeding: current status and recent advancements in genomic selection and genome editing. Biotechnol. Biotechnol. Equip. 32 (2), 261–285. https://doi.org/10.1080/13102818.2017.1400401.

Naert, T., Tulkens, D., Edwards, N.A., Carron, M., Shaidani, N.-I., Wlizla, M., Boel, A., Demuynck, S., Horb, M.E., Paul, C., Willaert, A., Zorn, A.M., Vleminckx, K., 2020. Maximizing CRISPR/Cas9 phenotype penetrance applying predictive modeling of editing outcomes in Xenopus and zebrafish embryos. Sci. Rep. 10 (1). https://doi.org/10.1038/s41598-020-71412-0.

Podevin, N., Davies, H.V., Hartung, F., Nogué, F., Casacuberta, J.M., 2013. Site-directed nucleases: a paradigm shift in predictable, knowledge-based plant breeding. Trends Biotechnol. 31 (6), 375–383. https://doi.org/10.1016/j.tibtech.2013.03.004.

Puchta, H., 2005. The repair of double-strand breaks in plants: mechanisms and consequences for genome evolution. J. Exp. Bot. 56 (409), 1–14. https://doi.org/10.1093/jxb/eri025.

Rafiq, S., Macha, M.A., Assad, A., 2025. Machine learning and deep learning for genomic data: a data-centric approach to CRISPR/Cas9 gene editing. Int. J. Data Sci. Anal. https://doi.org/10.1007/s41060-025-00773-4.

Roth, N., Klimesch, J., Schulze, S.D.-, Pacher, M., Mannuss, A., Puchta, H., 2012. The requirement for recombination factors differs considerably between different pathways of homologous double-strand break repair in somatic plant cells. Plant J. 72 (5), 781–790. https://doi.org/10.1111/j.1365-313x.2012.05119.x.

Sadoine, M., Ishikawa, Y., Kleist, T.J., Wudick, M.M., Nakamura, M., Grossmann, G., Frommer, W.B., Ho, C.-H., 2021. Designs, applications, and limitations of genetically encoded fluorescent sensors to explore plant biology. Plant Physiol. 187 (2), 485–503. https://doi.org/10.1093/plphys/kiab353.

Sayyed, R.Z., Mujumdar, S., 2025. Microbial Surfactants in Pharmaceuticals and Cosmetics Microbial Surfactants in Pharmaceuticals and Cosmetics. CRC Pres, Oman, pp. 1–501. https://doi.org/10.1201/9781003424864.

Schenke, D., 2020. CRISPR/Cas9 or prime editing? – It depends on. Open Life Sci. 15 (1), 868–870. https://doi.org/10.1515/biol-2020-0109.

Sfeir, A., Symington, L.S., 2015. Microhomology-mediated end joining: a back-up survival mechanism or dedicated pathway? Trends Biochem. Sci. 40 (11), 701–714. https://doi.org/10.1016/j.tibs.2015.08.006.

Smirnov, A.V., Yunusova, A.M., 2025. Novel CRISPR/Cas9-Based approaches for quantitative study of DSB repair mechanics. Biochemistry (Mosc.) 90 (4), 437–456. https://doi.org/10.1134/S0006297924601813RussianFederation.

Sternberg, S.H., Lafrance, B., Kaplan, M., Doudna, J.A., 2015. Conformational control of DNA target cleavage by CRISPR-Cas9. Nature 527 (7576), 110–113. https://doi.org/10.1038/nature15544.

Sung, P., Klein, H., 2006. Mechanism of homologous recombination: mediators and helicases take on regulatory functions. Nat. Rev. Mol. Cell Biol. 7 (10), 739–750. https://doi.org/10.1038/nrm2008.

Umar Ibrahim, A., Coston Pwavodi, P., Ozsoz, M., Al-Turjman, F., Galaya, T., Johnson Agbo, J., 2023. CRISPR biosensing and AI driven tools for detection and prediction of Covid-19. J. Exp. Theor. Artif. Intell. 35 (4), 489–505. https://doi.org/10.1080/0952813x.2021.1952652.

Valieris, R., Amaro, L., , Cynthia Aparecida Bueno de Toledo Osório, Bueno, A.P., Mitrowsky, R.A.R., Carraro, D. M., Nunes, D.N., Dias-Neto, E., Silva, I.T.da, 2020. Deep learning predicts underlying features on pathology images with therapeutic relevance for breast and gastric cancer. Cancers 12 (12), 3687. https://doi.org/10.3390/cancers12123687.

Van Vu, T., Doan, D.T.H., Kim, J., Sung, Y.W., Tran, M.T., Jong Song, Y., Das, S., Kim, J.-Y., 2021. CRISPR/Cas-based precision genome editing via microhomology-mediated end joining. Plant Biotechnol. J. 19 (2), 230–239. https://doi.org/10.1111/pbi.13490.

Vogt, A., He, Y., 2023. Structure and mechanism in non-homologous end joining. DNA Repair 130, 103547. https://doi.org/10.1016/j.dnarep.2023.103547.

Wang, G., Liu, X., Wang, A., Wen, J., Kim, P., Song, Q., Liu, X., Zhou, X., 2025. CRISPRoffT: comprehensive database of CRISPR/Cas off-targets. Nucleic Acids Res. 53 (D1), D914. https://doi.org/10.1093/nar/gkae1025.

Waterworth, W.M., Drury, G.E., Bray, C.M., West, C.E., 2011. Repairing breaks in the plant genome: the importance of keeping it together. New Phytol. 192 (4), 805–822. https://doi.org/10.1111/j.1469-8137.2011.03926.x.

Xue, C., Greene, E.C., 2021. DNA repair pathway choices in CRISPR-Cas9-mediated genome editing. Trends Genet. 37 (7), 639–656. https://doi.org/10.1016/j.tig.2021.02.008.

Yuan, T., Wu, L., Li, S., Zheng, J., Li, N., Xiao, X., Zhang, H., Fei, T., Xie, L., Zuo, Z., Li, D., Huang, P., Feng, H., Cao, Y., Yan, N., Wei, X., Shi, L., Sun, Y., Wei, W., Sun, Y., Zuo, E., 2024. Deep learning models incorporating endogenous factors beyond DNA sequences improve the prediction accuracy of base editing outcomes. Cell Discov. 10 (1). https://doi.org/10.1038/s41421-023-00624-1.

Zhang, F., Xiang Neik, T., Thomas, W.J.W., Batley, J., 2023. CRISPR-based genome editing tools: an accelerator in crop breeding for a changing future. Int. J. Mol. Sci. 24 (10), 8623. https://doi.org/10.3390/ijms24108623.

Zhang, J.P., Li, X.L., Li, G.H., Chen, W., Arakaki, C., Botimer, G.D., Baylink, D., Zhang, L., Wen, W., Fu, Y.W., Xu, J., Chun, N., Yuan, W., Cheng, T., Zhang, X.B., 2017. Efficient precise knockin with a double cut HDR donor after CRISPR/Cas9-mediated double-stranded DNA cleavage. Genome Biol. 18 (1). https://doi.org/10.1186/s13059-017-1164-8.

Zhang, X.H., Tee, L.Y., Wang, X.G., Huang, Q.S., Yang, S.H., 2015. Off-target effects in CRISPR/Cas9-mediated genome engineering. Mol. Ther. Nucleic Acids 4 (11), e264. https://doi.org/10.1038/mtna.2015.37.

Zhou, Z., Liu, F., Chen, S., Li, Z., 2020. A truthful and efficient incentive mechanism for demand response in green datacenters. IEEE Trans. Parallel Distr. Syst. 31 (1), 1–15. https://doi.org/10.1109/TPDS.2018.2882174.

CHAPTER

13

AI and CRISPR: Studying gene functions and regulatory networks in plants

Kamel A. Abd-Elsalam[1], *Rawan K. Hassan*[2], *Mohamed A. Aref*[3], *Adham M. AlDakrany*[3], *and Toka E. Abdelkhalek*[4]

[1]Plant Pathology Research Institute, Agricultural Research Center, Giza, Egypt; [2]Smart Engineering Systems Research Centre (SESC), Nile University, Cairo, Egypt; [3]School of Biotechnology, Nile University, Giza, Egypt; [4]Biotechnology English Program, Faculty of Science, Cairo University, Giza, Egypt

13.1 Introduction

The field of modern plant science is being transformed by genetic engineering and computer science with implications that are exceptionally broad in terms of application to agriculture, food security, and environmental stewardship (Ai et al., 2023). Indeed, the CRISPR–Cas systems, as well as AI, are among the most exciting technologies that have acquired previously unmatched performance both separately and, more importantly, together. CRISPR–Cas9 systems, specifically, have offered scientists an efficient, low-cost, and versatile way for targeted genome editing (Lee, 2023). It can give researchers the ability to insert, delete, or modify certain sequences of plant genomes and, in the process, learn more about genes, understand how they are regulated, and develop the desired traits, such as pest resistance, drought resistance, and higher crop yields. Artificial intelligence (AI) has drawn attention as a significant technology to assist in data-driven discovery in genomics and systems biology (Sun et al., 2024). High-throughput sequencing, transcriptomics, proteomics, and metabolomics will be more the norm, and scientists will be in the vanguard of a flood of multidimensional biological data and information (Panahi et al., 2025). The analysis, interpretation, and modeling of these complexes of data are provided by the computational

Intelligent Gene Editing in Plants
https://doi.org/10.1016/B978-0-443-44079-3.00004-8

mechanisms of machine learning (ML) in general and deep learning algorithms (DL) in particular. Prediction of gene functions, identification of regulatory motifs, identification of gene regulatory networks (GRN), and optimization of experimental design are also among the uses of AI in plant biology (Gerakari et al., 2025). Plant genomics research could be viewed as a poster child of the CRISPR and AI union (Abbas et al., 2024). CRISPR offers the capability to directly intervene with the genome, and AI offers the prediction and analysis platform to be more informed about the actions and apply them more efficiently. As an example, AI may be employed to preempt off-target gRNA action, to screen omics data in order to locate gene targets, and to optimize multiplexed CRISPR experiment design (Anwar et al., 2025). Moreover, artificial intelligence (AI) can enable the incorporation of all sorts of biological data (genomic, transcriptomic, epigenomic, and environmental data) into integrative models that can guide the development of precision gene editing practices. Moreover, this form of integrative synergy enables us not only to more effectively comprehend plant gene activity and regulatory programs but also to accelerate our process of intelligent breeding systems in crops (Mansoor and Kim, 2024). CRISPR and AI in plant genomics, emphasizing how the combination of these two tools can be used to examine the functions and regulation of genes and signaling pathways. The genomics AI usage is also applied to predict gene functions, network inference, and multi-omics integration. Special attention is given to experimental design and validation of AI-based tools based on CRISPR (Soomro et al., 2025). This chapter seeks to provide a review of high-throughput screening technologies, computational reconstruction of gene regulatory networks based on artificial intelligence (AI), and its applications in crop improvements. Lastly, we give the perspective of the future of this multidisciplinary discipline, paying special attention to the ethical concerns, information restrictions, and prospects of collaboration between computer scientists and plant scientists.

13.2 CRISPR–Cas9: A revolutionary tool for plant genomics

The CRISPR-associated protein (Cas) technology and clustered regularly interspaced short palindromic repeats (CRISPR) technology have brought about a novel stage in the specific manipulation of plant genomes, increased understanding of biological functions of plant genes, and created novel opportunities in crop characteristic breeding (Uranga et al., 2025). Recent findings indicate that it is possible to have nutritional security and athletic performance needs through the application of CRISPR (Bu et al., 2025). note that the content of all the necessary nutrients in quinoa seed has increased, and lysine has increased by 35%, zinc by 43%, vitamin C by 114%, and vitamin E by 45%. It is the findings that show that CRISPR edits made the nutritional changes wanted, and it thus can be applied in the plant-based athlete recovery diet in addition to preventing malnutrition in the world (Bu et al., 2025). Besides improving their nutritional status, CRISPR–Cas systems have proved to be effective in diagnosing plant diseases because they are quicker and more accurate. Nevertheless, the use of CRISPR/Cas in disease diagnosis of plants has not been thoroughly studied in systematic research, and it is this fact that somewhat complicates the implementation (Gong et al., 2025). In addition to CRISPR/Cas systems, many other modern developments are also getting considerable attention on genome editing as a preservation system. The most recent technology is engineered homing endonucleases, meganucleases, zinc finger nucleases (ZFNs), transcription activator-like effector nucleases (TALENs), and CRISPR-associated protein systems (Afzal et al., 2020). These technologies, and specifically CRISPR/Cas9, can enable the regulation of genes in monocot and dicot plants. CRISPR/Cas9 has already demonstrated a positive role in intensifying vulnerability to a great variety of infections (fungal, viral, or bacterial), improving abiotic tolerance (chill, heat, drought), improving nutritional value and grain quality, extending shelf life, haploid seed formation, and yielding high agricultural productivity (Jeyaraj et al., 2024). CRISPR/Cas9 is easy and cheap and can be scaled for all sorts of applications. It has been of significant interest due to its ease of manipulation and the ability to find extensive applications in gene mutation and transcriptional regulation of plants. The initial application of CRISPR/Cas9 technology was in rice, sorghum, and wheat, and now more than ever before, it has been found that the number of studies on plant genome editing with CRISPR/Cas9-based technology has dramatically increased. However, CRISPR/Cas9 can potentially edit the genome of plants, but it is not likely to be used fully (Pillay and M., 2020).

13.2.1 Mechanism of CRISPR–Cas9

CRISPR–Cas systems groupings and classifications have been continuously heterogeneous and altered to fit the purposes and localizations (Koonin and Makarova, 2019). Numerous systems exist, including two overlapping categories of the tomato CRISPR/Cas system, specifically classes 1 and 2, which exhibit significant variations. The class 1 CRISPR/Cas system provides a multi-subunit crRNA effector complex, whereas class 2 delivers a cluster of single crRNA effector complexes. Class 1 is further subdivided into three subtypes: type I, type III, and type IV. It subdivides into subtypes, which are subsequently categorized into additional classes, as illustrated in Fig. 13.1. Likewise, class 2 encompasses types II, V, and VI, incorporating further classes within the subclassifications (Chaudhuri et al., 2022). They are also further subdivided into subcategories, and they are a component of different subtypes depending on more heterogeneous functionalities and the encoding of Cas proteins in different subtypes (Koonin and Makarova, 2019).

The class 2 CRISPR/Cas system comprises a singular, big multi-domain protein within its effector complex, suggesting a more sophisticated organization compared to the class 1 system. The class 2 CRISPR/Cas system has been further categorized into three subtypes: type II, type V, and type VI, as seen in (Fig. 13.2)

13.2.2 Simple gene editing: SgRNA, Cas9 protein, and repair processes

CRISPR/Cas9 has established a novel standard for the precise manipulation of plant genomes, enhancing comprehension of the biological roles of plant genes and offering new prospects in crop trait breeding (Uranga et al., 2025). Recent findings indicate that nutritional security and athletic performance needs can be attained using CRISPR (Bu et al., 2025). Also stated that all the necessary nutrients in quinoa seeds have increased in value, with lysine increasing by 35%, zinc by 43%, vitamin C by 114%, and vitamin E by 45%. These

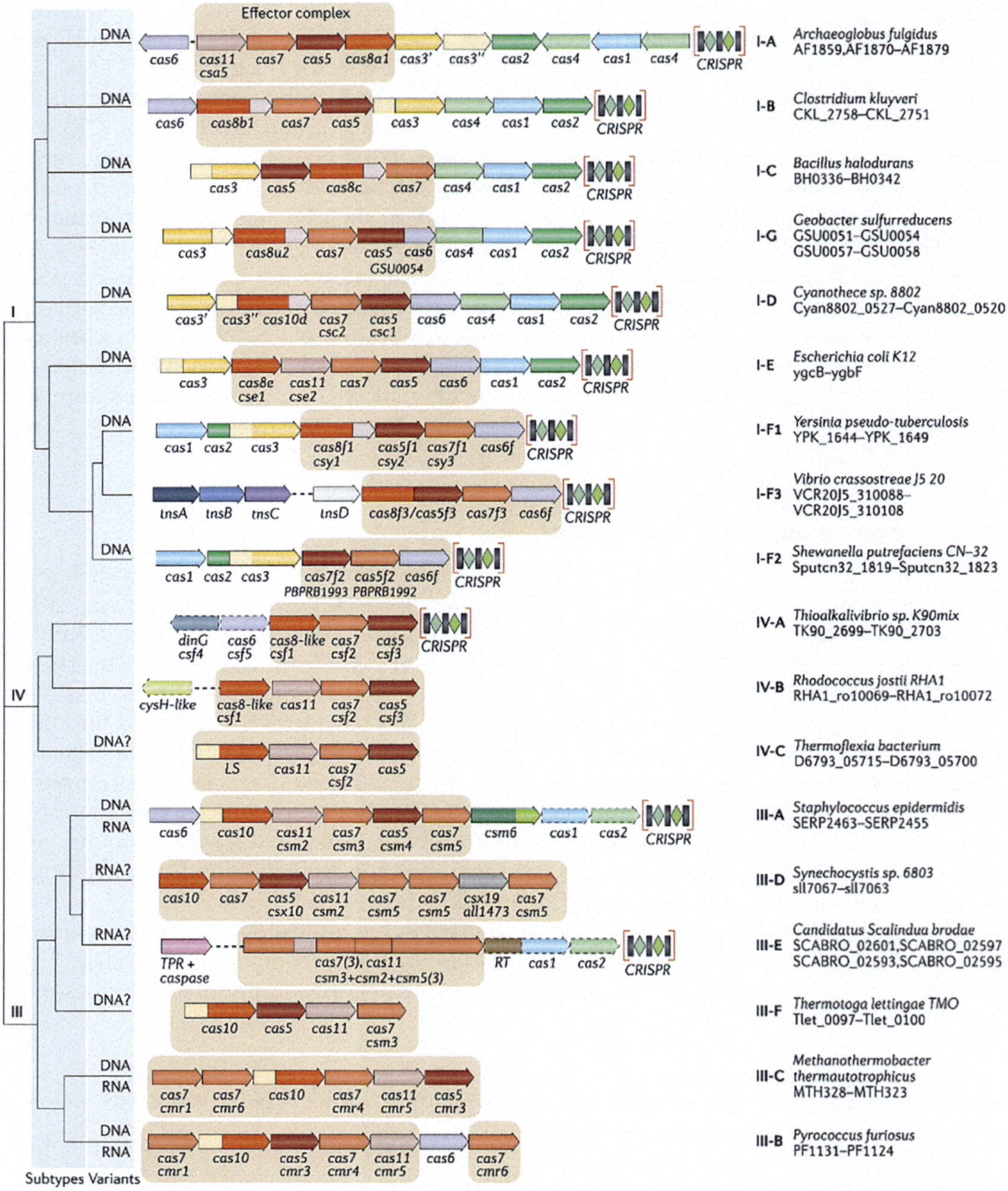

FIGURE 13.1 CRISPR–Cas loci class 1 domain structures of the effector molecules and target nucleic acids and PAM/PFS sequences; SS small subunit. The dendrogram shows the relationships between different types and subtypes as it was proposed before. *Chaudhuri, A., Halder, K. and Datta, A., 2022. Classification of CRISPR/Cas system and its application in tomato breeding. Theor. Appl. Genet., 135(2), pp.367-387.*

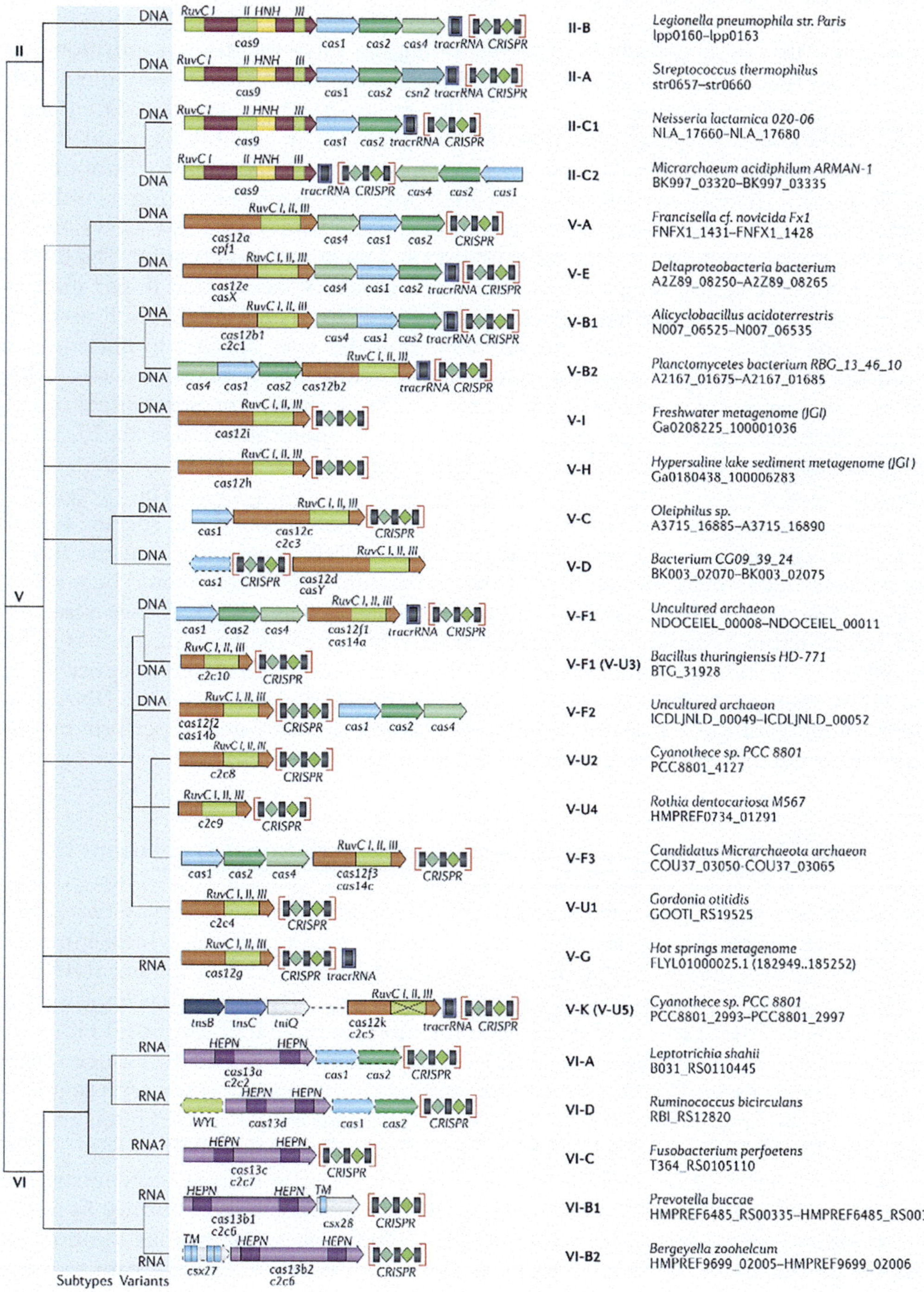

FIGURE 13.2 Class-2 CRISPR/Cas system classification according to the structure of the CRISPR/Cas loci, domain organization of the effector proteins and the target (predicted) nucleic acid, and PAM/PFS sequences; TM predicted transmembrane domain; HEPN higher eukaryotes and prokaryotes nucleotide binding. The color-diagnosed arrows show the respective genes of the class 2 system. The class 2 CRISPR/Cas effector modules consist of large single-domain proteins, which simplify and better organize them than class I systems. *Chaudhuri, A., Halder, K. and Datta, A., 2022. Classification of CRISPR/Cas system and its application in tomato breeding. Theor. Appl. Genet., 135(2), pp.367-387.*

are the findings that demonstrate that CRISPR mutations gave the desired nutritional outcomes, and therefore, it can be incorporated in the plant-based athlete recovery diet in addition to preventing malnutrition in the world (Bu et al., 2025). In addition to enhancing nutrition, CRISPR–Cas systems have demonstrated efficacy in identifying plant illnesses due to their speed and precision. Nonetheless, systematic investigations into the application of CRISPR/Cas for diagnosing plant diseases are inadequate, hence hindering its utilization (Gong et al., 2025). Other contemporary innovations, aside from CRISPR/Cas systems, are also receiving significant interest in genome editing as a means of preservation. The ultimate strategies include designed homing endonucleases, meganucleases, zinc finger nucleases (ZFNs), transcription activator-like effector nucleases (TALENs), and CRISPR-associated protein systems (Afzal et al., 2020). Researchers could be very specific in altering some DNA sequences. Transcription of the repeat-spacer array into a long precursor-crRNA (pre-crRNA) is the normal response of activation of CRISPR immunity. Subsequent processing of the crRNA molecules results in the shortening of the molecules (Liu et al., 2023a). The CRISPR–Cas ribonucleoproteins can recognize their targets, that is, DNA or RNA, during the interference stage to cleave them through an RNA-guided system. The protein Cas9 codes for *Streptococcus pyogenes,* which recognizes DNA molecular targets specific to guide RNAs to cleavage sites to initiate the native cellular DNA repair mechanisms that are already present. In contrast, without type II specificity dual-RNA, there would be no activation of Cas9. The complex between Cas9 and tracrRNA, as well as crRNA, causes a change to occur in the structure to enable Cas9 to bind to their target DNA through the two guides. Finally, when the bonding or nucleation has been successful to effectively identify a PAM on their DNA, then guide RNA binds to its target on the seed sequence on the DNA. Later, the HNH domain and RuvC domain of the Cas9 nuclease are activated to perform target or non-target cleavage of DNA strands (Plagens et al., 2015).

13.3 Gene knockout, knock-in, and transcriptional regulation

After a double-strand break (DSB) is created by CRISPR–Cas9, cells employ two broadly applicable mechanisms for repair: homology-directed repair (HDR), which introduces desired alleles from a donor DNA template, and non-homologous end joining (NHEJ), by which the cell introduces or deletes coding bits of its own. These two mechanisms have different outcomes: NHEJ can inhibit the gene functions, while HDR can result either way and retain a specific edit (Farag et al., 2025). NHEJ is commonly used to induce random indels to destroy genes, and HDR uses a donor template to replace the genes with sequences. Although the HDR generates controlled and precise knock-in models, NHEJ often generates knockout and knock-in models at the site of the cutting, as demonstrated in Fig. 13.3. The benefit of HDR compared to NHEJ is that it allows the precise editing of genes with the insertion of a single- or two-strand template of DNA at the target site, contrary to random indels and gene disruption (Collonnier et al., 2017). The most advantageous location for a gene knockout cleavage is an exon adjacent to either end codon (or downstream of the terminal codon ATG, etc.), likely resulting in a shift in the protein sequence, generally termed an out-of-frame score. Knock-ins possess fewer restrictions regarding targeting, as

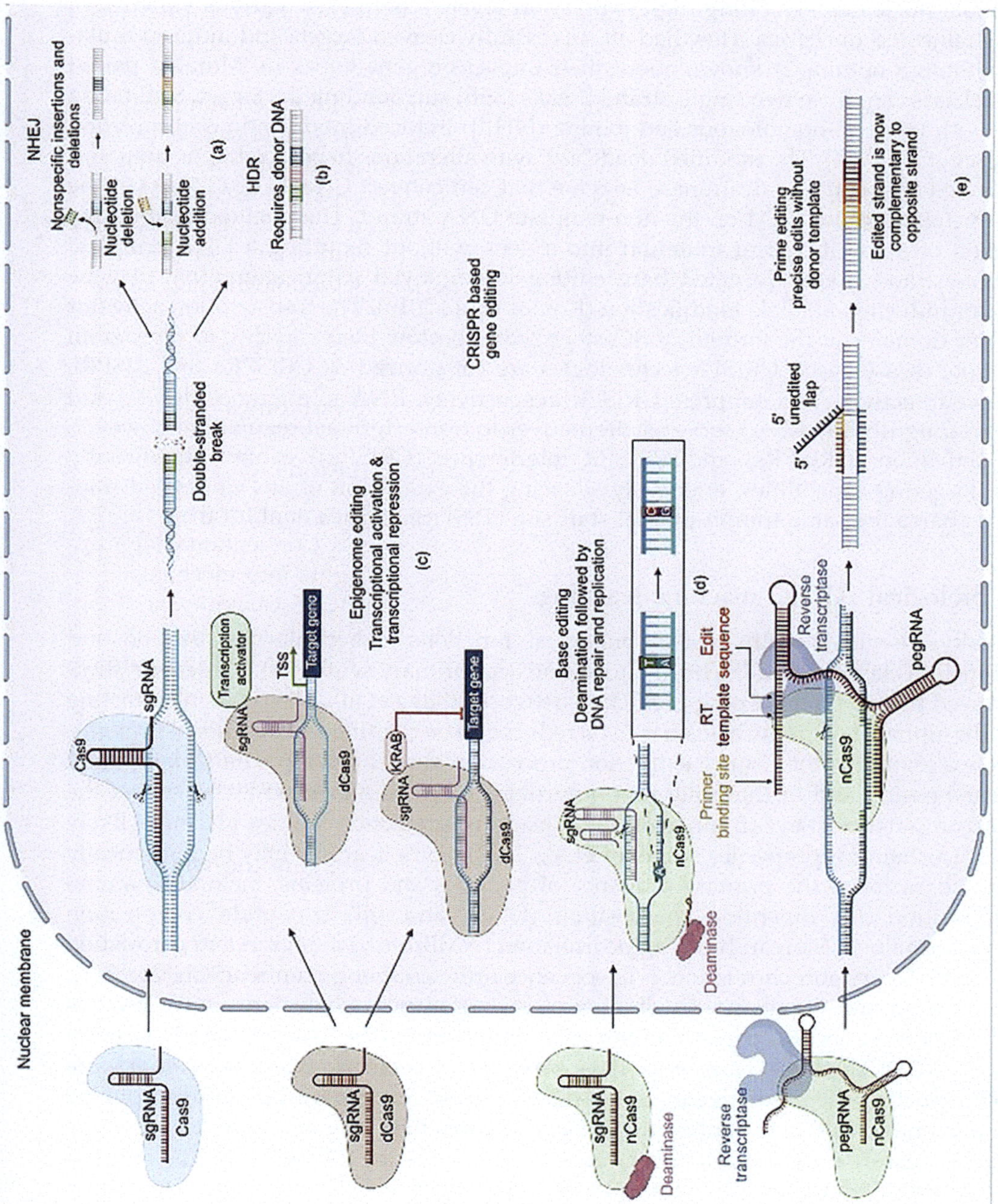

FIGURE 13.3 Various forms of CRISPR mechanisms and repair mechanisms in the cell nucleus (Veronica et al., 2025).

integration requires simply a PAM-proximal site. An effective cleavage of the target site is crucial for CRISPR-mediated gene knock-in. Optimal guide RNAs necessitate interaction with suitable nucleases exhibiting high expression levels. There exist various varieties of Cas and engineered nucleases. They had all successfully cleaved targets and induced mutagenesis, although nothing is known about their impact on gene knock-in. Mutated paired SpCas9 nickases can cause two single-strand breaks (SSB) surrounding the target, facilitating knock-in with less non-homologous end-joining (NHEJ)-induced modifications at insertion sites (Fauser et al., 2014). The modified deadCas9, with alterations to both catalytic domains, may be linked to a cytidine deaminase enzyme that can convert C to T (or G to A) at the location upstream of the PAM on the non-template DNA strand. This method enables the introduction of a specific point mutation into a gene without requiring a DNA template or a double-strand break. Targeted base editing is employed as an alternative to gene knock-in for inducing an allele modification (Komor et al., 2016). The transcription activator or repressor domains at the termini of dCas9 enzymes bestow their capacity to reprogram transcription. dCas9-based CRISPR technologies are categorized as CRISPRa or CRISPRi, which serve to activate and suppress CRISPR, respectively. RNA interference (RNAi) and small interfering RNA (siRNA) serve as alternatives to transcriptional regulation; however, CRISPR activation (CRISPRa) and CRISPR interference (CRISPRi) exhibit significantly restricted targeting capabilities, effectively silencing the expression of any undesired transcript that shares the same transcriptional start site (TSS) (Shakirova et al., 2020).

13.3.1 Biological AI and machine learning

Essentially associated with protein biological functions, subcellular localization and protein-function classification (according to amino acid primary sequence) is a generic strategy employed by machine learning predictive networks (Silva et al., 2019). When extracting features, the alphabet of amino acids has to be reduced to be classified as functional proteins. The priceless feature is the feature extraction process because it enhances data exploration and comprehension and at the same time improves forecasting precision and lowers the computer computation. Several other machine learning prediction studies are more likely to use physicochemical properties (Greener et al., 2022). Some features may be numerically defined to characterize the primary sequence of peptides and proteins, include (i) amino acid composition, (ii) dipeptide composition (DPC), and (iii) tripeptide composition (TPC). (iv) normalized Moreau Broto autocorrelation (NMBroto), (v) Moran autocorrelation (AAindex), (vi) Geary autocorrelation, (vii) sequence order coupling number (SOCNumber), (viii) composition, transition, and distribution of various structural elements, (vii) physicochemical properties, and (x) pseudo amino acid composition (PseAAC) descriptors (Lombardo et al., 2021). The PAAC attributes have been used to compile the datasets that were used in constructing the predictive model of disease resistance genes in plants. In machine learning, attribute selection is applied to discover features that vary in a minimal way. The advantages of feature selection prior to the data modeling are the following: (i) The reduced levels of overfitting translate to the reduced data redundancy and the opportunity to make a decision under the impact of noise and fake data; (ii) the reduced training time leads to the efficient algorithm due to the reduced volume of data (Dhudum et al., 2024).

13.3.1.1 *Sequence analysis deep learning*

Sequence analysis deep learning (SADL) is the use of deep neural networks to analyze ordered data such as DNA, RNA, proteins, language, or time-series signals. By automatically learning complex dependencies and hidden patterns within sequences, SADL enables accurate classification, prediction, and feature extraction across diverse scientific domains, including bioinformatics, natural language processing, and signal processing (LeCun et al., 2015). Deep learning (DL) has become a core methodology in computational biology and medicine, driving advances in protein classification, protein–protein interaction prediction, and protein structure/function inference. It also supports large-scale genomics and gene expression regulation studies. Beyond molecular applications, deep learning is widely implemented in biomedical imaging and signal analysis, enabling accurate detection of pathological features and contributing to early disease diagnosis (Maleki Varnosfaderani and Forouzanfar, 2024). The use of DL in plant science has been effectively implemented to measure biotic and abiotic stresses, to genetically enhance the major crop species, and to identify plant diseases through leaf images (Wang et al., 2023). Simply, DL is a group of numerous artificial neural networks with numerous layers of nonlinear functions; hence, it is an improvement of traditional neural networks. The most prominent technologies in the industry of bioinformatics are deep neural networks (DNNs), recurrent neural networks (RNNs), and convolutional neural networks (CNNs) (Ma et al., 2024). Deep neural networks share the same design as multilayer perceptron neural networks, comprising an input layer, many hidden layers, and an output layer. The values are subsequently computed in succession across the distinct network layers upon the provision of input data to the DNNs, utilizing a nonlinear function, which may include hyperbolic, sigmoid, tangent, or rectified linear unit (ReLU) functions. Deep neural networks (DNNs) have been employed in the analysis of high-dimensional biological data, enhancing computational efficiency and facilitating the examination of systems biology and bioinformatics research (Fang et al., 2023).

13.3.1.2 *AI for omics data analysis*

The omics approaches used in plant defense are quite diverse in nature and provide varying information regarding biochemical functions of plants in response to pathogens and other environmental issues. The next sections will give a short description of the key types of data that each omics method generates in the field of plant stress research application (Banh and Strobel, 2023). Plants are immobile organisms, and as such, a complex system of defense has developed against diverse stress factors, such as biotic and abiotic factors. The integration of genomics and artificial intelligence has also proven to be an effective mechanism of elucidating the genomic basis of plant defense. AI-assisted genomics in learning biotic and abiotic stress responses in plants (Chang et al., 2024). It entails the identification of genes that resist, the study of resistance mechanisms, and the enhancement of stress resistance. It is also discussed here in terms of the issues, ethics, and future developments of this emerging field. Genome-wide association studies (GWAS) are needed as genomic approaches to make crops more resistant to abiotic stress. GWAS value in establishing the effects of abiotic stress on the oil content of sunflower. Other past research has determined Quantitative Trait Loci (QTLs) linked to maize yield in reaction to thermal and hydric strain. Environmental GWAS research identified Single Nucleotide Polymorphisms (SNPs) that correlated with sorghum drought stress with 213 genomic locations of drought tolerance. Combined with genomics is epigenetics, which involves inheritable changes outside of the DNA sequences, into the newly developed field of epigenomics. In this unification, genetic control is revealed in cell response to stress and epigenomic response to environmental condition and stress. Genome-level analysis is essential throughout the stages of development to analyze these phenomena in the context of the genome or determine the changes brought about by plant diseases (Bekbolatova et al., 2024). The CRISPR–Cas9 technology based on bacterial viral defense has transformed genome editing and is a breakthrough technology. CRISPR–Cas systems are effective in editing the genomes of eukaryotic cells and provide a chance to design crop plants to increase their tolerance to abiotic and biotic stresses (Khalifa and Albadawy, 2024). The omics approaches in the study of plant defense are a broad-based research field and provide varying information on the processes by which plants respond to pathogens and other environmental stressors. Each of the omics methods generated some of the most significant data types to be used in the study of plant stress, which will be briefly described in subsequent sections (Banh and Strobel, 2023). Plants are immobile organisms; they have developed elaborate defense mechanisms to cope with many stressful environmental factors, such as biotic and abiotic factors. Integrating genomics and artificial intelligence has proven to be a powerful application in describing the genomic foundations of plant defense. It is an overview of the existing and emerging applications of AI-aided genomics in the study of biotic and abiotic stress responses in plants (Chang et al., 2024). It encompasses the identification of resistance genes, investigation of defensive mechanisms, and augmentation of stress resistance. This nascent discipline also examines the challenges, ethical considerations, and prospective trajectories of the field. Genome-wide association studies (GWAS) are essential for formulating genetic strategies aimed at enhancing crop tolerance to abiotic stress. The efficacy of GWAS in elucidating the impact of abiotic stress on sunflower oil content (Bekbolatova et al., 2024). CRISPR–Cas systems are effective in editing the eukaryotic genomes and offer a chance to engineer crop plants to be more resistant to both the abiotic and biotic stress (Khalifa and Albadawy, 2024).

13.4 Synergizing AI and CRISPR for plant research

CRISPR as a gene-editing technology has transformed the agriculture sector due to its applicability, which is exceptionally wide, as depicted in Fig. 13.4 CRISPR has transformed the crops with the ability to edit genes at a point (Zhang et al., 2025). Artificial intelligence (AI) is a young discipline that incorporates language processing, machine learning, or computer vision (Zeng et al., 2019). The two technologies can be united to solve the problem of food security and sustainability. Crops with improved characteristics can be developed because of the accuracy of gene editing by using CRISPR (Mok et al., 2020). This would be enhanced through the assistance of AI that will provide it with enhanced data analysis, prediction models, and optimization that will allow CRISPR to be used more efficiently and effectively. CRISPR and AI are not only the solution to the shortage of food in the immediate future, but also both technologies are the gateway to more environmentally friendly farming

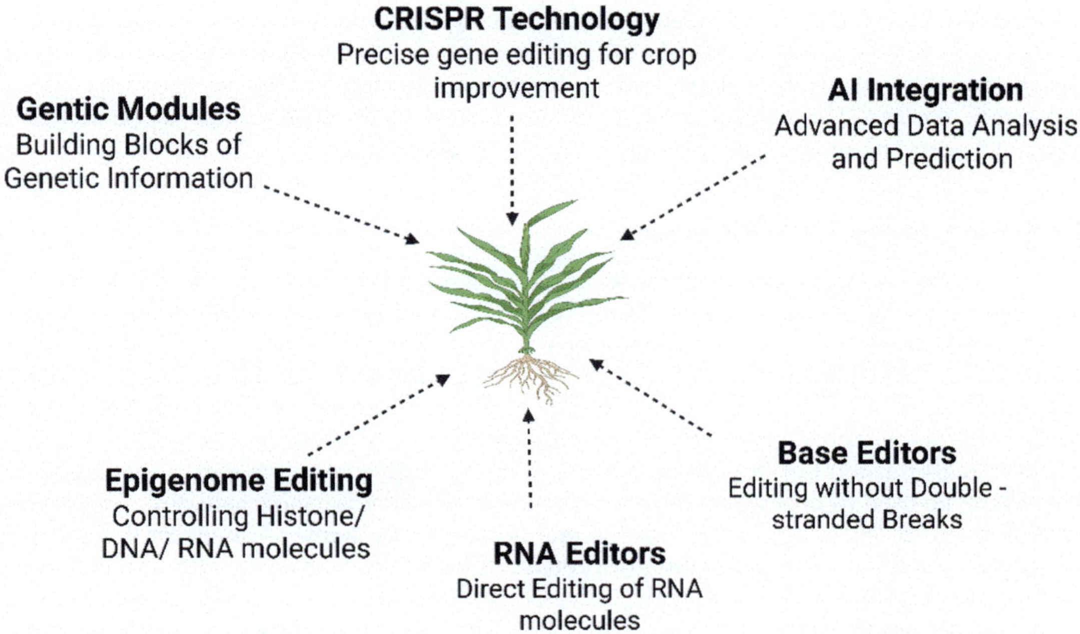

FIGURE 13.4 Conceptual diagram the interplay of enhanced biotechnological and computational tools to create innovation in genetic studies and crop enhancement. These elements constitute a synergistic platform of next-generation agriculture incorporating molecular accuracy and computational intelligence to improve crop resilience, productivity, and sustainability.

practices. The CRISPR-guided gene editing is a genetic engineering technology that has been found to be valuable in the production of crop yield traits and resistance (Zheng et al., 2024). Non-DNA-breaking editors like Cas9-NG, xCas9, nCas9, dCas9 (deactivated Cas9), and cytosine or adenosine deaminase domains are applied to editing without generating any double-strand break. RNA editors with dCas13 also can edit RNA molecules directly, and mitochondrial base editing is also possible with base editors based on DddA (Yu et al., 2021). CRISPR–Cas9 is also an effective variant that can be used to target mutagenesis to create more useful engineered proteins. CRISPR–Cas12 has a giant potential in crop enhancement with accurate and targeted genome editing of low off-target effects and high efficiency and rapidity, increasing crop generation with characteristics of interest (Palluk et al., 2018). The editing of the epigenome enhanced the traits of crops by controlling histone/DNA/RNA.

The protein transduced into rice and potatoes, depending on the editing of the epigenome, led to enhanced root systems and resistance to drought.The problem, however, persists in improving the effectiveness of changeover of various crops to enable the production of vast pools of gene variation. Genetic modules should be synthesized using a high-efficiency and accuracy editing process to obtain various genomes (Tian et al., 2021). The directed modification of DNA, RNA, and protein modules forms the input components, and a genetic module consists of most of them. Nucleic acids are carrying hereditary information. The primary component of the de novo genetic modules is the formation of nucleases and proteins (Çakmakçi et al., 2023). A linker molecule covalently joins a molecule of deoxyribonucleoside triphosphate with a template-independent deoxynucleotidyl transferase polymerase. This allows coupled deoxyribonucleoside triphosphate to be attached to the 3′ terminal end of de novo DNA primers (Xu et al., 2022).

13.4.1 AI-guided CRISPR design

The methods of AI include machine learning, learning algorithms with the data and making decisions; natural language processing, reading and writing human language; computer vision, working with the visual information, etc. In the pre-AI era, only a single amino acid substitution could hardly be predicted to have what impact it would impose on protein properties, and such an impact had to be measured directly, which was both time-consuming and labor-intensive (Krishnappa et al., 2021). Machine learning methods can perform high-throughput screening and selection of engineered enzymes or proteins, as well as can be used to provide a platform to produce large datasets to guide directed protein evolution (Voss-Fels et al., 2019). Directed engineering of proteins is usually in three steps. R1 (protein modeling and generation of protein variants), R2 (screening and refining feedback of the variants), and R3 (selecting targeted engineered proteins based on the findings of R1 and R2) are the steps (Trachsel et al., 2019). The focus in all these three steps is on AI. Deep learning algorithms, such as ProteinMPNN, are an effective technique to assemble the candidate amino acid sequences to represent a monomeric protein backbone. ProteinMPNN is more realistic and accurate than Rosetta and their protocols, which rely on physics (Sarker and I. H., 2021). Protein design is also of relevance in novel protein or enzyme design, based on the structure. An example is the homology model of the multi-functional P450 enzyme CYP8D2, which was created based on the crystal structure of the *Saccharomyces cerevisiae* P450 enzyme CYP51 (Li et al., 2019). The structures of natural proteins are small and variability. Therefore, protein-directed evolution will have to be complemented with existing protein structures. AlphaFold, a neural model, can predict the protein structure of a protein irrespective of the similarity of the structure that has been previously performed using a deep learning algorithm (Zhang et al., 2023a, 2023b). Plant protein engineering and directed evolution are methods of heterologous expression of exogenous or plant-derived engineered proteins. The context dependence of protein stability must be factored in on heterologous expression in plants (Siedhoff et al., 2020). The three-dimensional structure influences the stability of proteins, while AlphaFold offers a trustworthy and accurate way of showing the structure of a protein. The emergence of the OpenAI generative pre-trained transformers, for instance, GPTs, has been very helpful for the intelligent breeding of crops. The commonly used platforms are BioGPT and ChatGPT, which are founded on the GPT model (van Dis et al., 2023). The newly introduced BiomedGPT model applies self-supervision to all information types for learning multi-modal data, such as images, text, and bounding boxes, which marks the first milestone in the pursuit of biomedicine. Scaling out the domain knowledge improvement of the LLM CRISPR-GPT would enable the optimization of CRISPR gene editing experiments to be conducted in a fully autonomous manner (Qu et al., 2026). LLMs can select CRISPR systems, guide RNA design, propose a cellular delivery approach, write protocols, and design experiments to verify the results of editing by employing their excellent reasonableness power. GPTs can reach the literature of the world and implement an overwhelming database of knowledge on breeding and biology and can facilitate the creation of a large atlas of biological knowledge (Jiang et al., 2022). Such a world biological atlas and world research atlas can help in crop breeding done to produce additional data and information and as a part of global research and the world atlas of biology. This model that is powered by big data and GPTs has the potential to create a linear system to advance biological breeding processes repeatedly (Ahmad et al., 2024). In agriculture, DeepG2P applies natural language processing and neural network architecture in processing multi-modal inputs, including genomics, environment, weather, soil, crop management, and interactions, to improve crop yield. Moreover, the other emerging technologies like Double-Haploid (DH) technology, AI, genome editing, and multi-omics technologies are suggested to be incorporated in the CropGPT project to create a model of biological big data (LLM/GPT) that can be applied best in precision breeding. Besides, AI and integrating epigenetics with SynBio will also result in the agricultural age (Zhu et al., 2024).

13.4.2 High-throughput screening

It consists of a spectrum from molecular biology scale to ecosystem-wide studies, with high-throughput phenotyping (HTP) techniques to fast phenotype property and scale crop yields in intelligent plant centers. HTP entails parameter-based measurements (readings on environmental parameters, e.g., software sensors) and hyperspectral imaging and pigment data, thermal imaging and water content, and fluorescence measurements and photosynthesis rate (Zhang et al., 2023a, 2023b). HTP can be described as the rapid and accurate determination of plant characteristics with the help of the newest technologies and equipment. This type of approach enables obtaining phenotypic data on numerous plants in a short period to decrease the time of breeding a generation and discover additional information about the association of genotypes and phenotypes (Sudhakar et al., 2016). The proposed HTP system will also include robotics and AI to supplement the efficiency of data collection and processing to improve the accuracy and productivity of the proposed system. The technology is used in greenhouse and growth chamber research to study plant responses to environmental factors and in field research to determine how various environmental factors influence plants. Next-generation imaging and AI/ML is a key component of HTP, allowing the measurement of physical and nonphysical attributes of plants and their physiological condition without invasive techniques (Ma et al., 2024). Some of the techniques employed include RGB imaging technology to take images of the visible light of different specified features of plants, such as plant height, leaf area, and canopy structure. The aerial RGB images are then geometrically and radiometrically rectified and processed to estimate vegetation indices to show the health and vigor of plants. But NIR imaging is exploiting the NIR light to measure the content of water and biomass, which helps in forecasting the drought resistance and water consumption efficiency (Tian et al., 2025). Alternatively, thermal imaging relies on the emitted infrared radiation of the plant surfaces to determine leaf temperature, which will give information on transpiration rates, heat stress reactions, and details of water status and stress tolerance of plants. Furthermore, fluorescence imaging measures chlorophyll fluorescence that can be used to measure the level of stress and photosynthetic performance in different light settings (Bongomin et al., 2024). Sensor technologies can be used to measure environmental and physiological conditions continuously and parallel to imaging, in addition to imaging. The other possible sensors are a soil moisture sensor to measure soil moisture status to gauge the rate of irrigation and drought tolerance, a light sensor to measure the intensity and quality of light to aid plant growth and development, and a CO_2 sensor to measure CO_2 concentration to determine the rate of photosynthesis and C assimilation ability (Ye et al., 2025). The transmission of HTP and sensor technologies requires a large-scale data analysis and management system since both technologies produce large volumes of data. These phenotypic data cannot be analyzed just by simple statistical tools; that is where ML and AI can be priceless to discover the relationships and correlations that are often hard to most accurately discern (Stuart et al., 2025). Furthermore, systems that facilitate advanced analytics and data management platforms for the organization, sharing, and retrieval of phenotypic data are also necessary. Databases and cloud storage have been utilized to facilitate certain processes. The challenges of data transparency and its compatibility with other projects are essential and can only be addressed by adhering to established protocols and metadata profiles (Kundu et al., 2024).

13.4.3 Regulatory network mapping

As described in Fig. 13.5, the most important tool of life processes maintenance is called the gene regulatory network (GRN) and controls biochemical reactions and balances the concentration of compounds, which plays a huge role in diverse organisms and systems (Zhao et al., 2021). GRN reconstruction may help us to understand the molecular mechanism of organisms and to illustrate the key principles of most of the biological processes and responses in organisms. There are also some very good network reconstruction algorithms working with limited assumptions that affect predictive accuracy to be tolerant to the variability of computation (Weidemüller et al., 2021). To find out the reason why one approach is more suitable to a research problem or data received during an experiment, we conduct research based on the model-based, information-based, and machine learning-based approaches classifications. It is clear that various types of computational machines can be developed to distinguish among GRNs. Gene expression regulation is a complicated process during which genes and products (proteins, etc.) form complexes (Marku and Pancaldi, 2023). When depicted in the form of lines, they will have a network form, i.e., a gene regulatory network (GRN). GRN is an important component of every organism and system, as it is the necessary element to support the process of life, balance the biochemical processes, the level of compounds, etc (Mao et al., 2023). That is why the reconstruction of GRN can be a significant way of cognizing the molecular mechanism of organisms and revealing the fundamental principles of most biological processes and reactions in organisms. The various types of data have a small number of approaches that mainly include steady-state and time-series data. The steady-state data are provided in the form of the expression level of genes under steady state in different experimental conditions, e.g., gene knockout, gene knockdown, etc (Zhao et al., 2022). The response of the network to perturbation and recovery

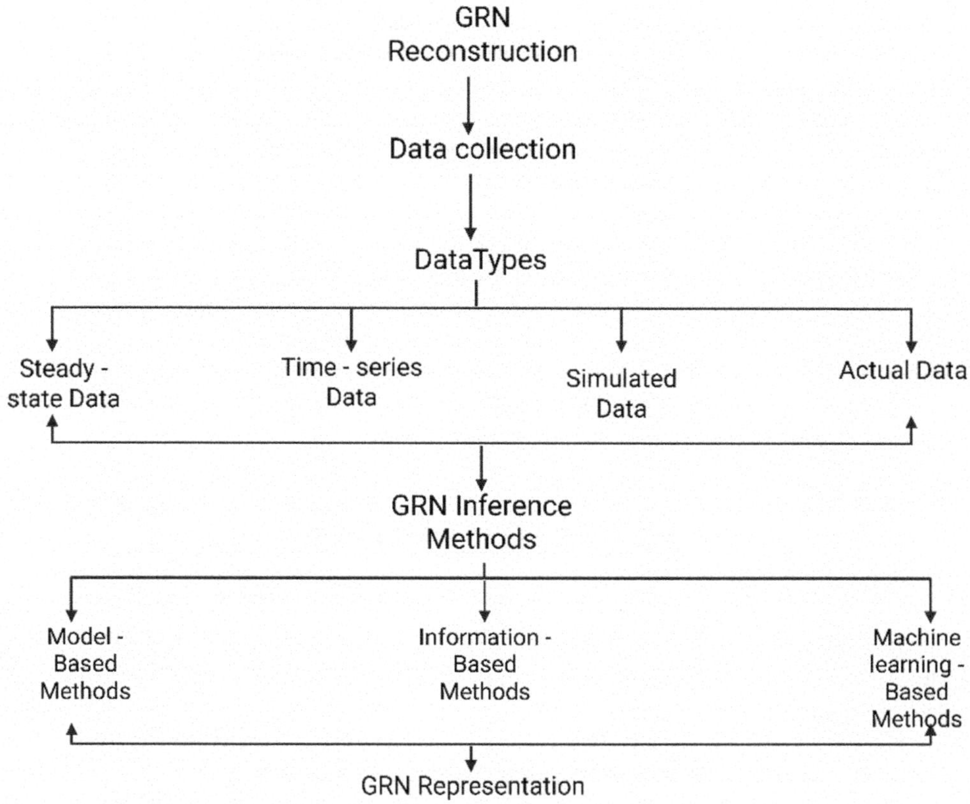

FIGURE 13.5 Flowchart outlines the stepwise process of reconstructing gene regulatory networks. It begins with data collection, branching into four data types: Steady-state, time-series, simulated, and actual biological data. These feed into GRN inference methods, categorized as model-based, information-based, and machine learning-based approaches. The final output is a structured GRN representation, enabling insights into gene interactions and regulatory mechanisms.

of the network after the perturbation are defined by the time-series data. It entails the amount of expression of every gene in different periods. Alternatively, based on the source of the data set, the input data can be divided into simulated data and actual data (Pušnik et al., 2022). The simulated data takes the synthetic network as the control network and simulates different conditions of the experiments mathematically. Data of this type do not merely provide steady-state and time-series data but can contain other parameters such as TFs to permit network inference. Most of the synthetic data are DREAM Challenges (DREAM4 and DREAM5 specifically), an open platform where biomedical competitions are contested. Biological experiments are also carried out on both typical network and expression level data to obtain actual data. In addition, these types of information may be cumulative steady-state information or time-series information of partial methodology. However, there are real regulatory relationships that cannot be tested by an experiment but are discovered in biological processes (Liu et al., 2023b). Consequently, in addition to established regulatory linkages, we must explore additional unexploited associations.

GRN inference techniques can employ standard statistical analysis methods to deduce necessary rules from partial gene expression data, discover relationships indicated by the levels of gene expression, and describe complex network interactions among regulators. The input data are evaluated against both actual and simulated expression data. The possible regulatory linkages can be ascertained using actual biological data, whereas simulation data is commonly employed to evaluate the efficacy of procedures (Ananda et al., 2024). Most of the simulation data is derived as part of the DREAM Challenges, an open and ongoing biological and pharmaceutical competition. The real data are mainly founded on GEO, ArrayExpress, and other databases. In this paper, we roughly divide such methods into three categories based on the nature and design underlying them (Chakraborty et al., 2022). Model-based approaches employ conventional models to delineate gene interactions, information theory-based approaches quantify intergenic correlations to express these interactions, and machine learning-based approaches utilize algorithms to discern the significant rules regulating gene expression levels (Wei et al., 2024).

13.5 Case studies and applications

13.5.1 CRISPR-AI in developing disease-resistant or climate-resilient crops

Recent developments with the CRISPR/Cas system enable targeted modification of more than a single gene within a genome with a high level of precision, as illustrated in Fig. 13.6. Various gRNAs may be co-expressed with a Cas nuclease and can bind to one or more genes simultaneously (Dara et al., 2024). The different gRNAs have been previously encoded in one expression cassette with an independent promoter or as a polycistronic message that is processed using indigenous nucleases, e.g., Cys4-based excision, tRNA-based gRNA cleavage, or ribozyme-flanked gRNA (Sollner and J. H., 2022). The multiplex CRISPR/Cas architecture has enabled researchers to target multiple DNA sequences at once and consequently permit a variety of applications, such as knockout of multiple genes, transcriptional regulation, base editing, and reconfigured plant metabolic pathways. Off targeting can be a critical problem in multiplex editing, but using online solutions during the design of gRNAs allows one to avoid it (Fernández et al., 2024). With the introduction of AI to CRISPR/Cas genome editing, the ideal architecture of gRNA can be designed, and off-target effects can be reduced by several algorithms, thus providing an opportunity to gene edit a wide variety of crops with enhanced characteristics (Pirali et al., 2025). How this mutation has been performed is not controlled, and the result of the mutation must be confirmed. Directional changes are also being investigated with advancing technology, one clip to another set base, but this mutation has not been effectively enhanced (Kanfer et al., 2021). Also, gene editing can be used to select mutants that foreign aid fragments have incorporated, which is not the case with the transgenic plant's breeders that researchers have been creating. When better and more accurate technology in mutant screening might be discovered without foreign aid fragment insertion and gene editing can be enhanced, it will be a revolutionary step to adopt gene editing technology (Rood et al., 2024). Transgenic technologies, including precision shearing, insertion, and substitution of multiple bases in specific genes, have been used to edit several genes in bulk to track alterations in phenotypes and select

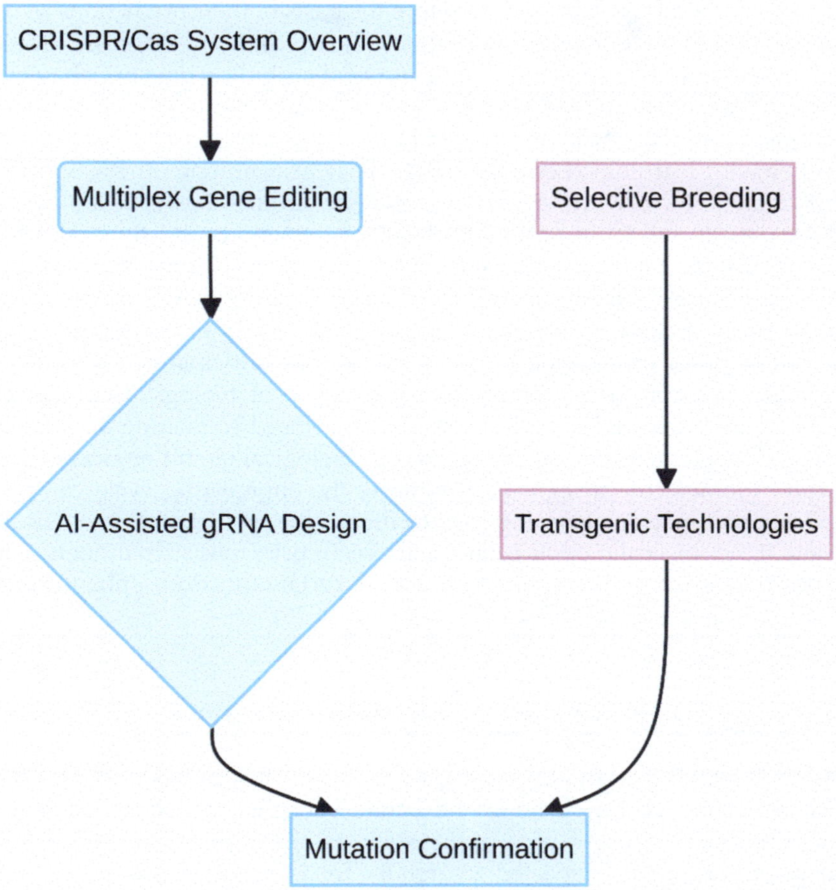

FIGURE 13.6 Flowchart outlines the sequential steps in CRISPR/Cas-mediated gene editing for agricultural enhancement. It begins with system introduction, followed by multiplex editing techniques for targeting multiple genes. AI integration supports guide RNA design and minimizes off-target effects. Mutation confirmation ensures edit accuracy, while transgenic technologies facilitate gene delivery. The process concludes with selective breeding to identify and propagate desirable crop traits.

variably phenotypic plants to use in future studies (Bock et al., 2022). The biodiversity development largely depends on random editing of plants. Selective breeding was also achieved by screening organisms with no exogenous fragments and crops that met breeding objectives (Schraivogel et al., 2022).

13.5.2 Designing synthetic gene circuits using AI predictions

The contemporary ML implementation, i.e., a contemporary artificial neural network model, will tend to require high volumes of data in order to be useful. It belongs to the issues that demand difficult-to-retrieve data that is expensive or even just limited, as in the majority of cases when assembling biological circuits (Regenwetter et al., 2022). Additionally, such models tend to be noninterpretable, and therefore the basis of their predictions as well is difficult to determine other than enhanced mechanistic biological knowledge. In addition to that, ML models do not extrapolate on extrapolative out-of-distribution data or behavior outside of the training data (Elahi et al., 2023). These models are usually trained to interpolate within the training data space, and their behavior may fail in unexpected circumstances that are not the focus of their behavior learnings (Lee et al., 2024). They also, among others, restrict the applications of ML models to forward engineering genetic circuits. Nevertheless, physics-based models may be used to model important molecular interactions and dynamics of biological processes. They offer the best design-based model of biological circuit engineering since parameters of the model are directly related to physical quantities that can be experimentally manipulated to fulfill a design purpose (Nemani et al., 2023). Consequently, theoretically, such models can facilitate prescribing behavior rather than anticipating it. Regrettably, these models frequently fail due to the profound dependence of biological circuit behavior on cellular, genetic, and extracellular settings, rendering it hard to include all interconnections among circuits. Machine learning can be applied in biological circuit engineering for genetic components and biomolecule design. Furthermore, the engineering cycle can be enhanced with the application of machine learning, as it facilitates the prediction of biological circuit activity in ways that biophysical models cannot achieve. In principle, such models would enable us to design novel gene circuit topologies to execute a certain function, a phenomenon already evident in other engineering disciplines (Wang et al., 2025).

13.6 Challenges and future directions

Transmission of genetic information of a gene is a dynamic process existing between transcription and translation, and the degree of fluctuation can be related to different functions. Suppose that the entire process of a gene, such as the replication of DNA and post-translational modification, is inspected in detail (Teixeira and Fussenegger, 2024). Then the role of the gene will be more apparent and closer to breeders. The most successful application of gene editing nowadays is to introduce non-regular alterations in dozens of cut pieces with the latest technology in the sector of molecular research (Palacios et al., 2025). As policy settings grow increasingly permissive, the biological implications of gene editing will expand significantly. Nonetheless, the advancement of gene editing enables a further reduction in the genetic variety of organisms. Certain potential paths may be disregarded due to breeders' selective breeding capabilities, stemming from their limited knowledge on the subject. Consequently, as breeding research swiftly integrates advanced information acquisition technologies such as gene sequencing and phenotype collection, an increasing volume of genetic, phenotypic, and environmental data is generated. The widespread implementation of artificial intelligence, information technology, and traditional breeding alongside biotechnology is catalyzing a transition from conventional hybrid breeding and field screening methods to data-driven intelligent breeding technology systems (Zhou et al., 2025). All the steps of data collection, data analysis, data storage, data management, and use of data in the breeding of the new breeding will be implemented with the help of information technology and intelligent equipment technology whereby computers will assist or even substitute the human brain work, and the human will be substituted with intelligent equipment (Qiu et al., 2025). The breeders will generate extensive genotype-phenotype-environment multidimensional data-driven precision breeding frameworks, while simultaneously advancing genome editing, synthetic biology, and other technological enhancements; facilitating efficient and selective development of new varieties; and fostering improved precision, intelligence, and a factory-based seed revolution. However, there are constraints to the advancement of vegetation, which breeders regard as a method. The unrestricted growing of crops in the wild constitutes another irreversible aspect of the research phenomena. The breeding direction is partially the responsibility of breeders, supported by applicable institutions and policies (Oesinghaus et al., 2024).

13.7 Conclusion

Artificial intelligence and CRISPR-mediated genome editing are transforming the field of plant functional genomics research because they can offer a resolution to some of the most pressing issues in the agricultural industry, including food security, climate resistance, and the sustainable increase in crop production. CRISPR reagents are stable, directed, and very versatile, thus enabling researchers to directly interrogate gene functions, create the features they want, and build artificial biological networks. In the meantime, AI, and specifically machine learning and deep learning algorithms, is a major enabler of high-throughput data analysis, predictive modeling, and smart experiment design.

Declaration

"The authors declare that no generative artificial intelligence tools were used in the writing or preparation of this chapter. The content reflects the authors' own scholarly work, and the authors assume full responsibility for all statements, interpretations, and conclusions presented".

References

Abbas, S., Li, Y., Lu, J., Hu, J., Zhang, X., Lv, X., Shahzad, A., Ao, D., Abbas, M., Wu, Y., Zhang, L., Fayyaz, M., 2024. Evaluation of stripe rust resistance and analysis of resistance genes in wheat genotypes from Pakistan and Southwest China. Front. Plant Sci. 15. https://doi.org/10.3389/fpls.2024.1494566.

Afzal, S., Sirohi, P., Singh, N.K., 2020. A review of CRISPR associated genome engineering: application, advances and future prospects of genome targeting tool for crop improvement. Biotechnol. Lett. 42 (9), 1611–1632.

Ahmad, I., Amelio, A., Merla, A., Scozzari, F., 2024. A survey on the role of artificial intelligence in managing Long COVID. Front. Artif. Intell. 6. https://doi.org/10.3389/frai.2023.1292466.

Ai, P., Xue, D., Wang, Y., Zeng, S., 2023. An efficient improved CRISPR mediated gene function analysis system established in Lycium ruthenicum Murr. Ind. Crop. Prod. 192, 116142. https://doi.org/10.1016/j.indcrop.2022.116142.

Ananda, R., Daud, K.M., Zainudin, S., 2024. A review of advances in integrating gene regulatory networks and metabolic networks for designing strain optimization. J. King Saud Univ., Comput. Inf. Sci. 36 (6), 102120.

Anwar, A., Tabassum, J., Ahmad, S., Ashfaq, M., Hussain, A., Ullah, M.A., Saad, N.S.B.M., Ghazy, A.I., Javed, M.A., 2025. Screening and assessment of genetic diversity of rice (Oryza sativa L.) Germplasm in response to soil salinity stress at germination stage. Agronomy 15 (2). https://doi.org/10.3390/agronomy15020376.

Banh, L., Strobel, G., 2023. Generative artificial intelligence. Electron. Mark. 33 (1), 63.

Bekbolatova, M., Mayer, J., Ong, C.W., Toma, M., 2024. Transformative potential of AI in healthcare: definitions, applications, and navigating the ethical landscape and public perspectives. Healthcare 12 (2), 125. https://doi.org/10.3390/healthcare12020125.

Bock, C., Datlinger, P., Chardon, F., Coelho, M.A., Dong, M.B., Lawson, K.A., Lu, T., Maroc, L., Norman, T.M., Song, B., Stanley, G., Chen, S., Garnett, M., Li, W., Moffat, J., Qi, L.S., Shapiro, R.S., Shendure, J., Weissman, J.S., Zhuang, X., 2022. High-content CRISPR screening. Nat. Rev. Methods Primers 2 (1). https://doi.org/10.1038/s43586-021-00093-4.

Bongomin, O., Lamo, J., Guina, J.M., Okello, C., Ocen, G.G., Obura, M., Alibu, S., Owino, C.A., Akwero, A., Ojok, S., 2024. UAV image acquisition and processing for high-throughput phenotyping in agricultural research and breeding programs. Plant Phenome Journal 7 (1). https://doi.org/10.1002/ppj2.20096.

Bu, G., Ma, X., Jin, H., Wang, L., 2025. CRISPR-Cas and RNA sequencing reveal nutrient enhancement pathways in quinoa for plant-based athlete recovery diets and future food security. Int. J. Biol. Macromol. 318 (Pt 4), 144785. https://doi.org/10.1016/j.ijbiomac.2025.144785.

Çakmakçı, R., Salık, M.A., Çakmakçı, S., 2023. Assessment and principles of environmentally sustainable food and agriculture systems. Agriculture 13 (5), 1073.

Chakraborty, S., Valdés-López, O., Stonoha-Arther, C., Ané, J.-M., 2022. Transcription factors controlling the rhizobium–legume symbiosis: integrating infection, organogenesis and the abiotic environment. Plant Cell Physiol. 63 (10), 1326–1343. https://doi.org/10.1093/pcp/pcac063.

Chang, Y., Wang, X., Wang, J., Wu, Y., Yang, L., Zhu, K., Chen, H., Yi, X., Wang, C., Wang, Y., Ye, W., Zhang, Y., Chang, Y., Yu, P.S., Yang, Q., Xie, X., 2024. A survey on evaluation of large language models. ACM Trans. Intell. Syst. Technol. 15 (3), 1–45. https://doi.org/10.1145/3641289.

Chaudhuri, A., Halder, K., Datta, A., 2022. Classification of CRISPR/Cas system and its application in tomato breeding. Theor. Appl. Genet. 135 (2), 367–387.

Collonnier, C., Guyon-Debast, A., Maclot, F., Mara, K., Charlot, F., Nogué, F., 2017. Towards mastering CRISPR-induced gene knock-in in plants: survey of key features and focus on the model Physcomitrella patens. Methods 121–122, 103–117. https://doi.org/10.1016/j.ymeth.2017.04.024.

Dara, M., Dianatpour, M., Azarpira, N., Omidifar, N., 2024. Convergence of CRISPR and artificial intelligence: a paradigm shift in biotechnology. Hum. Genet. 41, 201297. https://doi.org/10.1016/j.humgen.2024.201297.

Dhudum, R., Ganeshpurkar, A., Pawar, A., 2024. Revolutionizing drug discovery: a comprehensive review of AI applications. Drugs and Drug Candidates 3 (1), 148–171.

Elahi, M., Afolaranmi, S.O., Martinez Lastra, J.L., Perez Garcia, J.A., 2023. A comprehensive literature review of the applications of AI techniques through the lifecycle of industrial equipment. Discover Artificial Intelligence 3 (1). https://doi.org/10.1007/s44163-023-00089-x.

Fang, W., Chen, Y., Ding, J., Yu, Z., Masquelier, T., Chen, D., Huang, L., Zhou, H., Li, G., Tian, Y., 2023. SpikingJelly: an open-source machine learning infrastructure platform for spike-based intelligence. Sci. Adv. 9 (40). https://doi.org/10.1126/sciadv.adi1480.

Farag, V.E., Devey, E.A., Leong, K.W., 2025. 基因编辑与再生医学的交叉界面. Engineering 46 (3), 73–100.

Fauser, F., Schiml, S., Puchta, H., 2014. Both CRISPR/C as-based nucleases and nickases can be used efficiently for genome engineering in A rabidopsis thaliana. Plant J. 79 (2), 348–359.

Fernández, J.P., Petersen, B., Hassel, P., Lucas Hahn, A., Kielau, P., Geibel, J., Kues, W.A., 2024. Comparison between electroporation at different voltage levels and microinjection to generate porcine embryos with multiple xenoantigen knock-outs. Int. J. Mol. Sci. 25 (22). https://doi.org/10.3390/ijms252211894.

Gerakari, M., Katsileros, A., Kleftogianni, K., Tani, E., Bebeli, P.J., Papasotiropoulos, V., 2025. Breeding of solanaceous crops using AI: machine learning and deep learning approaches—a critical review. Agronomy 15 (3). https://doi.org/10.3390/agronomy15030757.

Gong, X.-Y., Wang, Z.-H., Bashir, M., Tang, T., Gan, X., Yang, W.-C., 2025. Recent application of CRISPR/Cas in plant disease detection. TrAC, Trends Anal. Chem. 189, 118251. https://doi.org/10.1016/j.trac.2025.118251.

Greener, J.G., Kandathil, S.M., Moffat, L., Jones, D.T., 2022. A guide to machine learning for biologists. Nat. Rev. Mol. Cell Biol. 23 (1), 40–55. https://doi.org/10.1038/s41580-021-00407-0.

Jeyaraj, G., Alphonse, V., Jayanthi, P., Neha Angelin, F., Swapna Geetanjali, A., Ganesan, G., 2024. Harnessing the potential of CRISPR/Cas system for enhancing virus resistance in plants: targets, strategies, and challenges. Physiol. Mol. Plant Pathol. 129, 102202. https://doi.org/10.1016/j.pmpp.2023.102202.

Jiang, S., Loomba, J., Sharma, S., Brown, D., 2022. Vital measurements of hospitalized COVID-19 patients as a predictor of long COVID: an EHR-based cohort study from the RECOVER program in N3C. In: 2022 IEEE International Conference on Bioinformatics and Biomedicine (BIBM). IEEE.

Kanfer, G., Sarraf, S.A., Maman, Y., Baldwin, H., Dominguez-Martin, E., Johnson, K.R., Ward, M.E., Kampmann, M., Lippincott-Schwartz, J., Youle, R.J., 2021. Image-based pooled whole-genome CRISPRi screening for subcellular phenotypes. JCB (J. Cell Biol.) 220 (2). https://doi.org/10.1083/JCB.202006180.

Khalifa, M., Albadawy, M., 2024. AI in diagnostic imaging: revolutionising accuracy and efficiency. Computer Methods and programs in biomedicine update 5, 100146.

Komor, A.C., Kim, Y.B., Packer, M.S., Zuris, J.A., Liu, D.R., 2016. Programmable editing of a target base in genomic DNA without double-stranded DNA cleavage. Nature 533, 420–424. https://doi.org/10.1038/nature17946.

Koonin, E.V., Makarova, K.S., 2019. Origins and evolution of CRISPR-Cas systems. Phil. Trans. Biol. Sci. 374 (1772), 20180087. https://doi.org/10.1098/rstb.2018.0087.

Krishnappa, G., Savadi, S., Tyagi, B.S., Singh, S.K., Mamrutha, H.M., Kumar, S., Mishra, C.N., Khan, H., Gangadhara, K., Uday, G., Singh, G., Singh, G.P., 2021. Integrated genomic selection for rapid improvement of crops. Genomics 113 (3), 1070–1086. https://doi.org/10.1016/j.ygeno.2021.02.007.

Kundu, S., Saini, D.K., Meena, R.K., Bahuguna, R.N., Jagadish, S.V.K., 2024. High-throughput phenotyping and AI technologies for deciphering crop resilience to heat stress. Plant Physiology Reports 29 (4), 699–715. https://doi.org/10.1007/s40502-024-00821-4.

LeCun, Y., Bengio, Y., Hinton, G., 2015. Deep learning. Nature 521 (7553), 436–444.

Lee, M., 2023. Deep learning in CRISPR-Cas systems: a review of recent studies. Front. Bioeng. Biotechnol. 11, 1226182.

Lee, D., Chen, W.(W.), Wang, L., Chan, Y.-C., Chen, W., 2024. Data-driven design for metamaterials and multiscale systems: a review. Adv. Mater. 36 (8). https://doi.org/10.1002/adma.202305254.

Li, D., Ma, Y., Zhou, Y., Gou, J., Zhong, Y., Zhao, L., Han, L., Ovchinnikov, S., Ma, L., Huang, S., Greisen, P., Shang, Y., 2019. A structural and data-driven approach to engineering a plant cytochrome P450 enzyme. Life Sci. 62 (7), 873–882. https://doi.org/10.1007/s11427-019-9538-3.

Liu, L., Helal, S.E., Peng, N., 2023a. CRISPR-Cas-based engineering of probiotics. BioDesign Research 5, 0017.

Liu, W., Yang, Y., Lu, X., Fu, X., Sun, R., Yang, L., Peng, L., 2023b. NSRGRN: a network structure refinement method for gene regulatory network inference. Briefings Bioinf. 24 (3). https://doi.org/10.1093/bib/bbad129.

Lombardo, T., Duquesnoy, M., El-Bouysidy, H., Årén, F., Gallo-Bueno, A., Jørgensen, P.B., Franco, A.A., 2021. Artificial intelligence applied to battery research: hype or reality? Chemical reviews 122 (12), 10899–10969.

Ma, X., Dai, X., Bai, Y., Wang, Y., Fu, Y., 2024. Rewrite the stars. In: Proceedings of the IEEE/CVF Conference on Computer Vision and Pattern Recognition.

Maleki Varnosfaderani, S., Forouzanfar, M., 2024. The role of AI in hospitals and clinics: transforming healthcare in the 21st century. Bioengineering 11 (4), 337.

Mansoor, S., Kim, I.J., 2024. Classical to modern biotechnology approaches, applications and future prospects in citrus breeding. Plant Biotechnology Reports 18 (7), 813–827.

Mao, G., Pang, Z., Zuo, K., Wang, Q., Pei, X., Chen, X., Liu, J., 2023. Predicting gene regulatory links from single-cell RNA-seq data using graph neural networks. Briefings Bioinf. 24 (6). https://doi.org/10.1093/bib/bbad414.

Marku, M., Pancaldi, V., 2023. From time-series transcriptomics to gene regulatory networks: a review on inference methods. PLoS Comput. Biol. 19 (8), e1011254.

Mok, B.Y., de Moraes, M.H., Zeng, J., Bosch, D.E., Kotrys, A.V., Raguram, A., Hsu, F.S., Radey, M.C., Peterson, S.B., Mootha, V.K., Mougous, J.D., Liu, D.R., 2020. A bacterial cytidine deaminase toxin enables CRISPR-free mitochondrial base editing. Nature 583 (7817), 631–637. https://doi.org/10.1038/s41586-020-2477-4.

Nemani, V., Biggio, L., Huan, X., Hu, Z., Fink, O., Tran, A., Wang, Y., Zhang, X., Hu, C., 2023. Uncertainty quantification in machine learning for engineering design and health prognostics: a tutorial. Mech. Syst. Signal Process. 205, 110796. https://doi.org/10.1016/j.ymssp.2023.110796.

Oesinghaus, L., Castillo-Hair, S., Ludwig, N., Keller, A., Seelig, G., 2024. Quantitative design of cell type-specific mRNA stability from microRNA expression data. bioRxiv. https://doi.org/10.1101/2024.10.28.620728.

Palacios, S., Collins, J.J., Del Vecchio, D., 2025. Machine learning for synthetic gene circuit engineering. Curr. Opin. Biotechnol. 92, 103263.

Palluk, S., Arlow, D.H., De Rond, T., Barthel, S., Kang, J.S., Bector, R., Baghdassarian, H.M., Truong, A.N., Kim, P. W., Singh, A.K., Hillson, N.J., Keasling, J.D., 2018. De novo DNA synthesis using polymerasenucleotide conjugates. Nat. Biotechnol. 36 (7), 645–650. https://doi.org/10.1038/nbt.4173.

Panahi, B., Hamid, R., Jalaly, H.M.Z., 2025. Deciphering plant transcriptomes: leveraging machine learning for deeper insights. Curr. Plant Biol. 41, 100432.

Pillay, M., 2020. Genome editing technologies for crop improvement. In: Quantitative Genetics, Genomics and Plant Breeding. CaBI, Wallingford UK, pp. 33–43.

Pirali, A., Jafarpour, F., Hajian, M., Hosseini Moghaddam, S.H., Moradi, R., Tanhaie-Vash, N., Rahimi Andani, M., Izadi, T., Shiralian-Esfahani, H., Safaeinejad, Z., Kues, W., Nasr-Esfahani, M.H., Eghbalsaied, S., 2025. Editing the CYP19 gene in goat embryos using CRISPR/Cas9 and somatic cell nuclear transfer techniques. Cell. Reprogr. 27 (2), 86–93. https://doi.org/10.1089/cell.2024.0109.

Plagens, A., Richter, H., Charpentier, E., Randau, L., 2015. DNA and RNA interference mechanisms by CRISPR-Cas surveillance complexes. FEMS (Fed. Eur. Microbiol. Soc.) Microbiol. Rev. 39 (3), 442–463. https://doi.org/10.1093/femsre/fuv019.

Pušnik, Ž., Mraz, M., Zimic, N., Moškon, M., 2022. Review and assessment of Boolean approaches for inference of gene regulatory networks. Heliyon 8 (8), e10222. https://doi.org/10.1016/j.heliyon.2022.e10222.

Qiu, X., Zhu, L., Wang, H., Xie, M., 2025. Biocomputing at the crossroad between emulating artificial intelligence and cellular supremacy. Curr. Opin. Biotechnol. 92, 103264. https://doi.org/10.1016/j.copbio.2025.103264.

Qu, Y., Huang, K., Yin, M., Zhan, H., Liu, D., Yan, D., Cousin, HC., et al., 2026. CRISPR-GPT for agentic automation of gene-editing experiments. Nat. Biomed. Eng. 10, 245–258. https://doi.org/10.1038/s41551-025-01463-z.

Regenwetter, L., Nobari, A.H., Ahmed, F., 2022. Deep generative models in engineering design: a review. J. Mech. Des. 144 (7), 071704.

Rood, J.E., Hupalowska, A., Regev, A., 2024. Toward a foundation model of causal cell and tissue biology with a Perturbation Cell and Tissue Atlas. Cell 187 (17), 4520–4545.

Sarker, I.H., 2021. Machine learning: algorithms, real-world applications and research directions. SN Comput. Sci. 2 (3), 160.

Schraivogel, D., Kuhn, T.M., Rauscher, B., Rodríguez-Martínez, M., Paulsen, M., Owsley, K., Middlebrook, A., Tischer, C., Ramasz, B., Ordoñez-Rueda, D., Dees, M., Cuylen-Haering, S., Diebold, E., Steinmetz, L.M., 2022. High-speed fluorescence image-enabled cell sorting. Science 375 (6578), 315–320. https://doi.org/10.1126/science.abj3013.

Shakirova, K.M., Ovchinnikova, V.Y., Dashinimaev, E.B., 2020. Cell reprogramming with CRISPR/Cas9 based transcriptional regulation systems. Front. Bioeng. Biotechnol. 8, 882.

Siedhoff, N.E., Schwaneberg, U., Davari, M.D., 2020. Machine learning-assisted enzyme engineering. Methods Enzymol. 643, 281–315.

Silva, J.C.F., Teixeira, R.M., Silva, F.F., Brommonschenkel, S.H., Fontes, E.P.B., 2019. Machine learning approaches and their current application in plant molecular biology: a systematic review. Plant Sci. 284, 37–47. https://doi.org/10.1016/j.plantsci.2019.03.020.

Söllner, J.H., 2022. An Experimental Approach of an in Vivo Pathogen Genome Targeting Strategy to Generate African Swine Fever Resistant Pigs. Tierärztliche Hochschule Hannover. Doctoral dissertation.

Soomro, S.R., Soomro, S.N., Altaf, M.T., Liaqat, W., Nadeem, M.A., Baloch, F.S., Aasim, M., Mohamed, H.I., 2025. Development of tetraploids in tissue culture: modern techniques and biotechnological innovations. Plant Cell Tissue Organ Cult. 160 (2). https://doi.org/10.1007/s11240-025-02994-8.

Stuart, L.A.G., Wells, D.M., Atkinson, J.A., Castle-Green, S., Walker, J., Pound, M.P., 2025. High-fidelity wheat plant reconstruction using 3D Gaussian splatting and neural radiance fields. GigaScience 14. https://doi.org/10.1093/gigascience/giaf022.

Sudhakar, P., Latha, P., Reddy, P.V., 2016. Phenotyping Crop Plants for Physiological and Biochemical Traits. Academic Press.

Sun, L., Lai, M., Ghouri, F., Nawaz, M.A., Ali, F., Baloch, F.S., Nadeem, M.A., Aasim, M., Shahid, M.Q., 2024. Modern plant breeding techniques in crop improvement and genetic diversity: from molecular markers and gene editing to artificial intelligence—a critical review. Plants 13 (19). https://doi.org/10.3390/plants13192676.

Teixeira, A.P., Fussenegger, M., 2024. Synthetic macromolecular switches for precision control of therapeutic cell functions. Nat. Rev. Bioeng. 2 (12), 1005–1022.

Tian, Z., Wang, J.W., Li, J., Han, B., 2021. Designing future crops: challenges and strategies for sustainable agriculture. Plant J. 105 (5), 1165–1178. https://doi.org/10.1111/tpj.15107.

Tian, Z., Nepomuceno, A.L., Song, Q., Stupar, R.M., Liu, B., Kong, F., Ma, J., Lee, S.H., Jackson, S.A., 2025. Soybean2035: a decadal vision for soybean functional genomics and breeding. Mol. Plant 18 (2), 245–271. https://doi.org/10.1016/j.molp.2025.01.004.

Trachsel, S., Dhliwayo, T., Perez, L.G., Lugo, J.A.M., Trachsel, M., Ali, J., 2019. Estimation of physiological genomic estimated breeding values (PGEBV) combining full hyperspectral and marker data across environments for grain yield under combined heat and drought stress in tropical maize (Zea mays L.). PLoS One 14 (3), e0212200. https://doi.org/10.1371/journal.pone.0212200.

Uranga, M., Martín-Hernández, A.M., Pasin, F., De Storme, N., 2025. CRISPR plants for studying functional genomics: methods and applications. In: CRISPR for Climate-Smart and Sustainable Agriculture. Academic Press., pp. 27–59.

van Dis, E.A.M., Bollen, J., Zuidema, W., van Rooij, R., Bockting, C.L., 2023. ChatGPT: five priorities for research. Nature 614 (7947), 224–226. https://doi.org/10.1038/d41586-023-00288-7.

Veronica, A., Reiprich, T.H., Pacaud, F., Sanders, J.S., Gattuzz, E., Yeung, M.C.H., Bulbul, E., Ghirardini, V., Liu, A., Mannes, C., Morelli, A., Ota, N., 2025. The SRG/eROSITA all-sky survey: large-scale view of the centaurus cluster. Astron. Astrophys. 694, A168. https://doi.org/10.1051/0004-6361/202450277.

Voss-Fels, K.P., Cooper, M., Hayes, B.J., 2019. Accelerating crop genetic gains with genomic selection. Theor. Appl. Genet. 132 (3), 669–686.

Wang, H., Fu, T., Du, Y., Gao, W., Huang, K., Liu, Z., Chandak, P., Liu, S., Katwyk, P.Van, Deac, A., Anandkumar, A., Bergen, K., Gomes, C.P., Ho, S., Kohli, P., Lasenby, J., Leskovec, J., Liu, T.-Y., Manrai, A., Marks, D., Ramsundar, B., Song, Le, Sun, J., Tang, J., Veličković, P., Welling, M., Zhang, L., Coley, C.W., Bengio, Y., Zitnik, M., 2023. Scientific discovery in the age of artificial intelligence. Nature 620 (7972), 47–60. https://doi.org/10.1038/s41586-023-06221-2.

Wang, Z., Liang, X., Li, M., Li, S., Liu, J., Zheng, L., 2025. Towards cognitive intelligence-enabled product design: the evolution, state-of-the-art, and future of AI-enabled product design. J. Ind. Inf. Integr. 43, 100759. https://doi.org/10.1016/j.jii.2024.100759.

Wei, P.J., Guo, Z., Gao, Z., Ding, Z., Cao, R.F., Su, Y., Zheng, C.H., 2024. Inference of gene regulatory networks based on directed graph convolutional networks. Briefings Bioinf. 25 (4). https://doi.org/10.1093/bib/bbae309.

Weidemüller, P., Kholmatov, M., Petsalaki, E., Zaugg, J.B., 2021. Transcription factors: bridge between cell signaling and gene regulation. Proteomics 21 (23–24), 2000034.

Xu, Y., Zhang, X., Li, H., Zheng, H., Zhang, J., Olsen, M.S., Varshney, R.K., Prasanna, B.M., Qian, Q., 2022. Smart breeding driven by big data, artificial intelligence, and integrated genomic-enviromic prediction. Mol. Plant 15 (11), 1664–1695. https://doi.org/10.1016/j.molp.2022.09.001.

Ye, K., Hu, G., Tong, Z., Xu, Y., Zheng, J., 2025. Key intelligent pesticide prescription spraying technologies for the control of pests, diseases, and weeds: a review. Agriculture 15 (1), 81. https://doi.org/10.3390/agriculture15010081.

Yu, Q., Liu, S., Yu, L., Xiao, Y., Zhang, S., Wang, X., Xu, Y., Yu, H., Li, Y., Yang, J., Tang, J., Duan, H.-C., Wei, L.-H., Zhang, H., Wei, J., Tang, Q., Wang, C., Zhang, W., Wang, Y., Song, P., Lu, Q., Zhang, W., Dong, S., Song, B., He, C., Jia, G., 2021. RNA demethylation increases the yield and biomass of rice and potato plants in field trials. Nat. Biotechnol. 39 (12), 1581–1588. https://doi.org/10.1038/s41587-021-00982-9.

Zeng, D., Li, X., Huang, J., Li, Y., Cai, S., Yu, W., Zhu, Q., 2019. Engineered Cas9 variant tools expand targeting scope of genome and base editing in rice. Plant Biotechnol. J. 18 (6), 1348.

Zhang, H., Wang, L., Jin, X., Bian, L., Ge, Y., 2023a. High-throughput phenotyping of plant leaf morphological, physiological, and biochemical traits on multiple scales using optical sensing. The crop journal 11 (5), 1303–1318.

Zhang, Z., Wu, Q.Y., Ge, Y., Huang, Z.Y., Hong, R., Li, A., Xu, J.H., Yu, H.L., 2023b. Hydroxylases involved in terpenoid biosynthesis: a review. Bioresources and Bioprocessing 10 (1). https://doi.org/10.1186/s40643-023-00656-1.

Zhang, D., Xu, F., Wang, F., Le, L., Pu, L., 2025. Synthetic biology and artificial intelligence in crop improvement. Plant Commun. 6 (2), 101220. https://doi.org/10.1016/j.xplc.2024.101220.

Zhao, M., He, W., Tang, J., Zou, Q., Guo, F., 2021. A comprehensive overview and critical evaluation of gene regulatory network inference technologies. Briefings Bioinf. 22 (5). https://doi.org/10.1093/bib/bbab009.

Zhao, M., He, W., Tang, J., Zou, Q., Guo, F., 2022. A hybrid deep learning framework for gene regulatory network inference from single-cell transcriptomic data. Briefings Bioinf. 23 (2), bbab568.

Zheng, Z., Liu, T., Chai, N., Zeng, D., Zhang, R., Wu, Y., Hang, J., Liu, Y., Deng, Q., Tan, J., Liu, J., Xie, X., Liu, Y.-G., Zhu, Q., 2024. PhieDBEs : a DBD -containing, PAM -flexible, high-efficiency dual base editor toolbox with wide targeting scope for use in plants. Plant Biotechnol. J. 22 (11), 3164–3174. https://doi.org/10.1111/pbi.14438.
Zhou, Y., Wei, Y., Li, L., Yan, T., Ye, H., 2025. Optogenetics in medicine: innovations and therapeutic applications. Curr. Opin. Biotechnol. 92, 103262. https://doi.org/10.1016/j.copbio.2025.103262.
Zhu, W., Han, R., Shang, X., Zhou, T., Liang, C., Qin, X., Chen, H., Feng, Z., Zhang, H., Fan, X., Li, W., Lin, L., 2024. The CropGPT project: call for a global, coordinated effort in precision design breeding driven by AI using biological big data. Mol. Plant 17 (2), 215–218. https://doi.org/10.1016/j.molp.2023.12.015.

CHAPTER

14

AI-driven gene editing for disease resistance in plants

Rajrishi Awadhesh Kumar[a], *Anshuman Mishra*[a], *and Siddhesh B. Ghag*

Department of Atomic Energy, Centre for Excellence in Basic Sciences, School of Biological Sciences, University of Mumbai, Mumbai, Maharashtra, India

14.1 Introduction

14.1.1 The global burden of phytopathogens

Global food security is under constant threat from plant diseases. Phytopathogenic microorganisms, including fungi, bacteria, viruses, and oomycetes, exert a relentless pressure on agricultural systems worldwide, leading to major economic, ecological, and social consequences. Plant diseases reduce global crop fields by as much as 20%–40% annually, taking a toll on the lives of billions of people across the globe, especially those in regions of extreme poverty and hunger (Savary et al., 2019). In the United States alone, crop losses every year due to a variety of invasive plant pathogens, especially fungi, lead to an estimated $21 billion financial loss (Rossman, 2009). The bacterial pathogen *Xylella fastidiosa*, since its emergence in Southern Italy in 2013, has led to the death of an estimated 21 million olive trees. It could lead to a staggering financial loss of around €5.5 billion annually if it spreads across the European Union (Sabelli, 2023; Xylella fastidiosa, 2024). This will have a severe impact on the lives of the people directly or indirectly associated with its production, affecting almost around 300,000 jobs (Xylella fastidiosa, 2024). The Irish Potato Famine in the mid-19th century, caused by the oomycete *Phytophthora infestans*, stands as a stark reminder of how the collapse of a staple crop due to disease can lead to mass starvation, high mortality, and political instability, leading to social unrest (Coomber et al., 2024).

The impact of plant diseases extends far beyond mere yield reduction. It influences the global economy, affecting international trade, food prices, and employment. This changes

[a] Equal Contribution.

Intelligent Gene Editing in Plants
https://doi.org/10.1016/B978-0-443-44079-3.00001-2

the behavior of the producers and consumers, affecting market prices and the processing industries at large. Neglecting plant health would result in loss of biodiversity, seriously disrupting environmental balances. This will lead to long-term instability in natural and agricultural ecosystems. Widespread monocropping practices due to growing international trade, as well as global climate change, are facilitating the rapid spread of plant pathogens. This major crisis urgently requires the development of novel strategies for crop protection and disease resistance in plants.

14.1.2 Limitations of conventional breeding

Conventional plant breeding has been the primary method for developing disease-resistant crop varieties across the world. It relies on the hybridization of plants with desirable traits and selections based on phenotypic characters. This has successfully introduced numerous resistance traits into elite cultivars (Ikram et al., 2025). However, conventional breeding is limited mainly by the time taken and the laborious effort required to develop a disease-resistant plant variety. Also, the dependence of conventional breeding exclusively on the existing gene pool highly limits the potential of developing disease-resistant cultivars (Ikram et al., 2025; Savary et al., 2019). Furthermore, the evolutionary arms race between the slow-growing plants and the rapidly adapting pathogens pose a serious hurdle on the pace of development of resistance mechanisms in plants. These limitations, hence, necessitate the adoption of new methods such as precision genome editing for the rapid development of disease resistance mechanisms in plants.

14.1.3 The advent of genome editing: a paradigm shift

A major advantage of using precision genome editing tools is the substantial reduction of breeding time. By directly engineering desirable traits into elite crop varieties, the development process can be shortened from over a decade to as low as 2–5 years (Erdoğan et al., 2023; Kumar et al., 2024). This allows breeders to mount a much more effective response to emerging plant diseases. Genome-editing techniques, especially the Clustered Regularly Interspaced Short Palindromic Repeats (CRISPR) and CRISPR-associated protein (Cas) system, have completely transformed conventional plant breeding practices (Erdoğan et al., 2023; Nisa et al., 2022). This technology provides the ability to make precise and targeted gene modifications in plants at a much higher efficacy as compared to the earlier genome editing tools, like zinc finger nucleases (ZFNs) and transcription activator-like effector nucleases (TALENs) (Borrelli et al., 2018). The CRISPR/Cas9 system uses a synthetic guide RNA (gRNA) sequence complementary to the desired gene of interest. This directs the Cas9 enzyme complex to make a double-strand break (DSB) at the target site. The cell's natural DNA repair mechanisms, such as nonhomologous end joining (NHEJ) and homology-directed repair (HDR) pathways, then resolve this DSB break. The error-prone NHEJ pathway often introduces small insertions or deletions (indels) at the cut site, effectively knocking out the gene's function. While the HDR pathway can also be utilized to introduce known changes in the plant's genome by providing a donor DNA as a template to modify specific DNA stretches during the repair process. To reuse the CRISPR system, one only needs to synthesize a new, short gRNA molecule without the need to engineer the entire protein. This reduces the overall cost and complexity of the genome editing process and makes it accessible to a vast community of researchers across the globe (Borrelli et al., 2018; Nisa et al., 2022).

14.2 Genome editing strategies for engineering disease resistance in plants

14.2.1 Targeting single loci for monogenic resistance

One of the most important applications of the CRISPR/Cas9 system for harboring disease resistance is the modification of the single genetic loci. These strategies are quite simple yet highly effective. They can provide strong and durable resistance by precisely altering host–pathogen interaction at the molecular level. Among these approaches, the most successful and widely adopted strategy involves the targeted inactivation of host susceptibility (*S*) genes (Erdoğan et al., 2023; Nisa et al., 2022). Pathogens have evolved to exploit proteins encoded by these host genes for infection, colonization, or nutrient acquisition. CRISPR/Cas9-mediated loss-of-function mutations introduced in *S* genes make the plant nonpermissive to the pathogen. This approach often results in a broad-spectrum resistance when compared to the traditional *R*-gene-based immunity. Knocking out *Mildew Locus O* (*MLO*) genes in a variety of crops like wheat, tomato, and grapevine provides durable resistance to powdery mildew fungi (Erdoğan et al., 2023; Jorgensen, 1992). An attempt to edit promoter regions of the *SWEET* (*Sugar Will Eventually be Exported Transporters*) gene in rice was made to prevent pathogen-mediated transcriptional induction. Pathogens such as *Xanthomonas oryzae* pv. *oryzae*, the causal agent of bacterial blight in rice, secretes transcription activator-like (TAL) effector proteins to induce expression of these *SWEET* genes in host plants. The plant's sugar transport system is then hijacked and repurposed for feeding the invading bacteria. By introducing targeted mutations in the effector binding sites within the *SWEET* gene promoters, this induction is prevented. This starves the pathogen and confers robust disease resistance to the host plants (Zhou et al., 2015). Similar success has been achieved by editing the promoter of the *Lateral Organ Boundaries 1* (*CsLOB1*) gene in citrus to enhance resistance to citrus canker (Jia et al., 2016). Similarly, CRISPR/Cas9-mediated mutations can be introduced in host resistance (*R*) genes, responsible for recognizing pathogenic effector molecules, expanding their recognition spectrum, and/or enhancing their expression, strengthening the plant's defense response against invading pathogens.

14.2.2 Multiplex genome editing (MGE) for durable, broad-spectrum resistance

Agronomic traits, including disease resistance, often depend on multiple genes exhibiting genetic redundancy. This is a major challenge, especially within polyploid crops like wheat and banana, having multiple sets of subgenomes. CRISPR has the ability for MGE, which can be used to tackle this problem and engineer highly resilient plant varieties. In particular cases, knocking out a homeoallele (single copy of the gene) won't show any observable phenotype because of the genetic compensation from the remaining allelic copies. MGE, however, allows for the simultaneous knockout of all homeoalleles, completely masking the genes' functions, providing us with the desired trait (Abdelrahman et al., 2021). MGE is also effective in pyramiding or layering multiple resistance mechanisms into a single elite cultivar. This strategy works by merging several distinct protective modes of action, presenting a multilayered defense against the pathogen, making it far more robust than monogenic resistance. This is achieved by simultaneously knocking out multiple unrelated *S* genes or by combining multiple *S*-gene knockouts with the enhancements of *R*-gene pathways. To combat diverse strains of the bacterial blight disease in rice, researchers used a multiplex CRISPR approach to simultaneously induce mutations in the promoter region of three different *SWEET* genes (*OsSWEET11*, *OsSWEET13*, and *OsSWEET14*), giving a stable genotype with a broad-spectrum resistance, which would be way more troublesome if done by the conventional breeding method (Oliva et al., 2019). The precision and efficacy of CRISPR-based genome editing experiments are totally dependent on the design of its components, mainly the gRNA. Selection of an optimal gRNA is extremely crucial and a multiparameter problem, which has boosted the development of various computational tools. The integration of machine learning (ML) and deep learning (DL) methods, as well as large language models (LLMs), into CRISPR-based genome workflows has changed the field from being an error-prone, hit-and-trial-based method into data-driven and predictive engineering.

14.3 Artificial intelligence and genome editing—a new era

AI has completely revolutionized the way we edit genomes. It seamlessly integrates into every step of the process, making it more accurate, efficient, and highly scalable. Modern-day AI-driven genome editing encompasses several domains, mainly.

1. Optimal gRNA design prediction
2. Predicting off-target effects of gRNA edits
3. High-throughput phenotyping and screening
4. Chat-based CRISPR tools for planning and designing genome editing workflows

Algorithms used in gene editing have evolved from rigid, rule-based, static systems to adaptive, data-driven, dynamic frameworks. These tools now even integrate LLMs, completely redefining the scope of the field. Drastic progress has been made from the manual error-prone workflows to an intelligent automated system, which is highly capable of handling complex genomic architecture. Primer3, for example, is a great tool designed to optimize PCR primer design by calculating melting temperatures (T_m) and GC content and avoiding the formation of primer dimers and hairpin looping. Its traditional approach, based on thermodynamic principles, though effective for pilot sequences, could not predict off-target effects or adapt to a large genomic dataset or a complex genome with high conservation (Untergasser et al., 2012). Similarly, CRISPRitz introduced hash-table indexing, which found potential genome-wide off-target sites by ranking hits using simple scoring functions

(Cancellieri et al., 2020). Though these approaches offered precision and accuracy, their performance dipped significantly when faced with very large genomic datasets, especially polyploidy. CRISPResso2, on the other hand, transformed the workflow of next-generation sequencing (NGS) experiments. It used alignment-based methods (e.g., Burrows–Wheeler Aligner or Needleman–Wunsch algorithms) to quantify indel percentage and editing efficacies, aiding in the interpretation of sequencing results (Clement et al., 2019). Even though vital for early progress, these computationally intensive and tedious methods necessitated the development of newer algorithms to account for the increasing complexity of genomic datasets.

The integration of ML and DL methods in the gene-editing workflow marks a significant shift. For instance, the development of tools such as CCLMoff and CRISPR-M significantly improved the accuracy of predicting off-target activity and indels. CCLMoff used a pretrained RNA-based LLM model to predict any off-target activity. This approach turned out to be very efficient and accurate across species in identifying even subtle sequence mismatches (Du et al., 2025). CRISPR-M, on the other hand, is an innovative algorithm that merged convolutional neural networks (CNNs) and bidirectional long short-term memory (LSTM) to accurately predict off-target effects at target sites containing indels and mismatches (Sun et al., 2024). Elektrum further enhanced this by integrating kinetically interpretable neural networks (KINNs) with standard models trained on in vitro kinetic data to predict any indel possibilities (Zhang et al., 2023). These tools demonstrated how ML algorithms can predict more subtle off-target effects and repair outcomes depending on the dynamic information they acquire over time.

For potential gRNA design, tools like CHOPCHOP, GuideMaker, and CRISPRware introduce ML-based scoring systems (Labun et al., 2019; Poudel et al., 2022; Malekos et al., 2024). CRISPRware presents an exciting development in designing efficient gRNA libraries at a genome-wide scale (Schmidt et al., 2025; Malekos et al., 2024). Whereas CROPSR is known to handle complex polyploid plant genomes, integrating interdisciplinary ML-based efficiency predictors with threshold filters (Müller Paul et al., 2022). Variant-aware Cas-OFFinder extends this further with off-target analysis by incorporating single nucleotide polymorphisms (SNPs) and indels for individual, haplotype-specific predictions (Bae et al., 2014). Though these tools significantly improved the specificity and scalability of genome editing workflows, they still, however, need human reliance, as they only work on labeled datasets with limited contextual reasoning, hindering fully automated end-to-end gRNA design.

The latest advancement in gene editing came with the introduction of LLMs, which combine the strengths of DL with natural language processing (NLP), enabling intelligent, self-optimizing experimental workflows. CRISPR-GPT, a great example of this shift, is fine-tuned on Llama3-instruct (8B parameters) with CRISPR discussion data. It uses retrieval-augmented generation (RAG) to retrieve information from literature and protocols for predictions and conclusions. Unlike earlier tools, which required manual input and interpretation at multiple stages, CRISPR-GPT can reason about complex biological systems, generate human-like text for experimental design, and troubleshoot ideas in real time. This multifactorial approach allows seamless transitions from planning to automated execution and validation of gene-editing experiments (Qu et al., 2025).

14.4 Scope of AI-mediated plant genome-editing workflows

With rapidly evolving AI algorithms and improving computational capabilities of modern-day computers, AI models can specifically be trained on enormous experimental datasets catering to the needs of modern agriculture. This would significantly diversify and improve their scope of applications. From designing efficient gRNA molecules to predicting the outcomes of the gene edits as well as the phenotypical characteristics of the crop, AI can be crucial at all these intermittent steps. This integration of AI into the experimental workflow would transform genome editing from a process of trial and error into a more predictive, design-oriented science, accelerating the entire crop improvement pipeline from lab to the field (Table 14.1).

14.4.1 Enhanced gRNA design and better off-target predictions

A significant area of research in the field of AI-driven genome editing is the ability to predict the precise outcome of the DNA repair process that follows a CRISPR-induced DSB. When a cell repairs a break via the error-prone NHEJ pathway, it often introduces indels at the cut site. For a long time, predicting the exact nature and frequency of these indels was a major challenge, making the outcome of a gene knockout experiment somewhat unpredictable. Now, AI algorithms can not only analyze large datasets to predict optimal gRNAs for CRISPR–Cas systems, but also assess large datasets of genomic sequences to identify potential off-target sites, improving on-target efficiency and reducing off-target effects. By training on massive datasets generated from the results of experimental genome editing, where the outcomes of thousands of different gRNAs have been sequenced, these algorithms can learn the complex and consistent rules that govern the DNA repair process. These models have revealed that repair outcomes are not random but are dependent on the local sequence context around the cleavage site. This allows them to predict the indel patterns that may arise from a specific cut. For example, tools such as "Pythia" and "Apindel" have been developed on similar principles, which can predict the outcomes of a DNA DSB repair process. This can improve the efficiency of designing gRNA molecules for the target gene (Liu et al., 2022; Naert et al., 2024, 2025).

14.4.2 Discovery and design of novel CRISPR systems

In addition, AI can aid in discovering and/or designing novel CRISPR systems, expanding the toolkit for genome editing. Instead of merely finding the best gRNA to use with a natural Cas9 enzyme, AI is now capable of designing better Cas9 enzymes with specific properties like enhanced specificity or reduced off-target cuts. For example, researchers at Profluent Bio used an LLM to generate millions of novel Cas9-like protein sequences. The best sequence, OpenCRISPR-1, demonstrated on-target editing efficiency in human cells similar to the widely used, naturally occurring *Streptococcus pyogenes* Cas9 (SpCas9) enzyme. Remarkably, it showed a 95% reduction in off-target editing activity (Ruffolo et al., 2025).

TABLE 14.1 List of software/databases relevant at different stages of the gene-editing workflow.

Workflow stage	Tool name	Core function	Underlying technology	Access	Reference
1. Target ID	KnetMiner Nova	Gene-trait knowledge discovery	Knowledge graph, RAG	Freemium SaaS	Hassani-Pak et al. (2021)
	LLM-GeneGraph	Extract gene interactions from literature	LLM, RAG	Open-source framework	Li et al. (2025b)
	DSP database	Repository of plant susceptibility genes	Curated database	Public web database	Kaur et al. (2023)
	PRGdb	Repository of plant resistance genes	Curated database	Public web database	Calle García et al. (2022)
2. gRNA design	DeepCRISPR	Predict on-target gRNA efficacy	Hybrid deep neural network	Open-source software	Chuai et al. (2018)
	CRISPRon	Predict on-target gRNA efficacy	Deep learning	Free web server	Anthon et al. (2022)
	DNABERT	Predict gRNA activity (fine-tunable)	Transformer (foundational model)	Open-source model	Ji et al. (2021)
	Benchling	Integrated gRNA design and workflow management	Web platform	Commercial SaaS	(Benchling)
	Synthego tool	Design knockout gRNAs	Web platform (Azimuth 2.0)	Free web tool	(Synthego)
	GenScript tool	Design gRNA and HDR templates	Web platform (Doench score)	Free web tool	(GenScript)
	CRISPOR	Design gRNAs with aggregated scores	Web platform	Free web server	Concordet and Haeussler (2018)
	CHOPCHOP	Design gRNAs for various applications	Web platform	Free web server	Labun et al. (2019)
3. Off-target	Cas-OFFinder	Identify potential off-target sites	Sequence alignment (GPU-accelerated)	Free web server/CLI	Bae et al. (2014)
	CRISPR-PLANT v2	Plant-specific gRNA design/off-target analysis	Precomputed database	Public web database	Minkenberg et al. (2019)
	CRISPR-net	Quantify off-target cleavage activity	Recurrent convolutional neural network	Open-source software	Lin et al. (2020)

(*Continued*)

TABLE 14.1 List of software/databases relevant at different stages of the gene-editing workflow.—cont'd

Workflow stage	Tool name	Core function	Underlying technology	Access	Reference
	Variant-aware Cas-OFFinder	Identify off-targets using personal genomes	Sequence alignment	Free web server/CLI	Mekonnen et al. (2025)
	CRISPRoffT	Database of validated off-target sites	Curated database	Public web database	Wang et al. (2025)
4. Validation	genoTYPER-NEXT™	High-throughput NGS validation of edits	NGS	Commercial service	admin (2024)
	Geneious prime	Analyze NGS/Sanger data for edits	Bioinformatics software	Commercial license	(Geneious)
	CRISPResso2	Analyze NGS data for edit outcomes	Software pipeline	Open-source software	Clement et al. (2019)
	ICE tool	Analyze Sanger data for indel frequency	Deconvolution algorithm	Free web tool	Conant et al. (2022)
5. Phenotyping	PlantPAD	Database and models for disease diagnosis	Deep learning, image database	Public web database	(PlantPAD)
	Plantix	Mobile app for disease/pest diagnosis	AI image recognition	Free mobile app	Siddiqua et al. (2022)
	Leaf doctor	Mobile app to quantify disease severity	Image analysis algorithm	Free mobile app	Pethybridge and Nelson (2015)

14.4.3 Integration with multiomics for better phenotypical predictions

It is necessary to incorporate AI in various omics-related data analysis workflows to get a comprehensive understanding of plant biology. This would help to predict phenotypic outcomes of any genetic modification in a crop, aiding in its design with desired traits. These technologies generate massive, high-dimensional, and heterogeneous datasets that are impossible to interpret with traditional statistical methods. This is where AI and ML algorithms are uniquely capable of identifying subtle patterns and correlations in these distinctive layers of data. By analyzing how changes at the genomic level affect the transcriptome, proteome, and metabolome to produce an observable trait, these algorithms would be helpful in deciphering the genetic components of even the most complex and difficult-to-distinguish traits. *R*-genes, for example, comprising an intracellular nucleotide-binding domain and leucine-rich repeat receptors (NLRs) and cell-surface pattern recognition receptors (PRRs) can now be annotated with high sensitivity using DL and protein language models trained on sequence and domain patterns (Liu et al., 2024). Similarly, *S*-genes that lack any unifying sequence signature are instead identified through correlational and expression-based approaches. Techniques such as genome-wide association studies (GWAS) and quantitative trail loci (QTL) mapping identify loci associated with susceptibility, whereas transcriptomic analysis identifies genes upregulated during infection. Homology searches, on the other hand, give an idea of functionally characterized *S*-genes across species (Liu et al., 2024; Thakur et al., 2025). Furthermore, the cis-regulatory elements (CREs) that control defense gene expression are decoded using motif databases and multiomics inference of gene regulatory networks (Rombauts et al., 2003). Methods like ConSReg integrate data from RNA-Seq, transcription factor (TF)-binding assays, and chromatin accessibility assays to predict condition-specific TF target relationships (Song et al., 2020). These do not work independently but form interconnected networks that provide a translational route to CRISPR-based tools. CRISPR/Cas9 knockouts enable the targeted disruption of *S*-genes for durable, broad-spectrum resistance, while precision edits to *R*-gene domains can refine pathogen recognition or manipulate expression. Future progress will depend on integrative AI models trained on multiomics datasets, which can help maximize durable resistance while minimizing negative agronomic trade-offs.

14.4.4 High-throughput phenotyping (HTP) for precision agriculture

Traditional phenotyping involves manually measuring plant phenotypic traits. It is often a major hurdle in breeding programs, as it is labor-intensive, slow, and highly subjective. HTP, on the other hand, enables rapid and nondestructive measurement of plant traits using advanced imaging technologies, sensor platforms, and unmanned aerial vehicles (UAVs). HTP platforms rely on phenotyping crops by using a range of sensors and cameras, such as the RGB (red, green, blue), thermal, and hyperspectral cameras, mounted on ground-based surveying systems or UAVs. These systems can aid in the phenotyping of crops across developmental stages rapidly and nondestructively. It is mainly used to monitor stress responses, plant growth, and yield potential under diverse environmental conditions (Gill et al., 2022; Sheikh et al., 2024). This has led to the development of high-throughput screening (HTS) practices.

The integration of AI into HTP workflow has greatly transformed crop-breeding practices. Computer vision and DL models are used to process huge imaging datasets to extract and quantify key phenotypic characters related to disease resistance. This helps in detecting subtle signs of disease symptoms before they are even visible to the human eye. These difficult-to-measure traits can be discovered and selected for by integrating these algorithms into the data analysis protocol. These models can also be trained on a variety of imaging datasets for accurately predicting plant responses to environmental stresses. This would help breeders to make informed decisions, assisting in the selection of resilient genotypes (Ghosal et al., 2018; Singh et al., 2016). AI hastens breeding cycles by automating the phenotyping practices, supporting the selection of superior genotypes, and designing efficient genetic crosses. Collectively, these advancements not only improve the accuracy and reliability of trait evaluation but also ensure that all the genetic modifications are effectively assessed for their phenotypes, including yield, stress tolerance, and resistance to biotic and abiotic challenges (Bai et al., 2023; Gill et al., 2022; Sheikh et al., 2024; Chivasa et al., 2020). Moreover, combining omics data with environmental information into these algorithms would improve the scalability and precision of these platforms. An overall analysis of the plants' performance is crucial to understanding how genetic modifications would influence complex traits. This would optimize resource utilization by prioritizing the most promising genotypes early on in the breeding programs, thereby reducing time and costs (Shawon et al., 2025).

Macro-scale adoption of HTP and HTS technologies in breeding practices would require inputs from various industries, including the semiconductor and IT industries. The overall workflow needs to be streamlined and simplified for a common farmer to be able to use it without much technical expertise. Also, building and managing large facilities to store the large amount of data that would be generated in the process poses critical challenges for its large-scale accessibility. In addition, framing laws to ensure ethical and judicious use of this data is extremely crucial. Addressing these hurdles through interdisciplinary collaboration and robust data management frameworks will be essential to fully realize the potential of AI in the development of precision agricultural practices.

14.5 Precision agricultural practices for developing disease resistance in plants

Integration of AI into genome editing workflows is reshaping modern agriculture. New and advanced genome editing tools equipped with efficient AI algorithms are increasingly being used to better optimize experimental strategies. The following are some of the interesting applications of these emerging new technologies in the field of agriculture to improve disease resistance and crop qualities.

14.5.1 Receptor engineering to improve recognition and immune response

A highly conserved PRR on the plant cell surface known as Flagellin-sensing 2 (FLS2) is known for recognizing a conserved 22-amino acid-long immunogenic peptide (flg22) from the bacterial flagellin protein. This epitope (flg22) is polymorphic in nature, while most of the FLS22 receptors have a narrow range of perception for these peptides. Using various bioinformatics approaches, including AlphaFold modeling, key amino acid residues responsible for a broad range of pathogen recognition were mapped to the concave surface of the FLS2 receptor. Furthermore, when used for predicting recognition of peptide variants by FLS2 homologs, AlphaFold3 predicted a complex structure that resembled that experimentally obtained and published results (Li et al., 2025a).

14.5.2 Citrus canker resistance in sweet orange

Citrus canker is a severe disease caused by the bacterial pathogen *Xanthomonas citri* subsp. *citri*. The symptoms are marked by lesions on the fruit, leaves, and stems of citrus trees. The *CsLOB1* (*Citrus sinensis* Lateral Organ Boundaries) gene is a susceptibility factor for citrus canker in grapefruit. It is induced by the binding of the pathogenicity factor PthA4. Using the CRISPR/Cas9 system, PthA4 effector binding elements (EBEs) in the *CsLOB1* promoter were modified to generate citrus-canker-resistant plants. Software such as CRISPR-P 2.0 and CasOFFinder were used for predicting off-target sites of gRNA edits in the study (Jia et al., 2016, 2024; Huang et al., 2022).

14.5.3 Citrus greening disease, or huanglongbing

Citrus greening disease, also known as Huanglongbing (HLB) disease, is caused by the bacterium *Candidatus* Liberibacter asiaticus (CLas). It is primarily transmitted by the Asian citrus psyllid (*Diaphorina citri*) (Qiu et al., 2022). Affected plants are small and deformed and produce green or "red nose" fruits, leading to a decline in the fruit yield and quality. These symptoms are complex and variable, which makes their diagnosis very difficult. The DL-based YOLOv5l algorithm was used to develop a software for early detection of HLB symptoms with high accuracy. Furthermore, a mobile software was developed that can be used by common farmers for efficient and early detection of diseased plants (Qiu et al., 2022). The MGA-UNet segmentation model was developed using multispectral imaging data collected from UAVs and deep CNNs. This improves the efficiency of diagnosis for HLB (Ye et al., 2025).

14.5.4 Rice Blast disease resistance

The CRISPR/Cas9 system was used to create small indels in the host susceptibility (*S*) genes, which include the ethylene-responsive element binding factor (ERF) transcription factor gene *OsERF922* and the histone deacetylase gene *OsHDT701*, to confer disease resistance against rice blast disease caused by the fungus *Magnaporthe oryzae* (Wang et al., 2016). Software such as UNAFold, based on mfold and DINAMelt algorithms, was used to design the vector system and assess secondary structures of gRNAs (Ma et al., 2015). In addition, UAV platforms were used to capture images and develop the RiceBlastYolo model for early detection of symptoms (Lin et al., 2023).

14.5.5 Powdery mildew resistance in wheat

The CRISPR/Cas9 system was used to create a large deletion at a mutant mildew resistance gene locus (*MLO-B1*) in wheat. This resulted in conferring strong and durable disease resistance to the plants without affecting the yield. This large deletion upregulated a nearby dormant gene, *TaTMT3B*, which rescued the plants from the negative side effects of *MLO-B1* knockout. This crucial evidence of rescue of plants from pleiotropic side effects resulting from knocking out susceptibility genes opens up a new possibility of identification of such dormant gene loci responsible for suppressing pleiotropic side effects using precision genome editing tools (Li et al., 2022).

14.5.6 Resistance to bacterial pathogens and improving tomato cultivars (*Solanum lycopersicum*)

Using a modified, noncutting Cas9 enzyme fused to transcriptional activators, known as CRISPR activation (CRISPRa), the expression of native defense genes such as *PATHOGENESIS-RELATED GENE 1* (*SlPR-1*) was upregulated, which successfully enhanced tomato plant defense against *Clavibacter michiganensis* infection (García-Murillo et al., 2023). In another study, CRISPR was used to edit the *SlHyPRP1* and *SlDEA1* genes, resulting in plants with reduced cell death in leaves and roots under biotic stress conditions (Saikia et al., 2024). Both these studies mention the use of software such as CRISPR-P 2.0 and CHOPCHOP in the design of gRNAs. CRISPR has also been used to improve traits such as fruit size and vitamin C content to improve common cultivars. For example, CRISPR-engineered tomato enriched with γ-aminobutyric acid (GABA), which is known for its health-promoting properties, was approved in Japan in 2021 for commercial use (Ezura, 2022).

14.5.7 Combating banana fusarium wilt (TR4)

Using precision genome editing tools, a genetically modified banana line known as QCAV-4 was created by transforming it with the *RGA2* resistance gene identified in a wild variety of banana (Dale et al., 2017). In 2024, this genetically engineered banana was approved for commercial cultivation in Australia, making it the world's first genetically modified banana approved for commercial production and consumption (Harding et al., 2025). Also, a recent study aimed to specifically introduce mutations in the effector gene (*SIX9*) of *Fusarium oxysporum* f.sp. *cubense race 1* used the Breaking-Cas web server for designing gRNAs. This marks the rising adaptability of these tools in precision agricultural practices (Villao et al., 2025).

In addition to the above examples, there are numerous other exciting cases of applications of precision genome editing. In another study, a Random Forest (RF) classifier analyzed UAV-derived spectral data to successfully classify maize varieties as resistant, moderately resistant, or susceptible to Maize Streak Virus (MSV) with 77.3% accuracy (Chivasa et al., 2020). Similarly, for wheat stripe rust, AI models have processed UAV-based hyperspectral imagery to quantify disease severity with high precision, such as by combining spectral vegetation indices with texture features predicting disease severity with an R^2 of 0.88, as well as by using automated machine learning (AutoML) framework to achieve 95.35% accuracy in another study, significantly outperforming standard DL models (Guo et al., 2021; Uzair et al., 2025). In cases where diseases can have visually similar symptoms, such as in soybeans, the development of advanced platforms combining a CNN with a graph neural network (GNN) utilizing MobileNetV2 for localized feature extraction and GraphSAGE for relational modeling achieved very high accuracy in soybean leaf disease detection (Goshika et al., 2024; Jahin et al., 2025).

14.6 Ethical and regulatory considerations in the application of gene editing

The rapid advancement of AI-driven gene editing presents serious ethical and regulatory challenges. The wide gap between the existing laws and the sociocultural influence of modern technology needs to be addressed critically. Key ethical concerns include ensuring the safety of edited crops by minimizing unintended off-target effects, promoting equitable access to the technology to avoid widening the gap between large corporations and small landholding farmers, and maintaining public trust through transparency. The global regulatory landscape for gene-edited crops is complex and fragmented. Different jurisdictions have adopted fundamentally different approaches.

- United States: The United States regulates biotechnology under the product-based Coordinated Framework for the Regulation of Biotechnology (1986), which involves the USDA, FDA, and EPA. A significant development occurred on December 2, 2024, when a federal court vacated USDA's 2020 SECURE rule (7 CFR part 340). As a result, USDA's regulatory process has reverted to the pre-2020 framework, in which oversight is triggered primarily if a genetically engineered organism was developed using DNA from a plant pest or is likely to pose a plant pest risk. Because many modern genome-edited crops do not use plant pest sequences in their construction, this older standard may result in less regulatory oversight for some newer products. Actions taken under the 2020 rule before the court's decision remain valid, but future submissions must follow the reinstated pre-2020 process. (Bickell, 2023).
- European Union (EU): The EU has traditionally used a strict, process-based approach to biotech regulation, meaning many gene-edited organisms are treated as genetically modified organisms (GMOs). However, a legislative proposal on new genomic techniques (NGTs) seeks to introduce a two-tiered system: NGT1 plants, which meet defined equivalence thresholds and are considered close to conventionally bred crops, would be largely exempt from the full GMO regulatory requirements (though subject to verification, seed labeling, and registration). NGT2 plants, which involve more extensive or novel genetic changes, would remain under a modified GMO framework requiring risk assessment, authorization, traceability, and labeling (Lin et al., 2023; COD, 2023).
- Japan: In Japan, the regulatory approach to genome editing is a hybrid model. Modifications made using the SDN-1 method (i.e., edits without employing an external DNA template) are excluded from regulation under the Japanese Cartagena Act and are not treated as "living modified organisms" (LMOs). However, developers are

expected to notify or consult with relevant ministries (e.g., Health, Environment, Agriculture) before marketing such products, and summaries of exempted edits are made publicly available, enabling transparency in oversight (Tsuda et al., 2019).

These diverse and somewhat conflicting regulations for GMOs across the world necessitate cooperation at an international forum. This would help in standardizing and guiding the regulatory frameworks in a concerted direction to tackle modern agricultural issues. Regulatory bodies can leverage AI to analyze large and complex datasets for more accurate risk assessments of new gene-edited crops. AI-driven simulation can also be used to model the potential long-term environmental impacts of a new crop and assess its potential risk of gene flow to wild relatives, providing crucial data for decision-making before any commercial release. However, this raises many new challenges, invariably due to the "black box" nature of many AI models. The proprietary nature of their source code makes it difficult to understand their backend operations.

14.7 Future directions, challenges, and innovations

The integration of AI and CRISPR is rapidly evolving. From simple optimization tools, it has grown to a fully self-autonomous engineering system for disease-resistant crops. One of the most transformative emerging trends is the development of "digital-twins" for plant–pathogen systems, which are virtual replicas of crops and their pathogens that simulate infection dynamics and resistance mechanisms in silico before any physical editing occurs (Escribà-Gelonch et al., 2024). These simulations allow testing thousands of potential CRISPR edits and predict their long-term durability under various climatic conditions. Another interesting trend is the application of models like PlantGPT, trained on vast plant genomic and protein data obtained from *Arabidopsis thaliana*, in synthesizing existing knowledge to generate testable hypotheses about disease resistance, appreciating the fundamental "language" of plant immunity (Zhang et al., 2025).

Nowadays, many computational biologists, plant pathologists, and robotic engineers are working together to innovate new tools for easy and fast diagnosis, treatment, and prevention of plant diseases. Also, many ambitious efforts are undertaken to combine synthetic plant biology with AI, so that crops themselves can act as built-in health monitors. Researchers are also developing CRISPR-based biosensors that detect specific pathogen biomarkers and trigger visible color changes in leaves. ML-based smartphone algorithms can then be used to detect disease symptoms with high accuracy.

Meanwhile, organizations like the International Plant Sentinel Network (IPSN), coordinated by Botanic Gardens Conservation International (BGCI), are already using global networks of botanic gardens to spot novel emerging pests. One of the most significant advantages of precision agriculture is the ability to create durable and robust resistance, which could drastically cut global pesticide use. This, in turn, would reduce environmental contamination and help protect nontarget organisms, especially essential pollinators. This technology is a boon for farmers and developing countries, who often suffer the most due to crop losses. Initiatives like the Vision for Adopted Crops and Soils (VACS) aim to implement these ideas in places like West Africa to enhance "opportunity crops" like cowpea. This

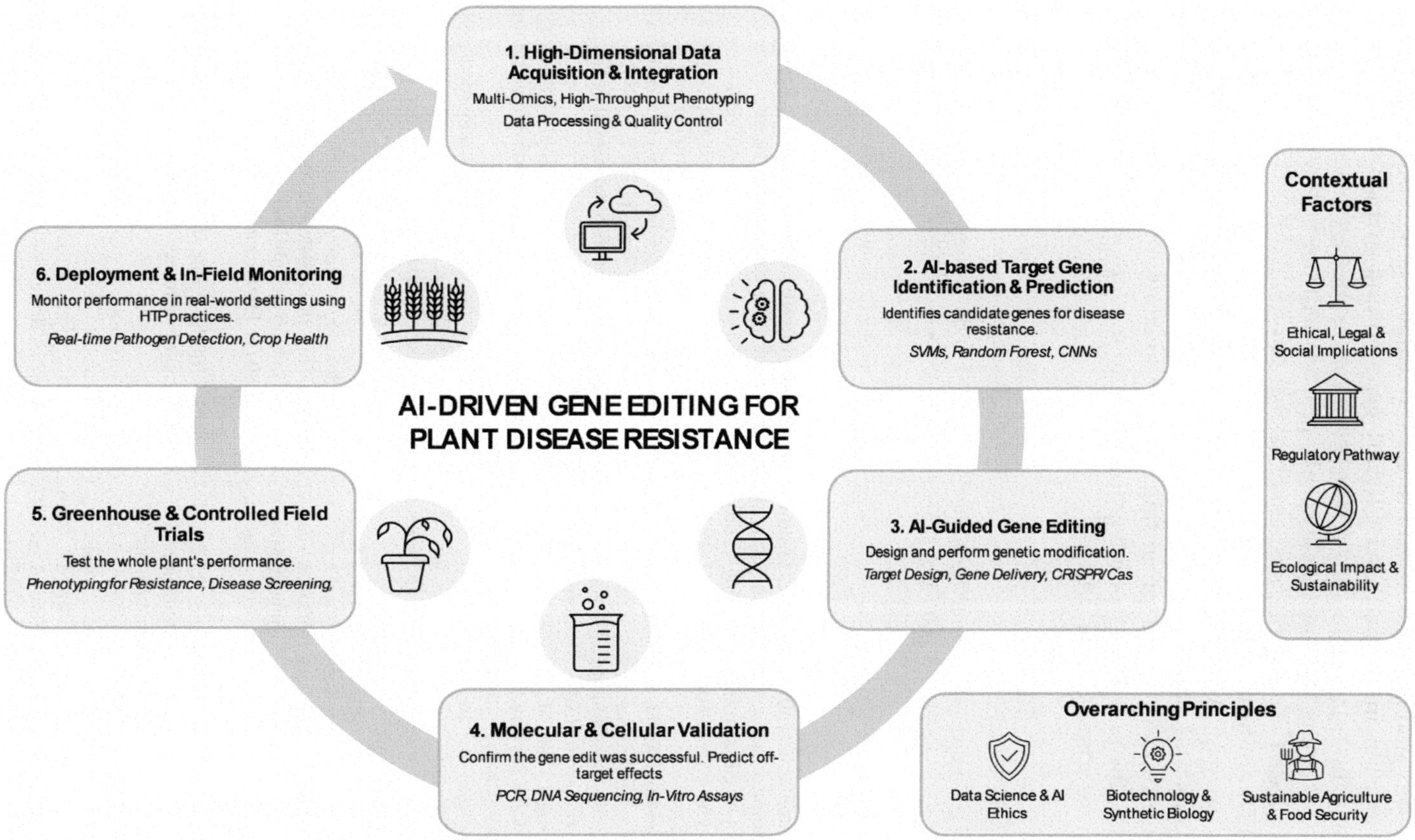

FIGURE 14.1 Schematic representation of the AI-driven gene editing workflow for developing disease-resistant plants.

would ensure food security and eradicate mass hunger (Fredenberg et al., 2024). However, significant issues still pertain in AI-mediated CRISPR gene editing in plants. The most fundamental challenge is the "plant data gap." Plant-specific outcomes are overwhelmingly underrepresented in AI training datasets. As a result, practical editing efficacy often falls short of what models predict. To fix this, systematic global efforts like the Plant Genome Editing Database (PGED) are formed to collect and organize plant-specific CRISPR results. Along with the plant data gap, AI models must also account for the complex evolutionary arms race between plants and pathogens. Furthermore, integrating climate datasets with plant biology and AI would open up promising broader avenues of "smart farming." Despite these difficulties, the potentiality of integrating AI into modern agronomic architecture is substantially transformative (Fig. 14.1).

Declaration

The authors declare that no generative artificial intelligence tools were used in the writing or preparation of this chapter. The content reflects the authors' own scholarly work, and the authors assume full responsibility for all statements, interpretations, and conclusions presented.

References

Abdelrahman, M., Wei, Z., Rohila, J.S., Zhao, K., 2021. Multiplex genome-editing technologies for revolutionizing plant biology and crop improvement. Front. Plant Sci. 12. https://doi.org/10.3389/fpls.2021.721203.

admin, 2024. High-Throughput Genotyping | CRISPR Validation | GENEWIZ from Azenta. GENEWIZ Azenta Life Sci. URL https://www.genewiz.com/public/services/next-generation-sequencing/high-throughput-genotyping (accessed 10.3.25).

Anthon, C., Corsi, G.I., Gorodkin, J., 2022. CRISPRon/off: CRISPR/Cas9 on- and off-target gRNA design. Bioinformatics 38, 5437–5439.

Bae, S., Park, J., Kim, J.-S., 2014. Cas-OFFinder: a fast and versatile algorithm that searches for potential off-target sites of Cas9 RNA-guided endonucleases. Bioinformatics 30, 1473–1475.

Bai, X., Fang, H., He, Y., Zhang, J., Tao, M., Wu, Q., Yang, G., Wei, Y., Tang, Y., Tang, L., Lou, B., Deng, S., Yang, Y., Feng, X., 2023. Dynamic UAV phenotyping for rice disease resistance analysis based on multisource data. Plant Phenomics 5, 0019. https://doi.org/10.34133/plantphenomics.0019.

Benchling Cloud-based platform for biotech R&D | Benchling, URL https://www.benchling.com (accessed 10.3.25).

Bickell, E.G., 2023. Gene-Edited Plants: Regulation and Issues for Congress, 2023. Congressional Research Service. https://crsreports.congress.gov.

Borrelli, V.M.G., Brambilla, V., Rogowsky, P., Marocco, A., Lanubile, A., 2018. The enhancement of plant disease resistance using CRISPR/Cas9 technology. Front. Plant Sci. 9. https://doi.org/10.3389/fpls.2018.01245.

Calle García, J., Guadagno, A., Paytuvi-Gallart, A., Saera-Vila, A., Amoroso, C.G., D'esposito, D., Andolfo, G., Aiese Cigliano, R., Sanseverino, W., Ercolano, M.R., Sanseverino, W., 2022. PRGdb 4.0: an updated database dedicated to genes involved in plant disease resistance process. Nucleic Acids Res. 50 (1), D1483. https://doi.org/10.1093/nar/gkab1087.

Cancellieri, S., Canver, M.C., Bombieri, N., Giugno, R., Pinello, L., 2020. CRISPRitz: rapid, high-throughput and variant-aware in silico off-target site identification for CRISPR genome editing. Bioinformatics 36 (7), 2001–2008. https://doi.org/10.1093/bioinformatics/btz867.

Chivasa, W., Mutanga, O., Biradar, C., 2020. UAV-based multispectral phenotyping for disease resistance to accelerate crop improvement under changing climate conditions. Remote Sens. 12, 2445.

Chuai, G., Ma, H., Yan, J., Chen, M., Hong, N., Xue, D., Zhou, C., Zhu, C., Chen, K., Duan, B., Gu, F., Qu, S., Huang, D., Jia, W., Qi, L., 2018. DeepCRISPR: optimized CRISPR guide RNA design by deep learning. Genome Biol. 19 (1). https://doi.org/10.1186/s13059-018-1459-4.

Clement, K., Rees, H., Canver, M.C., Gehrke, J.M., Farouni, R., Hsu, J.Y., Cole, M.A., Liu, D.R., Joung, J.K., Bauer, D. E., Pinello, L., 2019. CRISPResso2 provides accurate and rapid genome editing sequence analysis. Nat. Biotechnol. 37 (3), 224–226. https://doi.org/10.1038/s41587-019-0032-3.

COD), 2023. Plants obtained by certain new genomic techniques and their food and feed. https://eur-lex.europa.eu.

Conant, D., Hsiau, T., Rossi, N., Oki, J., Maures, T., Waite, K., Yang, J., Joshi, S., Kelso, R., Holden, K., Enzmann, B. L., Stoner, R., 2022. Inference of CRISPR edits from Sanger trace data. CRISPR J. 5 (1), 123–130. https://doi.org/10.1089/crispr.2021.0113.

Concordet, J.P., Haeussler, M., 2018. CRISPOR: intuitive guide selection for CRISPR/Cas9 genome editing experiments and screens. Nucleic Acids Res. 46 (1), W242. https://doi.org/10.1093/nar/gky354.

Coomber, A., Saville, A., Ristaino, J.B., 2024. Evolution of Phytophthora infestans on its potato host since the Irish potato famine. Nat. Commun. 15, 6488.

Dale, J., James, A., Paul, J.Y., Khanna, H., Smith, M., Peraza-Echeverria, S., Garcia-Bastidas, F., Kema, G., Waterhouse, P., Mengersen, K., Harding, R., 2017. Transgenic Cavendish bananas with resistance to Fusarium wilt tropical race 4. Nat. Commun. 8 (1). https://doi.org/10.1038/s41467-017-01670-6.

Du, W., Zhao, L., Diao, K., Zheng, Y., Yang, Q., Zhu, Z., Zhu, X., Tang, D., 2025. A versatile CRISPR/Cas9 system off-target prediction tool using language model. Commun. Biol. 8 (1). https://doi.org/10.1038/s42003-025-08275-6.

Erdoğan, İ., Cevher-Keskin, B., Bilir, Ö., Hong, Y., Tör, M., 2023. Recent developments in CRISPR/Cas9 genome-editing technology related to plant disease resistance and abiotic stress tolerance. Biology 12 (7). https://doi.org/10.3390/biology12071037.

Escribà-Gelonch, M., Liang, S., van Schalkwyk, P., Fisk, I., Long, N.V.D., Hessel, V., 2024. Digital twins in agriculture: orchestration and applications. J. Agric. Food Chem. 72 (19), 10737–10752. https://doi.org/10.1021/acs.jafc.4c01934.

Ezura, H., 2022. Letter to the editor: the world's first CRISPR tomato launched to a Japanese market: the social-economic impact of its implementation on crop genome editing. Plant Cell Physiol. 63 (6), 731–733. https://doi.org/10.1093/pcp/pcac048.

Fredenberg, E., Karl, K., Passarelli, S., Porciello, J., Rattehalli, V., Auguston, A., Chimwaza, G., Grande, F., Holmes, B., Kozlowski, N., Laborde, D., MacCarthy, D., Masikati, P., McMullin, S., Méndez Leal, E., Valdivia, R., Van Deynze, A., van Zonneveld, M., 2024. Vision for Adapted Crops and Soils (VACS) Research in Action: Opportunity Crops for Africa.

García-Murillo, L., Valencia-Lozano, E., Priego-Ranero, N.A., Cabrera-Ponce, J.L., Duarte-Aké, F.P., Vizuet-de-Rueda, J.C., Rivera-Toro, D.M., Herrera-Ubaldo, H., de Folter, S., Alvarez-Venegas, R., 2023. CRISPRa-mediated transcriptional activation of the SlPR-1 gene in edited tomato plants. Plant Sci. 329. https://doi.org/10.1016/j.plantsci.2023.111617.

Geneious | Bioinformatics Software for Sequence Data Analysis. Geneious. https://www.geneious.com/(accessed 10.3.25).

GenScript - Scripting Possibilities - The leader in molecular cloning and gene synthesis, peptide synthesis, protein and antibody engineering. https://www.genscript.com/(accessed 10.3.25).

Ghosal, S., Blystone, D., Singh, A.K., Ganapathysubramanian, B., Singh, A., Sarkar, S., 2018. An explainable deep machine vision framework for plant stress phenotyping. Proc. Natl. Acad. Sci. U. S. A 115 (18), 4613–4618. https://doi.org/10.1073/pnas.1716999115.

Gill, T., Gill, S.K., Saini, D.K., Chopra, Y., de Koff, J.P., Sandhu, K.S., 2022. A comprehensive review of high throughput phenotyping and machine learning for plant stress phenotyping. Phenomics 2 (3), 156–183. https://doi.org/10.1007/s43657-022-00048-z.

Goshika, S., Meksem, K., Ahmed, K.R., Lakhssassi, N., 2024. Deep learning model for classifying and evaluating soybean leaf disease damage. Int. J. Mol. Sci. 25 (1). https://doi.org/10.3390/ijms25010106.

Guo, A., Huang, W., Dong, Y., Ye, H., Ma, H., Liu, B., Wu, W., Ren, Y., Ruan, C., Geng, Y., 2021. Wheat yellow rust detection using UAV-based hyperspectral technology. Remote Sens. 13 (1), 123. https://doi.org/10.3390/rs13010123.

Harding, R., Paul, J.Y., James, A., Smith, M., Kleidon, J., Shekhawat, U., Phillips, A., Kidanemariam, D., Dawit, A., Dale, J., 2025. QCAV-4, the first genetically modified Cavendish (cv. Grand Nain) banana resistant to Fusarium wilt tropical race 4 approved for commercial production and consumption. Plant Biotechnol. J. 23 (9), 3628–3637. https://doi.org/10.1111/pbi.70178.

Hassani-Pak, K., Singh, A., Brandizi, M., Hearnshaw, J., Parsons, J.D., Amberkar, S., Phillips, A.L., Doonan, J.H., Rawlings, C., 2021. KnetMiner: a comprehensive approach for supporting evidence-based gene discovery and complex trait analysis across species. Plant Biotechnol. J. 19 (8), 1670–1678. https://doi.org/10.1111/pbi.13583.

Huang, X., Wang, Y., Wang, N., 2022. Highly efficient generation of canker-resistant sweet orange enabled by an improved CRISPR/Cas9 system. Front. Plant Sci. 12.

Ikram, A.U., Khan, M.S.S., Islam, F., Ahmed, S., Ling, T., Feng, F., Sun, Z., Chen, H., Chen, J., 2025. All roads lead to Rome: pathways to engineering disease resistance in plants. Adv. Sci. 12 (5). https://doi.org/10.1002/advs.202412223.

Jahin, M.A., Soudeep, S., Mridha, M.F., Dey, N., 2025. Soybean disease detection via interpretable hybrid CNN-GNN: integrating MobileNetV2 and GraphSAGE with cross-modal attention. arXiv. https://doi.org/10.48550/arXiv.2503.01284.

Ji, Y., Zhou, Z., Liu, H., Davuluri, R.V., 2021. DNABERT: pre-trained bidirectional encoder representations from transformers model for DNA-language in genome. Bioinformatics 37 (15), 2112–2120. https://doi.org/10.1093/bioinformatics/btab083.

Jia, H., Orbovic, V., Jones, J.B., Wang, N., 2016. Modification of the PthA4 effector binding elements in Type I CsLOB1 promoter using Cas9/sgRNA to produce transgenic Duncan grapefruit alleviating XccΔpthA4: DCsLOB1.3 infection. Plant Biotechnol. J. 14 (5), 1291–1301. https://doi.org/10.1111/pbi.12495.

Jia, H., Omar, A.A., Xu, J., Dalmendray, J., Wang, Y., Feng, Y., Wang, W., Hu, Z., Grosser, J.W., Wang, N., 2024. Generation of transgene-free canker-resistant Citrus sinensis cv. Hamlin in the T0 generation through Cas12a/CBE co-editing. Front. Plant Sci. 15. https://doi.org/10.3389/fpls.2024.1385768.

Jorgensen, J.H., 1992. Discovery, characterization and exploitation of Mlo powdery mildew resistance in barley.

Kaur, S., Bishnoi, R., Priyadarshini, P., Singla, D., Chhuneja, P., 2023. DSP: database of disease susceptibility genes in plants. Funct. Integr. Genom. 23 (3). https://doi.org/10.1007/s10142-023-01132-x.

Kumar, R., Das, S.P., Choudhury, B.U., Kumar, A., Prakash, N.R., Verma, R., Chakraborti, M., Devi, A.G., Bhattacharjee, B., Das, R., Das, B., Devi, H.L., Das, B., Rawat, S., Mishra, V.K., 2024. Advances in genomic tools for plant breeding: harnessing DNA molecular markers, genomic selection, and genome editing. Biol. Res. 57 (1). https://doi.org/10.1186/s40659-024-00562-6.

Labun, K., Montague, T.G., Krause, M., Torres Cleuren, Y.N., Tjeldnes, H., Valen, E., 2019. CHOPCHOP v3: expanding the CRISPR web toolbox beyond genome editing. Nucleic Acids Res. 47 (1), W171. https://doi.org/10.1093/nar/gkz365.

Li, S., Lin, D., Zhang, Y., Deng, M., Chen, Y., Lv, B., Li, B., Lei, Y., Wang, Y., Zhao, L., Liang, Y., Liu, J., Chen, K., Liu, Z., Xiao, J., Qiu, J.L., Gao, C., 2022. Genome-edited powdery mildew resistance in wheat without growth penalties. Nature 602 (7897), 455–460. https://doi.org/10.1038/s41586-022-04395-9.

Li, T., Jarquin Bolaños, E., Stevens, D.M., Sha, H., Prigozhin, D.M., Coaker, G., 2025a. Unlocking expanded flagellin perception through rational receptor engineering. Nat. Plants 11, 1628–1641.

Li, B., Ritchotte, M.Z., He, S., Nabavi, S., 2025b. A Zero-shot LLM-based Framework for Descriptive Gene-gene Interaction Network Generation.

Lin, J., Zhang, Z., Zhang, S., Chen, J., Wong, K.C., 2020. CRISPR-net: a recurrent convolutional network quantifies CRISPR off-target activities with mismatches and indels. Adv. Sci. 7 (13). https://doi.org/10.1002/advs.201903562.

Lin, S., Li, J., Huang, D., Cheng, Z., Xiang, L., Ye, D., Weng, H., 2023. Early detection of rice Blast using a semi-supervised contrastive unpaired translation iterative network based on UAV images. Plants 12 (21), 3675. https://doi.org/10.3390/plants12213675.

Liu, X., Wang, S., Ai, D., 2022. Predicting CRISPR/Cas9 repair outcomes by attention-based deep learning framework. Cells 11, 1847.

Liu, Z., Wang, X., Cao, S., Lei, T., Chenzhu, Y., Zhang, M., Liu, Z., Lu, J., Ma, W., Su, B., Wang, Y., Zhou, Y., 2024. Deep learning facilitates precise identification of disease-resistance genes in plants. bioRxiv. https://doi.org/10.1101/2024.09.26.615248.

Ma, X., Zhang, Q., Zhu, Q., Liu, W., Chen, Y., Qiu, R., Wang, B., Yang, Z., Li, H., Lin, Y., Xie, Y., Shen, R., Chen, S., Wang, Z., Chen, Y., Guo, J., Chen, L., Zhao, X., Dong, Z., Liu, Y.-G., 2015. A robust CRISPR/Cas9 system for convenient, high-efficiency multiplex genome editing in monocot and dicot plants. Mol. Plant 8 (8), 1274–1284. https://doi.org/10.1016/j.molp.2015.04.007.

Malekos, E., Montano, C., Carpenter, S., 2024. CRISPRware: an efficient method for contextual gRNA library design.

Mekonnen, A.M., Seong, K., Kim, H., Park, J., 2025. Variant-aware Cas-OFFinder: web-based in silico variant-aware potential off-target site identification for genome editing applications. Nucleic Acids Res. 53 (1), W118. https://doi.org/10.1093/nar/gkaf389.

Minkenberg, B., Zhang, J., Xie, K., Yang, Y., 2019. CRISPR - PLANT v2: an online resource for highly specific guide RNA spacers based on improved off-target analysis. Plant Biotechnol. J. 17 (1), 5–8. https://doi.org/10.1111/pbi.13025.

Müller Paul, H., Istanto, D.D., Heldenbrand, J., Hudson, M.E., 2022. CROPSR: an automated platform for complex genome-wide CRISPR gRNA design and validation. BMC Bioinf. 23 (1). https://doi.org/10.1186/s12859-022-04593-2.

Naert, T., Yamamoto, T., Han, S., Horn, M., Bethge, P., Vladimirov, N., Voigt, F.F., Figueiro-Silva, J., Bachmann-Gagescu, R., Helmchen, F., Lienkamp, S.S., 2024. Pythia: Non-random DNA repair allows predictable CRISPR/Cas9 integration and gene editing.

Naert, T., Yamamoto, T., Han, S., Röck, R., Horn, M., Bethge, P., Vladimirov, N., Voigt, F.F., Figueiro-Silva, J., Bachmann-Gagescu, R., Vleminckx, K., Helmchen, F., Lienkamp, S.S., 2025. Precise, predictable genome integrations by deep-learning-assisted design of microhomology-based templates. Nat. Biotechnol. https://doi.org/10.1038/s41587-025-02771-0.

Nisa, R.T., Jan, S.K., Ahmad Bhat, F., Rasool Rather, T., Nabi, A., Altaf, A.W., Rehana, A., Nahida, A., Bashir, M., Ishfaq Majeed, S., Zohra, S., 2022. Review on "CRISPR-Cas9- A genome editing tool for plant disease management. Plant Cell Biotechnol. Mol. Biol. 1–14. https://doi.org/10.56557/pcbmb/2022/v23i15-167619.

Oliva, R., Ji, C., Atienza-Grande, G., Huguet-Tapia, J.C., Perez-Quintero, A., Li, T., Eom, J.S., Li, C., Nguyen, H., Liu, B., Auguy, F., Sciallano, C., Luu, V.T., Dossa, G.S., Cunnac, S., Schmidt, S.M., Slamet-Loedin, I.H., Vera Cruz, C., Szurek, B., Frommer, W.B., White, F.F., Yang, B., 2019. Broad-spectrum resistance to bacterial blight in rice using genome editing. Nat. Biotechnol. 37 (11), 1344–1350. https://doi.org/10.1038/s41587-019-0267-z.

Pethybridge, S.J., Nelson, S.C., 2015. Leaf doctor: a new portable application for quantifying plant disease severity. Plant Dis. 99 (10), 1310–1316. https://doi.org/10.1094/PDIS-03-15-0319-RE.

PlantPAD https://plantpad.samlab.cn/index.html (accessed 10.3.25).

Poudel, R., Rodriguez, L.T., Reisch, C.R., Rivers, A.R., 2022. GuideMaker: software to design CRISPR-Cas guide RNA pools in non-model genomes. GigaScience 11. https://doi.org/10.1093/gigascience/giac007.

Qiu, R.Z., Chen, S.P., Chi, M.X., Wang, R.B., Huang, T., Fan, G.C., Zhao, J., Weng, Q.Y., 2022. An automatic identification system for citrus greening disease (Huanglongbing) using a YOLO convolutional neural network. Front. Plant Sci. 13. https://doi.org/10.3389/fpls.2022.1002606.

Qu, Y., Huang, K., Yin, M., Zhan, K., Liu, D., Yin, D., Cousins, H.C., Johnson, W.A., Wang, X., Shah, M., Altman, R. B., Zhou, D., Wang, M., Cong, L., 2025. CRISPR-GPT for agentic automation of gene-editing experiments. Nat. Biomed. Eng. https://doi.org/10.1038/s41551-025-01463-z.

Rombauts, S., Florquin, K., Lescot, M., Marchal, K., Rouzé, P., Van de Peer, Y., 2003. Computational approaches to identify promoters and cis-regulatory elements in plant genomes. Plant Physiol. 132 (3), 1162–1176. https://doi.org/10.1104/pp.102.017715.

Rossman, A.Y., 2009. The impact of invasive fungi on agricultural ecosystems in the United States. In: Langor, D.W., Sweeney, J. (Eds.), Ecological Impacts of Non-native Invertebrates and Fungi on Terrestrial Ecosystems. Springer, Netherlands, Dordrecht, pp. 97–107.

Ruffolo, J.A., Nayfach, S., Gallagher, J., Bhatnagar, A., Beazer, J., Hussain, R., Russ, J., Yip, J., Hill, E., Pacesa, M., Meeske, A.J., Cameron, P., Madani, A., 2025. Design of highly functional genome editors by modelling CRISPR–Cas sequences. Nature 645 (8080), 518–525. https://doi.org/10.1038/s41586-025-09298-z.

Sabelli, C., 2023. Deadly olive tree pathogen came by road and rail. Nature Italy. https://doi.org/10.1038/d43978-023-00118-4.

Saikia, B., Remya, S., Debbarma, J., Maharana, J., Sastry, G.N., Chikkaputtaiah, C., 2024. CRISPR/Cas9-based genome editing and functional analysis of SlHyPRP1 and SlDEA1 genes of Solanum lycopersicum L. in imparting genetic tolerance to multiple stress factors. Front. Plant Sci. 15. https://doi.org/10.3389/fpls.2024.1304381.

Savary, S., Willocquet, L., Pethybridge, S.J., Esker, P., McRoberts, N., Nelson, A., 2019. The global burden of pathogens and pests on major food crops. Nat. Ecol. Evol. 3 (3), 430–439. https://doi.org/10.1038/s41559-018-0793-y.

Schmidt, H., Zhang, M., Chakarov, D., Bansal, V., Mourelatos, H., Sánchez-Rivera, F.J., Lowe, S.W., Ventura, A., Leslie, C.S., Pritykin, Y., 2025. Genome-wide CRISPR guide RNA design and specificity analysis with GuideScan2. Genome Biol. 26 (1). https://doi.org/10.1186/s13059-025-03488-8.

Shawon, S.M., Ema, F.B., Mahi, A.K., Niha, F.L., Zubair, H.T., 2025. Crop yield prediction using machine learning: an extensive and systematic literature review. Smart Agric. Technol. 10. https://doi.org/10.1016/j.atech.2024.100718.

Sheikh, M., Iqra, F., Ambreen, H., Pravin, K.A., Ikra, M., Chung, Y.S., 2024. Integrating artificial intelligence and high-throughput phenotyping for crop improvement. J. Integr. Agric. 23 (6), 1787–1802. https://doi.org/10.1016/j.jia.2023.10.019.

Siddiqua, A., Kabir, M.A., Ferdous, T., Ali, I.B., Weston, L.A., 2022. Evaluating plant disease detection mobile applications: quality and limitations. Agronomy 12 (8), 1869. https://doi.org/10.3390/agronomy12081869.

Singh, A., Ganapathysubramanian, B., Singh, A.K., Sarkar, S., 2016. Machine learning for high-throughput stress phenotyping in plants. Trends Plant Sci. 21 (2), 110–124. https://doi.org/10.1016/j.tplants.2015.10.015.

Song, Q., Lee, J., Akter, S., Rogers, M., Grene, R., Li, S., 2020. Prediction of condition-specific regulatory genes using machine learning. Nucleic Acids Res. 48 (11), e62. https://doi.org/10.1093/nar/gkaa264.

Sun, J., Guo, J., Liu, J., Sinha, S., Crispr, M., 2024. Predicting sgRNA off-target effect using a multi-view deep learning network. PLoS Comput. Biol. 20 (3), e1011972. https://doi.org/10.1371/journal.pcbi.1011972.

Synthego, https://tools.synthego.com/#/(accessed 10.3.25).

Thakur, S., Kaur, S., Adhikari, S., Sabharwal, P., Fu, Y., Meru, G., 2025. Turning susceptibility into strength: a new era of durable resistance in plants through genome editing. Plants 14 (19). https://doi.org/10.3390/plants14193080.

Tsuda, M., Watanabe, K.N., Ohsawa, R., 2019. Regulatory status of genome-edited organisms under the Japanese Cartagena act. Front. Bioeng. Biotechnol. 7.

Untergasser, A., Cutcutache, I., Koressaar, T., Ye, J., Faircloth, B.C., Remm, M., Rozen, S.G., 2012. Primer3—new capabilities and interfaces. Nucleic Acids Res. 40 (15), e115. https://doi.org/10.1093/nar/gks596.

Uzair, M., ElShawi, R., Tomasiello, S., 2025. Context-Aware AutoML for Accurate Wheat Disease Detection.

Villao, L., Vargas, J., Diez, N., Magdama, F., Santos-Ordóñez, E., 2025. Optimization of a CRISPR-Cas9 in vitro protocol for targeting the SIX9 gene of Fusarium oxysporum f.sp. cubense race 1 associated with banana Fusarium wilt. Front. Plant Sci. 16. https://doi.org/10.3389/fpls.2025.1523884.

Wang, F., Wang, C., Liu, P., Lei, C., Hao, W., Gao, Y., Liu, Y.-G., Zhao, K., Wilson, R.A., 2016. Enhanced rice Blast resistance by CRISPR/Cas9-Targeted mutagenesis of the ERF transcription factor gene OsERF922. PLoS One 11 (4), e0154027. https://doi.org/10.1371/journal.pone.0154027.

Wang, G., Liu, X., Wang, A., Wen, J., Kim, P., Song, Q., Liu, X., Zhou, X., 2025. CRISPRoffT: comprehensive database of CRISPR/Cas off-targets. Nucleic Acids Res. 53 (D1), D914. https://doi.org/10.1093/nar/gkae1025.

Xylella fastidiosa - Food Safety - European Commission https://food.ec.europa.eu/plants/plant-health-and-biosecurity/plant-health-rules/control-measures/xylella-fastidiosa_en (accessed 10.3.25).

Xylella fastidiosa | EFSA, 2024. https://www.efsa.europa.eu/en/topics/topic/xylella-fastidiosa (accessed 10.3.25).

Ye, N., Mai, W., Qin, F., Yuan, S., Liu, B., Li, Z., Liu, C., Wan, F., Qian, W., Wu, Z., Qiao, X., 2025. Early detection of Citrus Huanglongbing by UAV remote sensing based on MGA-UNet. Front. Plant Sci. 16. https://doi.org/10.3389/fpls.2025.1503645.

Zhang, Z., Lamson, A.R., Shelley, M., Troyanskaya, O., 2023. Interpretable neural architecture search and transfer learning for understanding CRISPR–Cas9 off-target enzymatic reactions. Nat. Comput. Sci. 3 (12), 1056–1066. https://doi.org/10.1038/s43588-023-00569-1.

Zhang, R., Wang, Y., Yang, W., Wen, J., Liu, W., Zhi, S., Li, G., Chai, N., Huang, J., Xie, Y., Xie, X., Chen, L., Gu, M., Liu, Y.G., Zhu, Q., 2025. PlantGPT: an Arabidopsis-based intelligent agent that answers questions about plant functional genomics. Adv. Sci. 12 (30). https://doi.org/10.1002/advs.202503926.

Zhou, J., Peng, Z., Long, J., Sosso, D., Liu, B., Eom, J.S., Huang, S., Liu, S., Vera Cruz, C., Frommer, W.B., White, F.F., Yang, B., 2015. Gene targeting by the TAL effector PthXo2 reveals cryptic resistance gene for bacterial blight of rice. Plant J. 82 (4), 632–643. https://doi.org/10.1111/tpj.12838UnitedStates.

CHAPTER

15

Artificial intelligence tools for predicting off-targets in plant genome editing

Fredy Davi Albuquerque Silva, Eulálio Gutenberg Dias dos Santos, and Elizabeth Pacheco Batista Fontes

Department of Biochemistry and Molecular Biology/Bioagro, National Institute of Science and Technology in Plant-Pest Interactions, Federal University of Viçosa, Viçosa, Brazil

15.1 Introduction

Advances in systems biology, in silico predictions, and gene editing have significantly boosted several aspects of plant biotechnology. Gene editing technology is designed to make precise changes to nucleic acid sequences within the target genome of any organism. Technologies using zinc finger nucleases (ZFNs) and protein fusions, including *FoK1*-type nucleases or transcription-activator-like effector nucleases (TALENs), have been widely used for genome editing (Bibikova et al., 2003; Cheng et al., 2015). ZFNs are fusion proteins of zinc finger transcription factor and *FoK1* endonucleases, while in TALENs, *Fok1* is fused to TALE. Both engineered endonucleases have been used for addressing effectors to a specific genomic region. However, their design requires a zinc finger or TALE motif that specifically binds to a genomic DNA region. Since ZFNs require a specific amino acid sequence as a target site, multiple designs are needed to identify different genomic regions simultaneously. Similarly, a TALEN motif identifies a single base pair; therefore, a TALEN matrix is needed to target more extended DNA sequences. In both approaches, specificity relies on protein-DNA binding, which restricts editing to a single DNA-binding site. In contrast, the specificity of clustered regularly interspaced short palindromic repeats associated with the Cas nuclease (CRISPR/Cas) technology relies on highly specific hydrogen bonds for base-

Intelligent Gene Editing in Plants
https://doi.org/10.1016/B978-0-443-44079-3.00009-7

pair annealing of complementary nucleic acid sequences on the target gene. These properties allow editing multiple traits with an unprecedented high precision.

The CRISPR/Cas technology in plant biotechnology has significantly improved genome engineering, enabling targeted gene editing in crops. Progress in developing more versatile CRISPR tools, including CRISPR nucleases, base and prime editors, has expanded the toolkit to induce insertions or deletions, nucleotide conversions, and a broader range of genetic variations in specific genomic regions (Zhang et al., 2025). Due to its variable mechanism, ease of assembly, and accuracy, CRISPR–Cas has become a valuable alternative for genome editing across multiple targets in plant biotechnology.

15.2 CRISPR–Cas systems and assembly

The CRISPR/Cas technology consists of an RNA complementary sequence to the target DNA sequence, designated the spacer or CRISPR-RNA, a scaffold RNA sequence, referred to as the trans-activator CRISPR RNA (tracrRNA), and the Cas nuclease (Deltcheva et al., 2011). CRISPR–Cas9 is the most widely employed system, which requires a single 20 nt gRNA guide and a spacer sequence that anchors and drives the Cas endonuclease to a specific targeted genomic region (Fig. 15.1). A protospacer adjacent motif (PAM), typically NGG, is required for Cas9 double-strand cleavage in the double-stranded DNA sequence (Doudna and Charpentier, 2014). Cas9 interacts with the sgRNA, inducing a conformational change that facilitates formation of the PAM interaction domain (PI domain), and the resulting Cas9-sgRNA complex then binds the target DNA. The PI domain physically separates the target DNA strands, initiating cleavage.

Specific amino acid residues in the Cas9 sequence bind to the GG motif of the PAM sequence (5′-NGG-3′). The gRNA pairs with the target DNA by the seed domain, and double-strand cleavage occurs by the Cas9's HNH and RuvC domains. In general, Cas9 cleavage does not generate homologous ends that are repaired by the host cell. This process leads to insertions or deletions (indels), which introduce frameshift mutations and a premature stop codon, resulting in a knockout (Anders et al., 2014). A recently generated PAM-less SpCas9 variant, exhibiting a more flexible multiplicity preference, has overcome some limitations of the traditional system by expanding the number of targetable sites and enabling multiplex targeting of both DNA and RNA genomes (Walton et al., 2020). Additional CRISPR–Cas variants, such as Cas3, Cas9, Cas12, Cas13, and Cas14, have been developed.

The CRISPR–Cas3 system can target multiple sites in double-stranded RNA (dsRNA) and RNA viruses through coordinated interactions between guide RNA and Cas3 protein (Kim et al., 2024). Cas3 exhibits single-stranded (ssDNA)-dependent ATPase activity, and helicase activity, which unwinds DNA/DNA and RNA/DNA duplexes as well as an ATP-independent nuclease activity with a strong preference for ssDNA substrates. The process starts with the transcription of pre-crRNA, which is processed into a mature crRNA containing sequences complementary to the target DNA. Cas3 interacts with the crRNA to form a ribonucleoprotein (RNP) complex that is guided to the target site. Upon target recognition, the Cas3 helicase separates the DNA strands, followed by exonucleolytic degradation of

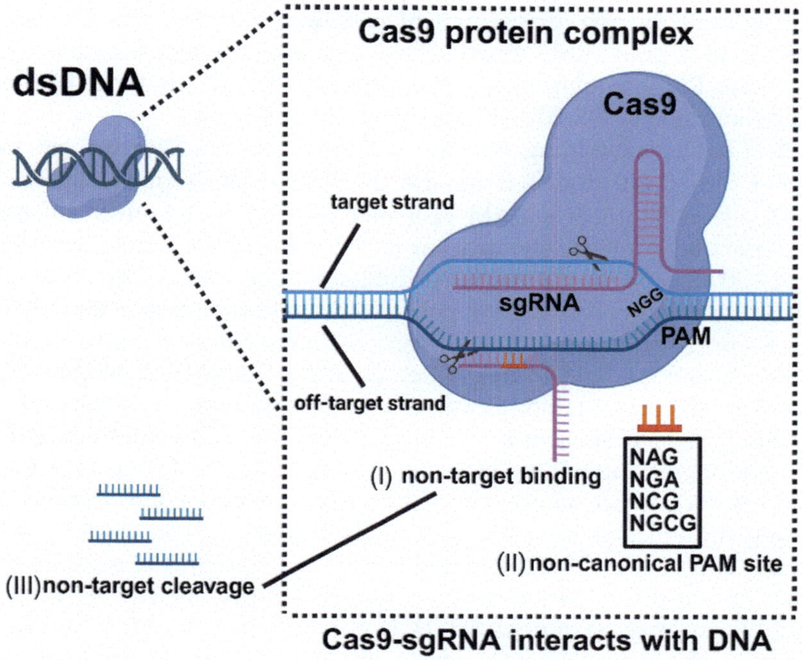

FIGURE 15.1 CRISPR-associated protein (CRISPR/Cas9) complex interacts specifically with target DNA. A functional Cas9 protein complex with small guide RNA (Cas9/sgRNA) binds to the plant genome (dsDNA) sequence, which is complementary to a short 20-nucleotide sequence. This process places the Cas9 endonuclease on the target strand of genomic DNA, where it acts as a molecular scissor. A canonical protospacer adjacent motif (NGG) is required. The Cas9 endonuclease introduces a double-strand break, leading to pinpoint mutations. Off-target edits in CRISPR can occur when the CRISPR/Cas9 binds to unintended non-target strand regions (I), or (II) recognizes the non-canonical PAM sites (NAG, NGA, NCG, or NGCG), and (III) creates non-target cleavage of DNA. *The Figure was created with BioRander.com.*

extensive regions surrounding the cleavage site. As a result, Cas3 is more effective than Cas9 at eliminating larger genomic regions (Nimkar and Anand, 2020).

The CRISPR–Cas13 system is a recent variant that acts on single-stranded RNA. Its mechanism involves a combination of guide RNA (crRNA) and Cas13 protein, which work together to recognize the RNA target (Abudayyeh et al., 2016). CRISPR–Cas13 exhibits ribonuclease activity and is guided by the crRNA, which carries the complementary sequence of the target RNA. After recognizing the target RNA, the Cas13 riboibonuclease activity hydrolyzes the RNA, selectively cleaving the RNA strand. A characteristic of Cas13 is its ability to cleave RNA even in regions lacking the PAM sequence, a feature that distinguishes it from other systems. The CRISPR–Cas13 system has been widely employed to generate immunity in plants against RNA viruses such as tomato mosaic virus (ToMV) and tobacco turnip mosaic virus (TuMV) (Aman et al., 2018). Cas13 also conferred protection against potato virus Y (PVY) in potato plants (Zhan et al., 2019). Unlike Cas9 and Cas12, which cleave DNA, Cas13 acts specifically on RNA and can alter gene expression without modifying the DNA sequence. Additionally, Cas13 does not cause DNA breaks, allowing it to be used as a gene silencing tool without the risk of permanent genetic mutations. The Cas13 also has the continuous ability to degrade RNA after cutting, due to its non-specific exonuclease activity that degrades other RNA sequences near the target. There are variations of Cas13, such as C2c6 and C2c7 (Smargon et al., 2017; Shmakov et al., 2017).

The CRISPR/Cas12 gene-editing system can target dsDNA. Unlike Cas9, Cas12 makes cuts in different forms and does not require the PAM-NGG sequence at its 3′ end. The CRISPR–Cas12 system requires a PAM sequence 5′-TTTV-3′, where V is either A or G. This sequence must be immediately adjacent to the target site for the Cas12 protein action (Karvelis et al., 2020). The Cas12 protein exhibits a nuclease activity that interacts with DNA to form the recognition complex. Upon recognition of the target site, the Cas12 endonuclease makes a precise cut, generating sticky ends that facilitate the insertion of additional DNA sequences during genome editing (Li et al., 2020). These sticky ends (overhangs) are useful for introducing desired modifications into the genome. A feature of Cas12 is its exonuclease activity, which can continue to degrade the DNA adjacent to the cut, amplifying the cut region and increasing editing efficiency (Deng et al., 2024). Cas12 includes variants such as Cas12a (Cpf1), which is effective on double-stranded DNA and generates sticky ends, facilitating the efficient insertion of Cas12b sequences. The CRISPR-MAD7 system comprises a class 2, type V-A CRISPR related to the Cas12a/Cpf1 but exhibits low sequence identity to the canonical Cas12a nuclease, making it a representative example of recently identified CRISPR nucleases (Lin et al., 2021). In contrast, the CRISPR/Cas14a system comprises a compact nuclease derived from archaea that targets single-stranded DNA (ssDNA). The activity of Cas14a is largely sequence-independent, and its cleavage mechanism is unrestricted (Harrington et al., 2018).

15.3 Impact of off-targets in the plant genome

Gene editing technology is designed to make precise changes to nucleic acid sequences within the target genome of any organism. The CRISPR/Cas9 system is widely valued for its high accuracy and efficiency. However, despite its precision, CRISPR/Cas9 can produce off-target effects by inducing unintended genetic alterations elsewhere in the genome. The Cas9 caspase protein cleaves DNA, generating a double-strand break (indels), and as an undesirable consequence, can edit off-target sequences, creating non-specific genetic modifications by engineered nucleases.

The generation of cleavages far from the target site can be affected by several factors, such as the configuration and structure of the gRNA, homology of the gRNA sequence and genome sequences, and the presence of adjacent motifs to the protospacer PAM, which can prevent Cas9 activation by changing the conformation of the HNH domain and initiating cleavage (Guo et al., 2023). Off-target edits in CRISPR can occur when the CRISPR/Cas system binds to unintended non-target strand regions or recognizes non-canonical PAM sites, leading to non-target DNA cleavage (Fig. 15.1) (Li et al., 2021). The generation of off targets can lead to unintended mutations with phenotypic changes and may impact the development.

Experimental strategies, such as selecting gRNAs with GC content (40%–60%) that increase on-target editing by stabilizing interactions with DNA and using guide sequences shorter than 20 nucleotides, decrease off-target effects (Mengstie et al., 2024). Some effort to minimize off-targets using CRISPR–Cas systems include structural modifications in Cas variants, such as eSpCas9 and Sp-Cas9-HF1, which significantly improve the editing specificity (Allemailem et al., 2023). For example, xCas9 3.7 recognizes an expanded range of PAMs (NGG, NG, GAT, GAA), with greater specificity and lower off-target activity (Hua et al., 2019). Other caspases, including enAsCas12a and enAsCas12a HF1, are optimized variants of Cas12a with high fidelity that reduce off-target effects (Kleinstiver et al., 2019). CRISPR–Cas systems are regulated by several mobile genetic elements that encode anti-CRISPR–Cas systems (ACRs) that inhibit CRISPR–Cas proteins, such as Cas3, Cas9, Cas12, and Cas13. Their mechanism involves binding to DNA and inhibition of crRNA loading and DNA cleavage. Employing ACRs to shut down CRISPR–Cas systems is an additional tool against the formation of unwanted off-target effects (Liu et al., 2020).

Rationally designed gRNAs are an additional approach for minimizing off-target effects in genome editing. In prime editing, a specialized guide RNA (pegRNA) directs a prime editor protein complex to a specific genomic locus and mediates precise genetic alterations, including point mutations or small insertions/deletions. The pegRNA comprises three functional components: (i) a spacer sequence that determines target specificity, (ii) a primer binding site (PBS) that initiates reverse transcription, and (iii) a reverse transcriptase template (RTT) encoding the desired edit. Upon target recognition, the Cas9 nickase introduces a single-strand break in the target DNA, allowing the reverse transcriptase (fused to Cas9) to synthesize the edited DNA sequence using the RTT as a template (Vu et al., 2024). This mechanism enables the introduction of highly specific genetic modifications and has the potential to enhance the precision and efficiency of crop breeding applications (Sretenovic and Qi, 2022).

15.4 The use of AI as a tool to predict off-targets and improve gene editing efficiencies

CRISPR–Cas-mediated gene editing has enabled increasingly precise genetic modifications across a wide range of organisms. However, off-target nuclease activity in non-target regions remains a common problem. Although guide RNAs (gRNAs) are designed to edit specific regions, Cas nucleases can tolerate mismatches between the gRNA and genomic DNA, resulting in cleavage of highly similar non-target sequences, leading to unintended mutations. These mutations can have unpredictable consequences, potentially harm the organism, and compromise the safety of the method used, since the target of interest is crops for which precision is critical to commercial viability (Mahfouz et al., 2014; Tsai et al., 2015). Therefore, predicting off targets becomes a complex process that relies on computational approaches using various statistical methods, including heuristics. In this scenario, AI approaches can process and analyze large volumes of genomic, transcriptomic, proteomic, and other data, enabling the identification of behavioral patterns and correlations that would be difficult with conventional analysis methods (Bradbury et al., 2025).

Artificial Intelligence (AI) is an approach that can be used to improve off-target prediction in several ways, as it is built on algorithms that learn from training data containing information about known processed on-target and off-target sites. This data, used for simulations, includes gRNA sequences, target DNA sequences, and potential off-target sequences, along with free energies, interactions, and cleavage efficiencies at each site. By learning from training and test data, AI models can develop the ability to predict the probability of off-target effects for new gRNA experimental designs (Xu et al., 2021). In addition to prediction, this approach allows optimization of gRNA construction through algorithms that explore candidate gRNA sequences, selecting those that maximize on-target cleavage efficiency while minimizing off-target cleavage by analyzing the target genome of the organism. This is done by considering multiple parameters, such as sequence specificity, pairing energy, chromatin accessibility, and the presence of PAM motifs at non-target sites (Jiang and Doudna, 2017).

In the context of AI applications in gene editing, integrated, optimized pipelines are already in use and can combine multiple analysis steps. Deep learning models, such as Convolutional Neural Networks (CNNs), are especially used to analyze gridded data, such as 2D or 3D images, Recurrent Neural Networks (RNNs), which are used when there is a sequential order, in which the results of one layer influence other analysis steps, creating a memory, and large language models (LLMs). These models can make decisions, decompose complex tasks into steps, and promote human-AI interaction. These methods have been widely explored to predict gRNA activity and identify off targets with high accuracy and precision. A notable tool is CRISPR-DIPOFF, which is a model developed based on deep learning that uses RNN models to predict potential off-targets, demonstrating significant and optimized improvements compared to methods before AI approaches (Toufikuzzaman et al., 2024). In addition to this tool, the CRISPR-GPT represents a new-generation model that automates the entire experimental design process, from selecting the most appropriate CRISPR system and designing gRNAs to reporting, data analysis, and building optimized LLM-based pipelines (Huang et al., 2024; Qu et al., 2025). These methods not only improve the safety of gene editing and save resources but also significantly increase process efficiency and reduce human-related errors, making CRISPR–Cas technology more accurate and accessible for plant biotechnology applications (Angelov et al., 2021).

15.5 The use of AI in plant gene editing

Artificial intelligence (AI) originates with the seminal work or Alan Turing in 1950, in which he proposed the first computational model capable of predicting encodings, and introduced the concept of machine learning and intelligent behavior (Turing, 1950). Since then, AI has shown great potential for application a wide range of human activities. These technologies requires a large amount of data, along with multiple variables and parameters, which are then organized for training and testing purposes, resulting in a learning model capable of classifying, predicting, and identifying patterns, among other functions (Angelov et al., 2021). Biotechnology is no different, providing highly effective tools for analyzing and interpreting large volumes of data and optimizing highly complex biological processes, such as the study of genomic sequences and genetic editing.

Plant breeding is a dynamic field within plant biotechnology, driven by the growing demand for food security and sustainability in a climate change scenario. It has benefited immensely from advances in gene editing. The capacity to precisely modify plant genomes to introduce desirable agronomic characters, such as higher productivity, resistance to pathogen, pests and abiotic stresses, is critical to the future of agriculture (Niazian and Niedbała, 2020). Gene improvement was initiated by Gregor Mendel between 1856 and 1863, observing how the phenotype of peas behaved in genetic crosses, evaluating genetic diversity, statistical proportions, and genotype-by-environment interactions, and attributing Mendelian genetics as an initial milestone for the study of plant genetic improvement (Weiling, 1991). Biotechnological advances over the following decades enabled the selection of genes with phenotypes of interest, improving plant nutritional value, resistance to biotic stresses (pathogens, insects), and tolerance to adverse climatic conditions (drought, salinity, high temperatures), among other characteristics (Niazian and Niedbała, 2020)). These demands have arisen due to climate change, coupled with increased food needs and limited resources resulting from population growth, and they require the development of science to provide short-term solutions.

Genetic editing systems introduce precise changes to plant genomes, whereas historically these modifications were induced randomly by mutagenic agents. Recent genetic editing tools, such as CRISPR–Cas, have improved genetic and breeding research by enabling precise and efficient DNA modification (Jiang and Doudna, 2017). The simplicity, efficiency, and accuracy of CRISPR–Cas systems have overcome several limitations of other methods, such as ZFNsTALENs (Mahfouz et al., 2014). The efficiency of CRISPR systems, due to targeted mutations, has enabled the generation of crops with agronomic traits and genotypes of interest in a faster, more controlled manner.

Given this context, AI assumptions applied to genetic editing emerge as an auxiliary tool in various contexts of plant genetic engineering, through the implementation of machine learning (ML) and deep learning (DL) approaches, intelligent systems capable of learning, resembling activities that require human intellect (Xu et al., 2021). These tools significantly enhance the capacity to process and analyze large volumes of data generated by omics technologies. ML algorithms are highly effective at identifying patterns, correlations, and classifications that were previously difficult or impossible to detect using traditional known tools. As a result, they are more effective for selecting genetic targets in plants, enabling the identification of candidate genes whose modifications can confer the desired characteristics. For example, ML algorithms can analyze gene expression data (RNA-seq) and protein-DNA interaction data to predict which genes are responsive to a given stress or in the biosynthesis of a metabolite of interest (Bhardwaj et al., 2022).

In addition to genomic analysis, AI has been used in high-throughput phenotyping in plants, since traditional phenotyping is a manual, time-consuming, and often subjective process. Deep learning (DL), especially CNNs, enables automated analysis of plant images (obtained by RGB, multispectral, thermal, or 3D cameras) to accurately quantify characteristics such as leaf area, height, biomass, disease symptoms, and root architecture (Xu et al., 2021). This capability improves the identification of superior genotypes in breeding programs by obtaining accurate phenotypic data that can be correlated with genetic modifications induced by gene editing. For example, CNNs can be trained to detect early-stage plant diseases in leaf images or to estimate crop yields from field images. This enables faster, more efficient evaluation of gene-editing results.

AI-based processes are fundamental for the integration and interpretation of omic data (genomics, transcriptomics, proteomics, metabolomics) generated from experiments involving genetic editing. By combining and analyzing these vast, heterogeneous datasets, AI can provide insights into the effects of genetic modifications on plant metabolism and development. This allows a more accurate assessment of the outcomes of genetic editing and the identification of any unwanted effects, contributing to the generation of safer, more effective genetically edited plants (Angelov et al., 2021). The combination of CRISPR/Cas9 with omics strategies has provided a powerful approach for accessing almost any gene sequence in the genome to assign gene function. Several research groups focused on crop improvement have identified several key related-resistance genes to various abiotic stresses (Razzaq et al., 2021). CRISPR-edited mutants of the *slmapk3*-activated protein kinase gene in *Solanum lycopersicum* conferred drought tolerance (Wang et al., 2017). An edited triple mutant of the CBFS transcription factors in Arabidopsis revealed the role of C-repeat binding factors under cold stress (Jia et al., 2016). In corn, CRISPR was used to alter *ARGOS8* expression, a gene involved in ethylene signaling, to improve tolerance to dought and increase productivity (Shi et al., 2017).

15.6 AI for prediction of off-target effects in plants

When using AI to improve predictive performance in plants, the first step is to properly collect and prepare the data. The second step would be to select the best predictive models compatible with the data to be analyzed. Therefore, some of the most common models can be tested to determine the best-performing model for the chosen dataset. Starting with regression functions, whether linear or logistic, which provide a reasonable estimate for continuous or binary data, assuming a relationship between the variables (Fahrmeir et al., 2022). Decision tree and Random Forest models are ensemble methods that construct multiple decision trees during training and produce the class with the highest frequency (classification) or the mean prediction (regression) across the individual trees. They are robust to overfitting and can handle large datasets, effective for off-target variations (Breiman, 2001; Hesami et al., 2021).

Other classic models include Support Vector Machines (SVMs), which are supervised learning algorithms applied in both classification (regardless of whether the site is off-target) and regression (cleavage probability prediction). SVMs are particularly effective in identifying the optimal hyperplane that maximally separates data classes (Noble, 2006). The K-Nearest Neighbors (KNN) algorithm is a non-parametric classification approach that clusters gRNAs based on sequence similarity without requiring prior model training. While well-suited for smaller datasets, KNN can be compromised by noise and reduced scalability in large-scale datasets (Guo et al., 2003). Naive Bayes, a fast and effective classifier for text-based data, may perform poorly when features are highly correlated, but it is still used in gRNA efficiency classification tasks (Bhardwaj et al., 2023).

In predictive unsupervised learning, the focus is on forming groups with similar characteristics. In this context, K-Means is an alternative, especially for grouping potential genomic regions as CRISPR targets. However, it requires prior specification of the number of clusters and is sensitive to data scale (Ahmed et al., 2020). The Hierarchical Clustering model, therefore, does not require a predefined number of clusters; by automating the selection of the number of clusters, it produces visually interpretable dendrograms (Cohen-Addad et al., 2019). Another tool for cluster formation is PCA, well-known in omics analysis. It reduces dimensionality and noise, normalizes to remove off-targets, but is often difficult to interpret from a biological perspective (Jafarzadegan et al., 2019).

Another class of algorithms is deep learning approaches, such as Artificial Neural Networks (ANNs), which are models derived from processing methods that mimic the human brain. ANNs can learn complex patterns in data by observing common behaviors. For off-target prediction, they can be used to classify sequences as off-target or not, as well as predict cleavage efficiency. Simpler neural networks can be effective for problems with fewer variables (Montesinos-López et al., 2022). CNNs are a type of DL particularly effective for sequence data, such as DNA sequences. Through training, CNNs can learn hierarchical features and local patterns in gRNA and target DNA sequences, identifying motifs that influence cleavage specificity. This model is widely used in bioinformatics for sequence analysis, as it employs a progressive, sequential approach to genetic analysis. Finally, RNNs and Long Short-Term Memory (LSTM) models, although less commonly used for off-target prediction than the previously discussed approaches, remain well-suited for sequence-processing tasks. These models can capture long-range dependencies within DNA sequences, which may be relevant for cleavage specificity (Hochreiter and Schmidhuber, 1997; Zargar, 2021). A summary of the most commonly used algorithm models is presented in Table 15.1.

TABLE 15.1 Main artificial intelligence-based algorithms for optimizing CRISPR–Cas analyses.

Algorithm	Advantage (Pros)	Disadvantage (Cons)	Application in CRISPR–Cas analysis
Linear regression	Simple, interpretable, fast	Sensitive to outliers, assumes linearity	Predicting continuous outcomes such as gene-editing efficiency scores
Logistic regression	Probabilistic, interpretable	Weak with non-linear boundaries	Classifying gRNA as efficient/ inefficient
Decision tree	Easy to interpret	Overfitting, unstable with small changes	Rule-based classification of on/off-target sites
Random forest	High accuracy, robust	Slower, less interpretable	Off-target prediction combining multiple genomic features
Gradient boosting	State-of-the-art accuracy, handles complex data	Overfitting, hard to interpret	Predicting CRISPR activity efficiency with high-dimensional features
SVM	Works in high dimensions, strong classification	Slow on large datasets, requires scaling	gRNA sequence classification for target recognition
KNN	Simple, no training required	Slow, noisy predictions, needs scaling	Grouping gRNAs based on sequence similarity

(*Continued*)

TABLE 15.1 Main artificial intelligence-based algorithms for optimizing CRISPR–Cas analyses.—cont'd

Algorithm	Advantage (Pros)	Disadvantage (Cons)	Application in CRISPR–Cas analysis
Naive bayes	Fast, good with text-like data	Fails with correlated features	Classifying gRNA efficiency using sequence-based features
K-means	Fast, easy clustering	Needs K, sensitive to scale	Clustering genomic regions for potential CRISPR target sites
Hierarchical clustering	No need for K, interpretable dendrograms	Memory-intensive, slow on large data	Grouping similar genes or targets for CRISPR design
PCA	Noise reduction, dimensionality reduction	Hard to interpret	Reducing features before CRISPR off-target prediction
Neural networks (MLP)	Learns complex patterns	Needs large data, non-linear	Predicting CRISPR efficiency with multiple biological features
CNN	Excellent for sequence/ structural patterns	High resource demand	Analyzing CRISPR target sequences (1D CNN for gRNA design)
RNN	Captures sequence dependencies	Vanishing gradient, long training	Modeling nucleotide sequence context for CRISPR target prediction
Transformer (BERT, GPT)	Context-aware, handles long sequences	Large data & compute needs	NLP-like analysis of DNA for CRISPR off-target discovery
Autoencoders	Effective dimensionality reduction, anomaly detection	Can overfit, black-box	Detecting unusual CRISPR editing patterns or rare off-targets
DBSCAN	Noise-tolerant, shape-flexible clustering	Fails with varying densities	Clustering CRISPR targets in genomic space based on density

To select the most efficient algorithm for classification, prediction, and clustering applied to CRISPR systems, especially for off-target prediction, a series of well-described steps and protocols helps build a robust experimental design. The first step is data collection and quality assessment, which involves obtaining genomic and experimental data (on-target and off-target) from reliable databases. The next step, curation, includes data cleaning to remove errors and inconsistencies, normalization, and handling missing data. This allows for standardization of formats, scales, and completion of missing data, since the quality of the input data directly impacts model performance. The second phase involves feature selection, in which raw DNA and gRNA sequence information is transformed into numerical vectors, which can be interpreted by algorithms (Sherkatghanad et al., 2023). Some of the most common types are One-Hot Encoding, in which the bases A, T, G, and C are converted into binary vectors, facilitating interpretation without loss of nucleotide information, and K-mer counting, used primarily for base-frequency counting (Li et al., 2025). Furthermore, pairing energy calculations are relevant, as they take into account interactions between the gRNA and the target DNA, assessing potential off-target effects and allowing prediction of complex stability and chromatin accessibility, including ATAC-seq data or other information on chromatin structure and PAM motifs, with information on the locations of PAM motifs in the target DNA (Noshay et al., 2023).

Model training begins with data preparation, followed by splitting the data set into training, validation, and test sets. The training set, which contains the largest volume of information, is used to optimize the model's parameters, allowing it to learn patterns during each training epoch. Throughout this phase, the model attempts to minimize a loss function that measures the difference between predicted and true values, repeating this process multiple times. Model validation and testing are tasks for adjusting hyperparameters (parameters that are not learned directly from the data, such as the learning rate or the number of layers in a neural network). This data is new to the model and is used to assess the final model's performance on unseen data, providing an unbiased estimate of its ability to generalize. Finally, model interpretability is essential for practical biological applications, where it is not enough to have a model; it is also necessary to understand the statistical basis for these predictions. Explainable AI (XAI) techniques can be used to identify the most relevant features, such as specific nucleotides, pairing patterns, and others, that are most important for decision-making. This can provide valuable biological insights into mechanisms of action and help refine gRNA design (Adadi and Berrada, 2018).

After selecting features and parameters for the data to be analyzed, training proceeds by splitting the data into three groups: training data for the selected model, a test set, and a third to validate the results from the previous steps. The training set is used to adjust the model's parameters, allowing it to learn patterns in the data at each processing epoch. The model attempts to minimize a loss function, such as false positives or false negatives, which measures the difference between predicted and true values. For validation and testing, hyperparameters are adjusted to achieve higher accuracy, including relevant measures that are not learned directly from the data and require human intervention, such as the learning rate or the number of layers in a neural network. Testing is also crucial, as previously unseen data is used to ensure the chosen model provides an unbiased estimate of the results. At the end, the algorithm is trained and ready to input biological data from new experiments.

Recently, several tools and computational platforms have been developed and are now available for use with embedded AI algorithms to predict off-targets. These artificial intelligence tools are described in Table 15.2 The methodologies listed include ML algorithms, DL architectures, and CNNc, RNNs, and language models. The programs can be used online or locally.

15.7 Perspectives

Artificial intelligence is revolutionizing gene editing via CRISPR by enabling more accurate, precise design of genetic models, making it a powerful tool for advancing plant biotechnology. By modeling more assertive gene constructs to minimize off-target effects, AI accelerates the generation of agronomic cultivars with desired traits, shortening the time to bring these innovations to the field and ensuring global food security.

TABLE 15.2 Artificial Intelligence tools for predicting off-target effects in plant genome editing.

Tool	Publication	Description	Reference
CRISPR-PLANT v2	2018	An online resource specifically designed for plants that provides precise guide RNA spacers based on off-target analysis for different plant models and crop genomes.	
CRISPR-net	2020	This tool, based on RNN-CNN, quantifies CRISPR off-target activities, including mismatches and indels, surpassing traditional scoring methods.	Lin et al. (2020)
Off-target-sgRNA-Predictor (OSP)	2023	This tool combines different sequence features to predict and classify off-targets, demonstrating how integrating multiple AI models can enhance prediction accuracy.	Sherkatghanad et al. (2023)
CRISPR-M	2024	This deep learning neural network algorithm and RNNs employ a novel coding model to predict off-target effects of sgRNA, aiming to overcome limitations of previous methods.	
CRISPR-DIPOFF	2024	Deep learning tool (CNN) for accurately predicting off-target locations, with calculations of cleavage efficiency at the target site of the CRISPR–Cas9 system, using guide RNA and target DNA sequences as input data.	Toufikuzzaman et al. (2024)
CCLMoff	2025	A deep learning model that has a pre-trained RNA algorithm to predict off-target CRISPR/Cas9 sites, to store and process mutual sequence data between sgRNAs and target sites.	Du et al. (2025)
CRISPR-embedding	2025	A deep learning model based on CNNs using 9 layers. This tool is designed to predict off-target activity using the k-mer coding model.	Tahsin et al. (2025)
CRISPR-GPT	2025	A large language model (LLM) based on agent system capable of enhancing CRISPR genome editing experiments, including customized off-target prediction and optimized guide RNA design.	Qu et al. (2025)

Optimized algorithmic models for gRNA design, coupled with molecular dynamics techniques such as ML and DL, have already demonstrated the ability to analyze vast genomic datasets and predict off-target effects with high accuracy. Several pipelines have been developed to minimize errors by integrating multiple steps, ensuring the creation of highly reliable models. CRISPR-GPT, a large language model (LLM), exemplifies this process by automating and improving CRISPR-based gene-editing workflows, from CRISPR system selection and gRNA design to data analysis and the construction of optimized pipelines. Prospects include integrating CRISPR systems analysis pipelines with omics data, enabling a more comprehensive evaluation of genetic modifications and their downstream effects on metabolism, gene expression, and plant development. The application of AI in plant gene editing represents a remarkable advance in agricultural biotechnology, holding the potential to develop more resilient, high-yielding cultivars to support sustainable food production.

Declaration

The authors declare that no generative artificial intelligence tools were used in the writing or preparation of this chapter. The content reflects the authors' own scholarly work, and the authors assume full responsibility for all statements, interpretations, and conclusions presented.

References

Abudayyeh, O.O., et al., 2016. C2c2 is a single-component programmable RNA-guided RNA-targeting CRISPR effector. Science 353 (6299), aaf5573. https://doi.org/10.1126/science.aaf5573.

Adadi, A., Berrada, M., 2018. Peeking inside the black-box: a survey on explainable artificial intelligence (XAI). IEEE Access 6, 52138–52160. https://doi.org/10.1109/ACCESS.2018.2870052.

Ahmed, M., Seraj, R., Islam, S.M.S., 2020. The k-means algorithm: a comprehensive survey and performance evaluation. Electronics 9 (8), 1295. https://doi.org/10.3390/electronics9081295.

Allemailem, K.S., Almatroodi, S.A., Almatroudi, A., et al., 2023. Recent advances in genome-editing technology with CRISPR/Cas9 variants and stimuli-responsive targeting approaches within tumor cells: a future perspective of cancer management. Int. J. Mol. Sci. 24, 7052. https://doi.org/10.3390/ijms24087052.

Aman, R., Ali, Z., Butt, H., et al., 2018. RNA virus interference via CRISPR/Cas13a system in plants. Genome Biol. 19, 1. https://doi.org/10.1186/s13059-017-1381-1.

Anders, C., Niewoehner, O., Duerst, A., Jinek, M., 2014. Structural basis of PAM-dependent target DNA recognition by the Cas9 endonuclease. Nature 513 (7519), 569–573. https://doi.org/10.1038/nature13579.

Angelov, P.P., Soares, E.A., Jiang, R., Arnold, N.I., Atkinson, P.M., 2021. Explainable artificial intelligence: an analytical review. Wiley Interdiscip. Rev.: Data Min. Knowl. Discovery 11 (5), e1424. https://doi.org/10.1002/widm.1424.

Bhardwaj, A., Kishore, S., Pandey, D.K., 2022. Artificial intelligence in biological sciences. Life 12 (9), 1430. https://doi.org/10.3390/life12091430.

Bhardwaj, A., Tomar, P., Nain, V., 2023. Identification and evaluation of machine learning classification algorithm to predict the efficacy of gRNA in CRISPR/Cas9 genome editing system using WEKA. AIP Conf. Proc. 2938 (1). https://doi.org/10.1063/5.0184638.

Bibikova, M., Beumer, K., Trautman, J.K., Carroll, D., 2003. Enhancing gene targeting with designed zinc finger nucleases. Science 300 (5620), 764. https://doi.org/10.1126/science.1079512.

Bradbury, A., Clapp, O., Biacsi, A.S., Kuo, P., Gaju, O., Hayta, S., et al., 2025. Integrating genome editing with omics, artificial intelligence, and advanced farming technologies to increase crop productivity. Plant Commun. 6 (7), 101248. https://doi.org/10.1016/j.xplc.2024.101248.

Breiman, L., 2001. Random forests. Mach. Learn. 45 (1), 5–32. https://doi.org/10.1023/A:1010933404324.

Cheng, X., Li, F., Cai, J., et al., 2015. Artificial TALE as a convenient protein platform for engineering broad-spectrum resistance to begomoviruses. Viruses 7 (8), 4772–4782. https://doi.org/10.3390/v7082843.

Cohen-Addad, V., Kanade, V., Mallmann-Trenn, F., Mathieu, C., 2019. Hierarchical clustering: objective functions and algorithms. J. ACM 66 (4), 1–42. https://doi.org/10.1145/3326284.

Deltcheva, E., Chylinski, K., Sharma, C.M., Gonzales, K., Chao, Y., Pirzada, Z.A., Eckert, M.R., Vogel, J., Charpentier, E., 2011. CRISPR RNA maturation by trans-encoded small RNA and host factor RNase III. Nature 471 (7340), 602–607. https://doi.org/10.1038/nature09886.

Deng, F., Li, Y., Yang, B., et al., 2024. Topological barrier to Cas12a activation by circular DNA nanostructures facilitates autocatalysis and transforms DNA/RNA sensing. Nat. Commun. 15, 1818. https://doi.org/10.1038/s41467-024-46001-8.

Doudna, J.A., Charpentier, E., 2014. The new frontier of genome engineering with CRISPR-Cas9. Science 346, 1258096. https://doi.org/10.1126/science.1258096.

Du, W., Zhao, L., Diao, K., Zheng, Y., Yang, Q., Zhu, Z., Tang, D., 2025. A versatile CRISPR/Cas9 system off-target prediction tool using language model. Commun. Biol. 8 (1), 882. https://doi.org/10.1038/s42003-025-08275-6.

Fahrmeir, L., Kneib, T., Lang, S., Marx, B.D., 2022. Regression models. In: Fahrmeir, L., et al. (Eds.), Regression: Models, Methods and Applications. Springer Berlin Heidelberg, Berlin, Heidelberg, pp. 23–84. https://doi.org/10.1007/978-3-662-64415-8_2.

Guo, C., Ma, X., Gao, F., Guo, Y., 2023. Off-target effects in CRISPR/Cas9 gene editing. Front. Bioeng. Biotechnol. 11, 1143157. https://doi.org/10.3389/fbioe.2023.1143157.

Guo, G., Wang, H., Bell, D., Bi, Y., Greer, K., 2003. KNN model-based approach in classification. In: Meersman, R., Tari, Z., Schmidt, D.C. (Eds.), OTM Confederated International Conferences "On the Move to Meaningful Internet Systems. Springer, Berlin, Heidelberg, pp. 986–996. https://doi.org/10.1007/978-3-540-39964-3_67.

Huang, Y., Wan, L.J., Ye, H., Jha, M., Wang, J., Li, Y., Zhang, X., Chen, D., 2024. Invited: new solutions on LLM acceleration, optimization, and application. In: Proceedings of the 61st ACM/IEEE Design Automation Conference, 369. Association for Computing Machinery, New York, NY, USA, pp. 1–4. https://doi.org/10.1145/3649329.3663517.

Harrington, L.B., et al., 2018. Programmed DNA destruction by miniature CRISPR Cas14 enzymes. Science 362, 839–842. https://doi.org/10.1126/science.aav4294.

Hesami, M., Yoosefzadeh Najafabadi, M., Adamek, K., Torkamaneh, D., Jones, A.M.P., 2021. Synergizing off target predictions for in silico insights of CENH3 knockout in cannabis through CRISPR/CAS. Molecules 26 (7), 2053. https://doi.org/10.3390/molecules26072053.

Hochreiter, S., Schmidhuber, J., 1997. Long short-term memory. Neural Comput. 9 (8), 1735–1780. https://doi.org/10.1162/neco.1997.9.8.1735.

Hua, K., Tao, X., Han, P., Wang, R., Zhu, J.K., 2019. Genome engineering in rice using Cas9 variants that recognize NG PAM sequences. Mol. Plant 12 (7), 1003–1014. https://doi.org/10.1016/j.molp.2019.03.009.

Jafarzadegan, M., Safi Esfahani, F., Beheshti, Z., 2019. Combining hierarchical clustering approaches using the PCA method. Expert Syst. Appl. 137, 1–10. https://doi.org/10.1016/j.eswa.2019.07.007.

Jia, Y., Ding, Y., Shi, Y., Zhang, X., Gong, Z., Yang, S., 2016. The cbfs triple mutants reveal the essential functions of CBFs in cold acclimation and allow the definition of CBF regulons in Arabidopsis. New Phytol. 212 (2), 345–353. https://doi.org/10.1111/nph.14004.

Jiang, F., Doudna, J.A., 2017. CRISPR–Cas9 structures and mechanisms. Annu. Rev. Biophys. 46, 505–529. https://doi.org/10.1146/annurev-biophys-062215-010822.

Karvelis, T., Bigelyte, G., Young, J.K., Hou, Z., Zedaveinyte, R., Budre, K., Paulraj, S., Djukanovic, V., Gasior, S., Silanskas, A., Venclovas, Č., Siksnys, V., 2020. PAM recognition by miniature CRISPR–Cas12f nucleases triggers programmable double-stranded DNA target cleavage. Nucleic Acids Res. 48 (9), 5016–5023. https://doi.org/10.1093/nar/gkaa208.

Kim, D.Y., Lee, S.Y., Ha, H.J., Park, H.H., 2024. Structural basis of Cas3 activation in type I-C CRISPR–Cas system. Nucleic Acids Res. 52 (17), 10563–10574. https://doi.org/10.1093/nar/gkae723.

Kleinstiver, B.P., Sousa, A.A., Walton, R.T., et al., 2019. Engineered CRISPR-Cas12a variants with increased activities and improved targeting ranges for gene, epigenetic and base editing. Nat. Biotechnol. 37 (3), 276–282. https://doi.org/10.1038/s41587-018-0011-0.

Li, C., Zou, Q., Li, J., Feng, H., 2025. Prediction of CRISPR-Cas9 on-target activity based on a hybrid neural network. Comput. Struct. Biotechnol. J. 27, 1–8. https://doi.org/10.1016/j.csbj.2025.05.001.

Li, J., Xu, R., Qin, R., Liu, X., Kong, F., Wei, P., 2021. Genome editing mediated by SpCas9 variants with broad non-canonical PAM compatibility in plants. Mol. Plant 14, 352–360. https://doi.org/10.1016/j.molp.2020.12.017.

Li, P., Zhang, L., Li, Z., Xu, C., Du, X., Wu, S., 2020. Cas12a mediates efficient and precise endogenous gene tagging via MITI: microhomology-dependent targeted integrations. Cellular and Molecular Life Science 19, 3875–3884. https://doi.org/10.1007/s00018-019-03396-8.

Lin, J., Zhang, Z., Zhang, S., Chen, J., Wong, K.C., 2020. CRISPR-net: a recurrent convolutional network quantifies CRISPR off-target activities with mismatches and indels. Adv. Sci. 7 (13), 1903562. https://doi.org/10.1002/advs.201903562.

Lin, Q., Zhu, Z., Liu, G., Sun, C., Lin, D., Xue, C., Li, S., Zhang, D., Gao, C., Wang, Y., Qiu, J.L., 2021. Genome editing in plants with MAD7 nuclease. J. Genet. Genomics 48 (6), 444–451. https://doi.org/10.1016/j.jgg.2021.04.003.

Liu, Q., Zhang, H., Huang, X., 2020. Anti-CRISPR proteins targeting the CRISPR–Cas system enrich the toolkit for genetic engineering. FEBS J. 287 (4), 626–644. https://doi.org/10.1111/febs.15139.

Mahfouz, M.M., Piatek, A., Stewart, J.,C.N., 2014. Genome engineering via TALENs and CRISPR/Cas9 systems: challenges and perspectives. Plant Biotechnol. J. 12 (8), 1006–1014. https://doi.org/10.1111/pbi.12219.

Mengstie, A.M., Azezew, M.T., Dejenie, A.T., Teshome, A.A., Admasu, F.T., Teklemariam, A.B., Mulu, A.T., Agidew, M.M., Adugna, D.G., Geremew, H., Abebe, E.C., 2024. Recent advancements in reducing the off-target effect of CRISPR-Cas9 genome editing. Biologics 18, 21–28. https://doi.org/10.2147/BTT.S429411.

Montesinos-López, O.,A., Montesinos-López, A., Bernal-Sandoval, D.A., Mosqueda-Gonzalez, B.A., Valenzo-Jiménez, M.A., Crossa, J., 2022. Multi-trait genome prediction of new environments with partial least squares. Front. Genet. 13, 966775. https://doi:10.3389/fgene.2022.966775.

Niazian, M., Niedbała, G., 2020. Machine learning for plant breeding and biotechnology. Agriculture 10 (10), 436. https://doi.org/10.3390/agriculture10100436.

Nimkar, S., Anand, B., 2020. Cas3/I-C mediated target DNA recognition and cleavage during CRISPR interference are independent of the composition and architecture of Cascade surveillance complex. Nucleic Acids Res. 48 (5), 2486–2501. https://doi.org/10.1093/nar/gkz1218.

Noble, W.S., 2006. What is a support vector machine? Nat. Biotechnol. 24 (12), 1565–1567. https://doi.org/10.1038/nbt1206-1565.

Noshay, J.M., Walker, T., Alexander, W.G., Klingeman, D.M., Romero, J., Walker, A.M., Prates, E., Eckert, C., Irle, S., Kainer, D., Jacobson, D.A., 2023. Quantum biological insights into CRISPR–Cas9 sgRNA efficiency from explainable AI-driven feature engineering. Nucleic Acids Res. 51 (19), 10147–10161. https://doi.org/10.1093/nar/gkad736.

Qu, Y., Huang, K., Yin, M., et al., 2025. CRISPR GPT for agentic automation of gene editing experiments. Nat. Biomed. Eng. https://doi.org/10.1038/s41551-025-01463-z.

Razzaq, M.K., Aleem, M., Mansoor, S., Khan, M.A., Rauf, S., Iqbal, S., Siddique, K.H., 2021. Omics and CRISPR-Cas9 approaches for molecular insight, functional gene analysis, and stress tolerance development in crops. Int. J. Mol. Sci. 22 (3), 1292. https://doi.org/10.3390/ijms22031292.

Sherkatghanad, Z., Abdar, M., Charlier, J., Makarenkov, V., 2023. Using traditional machine learning and deep learning methods for on-and off-target prediction in CRISPR/Cas9: a review. Briefings Bioinf. 24 (3), bbad131.

Shi, J., Gao, H., Wang, H., Lafitte, H.R., Archibald, R.L., Yang, M., Hakimi, S.M., Mo, H., Habben, J.E., 2017. ARGOS8 variants generated by CRISPR-Cas9 improve maize grain yield under field drought stress conditions. Plant Biotechnol. J. 2, 207–216. https://doi.org/10.1111/pbi.12603.

Shmakov, S., Smargon, A., Scott, D., et al., 2017. Diversity and evolution of class 2 CRISPR Cas systems. Nat. Rev. Microbiol. 15, 169–182. https://doi.org/10.1038/nrmicro.2016.184.

Smargon, A.A., Cox, D.B.T., Pyzocha, N.K., et al., 2017. Cas13b is a type VI B CRISPR associated RNA guided RNase differentially regulated by accessory proteins Csx27 and Csx28. Mol. Cell 65, 618–630.e7. https://doi.org/10.1016/j.molcel.2016.12.023.

Sretenovic, S., Qi, Y., 2022. Plant prime editing goes prime. Nat. Plants 8, 20–22. https://doi.org/10.1038/s41477-021-01047-0.

Tahsin, A., Tasnim, Z., Chowdhury, M., Hassin, K., Meraz, G.H., Aziz, K.F., Shatabda, S., 2025. CRISPR-Embedding: CRISPR/Cas9 off-target activity prediction using DNA k-mer embedding. Computational and Structural Biotechnology Reports, 100043. https://doi.org/10.1016/j.csbr.2025.100043.

Toufikuzzaman, M., Hassan Samee, M.A., Sohel Rahman, M., 2024. Crispr DIPOFF: an interpretable deep learning approach for CRISPR Cas 9 off target prediction. Briefings Bioinf. 25 (2), bbad530. https://doi.org/10.1093/bib/bbad530.

Tsai, S.Q., Zheng, Z., Nguyen, N.T., Liebers, M., Topkar, V.V., Thapar, V., et al., 2015. GUIDE seq enables genome wide profiling of off target cleavage by CRISPR Cas nucleases. Nat. Biotechnol. 33 (2), 187–197. https://doi.org/10.1038/nbt.3117.

Turing, A.M., 1950. Mind. Mind 59 (236), 433–460.

Vu, T.V., Nguyen, N.T., Kim, J., Hong, J.C., Kim, J.Y., 2024. Prime editing: mechanism insight and recent applications in plants. Plant Biotechnol. J. 22, 19–36. https://doi.org/10.1111/pbi.14188.

Walton, R.T., Christie, K.A., Whittaker, M.N., Kleinstiver, B.P., 2020. Unconstrained genome targeting with near PAMless engineered CRISPR Cas9 variants. Science 368 (6488), 290–296. https://doi.org/10.1126/science.aba8853.

Wang, L., Chen, L., Li, R., Zhao, R., Yang, M., Sheng, J., Shen, L., 2017. Reduced drought tolerance by CRISPR/Cas9-Mediated SlMAPK3 mutagenesis in tomato plants. Journal Agricultural and Food Chemistry. Oct 4 65 (39), 8674–8682. https://doi.org/10.1021/acs.jafc.7b02745.

Weiling, F., 1991. Historical study: johann gregor Mendel 1822 1884. Am. J. Med. Genet. 40 (1), 1–25. https://doi.org/10.1002/ajmg.1320400103.

Xu, Y., Liu, X., Cao, X., Huang, C., Liu, E., Qian, S., et al., 2021. Artificial intelligence: a powerful paradigm for scientific research. Innovation 2 (4). https://doi.org/10.1016/j.xinn.2021.100143.

Zargar, S., 2021. Introduction to Sequence Learning Models: RNN, LSTM, GRU. Department of Mechanical and Aerospace Engineering. North Carolina State University. https://repository.lib.ncsu.edu/handle/1840.20/37988.

Zhan, X., Zhang, F., Zhong, Z., Chen, R., Wang, Y., Chang, L., Bock, R., Nie, B., Zhang, J., 2019. Generation of virus resistant potato plants by RNA genome targeting. Plant Biotechnol. J. 17 (9), 1814–1822. https://doi.org/10.1111/pbi.13102.

Zhang, W., Zhao, Y., Zhang, J., Ning, Y., Wang, G.-L., Ye, G., He, Z., 2025. Genetic engineering, including genome editing, for enhancing broad spectrum disease resistance in crops. Plant Commun. 6 (2), 101195. https://doi.org/10.1016/j.xplc.2024.101195.

CHAPTER

16

AI and multiplex gene editing: The future of gene editing in plants

Benoit Mermaz[1], *Alessandra Alves de Souza*[2], and *Cintia H.D. Sagawa*[3]

[1]Department of Molecular, Cellular, and Developmental Biology, Yale University, New Haven, CT, United States; [2]Citrus Research Center "Sylvio Moreira", Agronomic Institute - IAC, Cordeirópolis, Brazil; [3]PreScouter Inc., Chicago, IL, United States

16.1 Introduction

Growing global food demand, combined with the rise of extreme climate events such as prolonged droughts, heavy rainfall, and unexpected flooding, is posing major challenges to agricultural productivity. These pressures highlight the urgent need to integrate new tools into production systems that must not only deliver high productivity on a scale but also ensure environmental sustainability. This shifting scenario signals a pivotal moment for agriculture. To meet the dual challenges of climate resilience and sustainability, biotechnology must take on a more prominent role in plant breeding. While conventional breeding laid the groundwork for today's agricultural systems, its limitations, such as long breeding cycles, low precision, and random genomic changes, make it less equipped to meet the demands of a rapidly changing world (Jeon et al., 2023). Moreover, traits once considered desirable through traditional breeding are no longer ideal. Take plant architecture, for example: while semi-dwarf stature has been a cornerstone of cereal yield improvement since the Green Revolution, new challenges, such as mechanized harvesting, higher planting densities, and stress tolerance, continue to drive the search for finely tuned combinations of height, root architecture, and shoot morphology that cannot be easily achieved by crossing alone. Achieving such multitrait improvements through traditional methods is often slow or even unattainable. In this context, CRISPR/Cas systems are emerging as transformative tools, enabling precise and efficient genome editing to accelerate crop improvement. Early applications of CRISPR in plants, however, were largely limited to single-gene targets (Lowder et al., 2015; Pacesa et al., 2024). The advent of multiplex gene editing, which allows the

Intelligent Gene Editing in Plants
https://doi.org/10.1016/B978-0-443-44079-3.00018-8

simultaneous modification of multiple genomic sites, now provides a way to overcome these bottlenecks and address more intricate agronomic challenges with higher precision.

Recent advances suggest that multiplex gene editing could dramatically accelerate crop improvement by allowing multiple agronomic traits, such as drought tolerance, disease resistance, and higher yield, to be introduced in a single generation. This potential has been highlighted in several recent studies (Abdelrahman et al., 2021) and put into practice with high-capacity CRISPR systems in rice, targeting nearly 50 sites at once (Wu et al., 2024). CRISPR/Cas technology is revolutionizing crop improvement. By making genetic changes more precise and efficient, CRISPR holds promise for strengthening the resilience, productivity, and nutritional value of key staple crops. Despite ongoing challenges, such as technical, ethical, and regulatory, this approach is increasingly viewed as a critical part of the solution to global food security (Chen et al., 2024a). Bringing these targeted breeding tools into mainstream agriculture could mark a major shift in how we develop crops for the 21st century, especially as the demand for climate-resilient, high-yielding varieties continues to grow.

Artificial intelligence (AI) is now set to transform the field of gene editing by improving every step of the workflow, from guide RNA (gRNA) design and off-target prediction to trait prioritization and phenotypic assessment. AI-driven platforms can analyze large multiomics datasets, model DNA and protein interactions, and predict editing outcomes with levels of accuracy that were not possible before. This helps reduce experimental workload and makes multiplex strategies more reliable (Chen et al., 2024a; Müller et al., 2022). Machine learning (ML) models that integrate genomic and epigenomic data have begun to outperform traditional heuristic methods. These rely on fixed criteria such as GC content, sequence motifs, or manually designed scoring functions in predicting gRNA efficiency and specificity. This is an essential step toward large-scale multiplex designs for complex traits (Xiang et al., 2021). Altogether, the convergence of computational intelligence and genome engineering promises to accelerate innovation in plant biotechnology while lowering costs and minimizing risks (Li et al., 2025b).

This chapter is structured to first provide an overview of the current capabilities and promise of multiplex gene editing in plants, followed by an in-depth examination of the technological advances and associated challenges in achieving efficient, precise multilocus edits. Subsequent sections explore the transformative role of AI in enhancing gRNA design, vector engineering, and trait discovery, alongside the data challenges that limit AI model performance in plant systems. Finally, the chapter looks ahead to the future integration of AI and multiplex CRISPR editing in sustainable agriculture, considering not only technical possibilities but also the ethical, legal, and societal dimensions of these emerging tools. Together, these discussions aim to offer a comprehensive view of how AI-powered multiplex gene editing could become a central driver of climate-smart crop development in the coming decade.

16.2 The role and promise of multiplex gene editing

Multiplex gene editing refers to the modification of multiple genomic loci within a single transformation event (Fig. 16.1a). Unlike traditional approaches that focus on editing one gene at a time, multiplexing allows researchers to study, target, and manipulate entire gene families, metabolic pathways, or regulatory networks. It is especially interesting in plants where traits such as yield, stress tolerance, development, and disease resistance tend to be the product of multiple genes acting together. Taking advantage of multiplex editing, scientists can improve crops faster, uncover new gene functions, and develop more robust varieties to face global agricultural challenges (Lowder et al., 2015).

The CRISPR–Cas system became the main system used to perform multiplex gene editing due to the ease of implementation and adaptation to most studied organisms. Cas9 remains the most common nuclease, often used in combination with multiple gRNAs expressed from polycistronic transcripts, tRNA spacers, or individual promoters (Fig. 16.1a). However, new CRISPR variants have been identified and are expanding possibilities.

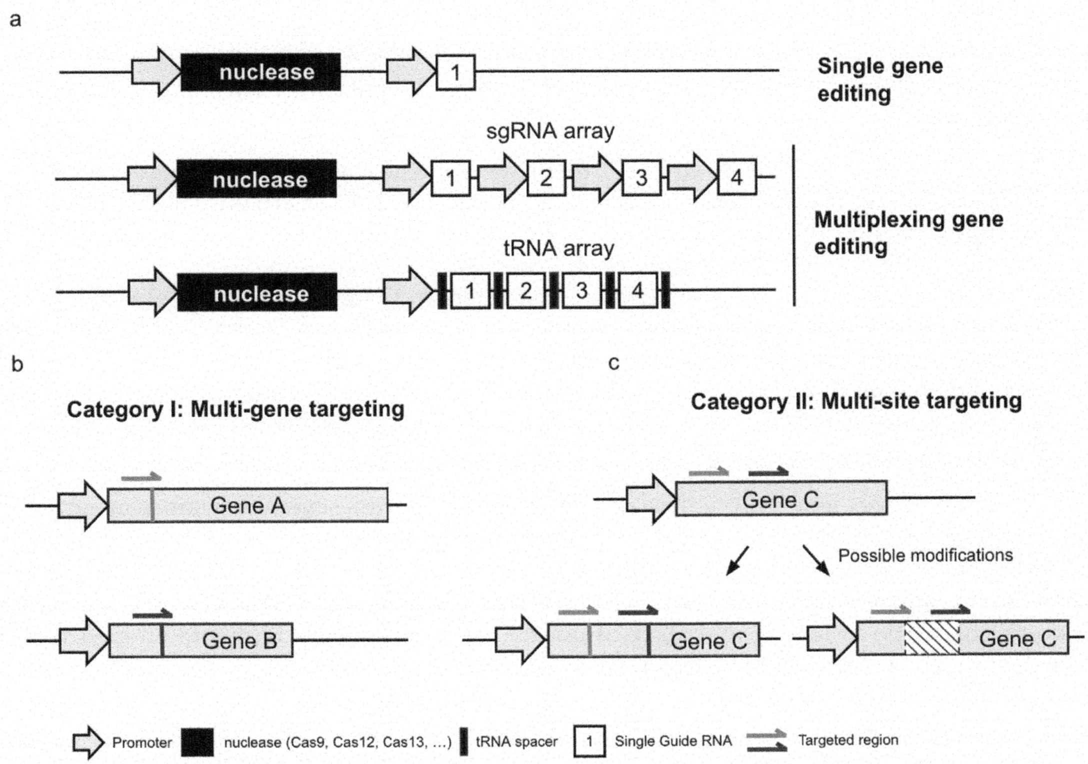

FIGURE 16.1 Comparing singleplex and multiplex gene editing strategies.

- Cas12a (Cpf1), for instance, simplifies multiplexing by processing its own crRNA array from a single transcript, reducing construct complexity (McCarty et al., 2020; Luo et al., 2025).
- Cas13 systems, which target RNA rather than DNA, offer powerful tools for virus resistance and transient gene regulation (Kavuri et al., 2022).
- Cas**Φ**, a hypercompact Cas protein, reduces construct size, improving delivery efficiency and enabling use in size-limited vectors such as certain plant viruses. (Li et al., 2023)
- Base editors and prime editors provide high-precision alternatives to traditional double-strand break (DSB)-based editing, enabling nucleotide substitutions or small insertions without inducing genome instability (Abdelrahman et al., 2021).
  a. Introducing single-gene editing and the different types of multiplexing gene editing, with an array of sgRNA, where each guide has its own promoter, and the tRNA array, where all guides are expressed under a single promoter.
  b. Present the multigene targeting that enables the simultaneous targeting and modification of multiple genes within the same plants. It is useful to improve traits, multigene families, or study complex pathways.
  c. Present the multisite targeting that enables targeting several locations of the same loci or promoter. This allows users to create various types of edits, either edits at each target site or create large deletions, truncated proteins, etc.

In plant research, the term "multiplexing" generally falls into two main categories (Fig. 16.1b and c).

(I) Multigene targeting refers to the editing of several loci within a single plant, whether the genes are part of the same pathway, a common family, or are unrelated. This approach enables the simultaneous modification of multiple traits or the study of complex pathways (Sagawa et al., 2024; Wu et al., 2024).

(II) Multisite targeting within a single gene that refers to the introduction of edits at multiple positions in the same gene to ensure complete knockout (Khadgi et al., 2025), generate large deletions (Zhou et al., 2014), create large genomic rearrangements (Lu et al., 2021; Samach et al., 2023), truncate protein coding sequences, or alter promoter regions to modulate gene expression (Li et al., 2021a). This is particularly useful to study specific domains of genes or promoters.

CRISPR multiplex gene editing has already been used across several key areas in agriculture, summarized in Table 16.1, which includes the metabolic engineering of tomatoes where several genes regulating color were simultaneously edited (Yang et al., 2022a, 2022b). Similarly, in rice and maize, multiplexing was used to knock out negative regulators of yield (Zhou et al., 2019; Waqas et al., 2025). Stress tolerance to drought and cold storage were achieved in wheat and potato by editing stress-related transcription factor families or receptors (Ly et al., 2023; Abdallah et al., 2022), which is promising to develop more climate-resilient crops; disease resistance was achieved in citrus and soybean by editing genes involved in the transport of sugar and flavonoid pathways, respectively (Khadgi et al., 2025; Zhang et al., 2022). Lastly, a new promising ultra-multiplex approach was developed

TABLE 16.1 Applications of CRISPR multiplex gene editing in plants.

Crop	Species	No. of guides	No. of targets	Goal(s)	References
Rice	Oryza sativa	49	49	Ultra-multiplex genome editing	Wu et al. (2024)
		4	4	Genomic rearrangement	Lu et al. (2021)
		4	4	Improved yield	Waqas et al. (2025)
		3	3	Improved yield	Zhou et al. (2019)
		2	1	High oleic acid production	Bahariah et al. (2021)
		2	1	Enhanced lysine content	Rastogi et al. (2025)
Citrus	Carrizo citrange, 'Duncan' grapefruit	6	3	Disease resistance	Khadgi et al. (2025)
		4	4	Induced mutations at multiple targets	Sagawa et al. (2024)
		3	1	Intron Cas9 to target PDS in different cultivars	Sarkar et al. (2024)
	Citrus sinensis	3	1	Disease resistance	de Souza-Neto et al. (2025)
Wheat	Triticum aestivum	4	4	Multiplex base editing	Ni et al. (2023)
		3	8	Improved CRISPR/Cas9 system for multiplex editing	Li et al. (2021b)
		3	6	Drought tolerance	Abdallah et al. (2022)
		1	1	Powdery mildew resistance, growth restoration	Li et al. (2022)
Tomato	Solanum lycopersicum	6	3	Fruit color	Yang et al. (2022a)
		2	1	Disruption of protein functional domains	Hashimoto et al. (2018)
		2	1	Increased GABA accumulation	Li et al. (2018)
Maize	Zea mays	4	1	Multiplex CRISPR/Cas9 system with genotyping approach	Rai et al. (2025)
		3	3	Utility of tRNA-processing–based multiplex strategy	Qi et al. (2016)
Potato	Solanum tuberosum	4	2	Reduced browning and acrylamide	Ly et al. (2023)
		2	2	Cold storage and bruising resistance	Massa et al. (2025)
Soybean	Glycine max	3	3	Disease resistance	Zhang et al. (2022)
		2	2	Oil content and fatty acid composition	Do et al. (2019)

(*Continued*)

TABLE 16.1 Applications of CRISPR multiplex gene editing in plants.—cont'd

Crop	Species	No. of guides	No. of targets	Goal(s)	References
Barley	Hordeum vulgare	7	7	Higher grain weight or dietary fiber	Yang et al. (2024)
Grapes	Vitis vinifera	6	3	Altered flavonoid accumulation	Ren et al. (2023)
Cassava	Manihot esculenta	4	2	Reduced cyanogen production	Gomez et al. (2023)
Cotton	Gossypium hirsutum	3	2	Plant architecture	Wang et al. (2024)
Plantain	–	3	9	Inactivation of eBSV, symptom-free under stress	Tripathi et al. (2019)
Walnut	Juglans sp.	3	1	JrWOX11 edits for rooting and vegetative growth	Chang et al. (2022)
Carrot	Daucus carota	2	1	Anthocyanin biosynthesis	Klimek-Chodacka et al. (2018)
Sugarcane	Saccharum spp. hybrid	2	1	Herbicide resistance	Oz et al. (2021)
Arabidopsis	Arabidopsis thaliana	1	3	Multi-gene ABA receptor edits	Zhang et al. (2022)
Avocado	Persea americana	1	2	Patent: PPO gene knockout to reduce browning	Yang et al. (2022b) (WO2022036074A2)

to target up to 49 loci in a single plant of rice (Wu et al., 2024), which opens a new scale of multiplexing. To continue, the following section will review the current advances and challenges in adapting multiplex editing for agricultural use.

16.3 Advances and challenges in multiplex gene editing technologies

The rapid expansion of multiplex gene editing in plants has been made possible with major innovations in several areas. These include advances in molecular design, delivery systems, and analytical tools. While these advances have substantially increased editing efficiency and precision, several technical challenges remain, including optimizing delivery methods, expression of the CRISPR constructs, guide optimization, and off-target effects. In addition, regulatory challenges continue to restrict the widespread use of multiplex editing in crop improvement.

The success of editing multiple genes at the same time depends on several factors. One of them is the expression and activity of multiple gRNAs within a single plant cell. Early strategies used to express each gRNA under its own promoter. With an increasing number of guides, this approach quickly resulted in large constructs that became difficult to introduce into a cell (McCarty et al., 2020). More recent advances in gRNA architecture now allow the expression of several gRNAs under the control of a single promoter, also known as polycistronic arrays, such as tRNA–gRNA systems. Next, the ribozyme-flanked transcript refers to an artificial gene that generates an RNA with a ribozyme that self-catalyzes to produce the desired gRNAs. Finally, the CRISPR arrays can be transcribed into a long RNA that is processed by Cas12a in order to release the gRNAs. Altogether, these advances allowed streamlining construct design, maintaining a balance of gRNA abundance and overall keeping the size of vectors to a minimum (Campa et al., 2019; Xie et al., 2015).

With the development of synthetic biology tools, it has become possible to design gRNAs computationally to improve their folding and secondary structure and reduce the chance of off-target activity. Some recent AI-based design platforms (Table 16.2) combine genome-wide off-target and target prediction and thermodynamic modeling to predict binding and cleavage and generate an activity score. This is useful to guide users to select guides with high cleavage potential while minimizing off-target effects (Wang et al., 2019; Haeussler et al., 2016). Another factor influencing the efficiency of multiplex gene editing is how the

TABLE 16.2 Overview of AI and machine learning tools for gene editing applications.

AI tool name	Role	Function	Tested in plants?	References
CRISPResso2	Post-editing analysis	AI-enhanced analysis of amplicon sequencing data, including a batch mode for analyzing multiple editing experiments.	Yes (cotton)	Clement et al. (2019)
OpenCRISPR-1	De novo gene editor	AI-designed CRISPR–Cas variant (OpenCRISPR-1) with broad editing capabilities (including base editing compatibility).	Potential use[a]	Ruffolo et al. (2025)
CRISPR-GPT	Experimental design	LLM-powered multi-agent system for CRISPR experiment design and analysis (assists in system selection, gRNA design, planning, data analysis, etc.).	Potential use[a]	Qu et al. (2025)
DeepCRISPR	gRNA design/ off-target	Deep learning platform to predict sgRNA on-target knockout efficiency and off-target profiles, integrating sequence and epigenetic features.	Potential use[a]	Chuai et al. (2018)
CRISPR-broad	Multi-target gRNA design	Flexible tool to identify and rank genomic regions that can be multi-targeted by few gRNAs, and to design gRNAs targeting broad/multiple regions.	Potential use[a]	Veluchamy et al. (2023)
Easy-Prime	Experimental design (prime editing)	Machine learning–based pegRNA/ngRNA design tool to optimize prime editing efficiency (trained on published PE datasets).	Potential use[a]	Li et al. (2021a)
OPED	pegRNA design (prime editing)	Transfer-learning nucleotide language model for optimized pegRNA design and efficiency prediction.	Potential use[a]	Liu et al. (2023)
BE-DICT	Base-edit outcome prediction	Attention-based deep learning algorithm to predict cytosine/adenine base editor outcomes on any target sequence.	Potential use[a]	Marquart et al. (2021)
PRIDICT	pegRNA outcome prediction	RNN model (attention-based) for predicting prime editing efficiency and outcomes (trained on 92k pegRNAs).	Potential use[a]	Mathis et al. (2023)

[a]*Potential use would require retraining on plant data for plant-specific tuning.*

constructs are delivered into plant cells. While Agrobacterium-mediated transformation remains the most widely used method for introducing constructs into plant cells, newer approaches have started to gain traction. These include viral vectors (Ma et al., 2020; Ellison et al., 2020), nanomaterials (Demirer et al., 2019; Kwak et al., 2019), protoplast transformation (Woo et al., 2015; Li et al., 2018), and particle bombardment (Svitashev et al., 2016; Liang et al., 2017). These strategies allow both transient and stable multiplex editing, even in species that are difficult to transform. For example, using plant viral replicons can drive high expression of several gRNAs at once, without having to assemble large constructs (Baltes et al., 2014; Čermák et al., 2015; Uranga et al., 2023; Cody et al., 2017).

One of the main challenges with multiplex editing is the off-target effect. This happens when the editing machinery binds and cuts at unintended genomic loci because of partial sequence similarity with the gRNA. In some cases, even a single mismatch can be enough to induce unwanted editing. The risk of off-target effects increases with the number of gRNAs used. To mitigate it, several approaches have emerged: high-fidelity Cas variants (e.g., SpCas9-HF1, eCas9, andHypaCas9) reduce nonspecific DNA cleavage (Kleinstiver et al., 2016); shortened gRNAs (17–18 nt) can improve selectivity between target and nontarget sites (Fu et al., 2014); in silico genome-wide screening now identifies gRNAs with minimal sequence similarity to nontarget loci prior to experimental validation (Doench et al., 2016); temporal control of nuclease activity, such as inducible promoters or heat/light-responsive systems, limits the exposure window for off-target cutting (Polstein and Gersbach, 2015). Together, these approaches enable higher gRNA multiplexing without increasing unintended edits, a critical factor for research integrity and regulatory approval, and AI can further assist by predicting off-target sites and optimizing gRNA design to maximize specificity and efficiency.

Despite substantial progress, plant multiplex gene editing still faces several biological and technical challenges: transformation efficiency to introduce the CRISPR/Cas machinery remains low or impossible in some crops, particularly when delivering large multiplex constructs (Ma et al., 2015); regeneration can be genotype-dependent and inefficient, limiting recovery of fully edited plants (Xing et al., 2014); somatic mosaicism and chimerism can produce heterogeneous edits complicating phenotype analysis and necessitating stabilization of edits by going through different generations when possible (Feng et al., 2014; Scintilla et al., 2022); edits at multiple targeted loci can also lead to pleiotropic effects impacting plant growth and fitness (Waqas et al., 2025); and finally, the costs and screening complexity increase with the number of targeted loci.

Research efforts are also focused on developing transformation systems that work independently of genotype, using meristematic promoters and transgene-free delivery approaches to get around existing limitations. In parallel, the drop in sequencing cost and improvements in the genotyping pipeline now make it possible to rapidly identify edited plants at multiple loci (Jones et al., 2023; Sagawa et al., 2024; Thomson et al., 2025). The use of AI-driven analytical tools is also starting to change how genome engineering workflows are done, helping to streamline construct design, predict editing outcomes, and analyze phenotypes more efficiently (Cheng et al., 2021).

Since multiplex editing can introduce several genetic changes within a single generation, it brings a number of regulatory and biosafety considerations. Rules differ across countries and will be subject to changes over time: some regulatory bodies exempt genome-edited plants that are free of transgenes from GMO regulations, while others apply the same strict biotechnology laws to all genome modification. The complexity of multiplex edits complicates molecular characterization and risk assessment, especially when large-scale chromosomal rearrangements may occur.

From a biosafety perspective, potential ecological impacts, such as altered metabolic pathways influencing nontarget organisms, must be carefully evaluated. Transparent communication of the technology's benefits, combined with rigorous molecular and phenotypic profiling, will be essential to ensure responsible deployment and societal acceptance.

16.4 AI-driven multiplexing strategies in gene editing

- Tools based on AI and ML hold great potential to improve the design, off-target prediction, and execution of genome editing in plants, as well as phenotyping evaluation. The development of strategies incorporating these technologies has steadily increased in recent years, particularly with the availability of large experimental and multiomics datasets that can be used to train models (Li et al., 2025a). According to Chen et al. (2024a, 2024b), there are three main ways AI/ML-driven approaches can help advance genome editing in plants: Increase efficiency and specificity by optimizing gRNA sequences and prioritizing Cas orthologs, engineered variants, and engineered fusion proteins designed for domain substitution;
- Predict edit outcomes by learning DNA repair pathway preferences from extensive outcome datasets;
- Speed target discovery and gRNA design by integrating transcriptomic, epigenomic, and chromatin-accessibility data to pinpoint genes and genomic sites.

Examples of AI/ML improving plant gRNA design at scale by addressing the complexity of plant genomes and enabling precise multiplex editing have been demonstrated (Müller et al., 2022; Das et al., 2023; Chen et al., 2024a, 2024b). CROPSR is an open-source platform built for complex crop genomes that performs genome-wide gRNA design and integrates PCR primer design for experimental validation; it introduces a linear support-vector-regression scoring model to avoid failures in repetitive, A/T-rich and polyploid regions, increasing both prediction accuracy and the number of usable guides, and was developed with crops such as soybean and Miscanthus in mind (and demonstrated on Miscanthus) (Müller et al., 2022). In addition to individual tools, ML models trained on integrated plant multiomics and cistrome datasets can effectively identify functional regulatory elements and critical nucleotide positions. This accelerates the discovery of genomic targets, ultimately informing precise gRNA design for trait-specific genome editing in crops (Chen et al., 2024a; Peleke et al., 2024). General deep-learning frameworks like DeepCRISPR (Chuai et al., 2018), which simultaneously predict on-target efficiency and genome-wide off-target effects, and CRISPR-Net (Lin et al., 2020), which evaluates gRNA–target interactions including mismatches and insertions/deletions, are becoming increasingly valuable for optimizing gRNA selection. These models can enhance both specificity and efficiency, making them directly applicable to genome editing workflows in plants. Recent studies also show that ML can predict gRNA efficiency and specificity across different SpCas9 variants, enabling the selection of the nuclease–gRNA combination best suited to the editing context, an advantage that is equally valuable in plant systems (Li et al., 2024). Beyond gRNA design, AI can help predict off-target effects, which is a major limitation in multiplex editing applications. Some of the factors involved in off-target cleavage include mismatch type and position, Protospacer Adjacent Motif (PAM) context, local chromatin state, and the energetics of R-loop formation. Those factors are difficult to capture with simple rules but can be learned from large, genome-wide datasets (Listgarten et al., 2018; Liu et al., 2019). ML models, such as Elevation (Listgarten et al., 2018), score guide target pairs and then aggregate genome-wide off-target risks for a given sgRNA, outperforming heuristic methods on held-out assays, which directly improves gRNA ranking in multiplex designs. The development of numerical scores (e.g., the cutting frequency determination "CFD" score) from large-scale experiments helps measure how likely CRISPR is to cut specific DNA sites (Doench et al., 2016). These scores capture how mismatches and PAM sequences affect cutting, and those are still widely used (Hesami et al., 2021; Berman et al., 2025) and could be used as key inputs in modern AI models that predict CRISPR accuracy. Recent studies also further integrate biophysical priors: thermodynamic/single-molecule evidence that gRNA activity and off-target tolerance track the free-energy landscape of R-loop nucleation and propagation has been translated into specificity scores (e.g., CRISPRspec) and hybrid AI models that fuse molecular-dynamics-derived interaction fingerprints with learning-based predictors (Corsi et al., 2022). This capability is especially important in crop species, where the genomes tend to be large, repetitive, and often polyploid. Such complexities increase the number of near-matching sequences and raise the risk of off-target edits (Graham et al., 2020). Simultaneously, AI models predict on- and off-target editing efficiency, enabling researchers to prioritize gRNAs that balance efficacy with specificity, a dual capacity that can represent a shift in gene editing technology (Xiang et al., 2021).

Advances in AI/ML are also accelerating synthetic-biology workflows that can support plant gene editing strategies (Zhang et al., 2025). Deep learning models trained on plant genomes and transcriptomes now can recover cis-regulatory sequences and prioritize candidate promoters and terminators, enabling tighter, tissue- or condition-specific control of transgene expression (e.g., convolutional neural networks [CNNs] predicting expression from flanking sequences and extracting interpretable motifs) in Arabidopsis, tomato, sorghum, and maize, and informing synthetic promoter design pipelines in plants (Yasmeen et al., 2023; Peleke et al., 2024). Codon-optimization engines powered by ML improve transgene output by learning host-specific translational preferences and sequence context. A BiLSTM-CRF model increased protein yields over vendor and index-based methods in wet-lab tests (Fu et al., 2020). In crops, codon choice strongly shapes in-planta performance: codon-optimized Bt vip3Aa11 variants in maize produced divergent outcomes, including robust insecticidal activity versus truncated, inactive protein, underscoring context-dependent effects and the need for data-driven design (Li et al., 2025a). Reviews in plant molecular farming document AI/ML integration for codon selection and host/vector engineering to raise yield and stability (Parthiban et al., 2023). AI is also being woven into plant genetic-circuit engineering: circuit components and behaviors are predicted with data-driven models, and ML-guided design/optimization is increasingly used to build programmable networks that integrate sensors, logic, and actuators for context-dependent control.

This is an approach reviewed and demonstrated across recent plant circuit studies (Vazquez-Vilar et al., 2023; Kong et al., 2025; Khan and Lister, 2025). Together, these advances show how AI links regulatory-element discovery, codon-level optimization, and synthetic-circuit design, paving the way for more precise and reliable multiplex editing systems in crops. In the context of multiplex gene editing, ML models like CNNs, support vector regression (SVR), and hybrid ensembles are proving especially valuable. These approaches can uncover patterns that traditional methods often miss, making it possible to generate more accurate predictions, which is an essential capability when designing gRNAs for simultaneously targeting multiple genomic sites with high precision and minimal off-target effects (Zhang et al., 2020; McCarty et al., 2020). One promising example is a hybrid model that combines CNNs with SVR to predict CRISPR/Cas9 gRNA activity. In this approach, CNNs are used to extract key sequence features, while the SVR component handles the regression needed to estimate gRNA efficiency (Zhang et al., 2020). More recently, deep learning frameworks have continued to advance the field by improving predictions of both on-target activity and off-target risks, directly enhancing the design of robust multiplex editing strategies (Lee, 2023).

Some case studies demonstrate the successful application of AI-guided editing pipelines; however, they have not yet been validated in empirical plant CRISPR experiments. Deep learning and ML predictors for gRNA activity and off-target risk, such as DeepCRISPR (Chuai et al., 2018) and CRISPR-Net (Lin et al., 2020), are now used upstream to rank guides, reducing wet-lab screening while improving hit rates and specificity. While the original DeepCRISPR and CRISPR-Net models were explicitly trained using human data, they could be theoretically easily extended to other Cas9 species or variants and other species, according to the authors (Chuai et al., 2018; Lin et al., 2020). To effectively adapt these models for plants, high-quality, large-scale data on empirically validated sgRNA efficacy, genome-wide off-target profiles, and plant-specific epigenetic features would be needed. Scored guides have been deployed with high-throughput multiplex vectors (tRNA–sgRNA arrays and intron-optimized Cas9) delivering efficient multilocus edits and shortening design–build–test cycles. In monocots, data-driven prime-editing design tools trained on rice experiments, such as PlantPegDesigner (Jin et al., 2023), and broader deep-learning frameworks for pegRNA optimization (Li et al., 2021a) translate into higher on-target efficiencies and fewer iterative rounds, accelerating trait programs. Together with crop-focused ML reviews documenting gRNA/target prioritization workflows (Chen et al., 2024a, 2024b), these examples link potential use of AI-assisted design to multiplex editing outcomes aligned with drought resilience, disease resistance, and yield improvement agendas in cereals and fruit crops. Table 2 provides an overview of some of the available AI/ML tools available for gene editing.

Overall, AI-driven multiplexing strategies represent a potential major advance in plant genome engineering by enabling the systematic prioritization of gene targets, the intelligent design of combinatorial editing tools, and the optimization of complex biological pathways. These capabilities can accelerate the generation of crop varieties with improved stress tolerance, disease resistance, and nutrient use efficiency. These can also reduce the time, cost, and uncertainty associated with multiplex gene editing experiments, ultimately supporting sustainable agricultural innovation through precise and scalable genome engineering.

16.5 Integrating AI for gene discovery and vector design

The advantages of the integration of AI into plant genome editing go far beyond gRNA optimization to support large-scale gene discovery and automated design pipelines. ML algorithms transform gene function prediction by leveraging multiomics datasets to identify links between genomic sequences and complex agricultural traits (Montesinos-López et al., 2021). By integrating evolutionary information, ML approaches that leverage phenotypic diversity and conserved transcriptome responses across species have further improved predictions of gene-to-phenotype relationships, as shown in Arabidopsis, rice, and maize for nitrogen use efficiency studies (Cheng et al., 2021).

Deep learning models in nonplant systems have demonstrated impressive accuracy in modeling gene regulatory mechanisms, both by predicting the expression pattern of cis-regulatory sequences and by identifying the structure of regulatory networks (Avsec et al., 2021; Yuan and Duren, 2025; Kamimoto et al., 2023). Similar approaches are now emerging in plants (Shoaib et al., 2023), though multiomics integrative AI models for constructing gene regulatory networks and prioritizing multiplex-editing targets remain an active area of development in crops. Advanced ML models have been applied to uncover and predict the genetic architecture of complex traits, such as grain yield under drought stress in maize. In particular, deep neural networks have been used to predict hybrid yield from genotype and environmental data (Khaki and Wang, 2019), and deep learning methods have significantly improved genomic prediction accuracy in wheat breeding programs (Montesinos-López et al., 2023).

Genomic prediction models based on ML approaches, including statistical ML tools for continuous, binary, and categorical phenotypes, are being democratized through user-friendly software platforms that facilitate the translation of computational predictions and practical breeding applications (Crossa et al., 2025). Another example is the development of models based on three ML-based techniques, artificial neural network (ANN), support vector machine (SVM), and random forest (RF), for the prediction of the CRISPR Cas9 cleavage sites (Das et al., 2023). AI and ML have rapidly developed in human research, and while their application in plant science is still emerging, the field is quickly catching up. In the coming years, we can expect more sophisticated AI-driven approaches to improve CRISPR target prediction and accelerate crop improvements. Importantly, as more plant genomic and phenotypic data become available, the performance and generalizability of these AI models will improve. However, this reliance on data also highlights one of the major challenges currently facing the field.

16.6 Data challenges and AI model training

Despite the promise of AI-driven approaches, their success is fundamentally limited by the quality and availability of data. One of the key obstacles in using AI to multiplex CRISPR gene editing in plants is the lack of availability of high-quality, plant-specific genomic datasets. While model organisms like *Arabidopsis thaliana* and major crops such as rice and maize have well-developed genomic resources, many other agriculturally important species still lack complete, well-annotated reference genomes and functional datasets (Abbasi et al., 2025). This gap presents a significant obstacle for multiplex gene editing, which depends on detailed insights into gene networks, regulatory elements, and noncoding regions in order to design gRNAs that can act on multiple targets at once (Pan et al., 2023). The challenge is made even more pressing by the fact that most available datasets have been developed with single-gene CRISPR applications in mind. As a result, they often fall short when applied to multiplex strategies, which require much broader coverage and more complex, combinatorial sequence data (Chen et al., 2024a, 2024b). Even when plant genomic datasets are available, those often suffer from quality issues, incomplete annotations, and insufficient phenotypic metadata, all of which constrain the predictive accuracy of AI models. Annotation gaps, such as missing or incorrectly predicted transcription start sites, untranslated regions, or noncoding RNAs, can introduce systematic errors into gRNA efficacy predictions (Das et al., 2023). Multiplex CRISPR applications are particularly sensitive to these issues, since off-target interactions between multiple gRNAs can amplify unintended edits if gene annotations are incomplete or erroneous (Lee, 2023). Another challenge lies in the wide phenotypic variability seen in plants, much of which comes from complex genotype-by-environment (G×E) interactions. This variability makes it harder to draw clear connections between specific genomic edits and observable traits, posing a real hurdle for AI models that need to perform reliably across diverse experimental conditions (Karimi et al., 2025). Without consistent, standardized phenotyping systems in place, predictions for multiplex gene editing often remain highly dependent on context and difficult to scale.

Recent reviews point to the growing role of transfer learning in agricultural AI, with promising results, especially in areas like plant disease detection (Hossen et al., 2025; Sambana et al., 2025). Although plant-specific data for genome editing remains limited, researchers are beginning to explore how techniques like transfer learning, active learning, and multitask modeling could help tackle key challenges, such as predicting targets, minimizing off-target effects, and managing the combinatorial complexity of multiplex CRISPR systems. That said, these approaches still need to be validated through further experimental work.

Finally, the development of collaborative platforms for data sharing and AI benchmarking is emerging as a critical driver of progress in this field. Public repositories hosting multiplex CRISPR datasets, such as gRNA activity libraries and high-throughput phenotyping results, provide the foundation for reproducible AI model development (Pan et al., 2023). Benchmarking platforms that standardize how data is formatted, how models are evaluated, and how results are validated make it possible for fair comparisons of competing AI architectures, accelerating the identification of optimal approaches (Abbasi et al., 2025). Equally important are cross-institutional collaborations that bring together plant biologists, computational scientists, and breeders. These partnerships help create diverse, high-quality datasets and support the development of AI tools specifically designed for the complex challenges of plant genome editing (Chen et al., 2024a, 2024b; Karimi et al., 2025). Without such open, interoperable ecosystems, AI model progress risks remaining fragmented and species-specific, limiting its translational impact in agricultural biotechnology.

16.7 Outlook: The future of AI and CRISPR in sustainable agriculture

Integrating AI into CRISPR/Cas9 multiplex genome editing approaches is expected to take plant breeding pipelines to the next level in the next few decades, particularly for the development of climate-smart crops and precision agriculture applications. Advances in multiplex CRISPR technologies, such as tRNA-based sgRNA arrays (Zhang et al., 2019; Sagawa et al., 2024) and ultra-multiplex Cas9 systems (Wu et al., 2024), now permit the simultaneous targeting of multiple genomic loci, enabling complex trait stacking in a single generation. For example, Cas12a-based systems have demonstrated the ability to edit several genomic regions simultaneously (Zhang et al., 2021), while RNA virus-based delivery approaches have made it feasible to perform whole-plant multiplex editing without the need for tissue culture (Uranga et al., 2021). When paired with AI-driven target design and predictive modeling, these breakthroughs could dramatically shorten breeding cycles and accelerate the development of improved crop varieties.

AI-assisted genome engineering is anticipated to help overcome current challenges in gRNA design, off-target prediction, and editing efficiency. Recent deep learning models, such as CRISPR-SGRU (Zhang et al., 2024) and CRISPR-BERT (Luo et al., 2024), have shown notable improvements in predicting CRISPR–Cas9 activity by accounting for mismatches, insertions, or deletions (indels), and the broader sequence context. Beyond enhancing prediction accuracy, these in silico approaches are helping to cut down the time and cost of experimental screening. They also make it possible to preselect gRNA combinations for multiplex editing systems that provide optimal genome coverage (Sun et al., 2024). As AI models continue to be trained on plant-specific genomic datasets, their capabilities are expected to grow—from simple sequence recognition to deeper insights into functional genomics. This shift could significantly speed up the identification of causal variants linked to complex traits such as climate resilience and resource-use efficiency.

Phenotyping remains a critical bottleneck in translating genome edits into field performance, and here AI is expected to play a transformative role. High-throughput phenotyping integrated with explainable AI has already been applied to marker-assisted selection for drought-tolerant wheat (Chang-Brahim et al., 2024) and multiomics-based horticultural breeding (Cembrowska-Lech et al., 2023). The combination of multiplex CRISPR editing for trait introduction and AI-driven phenotyping for trait validation creates a closed feedback loop in breeding pipelines, enabling faster genotype-to-phenotype mapping and accelerating cultivar deployment in response to climate stressors. Advances in remote sensing and deep learning are projected to further reduce phenotyping costs and increase selection accuracy, particularly in variable field conditions (Murphy et al., 2024).

Looking ahead, the intersection of AI and CRISPR brings not only exciting scientific opportunities but also important ethical, legal, and societal considerations (ELSI) that will influence how these technologies are adopted. Regulatory frameworks continue to differ widely across regions, with varying levels of oversight for gene-edited versus transgenic crops (Caradus et al., 2023). Notably, the European Union has moved toward a regulatory distinction between simple genome-edited crops (those carrying modifications that could also arise naturally or through conventional breeding) and more complex edits that remain subject to full GMO legislation, with a provisional agreement reached between the Council and the European Parliament in December 2025 and formal adoption by the Council in April 2026 (European Parliament, 2025). If enacted as expected in mid-2028, this would represent the most significant regulatory shift for gene-edited crops in Europe since the 2018 ruling of the Court of Justice of the EU, and could substantially affect the global commercialization landscape for CRISPR-improved varieties. Public acceptance often depends on perceptions of safety and the strength of supporting evidence (McFadden et al., 2024), while broader social and cultural values will ultimately shape how readily these innovations are embraced in the marketplace (Henderson et al., 2024).

Here, however, we return to the starting point of this chapter, the importance of considering how society accepts new and safe technologies as a key part of securing food supplies. In the case of transgenic crops, strict regulatory frameworks in many countries limit their use to major companies able to cover the high cost of compliance. As a result, transgenics were mainly introduced in a few major crops where production volumes justified the expense, leaving many others in the back.

By contrast, CRISPR technology represents a potential turning point. Because many of its applications do not involve transgenesis, it offers the prospect of extending these advances to a broader range of essential food crops. This shift could not only strengthen agricultural resilience to climate challenges but also promote more sustainable and equitable food systems worldwide. For AI-CRISPR breeding pipelines to achieve their full potential in sustainable agriculture, transparent governance frameworks, inclusive stakeholder engagement, and equitable access to technology will be as essential as the technical innovations themselves. Fig. 16.2 illustrates the main steps in a pipeline for developing and implementing multiplex AI combined with multiplex CRISPR/Cas platforms in next-generation plant breeding.

16.8 Disclaimer

The opinions expressed in this chapter are solely from the authors and do not represent any opinion or official stance of the institutions to which they are affiliated. The authors have reviewed the available literature, and some of the content in this chapter was produced with the assistance of AI tools, specifically ChatGPT-4o and ChatGPT-5o, and manually verified. The authors declare no conflict of interest and intend to provide a general outlook of available AI tools for gene editing multiplexing in plants.

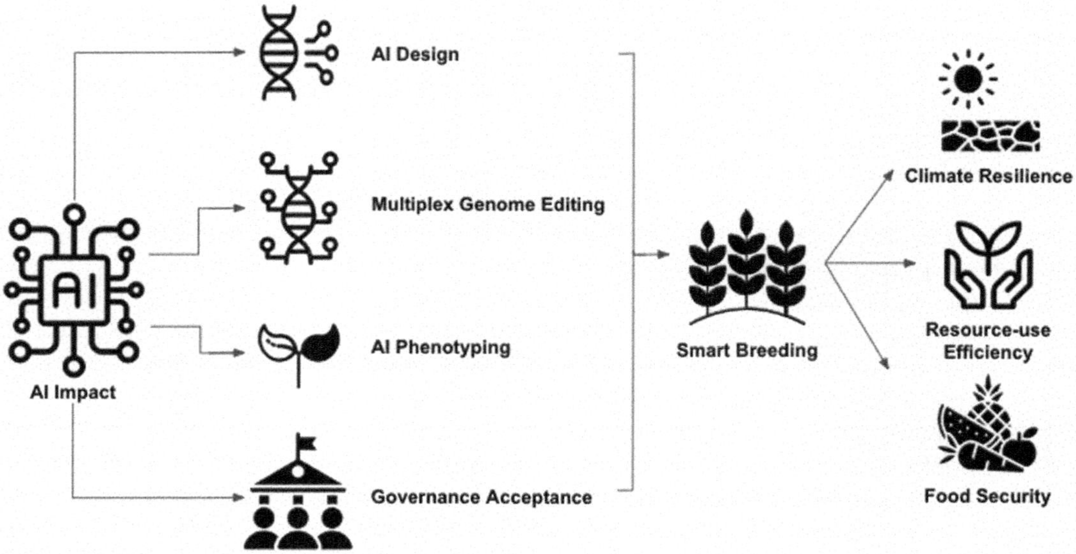

FIGURE 16.2 Integration of AI and multiplex CRISPR/Cas platforms in next-generation plant breeding pipelines. *AI Design*: Predicts optimal gRNAs, minimizes off-target effects, and selects multiplex targets using plant genomic and multiomics data for precise trait editing; *Multiplex Genome Editing:* Advanced CRISPR systems (e.g., tRNA-sgRNA arrays, Cas12a/Cas9) enable simultaneous multigene editing, facilitating trait stacking and shortening breeding cycles; *AI Phenotyping:* High-throughput, AI-driven imaging and sensor analyses connect genome edits to field performance, creating a feedback loop for improved accuracy and efficiency; *Governance Acceptance:* Transparent, inclusive frameworks ensure safe, equitable, and socially responsible implementation of AI–CRISPR technologies. Together, these steps create an intelligent pipeline for accelerating the development of resilient, sustainable smart crops that can withstand biotic and abiotic stresses while contributing to global food security. Designed with icons from Flaticon.

References

Abbasi, A.F., Asim, M.N., Dengel, A., 2025. Transitioning from wet lab to artificial intelligence: a systematic review of AI predictors in CRISPR. J. Transl. Med. 23, 60. https://doi.org/10.1186/s12967-024-06013-w.

Abdallah, N.A., Elsharawy, H., Abulela, H.A., Thilmony, R., Abdelhadi, A.A., Elarabi, N.I., 2022. Multiplex CRISPR/Cas9-mediated genome editing to address drought tolerance in wheat. GM Crops Food 16 (1), 1–17. https://doi.org/10.1080/21645698.2022.2120313. PMID:36200515; PMCID:PMC11702957.

Abdelrahman, M., Wei, Z., Rohila, J.S., Zhao, K., 2021. Multiplex genome-editing technologies for revolutionizing plant biology and crop improvement. Front. Plant Sci. 12. https://doi.org/10.3389/fpls.2021.721203.

Avsec, Ž., Agarwal, V., Visentin, D., Ledsam, J.R., Grabska-Barwinska, A., Taylor, K.R., Assael, Y., Jumper, J., Kohli, P., Kelley, D.R., 2021. Effective gene expression prediction from sequence by integrating long-range interactions. Nat. Methods 18 (10), 1196–1203. https://doi.org/10.1038/s41592-021-01252-x.

Bahariah, B., Masani, M.Y.A., Rasid, O.A., Parveez, G.K.A., 2021. Multiplex CRISPR/Cas9-mediated genome editing of the FAD2 gene in rice: a model genome editing system for oil palm. J. Genet. Eng. Biotechnol. 19 (1), 86. https://doi.org/10.1186/s43141-021-00185-4. PMID:34115267; PMCID:PMC8196110.

Baltes, N.J., Gil-Humanes, J., Cermak, T., Atkins, P.A., Voytas, D.F., 2014. DNA replicons for plant genome engineering. Plant Cell 26 (1), 151–163. https://doi.org/10.1105/tpc.113.119792.

Berman, A., Su, N., Li, Z., Landau, U., Chakraborty, J., Gerbi, N., Liu, J., Qin, Y., Yuan, B., Wei, W., Yanai, O., Mayrose, I., Zhang, Y., Shani, E., 2025. Construction of multi-targeted CRISPR libraries in tomato to overcome functional redundancy at genome-scale level. Nat. Commun. 16 (1). https://doi.org/10.1038/s41467-025-59280-6.

Campa, C.C., Weisbach, N.R., Santinha, A.J., Incarnato, D., Platt, R.J., 2019. Multiplexed genome engineering by Cas12a and CRISPR arrays encoded on single transcripts. Nat. Methods 16 (9), 887–893. https://doi.org/10.1038/s41592-019-0508-6.

Caradus, J.R., Fennessy, P.F., George, D.G., 2023. Processes for regulating genetically modified and gene-edited plants differ in outcomes. GM Crops Food 14 (1), 1–15. https://doi.org/10.1080/21645698.2023.2252947.

Cembrowska-Lech, D., Wójtowicz, T., Niemira, M., Szot, I., 2023. An integrated multi-omics and artificial intelligence framework for horticultural research and breeding. Hortic. Res. 10, uhac267. https://doi.org/10.1093/horticulture/uhac267.

Čermák, T., Baltes, N.J., Čegan, R., Zhang, Y., Voytas, D.F., 2015. High-frequency, precise modification of the tomato genome. Genome Biol. 16 (1). https://doi.org/10.1186/s13059-015-0796-9.

Chang, Y., Song, X., Zhang, Q., Zhang, P., Lei, X., Pei, D., 2022. Robust CRISPR/Cas9-mediated gene editing of JrWOX11 manipulated adventitious rooting and vegetative growth in a nut tree species of walnut. Sci. Hortic. 303, 111199. https://doi.org/10.1016/j.scienta.2022.111199.

Chang-Brahim, I., Koppensteiner, L.J., Beltrame, L., Bodner, G., Saranti, A., Salzinger, J., Fanta-Jende, P., Sulzbachner, C., Bruckmüller, F., Trognitz, F., Samad-Zamini, M., Zechner, E., Holzinger, A., Molin, E.M., 2024. Reviewing the essential roles of remote phenotyping, GWAS and explainable AI in practical marker-assisted selection for drought-tolerant winter wheat breeding. Front. Plant Sci. 15. https://doi.org/10.3389/fpls.2024.1319938.

Chen, F., Jiang, L., Xiao, Y., Li, Z., Du, Y., 2024a. Recent advances of CRISPR-based genome editing for sustainable crop improvement. Front. Plant Sci. 15, 1478398. https://doi.org/10.3389/fpls.2024.1478398.

Chen, L., Liu, G., Zhang, T., 2024b. Integrating machine learning and genome editing for crop improvement. aBIOTECH 5 (2), 262–277. https://doi.org/10.1007/s42994-023-00133-5. PMID:38974863; PMCID:PMC11224061.

Cheng, C.Y., Li, Y., Varala, K., Bubert, J., Huang, J., Kim, G.J., Halim, J., Arp, J., Shih, H.J.S., Levinson, G., Park, S. H., Cho, H.Y., Moose, S.P., Coruzzi, G.M., 2021. Evolutionarily informed machine learning enhances the power of predictive gene-to-phenotype relationships. Nat. Commun. 12 (1). https://doi.org/10.1038/s41467-021-25893-w.

Chuai, G., Ma, H., Yan, J., Chen, M., Hong, N., Xue, D., Zhou, C., Zhu, C., Chen, K., Duan, B., Gu, F., Qu, S., Huang, D., Jia, W., Qi, L., 2018. DeepCRISPR: optimized CRISPR guide RNA design by deep learning. Genome Biol. 19 (1). https://doi.org/10.1186/s13059-018-1459-4.

Clement, K., Rees, H., Canver, M.C., Gehrke, J.M., Farouni, R., Hsu, J.Y., Cole, M.A., Liu, D.R., Joung, J.K., Bauer, D. E., Pinello, L., 2019. CRISPResso2 provides accurate and rapid genome editing sequence analysis. Nat. Biotechnol. 37 (3), 224–226. https://doi.org/10.1038/s41587-019-0032-3. PMID:30809026; PMCID:PMC6533916.

Cody, W.B., Scholthof, H.B., Mirkov, T.E., 2017. Multiplexed gene editing and protein overexpression using a tobacco mosaic virus viral vector. Plant Physiol. 175, 23–35. https://doi.org/10.1104/pp.17.00701.

Corsi, G.I., Qu, K., Alkan, F., Pan, X., Luo, Y., Gorodkin, J., 2022. CRISPR/Cas9 gRNA activity depends on free energy changes and on the target PAM context. Nat. Commun. 13 (1). https://doi.org/10.1038/s41467-022-30515-0.

Crossa, J., Martini, J.W.R., Vitale, P., Pérez-Rodríguez, P., Costa-Neto, G., Fritsche-Neto, R., Runcie, D., Cuevas, J., Toledo, F., Li, H., De Vita, P., Gerard, G., Dreisigacker, S., Crespo-Herrera, L., Saint Pierre, C., Bentley, A., Lillemo, M., Ortiz, R., Montesinos-López, O.A., Montesinos-López, A., 2025. Expanding genomic prediction in plant breeding: harnessing big data, machine learning, and advanced software. Trends Plant Sci. 30 (7), 756–774. https://doi.org/10.1016/j.tplants.2024.12.009.

Das, J., Paul, S., Das, S., et al., 2023. Machine learning in the estimation of CRISPR-Cas9 efficiency and specificity in plant systems. Front. Genet. 13, 1085332. https://doi.org/10.3389/fgene.2022.1085332.

de Souza-Neto, R.R., Cavalcante, L.N., Carvalho, I.G.B., Curtolo, M., Benedetti, C.E., Takita, M.A., Wang, N., de Souza, A.A., 2025. CRISPR/Cas9-mediated disruption of *CsLIEXP1* reveals expansin as a key susceptibility factor for citrus canker disease. Mol. Plant Microbe Interact. PMID:40173114. https://doi.org/10.1094/MPMI-12-24-0151-R.

Demirer, G.S., Zhang, H., Matos, J.L., Goh, N.S., Cunningham, F.J., Sung, Y., Chang, R., Aditham, A.J., Chio, L., Cho, M.J., Staskawicz, B., Landry, M.P., 2019. High aspect ratio nanomaterials enable delivery of functional genetic material without DNA integration in mature plants. Nat. Nanotechnol. 14 (5), 456–464. https://doi.org/10.1038/s41565-019-0382-5.

Do, P.T., Nguyen, C.X., Bui, H.T., Tran, L.T.N., Stacey, G., Gillman, J.D., Zhang, Z.J., Stacey, M.G., 2019. Demonstration of highly efficient dual gRNA CRISPR/Cas9 editing of the homeologous GmFAD2-1A and GmFAD2-1B genes to yield a high oleic, low linoleic and α-linolenic acid phenotype in soybean. BMC Plant Biol 19 (1), 311. https://doi.org/10.1186/s12870-019-1906-8. PMID:31307375; PMCID:PMC6632005.

Doench, J.G., Fusi, N., Sullender, M., Hegde, M., Vaimberg, E.W., Donovan, K.F., Smith, I., Tothova, Z., Wilen, C., Orchard, R., Virgin, H.W., Listgarten, J., Root, D.E., 2016. Optimized sgRNA design to maximize activity and minimize off-target effects of CRISPR-Cas9. Nat. Biotechnol. 34 (2), 184–191. https://doi.org/10.1038/nbt.3437.

Ellison, E.E., Nagalakshmi, U., Gamo, M.E., Huang, P.j., Dinesh-Kumar, S., Voytas, D.F., 2020. Multiplexed heritable gene editing using RNA viruses and mobile single guide RNAs. Nat. Plants 6 (6), 620–624. https://doi.org/10.1038/s41477-020-0670-y.

European Parliament, 2025. New Genomic Techniques: Council and Parliament Strike Deal to Boost the Competitiveness and Sustainability of Our food systems. Available at: https://www.consilium.europa.eu/en/press/press-releases/2025/12/04/new-genomic-techniques-council-and-parliament-strike-deal-to-boost-the-competitiveness-and-sustainability-of-our-food-systems/.

Feng, Z., Mao, Y., Xu, N., Zhang, B., Wei, P., Yang, D.-L., Wang, Z., Zhang, Z., Zheng, R., Yang, L., Zeng, L., Liu, X., Zhu, J.-K., 2014. Multigeneration analysis reveals the inheritance, specificity, and patterns of CRISPR/Cas-induced gene modifications in Arabidopsis. Proc. Natl. Acad. Sci. 111 (12), 4632–4637. https://doi.org/10.1073/pnas.1400822111.

Fu, H., Liang, Y., Zhong, X., Pan, Z.L., Huang, L., Zhang, H.L., Xu, Y., Zhou, W., Zhong, L., 2020. Codon optimization with deep learning to enhance protein expression. Sci. Rep. 10 (1). https://doi.org/10.1038/s41598-020-74091-z.

Fu, Y., Sander, J.D., Reyon, D., Cascio, V.M., Joung, J.K., 2014. Improving CRISPR-Cas nuclease specificity using truncated guide RNAs. Nat. Biotechnol. 32 (3), 279–284. https://doi.org/10.1038/nbt.2808. Epub 2014 Jan 26. PMID: 24463574; PMCID: PMC3988262.

Gomez, M.A., Berkoff, K.C., Gill, B.K., Iavarone, A.T., Lieberman, S.E., Ma, J.M., Schultink, A., Karavolias, N.G., Wyman, S.K., Chauhan, R.D., Taylor, N.J., Staskawicz, B.J., Cho, M.J., Rokhsar, D.S., Lyons, J.B., 2023. CRISPR-Cas9-mediated knockout of CYP79D1 and CYP79D2 in cassava attenuates toxic cyanogen production. Front. Plant Sci. 13, 1079254. https://doi.org/10.3389/fpls.2022.1079254. PMID:37007603; PMCID: PMC10064795.

Graham, N., Patil, G.B., Bubeck, D.M., Dobert, R.C., Glenn, K.C., Gutsche, A.T., Kumar, S., Lindbo, J.A., Maas, L., May, G.D., Vega-Sanchez, M.E., Stupar, R.M., Morrell, P.L., 2020. Plant genome editing and the relevance of off-target changes. Plant Physiol. 183 (4), 1453–1471. https://doi.org/10.1104/pp.19.01194.

Haeussler, M., Schönig, K., Eckert, H., Eschstruth, A., Mianné, J., Renaud, J.B., Schneider-Maunoury, S., Shkumatava, A., Teboul, L., Kent, J., Joly, J.S., Concordet, J.P., 2016. Evaluation of off-target and on-target scoring algorithms and integration into the guide RNA selection tool CRISPOR. Genome Biol. 17 (1). https://doi.org/10.1186/s13059-016-1012-2.

Hashimoto, R., Ueta, R., Abe, C., Osakabe, Y., Osakabe, K., 2018. Efficient multiplex genome editing induces precise and self-ligated type mutations in tomato plants. Front. Plant Sci. 9, 916. https://doi.org/10.3389/fpls.2018.00916. PMID:30018630; PMCID:PMC6037947.

Henderson, K., Lang, B., Kemper, J., Conroy, D., 2024. Exploring diverse food system actor perspectives on gene editing: a systematic review of socio-cultural factors influencing acceptability. Agric. Hum. Val. 41 (2), 883–907. https://doi.org/10.1007/s10460-023-10523-6.

Hesami, M., Yoosefzadeh Najafabadi, M., Adamek, K., Torkamaneh, D., Jones, A.M.P., 2021. Synergizing off-target predictions for in silico insights of cenh3 knockout in cannabis through CRISPR/Cas. Molecules 26 (7). https://doi.org/10.3390/molecules26072053.

Hossen, M.I., Awrangjeb, M., Pan, S., Mamun, A.A., 2025. Transfer learning in agriculture: a review. Artif. Intell. Rev. 58 (4). https://doi.org/10.1007/s10462-024-11081-x.

Jeon, D., Kang, Y., Lee, S., Choi, S., Sung, Y., Lee, T.H., Kim, C., 2023. Digitalizing breeding in plants: a new trend of next-generation breeding based on genomic prediction. Front. Plant Sci. 14. https://doi.org/10.3389/fpls.2023.1092584.

Jin, S., et al., 2023. Optimized prime editing in monocot plants using data-driven pegRNA design (PlantPegDesigner). Nat. Protoc. 18, 2749–2781. https://doi.org/10.1038/s41596-023-00851-0.

Kamimoto, K., Stringa, B., Hoffmann, C.M., Jindal, K., Solnica-Krezel, L., Morris, S.A., 2023. Dissecting cell identity via network inference and in silico gene perturbation. Nature 614 (7949), 742–751. https://doi.org/10.1038/s41586-022-05688-9.

Karimi, R., Mohammadi, R., Tavakoli, A., et al., 2025. Integrating AI and CRISPR Cas13a for rapid detection of plant pathogens. Sci. Rep. 15, 11405. https://doi.org/10.1038/s41598-025-11405-z.

Kavuri, N.R., Ramasamy, M., Qi, Y., Mandadi, K., 2022. Applications of CRISPR/Cas13-Based RNA editing in plants. Cells 11 (17). https://doi.org/10.3390/cells11172665.

Khadgi, A., Zayed, O., Sagawa, C.H.D., Zhang, F., Seymour, D.K., Irish, V.F., 2025. Mutations in the SWEET15 sugar transporter gene affect response of citrus to huanglongbing disease and citrus canker. Mol. Plant Pathol. 26 (5). https://doi.org/10.1111/mpp.70094.

Khaki, S., Wang, L., 2019. Crop yield prediction using deep neural networks. Front. Plant Sci. 10. https://doi.org/10.3389/fpls.2019.00621.

Khan, A., Lister, R., 2025. Synthetic gene circuits in plants: recent advances and challenges. Quantitative Plant Biology 6. https://doi.org/10.1017/qpb.2025.3.

Kleinstiver, B.P., Pattanayak, V., Prew, M.S., Tsai, S.Q., Nguyen, N.T., Zheng, Z., Joung, J.K., 2016. High-fidelity CRISPR-Cas9 nucleases with no detectable genome-wide off-target effects. Nature 529 (7587), 490–495. https://doi.org/10.1038/nature16526.

Klimek-Chodacka, M., Oleszkiewicz, T., Lowder, L.G., Qi, Y., Baranski, R., 2018. Efficient CRISPR/Cas9-based genome editing in carrot cells. Plant Cell Rep. 37 (4), 575–586. https://doi.org/10.1007/s00299-018-2252-2. PMID:29332168; PMCID:PMC5852178.

Kong, C., Yang, Y., Qi, T., Zhang, S., 2025. Predictive genetic circuit design for phenotype reprogramming in plants. Nat. Commun. 16 (1). https://doi.org/10.1038/s41467-025-56042-2.

Kwak, S.Y., Lew, T.T.S., Sweeney, C.J., Koman, V.B., Wong, M.H., Bohmert-Tatarev, K., Snell, K.D., Seo, J.S., Chua, N.H., Strano, M.S., 2019. Chloroplast-selective gene delivery and expression in planta using chitosan-complexed single-walled carbon nanotube carriers. Nat. Nanotechnol. 14 (5), 447–455. https://doi.org/10.1038/s41565-019-0375-4.

Lee, M., 2023. Deep learning in CRISPR-Cas systems: a review of recent studies. Front. Bioeng. Biotechnol. 11. https://doi.org/10.3389/fbioe.2023.1226182.

Li, R., Li, R., Li, X., Fu, D., Zhu, B., Tian, H., Luo, Y., Zhu, H., 2018. Multiplexed CRISPR/Cas9-mediated metabolic engineering of γ-aminobutyric acid levels in Solanum lycopersicum. Plant Biotechnol. J. 16 (2), 415–427. https://doi.org/10.1111/pbi.12781. PMID:28640983; PMCID:PMC5787826.

Li, J., Zhang, S., Zhang, R., Gao, J., Qi, Y., Song, G., Li, W., Li, Y., Li, G., 2021a. Efficient multiplex genome editing by CRISPR/Cas9 in common wheat. Plant Biotechnol. J. 19 (3), 427–429. https://doi.org/10.1111/pbi.13508. PMID:33150679; PMCID:PMC7955872.

Li, Y., Chen, J., Tsai, S.Q., Cheng, Y., 2021b. Easy-Prime: a machine learning–based prime editor design tool. Genom. Biol. 22 (1), 235. https://doi.org/10.1186/s13059-021-02458-0. PMID:34407862; PMCID:PMC8377858.

Li, Z., Zhong, Z., Wu, Z., Pausch, P., Al-Shayeb, B., Amerasekera, J., Doudna, J.A., Jacobsen, S.E., 2023. Genome editing in plants using the compact editor CasΦ. Proc. Natl. Acad. Sci. U.S.A. 120 (4). https://doi.org/10.1073/pnas.2216822120.

Li, J., Wu, P., Cao, Z., Huang, G., Lu, Z., Yan, J., Zhang, H., Zhou, Y., Liu, R., Chen, H., Ma, L., Luo, M., 2024. Machine learning-based prediction models to guide the selection of Cas9 variants for efficient gene editing. Cell Rep. 43 (2), 113765. https://doi.org/10.1016/j.celrep.2024.113765.

Li, S., Lin, D., Zhang, Y., Deng, M., Chen, Y., Lv, B., Li, B., Lei, Y., Wang, Y., Zhao, L., Liang, T., Liu, J., Chen, K., Xia, B., Jia, H., Wang, N., Yang, J., Gao, C., Qiu, J.L., 2022. Genome-edited powdery mildew resistance in wheat without growth penalties. Nature 602, 455–460. https://doi.org/10.1038/s41586-022-04395-9. PMID:35140402.

Li, S., Wen, N., Lv, W., Zhang, M., Wang, Y., Lang, Z., 2025a. Impact of codon optimization on vip3Aa11 gene expression and insecticidal efficacy in maize. Front. Plant Sci. 16, 1579465. https://doi.org/10.3389/fpls.2025.1579465.

Li, X., Gao, Y., Zhang, Z., Deng, W., Cao, W., Wei, X., Gao, Z., Yao, L., Wang, S., Xie, Y., Li, M., 2025b. Biosafety considerations triggered by genome-editing technologies. Biosafety and Health 7 (3), 141–151. https://doi.org/10.1016/j.bsheal.2025.05.003.

Liang, Z., Chen, K., Li, T., Zhang, Y., Wang, Y., Zhao, Q., Liu, J., Zhang, H., Liu, C., Ran, Y., Gao, C., 2017. Efficient DNA-free genome editing of bread wheat using CRISPR/Cas9 ribonucleoprotein complexes. Nat. Commun. 8. https://doi.org/10.1038/ncomms14261.

Lin, J., Zhang, Z., Zhang, S., Chen, J., Wong, K.C., 2020. CRISPR-net: a recurrent convolutional network quantifies CRISPR off-target activities with mismatches and indels. Adv. Sci. 7 (13). https://doi.org/10.1002/advs.201903562.

Listgarten, J., Weinstein, M., Kleinstiver, B.P., Sousa, A.A., Joung, J.K., Crawford, J., Gao, K., Hoang, L., Elibol, M., Doench, J.G., Fusi, N., 2018. Prediction of off-target activities for the end-to-end design of CRISPR guide RNAs. Nat. Biomed. Eng. 2 (1), 38–47. https://doi.org/10.1038/s41551-017-0178-6.

Liu, F., Huang, S., Hu, J., et al., 2023. Design of prime-editing guide RNAs with deep transfer learning. Nat. Mach. Intell. 5, 1261–1274. https://doi.org/10.1038/s42256-023-00739-w.

Liu, G., Zhang, Y., Zhang, T., 2019. Computational approaches for effective CRISPR guide RNA design and evaluation. Comput. Struct. Biotechnol. J. 18, 35–44. https://doi.org/10.1016/j.csbj.2019.11.006.

Lowder, L.G., Zhang, D., Baltes, N.J., Paul, J.W., Tang, X., Zheng, X., Voytas, D.F., Hsieh, T.F., Zhang, Y., Qi, Y., 2015. A CRISPR/Cas9 toolbox for multiplexed plant genome editing and transcriptional regulation. Plant Physiol. 169 (2), 971–985. https://doi.org/10.1104/pp.15.00636.

Lu, Y., Wang, J., Chen, B., Mo, S., Lian, L., Luo, Y., Ding, D., Ding, Y., Cao, Q., Li, Y., Li, Y., Liu, G., Hou, Q., Cheng, T., Wei, J., Zhang, Y., Chen, G., Song, C., Hu, Q., Sun, S., Fan, G., Wang, Y., Liu, Z., Song, B., Zhu, J. K., Li, H., Jiang, L., 2021. A donor-DNA-free CRISPR/Cas-based approach to gene knock-up in rice. Nat. Plants 7 (11), 1445–1452. https://doi.org/10.1038/s41477-021-01019-4.

Luo, Y., Chen, Y., Xie, H.Z., Zhu, W., Zhang, G., 2024. Interpretable CRISPR/Cas9 off-target activities with mismatches and indels prediction using BERT. Comput. Biol. Med. 169. https://doi.org/10.1016/j.compbiomed.2024.107932.

Luo, W., Ors, S., Fang, H., Zeng, M., Cheng, Y., Sidawy, A.E., Qi, Y., 2025. Efficient genome editing by CRISPR-Cas12a in common wheat via shoot apical meristem delivery. Plant Physiol. 199 (4), kiaf611. https://doi.org/10.1093/plphys/kiaf611.

Ly, D.N.P., Iqbal, S., Fosu-Nyarko, J., Milroy, S., Jones, M.G.K., 2023. Multiplex CRISPR-Cas9 gene-editing can deliver potato cultivars with reduced browning and acrylamide. Plants 12 (2). https://doi.org/10.3390/plants12020379.

Ma, X., Zhang, Q., Zhu, Q., Liu, W., Chen, Y., Qiu, R., Wang, B., Yang, Z., Li, H., Lin, Y., Xie, Y., Shen, R., Chen, S., Wang, Z., Chen, Y., Guo, J., Chen, L., Zhao, X., Dong, Z., Liu, Y.G., 2015. A robust CRISPR/Cas9 system for convenient, high-efficiency multiplex genome editing in monocot and dicot plants. Mol. Plant 8 (8), 1274–1284. https://doi.org/10.1016/j.molp.2015.04.007.

Ma, X., Zhang, X., Liu, H., Li, Z., 2020. Highly efficient DNA-free plant genome editing using virally delivered CRISPR–Cas9. Nat. Plants 6 (7), 773–779. https://doi.org/10.1038/s41477-020-0704-5.

Marquart, K.F., Allam, A., Janjuha, S., et al., 2021. Predicting base editing outcomes with an attention-based deep learning algorithm trained on high-throughput target library screens. Nat. Commun. 12, 5114. https://doi.org/10.1038/s41467-021-25375-z.

Massa, G.A., Décima Oneto, C.A., González, M.N., Poulsen Hornum, A., Arizmendi, A., Sucar, S., Divito, S.B., Feingold, S.E., 2025. CRISPR/Cas9-mediated development of potato varieties with long-term cold storage and bruising resistance. Biology 14 (4), 445. https://doi.org/10.3390/biology14040445. Erratum in: Biology, 14(7), 768. https://doi.org/10.3390/biology14070768. PMID:40282310; PMCID:PMC12024581.

Mathis, N., Allam, A., Kissling, L., Marquart, K.F., Schmidheini, L., Solari, C., Balázs, Z., Krauthammer, M., Schwank, G., 2023. Predicting prime editing efficiency and product purity by deep learning. Nat. Biotechnol. 41 (8), 1151–1159. https://doi.org/10.1038/s41587-022-01613-7. Epub 2023 Jan 16. PMID:36646933; PMCID: PMC7614945.

McCarty, N.S., Graham, A.E., Studená, L., Ledesma-Amaro, R., 2020. Multiplexed CRISPR technologies for gene editing and transcriptional regulation. Nat. Commun. 11 (1). https://doi.org/10.1038/s41467-020-15053-x.

McFadden, B.R., Rumble, J.N., Stofer, K.A., Folta, K.M., 2024. U.S. public opinion about the safety of gene editing in the agriculture and medical fields and the amount of evidence needed to improve opinions. Front. Bioeng. Biotechnol. 12. https://doi.org/10.3389/fbioe.2024.1340398.

Montesinos-López, O.A., Montesinos-López, A., Mosqueda-González, B.A., Bentley, A.R., Lillemo, M., Varshney, R. K., Crossa, J., 2021. A new deep learning calibration method enhances genome-based prediction of continuous crop traits. Front. Genet. 12, 798840. https://doi.org/10.3389/fgene.2021.798840. PMID:34976026; PMCID: PMC8718701.

Montesinos-López, O.A., Saint Pierre, C., Gezan, S.A., Bentley, A.R., Mosqueda-González, B.A., Montesinos-López, A., van Eeuwijk, F., Beyene, Y., Gowda, M., Gardner, K., Gerard, G.S., Crespo-Herrera, L., Crossa, J., 2023. Optimizing sparse testing for genomic prediction of plant breeding crops. Genes 14 (4). https://doi.org/10.3390/genes14040927.

Müller, P.H., Istanto, D.D., Heldenbrand, J., et al., 2022. CROPSR: an automated platform for complex genome-wide CRISPR gRNA design and validation. BMC Bioinf. 23, 74. https://doi.org/10.1186/s12859-022-04593-2.

Murphy, K.M., Ludwig, E., Gutierrez, J., Gehan, M.A., 2024. Deep learning in image-based plant phenotyping. Annu. Rev. Plant Biol. 75 (1), 771–795. https://doi.org/10.1146/annurev-arplant-070523-042828.

Ni, P., Zhao, Y., Zhou, X., Liu, Z., Huang, Z., Ni, Z., Sun, Q., Zong, Y., 2023. Efficient and versatile multiplex prime editing in hexaploid wheat. Genome Biol. 24 (1), 156. https://doi.org/10.1186/s13059-023-02990-1. PMID:37386475; PMCID:PMC10308706.

Oz, M.T., Altpeter, A., Karan, R., Merotto, A., Altpeter, F., 2021. CRISPR/Cas9-mediated multi-allelic gene targeting in sugarcane confers herbicide tolerance. Front. Genom. Ed. 3, 673566. https://doi.org/10.3389/fgeed.2021.673566. PMID:34713261; PMCID:PMC8525412.

Pacesa, M., Kampmann, M., Corn, J.E., 2024. Past, present, and future of CRISPR genome editing. Cell 187 (10), 2390–2415. https://doi.org/10.1016/j.cell.2024.03.026.

Pan, C., Li, G., Bandyopadhyay, A., Qi, Y., 2023. Guide RNA library-based CRISPR screens in plants: opportunities and challenges. Curr. Opin. Biotechnol. 79, 102883. https://doi.org/10.1016/j.copbio.2022.102883.

Parthiban, S., Vijeesh, T., Gayathri, T., et al., 2023. Artificial intelligence-driven systems engineering for next-generation plant-derived biopharmaceuticals. Front. Plant Sci. 14, 1252166. https://doi.org/10.3389/fpls.2023.1252166.

Peleke, F.F., Zumkeller, S.M., Gültas, M., Schmitt, A., Szymański, J., 2024. Deep learning the cis-regulatory code for gene expression in selected model plants. Nat. Commun. 15 (1). https://doi.org/10.1038/s41467-024-47744-0.

Polstein, L.R., Gersbach, C.A., 2015. A light-inducible CRISPR-Cas9 system for control of endogenous gene activation. Nat. Chem. Biol. 11 (3), 198–200. https://doi.org/10.1038/nchembio.1753.

Qi, W., Zhu, T., Tian, Z., Li, C., Zhang, W., Song, R., 2016. High-efficiency CRISPR/Cas9 multiplex gene editing using the glycine tRNA-processing system-based strategy in maize. BMC Biotechnol. 16 (1), 58. https://doi.org/10.1186/s12896-016-0289-2. PMID:27515683; PMCID:PMC4982333.

Qu, Y., Huang, K., Yin, M., Zhan, K., Liu, D., Yin, D., Cousins, H.C., Johnson, W.A., Wang, X., Shah, M., Altman, R. B., Zhou, D., Wang, M., Cong, L., 2025. CRISPR-GPT for agentic automation of gene-editing experiments. Nat. Biomed. Eng. 9 (Epub ahead of print, July 30). https://doi.org/10.1038/s41551-025-01463-z. PMID:40738974.

Rai, M.N., Rhodes, B., Jinga, S., Kanchupati, P., Ross, E., Carlson, S.R., Moose, S.P., 2025. Efficient mutagenesis and genotyping of maize inbreds using biolistics, multiplex CRISPR/Cas9 editing, and Indel-Selective PCR. Plant Methods 21 (1), 43. https://doi.org/10.1186/s13007-025-01365-w. PMID:40128730; PMCID:PMC11934539.

Rastogi, K., Mankar, S.P., Ibarra, O., Molina-Risco, M., Faion-Molina, M., Bellinatti-Della Gracia, M., Rathore, K.S., Thomson, M.J., Septiningsih, E.M., 2025. CRISPR/Cas9 multiplex genome editing to enhance grain lysine concentration in a US rice cultivar. Curr. Plant Biol. 42, 100490. https://doi.org/10.1016/j.cpb.2025.100490.

Ren, C., Gathunga, E.K., Li, X., Li, H., Kong, J., Dai, Z., Liang, Z., 2023. Efficient genome editing in grapevine using CRISPR/LbCas12a system. Mol. Hortic. 3 (1), 21. https://doi.org/10.1186/s43897-023-00069-w. Erratum in: Molecular Horticulture, 4(1), 44. https://doi.org/10.1186/s43897-024-00133-z. PMID:37853418; PMCID: PMC10583370.

Ruffolo, J.A., Nayfach, S., Gallagher, J., Bhatnagar, A., Beazer, J., Hussain, R., Russ, J., Yip, J., Hill, E., Pacesa, M., Meeske, A.J., Cameron, P., Madani, A., 2025. Design of highly functional genome editors by modelling CRISPR-Cas sequences. Nature 645 (8080), 518–525. https://doi.org/10.1038/s41586-025-09298-z. Epub 2025 Jul 30. PMID:40739342; PMCID:PMC12422970.

Sagawa, C.H.D., Thomson, G., Mermaz, B., Vernon, C., Liu, S., Jacob, Y., Irish, V.F., 2024. An efficient multiplex approach to CRISPR/Cas9 gene editing in citrus. Plant Methods 20 (1). https://doi.org/10.1186/s13007-024-01274-4.

Samach, A., Mafessoni, F., Gross, O., Melamed-Bessudo, C., Filler-Hayut, S., Dahan-Meir, T., Amsellem, Z., Pawlowski, W.P., Levy, A.A., 2023. CRISPR/Cas9-induced DNA breaks trigger crossover, chromosomal loss, and chromothripsis-like rearrangements. Plant Cell 35 (11), 3957–3972. https://doi.org/10.1093/plcell/koad209.

Sambana, B., Nnadi, H.S., Wajid, M.A., Fidelia, N.O., Camacho-Zuñiga, C., Ajuzie, H.D., Onyema, E.M., 2025. An efficient plant disease detection using transfer learning approach. Sci. Rep. 15 (1). https://doi.org/10.1038/s41598-025-02271-w.

Sarkar, P., Santiago Vazquez, J., Zhou, M., Levy, A., Mou, Z., Orbović, V., 2024. Multiplexed gene editing in citrus by using a multi-intron containing Cas9 gene. Transgenic Res 33 (1–2), 59–66. https://doi.org/10.1007/s11248-024-00380-2. PMID:38564120.

Scintilla, S., Salvagnin, U., Giacomelli, L., Zeilmaker, T., Malnoy, M.A., Rouppe van der Voort, J., Moser, C., 2022. Regeneration of non-chimeric plants from DNA-free edited grapevine protoplasts. Front. Plant Sci. 13. https://doi.org/10.3389/fpls.2022.1078931.

Shoaib, M., Shah, B., Ei-Sappagh, S., Ali, A., Ullah, A., Alenezi, F., Gechev, T., Hussain, T., Ali, F., 2023. An advanced deep learning models-based plant disease detection: a review of recent research. Front. Plant Sci. 14, 1158933. https://doi.org/10.3389/fpls.2023.1158933. Erratum in: Frontiers in Plant Science, 14, 1282443. doi:10.3389/fpls.2023.1282443. PMID:37025141; PMCID:PMC10070872.

Sun, J., Guo, J., Liu, J., Sinha, S., 2024. CRISPR-M: predicting sgRNA off-target effect using a multi-view deep learning network. PLoS Comput. Biol. 20 (3), e1011972. https://doi.org/10.1371/journal.pcbi.1011972.

Svitashev, S., Schwartz, C., Lenderts, B., Young, J.K., Mark Cigan, A., 2016. Genome editing in maize directed by CRISPR-Cas9 ribonucleoprotein complexes. Nat. Commun. 7. https://doi.org/10.1038/ncomms13274.

Thomson, G., Mermaz, B., Sagawa, C.H.D., Lin, C.Y.I., Tachev, M., Joly, V., Irish, V.F., Jacob, Y., 2025. Enzymatic depletion of transposable elements in sequencing libraries and its application for genotyping multiplexed CRISPR-edited plants. Plant J. 124 (1). https://doi.org/10.1111/tpj.70501.

Uranga, M., Vazquez-Vilar, M., Orzáez, D., Daròs, J.A., 2021. CRISPR-Cas12a genome editing at the whole-plant level using two compatible RNA virus vectors. CRISPR J. 4 (5), 761–769. https://doi.org/10.1089/crispr.2021.0049.

Tripathi, J.N., Ntui, V.O., Ron, M., Muiruri, S.K., Britt, A., Tripathi, L., 2019. CRISPR/Cas9 editing of endogenous banana streak virus in the B genome of *Musa* spp. overcomes a major challenge in banana breeding. Commun. Biol. 2, 46. https://doi.org/10.1038/s42003-019-0288-7. PMID:30729184; PMCID:PMC6355771.

Uranga, M., Aragonés, V., Daròs, J.-A., Pasin, F., 2023. Heritable CRISPR-Cas9 editing of plant genomes using RNA virus vectors. STAR Protoc. 4 (1), 102091. https://doi.org/10.1016/j.xpro.2023.102091.

Vazquez-Vilar, M., Selma, S., Orzaez, D., 2023. The design of synthetic gene circuits in plants: new components, old challenges. J. Exp. Bot. 74 (13), 3791–3805. https://doi.org/10.1093/jxb/erad167.

Veluchamy, A., Teles, K., Fischle, W., 2023. CRISPR-broad: combined design of multi-targeting gRNAs and broad, multiplex target finding. Sci. Rep. 13 (1), 19717. https://doi.org/10.1038/s41598-023-46212-x. PMID:37953351; PMCID:PMC10641073.

Wang, G., Wang, F., Xu, Z., Wang, Y., Zhang, C., Zhou, Y., Hui, F., Yang, X., Nie, X., Zhang, X., Jin, S., 2024. Precise fine-tuning of GhTFL1 by base editing tools defines ideal cotton plant architecture. Genome Biol 25 (1), 59. https://doi.org/10.1186/s13059-024-03189-8. PMID:38409014; PMCID:PMC10895741.

Wang, D., Zhang, C., Wang, B., Li, B., Wang, Q., Liu, D., Wang, H., Zhou, Y., Shi, L., Lan, F., Wang, Y., 2019. Optimized CRISPR guide RNA design for two high-fidelity Cas9 variants by deep learning. Nat. Commun. 10 (1). https://doi.org/10.1038/s41467-019-12281-8.

Waqas, M.A.B., Awan, M.J.A., Amin, I., Arif, M., Mukhtar, Z., Mansoor, S., 2025. Engineering high yield basmati rice by editing multiple negative regulators of yield. Mol. Biol. Rep. 52 (1). https://doi.org/10.1007/s11033-025-10660-7.

Woo, J.W., Kim, J., Kwon, S.I., Corvalán, C., Cho, S.W., Kim, H., Kim, S.G., Kim, S.T., Choe, S., Kim, J.S., 2015. DNA-free genome editing in plants with preassembled CRISPR-Cas9 ribonucleoproteins. Nat. Biotechnol. 33 (11), 1162–1164. https://doi.org/10.1038/nbt.3389.

Wu, S., Kyaw, H., Tong, Z., Yang, Y., Wang, Z., Zhang, L., Deng, L., Zhang, Z., Xiao, B., Quick, W.P., Lu, T., Xiao, G., Qin, G., Cui, X., 2024. A simple and efficient CRISPR/Cas9 system permits ultra-multiplex genome editing in plants. Crop Journal 12 (2), 569–582. https://doi.org/10.1016/j.cj.2024.01.010.

Xiang, X., Corsi, G.I., Anthon, C., Qu, K., Pan, X., Liang, X., Han, P., Dong, Z., Liu, L., Zhong, J., Ma, T., Wang, J., Zhang, X., Jiang, H., Xu, F., Liu, X., Xu, X., Wang, J., Yang, H., Bolund, L., Church, G.M., Lin, L., Gorodkin, J.,

Luo, Y., 2021. Enhancing CRISPR-Cas9 gRNA efficiency prediction by data integration and deep learning. Nat. Commun. 12 (1). https://doi.org/10.1038/s41467-021-23576-0.

Xie, K., Minkenberg, B., Yang, Y., 2015. Boosting CRISPR/Cas9 multiplex editing capability with the endogenous tRNA-processing system. Proc. Natl. Acad. Sci. U.S.A. 112, 3570–3575. https://doi.org/10.1073/pnas.1420294112.

Xing, H.L., Dong, L., Wang, Z.P., Zhang, H.Y., Han, C.Y., Liu, B., Wang, X.C., Chen, Q.J., 2014. A CRISPR/Cas9 toolkit for multiplex genome editing in plants. BMC Plant Biol. 14 (1). https://doi.org/10.1186/s12870-014-0327-y.

Yang, T., Ali, M., Lin, L., Li, P., He, H., Zhu, Q., Sun, C., Wu, N., Zhang, X., Huang, T., Li, C.-B., Li, C., Deng, L., 2022a. Recoloring tomato fruit by CRISPR/Cas9-mediated multiplex gene editing. Hortic. Res. 10 (1), uhac214. https://doi.org/10.1093/hr/uhac214. PMID:36643741; PMCID:PMC9832834.

Yang, L., Zhang, Y. and Gao, C. (2022b) Intron-based CRISPR/Cas9 System for Efficient Gene Editing in Plants. World Intellectual Property Organization Patent WO2022036074A2.

Yang, Q., Ral, J.P., Wei, Y., Zheng, Y., Li, Z., Jiang, Q., 2024. Genome editing of five starch synthesis genes produces highly resistant starch and dietary fibre in barley grains. Plant Biotechnol. J. 22 (7), 2051–2053. https://doi.org/10.1111/pbi.14324. PMID:38415905; PMCID:PMC11182577.

Yasmeen, E., Wang, J., Riaz, M., Zhang, L., Zuo, K., 2023. Designing artificial synthetic promoters for accurate, smart, and versatile gene expression in plants. Plant Commun. 4 (4), 100558. https://doi.org/10.1016/j.xplc.2023.100558.

Yuan, Q., Duren, Z., 2025. Inferring gene regulatory networks from single-cell multiome data using atlas-scale external data. Nat. Biotechnol. 43 (2), 247–257. https://doi.org/10.1038/s41587-024-02182-7.

Zhang, Y., Wang, J., Wang, Z., Zhang, Y., Shi, S., Nielsen, J., Liu, Z., 2019. A gRNA-tRNA array for CRISPR-Cas9 based rapid multiplexed genome editing in Saccharomyces cerevisiae. Nat. Commun. 10 (1). https://doi.org/10.1038/s41467-019-09005-3.

Zhang, P., Du, H., Wang, J., Pu, Y., Yang, C., Yan, R., Yang, H., Cheng, H., Yu, D., 2020. Multiplex CRISPR/Cas9-mediated metabolic engineering increases soya bean isoflavone content and resistance to soya bean mosaic virus. Plant Biotechnol. J. 18 (6), 1384–1395. https://doi.org/10.1111/pbi.13302.

Zhang, Y., Ren, Q., Tang, X., Liu, S., Malzahn, A.A., Zhou, J., Wang, J., Yin, D., Pan, C., Yuan, M., Huang, L., Yang, H., Zhao, Y., Fang, Q., Zheng, X., Tian, L., Cheng, Y., Le, Y., McCoy, B., Franklin, L., Selengut, J.D., Mount, S.M., Que, Q., Zhang, Y., Qi, Y., 2021. Expanding the scope of plant genome engineering with Cas12a orthologs and highly multiplexable editing systems. Nat. Commun. 12 (1). https://doi.org/10.1038/s41467-021-22330-w.

Zhang, Z., Wang, W., Ali, S., Luo, X., Xie, L., 2022. CRISPR/Cas9-Mediated multiple knockouts in abscisic acid receptor genes reduced the sensitivity to ABA during soybean seed germination. Int. J. Mol. Sci. 23 (24), 16173. https://doi.org/10.3390/ijms232416173.

Zhang, G., Luo, Y., Xie, H., Dai, Z., 2024. Crispr-SGRU: prediction of CRISPR/Cas9 off-target activities with mismatches and indels using stacked BiGRU. Int. J. Mol. Sci. 25 (20), 10945. https://doi.org/10.3390/ijms252010945.

Zhang, D., Xu, F., Wang, F., Le, L., Pu, L., 2025. Synthetic biology and artificial intelligence in crop improvement. Plant Commun. 6 (2), 101220. https://doi.org/10.1016/j.xplc.2024.101220.

Zhou, H., Liu, B., Weeks, D.P., Spalding, M.H., Yang, B., 2014. Large chromosomal deletions and heritable small genetic changes induced by CRISPR/Cas9 in rice. Nucleic Acids Res. 42 (17), 10903–10914. https://doi.org/10.1093/nar/gku806.

Zhou, J., Xin, X., He, Y., Chen, H., Li, Q., Tang, X., Zhong, Z., Deng, K., Zheng, X., Akher, S.A., Cai, G., Qi, Y., Zhang, Y., 2019. Multiplex QTL editing of grain-related genes improves yield in elite rice varieties. Plant Cell Rep. 38 (4), 475–485. https://doi.org/10.1007/s00299-018-2340-3.

Further reading

Jones, A., Stanley, D., Ferguson, S., Schwessinger, B., Borevitz, J., Warthmann, N., Perovic, D., 2023. Cost-conscious generation of multiplexed short-read DNA libraries for whole-genome sequencing. PLoS One 18 (1), e0280004. https://doi.org/10.1371/journal.pone.0280004.

Lin, C.S., Hsu, C.T., Yang, L.H., Lee, L.Y., Fu, J.Y., Cheng, Q.W., Wu, F.H., Hsiao, H.C., Zhang, Y., Zhang, R., Chang, W.C., Yu, C.T., Wang, W., Liao, L.J., Gelvin, S.B., Shih, M.C., 2018. Application of protoplast technology to CRISPR/Cas9 mutagenesis: from single-cell mutation detection to mutant plant regeneration. Plant Biotechnol. J. 16, 1295–1310. https://doi.org/10.1111/pbi.12870.

Li, G., An, L., Yang, W., Yang, L., Wei, T., Shi, J., Wang, J., Doonan, J.H., Xie, K., Fernie, A.R., Lagudah, E.S., Wing, R.A., Gao, C., 2025c. Integrated biotechnological and AI innovations for crop improvement. Nature 643 (8073), 925–937. https://doi.org/10.1038/s41586-025-09122-8.

Lin, Q., et al., 2021. High-efficiency prime editing with optimized, paired pegRNAs in plants. Nat. Biotechnol. 39, 923–927. https://doi.org/10.1038/s41587-021-00844-w.

Liu, L., Gallagher, J., Arevalo, E.D., et al., 2021. Enhancing grain-yield-related traits by CRISPR–Cas9 promoter editing of maize CLE genes. Nat. Plants 7, 287–294. https://doi.org/10.1038/s41477-021-00858-5.

Subramanian, P., Vijeesh, T., Gayathri, T., Shanmugaraj, B., Sharma, A., Sathishkumar, R., 2023. Artificial intelligence-driven systems engineering for next-generation plant-derived biopharmaceuticals. Front. Plant Sci. 14. https://doi.org/10.3389/fpls.2023.1252166.

CHAPTER

17

Artificial intelligence and CRISPR technologies for enhancing food security

Nabarun Roy, Shalini Singh, and Hari Shankar Gaur
School of Agriculture, Galgotias University, Greater Noida, Uttar Pradesh, India

17.1 Introduction

Climate change, fast growing populations, and persistent environmental damage have generated an urgent need for novel techniques to support sustainable agriculture and long-term nutritional security (Ahmad, 2023). Between 1960 and 2000, worldwide agricultural productivity tripled, owing primarily to green revolution (GR) which included enhanced germplasm and widespread use of better infrastructure, mechanization, fertilizers, pesticides, and supportive policies. While these advancements considerably enhanced food security and helped the globe avoid expected shortages, the benefits were not evenly distributed. The yield gain was less for many countries as compared to others (Chavhan et al., 2025). This was due to the reason that GR relied heavily on conventional/traditional breeding and resource-intensive farming practices, which is generally slow and often takes decades to develop new varieties making them inefficient for today's rapidly evolving food security and environmental challenges (Atia et al., 2024).

This is where artificial intelligence (AI) and Clustered Regularly Interspaced Short Palindromic Repeats (CRISPR) has shown groundbreaking potential in tackling the challenges mentioned above. AI tools can estimate yields, optimize farm inputs, and manage risks associated with climate instability, hence improving productivity and long-term sustainability (Dhal and Kar, 2025). Simultaneously, cutting-edge gene-editing tools like CRISPR enable scientists to precisely alter plant genomes to improve yield, stress tolerance, nutrient content, and resilience to climate stressors (Manzoor et al., 2024). Together, the combined power of AI and CRISPR offers a promising path towards creating climate-resilient high-yielding crops (Faizal et al., 2024).

Intelligent Gene Editing in Plants
https://doi.org/10.1016/B978-0-443-44079-3.00026-7

17.2 Current state of food security

The world population is predicted to reach 9.7 billion by 2050. This will put more and more burden on agricultural systems that are already dealing with unpredictable weather patterns and climate change (Shaheen et al., 2023). Modern agriculture relies on the practice of monoculture and enormous amounts of chemical inputs. The use of pesticide, fungicide and fertilizer has surged by more than 850% over the last 50 years. Despite extensive pesticide and fungicide use, pests continue to cause significant yield losses in wheat (21.5%), rice (30.3%), and maize (22.6%), whereas plant diseases account for an estimated loss of US$220 billion yearly (Ristaino et al., 2021). Excess fertilizer use makes crops more vulnerable to pests and poor irrigation practices can cause soil salinization, ion toxicity, and nutritional imbalances. Although intercropping has ecological benefits, many farmers continue to rely on monocropping due to market pressures, automation constraints, and policy limitations. All these problems have led to yield stagnation of crops, soil degradation, biodiversity loss, and financial instability (Léonidas et al., 2024).

Although the GR once boosted production, progress has slowed due to inefficient input use and poor agronomic practices. Globally, yield improvements in major staples like maize, rice, wheat, and soybean are less than 2.4% yearly growth required to fulfill future food demand. Nearly 75% of plant genetic diversity has been lost because of habitat destruction, the replacement of traditional landraces with modern HYVs (High Yielding Varieties) and climate impacts, which limits the ability of breeders to incorporate traits for disease resistance and climate resilience (Salgotra and Chauhan, 2023; Bohra et al., 2022).

Mutation and radiation breeding increase variation through triggered mutagenesis, whereas TILLING (Targeting Induced Local Lesions in Genomes) allows for high-throughput identification of point mutations in coding and non-coding areas (Singh et al., 2024). While these procedures provide valuable variation, they frequently result in many unwanted changes, requiring prolonged backcrossing. Meeting future food demands will necessitate a combination of advanced breeding technology, improved agronomic practices, and precision farming strategies. Emerging techniques such as AI and CRISPR/Cas genome editing (GE) provide unprecedented prospects to speed the development of high-yielding, stress-tolerant, and resource-efficient crops that will form the foundation of climate-smart agriculture (Zhan et al., 2024).

17.3 Fundamentals of CRISPR technology and its variants

GE is a highly advanced and precise technique that alters specific Deoxyribonucleic Acid (DNA) sequences using site-specific nucleases (SSNs) without introducing foreign genes. Among the various GE techniques, CRISPR/Cas9, first found as an immunological defense system in *Streptococcus pyogenes*, has become the most commonly utilized because of its reliability, efficiency, and fairly low cost (Rai et al., 2023). CRISPR/Cas9 works via a guide RNA (gRNA), composed up of CRISPR Ribonucleic Acid (crRNA) and trans-activating crRNA (tracrRNA), that guides the Cas9 enzyme to a complementary 20-nt sequence of DNA and generates a double-strand break (DSB) (Chawla et al., 2023). This cut allows researchers to knock out genes, insert new sequences, or replace existing DNA (Ebrahimi and Hashemi, 2024). The cell then repairs the DSBs via either non-homologous end joining (NHEJ) or homology-directed repair (HDR) (Wang and Doudna, 2023). Previously, editing methods depended mainly on SSNs, but as the concept of GE progressed, the Site-Directed Nuclease (SDN) classification system (SDN-1/2/3) became the standard for identifying editing categories based on donor DNA use, repair routes, and foreign DNA integration (Munaweera et al., 2022; EFSA GMO Panel et al., 2020). This system is now widely used to differentiate GE crops from traditional Genetically modified Organism (GMOs) (Abdallah et al., 2021).

The CRISPR/Cas system itself originates from a bacterial immune process in which fragments of viral DNA, known as "spacers", are incorporated into the bacterial genome to provide a molecular memory of past infections (Mamrutha et al., 2024). When the same virus reappears, Cas proteins quickly recognize and destroy its genetic material. SpCas9, the most commonly used Cas9 enzyme, recognizes an NGG Protospacer Adjacent Motif (PAM) sequence and generates blunt-end DSBs through its HNH and RuvC nuclease domains (Zaidi et al., 2020). To expand the range of editable sites, many Cas9 orthologues such as St1Cas9, SaCas9, ScCas9, LrCas9, FrCas9, FnCas9, and Nm/Nme2Cas9 have been developed, each with distinct PAM requirements (Zhang et al., 2021a). Engineered variants like SpCas9-VQR, EQR, VRER, NRRH, NRTH, NRCH, SpCas9-NG, xCas9, SpG, and iSpy-MacCas9 further increase flexibility across plant genomes (Bradbury et al., 2025).

Beyond Cas9, Cas12 (Cas12a/Cas12b) and Cas13 systems (see Table 17.1) broaden GE possibilities, including RNA-level targeting (Sharma et al., 2021). Cas12a uses only a crRNA, recognizes T-rich PAMs (TTTV), and generates staggered DSBs in DNA (Hussen et al., 2024).

TABLE 17.1 Comparative overview of CRISPR/Cas systems used in plant GE.

CRISPR system	Nuclease Domain (s)	PAM requirement	Target molecule	Major orthologues	Key features/Applications	References
Cas9	HNH + RuvC	5′-NGG (G-rich)	DNA	SpCas9, SaCas9, St1Cas9, FnCas9, NmCas9	First and most widely used system; induces blunt DSBs; multiple engineered variants (SpCas9-NG, SpRY) broaden PAM scope	Tuncel et al. (2023)
Cas12a (Cpf1)	RuvC-like	5′-TTTV (T-rich)	DNA	AsCas12a, LbCas12a, FnCas12a, Mb2Cas12a	Requires only crRNA; produces staggered DSBs with sticky ends; suitable for multiplex editing	Zhai et al. (2022)
Cas12b	RuvC-like	VTTV	DNA	AaCas12b, AacCas12b, BthCas12b	Requires crRNA + tracrRNA; high-temperature tolerance; smaller size suitable for viral vectors	Lei et al. (2024)
Cas13	HEPN	PAM-less (targets single stranded ribonucleic acid)	RNA	LwaCas13a, CasRx, Cas13d	Enables RNA knockdown; transient and reversible; potential for virus resistance	Kim et al. (2022)

Key Cas12a orthologues include AsCas12a, LbCas12a, FnCas12a, Mb2Cas12a, Ev1Cas12a, and Hs1Cas12a (Zhai et al., 2022). Engineered versions such as LbCas12a-RR, RVR, RVQ, and Mb2Cas12a-RVR further relax PAM restrictions and improve editing efficiency (Mangin et al., 2024). Cas12b variants like AaCas12b, AacCas12b, and BthCas12b also function effectively in plants and share similar T-rich PAM preferences. Other expanding Cas families including Cas12i, CasΦ/12j, and Cas12f continue to widen editing options (Lei et al., 2024). Cas13, which targets single-stranded RNA and does not require a PAM, is still in early development for plant applications (Kim et al., 2022).

17.4 CRISPR reagents

CRISPR/Cas GE depends on several core components; each supplied in forms designed for stability and reliable performance. Cas nucleases such as Cas9, Cas12a, and Cas13 are usually provided in liquid storage buffers (pH 7.5–8.0) containing Tris-HCl, salts, glycerol, and reducing agents, and are kept frozen at −20°C or −80°C (Aksoy et al., 2022). Guide-RNAs typically come as lyophilized powders that are rehydrated with RNase-free water or Tris–Ethylenediaminetetraacetic acid (EDTA) buffer (pH 7.0–8.0) and stored at −20°C. Donor DNA templates may be supplied as dry ssODNs (single-stranded oligodeoxynucleotides), lyophilized double-stranded DNA, or liquid plasmid DNA suspended in Tris–EDTA (pH 8.0), also stored at −20°C. Delivery tools such as lipid nanoparticles, viral vectors, electroporation buffers, and *Agrobacterium*-based constructs are generally provided in liquid form. Additional reagents used for DNA isolation, cloning, amplification, and assembly are commonly prepared as liquid enzyme formulations stabilized with glycerol and kept at −20°C. Materials like agarose, nucleic acid stains, loading dyes, and nucleotides are distributed either as powders or concentrated liquids that are later dissolved or diluted in neutral-pH buffers (Laforest and Nadakuduti, 2022).

Growth media constituents for CRISPR-based procedures such as MS (Murashige and Skoog) basal salts, sucrose, and plant hormones are generally powdered or concentrated stock solutions which are normally maintained at 4°C. Enzymes used for protoplast isolation are provided in lyophilized form and are reconstituted in buffers with pH values between 5.5 and 6.0. Screening reagents including PCR (Polymerase Chain Reaction) primers, mismatch-detection enzymes like T7 Endonuclease I (T7E1), restriction enzyme systems, and sequencing kits are usually supplied as liquid formulations, function within pH ranges of 7.0–8.0, and are stored at −20°C (Sandhya et al., 2020).

17.5 Methods for delivering genome editing tools into plant cells

17.5.1 *Agrobacterium*-mediated delivery

Agrobacterium-mediated transformation remains the most widely used approach for delivering GE components into plants because it is efficient, reproducible, and supported by well-established protocols (Azizi-Dargahlou and pouresmaeil, 2024) (see Fig. 17.1). The process operates through two main pathways: stable transformation, where Transfer

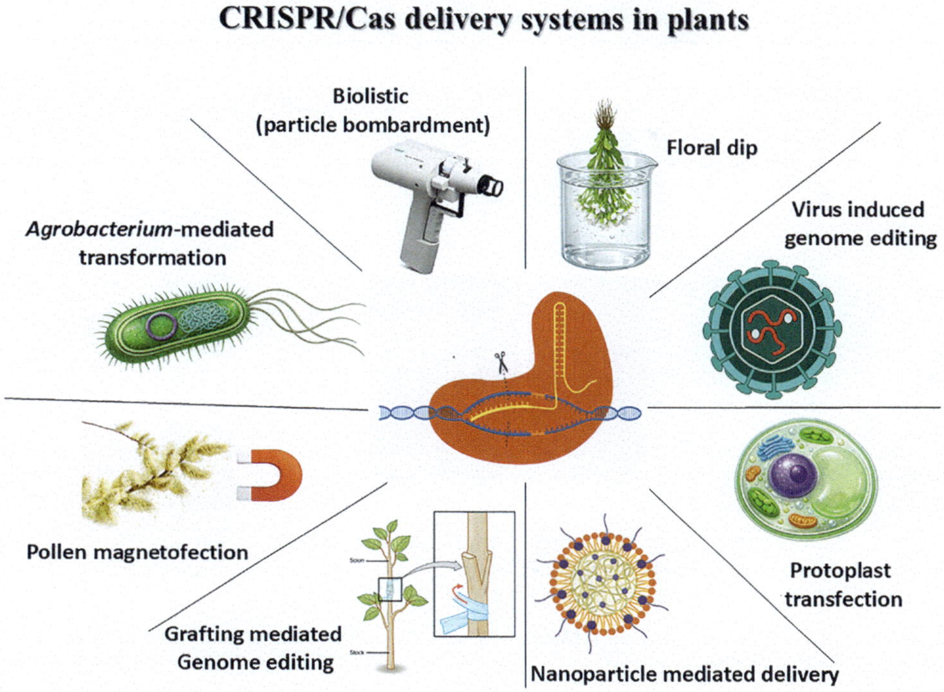

FIGURE 17.1 Different methods of delivery of CRISPR system in plants.

Deoxyribonucleic Acid (T-DNA) integrates into the genome and is passed on to future generations through co-cultivation or floral dip methods, and transient transformation, which allows temporary expression of introduced genes without integration. Although stable transformation is reliable, it is often genotype-dependent, labor-intensive, and makes it difficult to produce transgene-free plants especially in perennial and vegetatively propagated species (Kaur et al., 2021).

To address these challenges, transient expression systems such as agroinfiltration and optimized derivatives like FAST (Fast Agro-mediated Seedling Transformation) and AGROBEST (*Agrobacterium*-mediated Enhanced Seedling Transformation) have been developed. These systems support short-term CRISPR/Cas activity without genomic integration and can achieve efficiencies up to 1000 times higher than traditional stable methods. Agroinfiltration, which uses simple syringe or vacuum infiltration, offers a rapid, low-cost, and tissue-culture-free technique well suited for functional genomics and recombinant protein studies. Because it does not produce transgenic plants, it is also more acceptable to the public (Mohamed et al., 2024).

Further improvements such as *ALS* gene co-editing to eliminate selectable markers and site-specific recombination systems for removing transgenes have increased both precision and regulatory compatibility. Additionally, co-expressing developmental regulators like *BABY BOOM (BBM)* and *WUSCHEL (WUS)* has greatly boosted transformation efficiency

TABLE 17.2 Delivery methods for CRISPR components in plants.

Delivery method	Principle/Mechanism	Key advantages	Limitations	References
Agrobacterium-mediated transformation	Transfer of T-DNA containing CRISPR components into plant genome	High efficiency, reproducibility, well-established protocols	Genotype-dependent, transgene integration	Mohamed et al. (2024)
Biolistic	High-velocity delivery of DNA/RNA/protein-coated particles	Transgene-free editing possible using RNPs; species-independent	Potential tissue damage; low efficiency	Rönspies et al. (2021)
Protoplast transfection	PEG/electroporation-mediated uptake of DNA/RNPs into cell-wall-free protoplasts	Transgene-free editing; rapid assay of gRNAs	Limited regeneration in monocots	Xu et al. (2022)
Virus induced genome editing (VIGE)	Plant viruses deliver gRNA/Cas via infection	Tissue-culture-free; systemic editing	Limited cargo capacity; host-range restricted	Uranga and Daròs (2023)
Grafting mediated editing (GEG)	Movement of mobile RNAs from edited rootstock to non-transgenic scion	Transgene-free, heritable edits	Low efficiency; species-specific	Zaman et al. (2024)
Nanotechnology-based delivery (NBD)	Use of nanocarriers (CNTs, MSNs, LNPs) for RNP transport	Tissue-culture-free; DNA-free; minimal cell damage	ROS toxicity, low uptake efficiency	Demirer et al. (2021)

and enabled tissue-culture-independent GE. These innovations expand the applicability of *Agrobacterium*-mediated CRISPR delivery across a broader range of plant species (Huang et al., 2023) (see Table 17.2).

17.5.2 Biolistic transformation/particle bombardment

Biolistic transformation is a commonly used physical method for delivering GE components into plant cells. In this technique, tiny gold or tungsten particles coated with DNA, RNA, or protein complexes are shot into plant tissues using a gene gun, allowing the materials to enter cells directly (Rönspies et al., 2021). A key recent advancement is the use of Ribonucleoprotein (RNP) complexes, which enable DNA-free GE by removing the need for plasmid construction and preventing foreign DNA from integrating into the genome. Successful RNP-based editing protocol has now been established in crops like maize and wheat. However, biolistic delivery still has some limitations like labor-intensive mutant screening, lower editing efficiency and potential genomic damage caused by particle impact (Ma et al., 2020).

Recent advancements in biolistics with *in planta* particle bombardment have expanded its usefulness, where CRISPR/Cas9 plasmids or RNPs can be delivered directly into seeds or shoot apical meristems. This type of approach has been effectively applied in crops such as wheat, soybean, rice, and chickpea. This allowed heritable, tissue-culture-free editing that has seem to work across genotypes (Bélanger et al., 2024). This method is technically demanding and is often less efficient than *Agrobacterium*-mediated transformation. Improvement opportunities like optimizing particle size and delivery pressure are enhancing the procedures' precision and consistency. Biolistic transformation thus serves as a valuable tool for plant GE (Chen et al., 2022).

17.5.3 Protoplast transformation

Protoplast transformation is another important method for delivering GE reagents into plant cells. Protoplasts can readily take up DNA, RNA, or protein complexes making them powerful system for applying CRISPR/Cas9. In earlier days, plasmid DNA delivery was done into protoplasts to drive transient CRISPR/Cas expression. But this method carried the risk of random DNA integration and possible insertional mutagenesis (Tuncel et al., 2025). The development of RNP delivery has since enabled precise, transgene-free editing, successfully demonstrated in crops such as lettuce, potato, and tomato (Xu et al., 2022).

Even with its precision, the approach is still limited by species-specific challenges, particularly the difficulty of regenerating protoplasts in major monocots like rice, maize, and wheat. The technique can also introduce somaclonal variation and typically requires highly refined culture and regeneration protocols. Advances in Polyethylene Glycol (PEG)-mediated delivery, electroporation, and improved culture media continue to enhance both transformation and regeneration success. With ongoing improvements, protoplast-based systems remain a promising, highly controlled, DNA-free platform for sophisticated plant GE applications (Lokya et al., 2025).

17.5.4 Virus-induced genome editing (VIGE)

Viral delivery systems have become an innovative and effective approach for introducing CRISPR/Cas components into plant cells (Uranga and Daròs, 2023). So far, at least 22 plant viruses including Tobacco rattle virus (TRV), Potato virus X (PVX), Foxtail mosaic virus (FoMV), and Barley stripe mosaic virus (BSMV) have been adapted for VIGE, offering strong potential for producing GE plants without introducing transgenes. However, challenges such as limited cargo capacity, narrow host range, vector instability, non-seed transmission and low CRISPR/Cas expression continue to restrict wider use (Mikhaylova, 2025).

To address size limitations, researchers have turned to Sonchus yellow net rhabdovirus (SYNV) and tomato spotted wilt virus (TSWV) which can carry larger inserts. Compact Cas proteins including Cas12f, Cas12j, and Cas14 also help reduce vector load (Gentzel et al., 2022). Achieving heritable edits has been made possible by attaching Flowering Locus T (FT) transcripts and tRNA-like sequences (TLSs) to gRNAs, which promotes long-distance movement of CRISPR components to reproductive tissues through plasmodesmata and the phloem (Tamilselvan-Nattar-Amutha et al., 2023). Seed-transmissible viruses such as TRV and BSMV show promise for passing edits to the next generation (Uranga et al., 2021). Thus, continuous progress in vector design, development of compact Cas proteins, and improvements in systemic viral movement are rapidly transforming VIGE into a flexible, efficient, and DNA-free GE platform suitable for many crop species (Li et al., 2023; Bai et al., 2025).

17.5.5 Grafting-mediated gene editing (GEG)

GEG is an emerging, tissue-culture-free and transgene-free strategy that takes advantage of the natural mobility of RNA molecules in plants to produce heritable genome edits (Kocsisova and Coneva, 2023; Li et al., 2024). In this approach, a transgenic rootstock engineered to express Cas9 and sgRNAs fused with tRNA-like sequences (TLSs) sends mobile CRISPR/Cas9 RNA molecules through the plant's vascular system. When a non-transgenic (wild-type) scion is grafted onto this rootstock, these mobile RNAs move into the scion, enabling GE in both its somatic and reproductive tissues. The seeds produced by the scion then inherit the edited alleles without any foreign DNA, resulting in regulation-compliant, truly transgene-free plants (Yang et al., 2023a; Zaman et al., 2024). Successful demonstrations in *Arabidopsis thaliana* and *Brassica rapa* show that this method can even produce homozygous mutants in a single generation, offering an attractive alternative to traditional transformation systems that rely on tissue culture.

Several challenges still limit the broader adoption of GEG. The relatively low frequency of heritable edits for the technique of GEG indicates that improvement is required to enhance the long-distance RNA transport and stability of CRISPR components across graft junctions. The requirement to create target-specific transgenic rootstocks also reduces scalability and complicates cross-species application, which requires the development of multi-target rootstocks capable of delivering several editing reagents at once (Hu and Gao, 2023; Antony Ceasar and Ignacimuthu, 2023). Pairing GEG with other transformation-free delivery systems such as virus-based vectors or nanoparticle-mediated transport could further accelerate the production of transgene-free GE plants (Simoni et al., 2025).

17.5.6 Cut-dip-budding (CDB) and floral dip-mediated GE

The development of CDB and floral dip has greatly broadened the possibilities for CRISPR/Cas-based GE in plants. These approaches eliminate the labor-intensive tissue-culture steps typically required in *Agrobacterium tumefaciens* or biolistic transformation systems (Li et al., 2025a). In the CDB method, stem or leaf cuttings are dipped into an *Agrobacterium rhizogenes* suspension carrying CRISPR reagents, which stimulates the formation of hairy roots, which in turn help GE shoots to regenerate. This technique has shown particularly strong success in species with high natural regenerative capacity and root-suckering abilities (Lu et al., 2024).

The floral dip technique widely used in *Arabidopsis thaliana* works by infiltrating floral tissues with *A. tumefaciens* carrying CRISPR components. This allows the reagents to directly target gametes and produce edited offspring without any tissue culture steps, which makes both the CDB and floral dip methods as cheap methods of transformation (Cao et al., 2023).

Despite these benefits, both CDB and floral dip remain limited by species-specific constraints. The effectiveness of CDB depends heavily on the ability of a plant to regenerate shoots from hairy roots (Hu and Liu, 2025). The floral dip, method still remains largely restricted to *Arabidopsis*(Ma et al., 2025). Expanding their use for other plants will require advancements that improve transformation rates, reduce genotype dependence, and increase delivery precision (Cao et al., 2024).

17.5.7 Nanotechnology-based delivery (NBD)

Nanotechnology-based delivery systems have emerged as a powerful technique for introducing CRISPR components into plant cells with high precision (Shende and Patil, 2025). A wide range of nanomaterials including carbon nanotubes, mesoporous silica nanoparticles (MSNs), lipid nanoparticles (LNPs), and poly (lactic-co-glycolic acid) (PLGA) carriers have been used to transport DNA, RNA, and Cas9–sgRNA RNPs across the plant cell wall without relying on tissue culture (Demirer et al., 2021; Mujtaba et al., 2021). These nanocarriers improve the stability of GE molecules, support transient and transgene-free editing.

Innovative approaches such as pollen magnetofection further highlight the flexibility of NBD, allowing genetic material to be delivered directly into pollen grains. This makes it possible to generate GE seeds in a genotype-independent, tissue-culture-free manner (Vats et al., 2022). Despite these promising advances, challenges like potential nanoparticle toxicity, unintended off-target effects and production of reactive oxygen species (ROS) can negatively affect plant health (Gupta et al., 2025). Continued improvements in nanoparticle composition, size, and charge along with refinement of CRISPR/Cas systems, will be crucial for achieving safe, efficient, and scalable NBD based GE in plants (Ahmadikhah et al., 2025; Yau et al., 2024).

17.6 CRISPR in crop improvement

CRISPR/Cas systems provide an excellent opportunity to tackle global food security challenges by boosting nutritional quality, increasing yields and enhancing resistance to both biotic and abiotic stresses thus offering sustainable strategies for strengthening crop production (see Fig. 17.2). Targeted genetic changes can produce plants with better photosynthetic efficiency or improved nutrient uptake, directly enhancing crop productivity (Tang et al., 2023). CRISPR can create disease-resistant crops by disabling susceptibility genes or introducing resistance pathways (Zhu et al., 2020). Additionally, CRISPR helps the fine-tuning of plant architecture and developmental processes, supporting more efficient resource use and increasing the amount of harvestable biomass. A plenty of review articles and book chapters from reputed publishers has highlighted the expanding role of CRISPR technology in advancing agricultural crop improvement (Ikram et al., 2024a; Li et al., 2025b; Pal et al., 2025; Sharma et al., 2025).

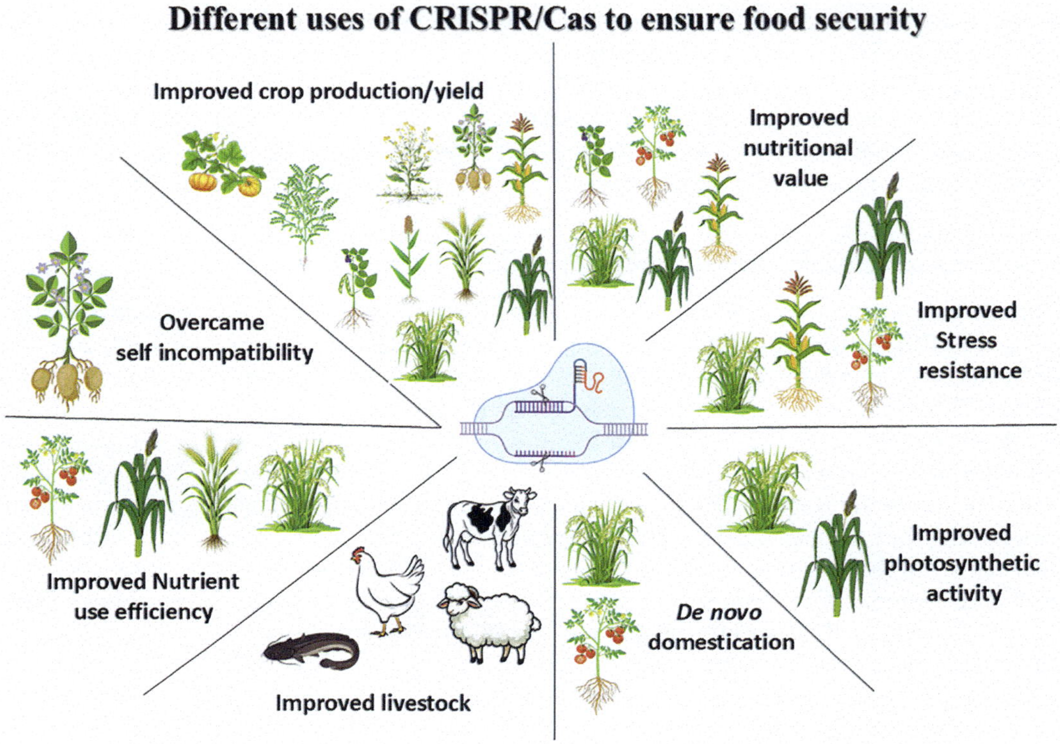

FIGURE 17.2 The various applications of CRISPR to ensure food security.

17.6.1 Improving crop production/yield

Enhancing crop yield remains a central priority for global food security, yet the traits governing yield are complex, polygenic, and largely regulated by quantitative trait loci (QTLs) that influence thousand-grain weight, grain number, floret development, and panicle architecture. CRISPR technology has accelerated progress in improving these traits across major food crops including rice, wheat, maize, barley, soybean, pulses, potato, brinjal, and tomato (Ukhatova et al., 2023; Chen et al., 2024; Kumar et al., 2024; Shraddha et al., 2024; Sood et al., 2025; Kaur et al., 2025). Since the first report of CRISPR/Cas editing in rice in 2013, the technology has enabled the rapid development of both high-yielding and climate-resilient rice varieties (Rengasamy et al., 2024). Key yield-associated genes and QTLs such as *OsGS3*, *OsGW2*, *OsGW5*, *OsGW6*, *OsGn1a*, *OsGL3.1*, *HTD1*, *IPA1*, *DEP1*, *OsCKX2*, *OsSPL14*, *OsHXK1*, *OsMADS14*, and *OsMADS15* have been edited to enhance grain size, panicle traits, cytokinin regulation, and overall plant architecture in rice (Usman et al., 2021; Zheng et al., 2021; Bhavya et al., 2024) (a).

In wheat, CRISPR/Cas9 editing of genes including *TaSPL13*, *TaRPK1*, *TaARF15-A1*, *TaD27*, and *TaSBEIIa* has led to improvements in grain size, grain number, tillering, and photosynthetic performance, contributing to higher yields under both optimal and stress conditions (Elsharawy and Refat, 2023; Feng, 2025). Similarly, in maize, editing of genes such as *ZmCEP1*,

ZmACO2, *ZmCLE7*, *ZmFCP1*, *ZmCLE1E5*, and *ZmRAVL1* have enhanced ear development, kernel traits, meristem activity, and overall ear length (Basu et al., 2023). Advances in multiplex and pyramiding strategies now allow several QTLs to be modified simultaneously, accelerating yield improvement within a single generation (Roy et al., 2023, 2025). Similar successes have been reported in sorghum (Namata et al., 2025), barley (Ahmar et al., 2024), soybean (Li et al., 2025c; Kim et al., 2025), and potato (Tang et al., 2022).

Plant structural architecture traits which include shoot branching, root development and leaf morphology is another critical determinant of yield (Gao, 2021). CRISPR has made it possible to fine-tune these architectural traits across a wide range of crops such as rice, maize, wheat, barley, soybean, rapeseed, mustard, chickpea, tomato, and pumpkin (Kořínková et al., 2022). Improving photosynthetic efficiency through enhanced CO_2 assimilation, optimized Rubisco function, and reduced photorespiration also contributes significantly to yield gains (Hilioti et al., 2024). For instance, CRISPR knockout of *NRP1* in rice increased photosynthetic capacity and biomass, while substituting native rice *RbcS* with the sorghum *RbcS* gene generated a more efficient hybrid Rubisco, highlighting the potential of CRISPR for crop improvement (Saini et al., 2025).

17.6.2 Improving stress resistance

Enhancing crop productivity depends not only on boosting yield potential but also on reducing the substantial losses caused by biotic and abiotic stresses. Plant pathogens including bacteria, fungi, and viruses remain major constraints on global agriculture, often decreasing yields through disease outbreaks, while insect pests inflict direct feeding damage and frequently act as vectors that spread additional pathogens (Sharma et al., 2023). At the same time, environmental challenges such as drought, heat stress, flooding, and soil salinity further restrict crop growth, and these pressures are becoming more severe with climate change, contributing to worldwide yield reductions estimated at 20%–40% (Huang et al., 2021). Because many stress-response pathways overlap at the molecular and physiological levels, CRISPR-based GE has become an essential tool for identifying, modifying, and improving genes associated with stress tolerance (Shelake et al., 2022).

Through targeted manipulation of susceptibility (*Su*) and sensitivity (*Se*) genes exploited by pathogens or environmental factors, CRISPR/Cas systems enable lasting resistance to diseases and improved tolerance to abiotic stress. In rice, for example, editing genes such as *OsSWEET11*, *OsSWEET13*, and *OsSWEET14* provides strong resistance to bacterial blight, while knocking out *MLO* in wheat enhances resistance to powdery mildew. Additionally, CRISPR editing of fungal virulence-related genes including *ALB1*, *RSY1*, *USTA*, and *UvSLT2* has been shown to bolster resistance against blast disease and false smut (Ahmad et al., 2021). The technology has also proven effective for combatting viral infections such as modifying *eIF4G* confers resistance to Rice tungro disease, and targeting the coat protein of Grapevine virus A (*GVA*) reduces viral load in infected plants (Jeyaraj et al., 2024; Zhan et al., 2024; Spencer et al., 2023).

CRISPR has also accelerated progress in abiotic stress tolerance. In rice, disruption of *OsMYB30* improves cold tolerance, while mutating *OsDST* enhances plant resilience to both drought and salinity. In maize, editing *ARGOS8* increases drought tolerance, and in tomato, modifications to genes such as *SlMAPK3*, *SlAGL6*, and *SlARF4* enhance heat and salinity tolerance (Benitez-Alfonso et al., 2023). Through these kinds of targeted genetic improvements, CRISPR has transformed breeding programs across many major crops including wheat, rice, maize, sorghum, soybean, tomato, and potato allowing the development of climate-resilient, adaptable varieties suited to rapidly changing environmental and biological pressures (Yuan et al., 2023; Phogat et al., 2024; Arif et al., 2025). As a result, CRISPR technology stands out as one of the most powerful tools available for strengthening global food security in the face of intensifying climate challenges and rising pest and disease threats.

17.6.3 Improving nutrient use efficiency

The GR dramatically boosted crop yields through heavy fertilizer application, but this progress also increased production costs and contributed to environmental pollution through nutrient leaching and runoff. As a result, improving the use efficiency of nitrogen (N), phosphorus (P), and potassium (K) has become essential for achieving sustainable productivity with better crop quality and improved stress tolerance (Sathee et al., 2022). CRISPR/Cas GE has emerged as a powerful tool for enhancing nutrient use efficiency (NUE, PUE, and KUE) by precisely modifying key transporters, transcription factors (TFs), and regulatory genes involved in nutrient uptake and signaling (Hazrati et al., 2025). In the case of nitrogen, CRISPR-based edits in rice, wheat, and barley have strengthened nitrate uptake, signaling pathways, and assimilation processes. Mutations in *NRT1.1B*, *SLR1 DELLA* base editing, *ARE1* knockouts, and *MIR396e/f* editing have all contributed to increased NUE and grain yield under reduced nitrogen conditions (Zhang et al., 2021b). Conversely, knockouts of *OsNLP3* and *OsNLP4* in rice were shown to decrease growth, grain yield, and NUE under nitrogen limitation (Wu et al., 2021). For phosphorus, CRISPR editing of genes such as *PHO1;1*, *PHT1*, *OsPHR3*, *MYB1*, and vacuolar Pi transporters (*Osvpe1*/*Osvpe2*) has improved P uptake and internal homeostasis. Examples include the *SlPHO1;1* knockout in tomato and *OsPHR3* overexpression in rice. However, excessive disruption of *PHR1*, *PHR2*, or *PHR3* negatively affected plant growth under both phosphorus-sufficient and phosphorus-deficient conditions (Karunarathne et al., 2022).

Similarly, CRISPR-mediated editing of potassium transporters and regulatory components including *OsHAK1*, *OsHAK3*, *OsHAK5*, *OsHAK16*, and *OsRUPO* has enhanced K^+ uptake and homeostasis. Notably, *OsRUPO* knockout increased K^+ accumulation in pollen and improved root-to-shoot translocation. Regulatory modules such as *CBL1/9–CIPK23–AKT1* and TFs like *AtARF2* also play critical roles in coordinating potassium transporter activity and represent promising targets for elevating KUE (Zhang et al., 2022). Overall, CRISPR/Cas GE offers a precise and flexible approach for improving nutrient acquisition, transport, and assimilation, leading to higher yield potential and greater stress resilience in rice, wheat, and barley. At the same time, careful targeting and modulation are necessary to prevent unintended developmental effects (Ceasar et al., 2022).

17.6.4 Improving nutrient quality/nutritional value

CRISPR/Cas GE offers a precise and highly effective strategy for biofortifying crops with essential nutrients while reducing anti-nutritional compounds such as phytic acid and steroidal glycoalkaloids. This approach enhances the bioavailability of vitamins, minerals, and key metabolites across cereals, vegetables, and oilseed crops. For example, carotenoid-rich "golden" varieties such as Golden Rice Kitaake which contains 7.9 μg/g dry weight of β-carotene in the endosperm have been developed through targeted, transgene-free CRISPR modifications (Zheng et al., 2021b). In barley, CRISPR/Cas9-mediated overexpression of *HvHPT* and *HvHGGT* has significantly increased vitamin E levels by elevating both tocopherol and tocotrienol content.

Mineral biofortification has also advanced through CRISPR technologies. Disruption of the *IPK1* gene in wheat has boosted grain iron concentration, while editing *TaIPK1* has improved zinc accumulation by lowering phytic acid levels, thereby increasing mineral bioavailability. Improvements in protein and oil quality further highlight the versatility of CRISPR/Cas systems: in maize, editing members of the 19 kDa *α-zein* gene family increased lysine levels by up to 30% (Hurst et al., 2024); in wheat, targeted α-gliadin gene knockouts reduced gluten content and decreased immunoreactivity; and in soybean, *FAD2-1A* and *FAD2-1B* editing has raised oleic acid content. CRISPR can also enhance aesthetic and nutraceutical qualities, as demonstrated by the insertion of the cauliflower mosaic virus 35S promoter upstream of the *ANT1* gene in tomato, resulting in purple tomatoes enriched with anthocyanins (Tyczewska et al., 2023). Collectively, these advances underscore the powerful potential of CRISPR/Cas GE to create nutrient-dense, health-promoting crops that contribute to sustainable improvements in global nutrition (Kumar et al., 2022).

17.6.5 Broadening the genetic diversity of crop germplasm

Over thousands of years of domestication, humans have mainly selected plants for traits related to high yield such as larger fruits and grains, improved plant architecture, better flavor, and reduced toxicity. While these efforts have transformed agriculture, they have also narrowed genetic diversity and, in some cases, reduced nutritional quality, making modern crops more vulnerable to environmental stresses and emerging diseases (Farooq et al., 2025). In contrast, wild relatives retain extensive genetic variation and naturally evolved stress tolerance. With the advent of GE, it is now possible to rapidly domesticate these wild species by precisely modifying key domestication genes (Zeinali et al., 2025). In *Solanum pimpinellifolium* (wild tomato), CRISPR/Cas9 editing of *SELF PRUNING*, *SP5G*, *CLAVATA3*, *WUSCHEL*, and *GGP* produced compact, day-length-independent plants with larger fruits and elevated ascorbic acid levels, while preserving the species' inherent salt and disease tolerance (Zsögön et al., 2025). Similar GE efforts have increased lycopene accumulation in tomato, boosted productivity in groundcherry, and improved agronomic performance in *Oryza alta*, a wild rice species (Yu et al., 2021). These achievements highlight the promise of GE in bridging wild and cultivated gene pools and accelerating the development of resilient, high-yielding crops (Varshney et al., 2021).

Another approach to expanding the genetic base of modern crops involves overcoming reproductive barriers such as self-incompatibility, which restricts natural recombination (Zhang et al., 2024b). In potato, CRISPR-mediated mutagenesis of the *S-RNase* gene, responsible for blocking self-pollen tube growth, has successfully converted self-incompatible wild lines into self-compatible plants capable of producing viable berries. This advancement enables sexual reproduction and crossbreeding in a crop historically propagated vegetatively, creating new opportunities to generate genetically diverse and improved potato varieties (Hou et al., 2023).

17.7 CRISPR for livestock enhancement

Farm animals including cattle, sheep, goats, poultry, and fish remain vital sources of meat, milk, and eggs that support global nutrition. Traditional breeding has contributed greatly to livestock improvement, but these methods are often slow, limited by existing genetic variation, and unable to precisely target traits of interest. The emergence of CRISPR/Cas9 GE has transformed livestock breeding by enabling accurate and efficient modification of specific genes to enhance productivity, disease resistance, and nutritional value. This technological shift supports more sustainable and ethical livestock production by improving animal health and strengthening resilience in global food systems (Jabbar et al., 2021).

In cattle, CRISPR/Cas9 has been used to generate lines resistant to major infectious diseases. For example, *NRAMP1* knock-in cattle show enhanced resistance to tuberculosis, while *IL10RA* knock-out animals demonstrate reduced susceptibility to Johne's disease. Edits targeting *CD46* have also been employed to protect against brucellosis and bovine viral diarrhea (Szillat et al., 2020). In sheep and goats, modification of genes such as *MSTN* and *CLPG1* increases muscle mass and meat yield, whereas *SOCS2* editing has improved both milk production and body weight (Zhou et al., 2022). Knockout of the β-lactoglobulin (*BLG*) gene reduces milk allergenicity and makes the milk composition more similar to human milk. Additional edits in *SCD1* and *ACAA2* enhance milk fat quality, and insertion of *ASMT* and *AANAT* genes has been shown to increase melatonin-rich milk, demonstrating how GE can be tailored toward specific nutritional or functional properties (Aboelhassan and Abozaid, 2024).

CRISPR/Cas9 applications in poultry have also advanced rapidly. *MSTN* knockout has led to improved muscle growth and lean meat production, while targeted edits in *chNHE1* and *chANP32A* confer resistance to avian leukosis virus (ALV-J) and avian influenza virus (AIV), strengthening overall flock health (Singh and Ali, 2021). In aquaculture, the disruption of MSTN in species such as catfish and Alburnus has resulted in substantial increases in body weight and muscle mass, highlighting the broad applicability of CRISPR across terrestrial and aquatic animals (Zheng et al., 2024). Collectively, these developments demonstrate that CRISPR/Cas9 is reshaping animal agriculture by accelerating genetic gains, enhancing disease resistance, and improving the nutritional quality of animal products. This progress marks a significant step toward more precise, sustainable, and ethically aligned livestock breeding systems.

17.8 Turbo crop improvement with multiplexing

Multiplex CRISPR editing has become a powerful strategy for improving polygenic traits that are essential for global food and nutrition security, particularly those tied to nutritional value, processing quality, and allergen reduction. In tomato, the simultaneous editing of *PSY1*, *MYB12*, and *SGR1* produced a wide range of fruit colors within a single year, demonstrating how coordinated gene modification can rapidly diversify both market traits and nutritional profiles (Yang et al., 2023b). Similar progress has been made in cereals: multiplex editing of starch branching enzymes (*SBE*) and starch synthases (*SS*) in rice, wheat, and barley has increased amylose and resistant starch, the key contributors to dietary fiber and lower glycemic index with the greatest improvements appearing in higher-order mutants (Biswas et al., 2023; Yang et al., 2024). These advances enhance the nutritional quality of staple crops while maintaining agronomic performance, supporting broader goals of global food and nutrition security.

Targeting multiple genes simultaneously has also delivered significant benefits for crop health and consumer safety, particularly for traits governed by large gene families. In wheat, coordinated disruption of *γ-*, *ω-*, and *α-gliadin* clusters resulted in 64%–97% reductions in gluten allergens, producing non-transgenic lines better suited for gluten-sensitive individuals (Yu et al., 2024). In cucumber, durable resistance to powdery mildew required simultaneous knockout of three *MLO* homologs viz. *Csmlo1*, *Csmlo8*, and *Csmlo11*—highlighting the necessity of multiplex approaches when gene redundancy limits single-gene edits (Yin and Tsai, 2025). These examples illustrate how multiplex CRISPR strengthens crop resilience and enhances food safety by addressing complex biological constraints in a single editing step.

Beyond trait improvement, multiplex GE has accelerated progress in crop protection and streamlined breeding. A landmark case is the inactivation of endogenous banana streak virus (eBSV) in plantain, where simultaneous editing of three viral ORFs produced virus-free lines in vegetatively propagated cultivars overcoming a major bottleneck in banana and plantain breeding (Tripathi et al., 2019). Multiplex editing of regulatory elements has likewise enabled rapid manipulation of agronomic traits, including modified fruit size in tomato, optimized amylose levels in rice, and improved tiller number, panicle size, and multiple-ear formation in maize (Kumar et al., 2025). By enabling coordinated improvement of several traits in a single generation, multiplex CRISPR approaches greatly accelerate breeding timelines and provide a robust platform for developing resilient, high-performing crops needed to support future food and nutrition demands.

17.9 Biosafety concerns surrounding CRISPR and global regulation

The rapid expansion of CRISPR/Cas-mediated GE has reshaped plant biotechnology by enabling precise, efficient, and heritable genetic modifications. At the same time, it has brought forward a range of regulatory, ethical, and biosafety challenges that require careful consideration (Munawar et al., 2024). Biosafety concerns remain at the forefront, particularly with respect to off-target edits and unintended genomic changes that could alter plant physiology, disrupt ecological interactions, or affect non-target species (Tariq et al., 2023). In addition, multiplex editing and potential gene-drive applications raise concerns about uncontrolled gene flow and ecological imbalance due to the biased inheritance of engineered traits (Movahedi et al., 2023). Given these complexities, robust risk-assessment frameworks, long-term monitoring strategies, and internationally harmonized biosafety regulations are essential for the responsible application of CRISPR technologies in agriculture.

Globally, regulatory oversight of GE crops varies widely. Countries such as the United States, Brazil, Canada, Australia, and India do not classify non-transgenic edits (SDN-1/2) as GMOs, which has facilitated innovation and commercialization (Atimango et al., 2024). In contrast, the European Union and Russia regulate all GE organisms under GMO legislation, imposing more stringent requirements. China and Japan have adopted intermediate policies that exempt transgene-free edits from full GMO procedures, while countries such as Argentina, Chile, and Brazil follow case-by-case assessments aligned with the biosafety principles of the Cartagena Protocol. These differing frameworks underscore the need for globally coherent, science-based regulation that clearly distinguishes GE from conventional transgenesis, enabling responsible innovation while safeguarding biosafety and maintaining public trust (Tuncel et al., 2023).

17.10 Responsible use of AI in agriculture

The integration of AI into agriculture marks a transformative shift toward sustainable, high-yield, and resource-efficient farming systems. Although adoption of AI is currently highest in technologically advanced countries, it offers immense potential for developing regions to enhance productivity and strengthen food security (Ahmad et al., 2024) (refer to Fig. 17.3). Emerged from defense and aerospace research, AI and Machine Learning (ML) now support precise farmland characterization, optimized input management, early detection of biotic and abiotic stresses, crop monitoring, high-throughput phenotyping, disease diagnosis, yield forecasting, weather prediction, post-harvest management thereby reducing trial-and-error approaches and improving global food system resilience. AI-enabled irrigation systems integrating Internet of Things (IoT) sensors and ML algorithms deliver precise amount of water for irrigation (Ben Ayed, Hanana and Khan, 2021), while deep learning platforms such as Agpest and PlantDoc achieve more than 95% accuracy in disease identification (Khan et al., 2022, 2025). Despite persistent limitations including restricted datasets, high computational requirements, and the need for trained personnel, AI-driven chatbots, predictive models, and diagnostic tools continue to support improved agricultural decision-making (Jarray et al., 2023; Zhang et al., 2024a).

The performance of GE crops increasingly relies on precision agriculture technologies that combine AI, robotics, and IoT. These integrated systems generate real-time insights into crop physiology and soil conditions, enabling data-driven decisions regarding irrigation scheduling, fertilizer application, and pest management (Nath et al., 2024; Pandey and Mishra, 2024; Naseem and Rizwan, 2025). IoT-based sensors and drones provide continuous measurements of moisture, pH, nutrient levels, and canopy health, while AI-powered models diagnose nutrient deficiencies, predict stress events, and optimize resource use. Decision support systems disseminate timely information on pest outbreaks, disease spread, and weather variability, improving accessibility and resilience for smallholder farmers (Assimakopoulos et al., 2025). Novel technologies such as the NASA–ISRO Synthetic Aperture Radar (NISAR) satellite are poised to revolutionize agricultural monitoring through dual-frequency radar that delivers cloud-penetrating, high-resolution imagery every 12 days.

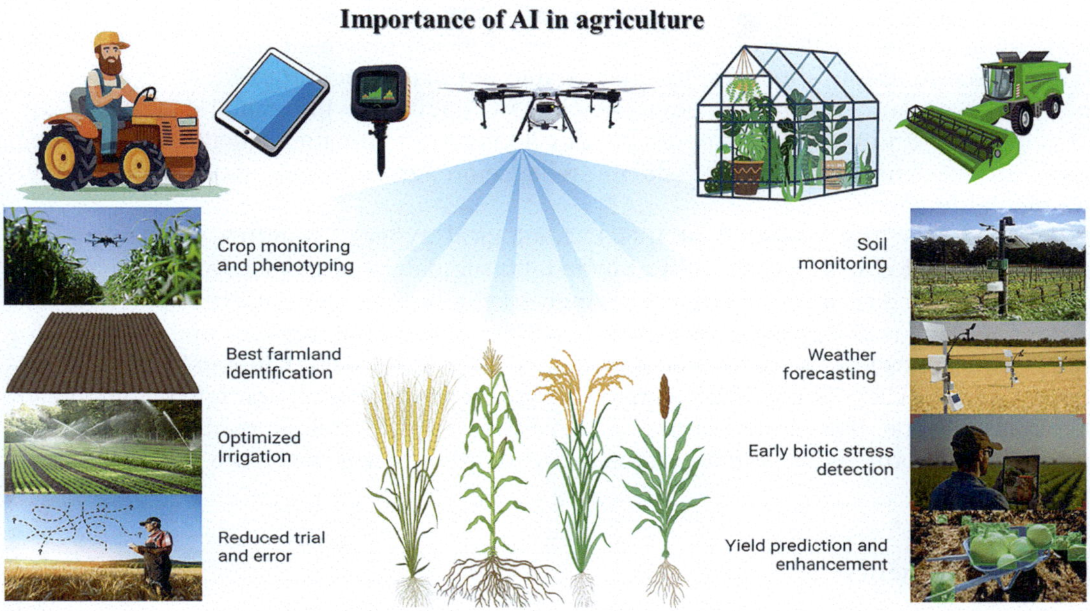

FIGURE 17.3 The different applications of AI in agriculture.

These capabilities enhance crop forecasting, soil moisture assessment, and long-term food security planning (Karanth et al., 2023; Thingujam et al., 2025). AI-integrated robotics and unmanned ground vehicles (UGVs) further strengthen field operations by enabling autonomous planting, crop surveillance, precision weeding, and harvesting (Rejeb et al., 2022; Jarroudi et al., 2024), supported by innovations such as AgriQ navigation systems, ecoRobotix precision weeders, and robotic harvesters (Ikram et al., 2024b).

As AI becomes increasingly embedded in global agriculture, responsible deployment is critical to ensure sustainability and public trust. Fragmented and biased datasets limit model accuracy and fairness, emphasizing the need for frameworks that promote transparency, accessibility, and ethical governance (Richards et al., 2023; Rudrakar and Rughani, 2024). Participatory innovation engaging farmers, scientists and policymakers can help align AI solutions with local socio-economic and environmental priorities. Ultimately, integrating responsible AI with robotics and GE represents a paradigm shift that enhances yield potential, reduces chemical inputs, improves climate resilience, and contributes to global food security and ecological balance (Tzachor et al., 2022; Javaid et al., 2023; Rugji et al., 2025; Shao and Marwa, 2025).

17.11 Conclusion

The convergence of AI and CRISPR-based GE represents a pivotal turning point in modern agriculture, offering unprecedented opportunities to reinforce global food security amid intensifying pressures such as climate change, population growth, and declining natural resources. CRISPR technologies provide precise, heritable, and transgene-free genetic modifications that significantly accelerate crop improvement, while AI-driven analytics enhance these advances by identifying optimal gene targets, forecasting crop performance, and managing environmental variability with remarkable accuracy. Together, these innovations support the development of climate-resilient, high-yielding, and nutrient-dense crops, while reducing reliance on agrochemicals and mitigating environmental impacts.

Realizing the full potential of this technological synergy requires confronting key challenges related to biosafety, ethical oversight, regulatory alignment, and equitable access. Globally consistent governance frameworks, strengthened data transparency, and expanded technical capacity particularly within developing regions are essential for ensuring responsible, inclusive, and sustainable implementation. Ultimately, the integration of AI with CRISPR-based GE signals the emergence of a new paradigm of "intelligent agriculture", one that unites biological precision with digital innovation to promote sustainable productivity, ecological balance, and long-term resilience of global food systems.

Acknowledgments

We thank all the faculty members of the School of Agriculture, Galgotias University, Greater Noida, India, with whom we discussed different aspects of this review. We also acknowledge that help of ChatGPT has been taken at a few places (PLEASE EXPLAIN THIS). The content reflects the authors' own scholarly work, and the authors assume full responsibility for all statements, interpretations, and conclusions presented.

References

Abdallah, N.A., Hamwieh, A., Radwan, K., Fouad, N., Prakash, C., 2021. Genome editing techniques in plants: a comprehensive review and future prospects toward zero hunger. GM Crops Food 12 (2), 601–615. https://doi.org/10.1080/21645698.2021.2021724.

Aboelhassan, D.M., Abozaid, H., 2024. Opportunities for CRISPR-Cas9 application in farm animal genetic improvement. Mol. Biol. Rep. 51 (1). https://doi.org/10.1007/s11033-024-10052-3.

Ahmad, M., 2023. Plant breeding advancements with "CRISPR-Cas" genome editing technologies will assist future food security. Front. Plant Sci. 14. https://doi.org/10.3389/fpls.2023.1133036.

Ahmad, S., Tang, L., Shahzad, R., Mawia, A.M., Rao, G.S., Jamil, S., Wei, C., Sheng, Z., Shao, G., Wei, X., Hu, P., Mahfouz, M.M., Hu, S., Tang, S., 2021. CRISPR-based crop improvements: a way forward to achieve zero hunger. J. Agric. Food Chem. 69 (30), 8307–8323. https://doi.org/10.1021/acs.jafc.1c02653.

Ahmad, A., Liew, A.X.W., Venturini, F., Kalogeras, A., Candiani, A., Di Benedetto, G., Ajibola, S., Cartujo, P., Romero, P., Lykoudi, A., De Grandis, M.M., Xouris, C., Lo Bianco, R., Doddy, I., Elegbede, I., D'Urso Labate, G.F., García del Moral, L.F., Martos, V., 2024. AI can empower agriculture for global food security: challenges and prospects in developing nations. Front. Artif. Intell. 7. https://doi.org/10.3389/frai.2024.1328530.

Ahmadikhah, A., Zarabizadeh, H., Nayeri, S., Sadegh Abbasi, M., 2025. Advancements in genome editing tools for genetic studies and crop improvement. Front. Plant Sci. 15. https://doi.org/10.3389/fpls.2024.1370675.

Ahmar, S., Usman, B., Hensel, G., Jung, K.H., Gruszka, D., 2024. CRISPR enables sustainable cereal production for a greener future. Trends Plant Sci. 29 (2), 179–195. https://doi.org/10.1016/j.tplants.2023.10.016.

Aksoy, E., Yildirim, K., Kavas, M., Kayihan, C., Yerlikaya, B.A., Çalik, I., Sevgen, İ., Demirel, U., 2022. General guidelines for CRISPR/Cas-based genome editing in plants. Mol. Biol. Rep. 49 (12), 12151–12164. https://doi.org/10.1007/s11033-022-07773-8.

Antony Ceasar, S., Ignacimuthu, S., 2023. CRISPR/Cas genome editing in plants: dawn of Agrobacterium transformation for recalcitrant and transgene-free plants for future crop breeding. Plant Physiol. Biochem. 196, 724–730. https://doi.org/10.1016/j.plaphy.2023.02.030.

Arif, M., Shamshad, J., Khalid, F., Kuzyakov, Y., Younas, A., Li, L., 2025. Biotechnological and sustainable approaches to climate-resilient agriculture. J. Agron. Crop Sci. 211 (4). https://doi.org/10.1111/jac.70105.

Assimakopoulos, F., Vassilakis, C., Margaris, D., Kotis, K., Spiliotopoulos, D., 2025. AI and related technologies in the fields of smart agriculture: a review. Information 16 (2), 100. https://doi.org/10.3390/info16020100.

Atia, M., Jiang, W., Sedeek, K., Butt, H., Mahfouz, M., 2024. Crop bioengineering via gene editing: reshaping the future of agriculture. Plant Cell Rep. 43 (4). https://doi.org/10.1007/s00299-024-03183-1.

Atimango, A.O., Wesana, J., Kalule, S.W., Verbeke, W., De Steur, H., 2024. Genome editing in food and agriculture: from regulations to consumer perspectives. Curr. Opin. Biotechnol. 87. https://doi.org/10.1016/j.copbio.2024.103127.

Azizi-Dargahlou, S., pouresmaeil, M., 2024. Agrobacterium tumefaciens-mediated plant transformation: a review. Mol. Biotechnol. 66 (7), 1563–1580. https://doi.org/10.1007/s12033-023-00788-x.

Bai, S., Cao, X., Hu, L., Hu, D., Li, D., Sun, Y., 2025. Engineering an optimized hypercompact CRISPR/Cas12j-8 system for efficient genome editing in plants. Plant Biotechnol. J. 23 (4), 1153–1164. https://doi.org/10.1111/pbi.14574.

Basu, U., Riaz Ahmed, S., Bhat, B.A., Anwar, Z., Ali, A., Ijaz, A., Gulzar, A., Bibi, A., Tyagi, A., Nebapure, S.M., Goud, C.A., Ahanger, S.A., Ali, S., Mushtaq, M., 2023. A CRISPR way for accelerating cereal crop improvement: progress and challenges. Front. Genet. 13. https://doi.org/10.3389/fgene.2022.866976.

Bélanger, J.G., Copley, T.R., Hoyos-Villegas, V., Charron, J.B., O'Donoughue, L., 2024. A comprehensive review of in planta stable transformation strategies. Plant Methods 20 (1). https://doi.org/10.1186/s13007-024-01200-8.

Ben Ayed, R., Hanana, M., Khan, R., 2021. Artificial intelligence to improve the food and agriculture sector. J. Food Qual. 2021, 1–7. https://doi.org/10.1155/2021/5584754.

Benitez-Alfonso, Y., Soanes, B.K., Zimba, S., Sinanaj, B., German, L., Sharma, V., Bohra, A., Kolesnikova, A., Dunn, J. A., Martin, A.C., Khashi u Rahman, M., Saati-Santamaría, Z., García-Fraile, P., Ferreira, E.A., Frazão, L.A., Cowling, W.A., Siddique, K.H.M., Pandey, M.K., Farooq, M., Varshney, R.K., Chapman, M.A., Boesch, C., Daszkowska-Golec, A., Foyer, C.H., 2023. Enhancing climate change resilience in agricultural crops. Curr. Biol. 33 (23), R1246. https://doi.org/10.1016/j.cub.2023.10.028.

Bhavya, G., Acharya, A., Abida, P.S., Preetha, R., Kiran, A.G., Melna Mary, C.J., Krishna, G.K., 2024. CRISPR/Cas9 driven targeted editing of grain number 1a gene: sgRNA constructs design for yield enhancement in rice (Oryza sativa L.). Journal of Advances in Biology & Biotechnology 27 (12), 308–324. https://doi.org/10.9734/jabb/2024/v27i121778.

Biswas, S., Ibarra, O., Shaphek, M., Molina-Risco, M., Faion-Molina, M., Bellinatti-Della Gracia, M., Thomson, M.J., Septiningsih, E.M., 2023. Increasing the level of resistant starch in 'Presidio' rice through multiplex CRISPR–Cas9 gene editing of starch branching enzyme genes. Plant Genome 16 (2). https://doi.org/10.1002/tpg2.20225.

Bohra, A., Kilian, B., Sivasankar, S., Caccamo, M., Mba, C., McCouch, S.R., Varshney, R.K., 2022. Reap the crop wild relatives for breeding future crops. Trends Biotechnol. 40 (4), 412–431. https://doi.org/10.1016/j.tibtech.2021.08.009.

Bradbury, A., Clapp, O., Biacsi, A.S., Kuo, P., Gaju, O., Hayta, S., Zhu, J.K., Lambing, C., 2025. Integrating genome editing with omics, artificial intelligence, and advanced farming technologies to increase crop productivity. Plant Commun. 6 (7). https://doi.org/10.1016/j.xplc.2025.101386.

Cao, X., Xie, H., Song, M., Lu, J., Ma, P., Huang, B., Wang, M., Tian, Y., Chen, F., Peng, J., Lang, Z., Li, G., Zhu, J.K., 2023. Cut–dip–budding delivery system enables genetic modifications in plants without tissue culture. Innovation 4 (1). https://doi.org/10.1016/j.xinn.2022.100345.

Cao, X., Xie, H., Song, M., Zhao, L., Liu, H., Li, G., Zhu, J.K., 2024. Simple method for transformation and gene editing in medicinal plants. J. Integr. Plant Biol. 66 (1), 17–19. https://doi.org/10.1111/jipb.13593.

Ceasar, S.A., Maharajan, T., Hillary, V.E., Ajeesh Krishna, T.P., 2022. Insights to improve the plant nutrient transport by CRISPR/Cas system. Biotechnol. Adv. 59. https://doi.org/10.1016/j.biotechadv.2022.107963.

Chavhan, R.L., Jaybhaye, S.G., Hinge, V.R., Deshmukh, A.S., Shaikh, U.S., Jadhav, P.K., Kadam, U.S., Hong, J.C., 2025. Emerging applications of gene editing technologies for the development of climate-resilient crops. Frontiers in Genome Editing 7. https://doi.org/10.3389/fgeed.2025.1524767.

Chawla, R., Poonia, A., Samantara, K., Mohapatra, S.R., Naik, S.B., Ashwath, M.N., Djalovic, I.G., Prasad, P.V.V., 2023. Green revolution to genome revolution: driving better resilient crops against environmental instability. Front. Genet. 14. https://doi.org/10.3389/fgene.2023.1204585.

Chen, Z., Debernardi, J.M., Dubcovsky, J., Gallavotti, A., 2022. Recent advances in crop transformation technologies. Nat. Plants 8 (12), 1343–1351. https://doi.org/10.1038/s41477-022-01295-8.

Chen, F., Chen, L., Yan, Z., Xu, J., Feng, L., He, N., Guo, M., Zhao, J., Chen, Z., Chen, H., Yao, G., Liu, C., 2024. Recent advances of CRISPR-based genome editing for enhancing staple crops. Front. Plant Sci. 15. https://doi.org/10.3389/fpls.2024.1478398.

Demirer, G.S., Silva, T.N., Jackson, C.T., Thomas, J.B., Ehrhardt, D.W., Rhee, S.Y., Mortimer, J.C., Landry, M.P., 2021. Nanotechnology to advance CRISPR–Cas genetic engineering of plants. Nat. Nanotechnol. 16 (3), 243–250. https://doi.org/10.1038/s41565-021-00854-y.

Dhal, S.B., Kar, D., 2025. Leveraging artificial intelligence and advanced food processing techniques for enhanced food safety, quality, and security: a comprehensive review. Discov. Appl. Sci. 7 (1). https://doi.org/10.1007/s42452-025-06472-w.

Ebrahimi, V., Hashemi, A., 2024. CRISPR-based gene editing in plants: focus on reagents and their delivery tools. Bioimpacts 15. https://doi.org/10.34172/bi.30019.

EFSA Panel on Genetically Modified Organisms (EFSA GMO Panel), Naegeli, H., Bresson, J.L., Dalmay, T., Dewhurst, I.C., Epstein, M.M., Rostoks, N., 2020. Applicability of the EFSA Opinion on site-directed nucleases type 3 for the safety assessment of plants developed using site-directed nucleases type 1 and 2 and oligonucleotide-directed mutagenesis. EFSA J. 18 (11), e06299.

Elsharawy, H., Refat, M., 2023. CRISPR/Cas9 genome editing in wheat: enhancing quality and productivity for global food security—a review. Funct. Integr. Genom. 23 (3). https://doi.org/10.1007/s10142-023-01190-1.

Faizal, A., Nugroho, S., Awal Sembada, A., Theda, Y., Komariyah, T., Rachmi Esyanti, R., 2024. Genome editing in future crop protection: utilizing CRISPR/Cas9 to improve crop resistance against diseases, pests, and weeds. Discover Agriculture 2 (1). https://doi.org/10.1007/s44279-024-00124-0.

Farooq, M., Frei, M., Zeibig, F., Pantha, S., Özkan, H., Kilian, B., Siddique, K.H.M., Pecchioni, N., 2025. Back into the wild: harnessing the power of wheat wild relatives for future crop and food security. J. Exp. Bot. https://doi.org/10.1093/jxb/eraf141.

Feng, X., 2025. CRISPR/Cas9-mediated editing of TaGW2 to enhance grain size in wheat. Triticeae Genomics and Genetics 16, 15.

Gao, C., 2021. Genome engineering for crop improvement and future agriculture. Cell 184 (6), 1621–1635. https://doi.org/10.1016/j.cell.2021.01.005.

Gentzel, I.N., Ohlson, E.W., Redinbaugh, M.G., Wang, G.L., 2022. VIGE: virus-induced genome editing for improving abiotic and biotic stress traits in plants. Stress Biology 2 (1). https://doi.org/10.1007/s44154-021-00026-x.

Gupta, I., Sharma, J.G., Kaul, T., 2025. Nanoparticle-driven CRISPR-Cas9 genome editing: a new frontier in crop improvement. Mol. Biol. Rep. 52 (1), 1–22.

Hazrati, R., Asghari-Zakaria, R., Zare, N., 2025. Crop genome editing for nutrient use efficiency. In: Omics and Genome Editing: Revolution in Crop Improvement for Sustainable Agriculture. Springer Nature Switzerland, Cham, pp. 87–106.

Hilioti, Z., Antunes, D., Kalaitzis, P., Merkouropoulos, G., 2024. Manipulation of plant architecture for crop production. Front. Plant Sci. 15, 1502833.

Hou, X., Guo, X., Zhang, Y., Zhang, Q., 2023. CRISPR/Cas genome editing system and its application in potato. Front. Genet. 14. https://doi.org/10.3389/fgene.2023.1017388.

Hu, J., Gao, C., 2023. CRISPR-edited plants by grafting. Nat. Biotechnol. 41 (7), 909–910. https://doi.org/10.1038/s41587-022-01516-7.

Hu, M., Liu, D., 2025. Unlocking the potential of genome editing in agriculture with tissue culture-free techniques. New Phytol. 246 (4), 1478–1484. https://doi.org/10.1111/nph.70046.

Huang, X., Hilscher, J., Stoger, E., Paul, C., Zhu, C., 2021. Modification of cereal plant architecture by genome editing to improve yields. Plant Cell Rep. 40 (6), 953–978. https://doi.org/10.1007/s00299-021-02668-7.

Huang, X., Jia, H., Xu, J., Wang, Y., Wen, J., Wang, N., 2023. Transgene-free genome editing of vegetatively propagated and perennial plant species in the T0 generation via a co-editing strategy. Nat. Plants 9 (10), 1591–1597. https://doi.org/10.1038/s41477-023-01520-y.

Hurst, J.P., Sato, S., Ferris, T., Yobi, A., Zhou, Y., Angelovici, R., Clemente, T.E., Holding, D.R., 2024. Editing the 19 kDa alpha-zein gene family generates non-opaque2-based quality protein maize. Plant Biotechnol. J. 22 (4), 946–959. https://doi.org/10.1111/pbi.14237.

Hussen, B.M., Najmadden, Z.B., Abdullah, S.R., Rasul, M.F., Mustafa, S.A., Ghafouri-Fard, S., Taheri, M., 2024. CRISPR/Cas9 gene editing: a novel strategy for fighting drug resistance in respiratory disorders. Cell Commun. Signal. 22 (1). https://doi.org/10.1186/s12964-024-01713-8.

Ikram, M., Rauf, A., Rao, M.J., Maqsood, M.F.K., Bakhsh, M.Z.M., Ullah, M., Tahira, M., 2024a. CRISPR-Cas9-based molecular breeding in crop plants: a review. Mol. Biol. Rep. 51 (1), 227–241.

Ikram, A., Mehmood, H., Arshad, M.T., Rasheed, A., Noreen, S., Gnedeka, K.T., 2024b. Applications of artificial intelligence (AI) in managing food quality and ensuring global food security. CYTA - Journal of Food 22 (1). https://doi.org/10.1080/19476337.2024.2393287.

Jabbar, A., Zulfiqar, F., Mahnoor, M., Mushtaq, N., Zaman, M.H., din, A.S.U., Khan, M.A., Ahmad, H.I., 2021. Advances and perspectives in the application of CRISPR-cas9 in livestock. Mol. Biotechnol. 63 (9), 757–767. https://doi.org/10.1007/s12033-021-00347-2.

Jarray, N., Abbes, A.B., Farah, I.R., 2023. Machine learning for food security: current status, challenges, and future perspectives. Artif. Intell. Rev. 56 (3), 3853–3876.

Jarroudi, M.E., Kouadio, L., Delfosse, P., Bock, C.H., Mahlein, A.K., Fettweis, X., Hamdioui, S., 2024. Leveraging edge artificial intelligence for sustainable agriculture. Nat. Sustain. 7 (7), 846–854.

Javaid, M., Haleem, A., Khan, I.H., Suman, R., 2023. Understanding the potential applications of artificial intelligence in agriculture sector. Advanced Agrochem 2 (1), 15–30. https://doi.org/10.1016/j.aac.2022.10.001.

Jeyaraj, G., Alphonse, V., Jayanthi, P., Neha Angelin, F., Swapna Geetanjali, A., Ganesan, G., 2024. Harnessing the potential of CRISPR/Cas system for enhancing virus resistance in plants: targets, strategies, and challenges. Physiol. Mol. Plant Pathol. 129, 102202. https://doi.org/10.1016/j.pmpp.2023.102202.

Karanth, S., Benefo, E.O., Patra, D., Pradhan, A.K., 2023. Importance of artificial intelligence in evaluating climate change and food safety risk. J. Agric. Food Res. 11. https://doi.org/10.1016/j.jafr.2022.100485.

Karunarathne, S.D., Han, Y., Zhang, X.Q., Li, C., 2022. CRISPR/Cas9 gene editing and natural variation analysis demonstrate the potential for HvARE1 in improvement of nitrogen use efficiency in barley. J. Integr. Plant Biol. 64 (3), 756–770. https://doi.org/10.1111/jipb.13214.

Kaur, M., Manchanda, P., Kalia, A., Ahmed, F.K., Nepovimova, E., Kuca, K., Abd-Elsalam, K.A., 2021. Agroinfiltration-mediated scalable transient gene expression in genome-edited crop plants. Int. J. Mol. Sci. 22 (19), 10882.

Kaur, N., Qadir, M., Francis, D.V., Alok, A., Tiwari, S., Ahmed, Z.F.R., 2025. CRISPR/Cas9: a sustainable technology to enhance climate resilience in major Staple Crops. Frontiers in Genome Editing 7. https://doi.org/10.3389/fgeed.2025.1533197.

Khan, M.H.U., Wang, S., Wang, J., Ahmar, S., Saeed, S., Khan, S.U., Xu, X., Chen, H., Bhat, J.A., Feng, X., 2022. Applications of artificial intelligence in climate-resilient smart-crop breeding. Int. J. Mol. Sci. 23 (19). https://doi.org/10.3390/ijms231911156.

Khan, G., Irshad, G., Ijaz, R., Javaid, S., Tahir, N., Mehmood, S., 2025. Emboldening food security for global sustainability yoking artificial intelligence. Discover Food 5 (2).

Kim, H.S., Kim, D.w., Kim, S., Choe, S., 2022. Biochemical characterization of the two novel mgCas12a proteins from the human gut metagenome. Sci. Rep. 12 (1). https://doi.org/10.1038/s41598-022-25227-w.

Kim, C.Y., Karthik, S., Kim, H., 2025. Soybean molecular breeding through genome editing tools: recent advances and future perspectives. Agronomy 15 (8), 1983.

Kocsisova, Z., Coneva, V., 2023. Strategies for delivery of CRISPR/Cas-mediated genome editing to obtain edited plants directly without transgene integration. Frontiers in Genome Editing 5. https://doi.org/10.3389/fgeed.2023.1209586.

Kořínková, N., Fontana, I.M., Nguyen, T.D., Pouramini, P., Bergougnoux, V., Hensel, G., 2022. Enhancing cereal productivity by genetic modification of root architecture. Biotechnol. J. 17 (7). https://doi.org/10.1002/biot.202100505.

Kumar, U., Dwivedi, D., Das, U., 2025. Advancements in CRISPR-mediated multiplex genome editing: transforming plant breeding for crop improvement and polygenic trait engineering. Biotechnol. J. 20 (11), e70148.

Kumar, D., Yadav, A., Ahmad, R., Dwivedi, U.N., Yadav, K., 2022. CRISPR-based genome editing for nutrient enrichment in crops: a promising approach toward global food security. Front. Genet. 13. https://doi.org/10.3389/fgene.2022.932859.

Kumar, V., Yadav, M., Awala, S.K., Valombola, J.S., Saxena, M.S., Ahmad, F., Saxena, S.C., 2024. Millets: a nutritional powerhouse for ensuring food security. Planta 260 (4). https://doi.org/10.1007/s00425-024-04533-9.

Laforest, L.C., Nadakuduti, S.S., 2022. Advances in delivery mechanisms of CRISPR gene-editing reagents in plants. Frontiers in Genome Editing 4. https://doi.org/10.3389/fgeed.2022.830178.

Lei, L., Pan, W., Shou, X., Shao, Y., Ye, S., Zhang, J., Kolliputi, N., Shi, L., 2024. Nanomaterials-assisted gene editing and synthetic biology for optimizing the treatment of pulmonary diseases. J. Nanobiotechnol. 22 (1). https://doi.org/10.1186/s12951-024-02627-w.

Léonidas, H., Toru, T., Fidèle, N., Watabaji, M.D., Gelete, T.B., Hubert, H., 2024. Monocropping vs. mixed cropping systems under a changing climate: smallholder farmers' perceptions and farm profitability in Eastern Rwanda. Environ. Sustain. Indic. 24, 100527.

Li, Z., Zhong, Z., Wu, Z., Pausch, P., Al-Shayeb, B., Amerasekera, J., Doudna, J.A., Jacobsen, S.E., 2023. Genome editing in plants using the compact editor CasΦ. Proc. Natl. Acad. Sci. 120 (4). https://doi.org/10.1073/pnas.2216822120.

Li, B., Sun, C., Li, J., Gao, C., 2024. Targeted genome-modification tools and their advanced applications in crop breeding. Nat. Rev. Genet. 25 (9), 603–622. https://doi.org/10.1038/s41576-024-00720-2.

Li, X.X., Gao, Y.H., Ma, H.W., Wang, Y.Q., Bu, T., Yin, W., Xia, X., Wang, H.L., 2025a. Non-tissue culture genetic modifications for plant improvement. Plant Mol. Biol. 115 (3). https://doi.org/10.1007/s11103-025-01594-6.

Li, L., Zhang, D., Zhang, Z., Zhang, B., 2025b. CRISPR/Cas: a powerful tool for designing and improving oil crops. Trends Biotechnol. 43 (4), 773–789.

Li, N., Yuan, X., Han, B., Guo, W., Chen, H., 2025c. CRISPR/Cas-mediated optimization of soybean shoot architecture for enhanced yield. Int. J. Mol. Sci. 26 (16), 7925.

Lokya, V., Singh, S., Chaudhary, R., Jangra, A., Tiwari, S., 2025. Emerging trends in transgene-free crop development: insights into genome editing and its regulatory overview. Plant Mol. Biol. 115 (4). https://doi.org/10.1007/s11103-025-01600-x.

Lu, J., Li, S., Deng, S., Wang, M., Wu, Y., Li, M., Dong, J., Lu, S., Su, C., Li, G., Lang, Z., Zhu, J.K., 2024. A method of genetic transformation and gene editing of succulents without tissue culture. Plant Biotechnol. J. 22 (7), 1981–1988. https://doi.org/10.1111/pbi.14318.

Ma, X., Zhang, X., Liu, H., Li, Z., 2020. Highly efficient DNA-free plant genome editing using virally delivered CRISPR–Cas9. Nat. Plants 6 (7), 773–779. https://doi.org/10.1038/s41477-020-0704-5.

Ma, F., Zheng, Q., Sun, Y., Zhang, N., Xu, W., 2025. Toward tissue-free plant engineering: emerging platforms for sustainable horticultural transformation. J. Agric. Food Chem. 73 (42), 26457–26471. https://doi.org/10.1021/acs.jafc.5c06803.

Mamrutha, H.M., Zeenat, W., Kapil, D., Budhagatapalli, N., Tikaniya, D., Rakesh, K., Krishnappa, G., Singh, G., Singh, G.P., 2024. Evidence and opportunities for developing non-transgenic genome edited crops using site-directed nuclease 1 approach. Crit. Rev. Biotechnol. 44 (6), 1140–1150. https://doi.org/10.1080/07388551.2023.2270581.

Mangin, A., Dion, V., Menzies, G., 2024. Developing small Cas9 hybrids using molecular modeling. Sci. Rep. 14 (1), 12482.

Manzoor, S., Nabi, S.U., Rather, T.R., Gani, G., Mir, Z.A., Wani, A.W., Ali, S., Tyagi, A., Manzar, N., 2024. Advancing crop disease resistance through genome editing: a promising approach for enhancing agricultural production. Frontiers in Genome Editing 6. https://doi.org/10.3389/fgeed.2024.1399051.

Mikhaylova, E., 2025. Virus-Induced genome editing (VIGE): one step away from an agricultural revolution. Int. J. Mol. Sci. 26 (10), 4599. https://doi.org/10.3390/ijms26104599.

Mohamed, H.I., Khan, A., Basit, A., 2024. CRISPR-Cas9 system-mediated genome editing technology: an ultimate tool to enhance abiotic stress in crop plants. J. Soil Sci. Plant Nutr. 24 (2), 1799–1822.

Movahedi, A., Aghaei-Dargiri, S., Li, H., Zhuge, Q., Sun, W., 2023. CRISPR variants for gene editing in plants: biosafety risks and future directions. Int. J. Mol. Sci. 24 (22), 16241. https://doi.org/10.3390/ijms242216241.

Mujtaba, M., Wang, D., Carvalho, L.B., Oliveira, J.L., Espirito Santo Pereira, A.D., Sharif, R., Jogaiah, S., Paidi, M.K., Wang, L., Ali, Q., Fraceto, L.F., 2021. Nanocarrier-mediated delivery of miRNA, RNAi, and CRISPR-cas for plant protection: current trends and future directions. ACS Agricultural Science and Technology 1 (5), 417–435. https://doi.org/10.1021/acsagscitech.1c00146.

Munawar, N., Faheem, M., Niamat, A., Munir, A., Habibullah Khan, S., Zahoor, M.K., Aslam, A., Ahmad, A., 2024. Regulatory, ethical, social, and biosafety concerns in genome-edited horticultural crops. In: CRISPRized Horticulture Crops: Genome Modified Plants and Microbes in Food and Agriculture. Elsevier, United Arab Emirates, pp. 421–438. https://doi.org/10.1016/B978-0-443-13229-2.00026-0.

Munaweera, T.I.K., Jayawardana, N.U., Rajaratnam, R., Dissanayake, N., 2022. Modern plant biotechnology as a strategy in addressing climate change and attaining food security. Agric. Food Secur. 11 (1). https://doi.org/10.1186/s40066-022-00369-2.

Namata, M.J., Xu, J., Habyarimana, E., Palakolanu, S.R., Wang, L., Li, J., 2025. Genome editing in maize and sorghum: a comprehensive review of CRISPR/Cas9 and emerging technologies. Plant Genome 18 (2). https://doi.org/10.1002/tpg2.70038.

Naseem, S., Rizwan, M., 2025. The role of artificial intelligence in advancing food safety: a strategic path to zero contamination. Food Control 175, 111292. https://doi.org/10.1016/j.foodcont.2025.111292.

Nath, P.C., Mishra, A.K., Sharma, R., Bhunia, B., Mishra, B., Tiwari, A., Nayak, P.K., Sharma, M., Bhuyan, T., Kaushal, S., Mohanta, Y.K., Sridhar, K., 2024. Recent advances in artificial intelligence towards the sustainable future of agri-food industry. Food Chem. 447. https://doi.org/10.1016/j.foodchem.2024.138945.

Pal, S., Krishna, R., Dedhia, L., Panwar, H.S., Karkute, S.G., Rai, N., Singh, A.K., 2025. CRISPR-mediated gene editing for economically important traits in horticultural crops: progress and prospects. Transgenic Res. 34 (1), 26.

Pandey, D.K., Mishra, R., 2024. Towards sustainable agriculture: harnessing AI for global food security. Artificial Intelligence in Agriculture 12, 72–84. https://doi.org/10.1016/j.aiia.2024.04.003.

Phogat, S., Poudel, A., Bharati, A., Kumar, A., Mandal, P.K., 2024. Recent Advances in CRISPR-Cas for Climate Resilient Agriculture in Cereals. Informa UK Limited, pp. 343–371. https://doi.org/10.1201/9781003455271-14.

Rai, G.K., Khanday, D.M., Kumar, P., Magotra, I., Choudhary, S.M., Kosser, R., Kalunke, R., Giordano, M., Corrado, G., Rouphael, Y., Pandey, S., 2023. Enhancing crop resilience to drought stress through CRISPR-cas9 genome editing. Plants 12 (12). https://doi.org/10.3390/plants12122306.

Rejeb, A., Rejeb, K., Zailani, S., Keogh, J.G., Appolloni, A., 2022. Examining the interplay between artificial intelligence and the agri-food industry. Artificial Intelligence in Agriculture 6, 111–128. https://doi.org/10.1016/j.aiia.2022.08.002.

Rengasamy, B., Manna, M., Thajuddin, N.B., Sathiyabama, M., Sinha, A.K., 2024. Breeding rice for yield improvement through CRISPR/Cas9 genome editing method: current technologies and examples. Physiol. Mol. Biol. Plants 30 (2), 185–198. https://doi.org/10.1007/s12298-024-01423-y.

Richards, C.E., Tzachor, A., Avin, S., Fenner, R., 2023. Rewards, risks and responsible deployment of artificial intelligence in water systems. Nat. Water 1 (5), 422–432. https://doi.org/10.1038/s44221-023-00069-6.

Ristaino, J.B., Anderson, P.K., Bebber, D.P., Brauman, K.A., Cunniffe, N.J., Fedoroff, N.V., Finegold, C., Garrett, K. A., Gilligan, C.A., Jones, C.M., Martin, M.D., MacDonald, G.K., Neenan, P., Records, A., Schmale, D.G., Tateosian, L., Wei, Q., 2021. The persistent threat of emerging plant disease pandemics to global food security. Proc. Natl. Acad. Sci. U. S. A 118 (23). https://doi.org/10.1073/pnas.2022239118.

Rönspies, M., Dorn, A., Schindele, P., Puchta, H., 2021. CRISPR–Cas-mediated chromosome engineering for crop improvement and synthetic biology. Nat. Plants 7 (5), 566–573. https://doi.org/10.1038/s41477-021-00910-4.

Roy, N., Verma, R.K., Chetia, S.K., Sharma, V., Sen, P., Modi, M.K., 2023. Molecular mapping of drought-responsive QTLs during the reproductive stage of rice using a GBS (genotyping-by-sequencing) based SNP linkage map. Mol. Biol. Rep. 50 (1), 65–76. https://doi.org/10.1007/s11033-022-08002-y.

Roy, N., Debnath, P., Gaur, H.S., 2025. Adoption of multi-omics approaches to address drought stress tolerance in rice and mitigation strategies for sustainable production. Mol. Biotechnol. 1–13.

Rudrakar, S., Rughani, P., 2024. IoT based agriculture (Ag-IoT): a detailed study on architecture, security and forensics. Inf. Process. Agric. 11 (4), 524–541. https://doi.org/10.1016/j.inpa.2023.09.002.

Rugji, J., Erol, Z., Taşçı, F., Musa, L., Hamadani, A., Gündemir, M.G., Karalliu, E., Siddiqui, S.A., 2025. Utilization of AI–reshaping the future of food safety, agriculture and food security–a critical review. Crit. Rev. Food Sci. Nutr. 65 (26), 5136–5180. https://doi.org/10.1080/10408398.2024.2430749.

Saini, H., Devrani, A., Synrem, G., Priyanka, K.F., Salem, M., 2025. Application of CRISPR technology in plant improvement: an update review. Advances in Agriculture 2025 (1). https://doi.org/10.1155/aia/4578877.

Salgotra, R.K., Chauhan, B.S., 2023. Genetic diversity, conservation, and utilization of plant genetic resources. Genes 14 (1). https://doi.org/10.3390/genes14010174.

Sandhya, D., Jogam, P., Allini, V.R., Abbagani, S., Alok, A., 2020. The present and potential future methods for delivering CRISPR/Cas9 components in plants. J. Genet. Eng. Biotechnol. 18 (1). https://doi.org/10.1186/s43141-020-00036-8.

Sathee, L., Jagadhesan, B., Pandesha, P.H., Barman, D., Adavi B, S., Nagar, S., Krishna, G.K., Tripathi, S., Jha, S.K., Chinnusamy, V., 2022. Genome editing targets for improving nutrient use efficiency and nutrient stress adaptation. Front. Genet. 13. https://doi.org/10.3389/fgene.2022.900897.

Shaheen, N., Ahmad, S., Alghamdi, S.S., Rehman, H.M., Javed, M.A., Tabassum, J., Shao, G., 2023. CRISPR-cas system, a possible "savior" of rice threatened by climate change: an updated review. Rice 16 (1). https://doi.org/10.1186/s12284-023-00652-1.

Shao, D., Marwa, N., 2025. Examining the role of artificial intelligence in post-harvest management for enhanced food security: a critical review. Sustain. Futures 10, 101286. https://doi.org/10.1016/j.sftr.2025.101286.

Sharma, S.K., Gupta, O.P., Pathaw, N., Sharma, D., Maibam, A., Sharma, P., Sanasam, J., Karkute, S.G., Kumar, S., Bhattacharjee, B., 2021. CRISPR-Cas-Led revolution in diagnosis and management of emerging plant viruses: new avenues toward food and nutritional security. Front. Nutr. 8. https://doi.org/10.3389/fnut.2021.751512.

Sharma, U., Nisha, Ray, A., 2025. An insight into the journey of CRISPR-Cas9 and its application in crop improvement. Discover Plants 2 (1), 266.

Sharma, P., Pandey, A., Malviya, R., Dey, S., Karmakar, S., Gayen, D., 2023. Genome editing for improving nutritional quality, post-harvest shelf life and stress tolerance of fruits, vegetables, and ornamentals. Frontiers in Genome Editing 5. https://doi.org/10.3389/fgeed.2023.1094965.

Shelake, R.M., Kadam, U.S., Kumar, R., Pramanik, D., Singh, A.K., Kim, J.Y., 2022. Engineering drought and salinity tolerance traits in crops through CRISPR-mediated genome editing: targets, tools, challenges, and perspectives. Plant Commun. 3 (6). https://doi.org/10.1016/j.xplc.2022.100417.

Shende, H., Patil, R.H., 2025. Harnessing nanotechnology for enhanced CRISPR-Cas plant genome editing. In: Nanotechnology in Agriculture: Pioneering Progress and Challenges. Springer Nature, pp. 265–296.

Shraddha, Bhardwaj, R.K., Shukla, Y.R., Akshay, D.A., Vashishat, R.K., 2024. Underexploited tropical and subtropical vegetable crops for diversification and nutritional security: a review. Agroecology and Sustainable Food Systems 48 (8), 1094–1114. https://doi.org/10.1080/21683565.2024.2348638.

Simoni, S., Fambrini, M., Pugliesi, C., Rogo, U., 2025. Genome editing by grafting. Int. J. Mol. Sci. 26 (19), 9294. https://doi.org/10.3390/ijms26199294.

Singh, P., Ali, S.A., 2021. Impact of CRISPR-cas9-based genome engineering in farm animals. Vet. Sci. 8 (7). https://doi.org/10.3390/vetsci8070122.

Singh, D., Chaudhary, P., Taunk, J., Singh, C.K., Chinnusamy, V., Sevanthi, A.M., Singh, V.J., Pal, M., 2024. Targeting Induced Local Lesions in Genomes (TILLING): advances and opportunities for fast tracking crop breeding. Crit. Rev. Biotechnol. 44 (5), 817–836. https://doi.org/10.1080/07388551.2023.2231630.

Sood, A., Sharma, P., Sharma, A., Chaudhary, D.R., 2025. Manisha, Genome editing in vegetable crops: a new era of sustainable agriculture. Mol. Biol. Rep. 52 (1). https://doi.org/10.1007/s11033-025-11063-4.

Spencer, K.P., Burger, J.T., Campa, M., 2023. CRISPR-based resistance to grapevine virus A. Front. Plant Sci. 14, 1296251.

Szillat, K.P., Koethe, S., Wernike, K., Höper, D., Beer, M., 2020. A CRISPR/Cas9-generated bovine CD46-knockout cell line: a tool to elucidate the adaptability of bovine viral diarrhea viruses (BVDV). Viruses 12 (8), 859.

Tamilselvan-Nattar-Amutha, S., Hiekel, S., Hartmann, F., Lorenz, J., Dabhi, R.V., Dreissig, S., Hensel, G., Kumlehn, J., Heckmann, S., 2023. Barley stripe mosaic virus-mediated somatic and heritable gene editing in barley (Hordeum vulgare L.). Front. Plant Sci. 14. https://doi.org/10.3389/fpls.2023.1201446.

Tang, D., Jia, Y., Zhang, J., Li, H., Lin, C., Wang, P., Bao, Z., Liu, Z., Feng, S., Zhu, X., Li, D., Zhu, G., Wang, H., Zhou, Y., Zhou, Y., Bryan, G.J., Robin Buell, C., Zhang, C., Huang, S., 2022. Genome evolution and diversity of wild and cultivated potatoes. Nature 606 (7914), 535–541. https://doi.org/10.1038/s41586-022-04822-x.

Tang, Q., Wang, X., Jin, X., Peng, J., Zhang, H., Wang, Y., 2023. CRISPR/Cas technology revolutionizes crop breeding. Plants 12 (17), 3119. https://doi.org/10.3390/plants12173119.

Tariq, A., Mushtaq, M., Yaqoob, H., Bhat, B.A., Zargar, S.M., Raza, A., Ali, S., Charagh, S., Mubarik, M.S., Zaman, Q. U., Prasad, P.V., Mir, R.A., 2023. Putting CRISPR-Cas system in action: a golden window for efficient and precise genome editing for crop improvement. GM Crops Food 14 (1), 1–27. https://doi.org/10.1080/21645698.2023.2219111.

Thingujam, D., Gouli, S., Cooray, S.P., Chandran, K.B., Givens, S.B., Gandhimeyyan, R.V., Tan, Z., Wang, Y., Patam, K., Greer, S.A., Acharya, R., Moseley, D.O., Osman, N., Zhang, X., Brooker, M.E., Tagert, M.L., Schafer, M.J., Jeong, C., Hoffseth, K.F., Bheemanahalli, R., Wyss, J.M., Wijewardane, N.K., Ham, J.H., Mukhtar, M.S., 2025. Climate-resilient crops: integrating AI, multi-omics, and advanced phenotyping to address global agricultural and societal challenges. Plants 14 (17). https://doi.org/10.3390/plants14172699.

Tripathi, J.N., Ntui, V.O., Ron, M., Muiruri, S.K., Britt, A., Tripathi, L., 2019. CRISPR/Cas9 editing of endogenous banana streak virus in the B genome of Musa spp. overcomes a major challenge in banana breeding. Commun. Biol. 2 (1). https://doi.org/10.1038/s42003-019-0288-7.

Tuncel, A., Pan, C., Sprink, T., Wilhelm, R., Barrangou, R., Li, L., Shih, P.M., Varshney, R.K., Tripathi, L., Van Eck, J., Mandadi, K., Qi, Y., 2023. Genome-edited foods. Nat. Rev. Bioeng. 1 (11), 799–816. https://doi.org/10.1038/s44222-023-00115-8.

Tuncel, A., Pan, C., Clem, J.S., Liu, D., Qi, Y., 2025. CRISPR–Cas applications in agriculture and plant research. Nat. Rev. Mol. Cell Biol. 26 (6), 419–441. https://doi.org/10.1038/s41580-025-00834-3.

Tyczewska, A., Twardowski, T., Woźniak-Gientka, E., 2023. Agricultural biotechnology for sustainable food security. Trends Biotechnol. 41 (3), 331–341.

Tzachor, A., Devare, M., King, B., Avin, S., hÉigeartaigh, S.Ó., 2022. Responsible artificial intelligence in agriculture requires systemic understanding of risks and externalities. Nat. Mach. Intell. 4 (2), 104–109. https://doi.org/10.1038/s42256-022-00440-4.

Ukhatova, Y.V., Erastenkova, M.V., Korshikova, E.S., Krylova, E.A., Mikhailova, A.S., Semilet, T.V., Tikhonova, N. G., Shvachko, N.A., Khlestkina, E.K., 2023. Improvement of crops using the CRISPR/cas system: new target genes. Mol. Biol. 57 (3), 375–397. https://doi.org/10.1134/S0026893323030135.

Uranga, M., Daròs, J.A., 2023. Tools and targets: the dual role of plant viruses in CRISPR–Cas genome editing. Plant Genome 16 (2). https://doi.org/10.1002/tpg2.20220.

Uranga, M., Vazquez-Vilar, M., Orzáez, D., Daròs, J.A., 2021. CRISPR-Cas12a genome editing at the whole-plant level using two compatible RNA virus vectors. CRISPR J. 4 (5), 761–769. https://doi.org/10.1089/crispr.2021.0049.

Usman, B., Zhao, N., Nawaz, G., Qin, B., Liu, F., Liu, Y., Li, R., 2021. CRISPR/Cas9-guided mutagenesis of grain size 3 confers increased rice (Oryza sativa L.) grain length by regulating cysteine proteinase inhibitor and ubiquitin-related proteins. Int. J. Mol. Sci. 22 (6), 3225.

Varshney, R.K., Bohra, A., Yu, J., Graner, A., Zhang, Q., Sorrells, M.E., 2021. Designing future crops: genomics-assisted breeding comes of age. Trends Plant Sci. 26 (6), 631–649. https://doi.org/10.1016/j.tplants.2021.03.010.

Vats, S., Kumawat, S., Brar, J., Kaur, S., Yadav, K., Magar, S.G., Jadhav, P.V., Salvi, P., Sonah, H., Sharma, S., Deshmukh, R., 2022. Opportunity and challenges for nanotechnology application for genome editing in plants. Plant Nano Biology 1. https://doi.org/10.1016/j.plana.2022.100001.

Wang, J.Y., Doudna, J.A., 2023. CRISPR technology: a decade of genome editing is only the beginning. Science 379 (6629). https://doi.org/10.1126/science.add8643.

Wu, J., Zhang, Z.S., Xia, J.Q., Alfatih, A., Song, Y., Huang, Y.J., Wan, G.Y., Sun, L.Q., Tang, H., Liu, Y., Wang, S.M., Zhu, Q.S., Qin, P., Wang, Y.P., Li, S.G., Mao, C.Z., Zhang, G.Q., Chu, C., Yu, L.H., Xiang, C.B., 2021. Rice NIN-LIKE PROTEIN 4 plays a pivotal role in nitrogen use efficiency. Plant Biotechnol. J. 19 (3), 448–461. https://doi.org/10.1111/pbi.13475.

Xu, Y., Li, R., Luo, H., Wang, Z., Li, M.W., Lam, H.M., Huang, C., 2022. Protoplasts: small cells with big roles in plant biology. Trends Plant Sci. 27 (8), 828–829. https://doi.org/10.1016/j.tplants.2022.03.010.

Yang, L., Machin, F., Wang, S., Saplaoura, E., Kragler, F., 2023a. Heritable transgene-free genome editing in plants by grafting of wild-type shoots to transgenic donor rootstocks. Nat. Biotechnol. 41 (7), 958–967.

Yang, T., Ali, M., Lin, L., Li, P., He, H., Zhu, Q., Deng, L., 2023b. Recoloring tomato fruit by CRISPR/Cas9-mediated multiplex gene editing. Hortic. Res. 10 (1), uhac214.

Yang, Q., Ral, J.P., Wei, Y., Zheng, Y., Li, Z., Jiang, Q., 2024. Genome editing of five starch synthesis genes produces highly resistant starch and dietary fibre in barley grains. Plant Biotechnol. J. 22 (7), 2051–2053. https://doi.org/10.1111/pbi.14324.

Yau, Y.Y., Easterling, M., Kumar, A., 2024. Advances in delivery of CRISPR–Cas reagents for precise genome editing in plants. In: Gene Editing in Plants: CRISPR-Cas and its Applications. Springer Nature Singapore, Singapore, pp. 543–569.

Yin, K., Tsai, C.J., 2025. Turbo-charging crop improvement: harnessing multiplex editing for polygenic trait engineering and beyond. Plant J. 124 (1), e70527.

Yu, H., Lin, T., Meng, X., Du, H., Zhang, J., Liu, G., Chen, M., Jing, Y., Kou, L., Li, X., Gao, Q., Liang, Y., Liu, X., Fan, Z., Liang, Y., Cheng, Z., Chen, M., Tian, Z., Wang, Y., Chu, C., Zuo, J., Wan, J., Qian, Q., Han, B., Zuccolo, A., Wing, R.A., Gao, C., Liang, C., Li, J., 2021. A route to de novo domestication of wild allotetraploid rice. Cell 184 (5), 1156. https://doi.org/10.1016/j.cell.2021.01.013.

Yu, Z., Yunusbaev, U., Fritz, A., Tilley, M., Akhunova, A., Trick, H., Akhunov, E., 2024. CRISPR -based editing of the ω- and γ-gliadin gene clusters reduces wheat immunoreactivity without affecting grain protein quality. Plant Biotechnol. J. 22 (4), 892–903. https://doi.org/10.1111/pbi.14231.

Yuan, Y., Ton, B.L., Thomas, W.J.W., Batley, J., Edwards, D., 2023. Supporting crop plant resilience during climate change. Crop Sci. 63 (4), 1816–1828. https://doi.org/10.1002/csc2.21019.

Zaidi, S.S.e.A., Mahas, A., Vanderschuren, H., Mahfouz, M.M., 2020. Engineering crops of the future: CRISPR approaches to develop climate-resilient and disease-resistant plants. Genome Biol. 21 (1). https://doi.org/10.1186/s13059-020-02204-y.

Zaman, Q.U., Raza, A., Lozano-Juste, J., Chao, L., Jones, M.G.K., Wang, H.F., Varshney, R.K., 2024. Engineering plants using diverse CRISPR-associated proteins and deregulation of genome-edited crops. Trends Biotechnol. 42 (5), 560–574. https://doi.org/10.1016/j.tibtech.2023.10.007.

Zeinali, M., Sabzehzari, M., Javid, S., Poczai, P., 2025. Crop wild relatives for breeding climate-smart and highly productive cultivars. In: Plant Breeding 2050: Next-Gen Crops. Springer Nature, Singapore, pp. 137–151.

Zhai, R., Zheng, C., Yang, Z., Li, T., Chen, J., Shen, X., 2022. Contribution of CRISPRable DNA to human complex traits. Commun. Biol. 5 (1). https://doi.org/10.1038/s42003-022-03969-7.

Zhan, X., Zhang, F., Li, N., Xu, K., Wang, X., Gao, S., Yin, Y., Yuan, W., Chen, W., Ren, Z., Yao, M., Wang, F., 2024. CRISPR/Cas: an emerging toolbox for engineering virus resistance in plants. Plants 13 (23), 3313. https://doi.org/10.3390/plants13233313.

Zhang, Y., Zhou, P., Bozorov, T.A., Zhang, D., 2021a. Application of CRISPR/Cas9 technology in wild apple (Malus sieverii) for paired sites gene editing. Plant Methods 17 (1). https://doi.org/10.1186/s13007-021-00769-8.

Zhang, J., Zhang, H., Li, S., Li, J., Yan, L., Xia, L., 2021b. Increasing yield potential through manipulation of an ARE1 ortholog related to nitrogen use efficiency in wheat by CRISPR/Cas9. J. Integr. Plant Biol. 63 (9), 1649–1663.

Zhang, X., Ibrahim, Z., Khaskheli, M.B., Raza, H., Zhou, F., Shamsi, I.H., 2024a. Integrative approaches to abiotic stress management in crops: combining bioinformatics educational tools and artificial intelligence applications. Sustainability 16 (17), 7651.

Zhang, D., Li, Y.Y., Zhao, X., Zhang, C., Liu, D.K., Lan, S., Liu, Z.J., 2024b. Molecular insights into self-incompatibility systems: from evolution to breeding. Plant Commun 5 (2), 100780.

Zhang, Z.S., Xia, J.Q., Alfatih, A., Song, Y., Huang, Y.J., Sun, L.Q., Wan, G.Y., Wang, S.M., Wang, Y.P., Hu, B.H., Zhang, G.H., Qin, P., Li, S.G., Yu, L.H., Wu, J., Xiang, C.B., 2022. Rice NIN-LIKE PROTEIN 3 modulates nitrogen use efficiency and grain yield under nitrate-sufficient conditions. Plant Cell Environ. 45 (5), 1520–1536. https://doi.org/10.1111/pce.14294.

Zheng, S., Ye, C., Lu, J., Liufu, J., Lin, L., Dong, Z., Li, J., Zhuang, C., 2021. Improving the rice photosynthetic efficiency and yield by editing oshxk1 via crispr/cas9 system. Int. J. Mol. Sci. 22 (17). https://doi.org/10.3390/ijms22179554.

Zheng, X., Kuijer, H.N., Al-Babili, S., 2021b. Carotenoid biofortification of crops in the CRISPR era. Trends Biotechnol. 39 (9), 857–860.

Zheng, J., Liu, S., Jiang, W., Li, F., Chi, M., Cheng, S., Liu, Y., 2024. CRISPR/Cas9-mediated mutation of Mstn confers growth performance in Culter alburnus juveniles. Aquaculture and Fisheries 9 (6), 900–907. https://doi.org/10.1016/j.aaf.2023.04.008.

Zhou, S., Kalds, P., Qi, L., Sun, K., Zhao, X., Gao, Y., Cai, B., Huang, S., Kou, Q., Petersen, B., Chen, Y., Ma, B., Wang, X., 2022. Optimized Cas9:sgRNA delivery efficiently generates biallelic MSTN knockout sheep without affecting meat quality. BMC Genom. 23 (1). https://doi.org/10.1186/s12864-022-08594-6.

Zhu, H., Li, C., Gao, C., 2020. Applications of CRISPR–Cas in agriculture and plant biotechnology. Nat. Rev. Mol. Cell Biol. 21 (11), 661–677.

Zsögön, A., Rodríguez Gargantini, L., Marfil, C.F., 2025. The power of de novo domestication of wild Solanaceae for a changing climate. In: Plant Breeding 2050: Next-Gen Crops. Springer Nature Singapore, Singapore, pp. 153–171.

CHAPTER

18

Intelligent gene editing for stress tolerance in plants

Jyotsana Tilgam[1], *Shbana Begam*[2], *Deepanshu Jayaswal*[3], *Kuldip Jayaswall*[3], *Joy Das*[4], *and Rakesh Kumar*[4]

[1]ICAR-National Bureau of Agriculturally Important Micro-organisms, Maunath Bhanjan, Uttar Pradesh, India; [2]ICAR-National Institute for Plant Biotechnology, New Delhi, Delhi, India; [3]ICAR-National Institute of Seed Science & Technology, Maunath Bhanjan, Uttar Pradesh, India; [4]ICAR-Central Institute for Cotton Research, Nagpur, Maharashtra, India

18.1 Introduction

Agriculture forms the foundation of human survival. However, challenges such as climate change, a rapidly growing population and shrinking agricultural land are posing serious threats to food security. Wild germplasm holds immense genetic diversity, offering valuable alleles that could be harnessed to strengthen existing cultivars against both biotic and abiotic stresses (Bohra et al., 2022). Yet, traditional breeding approaches often face limitations like crossing incompatibility, which restricts the direct use of wild germplasm. In addition, traditional breeding methods may unintentionally miss other important genes or QTLs that are vital for stress tolerance. In recent years, translational research and modern technologies have opened new avenues for crop improvement. Among these, molecular breeding has emerged as a powerful tool, enabling precise trait selection and integration of desirable genes into elite cultivars (Hasan et al., 2021). This approach has already shown remarkable success in improving key crops such as rice, maize and millets, offering hope for developing varieties that are more productive, resilient and sustainable in the face of global challenges. Molecular breeding, while powerful, comes with its own set of challenges such as high initial costs, the need for well-equipped laboratories and trained personnel (Sinha et al., 2023). Following this, the transgenic approach has shown great

Intelligent Gene Editing in Plants
https://doi.org/10.1016/B978-0-443-44079-3.00025-5

success, with several improved crops already commercialized. However, public concerns, fear and ongoing controversies have led to bans on transgenic products in many countries, limiting its widespread adoption. In this context, genome editing has emerged as the most promising alternative. This technology allows precise targeting of specific genes to introduce desirable traits in crops (Avci and Sipahi, 2023). Editing monogenic traits has proven relatively simple and highly effective. On the other hand, improving complex traits controlled by polygenic traits remains a challenge, as altering them without disturbing other important crop features is much more difficult (Venezia and Krainer, 2021). Genome sequencing and accurate annotation to identify the genes or QTLs responsible for important traits are often expensive, especially in crops with large and complex genomes (polyploidy), such as wheat, onion and sugarcane. Adding to this challenge, transposons can affect the stability of desirable traits, making crop improvement even more difficult. The continuous research efforts in genome editing technologies are steadily advancing, with each new version overcoming the shortcomings of the previous ones. Looking ahead, smarter approaches like artificial intelligence (AI) and machine learning (ML) hold great promise. These emerging technologies can help address many of the current challenges in crop improvement and should be prioritized for future research (Farooq et al., 2024). Datasets generated from various crops using both classical and modern biological techniques can be effectively analyzed with the help of AI and machine learning tools. However, the large and complex datasets produced by transcriptomic, proteomic and metabolomic studies are difficult and time-consuming to process using traditional software and methods (Yetgin, 2025). Extracting accurate and meaningful information from these large datasets remains a significant challenge. Additionally, many advanced software tools are often complex and require highly skilled operators, making them less accessible for routine use. Machine learning helps uncover patterns in gene expression that might be missed by traditional analysis. The cis-regulatory elements that turn genes on or off under stress can be predicted by AI models trained on promoter sequences. These techniques can identify important genetic variants associated with stress tolerance when paired with phenotypic data, such as yields affected by drought. Instead of testing thousands of genes randomly, AI prioritizes the most promising candidates for precise targeting, saving time and resources (Gamal et al., 2025). It can even simulate how gene modifications affect other pathways, helping avoid unintended consequences. Together, AI and ML are guiding smarter, faster, and more efficient strategies for developing stress-resilient crops. Artificial intelligence (AI) is transforming crop breeding by integrating multi-trait analysis with genome editing to rapidly develop stress-tolerant varieties. Machine learning tools like deep learning and random forests analyze large phenotypic and genomic datasets to identify key stress-related genes, improving CRISPR/Cas9 target design and minimizing off-target effects. AI also models gene interactions to optimize multiple traits such as yield, drought, and salinity tolerance simultaneously (Liu et al., 2020). Combined with high-throughput phenotyping and imaging (thermal, hyperspectral), this creates a closed-loop system that quickly validates gene edits under stress. Recent advances in multiplex CRISPR enable simultaneous editing of several genes in stress-response networks (e.g., ABA pathway). This integrated AI–CRISPR–phenotyping framework accelerates breeding cycles, enhances precision, and supports climate-resilient agriculture.

18.2 Concept of intelligent gene editing

The main goal of intelligent gene editing is to make targeted and desirable modifications to the genome with greater precision, efficiency and accuracy by integrating genome editing technologies like CRISPR with the power of AI and ML. Editing the genome is a complex task because of its large size and complex structure, which includes introns, repetitive sequences and codon redundancy. These features often create obstacles and increase the risk of off-target effects, undesirable edits that can lead to harmful mutations. Traditional gene editing methods can struggle to pinpoint the exact editing sites and may unintentionally affect similar sequences elsewhere in the genome (Chehelgerdi et al., 2024). This is where AI becomes a powerful tool. It can analyze vast genomic datasets quickly and identify the most suitable and unique target locations for editing. Since the genome may contain multiple regions with similar sequences, AI can scan the entire genome in a short time to detect these areas, helping researchers avoid unintentional edits and reducing the risk of failure (Sun et al., 2020; Abbaszadeh and Shahlai, 2025). An easy way to understand this is to imagine delivering a parcel to a specific house on a street full of similar-looking houses. In this analogy, CRISPR is the delivery tool, DNA is the street, the target gene is the correct house and an off-target edit is like delivering the parcel to the wrong house. Here, AI acts like a GPS, using its past learning to recognize subtle differences between houses (genes) and guide the delivery to the right house. In gene editing, the guide RNA (gRNA) is like the GPS signal that directs the CRISPR–Cas9 enzyme to the correct location in the genome. If the gRNA isn't designed properly, CRISPR might cut the wrong part of the DNA or miss the target entirely (Manghwar et al., 2020). This is where machine learning becomes crucial. ML models can be trained using large datasets from previous CRISPR experiments to learn which gRNA designs are more effective, which sequences are easier to edit and which approaches are less likely to cause failure (Konstantakos et al., 2022). These models can analyze various factors such as DNA structure, GC content, off-target risks and expected outcomes and then suggest optimized gRNA designs (Kim et al., 2025). Beyond gRNA design, AI and ML can also be used to read and compare gene sequences before and after editing, detect any mistakes and recommend better alternatives. As these systems are exposed to more data, their predictions become more accurate, helping researchers avoid high-risk edits and focus on strategies that are likely to succeed. Traditional gene editing methods are often time-consuming, expensive and prone to trial and error. With AI simulations, researchers can test potential edits virtually, quickly identifying the most promising ones and saving months of experimental work.

18.2.1 Using AI/ML to design efficient gRNAs for CRISPR/Cas systems

Quality of the guide RNA (gRNA) plays a significant role in CRISPR-based genome editing. The wrong selection of gRNA increases the chance of undesirable changes in the genome. The successful desirable change in the genome depends on the high-activity guide RNA, i.e., its ability to carry the Cas proteins to the specific target efficiently. However, the major intricacy lies in the complex relationship of sequence-activity in the gRNAs (Doench et al., 2016). The available and routinely used conventional analytical methods are unable to analyze large datasets containing complex sequences and signals, elevating the role of advanced computational tools. AI and ML have the promising capacity to mine large experimental datasets (position-specific effects, sequence motifs), which have measured gRNA activities and validated off-target events (Sherkatghanad et al., 2023). The traditional ML considers local sequence features, GC content, nucleotide positions, etc., whereas deep learning with several predictors learns informative patterns directly from unrefined sequence data. Whereas, hybrid models with more accuracy integrate sequence-based architectures with additional genomic context such as chromatin accessibility, DNA methylation, and nucleotide composition. In the deep learning models, algorithms are designed with the ability to learn complex and nonlinear relationships from raw data (Kyriazos and Poga, 2024). To do so, the deep learning models are trained on datasets that include gRNA sequences along with their experimentally measured activities, allowing them to learn the patterns that determine gRNA efficiency. The AI-based models usually consider multiple signals while designing gRNA, such as base composition, positional preferences, motifs, different Cas enzymes and their PAM, mismatched sites, chromatin structure, and epigenetic modifications (Lee, 2023; Dixit et al., 2024; Khammampalli and Vindal, 2025). Several researchers have reported AI-based approaches for designing specific and precise gRNAs tailored to different Cas variants (Kim et al., 2020; Feng et al., 2021; Baisya et al., 2022). The deep learning architecture, like convolutional neural networks (CNNs) and recurrent neural networks (RNNs), to predict the gRNA activities and their sequence-activity relationships had been applied to study the CRISPR–Cas systems (Chen et al., 2018; Yin et al., 2019; Ding et al., 2023). The potential of machine learning models was evaluated by analyzing the on-target activities due to the specificity of the designed optimal gRNA. The widespread applications of deep learning models across diverse areas of CRISPR technology highlight their growing significance. However, there are a few things that need to be taken care of while using AI and ML to design efficient gRNA. There are different models that have been designed and fed with different datasets and probabilistic scores and have different regulatory compliance. Additionally, current machine learning models for gene editing can sometimes be affected by biases in their training datasets, which can limit their broader applicability. Therefore, their practical consideration is highly important. There should be a focus on developing improved models and exploring the application of machine learning in other areas of gene editing. AI and ML are the powerful tools that save time and effort, but small validation or experiments must be done before the large-scale experiment is totally dependent on the AI, especially to avoid risk in the biological systems.

18.2.2 Prediction of off-target effects and editing efficiency

One of the key challenges in CRISPR-based gene editing is ensuring high precision. While CRISPR–Cas systems are celebrated for their accuracy, unintended cleavage at genomic sites resembling the target sequence, known as off-target effects that can compromise both the safety and effectiveness of edited plants. For agricultural applications, where gene edits directly influence traits such as stress tolerance, minimizing off-targets is essential to avoid detrimental phenotypes or regulatory complications (Jiang et al., 2025). Factors influencing off-target activity include sequence similarity between guide RNAs (gRNAs) and non-target sites, chromatin accessibility, DNA methylation status, and the secondary structure of gRNAs. Similarly, editing efficiency is shaped by target sequence composition (e.g., GC content, presence of protospacer adjacent motifs or PAMs), Cas nuclease variant, and delivery method used in plant systems.

To address these challenges, computational tools powered by ML, have emerged as indispensable (Table 18.1). AI models trained on experimentally validated datasets can predict gRNA binding affinity, cleavage likelihood, and editing outcomes. Examples include DeepCRISPR, CRISPR-P 2.0, and Elevated CRISPR Learner, which integrate genomic sequence features with epigenetic and structural parameters (Chuai et al., 2018). These tools not only rank potential gRNAs for efficiency but also highlight probable off-target regions. In the plant context, genome complexity poses unique difficulties. Polyploid crops (e.g., wheat, sugarcane, potato) harbor multiple homologous sequences, raising the risk of unintended edits. AI-based models are increasingly adapted to incorporate plant-specific genomic features. By integrating plant pangenomes, stress-responsive transcriptional profiles, and chromatin landscapes, predictions become context-aware, enabling more reliable editing decisions. An emerging frontier is multi-objective optimization, where AI simultaneously maximizes editing efficiency while minimizing off-targets. Coupled with in silico simulations and validation pipelines, researchers can prioritize editing targets with greater confidence before moving into wet-lab experiments (Lambrinidis and Tsantili-Kakoulidou, 2021). Prediction of off-target effects and editing efficiency has become central to intelligent gene editing. By combining CRISPR design algorithms with AI-driven prediction models, plant scientists can engineer stress-tolerant crops with unprecedented precision. This convergence reduces experimental costs, accelerates trait development, and ensures safer adoption of gene-edited plants in agriculture.

TABLE 18.1 Applications of artificial intelligence and machine learning in CRISPR–Cas technology.

References	AI/ML model used	Target	Description
Xue et al. (2018)	DeepCas9 (1D CNN)	On-target	First Deep learning technique to recognize CRISPR–Cas9 sgRNA activity directly from genomic sequences without feature input information
Lin and Wong (2018)	CNN, Random Forest, GBTs and LR	Off-target	Use of deep CNN for accurately predicting off target mutations in CRISPR–Cas9 genome editing
Liu et al. (2019)	SeqCrispr (RNN, CNN and transfer learning)	On-target	Deep learning model that integrates gene network feature specific to a asigned context into the model
Wang and Zhang (2019)	DeepHF (RNN)	On-target	This model extract features using a Bi-LSTM and combine it with biological features.
Aktas et al. (2019)	CNN, MLP	On-target &Off-target	Deep Learning used to Estimate the sgRNA target for CRISPR–Cas9
Liu et al. (2020)	LSTM and CNN	Off-target	Predict the off-target tendency of sgRNA at specific DNA fragments
Kim et al. (2019)	SVM and Random Forest	On-target & Off-target	

18.2.3 Integrating phenotypic data with editing targets

While genomic information identifies candidate genes, true biological value emerges when these targets are linked to observable plant traits. For stress tolerance, phenotypic data such as drought survival rates, leaf wilting indices, root architecture, photosynthetic efficiency, or ion accumulation under salinity, provides crucial context. Integrating such phenotypic information with gene editing strategies ensures that editing targets are not only biologically relevant but also agriculturally meaningful. Traditionally, gene selection relied heavily on transcriptomic and genomic datasets, often overlooking the direct relationship with plant performance under stress. However, phenotype-to-genotype integration offers a more holistic approach. High-throughput phenotyping platforms, including drone-based imaging, multispectral cameras, chlorophyll fluorescence sensors, and automated growth chambers, now generate rich datasets describing plant stress responses in real time. These datasets, when combined with genomic information, enable researchers to identify key genes underpinning complex stress tolerance traits.

Artificial intelligence plays a pivotal role in this integration. ML algorithms such as random forests, support vector machines, and deep learning networks can correlate phenotypic patterns with specific genomic variants or gene expression modules. For example, by linking drought-related root architecture traits with genomic loci, AI can highlight potential CRISPR targets in transcription factors regulating root development. Similarly, combining salt-stress phenotypes with metabolomic data can refine editing targets to transporter genes crucial for ion homeostasis (Dixit et al., 2024). Another powerful approach is genome-to-phenome modeling, where integrated datasets guide iterative cycles of hypothesis, editing, and validation. AI-driven pipelines not only prioritize targets but also simulate the likely impact of edits on phenotype. For instance, predictive models can estimate how editing a heat shock protein gene might alter plant growth under high temperature, before committing resources to experimental trials. Integrating phenotypic data also enhances precision in polygenic traits, where stress tolerance results from the combined action of multiple genes. By mapping complex trait networks, researchers can design multiplex CRISPR strategies targeting several loci simultaneously, guided by phenotype-linked evidence.

18.3 AI and machine learning for plant stress response and resilience

Understanding and enhancing plant stress tolerance is inherently complex, as it involves multiple genes, pathways, and environmental interactions. AI and ML provide the computational power to decode these complexities, offering predictive and adaptive frameworks for intelligent gene editing. By establishing a workflow that spans from data acquisition to model-driven genome editing, researchers can accelerate the design of stress-resilient plants (Fig. 18.1).

18.3.1 Step 1: Data acquisition

The foundation lies in collecting diverse datasets.

- Genomic data: reference genomes, pangenomes, and sequence variants

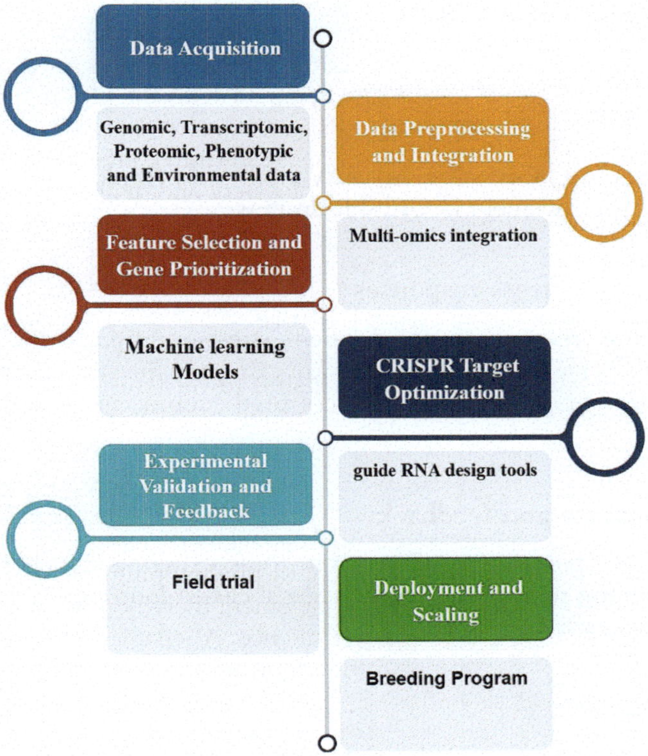

FIGURE 18.1 Workflow of plant stress response and resilience using AI.

- Transcriptomic and proteomic data: gene expression under stress conditions
- Phenotypic data: high-throughput stress assays, imaging, and field performance
  Environmental data: soil conditions, weather, and stress exposure patterns

18.3.2 Step 2: Data preprocessing and integration

Data must be standardized, cleaned, and integrated. Multi-omics integration enables linking gene expression patterns to stress phenotypes. AI algorithms such as data fusion models combine heterogeneous datasets, ensuring that stress responses are contextualized in both genetic and environmental dimensions.

18.3.3 Step 3: Feature selection and gene prioritization

Machine learning methods identify the most relevant genes or pathways:

- Random forests highlight key stress-responsive features
- LASSO regression pinpoints regulatory genes
- Deep learning networks capture non-linear gene–trait interactions This stage outputs prioritized editing targets with high relevance to stress resilience.

18.3.4 Step 4: CRISPR target optimization

AI tools optimize guide RNA design:

- Predicting on-target cleavage efficiency
- Minimizing off-target activity
- Recommending suitable Cas variants (e.g., Cas9, Cas12a). Advanced tools employ reinforcement learning for iterative improvement of CRISPR designs.

18.3.5 Step 5: Predictive modeling of stress responses

AI simulates how edited plants may respond under different stress conditions. For instance, models can forecast drought survival probabilities after knocking out specific negative regulators. Integration with environmental models enhances prediction accuracy for field conditions.

18.3.6 Step 6: Experimental validation and feedback

Edited plants undergo laboratory and field testing. Data from these experiments feed back into AI models, continuously refining predictions. This creates a closed-loop pipeline where machine learning improves with each experimental cycle.

18.3.7 Step 7: Deployment and scaling

Validated AI-assisted editing strategies can be scaled for crop improvement programs. Integration with breeding pipelines ensures that gene-edited lines are combined with elite varieties, accelerating adoption in agriculture.

18.4 Advances in phenotypic data collection and smart analytical tools

High-throughput phenotyping (HTP) has emerged as a transformative approach in modern plant science, offering unprecedented opportunities to dissect plant growth, development, and stress responses under both controlled and field conditions. Central to this progress is the development of non-invasive imaging techniques, which enable dynamic monitoring of plant traits without destructive sampling. These methods allow the visualization of plant tissues and physiological processes at scale, generating highly detailed datasets that support complex biological interpretations of plant performance (Walter et al., 2010; Chen et al., 2014). In stress-resistance research, accurate phenotypic data collection and advanced analytical tools are essential for evaluating crop performance following gene editing or other genetic interventions. Conventional phenotyping methods rely heavily on manual scoring of traits such as leaf wilting, chlorosis, or plant height. While informative, these approaches are time-consuming, prone to observer bias, and often lack reproducibility and precision—particularly when applied to large breeding populations. By contrast, modern platforms integrate advanced sensors with automated imaging systems, enabling fast, objective, and high-resolution acquisition of stress-related phenotypic traits.

A diverse set of imaging modalities has been successfully applied in HTP platforms (Table 18.2). Thermal imaging provides insights into canopy temperature and water status (Oerke et al., 2006); chlorophyll fluorescence reveals photosynthetic efficiency and photoinhibition responses (Chaerle et al., 2006; Zarco-Tejada et al., 2009); digital imaging enables the quantification of morphological and growth traits (Jensen et al., 2007); and spectroscopic imaging captures biochemical signatures related to pigment composition, nutrient status, and stress responses (Montes et al., 2006). More advanced tools such as hyperspectral and multispectral imaging, X-ray computed tomography (X-ray CT), and LiDAR (Light Detection and Ranging) have further expanded the capacity of HTP systems to capture a wide range of non-invasive phenotypes (Zhang et al., 2024). These tools deliver real-time insights into photosynthetic activity, canopy architecture, leaf pigmentation, water status, plant height, leaf area index (LAI), and root system architecture, providing a comprehensive view of plant physiology (Atkinson et al., 2019; Khadka et al., 2020). Crucially, they enable precise characterization of stress responses, such as leaf wilting under drought, growth inhibition under salinity, or discoloration due to nutrient deficiencies, thus generating reliable datasets for evaluating stress-tolerance traits (Su et al., 2025). The power of these imaging technologies lies in their integration with advanced image-processing algorithms. High-resolution computational models can extract traits such as leaf water content, photosynthetic performance, canopy structure, and growth trajectories under diverse environmental conditions. Recognition models further allow the detection and quantification of subtle stress indicators often overlooked by human observation. For instance, computer vision systems can automatically identify drought-induced lesions or salt stress spots on leaves and convert them into quantitative parameters. Similarly, thermal infrared imaging (TIR) has been widely applied to monitor canopy temperature (García-Tejero et al., 2018), offering indirect but reliable estimates of stomatal conductance and transpiration rates, particularly under salinity stress (Sirault et al., 2009). Such automated systems substantially accelerate the identification of stress-tolerant genotypes, reducing both time and resource requirements in breeding programs. Ground-based phenotyping platforms extend HTP from controlled laboratory environments to field conditions, thereby bridging the gap between experimental precision and real-world variability. These platforms enable in situ trait measurements while accounting for environmental fluctuations typical of field trials. By facilitating large-scale data collection across multiple growing seasons, they provide valuable insights into plant responses under realistic and dynamic conditions (Harris et al., 2021). Complementary to these, remote sensing technologies, such as drones, unmanned aerial vehicles (UAVs), and satellites enable large-scale phenotypic assessments across extensive crop fields. They offer rapid, scalable, and non-invasive data collection across varied terrains and environmental conditions, with particular strength in capturing spatial and temporal variations. Remote sensing is especially relevant for large breeding programs that target complex traits such as abiotic stress tolerance (Maes and Steppe, 2019). Together, ground-based and aerial platforms expand the scope of phenotyping, offering scalable solutions that strengthen the genotype–phenotype–environment nexus (Angidi et al., 2025).

Despite these advances, a major challenge of HTP is the enormous volume and complexity of the data it generates. Time-series imaging, hyperspectral profiles, and spatial

TABLE 18.2 High-throughput phenotypic data acquisition methods for plant stress studies.

Technology	Traits captured	Scientific insight	Platform	References
RGB/Visible Light Imaging (VIS: 400–700 nm)	Photosynthetic activity, Plant height, leaf area, canopy coverage, biomass (2D/3D), root architecture.	Morphological and Growth Metrics. Provides the most basic and visible stress symptoms (e.g., growth reduction, leaf color).	Ground-based platforms, benchtop, UAVs (unmanned aerial vehicles).	Lichtenthaler et al. (2005)
Near-Infrared (NIR) (0.7–1.3 μm) & Multispectral Imaging	Normalized Difference Vegetation Index (NDVI), water content, plant structure, pigment content.	Vegetation Vigor & Photosynthetic Health. Detects early signs of stress before they are visible to the human eye, based on changes in chlorophyll and cell structure.	Ground-based, UAVs, Satellites.	Blackford (2010)
Hyperspectral Imaging	Comprehensive spectral signatures across many narrow bands, unique to specific biochemical compounds.	Biochemical & Physiological Status. Allows for the estimation of complex traits like chlorophyll, nitrogen, and protein content, revealing the plant's metabolic response to stress.	Ground-based, UAVs.	Mertens et al. (2023)
Thermal Infrared (TIR: 8–14 μm) Imaging (Long-Wave IR)	Canopy temperature, photosynthetic efficiency and photoinhibition responses	Water Status & Stomatal Conductance. Increased canopy temperature is a proxy for reduced transpiration (stomatal closure), an early response to drought or heat stress.	Ground-based, UAVs.	Oerke et al. (2006); Chaerle et al. (2006); Zarco-Tejada et al. (2009); Mertens et al. (2023)
LiDAR (0.7–1.55 μm)/3D Imaging	Accurate plant volume, canopy structure, organ-level feature (e.g., leaf area index, leaf tilt index) extraction.	Architectural Metrics. Crucial for precise biomass estimation and detailed structural changes under stress.	Ground-based scanners, UAVs.	Lefsky et al. (2002); Khadka et al. (2020)
X-ray Computed Tomography (CT)	Root System Architecture (RSA) in intact soil (3D/4D), tuber growth, soil structure (porosity, water distribution).	Subterranean Stress Response. Non-destructive, high-resolution visualization of root growth dynamics (length, volume, branching angles) and how roots interact with compacted or stressed soil environments. Critical for phenotyping traits like drought/salinity tolerance.	Controlled-environment scanners.	Atkinson et al. (2019); Zhang et al. (2024)
Nanosensors	Real-time signaling molecules (e.g., H2O2) within plant organs/cells.	Rapid, Cellular-level Stress Signaling. Detects internal stress responses at short time-scales (seconds to minutes), providing insights into defense pathways.	Species-independent analytical tools.	Lew et al. (2020); Shaw and Honeychurch (2022)

datasets can reach terabyte scales, demanding computational tools beyond conventional statistical approaches. Artificial intelligence (AI), particularly machine learning (ML) and deep learning (DL), has emerged as a critical enabler in this context. ML methods such as support vector machines (SVMs) and random forests, alongside DL architectures like convolutional neural networks (CNNs), are widely applied to interpret multidimensional datasets. CNNs are especially powerful in image-based analysis, automatically extracting features (e.g., leaf edges, lesions, or color changes) from raw pixels without manual programming. This capability underpins the development of highly accurate stress-detection models. Trained on large-scale phenotypic and environmental data, these models reveal hidden associations between stress conditions and plant responses. For instance, DL approaches can identify adaptive traits under drought, salinity, or heat stress and even predict how specific genetic modifications may influence crop performance. While DL excels at feature discovery, ML methods remain widely used for detection, classification, quantification, and prediction of plant diseases and stress symptoms. Such predictive and diagnostic power not only validates gene-editing outcomes but also supports the design of environmental adaptability models, thereby guiding more targeted breeding strategies.

The integration of sensor-based phenotyping, image processing, and AI-driven analysis has fundamentally reshaped stress-resistance evaluation. Multi-mode analytics (MMA) approaches, in particular, have shown substantial improvements in the accuracy and reliability of early stress detection, outperforming traditional single-mode methods by capturing complex interactions among multiple abiotic stressors (Zandi et al., 2025). A cornerstone of MMA is data fusion, where information from different sensors is combined using low-, mid-, or high-level fusion techniques. For example, spatial data from LiDAR, spectral data from hyperspectral imaging, and temporal data from repeated measurements can be integrated to enhance the robustness of stress-detection models. To handle the vast data streams, dimensionality reduction techniques such as Principal Component Analysis (PCA) are employed to reduce variable complexity while retaining essential information—an especially critical step when analyzing hundreds of spectral bands from hyperspectral sensors. Practical applications demonstrate the value of such integration: for instance, combining hyperspectral and thermal imaging has been shown to improve the precision of drought stress assessment and transpiration monitoring (Mertens et al., 2023; Zandi et al., 2025). By uniting sensor-based data acquisition, advanced analytics, and AI-driven modeling, these approaches significantly enhance the accuracy, efficiency, and scalability of trait measurement. More importantly, they deepen our understanding of genotype–phenotype interactions under stress conditions. Ultimately, this integrated framework accelerates the identification and optimization of stress-resistance traits, informs precise breeding targets, and drives the development of resilient crop varieties capable of thriving in the face of climate variability and environmental challenges.

18.5 Integrating intelligent gene editing and phenotypic analysis

Artificial intelligence (AI) is revolutionizing crop breeding by linking multi-trait analysis with precise genome editing, enabling faster, more accurate development of stress-tolerant varieties. Advanced algorithms such as deep learning, random forests, and support vector machines enable rapid analysis of large-scale phenotypic, genomic, and environmental datasets to identify key genes linked to traits like drought resistance, salinity tolerance, and disease resistance. This accelerates candidate gene discovery and provides a robust foundation for genome editing (Su et al., 2025). In genome editing, AI improves CRISPR/Cas9 target design, minimizing off-target effects and predicting the performance of gene combinations under specific environmental conditions. By modeling synergistic interactions among edits, AI reduces experimental phases, improves accuracy, and enables multi-objective optimization that balances trade-offs among yield, stress tolerance, and grain quality. This shift breeding strategies from single-trait improvement to multi-trait balance optimization, ensuring better adaptability and resilience under variable climates (Kim et al., 2025; Abbaszadeh and Shahlai, 2025). Integrating gene editing with high-throughput phenotyping (HTP) creates a closed-loop system for improving stress-resistant crops (Su et al., 2025). CRISPR/Cas9 modifications can be evaluated rapidly through non-invasive imaging methods—such as hyperspectral, multispectral, and thermal imaging—that monitor physiological responses to stress. For instance, thermal imaging detects canopy temperature differences linked to stomatal regulation, while hyperspectral imaging captures pigment shifts associated with photosynthesis (Mertens et al., 2023). Targeted edits, such as those in drought-responsive DREB transcription factors, can thus be validated by tracking leaf wilting, chlorosis, or photosynthetic efficiency (Lata and Prasad, 2011). AI-driven image analysis further strengthens this loop. Convolutional neural networks (CNNs) can extract subtle stress indicators, such as early drought lesions or salinity-induced growth inhibition, often invisible to human observation. Combined with large-scale phenotypic data, these models predict the impact of genetic modifications across diverse environmental conditions (Danilevicz et al., 2022).

An important advancement in this integration is the shift from single-gene edits to combinatorial and network-level editing. Many stress-tolerance traits are polygenic, requiring coordinated regulation of multiple pathways. CRISPR multiplexing systems now enable simultaneous modification of several genes. For instance, edits targeting genes in the ABA signaling pathway (e.g., PYL receptors and SnRK2 kinases) could enhance drought and salinity tolerance by improving stomatal regulation and osmotic adjustment (Weiner et al., 2010; Dalal and Chinnusamy 2015). When these edits are assessed through multi-stress phenotyping platforms, which simulate drought, heat, and salinity stresses together, researchers can identify stable trait combinations that ensure resilience under fluctuating environments (Liu et al., 2020). This closed-loop framework comprising gene editing, rapid phenotyping, AI-driven analytics, and iterative optimization which represents a paradigm shift in crop improvement. It reduces breeding cycles from years to seasons, enhances precision by validating edits against real stress responses, and ensures scalability through remote sensing and automation. Moreover, the integration of multi-stress phenotyping aligns with future agricultural challenges, where crops must withstand simultaneous exposure to heat, drought, salinity, and emerging pathogens. However, most published studies currently integrate only parts of this pipeline: AI/ML for candidate discovery (GWAS, omics, phenomics), genome editing of selected targets, and greenhouse or limited field evaluation. While AI is widely applied for sgRNA design and target prioritization, few reports demonstrate a fully iterative AI→edit→field→AI feedback loop with multi-environment validation. As highlighted by Chen et al. (2024), major bottlenecks include limited large-scale genotype–phenotype–environment datasets, the complexity of gene × environment × management interactions, and the long timelines required for regulatory approval and field trials. Addressing these challenges will be critical to realizing the full potential of intelligent genome editing in agriculture.

18.6 Case studies on genome-editing outcomes for stress tolerance

18.6.1 Case study 1: Drought tolerance in maize (Shi et al., 2017)

One of the most cited examples of genome editing for abiotic stress tolerance is the modification of the *ARGOS8* gene in maize. Using CRISPR/Cas9, researchers created promoter variants that enhanced *ARGOS8* expression, a negative regulator of ethylene responses. Field evaluations under flowering-stage drought stress demonstrated a significant increase in grain yield up to five bushels per acre and without any yield penalty under well-watered conditions. This study represents a rare case where edited alleles were tested under real agricultural conditions, confirming both stress tolerance and agronomic value.

18.6.2 Case study 2: Salinity tolerance in rice (Zhang et al., 2019a)

In rice, the *OsRR22* gene, a type-B response regulator in cytokinin signaling, was targeted for mutagenesis using CRISPR/Cas9. Knockout lines carrying *OsRR22* mutations showed enhanced survival and growth under high salinity conditions compared with wild-type plants. Importantly, these mutations were heritable, and T1 segregants free of the transgene retained the tolerance trait, highlighting the feasibility of developing non-transgenic, edited rice lines with improved salt resilience.

18.6.3 Case study 3: Drought survival in Rapeseed (Wang et al., 2024)

A recent study in *Brassica napus* integrated genome-wide association (GWAS)-based gene discovery with functional editing to validate the role of *BnaA9.NF-YA7*, a nuclear transcription factor. Natural variation in this gene correlated with drought survival. CRISPR/Cas9-mediated knockout and RNAi suppression of *BnaA9.NF-YA7* significantly improved water-use efficiency, reduced leaf water loss, and enhanced seedling recovery after drought stress. A specific amino acid substitution (M63I) further illustrated allelic diversity contributing to drought adaptation, providing a strong candidate locus for crop improvement.

18.6.4 Case study 4: Drought resilience in potato (Decima Oneto et al., 2025)

In tetraploid potato (Solanum tuberosum), CRISPR/Cas9 was used to edit the *StCBP80* gene, which encodes a cap-binding protein involved in RNA metabolism and stress response regulation. Edited lines exhibited improved drought tolerance, manifested as enhanced stomatal control, altered expression of drought-responsive genes, and lower yield penalties under water deficit. These results provide one of the first demonstrations of targeted genome editing in a vegetatively propagated, polyploid crop leading to improved abiotic stress performance.

18.6.5 Case study 5: Multi-stress tolerance in tomato (Tran et al., 2023)

In tomato, the hybrid proline-rich protein gene *SlHyPRP1* was edited using CRISPR–Cas9 to examine its role in stress resilience. The edited lines showed enhanced tolerance not only to salinity stress but also to drought and high temperature, alongside improved resistance to certain pathogens. Physiological assessments indicated better osmotic adjustment and stress-associated gene regulation, positioning *SlHyPRP1* as a multifunctional target for engineering broad-spectrum tolerance in tomato and related crops.

18.6.6 Case study 6: Disease resistance in wheat (Liu et al., 2024a)

In wheat, the role of *MAP kinase phosphatase1 (TaMKP1)* in immunity was investigated using CRISPR/Cas9 knock-down mutants. Loss of *TaMKP1* enhanced resistance against *Puccinia striiformis* f. sp. *tritici* (stripe rust) and *Blumeria graminis* f. sp. *tritici* (powdery mildew), while unexpectedly also increasing yield compared with wild-type plants. Mechanistic studies revealed that *TaMKP1* acts as a negative regulator of MAPK-mediated defense by interacting with *TaMPK3/4/6*, providing insights into plant immunity and valuable germplasm for breeding high-yield, disease-resistant wheat.

18.7 Future prospects

18.7.1 Future prospects of intelligent genome editing for multi-stress tolerance

The future of intelligent gene editing for various biotic and abiotic stress tolerance in plants is based on the integration of system biology, artificial intelligence, and precise molecular biology tools. Amid extreme climate change, there is an urgent need to move beyond single-gene genome editing to integrated polygenic multiple-trait genome editing (Pacesa et al., 2024). Various genome editing tools, such as base editors, prime editors, and epigenome modifiers, can enable heritable changes at single-nucleotide resolution (Cho et al., 2013). Meanwhile, other advanced CRISPR machinery can simultaneously edit multiple loci to improve integrated biotic and abiotic stress tolerance. Similarly, the artificial intelligence-guided integration of pangenome data, high-throughput genotyping & phenotyping, and machine learning would enable precise target gene selection, reduce off-target risk and minimize negative trade-offs with quality and yield (Knott and Doundna, 2018).

Identifying a particular allele responsible for plant disease is a key focus of agricultural research. High-throughput sequencing and system biology have advanced our ability to identify mutations responsible for disease. However, to validate the identity allele responsible for the disease, experimental validation through reverse genetics is essential (Boycott et al., 2013). Hence, genome editing tools are very crucial to enabling researchers to introduce variants in the wild type or convert the mutation of a diseased plant into to wild type. A CRISPR-engineered plant, which can produce pest-toxic protein, could be used to protect crops from pest infestation. A major advantage of CISPR genome editing is the possibility to create a multiplex edit by targeting multiple loci responsible for multiple traits (McCarty et al., 2020). This approach has enabled us to make a crop with more than one mutation targeting multiple trait improvements simultaneously (McCarty et al., 2020, Zhu et al., 2020; Zhang et al., 2019b). CRISPR genome editing, enabling systematic perturbation of regulatory elements and genes associated with seed longevity, multiple biotic and abiotic stress tolerance, and novel traits. High-throughput CRISPR screens in plant cells or tissues delivered via protoplast transformation, viral vectors, or Agrobacterium allow identification of genes whose modification enhances climate resilience by tracking sgRNA enrichment or depletion under selective stress conditions (Xue and Greene, 2021). CRISPR can be used to knock out negative regulators of stress response genes, activate stress protective genes, or introduce a favorable allele for seed longevity enhancement, biotic and abiotic stress tolerance. Edited plant callus can be selected under stress conditions in culture media, and only the callus showing the desired phenotype would be regenerated into plants. Advanced approaches like genome-edited synthetic promoters or an inducible system to confer stress memory, enabling the plant to remember past stress and respond more robustly (Yang et al., 2013). Several studies have also shown the possibilities of in vivo editing. For example, the CRISPR–Cas9 system has been introduced in the shoot apical meristem using viral vectors or nanocarrier-mediated delivery, enabling heritable genome editing without transformation (Parvez et al., 2021). Another promising approach involves direct injection or biolistic delivery of CRISPR components into reproductive tissue. Further vascular bundles provide an excellent target for in vivo gene editing due to their natural transport. Lipid nanoparticle, DNA nanostructure, and cell-penetrating peptide could be optimized to deliver Cas9 mRNA and sgRNAs (Kasiewicz et al., 2023).

18.7.2 Emerging molecular tools and delivery systems in plant genome editing

Now emerging third-generation tools hold the potential to overcome long-standing barriers in crop improvement, such as efficient delivery systems, precise large fragment insertion, and multiplex editing across diverse plant genomes. One of the major challenges is having a stable insertion of a long fragment of DNA, especially in species and tissues where homology-directed repair is inefficient. Hence, CRISPR-guided recombinases, integrases, and transposon-based systems hold a promising new class of tools for targeted gene insertion (Meers et al., 2023). The search for novel RNA-guided nucleases is important, where editing efficiency and delivery remain constrained. Compact nucleases with reduced size can greatly facilitate fusion with base editors, prime editors, and can improve their delivery through viral vectors, nanoparticles, or plant transformation.

Unlike prime editing, which relies on a Cas 9 fused with reverse transcriptase and uses RNA as a template, various novel approaches are emerging that exploit DNA polymerases for precise genome editing (Halperin et al., 2018). DNA polymerase-based genome editing provides a wider spectrum of genetic alterations, greater control over mutagenesis, the ability to generate high diversity allelic variants, and reduces the risk of unintended insertions or rearrangements, which is crucial for regulatory acceptance of genome-edited plants (Liu et al., 2024b). The efficiency of homologous recombination-mediated insertions decreases sharply as the size of the DNA fragment increases, which restricts the ability to introduce long DNA sequences or entire genes into the plant genome. A recent study combined prime editing with serine recombinase or integrases, where the prime editor mediated the installation of recombinase recognition, allowing subsequent integration of large DNA sequences by recombinases (Durrant et al., 2023). Hence, characterization of novel plant-compatible recombinases holds promise for enhancing the efficiency, precision, and stability of large fragment DNA insertion. Transposons provide a promising alternative, as they can automatically catalyze the insertion of large fragments independent of double-strand break induction and repair pathways. Retroelements are a promising tool for inserting long DNA sequences into the plant genome with high specificity and stability. It mediates insertion through target primed reverse transcriptase, a process in which Target DNA is nicked and the exposed 3′ end is used to prime reverse transcription of the retroelement RNA. Hence, by integrating the retroelement with Cas 9 nickase, researchers could target insertions in the plant genome (Tang and Sternberg, 2023).

18.7.3 Reversible and precise gene regulation through epigenome and RNA editing for stress resilience

Permanent genetic modification in plants raises biosafety, ecological, and regulatory concerns due to unintended mutation, off-target effects, and heritable changes. Hence alternative strategy, epigenome editing, modifies chromatin states without altering DNA sequences. CRISPR-based epigenome editors using nuclease-dead CAS9 fused to DNA or histone-modifying enzymes allow locus-specific regulation of gene expression (Nuñez et al., 2021). Hence, employing DNA methyltransferases or histone acetyltransferases, or histone deacetylases enables fine-tune activation or repression of genes involved in stress tolerance, seed longevity, or secondary metabolite production. Similar to epigenome editing, RNA editing provides a reversible and transient approach to regulate gene expression (Kuscu et al., 2014). In plants particularly C to U editing in chloroplasts and mitochondria already plays a key regulatory role in stress tolerance. Hence, harnessing CRISPR-based RNA editors expands these possibilities in the nuclear transcriptome. CAS7-11, a single protein RNA-guided nuclease and engineered Cas 13 variant, enables precise mRNA knockdown or transient modification with minimal collateral activity (Özcan et al., 2021). These systems could be applied for transient suppression of susceptibility genes, dynamic regulation of stress-responsive transcription factors, fine-tuning seed storage protein and oil biosynthesis genes, and temporal modulation of growth regulators. Because RNA molecules have a short life span, RNA editing a particularly suited for transient traits without permanent ecological or evolutionary consequences.

In the future, AI could enable tailored genome editing strategies in plants, such as designing crop-specific CRISPR tools. Beyond guide RNA design, AI is being increasingly used in computational protein engineering, accelerating the creation of novel Cas variants, base editors, and efficient prime editors. Despite these advances, the quality and diversity of plant genome datasets are uneven across species, which limits the use of AI (Abudayyeh et al., 2019). Nevertheless, the potential of AI in plant genome editing is immense. Its integration with epigenome editing, RNA editing, and system biology could drive the development of next-generation climate-resilient crops with programmable stress memory, higher seed vigor, improved nutritional quality, and enhanced adaptability to climate change.

18.7.4 Advancing crop improvement through AI and precision breeding

AI is poised to transform precision breeding by integrating large-scale genomic, phenotypic, and environmental data to identify functional genes influencing stress resistance, yield, and quality. Deep learning and neural networks enable faster, more accurate gene selection compared to traditional methods, significantly reducing breeding cycles. When combined with gene editing tools like CRISPR/Cas9 and TALEN, AI can optimize target selection, predict pleiotropic effects, and simulate long-term crop performance under diverse conditions, ensuring safe and effective edits (Chen et al., 2024). Future AI-driven breeding platforms will incorporate real-time environmental data through smart sensors and remote sensing, allowing dynamic feedback to adjust breeding strategies, enhance crop adaptability, and provide tailored solutions for diverse climates and seasons (Su et al., 2025).

18.8 Challenges and ethical aspects

The primary concern in crops is that the heritable nature of edits can permanently alter the plant genome, hence affecting the ecosystem, biodiversity, and seed system. Although genome editing technologies have remarkable precision, uncertainties regarding off-target effects, unintended ecological consequences, and gene flow to wild relatives may impact natural plant populations. Although unequal access, high cost, and intellectual property restrictions are challenges (Turocy et al., 2021). Genome editing, particularly via CRISPR–Cas9, offers transformative potential for addressing climate change, food insecurity, and agricultural sustainability. However, its responsible deployment requires transparent governance, global collaboration, and inclusive societal dialog. Ethical considerations encompass autonomy, beneficence, non-maleficence, accountability, transparency, fairness, privacy, and global governance (Gupta et al., 2025). Autonomy emphasizes informed decision-making, while beneficence and non-maleficence highlight maximizing benefits and minimizing harm. Equitable access, data confidentiality, and traceability of AI-driven decisions remain critical for responsible use. Mathematical and socio-ethical frameworks provide structured approaches to define acceptable boundaries of human-directed genetic modification, while global governance ensures uniform ethical standards. AI-integrated CRISPR applications illustrate both the promise and ethical complexities of genome editing. Quantitative modeling combined with ethics-informed public engagement enhances accountability, enabling genome editing to navigate complex moral landscapes (Gupta et al., 2025). Through rigorous evaluation and ethical frameworks, AI-augmented genome editing can achieve higher standards of responsibility, fairness, and societal benefit, ensuring that technological innovation advances both scientific progress and global good.

18.9 Conclusion

First-generation CRISPR technologies, which rely on double-strand breaks and endogenous DNA repair, have enabled precise mutagenesis and targeted trait modification in crops. Second-generation systems, such as base and prime editing, introduced safer, more predictable nucleotide substitutions and insertions without causing double-strand breaks. Emerging third-generation tools now focus on precise large-fragment integration, epigenome editing, and improved delivery systems. Although CRISPR and related RNA-guided systems currently dominate plant genome editing, further refinements are needed to enhance efficiency, tissue specificity, and transient delivery methods that avoid transgene integration and simplify regulatory approval. At the same time, artificial intelligence (AI) is revolutionizing plant biotechnology by accelerating gene discovery, trait prediction, and phenotype analysis for stress resilience and adaptability. The integration of AI with CRISPR-based editing has created a powerful closed-loop system linking gene identification to phenotype optimization, facilitating multi-trait improvement for yield, stress tolerance, and nutritional quality. Despite challenges in delivery, regulation, data integration, and model interpretability, the convergence of AI, systems biology, and advanced genome editing holds immense promise. Together, these innovations are poised to redefine crop breeding—enabling the development of resilient, high-yielding, and climate-adaptive varieties that will underpin sustainable agriculture and global food security in the era of climate change.

Acknowledgments

The authors thank the Indian Council of Agricultural Research, Department of Agricultural Research and Education, Government of India, for financial support for genome editing research under the ICAR EFC on the genome editing Scheme: Crop Science for Food and Nutritional Security Sub-Scheme 10: Enhancing climate resilience and ensuring food security with genome editing tools.

Data Availability Statement

Not applicable.

Declaration

"The authors declare that no generative artificial intelligence tools were used in the writing or preparation of this chapter. The content reflects the authors' own scholarly work, and the authors assume full responsibility for all statements, interpretations, and conclusions presented."

Ethical Approval

Not applicable.

Conflicts of Interest

The authors declare that they have no conflicting or competing interests.

References

Abbaszadeh, A., Shahlai, A., 2025. Artificial Intelligence for CRISPR Guide RNA Design: Explainable Models and Off-Target Safety. arXiv preprint arXiv:2508.20130.

Abudayyeh, O.O., Gootenberg, J.S., Franklin, B., Koob, J., Kellner, M.J., Ladha, A., Joung, J., Kirchgatterer, P., Cox, D.B.T., Zhang, F., 2019. A cytosine deaminase for programmable single-base RNA editing. Science 365 (6451), 382–386. https://doi.org/10.1126/science.aax7063.

Aktas, Ö., Dogan, E., Ensari, T., 2019. CRISPR/Cas9 target prediction with deep learning. In: 2019 Scientific Meeting on Electrical-Electronics & Biomedical Engineering and Computer Science (EBBT). IEEE, Istanbul, Turkey, pp. 1–5.

Angidi, S., Madankar, K., Tehseen, M.M., Bhatla, A., 2025. Advanced high-throughput phenotyping techniques for managing abiotic stress in agricultural crops—a comprehensive review. Multidisciplinary digital publishing Institute (MDPI). United States Crops 5 (2). https://doi.org/10.3390/crops5020008.

Atkinson, J.A., Pound, M.P., Bennett, M.J., Wells, D.M., 2019. Uncovering the hidden half of plants using new advances in root phenotyping. Curr. Opin. Biotechnol. 55, 1–8. https://doi.org/10.1016/j.copbio.2018.06.002.

Avci, U., Sipahi, H., 2023. Genome editing of gene families for crop improvement. In: A Roadmap for Plant Genome Editing. Springer Nature, Turkey, pp. 3–15. https://doi.org/10.1007/978-3-031-46150-7_1.

Baisya, D., Ramesh, A., Schwartz, C., Lonardi, S., Wheeldon, I., 2022. Genome-wide functional screens enable the prediction of high activity CRISPR-Cas9 and -Cas12a guides in Yarrowia lipolytica. Nat. Commun. 13 (1). https://doi.org/10.1038/s41467-022-28540-0.

Blackford, J.C., 2010. Predicting the impacts of ocean acidification: challenges from an ecosystem perspective. J. Marine Syst. 81 (1–2), 12–18. https://doi.org/10.1016/j.jmarsys.2009.12.016.

Bohra, A., Kilian, B., Sivasankar, S., Caccamo, M., Mba, C., McCouch, S.R., Varshney, R.K., 2022. Reap the crop wild relatives for breeding future crops. Trends Biotechnol. 40 (4), 412–431. https://doi.org/10.1016/j.tibtech.2021.08.009.

Boycott, K.M., Vanstone, M.R., Bulman, D.E., MacKenzie, A.E., 2013. Rare-disease genetics in the era of next-generation sequencing: discovery to translation. Nat. Rev. Genet. 14 (10), 681–691. https://doi.org/10.1038/nrg3555.

Chaerle, L., Pineda, M., Romero-Aranda, R., Van Der Straeten, D., Barón, M., 2006. Robotized thermal and chlorophyll fluorescence imaging of pepper mild mottle virus infection in Nicotiana benthamiana. Plant Cell Physiol. 47 (9), 1323–1336. https://doi.org/10.1093/pcp/pcj102.

Chehelgerdi, M., Chehelgerdi, M., Khorramian-Ghahfarokhi, M., Shafieizadeh, M., Mahmoudi, E., Eskandari, F., Mokhtari-Farsani, A., 2024. Comprehensive review of CRISPR-based gene editing: mechanisms, challenges, and applications in cancer therapy. Mol. Cancer 23, 9.

Chen, D., Neumann, K., Friedel, S., Kilian, B., Chen, M., Altmann, T., Klukas, C., 2014. Dissecting the phenotypic components of crop plant growth and drought responses based on high-throughput image analysis. Plant Cell 26, 4636–4655.

Chen, J.S., Ma, E., Harrington, L.B., Da Costa, M., Tian, X., Palefsky, J.M., Doudna, J.A., 2018. CRISPR-Cas12a target binding unleashes indiscriminate single-stranded DNase activity. Science 360 (6387), 436–439. https://doi.org/10.1126/science.aar6245.

Chen, L., Liu, G., Zhang, T., 2024. Integrating machine learning and genome editing for crop improvement. aBIOTECH 5, 262–277.

Cho, S.W., Kim, S., Kim, J.M., Kim, J.S., 2013. Targeted genome engineering in human cells with the Cas9 RNA-guided endonuclease. Nature Biotechnol. 31 (3), 230–232. https://doi.org/10.1038/nbt.2507.

Chuai, G., Ma, H., Yan, J., Chen, M., Hong, N., Xue, D., Liu, Q., 2018. DeepCRISPR: optimized CRISPR guide RNA design by deep learning. Genome Biol. 19, 80.

Dalal, M., Chinnusamy, V., 2015. ABA receptors: prospects for enhancing biotic and abiotic stress tolerance of crops. In: Pandey, G.K. (Ed.), Elucidation of Abiotic Stress Signaling in Plants: Functional Genomics Perspectives, 1. Springer New York, pp. 271–298.

Danilevicz, M.F., Gill, M., Anderson, R., Batley, J., Bennamoun, M., Bayer, P.E., Edwards, D., 2022. Plant genotype to phenotype prediction using machine learning. Front. Genet. 13, 822173.

Decima Oneto, C., Massa, G.A., Echarte, L., Rey Burusco, M.F., González, M.N., Alfonso, C., Feingold, S., 2025. CRISPR/Cas9 editing of CBP80 enhances drought tolerance in potato (*Solanum tuberosum*). Front. Plant Sci. 16, 1598947.

Ding, L., Wu, Y., Liu, L.E., He, L., Yu, S., Effah, C.Y., Wu, Y., 2023. Universal DNAzyme walkers-triggered CRISPR-Cas12a/Cas13a bioassay for the synchronous detection of two exosomal proteins and its application in intelligent diagnosis of cancer. Biosens. Bioelectron. 219, 114827.

Dixit, S., Kumar, A., Srinivasan, K., Vincent, P.D.R., Ramu Krishnan, N., 2024. Advancing genome editing with artificial intelligence: opportunities, challenges, and future directions. Front. Bioeng. Biotechnol. 11, 1335901.

Doench, J.G., Fusi, N., Sullender, M., Hegde, M., Vaimberg, E.W., Donovan, K.F., Root, D.E., 2016. Optimized sgRNA design to maximize activity and minimize off-target effects of CRISPR-Cas9. Nat. Biotechnol. 34, 184–191.

Durrant, M.G., Fanton, A., Tycko, J., Hinks, M., Chandrasekaran, S.S., Perry, N.T., Hsu, P.D., 2023. Systematic discovery of recombinases for efficient integration of large DNA sequences into the human genome. Nat. Biotechnol. 41, 488–499.

Farooq, M.A., Gao, S., Hassan, M.A., Huang, Z., Rasheed, A., Hearne, S., Li, H., 2024. Artificial intelligence in plant breeding. Trends Genet. 40, 891–908.

Feng, H., Guo, J., Wang, T., Zhang, C., Xing, X.H., 2021. Guide-target mismatch effects on dCas9–sgRNA binding activity in living bacterial cells. Nucleic Acids Res. 49, 1263–1277.

Gamal, H., Shoeib, E.M., Hajjaj, A., Abdullah, H.E.A., Elramy, E.H., Abd Ellah, D.A., Khder, M.F., 2025. Incorporating AI, in silico, and CRISPR technologies to uncover the potential of repurposed drugs in cancer therapy. RSC Pharmaceut.

García-Tejero, I.F., Gutiérrez-Gordillo, S., Ortega-Arévalo, C., Iglesias-Contreras, M., Moreno, J.M., Souza-Ferreira, L., Durán-Zuazo, V.H., 2018. Thermal imaging to monitor the crop-water status in almonds by using the non-water stress baselines. Sci. Hortic. 238, 91–97.

Gupta, U., Pranav, A., Modi, R.K., Dubey, A., 2025. Ethical implications of AI in CRISPR: responsible genome editing. In: Bioinformatics and beyond. CRC Press, pp. 256–275.

Halperin, S.O., Tou, C.J., Wong, E.B., Modavi, C., Schaffer, D.V., Dueber, J.E., 2018. CRISPR-guided DNA polymerases enable diversification of all nucleotides in a tunable window. Nature 560 (7717), 248–252.

Harris, Z.N., Awale, M., Bhakta, N., Chitwood, D.H., Fennell, A., Frawley, E., Klein, L.L., 2021. Multi-dimensional leaf phenotypes reflect root system genotype in grafted grapevine over the growing season. GigaScience 10, giab087.

Hasan, N., Choudhary, S., Naaz, N., Sharma, N., Laskar, R.A., 2021. Recent advancements in molecular marker-assisted selection and applications in plant breeding programmes. J. Genet. Eng. Biotechnol. 19, 128.

Jensen, T., Apan, A., Young, F., Zeller, L., 2007. Detecting the attributes of a wheat crop using digital imagery acquired from a low-altitude platform. Comput. Electron. Agric. 59, 66–77.

Jiang, B., An, Z., Niu, L., Qin, D., 2025. Precise genome editing process and its applications in plants driven by AI. Funct. Integr. Genom. 25, 109.

Kasiewicz, L.N., Biswas, S., Beach, A., Ren, H., Dutta, C., Mazzola, A.M., Bellinger, A.M., 2023. GalNAc-lipid nanoparticles enable non-LDLR dependent hepatic delivery of a CRISPR base editing therapy. Nat. Commun. 14, 2776.

Khadka, K., Earl, H.J., Raizada, M.N., Navabi, A., 2020. A physio-morphological trait-based approach for breeding drought tolerant wheat. Front. Plant Sci. 11, 715.

Khammampalli, S., Vindal, V., 2025. Artificial intelligence in CRISPR-Cas systems: a review of tool applications. In: Artificial Intelligence (AI) in Cell and Genetic Engineering, pp. 243–257.

Kim, H.K., Kim, Y., Lee, S., Min, S., Bae, J.Y., Choi, J.W., Kim, J.S., 2019. SpCas9 activity prediction by DeepSpCas9, a deep learning–based model with high generalization performance. Sci. Adv. 5, eaax9249. https://doi.org/10.1126/sciadv.aax9249.

Kim, H.K., Lee, S., Kim, Y., Park, J., Min, S., Choi, J.W., Kim, H.H., 2020. High-throughput analysis of the activities of xCas9, SpCas9-NG and SpCas9 at matched and mismatched target sequences in human cells. Nat. Biomed. Eng. 4, 111–124.

Kim, M.G., Go, M.J., Kang, S.H., Jeong, S.H., Lim, K., 2025. Revolutionizing CRISPR technology with artificial intelligence. Exp. Mol. Med. 1–13.

Knott, G.J., Doudna, J.A., 2018. CRISPR-Cas guides the future of genetic engineering. Science 361 (6405), 866–869.

Konstantakos, V., Nentidis, A., Krithara, A., Paliouras, G., 2022. CRISPR–Cas9 gRNA efficiency prediction: an overview of predictive tools and the role of deep learning. Nucleic Acids Res. 50, 3616–3637.

Kuscu, C., Arslan, S., Singh, R., Thorpe, J., Adli, M., 2014. Genome-wide analysis reveals characteristics of off-target sites bound by the Cas9 endonuclease. Nat. Biotechnol. 32, 677–683.

Kyriazos, T., Poga, M., 2024. Application of machine learning models in social sciences: managing nonlinear relationships. Encyclopedia 4, 1790–1805.

Lambrinidis, G., Tsantili-Kakoulidou, A., 2021. Multi-objective optimization methods in novel drug design. Expert Opin. Drug Discov. 16, 647–658.

Lata, C., Prasad, M., 2011. Role of DREBs in regulation of abiotic stress responses in plants. J. Exp. Bot. 62, 4731–4748.

Lee, M., 2023. Deep learning in CRISPR-Cas systems: a review of recent studies. Front. Bioeng. Biotechnol. 11, 1226182.

Lefsky, M.A., Cohen, W.B., Harding, D.J., Parker, G.G., Acker, S.A., Gower, S.T., 2002. Lidar remote sensing of above-ground biomass in three biomes. Global Ecol. Biogeogr. 11, 393–399.

Lew, T.T.S., Sarojam, R., Jang, I.C., Park, B.S., Naqvi, N.I., Wong, M.H., Strano, M.S., 2020. Species-independent analytical tools for next-generation agriculture. Nat. Plants 6, 1408–1417.

Lichtenthaler, H.K., Langsdorf, G., Lenk, S., Buschmann, C., 2005. Chlorophyll fluorescence imaging of photosynthetic activity with the flash-lamp fluorescence imaging system. Photosynthetica 43, 355–369.

Lin, J., Wong, K.C., 2018. Off-target predictions in CRISPR-Cas9 gene editing using deep learning. Bioinformatics 34, i656–i663. https://doi.org/10.1093/bioinformatics/bty554.

Liu, Q., He, D., Xie, L., 2019. Prediction of off-target specificity and cell-specific fitness of CRISPR-Cas system using attention boosted deep learning and network-based gene feature. PLoS Comput. Biol. 15, e1007480. https://doi.org/10.1371/journal.pcbi.1007480.

Liu, Q., Cheng, X., Liu, G., Li, B., Liu, X., 2020. Deep learning improves the ability of sgRNA off-target propensity prediction. BMC Bioinf. 21, 51. https://doi.org/10.1186/s12859-020-3395-z.

Liu, B., Dong, X., Zheng, C., Keener, D., Chen, Z., Cheng, H., Sontheimer, E.J., 2024a. Targeted genome editing with a DNA-dependent DNA polymerase and exogenous DNA-containing templates. Nat. Biotechnol. 42, 1039–1045.

Liu, S., Zhang, F., Su, J., Fang, A., Tian, B., Yu, Y., Yang, Y., 2024b. CRISPR-targeted mutagenesis of mitogen-activated protein kinase phosphatase 1 improves both immunity and yield in wheat. Plant Biotechnol. J. 22, 1929–1941.

Maes, W.H., Steppe, K., 2019. Perspectives for remote sensing with unmanned aerial vehicles in precision agriculture. Trends Plant Sci. 24, 152–164.

Manghwar, H., Li, B., Ding, X., Hussain, A., Lindsey, K., Zhang, X., Jin, S., 2020. CRISPR/Cas systems in genome editing: methodologies and tools for sgRNA design, off-target evaluation, and strategies to mitigate off-target effects. Adv. Sci. 7, 1902312.

McCarty, N.S., Graham, A.E., Studená, L., Ledesma-Amaro, R., 2020. Multiplexed CRISPR technologies for gene editing and transcriptional regulation. Nat. Commun. 11, 1281.

Meers, C., Le, H.C., Pesari, S.R., Hoffmann, F.T., Walker, M.W., Gezelle, J., Sternberg, S.H., 2023. Transposon-encoded nucleases use guide RNAs to promote their selfish spread. Nature 622 (7984), 863–871.

Mertens, S., Verbraeken, L., Sprenger, H., De Meyer, S., Demuynck, K., Cannoot, B., Wuyts, N., 2023. Monitoring of drought stress and transpiration rate using proximal thermal and hyperspectral imaging in an indoor automated plant phenotyping platform. Plant Methods 19, 132.

Montes, J.M., Utz, H.F., Schipprack, W., Kusterer, B., Muminovic, J., Paul, C., Melchinger, A.E., 2006. Near-infrared spectroscopy on combine harvesters to measure maize grain dry matter content and quality parameters. Plant Breed. 125, 591–595.

Nuñez, J.K., Chen, J., Pommier, G.C., Cogan, J.Z., Replogle, J.M., Adriaens, C., Weissman, J.S., 2021. Genome-wide programmable transcriptional memory by CRISPR-based epigenome editing. Cell 184, 2503–2519.

Oerke, E.C., Steiner, U., Dehne, H.W., Lindenthal, M., 2006. Thermal imaging of cucumber leaves affected by downy mildew and environmental conditions. J. Exp. Bot. 57, 2121–2132.

Özcan, A., Krajeski, R., Ioannidi, E., Lee, B., Gardner, A., Makarova, K.S., Gootenberg, J.S., 2021. Programmable RNA targeting with the single-protein CRISPR effector Cas7-11. Nature 597 (7878), 720–725.

Pacesa, M., Pelea, O., Jinek, M., 2024. Past, present, and future of CRISPR genome editing technologies. Cell 187, 1076–1100.

Parvez, S., Herdman, C., Beerens, M., Chakraborti, K., Harmer, Z.P., Yeh, J.R.J., Peterson, R.T., 2021. MIC-Drop: a platform for large-scale in vivo CRISPR screens. Science 373 (6559), 1146–1151.

Shaw, D.S., Honeychurch, K.C., 2022. Nanosensor applications in plant science. Biosensors 12, 675.
Sherkatghanad, Z., Abdar, M., Charlier, J., Makarenkov, V., 2023. Using traditional machine learning and deep learning methods for on-and off-target prediction in CRISPR/Cas9: a review. Briefings Bioinf. 24, bbad131.
Shi, J., Gao, H., Wang, H., Lafitte, H.R., Archibald, R.L., Yang, M., Habben, J.E., 2017. ARGOS 8 variants generated by CRISPR-Cas9 improve maize grain yield under field drought stress conditions. Plant Biotechnol. J. 15, 207–216.
Sinha, D., Maurya, A.K., Abdi, G., Majeed, M., Agarwal, R., Mukherjee, R., Chen, J.T., 2023. Integrated genomic selection for accelerating breeding programs of climate-smart cereals. Genes 14, 1484.
Sirault, X.R., James, R.A., Furbank, R.T., 2009. A new screening method for osmotic component of salinity tolerance in cereals using infrared thermography. Funct. Plant Biol. 36, 970–977.
Su, X., Xu, J., Li, C., Wei, M., 2025. The research progress on crop stress resistance and environmental adaptability improvement based on artificial intelligence. Adv. Resour. Res. 5, 1589–1610.
Sun, J., Zhang, Y., Wang, M., Guan, Q., Yang, X., Ou, J.X., Zhang, Z., 2020. The biological significance of multi-copy regions and their impact on variant discovery. Genom. Proteom. Bioinform. 18, 516–524.
Tang, S., Sternberg, S.H., 2023. Genome editing with retroelements. Science 382 (6669), 370–371.
Tran, M.T., Son, G.H., Song, Y.J., Nguyen, N.T., Park, S., Thach, T.V., Kim, J.Y., 2023. CRISPR-Cas9-based precise engineering of SlHyPRP1 protein towards multi-stress tolerance in tomato. Front. Plant Sci. 14, 1186932.
Turocy, J., Adashi, E.Y., Egli, D., 2021. Heritable human genome editing: research progress, ethical considerations, and hurdles to clinical practice. Cell 184, 1561–1574.
Venezia, M., Creasey Krainer, K.M., 2021. Current advancements and limitations of gene editing in orphan crops. Front. Plant Sci. 12, 742932.
Walter, T., Shattuck, D.W., Baldock, R., Bastin, M.E., Carpenter, A.E., Duce, S., Hamilton, N., 2010. Visualization of image data from cells to organisms. Nat. Methods 7 (Suppl. 3), S26–S41.
Wang, L., Zhang, J., 2019. Prediction of sgRNA on-target activity in bacteria by deep learning. BMC Bioinf. 20, 1–14.
Wang, J., Mao, L., Li, Y., Lu, K., Qu, C., Tang, Z., Liu, L., 2024. Natural variation in BnaA9. NF-YA7 contributes to drought tolerance in *Brassica napus* L. Nat. Commun. 15, 2082.
Weiner, J.J., Peterson, F.C., Volkman, B.F., Cutler, S.R., 2010. Structural and functional insights into core ABA signaling. Curr. Opin. Plant Biol. 13, 495–502.
Xue, C., Greene, E.C., 2021. DNA repair pathway choices in CRISPR-Cas9-mediated genome editing. Trends Genet. 37, 639–656.
Xue, L., Tang, B., Chen, W., Luo, J., 2018. Prediction of CRISPR sgRNA activity using a deep convolutional neural network. J. Chem. Inf. Model. 59, 615–624.
Yang, H., Wang, H., Shivalila, C.S., Cheng, A.W., Shi, L., Jaenisch, R., 2013. One-step generation of mice carrying reporter and conditional alleles by CRISPR/Cas-mediated genome engineering. Cell 154, 1370–1379.
Yetgin, A., 2025. Revolutionizing multi-omics analysis with artificial intelligence and data processing. Quant. Biol. 13, e70002.
Yin, H., Xue, W., Anderson, D.G., 2019. CRISPR–Cas: a tool for cancer research and therapeutics. Nat. Rev. Clin. Oncol. 16, 281–295.
Zandi, A., Hosseinirad, S., Kashani Zadeh, H., Tavakolian, K., Cho, B.K., Vasefi, F., Tavakolian, P., 2025. A systematic review of multi-mode analytics for enhanced plant stress evaluation. Front. Plant Sci. 16, 1545025.
Zarco-Tejada, P.J., Berni, J.A.J., Suárez, L., Sepulcre-Cantó, G., Morales, F., Miller, J.R., 2009. Imaging chlorophyll fluorescence with an airborne narrow-band multispectral camera for vegetation stress detection. Remote Sens. Environ. 113, 1262–1275.
Zhang, A., Liu, Y., Wang, F., Li, T., Chen, Z., Kong, D., Luo, L., 2019a. Enhanced rice salinity tolerance via CRISPR/Cas9-targeted mutagenesis of the OsRR22 gene. Mol. Breed. 39, 47.
Zhang, Y., Wang, J., Wang, Z., Zhang, Y., Shi, S., Nielsen, J., Liu, Z., 2019b. A gRNA-tRNA array for CRISPR-Cas9 based rapid multiplexed genome editing in *Saccharomyces cerevisiae*. Nat. Commun. 10, 1053.
Zhang, Q., Luan, R., Wang, M., Zhang, J., Yu, F., Ping, Y., Qiu, L., 2024. Research progress of spectral imaging techniques in plant phenotype studies. Plants 13, 3088.
Zhu, H., Li, C., Gao, C., 2020. Applications of CRISPR–Cas in agriculture and plant biotechnology. Nat. Rev. Mol. Cell Biol. 21, 661–677.

CHAPTER

19

AI-guided CRISPR in crop yield enhancement

Elshan Musazade[1], *Isack Ibrahim Mrisho*[2], *Diming Zhang*[1,3], *Jinshan Gao*[1,3], *Zhiyuan Cheng*[1,3], *and Xianzhong Feng*[1,3]

[1]Key Laboratory of Soybean Molecular Design Breeding, Northeast Institute of Geography and Agroecology, Chinese Academy of Sciences, Changchun, PR China; [2]College of Chinese Medicinal Material, Jilin Agricultural University, Changchun, PR China; [3]University of Chinese Academy of Sciences, Beijing, PR China

19.1 Introduction

The world's population is projected to reach about 10 billion by 2050, creating unprecedented pressure on the world's agricultural systems to feed more people. This issue is intensified by the fact that higher average incomes increase the difference between food production and consumption, particularly in developing countries (Lenaerts et al., 2019). At the same time, climate-induced extreme weather events, in conjunction with rampant land degradation, shortages in water resources, all of which are exacerbated by careless application of pesticides and fertilizers and deforestation on a massive scale, are worsening the world food security crisis. These complex pressures establish a requirement for revolutionary and new agricultural strategies. Therefore, the improvement in both the efficiency of crop production and the improvements in resistance to pests, diseases, and abiotic stress factors is the most critical aim for worldwide scientists and agronomists (Farhaoui et al., 2025).

The establishment of advanced technologies for crop improvement is the cornerstone of "the third green revolution" in providing solutions to the present difficulties in agriculture (Tong and Nikoloski, 2021). Traditional plant breeding had for many years been devoted to the principles of improvement of yield and quality of the crops which they grew, but because of the slow extent of the progress made and the necessary labor and expense involved it was found that the constant demands for the production of high-yielding and high-quality crops and for the necessities of climatic resistance could not be satisfied. The gap thus created was filled by some of the new techniques of gene editing which were so

Intelligent Gene Editing in Plants
https://doi.org/10.1016/B978-0-443-44079-3.00005-X

readily at hand, such, for instance, as the zinc-finger nucleases (ZFNs) and the transcription activator-like effector nucleases (TALENs), which were supposed to fulfil the required conditions. The introduction of CRISPR–Cas9 has been a reforming spectacle, the surprise being that these techniques possess exceptional ease, versatility, and efficiency which are revolutionizing not only the scope but also the rapidity of plant breeding (Sander and Joung, 2014).

The CRISPR mechanism, found in about half of all bacterial genomes and up to 90% of all archaeal genomes, is based on the incorporation of segments of invading DNA into "memories" stored in the CRISPR locus which will then allow them to be recognized in subsequent encounters (Bortesi and Fischer, 2015). In the event of re-infection, a short guide RNA (gRNA), made from the CRISPR locus, will deliver Cas9 endonuclease to cognate foreign DNA sequences, which will be cleaved with remarkable specificity (Sander and Joung, 2014). A key feature of this mechanism is a PAM site that is responsible for the great discrimination between foreign DNA and host DNA allowing targeted and efficient defense (Terns and Terns, 2011) (Fig. 19.1). These advances notwithstanding, current methods of plant breeding are still challenged by abnormalities in the plant genome and the immense dimensionality of multi-omic data making the identification of causative variants, regulatory networks, and connections between genotype and phenotype (Wu et al., 2024).

At the same time, artificial intelligence (AI), sometimes referred to as the "new electricity" of the 21st century, is revolutionizing genomic data analysis by allowing faster and better informed biological decision making using genetic data (Grover et al., 2022). One of the underlying domains of AI is machine learning (ML), and its extension deep learning (DL),

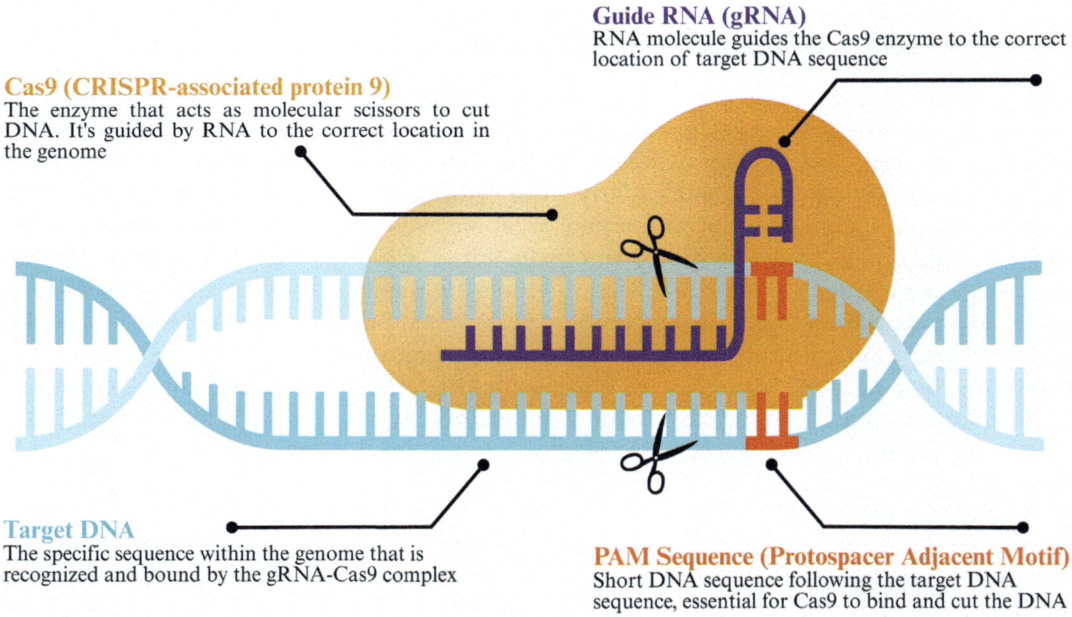

FIGURE 19.1 Major components of the CRISPR–Cas9 system. The Cas9 endonuclease (orange) acts as scissors to cut DNA. The gRNA (purple) is responsible for directing Cas9 to the complementary target DNA sequence (blue). Recognition is aided by a PAM (red) that binds next to the target and is required for Cas9 binding and cleavage.

which enables the training of AI models using very large databases to detect complex patterns and generate accurate predictive models. These functions are increasingly being employed to predict molecular phenotypes from DNA sequences, identifying key agronomic traits, and directing the selection of the best genotypes, thus greatly enhancing data-driven crop breeding (Eraslan et al., 2019).

The combination of AI and gene editing represents a strong synergy that can be directed toward the breeding of new crop cultivars with better yield and better resistance to biotic and abiotic stress. Using predictive modeling, AI accelerates the interpretation of genomic and multi-omics data, allowing the identification of causal variants and gene-trait interactions underlying those agronomic traits of economic importance (Farooq et al., 2024). Such discoveries can be observed in practice with fine CRISPR-based editing complemented by the AI-assisted optimized gRNA design and off-target cleavage (Kim et al., 2025). These developments are underpinned by ML, the predictive models of which are widely used in plant sciences, from breeding to pathogen detection (van Dijk et al., 2021). As such, these models in conjunction with CRISPR will facilitate the discovery of functional genomic elements and precision editing of genomes. Overall, this important synergy can drive breeding cycles while reinforcing cost-effectiveness and reliability, in an era of climate change and limited resources, contributing to world food security (Lenaerts et al., 2019).

This chapter provides a comprehensive investigation of AI- and CRISPR-based synthetic gene editing technologies to improve crops, integrating key principles with actual developments. The chapter begins by summarizing CRISPR as a revolutionary tool and then highlights the role of AI in the evaluation of genomic data and editing approaches. The main section of the paper then describes the work of AI in enhancing CRISPR efficiency based on performing four essential tasks, for example, predicting gRNA efficiency, the selection of appropriate Cas forms, the filtering of one or more specific targets and the capture of useful genes and regulatory elements. Important issues related to quality of data, integration of multi-omics, interpretability, computation and the need for collaboration between disciplines are discussed. The future of agricultural AI products can be visualized in this chapter, and from subsequent suggestions if emphasis is placed upon personalized breeding approaches and the development of multi-omical rooted products greater impetus to both computer biologists and agriculturists will be present. In the summary, this review heralds the arrival of AI-guided CRISPR as a game-changer in crop breeding designed to produce crops with a greater yield potential, enhanced tolerance to aggression and increased use of resources that will develop more sustainable agriculture systems.

19.2 CRISPR technology in modern gene editing

Gene editing is a new tool to make precise changes in the genome of an organism through the addition, removal, or alteration of the DNA at a given locus in the genome. This powerful technology facilitates crop improvement by providing an efficient new tool to conventionally time-consuming breeding methods such as crossing or backcrossing. Because of its ability to create mutant alleles in elite cultivars, gene editing facilitates the rapid generation of new and improved lines that have been engineered for a specific agronomic trait, and thus reduces the generation time for an improved line and promotes the earlier introduction of superior crops into fields (Wulff and Dhugga, 2018). In the past 10 years, rapid developments of gene-editing tools have fostered innovations in agricultural biotechnology, especially those associated with improving yield potential, stress tolerance, and sustainable food production (Dhugga, 2022; Wang et al., 2020).

Several tools for gene editing have been developed to create site-specific alterations, including meganucleases, ZFNs, TALENs, and the latest CRISPR–Cas (Clustered Regularly Interspersed Short Palindromic Repeats and CRISPR-associated proteins) systems (Montecillo et al., 2020). By far the most promising of these platforms is CRISPR–Cas9, which is less expensive, easy to use, programmable, and efficient as compared to previous gene editing systems (Afzal et al., 2020). Unlike protein engineering-intensive ZFNs and TALENs, for each new target site design, CRISPR–Cas9 leverages the simple design of a gRNA that is complementary to the DNA sequence of interest, making editing available across numerous agricultural applications (Montecillo et al., 2020).

An important feature of CRISPR–Cas9 is its ability to simultaneously edit multiple genes. By designing of different gRNA, it is thus possible for researchers to target multiple loci simultaneously even within a generation, in order to stack traits and engineer complex trait features such as yield increase, abiotic stress tolerance, and disease resistance (Yang et al., 2017). This scalability paves the way for further exploitation of CRISPR beyond simple gene changes to reformatting entire genetic networks for more robust and efficient crop systems. In combination, these traits mean that CRISPR–Cas technologies are an important part of plant breeding in the present time. CRISPR–Cas9 is an indispensable technology for the development of cultivars that have improved yield potential, stress tolerance, and nutritional value, making it an indispensable platform for the production of food in increased demand worldwide, considering the requirement for the development of sustainable agriculture (Muha-Ud-Din et al., 2024) (Table 19.1).

CRISPR–Cas9 was originally described as an adaptive immune system in bacteria and archaea, where it provides protection against infection by foreign viruses through the recognition and cleavage of foreign viral DNA by sequence-specific recognition. In these organisms, CRISPR loci contain Cas-encoding nucleolytic genes and arrays of short repeats interspersed with spacers which are remnants of foreign invaders. Upon infection of a known host by its foreign invader, the spacers direct the action of the associated Cas proteins to identify and destroy the invader (Barrangou et al., 2007). In eukaryotes, this system has been adapted to provide a molecular tool for gene editing by reprogramming of the Cas9 protein to interact with an arbitrarily synthesized gRNA which together generates mutations in the shape of site-specific double-strand breaks (DSB) in the genomic targets selected. The fate of the DSBs can now be modified by endogenous repair systems including homology directed repair (HDR) which allows for the accurate replacement of genes or targeting insertions if a donor template is provided or non-homologous end-joining (NHEJ) which usually produces insertions and deletions resulting in loss of a gene function (Sander and Joung, 2014).

Streptococcus pyogenes Cas9 (SpCas9) has become a widely used gene editing tool among the various CRISPR systems. It consists of a 20-nucleotides (nt) gRNA and the Cas9 nuclease. Cas9 is specific to the gRNA-complementary sequences when a PAM is available. In the case of SpCas9, the canonical PAM is "NGG," which is plentiful in genomes and makes for a large

TABLE 19.1 CRISPR-edited genes in different crops for yield enhancement.

Crop type	Trait modified	Targeted gene/s	Results	Reference
Barley	Development and yield	*HvCKX1*	Increased grain yield	Zhang et al. (2022)
Rice	Grain number	*OsGn1a*	Increased grain number	Li et al. (2016)
	Grain size	*OsGS3*	Increased grain size	Yuyu et al. (2020)
	Grain size	*OsGS2/GRF4*	Boosted grain number per spikelet	Wang et al. (2022)
	Grain width and weight	*OsGW2*, *OsGW5*, and *TGW6*	Increased grain width and weight	Xu et al. (2016)
	Grain yield	*OsDEP1*	Enhanced overall yield	Yang et al. (2024a)
	Grain yield	*OsPIN5b*	Increased tiller number contributing to yield gains	Zeng et al. (2020)
	Grain yield	*OsLOGL5*	Enlarged kernel size and yield	Wang et al. (2020)
	Plant growth and grain yield	*PYL1-PYL6*, *PYL12* (*gp-1*), *PYL7-PYL11* and *PYL13* (*gp-2*)	Improved grain productivity and plant growth	Miao et al. (2018)
	Tiller number	*OsAAP3*	Enhanced fruit size and yield	Lu et al. (2018)
Tomato	Fruit size	*SlCLV3*/Promoter	Strengthened yield-related components	Rodríguez et al. (2011)
	Fruit size	*SlENO*	Increased grain yield	Yuste-Lisbona et al. (2020)
Wheat	Grain size	*TaGS5-3A*	Increased grain size	Ma et al. (2016b)
	Grain number	*TaCKX2-D1*	Increased grain width and weight	Zhang et al. (2019)
	Grain yield	*TaDEP1*, *TaGASR7*, *TaLOX2*, *TaNAC2*, *TaPIN1*, *TaGW2*, *TaGLW7*, *TaGW2*, and *TaGW8*	Increased grain number	Zhang et al. (2016)
	Grain yield	*TaRPK1*	Enhanced overall yield	Abdul Rahim et al. (2024)

number of target sites (Rabinowitz et al., 2020). Cas9 induces a DSB about three base pairs upstream of the PAM (Chen et al., 2024). The gRNA can exist as either a dual-RNA complex, comprising CRISPR RNA (crRNA) and trans-activating CRISPR RNA (tracrRNA), or as a single-guide RNA (sgRNA) that combines both functions, an improvement that has significantly enhanced usability and efficiency (Alipanahi et al., 2023).

The inherent traits of CRISPR–Cas9 that allow for its versatility have their limitations. The PAM requirement of the system limits the sites of the genome that may be edited and the occurrence of off-target effects occurs when the gRNAs target those partially complementary sequences. For example, the canonical 5′-NGG-3′ PAM for *S. pyogenes* Cas9 (SpCas9) is found approximately every 42 base pairs in the human genome, indicating the potential for off-target cleavage events (Duan et al., 2021). Thus, efficient editing depends on the rational design of gRNAs and on nuclease variants with appropriate PAM specificities and heightened cleavage properties (Manibalan et al., 2020).

To overcome the limitations of the standard CRISPR–Cas9 tools, many variants and alternative nucleases have been developed. Notably, Cas12a recognizes a different PAM sequence of "TTTV" and produces staggered DSBs expanding the range over which genomic targets can be changed, giving more flexible experimental design (Zetsche et al., 2015). These advances have led to entirely different CRISPR-based tools that do not rely on DSB-mediated changes. For example, base editors (BEs) use a Cas9 nickase (nCas9) or a nuclease-inactive dCas9 fusion protein combined with a nucleobase deaminase to directly modify single nucleotides, allowing a specific mutation to be made without the induction of insertions and deletions (indels) during DSB repair processes (Komor et al., 2016). Prime editors (PEs) also combine reverse transcriptases and dCas9 under the guidance of prime editing guide RNAs (pegRNAs) containing both target and edited sequences in close proximity in a single molecule, leaving only single-strand breaks as intermediates for a wider range of base changes without generating DSBs (Anzalone et al., 2019). The successful application of base and prime editing in plants has opened up new opportunities for the engineering of complex traits such as yield improvement, abiotic tolerance, and quality traits (Li et al., 2018; Lin et al., 2020).

However, there are still many challenges to overcome in agriculture. Drawbacks arise from less precision and biosafety from off-target effects, the malaise of weak activity in the editing process itself, or the problems which arise in the attempts to edit in plants with large, complex or polyploid genomes where the interaction of many genes leads to a generalized effect (de Lange et al., 2022). These bottlenecks will require a better design of gRNAs, improved nucleases and the use of the mathematical tools which will provide a more efficient solution to the problems involved. One of the most powerful aids in this field is AI. AI programs are able to predict and optimize gRNA activities and specificity with the maximum on-target activity and minimum off-target events by utilizing the large amounts of data available to them on gRNA sequences and experimental results. The actual verification of the activity of gRNAs is still the gold standard for efficacy but is time-consuming, expensive, and impractical for applications needing high throughput (Yan et al., 2018).

By contrast, ML models on empirical data provide a generalized, inexpensive, and scalable solution for the assessment of thousands of candidate gRNAs. While previous models suffered from quality of data, its heterogeneity, and the lack of cross-species portability, improvements in the voluminous CRISPR databases and in algorithmic frameworks employed lately have markedly enhanced prediction performance. Further increases in curated training datasets, optimization of computational models, and experimental validation consistently in cross-species studies will be necessary to improve these deficiencies in the future.

In conclusion, CRISPR–Cas9 is a universally applicable and far better biotechnological platform for plant breeding than any other biotechnology. The integration of CRISPR–Cas9 with AI-based approaches will provide greater precision, controllability, and scalability. This is achieved through the design of highly efficient gRNAs, an increased range of loci that can be targeted, and the ability to manipulate traits that are multigenic in nature. The interaction between CRISPR and AI is therefore predicted to accelerate the development of high-yielding, stress-resistant, nutritious crop genotypes. Thus, these advances provide the evidence to allow the CRISPR–Cas systems to move from being a collection of technique with promise for crop improvement to radical technologies which are necessary for sustainable agriculture and food security during the rest of this century.

19.3 AI applications in gene editing

AI has emerged as a driving force in scientific development, discovery, and innovation, delivering incredible computational techniques for complex problems via simulation of the human thought processes. AI consists of several important domains, namely ML, DL, and artificial neural networks. ML is the avenue that allows systems to learn automatically about the patterns that may exist in large data repositories that can drive modelling predictions for both new and unseen data. ML can be classified into four broad classes, supervised, unsupervised, semi-supervised, and reinforcement learning (Yoosefzadeh Najafabadi et al., 2023). Among these methods, supervised learning is the most well-known and depends upon the knowledge of input–output for learning from labeled databases. This methodology has been applied to classification work including the detection of disease in crops (Jiang et al., 2020) and regression, which is used to predict the yield of crops (Pantazi et al., 2016), which is very important in promoting agricultural biotechnology-based strategies toward enhancement.

However, unsupervised learning utilizes unlabeled data to reveal hidden structures or trends. Typical methods here are clustering (which groups similar data points) and dimensionality reduction (which reduces the complexity of a dataset while retaining the essential characteristics of the data). Semi-supervised learning is an exceptionally important field of biological research, since, due to the rarity and expense of complete, annotated datasets, it takes a few labeled examples and combines them with a large quantity of unlabeled data. On the other hand, reinforcement learning allows the model to improve its decision-making through iterative trial and error experiences with a situation strengthened by rewards and penalties. This can be very well applied in agriculture for dynamic management systems (Chen et al., 2024).

DL, a sophisticated discipline of ML, is based on multilayer artificial neural networks (ANN), which are inspired by cognitive functions of the human brain. It is based on architectures such as convolutional neural networks (CNNs), recurrent neural networks (RNNs), transformers, and long short-term memory networks (LSTMs), and DL learns automatically to recognize clearly the hierarchical and abstract representations of raw data in progressive layer-wise modeling of the data pattern. This capability to recognize complex structures from multidimensional data makes it especially suited to the areas of genomics and other complex biological fields (Chen et al., 2024).

AI-based approaches are revolutionizing the gene-editing of crops in an innovative data-driven paradigm that promotes targeting design and screening of genes, while reducing human bias (Jiang et al., 2022). AI algorithms having the capability to analyze vast genomic data, such as those associated with important agronomic traits such as resistance to disease, drought and nutritional improvement, thereby allowing researchers to rapidly screen and prioritize gene editing candidates that hasten the improvement of crops (Tang et al., 2024).

AI not only increases exploratory potential, but it also refines target validation through the modeling of gene functions and interactions to predict effects of certain edits, thereby increasing confidence in experimental design and success of validation. An especially transformative aspect of AI is its ability to incorporate multi-omics data (proteomics, genomics, transcriptomic and metabolomic) in a common analytic platform. Such integrated studies allow for enhanced insights into complex biological phenomena. It also allows for the fine mapping of key control genes, laying a firm basis for possible and efficient gene editing strategies (Tang et al., 2024).

ML provides an important technique used for plant multi-omics and can unleash the potential power of plant genomes to elucidate their enigma. Although the decreasing cost of sequencing has resulted in better assemblies of genomes, is still a substantial problem to elucidate the functional regions of genomes. ML algorithms have been effectively used very successfully in this area, for example, for the prediction of protein coding genes (PCGs), like in their implementation in AUGUSTUS, which uses Support Vector Machines (SVMs) and Hidden Markov Models (HMMs) for high accuracy ab initio annotation (Schweikert et al., 2009; Stanke et al., 2006), or the prediction of cis-regulatory elements (CREs) in which case also CNNs can do predictions correctly (Umarov and Solovyev, 2017). Furthermore, ML has been used to design gene regulatory networks, and predict gene/gene interactions to obtain meaningful molecular information about the underlying mechanisms of complex traits (Haque et al., 2019). The combination of the above data-driven analytical approaches with AI and gene-editing technology represents an un-precedented opportunity for advancing the pipeline of crop breeding and to enhance the development of new climate-resilient crops that will help us meet our global quest for food security.

19.4 Integrating AI with CRISPR for crop breeding

The combination of AI and CRISPR systems represents a revolutionary step in modern crop genetics, affording new levels of accuracy, scalability, and efficiency in gene-editing applications. The algorithms of AI do more than hasten target searches for editing: they also enhance the accuracy of gRNA design, as well as determining which Cas variant to use and optimizing workflows, all of which will enhance efforts to develop stress-resistant high-yielding crop types (Fig. 19.2 and Table 19.2).

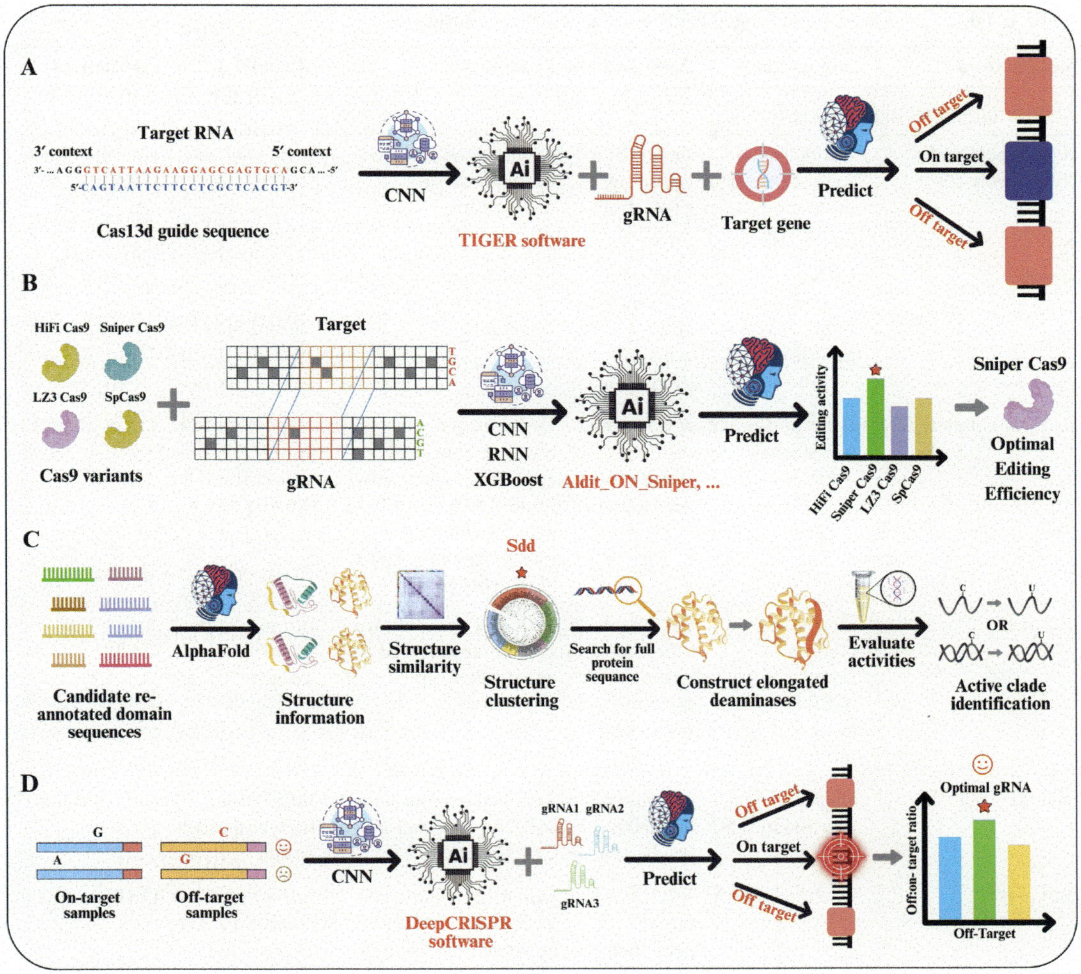

FIGURE 19.2 The multifaceted applications of AI in optimizing CRISPR: From gRNA design to precise targeting. (A) On-target and off-target activity prediction: The TIGER model (targeted inhibition of gene expression using gRNA design, modeled with CNNs) evaluates gRNAs by predicting their effectiveness based on their genomic context for estimating on-target activity and estimating the risks of off-target activity in parallel. This model produces compartmentalizations, and offers a great potential to increase the safety, efficiency, and accuracy of gene editing (Wessels et al., 2024). (B) AI-driven selection of Cas9 variants: The prediction models based on ML offer the possibility of identifying new homologous nucleases and offer a tool for the rational design of Cas9 variants (e.g., Versions of Sniper-Cas9). Furthermore, these platforms offer the possibility to optimally design the mutants and accelerate the optimization of Cas9 variants with superior editing activity using combinatorial sgRNA-target high throughput synthetic libraries (Li et al., 2024b). (C) AI-driven deaminase engineering: AlphaFold mediated structural prediction correctly offers the possibility to predict the localization of the functional domains of deaminases, based on their amino acid sequence and their classification, based on structural similarity, to represent different families of deaminases. Within the SCP1.201 clade for instance, enzymes such as Sdd (ssDNA deaminase) showed a very strong ssDNA deamination activity and they showed experimentally, their potentials, to be used as a scaffold for the development of the next generation of programmable BEs (Huang et al., 2023). (D) DeepCRISPR for gRNA optimization: The architecture of DeepCRISPR is based on CNN and trained on paired activity databases, is capable of predicting both the on-target efficacy and off-target efficiencies. Thus allowing the data-driven selection of high accuracy gRNAs as well as supporting the efficiency and greater reliability of the applications of CRISPR (Chuai et al., 2018).

TABLE 19.2 AI-based tools for CRISPR gene editing applications.

Tool/Method	Algorithm	Applications	Limitations	Website	References
On- and off-target activity prediction					
ABEdeepoff and CBEdeepoff	Bidirectional long short-term memory (BiLSTM)	Base editor off-target prediction	Excludes sequences lacking editable nucleotides.	http://www.deephf.com/#/bedeep/bedeepoff	Zhang et al. (2023)
Apindel	GloVe embeddings + BiLSTM with attention	On-target efficiency and editing outcome prediction	Restricted to fixed classes of deletions and insertions.	https://github.com/MoonLBH/Apindel	Liu et al. (2022)
AttnToMismatch_CNN	Attention-based transformer + CNN	sgRNA on-/off-target specificity prediction	Depending on the training data, this approach is generally limited in terms of generalization, interpretability, and risk of overfitting.	https://github.com/qiaoliuhub/AttnToCrispr	Liu et al. (2019)
BE-Hive	Gradient boosted regression trees (GBRT) + deep autoregressive model	On-target efficiency and editing outcome prediction	Reduced accuracy for rare or complex editing events; cell-type dependent.	https://www.crisprbehive.design/	Arbab et al. (2020)
CINDEL	Logistic regression (LR)	On-target efficiency prediction	Reduced accuracy for sgRNAs with extreme GC content.	http://big.hanyang.ac.kr/cindel	Kim et al. (2017)
CnnCrispr	CNN + BiLSTM	sgRNA off-target propensity prediction	Due to its reliance on large datasets, this approach has limited generalizability, high computational costs, and interpretability challenges.	https://github.com/LQYoLH/CnnCrispr	Liu et al. (2020)
CrisprBERT	BERT (bidirectional encoder representations from transformers) + BiLSTM	Off-target effect prediction	Trained on a single dataset; does not consider indel mismatches.	https://github.com/OSsari/CrisprBER	Sari et al. (2024)
CRISPR-BERT	BERT + CNN + BiGRU (bidirectional gated recurrent unit)	Off-target activity prediction	Modest performance gains are observed; however, more diverse datasets are required.	https://github.com/BrokenStringx/CRISPR-BERT	Luo et al. (2024)

TABLE 19.2 AI-based tools for CRISPR gene editing applications.—cont'd

Tool/Method	Algorithm	Applications	Limitations	Website	References
CRISPR-DIPOFF	LSTM	CRISPR/Cas9 off-target effect prediction	Trained on a single dataset; does not consider indel mismatches.	https://github.com/tzpranto/CRISPR-DIPOFF	Toufikuzzaman et al. (2024)
CrisprDNT	CNN + LSTM + transformer	Off-target activity prediction	Limited generalization due to data bias and scalability issues.	http://bioinfolab.miamioh.edu/CRISPR-DT	Guan and Jiang (2023)
CRISPRLearner	CNN	sgRNA cleavage efficiency prediction	Model bias restricts scalability across diverse datasets.	https://github.com/gzrgzx/CrisprDNT	Dimauro et al. (2019)
CRISPR-M	CNN + BiLSTM	sgRNA off-target effect prediction	Reduced accuracy under epigenetic modifications; poor generalization.	https://github.com/lyotvincent/CRISPR-M	Sun et al. (2024)
CRISPR-MCA	CNN	Off-target effect prediction	Dataset imbalance complicates off-target prediction.	https://github.com/Yang-k955/CRISPR-MCA	Yang et al. (2024b)
CRISPR-OTE	CNN + BiLSTM	gRNA efficiency prediction	Requires high-quality labels; less effective for rare off-targets.		Xie et al. (2023)
CRISPRscan	LR	On-target efficiency prediction	Optimized for zebrafish; limited accuracy in other species.	https://www.crisprscan.org/	Moreno-Mateos et al. (2015)
Crispr-SGRU	CNN + BiGRU	Off-target activity prediction	Requires large datasets; computationally intensive.		Zhang et al. (2024a)
CrnnCrispr	CNN	sgRNA on-target activity prediction	Relies solely on the sgRNA sequence; ignores the broader biological context.		Zhu et al. (2024)
CROTON	CNN + NAS (neural architecture search)	On-target efficiency & editing outcome prediction	Restricted to limited events (deletions, small indels).	https://github.com/vli31/CROTON	Li et al. (2021a)
DeepCpf1	CNN	On-target efficiency prediction	Relies on chromatin accessibility data, but has limited accuracy in specific cell lines.	https://github.com/kipoi/models/issues/22	Kim et al. (2018)

(*Continued*)

TABLE 19.2 AI-based tools for CRISPR gene editing applications.—cont'd

Tool/Method	Algorithm	Applications	Limitations	Website	References
DeepCRISTL	LSTM	CRISPR/ Cas9 on-target efficiency prediction	Limited generalization, computationally costly, and weak correlation with biological data.	https://github.com/OrensteinLab/DeepCRISTL	Elkayam et al. (2024)
DeepHF	RNN + BiLSTM	gRNA efficiency prediction	Benchmarking is limited by data scarcity across biological contexts.	http://www.DeepHF.com/	Wang et al. (2019)
DeepMEns	CNN + transformer + LSTM + attention	sgRNA on-target activity prediction	Limited by data heterogeneity and experimental variability.	https://github.com/Dingshumei/DeepMEns	Li et al. (2025)
DeepPE	CNN	On-target efficiency prediction	Poor generalization beyond training data; weak performance on external datasets.		Kim et al. (2021)
DL-CRISPR	CNN	Off-target activity prediction	High computational complexity and dataset dependency; prone to overfitting.	https://github.com/yuuuuzhang/DL-CRISPR_offtarget_prediction	Zhang et al. (2020)
FORECasT-BE	GBRT	On-target efficiency prediction	Susceptible to unintended edits at non-targeted cytosines/ adenines.	https://elixir.ut.ee/forecast-be/	Pallaseni et al. (2022)
OpenCRISPR-1	Large language models (LLMs)	Accurate on-target prediction	Accuracy varies depending on the experimental conditions and delivery methods.	https://github.com/Profluent-AI/OpenCRISPR	Ruffolo et al. (2024)
PE_type & PE_Position	RF	On-target efficiency prediction	Biased by sequence preferences, variable efficiency predictions.		Kim et al. (2021)
PREDICT	Attention-based BiRNN (bidirectional RNN)	On-target efficiency & editing outcome prediction	Performance declines under variable chromatin or experimental contexts.	https://www.pridict.it/	Mathis et al. (2023)
SeqCrispr	RNN + CNN + transfer learning (TL)	On-target efficiency prediction	Impacted by an incomplete understanding of gene-phenotype relationships.	https://github.com/qiaoliuhub/AttnToCrispr	Liu et al. (2019)

TABLE 19.2 AI-based tools for CRISPR gene editing applications.—cont'd

Tool/Method	Algorithm	Applications	Limitations	Website	References
sgRNA design and optimization					
AttCRISPR	CNN + RNN	sgRNA on-target activity prediction	Requires high-quality labels, has limited generalizability, and is computationally costly.	https://github.com/South-Walker/AttCRISPR	Xiao et al. (2021)
DeepCas9	CNN	sgRNA activity identification	Focused on clinical assays; limited generalization.	https://github.com/lje00006/DeepCas9	Xue et al. (2019)
DeepCRISPR	Deep convolutional deep neural network (DCDNN)	On-/off-target efficiency prediction	The size of the dataset limits the ability to learn features.	http://www.deepcrispr.net/	Chuai et al. (2018)
DeepSgRNA	CNN	sgRNA efficiency scoring	Does not account for off-target effects.		Shrawgi and Sisodia (2019)
EXPosition	CNN + LSTM + SVM	sgRNA knockout efficiency enhancement	Biased toward coding regions; high computational demand.	https://www.cs.tau.ac.il/~tamirtul/EXPosition/	Cohen et al. (2024)
TransCrispr	Transformer + CNN	sgRNA cleavage efficiency prediction	Computationally heavy with low interpretability.	https://transcrispr.igcz.poznan.pl/transcrispr/	Wan and Jiang (2023)
Editing efficiency prediction					
BE_Seq & BE_Endo	CNN	ABE/CBE editing efficiency prediction	Accuracy is affected by endogenous factors, leading to poor generalization.	http://www.sunlab.fun:3838/BE_Endo/	Zarate et al. (2022)
BEguider	CNN + BiLSTM	Near-PAMless base editor outcome prediction	Reduced accuracy for consecutive A/C sites; poor cross-cell generalization.	http://beguider.bmicc.org	Zhou et al. (2024)
BoostMEC	LightGBM (light gradient boosting machine)	Cleavage efficiency prediction	Validated only for wild-type Cas9; limited variant testing.	https://github.com/oazarate/BoostMEC	Zarate et al. (2022)
DeepIndel	BERT + deep SHAP (SHapley additive exPlanations)	Editing outcome prediction	Weak performance under extreme chromatin conditions; poor cross-cell generalization.		Zhang et al. (2024b)

(*Continued*)

TABLE 19.2 AI-based tools for CRISPR gene editing applications.—cont'd

Tool/Method	Algorithm	Applications	Limitations	Website	References
igRNA-PS	CNN + TL	Near-PAMless ABE outcome prediction	Limited generalization beyond HeLa cell lines.	https://github.com/YQLiCAS/igRNAs	Li et al. (2024b)
panCRISPR toolbox	GBR + SVM	CRISPR experimental improvement	Poor cross-species generalization; lacks a user-friendly interface.	http://github.com/maximilianh/crisporPaper/tree/master/effData	Yasafova et al. (2022)
TEEP	CNN + RNN	TnpB RNA activity prediction	Dependent on the scale of the experimental dataset, it lacks in vivo validation.	https://www.tnpb.app	Marquart et al. (2024)
Cas9 and CRISPR activity prediction					
AcrTransAct	Transformer DNN	Acr-mediated CRISPR–Cas inhibition prediction	Limited to type I CRISPR systems, a small dataset, and a risk of overfitting.	https://github.com/USask-BINFO/AcrTransAct	Hasani et al. (2023)
DeepAcr	CNN + DNN	Anti-CRISPR prediction	Small dataset, high computational cost of AlphaFold, and poor interpretability.	https://proj.cse.cuhk.edu.hk/aihlab/AcrNET/	Li et al. (2023)
DeepSpCas9	DNN + CNN	SpCas9 activity prediction	Training data is insufficient.	https://deepcrispr.info/DeepSpCas9/	Kim et al. (2019)

19.5 AI-enhanced CRISPR target selection

The use of AI in combination with the workflow of CRISPR–Cas9 systems has changed the whole picture of gene editing, because of the greatly improved efficiency and accuracy of gRNA target selection. Thus, in the past gRNA designing was an arduous task because it was based mainly on time-consuming trial-and-error laboratory experimentation combined with empirical knowledge. But now, by means of ML and DL algorithms, AI can investigate large genome data banks with great speed and automate the searching and tuning of the best gRNA targets in the genome. This method is thus greatly increasing the work efficacy (Guo et al., 2022; Tang et al., 2024).

AI models, based on large datasets, are able to accurately predict candidate target sites and simultaneously their editing efficacy and off-target risks. For example, CNNs can investigate entire genomes for gRNA templates close to perfect matches to target genes that meet performance criteria and this may improve and quicken selection (Guo et al., 2022; Tang et al., 2024). A significant advantage of this strategy is the likely reduction in off-target effects that is a concern for other forms of CRISPR/Cas9 genome editing since erroneous binding of gRNAs with non-target sequences is known to result in unwanted genetic and biosafety changes. With the integration of genomic context, gRNA sequence features, and high-order interactions between them, AI models will predict potential off-target site (s) which may be inadvertently edited. AI can also predict gRNA and target binding patterns enabling selection of guides with stable bindings and a higher predicted active element. This integrated design not only improves design speed but also improves the efficiency and accuracy of gene editing, which gives a stronger base for improved ability to modify crop traits (Kim et al., 2025).

The integration of AI into CRISPR pipelines is a very rational approach where gains can be made by increasing efficiencies, decreasing off target events and also maximizing the accuracy with which design targets are identified. This technology leads to the rapid employment of CRISPR technology. Thus, food sciences move near its envisaged future of high productivity crop plant varieties which are sustainable and resilient being realized.

19.6 Improving editing efficiency

19.6.1 Predicting gRNA efficiency

There is huge variability in the efficiency of gene editing by CRISPR/Cas, and multiple studies have explored the factors responsible for this. Properties of both the gRNA and the target sequence, such as GC content, the tendency of the RNA to fold with itself, the composition of the seed sequence, and the local epigenetic configuration, have been shown to be important determinants of editing efficiency (Doench et al., 2016). However, although the interaction of the variables concerned is intricate and not all fully quantifiable experimentally, the rational design of efficient sgRNAs for specific targets has remained a problem. In these circumstances, ML models have been shown to be useful tools in the development of predictive models that can accommodate various sequence and structural characteristics and lead to better predictions of the efficiency of sgRNAs and the identification of highly successful sgRNA candidates (Chen et al., 2024).

The Rule Set models first introduced by Doench et al. greatly expanded and enhanced the list of computational tools available for designing gRNAs, being some of the first and most powerful ones, including the systematic characterization of gRNA performance by genome-wide pooled library screening in human and mouse cells. Their first predictive model included Rule Set one which comprised SVMs along with supplementary linear regression to distinguish gRNAs according to characteristics derived from the gRNA sequence which had a strong correlation with experimental data (Doench et al., 2014). Rule Set two was built upon the Rule Set one framework through the incorporation of greater sequence-contextual and thermodynamic features increasing its predictive power (Doench et al., 2016). Thus rendering the Rule Set two model appropriate for not only CRISPR-KO, but activation (CRISPRa) or interference (CRISPRi). Due mainly to the enhanced availability of greater datasets and feature sets, the group has more recently developed the Rule Set three through the use of a gradient boosting framework-based model which included elements such as tracrRNA identity and more general structural and sequence features greatly improving on their earlier models (DeWeirdt et al., 2022).In addition to these advances, the sgRNA Designer of the Broad Institute has even simplified the gRNA predictions, giving efficient access to all experimental gene editing by powerful computational algorithms (Pallarès Masmitjà et al., 2019). Collectively, these developments represent the gaining evolutionary recurrence of ML-based tools for the design of sgRNAs, but also point out the assistance given by CRISPR of existing precision editing and powerful applications.

In addition to rule-based methods, Li and Lin (2010) developed Elastic Net regression as an ML approach for predicting gRNA's utilizing a regularized regression model. A number of applications have utilized this versatile method for developing predictive models for CRISPRa/CRISPRi and CRISPR-KO, including Spacer Scoring for CRISPR (SSC) (Qin et al., 2019). The real-world utility of Elastic Net can be confirmed by the fact that its use has been incorporated into various design platforms (e.g., CHOPCHOP (Labun et al., 2019), CRISPR-FOCUS (Cao et al., 2017) and E-CRISP (Heigwer et al., 2014). For instance, E-CRISP has an integrated model that includes other models in addition to its use of DNA sequence and secondary structure as features to classify using SVM. These include Rule Set one and the Elastic Net-based SSC. The output from each of these models is used as part of a composite score called Specificity, Annotation, and Efficacy (SAE) to produce a well-rounded and reliable evaluation of potential gRNA candidates (Alipanahi et al., 2023).

Another ML-based approach is WU-CRISPR, which combines features derived from the sequences (depletion of adenine and enrichment of guanine) with other new features to create a more accurate prediction model (Wong et al., 2015). At the same time, SgRNAScorer uses an SVM framework to predict the on-target activity of gRNAs and has been updated in later versions for various Cas systems such as SaCas9 (Qin et al., 2019) and AsCpf1 (Safari et al., 2019). Thus, CRISPR-DO integrates rule-based guidelines with DL strategies, as well as chromatin accessibility information to enhance site selection in the target region and predict on-target and off-target activities (Ma et al., 2016a). Similarly, sgRNAcas9 uses support vector regression to predict the target efficiency by combining sequence feature, positional context, and secondary structure (Xie et al., 2014). Building on these works, DeepCRISPR uses CNNs to capture the complex sequence patterns in an end-to-end manner, which are able to automatically learn and extract informative features for gRNA activity prediction with satisfactory performance and ranking of productive candidates (Chuai et al., 2018). In addition to predicting gRNA activity, some specific computational models have been created for enhancing editing specificity. Elevation employs gradient boosting to address sequence composition, secondary structure, and epigenetic context for off-target predictions (Listgarten et al., 2018).

Although most prediction methods have been prepared mainly with Cas9 in mind, the emergence of new nucleases, including Cpf1 (Cas12a), has led to the development of nucleases-specific models. In addition, the CRISPR/Cpf1 system with a high efficiency will find widespread use in dicots and monocots, such as gene insertion/deletion, base editing, and locus tag. It has been found functional in such important crops as tobacco, rice, soybean, and wheat, with fewer off-target effects than Cas9 (Alok et al., 2020). Based on these improvements, DeepCpf1 was proposed as the first DLM-based model capable of accurately predicting activity of Cpf1 (Kim et al., 2018). With the incorporation of chromatin accessibility into the model framework, DeepCpf1 displayed much improved performance compared to the previous sequence-based methods, providing a basis for predictive modeling of CRISPR-Cpf1 activities based on system-oriented methods. CRISPR-DT is another Cpf1 target efficiency predictor based on SVM which has been recently developed (Zhu and Liang 2019). CRISPR-DT is also shown to possess better performance than some DLM-based models (e.g., DeepCpf1) in the optimization of Cpf1-mediated gene editing, emphasizing further the continuing merit of diverse algorithmic approaches.

Besides these system-specific tools, researchers have also applied state-of-the-art DL architectures, such as CNNs, for improved performance across a wider variety of CRISPR systems. For example, Seq-deepCpf1 learns informative features from input gRNA sequences through the use of CNNs to improve the accuracy of Cpf1 activity predictions (Hui Kwon et al., 2019). Also recently reported DeepCRISPR incorporates CNN-based modeling to predict both on-target knockout efficiency and whole-genome off-target patterns for CRISPR/Cas9 systems. DeepCas9 also takes advantage of CNN to automatically learn sequence determinants and predict gRNA activities on genomes of additional species, extending the cross-species applicability (Bhagwat and Khuri, 2021).

In addition to CNNs, several other architectures have been investigated for embedding more complex sequence relationships. Deep-Bind also uses an LSTM to learn the dependencies between sequence features for improved prediction of protein binding specificity (Hassanzadeh and Wang, 2016). In addition, C-RNNCrispr concatenates CNNs with bidirectional gated recurrent units (BGRU) to predict sgRNA cleavage efficacy, a measure of hybrid architectures' strength for improving predictive performance (Sledzinski et al., 2020).

The majority of the research to date has been centered around the optimization of both CRISPR–Cas9 and Cas12a, but it appears there is a growing interest in next-generation gene-editing tools, whereby some tools, such as BEs and PEs, have editing efficiencies, also varying somewhat significantly. This has led to the development of dedicated prediction tools for these classes of tool, for example, FORECasT-BE for base editing (Pallaseni et al., 2022) and DeepPE for prime editing (Kim et al., 2021). These tools are feature-rich systems-specific models, incorporating high-throughout efficiency data to improve the outcomes of predictions. Analysis and comparison between different prediction methods demonstrates that different prediction methods show different levels of accuracy when tested on the same experimental datasets as a function of selecting different features, the content of the dataset and the algorithm employed (Haeussler et al., 2016).

In conclusion, these developments suggest that ML-based models are important for improvement of predictability of CRISPR-editing outcomes. The vast number of emerging large-scale experimental datasets, as well as the definition of system-specific complexes and subsequent optimization of algorithmic approaches will greatly improve the ability to fully predict proficient and specific gRNAs for use in different editing platforms. Several DL methods have been specifically developed for the possible design of gRNAs, that consider local sequence motifs and long-range dependencies, thereby producing better and greater accurate predictions of gRNA activity and specificity.

19.7 Enhancing editing specificity

19.7.1 Designing high-specificity gRNA

Off-target effects are a major challenge in CRISPR/Cas-based gene editing and happen when gRNAs bind to non-intended genomic locations because of small sequence mismatches, in a similar fashion as unspecific primer binding in PCR, leading to incorrect cleavage by the Cas nuclease. Although the use of gRNA and PAM leads to greater specificity, these off-target effects may mean that mutations occur which lead to possible detrimental mutations altering the fidelity and safety of gene edits (Hsu et al., 2013). Thus, prediction and reduction of potential off-target activity is an important area of study within gene-editing. Ways in which this is achieved is the careful selection of gRNA's and placement of targets, delivery of CRISPR systems and increased use of computational resources (Zetsche et al., 2020). An example is the work by (Das et al., 2023) whereby ML models are constructed to predict cleavage sites using ANNs, SVMs and RF, with the RF model performing best. These results demonstrate the need to carefully balance the on-target efficacy and specificity genome-wide, in order to achieve gene-editing which is efficacious yet safe as mediated by predictive modelling (Das et al., 2023).

To address the off-target effect problem, computational prediction methods has improved from predicting individual models into using a systematic approach to predict model results. These systematic approaches can be viewed as a three-step process. First, for a particular gRNA sequence, the genome is searched intensively to find all sites that correspond to the guide within a certain number of mismatches (Afzal et al., 2020). Second, a predicted cleavage probability score is generated for each off-target site based on sequence similarity, location of mismatch(es), and PAM compatibility (Jacquin et al., 2019). Lastly, these scores are then combined to generate the total off-target risk score for a particular gRNA. The variation between predictive models is substantial and there is considerable influence over the sensitivity and specificity of predictions generated through various algorithmic parameters (e.g., maximum number of mismatches, length of PAM sequence evaluated, and search strategy/algorithm). The large amount of variability generated by predictive models illustrates the significance of selecting and customizing computational algorithms carefully to provide the most accurate and reliable off-target predictions during gRNA design (Alipanahi et al., 2023).

There are a variety of computational tools to predict off-target CRISPR activity, which can be divided conceptually into alignment-based and hypothesis-driven methods. Alignment-based tools, for example, CasOT (Xiao et al., 2014), Cas-OFFinder (Bae et al., 2014), and CRISPR/Cas9 Target Online Predictor (CCTop) (Stemmer et al., 2015), identify potential off-target sites for the searches for complementary sequences in the genome. CCTop, for example, operates in the following manner. (i) The input sequence searches first for the presence of PAM sequences. (ii) A modified version of Bowtie is used to match sequences of gRNA returning up to five mismatches where the target gRNA is present (iii) The gRNAs identified receive an integrated specificity score where this includes a CRISPRater on-target efficiency is available. This combined approach allows prediction of activity, specificity, and risk of off-target mutagenesis to be taken into consideration comprehensively for evaluating CRISPR efficacy and specificity of target gene modification while preferring gRNAs with lower cleavage propensity to maximize gene-targeting efficacy (Stemmer et al., 2015). Alternatively, there are technologies based on the creation of hypotheses aided by tools such CRISPOR (Concordet and Haeussler, 2018) or CHOPCHOP (Labun et al., 2016) based on rules created from experimentally validated hypotheses. There are drawbacks to the alternative approaches as unfortunately there are assumptions contained in the scoring functions of the programs which could in some cases not take into account some of the complex nature of the sequence, and the function relationship.

To address the shortcomings noted above, a category of learning techniques has been recently developed that utilize data from extensive experimental datasets to develop predictive models from training data rather than relying on pre-determined rules. This emphasis is reflected in the tools CROP-IT (Bao et al., 2021), CRISPR-FOCUS (Cao et al., 2017), CRISPRpred (Hwang et al., 2021), DeepSpCas9 (Kim et al., 2019) and SgRNAScorer (Chari et al., 2017), which increasingly utilize statistical and ML techniques to define sequence features and gRNA activity. However, their prediction suffers from a limitation due to the small diversity of types of cells or species represented in the training datasets. This limitation is particularly significant in plant breeding, since polyploidy, repeats, and extensive structural variation provide problems for offtarget prediction, and the necessity for more diverse and abundant genomic resources if the predictions are to be generalized to other species. More recently, learning-based frameworks such as Elevation (Listgarten et al., 2018), DeepCRISPR (Chuai et al., 2018), and CnnCrispr (Liu et al., 2020) have explored involving combinations of multiple sequence and structural descriptors to depict the complex biological relations demonstrated extensively in large datasets, and better predictive performance than the preceding alignment- or hypothesis-based methods. In general, however, such models consist of three major elements: the prediction of putative off-target sites opportunistically in the genome, the scoring of the individual candidates out of the whole for sequence composition or mismatch sign and, if applicable, chromatin accessibility, followed by integration into a general metric of specificity. Although chromatin accessibility is a valuable asset in mammalian systems, it is of limited applicability in plants due to the lack of adequate large scale epigenomic datasets. For example, Elevation employs a two-layer regression technique (seed and extension gradient boosted regression tree) to screen the near matching CRISPR–Cas9 off-target binding sites, whose scores are employed in conjunction with others in the overall specificity prediction (Elevation-aggregate) (Listgarten et al., 2018).

Although significant progress has been made in this area, the limited capacities of ML frameworks to fully learn gRNA–DNA interactions has led to greater interest in the use of DL solutions. In particular, DL models using CNNs are superior as they allow the direct automated learning of complex and hierarchical features from genetic sequences and therefore are able to learn fine level dependencies that traditional ML are likely to miss. Since they deal with large and complex data, DL approaches can produce better predictive performance levels, such as DeepCRISPR, which is an approach developed to integrate together the on- and off-target predictions for gRNA assessment (Chuai et al., 2018). Compared to methods based on hand-designed feature sets, which are used in traditional methods, in DeepCRISPR the model learns what the sequence and flanking context-associated features should look like in the course of training and does not require that all gRNA determinants should be beforehand defined. It incorporates deep neural networks and large scale datasets, can learn complex sequence patterns as well as the epigenetic determinants of knockout (KO) efficacy and is able to learn for improved prediction levels diverse biological variability. All in all, DL-based CRISPR frameworks including DeepCRISPR progress encouragingly toward computational gRNA design tools for efficient and specific gene editing, with the further evidence of the strength of these tools being ISA or more generally, for exploitation of gene-edited phenotypes as an important tool (Alipanahi et al., 2023).

Collectively, these ML- and DL-driven approaches are transforming off-target prediction from heuristic estimates into highly accurate, data-driven models, with particular promise in plants where they can minimize unintended edits and enhance the precision and reliability of CRISPR-based genome engineering.

19.8 AI-driven functional and regulatory gene editing

Functional gene identification and precise editing are central to present applications of genome-editing technologies in agriculture; through its aid, a major role can be played by AI. Using the high technology of big-data analysis available, AI is now able to efficiently predict from complex genomic data sets the functional genes, which promotes efficiency in research and also decreases the reliance upon the traditional method of experimental trial and error. By the conversion of gene expression profiles, genetic polymorphism and phenotypes, AI quickly finds candidates in relation to agronomic traits such as yield, disease resistance or stress tolerance (Liu and Cheng, 2024).

Predictive models have potentials to be used in a more precise way through the use of targeted gene editing to make specific changes to DNA in terms of either the coding region of genes or to the regulatory regions surrounding genes. Additionally, the use of AI-based models can enhance the precision and rapidity of tools for editing proteins like TALENs and CRISPR–Cas9, by modeling how these editing proteins will interact with the desired target sequence. These models are useful for selecting the most optimal gRNAs, reducing off-target effects and increasing the efficiency of editing. Furthermore, predictive models can also predict the likelihood of the intended biological effect of an edit, thereby ensuring that they can be safely and effectively applied in plant breeding (Saini et al., 2025). AI-based functional prediction and engineering can also be applied to the CREs including promoters and enhancers. The integrated approach of using AI to predict protein function while analyzing regulatory networks of genes highlights the potential of AI to provide a very strong and effective means of controlling crop traits and developing new tools for modern plant breeding.

19.9 Optimizing editing proteins

The realization of crop breeding using gene-editing tools still calls for plans to overcome off-target effects and to further improve specificity. Besides excellent gRNA design for specificity, a central strategy is to search and modify novel editing proteins. Several bioinformatic and evolutionary approaches have delivered novel Cas proteins such as the recently characterized Cas13X and Cas13Y (Xu et al., 2021) and LrCas9 (Zhong et al., 2023). Similarly, classical protein engineering methods such as rational design and directed evolution have also produced improved versions of the naturally existing nucleases with high fidelity variations such as SpCas9-NG (Nishimasu et al., 2018), VQR-Cas9 (Kleinstiver et al., 2015), xCas9 (Hu et al., 2018), SpRY (Walton et al., 2020), eSpCas9 (Slaymaker et al., 2016), SaKKH-Cas9 (Kleinstiver et al., 2015), and LZ3 Cas9 (Schmid-Burgk et al., 2020). These novel or engineered proteins generally possess one or more properties (i.e., smaller size, relaxed and/or enlarged PAM requirements, improved efficacy and/or specificity). Some have been optimized for plant gene editing with successful results, indicating their potential to improve the precision and efficacy of crop breeding (Li et al., 2021b; Hua et al., 2019).

However, traditional protein engineering is often a laborious or expensive process, as they generally are conducted using some form of pre-existing structural or functional information to guide mutation. Recent advances in the field of ML and protein structure prediction such as AlphaFold (Senior et al., 2020), AlphaFold2 (Jumper et al., 2021), and RoseTTAFold (Baek et al., 2021), have greatly accelerated the process. These tools now allow accurate prediction of protein structure and functions, providing a framework for the identification of potential mutational hotspots related to substrate-protein interaction. For example AlphaFold2 was employed to elucidate the conformational dynamics of wild-type Cas12a and its variants and elucidate the structural mechanisms underlying mutational effects (Ma et al., 2022). The power of the AI-empowered structural prediction is clearly seen through this work in facilitating the development of novel high specificity Cas variants.

In base editing, off-target mutations can occur when the deaminase enzyme is active even without binding to the sgRNA, this activity can lead to unwanted genomic changes (Sretenovic et al., 2023). For example, cytosine base editors (CBEs), which use rAPOBEC1 deaminase have been shown to create numerous off-target mutations throughout the rice genome (Jin et al., 2019). ML can be used to speed up the identification of new deaminases with low off target activity. A recent study by Huang et al. (2023) demonstrated the ability to predict deaminase domain structure by using AlphaFold2. The authors then grouped several deaminases into clusters based on their structural similarities. Several proteins in the SCP1.201 clade were found to have substantially lower levels of off target activity than the wild type rAPOBEC1. These results demonstrate how ML can facilitate the rapid identification and application of novel precision deaminases that can provide safer and more efficient gene editing capabilities for future applications in plant breeding (Huang et al., 2023).

Collectively, the interplay of protein engineering, structure-based predictions, and ML-enabled enzyme discovery is revolutionizing the area of gene editing. These developments are broadening the range of editing tools available, fine-tuning their precision and efficacy, and setting the stage for new generations of editing platforms tailored to the specific needs of breeding in crops.

19.10 Predicting gene editing outcomes

The DSBs generated by the Cas nuclease during the traditional CRISPR/Cas9 gene editing procedure are primarily repaired by non-template-dependent repair pathways that include NHEJ and microhomology-mediated end joining (MMEJ). Although these repair pathways are error-prone, causing deletions and insertions, their end products are non-random and can often be predicted based on various sequence characteristics, such as GC content, length of microhomologies, and the location of a motif in the sequence (Molla and Yang, 2020). This foundation has laid the groundwork for developing a series of ML models that predict the range of gene editing outcomes in human cells with a high degree of accuracy. Examples of these models include FORECasT (Allen et al., 2019); SPROUT (Leenay et al., 2019); and inDelphi (Shen et al., 2018).

Nevertheless, although these predictive models may find application in plant systems, uncertainty regarding the plant-specific DNA damage response to CRISPR/Cas9-induced editing inhibits the identification of the features which are required in order to construct the model or models. To this end, Li et al. (2021a) here proposed a DL framework CROTON by a process using deep multitask CNNs in conjunction with NAS. Aside from enabling better prediction of editing results from raw sequence data, it has the additional benefit of enabling automatic feature extraction and model optimization. It is therefore possible that by these and other AI methods, repair outcome predictions for new targets will be made with greater accuracy, optimizing CRISPR designs and helping downstream validation, in short streamlining plant genome engineering.

19.11 Identifying cis-regulatory elements and key sites

Recently, researchers have discovered potential in combining ML with CRISPR-based gene editing to develop methods to finely adjust multiple genetic factors (polygenes) through regulatory manipulation based on their structure. For example (Zhou et al., 2023), developed a sophisticated computational model using an integrated approach of both ML with empirical knowledge to predict how site-specific modifications to promoters can influence gene expression. They found several key nucleotide positions that were significantly associated with transcriptional regulation. Then, they combined this prediction model with the CRISPR–Cas12a promoter editing system to generate multiple different levels of quantitative trait variation in crops, and thereby produce novel engineered alleles (Zhou et al., 2023). Once the genomic targets have been established, different generations of newer CRISPR-related tools, such as base editing and prime editing, are used by breeders to make exactly targeted sequence changes. Both of these techniques have been used to enhance many agronomic traits in numerous crop species (Butt et al., 2020; Mao et al., 2019). The application of ML enhances the precision and speed of those design processes by predicting edit efficiencies and outcomes and ultimately matching any genomic changes to desired improvement in traits.

19.12 AI-optimized gene-editing workflows

In the context of gene-editing experiments, AI evolves from a predictive model to an integrative optimizer of the entire workflow of gene-editing experiments. Specifically, it employs sophisticated computer algorithms to determine the optimal strategies for gene editing by selecting optimal CRISPR–Cas9 target sites. The latter enable a parallel characterization of both on-target efficiency and potential off-target events, leading to very accurate gRNA design. Furthermore, in silico prediction of editing outcomes using AI, as well as prediction of any associated risks, helps researchers explore the feasibility of different strategies before getting into the lab. This mode of virtual pre-validation allows for a more rational decision making, less "trial and error" in the empirical situation and gives information as to which is the best strategy. Therefore this approach of using AI to help believe, design and validate not only saves time and resources, but considerably increases the efficiency, safety and success of gene-editing applications (Dixit et al., 2024).

19.13 Challenges and limitations

The intersection of AI and genetic plant or crop improvement is transformational. Nevertheless, some of the related problems, including poor quality of data, a lack of uniformity of data, inadequate performance of multi-omic integration methods, non-readable models, computing intensive methodologies and lack of an interdisciplinary skilled workforce are yet to be addressed. These problems are also essential to fully realize the power of AI as a means to attain precision breeding and sustainable agricultural innovation. A major problem lies in the inability to utilize the very large, often low-quality, genomic data generated by the high-technology sequencing methods. Those raw data often in terabytes or petabytes are subject frequently to noise, missing, inconsistencies, which impinge on the reproducibility and accuracy of the AI (Li et al., 2024c). The high dimensionality of the data meant that its computing and storage requirements are high. To cater for this improved sequencing accuracy, bioinformatics AI pipelines, and data curation were required. It was important to have standardized and interoperable data repositories. The improved quality of data and its harmonization will lead to improved accuracy of the models used and as a result improved gene and trait discoveries and improved precision breeding activity.

Along with data quality, integrating multiple layers of omics, including genomics, transcriptomics, proteomics and metabolomics, remains a challenging task due to lack of standardized data formats, resolution differences, and the complex inter-layer biological modelling hinder AI from capturing the interaction factors underpinning complex traits (Xu et al., 2022). This requires the close integration of all the areas discussed above, including bioinformatics, computer science and statistics, to generate global methodologies for multi-omics data fusion that support the overall understanding of genes functions and networks of interactions.

Another major disadvantage is that the model is not very interpretable or transparent. Even though the predictive performance is very high, DL architectures are "black boxes", meaning that the rationale for predictions is unclear, undermining scientific trust and reproducibility (Ali et al., 2023). In the context of CRISPR-related tasks, it is necessary for AI not only to accurately predict CRISPR-targeting sites, but also to explain mechanisms of editing and off-target effects. Explainable AI (XAI) such as attention mechanisms, feature attribution techniques as well as interpretable models is an up-and-coming field of research with large potential to ease the interpretability and biological relevance of systems (Mohan et al., 2025).

The computational challenges associated with the incorporation of AI into genomics are equally important. The scale of these models which may have millions or billions of parameters means that they commonly need to be trained on large scale high performance computing (HPC) infrastructure namely GPU or TPU clusters that are inaccessible to a large majority of research institutions (Dixit et al., 2024). The complexity of the model also leads to increased overfitting risk and decreased explainability. To overcome these difficulties, pipelined and scalable approaches should be employed that improve the limitations of storage bandwidth and extensive computational costs through transfer learning, federated learning, and optimized training.

Finally, the development of AI-for-agriculture will need prudent interdisciplinary collaboration and talent development. Successful crop improvement needs the integration of capabilities in biology, gene editing, agricultural science, computer science, and statistics. However, lack of inter-disciplinary experts hinders the realization of AI-based breeding (Dixit et al., 2024). The development of researchers skilled in both computational and biological sciences, which will be facilitated by proper education, training, and collaborative practices will be critical in ensuring sustainability, equitability and globalization of the yields of AI-driven innovations in agriculture.

19.14 Future perspectives

The application of AI and gene-editing technologies together should be beneficial in the revolution of crop improvement. Due to newer and better computational, algorithmic and data resources, AI will do better in modeling the complex relationships among genes and phenotypes, which is necessary for more accurate and cheaper genome editing. It has not escaped notice that a variety of gene editing systems have been developed, such as base editing and non-cutting changes, in the end improving the precision of on-targets, reducing the number of off-targets and improving the efficiency of both on- and off-target hits. These results will combine together to improve the pace for improvement of crop traits and to demonstrate a sustainable, innovative and open access system for fundamentally addressing global agricultural problems (Dublino and Ercolano, 2025).

AI is furthermore achieving great breakthroughs among personalized plant breeding and precision agriculture. Scouring immense genomic databases can also contribute to identify the genes responsible for having desirable agronomic traits and adapt breeding programs aim at a local climate or agricultural method. We all seem to have the same ambition of proceeding to a more directed approach in order that be obtained more yield, better quality and more food resistant to climate problems to supply definite answers to difficult problems locally. Meanwhile, AI for precision agriculture is making real time monitoring and predictive analytics for maximizing inputs and productivity by means of sustainable management of resources which, should it give an unprecedented and solid backing to the food security in the world (Dublino and Ercolano, 2025).

One of the major lines that should be considered for future research is to integrate more widely the multi-omic data using alternatives based on AI. Genomics, transcriptomics, proteomics, epigenomis, and metabolomics or phenomics provide different levels of information but combination of these is absolutely necessary for giving an extensive knowledge of the functions of genes the communication of the regulations of genes or training of genes and environment in connection with the agronomical or other traits (Fan et al., 2025). AI, particularly multi-modal learning specifically suited for developing integrated framework systems, is especially well placed for syntactically integrating heterogeneous datasets and extracting biologically meaningful patterns of growth parameters, stress tolerance and yield (Farooq et al., 2024). As the sequencing technologies and computational implementation develop offering automated, traceable and integrated datasets, the integration of syntactically integrated datasets for explaining the basis of predictions will improve reproducibility and transparency in applicable sections, utilizing structural bridges from empirical agriculture into smart, data-based crop systems facilitating alterations for climate resilience and productivity sustainability.

Beyond the potential hosted within genomics, epigenetics is another area with much scope for innovation. Epigenetic mechanisms mediated through other processes, such as histone alterations, DNA methylation and non-coding RNAs, are also crucial for gene regulation particularly without changing the nucleic basis of gene sequences (Berger et al., 2009). Epigenome editing using CRISPR, employing dCas9 that has been rendered catalytically inactive coupled to enzymes for chromatin modification, provides a gene specific control over transcriptional activity (Konermann et al., 2015; Zalatan et al., 2015). Such strategies have been demonstrated to successfully re-program flowering time and stress responses in Arabidopsis (Papikian et al., 2019). These systems can be improved by the inclusion of ML for optimization of gRNA design and prediction of editing (Yang et al., 2023). With the introduction of such tools beyond model plants, coupling CRISPR-based epigenome editing with ML would provide further understanding of epigenetic mechanisms and data-driven breeding approaches.

Likewise, single-cell methods are also transforming functional genomics due to their high resolution analyses of cellular heterogeneity. Approaches such as Perturb-seq, CRISP-seq, and CROP-seq integrate pooled CRISPR screening with single-cell RNA sequencing to allow for the characterization of gene function and regulatory networks with an unprecedented level of accuracy (Datlinger et al., 2017; Dixit et al., 2016; Rubin et al., 2019). Their application in plants is limited however by considerable variation in size and thick cell walls which make isolation of single cells difficult (Shulse et al., 2019); however, ML is proving to be a useful way to uncover important regulatory patterns from limited amounts of data and to improve interpretive accuracy (Dixit et al., 2016). Gradual advances in plant CRISPR screening and single cell sequencing are establishing a dynamic foundation for their deeper integration with the functional genomics, systems biology and crop improvement (Lu et al., 2017; Tao et al., 2022).

De novo domestication of wild relatives also shows potential, involving targeted gene editing to retain ecological resilience, as well as improve agronomic traits (Zsögön et al., 2017). Applications that show positive promise include enhancements to wild-type allotetraploid rice by gene editing of homologous genes identified in diploid forms (Yu et al., 2021). ML technologies may expedite these approaches through identification of genes involved in domestication processes, simulating gene interactions, predicting phenotypic consequences of alterations in combinations of alleles, thus refining targeted editing strategies. AI analytics collaborating with targeted gene editing leads on to predictive data-driven agriculture as well as domestication of productive, resilient, and sustainable cultivars for future agriculture (Chen et al., 2024).

In summary, the successful advance of AI and gene editing to agriculture depend on inter-disciplinary co-operation and talent fostering of the kind necessary for winter wheat improvement involves the biology and the agronomy, computer science and statistics, and it is maintenance of a sound inter-disciplinary co-ordination that is the sine qua non for the translation of the product of research into practice (Zhang et al., 2025). Education and training must consider the changing need for cross-disciplinary skills and creative problem-solving in creating the workforce responsible for innovating in the area of AI-breeding. Improving how scientific disciplines coordinate with each other and providing manpower for talent enhancement can show how cross-pollination between AI and gene-editing will effect continued innovation and globalization in crop genomics in the improvement of climate-resilient crop varieties as sustainable food systems worldwide in the future.

19.15 Conclusion

The merging of AI and CRISPR gene editing will revolutionize the manner in which crops are bred, changing it to a predictive data-driven science. The beauty of this synergy is the ability to incorporate high-throughput genomic and phenotypic data to ascertain the regulome(s), predict candidate editing targets, and rationally design intelligent precision genome modification strategies. This invites the rapid development of new crop varieties characterized by desirable traits, such as higher yield, resistance to stress, and/or disease, which meet global challenges in the arena of climate change and food insecurity. However, AI should not replace classical breeding techniques, but rather enhance and amplify their efficacy.

However fundamental limitations exist in terms of the potential use of AI-directed CRISPR widely. One main limitation is the reliance of many ML's on restricted sets of training data typically from non-plant systems. This problem, together with the genomic variability of the different crops but also the complexity of epigenetic regulation and diversity of data created by differing platforms, means that existing predictive models cannot be readily exploited. The success of increasing plant-specific multi-omics reference resources and improving next generation sequencing assays is important to avoid these problems, but these aims are entirely dependent on collaborative engagement between computational scientists, molecular biologists, and plant breeders.

The incorporation of AI, CRISPR, and multi-omics technologies will transform the arena of agricultural research. This triad of power promises rapid and precise genetic improvements developed to fit particular agro-ecological conditions. Aided by the shortened time periods of crop production modification, and reduced resource allocation, the AI lead genome editing process is undoubtedly a means of achieving sustainability in agriculture. When used effectively, this will lead to momentous changes in the world food system and new frontiers in agricultural innovation.

Author contributions

E.M. and X.F. conceptualized the manuscript; E.M. prepared the initial draft; E.M., I.I.M., D.Z., J.G., Z.C., and X.F. reviewed and revised the manuscript. All authors have read and agreed to the published version of the manuscript.

Funding

This research was supported by the Biological Breeding-National Science and Technology Major Project of China (2023ZD0403203) and the National Natural Science Foundation of China (U21A20215 and 32,488,102).

Acknowledgments

We thank the editors and reviewers for their critical comments on the manuscript.

Declaration

"The authors declare that no generative artificial intelligence tools were used in the writing or preparation of this chapter. The content reflects the authors' own scholarly work, and the authors assume full responsibility for all statements, interpretations, and conclusions presented".

References

Abdul Rahim, A., Uzair, M., Rehman, N., Fiaz, S., Attia, K.A., Abushady, A.M., Yang, S.H., Khan, M.R., 2024. CRISPR/Cas9 mediated TaRPK1 root architecture gene mutagenesis confers enhanced wheat yield. J. King Saud Univ. Sci. 36 (2), 103063. https://doi.org/10.1016/j.jksus.2023.103063.

Afzal, S., Sirohi, P., Singh, N.K., 2020. A review of CRISPR associated genome engineering: application, advances and future prospects of genome targeting tool for crop improvement. Biotechnol. Lett. 42, 1611–1632. https://doi.org/10.1007/s10529-020-02950-w.

Ali, S., Abuhmed, T., El-Sappagh, S., Muhammad, K., Alonso-Moral, J.M., Confalonieri, R., Guidotti, R., Ser, J.Del, Díaz-Rodríguez, N., Herrera, F., 2023. Explainable artificial intelligence (XAI): what we know and what is left to attain trustworthy artificial intelligence. Inf. Fusion 99, 101805. https://doi.org/10.1016/j.inffus.2023.101805.

Alipanahi, R., Safari, L., Khanteymoori, A., 2023. CRISPR genome editing using computational approaches: a survey. Front Bioinforma 2. https://doi.org/10.3389/fbinf.2022.1001131.

Allen, F., Crepaldi, L., Alsinet, C., Strong, A.J., Kleshchevnikov, V., De Angeli, P., Páleníková, P., Khodak, A., Kiselev, V., Kosicki, M., Bassett, A.R., Harding, H., Galanty, Y., Muñoz-Martínez, F., Metzakopian, E., Jackson, S.P., Parts, L., 2019. Predicting the mutations generated by repair of Cas9-induced double-strand breaks. Nat. Biotechnol. 37 (1), 64–72. https://doi.org/10.1038/nbt.4317.

Alok, A., Sandhya, D., Jogam, P., Rodrigues, V., Bhati, K.K., Sharma, H., Kumar, J., 2020. The rise of the CRISPR/Cpf1 system for efficient genome editing in plants. Front. Plant Sci. 11. https://doi.org/10.3389/fpls.2020.00264.

Anzalone, A.V., Randolph, P.B., Davis, J.R., Sousa, A.A., Koblan, L.W., Levy, J.M., Chen, P.J., Wilson, C., Newby, G. A., Raguram, A., Liu, D.R., 2019. Search-and-replace genome editing without double-strand breaks or donor DNA. Nature 576 (7785), 149–157. https://doi.org/10.1038/s41586-019-1711-4.

Arbab, M., Shen, M.W., Mok, B., Wilson, C., Matuszek, Ż., Cassa, C.A., Liu, D.R., 2020. Determinants of base editing outcomes from target library analysis and machine learning. Cell 182 (2), 463. https://doi.org/10.1016/j.cell.2020.05.037.

Bae, S., Park, J., Kim, J.-S., 2014. Cas-OFFinder: a fast and versatile algorithm that searches for potential off-target sites of Cas9 RNA-guided endonucleases. Bioinformatics 30, 1473–1475. https://doi.org/10.1093/bioinformatics/btu048.

Baek, M., DiMaio, F., Anishchenko, I., Dauparas, J., Ovchinnikov, S., Lee, G.R., Wang, J., Cong, Q., Kinch, L.N., Dustin Schaeffer, R., Millán, C., Park, H., Adams, C., Glassman, C.R., DeGiovanni, A., Pereira, J.H., Rodrigues, A.V., Van Dijk, A.A., Ebrecht, A.C., Opperman, D.J., Sagmeister, T., Buhlheller, C., Pavkov-Keller, T., Rathinaswamy, M.K., Dalwadi, U., Yip, C.K., Burke, J.E., Christopher Garcia, K., Grishin, N.V., Adams, P.D., Read, R.J., Baker, D., 2021. Accurate prediction of protein structures and interactions using a three-track neural network. Science 373 (6557), 871–876. https://doi.org/10.1126/science.abj8754.

Bao, X.R., Pan, Y., Lee, C.M., Davis, T.H., Bao, G., 2021. Tools for experimental and computational analyses of off-target editing by programmable nucleases. Nat. Protoc. 16 (1), 10–26. https://doi.org/10.1038/s41596-020-00431-y.

Barrangou, R., Fremaux, C., Deveau, H., Richards, M., Boyaval, P., Moineau, S., Romero, D.A., Horvath, P., 2007. CRISPR provides acquired resistance against viruses in prokaryotes. Science 315 (5819), 1709–1712. https://doi.org/10.1126/science.1138140.

Berger, S.L., Kouzarides, T., Shiekhattar, R., Ali, S., 2009. An operational definition of epigenetics: Figure 1. Gene Dev. 23 (7), 781–783. https://doi.org/10.1101/gad.1787609.

Bhagwat, N., Khuri, N., 2021. Predicting Targets for Genome Editing with Long Short Term Memory Networks. Springer Science and Business Media LLC, pp. 657–670. https://doi.org/10.1007/978-3-030-71051-4_52.

Bortesi, L., Fischer, R., 2015. The CRISPR/Cas9 system for plant genome editing and beyond. Biotechnol. Adv. 33 (1), 41–52. https://doi.org/10.1016/j.biotechadv.2014.12.006.

Butt, H., Rao, G.S., Sedeek, K., Aman, R., Kamel, R., Mahfouz, M., 2020. Engineering herbicide resistance via prime editing in rice. Plant Biotechnol. J. 18 (12), 2370–2372. https://doi.org/10.1111/pbi.13399.

Cao, Q., Ma, J., Chen, C.-H., Xu, H., Chen, Z., Li, W., Liu, X.S., Zhang, J., 2017. CRISPR-FOCUS: a web server for designing focused CRISPR screening experiments. PLoS One 12 (9), e0184281. https://doi.org/10.1371/journal.pone.0184281.

Chari, R., Yeo, N.C., Chavez, A., Church, G.M., 2017. SgRNA scorer 2.0: a species-independent model to predict CRISPR/Cas9 activity. ACS Synth. Biol. 6 (5), 902–904. https://doi.org/10.1021/acssynbio.6b00343.

Chen, L., Liu, G., Zhang, T., 2024. Integrating machine learning and genome editing for crop improvement. aBIOTECH 5, 262–277. https://doi.org/10.1007/s42994-023-00133-5.

Chuai, G., Ma, H., Yan, J., Chen, M., Hong, N., Xue, D., Zhou, C., Zhu, C., Chen, K., Duan, B., Gu, F., Qu, S., Huang, D., Jia, W., Qi, L., 2018. DeepCRISPR: optimized CRISPR guide RNA design by deep learning. Genome Biol. 19 (1). https://doi.org/10.1186/s13059-018-1459-4.

Cohen, S., Bergman, S., Lynn, N., Tuller, T., 2024. A tool for CRISPR-Cas9 sgRNA evaluation based on computational models of gene expression. Genome Med. 16 (1). https://doi.org/10.1186/s13073-024-01420-6.

Concordet, J.P., Haeussler, M., 2018. CRISPOR: intuitive guide selection for CRISPR/Cas9 genome editing experiments and screens. Nucleic Acids Res. 46 (1), W242. https://doi.org/10.1093/nar/gky354.

Das, J., Kumar, S., Mishra, D.C., Chaturvedi, K.K., Paul, R.K., Kairi, A., 2023. Machine learning in the estimation of CRISPR-Cas9 cleavage sites for plant system. Front. Genet. 13. https://doi.org/10.3389/fgene.2022.1085332.

Datlinger, P., Rendeiro, A.F., Schmidl, C., Krausgruber, T., Traxler, P., Klughammer, J., Schuster, L.C., Kuchler, A., Alpar, D., Bock, C., 2017. Pooled CRISPR screening with single-cell transcriptome readout. Nat. Methods 14 (3), 297–301. https://doi.org/10.1038/nmeth.4177.

de Lange, J., Lanier Nalley, L., Yang, W., Shew, A., de Steur, H., 2022. The future of CRISPR gene editing according to plant scientists. iScience 25 (9), 105012. https://doi.org/10.1016/j.isci.2022.105012.

DeWeirdt, P.C., McGee, A.V., Zheng, F., Nwolah, I., Hegde, M., Doench, J.G., 2022. Accounting for small variations in the tracrRNA sequence improves sgRNA activity predictions for CRISPR screening. Nat. Commun. 13 (1). https://doi.org/10.1038/s41467-022-33024-2.

Dhugga, K.S., 2022. Gene editing to accelerate crop breeding. Front. Plant Sci. 13. https://doi.org/10.3389/fpls.2022.889995.

Dimauro, G., Colagrande, P., Carlucci, R., Ventura, M., Bevilacqua, V., Caivano, D., 2019. CRISPRLearner: a deep learning-based system to predict CRISPR/Cas9 sgRNA on-target cleavage efficiency. Electronics 8 (12), 1478. https://doi.org/10.3390/electronics8121478.

Dixit, A., Parnas, O., Li, B., Chen, J., Fulco, C.P., Jerby-Arnon, L., Marjanovic, N.D., Dionne, D., Burks, T., Raychowdhury, R., Adamson, B., Norman, T.M., Lander, E.S., Weissman, J.S., Friedman, N., Regev, A., 2016. Perturb-seq: Dissecting molecular Circuits with scalable single-cell RNA profiling of pooled genetic screens. Cell 167 (7), 1853. https://doi.org/10.1016/j.cell.2016.11.038.

Dixit, S., Kumar, A., Srinivasan, K., Durai Raj Vincent, P.M., Krishnan, N.R., 2024. Advancing genome editing with artificial intelligence: opportunities, challenges, and future directions. Front. Bioeng. Biotechnol. 11. https://doi.org/10.3389/fbioe.2023.1335901.

Doench, J.G., Hartenian, E., Graham, D.B., Tothova, Z., Hegde, M., Smith, I., Sullender, M., Ebert, B.L., Xavier, R.J., Root, D.E., 2014. Rational design of highly active sgRNAs for CRISPR-Cas9-mediated gene inactivation. Nat. Biotechnol. 32 (12), 1262–1267. https://doi.org/10.1038/nbt.3026.

Doench, J.G., Fusi, N., Sullender, M., Hegde, M., Vaimberg, E.W., Donovan, K.F., Smith, I., Tothova, Z., Wilen, C., Orchard, R., Virgin, H.W., Listgarten, J., Root, D.E., 2016. Optimized sgRNA design to maximize activity and minimize off-target effects of CRISPR-Cas9. Nat. Biotechnol. 34 (2), 184–191. https://doi.org/10.1038/nbt.3437.

Duan, L., Ouyang, Kan, Xu, X., Xu, L., Wen, C., Zhou, X., Qin, Z., Xu, Z., Sun, W., Liang, Y., 2021. Nanoparticle delivery of CRISPR/Cas9 for genome editing. Front. Genet. 12. https://doi.org/10.3389/fgene.2021.673286.

Dublino, R., Ercolano, M., 2025. Artificial intelligence redefines agricultural genetics by unlocking the enigma of genomic complexity. The Crop Journal 13 (5), 1350–1362. https://doi.org/10.1016/j.cj.2025.05.008.

Elkayam, S., Tziony, I., Orenstein, Y., Valencia, A., 2024. DeepCRISTL: deep transfer learning to predict CRISPR/Cas9 on-target editing efficiency in specific cellular contexts. Bioinformatics 40 (8). https://doi.org/10.1093/bioinformatics/btae481.

Eraslan, G., Avsec, Ž., Gagneur, J., Theis, F.J., 2019. Deep learning: new computational modelling techniques for genomics. Nat. Rev. Genet. 20 (7), 389–403. https://doi.org/10.1038/s41576-019-0122-6.

Fan, B.-L., Chen, L.-H., Chen, L.-L., Guo, H., 2025. Integrative multi-omics approaches for identifying and characterizing biological elements in crop traits: current progress and future prospects. Int. J. Mol. Sci. 26 (4), 1466. https://doi.org/10.3390/ijms26041466.

Farhaoui, A., Kouighat, M., Khadiri, M., Boutagayout, A., Assouguem, A., El Jarroudi, M., Lahlali, R., 2025. Innovative Approaches to Alleviate Climate Stress in Crop Production. Springer Science and Business Media LLC, pp. 271–307. https://doi.org/10.1007/978-3-031-84802-5_13.

Farooq, M.A., Gao, S., Hassan, M.A., Huang, Z., Rasheed, A., Hearne, S., Prasanna, B., Li, X., Li, H., 2024. Artificial intelligence in plant breeding. Trends Genet. 40 (10), 891–908. https://doi.org/10.1016/j.tig.2024.07.001.

Grover, P., Kar, A.K., Dwivedi, Y.K., 2022. Understanding artificial intelligence adoption in operations management: insights from the review of academic literature and social media discussions. Ann. Oper. Res. 308, 177–213. https://doi.org/10.1007/s10479-020-03683-9.

Guan, Z., Jiang, Z., 2023. Transformer-based anti-noise models for CRISPR-Cas9 off-target activities prediction. Briefings Bioinf. 24 (3). https://doi.org/10.1093/bib/bbad127.

Guo, W., Li, D., Xie, S., Yang, L., Li, C., Tian, J., et al., 2022. Artificial Intelligence Accelerates Epigenetics and Plant Breeding. https://doi.org/10.13304/j.nykjdb.2022.1031.

Haeussler, M., Schönig, K., Eckert, H., Eschstruth, A., Mianné, J., Renaud, J.B., Schneider-Maunoury, S., Shkumatava, A., Teboul, L., Kent, J., Joly, J.S., Concordet, J.P., 2016. Evaluation of off-target and on-target scoring algorithms and integration into the guide RNA selection tool CRISPOR. Genome Biol. 17 (1). https://doi.org/10.1186/s13059-016-1012-2.

Haque, S., Ahmad, J.S., Clark, N.M., Williams, C.M., Sozzani, R., 2019. Computational prediction of gene regulatory networks in plant growth and development. Curr. Opin. Plant Biol. 47, 96–105. https://doi.org/10.1016/j.pbi.2018.10.005.

Hasani, M., Trost C, N., Timmerman, N., Jin, L., 2023. AcrTransAct: pre-trained protein transformer models for the detection of type I anti-CRISPR activities. In: Proc. 14th ACM Int. Conf. Bioinformatics, Comput. Biol. Heal. Informatics, 169. ACM, New York, NY, USA, pp. 1–6. https://doi.org/10.1145/3584371.3613007.

Hassanzadeh, H.R., Wang, M.D., 2016. DeeperBind: enhancing prediction of sequence specificities of DNA binding proteins. 2016 IEEE Int. Conf. Bioinforma. Biomed. 2016, 178–183. https://doi.org/10.1109/BIBM.2016.7822515. IEEE.

Heigwer, F., Kerr, G., Boutros, M., 2014. E-CRISP: fast CRISPR target site identification. Nat. Methods 11, 122–123. https://doi.org/10.1038/nmeth.2812.

Hsu, P.D., Scott, D.A., Weinstein, J.A., Ran, F.A., Konermann, S., Agarwala, V., Li, Y., Fine, E.J., Wu, X., Shalem, O., Cradick, T.J., Marraffini, L.A., Bao, G., Zhang, F., 2013. DNA targeting specificity of RNA-guided Cas9 nucleases. Nat. Biotechnol. 31 (9), 827–832. https://doi.org/10.1038/nbt.2647.

Hu, J.H., Miller, S.M., Geurts, M.H., Tang, W., Chen, L., Sun, N., Zeina, C.M., Gao, X., Rees, H.A., Lin, Z., Liu, D.R., 2018. Evolved Cas9 variants with broad PAM compatibility and high DNA specificity. Nature 556 (7699), 57–63. https://doi.org/10.1038/nature26155.

Hua, K., Tao, X., Zhu, J., 2019. Expanding the base editing scope in rice by using Cas9 variants. Plant Biotechnol. J. 17, 499–504. https://doi.org/10.1111/pbi.12993.

Huang, J., Lin, Q., Fei, H., He, Z., Xu, H., Li, Y., Qu, K., Han, P., Gao, Q., Li, B., Liu, G., Zhang, L., Hu, J., Zhang, R., Zuo, E., Luo, Y., Ran, Y., Qiu, J.-L., Zhao, K.T., Gao, C., 2023. Discovery of deaminase functions by structure-based protein clustering. Cell 186 (15), 3182. https://doi.org/10.1016/j.cell.2023.05.041.

Hwang, G.-H., Song, B., Bae, S., 2021. Current widely-used web-based tools for CRISPR nucleases. Base Editors, and Prime Editors. Gene Genome 1, 100004. https://doi.org/10.1016/j.ggedit.2021.100004.

Jacquin, A.L.S., Odom, D.T., Lukk, M., 2019. Crisflash: open-source software to generate CRISPR guide RNAs against genomes annotated with individual variation. Bioinformatics 35, 3146–3147. https://doi.org/10.1093/bioinformatics/btz019.

Jiang, F., Lu, Y., Chen, Y., Cai, Di, Li, G., 2020. Image recognition of four rice leaf diseases based on deep learning and support vector machine. Comput. Electron. Agric. 179, 105824. https://doi.org/10.1016/j.compag.2020.105824.

Jiang, M., Fu, S.T., Wang, X.F., 2022. Current status and prospects of CRISPR/Cas9 genome editing technology in tomato. Northwest J Bot 42, 348–360.

Jin, S., Zong, Y., Gao, Q., Zhu, Z., Wang, Y., Qin, P., Liang, C., Wang, D., Qiu, J.-L., Zhang, F., Gao, C., 2019. Cytosine, but not adenine, base editors induce genome-wide off-target mutations in rice. Science 364 (6437), 292–295. https://doi.org/10.1126/science.aaw7166.

Jumper, J., Evans, R., Pritzel, A., Green, T., Figurnov, M., Ronneberger, O., Tunyasuvunakool, K., Bates, R., Žídek, A., Potapenko, A., Bridgland, A., Meyer, C., Kohl, S.A.A., Ballard, A.J., Cowie, A., Romera-Paredes, B., Nikolov, S., Jain, R., Adler, J., Back, T., Petersen, S., Reiman, D., Clancy, E., Zielinski, M., Steinegger, M., Pacholska, M., Berghammer, T., Bodenstein, S., Silver, D., Vinyals, O., Senior, A.W., Kavukcuoglu, K., Kohli, P., Hassabis, D., 2021. Highly accurate protein structure prediction with AlphaFold. Nature 596 (7873), 583–589. https://doi.org/10.1038/s41586-021-03819-2.

Kim, H.K., Song, M., Lee, J., Menon, A.V., Jung, S., Kang, Y.M., Choi, J.W., Woo, E., Koh, H.C., Nam, J.W., Kim, H., 2017. In vivo high-throughput profiling of CRISPR-Cpf1 activity. Nat. Methods 14 (2), 153–159. https://doi.org/10.1038/nmeth.4104.

Kim, H.K., Min, S., Song, M., Jung, S., Choi, J.W., Kim, Y., Lee, S., Yoon, S., Kim, H., 2018. Deep learning improves prediction of CRISPR-Cpf1 guide RNA activity. Nat. Biotechnol. 36 (3), 239–241. https://doi.org/10.1038/nbt.4061.

Kim, H.K., Kim, Y., Lee, S., Min, S., Bae, J.Y., Choi, J.W., Park, J., Jung, D., Yoon, S., Kim, H.H., 2019. SpCas9 activity prediction by DeepSpCas9, a deep learning–based model with high generalization performance. Sci. Adv. 5 (11). https://doi.org/10.1126/sciadv.aax9249.

Kim, H.K., Yu, G., Park, J., Min, S., Lee, S., Yoon, S., Kim, H.H., 2021. Predicting the efficiency of prime editing guide RNAs in human cells. Nat. Biotechnol. 39 (2), 198–206. https://doi.org/10.1038/s41587-020-0677-y.

Kim, M.-gyeong, Go, M.-ji, Kang, S.-H., Jeong, S.-hwan, Lim, K., 2025. Revolutionizing CRISPR technology with artificial intelligence. Exp. Mol. Med. 57 (7), 1419–1431. https://doi.org/10.1038/s12276-025-01462-9.

Kleinstiver, B.P., Prew, M.S., Tsai, S.Q., Topkar, V.V., Nguyen, N.T., Zheng, Z., Gonzales, A.P.W., Li, Z., Peterson, R. T., Yeh, J.R.J., Aryee, M.J., Joung, J.K., 2015. Engineered CRISPR-Cas9 nucleases with altered PAM specificities. Nature 523 (7561), 481–485. https://doi.org/10.1038/nature14592.

Komor, A.C., Kim, Y.B., Packer, M.S., Zuris, J.A., Liu, D.R., 2016. Programmable editing of a target base in genomic DNA without double-stranded DNA cleavage. Nature 533, 420–424. https://doi.org/10.1038/nature17946.

Konermann, S., Brigham, M.D., Trevino, A.E., Joung, J., Abudayyeh, O.O., Barcena, C., Hsu, P.D., Habib, N., Gootenberg, J.S., Nishimasu, H., Nureki, O., Zhang, F., 2015. Genome-scale transcriptional activation by an engineered CRISPR-Cas9 complex. Nature 517 (7536), 583–588. https://doi.org/10.1038/nature14136.

Kwon, K.H., Seonwoo, M., Myungjae, S., Soobin, J., Woo, C.J., Younggwang, K., Sangeun, L., Sungroh, Y., Henry, K. H., 2019. DeepCpf1: deep learning-based prediction ofCRISPR-Cpf1 activity at endogenous sites. Proceedings for Annual Meeting of The Japanese Pharmacological Society 92 (0), JKL–05. https://doi.org/10.1254/jpssuppl.92.0_jkl-05.

Labun, K., Montague, T.G., Gagnon, J.A., Thyme, S.B., Valen, E., 2016. CHOPCHOP v2: a web tool for the next generation of CRISPR genome engineering. Nucleic Acids Res. 44 (1), W272. https://doi.org/10.1093/NAR/GKW398.

Labun, K., Montague, T.G., Krause, M., Torres Cleuren, Y.N., Tjeldnes, H., Valen, E., 2019. CHOPCHOP v3: expanding the CRISPR web toolbox beyond genome editing. Nucleic Acids Res. 47 (1), W171. https://doi.org/10.1093/nar/gkz365.

Leenay, R.T., Aghazadeh, A., Hiatt, J., Tse, D., Roth, T.L., Apathy, R., Shifrut, E., Hultquist, J.F., Krogan, N., Wu, Z., Cirolia, G., Canaj, H., Leonetti, M.D., Marson, A., May, A.P., Zou, J., 2019. Large dataset enables prediction of repair after CRISPR–Cas9 editing in primary T cells. Nat. Biotechnol. 37 (9), 1034–1037. https://doi.org/10.1038/s41587-019-0203-2.

Lenaerts, B., Collard, B.C.Y., Demont, M., 2019. Review: improving global food security through accelerated plant breeding. Plant Sci. 287, 110207. https://doi.org/10.1016/j.plantsci.2019.110207.

Li, Q., Lin, N., 2010. The Bayesian elastic net. Bayesian Anal 5, 151–170. https://doi.org/10.1214/10-BA506.

Li, M., Li, X., Zhou, Z., Wu, P., Fang, M., Pan, X., Lin, Q., Luo, W., Wu, G., Li, H., 2016. Reassessment of the four yield-related genes Gn1a, DEP1, GS3, and IPA1 in rice using a CRISPR/Cas9 system. Front. Plant Sci. 7 (2016). https://doi.org/10.3389/fpls.2016.00377.

Li, C., Zong, Y., Wang, Y., Jin, S., Zhang, D., Song, Q., Zhang, R., Gao, C., 2018. Expanded base editing in rice and wheat using a Cas9-adenosine deaminase fusion. Genome Biol. 19 (1). https://doi.org/10.1186/s13059-018-1443-z.

Li, V.R., Zhang, Z., Troyanskaya, O.G., 2021a. CROTON: an automated and variant-aware deep learning framework for predicting CRISPR/Cas9 editing outcomes. Bioinformatics 37, i342–i348. https://doi.org/10.1093/bioinformatics/btab268.

Li, J., Xu, R., Qin, R., Liu, X., Kong, F., Wei, P., 2021b. Genome editing mediated by SpCas9 variants with broad non-canonical PAM compatibility in plants. Mol. Plant 14 (2), 352–360. https://doi.org/10.1016/j.molp.2020.12.017.

Li, Y., Wei, Y., Xu, S., Tan, Q., Zong, L., Wang, J., Wang, Y., Chen, J., Hong, L., Li, Y., Luigi Martelli, P., 2023. Acr-NET: predicting anti-CRISPR with deep learning. Bioinformatics 39 (5). https://doi.org/10.1093/bioinformatics/btad259.

Li, B., Zhu, X., Zhao, D., Li, Y., Yang, Y., Li, J., Bi, C., Zhang, X., 2024b. igRNA prediction and selection AI models (igRNA-PS) for bystander-less ABE base editing. J. Mol. Biol. 436 (18), 168714. https://doi.org/10.1016/j.jmb.2024.168714.

Li, J., Xu, M., Xiang, L., Chen, D., Zhuang, W., Yin, X., Li, Z., 2024c. Foundation models in smart agriculture: basics, opportunities, and challenges. Comput. Electron. Agric. 222, 109032. https://doi.org/10.1016/j.compag.2024.109032.

Li, Z., Khan, W.U., Bai, G., Dong, C., Wang, J., Zhang, Y., et al., 2025. From code to life: the AI-driven revolution in genome editing. Adv. Sci., e17029

Lin, Q., Zong, Y., Xue, C., Wang, S., Jin, S., Zhu, Z., Wang, Y., Anzalone, A.V., Raguram, A., Doman, J.L., Liu, D.R., Gao, C., 2020. Prime genome editing in rice and wheat. Nat. Biotechnol. 38 (5), 582–585. https://doi.org/10.1038/s41587-020-0455-x.

Listgarten, J., Weinstein, M., Kleinstiver, B.P., Sousa, A.A., Joung, J.K., Crawford, J., Gao, K., Hoang, L., Elibol, M., Doench, J.G., Fusi, N., 2018. Prediction of off-target activities for the end-to-end design of CRISPR guide RNAs. Nat. Biomed. Eng. 2 (1), 38–47. https://doi.org/10.1038/s41551-017-0178-6.

Liu, Y., Cheng, H., 2024. Research on the optimization of disease resistance gene editing targets in corn and future development directions. Geogr Res Bull 3, 373–376.

Liu, Q., He, Di, Xie, L., Segata, N., 2019. Prediction of off-target specificity and cell-specific fitness of CRISPR-Cas System using attention boosted deep learning and network-based gene feature. PLoS Comput. Biol. 15 (10), e1007480. https://doi.org/10.1371/journal.pcbi.1007480.

Liu, Q., Xiang, C., Liu, G., Li, B., Liu, X., 2020. Deep learning improves the ability of sgRNA off-target propensity prediction. BMC Bioinf. 21 (1). https://doi.org/10.1186/s12859-020-3395-z.

Liu, X., Wang, S., Ai, D., 2022. Predicting CRISPR/Cas9 repair outcomes by attention-based deep learning framework. Cells 11, 1847. https://doi.org/10.3390/cells11111847.

Lu, Y., Ye, X., Guo, R., Huang, J., Wang, W., Tang, J., Tan, L., Zhu, J.-kang, Chu, C., Qian, Y., 2017. Genome-wide targeted mutagenesis in rice using the CRISPR/Cas9 system. Mol. Plant 10 (9), 1242–1245. https://doi.org/10.1016/j.molp.2017.06.007.

Lu, K., Wu, B., Wang, J., Zhu, W., Nie, H., Qian, J., Huang, W., Fang, Z., 2018. Blocking amino acid transporter Os AAP 3 improves grain yield by promoting outgrowth buds and increasing tiller number in rice. Plant Biotechnol. J. 16 (10), 1710–1722. https://doi.org/10.1111/pbi.12907.

Luo, Y., Chen, Y., Xie, H.Z., Zhu, W., Zhang, G., 2024. Interpretable CRISPR/Cas9 off-target activities with mismatches and indels prediction using BERT. Comput. Biol. Med. 169, 107932. https://doi.org/10.1016/j.compbiomed.2024.107932.

Ma, J., Köster, J., Qin, Q., Hu, S., Li, W., Chen, C., et al., 2016a. CRISPR-DO for genome-wide CRISPR design and optimization. Bioinformatics 32, 3336–3338. https://doi.org/10.1093/bioinformatics/btw476.

Ma, L., Li, T., Hao, C., Wang, Y., Chen, X., Zhang, X., 2016b. Ta GS 5-3A, a grain size gene selected during wheat improvement for larger kernel and yield. Plant Biotechnol. J. 14, 1269–1280. https://doi.org/10.1111/pbi.12492.

Ma, E., Chen, K., Shi, H., Stahl, E.C., Adler, B., Trinidad, M., Liu, J., Zhou, K., Ye, J., Doudna, J.A., 2022. Improved genome editing by an engineered CRISPR-Cas12a. Nucleic Acids Res. 50 (22), 12689–12701. https://doi.org/10.1093/nar/gkac1192.

Manibalan, S., Thirukumaran, K., Varshni, M., Shobana, A., Achary, A., 2020. Report on biopharmaceutical profile of recent biotherapeutics and insilco docking studies on target bindings of known aptamer biotherapeutics. Biotechnol. Genet. Eng. Rev. 36 (2), 57–80. https://doi.org/10.1080/02648725.2020.1858395.

Mao, Y., Botella, J.R., Liu, Y., Zhu, J.K., 2019. Gene editing in plants: progress and challenges. Natl. Sci. Rev. 6 (3), 421–437. https://doi.org/10.1093/nsr/nwz005.

Marquart, K.F., Mathis, N., Mollaysa, A., Müller, S., Kissling, L., Rothgangl, T., Schmidheini, L., Kulcsár, P.I., Allam, A., Kaufmann, M.M., Matsushita, M., Haenggi, T., Cathomen, T., Kopf, M., Krauthammer, M., Schwank, G., 2024. Effective genome editing with an enhanced ISDra2 TnpB system and deep learning-predicted ωRNAs. Nat. Methods 21 (11), 2084–2093. https://doi.org/10.1038/s41592-024-02418-z.

Mathis, N., Allam, A., Kissling, L., Marquart, K.F., Schmidheini, L., Solari, C., Balázs, Z., Krauthammer, M., Schwank, G., 2023. Predicting prime editing efficiency and product purity by deep learning. Nat. Biotechnol. 41 (8), 1151–1159. https://doi.org/10.1038/s41587-022-01613-7.

Miao, C., Xiao, L., Hua, K., Zou, C., Zhao, Y., Bressan, R.A., Zhu, J.K., 2018. Mutations in a subfamily of abscisic acid recepto genes promote rice growth and productivity. Proc. Natl. Acad. Sci. U. S. A 115 (23), 6058–6063. https://doi.org/10.1073/pnas.1804774115.

Mohan, R.N.V.J., Rayanoothala, P.S., Sree, R.P., 2025. Next-gen agriculture: integrating AI and XAI for precision crop yield predictions. Front. Plant Sci. 15. https://doi.org/10.3389/fpls.2024.1451607.

Molla, K.A., Yang, Y., 2020. Predicting CRISPR/Cas9-Induced mutations for precise genome editing. Trends Biotechnol. 38 (2), 136–141. https://doi.org/10.1016/j.tibtech.2019.08.002.

Montecillo, J.A.V., Chu, L.L., Bae, H., 2020. CRISPR-Cas9 system for plant genome editing: current approaches and emerging developments. Agronomy 10, 1033. https://doi.org/10.3390/agronomy10071033.

Moreno-Mateos, M.A., Vejnar, C.E., Beaudoin, J.D., Fernandez, J.P., Mis, E.K., Khokha, M.K., Giraldez, A.J., 2015. CRISPRscan: designing highly efficient sgRNAs for CRISPR-Cas9 targeting in vivo. Nat. Methods 12 (10), 982–988. https://doi.org/10.1038/nmeth.3543.

Muha-Ud-Din, G., Ali, F., Hameed, A., Naqvi, S.A.H., Muhammad Nizamani, M., Jabran, M., Sarfraz, S., Yong, W., 2024. CRISPR/Cas9-based genome editing: a revolutionary approach for crop improvement and global food security. Physiol. Mol. Plant Pathol. 129, 102191. https://doi.org/10.1016/j.pmpp.2023.102191.

Nishimasu, H., Shi, X., Ishiguro, S., Gao, L., Hirano, S., Okazaki, S., Noda, T., Abudayyeh, O.O., Gootenberg, J.S., Mori, H., Oura, S., Holmes, B., Tanaka, M., Seki, M., Hirano, H., Aburatani, H., Ishitani, R., Ikawa, M., Yachie, N., Zhang, F., Nureki, O., 2018. Engineered CRISPR-Cas9 nuclease with expanded targeting space. Science 361 (6408), 1259–1262. https://doi.org/10.1126/science.aas9129.

Pallarès Masmitjà, M., Knödlseder, N., Güell, M., 2019. CRISPR-gRNA design. Methods Mol. Biol. 1961, 3–11. https://doi.org/10.1007/978-1-4939-9170-9_1. Humana Press.

Pallaseni, A., Peets, E.M., Koeppel, J., Weller, J., Vanderstichele, T., Ho, U.L., Crepaldi, L., Van Leeuwen, J., Allen, F., Parts, L., 2022. Predicting base editing outcomes using position-specific sequence determinants. Nucleic Acids Res. 50 (6), 3551–3564. https://doi.org/10.1093/nar/gkac161.

Pantazi, X.E., Moshou, D., Alexandridis, T., Whetton, R.L., Mouazen, A.M., 2016. Wheat yield prediction using machine learning and advanced sensing techniques. Comput. Electron. Agric. 121, 57–65. https://doi.org/10.1016/j.compag.2015.11.018.

Papikian, A., Liu, W., Gallego-Bartolomé, J., Jacobsen, S.E., 2019. Site-specific manipulation of Arabidopsis loci using CRISPR-Cas9 SunTag systems. Nat. Commun. 10 (1). https://doi.org/10.1038/s41467-019-08736-7.

Qin, R., Li, J., Li, H., Zhang, Y., Liu, X., Miao, Y., Zhang, X., Wei, P., 2019. Developing a highly efficient and wildly adaptive CRISPR - Sa Cas9 toolset for plant genome editing. Plant Biotechnol. J. 17 (4), 706–708. https://doi.org/10.1111/pbi.13047.

Rabinowitz, R., Almog, S., Roy, D., Offen, D., 2020. CrisPam: SNP-derived PAM analysis tool for allele-specific targeting of genetic variants using CRISPR-Cas systems. Front. Genet. 11. https://doi.org/10.3389/fgene.2020.00851.

Rodríguez, G.R., Muños, S., Anderson, C., Sim, S.-C., Michel, A., Causse, M., McSpadden Gardener, B.B., Francis, D., van der Knaap, E., 2011. Distribution of SUN, OVATE, LC , and FAS in the tomato germplasm and the relationship to fruit shape diversity. Plant Physiol. 156 (1), 275–285. https://doi.org/10.1104/pp.110.167577.

Rubin, A.J., Parker, K.R., Satpathy, A.T., Qi, Y., Wu, B., Ong, A.J., Mumbach, M.R., Ji, A.L., Kim, D.S., Cho, S.W., Zarnegar, B.J., Greenleaf, W.J., Chang, H.Y., Khavari, P.A., 2019. Coupled single-cell CRISPR screening and epigenomic profiling reveals causal gene regulatory networks. Cell 176 (1–2), 361. https://doi.org/10.1016/j.cell.2018.11.022.

Ruffolo, J.A., Nayfach, S., Gallagher, J., Bhatnagar, A., Beazer, J., Hussain, R., et al., 2024. Design of highly functional genome editors by modeling the universe of CRISPR-Cas sequences. Nature 645 (8080), 518–525. https://doi.org/10.1038/s41586-025-09298-z.

Safari, F., Zare, K., Negahdaripour, M., Barekati-Mowahed, M., Ghasemi, Y., 2019. CRISPR Cpf1 proteins: structure, function and implications for genome editing. Cell Biosci. 9 (1). https://doi.org/10.1186/s13578-019-0298-7.

Saini, H., Johar, V., Mondal, R., Vyas, S., Bisht, V., 2025. Integration of artificial intelligence with genome editing: a new era for molecular plant breeding. Physiol. Mol. Plant Pathol. 140, 102952. https://doi.org/10.1016/j.pmpp.2025.102952.

Sander, J.D., Joung, J.K., 2014. CRISPR-Cas systems for editing, regulating and targeting genomes. Nat. Biotechnol. 32 (4), 347–355. https://doi.org/10.1038/nbt.2842.

Sari, O., Liu, Z., Pan, Y., Shao, X., 2024. Predicting CRISPR-Cas9 off-target effects in human primary cells using bidirectional LSTM with BERT embedding. Bioinform. Adv. 5, vbae184–NA. https://doi.org/10.1093/bioadv/vbae184.

Schmid-Burgk, J.L., Gao, L., Li, D., Gardner, Z., Strecker, J., Lash, B., Zhang, F., 2020. Highly parallel profiling of Cas9 variant specificity. Mol. Cell 78 (4), 794. https://doi.org/10.1016/j.molcel.2020.02.023.

Schweikert, G., Zien, A., Zeller, G., Behr, J., Dieterich, C., Ong, C.S., Philips, P., De Bona, F., Hartmann, L., Bohlen, A., Krüger, N., Sonnenburg, S., Rätsch, G., 2009. mGene: accurate SVM-based gene finding with an application to nematode genomes. Genome Res. 19 (11), 2133–2143. https://doi.org/10.1101/gr.090597.108.

Senior, A.W., Evans, R., Jumper, J., Kirkpatrick, J., Sifre, L., Green, T., Qin, C., Žídek, A., Nelson, A.W.R., Bridgland, A., Penedones, H., Petersen, S., Simonyan, K., Crossan, S., Kohli, P., Jones, D.T., Silver, D., Kavukcuoglu, K., Hassabis, D., 2020. Improved protein structure prediction using potentials from deep learning. Nature 577 (7792), 706–710. https://doi.org/10.1038/s41586-019-1923-7.

Shen, M.W., Arbab, M., Hsu, J.Y., Worstell, D., Culbertson, S.J., Krabbe, O., Cassa, C.A., Liu, D.R., Gifford, D.K., Sherwood, R.I., 2018. Predictable and precise template-free CRISPR editing of pathogenic variants. Nature 563 (7733), 646–651. https://doi.org/10.1038/s41586-018-0686-x.

Shrawgi, H., Sisodia, D.S., 2019. Convolution neural network model for predicting single guide RNA efficiency in CRISPR/Cas9 system. Chemometr. Intell. Lab. Syst. 189, 149–154. https://doi.org/10.1016/j.chemolab.2019.04.008.

Shulse, C.N., Cole, B.J., Ciobanu, D., Lin, J., Yoshinaga, Y., Gouran, M., Turco, G.M., Zhu, Y., O'Malley, R.C., Brady, S.M., Dickel, D.E., 2019. High-throughput single-cell transcriptome profiling of plant cell types. Cell Rep. 27 (7), 2241. https://doi.org/10.1016/j.celrep.2019.04.054.

Slaymaker, I.M., Gao, L., Zetsche, B., Scott, D.A., Yan, W.X., Zhang, F., 2016. Rationally engineered Cas9 nucleases with improved specificity. Science 351 (6268), 84–88. https://doi.org/10.1126/science.aad5227.

Sledzinski, P., Nowaczyk, M., Olejniczak, M., 2020. Computational tools and resources supporting CRISPR-Cas experiments. Cells 9, 1288. https://doi.org/10.3390/cells9051288.

Sretenovic, S., Green, Y., Wu, Y., Cheng, Y., Zhang, T., Van Eck, J., Qi, Y., 2023. Genome- and transcriptome-wide off-target analyses of a high-efficiency adenine base editor in tomato. Plant Physiol. 193 (1), 291–303. https://doi.org/10.1093/plphys/kiad347.

Stanke, M., Keller, O., Gunduz, I., Hayes, A., Waack, S., Morgenstern, B., 2006. AUGUSTUS: ab initio prediction of alternative transcripts. Nucleic Acids Res. 34 (Web Server), W435. https://doi.org/10.1093/nar/gkl200.

Stemmer, M., Thumberger, T., Keyer, M.del S., Wittbrodt, J., Mateo, J.L., Maas, S., 2015. CCTop: an intuitive, flexible and reliable CRISPR/Cas9 target prediction tool. PLoS One 10 (4), e0124633. https://doi.org/10.1371/journal.pone.0124633.

Sun, J., Guo, J., Liu, J., Sinha, S., Crispr-M, 2024. Predicting sgRNA off-target effect using a multi-view deep learning network. PLoS Comput. Biol. 20 (3), e1011972. https://doi.org/10.1371/journal.pcbi.1011972.

Tang, D., Zhu, Y., Hu, X., 2024. The AI-driven potato gene editing: a new perspective on target identification. Geogr Res Bull 3, 370–372.

Tao, S., Liu, P., Shi, Y., Feng, Y., Gao, J., Chen, L., Zhang, A., Cheng, X., Wei, H., Zhang, T., Zhang, W., 2022. Single-cell transcriptome and network analyses unveil key transcription factors regulating mesophyll cell development in maize. Genes 13 (2), 374. https://doi.org/10.3390/genes13020374.

Terns, M.P., Terns, R.M., 2011. CRISPR-based adaptive immune systems. Curr. Opin. Microbiol. 14 (3), 321–327. https://doi.org/10.1016/j.mib.2011.03.005.

Tong, H., Nikoloski, Z., 2021. Machine learning approaches for crop improvement: leveraging phenotypic and genotypic big data. J. Plant Physiol. 257, 153354. https://doi.org/10.1016/j.jplph.2020.153354.

Toufikuzzaman, M., Hassan Samee, M.A., Sohel Rahman, M., 2024. CRISPR-DIPOFF: an interpretable deep learning approach for CRISPR Cas-9 off-target prediction. Briefings Bioinf. 25. https://doi.org/10.1093/bib/bbad530. NA-NA.

Umarov, R.K., Solovyev, V.V., 2017. Recognition of prokaryotic and eukaryotic promoters using convolutional deep learning neural networks. PLoS One 12, e0171410. https://doi.org/10.1371/journal.pone.0171410.

van Dijk, A.D.J., Kootstra, G., Kruijer, W., de Ridder, D., 2021. Machine learning in plant science and plant breeding. iScience 24 (1), 101890. https://doi.org/10.1016/j.isci.2020.101890.

Walton, R.T., Christie, K.A., Whittaker, M.N., Kleinstiver, B.P., 2020. Unconstrained genome targeting with near-PAMless engineered CRISPR-Cas9 variants. Science 368 (6488), 290–296. https://doi.org/10.1126/science.aba8853.

Wan, Y., Jiang, Z., 2023. TransCrispr: transformer based hybrid model for predicting CRISPR/Cas9 single guide RNA cleavage efficiency. IEEE ACM Trans. Comput. Biol. Bioinf 20 (2), 1518–1528. https://doi.org/10.1109/tcbb.2022.3201631.

Wang, T., Li, B., Nelson, C.E., Nabavi, S., 2019. Comparative analysis of differential gene expression analysis tools for single-cell RNA sequencing data. BMC Bioinf. 20 (1). https://doi.org/10.1186/s12859-019-2599-6.

Wang, H., Cimen, E., Singh, N., Buckler, E., 2020. Deep learning for plant genomics and crop improvement. Curr. Opin. Plant Biol. 54, 34–41. https://doi.org/10.1016/j.pbi.2019.12.010.

Wang, C., Wang, G., Gao, Y., Lu, G., Habben, J.E., Mao, G., Chen, G., Wang, J., Yang, F., Zhao, X., Zhang, J., Mo, H., Qu, P., Liu, J., Greene, T.W., 2020. A cytokinin-activation enzyme-like gene improves grain yield under various field conditions in rice. Plant Mol. Biol. 102 (4–5), 373–388. https://doi.org/10.1007/s11103-019-00952-5.

Wang, W., Wang, W., Pan, Y., Tan, C., Li, H., Chen, Y., Liu, X., Wei, J., Xu, N., Han, Y., Gu, H., Ye, R., Qi, D., Ma, C., 2022. A new gain-of-function OsGS2/GRF4 allele generated by CRISPR/Cas9 genome editing increases rice grain size and yield. The Crop Journal 10 (4), 1207–1212. https://doi.org/10.1016/j.cj.2022.01.004.

Wessels, H.H., Stirn, A., Méndez-Mancilla, A., Kim, E.J., Hart, S.K., Knowles, D.A., Sanjana, N.E., 2024. Prediction of on-target and off-target activity of CRISPR–Cas13d guide RNAs using deep learning. Nat. Biotechnol. 42 (4), 628–637. https://doi.org/10.1038/s41587-023-01830-8.

Wong, N., Liu, W., Wang, X., 2015. Wu-CRISPR: characteristics of functional guide RNAs for the CRISPR/Cas9 system. Genome Biol. 16, 218. https://doi.org/10.1186/s13059-015-0784-0.

Wu, C., Luo, J., Xiao, Y., 2024. Multi-omics assists genomic prediction of maize yield with machine learning approaches. Mol. Breed. 44, 14. https://doi.org/10.1007/s11032-024-01454-z.

Wulff, B.B.H., Dhugga, K.S., 2018. Wheat——the cereal abandoned by GM. Science 361 (6401), 451–452. https://doi.org/10.1126/science.aat5119.

Xiao, A., Cheng, Z., Kong, L., Zhu, Z., Lin, S., Gao, G., Zhang, B., 2014. CasOT: a genome-wide Cas9/gRNA off-target searching tool. Bioinformatics 30 (8), 1180–1182. https://doi.org/10.1093/bioinformatics/btt764.

Xiao, L.-M., Wan, Y.-Q., Jiang, Z.-R., 2021. AttCRISPR: a spacetime interpretable model for prediction of sgRNA on-target activity. BMC Bioinf. 22, 589. https://doi.org/10.1186/s12859-021-04509-6.

Xie, S., Shen, B., Zhang, C., Huang, X., Zhang, Y., Khodursky, A.B., 2014. sgRNAcas9: a software package for designing CRISPR sgRNA and evaluating potential off-target cleavage sites. PLoS One 9 (6), e100448. https://doi.org/10.1371/journal.pone.0100448.

Xie, J., Liu, M., Zhou, L., 2023. CRISPR-OTE: prediction of CRISPR on-target efficiency based on multi-dimensional feature fusion. IRBM 44, 100732. https://doi.org/10.1016/j.irbm.2022.07.003.

Xu, R., Yang, Y., Qin, R., Li, H., Qiu, C., Li, L., Wei, P., Yang, J., 2016. Rapid improvement of grain weight via highly efficient CRISPR/Cas9-mediated multiplex genome editing in rice. J. Genet. Genomics 43 (8), 529–532. https://doi.org/10.1016/j.jgg.2016.07.003.

Xu, C., Zhou, Y., Xiao, Q., He, B., Geng, G., Wang, Z., Cao, B., Dong, X., Bai, W., Wang, Y., Wang, X., Zhou, D., Yuan, T., Huo, X., Lai, J., Yang, H., 2021. Programmable RNA editing with compact CRISPR–Cas13 systems from uncultivated microbes. Nat. Methods 18 (5), 499–506. https://doi.org/10.1038/s41592-021-01124-4.

Xu, Y., Zhang, X., Li, H., Zheng, H., Zhang, J., Olsen, M.S., Varshney, R.K., Prasanna, B.M., Qian, Q., 2022. Smart breeding driven by big data, artificial intelligence, and integrated genomic-enviromic prediction. Mol. Plant 15 (11), 1664–1695. https://doi.org/10.1016/j.molp.2022.09.001.

Xue, L., Tang, B., Chen, W., Luo, J., 2019. Prediction of CRISPR sgRNA activity using a deep convolutional neural network. J. Chem. Inf. Model. 59 (1), 615–624. https://doi.org/10.1021/acs.jcim.8b00368.

Yan, J., Chuai, G., Zhou, C., Zhu, C., Yang, J., Zhang, C., Gu, F., Xu, H., Jia, W., Qi, L., 2018. Benchmarking CRISPR on-target sgRNA design. Briefings Bioinf. 19 (4), 721–724. https://doi.org/10.1093/bib/bbx001.

Yang, H., Wu, J.J., Tang, T., Liu, K.D., Dai, C., 2017. CRISPR/Cas9-mediated genome editing efficiently creates specific mutations at multiple loci using one sgRNA in Brassica napus. Sci. Rep. 7 (1). https://doi.org/10.1038/s41598-017-07871-9.

Yang, Q., Wu, L., Meng, J., Ma, L., Zuo, E., Sun, Y., EpiCas-Dl, 2023. Predicting sgRNA activity for CRISPR-mediated epigenome editing by deep learning. Comput. Struct. Biotechnol. J. 21, 202–211. https://doi.org/10.1016/j.csbj.2022.11.034.

Yang, Q., Tang, X., Wu, Y., Zhu, W., Zhang, T., Zhang, Y., 2024a. CRISPR-based modulation of uORFs in DEP1 and GIF1 for enhanced rice yield traits. Rice 17, 67. https://doi.org/10.1186/s12284-024-00743-7.

Yang, Y., Zheng, Y., Zou, Q., Li, J., Feng, H., 2024b. Overcoming CRISPR-Cas9 off-target prediction hurdles: a novel approach with ESB rebalancing strategy and CRISPR-MCA model. PLoS Comput. Biol. 20, e1012340. https://doi.org/10.1371/journal.pcbi.1012340.

Yasafova D, Gankin D, Sharapov Y, Bright C, Driss F, Campi F. panCRISPR Toolbox—a deep learning approach to improve CRISPR/Cas experiments 2022.

Yoosefzadeh Najafabadi, M., Hesami, M., Eskandari, M., 2023. Machine learning-assisted approaches in modernized plant breeding programs. Genes 14, 777. https://doi.org/10.3390/genes14040777.

Yu, H., Lin, T., Meng, X., Du, H., Zhang, J., Liu, G., Chen, M., Jing, Y., Kou, L., Li, X., Gao, Q., Liang, Y., Liu, X., Fan, Z., Liang, Y., Cheng, Z., Chen, M., Tian, Z., Wang, Y., Chu, C., Zuo, J., Wan, J., Qian, Q., Han, B., Zuccolo, A., Wing, R.A., Gao, C., Liang, C., Li, J., 2021. A route to de novo domestication of wild allotetraploid rice. Cell 184 (5), 1156. https://doi.org/10.1016/j.cell.2021.01.013.

Yuste-Lisbona, F.J., Fernández-Lozano, A., Pineda, B., Bretones, S., Ortíz-Atienza, A., García-Sogo, B., Müller, N.A., Angosto, T., Capel, J., Moreno, V., Jiménez-Gómez, J.M., Lozano, R., 2020. ENO regulates tomato fruit size through the floral meristem development network. Proc. Natl. Acad. Sci. U. S. A 117 (14), 8187–8195. https://doi.org/10.1073/pnas.1913688117.

Yuyu, C., Aike, Z., Pao, X., Wen, X., Cao, Y., Wang, B., Zhang, Y., Shah, L., Cheng, S., Cao, L., Zhang, Y., 2020. Effects of GS3 and GL3.1 for grain size editing by CRISPR/Cas9 in rice. Rice Sci. 27 (5), 405–413. https://doi.org/10.1016/j.rsci.2019.12.010.

Zalatan, J.G., Lee, M.E., Almeida, R., Gilbert, L.A., Whitehead, E.H., La Russa, M., Tsai, J.C., Weissman, J.S., Dueber, J.E., Qi, L.S., Lim, W.A., 2015. Engineering complex synthetic transcriptional programs with CRISPR RNA scaffolds. Cell 160 (1–2), 339–350. https://doi.org/10.1016/j.cell.2014.11.052.

Zarate, O.A., Yang, Y., Wang, X., Wang, J.P., 2022. BoostMEC: predicting CRISPR-Cas9 cleavage efficiency through boosting models. BMC Bioinf. 23 (1). https://doi.org/10.1186/s12859-022-04998-z.

Zeng, Y., Wen, J., Zhao, W., Wang, Q., Huang, W., 2020. Rational improvement of rice yield and cold tolerance by editing the three genes OsPIN5b, GS3, and OsMYB30 with the CRISPR–cas9 system. Front. Plant Sci. 10. https://doi.org/10.3389/fpls.2019.01663.

Zetsche, B., Gootenberg, J.S., Abudayyeh, O.O., Slaymaker, I.M., Makarova, K.S., Essletzbichler, P., Volz, S.E., Joung, J., Van Der Oost, J., Regev, A., Koonin, E.V., Zhang, F., 2015. Cpf1 is a single RNA-guided endonuclease of a Class 2 CRISPR-Cas system. Cell 163 (3), 759–771. https://doi.org/10.1016/j.cell.2015.09.038.

Zetsche, B., Strecker, J., Abudayyeh, O.O., Gootenberg, J.S., Scott, D.A., Zhang, F., 2020. A survey of genome editing activity for 16 CAS12A orthologs. Keio J. Med. 69 (3), 59–65. https://doi.org/10.2302/kjm.2019-0009-OA.

Zhang, Y., Liang, Z., Zong, Y., Wang, Y., Liu, J., Chen, K., Qiu, J.-L., Gao, C., 2016. Efficient and transgene-free genome editing in wheat through transient expression of CRISPR/Cas9 DNA or RNA. Nat. Commun. 7 (1). https://doi.org/10.1038/ncomms12617.

Zhang, Z., Hua, L., Gupta, A., Tricoli, D., Edwards, K.J., Yang, B., Li, W., 2019. Development of an Agrobacterium-delivered CRISPR/Cas9 system for wheat genome editing. Plant Biotechnol. J. 17 (8), 1623–1635. https://doi.org/10.1111/pbi.13088.

Zhang, Y., Long, Y., Yin, R., Kwoh, C.K., 2020. DL-CRISPR: a deep learning method for off-target activity prediction in CRISPR/Cas9 with data augmentation. IEEE Access 8, 76610–76617. https://doi.org/10.1109/ACCESS.2020.2989454.

Zhang, Y., Guo, W., Chen, L., Shen, X., Yang, H., Fang, Y., Ouyang, W., Mai, S., Chen, H., Chen, S., Qingnan, H., Yuan, S., Zhang, C., Huang, Y., Shan, Z., Yang, Z., Qiu, D., Zhou, X., Cao, D., Xia, L., Jiao, Y., 2022. CRISPR/Cas9-Mediated targeted mutagenesis of GmUGT enhanced soybean resistance against leaf-chewing insects through flavonoids biosynthesis. Front. Plant Sci. 13. https://doi.org/10.3389/fpls.2022.802716.

Zhang, C., Yang, Y., Qi, T., Zhang, Y., Hou, L., Wei, J., Yang, J., Shi, L., Ong, S.-G., Wang, H., Wang, H., Yu, B., Wang, Y., 2023. Prediction of base editor off-targets by deep learning. Nat. Commun. 14 (1). https://doi.org/10.1038/s41467-023-41004-3.

Zhang, G., Xie, H., Dai, X., 2024a. DeepIndel: an interpretable deep learning approach for predicting CRISPR/Cas9-Mediated editing outcomes. Int. J. Mol. Sci. 25, 10928. https://doi.org/10.3390/ijms252010928.

Zhang, G., Luo, Y., Xie, H., Dai, Z., 2024b. Crispr-SGRU: prediction of CRISPR/Cas9 off-target activities with mismatches and indels using stacked BiGRU. Int. J. Mol. Sci. 25 (20), 10945. https://doi.org/10.3390/ijms252010945.

Zhang, Y., Huang, G., Zhao, Y., Lu, X., Wang, Y., Wang, C., Guo, X., Zhao, C., 2025. Revolutionizing crop breeding: next-generation artificial intelligence and big data-driven intelligent design. Engineering 44 (1), 245–255. https://doi.org/10.1016/j.eng.2024.11.034.

Zhong, Z., Liu, G., Tang, Z., Xiang, S., Yang, L., Huang, L., He, Y., Fan, T., Liu, S., Zheng, X., Zhang, T., Qi, Y., Huang, J., Zhang, Y., 2023. Efficient plant genome engineering using a probiotic sourced CRISPR-Cas9 system. Nat. Commun. 14 (1). https://doi.org/10.1038/s41467-023-41802-9.

Zhou, J., Liu, G., Zhao, Y., Zhang, R., Tang, X., Li, L., Jia, X., Guo, Y., Wu, Y., Han, Y., Bao, Y., He, Y., Han, Q., Yang, H., Zheng, X., Qi, Y., Zhang, T., Zhang, Y., 2023. An efficient CRISPR–Cas12a promoter editing system for crop improvement. Nat. Plants 9 (4), 588–604. https://doi.org/10.1038/s41477-023-01384-2.

Zhou, X., Gao, J., Luo, L., Huang, C., Wu, J., Wang, X., 2024. Comprehensive evaluation and prediction of editing outcomes for near-PAMless adenine and cytosine base editors. Commun. Biol. 7 (1). https://doi.org/10.1038/s42003-024-07078-5.

Zhu, H., Liang, C., 2019. CRISPR-DT: designing gRNAs for the CRISPR-Cpf1 system with improved target efficiency and specificity. Bioinformatics 35, 2783–2789. https://doi.org/10.1093/bioinformatics/bty1061.

Zhu, W., Xie, H., Chen, Y., Zhang, G., 2024. CrnnCrispr: an interpretable deep learning method for CRISPR/Cas9 sgRNA on-target activity prediction. Int. J. Mol. Sci. 25 (8), 4429. https://doi.org/10.3390/ijms25084429.

Zsögön, A., Cermak, T., Voytas, D., Peres, L.E.P., 2017. Genome editing as a tool to achieve the crop ideotype and de novo domestication of wild relatives: Case study in tomato. Plant Sci. 256, 120–130. https://doi.org/10.1016/j.plantsci.2016.12.012.

Further reading

Höijer, I., Johansson, J., Gudmundsson, S., Chin, C.-S., Bunikis, I., Häggqvist, S., et al., 2020. Amplification-free long-read sequencing reveals unforeseen CRISPR-Cas9 off-target activity. Genome Biol. 21, 290. https://doi.org/10.1186/s13059-020-02206-w.

Li, J., Wu, P., Cao, Z., Huang, G., Lu, Z., Yan, J., et al., 2024a. Machine learning-based prediction models to guide the selection of Cas9 variants for efficient gene editing. Cell Rep. 43, 113765. https://doi.org/10.1016/j.celrep.2024.113765.

CHAPTER

20

AI-assisted regulation of gene-edited plants: Opportunities and challenges

Muhammad Riaz[1], *Hamad Khan*[2], *Muhammad Faheem*[3], *Ahmed Al-Harrasi*[2], *Muhammad Umer Masood*[4], *Nayla Munawar*[5], *Paulo A. Zaini*[6], *Muhammad Rasheed*[7], *Stanislaus Antony Ceasar*[8], *and Aftab Ahmad*[2]

[1]Yale University, New Haven, CT, United States; [2]Natural and Medical Sciences Research Center (NMSRC), University of Nizwa, Nizwa, Oman; [3]Department of Biochemistry, University of Agriculture, Faisalabad, Pakistan; [4]Department of Plant Sciences and Plant Pathology, Montana State University, Bozeman, MT, United States; [5]Department of Chemistry, United Arab Emirates University, Al-Ain, United Arab Emirates; [6]Department of Plant Sciences, University of California, Davis, CA, United States; [7]Department of Physiology, School of Basic Medical Science, Chongqing Medical University, Chongqing, China; [8]Department of Biotechnology, Karunya Institute of Technology and Sciences, Coimbatore, Tamil Nadu, India

20.1 Introduction

The advent of gene-editing technologies has transformed modern plant biotechnology and created unprecedented opportunities to improve crops that were not possible with conventional genetic engineering. Among these technologies, CRISPR-Cas systems have been the most widely adopted for their accuracy, efficiency, accessibility, and versatility (Chen et al., 2019). Unlike conventional transgenic approaches, which often involve the permanent integration of foreign DNA, gene editing enables targeted modifications of endogenous genes to develop crops with improved yields, pest resistance, stress tolerance, nutritional quality, and resource efficiency (Zhu et al., 2020). As a result, gene-edited crops are increasingly integrated into agricultural innovation pipelines worldwide.

Intelligent Gene Editing in Plants
https://doi.org/10.1016/B978-0-443-44079-3.00020-6

Several gene-edited crops, particularly those edited with site-directed nuclease-1 (SDN-1), have been approved for commercialization, thereby exempting them from conventional GMO regulations. Examples include high-γ-aminobutyric acid (GABA) tomato, disease-resistant rice varieties developed through targeted modification of susceptible genes, high-oleic soybeans, waxy corn, and nonbrowning mushrooms (FAO., 2023; Oliva et al., 2019; Polidoros et al., 2024). These developments demonstrate the practical importance of CRISPR-Cas technology in addressing global food security challenges in the face of climate change, population growth, emerging pests and diseases, and the growing demand for sustainable food production systems (FAO., 2023; Zhu et al., 2020). However, the rapid development of CRISPR-edited crops and the divergent regulatory frameworks governing them have raised regulatory challenges. Gene editing in crops using advanced CRISPR systems, such as CRISPR-Cas9, CRISPR-Cas12a, CRISPRi, CRISPRa, and prime editing, is producing increasingly complex data sets, including repair outcomes, exogenous DNA analyses, global off-target effects, transcriptomics, proteomics, metabolomics, environmental risk assessment, and post-marketing monitoring (Bodin et al., 2021; Kuzma et al., 2023). Regulatory authorities must evaluate these forms of evidence while safeguarding food safety, protecting the environment, ensuring transparency, and maintaining public confidence (Friedrichs et al., 2019). Traditional regulatory frameworks were largely developed in the era of transgenic GM crops and often rely on extensive manual reviews of scientific documentation, literature, and risk assessment data. As a result, regulators are increasingly faced with the challenge of assessing the vast amount of biological information generated by rapidly evolving gene-editing techniques (FAO, 2023; Kuzma et al., 2023).

The adoption of mosaic regulatory approaches across jurisdictions further complicates the governance of gene-edited crops (Tachikawa and Matsuo, 2023). The United States (US) generally uses a product-oriented regulatory framework, under which certain gene-edited crops (SDN-1 and SDN-2) without transgenes or exogenous DNA can be exempted from conventional GMO regulations (Animal, 2020, 2023). By contrast, the European Union (EU) continues to regulate most gene-edited crops under its GMO legislation. Other countries, such as Canada, Argentina, Brazil, Australia, Japan, China, and the UK, use different combinations of product, process, and novelty triggers to regulate gene-edited crops (Canada, 2022; Tachikawa and Matsuo, 2023; Whelan and Lema, 2015). These differences create regulatory uncertainty, increase regulatory costs, complicate international trade, and contribute to inconsistent risk assessments and product approvals. In this context, AI is emerging as a potentially transformational tool for the regulatory science of gene-edited crops. AI uses machine learning (ML), deep learning (DL), natural language processing (NLP), predictive analytics, and data integration to help scientists extract meaningful information from large and complex datasets (Authority, 2025; Bersani et al., 2022). Originally developed to support biological discovery and gene-editing applications, AI is now being explored for regulatory purposes in biotechnology. In the context of gene-edited crops, AI-assisted models can facilitate the synthesis of evidence, analyze molecular characterization data, anticipate potential off-target effects, assess environmental risk factors, integrate multi-omics data sets, automate dossier review, and support post-market monitoring activities (Sherkatghanad et al., 2023).

AI platforms such as DeepCRISPR, DeepSpCas9, CRISPRon, and related ML tools can support guide-RNA (gRNA) design, improve predictions of editing efficiency, predict repair outcomes, and help identify potential off-target risks in CRISPR studies (Chuai et al., 2018; Kim et al., 2019; Xiang et al., 2021). These advances are generating more structured, interpretable datasets that may support evidence synthesis, molecular characterization, and risk-assessment processes relevant to future regulatory evaluation of gene-edited crops. Beyond laboratory research, regulatory authorities are increasingly evaluating AI tools to streamline biotechnology regulatory frameworks. For example, the European Food Safety Authority (EFSA) has been exploring ML for evidence synthesis, hazard identification, predictive toxicology, and food safety assessment (Authority, 2025; Bersani et al., 2022). Similarly, other regulatory agencies, including the US Food and Drug Administration (FDA), the USDA, and the Canadian Food Inspection Agency (CFIA), are developing AI strategies, digital infrastructures, predictive modeling capacity, and data-modernization initiatives that may support more data-driven regulatory oversight in the future (CFIA, 2026; Food and Administration, 2025; USDA, 2025).

The importance of AI in regulatory systems extends beyond improving efficiency and specificity in CRISPR-Cas systems. Regulatory decisions on gene-edited crops increasingly require integrated genomic, environmental, agronomic, toxicological, ecological, ethical, and socio-economic information (Bodin et al., 2021; FAO, 2023). AI-based approaches can improve the consistency, transparency, reproducibility, and scalability of regulatory assessments of gene-edited crops (AI, 2023). ML algorithms can support comparative genomics, substantial equivalence assessment, environmental risk assessment, gene flow predictions, dossier and literature reviews, and post-marketing monitoring (EFSA, 2022; Iacono et al., 2022; Kuiper et al., 2002). These capabilities are particularly important as regulators face an increasing number of gene-edited products. Moreover, gene-edited crops with complex trait modifications, including multiplex gene-editing, are emerging (Polidoros et al., 2024; Xie et al., 2022). At the same time, integrating AI into regulatory systems raises important issues of transparency, accountability, clarity, cybersecurity, data management, algorithmic bias, and public confidence (AI, 2023; Smuha, 2025; Yeung, 2020). Regulatory decisions that affect food systems and the environment must remain scientifically and legally sound and socially acceptable. Therefore, AI should be seen as a decision-support technology operating within human-centered regulatory frameworks, rather than as a self-regulatory decision-maker (Authority, 2025; Yeung, 2020).

This chapter explores the growing convergence between artificial intelligence (AI) and the regulation of gene-edited crops. It examines how AI can support molecular characterization, off-target assessment, comparative genomics, risk assessment, dossier review, post-market monitoring, regulatory harmonization, and policy development. This chapter also assesses ethical, legal, and governance issues related to AI-assisted regulation and discusses future opportunities to develop adaptive, transparent, and evidence-based regulatory frameworks that can keep pace with advances in genome-editing technologies. This integration can ultimately contribute to more effective, scientifically sound, and socially responsible surveillance of gene-edited crops in the decades to come.

20.2 The significance of AI integration in the regulation of gene-edited crops

The increasing use of CRISPR-Cas to develop gene-edited crops, coupled with the exemption of certain CRISPR-edited products (SDN-1 and SDN-2) in some jurisdictions, has introduced new challenges for regulatory and governance authorities responsible for evaluating the safety, environmental impact, and social implications of these crops. Advanced CRISPR platforms can produce increasingly precise and complex genetic modifications, while generating a large volume of genomic, molecular, environmental, and agronomic data that require comprehensive evaluation for regulatory approval (FAO., 2023; Iacono et al., 2022). Conventional regulatory approaches, which often rely on extensive manual analysis of scientific evidence, may face increasing challenges in processing these data in an effective, transparent, and consistent manner (Bersani et al., 2022). AI offers a promising approach to support regulatory decision-making by facilitating the synthesis of evidence, the assessment of risks, and the interpretation of data (Cavalli et al., 2022). For example, ML algorithms can help with genomic sequencing analysis, the prediction of possible unintended cleavages, the integration of multi-omics data sets, and the identification of patterns associated with food safety or environmental hazards (Sherkatghanad et al., 2023). These capabilities are particularly relevant to gene-edited crops, for which regulatory authorities may need to evaluate data for molecular characterization, environmental risk assessment, and post-market monitoring (Bodin et al., 2021).

AI can contribute to improving consistency in regulatory assessments by enabling more standardized analysis of scientific evidence within regulatory workflows (Authority, 2025). For example, organizations such as the EFSA are exploring AI-enabled tools for literature monitoring, hazard identification, and evidence synthesis, while regulatory agencies such as the FDA, USDA, and CFIA are investing in digital infrastructure to support more data-driven regulatory processes (FDA, 2025; Authority, 2025; CFIA, 2026; USDA, 2025). As the number of gene-edited crops entering research pipelines and the commercialization process continues to grow, the ability to analyze large, complex data sets in a transparent, consistent, and scientifically sound manner is increasingly important (AI, 2023; Polidoros et al., 2024). AI should not be viewed as a substitute for regulatory expertise, but as a complementary tool that can strengthen evidence-based monitoring, improve regulatory effectiveness, and promote the safe and responsible use of gene-edited crops (Yeung, 2020).

Biotechnology regulation has traditionally been a paper-intensive, time-consuming process characterized by sequential reviews, established regulatory frameworks, and a lengthy approval process. In the early 2000s, approval of transgenic crops often required years of extensive paperwork, field trials, and stakeholder consultations (Hoffman, 2021; Smart et al., 2017). While these protocols were intended to enhance safety, they often created bottlenecks that could slow down innovation and delay the deployment of new technologies (Phillips and Williams, 2022). The transition toward digital regulation began as many regulatory agencies adopted electronic processes for application submission and review. However, the most recent developments have focused on exploring the use of AI to support risk assessment, analyze factors influencing social acceptance, and evaluate the potential ecological impact of gene-edited crops (Bersani et al., 2022). This transition from traditional to digital regulation highlights the growing need for more universal, adaptive, transparent, and data-based regulatory frameworks for gene-edited plants (FAO, 2023). This transition also provides a foundation for integrating AI into established regulatory frameworks and the governance of gene-edited plants, potentially enhancing regulators' capacity to address the challenges posed by the increasing pace and complexity of gene-edited crop development.

20.3 Present regulatory triggers for gene-edited plants

20.3.1 Product-based trigger

Product-triggered regulation is based on the characteristics of the final organism rather than the process used to develop it. This approach focuses on the presence of novel traits and their potential risks, regardless of whether they were developed through traditional breeding, mutagenesis, genetic engineering, or gene editing (Canada, 2022; CFIA, 2026). Under this approach, some jurisdictions, such as Argentina and the US, may regulate certain gene-edited crops similarly to conventionally bred crops when they do not contain exogenous DNA, do not contain a novel combination of genetic material, or fall within the scope of plant-pest-risk-based biotechnology regulation (Goberna et al., 2022; Whelan and Lema, 2015). However, some jurisdictions, such as Canada, focus on novelty rather than the presence or absence of foreign DNA (CFIA, 2026). The US regulatory model has historically reflected a product- and risk-oriented approach. Under the 2020 USDA SECURE Rule, certain gene-edited crops with modifications achievable through conventional breeding were exempted from USDA biotechnology regulation. However, a federal court vacated this rule in December 2024, after which Animal and Plant Health Inspection Service (APHIS) restored the pre2020 regulatory framework (Hoffman, 2021; USDA, 2025). Under this policy, CRISPR-edited crops such as nonbrowning mushrooms and waxy corn were exempt from certain USDA biotechnology requirements (Waltz, 2016). The product model is generally considered flexible because it focuses on the product rather than the breeding method, thereby minimizing regulatory requirements for certain CRISPR-edited crops and facilitating their commercialization. However, critics argue that this may not account for the minor risks associated with the gene-editing process itself. These concerns have prompted some countries to adopt a process-based approach, focusing on the method used to develop the organism rather than solely on the final characteristics (Agapito-Tenfen et al., 2018; Ahmad et al., 2021).

20.3.2 Process-based trigger

The process-based trigger determines regulation based on how an organism is developed, emphasizing the method used to develop the product, such as gene editing, transgenesis, or RNA interference. Under this system, plants developed through gene editing are regulated under the existing GM framework, irrespective of whether the final product is indistinguishable from conventionally bred crops (McHughen, 2016). The EU and New Zealand remain the most prominent examples of the process-based framework. In 2018, the Court of Justice of the European Union (CJEU) ruled in Case C-528/16 that all organisms developed through new mutagenesis technologies, including gene-editing, fall within the scope of the EU GMO Directive (Directive 2001/18/EC). Therefore, plants developed through CRISPR-Cas are subject to the same regulatory requirements as GMOs under the EU legislation (Dederer, 2019; Eckerstorfer et al., 2023; Vives-Vallés and Collonnier, 2020). Consequently, CRISPR-edited crops, including wheat and grapevine, developed in EU member states such as France and Italy, are generally subject to the EU regulatory framework, including environmental risk assessment, traceability, and labeling requirements, even when the final product contains no foreign genetic material (Dima et al., 2020).

20.3.3 Novelty-based trigger

Under this framework, regulatory oversight depends on whether a newly developed plant possesses a novel trait significantly different from those already present in cultivated plants or wild relatives (CFIA, 2026). Consequently, assessment focuses on the risks and benefits associated with novel traits, irrespective of the method used to develop them (Ahmad et al., 2021). The novelty-based trigger has been proposed as a potentially risk-proportionate and adaptive regulatory model that could contribute to greater international alignment; however, harmonization remains challenging because jurisdictions differ in legal definitions, regulatory mandates, and public-policy priorities (Tachikawa and Matsuo, 2023).

20.3.4 Hybrid regulatory frameworks

In 2019, Australia amended its Gene Technology Regulations to exempt SDN-1 gene-edited organisms that do not contain foreign DNA in their genome. This exemption generally covers most gene-editing outcomes that could also be achieved through conventional breeding (Zhang et al., 2021). Although Australia generally follows a process-based regulatory framework, the Office of the Gene Technology Regulator has exempted certain SDN-1-modified plants with no foreign DNA in their genomes from regulation under the Gene Technology Act (Tachikawa and Matsuo, 2023). Japan follows a product-focused, case-by-case framework, allowing certain CRISPR-modified crops, such as high-GABA tomato, to enter the market without premarket safety approval, provided no transgene or foreign DNA is present in the final product (Ishii, 2025; Waltz, 2022). However, developers must submit reports (notification and molecular characterization) confirming the removal of any foreign gene. Similarly, Argentina and Brazil use a product-based framework with case-by-case regulation of gene-edited crops. Under these frameworks, plants that do not contain novel combinations of genetic material may be excluded from GMO-specific regulations, depending on the characteristics of the final plant (Kuiken and Kuzma, 2021; Nepomuceno et al., 2020; Whelan and Lema, 2015).

20.4 Understanding site-directed nucleases classification

20.4.1 Site-directed nuclease-1: Minor edits

SDN-1 is a gene-editing method in which a double-stranded DNA break is induced at a specific site and then repaired by the cell's native repair machinery (nonhomologous end joining), without the use of external or donor DNA (Shillito et al., 2021). This process results in minor insertions, deletions, or point mutations that also occur naturally or can be achieved through conventional mutagenesis (Entine et al., 2021; Organisms et al., 2020). Many countries have exempted, or are considering exempting, SDN-1-edited plants from existing regulatory requirements for GMOs, particularly when they are free of transgene or exogenous DNA (Buchholzer and Frommer, 2023; Entine et al., 2021). For example, Argentina's Resolution 173/2015 established a case-by-case regulatory framework for gene-edited crops under which SDN-1-edited crops may be classified as nonGM products when they do not contain a novel combination of genetic material, thereby falling outside the scope of GMO regulatory requirements (Kuiken and Kuzma, 2021; Whelan and Lema, 2015). Similarly, the USDA did not subject Calyxt's high-oleic soybean—developed through TALEN-mediated SDN-1 editing of the fatty acid biosynthesis pathway—to extensive GMO regulation, because no transgene was introduced and it was not considered subject to USDA-APHIS regulatory oversight (Demorest et al., 2016; Food and Administration, 2019). This product successfully entered the market under a different regulatory oversight than conventional GMOs, representing an early example of the commercialization of gene-edited crops (Nagamine and Ezura, 2022). SDN-1 edited plants are often considered low-risk products because they do not contain foreign DNA and could arise from natural mutations or conventional breeding methods. As a result, several jurisdictions, such as the USDA, Argentina, Australia, and Brazil, have adopted proportionate, risk-based regulation for certain SDN-1-edited crops that do not contain a novel combination of genetic material or exogenous DNA (Organisms et al., 2020). However, regulatory oversight becomes more complex when a repair template is used to introduce predetermined changes, as occurs in SDN-2- and SDN-3-edited plants (Friedrichs et al., 2019).

20.4.2 SDN-2 small edits using a repair template

SDN-2 modifications involve creating a targeted double-stranded break (DSB) in the genome, followed by DNA repair using a homologous donor template to introduce predetermined small modifications, such as insertions, deletions, or substitutions of a single or a few nucleotides. This enables more precise genetic modifications, such as targeted base substitutions, but it requires an exogenous repair template during the editing process (Organisms et al., 2020). The regulatory oversight of SDN-2-edited crops differs significantly across jurisdictions, reflecting variations in legal definitions, regulatory triggers, and risk assessment. For example, Japan applies a case-by-case, product-based approach to regulate SDN-2-edited crops. Certain SDN-2-edited products may be exempt from the GMO regulatory framework if the final product contains no transgene, subject to notification and evaluation by the relevant regulatory authority (Kondo and Taguchi, 2022; Tachikawa and Matsuo, 2023). In contrast, Australia follows a process-based approach and regulates SDN-2 edited crops under existing GMO regulations rather than exempting them from regulatory oversight. Similarly, the EU follows a process-based regulatory approach and subjects all CRISPR-edited crops, including SDN-1 and SDN-2 edited crops, to GMO regulatory oversight based on the interpretation of the 2018 CJEU judgment (Vives-Vallés and Collonnier, 2020). This reflects the EU's process-based regulatory approach under which gene-edited crops generally fall under the scope of GMO legislation. However, critics argue that regulating GMOs and gene-edited crops under the same regulatory framework may hinder the timely adoption of gene-edited crops and constrain scientific innovation in the member states, despite growing calls for reform within the EU (Smart et al., 2017). Overall, the regulatory status of SDN-2-modified organisms remains uncertain in many jurisdictions: they are subject to more stringent regulation than SDN-1-edited modifications, yet are often differentiated from transgenics because they involve smaller and targeted modifications. However, when entire genes or long sequences are introduced, regulatory oversight becomes stricter under SDN-3-approaches (Organisms et al., 2020).

20.4.3 SDN-3 site-specific integration of a long DNA sequence or a gene

The SDN-3 approach involves the targeted insertion of one or more genes or a long stretch of DNA into the host genome, often derived from the same species, sexually compatible species, or noncrossable species. This approach resembles conventional genetic engineering, and the resulting products are generally regulated under GMO frameworks in most jurisdictions (Entine et al., 2021). Most SDN-3-modified organisms contain introduced genes or exogenous DNA sequences and are therefore subject to GMO regulatory oversight in the US by USDA-APHIS and, where applicable, by the FDA and EPA (Fernandez-Cornejo et al., 2014). For example, Bt cotton, developed by introducing genes from *Bacillus thuringiensis* into the cotton genome to confer insect/pest resistance, was regulated under GMO oversight before being approved for commercialization (Mendelsohn et al., 2003). Similarly, in Brazil and India, SDN-3-modified crops are generally subject to GMO oversight, including environmental risk assessments (ERAs), food safety evaluations, and, in some cases, additional regulatory assessment before commercialization (Apple A. 2022; Molinari et al., 2021). Introducing foreign DNA into SDN-3 products may raise concerns about unintended effects and gene flow to related species (Entine et al., 2021). Therefore, SDN-3-modified crops may face stricter regulatory oversight and risk assessment than SDN-1- or SDN-2-modified crops. Nonetheless, AI-driven resources are being explored to support risk assessments, molecular characterization, and evaluation of the potential long-term environmental impact of gene-edited organisms (Bersani et al., 2022). In the future, AI and ML tools are expected to play an increasingly important role in improving the consistency, efficiency, and scientific rigor of regulatory decisions across SDN-1, SDN-2, and SDN-3 gene-edited crops (Bersani et al., 2022). AI-assisted tools may enable faster molecular characterization, biosafety assessment, detection of foreign DNA, allergenicity prediction, and environmental impact analysis, thereby supporting more evidence-based and scalable regulatory decision-making (Nedyalkova et al., 2023). Therefore, as gene-edited crops become more diverse and complex, AI is expected to contribute to greater regulatory coherence while maintaining safety, transparency, and evidence-based decision-making (AI, 2023).

20.5 AI integration in CRISPR

20.5.1 Improved accuracy and on-target cleavage

AI tools are increasingly used to improve the precision and efficiency of CRISPR-based gene editing by selecting appropriate gRNAs, predicting unique target sites, and reducing off-target effects. While CRISPR has transformed gene editing, conventional Cas nucleases have often been limited by variable and unpredictable editing efficiency. AI can help reduce trial-and-error by predicting target sites, optimal gRNAs, and repair outcomes, while minimizing off-target effects (Chuai et al., 2018; Kim et al., 2019). Several AI and ML tools have been developed to improve the specificity and precision of CRISPR-Cas gene editing. For instance, CRISPR-Net uses a recurrent convolutional neural network to quantify CRISPR off-target activities associated with sequence mismatches and indels (Lin et al., 2020). In addition, DeepCRISPR, developed by Chuai and colleagues, uses deep learning to predict gRNA activity and unintended cleavage from large experimental datasets, thereby supporting the selection of gRNAs and target sites with improved efficiency and specificity (Chuai et al., 2018). Similarly, DeepSpCas9, uses deep neural networks trained on large datasets of SpCas9 activity to predict on-target cleavage with high accuracy (Kim et al., 2019). Moreover, DeepHF predicts the activity of wild-type SpCas9 and high-fidelity Cas9 variants, including eSpCas9(1.1) and SpCas9-HF1, thereby supporting the design of efficient gRNAs for high-fidelity genome editing (Wang et al., 2019). Together, these AI tools can improve the efficiency, specificity, and predictability of gene-editing, reducing the cost of experimentation and supporting the development of more precise gene-edited crops (Konstantakos et al., 2022). Improved editing efficiency and accuracy reduce the time and cost of gene editing, and support greater confidence in risk evaluation and regulatory approval of gene-edited crops (Dixit et al., 2024). Furthermore, AI can support the development of improved CRISPR-Cas systems and help researchers identify appropriate gene targets associated with desired traits. A list of important AI tools developed to predict on-target cleavage is summarized in Table 20.1.

20.5.2 AI and forecasting the best target site for CRISPR-Cas

Identifying the optimal genomic site for introducing precise genetic modifications is one of the most critical tasks in CRISPR experiments. AI models facilitate this process by analyzing sequence features, chromatin-related information, and predicted gRNA activity to identify effective target sites. For example, DeepSpCas9 was trained on a large CRISPR-Cas dataset to predict the on-target cleavage efficiency of gRNAs, helping scientists identify target sites with predictable editing efficiency. Similarly, DeepCRISPR uses sequence context and epigenetic features to predict both gRNA efficiency and off-target effects in gene-editing studies (Chuai et al., 2018; Kim et al., 2019). By enabling prediction of gRNA efficiency and off-target effects, AI may improve the reliability and specificity of CRISPR-mediated gene editing. For example, DeepGuide uses deep learning to predict highly active CRISPR-Cas9 and Cas12a gRNAs in *Yarrowia lipolytica*, although its direct applications in plants would require organism-specific validation or retraining (Baisya et al., 2022). Recent AI tools have further improved the identification and selection of optimal CRISPR target sites. For example, CCLMoff uses a pretrained RNA language model from RNAcentral and deep learning to predict CRISPR-Cas9 off-target effects across diverse NGS-based datasets (Du et al., 2025). In addition, CRISPR_HNN combines a hybrid neural-network architecture to predict sgRNA activity in CRISPR-Cas9 gene editing (Li et al., 2025). Similarly, DNABERT-Epi integrates a pretrained DNA model with epigenetic features to improve CRISPR-Cas9 off-target prediction (Kimata and Satou, 2025). Following the selection of an

TABLE 20.1 AI tools for predicting on-target cleavage of CRISPR.

AI tool	Website/Tool access	AI/ML approach	Purpose	Key features	References
DeepCRISPR	https://github.com/bm2-lab/DeepCRISPR (Last access: June 17, 2026)	Deep learning (CNN + autoencoder)	Predict on-target activity and off-target potential	Integrates sequence and epigenetic features	Chuai et al. (2018)
DeepSpCas9	https://deepcrispr.info/workflows/DeepSpCas9 (Last access: June 17, 2026)	Deep convolutional neural network	Predict SpCas9 cleavage efficiency	High predictive accuracy across datasets	Kim et al. (2019)
DeepHF	http://www.deephf.com/#/cas9 (Last access: June 17, 2026)	Deep learning	Predict activity of high-fidelity Cas9 variants	Balances efficiency and specificity	Wang et al. (2019)
CRISPRon	https://rth.dk/resources/crispr/crispron/ (Last access: June 17, 2026)	Machine learning/Deep learning	Predict gRNA on-target efficiency	Ranks guides using validated datasets	Xiang et al. (2021)
Azimuth	https://github.com/MicrosoftResearch/Azimuth (Last access: June 17, 2026)	Gradient boosted regression trees	Predict sgRNA activity	Widely used guide design model	Doench et al. (2016)
Elevation–Azimuth framework	https://github.com/microsoft/Elevation (Last access: June 17, 2026)	Machine learning ensemble models	Predict on-target and off-target CRISPR-Cas9 activity	Combines azimuth on-target prediction with elevation off-target prediction for end-to-end guide RNA design	Listgarten et al. (2018)
DeepGuide	https://github.com/dDipankar/DeepGuide (Last access: June 17, 2026)	Deep learning	Predict CRISPR guide activity	Useful for nonmodel organisms and plants	Baisya et al. (2022)

optimal target site, the next critical step is to select the most suitable gRNA to direct the relevant endonuclease, such as Cas9, to the desired target site in the genome.

20.5.3 AI role in selecting an optimal gRNA

The success of CRISPR-based gene editing largely depends on the efficiency and specificity of the gRNA, which guides the endonuclease, such as Cas9, to the target DNA site in the genome. AI and ML tools can facilitate both the design and prioritization of gRNAs by integrating assessments of sequence context, off-target potential, and epigenetic features

(Chari et al., 2017; Chuai et al., 2018). For example, DeepCRISPR employs deep learning to predict both gRNA activity and off-target effects, thereby facilitating the identification and selection of effective gRNAs in genome-editing experiments (Chuai et al., 2018). Similarly, CRISPRon facilitates the selection of candidate gRNAs based on the predicted on-target activity (Xiang et al., 2021). CRISPR-IP uses deep neural networks to analyze sgRNA–DNA sequence pairs and predict off-target activity, thereby supporting the selection of gRNAs with lower predicted off-target potential (Zhang and Jiang, 2022). Collectively, these tools improve gRNA selection, reduce screening requirements, and contribute to the specificity, efficiency, and safety of CRISPR-Cas-based gene editing.

20.5.4 Prediction of off-target effects

Although CRISPR is generally considered a precise gene-editing technology, off-target cleavage resulting from nonspecific gRNA binding at nontarget sites in the genome remains a recognized biosafety concern. Off-targets remain an important consideration, particularly in health and therapeutic applications of CRISPR-Cas, where a single unintended mutation may have significant biological and clinical implications. Similarly, the presence of off-target effects in food crops intended for human and animal consumption remains a significant concern in safety assessment. However, multiple studies have demonstrated that off-targets may pose relatively low risk in plant gene editing, as unintended mutations can often be identified through sequencing and removed through selection and crossing in subsequent generations (Graham et al., 2020; Hahn and Nekrasov, 2019). Nevertheless, minimizing off-targets in CRISPR remains important to ensure genome integrity, facilitate regulatory approval, and improve social acceptance of gene-edited crops (Modrzejewski et al., 2020).

AI tools are increasingly used to predict off-target effects in CRISPR experiments, improving the efficiency and specificity of genome editing in plants. AI-based screening for off-target effects is an important application of ML for identifying potential unintended mutations, enabling scientists to mitigate these risks before regulatory review and field trials. For example, Elevation, developed by Microsoft and the Broad Institute, integrates genome-wide data to predict off-target activity in CRISPR-based gene editing (Listgarten et al., 2018). Elevation has been used to predict off-target effects in gene-editing studies, helping researchers assess specificity and potential biosafety concerns. Another AI-based platform, CRISPR-Net, uses deep neural networks to predict potential off-target cleavage sites by modeling complex patterns of mismatches, insertions, and deletions between the gRNA and the target site (Lin et al., 2020). Similarly, the model-based off-target framework (MOFF) uses ML algorithms to quantify the impact of multiple mismatches on the specificity of CRISPR-Cas, thereby facilitating more accurate prediction of off-target effects (Fu et al., 2022). CRISTA is another ML tool that integrates sequence, thermodynamic, and genomic properties to identify potential off-target effects in CRISPR-based gene editing (Abadi et al., 2017). Major computational and ML tools for off-target predictions have been summarized in Table 20.2. By predicting and minimizing off-target effects in gene editing, AI can contribute to the biosafety assessment of gene-edited crops, a factor critical for both regulatory approval and public acceptance.

TABLE 20.2 Computational and AI-based tools for off-target predictions.

Tool	Website	Computational or AI approach	Function	Key features	References
Elevation	https://github.com/microsoft/Elevation (Last access: June 17, 2026)	Machine learning ensemble model	Predict CRISPR-Cas9 off-target activity	Integrates sequence mismatch features and genome-wide data to prioritize off-target sites	Listgarten et al. (2018)
DeepCRISPR	https://github.com/bm2-lab/DeepCRISPR (Last access: June 17, 2026)	Deep learning (CNN + autoencoder)	Predict on-target and off-target activity	Combines sequence and epigenetic information to improve specificity prediction	Chuai et al. (2018)
CRISPR-Net	https://github.com/JasonLinjc/CRISPR-Net (Last access: June 17, 2026)	Deep neural network	Off-target prediction	Captures complex mismatch, insertion, and deletion patterns between gRNA and DNA targets	Lin et al. (2020)
DeepHF	http://www.deephf.com/#/home (Last access: June 17, 2026)	Deep learning	Predict activity of high-fidelity Cas9 variants	Supports the selection of guides with reduced off-target effects	Wang et al. (2019)
DeepCpf1	https://github.com/lje00006/DeepCpf1 (Last access: June 17, 2026)	Deep learning	Predict Cas12a (Cpf1) activity and specificity	Optimizes guide design and minimizes unintended editing for Cas12a systems	(Kim et al., 2019; Luo et al., 2019)
CRISTA	http://crista.tau.ac.il/ Original server inactive/ No official replacement found	Random forest machine learning	Off-target prediction	Integrates sequence, thermodynamic, and genomic features	Abadi et al. (2017)
MOFF (model-based off-target framework)	https://github.com/MDhewei/MOFF (Last access: June 17, 2026)	Machine learning	Quantitative off-target prediction	Models combinatorial effects of multiple mismatches on Cas9 specificity	Fu et al. (2022)
AttnToMismatch_CNN	https://github.com/qiaoliuhub/AttnToCrispr (Last access: June 17, 2026)	Attention-based convolutional neural network	Off-target prediction	Uses attention mechanisms to improve recognition of mismatch patterns	Liu et al. (2019)
CFD (cutting frequency determination)	https://portals.broadinstitute.org/gppx/crispick/public (Last access: June 17, 2026)	Data-driven predictive model	Off-target scoring	Estimates the likelihood of cleavage at mismatched genomic sites	Doench et al. (2016)

TABLE 20.2 Computational and AI-based tools for off-target predictions.—cont'd

Tool	Website	Computational or AI approach	Function	Key features	References
CROP-IT	http://www.adlilab.org/CROP-IT/homepage.html Original server inactive/No official replacement found	Machine learning	Off-target prediction	Ranks candidate off-target sites using sequence-based features	Singh et al. (2015)
CCTop	https://cctop.cos.uni-heidelberg.de/ (Last access: June 17, 2026)	Computational scoring framework	Off-target identification	Genome-wide screening of potential off-target loci	Stemmer et al. (2015)
Cas-OFFinder	https://github.com/snugel/cas-offinder (Last access: June 17, 2026)	Genome-wide search algorithm	Off-target identification	Rapid identification of candidate off-target sites across large genomes	Bae et al. (2014)

20.5.5 Smart forecasting of repair outcomes

Predicting the biological effects of genome editing is a key part of the safety and regulatory assessment of gene-edited crops. AI can support this process by integrating genomic, transcriptomic, proteomic, metabolic, and phenotypic datasets to predict potential repair outcomes of targeted genetic modifications. These analyses can help scientists and regulators better understand the expected effects of modifications before large-scale greenhouse and field evaluations are undertaken (Chen et al., 2024). AI and ML models can also help predict how specific genetic modifications may influence agronomic characteristics such as yield, stress tolerance, disease resistance, and nutrient efficiency (Jubair and Domaratzki, 2022; Washburn et al., 2021). For example, DeepGS uses deep learning to predict complex phenotypic traits from genomic information, providing insight into the potential consequences of genetic modification and helping to prioritize candidate gene-editing targets for further evaluation (Ma et al., 2018). Similarly, Bayesian genomic genotypic x environmental models integrate genomic and environmental information to estimate crop performance under diverse environmental conditions (Granato et al., 2018).

From a regulatory perspective, such predictive approaches can provide additional evidence for molecular characterization and risk assessment. AI-assisted prediction can help identify potential unintended effects, assess trait stability across environments, and prioritize areas requiring further experimental validation (Organisms et al., 2022). This information can inform environmental risk assessment, food and feed safety, and the development of post-market monitoring strategies. Importantly, AI predictions do not replace laboratory experiments, greenhouse studies, or field trials; they serve as complementary tools to guide data collection, generate hypotheses, and support evidence-based regulatory decisions. As predictive models continue to improve, they may become an increasingly important part of the regulatory assessment framework for gene-edited crops.

20.6 Potential applications of AI in regulatory science for gene-edited crops

20.6.1 AI and risk prediction for CRISPR-edited crops

AI is increasingly integrated into risk assessment frameworks for gene-edited crops. Unlike traditional approaches, which often assess datasets separately, ML algorithms can integrate genomic, environmental, climate, and agronomic data to identify patterns relevant to potential safety issues and environmental impacts (Mmbando, 2024). This is particularly valuable because regulatory agencies must assess the increasingly complex molecular and biological datasets generated during the development of gene-edited crops (Radio et al., 2024).

Food and feed safety assessment is an important application area for AI and ML. Computing platforms such as AllergenFP and AlgPred 2.0 use ML algorithms to assess the potential allergenicity of newly expressed proteins by comparing their physicochemical and genomic properties with those of known allergens (Dimitrov et al., 2014; Sharma et al., 2021). Similarly, predictive toxicology frameworks such as Tox21 and ToxCast use large biological data sets and computational modeling to identify molecular signatures associated with toxicity, thereby supporting early-stage hazard identification (Thomas et al., 2019). These tools do not replace experimental testing but may help regulators and developers to prioritize proteins and properties that require further investigation.

Several commercially important gene-edited crops illustrate how AI-assisted risk assessment could contribute to regulatory assessments. For example, in Japan, CRISPR technology has been used to develop tomatoes with high levels of the bioactive substance GABA (Waltz, 2022). In such cases, AI-assisted analysis may help detect unintended changes in metabolic pathways or protein expression profiles that warrant further investigation (Organisms et al., 2022; Radio et al., 2024). Similarly, blast-resistant rice produced by targeted modification of susceptible genes could benefit from AI-based analysis of environmental interactions to evaluate resistance performance under different climatic and agronomic conditions (Huang et al., 2024). For high-oleic soybeans and waxy corn, AI and ML approaches can support comparison of compositional, nutritional, and molecular characteristics with conventional varieties, thereby providing evidence relevant to substantial equivalence assessment (Canada, 2022; Demorest et al., 2016).

Another area where AI can provide useful decision support is environmental risk assessment. Gene flow, persistence of altered characteristics, and interactions with surrounding ecosystems are often difficult to assess using short-term experimental studies alone (OECD, 2023). By integrating datasets from ecology, geography, climate, and reproductive biology, AI and ML models can help estimate the probability of pollen-mediated gene flow, hybridization with wild relatives, and the persistence of modified traits in agricultural landscapes (Chandler and Dunwell, 2008). Such analyses may provide additional evidence to support decisions on appropriate containment measures, isolation distances, and post-release monitoring requirements. Moreover, predictive ecological models can be used to investigate potential impacts on biodiversity and species interactions under different environmental scenarios (Pichler et al., 2020).

Recent advances in multi-omics technologies have further expanded the possibilities for AI-supported risk assessment. Modern ML approaches can analyze genomics, transcriptomics, proteomics, metabolomics, and phenomics data sets to identify unexpected changes associated with targeted genetic modification (Picard et al., 2021). Such integrated analyses may reveal subtle changes in gene expression networks, metabolic pathways, or protein abundances that conventional compositional analysis may not readily detect (Ricroch et al., 2011). Consequently, multi-omics integration provides an additional layer of evidence that can enhance molecular characterization and biosafety assessment (Benevenuto et al., 2023). Regulatory authorities have also begun exploring the applications of AI in evidence synthesis and hazard identification. For example, the EFSA has evaluated the potential of AI approaches for scientific literature screening, emerging risk identification, and support for environmental and food safety assessments (Bersani et al., 2022). This development reflects a broader shift toward data-driven regulatory science that can cope with the increased volume and complexity of information generated by modern biotechnology.

Despite their significant potential, the quality and representativeness of available datasets remain key limitations for AI-assisted risk assessment. Many predictive models have been developed using a limited number of crop species, traits, or environmental conditions, which may limit their applicability across diverse crop systems (Jubair and Domaratzki, 2022). In addition, predictive accuracy may be affected by incomplete training data, algorithmic bias, and variable data quality. For these reasons, AI outputs should be treated as supporting evidence rather than as final regulatory conclusions (Bersani et al., 2022). Experimental validation, field evaluation, and peer review remain essential components of the bio-safety assessment and regulatory decision-making process. As gene-editing technologies continue to evolve, AI-assisted approaches can provide regulators with additional tools to integrate molecular, environmental, and ecological information into existing risk-assessment frameworks (Organisms et al., 2022; Radio et al., 2024). A detailed AI-assisted regulatory design indicating how AI can support the regulation of gene-edited crops is shown in Fig. 20.1. Their greatest value likely lies in improving the efficiency, consistency, and transparency of the evaluation process while maintaining the scientific rigor required for evidence-based gene-editing regulation.

20.6.2 AI and gene flow modeling

Gene flow, defined as the transfer of genetic material from cultivated crops to sexually compatible species via pollen or seed dispersal, remains an important consideration in the environmental risk assessment of gene-edited crops (Ellstrand, 2018). Regulators often assess the likelihood of pollen-mediated transfer of modified traits to cultivated varieties, hybrids, or wild relatives, especially when compatible species are present in the receiving environment (Chandler and Dunwell, 2008). AI can support these assessments by integrating diverse datasets, including crop distribution, reproductive biology, flowering synchrony, pollen dispersal patterns, climatic conditions, and landscape features. ML and predictive modeling approaches can help researchers estimate the likelihood of gene flow across environmental and agronomic scenarios (Walklate et al., 2004). By integrating historical environmental data and environmental variables, AI models can help identify locations and conditions where hybridization events are more likely to occur (Lu et al., 2019). Such analyses may provide additional evidence to support decisions regarding containment

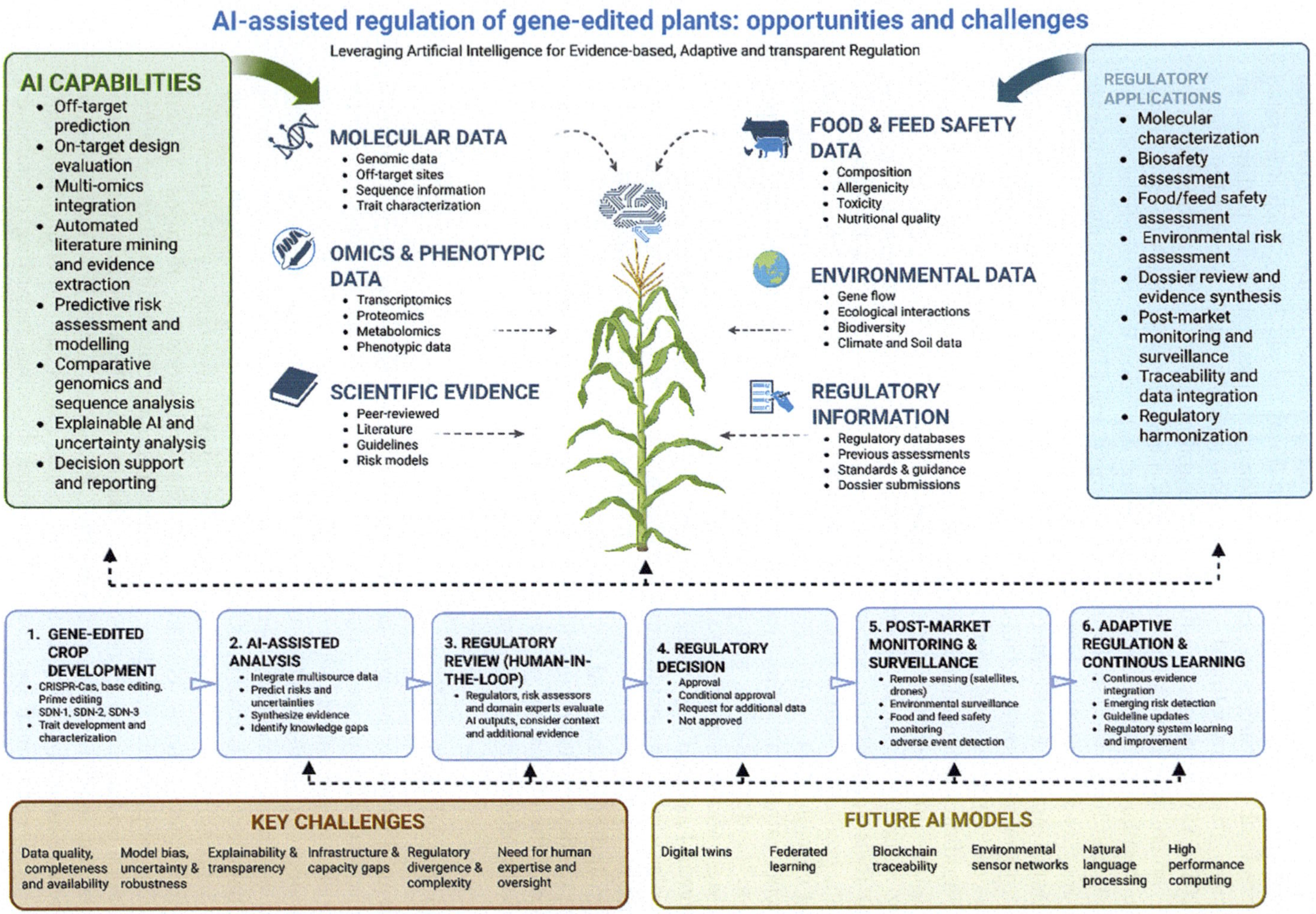

FIGURE 20.1 AI-assisted regulation of gene-edited crops. This figure summarizes how AI can support the regulation of gene-edited plants by integrating CRISPR design, multi-omics data, comparative genomics, risk prediction, automated literature review, and decision support systems. AI-enabled analysis can strengthen key regulatory processes, including biosafety assessment, food and feed safety assessment, environmental risk assessment, dossier review, post-market monitoring, traceability, and regulatory harmonization. The main implementation challenges include data quality and availability, model bias and uncertainty, clarity, infrastructure gaps, regulatory divergences, ethical and governance issues, privacy, and the ongoing need for human oversight.

measures, isolation distances, and monitoring requirements for field trials and commercial cultivation (Messeguer, 2003).

Gene flow modeling is especially important for crops with sexually compatible wild relatives or related cultivated species. For example, the environmental risk assessment of oilseed rape (canola), sugar beet, and rice has traditionally considered the possibility of pollen-mediated gene transfer to related species (OECD, 2022). Similar analytical frameworks could be used for gene-edited crops to assess the environmental behavior of modified traits under real-world farming conditions (Organisms et al., 2022). AI-based models can also support the assessment of trait persistence, the emergence of volunteer populations, and potential ecological interactions following gene transfer (Chandler and Dunwell, 2008). Importantly, AI predictions should complement, rather than replace, field studies and environmental assessments. Regulatory decisions regarding gene flow require multiple lines of evidence, including biological, environmental, and agronomic data. When combined with conventional approaches, AI-assisted environmental risk assessment can improve the effectiveness, consistency, and scientific basis for the regulatory assessment of gene-edited crops, while supporting biosafety oversight (OECD, 2022, 2023; Organisms et al., 2022).

20.6.3 AI and post-market monitoring of gene-edited crops

Post-market monitoring is an important part of regulatory oversight in jurisdictions where the evaluation of approved biotechnological products continues. Once a gene-edited crop is approved for commercial cultivation, regulatory authorities may need to verify that its agronomic performance, environmental behavior, and safety profile remain consistent with those assessed before commercialization (Organisms, 2015). AI and ML models can support these activities by analyzing large volumes of environmental, phenotypic, and agronomic data collected from commercially cultivated gene-edited crops (Kamilaris and Prenafeta-Boldú, 2018). Remote sensing platforms, coupled with AI and incorporating satellite imagery, unmanned aerial vehicles , and field sensors, can facilitate large-scale monitoring of crop growth, stress responses, disease occurrence, and geographical distribution (Kamilaris and Prenafeta-Boldu, 2018). ML algorithms can detect unusual patterns that may warrant further investigation, such as unexpected phenotypic responses, altered environmental interactions, or deviations from expected outcomes. Such systems can help regulatory authorities prioritize inspections and focus resources on areas of potential concern (Organisms, 2011).

AI can also support environmental monitoring by integrating weather, soil, land-use, and environmental datasets to identify trends relevant to post-market biosafety monitoring (Benos et al., 2021). For example, predictive analyses can help monitor the persistence of edited traits, track potential gene flow to sexually compatible relatives, or assess long-term environmental performance across diverse growing conditions. Similar data-driven monitoring approaches are already widely used in precision agriculture and environmental monitoring systems, and could provide valuable input for regulatory oversight of gene-editing crops (Mazzia et al., 2020).

20.6.4 Automated dossier review and decision support systems

Regulatory assessment of gene-edited crops requires analysis of extensive scientific documentation, including molecular characterization data, off-target evaluation, environmental risk studies, food and feed safety assessments, and post-market monitoring. As the number and complexity of gene-editing products continue to grow, regulatory agencies face an increasing challenge in reviewing these large, highly technical datasets (Radio et al., 2024). AI models, particularly NLP, can support this process by rapidly retrieving relevant information from scientific reports, identifying inconsistencies, summarizing evidence, and facilitating literature review (Blümmel et al., 2024). Rather than replacing peer review, NLP tools can help regulators manage the growing volume of information while improving efficiency and coherence (Bersani et al., 2022). Regulatory agencies have already begun exploring AI tools for evidence synthesis and document analysis. For example, the European Food Safety Authority (EFSA) has explored applications of AI in literature monitoring, evidence synthesis, and hazard identification to support food and environmental risk assessment (Cavalli et al., 2022). Similarly, the FDA has been exploring NLP and ML approaches to process large regulatory datasets and safety reports (FDA, 2025). Although these initiatives are not specific to gene-edited crops, they provide practical examples of how AI-supported document review could be integrated into future regulatory frameworks for gene-edited crops.

In addition to document review, AI-assisted decision support systems can help regulators assess the complex datasets generated during the evaluation of gene-edited crops (Organisms et al., 2022; Radio et al., 2024). By combining ML approaches, evidence synthesis, predictive risk assessment, and historical regulatory information, these systems can help identify key risk factors, prioritize files for further examination, and highlight areas where supporting evidence may be insufficient (Bersani et al., 2022; Cavalli et al., 2022). AI may also make it easier to compare newly submitted products with previously assessed gene-edited crops, thereby improving consistency across regulatory decisions. If implemented with appropriate transparency, validation, and human supervision, AI-assisted dossier review and decision support systems have the potential to improve the effectiveness, consistency, and scientific rigor of gene-edited regulatory assessment (Council, 2024).

20.6.5 AI-assisted regulatory harmonization

Differences in regulatory processes for gene-edited crops across jurisdictions continue to pose a major challenge for product authorization, risk assessment, and international trade (Ahmad et al., 2021; Fernández Ríos et al., 2025; Ishii, 2025). Countries currently use product-based, process-based, or hybrid regulatory frameworks, which often lead to different outcomes for similar genome-editing products. AI can support greater regulatory harmonization by facilitating the integration, analysis, and interpretation of complex datasets generated during the evaluation of gene-edited crops (Radio et al., 2024). With AI support, evidence synthesis tools can help regulatory agencies identify common scientific findings across jurisdictions, compare regulatory decisions, and assess the consistency of risk assessment approaches (Bersani et al., 2022). ML systems can also analyze large volumes of scientific literature and regulatory submissions, improving access to relevant evidence and promoting more transparent decision-making (Cavalli et al., 2022). Moreover, AI-enabled data-sharing platforms could facilitate cooperation among regulators by promoting standardized data formats, common assessment criteria, and interoperable assessment frameworks (Wilkinson et al., 2016).

International organizations such as the FAO, WHO, and the Codex Alimentarius Commission have long promoted science-based approaches to food safety assessment and international cooperation (FAO/WHO, 2009). Although these organizations do not directly regulate gene-edited crops, their guidance provides an important basis for harmonized risk assessment. AI can support this effort by improving evidence synthesis, promoting cross-jurisdictional comparability, and facilitating the exchange of scientific information among regulators (Cavalli et al., 2022; Radio et al., 2024). Although a globally harmonized regulatory framework for gene-edited crops is unlikely in the near future, AI-assisted collaborative evaluation systems could help reduce duplication in scientific evaluation, improve regulatory effectiveness, increase transparency, and foster greater consistency in biosafety oversight. This can ultimately facilitate international trade while maintaining high food, feed, and environmental safety standards (OECD, 2022).

20.6.6 AI and ecological interaction modeling

The era of gene-edited plants requires a comprehensive understanding of how novel and modified traits interact with the environment. Traditionally, this information has been generated through field trials and ecological research, which are often time-consuming and resource-intensive (Organisms, 2010; Organisms et al., 2022). AI offers an effective complementary approach by modeling potential ecological interactions using genomic, environmental, and biodiversity datasets, and simulation methodologies (Pichler et al., 2020; Shahhosseini et al., 2021). AI-based ecological modeling provides regulators with predictive information, supporting the planning of field trials and contributing to environmental risk assessment. Given the highly dynamic nature of ecological interactions, AI can provide valuable support for managing complex interaction networks; however, it is only effective when combined with accurate detection and characterization of unintended modifications, including off-target mutations (Modrzejewski et al., 2020).

20.6.7 AI for comparative genomics and substantial equivalence

Comparative genomics is a key component of the molecular characterization and regulatory assessment of gene-edited crops. Regulatory authorities often require evidence that the intended modifications are well characterized and that no biologically relevant unintended genomic modifications have occurred during the editing process (Sturme et al., 2022). The growing availability of whole-genome sequencing data has enabled researchers to assess genome integrity, but interpreting these large, complex datasets remains challenging (Graham et al., 2020). AI and ML tools can support comparative genomic analysis by identifying, classifying, and interpreting genomic variation between edited and nonedited plant genomes (Poplin et al., 2018). ML algorithms can help distinguish the proposed modifications from naturally occurring polymorphisms, background genetic variation, and sequencing errors. AI-supported annotation tools can further evaluate the potential biological relevance of detected mutations by integrating genomic, protein functional, and evolutionary information (Cheng et al., 2023; Cingolani et al., 2012). Such analyses can help identify candidates with off-target modifications that merit further investigation and provide additional evidence of the specificity of the gene-editing events. For regulatory assessment, AI-based comparative genomics can improve molecular characterization by evaluating genome integrity, editing accuracy, and off-target effects (Modrzejewski et al., 2020; Sturme et al., 2022).

Substantial equivalence remains a widely recognized principle in the safety assessment of novel crops. In this approach, the gene-edited crop is compared with a suitable conventional counterpart to determine whether there are biologically significant differences in composition, nutritional characteristics, agronomic performance, and other parameters relevant to safety (Organisms et al., 2022). These evaluations typically require extensive molecular, biochemical, nutritional, and phenotypic analyses of the edited plants. AI can support a substantial equivalence assessment by integrating multiple evidence sources, such as genomics, transcriptomics, proteomics, metabolomics, compositional analysis, and phenotypic data (Authority et al., 2018). ML approaches can detect patterns, correlations, and deviations that might not be easily detected by conventional statistical analysis (Eicher et al., 2020; Sanches et al., 2024). For example, AI and ML-based analyses of metabolomic and compositional datasets can help detect unexpected changes that require further investigation, while distinguishing normal biological variations from potentially relevant differences (Authority et al., 2018; Cellini et al., 2004). From a regulatory point of view, AI does not independently determine the existence of substantial equivalence; however, it provides decision support tools to facilitate data integration, improve the consistency of interpretation, and ensure evidence-based assessment (Bersani et al., 2022). These capabilities may be particularly valuable as gene-edited crops become more diverse and generate large multi-omics datasets (Radio et al., 2024). By promoting comprehensive comparisons between modified and conventional crops, AI-assisted comparative genomics and substantial equivalence assessments can contribute to more transparent, reproducible, and scientifically sound regulatory assessments, while maintaining established food, feed, and environmental safety standards.

20.7 AI and the social acceptance of gene-edited crops

Public perception remains a key factor in the acceptance and commercialization of gene-edited crops (Kato-Nitta et al., 2019; Spök et al., 2022). Beyond scientific evidence and regulatory approval, public confidence in the approval process often determines social acceptance of new agricultural technologies, including gene editing. As AI is increasingly integrated into regulatory science, its potential impact on public confidence in the regulatory process warrants careful consideration (Bersani et al., 2022). AI can support public trust in regulatory assessments by enhancing the coherence, transparency, and reproducibility of risk assessments. ML approaches can help regulators analyze large, complex datasets, promote greater consistency in interpreting evidence, and support systematic decision-making (Cavalli et al., 2022). However, public concerns may arise if AI systems are perceived as replacing human expertise or making decisions through opaque processes that are inadequately explained or difficult to verify independently (Barredo et al., 2020).

To address these concerns, researchers and policy-makers are increasingly promoting the use of explainable AI (XAI), which emphasizes transparent methodologies, interpretable outputs, and clear documentation of how those outputs are generated (Vilone and Longo, 2020). Human oversight remains important in this process. Regulatory decisions on food safety, environmental risks, and biotechnology should be made by qualified experts who can assess AI-generated evidence within the broader scientific, ethical, and societal context (Council, 2024). Transparent communication is crucial to maintaining public confidence (Kato-Nitta et al., 2019; Spök et al., 2022). Publicly available risk assessment reports, stakeholder engagement initiatives, and clear explanations of how AI contributes to regulatory assessments can help improve understanding and acceptance of both AI-assisted regulation and gene-edited crops. When implemented responsibly, AI can enhance evidence-based decision-making while promoting transparency, accountability, and public participation (Council, 2024). These factors are likely to play an increasingly important role in strengthening public confidence in the regulatory oversight of gene-edited crops.

20.8 AI and governance of gene-edited crops

The integration of AI into regulatory frameworks for CRISPR-edited crops now extends beyond technical risk assessment and off-target evaluation, to support broader policy making (Veale et al., 2023). These tools can help policymakers assess climate resilience, forecast economic impacts, identify research priorities, and estimate trade-offs among food security, sustainability, and environmental goals (Fernández Ríos et al., 2025). As a result, AI is emerging as a valuable decision-support tool for both regulatory oversight and policy planning. The growing role of AI models in regulatory and policy frameworks for gene-edited crops underscores the need for inclusive, ethically informed AI governance that balances scientific rigor, legal responsibility, and social values (UNESCO, 2021). With advances in AI, gene-editing governance may benefit from transparent, evidence-based, and accountable regulation, ensuring that AI serves the public interest beyond mere technical efficiency (AI, 2023; UNESCO, 2021).

20.9 Ethical use of AI in policy decision-making

20.9.1 Human-in-the-loop frameworks

As AI plays a role in the regulatory and policy frameworks for gene-edited crops, ethical considerations and governance oversight are increasingly important. The human-in-the-loop (HITL) approach is a key principle of AI governance, ensuring that human professionals are directly engaged in interpreting, validating, and even overriding AI recommendations (Amershi et al., 2014; Kamar, 2016). Under this framework, AI serves as a decision-support tool rather than an independent decision-maker. HITL systems preserve professional judgment, minimize overreliance on algorithms, and ensure that data-driven decisions account for broader scientific, ethical, and social factors (Kamar, 2016). Such oversight is essential for maintaining public trust in regulatory systems and for ensuring the responsible use of AI within the frameworks for gene-edited crops (Bersani et al., 2022; Council, 2024).

The broader concept of responsible AI is increasingly shaping discussions about AI-assisted regulation of gene-edited crops. Responsible AI emphasizes transparency, fairness, security, accountability, privacy protection, and human oversight throughout the AI system lifecycle (AI, 2023; UNESCO, 2021). In regulatory decision-making, these principles help ensure that AI-assisted assessments remain scientifically robust while respecting ethical, legal, and societal considerations (Cavalli et al., 2022; Council, 2024). Rather than focusing solely on efficiency, responsible AI frameworks seek to reconcile innovation with public benefit, environmental sustainability, and regulatory integrity.

The AI principles adopted by the Organization for Economic Co-operation and Development (OECD) in 2019 provide an internationally recognized framework for the use of AI. These principles emphasize inclusive growth, human values, transparency, resilience, security, and accountability. Although not developed specifically for agricultural biotechnology, they offer useful guidance on the responsible use of AI within the regulatory frameworks governing gene-edited plants. Their emphasis on transparency, traceability, and human supervision aligns with the requirements of evidence-based bio-safety regulation and risk assessment (Bersani et al., 2022; Council, 2024; Organisms et al., 2022).

20.9.2 Social license to operate and public trust

Several studies have reported that public trust in AI-based policymaking is often limited, particularly regarding food, health, and environmental issues (Burrell, 2016). To build greater public confidence in integrating AI systems into the regulatory framework for gene-edited crops, it is necessary to ensure that AI-assisted regulatory frameworks are just, transparent, consistent, and aligned with societal principles (Kato-Nitta et al., 2019; Spök et al., 2022). Otherwise, AI can be perceived as a technocratic tool that lacks transparency, accountability, and public trust. When implemented responsibly, AI can support transparent, data-driven, and participatory decision-making that improves social acceptance of gene-edited crops (Dare et al., 2014; Spök et al., 2022). Looking ahead, ethics, transparency, and accountability will be key components of the governance of AI-assisted regulatory frameworks for gene-edited crops. These principles can support responsible innovation while enabling the public to understand, benefit from, and participate in the governance and regulatory frameworks for gene-edited crops and AI (Council, 2024; Wachter et al., 2017).

20.9.3 AI and black-box problem

The black box problem, in which an AI model's internal processes are too complex or opaque for its developer to understand, is a pressing issue for integrating AI into regulatory systems. Regulatory decisions must be transparent, evidence-based, and capable of withstanding legal scrutiny (Council, 2024; Wachter et al., 2017). Lack of interpretability can raise questions within a regulatory framework about transparency, consistency, and legality (Burrell, 2016). Without sufficient transparency, regulators may be reluctant to rely on

AI systems that cannot provide understandable explanations of their outputs (Doshi-Velez and Kim, 2017). This concern has fueled growing interest in XAI, which aims to make AI-generated predictions more understandable and transparent. Techniques such as feature-importance analysis, attention mapping, and model explanation frameworks can help identify the factors influencing AI predictions and increase confidence in regulatory assessments (Ribeiro et al., 2016). As interest in AI for the regulation of gene-edited crops grows, ensuring transparency, accountability, and audibility will be crucial to maintaining public confidence and the credibility of regulatory frameworks (AI, 2023).

20.9.4 Right to explanation and auditability

Democratic, science-based regulatory environments require that AI-assisted regulatory assessments be transparent, evidence-based, and traceable for all stakeholders, including crop developers, regulators, and the public (Goodman and Flaxman, 2017). Therefore, the need for explainable and transparent AI has become a recent focus of AI- assisted regulatory initiatives, including the EU AI Act, which emphasizes transparency, human oversight, and auditability for high-risk AI applications (Nastoska et al., 2025). Such transparency and auditability help ensure that regulatory decisions can be thoroughly reviewed, appealed, or amended when new evidence becomes available, thereby supporting regulatory credibility and fostering public trust (AI, 2023; Wachter et al., 2017). Transparent systems not only safeguard democratic oversight but also help build public support for AI-assisted regulatory frameworks. This is particularly applicable when the subject matter involves sensitive issues such as food security, environmental impacts, and gene editing (Kato-Nitta et al., 2019; Spök et al., 2022).

Recent regulatory developments have underscored the importance of transparency and oversight. The EU AI Act sets out risk-based requirements for AI systems, with particular emphasis on transparency, human oversight, traceability, and accountability (Council, 2024). The principles outlined in the EU AI Act provide a useful reference for future governance frameworks that will include the AI-assisted regulation of gene-edited crops. Beyond transparency, algorithmic responsibility is becoming an integral component of governance frameworks for AI-assisted regulatory systems (AI, 2023). Accountability requires that organizations using AI systems remain responsible for the results these systems produce, irrespective of the level of automation involved. Clear documentation of data sets, model architecture, validation procedures, performance constraints, and decision-making pathways can help ensure accountability and transparency throughout the regulatory process. These mechanisms help ensure that AI-supported recommendations remain open to scientific peer review and independent scrutiny (AI, 2023; Bersani et al., 2022; Cavalli et al., 2022).

20.9.5 The risks of sets trained using biased data

Despite AI's significant potential to support the regulation of gene-edited crops, the effectiveness of the learning process depends heavily on the quality, completeness, and representativeness of the training data (Williamson et al., 2023). Training datasets that are biased, incomplete, or unrepresentative of specific regions, crops, or traits can lead to misleading predictions that may affect regulatory assessments (Mehrabi et al., 2021; Suresh and Guttag, 2021). For example, an ML model designed to assess the allergenicity of gene-edited crops such as maize and soybean may not perform well when applied to orphan crops such as sorghum and millet. Such limitations may increase the rate of false-positive predictions, leading to unnecessary testing, delays, and increased resource requirements for developers. These examples demonstrate that biased models can increase global inequalities, impede innovation, and decrease confidence in AI-assisted regulatory systems (Council, 2024; Mehrabi et al., 2021; Suresh and Guttag, 2021). These challenges highlight the importance of developing XIA systems for biotechnology applications and using high-quality datasets that accurately represent global agricultural diversity (MacNish et al., 2025; Williamson et al., 2023).

20.9.6 Data sovereignty

Another emerging issue is data sovereignty, especially when genomic, environmental, agricultural, and biodiversity data originate in different countries. Many countries increasingly view biological and genomic data as strategic resources that require proper management and protection (Montreal, 2011). International AI systems trained on plant genomic information may therefore need to address data ownership, access rights, consent, cross-border data sharing, and benefit-sharing arrangements (Wilkinson et al., 2016). These considerations are particularly relevant for crop species that are important for food security in developing countries and in regions with high genetic diversity (Williamson et al., 2023).

20.9.7 Creating representativeness in plant data

One approach to developing a broadly applicable, transparent, and scalable regulatory system is to ensure that AI systems are trained on representative, inclusive datasets. This requires incorporating not only the wide diversity of crop genomes but also a broad range of environmental conditions, trait combinations, and agricultural practices across different geographical regions (MacNish et al., 2025; Williamson et al., 2023). This diversity will help reduce algorithmic bias and improve the reliability of AI predictions for crops and environments. AI models trained on diverse genomics and multi-omics datasets have demonstrated improved capabilities for predicting gene functions, editing outcomes, and the biosafety risks associated with model and crop plants (MacNish et al., 2025; van Dijk et al., 2021; Williamson et al., 2023). High-quality, representative data will not only minimize bias but also enhance the credibility of AI tools in the international regulatory context for gene-edited crops.

The increasing use of digital infrastructure also poses significant security risks. AI-assisted regulatory systems may rely on interconnected databases containing genomic sequences, risk assessment documents, environmental monitoring data, and proprietary trade information (Williamson et al., 2023). Unauthorized access, data manipulation, or cyberattacks could jeopardize regulatory decisions and undermine public trust. Therefore, deploying AI in regulatory frameworks should be accompanied by robust security measures, including secure data storage, access controls, encryption, audit trails, and regular system validation. Ensuring data integrity and system security will be crucial to maintaining confidence in AI-assisted regulation of gene-edited crops (AI, 2023; Bersani et al., 2022).

20.10 Case studies of potential applications of AI in regulatory frameworks

20.10.1 EFSA exploration of AI for food and feed risk assessment

The EFSA has been among the most active regulatory authorities exploring potential applications of AI in food and feed safety assessments. The EFSA has explored ML approaches for evidence synthesis, hazard identification, predictive toxicology, and environmental risk assessment (Bersani et al., 2022; Blümmel et al., 2024; Cavalli et al., 2022). In 2024, EFSA published reports and activities highlighting the opportunities and limitations of AI in food safety assessment, emphasizing applications such as automated literature searches, extraction of scientific evidence, and detection of emerging risks. Although these tools are not currently used for autonomous regulatory decisions, they demonstrate how AI can support future gene-editing evaluation by assisting in the analysis of large datasets of genomics, toxicology, allergenicity, and environmental data (Radio et al., 2024). The work of EFSA represents one of the most prominent examples of a regulatory authority exploring AI approaches to support regulatory oversight of emerging agricultural biotechnologies (Authority, 2025; Bersani et al., 2022).

20.10.2 FDA ML applications in postmarket monitoring

The FDA is increasingly exploring ML and AI approaches for postmarket monitoring and safety assessment of regulated products (Ball et al., 2024). AI-assisted systems have been evaluated for detecting safety signals, identifying adverse event patterns, and processing large volumes of regulatory information. Although these initiatives primarily focus on pharmaceuticals and medical products, they offer a useful regulatory model for AI-assisted oversight of gene-edited crops and food products (Ball et al., 2024; FDA, 2023). The ML system's ability to detect unusual patterns in large datasets shows how similar approaches could be adapted to monitor the performance, safety, and environmental impact of gene-edited products after commercialization. These developments demonstrate the growing role of AI as a tool for regulatory support rather than as a substitute for scientific peer review (Organisms, 2011).

20.10.3 USDA digital transformation and biotechnology regulatory oversight

In particular, the USDA, through the APHIS, has modernized the regulation of emerging biotechnologies through electronic submission systems, digital databases, and data-driven review procedures (USDA, 2026). The implementation of the SECURE Regulation has introduced an easy, product-based framework for the regulatory approval of gene-edited crops. Similarly, the "AM I regulated?" platform represents a digital transformation aimed at improving transparency and facilitating communication between product developers and regulators. Although the USDA does not currently use fully autonomous AI systems for regulatory approval, it has developed a digital infrastructure capable of handling large genomic and regulatory data sets (USDA, 2026). These platforms provide the basis for future integration of AI-assisted analyses, including automated comparison of genomic sequences, evaluation of molecular characterization data, and identification of potential regulatory issues. The USDA's digital transformation illustrates how regulatory agencies are reshaping the infrastructure necessary for future AI-assisted regulatory oversight of gene-edited plants.

20.10.4 FAO/WHO initiatives on AI-supported food safety and risk assessment

The role of AI in food safety management and risk assessment is increasingly examined by the Food and Agriculture Organization (FAO) of the United Nations , particularly in predictive modeling, regulatory assessment, and evidence-based decision-making (van Meer et al., 2025). FAO has also hosted international discussions on AI for food safety, exploring practical applications, benefits, limitations, and regulatory considerations for competent authorities. Joint FAO/WHO scientific advice and Codex Alimentarius guidance continue to provide the broader international framework for food safety risk assessment, including foods derived from modern biotechnology (FAO/WHO, 2009). Although FAO and WHO are not directly involved in the regulatory approval of gene-edited crops, their work provides an important framework for international cooperation, scientific advice, and capacity building. This is particularly important in countries with limited regulatory resources, where AI-enabled tools may ultimately support food safety, environmental risk, and biosafety considerations related to gene-edited plants (FAO/WHO, 2009).

Taken together, these examples show that AI is gradually integrating into the regulatory sciences through evidence synthesis, risk identification, digital monitoring, predictive analytics, and data collection (Bersani et al., 2022; Cavalli et al., 2022). While AI-assisted regulation of gene-edited crops remains a future objective, current initiatives by EFSA, FDA, USDA, FAO, and WHO demonstrate that regulators and international regulatory authorities are building the scientific and digital foundations needed to oversee emerging biotechnology products effectively, transparently, and in an evidence-based manner.

20.11 Future of regulatory frameworks: AI-enabled models

Conventional regulatory frameworks for gene-edited crops are generally based on fixed assessment procedures that may struggle to keep pace with the rapid development of CRISPR-Cas and the increasing complexity of gene-editing techniques and products (Organisms et al., 2022). By contrast, AI-enabled adaptive regulatory systems can continuously incorporate new scientific evidence, including information from premarket analyses, postmarket monitoring, environmental assessments, and emerging research findings (FDA, 2025; Organisms et al., 2022). Such systems could support more dynamic and evidence-based regulatory decision-making, while maintaining established biosafety and transparency standards (Bersani et al., 2022). One emerging concept is the development of digital twins, which are virtual representations of biological systems generated using genomic, environmental, agronomic, and phenotypic data (Escribà-Gelonch et al., 2024; Kim et al., 2019). Although still largely experimental in agricultural biotechnology, digital twins could eventually allow regulators and developers to simulate the performance of gene-edited crops under a variety of environmental conditions before they are scaled up (Escribà-Gelonch et al., 2024; Kim et al., 2019). Such simulations can support the evaluation of trait stability, environmental interactions, and potential environmental impacts, complementing evaluations in the greenhouse and field.

Another promising approach is to use regulatory sandboxes, which provide a controlled environment for evaluating technologies under oversight (Council, 2024). Regulatory sandboxes have already been explored in several areas of digital governance and may offer a useful framework for testing AI-based risk assessment tools for gene-edited crops (OECD, 2023). By enabling regulators, developers, and researchers to evaluate new technologies in real-world settings, sandboxing can facilitate innovation while preserving regulatory oversight and public confidence. AI can also streamline regulatory workflows by integrating and analyzing evidence in real time. Instead of relying solely on periodic reviews, future regulatory systems could integrate new information from scientific publications, environmental monitoring programs, and postmarket monitoring activities on an ongoing basis (Authority, 2025; Bersani et al., 2022). While fully automated real-time approval is not currently in practice or desired for high-stakes regulatory decisions, continuous updating of evidence could allow regulators to identify emerging concerns more quickly and revise guidelines if needed (AI, 2023).

The increasing availability of molecular, biochemical, and agronomic datasets from multiple institutions has spurred interest in federated learning. This ML approach enables training AI models on distributed datasets without transferring sensitive information to a central database (Calvino et al., 2024). For regulatory purposes involving genomic, environmental, and agricultural data, federated learning can facilitate international cooperation while addressing data privacy, confidentiality, and data sovereignty concerns (Mc Cartney et al., 2022). Improved traceability can also be an important element of future regulatory oversight. Blockchain-enabled traceability is proposed as a mechanism to maintain secure and transparent records throughout the development, regulatory review, production, and distribution of agricultural products. When combined with AI-enabled monitoring systems, blockchain technology could enhance documentation of regulatory decision-making, facilitate audits, and increase transparency in the supply chains for gene-edited crops (Ellahi et al., 2023; Kamilaris et al., 2019).

Advances in remote sensing, environmental monitoring, and ML may enable the development of autonomous monitoring systems (Kamilaris and Prenafeta-Boldú, 2018). Satellites, drones, environmental sensors, and AI-driven analysis can support the collection and interpretation of large-scale field data following the commercialization of gene-edited products (Mazzia et al., 2020). These systems can help regulatory bodies identify unexpected agronomic responses, monitor environmental performance, and detect deviations from expected outcomes. In addition, they could improve postmarket monitoring while reducing the logistical burden of large-scale monitoring programs (Bersani et al., 2022). For example, EFSA has already begun using AI-assisted tools for literature monitoring, hazard identification, and evidence synthesis to support risk assessment practices (Cavalli et al., 2022).

Collectively, emerging technologies such as digital twins, regulatory sandboxing, federated learning, blockchain-enabled traceability, and autonomous monitoring systems highlight the potential of AI to develop adaptive regulatory frameworks for gene-edited crops (Calvino et al., 2024). While these approaches are still in the early stages of development, they offer significant potential to transform regulatory frameworks into more responsive, transparent, and scientifically informed systems capable of adapting to future advances in gene-editing technologies (AI, 2023).

20.12 Opportunities and challenges

AI applications within global regulatory frameworks for gene-edited crops hold significant potential to improve the effectiveness of regulatory processes by facilitating data integration, risk prediction, literature review, and evidence synthesis (Bersani et al., 2022). AI-enabled platforms can reduce administrative burden and facilitate certain regulatory review phases through automated, intelligent data analysis (FDA, 2025). Different regulatory agencies are exploring AI applications in dossier evaluation, scientific evidence synthesis, and the modernization of biotechnology regulations. Moreover, AI can help analyze complex datasets, including multi-omics, environmental, and ecological data, and identify patterns that are difficult to detect with conventional approaches (Radio et al., 2024; Williamson et al., 2023). These capabilities can improve risk assessments, ecological impact assessments, and biosafety evaluations, while supporting evidence-based and shared regulatory standards across jurisdictions. In the face of rapid CRISPR developments and increasing workloads, AI can become a valuable tool for facilitating evidence synthesis, supporting decision-making, and supporting regulators in scientific evaluation, public engagement, and policy making for gene-edited crops (FDA, 2025; Bersani et al., 2022).

Despite the considerable potential of AI-assisted regulatory frameworks for gene-editing crops, several challenges may limit the implementation and effectiveness of this transformative technology in developing countries (Unctad, 2021). Limited computing resources, internet connectivity, and access to multi-omics databases are major concerns across regulatory agencies in developing nations. Limited access to these resources may contribute to uneven regulatory capacity across countries and to the successful implementation of AI in regulatory frameworks. Moreover, integrating AI into the regulatory framework for gene-edited crops would require trained experts in both data science and biotechnology, and a lack of interdisciplinary expertise would significantly hinder progress in implementation (Williamson et al., 2023). Therefore, trained AI and biotechnology professionals, access to advanced databases, and investment in AI infrastructure are necessary to realize the full potential of AI-assisted gene-editing. Moreover, sensitive information, databases, and risk assessment information would be vulnerable to cybersecurity threats, intellectual property rights issues, and adverse manipulation (AI, 2023). Strong cybersecurity measures, transparent data governance frameworks, and secure data-sharing mechanisms will therefore be essential to maintain confidence, regulatory integrity, and international cooperation in AI-assisted regulation of gene-edited crops.

Declaration

"The chapter was professionally edited for the English language, and Grammarly Pro was also used to refine the English. After using this tool, the author(s) reviewed and edited the content as needed and take full responsibility for the publication's content."

References

Abadi, S., Yan, W.X., Amar, D., Mayrose, I., 2017. A machine learning approach for predicting CRISPR-Cas9 cleavage efficiencies and patterns underlying its mechanism of action. PLoS Comput. Biol. 13 (10), e1005807.

Agapito-Tenfen, S.Z., Okoli, A.S., Bernstein, M.J., Wikmark, O.-G., Myhr, A.I., 2018. Revisiting risk governance of GM plants: the need to consider new and emerging gene-editing techniques. Front. Plant Sci. 9, 1874.

Ahmad, A., Munawar, N., Khan, Z., Qusmani, A.T., Khan, S.H., Jamil, A., Ashraf, S., Ghouri, M.Z., Aslam, S., Mubarik, M.S., 2021. An outlook on global regulatory landscape for genome-edited crops. Int. J. Mol. Sci. 22 (21), 11753.

AI, N, 2023. Artificial Intelligence Risk Management Framework (AI RMF 1.0), pp. 100–101. https://nvlpubs.nist.gov/nistpubs/ai/nist.ai.

Amershi, S., Cakmak, M., Knox, W.B., Kulesza, T., 2014. Power to the people: the role of humans in interactive machine learning. AI Mag. 35 (4), 105–120.

Animal, U, 2020. Movement of certain genetically engineered organisms. In: Plant Health Inspection Service. 7 CFR Parts 330, 340, and 372. Final Rule. Federal Register.

Animal, U, 2023. Federal regulations governing the spongy moth quarantine found in, 7. In: Plant Health Inspection Service. 2023. ". Spongy Moth Program Manual.

Apple, A. F, 2022. Standard Operating Procedures for Regulatory Review of Genome Edited Plants under SDN-1 and SDN-2 Categories.

Authority, E. F. S, Aguilera, J., Aguilera-Gomez, M., Barrucci, F., Cocconcelli, P.S., Davies, H., Denslow, N., Lou Dorne, J., Grohmann, L., Herman, L., 2018. EFSA Scientific Colloquium 24–'omics in risk assessment: state of the art and next steps. EFSA Support. Publ. 15 (11), 1512E.

Authority, E. F. S, 2025. AI@EFSA.

Bae, S., Park, J., Kim, J.-S., 2014. Cas-OFFinder: a fast and versatile algorithm that searches for potential off-target sites of Cas9 RNA-guided endonucleases. Bioinformatics 30 (10), 1473–1475.

Baisya, D., Ramesh, A., Schwartz, C., Lonardi, S., Wheeldon, I., 2022. Genome-wide functional screens enable the prediction of high activity CRISPR-Cas9 and-Cas12a guides in Yarrowia lipolytica. Nat. Commun. 13 (1), 922.

Ball, R., Talal, A.H., Dang, O., Muñoz, M., Markatou, M., 2024. Trust but verify: lessons learned for the application of AI to case-based clinical decision-making from postmarketing drug safety assessment at the US Food and Drug Administration. J. Med. Internet Res. 26, e50274.

Barredo, A.A., Del Ser, J., Gil-Lopez, S., Díaz-Rodríguez, N., Bennetot, A., Chatila, R., Tabik, S., Garcia, S., Molina, D., Herrera, F., 2020. Explainable explainable artificial intelligence (XAI): concepts, taxonomies, opportunities and challenges toward responsible AI. Inf. Fusion 58, 82–115.

Benevenuto, R.F., Zanatta, C.B., Waßmann, F., Eckerstorfer, M.F., Agapito-Tenfen, S.Z., 2023. Integration of omics analyses into GMO risk assessment in Europe: a case study from soybean field trials. Environ. Sci. Eur. 35 (1), 14.

Benos, L., Tagarakis, A.C., Dolias, G., Berruto, R., Kateris, D., Bochtis, D., 2021. Machine learning in agriculture: a comprehensive updated review. Sensors 21 (11), 3758.

Bersani, C., Codagnone, J., David, L., Foiniotis, A., Galasso, G., Mancini, S., Michieletti, R., Orphanidou, C., Pellegrino, M., 2022. Roadmap for actions on artificial intelligence for evidence management in risk assessment. EFSA Support. Publ. 19 (5).

Blümmel, T., Rehn, J., Mereu, C., Graf, F., Bazing, F., Kneuer, C., Sonnenburg, A., Wittkowski, P., Padberg, F., Bech, K., 2024. Exploring the use of Artificial Intelligence (AI) for extracting and integrating data obtained through New Approach Methodologies (NAMs) for chemical risk assessment. EFSA Support. Publ. 21 (1), 8567E.

Bodin, J.E., Thorstensen, T., Alsheikh, M.K., Basic, D., Edvardsen, R.B., Dalen, K.T., Duale, N., Eklo, O.M., Ergon, Å. G., Jevnaker, A.-M.G., 2021. Genome editing in food and feed production–implications for risk assessment. In:

Scientific Opinion of the Scientific Steering Committee of the Norwegian Scientific Committee for Food and Environment.

Buchholzer, M., Frommer, W.B., 2023. An increasing number of countries regulate genome editing in crops. New Phytol. 237 (1).

Burrell, J., 2016. How the machine 'thinks': understanding opacity in machine learning algorithms. Big Data Soc. 3 (1), 2053951715622512.

Calvino, G., Peconi, C., Strafella, C., Trastulli, G., Megalizzi, D., Andreucci, S., Cascella, R., Caltagirone, C., Zampatti, S., Giardina, E., 2024. Federated learning: breaking down barriers in global genomic research. Genes 15 (12), 1650.

Canada, H., 2022. Scientific Opinion on the Regulation of Gene-Edited Plant Products Within the Context of Division 28 of the Food and Drug Regulations (Novel Foods).

Cavalli, E., Verloo, D., Paraskevopoulos, K., Kleiner, J., Heppner, C., Hugas, M., 2022. Theme (concept) paper-Artificial Intelligence in risk assessment. EFSA Support. Publ. 19 (5).

Cellini, F., Chesson, A., Colquhoun, I., Constable, A., Davies, H., Engel, K.H., Gatehouse, A.M., Kärenlampi, S., Kok, E.J., Leguay, J.-J., 2004. Unintended effects and their detection in genetically modified crops. Food Chem. Toxicol. 42 (7), 1089–1125.

Chandler, S., Dunwell, J.M., 2008. Gene flow, risk assessment and the environmental release of transgenic plants. Crit. Rev. Plant Sci. 27 (1), 25–49.

Chari, R., Yeo, N.C., Chavez, A., Church, G.M., 2017. sgRNA Scorer 2.0: a species-independent model to predict CRISPR/Cas9 activity. ACS Synth. Biol. 6 (5), 902–904.

Chen, K., Wang, Y., Zhang, R., Zhang, H., Gao, C., 2019. CRISPR/Cas genome editing and precision plant breeding in agriculture. Annu. Rev. Plant Biol. 70 (1), 667–697.

Chen, L., Liu, G., Zhang, T., 2024. Integrating machine learning and genome editing for crop improvement. aBIOTECH 5 (2), 262–277.

Cheng, J., Novati, G., Pan, J., Bycroft, C., Žemgulytė, A., Applebaum, T., Pritzel, A., Wong, L.H., Zielinski, M., Sargeant, T., 2023. Accurate proteome-wide missense variant effect prediction with AlphaMissense. Science 381 (6664).

Chuai, G., Ma, H., Yan, J., Chen, M., Hong, N., Xue, D., Zhou, C., Zhu, C., Chen, K., Duan, B., 2018. DeepCRISPR: optimized CRISPR guide RNA design by deep learning. Genome Biol. 19 (1), 80.

Cingolani, P., Platts, A., Wang, L.L., Coon, M., Nguyen, T., Wang, L., Land, S.J., Lu, X., Ruden, D.M., 2012. A program for annotating and predicting the effects of single nucleotide polymorphisms, SnpEff: SNPs in the genome of *Drosophila melanogaster* strain w1118; iso-2; iso-3. Fly 6 (2), 80–92.

Council, E. P. a, 2024. Regulation (EU) 2024/1689 laying down harmonised rules on artificial intelligence.

Dare, M., Schirmer, J., Vanclay, F., 2014. Community engagement and social licence to operate. Impact Assess. Proj. Apprais. 32 (3), 188–197.

Dederer, H.-G., 2019. Confédération Paysanne and others v. Premier Ministre and Ministre De L'agriculture, De L'Agroalimentaire et De La Forêt (CJEU). Int. Leg. Mater. 58 (6), 1281–1298.

Demorest, Z.L., Coffman, A., Baltes, N.J., Stoddard, T.J., Clasen, B.M., Luo, S., Retterath, A., Yabandith, A., Gamo, M.E., Bissen, J., 2016. Direct stacking of sequence-specific nuclease-induced mutations to produce high oleic and low linolenic soybean oil. BMC Plant Biol. 16 (1), 225.

Dima, O., Bocken, H., Custers, R., Inzé, D., Puigdomenech, P., 2020. Genome editing for crop improvement. Symposium summary. In: ALLEA.

Dimitrov, I., Naneva, L., Doytchinova, I., Bangov, I., 2014. AllergenFP: allergenicity prediction by descriptor fingerprints. Bioinformatics 30 (6), 846–851.

Dixit, S., Kumar, A., Srinivasan, K., Vincent, P., Ramu Krishnan, N., 2024. Advancing genome editing with artificial intelligence: opportunities, challenges, and future directions. Front. Bioeng. Biotechnol. 11, 1335901.

Doench, J.G., Fusi, N., Sullender, M., Hegde, M., Vaimberg, E.W., Donovan, K.F., Smith, I., Tothova, Z., Wilen, C., Orchard, R., 2016. Optimized sgRNA design to maximize activity and minimize off-target effects of CRISPR-Cas9. Nat. Biotechnol. 34 (2), 184–191.

Doshi-Velez, F., Kim, B., 2017. Towards a Rigorous Science of Interpretable Machine Learning arXiv preprint arXiv:1702.08608.

Du, W., Zhao, L., Diao, K., Zheng, Y., Yang, Q., Zhu, Z., Zhu, X., Tang, D., 2025. A versatile CRISPR/Cas9 system off-target prediction tool using language model. Commun. Biol. 8 (1), 882.

Eckerstorfer, M.F., Dolezel, M., Engelhard, M., Giovannelli, V., Grabowski, M., Heissenberger, A., Lener, M., Reichenbecher, W., Simon, S., Staiano, G., 2023. Recommendations for the assessment of potential environmental effects of genome-editing applications in plants in the EU. Plants 12 (9), 1764.

Eicher, T., Kinnebrew, G., Patt, A., Spencer, K., Ying, K., Ma, Q., Machiraju, R., Mathé, E.A., 2020. Metabolomics and multi-omics integration: a survey of computational methods and resources. Metabolites 10 (5), 202.

Ellahi, R.M., Wood, L.C., Bekhit, A.E.-D.A., 2023. Blockchain-based frameworks for food traceability: a systematic review. Foods 12 (16), 3026.

Ellstrand, N.C., 2018. "Born to run"? Not necessarily: species and trait bias in persistent free-living transgenic plants. Front. Bioeng. Biotechnol. 6, 88.

Entine, J., Felipe, M.S.S., Groenewald, J.-H., Kershen, D.L., Lema, M., McHughen, A., Nepomuceno, A.L., Ohsawa, R., Ordonio, R.L., Parrott, W.A., 2021. Regulatory approaches for genome edited agricultural plants in select countries and jurisdictions around the world. Transgenic Res. 30 (4), 551–584.

Escribà-Gelonch, M., Liang, S., van Schalkwyk, P., Fisk, I., Long, N.V.D., Hessel, V., 2024. Digital twins in agriculture: orchestration and applications. J. Agric. Food Chem. 72 (19), 10737–10752.

FAO, 2023. Gene Editing and Food Safety–Technical Considerations and Potential Relevance to the Work of Codex Alimentarius. Rome.

FAO/WHO, 2009. Foods Derived from Modern Biotechnology.

FDA, U, 2023. Using Artificial Intelligence & Machine Learning in the Development of Drug and Biological Products. Discussion paper.

FDA, 2025. Artificial Intelligence for Drug Development.

Fernández Ríos, D., Quintana, S.A., Gómez Paniagua, P., Arrúa, A.A., Brozón, G.R., Hicar, B., Castro Alegría, A., Goberna, M.F., 2025. Regulatory challenges and global trade implications of genome editing in agriculture. Front. Bioeng. Biotechnol. 13, 1609110.

Fernandez-Cornejo, J., Wechsler, S., Livingston, M., Mitchell, L., 2014. Genetically engineered crops in the United States. In: USDA-ERS Economic Research Report(162).

Food, U., & Administration, D, 2019. Biotechnology notification file No. 000164: high oleic acid soybean. In: FAD2KO (FDA, 2019).

Food, U., & Administration, D, 2025. FDA Announces Completion of First AI-Assisted Scientific Review Pilot and Aggressive Agency-Wide AI Rollout Timeline.

Friedrichs, S., Takasu, Y., Kearns, P., Dagallier, B., Oshima, R., Schofield, J., Moreddu, C., 2019. Meeting report of the OECD conference on "genome editing: applications in agriculture—implications for health, environment and regulation". Transgenic Res. 28 (3), 419–463.

Fu, R., He, W., Dou, J., Villarreal, O.D., Bedford, E., Wang, H., Hou, C., Zhang, L., Wang, Y., Ma, D., 2022. Systematic decomposition of sequence determinants governing CRISPR/Cas9 specificity. Nat. Commun. 13 (1), 474.

Goberna, M.F., Whelan, A.I., Godoy, P., Lewi, D.M., 2022. Genomic editing: the evolution in regulatory management accompanying scientific progress. Front. Bioeng. Biotechnol. 10, 835378.

Goodman, B., Flaxman, S., 2017. European Union regulations on algorithmic decision-making and a "right to explanation". AI Mag. 38 (3), 50–57.

Graham, N., Patil, G.B., Bubeck, D.M., Dobert, R.C., Glenn, K.C., Gutsche, A.T., Kumar, S., Lindbo, J.A., Maas, L., May, G.D., 2020. Plant genome editing and the relevance of off-target changes. Plant Physiol. 183 (4), 1453–1471.

Granato, I., Cuevas, J., Luna-Vázquez, F., Crossa, J., Montesinos-López, O., Burgueño, J., Fritsche-Neto, R., 2018. BGGE: a new package for genomic-enabled prediction incorporating genotype× environment interaction models. G3: Genes, Genomes, Genet. 8 (9), 3039–3047.

Hahn, F., Nekrasov, V., 2019. CRISPR/Cas precision: do we need to worry about off-targeting in plants? Plant Cell Rep. 38 (4), 437–441.

Hoffman, N.E., 2021. Revisions to USDA biotechnology regulations: the SECURE rule. Proc. Natl. Acad. Sci. 118 (22) e2004841118.

Huang, Z., Jiang, M., Sun, L., Zhang, J., Jiang, C., Lu, A., Li, D., Nan, N., Sun, W., 2024. Exploiting susceptibility genes in rice: from molecular mechanism to application. Phytopathol. Res. 6 (1), 67.

Iacono, G., Guerra, B., Kass, G., Paraskevopoulos, K., Kleiner, J., Heppner, C., Hugas, M., 2022. Theme (concept) paper-application of OMICS and BIOINFORMATICS approaches: towards next generation risk assessment. EFSA Support. Publ. 19 (5).

Ishii, T., 2025. Consumer choices regarding genome-edited food crops: lessons from Japan. Front. Genome Edit. 7, 1672358.

Jubair, S., Domaratzki, M., 2022. Crop genomic selection with deep learning and environmental data: a survey. Front. Artif. Intell. 5 (1040295).

Kamar, E., 2016. Directions in hybrid intelligence: complementing AI systems with human intelligence. IJCAI.

Kamilaris, A., Prenafeta-Boldú, F.X., 2018. Deep learning in agriculture: a survey. Comput. Electron. Agric. 147, 70–90.

Kamilaris, A., Fonts, A., Prenafeta-Boldú, F.X., 2019. The rise of blockchain technology in agriculture and food supply chains. Trends Food Sci. Technol. 91, 640–652.

Kato-Nitta, N., Maeda, T., Inagaki, Y., Tachikawa, M., 2019. Expert and public perceptions of gene-edited crops: attitude changes in relation to scientific knowledge. Palgrave Commun. 5 (1).

Kim, H.K., Kim, Y., Lee, S., Min, S., Bae, J.Y., Choi, J.W., Park, J., Jung, D., Yoon, S., Kim, H.H., 2019. SpCas9 activity prediction by DeepSpCas9, a deep learning–based model with high generalization performance. Sci. Adv. 5 (11).

Kimata, K., Satou, K., 2025. Improved CRISPR/Cas9 off-target prediction with DNABERT and epigenetic features. PLoS One 20 (11).

Kondo, K., Taguchi, C., 2022. Japanese regulatory framework and approach for genome-edited foods based on latest scientific findings. Food Safety 10 (4), 113–128.

Konstantakos, V., Nentidis, A., Krithara, A., Paliouras, G., 2022. CRISPR–Cas9 gRNA efficiency prediction: an overview of predictive tools and the role of deep learning. Nucleic Acids Res. 50 (7), 3616–3637.

Kuiken, T., Kuzma, J., 2021. Genome Editing in Latin America: regional Regulatory Overview.

Kuiper, H.A., Kleter, G.A., Noteborn, H.P., Kok, E.J., 2002. Substantial equivalence—an appropriate paradigm for the safety assessment of genetically modified foods? Toxicology 181, 427–431.

Kuzma, J., Grieger, K., Cimadori, I., Cummings, C.L., Loschin, N., Wei, W., 2023. Parameters, practices, and preferences for regulatory review of emerging biotechnology products in food and agriculture. Front. Bioeng. Biotechnol. 11, 1256388.

Li, Y., Liu, Z., Gmitter Jr., F.G., Deng, Z., Cheng, B., Duan, H., Li, Y., 2025. Substantial enhancement of Agrobacterium-mediated transgene-free genome editing via short-term chemical selection using citrus as a model plant. Hortic. Res. 12 (9) uhaf153.

Lin, J., Zhang, Z., Zhang, S., Chen, J., Wong, K.C., 2020. CRISPR-net: a recurrent convolutional network quantifies CRISPR off-target activities with mismatches and indels. Adv. Sci. 7 (13), 1903562.

Listgarten, J., Weinstein, M., Kleinstiver, B.P., Sousa, A.A., Joung, J.K., Crawford, J., Gao, K., Hoang, L., Elibol, M., Doench, J.G., 2018. Prediction of off-target activities for the end-to-end design of CRISPR guide RNAs. Nat. Biomed. Eng. 2 (1), 38–47.

Liu, Q., He, D., Xie, L., 2019. Prediction of off-target specificity and cell-specific fitness of CRISPR-Cas System using attention boosted deep learning and network-based gene feature. PLoS Comput. Biol. 15 (10) e1007480.

Lu, H., Lu, J., He, L., 2019. Modeling and estimation of pollen-mediated gene flow at the landscape scale. Ecol. Indic. 106, 105500.

Luo, J., Chen, W., Xue, L., Tang, B., 2019. Prediction of activity and specificity of CRISPR-Cpf1 using convolutional deep learning neural networks. BMC Bioinf. 20 (1), 332.

Ma, W., Qiu, Z., Song, J., Li, J., Cheng, Q., Zhai, J., Ma, C., 2018. A deep convolutional neural network approach for predicting phenotypes from genotypes. Planta 248 (5), 1307–1318.

MacNish, T.R., Danilevicz, M.F., Bayer, P.E., Bestry, M.S., Edwards, D., 2025. Application of machine learning and genomics for orphan crop improvement. Nat. Commun. 16 (1), 982.

Mazzia, V., Comba, L., Khaliq, A., Chiaberge, M., Gay, P., 2020. UAV and machine learning based refinement of a satellite-driven vegetation index for precision agriculture. Sensors 20 (9), 2530.

Mc Cartney, A.M., Anderson, J., Liggins, L., Hudson, M.L., Anderson, M.Z., TeAika, B., Geary, J., Cook-Deegan, R., Patel, H.R., Phillippy, A.M., 2022. Balancing openness with Indigenous data sovereignty: an opportunity to leave no one behind in the journey to sequence all of life. Proc. Natl. Acad. Sci. 119 (4) e2115860119.

McHughen, A., 2016. A critical assessment of regulatory triggers for products of biotechnology: product vs. process. GM Crops Food 7 (3–4), 125–158.

Mehrabi, N., Morstatter, F., Saxena, N., Lerman, K., Galstyan, A., 2021. A survey on bias and fairness in machine learning. ACM Comput. Surv. 54 (6), 1–35.

Mendelsohn, M., Kough, J., Vaituzis, Z., Matthews, K., 2003. Are Bt crops safe? Nat. Biotechnol. 21 (9), 1003–1009.

Messeguer, J., 2003. Gene flow assessment in transgenic plants. Plant Cell Tissue Organ Cult. 73 (3), 201–212.

Mmbando, G.S., 2024. The use of artificial intelligence in the production of genetically modified (GM) crops: a recent promising strategy for enhancing the acceptability of GM products? Discov. Appl. Sci. 6 (11), 581.

Modrzejewski, D., Hartung, F., Lehnert, H., Sprink, T., Kohl, C., Keilwagen, J., Wilhelm, R., 2020. Which factors affect the occurrence of off-target effects caused by the use of CRISPR/Cas: a systematic review in plants. Front. Plant Sci. 11, 574959.

Molinari, H., Vieira, L., Freitas, N., Justen, F., de Jesus, V., de Oliveira, B., 2021. Regulatory framework of genome editing in Brazil and worldwide. In: CRISPR Technology in Plant Genome Editing: Biotechnology Applied to Agriculture, pp. 169–195.

Montreal, S. o. t. C. o. B. D, 2011. Nagoya Protocol on Access to Genetic Resources and the Fair and Equitable Sharing of Benefits Arising from their Utilization.

Nagamine, A., Ezura, H., 2022. Genome editing for improving crop nutrition. Front. Genome Edit. 4, 850104.

Nastoska, A., Jancheska, B., Rizinski, M., Trajanov, D., 2025. Evaluating trustworthiness in AI: risks, metrics, and applications across industries. Electronics 14 (13), 2717.

Nedyalkova, M., Vasighi, M., Azmoon, A., Naneva, L., Simeonov, V., 2023. Sequence-based prediction of plant allergenic proteins: machine learning classification approach. ACS Omega 8 (4), 3698–3704.

Nepomuceno, A.L., Fuganti-Pagliarini, R., Felipe, M.S.S., Molinari, H.B.C., Velini, E.D., de Campos Pinto, E.R., Dagli, M.L.Z., Andrade Filho, G., Fernandes, P.M.B., 2020. Brazilian biosafety law and the new breeding technologies. Eng. Agri. 7 (2), 204.

OECD, 2022. Safety Assessment of Transgenic Organisms in the Environment, 9. OECD.

OECD, 2023. Consensus Document on Environmental Considerations for Risk/Safety Assessment for the Release of Transgenic Plants.

Oliva, R., Ji, C., Atienza-Grande, G., Huguet-Tapia, J.C., Perez-Quintero, A., Li, T., Eom, J.-S., Li, C., Nguyen, H., Liu, B., 2019. Broad-spectrum resistance to bacterial blight in rice using genome editing. Nat. Biotechnol. 37 (11), 1344–1350.

Organisms, E.P.o.G.M., Naegeli, H., Bresson, J.L., Dalmay, T., Dewhurst, I.C., Epstein, M.M., Firbank, L.G., Guerche, P., Hejatko, J., Moreno, F.J., 2020. Applicability of the EFSA Opinion on site-directed nucleases type 3 for the safety assessment of plants developed using site-directed nucleases type 1 and 2 and oligonucleotide-directed mutagenesis. EFSA J. 18 (11) e06299.

Organisms, E. P. o. G. M, Naegeli, H., Bresson, J.L., Dalmay, T., Dewhurst, I.C., Epstein, M.M., Firbank, L.G., Guerche, P., Hejatko, J., Moreno, F.J., 2022. Criteria for risk assessment of plants produced by targeted mutagenesis, cisgenesis and intragenesis. EFSA J. 20 (10) e07618.

Organisms, E. P. o. G. M, 2010. Guidance on the environmental risk assessment of genetically modified plants. EFSA J. 8 (11), 1879.

Organisms, E. P. o. G. M, 2011. Guidance on the post-market environmental monitoring (PMEM) of genetically modified plants. EFSA J. 9 (8), 2316.

Organisms, E. P. o. G. M, 2015. Guidance for renewal applications of genetically modified food and feed authorised under Regulation (EC) No 1829/2003. EFSA J. 13 (6), 4129.

Phillips, P., Williams, A., 2022. A meta-analysis of GM crops regulatory approval costs. In: Working Paper Series 2023-1. Johnson-Shoyama Graduate School of Public Policy, University of Saskatchewan.

Picard, M., Scott-Boyer, M.-P., Bodein, A., Périn, O., Droit, A., 2021. Integration strategies of multi-omics data for machine learning analysis. Comput. Struct. Biotechnol. J. 19, 3735–3746.

Pichler, M., Boreux, V., Klein, A.M., Schleuning, M., Hartig, F., 2020. Machine learning algorithms to infer trait-matching and predict species interactions in ecological networks. Methods Ecol. Evol. 11 (2), 281–293.

Polidoros, A., Nianiou-Obeidat, I., Tsakirpaloglou, N., Petrou, N., Deligiannidou, E., Makri, N.-M., 2024. Genome-editing products line up for the market: will Europe harvest the benefits from science and innovation? Genes 15 (8), 1014.

Poplin, R., Chang, P.-C., Alexander, D., Schwartz, S., Colthurst, T., Ku, A., Newburger, D., Dijamco, J., Nguyen, N., Afshar, P.T., 2018. A universal SNP and small-indel variant caller using deep neural networks. Nat. Biotechnol. 36 (10), 983–987.

Radio, S., Di Marsico, M., Bersani, C., Malinverni, R., Casacuberta, J., Corpetti, C., Cigliano, R.A., Sanseverino, W., 2024. Development of a roadmap for action on the application of omics and associated bioinformatics approaches in risk assessment. EFSA Support. Publ. 21 (10), 9086E.

Ribeiro, M.T., Singh, S., Guestrin, C., 2016. "Why should i trust you?" Explaining the predictions of any classifier. In: Proceedings of the 22nd ACM SIGKDD International Conference on Knowledge Discovery and Data Mining.

Ricroch, A.E., Bergé, J.B., Kuntz, M., 2011. Evaluation of genetically engineered crops using transcriptomic, proteomic, and metabolomic profiling techniques. Plant Physiol. 155 (4), 1752–1761.

Sanches, P.H.G., de Melo, N.C., Porcari, A.M., de Carvalho, L.M., 2024. Integrating molecular perspectives: strategies for comprehensive multi-omics integrative data analysis and machine learning applications in transcriptomics, proteomics, and metabolomics. Biology 13 (11), 848.

Shahhosseini, M., Hu, G., Huber, I., Archontoulis, S.V., 2021. Coupling machine learning and crop modeling improves crop yield prediction in the US Corn Belt. Sci. Rep. 11 (1), 1606.

Sharma, N., Patiyal, S., Dhall, A., Pande, A., Arora, C., Raghava, G.P., 2021. AlgPred 2.0: an improved method for predicting allergenic proteins and mapping of IgE epitopes. Briefings Bioinf. 22 (4) bbaa294.

Sherkatghanad, Z., Abdar, M., Charlier, J., Makarenkov, V., 2023. Using traditional machine learning and deep learning methods for on-and off-target prediction in CRISPR/Cas9: a review. Briefings Bioinf. 24 (3) bbad131.

Shillito, R.D., Whitt, S., Ross, M., Ghavami, F., De Vleesschauwer, D., D'Halluin, K., Van Hoecke, A., Meulewaeter, F., 2021. Detection of genome edits in plants—from editing to seed. In Vitro Cell. Dev. Biol. Plant 57 (4), 595–608.

Singh, R., Kuscu, C., Quinlan, A., Qi, Y., Adli, M., 2015. Cas9-chromatin binding information enables more accurate CRISPR off-target prediction. Nucleic Acids Res. 43 (18) e118–e118.

Smart, R.D., Blum, M., Wesseler, J., 2017. Trends in approval times for genetically engineered crops in the United States and the European Union. J. Agric. Econ. 68 (1), 182–198.

Smuha, N.A., 2025. Regulation 2024/1689 of the Eur. Parl. & Council of June 13, 2024 (eu artificial intelligence act). Int. Leg. Mater. 64 (5), 1234–1381.

Spök, A., Sprink, T., Allan, A.C., Yamaguchi, T., Dayé, C., 2022. Towards social acceptability of genome-edited plants in industrialised countries? Emerging evidence from Europe, United States, Canada, Australia, New Zealand, and Japan. Front. Genome Edit. 4, 899331.

Stemmer, M., Thumberger, T., del Sol Keyer, M., Wittbrodt, J., Mateo, J.L., 2015. CCTop: an intuitive, flexible and reliable CRISPR/Cas9 target prediction tool. PLoS One 10 (4) e0124633.

Sturme, M.H., van der Berg, J.P., Bouwman, L.M., De Schrijver, A., de Maagd, R.A., Kleter, G.A., Battaglia-de Wilde, E., 2022. Occurrence and nature of off-target modifications by CRISPR-Cas genome editing in plants. ACS Agri. Sci. Technol. 2 (2), 192–201.

Suresh, H., Guttag, J., 2021. A framework for understanding sources of harm throughout the machine learning life cycle. In: Proceedings of the 1st ACM Conference on Equity and Access in Algorithms, Mechanisms, and Optimization.

Tachikawa, M., Matsuo, M., 2023. Divergence and convergence in international regulatory policies regarding genome-edited food: how to find a middle ground. Front. Plant Sci. 14, 1105426.

The Canadian Food Inspection Agency's 2026 to 2027 Departmental Plan, 2026.

Thomas, R.S., Bahadori, T., Buckley, T.J., Cowden, J., Deisenroth, C., Dionisio, K.L., Frithsen, J.B., Grulke, C.M., Gwinn, M.R., Harrill, J.A., 2019. The next generation blueprint of computational toxicology at the US Environmental Protection Agency. Toxicol. Sci. 169 (2), 317–332.

Unctad, 2021. Technology and Innovation Report 2021: Catching Technological Waves-Innovation With Equity. UN.

UNESCO, 2021. Ethics of Artificial Intelligence.

USDA, 2025. Fiscal Year 2025–2026 AI Strategy, United States Department of Agriculture.

USDA, 2026. Biotechnology Regulatory Services.

van Dijk, A.D.J., Kootstra, G., Kruijer, W., de Ridder, D., 2021. Machine learning in plant science and plant breeding. iScience 24 (1).

van Meer, F., van der Velden, B., Takeuchi, M., 2025. Artificial Intelligence for Food Safety: a Literature Synthesis, Real-World Applications and Regulatory Frameworks.

Veale, M., Matus, K., Gorwa, R., 2023. AI and global governance: modalities, rationales, tensions. Annu. Rev. Law Soc. Sci. 19 (1), 255–275.

Vilone, G., Longo, L., 2020. Explainable artificial intelligence: a systematic review arXiv preprint arXiv:2006.00093.

Vives-Vallés, J.A., Collonnier, C., 2020. The judgment of the CJEU of 25 July 2018 on mutagenesis: interpretation and interim legislative proposal. Front. Plant Sci. 10, 1813.

Wachter, S., Mittelstadt, B., Russell, C., 2017. Counterfactual explanations without opening the black box: automated decisions and the GDPR. Harv. JL & Tech. 31, 841.

Walklate, P., Hunt, J., Higson, H., Sweet, J., 2004. A model of pollen–mediated gene flow for oilseed rape. Proc. Roy. Soc. Lond. B Biol. Sci. 271 (1538), 441–449.

Waltz, E., 2016. Gene-Edited CRISPR Mushroom Escapes US Regulation. Nature Publishing Group, UK London.

Waltz, E., 2022. GABA-enriched tomato is first CRISPR-edited food to enter market. Nat. Biotechnol. 40 (1), 9–11.

Wang, D., Zhang, C., Wang, B., Li, B., Wang, Q., Liu, D., Wang, H., Zhou, Y., Shi, L., Lan, F., 2019. Optimized CRISPR guide RNA design for two high-fidelity Cas9 variants by deep learning. Nat. Commun. 10 (1), 4284.

Washburn, J.D., Cimen, E., Ramstein, G., Reeves, T., O'Briant, P., McLean, G., Cooper, M., Hammer, G., Buckler, E. S., 2021. Predicting phenotypes from genetic, environment, management, and historical data using CNNs. Theor. Appl. Genet. 134 (12), 3997–4011.

Whelan, A.I., Lema, M.A., 2015. Regulatory framework for gene editing and other new breeding techniques (NBTs) in Argentina. GM Crops Food 6 (4), 253–265.

Wilkinson, M.D., Dumontier, M., Aalbersberg, I.J., Appleton, G., Axton, M., Baak, A., Blomberg, N., Boiten, J.-W., da Silva Santos, L.B., Bourne, P.E., Bouwman, J., Brookes, A.J., Clark, T., Crosas, M., Dillo, I., Dumon, O., Edmunds, S., Evelo, C.T., Finkers, R., Mons, B., 2016. The FAIR Guiding Principles for scientific data management and stewardship. Sci. Data 3 (1). https://doi.org/10.1038/sdata.2016.18, 160018.

Williamson, H.F., Brettschneider, J., Caccamo, M., Davey, R.P., Goble, C., Kersey, P.J., May, S., Morris, R.J., Ostler, R., Pridmore, T., 2023. Data management challenges for artificial intelligence in plant and agricultural research. F1000Research 10, 324.

Xiang, X., Corsi, G.I., Anthon, C., Qu, K., Pan, X., Liang, X., Han, P., Dong, Z., Liu, L., Zhong, J., 2021. Enhancing CRISPR-Cas9 gRNA efficiency prediction by data integration and deep learning. Nat. Commun. 12 (1), 3238.

Xie, Y., Haq, S.I.U., Jiang, X., Zheng, D., Feng, N., Wang, W., He, J.S., Qiu, Q.S., 2022. Plant genome editing: CRISPR, base editing, prime editing, and beyond. Grassland Res. 1 (4), 234–243.

Yeung, K., 2020. Recommendation of the council on artificial intelligence (OECD). Int. Leg. Mater. 59 (1), 27–34.

Zhang, Z.-R., Jiang, Z.-R., 2022. Effective use of sequence information to predict CRISPR-Cas9 off-target. Comput. Struct. Biotechnol. J. 20, 650–661.

Zhang, Y., Restall, J., Crisp, P., Godwin, I., Liu, G., 2021. Current status and prospects of plant genome editing in Australia. In Vitro Cell. Dev. Biol. Plant 57 (4), 574–583.

Zhu, H., Li, C., Gao, C., 2020. Applications of CRISPR–Cas in agriculture and plant biotechnology. Nat. Rev. Mol. Cell Biol. 21 (11), 661–677.

Further reading

Biotechnology Permits and Notifications, 2026.

EFSA, 2022. Application of OMICS and BIOINFORMATICS Approaches: Towards Next Generation.

Overview – Status of Organisms Modified using Gene Editing and Other New Technologies, 2025.

Principles for the Risk Analysis of Foods Derived from Modern Biotechnology (CAC/GL 44-2003), 2003.

Risk Assessment. (FDA), T. U. S. F. a. D. A., 2025. FDA Launches Agency-Wide AI Tool to Optimize Performance for the American People.

CHAPTER

21

AI and CRISPR: Biosafety, governance, and public perception of gene-edited organisms

Leire Escajedo San-Epifanio, Javier Uncetabarrenechea Larrabe and Igor Filibi

EU-Frames TransR. Department of Public Law, Historical-Legal Sciences and History of Political Thought, University of the Basque Country, Bilbao, Spain

21.1 Introduction: Between innovation and governance—CRISPR, AI and the future of plant biotechnology

In the global context, the regulatory situation of the European Union is the one that generates the greatest uncertainty. For more than 3 decades—that is, practically since the first Community legal instruments applicable to the use of genetically modified organisms (GMOs) in crops—disagreements among Member States have resulted in a permanent decision-making deadlock. De iure, since the 1990s, there has existed a premarket approval system for the use of GM crops, very demanding in terms of environmental, human, and animal health risks. De facto, however, even when all these requirements are met, applicants for authorization have no guarantee that the European Union will provide a favorable response within a reasonable period (Escajedo San-Epifanio et al., 2023).

The European Green Deal and the Farm to Fork strategy (European Commission, 2021, 2023) set out ambitious agri-food objectives that will be difficult to achieve without innovations in genome editing, yet regulatory uncertainty acts as a deterrent to investment. The problem lies not only in gaps in the substantive content of the rules but also in persistent obstacles within decision-making procedures. In certain comitology processes, for instance, Member States remain sharply divided, and public opinion remains cautious. Meanwhile, other regions are advancing with greater flexibility (Uncetabarrenechea, 2025).

Specifically, regulatory frameworks have not yet paid systematic attention to the interaction between CRISPR technologies, the digitalization of biology, and the use of artificial

Intelligent Gene Editing in Plants
https://doi.org/10.1016/B978-0-443-44079-3.00016-4

intelligence, such interaction is growing exponentially. The absence of specific regulatory attention is because this new need has arisen at a time when regulatory systems—particularly in regions such as the EU—have not yet consolidated their stance on plant genome editing. Among other issues, it remains necessary to clarify the legal status of digital sequence information (DSI), which until recently was regarded merely as data.

At present, AI-driven design and synthetic biology are blurring the boundary that previously separated the physical sample from innovation as a form of human intellectual activity—categories that were legally used to define, on the one hand, the status of genetic resources and, on the other, the recognition of intellectual property rights. This interaction undoubtedly offers considerable potential for agri-biotechnological innovation in terms of speed, scope, and reproducibility. However, the regulatory framework—already fragmented into silos—is becoming increasingly entangled and fails to provide sufficient legal certainty for investors, operators, and future consumers.

The context in which this tension around digital sequences is most clearly visible is the review of access and benefit-sharing frameworks under the Nagoya Protocol and the International Treaty on Plant Genetic Resources (Scholz et al., 2022). Questions such as who controls the data, under what conditions, and how these new value chains might exacerbate inequalities raise challenges that can only be addressed through robust legal frameworks (Oldham et al., 2021; Ruder, 2025). Some countries are seeking to ensure that such decisions are taken within the framework of the United Nations Convention on Biological Diversity or the FAO's International Treaty on Plant Genetic Resources for Food and Agriculture.

These two overarching issues—the legal status of plant-breeding techniques involving genome editing (including CRISPR) and that of synthetic biology using AI-mediated digital genetic sequences—are undoubtedly the most challenging in the medium and long term. Nevertheless, it should also be noted that there are other aspects of agri-biotechnologies which, although not new, continue to receive uneven attention depending on the region or country. Among these are the management of public information, social perception, public participation, and labeling, another area that deserves particular consideration.

21.2 Biosafety of AI-assisted CRISPR-edited crops: what still fits, what no longer does

The global regulatory architecture for biosafety began to take shape in the late 1980s and early 1990s, expanding particularly during the early 2000s. This architecture was—and still is—primarily centered on the use of recombinant DNA, especially in transgenic crops intended for deliberate release into the environment. In several parts of the world, particularly within the European Union, there has been little or no transition toward a more holistic framework encompassing the full range of modern plant-breeding techniques.

There are therefore no normative elements or planned updates capable of addressing the specific features that distinguish CRISPR from other plant-breeding technologies, or of regulating new aspects such as synthetic biology, digital sequence information (DSI), and the use of artificial intelligence in crop improvement. Broadly speaking, the regulatory frameworks currently applied to genetically modified organisms (GMOs) share a common rationale and structure, differing mainly in how risk assessment and decision-making are approached.

These commonalities stem from a series of global milestones that led many countries, almost simultaneously, to build their regulatory systems during the 1990s and early 2000s (Escajedo San-Epifanio, 2007). The first of these milestones is usually identified as the Asilomar moratorium (1975), followed by the OECD's first guideline (1986), the United States Coordinated Framework of the same year, and, from 2000 onwards, the Codex Alimentarius principles for food safety (FAO/WHO, 2000; Codex, 2003/2008).

The most influential and legally binding instrument is, without doubt, the Convention on Biological Diversity (CBD), which laid down a set of principles and values that have since inspired wide international collaboration. More specifically, in relation to biosafety, the Cartagena Protocol on Biosafety (2000) derived from the CBD and established the reference framework for the safe handling and transboundary movement of living modified organisms.

Taken together, these instruments consolidated a model characterized by broadly similar features, although each State implemented them in its own way. The key elements include science-based, case-by-case assessments founded on the concepts of familiarity and substantial equivalence; competent authorities and tiered assessments; requirements for public information and, in some regions, traceability and postmarket monitoring. Across systems, these frameworks draw inspiration from the precautionary principle, albeit with divergent projections: the European Union grants it strong legal force—including shifts in the burden of proof and lower thresholds for regulatory intervention—whereas other systems apply it more narrowly as a risk-management tool constrained by proportionality and available evidence.

In recent years, as some national frameworks have begun to update their rules to address genome editing, divergences among regulatory approaches have deepened, even though most countries remain broadly consistent with the rationale developed in the late 1990s. For example, Canada has adopted a product-based approach, focusing on the characteristics and risks of the final product rather than the techniques used to obtain it. The United States, in turn, introduced the SECURE Rule in 2020, which streamlines authorization for certain genome-edited plants that contain no foreign DNA and pose no additional plant-pest risks. This model seeks to rationalize evaluation procedures by reducing administrative burdens and prioritizing risk proportionality. By contrast, the European Union, South Africa, and New Zealand continue to operate under process-based frameworks, where the method employed—rather than the resulting phenotype—determines the legal status of the plant. This has fueled extensive debate over whether plants obtained through targeted mutagenesis should continue to be treated as genetically modified organisms, given that many exhibit traits indistinguishable from those produced by conventional breeding or natural mutation.

Between these two poles, hybrid regimes—such as those of Japan, the Philippines, and India—have emerged, seeking to balance flexibility with safety through notification systems or conditional exemptions for certain categories of edits. However, this diversity of models is producing an increasingly complex regulatory mosaic, with tangible implications for international trade, innovation, and the scientific coherence of policy.

These diverging trajectories are examined in greater detail in Section 3, which analyses the three main regulatory paradigms—product-based, process-based, and hybrid—and their implications for the global governance of CRISPR-AI. In this section, as already outlined—albeit only briefly—in the introduction, and leaving aside the existing gaps or inconsistencies of certain regulatory systems (notably the European one), which would clearly benefit from structural revision, it is evident that the interaction between CRISPR and artificial intelligence exceeds the boundaries of the current regulatory paradigm in at least four distinct ways. These overflows or "points of mismatch" reveal the growing tension between traditional biosafety logic and the data-driven nature of the new technologies. Let us examine them one by one:

a) Profound changes in detection and applicability compared with previous breeding techniques.
Many genome edits of the SDN-1 and SDN-2 types do not leave unique molecular signatures. When performed using ribonucleoprotein (RNP) complexes without DNA, they become virtually invisible to current detection tools, even though process-based regulations continue to attempt to capture them. This weakens enforceability and complicates postmarket monitoring systems that rely on event-specific markers (Jones et al., 2022). Regulators in Japan and the Philippines have responded with notification schemes or "certificates of non-coverage", implicitly acknowledging this detection gap.

b) Shifts in the scale of risk assessment and the expansion of the design space.
Earlier regulatory dossiers assumed single, isolated modification events. Today's reality is different: multiplex editing, iterative editing cycles, and AI-generated candidate lists vastly expand the combinatorial design space. Consequently, risk assessment must be reorganized around traits, intended uses, and environmental context rather than methodological labels, and must explicitly address the model uncertainty introduced by AI itself (Graham et al., 2020).

c) The emergence of data governance as a component of biosafety.
Algorithmic predictions depend on the quality and representativeness of their training data. Bias, data drift, or weak external validity can distort outcomes—sometimes producing false confidence, at other times concealing latent risks. For this reason, documentation, transparency, and auditability of AI models are not optional add-ons but integral parts of the safety dossier. As scholars of science, technology, and society have noted, the new predictive power of technology is accompanied by new blind spots—the so-called "ambivalence of technology" (Clapp and Ruder, 2025).

d) Particular challenges arising from the dematerialization of genetic resources.
With the increasing use of digital sequence information (DSI) and AI-based biological design, regulatory oversight must now follow data flows as well as physical samples. This shift destabilizes benefit-sharing systems originally conceived for tangible genetic resources and moves long-standing debates on sovereignty and equity into the digital realm (Scholz et al., 2022; Oldham et al., 2021; Abkallo et al., 2024).

Taken together, these developments expose a structural gap in the regulatory architecture: legal triggers still depend on how an organism was made, even when functionally equivalent products circulate freely elsewhere. Triggers still hinge on "how it was made," even as functionally equivalent products circulate freely elsewhere. Computational evidence is still treated as supportive, not central, despite AI shaping trait choice and gRNA design. Monitoring tools remain geared to traceable events rather than undetectable edits. And international coordination is struggling to keep up. While the U.S., Argentina, Brazil, Japan, the Philippines, and Kenya have mapped clear routes for transgene-free edits (Entine et al., 2021; Jones et al., 2022; NBA, 2022), the EU is still hampered by process triggers and bottlenecks-despite EFSA's recent guidance and the Commission's attempt to legislate for "new genomic techniques" (European Commission, 2023).

So, what would a credible update entail? A credible biosafety refresh should build on what already works. Trait- and use-centered assessment, long applied de facto in food safety, should become the formal entry point. Computational pipelines should be documented, versioned, and auditable, with uncertainty reported alongside wet-lab results. Post-authorization oversight should shift from event-specific detection to outcome-based monitoring when edits are indistinguishable, combining molecular checks where possible with agronomic and ecological indicators (OECD, 2024; Zischewski et al., 2022).

And because CRISPR is now inseparable from data, biosafety and benefit-sharing need to meet halfway. Discussions under the CBD about DSI must avoid closing off open databases essential for reproducibility, while still ensuring fair contributions from commercial users. In short, the science of regulation built for rDNA crops remains a strong foundation, but it needs refitting for a world where edits can be subtle, designs computational, and oversight must follow the trait rather than the technique.

21.3 Regulatory paradigms and global fragmentation

CRISPR-edited crops, at the global level, are not currently under a single and coherent order. Even across different regions of the world, we see that regulatory frameworks tend to be a patchwork, an overlay of older GMO laws, national risk cultures, and different mechanisms of recognition of intellectual property rights (sometimes complemented by the recognition of biodiversity custodians and traditional knowledge). To this complex framework is now added the debate on the regulatory framework needed to address the ethical, social, and political challenges implied by the computational component. With regard to biotechnological plant breeding, including GMOs, three major regulatory logics or approaches can be distinguished—product-based, process-based, and hybrid. This allows us to present them in broad strokes, but it must be noted that there is a risk of oversimplification in the explanation. It should be borne in mind that each approach, and the way in which it is articulated, shows variations depending on political traditions and trade strategies. The result is less a universal system than a plural and shifting order (Ahmad et al., 2021; Turnbull et al., 2021; Jones et al., 2022).

21.3.1 Product-based approaches

Systems organized around a product-based approach start from the question: what the plant is and what it does. The particularity of each approach is seen above all in practical application. Thus, following this approach, in the United States, the SECURE rule of May 2020 (85 FR 29790) exempts many edited plants if they do not represent a plant-pest risk and lack foreign DNA. Canada, for its part, applies a similar novelty principle, where traits, not methods, trigger oversight. Argentina, Brazil, Chile, Japan, and Kenya have also opened pathways for SDN-1/2 edits, which often conclude with a regulator's letter confirming that the product falls outside GMO law (Entine et al., 2021; Jones et al., 2022).

Under regulatory frameworks that apply this approach, tangible results have already been produced: high-oleic soybean oil in the US, the nonbrowning mushroom, or the GABA tomato in Japan, marketed for the first time in 2021. Lower transaction costs have also made it viable to invest in traits such as drought tolerance or nutritional improvements that would otherwise not be economically feasible. Normatively, these systems rely on Codex concepts of comparative safety and WTO proportionality rules. And given that dossiers already include in silico predictions—off-target maps, phenotypic simulations—this type of development integrates naturally with AI workflows.

21.3.2 Process-based approaches

The EU took the opposite route: an approach fundamentally based on process. The regulatory framework to be applied to each plant variety does not depend so much on the final product as on the specific technique that has been applied to obtain it. Thus, since the 1990s EU law has distinguished three major groups of techniques or breeding processes: (i) those techniques that do not give rise to GMOs; (ii) those techniques that give rise to GMOs but for which, prior to the 1990s, there was a history of safe use, such as classical mutagenesis; (iii) and the techniques that give rise to GMOs and which, in the 1990s, did not have a history of safe use. At first this last group mainly included transgenics, but over time, most NGTs have fallen into that category. This provoked much debate and some doubts, but everything was clarified by the CJEU. The 2018 judgment of the Court of Justice (Case C-528/16) interpreted that genome-edited plants are GMOs under Directive 2001/18, unless they fall within the narrow mutagenesis exemption. South Africa adopted a similar position in 2021, and New Zealand continues to regulate genome-edited organisms as GMOs under the HSNO Act.

These frameworks prioritize precaution and traceability, but also generate legal uncertainty for those developing agri-biotechnological innovations. Clearly, the scientific basis is one of the elements for the decision to authorize a modified variety, but it is not the only basis. Moreover, doubts arise when a process generates, in scientific terms, modifications very similar to those that occur in nature, including in terms of risk. How are process-based triggers applied when SDN-1 edits cannot be distinguished from natural mutations? Critics argue that such rigidity drives innovation abroad and locks Europe into a political stalemate (Wasmer, 2019; Eriksson et al., 2020). The EU has begun to debate a legislative reform for "new genomic techniques" (European Commission, 2023), notable, but as of the date of this writing, 2025, it has not yet been approved to enter into force. Perhaps 3 decades of conflict around GMOs make agreement difficult to reach.

21.3.3 Hybrid approaches and countries whose regulatory framework is in transition

Between these poles, hybrid regimes try to narrow the process-based trigger while at the same time maintaining certain safety backstops. Japan, for example, exempts SDN-1 and some SDN-2 edits, but only after notification, including molecular data and off-target checks. The Philippines went further in 2022, introducing a "Certificate of Non-Coverage" with a precise threshold: small insertions up to 19–20 base pairs. India's 2022 memorandum deregulated SDN-1/2 without foreign DNA, and Kenya's 2022 guidance is based on case-by-case determinations.

These models seek a certain flexibility, and are less rigid than those described above, but they generate new gray zones. Where exactly is the line drawn between SDN-2 and SDN-3? Moreover, their application will imply giving regulators a high technical capacity: they must be able to credibly assess AI-derived evidence. Without transparency in computational tools, confidence will remain limited (Smuha, 2021).

21.3.4 Practical consequences of fragmentation

It is difficult to fully capture the practical consequences of regulatory fragmentation and its uneven effects on the actors involved, because their interactions are deeply interlinked. A simple example illustrates this complexity: a soybean considered conventional in Argentina may be classified as a GMO in the European Union. Its importation—particularly for feed—may be authorized as safe, while at the same time the cultivation of that same soybean within the EU remains prohibited. Such inconsistencies trigger trade disputes under the SPS and TBT agreements and complicate bulk shipments when low-level presence of edited or transgenic material is detected (Pollack and Shaffer, 2009; Purnhagen, 2019).

Beyond the logistical aspects of trade, regulatory fragmentation also affects innovation dynamics. Developers tend to release new products in jurisdictions where approval pathways are more predictable—such as Argentina, Brazil, the United States, or Japan—while more restrictive markets risk falling behind and becoming dependent on the importation of innovations developed elsewhere. This asymmetry shapes global value chains and influences both business strategies and the capacity of countries to benefit from scientific advances in plant biotechnology.

21.3.5 And into this complex scenario came the CRISPR–AI interaction

The CRISPR–AI factor adds another layer. The fault line is no longer limited to whether an edit counts or not as a GMO, but also to whether in silico knowledge counts as legitimate evidence. Product-based systems already integrate computational predictions as standard; process-based regimes tend to downplay them. Hybrids, increasingly, require algorithmic transparency—training data, assumptions, uncertainty analysis—as part of the dossier.

Two changes seem inevitable. Everything points to AI pipelines needing to be documented and auditable to count as credible evidence. And data governance—bias, provenance, access—will have to be recognized as a biosafety issue in its own right (Jones et al., 2022). The point is that, for the moment, it is not clear how such auditability will be regulated.

It should be borne in mind, let us not forget, that fragmentation is not a temporary inconvenience but a structural condition, rooted in divergent legal triggers, political cultures, and now computational epistemologies. Any move toward interoperability—mutual recognition, trait-tiered standards, or DSI benefit-sharing—will have to speak both languages: the trait-and-use logic of product-based regimes and the process-and-provenance logic of precautionary systems.

21.4 Global governance challenges: DSI, trade, IP, and equity

The combined use of CRISPR–AI does more than transform plant genomes, because biology becomes datafied, and from that datafication, new biological material is generated again. The regulatory frameworks we have been using up to now, the categories they handle, their approaches are being overwhelmed in many ways.

21.4.1 How digital sequence information strains the regulatory framework of access/benefit-sharing

We cannot affirm, it is true, that the practical application of the principle of access and benefit sharing provided in the CBD has been particularly successful. Both the ITPGRFA and, almost a decade later, the Nagoya Protocol are seeking formulas that may translate in reality into better conservation and access to plant diversity, as well as into a benefit-sharing model that is fair to all actors. Without this having been completely resolved, regulators now face a series of scientific and technical advances that generate great perplexity. With DSI, researchers can already design edits without ever touching a seed or a leaf. The genome has become dematerialized, and may be accessible without accessing the original biological material. This revolution dramatically challenges both the Nagoya Protocol and the International Treaty on Plant Genetic Resources for Food and Agriculture, both drafted around physical resources under sovereign control (Scholz et al., 2022).

Biodiversity-rich countries fear that open access to sequence data will hollow out benefit-sharing obligations. Others argue that strong restrictions would strangle research and even food security (Oldham et al., 2021). Compounding this, AI increases the likelihood that, once sequences enter global databases, they can be recombined endlessly. If controlling flows at physical borders was not always possible (especially because it is perfectly legal to buy seeds or fruits with seeds in third countries, even if their use in research or breeding is not allowed), physical borders become less relevant. Who will control digital flows, and how can equity in access and benefit-sharing be guaranteed?.

21.4.2 The impact of regulatory fragmentation on international agri-food trade

The trade in agri-food products is already under scrutiny. A genetically modified soybean regarded as "conventional" in Argentina or the United States may be blocked as a genetically modified organism (GMO) in the European Union or South Africa. This lack of correspondence fuels disputes within the framework of the World Trade Organization (WTO), particularly under the rules on Sanitary and Phytosanitary Measures (SPS) and Technical Barriers to Trade (TBT) (Pollack and Shaffer, 2009; Purnhagen, 2019). The Global Low-Level Presence

Initiative was established to manage the accidental presence of traces of biotechnological crops, but CRISPR has complicated this problem, as many genome edits are indistinguishable from natural mutations (Zischewski et al., 2022).

The Asia–Pacific region provides a particularly clear example of the extent of this regulatory divergence. Australia excludes SDN-1 edits from its GMO legislation, considering them equivalent to the results of conventional breeding. Japan, for its part, also exempts SDN-1 and certain SDN-2 edits, although it maintains a system of voluntary notification intended to ensure transparency. At the opposite end of the spectrum, China and New Zealand adopt a much more restrictive approach, regulating almost all gene-edited plants as genetically modified organisms and subjecting them to the same authorization procedures as transgenic crops.

This regulatory patchwork, within a region characterized by frequent trade interactions, creates significant practical obstacles. Thus, a genome-edited crop approved in one country may be rejected or subjected to additional controls in another, introducing uncertainty and extra costs for exporters. Differences in assessment and labeling requirements further complicate the movement of agri-food products in one of the most economically interdependent regions of the world. For this reason, several ASEAN Member States have stated openly that their final position on the regulation of new genomic techniques may depend on the decisions taken by the European Union, whose rules continue to exert a decisive influence on global market access conditions (Jones et al., 2022; Uncetabarrenechea, 2025). The resulting fragmentation functions as a trade barrier that shapes both the geography of innovation and market access.

21.4.3 Intellectual property and data ownership

The CRISPR patent landscape was already a battlefield. Adding AI multiplies disputes: who owns the algorithms? And the training datasets or the resulting predictions? If it is said of CRISPR, because of its more affordable price, that it puts gene editing in more hands than transgenesis, the truth is that its convergence with AI points in the opposite direction. Only large companies, such as Bayer, Corteva, or Syngenta, and entities with large infrastructures will be able to adequately benefit from the large-scale management of digitized data useful for plant breeding. This vertical integration will allow them to reinforce their dominant position with respect to the entire chain, as Jennifer Clapp points out, from the seed to the data (Clapp, 2018; Clapp and Ruder, 2025). Farmers' data feed proprietary models, often with little return. Critics describe this as "data colonialism" (Bronson, 2022; Hackfort et al., 2024; Ruder, 2025).

The literature presents and discusses interesting alternatives—open-source CRISPR, humanitarian licensing, grassroots data cooperatives—but so far, they remain marginal compared to the corporate mainstream (Ruder and Wittman, 2025).

21.4.4 Equity, legitimacy, and food sovereignty

This line of reflection is easier to pose as a question than to answer. The question is the following: who benefits from AI-optimized plant breeding—multinationals, public breeders, or small farmers? The historical evolution of other agri-food innovations suggests that those who do not have adequate access to infrastructures or who cannot preserve the sovereignty of their data run the risk of deepening their economic dependency (Rock and Schurman, 2020). Although the way in which the spread of these technologies will evolve is not yet easy to predict.

In Africa and parts of Asia, debates about genome editing are and will continue to be inseparable from food sovereignty and postcolonial politics (Abkallo et al., 2024; Runo et al., 2024). Here, legitimacy refers not only to biosafety, but also to whose values define what is "sustainable." But in its own way, the EU is also concerned about its strategic sovereignty—including food sovereignty—and its strict sustainability policy.

Debates surrounding CRISPR-AI also reveal that the very notion of sustainability is deeply contested. Its proponents present these technologies as climate-smart tools capable of reducing agricultural inputs, strengthening crop resilience, and contributing to global food security (Santos et al., 2022; Balafoutis et al., 2017). Critics, by contrast, argue that much of current corporate research and development remains focused on herbicide-tolerant or high-yield traits, thereby reinforcing intensive, input-dependent production models. Between these positions lies a broader governance dilemma: sustainability cannot be assessed solely through metrics of biosafety or productivity, but rather in terms of the kind of agricultural future these innovations promote. Accordingly, regulatory frameworks should evaluate not only the safety of CRISPR-AI crops, but also the social and ecological models they institutionalize—whether they foster diversification, inclusion, and regeneration, or instead perpetuate concentration and dependency (Clapp and Ruder, 2025).

21.5 Public perception, legitimacy, and societal debates

The governance of CRISPR-AI in plant biotechnology is not determined solely by the technical assessment of risk. It is also shaped by imaginaries of progress, political cultures of risk, and, above all, by the delicate fabric of public trust. One notable difference compared with earlier biotechnologies is that, while GMOs provoked intense public mobilization by certain groups from the late 1990s onwards (Escajedo San-Epifanio et al., 2023), the controversy surrounding CRISPR-edited plants has been comparatively quieter and more confined to expert and policy circles (Ishii and Araki, 2016; Lassoued et al., 2021). The relative absence of public controversy should not, however, be interpreted as evidence of broad consensus. So far, debate on CRISPR-edited crops has been less intense than the earlier disputes surrounding GMOs, largely because discussion has remained confined to specialized and policy domains.

Nevertheless, the types of genetic modifications enabled by the new genomic techniques (NGTs)—and their associated risk profiles—are often comparable to those obtained through conventional breeding or even spontaneous mutation (Custers/Dima, 2022). This similarity calls into question the adequacy of existing regulatory categories and highlights the need to establish more precise scientific and legal criteria than those developed for first-generation GMOs. Consequently, the governance landscape remains uncertain: the social acceptance of CRISPR-AI applications will depend not only on empirical safety assessments, but also on the clarity of their objectives, the transparency of public communication, and the credibility of the institutions responsible for their oversight.

Thus, both continuities and discontinuities with the GMO debate are visible. In Europe, for example, some civil society groups tend to describe genome editing as a continuation of corporate biotech projects, raising concerns about ecological uncertainty, concentration of power, and ethical boundaries of human intervention in plant genomes (Wasmer, 2019; Eriksson et al., 2020). But it is not an opinion as widespread as was, in its day, the opposition to transgenics. Compared with the antiGMO campaigns of the 1990s, the echo has been smaller (see Escajedo San-Epifanio et al., 2023), less visible in mainstream public arenas. Protests and boycotts have not returned on a large scale, and debates remain concentrated among scientists, regulators, and activist networks (Ruder and Kandlikar, 2023).

In Africa and Latin America, as already mentioned, resistance has another profile: it is frequently interwoven with demands for food sovereignty and critiques of dependence on external inputs (Rock and Schurman, 2020).

Where survey data exist, they suggest a more favorable baseline than in the case of GMOs. In Asia, North America, and parts of Africa, CRISPR-edited crops are perceived more positively—especially when presented as "precision breeding" without foreign DNA (Ishii and Araki, 2016; Lassoued et al., 2021). Acceptance grows when traits are linked to public goods such as drought resistance, reduced pesticide use, or nutritional improvements (Zaidi et al., 2019). But when edits are presented as cosmetic or market-driven, skepticism increases. The empirical base, however, remains very limited: only a few large-scale surveys exist, often restricted to specific regions or traits (Qaim, 2020). Compared to the vast body of research on GMO perception, the social science of genome editing is still in its beginnings.

It is difficult to foresee the impact that the introduction of AI will have on this perception, because, beyond plant breeding, AI itself has generated an unprecedented regulatory challenge globally, affecting multiple sectors of the economy and even social life. If CRISPR raises questions about safety and ethics in the use of biotechnologies, algorithms, at the very least, raise issues of transparency and accountability. Who checks the accuracy of off-target predictions in proprietary models? What happens if regulators rely on algorithms trained with biased or incomplete datasets? Among the dangers warned of by critics is the fact that AI is, in sum, a process regulated through "a black box," a tool where decisions move away from public scrutiny (Smuha, 2021; Ruder, 2025). Public trust will depend not only on the safety of edited crops but also on whether algorithmic tools are explainable, auditable, and open to scrutiny.

Public narratives around CRISPR-AI remain limited for the moment, although an incipient polarisation is already apparent. Supporters present the technology as a cornerstone of climate-smart, data-driven agriculture, capable of producing more resilient crops, reducing chemical inputs, and strengthening global food security (Balafoutis et al., 2017; Santos et al., 2022). Critics, by contrast, regard it as a continuation of industrial agriculture models, characterized by corporate concentration, ecological risk, and what some authors in the literature describe as new forms of "data colonialism," in which genomic and agronomic information is controlled by a small number of dominant actors (Clapp and Ruder, 2025). Substantive differences in perspective remain between these positions, and the way governance frameworks address key issues such as transparency, public participation, and accountability will be decisive for the future legitimacy of these technologies. Without inclusive dialog and credible institutional oversight, legitimacy gaps could widen, reproducing the patterns of controversy that once characterized the European debate on GMOs.

21.6 Future scenarios for CRISPR–AI governance in agriculture

It is unlikely that the evolution of plant biotechnology—and, in particular, of CRISPR-AI applications—will follow a uniform or linear trajectory. At the global level, regulatory and policy pathways remain heterogeneous and are shaped by multiple factors: the adaptability of legal frameworks, the configuration of international trade, the level of public trust, and broader geopolitical conditions.

The previous experience with the implementation of regulatory frameworks for genetically modified organisms (GMOs) shows that expectations of rapid technological progress have often been constrained by social, institutional, and legal factors. Regardless of the requirements set for premarket approval or the different risk assessment models, regulatory systems have, in practice, offered operators very uneven levels of legal certainty and predictability for investment. In the case of CRISPR-AI, future developments will largely depend on how governance mechanisms manage to reconcile scientific innovation with regulatory legitimacy and public accountability.

In previous work, we have described how the European Union exemplifies this uncertainty (Escajedo San-Epifanio et al., 2023). At one end of the spectrum lies a scenario of cautious convergence, in which Europe reforms its rules and moves toward proportionate, risk-based regulation. Low-risk modifications (SDN-1/2 without foreign DNA) would be treated as equivalent to conventional breeding outcomes, and AI-assisted testing would become part of regulatory dossiers, provided that such processes are transparent and auditable.

Just as there are international fora currently debating the legal status of DSI uses in plant breeding, it would be desirable to establish forums for dialog and convergence on premarket approval of genome-edited plants and their scientific models of risk evaluation. As there was never full harmonization regarding GMOs, it is unlikely that there will be complete harmonization in the governance of CRISPR-AI interactions. Nevertheless, a concerted effort toward minimum coordination—particularly in the areas of biosafety and data governance—would be valuable. Such baseline agreements could help to reduce the current degree of regulatory fragmentation and prevent it from deepening in the future. It is true that the EU now faces a particularly delicate moment: it must make strategic decisions not only regarding the governance of its internal market but also its global economic and geopolitical positioning.

Even so, the European Union remains the governance framework where anticipatory governance is practiced most clearly. Although such consultations are not binding, the process of developing the new regulatory framework for genome-edited plants—and for their use in agriculture and food—has involved broad participation from producers, regulators, farmers, consumers, and civil society organizations. The EU proceeds from the conviction that such efforts enhance public trust in regulatory outcomes and that, conversely, when transparency is limited, or decision-making is hindered by institutional gridlock, disparities between jurisdictions are likely to widen and trust in supervisory authorities to erode.

In the medium term, the most plausible outlook is one of regulated pluralism, in which diverse models of risk assessment and governance continue to coexist according to distinct legal traditions and political priorities.

21.7 Governing CRISPR–AI: Lessons, dilemmas, and paths ahead

It may seem premature to consider how a coherent global regulatory framework for CRISPR-AI could be built, given the high degree of fragmentation that still characterizes existing regimes. Governance of biotechnology, intellectual property in plant innovation, and seed systems remains organized in what may be described as "regulatory silos"—that is, sectoral frameworks operating independently and with limited coordination between them. Many of these frameworks were designed decades ago and have not been updated to reflect the current convergence between digital and biological technologies. The coexistence of outdated legal instruments—ranging from GMO legislation to plant variety protection and data ownership regimes—makes it difficult to formulate an integrated approach capable of addressing the combined challenges of genome editing and artificial intelligence.

Both CRISPR and artificial intelligence, each transformative in its own domain, are reshaping the normative foundations of biotechnology and digital innovation alike. Their interaction raises not only new technical and scientific questions, but also ethical, social, and legal dilemmas that extend beyond the scope of traditional biosafety regulation. Addressing these challenges will require adapting existing regulatory frameworks and promoting controlled experimentation, for example, through pilot programs or regulatory sandboxes, which allow new approaches to be tested without compromising the core principles underpinning governance in each jurisdiction.

Unlike the first generation of GMOs, which emerged within a largely uncharted regulatory landscape, the interaction between CRISPR and AI is developing in an environment already shaped by longstanding disputes over legitimacy, intellectual property, and the limits of the precautionary principle. The present task is not to design an entirely new framework, but to restore coherence, transparency, and public trust in regulatory systems that have struggled to keep pace with the rapid advance of science and technology—and, once these foundations are reinforced, to incorporate specific measures to address the novelties introduced by the convergence of CRISPR and AI.

21.7.1 Lessons from the past, unfinished tasks in the present

Governance is often criticized for usually reacting only once conflicts have already erupted. But it so happens that regulation is precisely a social tool for conflict resolution, and, therefore, it is very difficult to try to resolve a conflict before it has manifested. It is undeniable, in any case, that few issues have required as much time (and effort) to achieve reasonable regulation as agribiotechnologies. The European Union, as has already been said, remains anchored in its (dis)union on agribiotechnology since the 1990s. The period

2019–25 is, without doubt, one of the times when the greatest effort has been made to overcome that situation, but for now, there are no successes to highlight—only progress in the right direction toward agreement. The Americas advanced faster with product-based systems, although that speed came accompanied by suspicions of underregulation. Hybrid regimes in Asia and Africa appear more pragmatic, yet they remain fragile, dependent on political consensus and sustained technical capacity (Jones et al., 2022; Abkallo et al., 2024; Runo et al., 2024). It remains to be seen when—and how—pending legislative work will be completed.

21.7.2 The dual challenge of CRISPR–AI

Even without AI, CRISPR already raised difficult questions: biosafety, benefit-sharing of digital sequence information, concentrated patents, and equity. Add AI, and the uncertainty doubles. Algorithms now guide trait discovery, risk assessment, and even the preparation of regulatory dossiers. Yet regulators cannot always verify their reliability. If the evidence comes from opaque or biased models, public trust is at risk. Governing CRISPR–AI therefore, means addressing molecular uncertainty and epistemic uncertainty at the same time (Smuha, 2021; Ruder, 2025).

21.7.3 Possible directions

Any discussion on the future governance of CRISPR–AI must begin from existing regulatory experience rather than from abstract normative ideals. The frameworks that currently govern biosafety, access and benefit-sharing, and agribiotechnological innovation—however imperfect—already reflect a set of values and principles articulated in the Convention on Biological Diversity (CBD): the precautionary approach, the fair and equitable sharing of benefits, respect for national sovereignty over genetic resources, and the pursuit of sustainability. These elements provide a conceptual and legal foundation on which new regulatory mechanisms for genome editing and artificial intelligence may be constructed.

From this perspective, certain characteristics could reasonably be expected of future regulatory frameworks. First, proportionate risk regulation should remain a cornerstone. Evaluation of CRISPR–AI applications ought to be anchored in the traits and intended uses of the resulting plants, rather than in the specific techniques used to obtain them. Developing trait-based assessment criteria, consistent with the principles of familiarity and substantial equivalence already recognized in international biosafety instruments, would improve the alignment between regulatory oversight and the real scale of potential risk or exposure, and would help to reduce inconsistencies across jurisdictions.

Second, the durability and legitimacy of governance will depend on its procedural robustness. Technical expertise is essential but not sufficient: regulatory decision-making requires transparent procedures, publicly accessible information, and mechanisms that allow for structured participation by relevant stakeholders, including farmers, consumers, and civil-society organizations. This procedural dimension reinforces both the credibility of regulatory outcomes and their enforceability, in line with the participatory ethos that underpins the CBD and related multilateral processes.

Finally, the increasing integration of artificial intelligence into plant breeding calls for a renewed emphasis on data governance. Because biosafety and data stewardship are now interdependent, future frameworks must ensure the traceability and auditability of AI systems used in genomic design, as well as equitable arrangements for access to and use of digital sequence information (DSI). Addressing these questions within the logic of the CBD—where fairness, transparency, and reproducibility are central—would help prevent new asymmetries of control over genetic and agronomic data.

Taken together, these elements suggest that credible governance for CRISPR–AI will combine scientific proportionality, procedural legitimacy, and data integrity. The challenge is not to depart from existing principles, but to reinterpret them in light of technological convergence, ensuring that the normative architecture inherited from the biodiversity regime remains capable of guiding innovation under conditions of accelerated scientific change.

21.7.4 Beyond coordination: The distance from a shared agenda

In the current context, it is difficult to speak of a truly collaborative global framework for the governance of CRISPR-AI in agriculture. Despite the rhetoric of convergence and openness, the management of plant genetic resources—particularly those used for food and agriculture—remains deeply fragmented. Longstanding disputes over intellectual property, access and benefit-sharing, and data control continue to hinder any meaningful coordination between jurisdictions. These tensions are not merely technical: they reflect a persistent asymmetry between regions that provide biological diversity and those that dominate technological and commercial capacity.

Invoking the need for international cooperation is therefore less an expression of optimism than a reminder of how far we still are from building a more equitable and inclusive global food system. Unlike other domains in which scientific coordination has proved possible—such as public health during the COVID-19 pandemic—agricultural innovation remains constrained by conflicting economic interests and by the lack of effective institutional mechanisms for joint governance.

The decisions—and indecisions—currently being made on the regulation of CRISPR-AI, the reform of benefit-sharing systems, and the definition of frameworks for digital sequence information will shape not only the trajectory of innovation but also our collective ability to progress toward the Sustainable Development Goals (SDGs). In this sense, moving beyond short-term regulatory disputes is essential if global governance is to avoid drifting further away from its stated principles of sustainability, inclusion, and equity.

21.7.5 Final reflection

Plant breeding directly affects some of humanity's most fundamental needs, regardless of where people live or the conditions in which they survive. Yet food is rarely governed from the perspective of its essential role in sustaining life; it is regulated in a fragmented manner across multiple governance arenas, usually reflecting the power relations that determine who controls food production and who depends on it (Escajedo San-Epifanio, 2022). It is therefore unsurprising that achieving a global regulatory framework that is both balanced and fair remains so difficult, particularly under the combined pressures of climate stress and growing geopolitical rivalry. The regulation of CRISPR-AI thus represents a crossroads: it may either deepen existing socioeconomic divides or serve as an opportunity to build a more just, sustainable, and cooperative agricultural future.

Acknowledgements

This work has been carried out within the framework of several research projects: VegEquity, Plant Breeding and European Integration (PID2021-123796OB), funded by the Spanish Ministry of Science and Innovation; the Research Group GIU24/007 EU-Frames, "Constructing Innovative Regulatory Frameworks in the EU," directed by L. Escajedo San-Epifanio and I. Filibi; and the Jean Monnet Module PlantSciLaw – European Law and Governance for the Agrosciences (Erasmus+), funded by the European Commission (grant code ERASMUS-JMO-2023-MODULE 101127121).

References

Abkallo, H.M., Arbuthnot, P., Auer, T.O., Berger, D.K., Burger, J., Chakauya, E., Wondji, C., 2024. Making genome editing a success story in Africa. Nat. Biotechnol. 42 (5), 551–554. https://doi.org/10.1038/s41587-024-02187-2.

Ahmad, A., Munawar, N., Khan, Z., Qusmani, A.T., Khan, S.H., Jamil, A., Ashraf, S., Ghouri, M.Z., Mubarik, M.S., Munir, A., Sultan, Q., Abd-Elsalam, K.A., Qari, S.H., 2021. An outlook on global regulatory landscape for genome-edited crops. Int. J. Mol. Sci. 22, 11753. https://doi.org/10.3390/ijms222111753.

Balafoutis, A., Beck, B., Fountas, S., Vangeyte, J., van der Wal, T., Soto, I., Gómez-Barbero, M., Barnes, A., Eory, V., 2017. Precision agriculture technologies positively contributing to GHG emissions mitigation, farm productivity and economics. Sustainability 9 (8), 1339. https://doi.org/10.3390/su9081339.

Bronson, K., 2022. The Immaculate Conception of Data: Agribusiness, Activists and Their Shared Politics of the Future. McGill-Queen's University Press.

Clapp, J., 2018. Mega-mergers on the menu: corporate concentration and the politics of sustainability in the global food system. Glob. Environ. Polit. 18 (2), 12–33. https://doi.org/10.1162/glep_a_00454.

Clapp, J., Ruder, S.-L., 2025. Tecnologías de precisión para la agricultura: Agricultura digital, cultivos editados genéticamente y la política de la sostenibilidad. En … [Referencia completa del libro cuando dispongas de los datos editoriales].

Entine, J., Eriksson, D., Felipe, M.S., Groenewald, J.H., Kershen, D.L., Lema, M., McHughen, A., Nepomuceno, A.L., Ohsawa, R., Ordonio, R.L., Parrott, W.A., Purnhagen, K.P., Quemada, H., Ramage, C., Slamet-Loedin, I., Smyth, S.J., Wasmer, M.S., Wray-Cahen, D., 2021. Regulatory approaches for genome edited agricultural plants in select countries and jurisdictions around the world. Transgenic Res. 30, 551–584. https://doi.org/10.1007/s11248-021-00257-8.

Eriksson, D., Custers, R., Björnberg, K.E., Hansson, S.O., Purnhagen, K., Qaim, M., Romeis, J., Schiemann, J., Schleissing, S., Tosun, J., 2020. Options to reform the European Union legislation on GMOs: scope and definitions. Trends Biotechnol. 38, 231–234. https://doi.org/10.1016/j.tibtech.2019.12.002.

Escajedo San-Epifanio, L., 2007. Towards a Sustainable and Healthy Development of Biotechnology. International Society of Bioethics, SIBI, Gijón.

Escajedo San-Epifanio, L., 2022. The right to food and the essential promotion of personal autonomy: the 'how' matters. In: Escajedo San-Epifanio, L., Rebato Ochoa, E.M. (Eds.), Ethics of Charitable Food: Dilemmas for Policy and Practice. Springer, pp. 61–74.

Escajedo San-Epifanio, L., Filibi, I., Lasa López, A., Puigdomènech, P., Uncetabarrenechea Larrabe, J., 2023. Possible EU futures for CRISPR-edited plants: little margin for optimism? Front. Plant Sci. 14, 1141455. https://doi.org/10.3389/fpls.2023.1141455.

European Commission, 2021. Study on the Status of New Genomic Techniques under Union Law and in Light of the Court of Justice Ruling in Case C-528/16 (SWD/2021/92 Final).

European Commission, 2023. Proposal for a Regulation on Plants Obtained by Certain New Genomic Techniques (COM/2023/411 Final) (Brussels).

Graham, N., Patil, G.B., Bubeck, D.M., et al., 2020. Plant genome editing and the relevance of off-target changes. Plant Physiol. 183, 1453–1471. https://doi.org/10.1104/pp.19.01194.

Hackfort, S., Marquis, S., Bronson, K., 2024. Harvesting value: corporate strategies of data assetization in agriculture and their socio-ecological implications. Big Data Soc. 11 (1), 1–15. https://doi.org/10.1177/20539517241234567.

Ishii, T., Araki, M., 2016. Consumer acceptance of food crops developed by genome editing. Plant Cell Rep. 35, 1507–1518. https://doi.org/10.1007/s00299-016-1974-2.

Jones, M.G.K., Fosu-Nyarko, J., Iqbal, S., Adeel, M., Romero-Aldemita, R., Arujanan, M., Kasai, M., Wei, X., Prasetya, B., Nugroho, S., Mewett, O., Mansoor, S., Awan, M.J.A., Ordonio, R.L., Rao, S.R., Poddar, A., Hundleby, P., Iamsupasit, N., Khoo, K., 2022. Enabling trade in gene-edited produce in Asia and Australasia: the developing regulatory landscape and future perspectives. Plants 11, 2538. https://doi.org/10.3390/plants11192538.

Lassoued, R., Smyth, S.J., Phillips, P.W.B., Hesseln, H., 2021. Consumer perceptions of genome editing in agriculture: a review of the literature. Trends Biotechnol. 39 (7), 634–646. https://doi.org/10.1016/j.tibtech.2020.12.002.

NBA (National Biosafety Authority, Kenya), 2022. Guidelines for Determining the Regulatory Process of Genome Edited Organisms and Products in Kenya.

OECD, 2024. Developments in delegations on the safety assessment of novel foods and feeds, may 2023–february 2024. In: Series on the Safety of Novel Foods and Feeds No. 38. ENV/CBC/MONO, 20.

Oldham, P., Hall, S., Forero, O., 2021. Digital sequence information: towards an inclusive science–policy interface. Plants People Planet 3, 505–514. https://doi.org/10.1002/ppp3.10202.

Pollack, M.A., Shaffer, G.C., 2009. When Cooperation Fails: The International Law and Politics of Genetically Modified Foods. Oxford University Press, Oxford.

Purnhagen, K., 2019. How to manage the Union's diversity of values in GMO regulation. Eur. J. Risk Regul. 10, 427–436. https://doi.org/10.1017/err.2019.45.

Qaim, M., 2020. Role of new plant breeding technologies for global food security. Appl. Econ. Perspect. Policy 42 (2), 129–150. https://doi.org/10.1002/aepp.13044.

Rock, J.S., Schurman, R., 2020. The complex choreography of agricultural biotechnology in Africa. Afr. Aff. 119 (477), 499–525. https://doi.org/10.1093/afraf/adaa021.

Ruder, S.-L., 2025. The "terms and conditions" of surveillance capitalism: theorizing agricultural data policy and governance. J. Peasant Stud. 54 (4), 725–750. https://doi.org/10.1080/03066150.2024.1234567.

Ruder, S.-L., Kandlikar, M., 2023. Governing gene-edited crops: risks, regulations, and responsibilities as perceived by agricultural genomics experts in Canada. J. Respons. Innov. 10 (1), 1–29. https://doi.org/10.1080/23299460.2023.2165678.

Ruder, S.-L., Wittman, H., 2025. Agricultural data governance from the ground up: exploring data justice with agri-food movements. Big Data Soc. 12 (1), 1–14. https://doi.org/10.1177/2053951725123456.

Runo, S., Karembu, M., Nan'gayo, F., Mekbib, F., Haileselassie, T., Tesfaye, K., Jones, T., 2024. Africanizing genome editing for food sustainability. Global Food Secur. 42, 100785. https://doi.org/10.1016/j.gfs.2024.100785.

Santos, R., Maier, O., Santos, R., 2022. Advancing tomorrow's innovations today. In: Bayer Crop Science Annual R&D Pipeline Update. https://www.bayer.com/sites/default/files/2022-02/Crop-Science-RD-Pipeline-Update-Webinar.pdf.

Scholz, A.H., Wellmann, C., Urnov, F., 2022. Digital sequence information: towards a policy for biodiversity governance. Nat. Biotechnol. 40, 1464–1470. https://doi.org/10.1038/s41587-022-01459-9.

Smuha, N.A., 2021. From a 'race to AI' to a 'race to AI regulation': regulatory competition for artificial intelligence. Law Innov. Technol. 13 (1), 57–84. https://doi.org/10.1080/17579961.2021.1898300.

Turnbull, C., Lillemo, M., Hvoslef-Eide, T.A., 2021. Global regulation of genetically modified crops amid the gene edited crop boom: a review. Front. Plant Sci. 12, 630396. https://doi.org/10.3389/fpls.2021.630396.

Uncetabarrenechea, J., 2025. La UE como actor en la gobernanza global de la biodiversidad agrícola y las agrobiotecnologías: desafíos y escenarios hacia el segundo cuarto del siglo XXI. En: Escajedo San-Epifanio, L., Filibi, I., Uncetabarrenechea, J. (Dirs.), Edición genómica de las plantas y desmaterialización de los recursos fitogenéticos. Dykinson, Madrid.

Wasmer, M., 2019. Roads forward for European GMO policy—uncertainties in wake of ECJ judgment have to be mitigated by regulatory reform. Front. Bioeng. Biotechnol. 7, 132. https://doi.org/10.3389/fbioe.2019.00132.

Zaidi, S.S., Vanderschuren, H., Qaim, M., Mahfouz, M.M., Kohli, A., Mansoor, S., Tester, M., 2019. New plant breeding technologies for food security. Science 363, 1390–1391.
Zischewski, J., Sprink, T., Hartung, F., 2022. Strategies to detect CRISPR/Cas-induced unintended modifications in plants. Plant Cell Tissue Organ Cult. 149, 459–472. https://doi.org/10.1007/s11240-022-02299-6.

Further reading

ALLEA, 2021. Genome Editing for Crop Improvement: Scientific Opportunities, Public Interests and Ethical Challenges. All European Academies.
Bostyn, S., 2019. Patentability of Plants.
Bradford, A., 2020. The Brussels effect: How the European Union Rules the World. Oxford University Press.
Brinegar, K., Yetisen, A., Choi, S., Vallillo, E., Ruiz-Esparza, G.U., Yun, S.H., 2017. The commercialization of genome-editing technologies. Crit. Rev. Biotechnol. 37 (7), 924–932.
Bronson, K., Knezevic, I., 2016. Big data in food and agriculture. Big Data Soc. 3 (1), 1–5.
CBAN (Canadian Biotechnology Action Network), 2020. Genome Editing in Food and Farming: risks and Unexpected Consequences. https://cban.ca/wp-content/uploads/Genome-Editing-Report-2020.pdf.
Chen, K., Wang, Y., Zhang, R., Zhang, H., Gao, C., 2019. CRISPR/Cas genome editing and precision plant breeding in agriculture. Annu. Rev. Plant Biol. 70, 667–697. https://doi.org/10.1146/annurev-arplant-050718-100049.
Clapp, J., 2025. Titans of Industrial Agriculture: How a Few Giant Corporations Came to Dominate the Farm Sector and Why it Matters. MIT Press.
Court of Justice of the European Union (CJEU), 2018. Case C-528/16, Confédération paysanne and Others v Premier ministre and Ministre de l'Agriculture, de l'Agroalimentaire et de la Forêt.
Crossroads at the crossroads between monopolizing nature and protecting technological innovation? J. World Intellect. Property 16 (3–4), 2013, 105–149.
Duarte Sagawa, C.H., de Almeida Barbosa Assis, R., Zaini, P.A., 2024. Regulatory framework of CRISPR-edited crops in the United States. In: Genome Modified Plants and Microbes in Food and Agriculture. Elsevier, pp. 167–195. https://doi.org/10.1016/B978-0-443-18444-4.00004-1.
Escajedo San-Epifanio, L., 2025. Edición genómica y gobernanza en África: marcos regulatorios, disputas sociopolíticas y trayectorias regionales. In: En: Escajedo San-Epifanio, L., Filibi, I., Uncetabarrenechea, J.(D.) (Eds.), Edición genómica de las plantas y desmaterialización de los recursos fitogenéticos: desafíos para la gobernanza global. Dykinson, Madrid.
NBMA (National Biosafety Management Agency, Nigeria), 2020. National Guidelines for the Regulation of Gene Editing. Abuja.
Rabitz, F., 2019. Institutional drift in international biotechnology regulation. Glob. Policy 10 (2), 227–237. https://doi.org/10.1111/1758-5899.12638.
Stilgoe, J., Owen, R., Macnaghten, P., 2013. Developing a framework for responsible innovation. Res. Policy 42 (9), 1568–1580. https://doi.org/10.1016/j.respol.2013.05.008.
Urzúa, N., 2025. Estado de la Regulación de las NGT en Latinoamérica. En: Escajedo San-Epifanio, L., Filibi, I., Uncetabarrenechea, J. (Dirs.), Edición genómica de las plantas y desmaterialización de los recursos fitogenéticos. Dykinson, Madrid.
USDA-APHIS, May 18, 2020. Movement of certain genetically engineered organisms (SECURE rule). In: Federal Register 85 FR 29790.
Waltz, E., 2018. With a free pass, CRISPR-edited plants reach market in record time. Nat. Biotechnol. 36, 6–7. https://doi.org/10.1038/nbt0118-6b.
WTO, 2006. European Communities — Measures Affecting the Approval and Marketing of Biotech Products (WT/DS291, WT/DS292, WT/DS293). Geneva.
WTO, 2015. Trade and Food Security: WTO Rules and Policy Options. Geneva.

CHAPTER

22

CRISPR meets AI: New frontiers in developing insect-resistant crops

Shivali Pandita[1,2], *Punam Chakraborty*[1,3], *and Praveen C. Verma*[1,3]

[1]Molecular Biology and Biotechnology Division, Council of Scientific and Industrial Research (CSIR)-National Botanical Research Institute, Lucknow, Uttar Pradesh, India; [2]Department of Zoology, University of Lucknow, Lucknow, Uttar Pradesh, India; [3]Academy of Scientific and Innovative Research (AcSIR), Ghaziabad, Uttar Pradesh, India

22.1 Introduction

Insects are the major biotic stressors, as they possess unique feeding behaviors and have the ability to transmit disease (Douglas, 2018). According to the Food and Agriculture Organization (FAO) of the United Nations, plant pests and diseases cause up to 40% loss of crop production (FAO, 2023). Plant diseases cause a loss of over $220 billion. Also, losses caused by invasive insects are estimated to be $70 billion each year (FAO, 2021). Modern techniques involve managing herbivorous pests, which can safeguard crops to increase crop yield. Most destructive pests are sap-sucking and chewing pests, which are the main reason for decreased crop production (Vanti et al., 2018). Though conventional pest management is dependent on chemical pesticides, they cause various environmental and health risks (Singh et al., 2022). Several pest species develop resistance due to improper use of these pesticides; therefore, there is a need for a sustainable control method (Sun et al., 2017). Recent improvements in the molecular groundworks of plant–insect associations. These and more environmentally friendly and specific methods have contributed to the development of an alternative biotechnological technique, for example, genome editing (Birkett and Pickett, 2014). In the long run, the use of genome-editing methods has extended dramatically. Recently, it has been found that nucleases hold massive potential for genome editing in both plants and insects (Taning et al., 2017; Islam, 2019). The most extensively used and effective technique of genome editing is CRISPR/Cas (Andolfo et al., 2016; Bhowmik and

https://doi.org/10.1016/B978-0-443-44079-3.00019-X

Islam, 2020). Using this system, several agronomic features, such as insect resistance, have been effectively created.

22.1.1 Introduction of pest-resistant trait

Developing insect pest-resistant varieties to sustain and increase crop yield is one of the major challenges faced in the current times. Modern agriculture includes host pest resistance against pests (Horgan et al., 2020; El-Dessouki et al., 2022). The heritable characteristics of a plant directly result in it sustaining less damage than a plant without those qualities. Following the landmark development of Hessian fly-resistant wheat cultivar published by Havens in 1782, many other insect-resistant cultivars have been developed by research centers and private firms using a combination of traditional breeding and modern conventional or biotechnological techniques (Jaiswal et al., 2018). So, to enhance the immunity of the crops, it is therefore important to enhance the resistance (R-genes). For example, using the CRISPR–Cas9 system, two *beta-1-3 glucanase* genes are altered in Golden Promise barley. These genes are involved in callus formation. Via editing of these genes, the callus formation reduces in sieve tubes, thus blocking the aphid (*Rhopalosiphum padi*) from accessing essential phloem sap nutrients (Bassett et al., 2013; Kim et al., 2020). The outward appearances of the host plant can also affect the insect's preferences, with color being the main factor. A study done on it has found that variations in plant color can influence an insect's host preferences. In a study done by Malone et al., the color of the red-leaf tobacco was changed by altering its anthocyanin pathway, which resulted in the prevention of *Helicoverpa armigera* and *Spodoptera litura* from feeding on the plant, as they were discouraged by the color. Therefore, by the genome editing of the color-producing gene, the plant was safeguarded from the pests' destruction (Malone et al., 2009). In another study done by Li et al., the calcium-dependent protein kinase *GmCDPK38* mutant with the Haplotype 3 (*Hap3*) deletion in *Glycine* max (soybeans) showed increased resistance to common cutworms (Li et al., 2022a, 2002b). Moreover, deletion of 1 bp and 33 bp region of the UDP-glycosyltransferase (*GmUGT*) gene resulted in resistance to *S. litura* and *H. armigera* (Zhang et al., 2022).

22.1.2 Silencing of susceptibility genes

By eliminating susceptibility (S) genes, insect-resistant plants can be produced. While plants with R (resistance) genes are more resistant to insect pests, those with S-genes are more susceptible to stress (Rathinam et al., 2019). CRISPR/Cas-mediated mutation of the tryptamine 5-hydroxylase-encoding *CYP71A1* gene, which converts tryptamine to serotonin in plants, can hamper plant hopper growth. Lu et al. explored how rice plants gain resistance to insect pests by preventing serotonin production. *CY71A1* mutated plants can be developed that can prev*ent* brown plant hoppers (*Nilaparvata lugens*) (Lu et al., 2018). This resulted in decreased eating, survival, and growth of brown plant hoppers because the transformation of tryptamine to serotonin is prevented, which is crucial for insect immunity and behavior (Table 22.1).

TABLE 22.1 Overview of genes edited or silenced via CRISPR/Cas systems to improve resistance against insect pests in various plant species.

Plant	Genes targeted	Insect pest resistance	Result	References
Oryza sativa (rice)	*OsLRR2* (a leucine-rich-repeat protein)	*Nilaparvata lugens*	Knockout of *OsLRR2* significantly reduced planthopper infestation and increased yield in untreated plots (~25%)	Kuai et al. (2024)
Oryza sativa (rice)	*CYP71A1* (Cytochrome P450 tryptamine 5-hydroxylase) knockout	*Nilaparvata lugens*, *Chilo suppressalis* (striped stem borer)	Knockout of *CYP71A1* suppressed serotonin synthesis, increased salicylic acid, and enhanced resistance to both pests	Lu et al. (2018)
(*Gossypium hirsutum* (cotton)	*GhMLP423* (major latex protein-like 423)	Piercing-sucking and chewing insects (e.g., *Helicoverpa armigera*—cotton bollworm)	Activated salicylic acid-mediated systemic acquired resistance (SAR), increased cytosolic Ca^{2+} flux and ROS production were observed, whereas knockout lines showed increased susceptibility	Sun et al. (2024).
Hordeum vulgare (barley)	Two β-1,3-glucanase genes *HvGlu1/1636* and *1639* (host-plant genes)	*Rhopalosiphum padi* (aphid)	The double-mutant lines had reduced β1,3-glucanase activity and more callose deposit after elicitor treatment, but no measurable increase in resistance to the aphid *Rhopalosiphum padi* (negative result)	Kim et al. (2020)
Glycine max (soybean)	*GmUGT* (UDP-glycosyltransferase)—1 bp and 33 bp deletions	*Spodoptera litura* (common cutworm), *Helicoverpa armigera*	Mutants showed enhanced insect resistance; editing disrupted the glycosylation pathway, affecting secondary metabolites used by insects	Zhang et al. (2022)
Glycine max (soybean)	*GmCDPK38* (calcium-dependent protein kinase)—Hap3 allele deletion	*Spodoptera litura*	*GmCDPK38* Hap3 mutant showed higher defense activation and increased resistance to feeding damage	Li et al. (2022a, 2002b)
Nicotiana tabacum (tobacco)	Anthocyanin biosynthesis pathway genes (altered color-producing genes)—gene editing	*Helicoverpa armigera*, *Spodoptera litura*	Altered leaf pigmentation (red-leaf phenotype) discouraged insect feeding—insects avoided the red leaves due to color preference	Malone et al. (2009)

22.1.3 Enhancing crop resistance via changing volatile organic compounds (VOCs)

CRISPR can be used to strengthen plants' natural defenses, making them less attractive or more resistant to pests. Numerous studies have made genome-edited plants resistant to various bacteria, viruses, and fungal-mediated diseases (Vats et al., 2019). However, regarding insects, this area has been less explored. Plants produce major chemical substances like VOCs, secondary metabolites, etc., which are crucial for insect behavior, immunity, and even development. Insects identify their host plants via the volatiles emitted by them, their structural features, phenotypes, visual signals, odor and taste cues, and oviposition sites, among others (Larsson et al., 2004). An insect picks a particular host for its oviposition site according to the availability of its favorite food for its young ones. Among the plant VOCs, which are blends of volatiles, insects can recognize only a few of them. A recent study demonstrated that a particular volatile mixture, which is a kairomone-mediated trap, can attract the predator *Nesidiocrois tenuis* to control the key tomato pest *Tuta absoluta* (Meyrick) and *Trialeurodes vaporariorum* (Westwood) (Ayelo et al., 2021). Alterations of these mixtures by genome editing can act as a fruitful strategy to manage insect pests (Wani et al., 2022).

22.1.4 CRISPR-based gene drives

CRISPR-based gene drives are an advanced genetic engineering technology designed to force a specific trait through a wild population by cheating the normal laws of inheritance. In standard reproduction, an offspring has a 50% chance of inheriting a gene from one parent. A CRISPR gene drive, however, ensures the modified trait is passed on to nearly 100% of the offspring (CRISPR-based gene drives are an advanced genetic engineering technology designed to force a specific trait through a wild population by cheating the normal laws of inheritance. In standard reproduction, an offspring has a 50% chance of inheriting a gene from one parent. A CRISPR gene drive, however, ensures the modified trait is passed on to nearly 100% of the offspring (CRISPR-based gene drives are an advanced genetic engineering technology designed to force a specific trait through a wild population by cheating the normal laws of inheritance. In standard reproduction, an offspring has a 50% chance of inheriting a gene from one parent. A CRISPR gene drive, however, ensures the modified trait is passed on to nearly 100% of the offspring (CRISPR-based gene drives are an advanced genetic engineering technology designed to force a specific trait through a wild population by cheating the normal laws of inheritance. In standard reproduction, an offspring has a 50% chance of inheriting a gene from one parent. A CRISPR gene drive, however, ensures the modified trait is passed on to nearly 100% of the offspring (Gantz and Bier, 2015). It works by embedding the CRISPR–Cas9 machinery directly into the organism's DNA. When a modified individual mates with a wild partner, the inherited CRISPR system acts like a molecular "copy-and-paste" tool: it cuts the wild, unmodified chromosome and forces the cell to repair it using the modified gene as a template (Esvelt et al., 2014). This converts the organism from heterozygous to homozygous for the trait. As this cycle repeats every generation, the engineered trait spreads rapidly through the entire population. While this technology could dramatically help suppress invasive agricultural pests or eradicate disease vectors like malaria, it also poses substantial ecological risks regarding its permanence and potential for unintended transboundary spread (National Academies of Sciences, 2016).

22.1.5 Targeting insect genes using the CRISPR/Cas system

CRISPR/Cas has emerged as a powerful tool in functional genomics and pest management. This allows researchers to precisely disrupt or modify specific genes linked to insect survival, reproduction, and behavior. Customized guide RNAs (gRNAs) are designed that direct the Cas9 endonuclease to a precise locus within the insect genome. This gRNA can introduce double-strand breaks that lead to gene knockouts through error-prone non-homologous end joining (NHEJ) (Taning et al., 2017). This approach is often used to silence important genes in destructive agricultural pests. For example, targeting the *doublesex* gene was used to destroy receptors involved in host-plant odor recognition to discourage feeding (Kyrou et al., 2018). Furthermore, targeting metabolic or detoxification genes such as cytochrome P450 monooxygenases, can significantly reverse pesticide resistance in targeted populations (Peng et al., 2017). Despite its specificity compared to traditional chemical control, its achievement relies heavily on overcoming challenges like off-target mutations, high genetic variation in wild populations, and delivery efficiency into insect embryos.

22.2 Emerging role of the CRISPR–Cas13 system for pest management

The CRISPR/Cas13 system, an RNA-guided editing tool, has emerged as a highly efficient and stable RNA editing technique. Unlike the CRISPR–Cas9 system targeting DNA, the programmable RNA-guided CRISPR–Cas13 editing system specifically recognizes and cleaves single-stranded RNA, enabling gene knockdown at the transcriptional level (Abudayyeh et al., 2016; Cox et al., 2017; Mahas et al., 2019; Smargon et al., 2017). A study conducted by Tang et al. successfully utilized the CRISPR–Cas13 system for RNA editing in the silkworm (*Bombyx mori*), making it a pioneer study in lepidopteran insects. Stable BmE cell lines expressing PspCas13b and CasRx efficiently downregulated both exogenous and endogenous target genes. Transgenic silkworms expressing CasRx and guide RNAs (gRNAs) targeting the *Sex combs reduced* (*Scr*) gene exhibited reduced *Scr* transcription and impaired larval growth. These results confirm that CRISPR–Cas13 functions effectively in silkworms, offering a promising tool for RNA manipulation in lepidopterans (Tang et al., 2024a, 2024b).

22.3 Challenges of CRISPR in pest resistance

CRISPR can be used to develop insect-resistant crops, though it has a few hindrances. One of the most significant problems is off-target mutations. These mutations create undesirable traits or reduced plant performance (Cong et al., 2013; Fu et al., 2013; Hsu et al., 2013; Jiang et al., 2013; Mali et al., 2013; Pattanayak et al., 2013; Zhang et al., 2019). Another disadvantage is that in the case of large gene families, plants have functional redundancy, which means removing a single susceptibility (S) gene would not be sufficient to produce long-lasting resistance because related homologs might make up for its loss (Hong et al., 2020; Zaidi et al., 2018). Choosing an efficient gRNA is important, as targeting of the nuclease to a particular locus relies on the gRNA's ability (Wright et al., 2016). These create a hurdle to apply in crop protection (Hsu et al., 2013; Zetsche et al., 2017). To create stable and long-lasting insect resistance, these problems should be overcome. Artificial intelligence (AI) can fulfill these barriers.

22.4 The role of AI in accelerating crop development

AI technologies can be used to innovate agricultural technology. AI can be used in industries, healthcare, manufacturing, and agriculture due to the presence of data processing and analytical knowledge (Yang et al., 2024a, 2024b, 2024c). AI not only analyses genomic data but also, by using algorithms, can improve gene editing efficiency (Sun and Wang, 2024; Hao and Li, 2024). AI can identify genes related to agronomic traits, enhance CRISPR targets, and also reduce off-target effects. With the help of multiomics data, AI can disclose metabolic pathways and gene regulatory networks (Chen et al., 2025). AI can assist with the following tasks:

22.4.1 Genomic data analysis

AI algorithms include advanced large models, deep learning (DL), and machine learning (ML). In particular, nonlinear interactions between genotypes and phenotypes can be captured by ML technology (Danilevicz et al., 2022). DL is a division of ML that employs artificial neural networks to fetch data. It is useful to extract unstructured data. One significant DL method of processing complicated, large-scale, and high-resolution images is the convolutional neural network (CNN) (Sarker, 2021). The term "big model" or "foundation model" usually refers to DL models that contain a high number of parameters. It allows for executing and solving different types of domains and challenging tasks (Myers et al., 2024). In conjunction with AI and ML, resistance genes can be graded with the help of weighted gene coexpression network analysis (WGCNA) (Lei et al., 2025). This ability greatly enhances CRISPR to select targets, reducing off-target effects via the identification of genes that show agronomic traits (Sun and Wang, 2024). One study was done by Lei et al., where ML and WGCNA were used. ML and WGCNA investigated the genomic data for important genes associated with Verticillium wilt, a fungal disease that is a major pest of *Gossypium hirsutum*. AI's capacity for pest resistance is confirmed by WGCNA, as it constructs gene coexpression networks, and ML, as it grades the vital genes responsible for resistant traits linked to resistance features (Lei et al., 2025).

22.4.2 Identifying target genes

AI technologies identify target genes. Traditional gene target identification is a time-consuming and complex analytical method. It can easily and precisely identify target genes linked to certain traits using large-scale genomic data processing and analysis (Liu et al., 2024). For example, a work by Gan et al. (2025) utilizes DL/ML algorithms, transcriptomics, plant phenomics, and quantitative genetics to perform genomic selection (GS) of pest resistance in grapevines. With a 95.7% precision rate in GS, they discovered 69 quantitative trait loci (QTLs) and 139 candidate genes for grapevine insect resistance. This proves AI can be utilized to develop pest-resistant crops (Gan and others, 2025). It can perform genome-wide association studies (GWAS) to recognize genetic markers linked to specific agronomic traits. This analytical approach combines large-scale genomic and phenotypic data to precisely locate target genes, thereby effectively guiding gene editing efforts. AI's application greatly enhances the efficiency and accuracy of gene target identification, providing robust support for the development of gene editing technologies (Liu et al., 2024).

22.4.3 Prediction of the outcomes of genome editing

Three major outcomes of gene editing are off-target consequences, specific repair results, and on-target proficiency. Through analyzing a large amount of genomic data, DL models like DeepCRISPR may reduce off-target activity by 50% by reducing the chances of unnecessary cleavage at nontarget locations (Liu et al., 2020). When the Cas9 nuclease creates a double-strand break, ML tools like InDelphi and FORECasT (**FO**rmulas for **R**epair **E**utcomes of **Cas9** **T**argeting) can simulate the cellular repair mechanisms. This aids investigators to expect the precise insertion and deletion (indel) patterns and also regulate whether a gene will be altered exactly or unintentionally knocked out (Dunka, 2022). Another platform called TIGER (**T**argeted **I**nhibition of **G**ene **E**xpression via **R**NA design) uses DL to predict the on- and off-target activity of the RNA-targeting CRISPR–Cas13 system, thus enabling accurate control of gene expression levels (Wessels et al., 2023), which ultimately enhances the accuracy, effectiveness, and safety of editing methods to generate vigorous crops.

22.5 Application of AI to identify crop stress resistance genes

22.5.1 Gene mining and analysis using big data

Big data technology enables scientists to process and analyze large-scale genomic data, unraveling the intricate relationship between disease resistance and genes involved. The datasets use sophisticated analytical techniques and technical tools to extract valuable information. In the field of agriculture, big data technology performs multiple aspects, including data collection, storage, processing, analysis, and visualization. Utilizing this big-data technology for GWAS and other gene association analysis results in key gene and locus identification associated with resistance to diseases. This technology opens new channels to gene mining as well as functional studies (Hou et al., 2016; Chen et al., 2021). With the integration of AI and advanced bioinformatics, gene mining for genes linked to diseases has rapidly grown. Researchers can now efficiently identify the novel resistance (R) genes and also the mechanisms of resistance from the large genomic datasets and vast transcriptomes. Tools like DL-based PRGminer can specifically forecast the R-genes that are involved in the defense mechanisms of plants. They do this with great precision. All these can increase the understanding of the molecular basis of disease and pest resistance (Duhan, N., and Kaundal, 2025). With the help of big data technology, early warning and precise control of pests and diseases can be achieved. By analyzing historical pest and disease data, meteorological data, and crop growth data, trends in pest and disease outbreaks and potential high-risk areas can be predicted. Additionally, image recognition technology and ML algorithms can automatically identify crop disease types, quickly diagnose causes, and provide targeted control measures. This approach effectively reduces the damage caused by pests and diseases, decreases pesticide use, and protects the ecological environment (Shi et al., 2019; Zhu et al., 2025).

22.5.2 Prediction of gene functions

Genome annotation is to recognize the genomic sequences that encompass functional regions. ML can detect polyadenylation sites (Gao et al., 2018), long noncoding RNAs (Sun et al., 2015), microRNAs (miRNAs), and protein-coding genes. These improvements can recognize various stress-resistant genes by merging genomic, transcriptomic, phenotypic, and environmental data. Under stress conditions, the ML model can predict stress-related genes. This was done via evaluating gene expression, phenotypic traits, and environmental data under various stress conditions. Support vector machine (SVM), random forests, k-nearest neighbors, and neural networks are frequently applied as ML methods (Cen et al., 2020; Zhou et al., 2018). SVM and random forests can predict stress resistance genes. This creates an easier way to validate the experiment (Su et al., 2025). In recent years, recurrent neural networks (RNNs) and CNNs have demonstrated remarkable performance in gene sequence and expression profile analysis. For example, RNNs can note gene expression changes over time, whereas CNNs can uncover important characteristics of stress-related genes. RNN is, therefore, a vital tool to investigate gene expression under different environmental conditions. DL can also forecast the biological reinforcements of stress-related characteristics (Su et al., 2025).

22.5.3 Construction and analysis of gene interaction networks

AI, for example, ML and DL, can help to build and analyze gene interaction networks. Gene interaction networks are created by gene interactions within the cell (Guan et al., 2025; Chen et al., 2016). AI can find the "commanders" that control the expression of the downstream genes during stress conditions. This network requires gene expression data, as well as environmental reactions. AI can extract linkages between genes, their locations, and their functional interactions using several analytical methodologies and data processing (Su et al., 2025). AI can create gene interaction networks using network analysis techniques. These techniques are WGCNA, graph neural networks, and Bayesian networks. WGCNA can organize genes into distinct modules and also recognize hub genes within each module. These hub genes play a major biological role in these networks. AI organizes these hub genes to target stress-resistant genes (Su et al., 2025).

AI is capable of dynamic analysis through deep learning architectures like RNNs and long short-term memory (LSTM) networks, which process temporal linkages to investigate variations in gene expression over time. By capturing these sequential shifts, these algorithms successfully construct dynamic gene interaction networks during complex stress response processes. Researchers are better able to examine gene activation, regulatory patterns, and connections at various levels of stress thanks to these dynamic network models. This contributes to a better understanding of how genes coordinate their reactions to environmental stressors (Su et al., 2025) (Fig. 22.1).

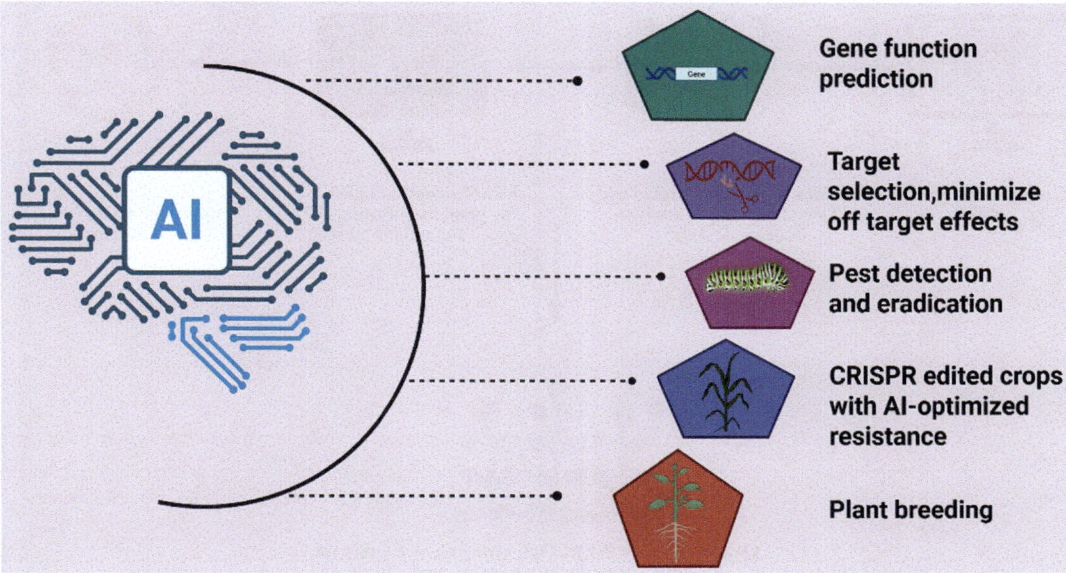

FIGURE 22.1 Applications of artificial intelligence (AI) in modern agriculture.

22.5.4 Identification and optimization of gene editing targets

22.5.4.1 AI optimization to select CRISPR target

AI can select a CRISPR target in gene editing technology to enhance efficiency. Traditional gRNA design includes laboratory trials and experience, which is a time-consuming process. By intercalating ML and DL algorithms, AI technology processes and analyzes vast amounts of genomic data to automatically recognize the most effective gRNA targets, thereby improving operational efficiency. AI models, like big data, can predict potential target sites' editing efficiency and also off-target risk (Guo et al., 2022; Tang et al., 2024a, 2024b). For instance, CNNs identify target selection by scanning the whole genome to recognize gRNA sequences that match the target gene. AI can reduce off-target effects, therefore enhancing gene editing. Also, AI can imitate the binding pattern between gRNA and the target sequence, which ensures the success rate and precision of gene editing. Overall, AI has great potential to optimize CRISPR target selection by enhancing editing efficiency, reducing off-target risks, and improving the accuracy of target identification (Chen et al., 2025) (Fig. 22.2).

In practical applications, AI can help breeders select crop varieties adapted to a particular environment. For example, AI models can predict crop performance under extreme conditions such as high temperatures, drought, or high humidity. Moreover, AI can also make agricultural management decisions, like irrigation, fertilization, and pest control strategies. With the assistance of AI technology, farmers and agricultural managers can make more accurate decisions, improve resource use efficiency, and reduce environmental costs, thereby driving progress toward sustainable agricultural development (Chen et al., 2025).

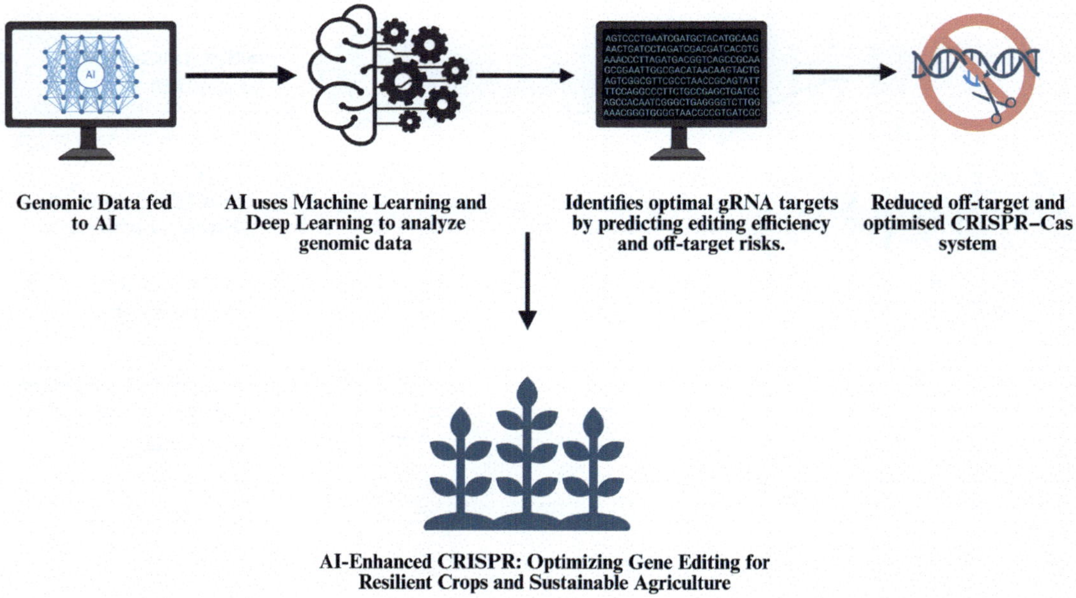

FIGURE 22.2 AI-enhanced CRISPR: Optimizing gene editing for resilient crops and sustainable agriculture.

22.6 CRISPR-edited crops with AI-optimized resistance

With the help of AI, a genetic sketch for an insecticidal maize plant is possible. This could be developed using new genomic techniques (NGTs). NGTs are defined as "techniques that are able to alter the genetic material of an organism, developed after the publication of EU Directive 2001/18/EC. NGTs can add sequence variations to the genome of an organism (a process called "genome editing"). Such variations may be entirely novel and may or may not occur already in other individuals of the species. NGTs may also be used to introduce new sequences derived from other species into an organism (Broothaerts et al., 2021). ChatGPT was used to design this (Testbiotech report / background).

The basic idea is to increase the gene that encodes serine protease inhibitor (SPI) of maize. This induces insect toxicity. SPIs are naturally occurring proteins that protect plants against pathogens. Upon pest infection, plants' natural defense responses are increased for a brief period of time (Chen et al., 2024a). When insects feed on these plants, these SPI proteins inhibit the insects from digesting plant material in the insect gut. This delays insect development and increases the risk of mortality. This can be seen in some lepidopteran insects. Some species of this order, like *Ostrinia* spp., are pest insects in maize (Chen et al., 2024b). Ongoing research is to increase the content of SPI in plants. Plants with an increased level of SPI have already been produced (Clemente et al., 2019). In this case, the target is to produce AI-designed insecticidal NGT maize with an increasing level of SPI without relocating any genes from other species.

22.6.1 Development of insecticidal NGT (new genomic technique) maize plants with inferences

In a first attempt, the suggestion of the AI was to replace the promoter that gives permanent expression instead of transient, and it will be a part of the maize plant. These plants were considered as cisgenic as the promoter belongs to the same species. In subsequent efforts by base editing, a few point mutations were introduced into the native promoter of the SPI gene. Cis-regulatory elements (CREs), which are a part of the native promoter, inhibit the expression of the SPI gene. The "knock-out" CRE can be presumed to increase expression of these genes (Testbiotech, 2025). Specific unification of several knock-outs may be needed to magnify the expression of knocking out repressive CRE (Xue et al., 2023). The role of AI was to select target gene sequences using the "deep search" mode of ChatGPT 4o.

22.6.2 Details of ChatGPT 4o in "deep research" modes

22.6.2.1 Development of NGT 1 plants with insecticidal properties

These NGT plants may not only be toxic to the targeted pest species, but also can pose serious risks to non-target insects, food webs, ecosystem functions and biodiversity. Under current EU legislative proposals, such plants could be considered as NGT-1 plants and could be released into the environment and marketed without prior environmental risk assessment.

To develop resistance against Lepidoptera, one approach was developed in NGT maize plants with insecticidal properties. Both classical and cis-regulatory genome editing techniques were considered, and also resistance factors were checked within the breeding pool. ChatGPT made an abstract of objectives, followed by a classification of the proposed project via classification related to the NGT1 criteria. It also suggested defending against insect pests via endogenous defense mechanisms of maize. Also recommended some genes to increase plant resistance to lepidoptera. For example, AI suggested insecticidal SPI genes. Several methods appeared in the program to enhance *SPI* gene expression. These approaches include exchanging the native SPI promoter with stronger or constitutive endogenous promoters ("Promoter-Swap") or inactivation of "repressors" like CREs of the *SPI* gene.

22.6.2.1.1 "Promoter-Swap": to create and design gRNA

Development of insecticidal NGT maize plant with improved *SPI* gene with increased expression in above-ground parts by replacing the native SPI promoter with either a string or a constitutive promoter. To develop a constitutive promoter, by the CRISPR/Cas9 technique, in "promoter swap" approach, a repair template from the donor ("Promoter-Swap" Cassette) to generate homology-directed repair (HDR), suitable "guide" RNAs and corresponding Cas9 cleavage sites were given.

22.6.2.1.2 Inactivation of CRE

CREs of the *SPI* gene of maize were identified. AI proposes point mutations via CRISPR–Cas9 to knock out repressive regulatory elements to give point mutations that increase *SPI* expression. Several maize "repressory" CREs of the *SPI* gene were identified as successful targets for inactivation. One of these target sites was chosen, as it contains binding sites for two different repressive transcription factors and the start codon of an upstream open reading frame (uORF). CRISPR/Cas (coupled with a deaminase) mutation was recommended to inactivate these three CREs.

Also, another approach was given to identify individual nucleotides of similar CREs, appropriate base editors, and suitable gRNAs. All recommended genomic sequences, binding sites, gRNAs, cleavage sites, etc., were further checked manually using NCBI (https://www.ncbi.nlm.nih.gov/gene), databases of the maize genome (https://maizegdb.org/), BLAST (https://blast.ncbi.nlm.nih.gov/Blast.cgi), and Clustal Omega (https://www.ebi.ac.uk/jdispatcher/msa/clustalo?stype=dna) to align sequences.

22.7 Multitrait integrated breeding technology with the assistance of AI

22.7.1 Multitrait selection and genome editing

Crop improvement increasingly demands the simultaneous enhancement of multiple traits, including yield, stress tolerance, nutrient efficiency, and pest resistance. Conventional breeding approaches face limitations due to the polygenic nature of these traits and issues such as linkage drag. Genome editing, especially CRISPR–Cas systems, has opened new possibilities for multitrait selection by enabling precise and simultaneous modifications at multiple loci (Su et al., 2025). Multitrait breeding is a selective breeding approach that aims to improve multiple traits simultaneously within a population of plants or animals. Contrasting with single-trait selection, which targets only one characteristic, multitrait breeding includes various traits such as yield, disease resistance, quality, adaptability, and environmental sustainability (Tollenaar and Lee, 2002). AI is an emerging technological innovation in multitrait breeding. AI significantly enhances breeding efficiency and improvement with the integration of multitrait selection (Yang et al., 2024a; 2024b; 2024c; Wang and Liu, 2024). As compared to traditional breeding, AI gives a scientific foundation for gene editing and also reduces time. For example, AI can help design accurate CRISPR/Cas9 targets minimizes off-target effects. With the help of AI, researchers can divine the synergistic effects and trait outcomes of multigene edits before performing experiments, thus optimizing experimental design. This reduces unnecessary trial phases and improves editing efficiency and accuracy (Su et al., 2025). Overall, AI-driven multitrait selection and genome editing enhance breeding precision and efficiency as well as ensure stress resistance improvements according to environmental and market needs (Su et al., 2025).

22.7.2 Multitrait breeding approaches in complex environments

Different types of stresses, like drought, salinity, high temperature, low temperature, and pest infections, have a great effect on plant growth and yield. These factors create difficulty and challenges in plant breeding. Multitrait breeding strategies based on AI have advantages in these cases. AI can rapidly integrate and analyze large-scale environmental, genomic, and phenotypic data to support precise breeding programs. Using ML and DL models, it can extract key genomic regions from complex datasets and also identify the most adaptive gene combinations under specific environmental conditions. This information allows theoretical and technical support for multitrait stress-resistant breeding (Wang et al., 2022; Liang, 2019).

Under extreme environmental conditions, AI-assisted breeding has a great effect. For example, an AI model can predict dynamic changes in gene expression under changing environments where low and high temperatures oscillate. AI with enhanced genomic improvement strategies by combining the effects of different environmental stresses. It improves crop adaptability and crop stability using these multitrait breeding practices. Overall, AI-supported multitrait breeding helps to understand complex environments and crop traits (Su et al., 2025).

22.8 Artificial intelligence applications and global efforts

22.8.1 Plantix

It is a movable crop application created by PEATit, which is a Berlin-based start-up. It can recognize pest damages, diseases, and nutrient deficiencies of different crops. It also gives recommendations for their control. Farmers can obtain knowledge about local weather and good agricultural assistance throughout the season and also acquire disease alerts when a disease is multiplying in their surroundings (Rahman and Ravi, 2022).

22.8.2 To monitor Fall Armyworm and early warning using the Nuru app

Fall Armyworm (FAW) first appeared in Africa in 2016. Many African farmers have heard about FAW, but they were incapable of recognizing it (Rahman and Ravi, 2022). FAMEWS (Fall Armyworm Monitoring and Early Warning Sign) and Nuru apps are important parts of FAO's sustainable management of the FAW program. They give information to farmers and extension workers so that they can eradicate pests and also give decision-makers' response that builds knowledge about pests to secure food (Bouroubi et al., 2018; Gao et al., 2019; Leafbyte, 2019; Apizoom, 2016; Bioleaf, 2017).

22.8.3 AI to save bees (via the Apizoom app)

This app is designed to manage colonies. Its slender and modern interface helps to track the Varroa's evolution. Features of the Apizoom app include controlling the apiaries and also precise actions like scanning, treatment, and laying boards. We can get results from a specific hive or in graphical form for all. Results can be extracted in CSV format (Rahman and Ravi, 2022).

22.8.4 Leaf byte app

It is a free and open-source mobile app to measure herbivory. It gives quick and accurate results. It measures total leaf area, consumed leaf area, and per cent consumed. It automatically saves data to Google Drive or on the phone. It also reads barcodes and saves them with data. It can be used nondestructively in the field (Rahman and Ravi, 2022).

22.9 Successful usage of AI in pest management

22.9.1 AI for Bharat (India)

Over 250 million people worldwide rely on cotton cultivation and processing as a livelihood. Around 99% of global farmers across 70 countries are small farm holders. In India, Telangana, Maharashtra, and Gujarat are the top cotton-producing states. It helps farmers to acquire knowledge about pink bollworm infestation in cotton fields so that they can provide protection. Professor Jayashankar Telangana State Agricultural University (PJTSAU) has manifested a memorandum of understanding (MoU) with Wadhwani Institute for Artificial Learning, Mumbai, to give AI-based interventions through the Cotton Ace application (Stevens Institute, 2020; SF Magazine, 2020; Optix, 2021; PJTSAU, 2021; Plantix, 2020; Wadhwani AI, 2020). As part of the agreement, the AI-based Cotton Ace app would identify the possibility of pink bollworm on cotton in advance and also distribute the information to farmers in the areas dedicated to the program. Therefore, farmers could take action early. CottonAce is an AI-powered early warning system accessible via an app on Android smartphones. It helps farmers to protect their crops by ascertaining the right time to spray pesticides through instant, localized advice (Rahman and Ravi, 2022).

22.9.2 Pest detection and identification

AI-powered drones are provided that include high-resolution cameras and multispectral, and hyperspectral sensors, which can collect a broad range of data from the crop environment. To process this data, computer vision and DL algorithms are applied. This permits the drone to identify and classify pest species with great precision. AI can learn from background data and improve its accuracy over time, further enhancing its pest detection capabilities (Verma, 2025).

For example, AI can be instructed to identify particular pests according to their shape, color, and movement patterns, also in complex agricultural environments. This potentiality allows for early pest detection, before they reach harmful levels, and also enables protective measures to be applied promptly (Verma, 2025).

22.9.3 Real-time data processing and decision-making

The major advantages of AI are the self-detected drone systems that process real-time data. Using algorithms, it can interpret the data while flying. It also determines the places where pest control must be implemented. An AI-driven drone can decide the ideal time and place for pest interference with the help of real-time data analysis in conjunction with other environmental data, for example, weather patterns, soil conditions, and crop health (Verma, 2025).

AI also surveys the pest distribution across a field and addresses their rate of growth. This information helps drones to analyze pest spread and find the target area susceptible to pest invasion. This confirms timely and cost-efficient pest control without the application of pesticides, thus minimizing ecological effects (Verma, 2025).

22.9.4 Pest control

AI-driven drones accurately control pests when their location is identified. The drones can apply pesticides, biopesticides, or even biological control agents like predators or pathogens directly to the infected areas. With the help of AI algorithms, the drones can measure the accurate quantity of pesticide or treatment needed. This level of clarity greatly decreases the amount of chemicals used, limits pesticide runoff, and reduces the likelihood of pest resistance. AI drones can be designed to release beneficial organisms, such as parasitoids, that target specific pests (Verma, 2025).

22.10 Conclusion

Integration of AI and CRISPR-mediated genome editing made a revolutionary change in the field of agriculture, as well as a sustainable pest management system. CRISPR delivers precision to introduce pest-resistant traits as well as suppresses the plant's own susceptibility genes via changing biochemical, morphological, and molecular pathways of the plant. It can also improve crop resilience by introducing the resistance genes that can offer immunity to pests. These properties can be detected by AI. AI also detects the resistance genes and can predict functional effects of these modifications with the help of big data analysis, gene function, and interaction network building. Integrating both AI and CRISPR can enhance pest resistance in crops. It can also open new paths to multitrait, stress-resistant crops with higher yields that need less or insignificant use of pesticides, thus preventing ecosystem damage. Though the CRISPR/Cas system has various drawbacks, such as off-target effects, functional redundancy in gene families, and choosing an efficient gRNA. These problems can be solved by the combined effects of these two techniques. Using AI, many international initiatives resolve pest management problems, which is a great concern for food security. Integration of these ground-breaking technologies in the agriculture field can make pest control more sustainable, eco-friendly, and targeted while also assuring long-term crop resilience against changing pest pressures.

Acknowledgments

PCV is thankful to the Director, CSIR-National Botanical Research Institute, Lucknow (U.P.), India, for the necessary support and facilities. Figures were created with BioRender.com.

Institutional review board statement

Approved for publication. Institutional Manuscript ID no. **CSIR-NBRI_MS/2026/05/10.**

Data availability statement

Not applicable.

Conflict of interest

The authors have no conflict of interest to declare.

Declaration

The authors declare that no generative artificial intelligence tools were used in the writing or preparation of this chapter. The content reflects the authors' own scholarly work, and the authors assume full responsibility for all statements, interpretations, and conclusions presented.

References

Abudayyeh, O.O., Gootenberg, J.S., Konermann, S., et al., 2016. C2c2 is a single-component programmable RNA-guided RNA-targeting CRISPR effector. Science 353 (6299), aaf5573.

Andolfo, G., Iovieno, P., Frusciante, L., Ercolano, M.R., 2016. Genome-editing technologies for enhancing plant disease resistance. Front. Plant Sci. 7, 1813.

Apizoom, 2016. https://www.apizoom.app/.

Ayelo, P.M., Yusuf, A.A., Pirk, C.W., Chailleux, A., Mohamed, S.A., Deletre, E., 2021. Terpenes from herbivore-induced tomato plant volatiles attract *Nesidiocoris tenuis* (Hemiptera: Miridae), a predator of major tomato pests. Pest Manag. Sci. 77 (11), 5255–5267.

Bassett, A.R., Tibbit, C., Ponting, C.P., Liu, J.L., 2013. Highly efficient targeted mutagenesis of *Drosophila* with the CRISPR/Cas9 system. Cell Rep. 4 (1), 220–228.

Bhowmik, P.K., Islam, M.T., 2020. CRISPR-Cas9-mediated gene editing in wheat: a step-by-step protocol. In: Islam, M.T., Bhowmik, P.K., Molla, K.A. (Eds.), CRISPR-cas Methods, vol. 1. Humana Press, pp. 203–222.

Bioleaf, 2017. https://www.quantitative-plant.org/software/bioleaf.

Birkett, M.A., Pickett, J.A., 2014. Prospects of genetic engineering for robust insect resistance. Curr. Opin. Plant Biol. 19, 59–67.

Bouroubi, Y., Bugnet, P., Nguyen-Xuan, T., Gosselin, C., Bélec, C., Longchamps, L., Vigneault, P., 2018. Pest detection on UAV imagery using a deep convolutional neural network. In: Proceedings of the International Society of Precision Agriculture, pp. 1–11.

Broothaerts, W., Jacchia, S., Angers, A., et al., 2021. New Genomic Techniques: State-of-the-Art Review (EUR 30430 EN). Publications Office of the European Union.

Cen, H., Zhu, Y., Sun, D., others, 2020. Application status and prospects of deep learning in plant phenotyping research. Trans. Chin. Soc. Agric. Eng. 36 (9), 1–16.

Chen, G., Hao, F., Sun, X., 2025. Artificial intelligence-driven gene editing and crop breeding: technological innovations and application prospects. Advances in Resources Research 5 (1), 235–254.

Chen, L., Duan, D., Xu, W., others, 2024a. Application and challenges of ground and aerial scale high-throughput phenotyping analysis in crops. Guangdong Agric. Sci. 51 (11), 149–162.

Chen, L., Fang, W., Si, H., others, 2016. Research on ontology construction for crop germplasm resources. Acta Agron. Sin. 42 (3), 407–414.

Chen, R., Peng, S., Wang, D., others, 2021. Genome-wide association analysis of rice blast resistance genes at the seedling stage. Acta Laser Biol. Sin. 30 (2), 163–169.

Chen, Y., Yang, S., Zeng, W., Zheng, X., Wang, P., Fu, H., Yang, F., 2024b. Salicylic acid inducing the expression of maize anti-insect gene SPI: a potential control strategy for *Ostrinia furnacalis*. BMC Plant Biol. 24 (1), 152.

Clemente, M., Corigliano, M.G., Pariani, S.A., Sánchez-López, E.F., Sander, V.A., Ramos-Duarte, V.A., 2019. Plant serine protease inhibitors: biotechnology application in agriculture and molecular farming. Int. J. Mol. Sci. 20 (6), 1345.

Cong, L., Ran, F.A., Cox, D., Lin, S., Barretto, R., Habib, N., others, 2013. Multiplex genome engineering using CRISPR/Cas systems. Science 339 (6121), 819–823.

Cox, D.B.T., Gootenberg, J.S., Abudayyeh, O.O., et al., 2017. RNA editing with CRISPR-Cas13. Science 358 (6366), 1019–1027.

Danilevicz, M.F., Gill, M., Anderson, R., Batley, J., Bennamoun, M., Bayer, P.E., Edwards, D., 2022. Plant genotype to phenotype prediction using machine learning. Front. Genet. 13, 822173.

Douglas, A.E., 2018. Strategies for enhanced crop resistance to insect pests. Annu. Rev. Plant Biol. 69, 637–660.

Duhan, N., Kaundal, R., 2025. PRGminer: harnessing deep learning for the prediction of resistance genes involved in plant defense mechanisms. Front. Plant Sci. 16, 1411525.

Dunka, V.K., 2022. AI-powered computational biology for genome editing: Developing deep learning models for CRISPR-Cas9 target site prediction, off-target effect minimization, and therapeutic efficacy enhancement. Artif. Intell. Mach. Learn. Autonom. Syst. 6 (1), 1–17.

El-Dessouki, W.A., Mansour, M.R.K., Eryan, N.L., 2022. Effects of certain weather, biotic factors and chemical components on the population of aphids in Egyptian wheat fields. Egyptian Academic Journal of Biological Sciences, A. Entomology 15 (1), 1–13.

Esvelt, K.M., Smidler, A.L., Catteruccia, F., Church, G.M., 2014. Concerning RNA-guided gene drives for the alteration of wild populations. Elife 3, e03401.

Food and Agriculture Organization of the United Nations, 2021. Climate Change Fans Spread of Pests and Threats Plants and Crops, New FAO Study. PreventionWeb. https://www.preventionweb.net/news/climate-change-fans-spread-pests-and-threats-plants-and-crops-new-fao-study.

Food and Agriculture Organization of the United Nations, 2023. Upto 40 Percent of Global Crop Production Is Lost Due to Pests and Diseases Every Year. FAO. https://www.fao.org/newsroom/detail/fao-puts-focus-on-importance-of-technology-for-international-day-of-plant-health/en.

Fu, Y., Foden, J.A., Khayter, C., Maeder, M.L., Reyon, D., Joung, J.K., others, 2013. High-frequency off-target mutagenesis induced by CRISPR-Cas nucleases in human cells. Nat. Biotechnol. 31 (9), 822–826.

Gan, Y., Liu, Z., Zhang, F., Xu, Q., Wang, X., Xue, H., Zhou, Y., 2025. Deep learning empowers genomic selection of pest-resistant grapevine. Hortic. Res. 12 (1), uhaf128.

Gao, P., Zhang, Y., Zhang, L., Noguchi, R., Ahamed, T., 2019. Development of a recognition system for spraying areas from unmanned aerial vehicles using a machine learning approach. Sensors 19 (2), 313.

Gantz, V.M., Bier, E., 2015. The mutagenic chain reaction: a method for converting heterozygous to homozygous mutations. Science 348 (6233), 442–444.

Gao, X., Zhang, J., Wei, Z., Hakonarson, H., 2018. DeepPolyA: a convolutional neural network approach for polyadenylation site prediction. IEEE Access 6, 24340–24349.

Guan, Z., Shan, Z., Li, T., others, 2025. Research on soybean breeding knowledge model and construction. Journal of Agricultural Library and Information Science 36 (11), 79–91.

Guo, W., Li, D., Xie, S., others, 2022. Artificial intelligence accelerates crop epigenetic design breeding. Bulletin of Chinese Academy of Agricultural Sciences 24 (12), 90–100.

Hao, J., Li, X., 2024. Future development directions and prospects of big data analysis in crop molecular breeding and intelligent breeding technologies. Geographical Research Bulletin 3, 259–262.

Hong, Y., others, 2020. CAFRI-Rice: CRISPR applicable functional redundancy inspector to accelerate functional genomics in rice. Plant J. 104 (6), 1771–1781.

Horgan, F.G., Garcia, C.P.F., Haverkort, F., de Jong, P.W., Ferrater, J.B., 2020. Changes in insecticide resistance and host range performance of planthoppers artificially selected to feed on resistant rice. Crop Prot. 127, 104963.

Hou, X., Wang, Y., Peng, S., et al., 2016. Dissection of the genetic architecture of rice resistance to the blast fungus Magnaporthe oryzae. Mol. Plant Pathol. 17 (6), 959–972.

Hsu, P.D., Scott, D.A., Weinstein, J.A., Ran, F.A., Konermann, S., Agarwala, V., Zhang, F., 2013. DNA targeting specificity of RNA-guided Cas9 nucleases. Nat. Biotechnol. 31 (9), 827–832.

Islam, T., 2019. CRISPR-Cas technology in modifying food crops. CAB Reviews 14, 1–16.

Jaiswal, D.K., Raju, S.V.S., Kumar, G.S., Sharma, K.R., Singh, D.K., Vennela, P.R., 2018. Biotechnology in plant resistance to insects. Indian Journal of Agricultural and Allied Sciences 4, 7–18.

Jiang, W., Bikard, D., Cox, D., Zhang, F., Marraffini, L.A., 2013. RNA-guided editing of bacterial genomes using CRISPR-Cas systems. Nat. Biotechnol. 31 (3), 233–239.

Kim, S.Y., Bengtsson, T., Olsson, N., 2020. Mutation in two aphid-regulated β-1,3-glucanase genes by CRISPR/Cas9 does not increase barley resistance to *Rhopalosiphum padi* L. Front. Plant Sci. 11, 1043.

Kuai, P., Lin, N., Ye, M., Ye, M., Chen, L., Chen, S., Zu, H., Hu, L., Gatehouse, A.M.R., Lou, Y., 2024. Identification and knockout of a herbivore susceptibility gene enhances planthopper resistance and increases rice yield. Nat. Food 5 (10), 846–859.

Kyrou, K., Hammond, A.M., Galizi, R., Kranjc, N., Burt, A., Beaghton, A.K., Nolan, T., Crisanti, A., 2018. A CRISPR–Cas9 gene drive targeting doublesex causes complete population suppression in caged Anopheles gambiae mosquitoes. Nat. Biotechnol. 36 (11), 1062–1066.

Larsson, M.C., Domingos, A.I., Jones, W.D., Chiappe, M.E., Amrein, H., Vosshall, L.B., 2004. Or83b encodes a broadly expressed odorant receptor essential for *Drosophila* olfaction. Neuron 43 (5), 703–714.

Leafbyte, 2019. https://zoegp.science/leafbytePest.

Lei, Y., Zhao, J., Hou, S., Xu, F., Zhang, C., Cai, D., Zhao, S., 2025. Integrative identification of key genes governing Verticillium wilt resistance in *Gossypium hirsutum* using machine learning and WGCNA. Front. Plant Sci. 16, 1621604.

Li, R., Maioli, A., Yan, Z., Bai, Y., Valentino, D., Milani, A.M., others, 2022a. CRISPR/Cas9-based knock-out of the PMR4 gene reduces susceptibility to late blight in two tomato cultivars. Int. J. Mol. Sci. 23 (23), 14542.

Li, X., Hu, D., Cai, L., Wang, H., Liu, X., Du, H., Yang, Z., Zhang, H., Hu, Z., Huang, F., others, 2022b. CALCIUM DEPENDENT PROTEIN KINASE38 regulates flowering time and common cutworm resistance in soybean. Plant Physiol. 190 (1), 480–499.

Liang, C., 2019. From crop genome analysis to integrated omics knowledge base construction. Hereditas 41 (9), 875–882.

Liu, B., Jiang, H., Li, L., 2024. The innovative applications and prospects of artificial intelligence in agricultural research: gene editing, crop improvement, and intelligent data analysis. Advances in Resources Research 4 (4), 624–653.

Liu, G., Zhang, Y., Zhang, T., 2020. Computational approaches for effective CRISPR guide RNA design and evaluation. Comput. Struct. Biotechnol. J. 18, 35–44.

Lu, H.P., Luo, T., Fu, H.W., Wang, L., Tan, Y.Y., Huang, J.Z., Wang, Q., Ye, G.Y., Gatehouse, A.M.R., Lou, Y.G., others, 2018. Resistance of rice to insect pests mediated by suppression of serotonin biosynthesis. Nat. Plants 4 (6), 338–344.

SF Magazine, 2020. https://sfmagazine.com/technotes/august-2020-bug-traps-with-ai-and-wi-fi/.

Mahas, A., Aman, R., Mahfouz, M., 2019. CRISPR-Cas13d mediates robust RNA virus interference in plants. Genome Biol. 20 (1), 263.

Mali, P., Yang, L., Esvelt, K.M., Aach, J., Guell, M., DiCarlo, J.E., others, 2013. RNA-guided human genome engineering via Cas9. Science 339 (6121), 823–826.

Malone, L.A., Barraclough, E.I., Lin-Wang, K., Stevenson, D.E., Allan, A.C., 2009. Effects of red-leaved transgenic tobacco expressing a MYB transcription factor on two herbivorous insects, *Spodoptera litura* and *Helicoverpa armigera*. Entomol. Exp. Appl. 133 (2), 117–127.

Myers, D., Mohawesh, R., Chellaboina, V.I., Sathvik, A.L., Venkatesh, P., Ho, Y.H., Jararweh, Y., 2024. Foundation and large language models: fundamentals, challenges, opportunities, and social impacts. Clust. Comput. 27 (1), 1–26.

PJTSAU, 2021. https://pjtsau.edu.in/mou-wadhani-doa-pjtsau-2021.html.

National Academies of Sciences, Medicine, Division on Earth, Life Studies, Board on Life Sciences, Committee on Gene Drive Research in Non-Human Organisms, & Recommendations for Responsible Conduct, 2016. Gene drives on the horizon: advancing science, navigating uncertainty, and aligning research with public values.

Optix, 2021. https://pestoptix.com/how-it-works/.

Pattanayak, V., Lin, S., Guilinger, J.P., Ma, E., Doudna, J.A., Liu, D.R., 2013. High-throughput profiling of off-target DNA cleavage reveals RNA-programmed Cas9 nuclease specificity. Nat. Biotechnol. 31 (9), 839–843.

Peng, L., Zhao, Y., Wang, H., Song, C., Shangguan, X., Ma, Y., Zhu, L., He, G., 2017. Functional study of cytochrome P450 enzymes from the brown planthopper (Nilaparvata lugens Stål) to analyze its adaptation to BPH-resistant rice. Front. Physiol. 8, 972.

Plantix, 2020. https://plantix.net/en/.

Rahman, S.M., Ravi, G., 2022. Role of artificial intelligence in pest management. Current Topics in Agricultural Sciences 7, 64–81.

Rathinam, M., Mishra, P., Mahato, A.K., Singh, N.K., Rao, U., Sreevathsa, R., 2019. Comparative transcriptome analyses provide novel insights into the differential response of pigeonpea (*Cajanus cajan* L.) and its wild relative (*Cajanus platycarpus* (Benth.) Maesen) to herbivory by *Helicoverpa armigera* (Hübner). Plant Mol. Biol. 101 (1–2), 163–182.

Sarker, I.H., 2021. Deep learning: a comprehensive overview on techniques, taxonomy, applications and research directions. SN Comput. Sci. 2 (6), 1–20.

Shi, D., Gao, D., Xue, W., others, 2019. Agricultural pest monitoring technology based on the Internet of Things and big data. J. Nanjing Agric. Univ. 42 (5), 967–974.

Singh, S., Rahangdale, S., Pandita, S., Saxena, G., Upadhyay, S.K., Mishra, G., Verma, P.C., 2022. CRISPR/Cas9 for insect pests management: a comprehensive review of advances and applications. Agriculture 12 (11), 1896.

Smargon, A.A., Cox, D.B.T., Pyzocha, N.K., et al., 2017. Cas13b is a type VI-B CRISPR-associated RNA-guided RNase differentially regulated by accessory proteins Csx27 and Csx28. Mol. Cell 65 (4), 618–630.

Stevens Institute, 2020. https://www.stevens.edu/news/pest-control-how-new-ai-driven-sensing-technologycould-help-save-our-forests.

Su, X., Xu, J., Li, C., Wei, M., 2025. The research progress on crop stress resistance and environmental adaptability improvement based on artificial intelligence. Adv. Resour. Res. 5 (3), 1589–1610. https://doi.org/10.50908/arr.5.3_1589.

Sun, D., Guo, Z., Liu, Y., Zhang, Y., 2017. Progress and prospects of CRISPR/Cas systems in insects and other arthropods. Front. Physiol. 8, 608.

Sun, L., Liu, H., Zhang, L., Meng, J., 2015. lncRScan-SVM: a tool for predicting long non-coding RNAs using support vector machine. PLoS One 10 (10), e0139654.

Sun, P., Han, X., Milne, R.J., Li, G., 2024. Trans-crop applications of atypical R genes for multipathogen resistance. Trends Plant Sci. 29 (10), 1103–1112.

Sun, W., Wang, S., 2024. The future research directions of artificial intelligence in crop growth prediction. Geographical Research Bulletin 3, 251–254.

Tang, D., Zhu, Y., Hu, X., 2024a. The AI-driven potato gene editing: a new perspective on target identification. Geographical Research Bulletin 3, 370–372.

Tang, Y.H., Zhang, X., Dai, Z.C., Li, H., Yang, Y., Zhao, T.J., Cheng, D.J., 2024b. CRISPR-Cas13-mediated RNA editing in the silkworm Bombyx mori. Zool. Res. 45 (6), 1249.

Taning, C.N.T., Eynde, B.V., Yu, N., Ma, S., Smagghe, G., 2017. CRISPR/Cas9 in insects: applications, best practices and biosafety concerns. J. Insect Physiol. 98, 245–257.

Testbiotech, 2025. 'Technical-tuned' NGT plants show: there is no magic threshold for the number of mutations to conclude on risks EU plans for the future legal framework for NGT plants need to be corrected. https://www.testbiotech.org/wp-content/uploads/2025/01/Technical-tuning-NGT-plants_v2.pdf.

Tollenaar, M., Lee, E.A., 2002. Yield potential, yield stability and stress tolerance in maize. Field Crops Res. 75 (2–3), 161–169.

Vanti, G.L., Katageri, I.S., Inamdar, S.R., Hiremathada, V., Swamy, B.M., 2018. Potent insect gut binding lectin from *Sclerotium rolfsii* imparts resistance to sucking and chewing type insects in cotton. J. Biotechnol. 278, 20–27.

Vats, S., Kumawat, S., Kumar, V., Patil, G.B., Joshi, T., Sonah, H., Sharma, T.R., Deshmukh, R., 2019. Genome editing in plants: exploration of technological advancements and challenges. Cells 8 (11), 1386.

Verma, S.K., 2025. Exploring the role of AI in autonomous drone technology for precision pest control in agriculture. International Research Journal of Engineering and Applied Sciences 13 (1).

Wadhwani AI, 2020. https://www.wadhwaniai.org/programs/pest-management/.

Wang, H., Lai, J., Wang, H., others, 2022. Crop intelligent design breeding: intelligent combination of natural variation and intelligent creation of artificial variation. Bulletin of Chinese Academy of Agricultural Sciences 24 (6), 1–8.

Wang, Z., Liu, J., 2024. Speciation research in the genomic era. Hereditas 47 (1), 71–100.

Wani, S.H., Choudhary, M., Barmukh, R., Bagaria, P.K., Samantara, K., Razzaq, A., Varshney, R.K., 2022. Molecular mechanisms, genetic mapping, and genome editing for insect pest resistance in field crops. Theor. Appl. Genet. 135 (11), 3875–3895.

Wessels, H.-H., others, 2023. Prediction of on-target and off-target activity of CRISPR–Cas13d guide RNAs using deep learning. Nat. Biotechnol. 41 (9), 1273–1280.

Wright, A.V., Nunez, J.K., Doudna, J.A., 2016. Biology and applications of CRISPR systems: harnessing nature's toolbox for genome engineering. Cell 164 (1–2), 29–44.

Xue, C., Qiu, F., Wang, Y., Li, B., Zhao, K.T., Chen, K., Gao, C., 2023. Tuning plant phenotypes by precise, graded downregulation of gene expression. Nat. Biotechnol. 41 (12), 1758–1764.

Yang, J., Wu, B., Wang, A., others, 2024a. Prediction of corn agronomic and quality traits based on multidimensional omics data. Acta Crops Sinica 50 (2), 373–382.

Yang, J., Yang, C., Zeng, Y., others, 2024b. Research progress on mining and utilization of blast resistance genes in rice. Chin. J. Rice Sci. 38 (6), 591–603.

Yang, Z., Zhao, X., Wang, C., 2024c. The future of crop molecular breeding and intelligent technologies: towards precision, personalization, and sustainable development. Geographical Research Bulletin 3, 247–250.

Zaidi, S.S.E.A., Mukhtar, M.S., Mansoor, S., 2018. Genome editing: targeting susceptibility genes for plant disease resistance. Trends Biotechnol. 36 (9), 898–906.

Zetsche, B., Heidenreich, M., Mohanraju, P., Fedorova, I., Kneppers, J., DeGennaro, E.M., Zhang, F., 2017. Multiplex gene editing by CRISPR–Cpf1 using a single crRNA array. Nat. Biotechnol. 35 (1), 31–34.

Zhang, Y., Guo, W., Chen, L., Shen, X., Yang, H., Fang, Y., Ouyang, W., Mai, S., Chen, H., Chen, S., others, 2022. CRISPR/Cas9-mediated targeted mutagenesis of GmUGT enhanced soybean resistance against leaf-chewing insects through flavonoids biosynthesis. Front. Plant Sci. 13, 802716.

Zhang, Y., Malzahn, A.A., Sretenovic, S., et al., 2019. The emerging and diverse applications of CRISPR-Cas toolkits in plant genome engineering. Nat. Plant. 5 (2), 135–145.

Zhou, J., Cheng, T., Zhu, Y., others, 2018. Plant phenomics: development, status, and challenges. J. Nanjing Agric. Univ. 41 (4), 580–588.

Zhu, L., Yu, S., Jin, Q., 2025. A review on the mining and functional research of rice disease resistance genes based on big data technology. Advances in Resources Research 5 (2), 793–816.

PART III

Emerging horizons of CRISPR

CHAPTER

23

Editing cis-regulatory elements to enhance desired traits in plants

Mariam Akhtar, Ifrah Imran, Muhammad Waseem Sajjad, Rubab Zahra Naqvi, and Imran Amin

National Institute for Biotechnology and Genetic Engineering, Faisalabad, Pakistan

23.1 Introduction

Modern agriculture is facing more challenges as the world's population rises, the environment becomes less stable, and the expectation of consumers also changes. The Green Revolution and traditional plant breeding have helped many staple crops grow, but now that growth has plateaued (Ray et al., 2013; Grassini et al., 2013). This goes with the burden of climate change, which increases the challenges, such as droughts, heatwaves, and more disease and pest susceptibility, all of which threaten the world's food security (Khatodia et al., 2016). Additionally, there is also more interest in agricultural features that are customized to consumers, such as less allergens, enhanced nutrition, and better environmental sustainability (Rodríguez-Leal et al., 2017). As a result, precision breeding has gotten more popular in the last 10 years, and genome editing with CRISPR has become a revolutionary method. CRISPR systems were initially used to edit protein-coding sequences or knock out genes. Now they can also be used to change noncoding genomic areas, such as cis-regulatory elements (CREs) (Saeed et al., 2022; Lu et al., 2017). These parts control the spatial, temporal, and quantitative regulation of gene expression, which makes them essential tools for changing complex characteristics in crops.

In the past, it was challenging to change CREs due to the lack of high-resolution functional genomics data and poor annotations. But now that epigenomic mapping, transcriptome data analyses, and chromatin accessibility profiling are becoming more popular, CREs are being routinely found in both model and crop species (Saeed et al., 2022). This has led to increased interest in using CRISPR techniques to target CREs, not to create permanent mutations in the DNA sequences, but to change how they are expressed with precision (Lu et al., 2017; Wan et al., 2025). Consequently, editing CREs through CRISPR has

Intelligent Gene Editing in Plants
https://doi.org/10.1016/B978-0-443-44079-3.00023-1

redefined the boundaries of plant genetic engineering by introducing novel methods to enhance production, resistance, and adaptability without altering protein sequences.

23.1.1 Why cis-regulation matters

Editing CREs alters the how, when, and where genes are expressed, while classical gene editing changes protein-coding regions and can disrupt essential functions. Gene expression patterns, not gene structure, control many critical agricultural traits, such as response to drought, flowering time, and accumulation of nutrients (Ludwig et al., 2023; Meyer and Purugganan, 2013). For example, the level of expression of a certain gene can be altered through modifying its enhancer or promoter region without changing the structure of the protein, enabling crops to adapt to environmental changes with greater precision.

Studies of natural variation have demonstrated that many traits that are important for domestication, including plants with reduced seed shattering or more compact shape, are generally caused by changes in noncoding regulatory sequences instead of protein-coding genes. So, CRE editing is not only a valuable alternative to gene knockouts, but it is also a proven way for evolution and crop improvement (Cui et al., 2023). Also, techniques like CRISPR activation (CRISPRa) and CRISPR interference (CRISPRi) enable accurate upregulation or downregulation of target genes. This makes gene expression regulation more predictable and reversible (Xu et al., 2023; Saeed et al., 2022). This is especially crucial for complicated traits that require gene regulation, specific to a particular tissue, stage of development, or environment.

This chapter examines how CRISPR gene editing and CRE engineering are integrated in plant biotechnology for genetic improvement of the plants. It is meant to provide readers with a complete and comprehensive picture of both the biology behind CREs and the practical ways to use them as precise regulatory tools to improve crops.

23.2 Fundamentals of CREs

23.2.1 Definition and classification of CREs

CREs are regulatory, noncoding DNA sequences that are on the same chromosome as the gene they regulate. They govern the patterns of gene expression, such as when, where, and the level of transcription of that gene. Regulation is done by bringing together transcription factors (TFs), chromatin remodelers, and cofactors (Marand et al., 2023). CREs are important potential targets for crop engineering because of their vital role in plant development, stress responses, and ability to adapt to diverse environments (Cui et al., 2023).

Promoters are the most basic form of CREs, and they are commonly located directly before the transcription start site (TSS). They have two functional zones: a core promoter that goes from about -50 to $+50$ base pairs around the TSS and has motifs like the TATA-box and initiator (Inr) that attract RNA polymerase II; and a proximal promoter that contains about 250 base pairs upstream and has TF-binding sites like G-boxes (CACGTG) and abscisic acid (ABA)-responsive elements (ABREs) that control the environmental and developmental responses (Villao-Uzho et al., 2023; O'Malley et al., 2016).

Enhancers, on the other hand, are distal CREs that are commonly found in introns or untranslated regions (UTRs). They can be up to a 1000 base pairs distant from the genes they control. They don't depend on position or how they are oriented; they loop chromatin to bring TFs and coactivators near gene promoters. In the last several years, high-throughput approaches like ATAC-seq and DNase-seq have discovered hundreds of plant enhancers in species including Arabidopsis, maize, and rice (Lu et al., 2019; Ricci et al., 2019; Tan et al., 2023).

Silencers and insulators are two additional types of CREs. Silencers halt transcription by using chromatin-modifying proteins or by adding repressive TFs. Insulators act as barriers to prevent the exchange of regulatory signals from neighboring genes. Recent analyses of the *Arabidopsis* genome have shown cohesin-mediated loops and insulator-like sequences that define regulatory regions, even though plants lack traditional insulator proteins like CTCF (Schmitz et al., 2022).

Researchers have also identified other types of CREs in plants that were previously unknown. In *Arabidopsis*, researchers have identified CREs called superenhancers (SEs) that are clusters of enhancers with a high number of TFs. These SEs may aid in the strong expression of stress-responsive genes (Zhao et al., 2022). Shadow enhancers give additional regulatory input that stabilizes expression in response to changes in the environment (Zhao et al., 2022; Cannavò et al., 2016). Enhancer RNAs (eRNAs), which come from active enhancer loci, help in chromatin looping and transcriptional regulation; however, researchers are still studying their role in plants (Sartorelli and Lauberth, 2020).

CREs come in a variety of forms, including insulators, promoters, enhancers, silencers, and more recent ones such as eRNAs, SEs, and shadow enhancers. These all interact to control how genes are expressed in plants. We are rapidly learning more about the functions and mechanisms of these elements thanks to high-resolution mapping and genome editing. Precision crop development greatly benefits from this.

23.2.2 Structural features of CREs

CREs are distinguished by some key structures, central to functioning in the regulation of plant genes. A primary characteristic is position in relation to the TSS.

Promoters are positioned directly upstream near the TSS and encompass core as well as proximal regions. The core promoter is usually between -50 and $+50$ base pairs surrounding the TSS and has conserved elements such as the TATA-box and Inr (initiator), which assist in bringing about RNA polymerase II and basal TFs for the initiation of transcription (Marand et al., 2023). The proximal promoter is more upstream with sequence-specific TF sites for binding in response to environmental and developmental stimuli for fine-tuning transcription initiation.

Compared to such positional specificity, enhancers are different in their positional flexibility. They may also reside upstream or downstream, or even within introns or UTRs of genes of interest. They play a regulatory role due to chromatin looping, which brings such distal elements into spatial proximity with promoter regions within the 3D nucleus architecture. Genome-wide chromatin accessibility profiling by technologies such as ATAC-seq and STARR-seq established the presence of active regions of enhancers within the intergenic as well as intronic regions with the help of high chromatin accessibility as well as histone acetylation within maize, rice, and pepper (Ricci et al., 2019; Cui et al., 2023). In contrast to promoters, enhancers work independently of their orientation so that they are effective irrespective of their direction to the target gene. In tobacco and maize, enhancers retained their functional effectiveness even if inverted, confirming that regulatory effectiveness is established to occur through spatial proximity using chromatin looping rather than sequential alignment (Lu et al., 2019).

The activity of CREs can also be affected by distinct epigenetic blueprints. Active promoters, generally enriched for H3K4me3 and have less DNA methylation, are suggestive of an open chromatin state permissive for transcription (Sun et al., 2019). Plant enhancers display heterogeneous changes depending on H3K27ac enrichment, low DNA methylation, and high chromatin accessibility. Compared with animal systems, H3K4me1 is irregularly enriched at plant enhancers, defining differences in epigenetic regulation between kingdoms (Ricci et al., 2019; Cui et al., 2023; Yang et al., 2025).

At the chromatin architecture level, the effect of CRE is controlled by 3D genome topology. Plant cells, although lacking typical CTCF proteins, create topologically associating domain (TAD)-like structures via cohesin-based complexes as well as plant-specific architectural proteins. In *Arabidopsis* and rice, Hi-C analyses showed that the interactions between promoter and enhancer are carried out by chromatin loops, mediated by such proteins. Disruption of cohesin-related proteins like PDS5 also disrupts domain boundaries, thereby changing gene expression patterns and stressing the regulatory capacity of genome topology (Shu et al., 2024; Göbel et al., 2024).

23.2.3 Molecular mechanisms of CREs

At the molecular level, CREs modulate gene expression by a series of biochemical interactions. Regulation starts with the recognition of TF motifs where TFs bind short, conserved DNA sequences in or near promoter or enhancer regions (Zhai and Li, 2019; Li et al., 2019). Transcriptional activation is reinforced by cooperative binding, where TFs support one another's binding through adjacent motifs. In contrast, competitive binding, where repressors displace activators from overlapping motifs, is a mechanism for repression or downregulation of gene expression. These interactions provide the precise modulation of transcriptional responses to various environmental or developmental stimuli. After TF engagement, CREs attract multiprotein complexes to facilitate either the activation or repression of transcription. The mediator complex serves as a channel to facilitate physical interactions between DNA-bound TFs and RNA polymerase II at promoter sites (Zhai and Li, 2019; Li et al., 2019; Hemsley et al., 2014).

In addition to the Mediator complex, chromatin remodelers such as SWI/SNF reposition nucleosomes to modify chromatin accessibility, while histone acetyltransferases (HATs) and deacetylases (HDACs) dynamically modify histones to affect transcriptional competence (Albini and Puri, 2010). Such combinatorial regulation at CREs allows plants to adjust gene expression finely in response to internal and external signals.

At highly active CRE hubs, such as SEs, a recently discovered mechanism showed the formation of phase-separated condensates. These membrane-less compartments recruit TFs,

Mediator, and RNA polymerase II and create localized transcriptional environments for promoting transcriptional efficiency. In plant nuclei, such biomolecular condensates and RNA granules induced by stress suggest a similar mechanism could result in the accumulation of regulatory machinery into active CREs (Zhang et al., 2020; Wang, He and Fang, 2023). CREs also control RNA polymerase II initiation and promoter-proximal pausing. When the promoter has attracted Pol II and transcription is established, the polymerase tends to pause just downstream of the TSS. The pausing is a regulatory checkpoint where the transcription complex seeks further indications for productive elongation.

Looping between the promoter and enhancer also has the effect of facilitating the release of Pol II into elongation, thereby controlling transcriptional output. Recent studies using CRISPR-mediated promoter motif mutagenesis in *Arabidopsis* established that it is sufficient to alter the structure of the CREs to elicit shifts in the dynamics of gene expression patterns (Zhu et al., 2018; Castel et al., 2019). Together, such molecular processes reveal the complex, multifaceted nature of CRE-mediated transcriptional regulation in plant species by integrating chromatin movement, transcriptional factors' networks, and functioning RNA polymerase II for effective gene expression regulation.

23.2.4 Functions of CREs

CREs are critical components in arranging transcriptional programs for defining plant development, environmental responses, and cascade signaling. By integrating temporal and spatial cues, CREs help plants to adjust the expression of genes on a highly context-dependent and dynamically fine-tuned level. CREs are required for many developmental transitions, responses to hormones, and adaptations to biotic as well as abiotic stress (Table 23.1). In developmental biology, CREs regulate significant events such as meristem maintenance, specification of organ identity, and flowering time. For instance, the *WUSCHEL* (*WUS*) gene in *Arabidopsis thaliana* that is responsible for maintaining the stem-cell niche is regulated by highly controlled enhancer modules located within its upstream region to enforce spatial restriction to the shoot apical meristem (Yadav et al., 2011). Similarly, the *LEAFY* gene for floral meristem identity is regulated by numerous CREs acting on environmental as well as endogenous flowering cues (Winter et al., 2011).

Hormone cascades of signaling also require CREs to attain specificity and feedback regulation. Auxin-responsive elements (AuxREs), including the conserved TGTCTC motif, mediate the induction of auxin-responsive genes by auxin response factors (ARFs) (Ulmasov et al., 1997). ABA-regulated transcription of genes during drought or salt stress is managed by ABREs recognized by bZIP TFs like ABI5 (Fujita et al., 2005). CREs also mediate the cross-talk between hormones; for example, the interaction between the salicylic acid (SA) and jasmonic acid (JA) cascades is often done through distinct but overlapping regulatory elements within the promoter regions (van Verk et al., 2011).

The CREs are the centers for generating proper transcriptional responses to plant stress. The heat shock elements (HSEs) are brought into close association with heat shock factors (HSFs), rapidly activating heat shock proteins (HSPs) to defend against thermal damage (Ohama et al., 2017). For defense against pathogens, the W-box motifs with WRKY-binding sites (TTGACC/T) are significant for activating the pathogenesis-related genes

TABLE 23.1 Examples that highlight cis-elements involved in hormonal regulation, development, and stress responses across diverse plant species.

Motif/CRE (consensus sequence)	Key TF(s)	Stimulus/ Condition	Canonical function	Example gene(s)	Species (characterized in)	
ABRE (e.g., CACGTGGC)	bZIP proteins (ABF/AREB family)	ABA, drought/ osmotic stress	ABA-mediated stress tolerance	*DREB2A*, *RD29B*	*A. thaliana*, rice	Choi et al. (2000)
DRE/CRT (RCCGAC)	AP2/ERF proteins (DREB/CBF family)	Cold, dehydration, salt	ABA-independent dehydration/cold response	*RD29A*, *COR15A*	*A. thaliana*, barley, rice	Yamamoto et al. (2007)
G-box (CACGTG)	bZIP (e.g., HY5, GBF)/bHLH factors	Light, hormones (GA/ABA), stress	Light-responsive and stress-regulated gene expression	*CHS* (chalcone synthase), *Lhcb*	*A. thaliana*, tobacco	Sheshadri et al. (2016)
W-box ((T) TGACC/T)	WRKY family TFs	Pathogen attack (SA), various stresses	Defense and stress-responsive regulation	*PR1*, *RD29A*	*A. thaliana*, tobacco	Sheshadri et al. (2016)
GCC-box (AGCCGCC)	AP2/ERF family (ERF TFs)	Ethylene, jasmonate (pathogen)	JA/ET-mediated defense gene activation	*PDF1.2*, *PR4*	*A. thaliana*, rice	Brown et al. (2003)
AuxRE (TGTCTC)	Auxin response factors (ARFs)	Auxin	Auxin-responsive growth regulation	*GH3* (soybean), *IAA* genes	Glycine max (soybean), *A. thaliana*	Ulmasov et al. (1995)
as-1 element (TGACG repeats)	bZIP (TGA subgroup)	Salicylic acid (pathogens)	SA-mediated defense/ SAR (systemic acquired resistance)	*PR-1*	*A. thaliana*, tobacco	Pape et al. (2010)
MBS/MYB-site (WAACCA)	MYB family (e.g., AtMYB2)	Drought, ABA	Drought-induced gene expression	*RD22*	*A. thaliana*	Yamamoto et al. (2007)
CArG-box (CC (A/T)*6 GG)	MADS-domain TFs (e.g., AP1, AG, SEP)	Development (flowering)	Floral organ identity and development	*AGAMOUS*, *SEP* genes	*A. thaliana*	Aerts et al. (2018)
HSE (nGAAnnTTCn repeats)	Heat shock factors (HSFs)	Heat shock	Heat shock protein gene activation	*HSP70*, *HSP101*	*A. thaliana*, rice	Jaeger et al. (2014)

against biotic stress (Pandey and Somssich, 2009). In response, drought- or cold-inducible elements like DRE/CRT are recognized by the DREB/CBF TFs to coordinate the sets of genes for drought- or cold tolerance (Nakashima et al., 2009).

The combinatorial and modular nature of CREs furthermore allows fine-tuning and robustness in gene expression. Shadow enhancers and redundant regulatory modules enable plants to have phenotypic robustness against diverse environmental conditions (Hong et al., 2008). With the advent of gene editing tools, CREs can now be targeted for manipulation to improve crop traits without introducing any foreign gene, as with CRISPR–Cas-mediated manipulation of promoter elements to fine-tune the expression of valuable yield or stress-related genes (Rodríguez-Leal et al., 2017).

23.3 Identification and validation pipelines

Cis-acting elements can alter the expression of genes at different levels. Because of their capacity to regulate gene expression, CREs, including various promoters and enhancers, have long been considered significant for detailed research and analysis (Wittkopp and Kalay, 2012). Accurately identifying and validating CREs requires a complex and cooperative fusion of experimental mapping, in which scientists contribute their roles, and computational predictions, which involve designing different databases and algorithms (Marand et al., 2023). Here are a few high-resolution, practical pipelines that are now employed for in-depth plant regulatory genomics analysis and research.

23.3.1 *In-silico* discovery

Coordinated motif databases are numerous for use, but the most widely used are PlantCARE, PLACE, and JASPAR, which gather numerous positioning weight matrices (PWMs). The PWMs are indicative of the binding of different types of plant TFs and are frequently the first point of reference for researchers to begin their work (Han et al., 2022). There are several examples of in-silico research and findings for further studies of plant TFs or cis-elements. ABRE, an ABA-responsive motif, was found when MEME-FIMO was used to scan 1 kb upstream areas of rice genes in different stress conditions (Ho and Geisler, 2019). PWM-driven predictive power was demonstrated by rescanning confirmed upstream promoter regions in *OsDREB2A*, which disrupted the motif and eliminated stress-induced expression (Matsukura et al., 2010).

Different effective CREs have also been reported through their conserved noncoding sequences (CNSs). A study compared the model plant *A. thaliana* to *Capsella grandiflora*, resulting in the identification of around 1000 CNSs (12–155 bp) with significant purifying selection pressure. Similarly, a clade-wide conserved motif workflow in the Cucurbitaceae family also identified preserved drought-responsive elements in cucumber and melon (Song et al., 2023; Haudry et al., 2013).

To anticipate further consideration for functional validations, different enhancers have been identified using sophisticated techniques like Enformer and EpiFormer. Additionally, combining sequences and different regulatory backgrounds has also been used to determine their further involvement and roles. For example, a convolutional neural network (CNN) developed using ATAC-seq and ChIP-seq profiles from rice root tissue was able to anticipate stress-inducible elements with over 80% accuracy, which was then verified experimentally through reporter studies. Long-range interconnections may now be integrated via transformers, improving the ability to identify complicated module syntax (Gao et al., 2022; Fang et al., 2024; Li et al., 2025).

23.3.2 Experimental mapping

One characteristic of functioning CREs is chromatin openness. Because of its minimal input requirements and high sensitivity, ATAC-seq has become one of the industry standards. To produce high-quality libraries in *Arabidopsis*, a methodology paired a commercial *Tn5* transposase element with nuclei separation (Lu et al., 2017). Regulatory mechanisms in the reproductive organs were revealed by ATAC-seq profiling across the ovary and spike regions in bread wheat, which produced around 20 GB of data per sample. Resources such as PlantCADB, which aggregates more than 15 million accessible chromatin regions (ACERs) from about 40 species, demonstrate how these tests show cis-regulatory network analyses throughout tissues and circumstances (Wang et al., 2022; Ding et al., 2023).

ATAC-seq, in conjunction with vital methodologies such as INTACT, has made cell-specific transcriptional monitoring possible in the present scenario (Maher et al., 2018). Cell-type-specific permeability alterations during heat stress were shown by scATAC-seq of over 50,000 nuclei in rice root tips, which correlated with the obtained transcriptome data. By identifying proliferative and stress-responsive CREs at the cellular level, different single-cell processes go beyond bulk averaging. Moreover, ChIP-seq, CUT&RUN, and DAP-seq are utilized to verify the interaction of different TFs (Tannenbaum et al., 2018). WRKY, one of the most studied TFs, in maize was profiled using CUT&RUN under fungal stimulation, and for further insights into regulatory pathways, binding at W-box areas associated with ATAC-seq spikes was also confirmed. Even for TFs without antibodies, DAP-seq screenings of about 450 TFs of the *Arabidopsis* species identified new sites throughout the genome (Wang et al., 2023b).

Moreover, Hi-C and Capture-C are used to depict long-range CRE–promoter interactions at different levels. Research on the development of maize embryos found different types of enhancer–promoter loops that were 9–46 kb apart. These loops were confirmed to be regulatorily dependent by promoter truncation tests. A database of more than 100 rice TFs was screened against 46–50 promoters using yeast one-hybrid (Y1H) assays. The test yielded approximately 30,000 types of different TF–promoter interactions. Enormous throughput protein–DNA characterization for regulated system rebuilding is made easier by DAP-seq-based libraries (Li et al., 2019; Lopato et al., 2006).

23.3.3 Functional validation

The gold standard for functional validations involving CREs is still reporter gene assays. Upstream regions from *OsSWEET11* that drive green fluorescent protein (GFP) during chilly conditions were momentarily incorporated into *Nicotiana benthamiana*. This resulted in cold-induced expression, which further disappeared by deleting an ABRE motif (Nicolas et al., 2014). Similarly, salt-responsive sites were validated by GFP promoter tests in rice protoplasts. Using different abridged constructs, a thorough promoter segmentation of the tomato

SlCLV3 gene identified an enhancer 290–490 kb upstream. A 70% decrease in promoter activity was seen when an MYB attachment site within this region was mutated (Zhang et al., 2023b).

A biochemical test method called electrophoretic mobility shift assay (EMSA) verified that WRKY binds to *ZmPR10* promoter W-box sequences with even minimal selectivity and specificity. Dynamical rate coefficients for bZIP TF adherence to ABRE regions ahead of *AtRD29A* were obtained via SPR studies (Ciolkowski et al., 2008). The amount of binding disruption caused by different minor or major alterations in the promoter has been measured using fluorescence polarization (FP) competition assays. Fabricated promoter libraries, following a combination of ABRE ×3 + minimum TATA, were barcoded and inserted into *Arabidopsis* protoplasts. Using 200 different designs in another study, dual-luciferase experiments showed that motif spacing and order might increase ABA reactivity by up to approximately five to six times. More than 460 promoters can be tested simultaneously according to different studies, thanks to barcoded RNA-seq (Zahm et al., 2024; Jores et al., 2021; Irfan et al., 2022) (Table 23.2).

23.4 CRE editing and synthetic reengineering strategies

The capacity to accurately alter CREs to fine-tune the spatial, temporal, and quantitative levels of gene expression in plants has been revolutionized by advances in gene editing technologies, especially those that employ the most widely used CRISPR–Cas systems. Experts can rewire and rearrange systems of gene regulation for intended agronomic results, such as stress tolerance, increased yield, or new metabolic routes, by specifically addressing the DNA regions that govern transcription. Site-directed nucleases, base and prime editors, regulatory modulators, and creative promoter technology are some of the growing tools for CRE editing (Tang and Zhang, 2023; Saeed et al., 2022).

23.4.1 Site-directed nuclease approaches

A single guide RNA (gRNA) directs the Cas9 to produce double-stranded breaks (DSBs) at a particular target site in the genome. Through nonhomologous end joining (NHEJ), these insertions or deletions introduced in CRE editing can cause minor mutations or additions, upset repressor regions, or eliminate repressive motifs. By skillfully resolving a crucial trade-off between the volume of tillers and the mass of the clusters, accurate knockout of small transcription factor binding (TFB) clusters has been achieved through intended mutations in the promoter region of the *IPA1* gene in rice, producing a 50 bp deletion. Similarly, random deletions were made using two gRNAs targeting tomato promoter regions. This resulted in a range of traits, from the size of fruit to inflorescence growth, illustrating how adaptable deletions in promoter functions act as a modifiable "dimmer" (Ludwig et al., 2024; Li, Sapkota and van der Knaap, 2020a).

Fruit size in tomatoes was altered by CRISPR–Cas9-mediated editing of regulatory elements to modify the promoter region of *SlWUS* and *SlCLV3*. Different site-dependent miniature deletions inside CArG boxes were produced that resulted in quantifiable trait variation

TABLE 23.2 Important experimental and in-silico tools for CRE validation and identification.

Tools	Principle	Output	Example/ Application	References	Link
PlantCARE	Promoter analysis database for cis-regulatory elements	Different cis-elements in plant promoters	Identification of different hormone-responsive elements	Lescot et al. (2002)	http://sphinx.rug.ac.be:8080/PlantCARE/
PLACE	Database of plant cis-acting elements	Position and sequence of motifs	Different light and hormone-responsive elements	Higo et al. (1999)	http://www.dna.affrc.go.jp/htdocs/PLACE/
JASPAR	TF binding profile based on PWMs	TF binding site prediction across genomes	Drought-responsive factors scanning like MYB	Sandelin et al. (2004)	http://jaspar.cgb.ki.se/
MEME suite	De novo motif discovery and scanning against known motifs	Location score and scanning of enriched motifs	Identification of conserved cis-elements	Bailey et al. (2009)	http://meme-suite.org/
PROMO	Estimating TFs and their binding sites using species-specific data	TF-target predictions	Detection of different stress-linked TFs	Farré et al. (2003)	http://www.lsi.upc.es/~alggen
RSAT	Regulatory sequence analysis tools	Combinatorial motif analysis	CRE analyses in different polyploid genomes	Thomas-Chollier et al. (2008)	http://rsat.ulb.ac.be/rsat/
PlantPAN3.0	Combination of TFs, cis-elements, and expression data	TF-gene network maps	Different stress-responsive regulatory network estimation	Chow et al. (2019)	http://plantpan.itps.ncku.edu.tw/
DeepCAPE	Use epigenomic features for TF prediction	Enhancer score predictions	Different enhancer identification in maize	Chen et al. (2021)	https://github.com/ShengquanChen/DeepCAPE
AGRIS	*Arabidopsis* gene regulatory information	TF-target interactions and motif data	Validation of heat stress CREs	Yilmaz et al. (2010)	http://arabidopsis.med.ohio-state.edu/
CNNProm	Classification of different promoters	Promoter strength classification	Prediction of promoter activity in crops like tomato	Umarov and Solovyev (2017)	https://github.com/solovictor/CNNPromoterData
GOMO	Link motifs to gene ontology terms	GO terms linked to cis motifs	Inferring functions of different motifs	Bailey et al. (2009)	https://meme-suite.org/meme/doc/gomo.html
TRANSFAC	TF database for both plants and animals	Matrix-based TF binding site prediction	Annotation of CREs in different plants	Matys et al. (2006)	http://www.gene-regulation.com/

TABLE 23.2 Important experimental and in-silico tools for CRE validation and identification.—cont'd

Tools	Principle	Output	Example/ Application	References	Link
EvoPrinter	Phylogenetic footprinting	CREs conserved across species	Identifying conserved leaf morphology enhancers	Odenwald et al. (2005)	https://evoprinter.ninds.nih.gov/evoprintprogramHD/evphd.html
SynergyFinder	Finds synergistic TF binding sites	Motif cooccurrence patterns	Analyzing CRE cooperation	Ianevski et al. (2022) (Li et al., 2022a, 2022b)	https://synergyfinder.fimm.fi/
ChIP-seq	Immunoprecipitation of DNA-bound proteins and their sequencing	TF occupancy peaks	Mapping of DREB21 binding during heat stress	Mardis (2007)	
CUT&RUN	Enzymatic cleavage for TF-DNA interaction	High-resolution TF binding profiles	Binding of TFs	Chow et al. (2019)	https://bitbucket.org/qzhudfci/cutruntools/
SPR (surface plasmon resonance)	Real-time biomolecular interaction	Binding affinity and kinetics	Measuring TF binding strength to mutated promoter elements	Liang et al. (2008)	
STARR-seq	Self-transcribing active regulatory region sequencing	Genome-wide enhancer activity maps	Identifying enhancers controlling secondary metabolism	Arnold et al. (2013)	

(Dong, 2024). By substituting a permissive sequence of the *GOS2* promoter for the native *ARGOS8* promoter in maize via CRISPR–Cas9 homology-directed repair (HDR), transcription of *ARGOS8* increased and led to significantly enhanced drought tolerance. This HDR-based promoter switch is a prime example of CRE-pragmatic technology (Zhan et al., 2021).

23.4.2 Use of base and prime editors

In the rice *OsSWEET11* promoter, transcription activator-like (TAL) effector binding sites were disrupted using a single base editor, such as cytidine base editing (C→T), to increase resistance to bacterial blight. This one-nucleotide alteration prevented *Xanthomonas* effectors from activating in rice without compromising the expression of the plants' defense genes. By slightly changing TF binding affinities, base editing of G-boxes and R6 motifs, the essential cis-elements in soybean and apple, resulted in regulated anthocyanin storage (Abdelrahman et al., 2021).

To target CREs, multiplex prime editing has also been used. Up to 21.8% editing accuracy was achieved through Prime Editing 3 (PE3) systems in rice protoplasts and further in transformed lines. Enhancement of PE2 in tobacco by a tomato-specific RPS5A promoter resulted in 6%–9% editing efficiency in *ALS2* and *PDS1* genes, with heritable features validated in seedlings of the next generation. A 25 bp cis-element was effectively incorporated into the *OsNAS2* promoter using prime editing, greatly boosting zinc relocation, vigor of the plant, and productivity. The specified study used XRF/ICP-OES phenotyping and the T7E1 test to confirm the alterations (Li et al., 2025; Wu et al., 2024).

23.4.3 Transcriptional modulation without DNA change

CRISPR interference (CRISPRi) uses a repressor domain, such as SRDX or KRAB, coupled to a chemically inactive catalytic Cas9 (dCas9) for gene suppression without introducing permanent editing in the genome. On the other hand, CRISPR activation (CRISPRa) makes use of dCas9 fused to stimulation domains such as p300-core, VP64, or p65 for activating transcription of the target gene. In maize and *Arabidopsis*, dCas9-KRAB/SRDX fused complexes have been used to selectively mute metabolic moderators and floral gene promoters, providing morphological transitions without disintegration. According to qPCR and phenotypic tests, rice and wheat exhibit higher abiotic stress tolerance as a result of the upregulation of indigenous genes as well as stress-responsive TFs (e.g., *DREB*, *NAC*), by dCas9-VP64 and dCas9-p300 activators. In models such as *Arabidopsis* protoplasts and *N. benthamiana* leaves, optogenetic dCas9-VP64 regulated by chemically inducible or light-responsive components enabled temporal regulation of CRE activation, allowing for brief but constrained stimulation frames (Li, Xiong and Li, 2020b; Ding et al., 2022).

23.4.4 Synthetic promoter and modular CRE design

A crucial component of plant genetic engineering is the construction of synthetic promoters. Synthetic promoters may be adjusted for accuracy, adaptive design, and transcriptional intensity by stacking many TF binding site motifs in predetermined orientations and positions. Synthetic circuits are used to confirm the rules for ideal motif separation, position, and complementarity that are established from investigations of natural promoters. In silico creation of synthetic promoters according to targeted expression patterns is made possible by tools such as Athena, CREAD, and PlantSynBio. Collections with coordinated CRE themes that are compatible with Gateway and Golden Gate cloning have been created for configurable replication. These databases include motifs that may be coupled to create promoters for particular uses. These motifs can be sensitive to light, hormonal substances, stress, or physiological cues (Pfotenhauer et al., 2022; Lemmon et al., 2018).

Arabidopsis protoplast-based tests have been successfully used to evaluate synthetic promoters that contain repetitive ABRE or DRE motifs with varied spacing. When contrasted with native promoters, a larger variety of evenly spaced ABREs enhanced stress-inducible production by around four times. ABRE, G-box, and R6 motif sections are included in Gateway and Golden Gate cloning libraries, which allow comprehensive CRE structure assessment in tomato and kiwi. Barcoded libraries provide multiplexed screening using RNA-seq and dual-luciferase assays (Nayak et al., 2025; Yasmeen et al., 2023).

MYB10-mediated activity was triggered in kiwifruit by HDR-mediated inclusion of the apple R6 motif across native regulators of anthocyanin cascade transcripts (F3′5′H, GGP), increasing the fruit's anthocyanin and vitamin C levels. Morphological and nutritional shifts were caused by the control of carotenoid biosynthesis regulation of genes in tomato and maize kernels by engineered promoters coupled with light-responsive and ABA-responsive elements. Instantaneous disease identification in plants was achieved by the fusion of synthetic promoters that respond to parasite elicitor patterns (such as W-box and WRKY-binding) to *LUC* or *GFP* transducers in tobacco and Arabidopsis (Yasmeen et al., 2023; Ali and Kim, 2019).

23.4.5 CRE-based de-novo domestication

Wild relative conversion in plants through CREs has been a focused aspect for de novo domestication. By concurrently adjusting many CREs to alter fruit size and inflorescence branching, tomato promoter editing produced novel trait combinations more quickly than breeding. Through CRE editing/promoter manipulation of *SP*, *CLV1*, and *ER* genes, the architecture of *Physalis pruinosa* was changed to resemble domesticated features, including bigger fruit, regulated branching, and increased yield. *CLE7* promoter editing changed the number of kernel rows in maize, whereas multiplex promoter editing changed floral timings in rice. These low-impact variations can add beneficial agronomic features without changing protein-coding sequences (Lemmon et al., 2018; Li et al., 2025; Vu et al., 2020).

23.5 Toward subtler genetics: The strategic move to CRE editing

23.5.1 Editing switches, not just the wires: Targeting CREs with CRISPR

CRISPR/Cas-mediated gene editing holds substantial potential for driving rapid advancements in crop breeding. Host machinery repairs DSBs caused by CRISPR/Cas9 via NHEJ to create InDels (insertion/deletion) or by HDR if a template of the donor is provided (Singh et al., 2023; Belhaj et al., 2015). To manipulate multiple genes simultaneously, large deletions can be induced among targets using multiplex genome editing. To enhance precision efficiency, base editors can be used where deaminase and nickase enzymes do not require a donor template or DSB (Minkenberg et al., 2017). A new gene editing technique, prime editing, enables genetic modifications with more precision in crop plants. This method facilitates the introduction of InDels and base substitutions with minimal off-target effect, utilizing a primer binding sequence and a short template guided by a reverse transcriptase fused to Cas9 nickase. Hence, CRE editing can provide an alternative approach that can be used when conventional overexpression and knockdown of functional genes cannot be manipulated easily. Moreover, since the CRISPR/Cas transgene insertion and modified target sites are inherited independently, it is easier to obtain progeny carrying the desired mutation but lacking the transgene (Anzalone et al., 2020). Desired functions of genes can be improved via CRE editing without disturbing coding sequences. Stimuli and tissue-specific expression can also be edited as desired using the CRE region, which can be difficult if done by targeting the coding region; hence, engineering of promoters through CRISPR/Cas-based CRE editing is promising (Singha et al., 2022).

23.5.2 Advancing agronomic traits via targeted editing of CREs

Many CREs have been edited using the CRISPR/Cas system to achieve desirable plant traits such as abiotic tolerance, yield, nutritional content improvement, and disease resistance in crop plants.

23.5.2.1 Drought tolerance

Plants respond to osmotic stress by activation of the responsible signaling pathways, thus leading to developmental and physiological adaptations. Genetically modified genes belonging to these pathways augment the drought resilience of plants. A positive regulator of drought-induced response is *AREB1/ABF2*, an ABRE-binding protein. A study used the CRISPR activation (CRISPRa) system to increase drought adaptability by targeting the promoter region of *AREB1/ABF2*. Using a chimeric dCAS9HAT system that encompasses histone acetyltransferase 1 (*AtHAT1*), they activated the endogenous promoter of *AREB1* by chromatin relaxation induction and transcriptional accessibility enhancement. Upregulation of upstream *AREB1* and downstream *RD29A* led to phenotypes with enhanced drought resilience, faster stomatal aperture, and high chlorophyll content (Roca Paixão et al., 2019).

In maize, *ARGOS8* negatively regulates ethylene responses, and its constitutive overexpression via a transgenic approach enhanced yield under water deficit conditions. They replaced the *ARGOS8* promoter with the *GOS2* promoter of maize or inserted the *GOS2* promoter in the 5′-UTR of the *ARGOS8* gene, which modified the tissue-specific expression into ubiquitous, resulting in high mRNA expression levels. Analysis of more than four generations demonstrated that *ARGOS8* variants were stably inherited. The novel variants improved grain yield under water scarcity conditions (Shi et al., 2017). Previously, Shi et al. reported the same results in *ARGOS8* overexpressing transgenic lines (Shi et al., 2015).

Transposable elements are dispersed widely throughout the genomes of plants and contribute significantly to plant evolution. Auxin plays a regulatory role in drought tolerance adaptation and root development. Zhao et al. identified a transposon in the *DRO1* gene, acting as an auxin-responsive promoter whose presence at different genome positions leads to physiological changes relevant to drought tolerance in rice. Auxin-mediated and high transcription of *DRO1* in roots is associated with better adaptation to drought by showing longer roots and deeper roots. An auxin response element is present in *INDITTO2*, which shows auxin-responsive promoter activity and thus enables auxin-regulated transcription of nearby genes. In the rice variety Acuce, which shows *DRO1*-mediated tolerance of drought, the *INDITTO2* transposon was identified within the promoter region of the *DRO1* gene. In the *DRO1* promoter of Nipponbare, a nonadapted variety of rice, a CRISPR/Cas-based insertion of the *INDITTO2* transposon efficiently promoted drought avoidance. This exemplifies how transposons can convey regulation to nearby genes by acting as promoters, thus improving overall plant health in response to various abiotic stresses (Zhao et al., 2021).

23.5.2.2 Salt tolerance

Soil salinization serves as a major obstacle to the progress of sustainable agriculture. Using ribosomal profiling and gene editing methodologies. Wang et al. reported the function of *SlRABGAP22*, a family member of the Rab GTPase-activating protein family, in salinity tolerance in tomatoes. Their finding suggests *SlRabGAP22* is a positive salt stress regulator and has four upstream open reading frames (uORFs), further divided into three types. Editing the functional uORFs, found to be negative regulators (type 1 out-of-frame overlapping type and type 3 in-frame overlapping type) alongside changing their distance, GC content, position, type, and length alleviated the negative effect on primary ORFs and also modulated gene expression due to translational repression of *SlRabGAP22*. Minimized toxicity of Na+, rummage of reactive oxygen species (ROS) species, and reduced water stress effects were associated with improved salt stress tolerance (Jia et al., 2024).

B13/VP1 (RAV) regulates abiotic stress tolerance in rice. Expression of *OsRAV2* is induced by salt in rice. Site-specific alterations and deletions identified the GT-1 element in its promoter, responsible for salt induction. Mutations in the GT-1 element of the promoter region of *OsRAV2* were installed by CRISPR–Cas9, and the mutants yielded abnormal phenotypes, hampering salinity-inducible response, while CRE-edited transgenic lines exhibited a wild-type phenotype (Duan et al., 2016).

23.5.2.3 Biotic resistance

Xanthomonas citri causes citrus canker and remains one of the most destructive diseases globally, and the development of citrus canker-resistant cultivars is regarded as the most eco-friendly and efficient control strategy. This pathogen induces the expression of susceptibility gene *LOB1* via a TAL effector, PthA4, which binds to a specific effector binding element (EBE) in the *LOB1* promoter. Earlier studies have reported mutations in the coding region of either *LOB1* or *EBE*, but biallelic or homozygous resistant grapefruit plants with two types of promoter sequences, *TIILOBP* and *LOBP*, have not been generated before for commercial use. Jia et al. used two sgRNAs to target both types of *EBE*. Expression of SpCas9 was derived from both the Yao and 35S promoters for modification of EBE_{PthA4}-*LOBP* in Duncan grapefruit epicotyls. One out of 19 CRISPR-edited lines, DunYao7, was biallelic and remained symptom-free (Jia et al., 2022).

In Africa and Asia, bacterial blight caused by *Xanthomonas pv.* Oryzae is an important disease in rice. The pathogen causes secretion of one or more out of six characterized transcription activator-like effectors (TALEs), which bind to the promoter region to activate expression of at least one of three sucrose transporter genes in rice, *SWEET11*, *SWEET13*, and *SWEET14*, that are essential for disease susceptibility. CRISPR/Cas9-mediated editing in all three promoters of *SWEET* genes by Olivia et al. All mutations in the promoter were introduced at once into the rice varieties, IR64, Kitaake, and Ciherang-Sub1. Field trials showed that edited promoters of *SWEET* confer strong resistance to bacterial blight (Oliva et al., 2019).

To protect cassava storage yields, an efficient weed control strategy is required. Expression of promoters responsible for the regulation of *EPSPS* genes was compared with many paired amino acid substitutions in a study by Hummel et al., and they discovered that persistent, high-strength expression is needed for glyphosate tolerance in cassava plants. Using NHEJ and homologous recombination (HR) DNA repair pathways, the best-performing allele was precisely inserted into the cassava genome by simultaneously substituting the original promoter and substituting dual amino acids at the *EPSPS* locus. Edited plants were resistant to a high dose of *EPSPS* and phenotypically normal (Hummel et al., 2018).

23.5.2.4 Yield improvement

Using the CRISPR editing system, Wang et al. mutated the conserved promoter region of *SlCLV3* in tomato, known to repress the proliferative activity of stem cells and influence fruit locule number. Individual mutations in the conserved promoter regions yielded a relatively weak effect on fruit locule number, while pairwise mutations induced a mild phenotypic change, showcasing amplifying and complementary interactions among conserved regions, reinforcing the idea of cis-regulatory interplay at higher complexity within the *SlCLV3* promoter. The same study states that a positive stem cell proliferation regulator, *SlWUS*, is repressed by *SlCLV3* and is resistant to promoter disruptions (Wang et al., 2021).

Another study identified a natural indel of 3 bp in the promoter of tomato *SlALMT9* that is responsible for the disruption of a WRKY TFBS, causing an increment in fruit malate content due to increased gene expression. This finding was validated by the CRISPR/Cas9 system using a dual sgRNA approach to create defined deletions (Ye et al., 2017).

Many agronomic yield traits selected during the enhancement and domestication of crops are related to an increase in meristem size, manipulated by peptide signals *CLE* in the CLAVATA-WUSCHEL pathway. Liu et al. genetically engineered quantitative trait variation affecting yield by CRISPR/Cas9 in maize by making promoters with reduced transcriptional strength of *CLE* genes *ZmCLE7* and *ZmFCP1*. This strategy improved many grain yield-associated traits in maize, thus emphasizing a substantial potential for agronomic improvement via editing cis-regulatory plant elements (Liu et al., 2021).

A study by Li et al. highlights the importance of fine-tuning gain-of-function alleles by CRISPR for positive regulators, such as *CYP78* family members, to achieve agronomic benefits like increased crop size and organ weight. They targeted a CRE within the *SlKLUH* promoter, located near an single nucleotide polymorphism (SNP) associated with fruit weight. Five out of 21 mutant alleles, m2+4bp, m3+1bp, m5–1bp, m13–8bp, and m14–9bp, with different insertions and deletions exhibited an increment in fruit weight and a reduction in the number of small fruits. Two mutants significantly boosted fruit weight by 10.7%–16.3%. This study highlights the fundamental role of CRISPR/Cas-based CRE editing in yield improvement (Li et al., 2022b).

Micronutrient deficiency is a major problem in developing countries, especially iron and zinc deficiency. The nicotineamine synthase (NOS) gene family is responsible for the regulation of micronutrient transport, such as Zn and Fe. Deletion in the *OsNAS2* promoter, especially the *ARR1AT* cis element at position −933, via CRISPR/Cas-based genome editing, increased Zn accumulation in grains per plant. This promoter editing also led to an increase in spikelet production per panicle and overall increased grain output, and all these attributes were inherited in homozygous transgene-free plant progenies (Ludwig et al., 2024).

A study shows the role of SEs in CRE editing in *A. thaliana*, where deletions of 131–157 bp in the upstream noncoding SE of the thalianol biosynthesis gene cluster via CRISPR/Cas led to transcriptional repression of all five genes, yielding a distinct phenotype. This shows that even a small edit within CRE can have large transcriptional consequences. Disruptions within T-DNA insertions of individual thalional genes changed the expression of respective genes only, while SE mutations affected the complete cluster as a regulatory control brought about by the shared TFB motif between promoters and SE of thalional genes (Zhao et al., 2022).

Genetic improvement of crops requires a complex balance of adjustments caused by linkage drag and gene pleiotropy. A traditional pleiotropic gene, *IPA1* (Ideal Plant Architecture 1), is responsible for increasing the number of grains per panicle but causes a reduction in the number of tillers. A study by Song et al. identified a cis-regulatory region of 54 bp in the *IPA1* gene through tiling deletion-based CRISPR/Cas9 screen. Upon deletion, the negative correlation between tiller number and panicle grain count ultimately boosts the plant's overall grain yield. The study shows that the deleted region is the potential target site for An-1 TF to repress the expression of *IPA1* in roots and panicles (Song et al., 2022).

Gene function divergence is an indication of evolution, but evaluating functional divergence across evolutionary timescales is challenging. For cross-species studies, the few available alleles have been insufficient to uncover the full range of gene functions, possibly masking their ancient and multifunctional roles. Hendelman et al. explored the divergence of *WOX9* (Wuschel Homeobox 9), known for its species-specific role in inflorescence and embryo development. They generated a thorough allelic series using the cis-regulatory editing system in tomato, which highlights the potential role of *WOX9* in pleiotropic roles. Conserved cis-regulatory sequences and accessible chromatin analysis identified the regions held responsible for pleiotropy, the roles that were conserved in ground cherry, a tomato lineage member. This study shows that targeting mutations in CRE can potentially reveal conserved gene functions for the reduction of unfavorable impacts on crop enhancement (Hendelman et al., 2021).

A study by Lu et al. highlights the contribution of structural variations (SV), including duplications and inversions, in agronomic trait improvement. SVs are known to be widespread and essential drivers of genetic diversity. They created an inversion of 911 kb on chromosome 1 via the CRISPR/Cas9 system, and it caused a promoter swap between *PPO1* and *CP12*, along with a duplication of 338 kb between *Ubq2* and *HPPD* on chromosome 2, causing a new gene cassette, $\text{promoter}_{\text{ubiquitin}}$:HPPD. The expression of *HPPD* and *PPO1* in homozygous SV allele-edited plants was amplified by 10-fold, and it resulted in increased herbicide tolerance without any detrimental effect on other necessary agronomic traits. Gene knock-out editing via the CRISPR/Cas system is considered challenging without inserting donor DNA as regulatory molecules, but this study provides a donor DNA-free method for the utilization of the CRISPR/Cas system in plant improvements (Lu et al., 2021).

In research by Yan et al., they optimized an RNA-guided Cas9 system in *A. thaliana* for efficient multiplex editing, and the reliability of the system was validated by knockout of various genes to create large fragment heritable deletions in the genome. To demonstrate feasibility, the second intron of the *AGAMOUS* gene of flower development was explored by the generation of Cas9-free mutant plants where that intron was excised from the genome. Upon deletion, gene expression was reduced by 40% without any alteration in gene splicing, which signifies the role of this regulatory region in the activation of *AGAMOUS* gene expression (Yan et al., 2016).

RPGE protein, a repressor of carotenoid biosynthesis, is encoded by DCAR_032,551. *DcRPGE1* interacts with *DcAPRR2* and suppresses activation of *DcPSY1*, *DcPSY2*, and *DcLCYE*, thus reducing carotenoid biosynthesis. This interaction between *DcRPGE1* and *DcAPRR2* prevents the binding of *DcAPRR2* with *RGATTY* promoter elements of *DcLCYE*, *DcPSY1*, and *DcPSY2*. Additionally, a mutation in the coding region of *DcRPGE1* was identified that causes alternatively spliced transcripts with truncated *DcRPGE1*, failing to repress carotenoid biosynthesis by being unable to interact with *DcAPRR2* (Wang et al., 2024) (Table 23.3).

23.6 Challenges and prospects

Conventional genetic transformation-based crop biotechnology mainly contributes to the refinement of desirable plant traits via loss-of-function or gain-of-function of beneficial genes. CRISPR/Cas-based genome editing of coding regions of repressor genes is also being used to achieve targeted crop traits for commercial deployment (Saeed et al., 2022). Mainly, CREs are present in the promoter region; their presence, absence, and variation regulate gene expression and can induce, reduce, and turn off gene expression. Any change in the CRE of the promoter can characterize variations in phenotypic attributes and gene expression, and many mutations caused by CRE editing can be beneficial over mutations in the coding region. There are still many challenges associated with characterization, discovery, identification of targets, and evaluation of regulatory elements that need to be overcome for their potential in plants (Swinnen et al., 2016).

CRE editing can significantly change CRE variants to yield desirable traits. For example, CRE editing may be used to make crops with improved agronomic characteristics if the region responsible for expression suppression is present in the promoter region of the gene. Especially, it has been seen that in some genes where a complete loss or gain of function mutation poses unintended adverse effects on multiple traits, CRE editing can be a potential way to deal with this problem (Saeed et al., 2022). In such cases, whole-genome resequencing of a wide range of cultivars will assist in targeting CREs of promoters for editing. Research has proven that editing flanking regions exerts unavoidable effects on promoter activity. So, to avoid detrimental impacts on the transcriptional pathway, a rigorous computational analysis before editing the CRE regions is required.

Methylation of DNA in the promoter region can suppress the corresponding gene's expression (Niederhuth and Schmitz, 2017). Methylation analysis of this CG island present in the promoter via DNA methylation global analysis is coupled with phenotypic characterization associated with genetic diversity present in that region. In this scenario, editing the methylated CG island might help in the activation of the expression of genes suppressed for the regulation of targeted agronomic traits (Cui et al., 2023; Saeed et al., 2022).

Hundreds of trait-associated SNPs have been discovered in promoters via genome-wide association studies (GWAS), but only a few of them are reported to play a significant role in crop breeding research. Mutations in the promoter region, including natural or induced, can aid in crop breeding, suggesting the potential of CRE editing to amplify agronomic quality, yield, and tolerance to multiple stress conditions (Rychel-Bielska et al., 2024; Liu et al., 2021;

TABLE 23.3 Engineered gene CREs via CRISPR/Cas-based gene editing in plants.

Crop	Target gene	Edited cis element	Output	References
Solanum lycopersicum	*SlCLV3, SlWUS*	Promoter deletion	Increased fruit size, flower number, and locule number	Wang et al. (2021)
Solanum lycopersicum	*SlALMT9*	Indel in promoter	High fruit malate accumulation	Ye et al. (2017)
Oryza sativa	*SWEET11/13/14*	Promoter mutations	Bacterial blight resistance	Oliva et al. (2019)
Oryza sativa	*OsSULTR3;6*	CRE disruption	Enhanced resistance to bacterial leaf streak	Xu et al. (2021)
Solanum lycopersicum	*SlKLUH*	Editing by SNP in the promoter	Increased fruit weight	Li et al. (2022b)
Oryza sativa	*OsRAV2*	Deletion and site-specific mutation in promoter	Improved salinity tolerance	Duan et al. (2016)
Oryza sativa	*OsNAS2*	Promoter deletion	Increased zinc uptake	Ludwig et al. (2024)
Oryza sativa	*OsTCP19*	Promoter deletion	N increase and grain yield	Tao et al. (2025)
Oryza sativa	*OsRFL*	Deletion in enhancer	Enhanced floral development in rice	Lou et al. (2017)
A. thaliana	*BGC*	Deletion in the superenhancer	Transcriptional alteration of five thalianol genes	Zhao et al. (2022)
Oryza sativa	*DRO1*	Insertion in the promoter	Increased drought tolerance and deeper roots	Zhao et al. (2021)
Triticum aestivum L.	*Pm41*	Insertion in promoter	Resistance to powdery mildew	Li et al. (2022a)
Citrus paradisi	*LOB1*	Promoter modification	Improved resistance to citrus canker	Jia et al. (2022)
Maize	*ARGOS8*	Promoter insertion	Improved growth under drought stress	Shi et al. (2017)
Oryza sativa	*SlRabGAP22*	Indels in uORFs	Improved salinity tolerance	Jia et al. (2024)
A. thaliana	AREB1	Promoter activation	Improved drought tolerance	Roca Paixão et al. (2019)
Zea Mays	*ZmCLE7, ZmFCP1*	Promoter variation	Yield trait improvement	Liu et al. (2021)
Solanum lycopersicum	*SlWOX9*	Promoter mutations	Uncovering conserved *WOX9* gene functions to minimize undesirable effects in crop improvement	Hendelman et al. (2021)

(*Continued*)

TABLE 23.3 Engineered gene CREs via CRISPR/Cas-based gene editing in plants.—cont'd

Crop	Target gene	Edited cis element	Output	References
Oryza sativa	*PPO1* and *CP12*	Promoter inversions	Herbicide tolerance	Lu et al. (2021)
A. thaliana	*TRY*	Deletion of intronic enhancer	Fine-tuned tissue and development-specific gene expression	Meng et al. (2021)
A. thaliana	*AGAMOUS*	Intron deletion	Change in gene expression	Yan et al. (2016)
Manihot esculenta	*EPSPS*	Promoter swap	Glyphosate tolerance	Hummel et al. (2018)
Daucus carota	*DcAPRR2*	Knockout	Decreased carotenoid accumulation, improved nutritional quality	Wang et al. (2024)
Oryza sativa	*OsIPA1*	Indel in promoter	Increased grain yield	Song et al. (2022)

Zhang et al., 2023a; Liang et al., 2025; Li et al., 2024; Poosapati et al., 2022). Despite an unavoidable delay between research and the application continuum, a large number of studies have demonstrated the utility of variants of CREs. Thus, CRE modifications via CRISPR/Cas have been recognized as a significant approach to tailor plant characteristics.

The discovery of condition and tissue-specific CREs is a major obstacle that can be solved by single-cell droplet assays and is thought to be powerful in the evaluation of CREs in the plant genome. Many research groups have employed single-cell RNA seq in plants; similarly, single-cell ATAC seq studies are reported in maize and Arabidopsis. Hi-C, single-cell ChIP-seq, and whole-genome bisulfite sequencing are the emerging methods in mammals but have not yet been used for plants (Farmer et al., 2021; Dorrity et al., 2021; Schmitz et al., 2022).

Addressing this knowledge gap is important, as only a limited number of silencers have been validated in plants so far. Additionally, it is important to define a set of markers or proteins that can reliably differentiate silencers from enhancers, as well as their activity. Considerable effort is required to achieve this; however, this should not be underestimated (Schmitz et al., 2022).

CRE gene pair detection is another challenge because many essential CREs regulate genes in addition to flanking immediate genes. Using single-cell 3C-based technology would be particularly challenging for this; however, single-cell ATAC-seq can be used along with single-cell RNA-seq to leverage coordinated accessibility of ACERs and expression levels of targeted genes across cells. Another complication associated with CRE and gene pair prediction is the opposite correlation for silencers and enhancers. Identification of cofactors and TFs bound to CREs is also challenging because ChIP-seq is labor-intensive and has a difficult implementation with low transformation efficiency (Marand et al., 2021; Ma et al., 2020; Schmitz et al., 2022). Plant scientists are in a discovery phase regarding CREs in plants, but in the future, a lot of attention will be given to this knowledge.

Declaration

The authors declare that no generative artificial intelligence tools were used in the writing or preparation of this chapter. The content reflects the authors' own scholarly work, and the authors assume full responsibility for all statements, interpretations, and conclusions presented.

References

Abdelrahman, M., Wei, Z., Rohila, J.S., Zhao, K., 2021. Multiplex genome-editing technologies for revolutionizing plant biology and crop improvement. Front. Plant Sci. 12. https://doi.org/10.3389/fpls.2021.721203.

Aerts, N., de Bruijn, S., van Mourik, H., Angenent, G.C., van Dijk, A.D.J., 2018. Comparative analysis of binding patterns of MADS-domain proteins in Arabidopsis thaliana. BMC Plant Biol. 18 (1). https://doi.org/10.1186/s12870-018-1348-8.

Albini, S., Puri, P.L., 2010. SWI/SNF complexes, chromatin remodeling and skeletal myogenesis: it's time to exchange. Exp. Cell Res. 316 (18), 3073–3080. https://doi.org/10.1016/j.yexcr.2010.05.023.

Ali, S., Kim, W.C., 2019. A fruitful decade using synthetic promoters in the improvement of transgenic plants. Front. Plant Sci. 10. https://doi.org/10.3389/fpls.2019.01433.

Anzalone, A.V., Koblan, L.W., Liu, D.R., 2020. Genome editing with CRISPR–Cas nucleases, base editors, transposases and prime editors. Nat. Biotechnol. 38, 824–844.

Arnold, C.D., Gerlach, D., Stelzer, C., Boryń, Ł.M., Rath, M., Stark, A., 2013. Genome-wide quantitative enhancer activity maps identified by STARR-seq. Science 339 (6123), 1074–1077. https://doi.org/10.1126/science.1232542.

Bailey, T.L., Boden, M., Buske, F.A., Frith, M., Grant, C.E., Clementi, L., Ren, J., Li, W.W., Noble, W.S., 2009. Meme SUITE: tools for motif discovery and searching. Nucleic Acids Res. 37 (Web Server), W202. https://doi.org/10.1093/nar/gkp335.

Belhaj, K., Chaparro-Garcia, A., Kamoun, S., Patron, N.J., Nekrasov, V., 2015. Editing plant genomes with CRISPR/Cas9. Curr. Opin. Biotechnol. 32, 76–84. https://doi.org/10.1016/j.copbio.2014.11.007.

Brown, R.L., Kazan, K., McGrath, K.C., Maclean, D.J., Manners, J.M., 2003. A role for the GCC-box in jasmonate-mediated activation of the PDF1.2 gene of Arabidopsis. Plant Physiol. 132 (2), 1020–1032. https://doi.org/10.1104/pp.102.017814.

Cannavò, E., Khoueiry, P., Garfield, D.A., Geeleher, P., Zichner, T., Gustafson, E.H., Ciglar, L., Korbel, J.O., Furlong, E.E.M., 2016. Shadow enhancers are pervasive features of developmental regulatory networks. Curr. Biol. 26 (1), 38–51. https://doi.org/10.1016/j.cub.2015.11.034.

Castel, B., Tomlinson, L., Locci, F., Yang, Y., Jones, J.D.G., Lai, E.-M., 2019. Optimization of T-DNA architecture for Cas9-mediated mutagenesis in arabidopsis. PLoS One 14 (1), e0204778. https://doi.org/10.1371/journal.pone.0204778.

Chen, S., Gan, M., Lv, H., Jiang, R., 2021. DeepCAPE: a deep convolutional neural network for the accurate prediction of enhancers. Genom. Proteom. Bioinform. 19 (4), 565–577. https://doi.org/10.1016/j.gpb.2019.04.006.

Choi, H.I., Hong, J.H., Ha, J.O., Kang, J.Y., Kim, S.Y., 2000. ABFs, a family of ABA-responsive element binding factors. J. Biol. Chem. 275 (3), 1723–1730. https://doi.org/10.1074/jbc.275.3.1723.

Chow, C.N., Lee, T.Y., Hung, Y.C., Li, G.Z., Tseng, K.C., Liu, Y.H., Kuo, P.L., Zheng, H.Q., Chang, W.C., 2019. Plantpan3.0: a new and updated resource for reconstructing transcriptional regulatory networks from chip-seq experiments in plants. Nucleic Acids Res. 47 (1), D1155. https://doi.org/10.1093/nar/gky1081.

Ciolkowski, I., Wanke, D., Birkenbihl, R.P., Somssich, I.E., 2008. Studies on DNA-binding selectivity of WRKY transcription factors lend structural clues into WRKY-domain function. Plant Mol. Biol. 68 (1–2), 81–92. https://doi.org/10.1007/s11103-008-9353-1.

Cui, Y., Cao, Q., Li, Y., He, M., Liu, X., 2023. Advances in cis-element- and natural variation-mediated transcriptional regulation and applications in gene editing of major crops. J. Exp. Bot. 74, 5441–5457.

Ding, X., Lu, Y., Chen, L., Li, Y., Zhang, J., Sheng, H., Ren, Z., Li, Y., Yu, X., Jin, S., Cao, J., 2022. Recent progress and future prospect of CRISPR/Cas-Derived transcription activation (CRISPRa) system in plants. Cells 11 (19), 3045. https://doi.org/10.3390/cells11193045.

Ding, K., Sun, S., Luo, Y., Long, C., Zhai, J., Zhai, Y., Wang, G., 2023. PlantCADB: a comprehensive plant chromatin accessibility database. Genom. Proteom. Bioinform. 21 (2), 311–323. https://doi.org/10.1016/j.gpb.2022.10.005.

Dong, H., 2024. Application of genome editing techniques to regulate gene expression in crops. BMC Plant Biol. 24 (1). https://doi.org/10.1186/s12870-024-04786-2.

Dorrity, M.W., Alexandre, C.M., Hamm, M.O., Vigil, A.L., Fields, S., Queitsch, C., Cuperus, J.T., 2021. The regulatory landscape of Arabidopsis thaliana roots at single-cell resolution. Nat. Commun. 12 (1). https://doi.org/10.1038/s41467-021-23675-y.

Duan, Y.B., Li, J., Qin, R.Y., Xu, R.F., Li, H., Yang, Y.C., Ma, H., Li, L., Wei, P.C., Yang, J.B., 2016. Identification of a regulatory element responsible for salt induction of rice OsRAV2 through ex situ and in situ promoter analysis. Plant Mol. Biol. 90 (1–2), 49–62. https://doi.org/10.1007/s11103-015-0393-z.

Fang, S., Ji, M., Zhu, T., Wang, Y., Tang, X., Zhu, X., Yang, Z., Xu, C., Wang, H., Li, P., 2024. Multi-omics analysis reveals the transcriptional regulatory network of maize roots in response to nitrogen availability. Agronomy 14 (7), 1541. https://doi.org/10.3390/agronomy14071541.

Farmer, A., Thibivilliers, S., Ryu, K.H., Schiefelbein, J., Libault, M., 2021. Single-nucleus RNA and ATAC sequencing reveals the impact of chromatin accessibility on gene expression in Arabidopsis roots at the single-cell level. Mol. Plant 14 (3), 372–383. https://doi.org/10.1016/j.molp.2021.01.001.

Farré, D., Roset, R., Huerta, M., Adsuara, J.E., Roselló, L., Albà, M.M., Messeguer, X., 2003. Identification of patterns in biological sequences at the ALGGEN server: PROMO and MALGEN. Nucleic Acids Res. 31 (13), 3651–3653. https://doi.org/10.1093/nar/gkg605.

Fujita, Y., Fujita, M., Satoh, R., Maruyama, K., Parvez, M.M., Seki, M., Hiratsu, K., Ohme-Takagi, M., Shinozaki, K., Yamaguchi-Shinozaki, K., 2005. AREB1 is a transcription activator of novel ABRE-dependent ABA signaling that enhances drought stress tolerance in Arabidopsis. Plant Cell 17 (12), 3470–3488. https://doi.org/10.1105/tpc.105.035659.

Gao, Y., Chen, Y., Feng, H., Zhang, Y., Yue, Z., 2022. RicENN: prediction of rice enhancers with neural network based on DNA sequences. Interdiscipl. Sci. Comput. Life Sci. 14 (2), 555–565. https://doi.org/10.1007/s12539-022-00503-5.

Göbel, A.M., Zhou, S., Wang, Z., Tzourtzou, S., Himmelbach, A., Zheng, S., Pradillo, M., Liu, C., Jiang, H., 2024. Mutations of PDS5 genes enhance TAD-like domain formation Arabidopsis thaliana. Nat. Commun. 15 (1). https://doi.org/10.1038/s41467-024-53760-x.

Grassini, P., Eskridge, K.M., Cassman, K.G., 2013. Distinguishing between yield advances and yield plateaus in historical crop production trends. Nat. Commun. 4, 2918.

Han, X., Rong, H., Tian, Y., Qu, Y., Xu, M., Xu, L.A., 2022. Genome-wide identification of PLATZ transcription factors in ginkgo biloba L. and their expression characteristics during seed development. Front. Plant Sci. 13. https://doi.org/10.3389/fpls.2022.946194.

Haudry, A., Platts, A.E., Vello, E., Hoen, D.R., Leclercq, M., Williamson, R.J., Forczek, E., Joly-Lopez, Z., Steffen, J. G., Hazzouri, K.M., Dewar, K., Stinchcombe, J.R., Schoen, D.J., Wang, X., Schmutz, J., Town, C.D., Edger, P.P., Pires, J.C., Schumaker, K.S., Jarvis, D.E., Mandáková, T., Lysak, M.A., Van Den Bergh, E., Schranz, M.E., Harrison, P.M., Moses, A.M., Bureau, T.E., Wright, S.I., Blanchette, M., 2013. An atlas of over 90,000 conserved noncoding sequences provides insight into crucifer regulatory regions. Nat. Genet. 45 (8), 891–898. https://doi.org/10.1038/ng.2684.

Hemsley, P.A., Hurst, C.H., Kaliyadasa, E., Lamb, R., Knight, M.R., De Cothi, E.A., Steele, J.F., Knight, H., 2014. The Arabidopsis mediator complex subunits MED16, MED14, and MED2 regulate mediator and RNA polymerase II recruitment to CBF-responsive cold-regulated genes. Plant Cell 26 (1), 465–484. https://doi.org/10.1105/tpc.113.117796.

Hendelman, A., Zebell, S., Rodriguez-Leal, D., Dukler, N., Robitaille, G., Wu, X., Kostyun, J., Tal, L., Wang, P., Bartlett, M.E., Eshed, Y., Efroni, I., Lippman, Z.B., 2021. Conserved pleiotropy of an ancient plant homeobox gene uncovered by cis-regulatory dissection. Cell 184 (7), 1724. https://doi.org/10.1016/j.cell.2021.02.001.

Higo, K., Ugawa, Y., Iwamoto, M., Korenaga, T., 1999. Plant cis-acting regulatory DNA elements (PLACE) database: 1999. Nucleic Acids Res. 27 (1), 297–300. https://doi.org/10.1093/nar/27.1.297.

Ho, C.L., Geisler, M., 2019. Genome-wide computational identification of biologically significant cis-regulatory elements and associated transcription factors from rice. Plants 8 (11). https://doi.org/10.3390/plants8110441.

Hong, J.-W., Hendrix, D.A., Levine, M.S., 2008. Shadow enhancers as a source of evolutionary novelty. Science 321, 1314-1314.

Hummel, A.W., Chauhan, R.D., Cermak, T., Mutka, A.M., Vijayaraghavan, A., Boyher, A., Starker, C.G., Bart, R., Voytas, D.F., Taylor, N.J., 2018. Allele exchange at the EPSPS locus confers glyphosate tolerance in cassava. Plant Biotechnol. J. 16 (7), 1275–1282. https://doi.org/10.1111/pbi.12868.

Ianevski, A., Giri, A.K., Aittokallio, T., 2022. SynergyFinder 3.0: an interactive analysis and consensus interpretation of multi-drug synergies across multiple samples. Nucleic Acids Res. 50, W739–W743.

Irfan, M., Kumar, P., Kumar, V., Datta, A., 2022. Fruit ripening specific expression of β-D-N-acetylhexosaminidase (β-Hex) gene in tomato is transcriptionally regulated by ethylene response factor SlERF.E4. Plant Sci. 323, 111380. https://doi.org/10.1016/j.plantsci.2022.111380.

Jaeger, A.M., Makley, L.N., Gestwicki, J.E., Thiele, D.J., 2014. Genomic heat shock element sequences drive cooperative human heat shock factor 1 DNA binding and selectivity. J. Biol. Chem. 289 (44), 30459–30469. https://doi.org/10.1074/jbc.M114.591578.

Jia, H., Omar, A.A., Orbović, V., Wang, N., 2022. Biallelic editing of the LOB1 promoter via CRISPR/Cas9 creates canker-resistant 'Duncan' grapefruit. Phytopathology® 112 (2), 308–314. https://doi.org/10.1094/phyto-04-21-0144-r.

Jia, C., Wang, J., Guo, B., Yang, T., Yang, H., Wang, B., Yu, Q., 2024. Editing and genome-wide analysis upstream open reading frames contributes to enhancing salt tolerance in tomato. Plant Biotechnol. J. 22 (12), 3312–3325. https://doi.org/10.1111/pbi.14450.

Jores, T., Tonnies, J., Wrightsman, T., Buckler, E.S., Cuperus, J.T., Fields, S., Queitsch, C., 2021. Synthetic promoter designs enabled by a comprehensive analysis of plant core promoters. Nat. Plants 7 (6), 842–855. https://doi.org/10.1038/s41477-021-00932-y.

Khatodia, S., Bhatotia, K., Passricha, N., Khurana, S.M.P., Tuteja, N., 2016. The CRISPR/Cas genome-editing tool: application in improvement of crops. Front. Plant Sci. 7 (2016). https://doi.org/10.3389/fpls.2016.00506.

Lemmon, Z.H., Reem, N.T., Dalrymple, J., Soyk, S., Swartwood, K.E., Rodriguez-Leal, D., Van Eck, J., Lippman, Z. B., 2018. Rapid improvement of domestication traits in an orphan crop by genome editing. Nat. Plants 4 (10), 766–770. https://doi.org/10.1038/s41477-018-0259-x.

Lescot, M., Déhais, P., Thijs, G., Marchal, K., Moreau, Y., Van De Peer, Y., Rouzé, P., Rombauts, S., 2002. PlantCARE, a database of plant cis-acting regulatory elements and a portal to tools for in silico analysis of promoter sequences. Nucleic Acids Res. 30 (1), 325–327. https://doi.org/10.1093/nar/30.1.325.

Li, E., Liu, H., Huang, L., Zhang, X., Dong, X., Song, W., Zhao, H., Lai, J., 2019. Long-range interactions between proximal and distal regulatory regions in maize. Nat. Commun. 10 (1). https://doi.org/10.1038/s41467-019-10603-4.

Li, Q., Sapkota, M., Van der Knaap, E., 2020a. Perspectives of CRISPR/Cas-mediated cis-engineering in horticulture: unlocking the neglected potential for crop improvement. Hortic. Res. 7.

Li, Z., Xiong, X., Li, J.-F., 2020b. The working dead: repurposing inactive CRISPR-associated nucleases as programmable transcriptional regulators in plants. aBIOTECH 1, 32–40.

Li, M., Dong, L., Zhu, K., Wu, Q., Chen, Y., Lu, P., Guo, G., Zhang, H., Zhang, P., Li, B., 2022a. Provoking a silent R gene in wheat genome confers resistance to powdery mildew. Plant Biotechnol. J. 20, 2039–2041.

Li, Q., Feng, Q., Snouffer, A., Zhang, B., Rodríguez, G.R., van der Knaap, E., 2022b. Increasing fruit weight by editing a cis-regulatory element in tomato KLUH promoter using CRISPR/Cas9. Front. Plant Sci. 13. https://doi.org/10.3389/fpls.2022.879642.

Li, Y.D., Liu, Y.C., Jiang, Y.X., Namisy, A., Chung, W.H., Sun, Y.H., Chen, S.Y., 2024. Analyzing genetic diversity in luffa and developing a Fusarium wilt-susceptible linked SNP marker through a single plant genome-wide association (sp-GWAS) study. BMC Plant Biol. 24 (1). https://doi.org/10.1186/s12870-024-05022-7.

Li, C., Liu, B., Dong, H., Yang, B., 2025. Enhancing resistance to bacterial blight in rice using CRISPR-based base editing technology. The Crop Journal 13 (1), 115–124. https://doi.org/10.1016/j.cj.2024.09.003.

Liang, X., Nazarenus, T.J., Stone, J.M., 2008. Identification of a consensus DNA-binding site for the Arabidopsis thaliana SBP domain transcription factor, AtSPL14, and binding kinetics by surface plasmon resonance. Biochemistry 47, 3645–3653.

Liang, Y.L., Yan, P.S., Li, W.X., Ding, W.N., Li, G.X., Xu, J.M., Ding, Z.J., Zheng, S.J., 2025. Bridging genomic insights to agricultural practice: GWAS applications in mining nutrient utilization efficiency genes for crop improvement. Plant Cell Environ. 48 (10), 7271–7283. https://doi.org/10.1111/pce.70027.

Liu, L., Gallagher, J., Arevalo, E.D., Chen, R., Skopelitis, T., Wu, Q., Bartlett, M., Jackson, D., 2021. Enhancing grain-yield-related traits by CRISPR–Cas9 promoter editing of maize CLE genes. Nat. Plants 7 (3), 287–294. https://doi.org/10.1038/s41477-021-00858-5.

Lopato, S., Bazanova, N., Morran, S., Milligan, A.S., Shirley, N., Langridge, P., 2006. Isolation of plant transcription factors using a modified yeast one-hybrid system. Plant Methods 2 (1). https://doi.org/10.1186/1746-4811-2-3Australia.

Lou, S., Chen, S., Zhao, X., Chen, L., Zhang, J., Fu, H., Liu, Y.G., Chen, Y., 2017. The far-upstream regulatory region of RFL is required for its precise spatial-temporal expression for floral development in rice. Plant Mol. Biol. 93 (1–2), 185–195. https://doi.org/10.1007/s11103-016-0556-6.

Lu, Z., Hofmeister, B.T., Vollmers, C., Dubois, R.M., Schmitz, R.J., 2017. Combining ATAC-seq with nuclei sorting for discovery of cis-regulatory regions in plant genomes. Nucleic Acids Res. 45, e41.

Lu, Z., Marand, A.P., Ricci, W.A., Ethridge, C.L., Zhang, X., Schmitz, R.J., 2019. The prevalence, evolution and chromatin signatures of plant regulatory elements. Nat. Plants 5 (12), 1250–1259. https://doi.org/10.1038/s41477-019-0548-z.

Lu, Y., Wang, J., Chen, B., Mo, S., Lian, L., Luo, Y., Ding, D., Ding, Y., Cao, Q., Li, Y., Li, Y., Liu, G., Hou, Q., Cheng, T., Wei, J., Zhang, Y., Chen, G., Song, C., Hu, Q., Sun, S., Fan, G., Wang, Y., Liu, Z., Song, B., Zhu, J.-K., Li, H., Jiang, L., 2021. A donor-DNA-free CRISPR/Cas-based approach to gene knock-up in rice. Nat. Plants 7 (11), 1445–1452. https://doi.org/10.1038/s41477-021-01019-4.

Ludwig, Y., Dueñas, C., Arcillas, E., Macalalad-Cabral, R.J., Kohli, A., Reinke, R., Slamet-Loedin, I.H., 2023. CRISPR-mediated promoter editing of a cis-regulatory element of OsNAS2 increases Zn uptake/translocation and plant yield in rice. Frontiers in Genome Editing 5. https://doi.org/10.3389/fgeed.2023.1308228.

Ludwig, Y., Dueñas, C., Arcillas, E., Macalalad-Cabral, R.J., Kohli, A., Russell, R., Slamet-Loedin, I.H., 2024. CRISPR-mediated promoter editing of a cis-regulatory element of OsNAS2 increases Zn uptake/translocation and plant yield in rice. Frontiers in Genome Editing 5. https://doi.org/10.3389/fgeed.2023.1308228.

Ma, S., Zhang, B., LaFave, L.M., Earl, A.S., Chiang, Z., Hu, Y., Ding, J., Brack, A., Kartha, V.K., Tay, T., Law, T., Lareau, C., Hsu, Y.C., Regev, A., Buenrostro, J.D., 2020. Chromatin potential identified by shared single-cell profiling of RNA and chromatin. Cell 183 (4), 1103. https://doi.org/10.1016/j.cell.2020.09.056.

Maher, K.A., Bajic, M., Kajala, K., Reynoso, M., Pauluzzi, G., West, D.A., Zumstein, K., Woodhouse, M., Bubb, K., Dorrity, M.W., Queitsch, C., Bailey-Serres, J., Sinha, N., Brady, S.M., Deal, R.B., 2018. Profiling of accessible chromatin regions across multiple plant species and cell types reveals common gene regulatory principles and new control modules. Plant Cell 30 (1), 15–36. https://doi.org/10.1105/tpc.17.00581.

Marand, A.P., Chen, Z., Gallavotti, A., Schmitz, R.J., 2021. A cis-regulatory atlas in maize at single-cell resolution. Cell 184 (11), 3041. https://doi.org/10.1016/j.cell.2021.04.014.

Marand, A.P., Eveland, A.L., Kaufmann, K., Springer, N.M., 2023. cis-Regulatory elements in plant development, adaptation, and evolution. Annu. Rev. Plant Biol. 74, 111–137.

Mardis, E.R., 2007. ChIP-seq: welcome to the new frontier. Nat. Methods 4 (8), 613–614. https://doi.org/10.1038/nmeth0807-613.

Matsukura, S., Mizoi, J., Yoshida, T., Todaka, D., Ito, Y., Maruyama, K., Shinozaki, K., Yamaguchi-Shinozaki, K., 2010. Comprehensive analysis of rice DREB2-type genes that encode transcription factors involved in the expression of abiotic stress-responsive genes. Mol. Genet. Genom. 283 (2), 185–196. https://doi.org/10.1007/s00438-009-0506-y.

Matys, V., Kel-Margoulis, O.V., Fricke, E., Liebich, I., Land, S., Barre-Dirrie, A., Reuter, I., Chekmenev, D., Krull, M., Hornischer, K., 2006. TRANSFAC® and its module TRANSCompel®: transcriptional gene regulation in eukaryotes. Nucleic Acids Res. 34, D108–D110.

Meng, F., Zhao, H., Zhu, B., Zhang, T., Yang, M., Li, Y., Han, Y., Jiang, J., 2021. Genomic editing of intronic enhancers unveils their role in fine-tuning tissue-specific gene expression in Arabidopsis thaliana. Plant Cell 33 (6), 1997–2014. https://doi.org/10.1093/plcell/koab093.

Meyer, R.S., Purugganan, M.D., 2013. Evolution of crop species: genetics of domestication and diversification. Nat. Rev. Genet. 14 (12), 840–852. https://doi.org/10.1038/nrg3605.

Minkenberg, B., Wheatley, M., Yang, Y., 2017. CRISPR/Cas9-enabled multiplex genome editing and its application. Prog. Mol. Biol. Transl. Sci. 149, 111–132.

Nakashima, K., Ito, Y., Yamaguchi-Shinozaki, K., 2009. Transcriptional regulatory networks in response to abiotic stresses in Arabidopsis and grasses. Plant Physiol. 149, 88–95.

Nayak, N., Mehrotra, S., Karamchandani, A.N., Santelia, D., Mehrotra, R., 2025. Recent advances in designing synthetic plant regulatory modules. Front. Plant Sci. 16. https://doi.org/10.3389/fpls.2025.1567659.

Nicolas, P., Lecourieux, D., Kappel, C., Cluzet, S., Cramer, G., Delrot, S., Lecourieux, F., 2014. The basic leucine zipper transcription factor ABSCISIC ACID RESPONSE ELEMENT-BINDING FACTOR2 is an important transcriptional regulator of abscisic acid-dependent grape berry ripening processes. Plant Physiol. 164 (1), 365–383. https://doi.org/10.1104/pp.113.231977.

Niederhuth, C.E., Schmitz, R.J., 2017. Putting DNA methylation in context: from genomes to gene expression in plants. Biochimica et Biophysica Acta - Gene Regulatory Mechanisms 1860 (1), 149–156. https://doi.org/10.1016/j.bbagrm.2016.08.009.

Odenwald, W.F., Rasband, W., Kuzin, A., Brody, T., 2005. EVOPRINTER, a multigenomic comparative tool for rapid identification of functionally important DNA. Proc. Natl. Acad. Sci. 102 (41), 14700–14705. https://doi.org/10.1073/pnas.0506915102.

Ohama, N., Sato, H., Shinozaki, K., Yamaguchi-Shinozaki, K., 2017. Transcriptional regulatory network of plant heat stress response. Trends Plant Sci. 22 (1), 53–65. https://doi.org/10.1016/j.tplants.2016.08.015.

Oliva, R., Ji, C., Atienza-Grande, G., Huguet-Tapia, J.C., Perez-Quintero, A., Li, T., Eom, J.S., Li, C., Nguyen, H., Liu, B., Auguy, F., Sciallano, C., Luu, V.T., Dossa, G.S., Cunnac, S., Schmidt, S.M., Slamet-Loedin, I.H., Vera Cruz, C., Szurek, B., Frommer, W.B., White, F.F., Yang, B., 2019. Broad-spectrum resistance to bacterial blight in rice using genome editing. Nat. Biotechnol. 37 (11), 1344–1350. https://doi.org/10.1038/s41587-019-0267-z.

O'Malley, R.C., Huang, S.-shan C., Song, L., Lewsey, M.G., Bartlett, A., Nery, J.R., Galli, M., Gallavotti, A., Ecker, J. R., 2016. Cistrome and epicistrome features shape the regulatory DNA landscape. Cell 165 (5), 1280–1292. https://doi.org/10.1016/j.cell.2016.04.038.

Pandey, S.P., Somssich, I.E., 2009. The role of WRKY transcription factors in plant immunity. Plant Physiol. 150 (4), 1648–1655. https://doi.org/10.1104/pp.109.138990.

Pape, S., Thurow, C., Gatz, C., 2010. Exchanging the as-1-like element of the PR-1 promoter by the as-1 element of the CaMV 35S promoter abolishes salicylic acid responsiveness and regulation by NPR1 and SNI1. Plant Signal. Behav. 5, 1669–1671.

Pfotenhauer, A.C., Occhialini, A., Nguyen, M.A., Scott, H., Dice, L.T., Harbison, S.A., Li, L., Reuter, D.N., Schimel, T.M., Stewart, C.N., Beal, J., Lenaghan, S.C., 2022. Building the plant SynBio toolbox through combinatorial analysis of DNA regulatory elements. ACS Synth. Biol. 11 (8), 2741–2755. https://doi.org/10.1021/acssynbio.2c00147.

Poosapati, S., Poretsky, E., Dressano, K., Ruiz, M., Vazquez, A., Sandoval, E., Estrada-Cardenas, A., Duggal, S., Lim, J.H., Morris, G., Szczepaniec, A., Walse, S.S., Ni, X., Schmelz, E.A., Huffaker, A., 2022. A sorghum genome-wide association study (GWAS) identifies a WRKY transcription factor as a candidate gene underlying sugarcane aphid (Melanaphis sacchari) resistance. Planta 255 (2). https://doi.org/10.1007/s00425-021-03814-x.

Ray, D.K., Mueller, N.D., West, P.C., Foley, J.A., 2013. Yield trends are insufficient to double global crop production by 2050. PLoS One 8 (6). https://doi.org/10.1371/journal.pone.0066428.

Ricci, W.A., Lu, Z., Ji, L., Marand, A.P., Ethridge, C.L., Murphy, N.G., Noshay, J.M., Galli, M., Mejía-Guerra, M.K., Colomé-Tatché, M., Johannes, F., Rowley, M.J., Corces, V.G., Zhai, J., Scanlon, M.J., Buckler, E.S., Gallavotti, A., Springer, N.M., Schmitz, R.J., Zhang, X., 2019. Widespread long-range cis-regulatory elements in the maize genome. Nat. Plants 5 (12), 1237–1249. https://doi.org/10.1038/s41477-019-0547-0.

Roca Paixão, J.F., Gillet, F.X., Ribeiro, T.P., Bournaud, C., Lourenço-Tessutti, I.T., Noriega, D.D., Melo, B.P.d., de Almeida-Engler, J., Grossi-de-Sa, M.F., 2019. Improved drought stress tolerance in Arabidopsis by CRISPR/dCas9 fusion with a Histone AcetylTransferase. Sci. Rep. 9 (1). https://doi.org/10.1038/s41598-019-44571-y.

Rodríguez-Leal, D., Lemmon, Z.H., Man, J., Bartlett, M.E., Lippman, Z.B., 2017. Engineering quantitative trait variation for crop improvement by genome editing. Cell 171, 470–480.

Rychel-Bielska, S., Bielski, W., Surma, A., Annicchiarico, P., Belter, J., Kozak, B., Galek, R., Harzic, N., Książkiewicz, M., 2024. A GWAS study highlights significant associations between a series of indels in a FLOWERING LOCUS T gene promoter and flowering time in white lupin (Lupinus albus L.). BMC Plant Biol. 24 (1). https://doi.org/10.1186/s12870-024-05438-1.

Saeed, S., Usman, B., Shim, S.H., Khan, S.U., Nizamuddin, S., Saeed, S., Shoaib, Y., Jeon, J.S., Jung, K.H., 2022. CRISPR/Cas-mediated editing of cis-regulatory elements for crop improvement. Plant Sci. 324. https://doi.org/10.1016/j.plantsci.2022.111435.

Sandelin, A., Alkema, W., Engström, P., Wasserman, W.W., Lenhard, B., 2004. JASPAR: an open-access database for eukaryotic transcription factor binding profiles. Nucleic Acids Res. 32, D91.

Sartorelli, V., Lauberth, S.M., 2020. Enhancer RNAs are an important regulatory layer of the epigenome. Nat. Struct. Mol. Biol. 27 (6), 521–528. https://doi.org/10.1038/s41594-020-0446-0.

Schmitz, R.J., Grotewold, E., Stam, M., 2022. Cis-regulatory sequences in plants: their importance, discovery, and future challenges. Plant Cell 34, 718–741.

Sheshadri, S., Nishanth, M., Simon, B., 2016. Stress-mediated cis-element transcription factor interactions interconnecting primary and specialized metabolism in planta. Frontiers in plant science 7, 1725.

Shi, J., Habben, J.E., Archibald, R.L., Drummond, B.J., Chamberlin, M.A., Williams, R.W., Renee Lafitte, H., Weers, B.P., 2015. Overexpression of ARGOS genes modifies plant sensitivity to ethylene, leading to improved drought tolerance in both Arabidopsis and maize. Plant Physiol. 169 (1), 266–282. https://doi.org/10.1104/pp.15.00780.

Shi, J., Gao, H., Wang, H., Lafitte, H.R., Archibald, R.L., Yang, M., Hakimi, S.M., Mo, H., Habben, J.E., 2017. ARGOS8 variants generated by CRISPR-Cas9 improve maize grain yield under field drought stress conditions. Plant Biotechnol. J. 15 (2), 207–216. https://doi.org/10.1111/pbi.12603.

Shu, J., Sun, L., Wang, D., Yin, X., Yang, M., Yang, Z., Gao, Z., He, Y., Calonje, M., Lai, J., Deng, X.W., He, H., Zhou, Y., 2024. EMF1 functions as a 3D chromatin modulator in Arabidopsis. Mol. Cell 84 (24), 4729. https://doi.org/10.1016/j.molcel.2024.10.031.

Singh, S., Chaudhary, R., Deshmukh, R., Tiwari, S., 2023. Opportunities and challenges with CRISPR-Cas mediated homologous recombination based precise editing in plants and animals. Plant Mol. Biol. 111 (1–2), 1–20. https://doi.org/10.1007/s11103-022-01321-5.

Singha, D.L., Das, D., Sarki, Y.N., Chowdhury, N., Sharma, M., Maharana, J., Chikkaputtaiah, C., 2022. Harnessing tissue-specific genome editing in plants through CRISPR/Cas system: current state and future prospects. Planta 255 (1). https://doi.org/10.1007/s00425-021-03811-0.

Song, X., Meng, X., Guo, H., Cheng, Q., Jing, Y., Chen, M., Liu, G., Wang, B., Wang, Y., Li, J., Yu, H., 2022. Targeting a gene regulatory element enhances rice grain yield by decoupling panicle number and size. Nat. Biotechnol. 40 (9), 1403–1411. https://doi.org/10.1038/s41587-022-01281-7.

Song, H., Wang, Q., Zhang, Z., Lin, K., Pang, E., 2023. Identification of clade-wide putative cis -regulatory elements from conserved non-coding sequences in Cucurbitaceae genomes. Hortic. Res. 10 (4). https://doi.org/10.1093/hr/uhad038.

Sun, J., He, N., Niu, L., Huang, Y., Shen, W., Zhang, Y., Li, L., Hou, C., 2019. Global quantitative mapping of enhancers in rice by STARR-seq. Genom. Proteom. Bioinform. 17 (2), 140–153. https://doi.org/10.1016/j.gpb.2018.11.003.

Swinnen, G., Goossens, A., Pauwels, L., 2016. Lessons from domestication: targeting cis-regulatory elements for crop improvement. Trends Plant Sci. 21, 506–515.

Tan, Y., Yan, X., Sun, J., Wan, J., Li, X., Huang, Y., Li, L., Niu, L., Hou, C., 2023. Genome-wide enhancer identification by massively parallel reporter assay in Arabidopsis. Plant J. 116 (1), 234–250. https://doi.org/10.1111/tpj.16373.

Tang, X., Zhang, Y., 2023. Beyond knockouts: fine-tuning regulation of gene expression in plants with CRISPR-Cas -based promoter editing. New Phytol. 239 (3), 868–874. https://doi.org/10.1111/nph.19020.

Tannenbaum, M., Sarusi-Portuguez, A., Krispil, R., Schwartz, M., Loza, O., Benichou, J.I.C., Mosquna, A., Hakim, O., 2018. Regulatory chromatin landscape in Arabidopsis thaliana roots uncovered by coupling INTACT and ATAC-seq. Plant Methods 14 (1). https://doi.org/10.1186/s13007-018-0381-9.

Tao, Y., Chen, Z., Xu, Y., Wang, F., Jiang, Y., Fan, F., Li, W., Zhu, J., Li, X., Wang, J., Zhu, Q.-H., Xu, G., Yang, J., 2025. CRISPR-mediated targeted mutagenesis for improving nitrogen use efficiency in japonica rice. Plant Commun. 6 (2), 101189. https://doi.org/10.1016/j.xplc.2024.101189.

Thomas-Chollier, M., Sand, O., Turatsinze, J.-V., Janky, R., Defrance, M., Vervisch, E., Brohee, S., van Helden, J., 2008. RSAT: regulatory sequence analysis tools. Nucleic Acids Res. 36 (Web Server), W119. https://doi.org/10.1093/nar/gkn304.

Ulmasov, T., Liu, Z.B., Hagen, G., Guilfoyle, T.J., 1995. Composite structure of auxin response elements. Plant Cell 7 (10), 1611–1623.

Ulmasov, T., Murfett, J., Hagen, G., Guilfoyle, T.J., 1997. Aux/IAA proteins repress expression of reporter genes containing natural and highly active synthetic auxin response elements. Plant Cell 9 (11), 1963–1971. https://doi.org/10.1105/tpc.9.11.1963.

Umarov, R.K., Solovyev, V.V., 2017. Recognition of prokaryotic and eukaryotic promoters using convolutional deep learning neural networks. PLoS One 12 (2). https://doi.org/10.1371/journal.pone.0171410.

van Verk, M.C., Bol, J.F., Linthorst, H.J., 2011. WRKY transcription factors involved in activation of SA biosynthesis genes. BMC Plant Biol. 11, 89.

Villao-Uzho, L., Chávez-Navarrete, T., Pacheco-Coello, R., Sánchez-Timm, E., Santos-Ordóñez, E., 2023. Plant promoters: their identification, characterization, and role in gene regulation. Genes 14 (6), 1226. https://doi.org/10.3390/genes14061226.

Vu, T.V., Das, S., Tran, M.T., Hong, J.C., Kim, J.Y., 2020. Precision genome engineering for the breeding of tomatoes: recent progress and future perspectives. Frontiers in Genome Editing 2. https://doi.org/10.3389/fgeed.2020.612137.

Wan, M., Xie, H., Guo, H., Jing, S., Zeng, D., Li, B., Zhu, B., Zeng, Z., 2025. Developing a pipeline for identification, characterization and molecular editing of cis-regulatory elements: a case study in potato. aBIOTECH 6 (1), 91–96. https://doi.org/10.1007/s42994-024-00185-1.

Wang, X., Aguirre, L., Rodríguez-Leal, D., Hendelman, A., Benoit, M., Lippman, Z.B., 2021. Dissecting cis-regulatory control of quantitative trait variation in a plant stem cell circuit. Nat. Plants 7 (4), 419–427. https://doi.org/10.1038/s41477-021-00898-x.

Wang, X., Chen, C., He, C., Chen, D., Yan, W., 2022. Mapping open chromatin by ATAC-seq in bread wheat. Front. Plant Sci. 13. https://doi.org/10.3389/fpls.2022.1074873.

Wang, Y., He, S., Fang, X., 2023a. Emerging roles of phase separation in plant transcription and chromatin organization. Curr. Opin. Plant Biol. 75, 102387.

Wang, Y., Li, W., Qu, J., Li, F., Du, W., Weng, J., 2023b. Genome-wide characterization of the maize (Zea mays L.) WRKY transcription factor family and their responses to Ustilago maydis. Int. J. Mol. Sci. 24 (19), 14916. https://doi.org/10.3390/ijms241914916.

Wang, Y.G., Zhang, Y.M., Wang, Y.H., Zhang, K., Ma, J., Hang, J.X., Su, Y.T., Tan, S.S., Liu, H., Xiong, A.S., Xu, Z.S., 2024. The y locus encodes a REPRESSOR of PHOTOSYNTHETIC GENES protein that represses carotenoid biosynthesis via interaction with APRR2 in carrot. Plant Cell 36 (8), 2798–2817. https://doi.org/10.1093/plcell/koae111.

Winter, C.M., Austin, R.S., Blanvillain-Baufumé, S., Reback, M.A., Monniaux, M., Wu, M.F., Sang, Y., Yamaguchi, A., Yamaguchi, N., Parker, J.E., Parcy, F., Jensen, S.T., Li, H., Wagner, D., 2011. LEAFY target genes reveal floral regulatory logic, cis motifs, and a link to biotic stimulus response. Dev. Cell 20 (4), 430–443. https://doi.org/10.1016/j.devcel.2011.03.019.

Wittkopp, P.J., Kalay, G., 2012. Cis-regulatory elements: molecular mechanisms and evolutionary processes underlying divergence. Nat. Rev. Genet. 13 (1), 59–69. https://doi.org/10.1038/nrg3095.

Wu, B., Luo, H., Chen, Z., Amin, B., Yang, M., Li, Z., Wu, S., Salmen, S.H., Alharbi, S.A., Fang, Z., 2024. Rice promoter editing: an efficient genetic improvement strategy. Rice 17 (1). https://doi.org/10.1186/s12284-024-00735-7.

Xu, X., Xu, Z., Li, Z., Zakria, M., Zou, L., Chen, G., 2021. Increasing resistance to bacterial leaf streak in rice by editing the promoter of susceptibility gene OsSULRT3; 6. Plant Biotechnol. J. 19 (6), 1101–1103. https://doi.org/10.1111/pbi.13602.

Xu, L., Sun, B., Liu, S., Gao, X., Zhou, H., Li, F., Li, Y., 2023. The evaluation of active transcriptional repressor domain for CRISPRi in plants. Gene 851, 146967. https://doi.org/10.1016/j.gene.2022.146967.

Yadav, R.K., Perales, M., Gruel, J., Girke, T., Jönsson, H., Venugopala Reddy, G., 2011. WUSCHEL protein movement mediates stem cell homeostasis in the Arabidopsis shoot apex. Genes Dev. 25 (19), 2025–2030. https://doi.org/10.1101/gad.17258511.

Yamamoto, Y.Y., Ichida, H., Matsui, M., Obokata, J., Sakurai, T., Satou, M., Seki, M., Shinozaki, K., Abe, T., 2007. Identification of plant promoter constituents by analysis of local distribution of short sequences. BMC Genom. 8. https://doi.org/10.1186/1471-2164-8-67.

Yan, W., Chen, D., Kaufmann, K., 2016. Efficient multiplex mutagenesis by RNA-guided Cas9 and its use in the characterization of regulatory elements in the AGAMOUS gene. Plant Methods 12, 1–9.

Yang, H., Yu, G., Lv, Z., Li, T., Wang, X., Fu, Y., Zhu, Z., Guo, G., He, H., Wang, M., Qin, G., Liu, F., Zhong, Z., Xue, Y., 2025. Epigenome profiling reveals distinctive regulatory features and cis-regulatory elements in pepper. Genome Biol. 26 (1). https://doi.org/10.1186/s13059-025-03595-6.

Yasmeen, E., Wang, J., Riaz, M., Zhang, L., Zuo, K., 2023. Designing artificial synthetic promoters for accurate, smart, and versatile gene expression in plants. Plant Commun. 4 (4), 100558. https://doi.org/10.1016/j.xplc.2023.100558.

Ye, J., Wang, X., Hu, T., Zhang, F., Wang, B., Li, C., Yang, T., Li, H., Lu, Y., Giovannoni, J.J., Zhang, Y., Ye, Z., 2017. An InDel in the promoter of AI-ACTIVATED MALATE TRANSPORTER9 selected during tomato domestication determines fruit malate contents and aluminum tolerance. Plant Cell 29 (9), 2249–2268. https://doi.org/10.1105/tpc.17.00211.

Yilmaz, A., Mejia-Guerra, M.K., Kurz, K., Liang, X., Welch, L., Grotewold, E., 2010. AGRIS: the Arabidopsis gene regulatory information server, an update. Nucleic Acids Res. 39 (Database), D1118. https://doi.org/10.1093/nar/gkq1120.

Zahm, A.M., Owens, W.S., Himes, S.R., Fallon, B.S., Rondem, K.E., Gormick, A.N., Bloom, J.S., Kosuri, S., Chan, H., English, J.G., 2024. A massively parallel reporter assay library to screen short synthetic promoters in mammalian cells. Nat. Commun. 15 (1). https://doi.org/10.1038/s41467-024-54502-9.

Zhai, Q., Li, C., 2019. The plant Mediator complex and its role in jasmonate signaling. J. Exp. Bot. 70, 3415–3424.

Zhan, X., Lu, Y., Zhu, J.K., Botella, J.R., 2021. Genome editing for plant research and crop improvement. J. Integr. Plant Biol. 63 (1), 3–33. https://doi.org/10.1111/jipb.13063.

Zhang, Y., Li, Z., Chen, N., Huang, Y., Huang, S., Chen, X., 2020. Phase separation of Arabidopsis EMB1579 controls transcription, mRNA splicing, and development. PLoS Biol. 18 (7), e3000782. https://doi.org/10.1371/journal.pbio.3000782.

Zhang, Y., Fan, X., Liang, Du, Guo, Q., Zhang, X., Nie, M., Li, C., Meng, S., Zhang, X., Xu, P., Guo, W., Wang, H., Liu, Q., Wu, Y., 2023a. The identification of a yield-related gene controlling multiple traits using GWAS in sorghum (Sorghum bicolor L.). Plants 12 (7), 1557. https://doi.org/10.3390/plants12071557.

Zhang, Y., Sun, H., Tian, L., Miao, Y., Hou, L., Sun, M., Qi, M., Li, T., 2023b. Transcriptome analysis identifies downstream genes of CLAVATA3 in tomato. Horticulturae 10, 11.

Zhao, Y., Wu, L., Fu, Q., Wang, D., Li, J., Yao, B., Yu, S., Jiang, L., Qian, J., Zhou, X., Han, L., Zhao, S., Ma, C., Zhang, Y., Luo, C., Dong, Q., Li, S., Zhang, L., Jiang, X., Li, Y., Luo, H., Li, K., Yang, J., Luo, Q., Li, L., Peng, S., Huang, H., Zuo, Z., Liu, C., Wang, L., Li, C., He, X., Friml, J., Du, Y., 2021. INDITTO2 transposon conveys auxin-mediated DRO1 transcription for rice drought avoidance. Plant Cell Environ. 44 (6), 1846–1857. https://doi.org/10.1111/pce.14029.

Zhao, H., Yang, M., Bishop, J., Teng, Y., Cao, Y., Beall, B.D., Li, S., Liu, T., Fang, Q., Fang, C., Xin, H., Nützmann, H.-W., Osbourn, A., Meng, F., Jiang, J., 2022. Identification and functional validation of super-enhancers in *Arabidopsis thaliana*. Proc. Natl. Acad. Sci. 119, e2215328119.

Zhu, J., Liu, M., Liu, X., Dong, Z., 2018. RNA polymerase II activity revealed by GRO-seq and pNET-seq in Arabidopsis. Nat. Plants 4 (12), 1112–1123. https://doi.org/10.1038/s41477-018-0280-0.

CHAPTER

24

AI integration with omics and CRISPR to decipher complex traits in plants

Ashwini Talakayala[1], *Kasi Rao Mediga*[2], *Sabavat Raju Naik*[1, 3], *and Mallikarjuna Garladinne*[4]

[1]**International Crops Research Institute for Semi-Arid Tropics, Hyderabad, Telangana, India;** [2]**Godrej Agrovet Limited, Khammam, Telangana, India;** [3]**Department of Genetics and Plant Breeding, Parbhani, Maharashtra, India;** [4]**Plantigen Agrisolutions, Pulivendula, Andhra Pradesh, India**

24.1 Introduction

Integration of artificial intelligence (AI) with "omics" technologies and CRISPR have greatly advanced plant science and agricultural innovation. Synergy within rapid development of crops with enhanced traits for disease resistance, drought tolerance, and increased nutrition (Varadharajan et al., 2025). These all directly determine crop productivity, stability, and resilient for variable climatic conditions. Several plant traits are controlled by multiple genes, epistatic interactions, and genotype through environment effects for their dissection and prediction is challenging (Holland, 2007; Mackay et al., 2009). However, QTL mapping and GWAS have identified key loci, however, these approaches often overlook thus, the full spectrum of genetic and regulatory factors underlying complex traits (Manolio et al., 2009).

24.1.1 Limitations of conventional breeding

Crop improvement through conventional breeding has limitations due phenotypic selection and marker-assisted techniques for complex traits, as well as long breeding cycles and gene-environment interactions. Developing resilient cultivars is crucial to adopt advanced predictive and integrative methods, such as Omics and CRISPR, alongside AI integration (Cooper et al., 2014).

Intelligent Gene Editing in Plants
https://doi.org/10.1016/B978-0-443-44079-3.00015-2

24.1.2 Emergence of omics technology, CRISPR, and AI as integrative tools

Integrating omics, CRISPR, and AI is transforming a complex plant trait by studying multiomics, which are significant (Roychowdhury et al., 2023). The CRISPR system enables gene validation and development of new alleles to improve plant ability (Zhu et al., 2020). AI enhances by analyzing complex omics data using ML and DL models, integrating multiomics datasets for trait predictions than statistical methods (Libbrecht and Noble, 2015; Montesinos-Lopez et al., 2019).

24.2 Omics approaches for complex traits

Omics technology serves as a key resource for crop improvement with growing omics data (Dai and Shen, 2022).

24.2.1 Genomics: Genome sequencing, QTL mapping, and GWAS

Genomics significantly understands genetic variation related to complex traits therefore, recent advances in NGS and long-read sequencing (e.g., PacBio, Nanopore) have enhanced computational capabilities and software applications, resulting in 3201 sequenced plant genomes (Ingvarsson and Street 2011; Khan et al., 2020). These resources support techniques such as QTL-Seq (Wang et al., 2022a, 2022b), MutMap (Abe et al., 2012), and GWAS, improving the analysis of genetic diversity regarding structural variants and SNPs. QTL mapping remains a popular method for identifying genetic regions tied to traits, despite its limitations in resolution (Collard et al., 2005). GWAS addresses these challenges by leveraging diverse population to detect small effect loci for yield and drought tolerance (Michael and VanBuren, 2020; Wellenreuther et al., 2019; Korte and Farlow, 2013). Emerging tools such as genomic prediction and genomic selection utilize genome marker data and statistical models for efficient breeding value prediction, expediting plant breeding efforts (Crossa et al., 2017).

24.2.1.1 Transcriptomics: expression profiling and regulatory networks

Transcriptome analysis reveals genome-wide gene activity and regulatory genes and pathways underlying complex plant traits (Tyagi et al., 2022). Techniques such as RNA sequencing and microarrays facilitates profiling of gene expression in various tissues, stages, and conditions (Wang et al., 2009), aiding identification of candidate genes for stress, yield, and nutrient use. DEGs uncovered by comparing tolerant with sensitive genotypes (Gelli et al., 2014). Time-series transcriptomics captures dynamic expression patterns under abiotic stresses (Mishra et al., 2025). Coexpression and gene regulatory network (GRN) analyses integrate transcription factors, signaling molecules, and target genes to elucidate trait regulation (Langfelder and Horvath, 2008). Furthermore, eQTL mapping and TWAS link genetic variation to expression and traits (Kremling et al., 2018).

24.2.1.2 Proteomics: protein interactions and posttranslational modifications

Proteomics is crucial for understanding plant traits by analyzing protein levels, interactions, and modifications (e.g., phosphorylation, acetylation) (Jorrín-Novo et al., 2015).

Techniques i.e., protein–protein interaction mapping (yeast two-hybrid, Co-IP/MS, AP-MS, BioID) applied in crops such as rice and maize (Braun et al., 2009; Ralser et al., 2006). These regulatory mechanisms that surpass transcriptomics, especially under stress environment, by identifying proteins for signaling pathways (Vu et al., 2021; Murmu et al., 2024). Proteomics is advanced further through AI and ML by predicting interactions, within integrating datasets, which aids in developing precise breeding strategies for improving stress resistance and yield.

24.2.1.3 *Metabolomics: metabolic pathways and biomarkers*

Metabolomics provides direct information at biochemical state by examining plants with primary and secondary metabolites, reflecting genetic and environmental interactions (Suruthi et al., 2025). With application of new methods including LC-MS, GC-MS, and NMR can quantify thousands of metabolites across tissues at its developmental stages (Fiehn, 2002; Ferinie and Schauer, 2009). By application it helps to identify key metabolic pathways involved in growth, nutrient-use-efficiency, and stress related (abiotic and biotic) response.

Plant metabolomes are composed by primary growth-related (e.g., sugars, amino acids) and secondary (i.e., defense) metabolites such as flavonoids or alkaloids. These metabolites are the biomarkers, which connect genotypic to phenotypic characteristics and also improve nutrient absorption and plant health through interactions with the microorganisms (Obata and Fernie, 2012). mQTL and mGWAS studies identify genome regions that control metabolite accumulation, which can be used for breeding and genome editing (Chen et al., 2014; Guo et al., 2020). AI application in metabolomics enhances biomarker discovery, genotype classification and multiomics integration to fast-track precision breeding and CRISPR-based crop improvement (Hamany Djande et al., 2020; Jamil et al., 2022).

24.2.1.4 *Epigenomics: DNA methylation, histone modifications, ncRNAs*

Epigenomics useful to know the heritable and flexible alterations in gene expression without changes in the DNA sequence instead by epigenetic modifications i.e., DNA methylation, histone modifications, and noncoding RNAs (ncRNAs) (Zhang et al., 2006). They control chromatin and gene access, affecting development, stress and traits (Springer and Schmitz 2017).

24.2.1.4.1 DNA methylation

Cytosine base methylation results in the suppression of transposable elements and ensures genome stability, as well as he regulation of gene expression following environmental stimuli (Zhou et al., 2025; Chmielowska-Bąk et al., 2023). WGBS and RRBS can map DNA methylation sites, identify variations of DNA methylation in crops such as rice, maize and wheat (Cokus et al., 2008; Secco et al., 2015; Wang et al., 2015). Methylation QTL (meQTL) studies also connect meQTL SNPs with methylation variation to discover regulatory loci (Li et al., 2025a, 2025b; Dubin et al., 2015).

24.2.1.4.2 Histone modifications

Histone are proteins that DNA is packed to undergoes acetylation, methylation, and phosphorylation, affects, where it associates with repressed gene expression, while looser structure for gene activation (Kajrolkar, 2025a, 2025b). They persist through cell division, building stress memory to acclimate its exposure and to transfer memory to their progeny (Gallusci et al., 2023). ChIP sequencing enables mapping its modifications such as H3K4me3 and H3K27me3, reveals actin and chromatin regions in rice, maize and barley (Zhang et al., 2009; Kawakatsu et al., 2016).

24.2.1.4.3 Noncoding RNA (nc-RNA)

miRNAs, siRNAs, and lncRNAs are different types of ncRNAs involved in regulating gene expression, therefore siRNAs control RNA-directed DNA methylation, particularly in plants. Small RNA-seq and degradome sequencing reveal their involvement in gene silencing, metabolism processes, and stress responses in wheat, and maize (Song et al., 2011; Budak et al., 2013). Integrative epiGWAS offers epigenetic biomarkers (Schmitz et al., 2013; Springer and Schmitz, 2017). AI and DL provide predictive modeling for epigenetic traits (Sperschneider, 2020; Wang et al., 2021).

24.2.1.5 Phenomics: high-throughput phenotyping platforms

Phenomics provide large scale, precise trait with data that complement with genomic and other omics resources for association studies in crop plants. Various advancing imaging, sensors, automation, and robotics for high-throughput phenotyping (HTTPs) (Zhao et al., 2019). Advanced usage of greenhouse-based platforms, including UAV/drone imaging, LiDAR, hyperspectral and thermal cameras, and automated scanners in greenhouse, to enable nondestructive monitoring for plant growth, biomass, canopy, temperature, water use efficiency, and stress responses (Araus and Cairns, 2014; Yang et al., 2020). Root phenotyping using rhizotrons, X-ray CT, and shovelomics with trait information for resilience breeding (Lynch, 2019; Tardieu et al., 2017). High dimensional phenotypic datasets are increasingly integrated with GWAS, QTL mapping and genomic prediction models to uncover trait-genotype associations and accelerate breeding (Cobb et al., 2013; Millet et al., 2019). AI and ML approaches, such as convolutional neural networks (CNNs), DL, and predictive modeling, are being applied for automated for trait prediction by knowing complex phenomic data into insights of breeding (Singh et al., 2016; Shakoor et al., 2017). Comprehensive integrative phenomics is driven in association with its application in maize, wheat, rice, sorghum, and barley, linking phenotypic plasticity with the underlying genetic architecture for improved stress and yield (Yang et al., 2020; Crain et al., 2021).

24.2.1.6 Microbiomics: Role of plant–microbe interactions

It emerged as important component associated with plant microbiome, encompassing with rhizosphere, phyllosphere, and endophytic microbial communities in crop improvement with significant role in nutrient acquisition, growth, stress resilience and for disease management (Compant et al., 2019; Vives-Peris et al., 2020). High-throughput sequencing in microbial diversity through meta-transcriptomics, metaproteomics and metabolomics with it functional insights into microbial-mediated. Omics that integrated by microbiome datasets with all omics application studies in enables the identification of microbial taxa or functional pathways related to productivity traits and stress adaptation (Awany et al., 2019; Wallace et al., 2018). AI and ML are increasingly used to decode plant–microbiome networks, to harness beneficial microbes for sustainable agriculture with the studies in maize, wheat, rice, and soybean supporting microbiome-based breeding and crop improvement (Trivedi et al., 2020; Chang et al., 2017). List of multiomics with applications are tabulated below.

24.2.2 Interactome: An interaction between biomolecules

The integration of AI with omics and CRISPR for transforming plants by enabling and decoding its complex traits through comprehensive interactome analysis. The interactome, encompassing protein–protein, protein–DNA, and signaling interactions, provides a systems-level understanding of cellular regulation. AI and machine learning helps in the prediction and reconstruction from multiomics datasets, regulatory nodes and pathways that control stress, growth, and yield (Kassem, 2025). Coupled with CRISPR-based functional validation, this integrated approach allows precise editing of candidate genes, which boosts climate resilient, high-yield crops (Table 24.1).

24.3 CRISPR tools for trait modification

Genome editing technology is an identified tools capable of programming of genome, with genome editing practices i.e., CRISPR–Cas technology is a simple and highly efficient tool, found with immune system of bacteria and archaea with defensive system against previously encountered viruses (Barrangou et al., 2007). CRISPR/Cas includes a Cas protein and a CRISPR array of repeats and genome targeting spacers. The Protospacer Adjacent Motif (PAM) is a targeting element with significance as a self- and a non-self-recognition system, from being targeted by CRISPR–Cas locus with Cas9 protein with target site with known crRNA-tracrRNA, i.e., single guide RNA (sgRNA) guides Cas9 to the PAM for precise DNA editing (Qi et al., 2013). CRISPR-based genome editing is revolutionizing plant breeding and functional genomics by enabling precise, efficient, and predictive trait modifications. CRISPR system including Cas9, Cas12a, and dCas9-based tools, which allows targets with single and multiple genes to enhance yield, stress tolerance, disease resistance, and nutrition (Jaganathan et al., 2018; Zhang et al., 2019). Complex traits in plants i.e., drought tolerance, nitrogen use efficiency yield, and disease resistance, are typically polygenic and are influenced by gene-gene environment interactions. Thus, CRISPR offers a precise, efficient, and versatile approach to functionally validate candidate genes, regulatory elements, and also pathways underlying complex traits.

24.3.1 Cas variants

SpCas9 and Cas12a (Cpf1) enable precise targeted knockouts or modifications of single or multiple genes associated with complex traits in plants. These variants signify multiplex genome editing, the loci to evaluate its epistatic effects. It showed its success application to edit drought-responsive genes, flowering regulators, and yield-related loci in crops i.e., rice, maize, and wheat (Jaganathan et al., 2018; Lowder et al., 2015; Zhang et al., 2019), b. dCas9-based: dCas9-based CRISPRa/CRISPRi are catalytically inactive when Cas9 fused to activators or repressor, which are enabled its modulation by gene expression without changes in DNA sequence. These variants allow fine-tuning gene networks that regulate complex traits targeting metabolic pathways, genes, for drought tolerance, nutrient uptake, and biomass accumulation (Piatek et al., 2015; Lowder et al., 2015).

TABLE 24.1 List of major multi-omics approaches used for crop improvement, each contributing unique insights into plant function and trait development.

S. No.	Name	Data type (input)	Key methods/analysis	Primary outputs	Example crops	References
Genomics						
1	SNP-based GWAS (array or sequencing)	SNP genotypes (arrays/WGS/GBS)	Mixed linear models (MLM, LMM), population structure & kinship correction	SNP–trait associations, candidate loci	Maize, rice, wheat, sorghum	Yu et al. (2006)
2	QTL mapping (biparental populations)	Linkage maps from RILs, DH, F2 + phenotypes	Interval mapping, composite interval mapping	QTL positions, effect sizes	Rice, wheat, soybean	Collard et al. (2005) Broman et al. (2003)
3	Sequence-based GWAS (WGS-GWAS)	Whole-genome sequencing variants (SNPs, indels, SVs)	GWAS with dense markers; haplotype/G×E models	High-resolution SNP/indel associations	Maize, rice, chickpea	Huang and Han (2014), Varshney et al. (2021)
4	Haplotype/Regional GWAS	Phased haplotypes or sliding-window SNP sets	Haplotype-based association	Haplotype blocks, allelic series	Rice, maize	Yano et al. (2016), Zhao et al. (2011)
5	Pan-genome/PAV GWAS	Gene presence/absence, structural variants	Association tests on PAVs, k-mer GWAS	PAVs, gene-level associations	Brassica, maize, wheat	Gao et al. (2019) Jayakodi et al. (2021)
6	eQTL mapping (genomics + transcriptomics)	SNPs + RNA-seq expression	cis/trans eQTL mapping	SNP–expression associations	Maize, *Arabidopsis*, rice	Albert and Kruglyak (2015);
7	TWAS (Transcriptome-wide association study)	GWAS summary + expression reference	PrediXcan, FUSION	Candidate genes mediated via expression	Maize, rice	Li et al. (2022), Liu et al. (2017c)
8	pQTL/mQTL mapping	Genotypes + proteomics/ metabolomics	QTL mapping or GWAS on molecular traits	pQTLs, mQTLs linking loci to proteins/metabolites	Rice, maize, wheat	Xu et al. (2025)
9	Genomic prediction/ selection (GS/GP)	Genome-wide markers + phenotypes	GBLUP, Bayesian models, ML/DL (RF, DL)	Genomic estimated breeding values (GEBVs)	Wheat, rice, maize	Meuwissen et al. (2001), (2025)
10	Multi-omics integrative GWAS	Genomics + transcriptomics + metabolomics	Multi-trait GWAS, ML integration	Multi-layer causal associations	Maize, rice, wheat	Riedelsheimer et al. (2012) Varshney et al. (2020) Li et al. (Unpublished)

Transcriptomics:						
1	Differential gene expression (DGE) analysis	RNA-seq or microarray expression profiles	DESeq2, edgeR, limma	Differentially expressed genes under conditions/genotypes	Rice, maize, wheat, sorghum	Wang et al. (2009), Zhang et al. (2010)
2	Co-expression network analysis	RNA-seq expression matrix	WGCNA, ARACNE, co-expression clustering	Gene modules, hub genes, regulatory networks	Rice, *Arabidopsis*, maize	Langfelder and Horvath (2008), Usadel et al. (2009)
3	Expression QTL (eQTL) mapping	Genotypes + RNA-seq expression	Linear models, MLM, Matrix eQTL	cis- and trans-eQTLs linking SNPs → gene expression	Maize, rice, *Arabidopsis*	Kremling et al. (2019), Albert and Kruglyak (2015)
4	Transcriptome-Wide Association Studies (TWAS)	GWAS summary stats + reference expression panel	PrediXcan, FUSION	Genes whose expression mediates trait variation	Maize, rice	Li et al. (2022), Liu et al. (2017c)
5	Time-series expression profiling	RNA-seq at multiple timepoints/ developmental stages	Clustering, PCA, spline regression	Temporal expression patterns, regulatory dynamics	Rice, *Arabidopsis*	Harb et al. (2010), Filichkin et al. (2010)
6	Alternative splicing/ isoform analysis	RNA-seq reads, transcript assemblies	rMATS, SUPPA, IsoformSwitchAnalyzeR	Differential isoform usage, splicing variants	Maize, rice, wheat	Wang et al. (2018a), 2018b; Thatcher et al. (2016)
7	Multi-omics integrative association	RNA-seq + genomics/ metabolomics	Multi-layer GWAS, ML/AI integration	Candidate genes, regulatory pathways, causal networks	Maize, rice, wheat	Riedelsheimer et al. (2012), Varshney et al. (2020)
Proteomics						
1	Protein–protein interaction (PPI) mapping	Proteins from tissue extracts, tagged proteins	Yeast two-hybrid, co-immunoprecipitation (Co-IP), affinity purification MS (AP-MS), BioID/TurboID	Protein interaction networks, hub proteins, regulatory modules	Rice, maize, *Arabidopsis*	Braun et al. (2009)
2	Posttranslational modification (PTM) profiling	Proteins from tissue/ stress conditions	Phosphoproteomics, ubiquitinomics, acetylomics via MS, enrichment strategies	PTM sites, modified proteins, regulatory pathways	Rice, maize, wheat	Vu et al. (2021)

(Continued)

TABLE 24.1 List of major multi-omics approaches used for crop improvement, each contributing unique insights into plant function and trait development.—cont'd

S. No.	Name	Data type (input)	Key methods/analysis	Primary outputs	Example crops	References
3	Differential protein expression analysis	Quantitative proteomics data (iTRAQ, TMT, label-free MS)	Statistical models, ANOVA, PCA, clustering	Differentially abundant proteins (DAPs) under stress or genotype differences	Rice, wheat, soybean	Jorrín-Novo et al. (2015)
4	ePPI/protein QTL mapping (pQTL)	Genotypes + protein abundance profiles	Linear models, mixed models, GWAS on protein traits	Protein-level QTLs linking loci → protein abundance	Maize, rice	Wen et al. (2014)
5	Integrative proteomics + AI/ML	Proteomics + transcriptomics + genomics + phenotypes	Machine learning, deep learning, network modeling	Predictive regulatory networks, candidate proteins, trait associations	Rice, maize, wheat	Sahito et al., 2024
6	Time-series proteomics	Proteomics at multiple developmental/ stress stages	Clustering, PCA, temporal modeling	Dynamic protein expression patterns, stress response pathways	Rice, *Arabidopsis*	Vu et al. (2021), Kosová et al. (2018)
Metabolomics						
1	Metabolite profiling (untargeted)	LC-MS, GC-MS, NMR metabolite spectra	Peak detection, PCA, PLS-DA, clustering	Metabolite fingerprints, biomarkers of stress/tolerance	Rice, wheat, maize	Fiehn (2002), Fernie and Schauer (2009)
2	Targeted metabolomics	Quantified metabolites (amino acids, sugars, lipids)	Targeted MS assays, isotope labeling	Specific metabolite quantification, pathway activity	Soybean, barley, rice	Saito and Matsuda (2010), Obata and Fernie (2012)
3	Metabolite QTL (mQTL) mapping	Genotypes + metabolite levels	Interval mapping, composite interval mapping	Genomic loci controlling metabolite variation	Tomato, maize, rice	Wen et al. (2015), Keurentjes et al. (2006)
4	Metabolite GWAS (mGWAS)	Genome-wide SNPs + metabolite data	Mixed linear models, population structure correction	SNP–metabolite associations, candidate genes	Rice, maize, wheat	Chen et al. (2014), Wen et al. (2014)

5	Metabolic pathway enrichment	Metabolite sets linked to KEGG/MetaCyc	Pathway enrichment, network-based analysis	Pathway-level insights into stress or nutrient metabolism	Rice, *Arabidopsis*	Tohge and Fernie (2010), Okazaki and Saito (2016)
6	Integrative metabolomics + genomics (systems biology)	Multi-omics: SNPs, transcriptome, metabolites	Network inference, machine learning, Bayesian modeling	Multi-layer networks linking genotype–metabolite–phenotype	Maize, chickpea, sorghum	Riedelsheimer et al. (2012) Gupta et al. (2021a, 2021b)
7	Biomarker discovery (AI/ML)	Large-scale metabolite datasets	Machine learning (RF, SVM, DL), feature selection	Predictive biomarkers for stress tolerance or yield	Wheat, rice, maize	Wang et al. (2022a), 2022b Jamil et al. (2022)
Epigenomics						
1	Whole-genome bisulfite sequencing (WGBS)	Cytosine methylation profiles	Bisulfite conversion + sequencing, DMR calling	Genome-wide DNA methylation maps, differentially methylated regions (DMRs)	Rice, maize, *Arabidopsis*	Cokus et al. (2008), Secco et al. (2015)
2	Reduced representation bisulfite sequencing (RRBS)	Subset of CpG-rich regions	Targeted methylome profiling	Methylation polymorphisms in specific genomic regions	Rice, wheat	Wang et al. (2015)
3	meQTL mapping (Methylation QTL)	SNPs + methylation variation	Association between SNPs and DNA methylation states	Genetic loci controlling methylation patterns	Maize, *Arabidopsis*	Li et al. (2014), Dubin et al. (2015)
4	ChIP-seq for histone modifications	Histone mark enrichment (H3K4me3, H3K27me3, etc.)	Chromatin immunoprecipitation + sequencing	Histone modification landscapes, active/repressive chromatin	Rice, maize, barley	Zhang et al. (2009), Kawakatsu et al. (2016)
5	ncRNA profiling (small RNA-seq, lncRNA-seq)	ncRNA sequences (miRNAs, siRNAs, lncRNAs)	Small RNA sequencing, target prediction	Regulatory RNAs controlling stress and development	Wheat, rice, maize	Song et al. (2011), Budak et al. (2013)
6	Epigenome-wide association studies (epiGWAS)	DNA methylation + histone/ncRNA layers	Association of epigenetic marks with traits	Epigenetic biomarkers, candidate regulatory regions	Rice, maize	Schmitz et al. (2013), Springer and Schmitz (2017)
7	AI/ML-enabled epigenomics	Multi-omics epigenetic datasets	Deep learning, CNNs, SVMs	Predictive modeling of epigenetic states and trait outcomes	Rice, maize	Sperschneider, 2020; Wang et al., 2013

(*Continued*)

TABLE 24.1 List of major multi-omics approaches used for crop improvement, each contributing unique insights into plant function and trait development.—cont'd

S. No.	Name	Data type (input)	Key methods/analysis	Primary outputs	Example crops	References
Phenomics						
1	Field-based HTP (e.g., drones, UAVs, tractor-mounted sensors)	RGB, multispectral, hyperspectral images; canopy temperature	Image analysis, vegetation indices (NDVI, PRI), thermal profiling	Growth dynamics, canopy temperature, water-use efficiency, yield prediction	Maize, wheat, sorghum	Araus and Cairns (2014), Singh et al. (2016)
2	Greenhouse/ controlled environment HTP platforms	Automated imaging (RGB, NIR, fluorescence, 3D scanning)	Growth modeling, time-series trait extraction	Shoot architecture, biomass, stress responses	Rice, *Arabidopsis*, barley	Granier and Vile (2014), Yang et al. (2020)
3	Root phenotyping (rhizotrons, X-ray CT, shovelomics)	Root images, 3D root architecture	Image segmentation, root trait quantification	Root depth, density, RSA traits linked to stress tolerance	Maize, wheat, soybean	Lynch (2019); Wilson et al., 2012
4	LiDAR and 3D imaging	Point clouds, 3D canopy structure	3D reconstruction, machine learning	Plant height, leaf area, canopy architecture	Rice, maize, sorghum	Paulus et al. (2013), Madec et al. (2017)
5	Thermal and hyperspectral imaging	Canopy temperature, spectral reflectance	Thermal indices, spectral unmixing	Heat/drought stress detection, pigment/nutrient profiling	Wheat, maize, barley	Berger et al. (2010), Furbank and Tester (2011)
6	AI/ML-enabled phenomics	Multi-sensor phenotyping datasets	Deep learning (CNNs, RNNs), trait prediction models	Automated feature extraction, predictive breeding values	Wheat, maize, rice	Singh et al. (2018), Shakoor et al. (2017)
Microbiomics						
1	Amplicon sequencing (16S rRNA, ITS)	Microbial community DNA (bacteria/fungi)	Amplicon sequencing, OTU/ASV clustering, diversity metrics	Microbial diversity, taxonomic composition	Maize, rice, wheat	Bulgarelli et al. (2013), Edwards et al. (2015)
2	Shotgun metagenomics	Total soil/plant-associated DNA	Whole-metagenome sequencing, assembly, functional annotation	Functional genes, pathways, microbial genome bins	Rice, soybean, barley	Xu et al. (2018)

3	Metatranscriptomics	Total RNA from plant–microbe systems	RNA-seq of microbial transcripts, co-expression with host	Active microbial functions, stress-response pathways	Maize, *Arabidopsis*	Turner et al. (2013)
4	Metaproteomics	Protein extracts from rhizosphere/endosphere	LC–MS/MS, protein identification	Functional proteins, enzyme activities, host–microbe signaling	Rice, wheat	Bastida et al. (2014), Hultman et al. (2015)
5	Metabolomics of root exudates & rhizosphere	Root exudates, soil metabolites	GC–MS, LC–MS, NMR	Microbial metabolites, signaling compounds	Maize, soybean	Vives-Peris et al. (2020), Hu et al. (2018)
6	Genome–microbiome-wide association studies (GMWAS)	Host SNPs + microbiome profiles	GWAS/QTL linking host genotype & microbiome traits	Host genetic control of microbiome composition	Maize, sorghum	Wallace et al. (2018)
7	AI/ML-enabled microbiome analysis	Multi-omics microbiome + host datasets	Machine learning, network inference, predictive modeling	Beneficial microbial taxa prediction, synthetic consortia design	Wheat, maize, rice	Chang et al. (2017) Trivedi et al. (2020)

24.3.2 Base and prime editing

Base Editors (BE) and Prime Editors (PE) are useful tools for allelic diversification and simulating natural variation to analyze causal variants underlying quantitative traits because they allow for precise nucleotide substitutions or minor insertions and deletions without causing double-strand breaks. In order to investigate the functional contributions of SNPs in yield-related and stress-responsive genes in plants, several methods have been used (Zong et al., 2017; Lin et al., 2020).

24.3.3 CRISPR–Cas13

CRISPR–Cas13 acts on RNA transcripts rather than DNA, enabling temporary modulation of gene expression. As noted by Abudayyeh et al. (2017) and Mahas and Mahfouz (2018), discussed that RNA-targeting enables precise analysis of plant stress and disease response networks.

24.3.4 Multiplex editing for polygenic traits in plants

It usually enables with the modifications of multiple genes or regulatory elements, for dissecting and improving polygenic traits i.e., yield, drought tolerance and for nutrient improvement. Using CRISPR technology, multiple guide RNAs (gRNAs) under one promoter allow targeting of complex gene networks (Xie et al., 2015; Lowder et al., 2015). Therefore, editing represents a transformative approach for precision breeding of polygenic traits to manipulate through conventional process (Wang et al., 2018a, 2018b; Zeng et al., 2020).

24.4 Role of AI in omics and genome editing

24.4.1 ML and DL for omics data integration

Machine learning drives an innovation and impact in crop improvement as well as in helping to work more sustainable on the farm (Chen et al., 2024). Genome editing has great potential in the field of precision breeding for yield, resistance to stress factors, pest control potential and nutrition quality (Li et al., 2024a, 2024b). ML is more usage than traditional methods in plant omics data analysis where it handles complex and larger datasets. Application usage of omics platforms i.e., high throughput sequencing technologies with an information, which is to explore to know it's association, decoding plant stress responses and uncovering genomic complexities (Singh et al., 2016). ML, an AI subset, builds models to analyze systems without explicating programming (Jordan and Mitchell, 2015), where DL (deep learning), an ML branch, uses layered neural networks to mimic human such as hierarchical learning (LeCun et al., 2015). Deep learning models, including convolutional neural networks (CNNs), recurrent neural networks (RNNs) and autoencoders, enable automated forms for extraction and nonlinear modeling of complex omics interactions (Ballard et al., 2024). With implication of heterogenous omics layers, ML/DL approaches enhance predictive accuracy and accelerate the identification of causal variants underlying complex traits (Wu et al., 2021). Merging the ML with editing tools such as CRISPR, allowing data within the target loci for crop improvement (Chen et al., 2024) (Fig. 24.1).

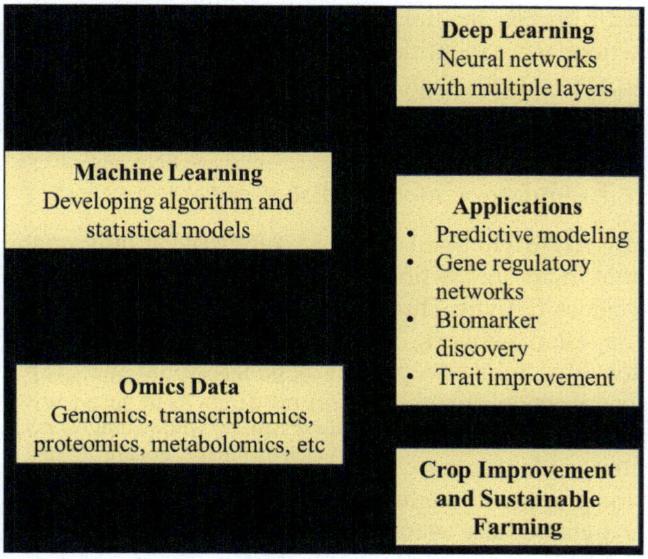

FIGURE 24.1 Integration of machine learning (ML) and deep learning (DL) with omics data and CRISPR for crop improvement. The flowchart illustrates how high-throughput omics datasets (genomics, transcriptomics, proteomics, metabolomics, and phenomics) are processed using ML & DL approaches such as convolutional neural networks (CNNs) and recurrent neural networks (RNNs) to decode complex trait–genotype associations. Extracted features and predictive models guide the design of editing strategies (CRISPR–Cas9, Cas12a, base/prime editors) for targeted trait modification. This integration enables the identification of causal variants, prediction of stress, and development of climate-resilient, high-yield, and nutrient-rich crops.

24.4.2 Predictive models for genotype–phenotype associations

Under crop improvement, predictive are increasingly applied to link genetic variation with various phenotypic traits i.e., yield, stress, nutrient use efficiency, and disease resistance (Bhardwaj et al., 2024a, 2024b). Omics and AI integration approaches performs its traditional quantitative genetics methods by handling high-dimensional, nonlinear, and interactive effects among loci. Wu and Xie (2025) demonstrated that AI driven multiomics has emerged as transformative form to decode the complex genotype–environment–phenotype (GxExP) relationship in crop plants, where the study combined omics models to extract hidden patterns and associations across biological scales. Montesionos-Lopez et al. (2019) reported in maize that multiomics datasets integrated with ML models, DL, which are used to predict yield under drought stress, with application of GxExP data for genomic selection accuracy and revealed that candidate genes are linked to its root architecture and water-use efficiency. In rice, with the combination genomic markers, metabolomics, and environmental data with deep neural networks (DNNs) an accurate prediction of grain with quality traits and stress tolerance connected, while another i.e., CRISPR editing was applied to target loci associated with salinity and drought resilience (Kajrolkar et al., 2025a). Juliana et al. (2019) used AI driven genomic prediction in wheat crop with a model that integrated genotypic, hyperspectral phenotyping and weather datasets with accuracy of grain yield, protein content under heat stress with transcriptome application under condition, which revealed candidate transcription factors regulating flowering and stress responses. In soybean, these omics tools were combined with AI-based models, which uncovers QTLs for drought, and nitrogen fixation efficiency, along with ML thus analysis of genomic-microbiome interactions were identified with applicable form with microbial taxa, which enhanced nutrient uptake (Xavier et al., 2018; Mohamedikbal et al., 2025) (Table 24.2).

24.4.3 Gene network inference and regulatory module discovery

Gene networking is crucial in applying AI/ML to plant systems biology using omics data to build gene regulatory networks that map interactions among transcription factors, signaling molecules, along with target genes (Huynh-Thu et al., 2014). Kassem (2025) discussed that ML approaches such as Bayesian network, random forests, and regression-based models help predict causal relationships, while deep learning (CNN,RNN, and graph neural networks) through genes into coexpression or coregulation modules associated with complex traits such as drought tolerance, nutrient uptake, or in flowering time therefore these modules serve as functional units for identifying candidate gene as regulators, biomarkers, and CRISPR-editable targets for precision breeding (Marbach et al., 2016; Kudapa et al., 2024). Some of the studies were explained here as follows RNA seq or microarrays to infer correlation-based networks, cluster genes usually to identify certain genes and trait-associated modules as reported in crops i.e., rice, maize, and soybean (Langfelder and Horvath, 2008; Sandhu and Raghuram, 2024). Bayesian networks and regression models are applied in *Arabidopsis* and wheat by integration of omics to perform graphical modeling

TABLE 24.2 Examples of AI-driven multi-omics integration in major crops for predictive modeling of genotype–phenotype associations. The table highlights the omics layers integrated, AI/ML approaches applied, key predicted traits or outcomes, and representative studies across maize, rice, wheat, and soybean.

S. No.	Crop	Omics integrated	AI/ML approach	Key outcomes/Traits predicted	References
1	Maize (*Zea mays*)	Genomics, transcriptomics, phenomics	Random forests, deep learning	Yield under drought stress; improved genomic selection accuracy; candidate genes for root traits, WUE	Montesinos-López et al. (2021)
2	Rice (*Oryza sativa*)	Genomics, metabolomics, environmental data	Deep neural networks	Grain quality traits; stress tolerance; CRISPR–Cas9 targeted loci for salinity & drought resilience	
3	Wheat (*Triticum aestivum*)	Genomics, transcriptomics, hyperspectral phenomics, weather data	Genomic prediction models, ML-based integration	Grain yield & protein content under heat stress; TFs for flowering & stress responses	Juliana et al. (2019), Krause et al. (2019)
4	Soybean (*Glycine max*)	Genomics, transcriptomics, phenomics, microbiomics	AI-enabled predictive models, ML	QTLs for drought & nitrogen fixation; beneficial microbial taxa for nutrient uptake	Xavier et al. (2018)

and causal inference, producing directional GRNs (Ajmal and Madden, 2023). Using a random forests and ensemble ML models leverage multiomics were applied for identifying key regulators and predictive trait-gene associations in maize and rice (Peignier et al., 2023; Cruz et al., 2020). DL learning includes CNNs, RNNs, and GNNs utilize transcriptomics, epigenomics and proteomics data with spatiotemporal regulation modules and intricate GRNs in rice, maize and wheat (Angermueller et al., 2016). Finally, multiomics integrative GRN discovery combines omic tools for using network integration, ML data to provide insights to complex trait in crop i.e., wheat, chickpea and maize (Marbach et al., 2016; Varshney et al., 2020) (Table 24.3).

24.4.4 AI in CRISPR target design and off-target minimization

AI and ML being applied to optimize CRISPR/Cas genome editing in plants where AI models are used to predict highly efficient sgRNAs for targeting specific genes in plants

TABLE 24.3 Overview of computational implication for inferring gene regulatory networks (GRNs) in plants. The table summarizes the type of input data, key analysis methods, primary outputs, example crop applications, and representative references for each approach. Methods range from correlation-based co-expression analysis and probabilistic Bayesian models to ML and DL frameworks, as well as integrative multi-omics strategies for identifying regulatory modules across genomics, transcriptomics, proteomics, and metabolomics layers.

S. No.	Approach/ Method	Data type (input)	Key methods/ analysis	Primary outputs	Example crops	References
1	Co-expression network analysis (WGCNA, ARACNe)	Transcriptomics (RNA-seq, microarrays)	Correlation-based network inference, module clustering	Co-expression modules, hub genes, trait-linked modules	Rice, maize, soybean	Langfelder and Horvath, (2008), Sandhu and Raghuram (2024)
2	Bayesian networks/ regression models	Genomics + transcriptomics	Probabilistic graphical modeling, causal inference	Directional gene regulatory networks, causal genes	*Arabidopsis*, wheat	Ajmal and Madden (2023)
3	Random forests/ ensemble ML models	Multi-omics (genome + transcript + pheno)	Feature importance, non-linear inference	Key regulators, predictive trait–gene associations	Maize, rice	Peignier et al. (2023), Cruz et al. (2020)
4	Deep learning (CNNs, RNNs, GNNs)	Transcriptomics, epigenomics, proteomics	Neural networks capturing non-linear patterns & context	Complex GRNs, spatiotemporal regulation modules	Rice, maize, wheat	Angermueller et al. (2016)
5	Multi-omics integrative GRN discovery	Genomics + transcriptomics + proteomics + metabolomics	Network integration, ML-based inference, clustering	Regulatory modules linking multiple omics layers	Wheat, chickpea, maize	Marbach et al. (2016), Varshney et al. (2020)

DL algorithm, including CNNs and RNNs were boosted models with datasets for experimentally validated sgRNAs to score and rank their targeting efficiency by using DeepCRISPR, CRISPR-P 2.0 by using ML to prioritize sgRNAs in plant genomes (Chen et al., 2024; Chuai et al., 2018). To predict and decrease off-target effects in plant by genome editing, where sequence similarity, chromatin accessibility, and local genomic context, allows ML to rank sgRNAs-based on-target and predict off-target risk (Bhardwaj et al., 2024a, 2024b). Furthermore, AI frameworks are adapted to form plant-specific genomic complexities such as polyploidy, repetitive sequences and structural variants, enhancing CRISPR accuracy in crops i.e., rice, wheat, maize and soybean (Li et al., 2025a; 2025b; Xie et al., 2015; Zhang et al., 2019).

24.4.5 AI-driven plant phenotyping and image-based trait analysis

AI, through ML and DL is transforming plant phenotyping by enabling precise, automated analysis. Image-based phenotyping platforms, which captures data with multiple sources, which are RGB cameras, hyperspectral imaging, thermal sensors, LiDAR, and multispectral sensors (Maraveas, 2024). AI algorithms process large, complex datasets to know quantitative phenotypic traits i.e., plant height, leaf area, biomass, chlorophyll content, stress and yield components (Kaya, 2025; Gill et al., 2022). Angidi et al. (2025) represents that CNNs and other DL architectures were used for image segmentation, object detection, with automated scoring of plant growth, disease symptoms, and physiological responses where ML integrate temporal and spatial data to monitor growth dynamics and predict stress tolerance in crops i.e., rice, wheat, maize, soybean, and tomato thus enabling breeders to accelerate selection, optimize resources use, and improve precision agriculture strategies (Singh et al., 2018; Ubbens and Stavness, 2017; Pound et al., 2017).

24.5 Integration of AI, omics, and CRISPR

Integration of AI with omics technologies and CRISPR-based editing is driving with new era for crop improvement (Kumar et al., 2021; Razzaq et al., 2021). Omics approach generates large scale, multilayered datasets that reveals regulatory networks, gene trait associates and stress responsive pathways. AI and ML are increasingly used to integrate these datasets, uncover hidden patterns, prioritize candidate genes within complex traits (Shelake et al., 2022). These insights with CRISPR/Cas systems, researchers can perform targeted and multiplex genome editing for traits i.e., drought tolerance, nitrogen use efficiency, yield enhancement, and disease resistance (Saini et al., 2025; Taheri et al., 2022).

24.5.1 Data-driven pipelines for trait dissection

In modern crop improvement, data-driven pipelines integrate multiomics datasets, HTTP (Yang et al., 2020), and advanced computational methods to dissect complex traits with combination of genomic (SNPs, structural variants), transcriptomic, proteomic, metabolomic (Crossa et al., 2017) and phenomic data with environmental parameters to identify key regulators with stress resilience, yield and quality traits in crops i.e., rice, wheat, maize, and chickpea, enabling the identification of quantitative trait loci (QTLs), candidate genes, and regulatory modules for drought tolerance, NUE and with yield improvement (Zhang et al., 2019; Varshney et al., 2021).

24.5.2 AI-guided selection of candidate genes for editing

In plant genomics integration of AI increased to accelerate the identification of candidate genes for genome editing (Sun et al., 2024). Both ML and DL models analyze large scale genomic, transcriptomic, and phenotypic datasets to uncover gene trait associations, prioritize functional variants for complex traits (Kassem, 2025). DL models used for identification of loci related with traits such as yield, stress tolerance, and nutrient use efficiency (Zhou and Troyanskaya, 2015; Montesinos-López et al., 2018a, 2018b; Wang et al., 2020). Plant specific omics and GWAS, with AI integration pipelines provide powerful tools to select high targets for CRISPR-mediated multiplex editing, thereby accelerates trait improvement in crops (Ma et al., 2015). Some of the studies demonstrated that AI guided pipelines are used in plant gene editing to prioritize candidate genes. Gupta et al. (2021a, 2021b) reported in rice that GRAiN network with QTG-Finder2 and ML identified *SD1*, thus editing reduced plant height with positive yield effects. Targeting promoter in rice *IPA1* (IPA1-Pro10 line) with editing to get improved tillering, grain number, and yield (Song et al., 2022). A hybrid ML framework combinedly shows CNNs and random forests identify genes, confirmed by RNA-seq and RT-PCR, highlighting new target for stress resilience (Dobránszki et al., 2025). In maize, high throughput CRISPR platform with datasets were enabled for rapid validation of candidate genes for trait dissection (Liu et al., 2020b, 2020c), in tomato large sgRNA libraries and phenotyping of ~1300 genome edited lines revealed genes that controlled fruit from nutrient and disease trait (Seed World, 2025). In wheat, omics technology with ML guided selection with susceptibility genes led to CRISPR edits improved among its agronomic traits and disease resistance (Zhou et al., 2023). In soybean, ML-integrated pipelines enabled transgene-free editing of architecture and stress-related genes, reducing breeding process (Tian et al., 2025). Ovesna et al. (2023) reported metabolomics-guided multiplex CRISPR editing achieved improved processing quality in potato. ML has also applied to engineer optimized CRISPR tool for precision editing, where cross species ML -CRISPR pipeline for orphan crops (MacNish et al., 2025) (Table 24.4).

24.5.3 Integrating multiomics datasets with AI/ML models for target prioritization in plants

Prioritizing candidate genes and regulatory elements that are helpful for genome editing is made possible by combining AI models with multiomics datasets (Murmu et al., 2024). Complex qualities such as drought tolerance, nutrient uptake, and yield potential can have their causal variations identified, gene–trait relationships predicted, and prioritized lists of viable targets generated using machine learning and deep learning frameworks (Yan et al., 2020; Yetgin 2025). AI speeds up the functional validation of integrated omics data, reducing with false positives and assisting in the development of precise CRISPR interventions for both model and nonmodel crops (Zhu et al., 2025).

TABLE 24.4 Examples of AI- and data-driven approaches guiding CRISPR-based gene editing in different crops. The table summarizes major studies across maize, tomato, wheat, soybean, potato, and orphan crops, highlighting how artificial intelligence (AI), machine learning (ML), and omics-informed pipelines were applied to prioritize candidate genes or regulatory loci, optimize sgRNA design, or customize CRISPR enzymes. Outcomes include streamlined trait dissection, improved agronomic performance, transgene-free edited lines, enhanced editing precision, and the extension of workflows to orphan and minor crops.

S. No.	Study/source (year)	Crop	AI/data-driven role	What was edited/ selected	Outcome/note	References
1	High-throughput CRISPR platform	Maize	High-throughput CRISPR screening combined with data pipelines to prioritize/ generate many edits	Multiplexed CRISPR targeting many gene families; platform enabled rapid target testing	Streamlined functional validation of many candidate genes and faster trait mapping in maize	Liu et al. (2020b), 2020c
2	Large-scale CRISPR libraries & phenotyping	Tomato	Design of large sgRNA libraries (data-driven) + high-throughput phenotyping to find causal genes	~15,000 sgRNAs across gene families; ~1300 edited lines screened	Identified mutants affecting fruit development, flavor, nutrient uptake and disease response example of massively parallel target selection + phenotyping.	Irum et al. (2025)
3	CRISPR reviews + wheat CRISPR	Wheat	Omics-informed target selection; AI/ML increasingly used in predicting regulatory elements & prioritizing S-genes	Candidate S-genes and regulatory loci edited to alter disease susceptibility and agronomic traits	Demonstrates integrating omics for target selection and use of transient editing systems for validation in wheat.	Zhou et al. (2023)
4	Soybean transgene-free editing pipelines + ML for selection	Soybean	Marker/ML-guided target selection and workflows for transgene-free edited plants	Architecture and stress-related genes edited; workflows to find and recover edited, transgene-free lines	Practical pipelines enabling trait edits with reduced breeding time and improved recovery of edited lines.	Teng (2024)
5	Multiplex CRISPR editing & candidate selection	Potato	Multiplex CRISPR with candidate selection informed by metabolomics/trait data	Edits to reduce processing-related compounds/ improve quality	Demonstrated multiplex editing to generate improved processing traits; candidate selection used trait and metabolite knowledge.	Ly et al. (2023)
6	AI for custom CRISPR enzymes and sgRNA optimization (protein engineering + ML)	Broad (tooling for many crops)	ML used to engineer/ customize Cas variants and to predict sgRNA efficiency/off-targets	Novel Cas variants and ML-ranked sgRNAs with improved specificity	Expands toolbox for crop editing by producing enzymes with tailored specificities and ML-optimized guides.	Chen et al. (2024)

TABLE 24.4 Examples of AI- and data-driven approaches guiding CRISPR-based gene editing in different crops. The table summarizes major studies across maize, tomato, wheat, soybean, potato, and orphan crops, highlighting how artificial intelligence (AI), machine learning (ML), and omics-informed pipelines were applied to prioritize candidate genes or regulatory loci, optimize sgRNA design, or customize CRISPR enzymes. Outcomes include streamlined trait dissection, improved agronomic performance, transgene-free edited lines, enhanced editing precision, and the extension of workflows to orphan and minor crops.—cont'd

S. No.	Study/source (year)	Crop	AI/data-driven role	What was edited/ selected	Outcome/note	References
7	AI/CRE discovery precise regulatory editing	Multi-crop/ conceptual	AI used to predict cis-regulatory elements (CREs) & editing plasticity to fine-tune expression	CREs edited to adjust gene expression levels rather than knockout	Highlights AI-guided precision regulatory edits next step beyond coding edits.	Qiu et al. (2025)
8	ML + CRISPR workflows applied to orphan/minor crops	Orphan crops (various)	ML used to prioritize favorable modifications and design accurate sgRNAs for non-model genomes	Candidate trait-improving edits proposed and tested in pilot pipelines	Shows transfer of ML + CRISPR approaches to crops with limited prior data by leveraging cross-species info and ML.	MacNish et al. (2025)

24.5.4 CRISPR validation and AI feedback loop

CRISPR technology through genome editing in plants depend on iterative cycles of its validation and computational refinement (Saini et al., 2025). Experimental purposes of AI and CRISPR result with critical feedback, which includes on-target efficiencies, observed off-target events, trait performance under various stress conditions and also with predictive frame works, refine sgRNA design, and improve prioritization of its targets (Saini et al., 2025). Thus, AI feedback loop creates a self-improving system where each round of genome editing enhances the predictive role of AI models by reducing experimental costs, minimizing unintended edits, and accelerating trait improvement. Therefore, integrative approaches are especially valuable for complex traits in crops such as rice, wheat, maize, and soybean, where multilayered regulation and polygenic effect have adaptive and data driven editing pipelines (Tsakirpaloglou et al., 2023).

24.6 Applications in plant trait improvement

The growing accessibility of cost-effective and precise genome-editing technologies, along with the relaxation of regulatory constraints on genome-edited crops, has shifted the research focus toward pinpointing suitable genetic targets for modification. Yield, being a complex polygenic trait, remains particularly challenging to enhance through manipulation of a single gene (Cao et al., 2020). Furthermore, crop improvement programs often face the dual challenge of simultaneously enhancing yield and stress tolerance two traits that frequently exhibit trade-offs. A comprehensive 20-year study by Corteva Agriscience examined with influence of 1671 genes on maize for yield, nitrogen utilization, and drought resistance, ultimately identifying 22 genes with key physiological roles (Simmons et al., 2021).

Genetic redundancy in polyploid crops, such as wheat, makes improvement efforts with more difficult by covering-off alleles that could leads to agronomic traits. More than 717,000 phenotypic data points were produced with recent multiomics study that examined 1035 wheat accessions, including both Watkins landraces and contemporary cultivars. It also found 8253 genetic associations, including 15 hitherto undiscovered loci associated with resistance to yellow rust (Cheng et al., 2024).

High-quality reference genome act as a vital tool for omics-based research and for genomics (Adamski et al., 2020; Yao et al., 2025). Whole genome sequences and annotations are made available for major crops such as rice, maize, soybean, and wheat (International Rice Genome Sequencing Project and Sasaki, 2005; Schnable et al., 2009; Schmutz et al., 2010; International Wheat Genome Sequencing Consortium et al., 2018). Although genomic resources are critical for researchers and plant breeders, they still contain regions composed of undefined sequences remained to specific chromosomes due to less sequence continuity.

Recent advances in long-read sequencing technologies have proven highly effective in bridging these assembly gaps (Liu et al., 2020a; Aury et al., 2022; Chen et al., 2023). Advances in this technology enable exploration of genetic variation across diverse germplasm, generating genomic datasets that reveal key loci for agronomic traits and guide targeted crop improvement (Li et al., 2020a; Shang et al., 2022; Zhang et al., 2022).

Plant phenomics, as defined by Furbank (2009), provides quantitative phenotypic data to link genetic determinants with agronomic traits and assess germplasm variability. Phenotyping systems may be ground-based or aerial, using manual, vehicle-mounted, or autonomous robotic platforms. Research institutions have developed phenotyping platforms for controlled and field settings. Notable examples include CropSight, an IoT-based open-source system for automated data collection, and INRA's ontology-driven Phenotyping Hybrid Information System for integrating phenotypic data (Sadok et al., 2007; Virlet et al., 2016).

Advances in phenotyping technologies have produced massive datasets requiring intensive processing. Robotics and autonomous systems are transforming agriculture (Pearson et al., 2022), yet managing large datasets from high-throughput phenotyping remains a key challenge despite the field's promise in linking gene discovery to practical application (Furbank, 2009). AI technologies link complex phenotypes with genotypes, predict gene functions, and forecast crop performance (Khan et al., 2022). AI-based sensors and imaging enable precise trait measurement, speeding up breeding selection.

ML and DL methods have demonstrated substantial potential with image-based phenotypic data (Khan et al., 2022; Poorter et al., 2023). Identifying quantitative trait loci associated with phenotypic diversity, and AlphaFold for predicting three-dimensional protein structures are driving major advances in computational biology (Alipanahi et al., 2015; Zhou and Troyanskaya, 2015; Jumper et al., 2021; Li et al., 2022).

The agronomic potential of genome-edited crops relies on integrating precision agriculture technologies using robotics, AI, and IoT to enhance sustainability and productivity (Sharma et al., 2023).

IoT networks i.e., sensors, drones, and analytical system to monitor environmental, soil, and crop conditions. Field sensors capture data on moisture, pH, and nutrients, which, combined with aerial and climate data, feed AI models to predict stress responses and optimize irrigation, fertilization, and pest management (Sharma et al., 2023).

AI-driven decision support and mobile technologies aid in informed farming management practices. Mobile-based applications have been particularly valuable in narrowing knowledge gaps, especially in rural areas where access to traditional information and communication infrastructure, i.e., computers, remains limited (Ayim et al., 2022). Recent advances use DL for early disease detection and digital platforms for crop management (Aslam et al., 2024; Puranik et al., 2024). Together, these tools empower smallholder farmers with accessible precision agriculture.

The NASA–ISRO NISAR mission (2025) will use dual-frequency radar to provide frequent, high-resolution crop monitoring, improving global forecasting and supporting food security. Publicly sharing these data and linking them with AI-driven tools and mobile apps can improve farm management, particularly where advanced monitoring is limited (Jet Propulsion Laboratory, 2025). As agriculture adapts to harsher climates, robotics will be key to managing resilient, genome-edited crops in diverse environments (Botta et al., 2022).

Advanced platforms AgriQ, terrain and GPS limitations (Botta and Cavallone, 2021). AI robots such as ecoRobotix enable precise weeding (Bykov, 2023), robotic harvesters improve efficiency and reduce losses in crops such as strawberries and tomatoes (Kim et al., 2022; Chang and Huang, 2024). Multiomics datasets offer a systems-level view of plant responses. Advancing AI applications in omics research can drive stress-tolerant crop development, improved management, and sustainable food (Arabnia and Tran, 2011; Yan and Wang, 2023).

Machine learning (ML) algorithms have become imperative tools for identifying genes associated with stress resistance, thereby assisting researchers and plant breeders for improving crop yield. Liang et al. (2011) employed a modified support vector machine (SVM), algorithm to the genes, which are linked to drought resistance in *Arabidopsis thaliana.*

The application of ML extends beyond drought studies. Wang et al. (2013) utilized an SVM-based predictive model for identifying genes conferring salt tolerance in rice. In another study, Ravari et al. (2015) evaluated Iranian wheat genotypes, which are salt tolerance and used artificial neural network (ANN) analysis to determine reliable indices for predicting salt-tolerant varieties. Moreover, Schwarz et al. (2020) applied ML techniques to decipher the cis-regulatory code underlying iron deficiency responses in *Arabidopsis* roots. SVM and other ML models effectively predict disease resistance proteins and pathogen effectors, as shown by Pal et al. (2016) and Sperschneider et al. (2016). ML also aids in identifying resistance and susceptibility genes for improved disease management (Yang and Guo, 2017). ML applications in single-cell genomics enable insights into cellular diversity, gene networks, and new cell types. Advances such as SIMLR and neural network models improve data accuracy and interpretation despite technical challenges (Silva et al., 2019; Raimundo et al., 2021; Wang et al., 2017; Lin et al., 2017).

Challenging the conventional view that abiotic and biotic stresses in plants operate through distinct signaling pathways, Shaik and Ramakrishna (2014) conducted a comprehensive study in rice that revealed significant overlap between the two response mechanisms. By performing a meta-analysis of multiple microarray datasets, they identified a set of genes exhibiting conserved expression under both stress conditions, with nearly 70% of the common differentially expressed genes showing similar regulatory behavior.

Meng et al. (2021) utilized ML models to identify genes exhibiting transcriptional responses to cold stress. Interestingly, genome assembly derived to show the only slight declines in predictive accuracy compared with those incorporating broader datasets models. Zhou et al. (2022) conducted a transcriptomic analysis of maize genotypes with thermal stress, revealing widespread changes in transcript abundance. Thus, study identified enriched sequence with motifs i.e., transcription start sites (TSSs) of stress-responsive genes.

Pradhan et al. (2023) applied artificial intelligence, particularly ML, for knowing the complex task of identifying long lncRNAs involved in plant responses to abiotic stress in recent times. Since stresses greatly affect crop productivity, identifying stress-responsive lncRNAs is critical for the development of resilient crop varieties. As miRNAs play a crucial role, the researchers developed a novel computational framework designed to predict abiotic stress associated lncRNAs using a dataset includes stress-responsive and nonresponsive sequences for predicting microRNAs (miRNAs) associated with distinct stresses, including cold, drought, heat, and salinity. AI regulatory role in plant stress responses, their identification is pivotal for developing stress-tolerant crop varieties. Employing machine learning, particularly SVM algorithms, the researchers established a computational framework to identify stress-responsive miRNAs.

Furthermore, Ahmed et al. (2023) proposed a deep learning framework employing a novel activation function, the Sigmoid-Integrated Exponential Linear Unit (SIELU), which outperformed conventional models in classifying uncharacterized abiotic stress-related protein sequences, achieving accuracies between 80.78% and 95.11%.

Liu et al. (2017a, 2017b) aimed a study to classifying 216 plant species using incomplete metabolite data. The researchers applied a network clustering algorithm to organize structurally similar metabolites, for each plant as a binary vector for analysis. Hierarchical clustering was then employed to categorize the plants. Remarkably, even with incomplete datasets, the method effectively grouped to establish evolutionary and metabolite profiles as reliable taxonomic indicators.

Fürtauer et al. (2018) used ML to analyze metabolic responses in *Arabidopsis* under cold and high-light stress, identifying a core network of 23 proteins—18 linked by interactions—revealing key mechanisms of stress adaptation. Deep learning (DL) has proven highly effective in crop management across species such as rice, wheat, tomato, and potato. Li et al. (2020) developed a DL-based video system for real-time pest and disease detection, while Mi et al. (2020) introduced C-DenseNet, achieving 97.99% accuracy in grading wheat stripe rust severity.

Wang and Liu (2021) achieved 94.13% F1 accuracy for early tomato leaf spot detection using MobileNetv2-YOLOv3. Polder et al. (2019) applied hyperspectral imaging with CNNs for potato virus detection, while Chen and Yuan (2019) used a CNN-RPN model for precise pest identification (precision = 0.93, miss rate = 0.10). Feng et al. (2020) combined deep learning and hyperspectral imaging to monitor salinity stress in okra. Their models accurately predicted multiple traits (r = 0.588–0.835), demonstrating the potential of DL and HSI for nondestructive crop stress assessment.

Supervised ML and DL are common in plant biology, but unsupervised and semisupervised methods are equally important for clustering, and gene network inference (Rai et al.,

2019). Hierarchical clustering remains a standard method for grouping genes for omic studies (Xu et al., 2012; Klepikova et al., 2016). Techniques such as t-SNE and OPTICS enhance the visualization of genotypic data, large-scale maize hybrid datasets (Yan et al., 2021). PCA, t-SNE, and OPTICS aid in visualizing omics and population data, while NMF and MDR uncover key metagenes and gene–environment interactions (Wilson et al., 2012). Algorithms such as UMAP, MAGIC, and PHATE address the complexity of single-cell data (Van Dijk et al., 2018; Pratapa et al., 2020; Wu and Zhang, 2020; Marand et al., 2021).

24.7 Challenges and future prospects

Rapid growth in biological data coupled with advances in machine learning (ML) has created new opportunities for deciphering complex biological systems. The application of ML in plant molecular research remains challenging, thus conventional molecular approaches, ML methods are highly dependent on experimental context. The surge in biological data and ML advances offers new insights into complex systems. However, ML in plant research remains challenging, as model performance depends on context, algorithms, and dataset variability.

Omics datasets contain rich biological data but often face noise, sparsity, and imbalance issues (Joyce and Palsson, 2006). Another common issue in ML-based analyses is data imbalance. To mitigate this, researchers often utilize resampling approaches, including sampling in excess (Maimon and Rokach, 2005). For example, Li et al. (2021) applied the Synthetic Minority Oversampling Technique (SMOTE) to accelerate the minority classes, thereby improving the detection of effector proteins. Overfitting, common in deep learning, reduces model generalization and is mitigated using regularization methods such as dropout (Scholz et al., 2004).

Dropout is a widely used in neural network training, wherein a random subset of neurons is temporarily deactivated during each iteration with an approach that mitigates the model's dependency on correlations within the training dataset, thereby improving its generalization capability to unseen data. By introducing this controlled randomness, dropout effectively reduces overfitting and fosters. Managing big data in plant systems characterized by volume, variety, veracity, and other major challenges. ML enables effective multiomics integration but faces issues such as sparsity, multicollinearity, and overfitting, requiring specialized analytical strategies and collaborative approaches.

Interpreting complex models, especially those derived from advanced ML techniques such as DL, remains a significant challenge due to their inherent "black-box" nature. In many cases, researchers place greater emphasis on elucidating the biological relevance of a model's predictions rather than solely focusing on its accuracy, necessitating thorough data analysis and alignment with established biological understanding.

Despite these challenges, AI has successful applications in plant omics. Robust, scalable algorithms and experimental validation to get reliable predictions. Collaboration among omics researchers, data scientists, and agricultural experts is vital to drive AI-based innovations in plant defense and agriculture.

24.8 Conclusion

Genome editing technologies have emerged as powerful tools for introducing novel traits in crops and its applications now extend beyond single-base modifications to the insertion of nucleotide sequences, with continued expansion driven by the discovery of new Cas orthologs with PAM specificities. The range of editable targets has also widened beyond the nuclear genome, as advancements in mitochondrial and chloroplast genome editing with manipulation of genes involved in photosynthesis and respiration that were previously inaccessible.

To achieve maximum benefits for crop improvement, genome editing should be integrated with recent technologies such as speed breeding, phenomics, AI, robotics, and satellite-based monitoring. With the commercialization of genome-edited crops, many countries are easing regulations, promoting wider agricultural adoption. Real advancement, however, requires integrated collaboration among scientists, breeders, engineers, farmers, and policymakers to translate AI and genome editing into practical, climate-resilient, high-yield solutions.

Declaration

The authors declare that no generative artificial intelligence tools were used in the writing or preparation of this chapter. The content reflects the authors' own scholarly work, and the authors assume full responsibility for all statements, interpretations, and conclusions presented

References

Abe, A., Kosugi, S., Yoshida, K., Natsume, S., Takagi, H., Kanzaki, H., Matsumura, H., Yoshida, K., Mitsuoka, C., Tamiru, M., Innan, H., 2012. Genome sequencing reveals agronomically important loci in rice using MutMap. Nat. Biotechnol. 30 (2), 174–178. https://doi.org/10.1038/nbt.2095.

Abudayyeh, O.O., Gootenberg, J.S., Essletzbichler, P., Han, S., Joung, J., Belanto, J.J., Verdine, V., Cox, D.B.T., Kellner, M.J., Regev, A., Lander, E.S., 2017. RNA targeting with CRISPR–Cas13. Nature 550 (7675), 280–284. https://doi.org/10.1038/nature24049.

Adamski, N.M., Borrill, P., Brinton, J., Harrington, S.A., Marchal, C., Bentley, A.R., Bovill, W.D., Cattivelli, L., Cockram, J., Contreras-Moreira, B., et al., 2020. A roadmap for gene functional characterisation in crops with large genomes: lessons from polyploid wheat. eLife 9, e55646. https://doi.org/10.7554/eLife.55646.

Ahmed, B., Haque, M.A., Iquebal, M.A., Jaiswal, S., Angadi, U.B., Kumar, D., et al., 2023. DeepAProt: deep learning-based abiotic stress protein sequence classification and identification tool in cereals. Front. Plant Sci. 13, 1008756. https://doi.org/10.3389/fpls.2022.1008756.

Ajmal, H.B., Madden, M.G., 2023. A review of Bayesian modelling methods for gene regulatory network inference. Bioinformatics 1.

Albert, F.W., Kruglyak, L., 2015. The role of regulatory variation in complex traits and disease. Nat. Rev. Genet. 16 (4), 197–212. https://doi.org/10.1038/nrg3891.

Alipanahi, B., Delong, A., Weirauch, M.T., Frey, B.J., 2015. Predicting the sequence specificities of DNA- and RNA-binding proteins by deep learning. Nat. Biotechnol. 33 (8), 831–838. https://doi.org/10.1038/nbt.3300.

Angermueller, C., Pärnamaa, T., Parts, L., Stegle, O., 2016. Deep learning for computational biology. Mol. Syst. Biol. 12 (7), 878. https://doi.org/10.15252/msb.20156651.

Angidi, S., Madankar, K., Tehseen, M.M., Bhatla, A., 2025. Advanced high-throughput phenotyping techniques for managing abiotic stress in agricultural crops—a comprehensive review. Crops 5 (2), 8. https://doi.org/10.3390/crops5020008.

Arabnia, H., Tran, Q.N., 2011. Software Tools and Algorithms for Biological Systems. Springer.

Araus, J.L., Cairns, J.E., 2014. Field high-throughput phenotyping: the new crop breeding frontier. Trends Plant Sci. 19 (1), 52–61. https://doi.org/10.1016/j.tplants.2013.09.008.

Aslam, T., Ghareeb, S., Mustafina, J., 2024. Agri Sage: a mobile application for agricultural disease detection, e-commerce, and real-time information systems. In: 17th International Conference on Development in eSystem Engineering (DeSE), pp. 84–88 (Khorfakkan, United Arab Emirates. IEEE).

Aury, J.M., Engelen, S., Istace, B., Monat, C., Lasserre-Zuber, P., Belser, C., Cruaud, C., Rimbert, H., Leroy, P., Arribat, S., et al., 2022. Long-read and chromosome-scale assembly of the hexaploid wheat genome achieves high resolution for research and breeding. GigaScience 11, giac034. https://doi.org/10.1093/gigascience/giac034.

Awany, D., Allali, I., Dalvie, S., Hemmings, S., Mwaikono, K.S., Thomford, N.E., Gomez, A., Mulder, N., Chimusa, E.R., 2019. Host and microbiome genome-wide association studies: current state and challenges. Front. Genet. 9, 637. https://doi.org/10.3389/fgene.2018.00637.

Ayim, C., Kassahun, A., Addison, C., Tekinerdogan, B., 2022. Adoption of ICT innovations in the agriculture sector in Africa: a review of the literature. Agric. Food Secur. 11, 22. https://doi.org/10.1186/s40066-022-00393-5.

Ballard, J.L., Wang, Z., Li, W., Shen, L., Long, Q., 2024. Deep learning-based approaches for multi-omics data integration and analysis. BioData Min. 17 (1), 38. https://doi.org/10.1186/s13040-024-00380-z.

Barrangou, R., Fremaux, C., Deveau, H., Richards, M., Boyaval, P., Moineau, S., Romero, D.A., Horvath, P., 2007. CRISPR provides acquired resistance against viruses in prokaryotes. Science 315 (5819), 1709–1712. https://doi.org/10.1126/science.1138140.

Bastida, F., Hernández, T., García, C., 2014. Metaproteomics of soils from semiarid environment: functional and phylogenetic information obtained with different protein extraction methods. J. Proteonomics 101, 31–42. https://doi.org/10.1016/j.jprot.2014.01.023.

Berger, B., Parent, B., Tester, M., 2010. High-throughput shoot imaging to study drought responses. J. Exp. Bot. 61 (13), 3519–3528. https://doi.org/10.1093/jxb/erq201.

Bhardwaj, A., Tomar, P., Nain, V., 2024a. Machine learning-driven prediction of CRISPR-Cas9 off-target effects and mechanistic insights. Environment 12, 13.

Bhardwaj, R., Khwairakpam, R., Negi, A.S., Rawat, B., 2024b. Implication of plant functional traits in crop improvement. In: Plant Functional Traits for Improving Productivity. Springer Nature Singapore, Singapore, pp. 23–39. https://doi.org/10.1007/978-981-99-1234-0_2.

Botta, A., Cavallone, P., 2021. Robotics used in precision agriculture: the sustainable Agri.q Rover case study. In: Quaglia, G., Gasparetto, A., Petuya, V., Carbone, G. (Eds.), Proceedings of the 14SDG Workshop 2021. Springer, p. 108.

Botta, A., Cavallone, P., Baglieri, L., Colucci, G., Tagliavini, L., Quaglia, G., 2022. A review of robots, perception, and tasks in precision agriculture. Applied Mechanics 3 (3), 830–854. https://doi.org/10.3390/applmech3030048.

Braun, P., Tasan, M., Dreze, M., Barrios-Rodiles, M., Lemmens, I., Yu, H., Sahalie, J.M., Murray, R.R., Roncari, L., De Smet, A.S., Venkatesan, K., 2009. An experimentally derived confidence score for binary protein–protein interactions. Nat. Methods 6 (1), 91–97. https://doi.org/10.1038/nmeth.1281.

Broman, K.W., Wu, H., Sen, Ś., Churchill, G.A., 2003. R/qtl: QTL mapping in experimental crosses. Bioinformatics 19 (7), 889–890. https://doi.org/10.1093/bioinformatics/btg112.

Budak, H., Kantar, M., Yucebilgili Kurtoglu, K., 2013. Drought tolerance in modern and wild wheat. Sci. World J. 2013 (1), 548246. https://doi.org/10.1155/2013/548246.

Bulgarelli, D., Schlaeppi, K., Spaepen, S., Van Themaat, E.V.L., Schulze-Lefert, P., 2013. Structure and functions of the bacterial microbiota of plants. Annu. Rev. Plant Biol. 64 (1), 807–838. https://doi.org/10.1146/annurev-arplant-050312-120106.

Bykov, S., 2023. World trends in the creation of robots for spraying crops. E3S Web of Conferences 380, 01011.

Cao, S., Xu, D., Hanif, M., Xia, X., He, Z., 2020. Genetic architecture underpinning yield component traits in wheat. Theor. Appl. Genet. 133, 1811–1823.

Chang, C.L., Huang, C.C., 2024. Design and implementation of an AI-based robotic arm for strawberry harvesting. Agriculture 14, 2057.

Chang, H.X., Haudenshield, J.S., Bowen, C.R., Hartman, G.L., 2017. Metagenome-wide association study and machine learning prediction of bulk soil microbiome and crop productivity. Front. Microbiol. 8, 519.

Chen, J., Wang, Z., Tan, K., Huang, W., Shi, J., Li, T., Hu, J., Wang, K., Wang, C., Xin, B., et al., 2023. A complete telomere-to-telomere assembly of the maize genome. Nat. Genet. 55, 1221–1231.

Chen, L., Yuan, Y., 2019. Agricultural disease image dataset for disease identification based on machine learning. In: Big Scientific Data Management: First International Conference, BigSDM 2018, Beijing, China, November 30–December 1, 2018, Revised Selected Papers. Springer International Publishing, pp. 263–274.

Chen, L., Liu, G., Zhang, T., 2024. Integrating machine learning and genome editing for crop improvement. aBIOTECH 5 (2), 262–277.

Chen, W., Gao, Y., Xie, W., Gong, L., Lu, K., Wang, W., Li, Y., Liu, X., Zhang, H., Dong, H., Zhang, W., 2014. Genome-wide association analyses provide genetic and biochemical insights into natural variation in rice metabolism. Nat. Genet. 46 (7), 714–721.

Cheng, S., Feng, C., Wingen, L.U., Cheng, H., Riche, A.B., Jiang, M., Leverington-Waite, M., Huang, Z., Collier, S., Orford, S., et al., 2024. Harnessing landrace diversity empowers wheat breeding. Nature 632, 823–831.

Chmielowska-Bąk, J., Searle, I.R., Wakai, T.N., Arasimowicz-Jelonek, M., 2023. The role of epigenetic and epitranscriptomic modifications in plants exposed to non-essential metals. Front. Plant Sci. 14, 1278185.

Chuai, G., Ma, H., Yan, J., Chen, M., Hong, N., Xue, D., Zhou, C., Zhu, C., Chen, K., Duan, B., Gu, F., 2018. DeepCRISPR: optimized CRISPR guide RNA design by deep learning. Genome Biol. 19 (1), 80.

Cobb, J.N., DeClerck, G., Greenberg, A., Clark, R., McCouch, S., 2013. Next-generation phenotyping: requirements and strategies for enhancing our understanding of genotype–phenotype relationships and its relevance to crop improvement. Theor. Appl. Genet. 126 (4), 867–887.

Cokus, S.J., Feng, S., Zhang, X., Chen, Z., Merriman, B., Haudenschild, C.D., Pradhan, S., Nelson, S.F., Pellegrini, M., Jacobsen, S.E., 2008. Shotgun bisulphite sequencing of the *Arabidopsis* genome reveals DNA methylation patterning. Nature 452 (7184), 215–219.

Collard, B.C., Jahufer, M.Z.Z., Brouwer, J.B., Pang, E.C.K., 2005. An introduction to markers, quantitative trait loci (QTL) mapping and marker-assisted selection for crop improvement: the basic concepts. Euphytica 142 (1), 169–196.

Compant, S., Samad, A., Faist, H., Sessitsch, A., 2019. A review on the plant microbiome: Ecology, functions, and emerging trends in microbial application. J. Adv. Res. 19, 29–37.

Cooper, M., Messina, C.D., Podlich, D., Totir, L.R., Baumgarten, A., Hausmann, N.J., Wright, D., Graham, G., 2014. Predicting the future of plant breeding: Complementing empirical evaluation with genetic prediction. Crop Pasture Sci. 65 (4), 311–336.

Crain, J., Wang, X., Lucas, M., Poland, J., 2021. Experiences of applying field-based high-throughput phenotyping for wheat breeding. In: High-Throughput Crop Phenotyping. Springer International Publishing, Cham, pp. 71–99.

Crossa, J., Martini, J.W., Vitale, P., Pérez-Rodríguez, P., Costa-Neto, G., Fritsche-Neto, R., Runcie, D., Cuevas, J., Toledo, F., Li, H., De Vita, P., 2025. Expanding genomic prediction in plant breeding: Harnessing big data, machine learning, and advanced software. Trends Plant Sci. 30 (7), 756–774.

Crossa, J., Pérez-Rodríguez, P., Cuevas, J., Montesinos-López, O., Jarquín, D., De Los Campos, G., Burgueño, J., González-Camacho, J.M., Pérez-Elizalde, S., Beyene, Y., Dreisigacker, S., 2017. Genomic selection in plant breeding: methods, models, and perspectives. Trends Plant Sci. 22 (11), 961–975.

Cruz, D.F., De Meyer, S., Ampe, J., Sprenger, H., Herman, D., Van Hautegem, T., De Block, J., Inzé, D., Nelissen, H., Maere, S., 2020. Using single-plant-omics in the field to link maize genes to functions and phenotypes. Mol. Syst. Biol. 16 (12), e9667.

Dai, X., Shen, L., 2022. Advances and trends in omics technology development. Front. Med. 9, 911861.

Dobránszki, J., Vassileva, V., Agius, D.R., Moschou, P.N., Gallusci, P., Berger, M.M., Farkas, D., Basso, M.F., Martinelli, F., 2025. Gaining insights into epigenetic memories through artificial intelligence and omics science in plants. J. Integr. Plant Biol. 67 (9), 2320–2349.

Dubin, M.J., Zhang, P., Meng, D., Remigereau, M.S., Osborne, E.J., Casale, F.P., Drewe, P., Kahles, A., Jean, G., Vilhjálmsson, B., Jagoda, J., 2015. DNA methylation in *Arabidopsis* has a genetic basis and shows evidence of local adaptation. eLife 4, e05255.

Edwards, J., Johnson, C., Santos-Medellín, C., Lurie, E., Podishetty, N.K., Bhatnagar, S., Eisen, J.A., Sundaresan, V., 2015. Structure, variation, and assembly of the root-associated microbiomes of rice. Proc. Natl. Acad. Sci. 112 (8), E911–E920.

Feng, X., Zhan, Y., Wang, Q., Yang, X., Yu, C., Wang, H., et al., 2020. Hyperspectral imaging combined with machine learning as a tool to obtain high-throughput plant salt-stress phenotyping. Plant J. 101 (6), 1448–1461. https://doi.org/10.1111/tpj.14597.

Fernie, A.R., Schauer, N., 2009. Metabolomics-assisted breeding: a viable option for crop improvement? Trends Genet. 25 (1), 39–48.

Fiehn, O., 2002. Metabolomics: the link between genotypes and phenotypes. Plant Mol. Biol. 48 (1), 155–171.

Filichkin, S.A., Priest, H.D., Givan, S.A., Shen, R., Bryant, D.W., Fox, S.E., Wong, W.K., Mockler, T.C., 2010. Genome-wide mapping of alternative splicing in *Arabidopsis thaliana*. Genome Res. 20 (1), 45–58.

Furbank, R.T., Tester, M., 2011. Phenomics–technologies to relieve the phenotyping bottleneck. Trends Plant Sci. 16 (12), 635–644.

Furbank, R.T., 2009. Plant phenomics: from gene to form and function. Funct. Plant Biol. 36 (1), 5–6.

Fürtauer, L., Pschenitschnigg, A., Scharkosi, H., Weckwerth, W., Nägele, T., 2018. Combined multivariate analysis and machine learning reveals a predictive module of metabolic stress response in *Arabidopsis thaliana*. Molecular Omics 14 (6), 437–449. https://doi.org/10.1039/C8MO00095F.

Gallusci, P., Agius, D.R., Moschou, P.N., Dobránszki, J., Kaiserli, E., Martinelli, F., 2023. Deep inside the epigenetic memories of stressed plants. Trends Plant Sci. 28 (2), 142–153.

Gao, L., Gonda, I., Sun, H., Ma, Q., Bao, K., Tieman, D.M., Burzynski-Chang, E.A., Fish, T.L., Stromberg, K.A., Sacks, G.L., Thannhauser, T.W., 2019. The tomato pan-genome uncovers new genes and a rare allele regulating fruit flavor. Nat. Genet. 51 (6), 1044–1051.

Gelli, M., Duo, Y., Konda, A.R., Zhang, C., Holding, D., Dweikat, I., 2014. Identification of differentially expressed genes between sorghum genotypes with contrasting nitrogen stress tolerance by genome-wide transcriptional profiling. BMC Genom. 15 (1), 179.

Gill, T., Gill, S.K., Saini, D.K., Chopra, Y., de Koff, J.P., Sandhu, K.S., 2022. A comprehensive review of high-throughput phenotyping and machine learning for plant stress phenotyping. Phenomics 2 (3), 156–183.

Granier, C., Vile, D., 2014. Phenotyping and beyond: modelling the relationships between traits. Curr. Opin. Plant Biol. 18, 96–102.

Guo, W., Chen, L., Herrera-Estrella, L., Cao, D., Tran, L.S.P., 2020. Altering plant architecture to improve performance and resistance. Trends Plant Sci. 25 (11), 1154–1170.

Gupta, D., Sharma, G., Saraswat, P., Ranjan, R., 2021a. Synthetic biology in plants: a boon for coming decades. Mol. Biotechnol. 63 (12), 1138–1154.

Gupta, M., Kumar, R., Singh, D., 2021b. Crop diseases detection and prevention using AI and machine learning techniques. In: Digital Twin Technology. CRC Press, pp. 61–82.

Hamany Djande, C.Y., Pretorius, C., Tugizimana, F., Piater, L.A., Dubery, I.A., 2020. Metabolomics: a tool for cultivar phenotyping and investigation of grain crops. Agronomy 10 (6), 831.

Harb, A., Krishnan, A., Ambavaram, M.M.R., Pereira, A., 2010. Molecular and physiological analysis of drought stress in Arabidopsis reveals early responses leading to acclimation in plant growth. Plant Physiol. 154 (3), 1254–1271.

Holland, J.B., 2007. Genetic architecture of complex traits in plants. Curr. Opin. Plant Biol. 10 (2), 156–161.

Hu, L., Robert, C.A., Cadot, S., Zhang, X.I., Ye, M., Li, B., Manzo, D., Chervet, N., Steinger, T., Van Der Heijden, M. G., Schlaeppi, K., 2018. Root exudate metabolites drive plant-soil feedbacks on growth and defense by shaping the rhizosphere microbiota. Nat. Commun. 9 (1), 2738.

Huang, X., Han, B., 2014. Natural variations and genome-wide association studies in crop plants. Annu. Rev. Plant Biol. 65 (1), 531–551.

Hultman, J., Waldrop, M.P., Mackelprang, R., David, M.M., McFarland, J., Blazewicz, S.J., Harden, J., Turetsky, M. R., McGuire, A.D., Shah, M.B., VerBerkmoes, N.C., 2015. Multi-omics of permafrost, active layer and thermokarst bog soil microbiomes. Nature 521 (7551), 208–212.

Huynh-Thu, V.A., Wehenkel, L., Geurts, P., 2014. Gene regulatory network inference from systems genetics data using tree-based methods. In: Gene Network Inference: Verification of Methods for Systems Genetics Data. Springer, Berlin, Heidelberg, pp. 63–85.

Ingvarsson, P.K., Street, N.R., 2011. Association genetics of complex traits in plants. New Phytol. 189 (4), 909–922.

International Rice Genome Sequencing Project, Sasaki, T., 2005. The map-based sequence of the rice genome. Nature 436, 793–800.

International Wheat Genome Sequencing Consortium (IWGSC), 2018. Shifting the limits in wheat research and breeding using a fully annotated reference genome. Science 361, eaar7191.

Irum, S., Rehman, N., Inam, S., Khan, M.Z.F., Uzair, M., Muhammad, A., Khan, M.R., 2025. CRISPR/Cas9 and beyond: genome editing for precision trait enhancement in tomatoes. In: Omics Approaches for Tomato Yield and Quality Trait Improvement. Springer Nature, Singapore, pp. 205–231.

Jaganathan, D., Ramasamy, K., Sellamuthu, G., Jayabalan, S., Venkataraman, G., 2018. CRISPR for crop improvement: an update review. Front. Plant Sci. 9, 985.

Jamil, F., Mukhtar, H., Fouillaud, M., Dufossé, L., 2022. Rhizosphere signaling: insights into plant–rhizomicrobiome interactions for sustainable agronomy. Microorganisms 10 (5), 899.

Jayakodi, M., Schreiber, M., Stein, N., Mascher, M., 2021. Building pan-genome infrastructures for crop plants and their use in association genetics. DNA Res. 28 (1), dsaa030.

Jet Propulsion Laboratory, 2025. NASA's New Satellite Could Change the Way We Farm Forever. SciTechDaily. https://scitechdaily.com/nasas-new-satellite-could-change-the-way-we-farm-forever/.

Jordan, M.I., Mitchell, T.M., 2015. Machine learning: trends, perspectives, and prospects. Science 349 (6245), 255–260.

Jorrín-Novo, J.V., Pascual, J., Sánchez-Lucas, R., Romero-Rodríguez, M.C., Rodríguez-Ortega, M.J., Lenz, C., Valledor, L., 2015. Fourteen years of plant proteomics reflected in Proteomics: moving from model species and 2DE-based approaches to orphan species and gel-free platforms. Proteomics 15 (5-6), 1089–1112.

Joyce, A.R., Palsson, B.Ø., 2006. The model organism as a system: integrating 'omics' data sets. Nat. Rev. Mol. Cell Biol. 7 (3), 198–210. https://doi.org/10.1038/nrm1857.

Juliana, P., Poland, J., Huerta-Espino, J., Shrestha, S., Crossa, J., Crespo-Herrera, L., Toledo, F.H., Govindan, V., Mondal, S., Kumar, U., Bhavani, S., 2019. Improving grain yield, stress resilience, and quality of bread wheat using large-scale genomics. Nat. Genet. 51 (10), 1530–1539.

Jumper, J., Evans, R., Pritzel, A., Green, T., Figurnov, M., Ronneberger, O., Tunyasuvunakool, K., Bates, R., Žídek, A., Potapenko, A., et al., 2021. Highly accurate protein structure prediction with AlphaFold. Nature 596, 583–589.

Kajrolkar, A., 2025a. Deep learning applications in plant biology: bridging genotype and phenotype. Plant Biol. 3, 100011.

Kajrolkar, A., 2025b. The role of epigenetics in plant development and adaptation: a comprehensive molecular and ecological perspective. DNA Sequence 4, 5.

Kassem, M.A., 2025. Harnessing artificial intelligence and machine learning for identifying quantitative trait loci (QTL) associated with seed quality traits in crops. Plants 14 (11), 1727.

Kawakatsu, T., Huang, S.S.C., Jupe, F., Sasaki, E., Schmitz, R.J., Urich, M.A., Castañon, R., Nery, J.R., Barragan, C., He, Y., Chen, H., 2016. Epigenomic diversity in a global collection of *Arabidopsis thaliana* accessions. Cell 166 (2), 492–505.

Kaya, C., 2025. Optimizing crop production with plant phenomics through high-throughput phenotyping and AI in controlled environments. Food Energy Secur. 14 (1), e70050.

Keurentjes, J.J., Fu, J., De Vos, C.R., Lommen, A., Hall, R.D., Bino, R.J., Van Der Plas, L.H., Jansen, R.C., Vreugdenhil, D., Koornneef, M., 2006. The genetics of plant metabolism. Nat. Genet. 38 (7), 842–849.

Khan, A.W., Garg, V., Roorkiwal, M., Golicz, A.A., Edwards, D., Varshney, R.K., 2020. Super-pangenome by integrating the wild side of a species for accelerated crop improvement. Trends Plant Sci. 25 (2), 148–158.

Khan, M.H.U., Wang, S., Wang, J., Ahmar, S., Saeed, S., Khan, S.U., Xu, X., Chen, H., Bhat, J.A., Feng, X., 2022. Applications of artificial intelligence in climate-resilient smart-crop breeding. Int. J. Mol. Sci. 23, 11156.

Kim, J., Pyo, H., Jang, I., Kang, J., Ju, B., Ko, K., 2022. Tomato harvesting robotic system based on Deep-ToMaToS: deep learning network using transformation loss for 6D pose estimation of maturity-classified tomatoes with side-stem. Comput. Electron. Agric. 201, 107300.

Klepikova, A.V., Kasianov, A.S., Gerasimov, E.S., Logacheva, M.D., Penin, A.A., 2016. A high-resolution map of the *Arabidopsis thaliana* developmental transcriptome based on RNA-seq profiling. Plant J. 88 (6), 1058–1070. https://doi.org/10.1111/tpj.13312.

Korte, A., Farlow, A., 2013. The advantages and limitations of trait analysis with GWAS: a review. Plant Methods 9 (1), 29.

Kosova, K., Vítámvás, P., Urban, M.O., Prášil, I.T., Renaut, J., 2018. Plant abiotic stress proteomics: the major factors determining alterations in cellular proteome. Frontiers Plant Sci. 9, 122.

Krause, M.R., González-Pérez, L., Crossa, J., Pérez-Rodríguez, P., Montesinos-López, O., Singh, R.P., Dreisigacker, S., Poland, J., Rutkoski, J., Sorrells, M., Gore, M.A., 2019. Hyperspectral reflectance-derived relationship matrices for genomic prediction of grain yield in wheat. G3: Genes, Genomes, Genet. 9 (4), 1231–1247.

Kremling, K.A., Chen, S.Y., Su, M.H., Lepak, N.K., Romay, M.C., Swarts, K.L., Lu, F., Lorant, A., Bradbury, P.J., Buckler, E.S., 2018. Dysregulation of expression correlates with rare-allele burden and fitness loss in maize. Nature 555 (7697), 520–523.

Kremling, K.A., Diepenbrock, C.H., Gore, M.A., Buckler, E.S., Bandillo, N.B., 2019. Transcriptome-wide association supplements genome-wide association in *Zea mays*. G3: Genes Genomes Genet. 9 (9), 3023–3033.

Kudapa, H., Ghatak, A., Barmukh, R., Chaturvedi, P., Khan, A., Kale, S., Fragner, L., Chitikineni, A., Weckwerth, W., Varshney, R.K., 2024. Integrated multi-omics analysis reveals drought stress response mechanism in chickpea (*Cicer arietinum* L.). Plant Genome 17 (1), e20337.

Kumar, A., Anju, T., Kumar, S., Chhapekar, S.S., Sreedharan, S., Singh, S., Choi, S.R., Ramchiary, N., Lim, Y.P., 2021. Integrating omics and gene editing tools for rapid improvement of traditional food plants for diversified and sustainable food security. Int. J. Mol. Sci. 22 (15), 8093.

Langfelder, P., Horvath, S., 2008. WGCNA: an R package for weighted correlation network analysis. BMC Bioinformat. 9 (1), 559.

LeCun, Y., Bengio, Y., Hinton, G., 2015. Deep learning. Nature 521 (7553), 436–444.

Li, Q., Eichten, S.R., Hermanson, P.J., Springer, N.M., 2014. Inheritance patterns and stability of DNA methylation variation in maize near-isogenic lines. Genetics 196 (3), 667–676.

Li, C., Xiang, X., Huang, Y., Zhou, Y., An, D., Dong, J., Zhao, C., Liu, H., Li, Y., Wang, Q., et al., 2020a. Long-read sequencing reveals genomic structural variations that underlie creation of quality protein maize. Nat. Commun. 11, 17.

Li, D., Wang, R., Xie, C., Liu, L., Zhang, J., Li, R., et al., 2020. A recognition method for rice plant diseases and pests video detection based on deep convolutional neural network. Sensors 20 (3), 578.

Li, G., An, L., Yang, W., Yang, L., Wei, T., Shi, J., Wang, J., Doonan, J.H., Xie, K., Fernie, A.R., Lagudah, E.S., 2025a. Integrated biotechnological and AI innovations for crop improvement. Nature 643 (8073), 925–937.

Li, J., Wei, L., Guo, F., Zou, Q., 2021. EP3: an ensemble predictor that accurately identifies type III secreted effectors. Briefings Bioinf. 22 (2), 1918–1928. https://doi.org/10.1093/bib/bbaa008.

Li, J., Wu, P., Cao, Z., Huang, G., Lu, Z., Yan, J., Zhang, H., Zhou, Y., Liu, R., Chen, H., Ma, L., 2024a. Machine learning-based prediction models to guide the selection of Cas9 variants for efficient gene editing. Cell Rep. 43 (2).

Li, K., Yu, Y., Yan, S., Li, W., Xu, J., Li, G., Li, W., Liu, J., Qi, X., Huang, W. and Zhang, Q. (unpublished). Multi-omics analyses reveal the flavor code of sweet corn.

Li, N., Li, G., Wang, D., Ma, L., Huang, X., Bai, Z., Wang, Y., Luo, M., Luo, Y., Zhu, Y., Cao, X., 2024b. Large-scale multi-omics analyses identified root-microbiome associations underlying plant nitrogen nutrition. bioRxiv 2024–02.

Li, X., Tong, X., Wang, B., Pu, M., Zheng, G., Wang, B., Li, J., He, X., Liu, Z., Ding, H., Zhang, Z., 2025b. Role of methyltransferase 1-dependent DNA methylation in affecting maize kernel development. Plant J. 121 (4), e17250.

Li, Z., Chen, X., Shi, S., Zhang, H., Wang, X., Chen, H., Li, W., Li, L., 2022. DeepBSA: a deep-learning algorithm improves bulked segregant analysis for dissecting complex traits. Mol. Plant 15, 1418–1427.

Liang, Y., Zhang, F., Wang, J., Joshi, T., Wang, Y., Xu, D., 2011. Prediction of drought-resistant genes in *Arabidopsis thaliana* using SVM-RFE. PLoS One 6 (7), e21750.

Libbrecht, M.W., Noble, W.S., 2015. Machine learning applications in genetics and genomics. Nat. Rev. Genet. 16 (6), 321–332.

Lin, C., Jain, S., Kim, H., Bar-Joseph, Z., 2017. Using neural networks for reducing the dimensions of single-cell RNA-Seq data. Nucleic Acids Res. 45 (17), e156. https://doi.org/10.1093/nar/gkx681.

Lin, Q., Zong, Y., Xue, C., Wang, S., Jin, S., Zhu, Z., Wang, Y., Anzalone, A.V., Raguram, A., Doman, J.L., Liu, D.R., 2020. Prime genome editing in rice and wheat. Nat. Biotechnol. 38 (5), 582–585.

Liu, H., Ding, Y., Zhou, Y., Jin, W., Xie, K., Chen, L.L., 2017a. CRISPR-P 2.0: an improved CRISPR-Cas9 tool for genome editing in plants. Mol. Plant 10 (3), 530–532.

Liu, H.J., Jian, L., Xu, J., Zhang, Q., Zhang, M., Jin, M., Peng, Y., Yan, J., Han, B., Liu, J., Gao, F., 2020b. High-throughput CRISPR/Cas9 mutagenesis streamlines trait gene identification in maize. Plant Cell 32 (5), 1397–1413.

Liu, J., Seetharam, A.S., Chougule, K., Ou, S., Swentowsky, K.W., Gent, J.I., Llaca, V., Woodhouse, M.R., Manchanda, N., Presting, G.G., et al., 2020a. Gapless assembly of maize chromosomes using long-read technologies. Genome Biol. 21, 121.

Liu, K., Abdullah, A.A., Huang, M., Nishioka, T., Altaf-Ul-Amin, M., Kanaya, S., 2017b. Novel approach to classify plants based on metabolite-content similarity. BioMed Res. Int. 2017. https://doi.org/10.1155/2017/5296729.

Liu, H., Luo, X., Niu, L., Xiao, Y., Chen, L., Liu, J., Wang, X., Jin, M., Li, W., Zhang, Q., Yan, J., 2017c. Distant eQTLs and non-coding sequences play critical roles in regulating gene expression and quantitative trait variation in maize. Mol. Plant 10 (3), 414–426.

Liu, M., Tan, X., Yang, Y., Liu, P., Zhang, X., Zhang, Y., Wang, L., Hu, Y., Ma, L., Li, Z., Zhang, Y., 2020c. Analysis of the genetic architecture of maize kernel size traits by combined linkage and association mapping. Plant Biotechnol. J. 18 (1), 207–221.

Lowder, L.G., Zhang, D., Baltes, N.J., Paul III, J.W., Tang, X., Zheng, X., Voytas, D.F., Hsieh, T.F., Zhang, Y., Qi, Y., 2015. A CRISPR/Cas9 toolbox for multiplexed plant genome editing and transcriptional regulation. Plant Physiol. 169 (2), 971–985.

Ly, D.N.P., Iqbal, S., Fosu-Nyarko, J., Milroy, S., Jones, M.G., 2023. Multiplex CRISPR-Cas9 gene-editing can deliver potato cultivars with reduced browning and acrylamide. Plants 12 (2), 379.

Lynch, J.P., 2019. Root phenotypes for improved nutrient capture: an underexploited opportunity for global agriculture. New Phytol. 223 (2), 548–564.

Ma, X., Zhang, Q., Zhu, Q., Liu, W., Chen, Y., Qiu, R., Wang, B., Yang, Z., Li, H., Lin, Y., Xie, Y., 2015. A robust CRISPR/Cas9 system for convenient, high-efficiency multiplex genome editing in monocot and dicot plants. Molecular plant 8 (8), 1274–1284.

Mackay, T.F., Stone, E.A., Ayroles, J.F., 2009. The genetics of quantitative traits: challenges and prospects. Nat. Rev. Genet. 10 (8), 565–577.

MacNish, T.R., Danilevicz, M.F., Bayer, P.E., Bestry, M.S., Edwards, D., 2025. Application of machine learning and genomics for orphan crop improvement. Nat. Commun. 16 (1), 982.

Madec, S., Baret, F., De Solan, B., Thomas, S., Dutartre, D., Jezequel, S., Hemmerlé, M., Colombeau, G., Comar, A., 2017. High-throughput phenotyping of plant height: comparing unmanned aerial vehicles and ground LiDAR estimates. Front. Plant Sci. 8, 2002.

Mahas, A., Mahfouz, M., 2018. Engineering virus resistance via CRISPR–Cas systems. Current Opin. Virol. 32, 1–8.

Maimon, O., Rokach, L. (Eds.), 2005. Data mining and knowledge discovery handbook, 2. Springer, New York.

Manolio, T.A., Collins, F.S., Cox, N.J., Goldstein, D.B., Hindorff, L.A., Hunter, D.J., McCarthy, M.I., Ramos, E.M., Cardon, L.R., Chakravarti, A., Cho, J.H., 2009. Finding the missing heritability of complex diseases. Nature 461 (7265), 747–753.

Marand, A.P., Chen, Z., Gallavotti, A., Schmitz, R.J., 2021. A cis-regulatory atlas in maize at single-cell resolution. Cell 184 (11), 3041–3055. https://doi.org/10.1016/j.cell.2021.04.014.

Maraveas, C., 2024. Image analysis artificial intelligence technologies for plant phenotyping: current state of the art. AgriEngineering 6 (3), 3375–3407.

Marbach, D., Lamparter, D., Quon, G., Kellis, M., Kutalik, Z., Bergmann, S., 2016. Tissue-specific regulatory circuits reveal variable modular perturbations across complex diseases. Nat. Methods 13 (4), 366–370.

Meng, X., Liang, Z., Dai, X., Zhang, Y., Mahboub, S., Ngu, D.W., et al., 2021. Predicting transcriptional responses to cold stress across plant species. Proc. Natl. Acad. Sci. 118 (10), e2026330118. https://doi.org/10.1073/pnas.2026330118.

Meuwissen, T.H., Hayes, B.J., Goddard, M., 2001. Prediction of total genetic value using genome-wide dense marker maps. Genetics 157 (4), 1819–1829.

Mi, Z., Zhang, X., Su, J., Han, D., Su, B., 2020. Wheat stripe rust grading by deep learning with attention mechanism and images from mobile devices. Front. Plant Sci. 11, 558126. https://doi.org/10.3389/fpls.2020.558126.

Michael, T.P., VanBuren, R., 2020. Building near-complete plant genomes. Curr. Opin. Plant Biol. 54, 26–33.

Millet, E.J., Kruijer, W., Coupel-Ledru, A., Alvarez Prado, S., Cabrera-Bosquet, L., Lacube, S., Charcosset, A., Welcker, C., van Eeuwijk, F., Tardieu, F., 2019. Genomic prediction of maize yield across European environmental conditions. Nat. Genet. 51 (6), 952–956.

Mishra, A., Bhat, A., Kumari, S., Sharma, R., Braynen, J., Tadesse, D., El Alaoui, S., Seaver, S., Grosjean, N., Ware, D., Xie, M., 2025. Time-series multi-omics analysis of micronutrient stress in *Sorghum bicolor* reveals iron and zinc crosstalk and regulatory network conservation. Plant Biol. https://doi.org/10.1111/plb.70038. Plb 70038.

Mohamedikbal, S., Al-Mamun, H.A., Bestry, M.S., Batley, J., Edwards, D., 2025. Integrating multi-omics and machine learning for disease resistance prediction in legumes. Theor. Appl. Genet. 138 (7), 1–21.

Montesinos-López, O.A., Montesinos-López, A., Crossa, J., Gianola, D., Hernández-Suárez, C.M., Martín-Vallejo, J., 2018a. Multi-trait, multi-environment deep learning modeling for genomic-enabled prediction of plant traits. G3: Genes, Genomes, Genet. 8 (12), 3829–3840.

Montesinos-López, O.A., Montesinos-López, A., Montesinos-López, J.C., Crossa, J., Luna-Vázquez, F.J., Salinas-Ruiz, J., 2018b. A Bayesian multiple-trait and multiple-environment model using the matrix normal distribution. Phys. Methods Stimul. Plant Mushroom Dev 19 (10.5772).

Montesinos-López, A., Montesinos-López, O.A., Gianola, D., Crossa, J., Hernández-Suárez, C.M., 2018c. Multi-environment genomic prediction of plant traits using deep learners with dense architecture. G3: Genes Genomes Genet. 8 (12), 3813–3828.

Montesinos-López, O.A., Martín-Vallejo, J., Crossa, J., Gianola, D., Hernández-Suárez, C.M., Montesinos-López, A., Juliana, P., Singh, R., 2019. A benchmarking between deep learning, support vector machine and Bayesian threshold best linear unbiased prediction for predicting ordinal traits in plant breeding. G3: Genes Genomes Genet. 9 (2), 601–618.

Montesinos-López, O.A., Montesinos-López, A., Pérez-Rodríguez, P., Barrón-López, J.A., Martini, J.W., Fajardo-Flores, S.B., Gaytan-Lugo, L.S., Santana-Mancilla, P.C., Crossa, J., 2021. A review of deep learning applications for genomic selection. BMC Genom. 22 (1), 19.

Murmu, S., Sinha, D., Chaurasia, H., Sharma, S., Das, R., Jha, G.K., Archak, S., 2024. A review of artificial intelligence-assisted omics techniques in plant defense: current trends and future directions. Front. Plant Sci. 15, 1292054.

Obata, T., Fernie, A.R., 2012. The use of metabolomics to dissect plant responses to abiotic stresses. Cell. Mol. Life Sci. 69 (19), 3225–3243.

Okazaki, Y., Saito, K., 2016. Integrated metabolomics and phytochemical genomics approaches for studies on rice. GigaScience 5 (1) s13742-016.

Ovesna, J., Kaminiaris, M.D., Tsiropoulos, Z., Collier, R., Kelly, A., De Mey, J., Pollet, S., 2023. Applicability of smart tools in vegetable disease diagnostics. Agronomy 13 (5), 1211.

Pal, T., Jaiswal, V., Chauhan, R.S., 2016. DRPPP: a machine-learning-based tool for prediction of disease resistance proteins in plants. Comput. Biol. Med. 78, 42–48. https://doi.org/10.1016/j.compbiomed.2016.09.008.

Paulus, S., Dupuis, J., Mahlein, A.K., Kuhlmann, H., 2013. Surface feature based classification of plant organs from 3D laserscanned point clouds for plant phenotyping. BMC Bioinform. 14 (1), 238.

Pearson, S., Camacho-Villa, T.C., Valluru, R., Gaju, O., Rai, M.C., Gould, I., Brewer, S., Sklar, E., 2022. Robotics and autonomous systems for net-zero agriculture. Current Robotics Reports 3, 57–64.

Peignier, S., Sorin, B., Calevro, F., 2023. Ensemble learning-based gene regulatory network inference. Int. J. Artif. Intell. Tools 32 (5), 2360005.

Piatek, A., Ali, Z., Baazim, H., Li, L., Abulfaraj, A., Al-Shareef, S., Aouida, M., Mahfouz, M.M., 2015. RNA-guided transcriptional regulation *in planta* via synthetic dCas9-based transcription factors. Plant Biotechnol. J. 13 (4), 578–589.

Polder, G., Blok, P.M., De Villiers, H.A., van der Wolf, J.M., Kamp, J., 2019. Potato virus Y detection in seed potatoes using deep learning on hyperspectral images. Front. Plant Sci. 10, 209.

Poorter, H., Hummel, G.M., Nagel, K.A., Fiorani, F., von Gillhaussen, P., Virnich, O., Schurr, U., Postma, J.A., van de Zedde, R., Wiese-Klinkenberg, A., 2023. Pitfalls and potential of high-throughput plant phenotyping platforms. Front. Plant Sci. 14, 1233794.

Pound, M.P., Atkinson, J.A., Townsend, A.J., Wilson, M.H., Griffiths, M., Jackson, A.S., Bulat, A., Tzimiropoulos, G., Wells, D.M., Murchie, E.H., Pridmore, T.P., 2017. Deep machine learning provides state-of-the-art performance in image-based plant phenotyping. GigaScience 6 (10), gix083. https://doi.org/10.1093/gigascience/gix083.

Pradhan, U.K., Meher, P.K., Naha, S., Rao, A.R., Gupta, A., 2023. ASLncR: a novel computational tool for prediction of abiotic stress-responsive long non-coding RNAs in plants. Funct. Integr. Genom. 23 (2), 113. https://doi.org/10.1007/s10142-023-01040-0.

Pratapa, A., Jalihal, A.P., Law, J.N., Bharadwaj, A., Murali, T.M., 2020. Benchmarking algorithms for gene regulatory network inference from single-cell transcriptomic data. Nat. Methods 17 (2), 147–154. https://doi.org/10.1038/s41592-019-0690-6.

Puranik, S.S., Hanamakkanavar, S.R., Bidargaddi, A.P., Ballur, V.V., Joshi, P.T., Meena, S.M., 2024. MobileNetV3 for mango leaf disease detection: an efficient deep learning approach for precision agriculture. In: Proceedings of the 5th International Conference for Emerging Technology (INCET). Belgaum, India, pp. 1–7.

Qi, L.S., Larson, M.H., Gilbert, L.A., Doudna, J.A., Weissman, J.S., Arkin, A.P., Lim, W.A., 2013. Repurposing CRISPR as an RNA-guided platform for sequence-specific control of gene expression. Cell 152 (5), 1173–1183. https://doi.org/10.1016/j.cell.2013.02.022.

Qiu, Y., Liu, L., Yan, J., Xiang, X., Wang, S., Luo, Y., Deng, K., Xu, J., Jin, M., Wu, X., Cheng, L., 2025. Precise engineering of gene expression by editing plasticity. Genome Biol. 26 (1), 51. https://doi.org/10.1186/s13059-025-03451-5.

Rai, A., Yamazaki, M., Saito, K., 2019. A new era in plant functional genomics. Curr. Opin. Syst. Biol. 15, 58–67. https://doi.org/10.1016/j.coisb.2019.03.005.

Raimundo, F., Meng-Papaxanthos, L., Vallot, C., Vert, J.P., 2021. Machine learning for single-cell genomics data analysis. Curr. Opin. Syst. Biol. 26, 64–71. https://doi.org/10.1016/j.coisb.2021.04.006.

Ralser, M., Querfurth, R., Warnatz, H.J., Lehrach, H., Yaspo, M.L., Krobitsch, S., 2006. An efficient and economic enhancer mix for PCR. Biochem. Biophys. Res. Commun. 347 (3), 747–751.

Ravari, S.Z., Dehghani, H., Naghavi, H., 2015. Evaluation of relationship between salinity stress tolerance indices and some physiological traits in bread wheat. Iran. J. Field Crop Sci. 46 (3), 423–432.

Razzaq, M.K., Aleem, M., Mansoor, S., Khan, M.A., Rauf, S., Iqbal, S., Siddique, K.H., 2021. Omics and CRISPR-Cas9 approaches for molecular insight, functional gene analysis, and stress tolerance development in crops. Int. J. Mol. Sci. 22 (3), 1292. https://doi.org/10.3390/ijms22031292.

Riedelsheimer, C., Czedik-Eysenberg, A., Grieder, C., Lisec, J., Technow, F., Sulpice, R., Altmann, T., Stitt, M., Willmitzer, L., Melchinger, A.E., 2012. Genomic and metabolic prediction of complex heterotic traits in hybrid maize. Nat. Genet. 44 (2), 217–220. https://doi.org/10.1038/ng.1033.

Roychowdhury, R., Das, S.P., Gupta, A., Parihar, P., Chandrasekhar, K., Sarker, U., Kumar, A., Ramrao, D.P., Sudhakar, C., 2023. Multi-omics pipeline and omics-integration approach to decipher plant's abiotic stress tolerance responses. Genes 14 (6), 1281. https://doi.org/10.3390/genes14061281.

Sadok, W., Naudin, P., Boussuge, B., Muller, B., Welcker, C., Tardieu, F., 2007. Leaf growth rate per unit thermal time follows QTL-dependent daily patterns in hundreds of maize lines under naturally fluctuating conditions. Plant Cell Environ. 30 (2), 135–146. https://doi.org/10.1111/j.1365-3040.2006.01612.x.

Sahito, J.H., Zhang, H., Gishkori, Z.G.N., Ma, C., Wang, Z., Ding, D., Zhang, X., Tang, J., 2024. Advancements and prospects of genome-wide association studies (GWAS) in maize. Int. J. Mol. Sci. 25 (3), 1918. https://doi.org/10.3390/ijms25031918.

Saini, H., Devrani, A., Synrem, G., Priyanka, 2025. Application of CRISPR technology in plant improvement: an updated review. Advances in Agriculture 2025 (1), 4578877. https://doi.org/10.1155/2025/4578877.

Saito, K., Matsuda, F., 2010. Metabolomics for functional genomics, systems biology, and biotechnology. Annu. Rev. Plant Biol. 61, 463–489. https://doi.org/10.1146/annurev.arplant.043008.092035.

Sandhu, K.S., Raghuram, N., 2024. Weighted gene co-expression network analysis of nitrogen (N)-responsive genes and the putative role of G-quadruplexes in N use efficiency (NUE) in rice. In: Nitrogen Use Efficiency: Plant Biology to Crop Improvement. Springer.

Schmitz, R.J., Schultz, M.D., Urich, M.A., Nery, J.R., Pelizzola, M., Libiger, O., Alix, A., McCosh, R.B., Chen, H., Schork, N.J., Ecker, J.R., 2013. Patterns of population epigenomic diversity. Nature 495 (7440), 193–198. https://doi.org/10.1038/nature11968.

Schmutz, J., Cannon, S.B., Schlueter, J., Ma, J., Mitros, T., Nelson, W., Hyten, D.L., Song, Q., Thelen, J.J., Cheng, J., et al., 2010. Genome sequence of the palaeopolyploid soybean. Nature 463 (7278), 178–183. https://doi.org/10.1038/nature08670.

Schnable, P.S., Ware, D., Fulton, R.S., Stein, J.C., Wei, F., Pasternak, S., Liang, C., Zhang, J., Fulton, L., Graves, T.A., et al., 2009. The B73 maize genome: complexity, diversity, and dynamics. Science 326 (5956), 1112–1115. https://doi.org/10.1126/science.1178534.

Scholz, M., Gatzek, S., Sterling, A., Fiehn, O., Selbig, J., 2004. Metabolite fingerprinting: Detecting biological features by independent component analysis. Bioinformatics 20 (15), 2447–2454. https://doi.org/10.1093/bioinformatics/bth270.

Schwarz, B., Azodi, C.B., Shiu, S.H., Bauer, P., 2020. Putative cis-regulatory elements predict iron deficiency responses in *Arabidopsis* roots. Plant Physiol. 182 (3), 1420–1439. https://doi.org/10.1104/pp.19.00760.

Secco, D., Wang, C., Shou, H., Schultz, M.D., Chiarenza, S., Nussaume, L., Ecker, J.R., Whelan, J., Lister, R., 2015. Stress-induced gene expression drives transient DNA methylation changes at adjacent repetitive elements. eLife 4, e09343. https://doi.org/10.7554/eLife.09343.

Seed World, 2025. CRISPR-Based Gene Editing Reveals Complex Multigenic Traits in Tomatoes.

Shaik, R., Ramakrishna, W., 2014. Machine learning approaches distinguish multiple stress conditions using stress-responsive genes and identify candidate genes for broad resistance in rice. Plant Physiol. 164 (1), 481–495. https://doi.org/10.1104/pp.113.225862.

Shakoor, N., Lee, S., Mockler, T.C., 2017. High throughput phenotyping to accelerate crop breeding and monitoring of diseases in the field. Curr. Opin. Plant Biol. 38, 184–192.

Shang, L., Li, X., He, H., Yuan, Q., Song, Y., Wei, Z., Lin, H., Hu, M., Zhao, F., Zhang, C., et al., 2022. A super pan-genomic landscape of rice. Cell Res. 32, 878–896.

Sharma, A., Sharma, A., Tselykh, A., Bozhenyuk, A., Choudhury, T., Alomar, M.A., Sánchez-Chero, M., 2023. Artificial intelligence and internet of things oriented sustainable precision farming: towards modern agriculture. Open Life Sci. 18, 20220713.

Shelake, R.M., Kadam, U.S., Kumar, R., Pramanik, D., Singh, A.K., Kim, J.Y., 2022. Engineering drought and salinity tolerance traits in crops through CRISPR-mediated genome editing: targets, tools, challenges, and perspectives. Plant Commun. 3 (6).

Silva, J.C.F., Teixeira, R.M., Silva, F.F., Brommonschenkel, S.H., Fontes, E.P., 2019. Machine learning approaches and their current application in plant molecular biology: a systematic review. Plant Sci. 284, 37–47. https://doi.org/10.1016/j.plantsci.2019.03.020.

Simmons, C.R., Lafitte, H.R., Reimann, K.S., Brugière, N., Roesler, K., Albertsen, M.C., Greene, T.W., Habben, J.E., 2021. Successes and insights of an industry biotech program to enhance maize agronomic traits. Plant Sci. 307, 110899.

Singh, A., Ganapathysubramanian, B., Singh, A.K., Sarkar, S., 2016. Machine learning for high-throughput stress phenotyping in plants. Trends Plant Sci. 21 (2), 110–124.

Singh, A.K., Ganapathysubramanian, B., Sarkar, S., Singh, A., 2018. Deep learning for plant stress phenotyping: trends and future perspectives. Trends Plant Sci. 23 (10), 883–898.

Song, Q.X., Liu, Y.F., Hu, X.Y., Zhang, W.K., Ma, B., Chen, S.Y., Zhang, J.S., 2011. Identification of miRNAs and their target genes in developing soybean seeds by deep sequencing. BMC Plant Biol. 11 (1), 5.

Song, X., Meng, X., Guo, H., Cheng, Q., Jing, Y., Chen, M., Liu, G., Wang, B., Wang, Y., Li, J., Yu, H., 2022. Targeting a gene regulatory element enhances rice grain yield by decoupling panicle number and size. Nat. Biotechnol. 40 (9), 1403–1411.

Sperschneider, J., 2020. Machine learning in plant–pathogen interactions: Empowering biological predictions from field scale to genome scale. New Phytol. 228 (1), 35–41.

Sperschneider, J., Gardiner, D.M., Dodds, P.N., Tini, F., Covarelli, L., Singh, K.B., et al., 2016. EffectorP: predicting fungal effector proteins from secretomes using machine learning. New Phytol. 210 (2), 743–761. https://doi.org/10.1111/nph.13794.

Springer, N.M., Schmitz, R.J., 2017. Exploiting induced and natural epigenetic variation for crop improvement. Nat. Rev. Genet. 18 (9), 563–575.

Sun, L., Lai, M., Ghouri, F., Nawaz, M.A., Ali, F., Baloch, F.S., Nadeem, M.A., Aasim, M., Shahid, M.Q., 2024. Modern plant breeding techniques in crop improvement and genetic diversity: from molecular markers and gene editing to artificial intelligence—a critical review. Plants 13 (19), 2676.

Suruthi, G.S., Begum, G.R.R., Vinothini, M., Balamuralikrishnan, B., Al-Dhabi, N.A., Arasu, M.V., Palani, V., Anand, A.V., 2025. High-throughput metabolomics for the exploration of plant secondary metabolism. In: High-Throughput Plant Metabolomics. GB: CABI, pp. 388–404.

Taheri, S., Gantait, S., Azizi, P., Mazumdar, P., 2022. Drought tolerance improvement in *Solanum lycopersicum*: an insight into "omics" approaches and genome editing. 3 Biotech 12 (3), 63.

Tardieu, F., Cabrera-Bosquet, L., Pridmore, T., Bennett, M., 2017. Plant phenomics, from sensors to knowledge. Curr. Biol. 27 (15), R770–R783.
Teng, C. (2024). Killing two birds with one stone: A breakthrough in transgene-free gene editing in soybean.
Thatcher, S.R., Danilevskaya, O.N., Meng, X., Beatty, M., Zastrow-Hayes, G., Harris, C., Van Allen, B., Habben, J., Li, B., 2016. Genome-wide analysis of alternative splicing during development and drought stress in maize. Plant Physiol. 170 (1), 586–599.
Tian, Z., Nepomuceno, A.L., Song, Q., Stupar, R.M., Liu, B., Kong, F., Ma, J., Lee, S.H., Jackson, S.A., 2025. Soybean2035: a decadal vision for soybean functional genomics and breeding. Mol. Plant 18 (2), 245–271.
Tohge, T., Fernie, A.R., 2010. Combining genetic diversity, informatics, and metabolomics to facilitate annotation of plant gene function. Nat. Protoc. 5 (6), 1210–1227.
Trivedi, P., Leach, J.E., Tringe, S.G., Sa, T., Singh, B.K., 2020. Plant–microbiome interactions: from community assembly to plant health. Nat. Rev. Microbiol. 18 (11), 607–621.
Tsakirpaloglou, N., Septiningsih, E.M., Thomson, M.J., 2023. Guidelines for performing CRISPR/Cas9 genome editing for gene validation and trait improvement in crops. Plants 12 (20), 3564.
Turner, T.R., James, E.K., Poole, P.S., 2013. The plant microbiome. Genome Biol. 14 (6), 209.
Tyagi, P., Singh, D., Mathur, S., Singh, A., Ranjan, R., 2022. Upcoming progress of transcriptomics studies on plants: an overview. Front. Plant Sci. 13, 1030890.
Ubbens, J.R., Stavness, I., 2017. Deep plant phenomics: a deep learning platform for complex plant phenotyping tasks. Front. Plant Sci. 8, 1190.
Usadel, B., Obayashi, T., Mutwil, M., Giorgi, F.M., Bassel, G.W., Tanimoto, M., Chow, A., Steinhauser, D., Persson, S., Provart, N.J., 2009. Co-expression tools for plant biology: opportunities for hypothesis generation and caveats. Plant Cell Environ. 32 (12), 1633–1651.
Van Dijk, D., Sharma, R., Nainys, J., Yim, K., Kathail, P., Carr, A.J., et al., 2018. Recovering gene interactions from single-cell data using data diffusion. Cell 174 (3), 716–729. https://doi.org/10.1016/j.cell.2018.05.061.
Varadharajan, V., Rajendran, R., Muthuramalingam, P., Runthala, A., Madhesh, V., Swaminathan, G., Murugan, P., Srinivasan, H., Park, Y., Shin, H., Ramesh, M., 2025. Multi-omics approaches against abiotic and biotic stress—a review. Plants 14 (6), 865.
Varshney, R.K., Bohra, A., Yu, J., Graner, A., Zhang, Q., Sorrells, M.E., 2021. Designing future crops: genomics-assisted breeding comes of age. Trends Plant Sci. 26 (6), 631–649.
Varshney, R.K., Sinha, P., Singh, V.K., Kumar, A., Zhang, Q., Bennetzen, J.L., 2020. 5Gs for crop genetic improvement. Curr. Opin. Plant Biol. 56, 190–196.
Virlet, N., Sabermanesh, K., Sadeghi-Tehran, P., Hawkesford, M.J., 2016. Field Scanalyzer: an automated robotic field phenotyping platform for detailed crop monitoring. Funct. Plant Biol. 44, 143–153.
Vives-Peris, V., De Ollas, C., Gómez-Cadenas, A., Pérez-Clemente, R.M., 2020. Root exudates: from plant to rhizosphere and beyond. Plant Cell Rep. 39 (1), 3–17.
Vu, L.D., Xu, X., Zhu, T., Pan, L., van Zanten, M., de Jong, D., Wang, Y., Vanremoortele, T., Locke, A.M., Van De Cotte, B., De Winne, N., 2021. The membrane-localized protein kinase MAP4K4/TOT3 regulates thermomorphogenesis. Nat. Commun. 12 (1), 2842.
Wallace, J.G., Kremling, K.A., Kovar, L.L., Buckler, E.S., 2018. Quantitative genetics of the maize leaf microbiome. Phytobiomes J. 2 (4), 208–224.
Wang, B., Zhu, J., Pierson, E., Ramazzotti, D., Batzoglou, S., 2017. Visualization and analysis of single-cell RNA-seq data by kernel-based similarity learning. Nat. Methods 14 (4), 414–416. https://doi.org/10.1038/nmeth.4207.
Wang, H., Cimen, E., Singh, N., Buckler, E., 2020. Deep learning for plant genomics and crop improvement. Curr. Opin. Plant Biol. 54, 34–41.
Wang, S., Lv, J., Zhang, L., Dou, J., Sun, Y., Li, X., Fu, X., Dou, H., Mao, J., Hu, X., Bao, Z., 2015. MethylRAD: a simple and scalable method for genome-wide DNA methylation profiling using methylation-dependent restriction enzymes. Open Biol. 5 (11), 150130.
Wang, W., Pan, Q., He, F., Akhunova, A., Chao, S., Trick, H., Akhunov, E., 2018a. Transgenerational CRISPR-Cas9 activity facilitates multiplex gene editing in allopolyploid wheat. CRISPR J. 1 (1), 65–74.
Wang, X., Liu, J., 2021. Multiscale parallel algorithm for early detection of tomato gray mold in a complex natural environment. Front. Plant Sci. 12, 620273.
Wang, X., Xu, Y., Hu, Z., Xu, C., 2018b. Genomic selection methods for crop improvement: current status and prospects. The Crop Journal 6 (4), 330–340.

Wang, Y., Zhang, P., Guo, W., Liu, H., Li, X., Zhang, Q., Du, Z., Hu, G., Han, X., Pu, L., Tian, J., 2021. A deep learning approach to automate whole-genome prediction of diverse epigenomic modifications in plants. New Phytol. 232 (2), 880–897.

Wang, Z., Gerstein, M., Snyder, M., 2009. RNA-Seq: a revolutionary tool for transcriptomics. Nat. Rev. Genet. 10 (1), 57–63.

Wang, J., Chen, L., Wang, Y., Zhang, J., Liang, Y., Xu, D., 2013. A computational systems biology study for understanding salt tolerance mechanism in rice. PLoS One 8 (6), e64929.

Wang, Z., Niu, Y., Vashisth, T., Li, J., Madden, R., Livingston, T.S., Wang, Y., 2022a. Nontargeted metabolomics-based multiple machine learning modeling boosts early accurate detection for citrus Huanglongbing. Hortic. Res. 9, uhac145.

Wang, Z., Yan, L., Chen, Y., Wang, X., Huai, D., Kang, Y., Jiang, H., Liu, K., Lei, Y., Liao, B., 2022b. Detection of a major QTL and development of KASP markers for seed weight by combining QTL-seq, QTL-mapping and RNA-seq in peanut. Theor. Appl. Genet. 135 (5), 1779–1795.

Wellenreuther, M., Mérot, C., Berdan, E., Bernatchez, L., 2019. Going beyond SNPs: the role of structural genomic variants in adaptive evolution and species diversification. Mol. Ecol. 28 (6), 1207–1220.

Wen, W., Li, D., Li, X., Gao, Y., Li, W., Li, H., Liu, J., Liu, H., Chen, W., Luo, J., Yan, J., 2014. Metabolome-based genome-wide association study of maize kernel leads to novel biochemical insights. Nat. Commun. 5 (1), 3438.

Wen, W., Li, K., Alseekh, S., Omranian, N., Zhao, L., Zhou, Y., Xiao, Y., Jin, M., Yang, N., Liu, H., Florian, A., 2015. Genetic determinants of the network of primary metabolism and their relationships to plant performance in a maize recombinant inbred line population. Plant Cell 27 (7), 1839–1856. https://doi.org/10.1105/tpc.15.00208.

Wilson, T.J., Lai, L., Ban, Y., Ge, S.X., 2012. Identification of metagenes and their interactions through large-scale analysis of Arabidopsis gene expression data. BMC Genom. 13 (1), 237. https://doi.org/10.1186/1471-2164-13-237.

Wu, D., Karhade, D.S., Pillai, M., Jiang, M.Z., Huang, L., Li, G., Cho, H., Roach, J., Li, Y., Divaris, K., 2021. Machine learning and deep learning in genetics and genomics. In: Machine Learning in Dentistry. Springer International Publishing, pp. 163–181.

Wu, Y., Xie, L., 2025. AI-driven multi-omics integration for multi-scale predictive modeling of genotype-environment-phenotype relationships. Comput. Struct. Biotechnol. J. 265–277.

Wu, Y., Zhang, K., 2020. Tools for the analysis of high-dimensional single-cell RNA sequencing data. Nat. Rev. Nephrol. 16 (7), 408–421. https://doi.org/10.1038/s41581-020-0262-0.

Xavier, A., Jarquin, D., Howard, R., Ramasubramanian, V., Specht, J.E., Graef, G.L., Beavis, W.D., Diers, B.W., Song, Q., Cregan, P.B., Nelson, R., 2018. Genome-wide analysis of grain yield stability and environmental interactions in a multiparental soybean population. G3: Genes, Genomes, Genet. 8 (2), 519–529.

Xie, K., Minkenberg, B., Yang, Y., 2015. Boosting CRISPR/Cas9 multiplex editing capability with the endogenous tRNA-processing system. Proc. Natl. Acad. Sci. 112 (11), 3570–3575.

Xu, F., Yoshida, H., Chu, C., Matsuoka, M., Sun, J., 2025. Seed dormancy and germination in rice: molecular regulatory mechanisms and breeding. Mol. Plant 18 (6), 960–977.

Xu, H., Zhang, W., Gao, Y., Zhao, Y., Guo, L., Wang, J., 2012. Proteomic analysis of embryo development in rice (*Oryza sativa*). Planta 235, 687–701. https://doi.org/10.1007/s00425-011-1535-4.

Xu, L., Naylor, D., Dong, Z., Simmons, T., Pierroz, G., Hixson, K.K., Kim, Y.M., Zink, E.M., Engbrecht, K.M., Wang, Y.I., Gao, C., 2018. Drought delays development of the sorghum root microbiome and enriches for monoderm bacteria. Proc. Natl. Acad. Sci. 115 (18), E4284–E4293.

Yan, J., Wang, X., 2023. Machine learning bridges omics sciences and plant breeding. Trends Plant Sci. 28 (2), 199–210. https://doi.org/10.1016/j.tplants.2022.08.018.

Yan, J., Xu, Y., Cheng, Q., Jiang, S., Wang, Q., Xiao, Y., et al., 2021. LightGBM: accelerated genomically designed crop breeding through ensemble learning. Genome Biol. 22, 1–24. https://doi.org/10.1186/s13059-021-02492-y.

Yan, J., Zou, D., Li, C., Zhang, Z., Song, S., Wang, X., 2020. SR4R: an integrative SNP resource for genomic breeding and population research in rice. Genom. Proteom. Bioinform. 18 (2), 173–185. https://doi.org/10.1016/j.gpb.2020.03.002.

Yang, W., Feng, H., Zhang, X., Zhang, J., Doonan, J.H., Batchelor, W.D., Xiong, L., Yan, J., 2020. Crop phenomics and high-throughput phenotyping: Past decades, current challenges, and future perspectives. Mol. Plant 13 (2), 187–214.

Yang, X., Guo, T., 2017. Machine learning in plant disease research. European Journal of Biomedical and Medical Research 3 (1), 6–9. https://doi.org/10.18088/ejbmr.3.1.2017.pp6-9.

Yano, K., Yamamoto, E., Aya, K., Takeuchi, H., Lo, P.C., Hu, L., Yamasaki, M., Yoshida, S., Kitano, H., Hirano, K., Matsuoka, M., 2016. Genome-wide association study using whole-genome sequencing rapidly identifies new genes influencing agronomic traits in rice. Nat. Genet. 48 (8), 927–934.

Yao, Y., Guo, W., Gou, J., Hu, Z., Liu, J., Ma, J., Zong, Y., Xin, M., Chen, W., Li, Q., et al., 2025. Wheat2035: integrating pan-omics and advanced biotechnology for future wheat design. Mol. Plant 18 (2), 272–297.

Yetgin, A., 2025. Revolutionizing multi-omics analysis with artificial intelligence and data processing. Quantitative Biology 13 (3), e70002.

Yu, J., Pressoir, G., Briggs, W.H., Vroh Bi, I., Yamasaki, M., Doebley, J.F., McMullen, M.D., Gaut, B.S., Nielsen, D.M., Holland, J.B., Kresovich, S., 2006. A unified mixed-model method for association mapping that accounts for multiple levels of relatedness. Nat. Genet. 38 (2), 203–208.

Zeng, D., Liu, T., Ma, X., Wang, B., Zheng, Z., Zhang, Y., Xie, X., Yang, B., Zhao, Z., Zhu, Q., Liu, Y.G., 2020. Quantitative regulation of *Waxy* expression by CRISPR/Cas9-based promoter and 5′UTR-intron editing improves grain quality in rice. Plant Biotechnol. J. 18 (12), 2385–2395.

Zhang, F., Xue, H., Dong, X., Li, M., Zheng, X., Li, Z., Xu, J., Wang, W., Wei, C., 2022. Long-read sequencing of 111 rice genomes reveals significantly larger pan-genomes. Genome Res. 32 (5), 853–863.

Zhang, X., Yazaki, J., Sundaresan, A., Cokus, S., Chan, S.W.L., Chen, H., Henderson, I.R., Shinn, P., Pellegrini, M., Jacobsen, S.E., Ecker, J.R., 2006. Genome-wide high-resolution mapping and functional analysis of DNA methylation in *Arabidopsis*. Cell 126 (6), 1189–1201.

Zhang, X., Bernatavichute, Y.V., Cokus, S., Pellegrini, M., Jacobsen, S.E., 2009. Genome-wide analysis of mono-, di- and trimethylation of histone H3 lysine 4 in *Arabidopsis thaliana*. Genome Biol. 10 (6), R62.

Zhang, Z., Ersoz, E., Lai, C.Q., Todhunter, R.J., Tiwari, H.K., Gore, M.A., Bradbury, P.J., Yu, J., Arnett, D.K., Ordovas, J.M., Buckler, E.S., 2010. Mixed linear model approach adapted for genome-wide association studies. Nat. Genet. 42 (4), 355–360.

Zhang, Y., Massel, K., Godwin, I.D., Gao, C., 2019. Correction to: Applications and potential of genome editing in crop improvement. Genome Biol. 20, 13. https://doi.org/10.1186/s13059-019-1622-6.

Zhao, C., Zhang, Y., Du, J., Guo, X., Wen, W., Gu, S., Wang, J., Fan, J., 2019. Crop phenomics: current status and perspectives. Front. Plant Sci. 10, 714.

Zhao, K., Tung, C.W., Eizenga, G.C., Wright, M.H., Ali, M.L., Price, A.H., Norton, G.J., Islam, M.R., Reynolds, A., Mezey, J., McClung, A.M., 2011. Genome-wide association mapping reveals a rich genetic architecture of complex traits in *Oryza sativa*. Nat. Commun. 2 (1), 467.

Zhou, J., Troyanskaya, O.G., 2015. Predicting effects of noncoding variants with deep learning–based sequence model. Nat. Methods 12 (10), 931–934.

Zhou, W., Wang, M., Wang, L., Liu, Y., Tian, Z., Xie, L., Wang, Y., 2025. Epigenetics in plant response to climate change. Biology 14 (6), 631.

Zhou, X., Zhao, Y., Ni, P., Ni, Z., Sun, Q., Zong, Y., 2023. CRISPR-mediated acceleration of wheat improvement: advances and perspectives. J. Genet. Genomics 50 (11), 815–834.

Zhou, Y., Zheng, H., Huang, X., Hao, S., Li, D., Zhao, J., 2022. Graph neural networks: Taxonomy, advances, and trends. ACM Trans. Intell. Syst. Technol. 13 (1), 1–54. https://doi.org/10.1145/3495161.

Zhu, H., Li, C., Gao, C., 2020. Applications of CRISPR–Cas in agriculture and plant biotechnology. Nat. Rev. Mol. Cell Biol. 21 (11), 661–677.

Zhu, W., Li, W., Zhang, H., Li, L., 2025. Big data and artificial intelligence-aided crop breeding: progress and prospects. J. Integr. Plant Biol. 67 (3), 722–739.

Zong, Y., Wang, Y., Li, C., Zhang, R., Chen, K., Ran, Y., Qiu, J.-L., Wang, D., Gao, C., 2017. Precise base editing in rice, wheat and maize with a Cas9-cytidine deaminase fusion. Nat. Biotechnol. 35 (5), 438–440.

CHAPTER

25

AI and next-generation gene editing tools

Muhammad Zawwad Raza[1,2], *Hira Khursheed*[1,2,3], *Qasim Ali*[4], *Yufeng Liu*[5], *Muhammad Riaz*[5], *Ahmed Al-Harrasi*[6], *and Aftab Ahmad*[6]

[1]Scholarious Scientific, Montreal, QC, Canada; [2]Mojoki BioPartners, Gujrat, Pakistan; [3]Clinical and Translational Sciences (CaTS) Lab, Douglas Research Center, McGill University, Montreal, QC, Canada; [4]Punjab Institute of Cardiology, Lahore, Pakistan; [5]Yale University, New Haven, CT, United States; [6]Natural and Medical Sciences Research Center (NMSRC), University of Nizwa, Nizwa, Oman

Abbreviations

3D Three-dimensional
AI Artificial Intelligence
CNN Convolutional Neural Network
CRISPR–Cas9 clustered regularly interspaced short palindromic repeats/CRISPR-associated protein nine
GNN Graph Neural Network
NN Neural Network
PAM Protospacer adjacent motif
RNN Recurrent Neural Networks
TALENs Transcription Activator-Like Effector Nucleases

25.1 Introduction

Artificial intelligence (AI) is an emerging field in computer science that focuses on developing systems and creating machines capable of functioning at a level similar to human intelligence (Hassani et al., 2020). Instead of mechanical parts, AI systems are equipped with machine learning models, algorithms, and neural networks that can learn, reason, and make decisions independently. These systems are designed to handle large volumes of data,

Intelligent Gene Editing in Plants
https://doi.org/10.1016/B978-0-443-44079-3.00011-5

identify hidden patterns, and perform complex or repetitive tasks that would take humans much longer to complete (Russell and Norvig, 2020). As a result, AI is beginning to transform nearly every field, including science and engineering, as well as environmental efforts. In particular, AI's predictive power is one of its greatest strengths, playing a key role in advancing precision medicine and agriculture (Gupta and Kumar Pal, 2025; Mesko, 2017).

Over the past decade, scientists have used AI's predictive capabilities to improve genome-editing techniques (Delfani et al., 2024). These techniques enable precise alterations to the DNA of living cells with substantial power, thereby removing faulty genes or inserting new, functional genes to improve cellular performance in both human and agricultural health care. The development of genome-editing tools has advanced from early simple techniques such as meganucleases and zinc-finger nucleases (ZFNs) to more advanced techniques like transcription activator-like effector nucleases (TALENs), culminating in a Nobel Prize-winning discovery of the clustered regularly interspaced short palindromic repeat (CRISPR)–Cas9 system. Of these techniques, meganucleases, ZFNs, and TALENs use their DNA-binding domains to identify the target DNA sequence, whereas CRISPR–Cas9 is guided by an RNA molecule complementary to the target DNA. For readers interested in the molecular mechanisms underlying these genome editing techniques, we refer them to Chapter 2 of our previously published book, *Gene-Edited Crops*, published by Taylor and Francis in 2025 (Ibrar et al., 2025). Here, we primarily focus on AI and its synergistic role in improving the specificity and accuracy of the two most versatile genome-editing tools, TALENs and CRISPR–Cas9, in crops to increase productivity and societal benefits.

Conventional methods of crop improvement, such as transgenesis, chemical and radiation mutagenesis, somaclonal variation from tissue culture, and marker-assisted selection (MAS), have been crucial for many decades in advancing agricultural productivity. But they are often limited by the complexity of polygenic traits, long breeding cycles, and unpredictable interactions between genotype and environment. In contrast, TALENs and CRISPR–Cas AI platforms enable targeted, site-specific nucleotide modifications (Gaj et al., 2016; Ruffolo et al., 2025). These precise tools have expedited both functional genomics and trait improvement across various crop species, leading to a new era of precision agriculture (Atia et al., 2024).

This chapter concisely explains how AI-powered computational methods improve the design, efficiency, and accuracy of TALEN- and CRISPR-mediated genome editing tools. It provides an in-depth summary of the core principles of these technologies, the algorithms and machine-learning techniques that underpin their development, and the broader impacts of AI-guided genome-wide editing on sustainable agriculture and global food security.

25.2 AI and TALENs gene editing

TALEs are a class of programmable DNA-binding proteins that have become key in genome editing. Initially discovered in 2009 in the phytopathogen *Xanthomonas oryzae*, a bacterium that infects plants, TALENs combine the customizable DNA-binding domain of TALEs with a DNA-cleaving enzyme, typically the *FokI* nuclease, which is located on each TALE in its monomeric form (Bogdanove and Voytas, 2011). The *FokI* domains dimerize when two TALENs bind to adjacent sites on opposite DNA strands, inducing a double-strand break at the target site (Bhardwaj and Nain, 2021). This break activates the naturally existing DNA repair mechanisms in living cells. Consequently, the broken DNA strands are repaired via error-prone non-homologous end-joining (NHEJ), allowing targeted gene disruption, insertion, or correction. TALEs operate on a remarkably elegant basic principle, which begins with identifying the patterns of DNA sequences at the target genomic region and generating TALE proteins that contain a series of nearly identical amino acid repeats, with each repeat responsible for recognizing a single DNA base. The specificity of this recognition is determined by a pair of amino acids known as the repeat-variable di-residue (RVD), within each repeat (Cermak et al., 2011; Wright et al., 2014). By assembling these repeats in a desired order, it is possible to engineer TALEs that can bind virtually any DNA sequence with high specificity (Fig. 25.1). This programmable nature and modular design make TALEs highly versatile tools, which, when combined with nucleases or other effectors, enable precise genomic changes, including gene regulation in plants with fine-tuned control (Bhardwaj and Nain, 2021; Tycko et al., 2016).

TALEN has been widely used in agricultural biology, mainly to improve traits such as disease resistance, drought tolerance, and nutrient efficiency (Piatek et al., 2018). However, despite its extensive applications, designing and building TALENs are labor-intensive, and achieving high specificity while reducing unintended off-target effects remains a major challenge (Mahfouz et al., 2014). Moreover, precision is critical, especially when editing complex plant genomes, where even minor errors can have cascading effects on growth, yield, or resistance (Bhardwaj and Nain, 2021; Wright et al., 2014; Zaman et al., 2025). AI has leveraged vast genomic datasets and machine learning predictive algorithms, enabling researchers to better assess the availability of editing-friendly DNA sequences that model TALE-DNA interactions more accurately, and forecast potential off-target sites to design more efficient TALE constructs with minimal unintended DNA changes post-cleavage (Zhang et al., 2025). In this context, the AI-powered tools have begun to enhance the TALEN gene-editing technique at multiple stages, making it versatile, operationally friendly, and easily accessible across the community at grassroots levels (Fig. 25.1).

25.3 AI enhances target genomic site selection

The first step in TALENs design is the selection of an appropriate genomic target site. This is a crucial step in TALENs design and has traditionally relied on well-established rule-based frameworks, developed through experiential observations of TALE-DNA interactions. These frameworks highlight specific design principles, including a thymine at the 5′ end of the target site (the “T0 rule”), the ideal spacer length between TALENs binding sites to facilitate *FokI* nuclease dimerization, and the avoidance of repetitive or highly GC-rich regions that may compromise binding efficiency. For instance, the TAL Effector-Nucleotide Targeter (TALE-NT) suite (Doyle et al., 2012), a framework developed by the Cornell University Institute for Biotechnology, includes tools such as the TALEN Targeter (TALEN-T), TAL Effector Targeter (TALEN-ET), and E-TALEN developed by the German Cancer Research Center (Heigwer et al., 2013).

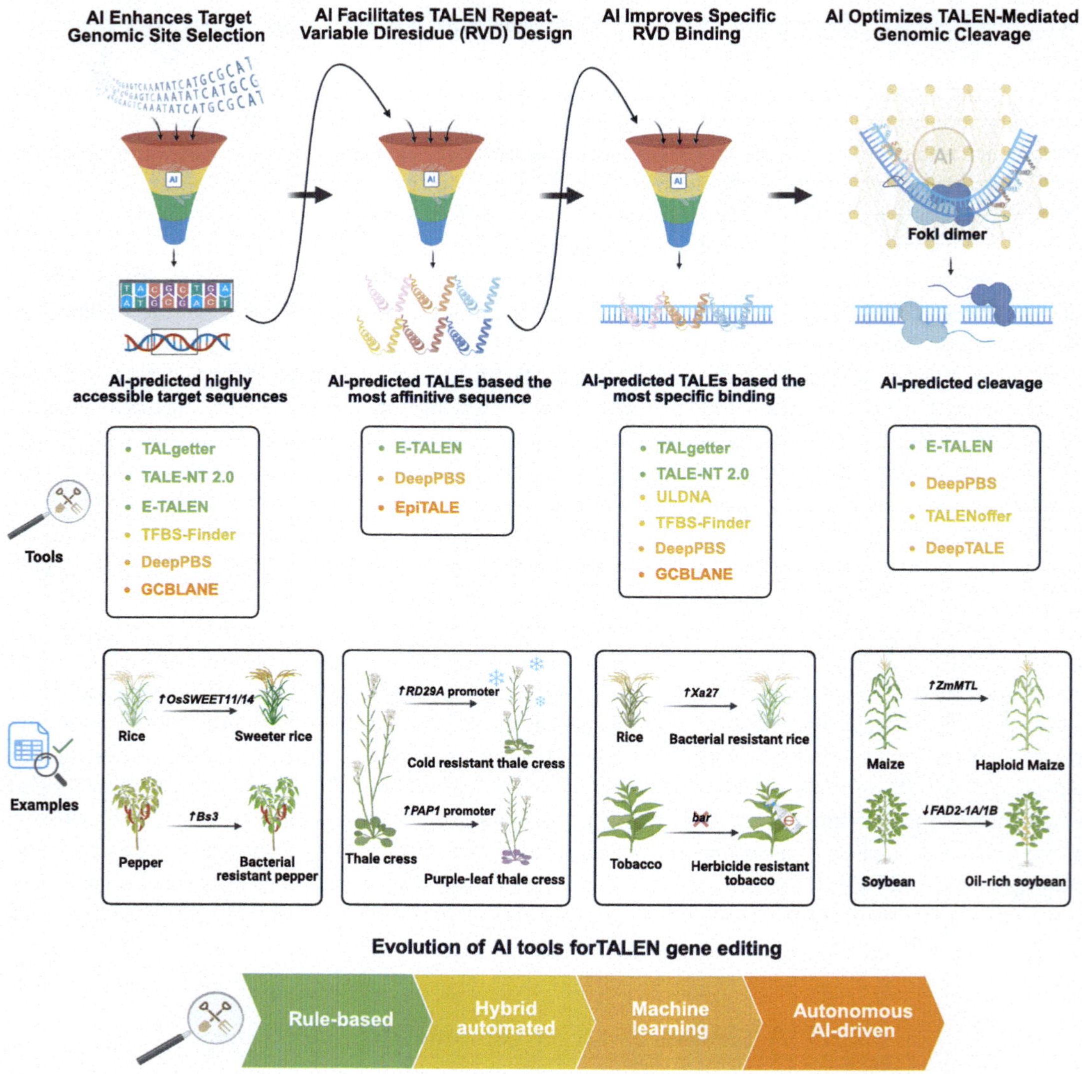

FIGURE 25.1 AI enhances TALEN design, specificity, and application in crops. Overview of how AI integration improves TALENs gene editing technology across design, prediction, and functional implementation. (Top) AI facilitates the optimization of repeat-variable di-residue (RVD) composition and enhances target genomic site selection by predicting highly accessible and high-affinity DNA sequences. Representative tools include DeepPBS, TFBS-Finder, GCBLANE, TALgetter, TALE-NT 2.0, and E-TALEN. (Middle) AI-based models such as EpiTALE, ULDNA, and TALENoffer refine TALEN-mediated genomic cleavage by predicting optimal *FokI* dimer configurations and minimizing off-target effects (Bottom; also see Table 25.1). Example applications illustrate improved editing precision and desirable phenotypes in major crops, including sweeter rice (↑*OsSWEET11/14*), bacterial-resistant rice (↑*Xa27*), bacterial-resistant pepper (↑*Bs3*), cold-resistant *Arabidopsis* (↑RD29A promoter), purple-leaf *Arabidopsis* (↑PAP1 promoter), haploid maize (↑*ZmMTL*), oil-rich soybean (↓*FAD2-1A/1B*), and herbicide-resistant tobacco (bar knockout). The bottom panel of the figure also depicts the evolution of AI tools from rule-based systems to machine learning, hybrid automated, and autonomous AI-driven frameworks implemented to improve TALEN design paradigms. The arrow indicates whether the gene of interest is upregulated or downregulated in the model crop.

These tools use sequence alignment algorithms to identify potential off-target sites by comparing the target sequence against the genome, thereby aiding in the selection of optimal target sites with minimal off-target activity. While these rule-based methods have provided a reliable foundation for TALEN construction, they largely overlook the broader genomic and epigenomic context, which can have a strong impact on accessibility and editing efficiency at the target site.

The AI boom and computational genomics have revolutionized different steps of TALENs editing design by introducing data-driven methodologies that overcome the limitations of traditional heuristics. AI models, particularly those based on deep learning and ensemble machine learning, can now be trained on extensive datasets of experimentally validated TALENs to infer complex, nonlinear relationships between DNA sequence features and nuclease activity (Wang et al., 2014; Grau et al., 2013) (Table 25.1). Through the integration of multidimensional genomic data, including nucleotide composition, local chromatin structure, methylation patterns, and histone modification states, AI systems can generate predictive rules that identify highly accessible sequences and low off-target sites. For instance, AI-driven tools such as *idTALE* (Li et al., 2012a), *TALE-NT 2.0* (Doyle et al., 2012), and *TALENoffer* (Grau et al., 2013; Moore et al., 2014) have been used to develop CNNs and transformer-based models capable of identifying high-level "AI rules". These include preferred nucleotide contexts, binding energy thresholds, and structural conformations that show strong correlations with successful TALEN binding. Importantly, these are dynamic models that adjust target predictions for each organism by incorporating differences in genome complexity, size, and polyploidy factors that are especially significant in plant genomes (Moore et al., 2014).

AI-based systems have also transformed off-target prediction and specificity assessment in TALEN design. While traditional methods relied on simple mismatch counting or sequence alignment, modern context-aware AI models are capable of learning how base mismatches, epigenetic modifications, and chromatin accessibility interact to influence nuclease binding affinity. These models employ AI-derived interpretive rules, for example, position-weighted mismatch penalties or accessibility-adjusted binding scores, to dynamically rank potential genomic sites by predicted specificity. Instead of merely copying existing biological knowledge, these systems reveal hidden dependencies between sequence motifs and structural DNA features, providing deeper insight into binding behavior. This capability is particularly valuable in polyploid crop genomes, where homologous and repetitive regions pose significant challenges to precise genome editing (Abdallah et al., 2015).

In agricultural biotechnology, integrating AI-based tools into TALEN design strategies can accelerate the development of improved crop varieties. Importantly, TALEN applications in creating agronomic traits have already demonstrated success, for example, editing the *OsSWEET14* promoter in rice to create bacterial blight resistance (Li et al., 2012b), disrupting the *TaMLO* alleles in wheat to achieve durable powdery mildew resistance (Wang et al., 2014), and modifying the *FAD2-1A/FAD2-1B* genes in soybean to produce high-oleic acid oil (Haun et al., 2014). The AI-assisted selection of target genomic sequences extends these achievements by integrating multi-omic and environmental datasets, such as transcriptomics, epigenomics, and stress-response profiles, to identify genomic loci most likely to yield stable, desirable phenotypes under field conditions (Abdallah et al., 2015; Venkataraman et al., 2025). Using AI-assisted TALENs design rules, frameworks, and predictive models can suggest target regions that are not only transcriptionally active in stress-responsive tissues but also epigenetically accessible and genetically stable across cultivars. This shift from static, rule-based heuristics to dynamic, AI-informed frameworks represents a pivotal evolution in TALEN technology. It marks a significant transformation of this technology into a predictive, context-aware tool for future precision breeding and sustainable agricultural innovation (Table 25.1).

25.4 AI facilitates TALENs repeat-variable di-residues design

The design of repeat-variable di-residues (RVDs) defines the success of TALENs specificity. Indeed, RVDs dictate the precise recognition of DNA bases by each TALE repeat. Historically, this process has been guided by a simple lookup table that maps individual RVDs to their corresponding nucleotides, NI to adenine (A), HD to cytosine (C), NN to guanine (G) or adenine (A), and NG to thymine (T) (Boch et al., 2009; Moscou and Bogdanove, 2009). This concept of "one RVD one base" has been established through early experimental studies in this domain. One of the main strengths of TALENs, as the most customizable gene-editing tool, stems from the idea that RVDs can be tailored using a modular approach to identify any target sequence. On the other side, this simplistic approach ignores other equally important factors in play. Studies have shown that neighboring repeats, base context, and three-dimensional DNA-protein interactions strongly influence this interaction (Mak et al., 2012). Experimental evidence increasingly demonstrates that RVD activity is not uniform across all genomic settings; instead, it can be significantly modulated by local sequence composition, GC content, and even the relative orientation of the TALEN monomer. Consequently, TALENs designed through simple lookup-based coding often fail in experimental settings and therefore require successive and constant experimental refinement to achieve optimal binding efficiency and nuclease activity. This trial-and-error process leads to substantial resource expenditure, including financial costs, time, and effort, highlighting the need for more predictive and context-aware design strategies.

Advances in AI and machine learning have offered great promise in this regard and have transformed TALEN RVD design from a static, lookup-based process into an adaptive, context-aware approach. AI-driven platforms are trained on extensive datasets of experimentally validated TALEN binding and cleavage assays and are therefore capable of modeling the complex interactions among sequence features, RVD composition, and local chromatin context. Furthermore, deep learning frameworks, including convolutional neural networks (CNNs) and transformer-based architectures, identify hidden dependencies that influence TALEN efficiency, enabling the derivation of AI-informed "rules" for dynamic RVD optimization (Dixit et al., 2023; Li et al., 2025). These predictive models estimate the relative binding affinities of different RVDs at specific nucleotide positions by taking into account neighboring repeat interactions, DNA conformational flexibility, and epigenetic features such as methylation and histone accessibility, thereby improving the odds of a successful TALEN-mediated genome editing (Morbitzer et al., 2011; Reyon et al., 2012). This integrative approach transforms RVD assignment from a rule-based task into a data-

TABLE 25.1 AI tools and TALENs genome editing.

Evolutionary point of view of TALENs tools	Tools	Core algorithm/ Architecture	Primary function and key features	Potential applications in agriculture	Representative references
Conventional rule-based and Probabilistic TALENs design tools	*TALgetter*	It is probabilistic model that integrates RVD-nucleotide binding and position-dependent weights	It predicts TAL effector and TALENs binding sites genome-wide and account for off-target	It offers genome-wide prediction of functional TALENs binding sites for editing stress- and disease-related genes.	Grau et al. (2013)
	TALE-NT 2.0	It is based in modular algorithm with TALENs targeter and assembly modules	It predicts and designs TAL effector binding sites through genome-wide scanning for RCD assembly	It is widely applicable for targeted gene modification.	(Doyle et al., 2012), Cermak et al. (2011), Mahfouz et al. (2014)
	E-TALEN	It based on rule-based algorithmic design	It uses GC contents to generate automated TALENs pair design and off-target analysis for integration of RVD-DNA binding.	It provides a potential for custom TALENs design for genes related to yield, resistance, or metabolism.	Heigwer et al. (2013)
Hybrid automated TALENs design tools	*TFBS-Finder*	It is based on machine learning with motif discovery and PWM modeling	It identifies transcription factor binding sites in regulatory regions and motives to integrates expression data.	It supports regulatory-level TALENs design for gene expression modulation.	Tsai et al. (2006)
	ULDNA	It is based on unsupervised learning (Autoencoder/ VAE/SOM)	It predicts DNA binding sites without labeled data and learn latent motives structures autonomously.	It is capable for de novo TALENs target discovery in under-characterized crop genomes.	Zhu et al. (2024)
	TALENoffer	It is based on mapping TALE repeats RVD to DNA bases, scoring binding sites.	It identifies TALENs binding sites by matching TALE repeats to DNA bases using the RVD code and predicts off-targets, and ranking candidates.	It predicts binding specificity, and gives the outputs for the best TALEN pairs with minimal off-target risk for crops	Grau et al. (2013)

(*Continued*)

TABLE 25.1 AI tools and TALENs genome editing.—cont'd

Evolutionary point of view of TALENs tools	Tools	Core algorithm/ Architecture	Primary function and key features	Potential applications in agriculture	Representative references
Machine learning based TALENs design tools	*DeepTALE*	It is based on deep learning (CNN + RNN hybrid)	It predicts TAL effector–DNA binding specificity in contextual sequence dependent manner and models non-linear RVD–DNA interactions.	It predicts high-affinity TALEN binding sites for genes related to yield and stress response related.	Grau et al. (2013)
	DeepPBS	It is based on geometric deep learning (graph neural network + spatial attention)	It predicts protein–DNA binding specificity based on 3D molecular structures for for RVD–DNA interaction modeling.	It guides structure-informed TALENs design for precise editing of yield and stress genes	Mitra et al. (2024)
Exclusively AI-driven TALENs design tools	*GCBLANE*	It is based on graph-enhanced CNN-BiLSTM with attention	It models protein–DNA binding with structural and sequential context through combination of GNNs and CNNs for local motifs and for BiLSTM for dependencies	It enables accurate TALENs target selection in polyploid and stress-tolerant crops	Ferrao et al. (2025)
	EpiTALE	It is based on utilization of genomic and epigenetic sequence data to predict TALE binding sites.	It predicts binding sites of TALEs in even methylated and chromatin non-accessible DNA sequence.	It develops TALENs array selection for epigenetic modification of regulatory genes in maize and wheat.	Erkes et al. (2021)

driven, adaptive design paradigm. As a result, TALENs are designed with enhanced accuracy and functional performance by precisely fine-tuning the genomic context (Table 25.1).

25.5 AI improves efficient RVD binding and specificity

Once the RVDs are designed, accurate prediction of their binding activity and specificity is equally critical for effective gene editing. Early methods have employed empirical scoring systems, in which each RVD-nucleotide interaction is assigned a fixed value derived from cleavage assays (Cermak et al., 2011). Tools such as TALE-NT 2.0 estimated overall binding affinity by summing these scores, assuming additive effects across the array. While applicable for initial design, this linear framework overlooked key contextual influences, as discussed in the previous section, and includes neighboring RVD interactions, DNA structural flexibility, and chromatin accessibility, leading to discrepancies between predicted and observed activities (Mak et al., 2012). This underscores the importance of contextual factors not only in RVDs' design but also in predicting their effectiveness and specificity.

The advent of AI has shifted this process from heuristic estimation to predictive, context-aware modeling. Deep learning frameworks now incorporate sequence composition, DNA topology, and epigenomic data to predict both on-target efficiency and off-target specificity. These models develop adaptive "AI rules" such as EpiTALE (Erkes et al., 2021) and TFBS-Finder (Tsai et al., 2006) that capture how RVD combinations and local genomic environments influence binding strength (Erkes et al., 2021; Wittmann et al., 2024). In agriculture, these tools enable the design of TALENs optimized for polyploid and epigenetically diverse crop genomes, improving precision in trait engineering (Bhardwaj and Nain, 2021; Erkes et al., 2021; Farooq et al., 2024) (Table 25.1). Therefore, the assessment of RVD binding activity has progressed from static scoring to AI-driven prediction, thus merging molecular understanding with computational intelligence for next-generation TALEN-based gene editing.

25.6 AI enhances TALENs-mediated cleavage at targeted genome sequences

Once RVDs are designed and their binding efficiency and specificity are predicted, the next crucial aspect is their functional efficiency, which involves determining TALENs-mediated cleavage at the targeted sequence site. Optimizing this step is essential for achieving high precision and consistency in gene editing. Early optimization strategies depended on empirical adjustments of spacer length, *FokI* nuclease configuration, and expression conditions (Moscou and Bogdanove, 2009). Although effective in specific cases, these methods often failed to account for the complex interactions among target sequence composition, chromatin accessibility, and nuclease conformational dynamics. As a result, cleavage optimization has transitioned from rule-based heuristics to AI-driven, data-informed modeling capable of interpreting complex biological relationships.

Modern AI and machine learning platforms, such as DeepTALE and TALENoffer (Grau et al., 2013) integrate sequence, structural, and epigenomic data to predict cleavage efficiency and specificity accurately. These systems employ deep learning architectures, including convolutional and attention-based neural networks, to identify optimal spacer lengths, nuclease dimer orientations, and target DNA and chromatin contexts that boost on-target activity while minimizing off-target effects (Bradbury et al., 2025) (Table 25.1). Bioinformatics frameworks such as DeepCRISPR and BE-Hive, although created for related systems, are being adapted to model TALEN cleavage kinetics using transfer learning methods (Arbab et al., 2020; Chuai et al., 2018; Xie et al., 2017). Overall, AI tools have transformed TALEN cleavage optimization into a predictive, automated, and biologically aware engineering process, connecting computational intelligence with molecular design. In agriculture and plant biotechnology, AI-guided tools enable context-specific TALENs optimization across complex, polyploid genomes, thereby supporting precision editing for traits such as disease resistance, stress tolerance, and yield improvement (Ding and Zhang, 2023; Wójcik-Gront et al., 2024; Zaman et al., 2025).

25.7 AI and CRISPR gene editing tool

The CRISPR–Cas9 system has revolutionized molecular biology and biotechnology by providing a significantly simplified and programmable approach to genome editing (Jinek et al., 2012). Despite its broad usefulness, the method remains far from perfect. Editing efficiency often varies between single-guide RNAs (sgRNAs). In addition, off-target activity, inconsistent repair events, and difficulties in delivering Cas9 components still limit its precision (Hsu et al., 2014). Significant advances have been made in devising AI-based approaches, particularly in deep learning and generative modeling, which have greatly facilitated addressing some of these problems by enabling data-driven design and predictive analysis within gene-editing workflows (Kim et al., 2025) (Fig. 25.2).

Trained on extensive experimental datasets, modern AI systems can now predict editing behavior at multiple stages of a CRISPR experiment: from choosing effective sgRNAs, refining Cas-protein variants, to simulating DNA-protein interactions, which were previously too complex to model. In agricultural research, the convergence of AI and CRISPR technology has supported breeders to produce crops with improved resilience, yield, and nutritional value (Lee, 2023; Alipanahi et al., 2025) (Table 25.2). The following sections will describe how AI contributes to these progresses and how its applications are spreading across diverse plant species.

25.8 AI improves sgRNA design for enhanced specificity

Choosing a unique gRNA for a target genomic site is a crucial step in CRISPR editing, as it affects both efficiency and specificity. Early scoring systems often only look at simple factors like GC content or the distance from PAMs, which offered useful guidance but couldn't fully capture the complexity of editing performance (Chuai et al., 2018; Labun et al., 2021). Recently, researchers have started using AI-based models, including DeepCRISPR (Chuai et al., 2018), CRISPR-Net (Lin et al., 2020), and Azimuth (Listgarten et al., 2018). These frameworks use deep-learning algorithms trained with thousands of experimentally tested

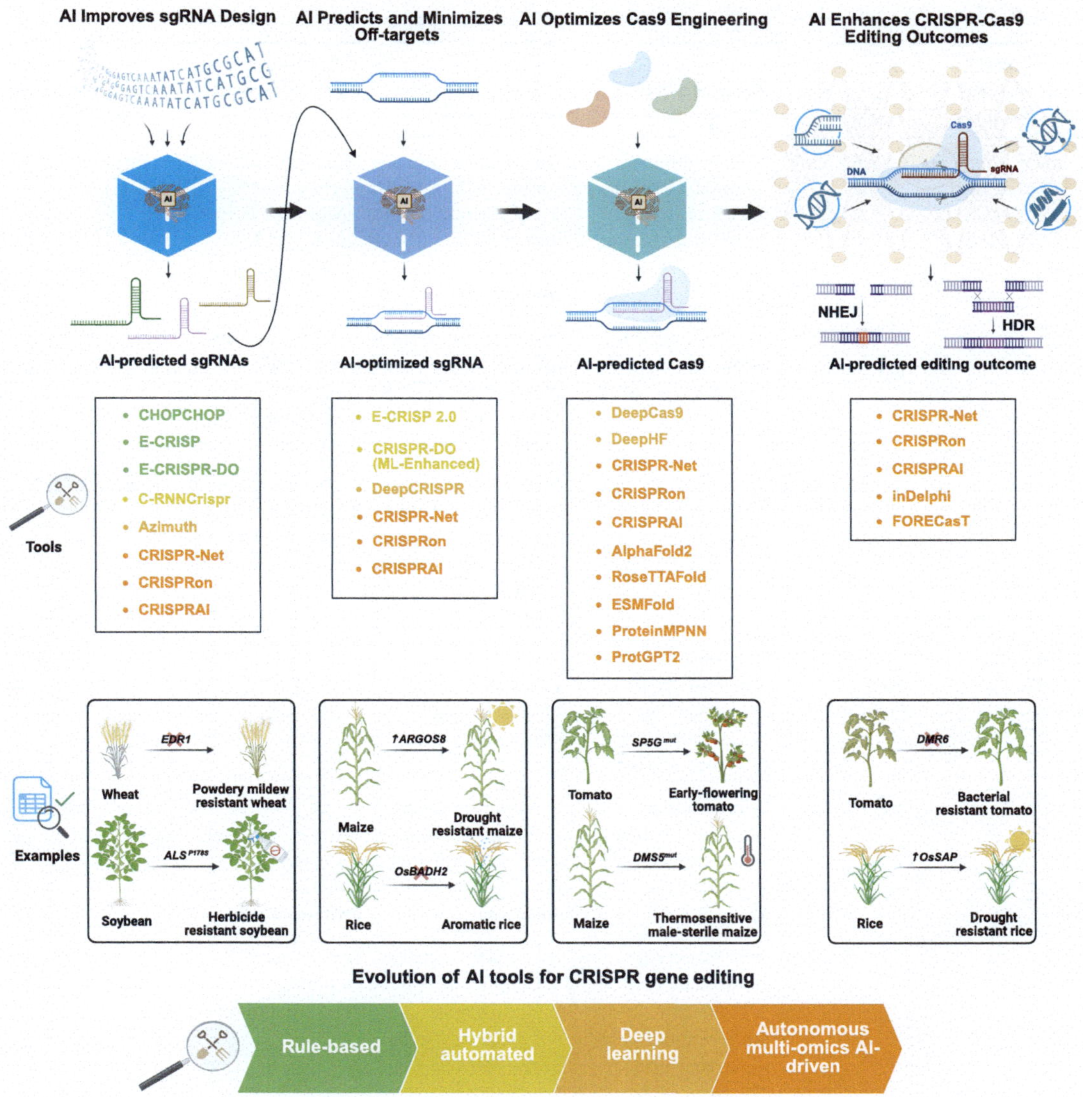

FIGURE 25.2 AI enhances CRISPR–Cas9 design, specificity, and application. Overview of how AI integration improves CRISPR–Cas9 technology across design, prediction, and functional implementation. (Top) AI facilitates the optimization of single-guide RNA (sgRNA) design by predicting highly efficient and specific targeting sequences while minimizing off-target effects. Representative models include DeepCas9, DeepHF, and DeepCRISPR. (Middle) AI-based platforms such as CRISPR-Net, CRISPRon, and CRISPRAI enhance editing efficiency prediction, guide selection, and outcome forecasting, while tools like CHOPCHOP, E-CRISP, and CRISPR-DO support improved Cas9 engineering and repair pathway choice (NHEJ vs. HDR) (see Table 25.2). (Bottom) Example applications demonstrate AI-assisted CRISPR editing achieving desirable agronomic traits, exemplified with powdery mildew-resistant wheat (EDR1 knockout), herbicide-resistant soybean (ALS P178S), thermosensitive male-sterile maize, drought-resistant maize (↑ARGOS8), aromatic rice (OsBADH2 knockout), bacterial-resistant tomato (DMR6 knockout), and early-flowering tomato (SP5G and DMS5 mutations). The bottom panel of the figure also depicts the evolution of CRISPR design paradigms from rule-based systems to deep learning, hybrid automated, and multi-omics AI-driven frameworks.

TABLE 25.2 AI tools and CRISPR genome editing.

Evolutionary point of view of TALENs tools	Tools	Core algorithm/ Architecture	Primary function and key features	Potential applications in agriculture	Representative references
Conventional rule-based and Probabilistic TALENs design tools	CHOPCHOP	It is rule-based algorithm; sequence alignment and empirical scoring	It identifies and ranks CRISPR and TALENs target sites and considers PAM and GC content mismatch tolerance.	It is applicable for early CRISPR design for rice, maize, and *Arabidopsis*.	Labun et al. (2021)
	E-CRISP	It is rule-based scoring; empirical off-target analysis	It integrates genome annotation data and automated CRISPR target site identification	It represents an early CRISPR technology in complex plant genomes.	(Heigwer et al., 2014)
	E-CRISPR-DO	It is also rule-based scoring; statistical regression	It predicts sgRNA efficiency and specificity through quantitative scoring if PAM and GC contents.	First genome wide CRISPR design for plant systems	Ma et al. (2016)
Hybrid automated CRISPR design systems	CRISPR-DO (ML-Enhanced)	It is based on machine learning (random Forest + Support vector machine (SVM)	It supports prediction of off-targets through empirical editing data	It is applicable for sgRNA design for crops with repetitive genomes	Kim et al. (2025)
	E-CRISP.2.0	It is based on statistical learning; semi-supervised optimization	It enhances CRISPR target and off-target prediction through multi-genome support	It has potential to improved target precision for complex polyploid crop genomes.	Heigwer et al. (2014)
	C-RNNCrispr	It is based on combination of CNN and bidirectional gated recurrent unit (BGRU)	It predicts on-target activity in the target sequence and epigenomic context in the form of regression score predicting sgRNA	It may be applicable to plants with large complex genomes	Zhang et al. (2020)
Deep Learning–enabled CRISPR systems	Azimuth	It is based on gradient boosted regression trees (GBRT)	It predicts sgRNA activity score (0–1) indicating cutting efficiency.	It integrates genomic and epigenomic features specific to plant like rice, maize, and wheat	Listgarten et al. (2018)

TABLE 25.2 AI tools and CRISPR genome editing.—cont'd

Evolutionary point of view of TALENs tools	Tools	Core algorithm/ Architecture	Primary function and key features	Potential applications in agriculture	Representative references
	DeepCas9	It combines CNN and attention mechanism.	It captures higher order sequence interaction to models Cas9 activity and PAM recognition.	It enhances Cas9 efficiency in complex crop genomes	Kim et al. (2019)
	DeepHF	It combines hybrid CNN-BiLSTM network	It predicts Cas9 variant fidelity and activity through variant-specific learning for enzyme optimization	It is applicable for sgRNA selection of Cas9 variants in plant systems	Pandey et al. (2025), Wang et al. (2021)
Autonomous, multi-omics and transformer-based AI frameworks	CRISPR-Net	It is based on graph neural network (GNN)	It predicts CRISPR activity in 3D chromatin through 3D genome topology and accessibility	It improves targeting precision in polyploid and epigenetically variable crops	Lin et al. (2020)
	CRISPRon	It is based on transformer model + transfer learning	It enables cross-species CRISPR efficiency prediction via domain adaptation using pretrained models.	It enables accurate CRISPR targeting in non-model crops	Anthon et al. (2022); Xiang et al. (2021)
	CRISPRAI	It is based on transformer-based multi-omics AI	It integrates genomic, transcriptomic, epigenomic, and phenotypic layers for regulatory and trait-level outcomes	It offers trait specific CRISPR applications for yield, stress tolerance and nutrient genes	Oberlin and McManus (2025); Truong et al. (2019)
	AlphaFold2	It is based on deep attention based NN + Evoformer processes for 3D structure	It predicts highly accurate 3D protein structures from amino acid sequences.	It helps provide functional annotation of crop proteomes, identifies target proteins for stress resistance and yield improvement.	Ruffolo et al. (2025)

(*Continued*)

TABLE 25.2 AI tools and CRISPR genome editing.—cont'd

Evolutionary point of view of TALENs tools	Tools	Core algorithm/ Architecture	Primary function and key features	Potential applications in agriculture	Representative references
	RoseTTAFold	It is transformer-based deep learning with attention networks trained on protein sequence and structure databases.	It predicts 3D protein structures and complexes.	It supports to engineer plant enzymes for modeling protein-pathogen interactions.	Baek et al. (2021)
	ESMFold	It is based on three-track NN and integrats sequence (1D), distance (2D), and coordinate (3D) information.	It predicts protein structure without MSAs (multiple sequence alignments).	It offers plant proteome modeling at large-scale to predict variant effect for trait improvement	Li et al. (2023)
	ProteinMPNN	It is transformer-based protein language model (Meta's ESM-2) trained on massive sequence datasets.	It designs protein sequences that fold into specified structures.	It helps designate stable plant enzymes and receptors for crop improvement.	Chen et al. (2025)
	ProtGPT2	It is based on graph NN of optimized amino acid sequences for a given 3D backbone.	It predicts de novo generation of protein sequences with natural-like properties.	It supports the creation of novel enzymes or antimicrobial peptides for agricultural biotechnology.	Ferruz et al. (2022)
	inDelphi	It is also transformer-based generative language model trained on protein sequences (similar to GPT architecture).	It predicts insertions and deletions (indels) after CRISPR editing.	It helps design predictable CRISPR edits in crops for knockout precision and trait engineering.	Shen et al. (2018)
	FORECasT	It is a machine learning model trained on CRISPR–Cas9 repair outcomes.	It predicts frequencies and patterns of CRISPR-induced indels.	It optimizes guide RNA design for crop genome editing for improving editing efficiency and precision in agricultural species.	Pallaseni et al. (2024)

gRNAs, enabling them to recognize subtle sequence patterns and contextual features that influence editing results. Unlike earlier heuristic tools, they incorporate a wide range of biological data, like chromatin accessibility, RNA secondary structure, and thermodynamic stability, to predict which guides are most likely to work well in a specific DNA environment. AI tools can now recommend highly effective sgRNAs across various species and genomic contexts. This has already reduced much of the trial-and-error that previously slowed down CRISPR research. In particular, in crops like wheat and maize, DeepCRISPR-derived gRNAs targeting drought-response genes showed significantly improved editing efficiency (Azameti and Dauda, 2021). Similarly, AI-optimized gRNAs have increased resistance to *Xanthomonas* infection in rice, helping to develop disease-resistant cultivars (Kim et al., 2020). Overall, these advances demonstrate that computational learning is making gRNA selection a more predictable, efficient, and flexible step in plant CRISPR-based genome editing (Fig. 25.2 and Table 25.2).

25.9 AI-mediated off-target prediction and minimization

Off-target mutations remain a major challenge in CRISPR–Cas9 applications, as unintended DNA cleavage can lead to unwanted genetic changes (Uddin et al., 2020). Conventional sequence-alignment programs can identify genomic regions resembling a target site, yet they often miss the biochemical subtleties that govern Cas9-DNA interactions. Accordingly, AI-powered systems, including CRISPRon, DeepSpCas9, DeepHF, sgDesigner, and TSAM (Chen et al., 2022), overcome this limitation by employing convolutional and transformer-based models trained on datasets such as GUIDE-seq and CIRCLE-seq (Tsai et al., 2015; Wang et al., 2025) (Fig. 25.2). These frameworks evaluate a broad range of factors, including sequence mismatches, base-pair bulges, PAM context, and chromatin accessibility. They can therefore build genome-wide specificity maps and help select guide RNAs with the lowest probability of off-target activity.

Plant genomes are even more challenging because they are often large, repetitive, and polyploid. In this context, AI-enhanced prediction tools have proven especially useful by increasing editing accuracy while safeguarding essential genes. Deep-learning-based models, for example, have decreased off-target events in soybean lines engineered for better oil quality (Das et al., 2022). In rice, the application DeepSpCas9 achieved similar benefits, reducing unwanted edits during multi-gene knockout experiments while preserving key agronomic traits (Narushima et al., 2021) (Table 25.2).

25.10 AI improves Cas9 protein engineering: OpenCRISPR-1 and DNA guided Cas12

The Cas9 nuclease is central to the CRISPR system, but its native Streptococcus pyogenes form has certain limitations, including strict PAM sequence requirements, large size, and potential immunogenicity (Nishimasu et al., 2018). In this regard, AI advances have improved protein engineering by allowing researchers to predict advantageous mutations without extensive trial-and-error screening. In particular, structural modeling platforms such as AlphaFold2 (Ruffolo et al., 2025), RoseTTAFold (Baek et al., 2021), and ESMFold (Li et al., 2023) now provide high-resolution 3D models of Cas enzymes by pinpointing amino acid residues that influence PAM recognition and catalytic activity (Fig. 25.2).

Newly developed AI tools, including ProteinMPNN (Chen et al., 2025) and ProtGPT2 (Ferruz et al., 2022), can design novel Cas protein sequences with improved fidelity, greater stability, or smaller size (Dauparas et al., 2022). Reinforcement learning algorithms refine these predicted designs through cycles of simulated performance testing, accelerating the discovery of optimized enzyme variants. Together, predictive and generative AI approaches have led to the development of versatile forms like xCas9 and SpCas9-NG, which expand the accessible editing range of the genome while improving precision. In crops, AI-engineered Cas proteins enable efficient genome modification in species with restrictive PAM constraints, such as barley and sugarcane (Xu et al., 2021). Compact Cas variants improve CRISPR component delivery via plant viral vectors, broadening the reach of gene-editing technologies to species that were previously difficult to manipulate (Table 25.2). Recently, researchers have developed plant-optimized AI-designed editor (PAiD) based on the AI-generated nuclease OpenCRISPR-1 and effectively demonstrated it for gene editing in rice plants. PAID has demonstrated genome editing efficiency comparable to, and in some cases superior to, conventional SPCas9, including wider editing ranges and improved efficiency. The study provides evidence that AI-generated Cas nucleases can function efficiently in plants, thereby expanding the future applications of CRISPR beyond the naturally occurring CRISPR systems (Das et al., 2026). In addition, researchers also developed pseudocode DNA (pseudo-guide DNA), a synthetic DNA-based guide that enables Cas12 to target RNA rather than DNA, overcoming the traditional requirement for RNA guides and extending Cas12's functionality beyond genome-editing. The authors demonstrated that the AsCas12a-ΨDNA system enables effective RNA detection, endogenous RNA knockdown (70%–95%), multiplex regulation of transcription, and simultaneous DNA and RNA editing with a single CRISPR effector, while maintaining significantly lower off-target effects than the Cas13 system. The work further demonstrated its therapeutic and diagnostic potential by achieving 100 percent detection of hepatitis C virus and enabling programmable RNA manipulation through fusions with RNA-degrading enzymes such as RNAase H1 and METTL3, as well as epitranscriptome editing (Orosco et al., 2026).

25.11 AI and prediction of CRISPR–Cas9 editing outcomes

Because of the stochastic nature of double-strand break repair pathways, predicting the outcome at the site has been difficult, especially whether it occurs through non-NHEJ or HDR (Allen et al., 2018). This challenge has been tackled by AI models such as Pythia (Naert et al., 2025), DeepIndel (Zhang et al., 2024), inDelphi (Shen et al., 2018), FORECasT (Pallaseni et al., 2024), and SPROUT (Leenay et al., 2019), which effectively use features like local sequence context and microhomology patterns to estimate the likelihood and distribution of insertions or deletions (Fig. 25.2). These tools allow researchers to anticipate repair outcomes in advance and, in some cases, to design editing strategies that favor the repair process toward a desired result. Section 3 of this chapter discusses these predictive models in greater detail.

In plant systems, this type of outcome forecasting has been particularly valuable for improving both precision and reproducibility. For example, AI-guided prediction has aided in creating targeted alleles in tomato, resulting in enhanced fruit quality (Bradbury et al., 2025). Likewise, in rice, the use of predictive models has boosted knock-in efficiency during nutrient biofortification efforts (Zhang et al., 2025) (Table 25.2). Overall, AI-assisted prediction of genome-editing outcomes thus improves the nutritional quality of crops, making them healthier for consumption by the growing population.

25.12 AI-aided functional and phenotypic changes after CRISPR editing

Linking specific genome edits to their phenotypic effects often involves analyzing large, diverse datasets, including multi-omic profiles and high-resolution imaging. AI is becoming essential in managing and interpreting these complexities. Models like scVI and scBERT can analyze single-cell RNA sequencing data to map transcriptional responses to specific CRISPR perturbations (Lopez et al., 2018; Yang et al., 2022). Meanwhile, Geneformer, a large transformer model trained on millions of gene expression profiles, can infer gene-gene interactions and predict likely phenotypic outcomes following targeted edits (Theodoris et al., 2023). In this context, multimodal AI pipelines are becoming more common for integrating expression data with imaging features, enabling researchers to identify beneficial trait changes more quickly in plants. AI-driven analysis of single-cell transcriptomics in rice is one of the best examples that highlight how CRISPR edits affect stress-response pathways (Zhang et al., 2021) (Table 25.2). Additionally, computer-vision systems such as CellProfiler have been used to detect subtle developmental or morphological differences in *Arabidopsis* and *Maize* seedlings that might otherwise go unnoticed (Chen et al., 2024). AI, therefore, has a profound ability to translate phenotypic features into genotypes and provide a dependable reflection of underlying genetics.

25.13 AI-aided experimental design and automation in CRISPR–Cas9 genetic editing

AI assists in optimizing both specific CRISPR components and the overall genome editing workflow. Active learning methods and Bayesian optimization tools, such as AutoCRISPR and CRISP-ML, leverage data from previous experiments to identify the most informative conditions to test next, helping reduce unnecessary iterations and overall costs. Moreover, when these computational frameworks are combined with laboratory automation platforms such as Benchling AI or LabMate, they facilitate closed-loop "design-build-test-learn" cycles that continuously refine genome-editing strategies through feedback (Carbonell et al., 2018). AI-guided automation has been helpful in supporting high-throughput mutagenesis and trait discovery, especially in major crop species. Large-scale screening pipelines in rice and maize are now being used to identify genes involved in drought tolerance, nutrient utilization, and flowering-time control (Kumar et al., 2023). In short, predictive modeling and automation have accelerated the translation of genomic data into practical crop improvement.

25.14 AI-based prediction of TALEN- or CRISPR–Cas9-mediated DNA repair outcomes

When TALENs or CRISPR–Cas9 induces a double-strand break, the cell activates DNA repair mechanisms that rejoin the broken DNA ends. The repair fate, however, is not uniform; it depends on features such as local sequence composition and the surrounding chromatin environment (Naert et al., 2025; van Overbeek et al., 2016). AI models trained on large datasets of editing results have permitted the detection of patterns in these variables, linking factors such as base context, microhomology, and chromatin accessibility to the mutation types being targeted (Shen et al., 2018; Allen et al., 2018). As a result, researchers can now more accurately anticipate the likelihood of repair outcomes than with traditional empirical prediction.

Deep-learning frameworks, including inDelphi and FORECasT, model how DNA structure, repair pathway choice, and even cell type influence editing outcomes (Shen et al., 2018; Liu et al., 2022). By evaluating properties such as GC content and microhomology motifs, these tools can determine whether repair will proceed via NHEJ, MMEJ, or HDR, and whether the resulting sequence change is more likely to be an insertion, deletion, or a precise correction. This predictive capacity can aid experimental planning, including the selection of gRNA targets and donor templates before gene editing. A newer set of machine-learning platforms even extends this approach to directly design repair templates, improving the reliability of precise genome modifications across both in vitro and in vivo systems (Naert et al., 2025).

Beyond outcome prediction, AI has also contributed to improving the fidelity and efficiency of gene editing. This is accomplished through AI-based identification of cellular states, chromatin accessibility maps that show high transcriptional activity, and DNA-repair factor data. Reinforcement-learning approaches further optimize nuclease engineering and delivery strategies to promote accurate repair (Kim et al., 2024; Li et al., 2024). In crop species such as rice, maize, and soybean, these computational tools have enabled more targeted and reproducible trait engineering, supporting efforts to improve stress resilience and agricultural productivity (Li et al., 2025). Together, these developments demonstrate how AI makes genome editing a more predictable, context-aware process for plants biotechnology.

25.15 AI-guided selection of efficient CRISPR cutters and modifiers

AI advancements are influencing how researchers select and engineer CRISPR components that are responsible for DNA cutting and modification. These components include SpCas9, SaCas9, and Cas12a, along with their catalytic variants used for plant genome editing (Li et al., 2023). Machine-learning models trained on datasets containing PAM specificities, off-target behavior, and structural parameters can now estimate which Cas variant is best suited for a particular genomic target (Li et al., 2024). These computational approaches reduce the need for extensive experiential screening and help design nucleases with improved fidelity, modified PAM recognition ranges, or smaller size for more efficient delivery (Li et al., 2025; Silverstein et al., 2025) (Fig. 25.3). Based on information from

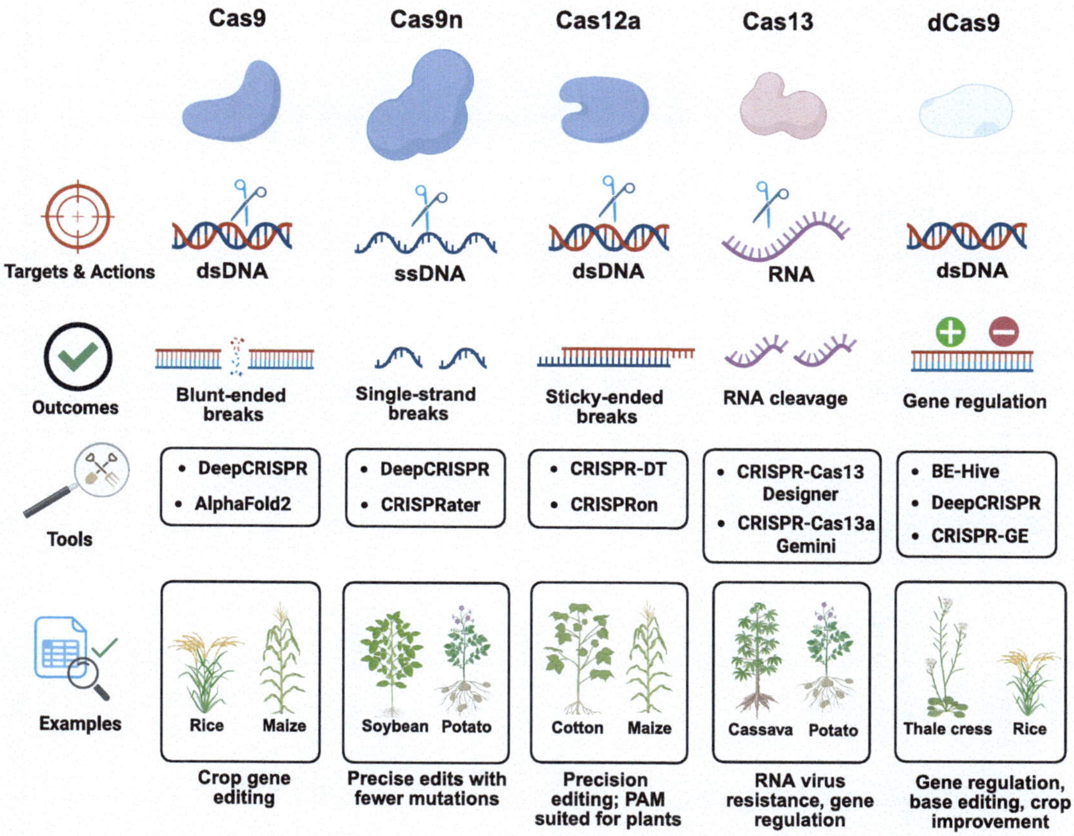

FIGURE 25.3 Comparison of CRISPR–Cas systems, target types, and editing outcomes. Schematic overview of five representative CRISPR–Cas platforms, Cas9, Cas9 nickase (Cas9n), Cas12a (Cpf1), Cas13, and dCas9, highlighting their distinct nucleic acid targets, cleavage types, and functional applications. The Cas9 Ind.uces blunt-ended double-strand breaks in DNA, while Cas9n produces single-strand (nick) breaks for more precise edits. Cas12a generates sticky-ended double-strand breaks and is advantageous for plant genome editing due to its T-rich PAM preference. Cas13 exclusively targets RNA for RNA virus resistance and transcript-level regulation. Catalytically inactive dCas9 enables transcriptional regulation, base editing, and epigenetic modification without cleavage. Representative bioinformatic tools for CRISPR design and prediction are listed (e.g., DeepCRISPR, CRISPRater, CRISPR-DT, CRISPRon, CRISPR–Cas13 Designer, BE-Hive, and CRISPR-GE). Example applications across major crop species, including rice, maize, potato, soybean, cotton, cassava, and thale cress, illustrate the breadth of genome-engineering outcomes achieved by different Cas systems (see also Table 25.3).

TABLE 25.3 AI tools and selection of Cas9 variants and genome-modifying proteins for genomic and epigenomic modifications.

Cutter/ Modifier Cas proteins	AI tools for selection	Target type and cutting mechanism	Guide RNA type and PAM requirement	Potential outcomes and off-target effects	Potential applications in agriculture	Representative references
Cas9	DeepCRISPR, AlphaFold2 to predict best Cas9 variant, guide RNA efficiency, minimize off-targets	It causes double-stranded DNA and blunt-ended double-strand breaks	Single guide RNA (sgRNA) with NGG PAM sequence	It offers gene knockout, insertion and editing with moderate off-target effect	It is applicable for precise editing of crops such as rice, wheat, maize, tomato	Chuai et al. (2018), Cancellieri et al. (2020)
Cas9n (Nickase)	DeepCRISPR, CRISPRater to optimize nickase guide design to improve specificity and reduce off-target cleavage	It promotes single-stranded DNA	Single guide RNA (sgRNA) with NGG PAM sequence	It induces precise edits with reduced off-targets as compared with Cas9	It is applicable for precise editing of crops with fewer mutations and suits for tomato, potato, soybean	Chuai et al. (2018), Labuhn et al. (2018)
Cas12a (Cpf1)	CRISPR-DT, CRISPRon to evaluate PAM compatibility, target accessibility, and off-target risk for Cas12a	It induces double-stranded DNA ands taggered (sticky-end) double-strand breaks	Single CRISPR RNA (crRNA) with TTTV (T-rich) PAM sequence	It leads to precise gene editing with alternative PAM and lead to lower off-target effect	It is applicable for precision editing and PAM identification for plants, including crops such as maize, rice, cotton	Xiang et al., 2021; Zhu and Liang, 2019
Cas13	Cas13-specific AI tools like CRISPR–Cas13 Designer and CRISPR–Cas13a Gemini System to optimize crRNA design for specificity and collateral activity	It induces single-stranded RNA cleavage	CRISPR RNA (crRNA) No PAM; requires as protospacer flanking site	It induces RNA knockdown, viral RNA detection with high RNA off-target activity	It is applicable for RNA virus resistance, gene regulation for plants like potato (virus resistance), cassava	Wessels et al. (2020), (Guo et al., 2021) uo et al. (2021), (Zhao et al., 2024)
dCas9 (dead Cas9)	BE-Hive, DeepCRISPR, CRISPR-GE to predict binding efficiency, optimize base editing outcomes, and model epigenetic impacts	No DNA cutting. It binds target DNA for transcriptional or epigenetic modulation	sgRNA with NGG PAM sequence is required	It performs epigenetic regulation and gene activation/repression (CRISPRa/i) as well as base editing with very low off-target effect	It is used for gene regulation and base editing for rice, arabidopsis, maize, potato	Arbab et al. (2020), Chuai et al. (2018), Xie et al. (2017)

mutational libraries, structure-function studies, and editing-performance datasets, AI frameworks can recommend the most appropriate Cas or dCas fusion for a given application. This capability is particularly valuable in plant species with large, polyploid, or otherwise restrictive genomes, where selecting an effective nuclease can be challenging (Wan et al., 2021; Illa-Berenguer et al., 2023). The variable nature of Cas9 cutters and modifiers, therefore, gives the CRISPR system modular capabilities to make broader changes to the crop genome for increased productivity while staying within ethical boundaries.

25.16 AI-powered smart Cas9 variants design

Cas9 functions as a 'molecular scissors' in the CRISPR system that cuts DNA at a specific target site. Different versions of Cas9 (e.g., SpCas9, SaCas9, Cas12a) differ in size, specificity, and cutting behavior (Li et al., 2023). The best choice depends on factors like the DNA sequence, the organism's genome, and the desired editing outcome. Trained on structural data, mutational libraries, and records from previous CRISPR experiments, models can now help identify which Cas9 variant is most likely to work effectively at a specific locus. By evaluating factors such as PAM compatibility, predicted off-target activity, and enzyme stability, these systems can suggest nuclease versions that minimize unintended edits while maintaining high on-target efficiency (Li et al., 2025; Silverstein et al., 2025). This computational guidance reduces trial-and-error screening and allows researchers to customize nucleases to specific experimental contexts, ranging from genome editing in mammalian cells to trait engineering in crop species, including rice, maize, potato, soybean, cotton, cassava, and thale cress (Fig. 25.3 and Table 25.3). In this way, AI offers direct assistance in improving the accuracy, safety, and applicability of CRISPR technologies across biological systems.

25.16.1 AI for developing efficient dCas9 complexes for epigenome editing

Dead Cas9 (dCas9) is a modified version of Cas9 created by inactivating its nucleolytic activity through a point mutation in the DNA sequence that encodes the cutting domain, which is then replaced with a functionally active domain (Sternberg et al., 2015). This dCas9 variant provides a platform for modifying epigenomic targets without causing double-stranded DNA breaks (Lin et al., 2023). A fully modified dCas9 complex includes factors that activate or repress genes, study chromatin structure, and directly remodel three-dimensional chromatin, as well as enzymes that perform base editing (Brezgin et al., 2019). AI-based tools help select optimal dCas9 variants and effector domains by considering guide RNA activity, chromatin accessibility, and DNA methylation context (Kissling et al., 2025; Pandey et al., 2025). Using AI-assisted design, dCas9 can be fused with regulatory domains such as DNMT, TET, KRAB, or p300 to target DNA methylation, demethylation, or transcriptional activation (Cai et al., 2023) (Fig. 25.3). These strategies have already proven useful in agriculture, where AI-guided dCas9 platforms enabled controlled methylation changes in Arabidopsis (Wang and Bart, 2025), activation of CBF4 to improve cold tolerance in grapevine (Ren et al., 2024), and precise cytosine base editing in phytopathogenic bacteria affecting multiple crops (Li et al., 2023). Overall, AI-assisted epigenome editing offers a targeted, reversible way to modify plant traits without altering the underlying DNA sequences, thus providing a versatile approach for sustainable crop improvement (Table 25.3).

25.16.2 AI optimizes TALEN and CRISPR–Cas9 components delivery to target tissues

Efficient delivery of TALEN and CRISPR components, including TALE proteins or plasmids expressing them, Cas proteins, gRNAs, and donor templates, remains a major technical challenge, particularly in plants, where rigid cell walls hinder protein or nucleic acid uptake (Sioson et al., 2021). AI now supports the rational design of delivery vehicles, including viral and nanoparticle-based vehicles. For instance, viral-capsid engineering structures such as AI-VAE-Capsid (Nisanov et al., 2025) and DeepCapsid employ generative models to predict capsid features relevant to tissue specificity, immunogenicity, and packaging capacity (Xu et al., 2021). Similarly, nanoparticle delivery strategies are being reoptimized through the integration of platforms such as NanoAI (Rauf et al., 2025), which utilizes machine learning to optimize formulation and targeting efficiency by leveraging physicochemical and biodistribution data (Chan et al., 2025; Kapoor et al., 2024). Many of these tools are used to facilitate the development of non-transgenic delivery systems for TALENs and CRISPR ribonucleoproteins. One prominent example is that AI-modeled nanoparticles have achieved efficient delivery in wheat and maize protoplasts, eliminating the need for traditional tissue culture (Liang et al., 2018; Namata et al., 2025). Such progress brings the possibility of direct, field-ready genome editing closer to practical reality.

25.17 Ethical considerations for AI-mediated gene editing in agriculture

While AI has offered a great deal of advantages to genome editing, it has also raised genuine ethical and societal concerns. Two main concerns in this regard are equity and access to advanced computational infrastructure and data, as AI tools are trained on massive, advanced datasets that are mostly concentrated in large research centers and private companies. This means that small farmers and low-resource agricultural systems are clearly at a disadvantage. Not only that, but it also raises serious concerns about transparency and accountability. Indeed, many AI models function as "black boxes" with no clear insight about the choice of a particular gene target and expected editing outcomes. As a result, it is very hard to assess potential bias, conflict of interests, ecological risks, off-target surveillance as well as regulatory compliance.

Additionally, environmental and societal implications are also major points of concern. Editing the sensitive gene responsible for regulatory or epigenetic pathways may lead to unwanted short-term or long-term effects. These include manipulation of plant-microbe interactions, competitive dynamics, and even biodiversity. Therefore, the use of AI-powered crops genome editing demands designing and strict compliance of comprehensive risk-assessment frameworks that integrate experimental validation. Finally, transparent and non-scientific communication with farmers and consumers is of pivotal importance to prevent misinformation and enhance trust. Addressing all these legitimate concerns in an adequate manner would set the stage for the responsible use of AI to enhance agricultural innovation at an unprecedented pace.

25.18 Emerging future of AI in agricultural genome editing

So far in this chapter, we have discussed in detail how AI has emerged as a key player in the field of agricultural genome editing. Considering its significant role in genome editing and improved outcomes in terms of efficiency, accuracy, and precision, it is safe to say that the future belongs to AI. However, in order to fully exploit the powers of AI and to make it more predictive, adaptive and environment-friendly, the research community needs to focus on developing plant-specific, multi-omics datasets integrating genomic, transcriptomic, epigenetic, phenotypic, and environmental factors. Indeed, AI feeds on such datasets to harness its predictive and adaptive power. On the other side, with rising ethical concerns, an improved transparency, accountability, and public trust is also warranted. A broader collaboration of all stakeholders such as farmers, breeders, computational scientists, ethicists, and policymakers can enhance the development of explainable AI systems, avoiding copyright infringement and formulation of strict regulations. Furthermore, a strict and long-term surveillance of off-target effects, trait monitoring, and environmental impact assessments should also be done in the modified crops. In short, regulation-compliant, ethical and safe use of AI holds immense potential to offer sustainable agricultural solutions by improving crops yield and nutritive values, resilience to changing climate, and ultimately offering food security and ecological balance.

References

Abdallah, N.A., Prakash, C.S., McHughen, A.G., 2015. Genome editing for crop improvement: challenges and opportunities. GM Crops Food 6 (4), 183–205. https://doi.org/10.1080/21645698.2015.1129937.

Alipanahi, R., Safari, L., Khanteymoori, A., December 9, 2025. Advancing CRISPR with deep learning: a comprehensive review of models and databases. Mol. Ther. Nucleic Acids 36 (4), 102691. https://doi.org/10.1016/j.omtn.2025.102691.

Allen, F., Crepaldi, L., Alsinet, C., Strong, A.J., Kleshchevnikov, V., De Angeli, P., Palenikova, P., Khodak, A., Kiselev, V., Kosicki, M., Bassett, A.R., Harding, H., Galanty, Y., Munoz-Martinez, F., Metzakopian, E., Jackson, S.P., Parts, L., 2018. Predicting the mutations generated by repair of Cas9-induced double-strand breaks. Nat. Biotechnol. https://doi.org/10.1038/nbt.4317.

Anthon, C., Corsi, G.I., Gorodkin, J., 2022. CRISPRon/off: CRISPR/Cas9 on- and off-target gRNA design. Bioinformatics 38, 5437–5439. https://doi.org/10.1093/bioinformatics/btac697.

Arbab, M., Shen, M.W., Mok, B., Wilson, C., Matuszek, Ż., Cassa, C.A., Liu, D.R., 2020. Determinants of base editing outcomes from target library analysis and machine learning. Cell 182 (2), 463. https://doi.org/10.1016/j.cell.2020.05.037.

Atia, M., Jiang, W., Sedeek, K., Butt, H., Mahfouz, M., 2024. Crop bioengineering via gene editing: reshaping the future of agriculture. Plant Cell Rep. 43 (4). https://doi.org/10.1007/s00299-024-03183-1.

Azameti, M.K., Dauda, W.P., 2021. Base editing in plants: applications, challenges, and future prospects. Front. Plant Sci. 12. https://doi.org/10.3389/fpls.2021.664997.

Baek, M., DiMaio, F., Anishchenko, I., Dauparas, J., Ovchinnikov, S., Lee, G.R., Wang, J., Cong, Q., Kinch, L.N., Dustin Schaeffer, R., Millán, C., Park, H., Adams, C., Glassman, C.R., DeGiovanni, A., Pereira, J.H., Rodrigues, A.V., Van Dijk, A.A., Ebrecht, A.C., Opperman, D.J., Sagmeister, T., Buhlheller, C., Pavkov-Keller, T.,

Rathinaswamy, M.K., Dalwadi, U., Yip, C.K., Burke, J.E., Christopher Garcia, K., Grishin, N.V., Adams, P.D., Read, R.J., Baker, D., 2021. Accurate prediction of protein structures and interactions using a three-track neural network. Science 373 (6557), 871–876. https://doi.org/10.1126/science.abj8754.

Bhardwaj, A., Nain, V., 2021. TALENs—an indispensable tool in the era of CRISPR: a mini review. J. Genet. Eng. Biotechnol. 19 (1), 125. https://doi.org/10.1186/s43141-021-00225-z.

Boch, J., Scholze, H., Schornack, S., Landgraf, A., Hahn, S., Kay, S., Lahaye, T., Nickstadt, A., Bonas, U., 2009. Breaking the code of DNA binding specificity of TAL-type III effectors. Science 326 (5959), 1509–1512. https://doi.org/10.1126/science.1178811.

Bogdanove, A.J., Voytas, D.F., 2011. TAL effectors: customizable proteins for DNA targeting. Science 333 (6051), 1843–1846. https://doi.org/10.1126/science.1204094.

Bradbury, A., Clapp, O., Biacsi, A.S., Kuo, P., Gaju, O., Hayta, S., Zhu, J.K., Lambing, C., 2025. Integrating genome editing with omics, artificial intelligence, and advanced farming technologies to increase crop productivity. Plant Commun. 6 (7). https://doi.org/10.1016/j.xplc.2025.101386.

Brezgin, S., Kostyusheva, A., Kostyushev, D., Chulanov, V., 2019. Dead Cas systems: types, principles, and applications. Int. J. Mol. Sci. 20 (23), 6041. https://doi.org/10.3390/ijms20236041.

Cai, R., Lv, R., Shi, X., Yang, G., Jin, J., 2023. CRISPR/dCas9 tools: epigenetic mechanism and application in gene transcriptional regulation. Int. J. Mol. Sci. 24 (19). https://doi.org/10.3390/ijms241914865.

Cancellieri, S., Canver, M.C., Bombieri, N., Giugno, R., Pinello, L., 2020. CRISPRitz: rapid, high-throughput and variant-aware in silico off-target site identification for CRISPR genome editing. Bioinformatics 36, 2001–2008. https://doi.org/10.1093/bioinformatics/btz867.

Carbonell, P., Jervis, A.J., Robinson, C.J., Yan, C., Dunstan, M., Swainston, N., Vinaixa, M., Hollywood, K.A., Currin, A., Rattray, N.J.W., Taylor, S., Spiess, R., Sung, R., Williams, A.R., Fellows, D., Stanford, N.J., Mulherin, P., Le Feuvre, R., Barran, P., Goodacre, R., Turner, N.J., Goble, C., Chen, G.G., Kell, D.B., Micklefield, J., Breitling, R., Takano, E., Faulon, J.L., Scrutton, N.S., 2018. An automated Design-Build-Test-Learn pipeline for enhanced microbial production of fine chemicals. Commun. Biol. 1 (1). https://doi.org/10.1038/s42003-018-0076-9.

Cermak, T., Doyle, E.L., Christian, M., Wang, L., Zhang, Y., Schmidt, C., Baller, J.A., Somia, N.V., Bogdanove, A.J., Voytas, D.F., 2011. Efficient design and assembly of custom TALEN and other TAL effector-based constructs for DNA targeting. Nucleic Acids Res. 39 (12), e82. https://doi.org/10.1093/nar/gkr218.

Chan, A., Kirtane, A.R., Qu, Q.R., Huang, X., Woo, J., Subramanian, D.A., Dey, R., Semalty, R., Bernstock, J.D., Ahmed, T., Honeywell, R., Hanhurst, C., Diaz Becdach, I., Prizant, L.S., Brown, A.K., Song, H., Law Cobb, J., DeRidder, L.B., Santos, B., Jimenez, M., Sun, M., Huang, Y., Byrne, C., Traverso, G., 2025. Designing lipid nanoparticles using a transformer-based neural network. Nat. Nanotechnol. 20 (10), 1491–1501. https://doi.org/10.1038/s41565-025-01975-4.

Chen, Y., Wang, X., Mathelier, A., 2022. Evaluation of efficiency prediction algorithms and development of ensemble model for CRISPR/Cas9 gRNA selection. Bioinformatics 38 (23), 5175–5181. https://doi.org/10.1093/bioinformatics/btac681.

Chen, L., Liu, G., Zhang, T., June 2024. Integrating machine learning and genome editing for crop improvement. aBIOTECH 5 (2), 262–277. https://doi.org/10.1007/s42994-023-00133-5.

Chen, G., Hou, L., Li, Z., Xie, B., Liu, Y., 2025. A new strategy for Cas protein recognition based on graph neural networks and SMILES encoding. Sci. Rep. 15 (1). https://doi.org/10.1038/s41598-025-99999-2.

Chuai, G., Ma, H., Yan, J., Chen, M., Hong, N., Xue, D., Zhou, C., Zhu, C., Chen, K., Duan, B., Gu, F., Qu, S., Huang, D., Wei, J., Liu, Q., 2018. DeepCRISPR: optimized CRISPR guide RNA design by deep learning. Genome Biol. 19 (1). https://doi.org/10.1186/s13059-018-1459-4.

Das, J., Kumar, S., Mishra, D.C., Chaturvedi, K.K., Paul, R.K., Kairi, A., 2022. Machine learning in the estimation of CRISPR-Cas9 cleavage sites for plant system. Front. Genet. 13, 1085332. https://doi.org/10.3389/fgene.2022.1085332.

Das, P., Saha, R., Panda, D., Ghosh, C., Avinash, S.P., Panda, S., Molla, K.A., 2026. AI-designed OpenCRISPR-1 performs robust knockout, base editing, and prime editing in rice. New Phytol.

Dauparas, J., Anishchenko, I., Bennett, N., Bai, H., Ragotte, R.J., Milles, L.F., Wicky, B.I.M., Courbet, A., de Haas, R. J., Bethel, N., Leung, P.J.Y., Huddy, T.F., Pellock, S., Tischer, D., Chan, F., Koepnick, B., Nguyen, H., Kang, A., Sankaran, B., Bera, A.K., King, N.P., Baker, D., 2022. Robust deep learning–based protein sequence design using ProteinMPNN. Science 378 (6615), 49–56. https://doi.org/10.1126/science.add2187.

Delfani, P., Thuraga, V., Banerjee, B., Chawade, A., 2024. Integrative approaches in modern agriculture: IoT, ML and AI for disease forecasting amidst climate change. Precis. Agric. 25 (5), 2589–2613. https://doi.org/10.1007/s11119-024-10164-7.

Ding, C., Zhang, Z., 2023. Effective omics tools are still lacking for improvement of stress tolerance in polyploid crops. Front. Plant Sci. 14. https://doi.org/10.3389/fpls.2023.1295528.

Dixit, S., Kumar, A., Srinivasan, K., Vincent, P.M.D.R., Ramu Krishnan, N., 2023. Advancing genome editing with artificial intelligence: opportunities, challenges, and future directions. Front. Bioeng. Biotechnol. 11. https://doi.org/10.3389/fbioe.2023.1335901.

Doyle, E.L., Booher, N.J., Standage, D.S., Voytas, D.F., Brendel, V.P., Vandyk, J.K., Bogdanove, A.J., 2012. TAL Effector-Nucleotide Targeter (TALE-NT) 2.0: tools for TAL effector design and target prediction. Nucleic Acids Res. 40 (1), W117. https://doi.org/10.1093/nar/gks608.

Erkes, A., Mücke, S., Reschke, M., Boch, J., Grau, J., 2021. Epigenetic features improve TALE target prediction. BMC Genom. 22 (1). https://doi.org/10.1186/s12864-021-08210-z.

Farooq, M.A., Gao, S., Hassan, M.A., Huang, Z., Rasheed, A., Hearne, S., Prasanna, B., Li, X., Li, H., 2024. Artificial intelligence in plant breeding. Trends Genet. 40 (10), 891–908. https://doi.org/10.1016/j.tig.2024.07.001.

Ferrao, C.J., Dias, D., Morajkar, S., Dessai, M.F.F., 2025. GCBLANE: A graph-enhanced convolutional BiLSTM attention network for improved transcription factor binding site prediction. arXiv:2503.12377 1–19. https://doi.org/10.48550/arXiv.2503.12377.

Ferruz, N., Schmidt, S., Hocker, B., 2022. ProtGPT2 is a deep unsupervised language model for protein design. Nat. Commun. 13 (1), 4348. https://doi.org/10.1038/s41467-022-32007-7.

Gaj, T., Sirk, S.J., Shui, S.L., Liu, J., 2016. Genome-editing technologies: principles and applications. Cold Spring Harbor Perspect. Biol. 8 (12). https://doi.org/10.1101/cshperspect.a023754.

Grau, J., Boch, J., Posch, S., November 15, 2013. TALENoffer: genome-wide TALEN off-target prediction. Bioinformatics 29 (22), 2931–2932. https://doi.org/10.1093/bioinformatics/btt501.

Guo, X., Rahman, J.A., Wessels, H.H., Mendez-Mancilla, A., Haro, D., Chen, X., Sanjana, N.E., 2021. Transcriptome-wide Cas13 guide RNA design for model organisms and viral RNA pathogens. Cell Genom. 1, 100001. https://doi.org/10.1016/j.xgen.2021.100001.

Gupta, G., Pal, S.K., 2025. Applications of AI in precision agriculture. Discover Agriculture 3 (1). https://doi.org/10.1007/s44279-025-00220-9.

Hassani, H., Silva, E.S., Unger, S., TajMazinani, M., Feely, S.Mac, 2020. Artificial intelligence (AI) or intelligence augmentation (IA): what is the future? AI 1 (2), 143–155. https://doi.org/10.3390/ai1020008.

Haun, W., Coffman, A., Clasen, B.M., Demorest, Z.L., Lowy, A., Ray, E., Retterath, A., Stoddard, T., Juillerat, A., Cedrone, F., Mathis, L., Voytas, D.F., Zhang, F., 2014. Improved soybean oil quality by targeted mutagenesis of the fatty acid desaturase 2 gene family. Plant Biotechnol. J. 12 (7), 934–940. https://doi.org/10.1111/pbi.12201.

Heigwer, F., Kerr, G., Boutros, M., 2014. E-CRISP: fast CRISPR target site identification. Nat. Methods 11 (2), 122–123. https://doi.org/10.1038/nmeth.2812.

Heigwer, F., Kerr, G., Walther, N., Glaeser, K., Pelz, O., Breinig, M., Boutros, M., 2013. E-TALEN: a web tool to design TALENs for genome engineering. Nucleic Acids Res. 41 (20), e190. https://doi.org/10.1093/nar/gkt789.

Hsu, P.D., Lander, E.S., Zhang, F., June 5, 2014. Development and applications of CRISPR-Cas9 for genome engineering. Cell 157 (6), 1262–1278. https://doi.org/10.1016/j.cell.2014.05.010.

Ibrar, M., Khurseed, H., Fatima, A., Raza, M.Z., Riaz, M., 2025. Genome-Editing Technologies: A Comprehensive Overview. Gene-Edited Crops 13–22. CRC Press.

Illa-Berenguer, E., LaFayette, P.R., Parrott, W.A., 2023. Editing efficiencies with Cas9 orthologs, Cas12a endonucleases, and temperature in rice. Front Genome 5, 1074641. https://doi.org/10.3389/fgeed.2023.1074641.

Jinek, M., Chylinski, K., Fonfara, I., Hauer, M., Doudna, J.A., Charpentier, E., 2012. A programmable dual-RNA-guided DNA endonuclease in adaptive bacterial immunity. Science 337 (6096), 816–821. https://doi.org/10.1126/science.1225829.

Kapoor, D.U., Sharma, J.B., Gandhi, S.M., Prajapati, B.G., Thanawuth, K., Limmatvapirat, S., Sriamornsak, P., 2024. AI-driven design and optimization of nanoparticle-based drug delivery systems. Science, Engineering and Health Studies 18. https://doi.org/10.69598/sehs.18.24010003.

Kim, H.K., Kim, Y., Lee, L., Min, S., Bae, J.Y., Choi, J.W., Park, J., Jung, D, Yoon, S., Kim, H.H., 2019. SpCas9 activity prediction by DeepCas9, a deep learning-based model withunparalleled generalization performance. Sci. Adv. 5 (11),1101/636472. https://doi.org/10.1126/sciadv.aax9249.

Kim, H.K., Lee, S., Kim, Y., Park, J., Min, S., Choi, J.W., Huang, T.P., Yoon, S., Liu, D.R., Kim, H.H., 2020. High-throughput analysis of the activities of xCas9, SpCas9-NG and SpCas9 at matched and mismatched target sequences in human cells. Nat. Biomed. Eng. 4 (1), 111–124. https://doi.org/10.1038/s41551-019-0505-1.

Kim, N., Choi, S., Kim, S., Song, M., Seo, J.H., Min, S., Park, J., Cho, S.R., Kim, H.H., 2024. Deep learning models to predict the editing efficiencies and outcomes of diverse base editors. Nat. Biotechnol. 42 (3), 484–497. https://doi.org/10.1038/s41587-023-01792-x.

Kim, M.G., Go, M.J., Kang, S.H., Jeong, S.H., Lim, K., 2025. Revolutionizing CRISPR technology with artificial intelligence. Exp. Mol. Med. 57 (7), 1419–1431. https://doi.org/10.1038/s12276-025-01462-9.

Kissling, L., Mollaysa, A., Janjuha, S., Mathis, N., Marquart, K.F., Weber, Y., Moon, W.J., Lin, P.J.C., Fan, S.H.Y., Muramatsu, H., Vadovics, M., Allam, A., Pardi, N., Tam, Y.K., Krauthammer, M., Schwank, G., 2025. Predicting adenine base editing efficiencies in different cellular contexts by deep learning. Genome Biol. 26 (1). https://doi.org/10.1186/s13059-025-03586-7.

Kumar, M., Prusty, M.R., Pandey, M.K., Singh, P.K., Bohra, A., Guo, B., Varshney, R.K., 2023. Application of CRISPR/Cas9-mediated gene editing for abiotic stress management in crop plants. Front. Plant Sci. 14. https://doi.org/10.3389/fpls.2023.1157678.

Labuhn, M., Adams, F.F., Ng, M., Knoess, S., Schambach, A., Charpentier, E.M., Schwarzer, A., Mateo, J.L., Klusmann, J.H., Heckl, D., 2018. Refined sgRNA efficacy prediction improves large- and small-scale CRISPR-Cas9 applications. Nucleic Acids Res. 46, 1375–1385. https://doi.org/10.1093/nar/gkx1268.

Labun, K., Krause, M., Torres Cleuren, Y., Valen, E., 2021. CRISPR genome editing made easy through the CHOP-CHOP website. Curr. Protoc. 1 (4). https://doi.org/10.1002/cpz1.46.

Lee, M., 2023. Deep learning in CRISPR-Cas systems: a review of recent studies. Front. Bioeng. Biotechnol. 11. https://doi.org/10.3389/fbioe.2023.1226182.

Leenay, R.T., Aghazadeh, A., Hiatt, J., Tse, D., Roth, T.L., Apathy, R., Shifrut, E., Hultquist, J.F., Krogan, N., Wu, Z., Cirolia, G., Canaj, H., Leonetti, M.D., Marson, A., May, A.P., Zou, J., 2019. Large dataset enables prediction of repair after CRISPR–Cas9 editing in primary T cells. Nat. Biotechnol. 37 (9), 1034–1037. https://doi.org/10.1038/s41587-019-0203-2.

Li, T., Liu, B., Spalding, M.H., Weeks, D.P., Yang, B., 2012a. High-efficiency TALEN-based gene editing produces disease-resistant rice. Nat. Biotechnol. 30 (5), 390–392. https://doi.org/10.1038/nbt.2199.

Li, L., Piatek, M.J., Atef, A., Piatek, A., Wibowo, A., Fang, X., Sabir, J.S.M., Zhu, J.K., Mahfouz, M.M., 2012b. Rapid and highly efficient construction of TALE-based transcriptional regulators and nucleases for genome modification. Plant Mol. Biol. 78 (4–5), 407–416. https://doi.org/10.1007/s11103-012-9875-4.

Li, T., Yang, Y., Qi, H., Cui, W., Zhang, L., Fu, X., He, X., Liu, M., Li, P.F., Yu, T., January 16, 2023. CRISPR/Cas9 therapeutics: progress and prospects. Signal Transduct Target Ther 8 (1), 36. https://doi.org/10.1038/s41392-023-01309-7.

Li, C., Wang, L., Cseke, L.J., Vasconcelos, F., Huguet-Tapia, J.C., Gassmann, W., Pauwels, L., White, F.F., Dong, H., Yang, B., 2023. Efficient CRISPR-Cas9 based cytosine base editors for phytopathogenic bacteria. Commun. Biol. 6 (1). https://doi.org/10.1038/s42003-023-04451-8.

Li, J., Wu, P., Cao, Z., Huang, G., Lu, Z., Yan, J., Zhang, H., Zhou, Y., Liu, R., Chen, H., Ma, L., Luo, M., 2024. Machine learning-based prediction models to guide the selection of Cas9 variants for efficient gene editing. Cell Rep. 43 (2), 113765. https://doi.org/10.1016/j.celrep.2024.113765.

Li, Z., Khan, W.U., Bai, G., Dong, C., Wang, J., Zhang, Y., Wang, C., Zhang, H., Wang, W., Luo, M., Chen, F., August 2025. From code to life: the AI-driven revolution in genome editing. Adv. Sci. (Weinh.) 12 (30), e17029. https://doi.org/10.1002/advs.202417029.

Li, G., An, L., Yang, W., Yang, L., Wei, T., Shi, J., Wang, J., Doonan, J.H., Xie, K., Fernie, A.R., Lagudah, E.S., Wing, R.A., Gao, C., 2025. Integrated biotechnological and AI innovations for crop improvement. Nature 643 (8073), 925–937. https://doi.org/10.1038/s41586-025-09122-8.

Liang, Z., Chen, K., Yan, Y., Zhang, Y., Gao, C., 2018. Genotyping genome-edited mutations in plants using CRISPR ribonucleoprotein complexes. Plant Biotechnol. J. 16 (12), 2053–2062. https://doi.org/10.1111/pbi.12938.

Lin, J., Zhang, Z., Zhang, S., Chen, J., Wong, K.C., 2020. CRISPR-net: a recurrent convolutional network quantifies CRISPR off-target activities with mismatches and indels. Adv. Sci. 7 (13). https://doi.org/10.1002/advs.201903562.

Lin, Z., Akin, H., Rao, R., Hie, B., Zhu, Z., Lu, W., Smetanin, N., Verkuil, R., Kabeli, O., Shmueli, Y., dos Santos Costa, A., Fazel-Zarandi, M., Sercu, T., Candido, S., Rives, A., 2023. Evolutionary-scale prediction of atomic-level protein structure with a language model. Science 379 (6637), 1123–1130. https://doi.org/10.1126/science.ade2574.

Listgarten, J., Weinstein, M., Kleinstiver, B.P., Sousa, A.A., Joung, J.K., Crawford, J., Gao, K., Hoang, L., Elibol, M., Doench, J.G., Fusi, N., 2018. Prediction of off-target activities for the end-to-end design of CRISPR guide RNAs. Nat. Biomed. Eng. 2 (1), 38–47. https://doi.org/10.1038/s41551-017-0178-6.

Liu, X., Wang, S., Ai, D., June 5, 2022. Predicting CRISPR/Cas9 repair outcomes by attention-based deep learning framework. Cells 11 (11). https://doi.org/10.3390/cells11111847.

Lopez, R., Regier, J., Cole, M.B., Jordan, M.I., Yosef, N., 2018. Deep generative modeling for single-cell transcriptomics. Nat. Methods 15 (12), 1053–1058. https://doi.org/10.1038/s41592-018-0229-2.

Ma, J., Köster, J., Qin, Q., Hu, S., Li, W., Chen, C., Cao, Q, et al., 2016. CRISPR-DO for genome-wide CRISPR design and optimization. Bioinformatics 32 (21), 3336–3338. https://doi.org/10.1093/bioinformatics/btw476.

Mahfouz, M.M., Piatek, A., Stewart Jr., C.N., October 2014. Genome engineering via TALENs and CRISPR/Cas9 systems: challenges and perspectives. Plant Biotechnol. J. 12 (8), 1006–1014. https://doi.org/10.1111/pbi.12256.

Mak, A.N.S., Bradley, P., Cernadas, R.A., Bogdanove, A.J., Stoddard, B.L., 2012. The crystal structure of TAL effector PthXo1 bound to its DNA target. Science 335 (6069), 716–719. https://doi.org/10.1126/science.1216211.

Mesko, B., 2017. The role of artificial intelligence in precision medicine. Expert Review of Precision Medicine and Drug Development 2 (5), 239–241. https://doi.org/10.1080/23808993.2017.1380516.

Mitra, R., Li, J., Sagendorf, J.M., Jiang, Y., Cohen, A.S., Chiu, T.P., Glasscock, C.J., Rohs, R., 2024. Geometric deep learning of protein-DNA binding specificity. Nat. Methods 21, 1674–1683. https://doi.org/10.1038/s41592-024-02372-w.

Moore, R., Chandrahas, A., Bleris, L., October 17, 2014. Transcription activator-like effectors: a toolkit for synthetic biology. ACS Synth. Biol. 3 (10), 708–716. https://doi.org/10.1021/sb400137b.

Morbitzer, R., Elsaesser, J., Hausner, J., Lahaye, T., 2011. Assembly of custom TALE-type DNA binding domains by modular cloning. Nucleic Acids Res. 39 (13), 5790–5799. https://doi.org/10.1093/nar/gkr151.

Moscou, M.J., Bogdanove, A.J., 2009. A simple cipher governs DNA recognition by TAL effectors. Science 326 (5959), 1501. https://doi.org/10.1126/science.1178817.

Naert, T., Yamamoto, T., Han, S., Röck, R., Horn, M., Bethge, P., Vladimirov, N., Voigt, F.F., Figueiro-Silva, J., Bachmann-Gagescu, R., Vleminckx, K., Helmchen, F., Lienkamp, S.S., 2025. Precise, predictable genome integrations by deep-learning-assisted design of microhomology-based templates. Nat. Biotechnol. https://doi.org/10.1038/s41587-025-02771-0.

Namata, M.J., Xu, J., Habyarimana, E., Palakolanu, S.R., Wang, L., Li, J., 2025. Genome editing in maize and sorghum: a comprehensive review of CRISPR/Cas9 and emerging technologies. Plant Genome 18 (2). https://doi.org/10.1002/tpg2.70038.

Narushima, J., Kimata, S., Shiwa, Y., Gondo, T., Akimoto, S., Soga, K., Yoshiba, S., Nakamura, K., Shibata, N., Kondo, K., 2021. Strategy for detecting off-target sites in genome-edited rice. bioRxiv. https://doi.org/10.1101/2021.05.28.446070.

Nisanov, A.M., Rivera de Jesus, J.A., Schaffer, D.V., May 7, 2025. Advances in AAV capsid engineering: integrating rational design, directed evolution and machine learning. Mol. Ther. 33 (5), 1937–1945. https://doi.org/10.1016/j.ymthe.2025.03.056.

Nishimasu, H., Shi, X., Ishiguro, S., Gao, L., Hirano, S., Okazaki, S., Noda, T., Abudayyeh, O.O., Gootenberg, J.S., Mori, H., Oura, S., Holmes, B., Tanaka, M., Seki, M., Hirano, H., Aburatani, H., Ishitani, R., Ikawa, M., Yachie, N., Zhang, F., Nureki, O., 2018. Engineered CRISPR-Cas9 nuclease with expanded targeting space. Science 361 (6408), 1259–1262. https://doi.org/10.1126/science.aas9129.

Oberlin, S., McManus, M.T., 2025. Decoding gene regulation with CRISPR perturbations. Nat. Biotechnol. 43, 304–305. https://doi.org/10.1038/s41587-024-02222-2.

Orosco, C., Huang, B., Rananaware, S.R., Bodin, A.P., Browning, I., Fang, A., Jain, P.K., 2026. DNA-guided CRISPR–Cas12 for cellular RNA targeting. Nat. Biotechnol. 1–12.

Pallaseni, A., Peets, E.M., Girling, G., Crepaldi, L., Kuzmin, I., Moor, M., Muñoz-Subirana, N., Schimmel, J., Serçin, Ö., Mardin, B.R., Tijsterman, M., Peterson, H., Kosicki, M., Parts, L., 2024. The interplay of DNA repair context with target sequence predictably biases Cas9-generated mutations. Nat. Commun. 15 (1). https://doi.org/10.1038/s41467-024-54566-7.

Pandey, S., Choudhari, J.K., Tripathi, A., Singh, A., Antony, A., Chouhan, U., 2025. Artificial intelligence-based genome editing in CRISPR/Cas9. Methods Mol. Biol. 2952, 273–282. https://doi.org/10.1007/978-1-0716-4690-8_16.

Piatek, A.A., Lenaghan, S.C., Stewart, C.N., 2018. Advanced editing of the nuclear and plastid genomes in plants. Plant Sci. 273, 42–45. https://doi.org/10.1016/j.plantsci.2018.02.025.

Rauf, M.A., Rao, A., Sivasoorian, S.S., Iyer, A.K., 2025. Nanotechnology-based delivery of CRISPR/Cas9 for cancer treatment: a comprehensive review. Cells 14 (15). https://doi.org/10.3390/cells14151136.

Ren, C., Gathunga, E.K., Li, X., Li, H., Kong, J., Dai, Z., Liang, Z., 2024. Correction: efficient genome editing in grapevine using CRISPR/LbCas12a system. Molecular Horticulture 4 (1). https://doi.org/10.1186/s43897-024-00133-z.

Reyon, D., Tsai, S.Q., Khgayter, C., Foden, J.A., Sander, J.D., Joung, J.K., 2012. FLASH assembly of TALENs for high-throughput genome editing. Nat. Biotechnol. 30 (5), 460–465. https://doi.org/10.1038/nbt.2170.

Ruffolo, J.A., Nayfach, S., Gallagher, J., Bhatnagar, A., Beazer, J., Hussain, R., Russ, J., Yip, J., Hill, E., Pacesa, M., Meeske, A.J., Cameron, P., Madani, A., 2025. Design of highly functional genome editors by modelling CRISPR–Cas sequences. Nature 645 (8080), 518–525. https://doi.org/10.1038/s41586-025-09298-z.

Russell, S., Norvig, P., 2020. Artificial intelligence. A Modern Approach, 4th ed. In: Part of: Pearson Series in Artifical Intelligence (1 books). Pearson.

Shen, M.W., Arbab, M., Hsu, J.Y., Worstell, D., Culbertson, S.J., Krabbe, O., Cassa, C.A., Liu, D.R., Gifford, D.K., Sherwood, R.I., 2018. Predictable and precise template-free CRISPR editing of pathogenic variants. Nature 563 (7733), 646–651. https://doi.org/10.1038/s41586-018-0686-x.

Silverstein, R.A., Kim, N., Kroell, A.S., Walton, R.T., Delano, J., Butcher, R.M., Pacesa, M., Smith, B.K., Christie, K.A., Ha, L.L., Meis, R.J., Clark, A.B., Spinner, A.D., Lazzarotto, C.R., Li, Y., Matsubara, A., Urbina, E.O., Dahl, G.A., Correia, B.E., Marks, D.S., Tsai, S.Q., Pinello, L., De Ravin, S.S., Liu, Q., Kleinstiver, B.P., 2025. Custom CRISPR–Cas9 PAM variants via scalable engineering and machine learning. Nature 643 (8071), 539–550. https://doi.org/10.1038/s41586-025-09021-y.

Sioson, V.A., Kim, M., Joo, J., August 2021. Challenges in delivery systems for CRISPR-based genome editing and opportunities of nanomedicine. Biomed. Eng. Lett. 11 (3), 217–233. https://doi.org/10.1007/s13534-021-00199-4.

Sternberg, S.H., Lafrance, B., Kaplan, M., Doudna, J.A., 2015. Conformational control of DNA target cleavage by CRISPR-Cas9. Nature 527 (7576), 110–113. https://doi.org/10.1038/nature15544.

Theodoris, C.V., Xiao, L., Chopra, A., Chaffin, M.D., Al Sayed, Z.R., Hill, M.C., Mantineo, H., Brydon, E.M., Zeng, Z., Liu, X.S., Ellinor, P.T., 2023. Transfer learning enables predictions in network biology. Nature 618 (7965), 616–624. https://doi.org/10.1038/s41586-023-06139-9.

Truong, V.A., Hsu, M.N., Kieu Nguyen, N.T., Lin, M.W., Shen, C.C., Lin, C.Y., Hu, Y.C., 2019. CRISPRai for simultaneous gene activation and inhibition to promote stem cell chondrogenesis and calvarial bone regeneration. Nucleic Acids Res. 47, e74. https://doi.org/10.1093/nar/gkz267.

Tsai, H.K., Huang, G.T.W., Chou, M.Y., Lu, H.H.S., Li, W.H., 2006. Method for identifying transcription factor binding sites in yeast. Bioinformatics 22 (14), 1675–1681. https://doi.org/10.1093/bioinformatics/btl160.

Tsai, S.Q., Zheng, Z., Nguyen, N.T., Liebers, M., Topkar, V.V., Thapar, V., Wyvekens, N., Khayter, C., Iafrate, A.J., Le, L.P., Aryee, M.J., Joung, J.K., 2015. GUIDE-seq enables genome-wide profiling of off-target cleavage by CRISPR-Cas nucleases. Nat. Biotechnol. 33 (2), 187–198. https://doi.org/10.1038/nbt.3117.

Tycko, J., Myer, V.E., Hsu, P.D., August 4, 2016. Methods for optimizing CRISPR-Cas9 genome editing specificity. Mol. Cell 63 (3), 355–370. https://doi.org/10.1016/j.molcel.2016.07.004.

Uddin, F., Rudin, C.M., Sen, T., 2020. CRISPR gene therapy: applications, limitations, and implications for the future. Front. Oncol. 10, 1387. https://doi.org/10.3389/fonc.2020.01387.

van Overbeek, M., Capurso, D., Carter, M.M., Thompson, M.S., Frias, E., Russ, C., Reece-Hoyes, J.S., Nye, C., Gradia, S., Vidal, B., Zheng, J., Hoffman, G.R., Fuller, C.K., May, A.P., 2016. DNA repair profiling reveals nonrandom outcomes at Cas9-mediated breaks. Mol. Cell 63 (4), 633–646. https://doi.org/10.1016/j.molcel.2016.06.037.

Venkataraman, S., Zaruk, D., Hefferon, K., 2025. Recent developments, challenges and opportunities in genome editing for crop science from a societal perspective. Front Genome 7, 1568072. https://doi.org/10.3389/fgeed.2025.1568072.

Wan, L., Wang, Z., Tang, M., Hong, D., Sun, Y., Ren, J., Zhang, N., Zeng, H., 2021. CRISPR-Cas9 gene editing for fruit and vegetable crops: strategies and prospects. Horticulturae 7 (7), 193. https://doi.org/10.3390/horticulturae7070193.

Wang, Z., Bart, R.S., 2025. Using targeted genome methylation for crop improvement. J. Exp. Bot. 76 (9), 2394–2404. https://doi.org/10.1093/jxb/eraf131.

Wang, Y., Cheng, X., Shan, Q., Zhang, Y., Liu, J., Gao, C., Qiu, J.L., 2014. Simultaneous editing of three homoeoalleles in hexaploid bread wheat confers heritable resistance to powdery mildew. Nat. Biotechnol. 32 (9), 947–951. https://doi.org/10.1038/nbt.2969.

Wang, G., Liu, X., Wang, A., Wen, J., Kim, P., Song, Q., Liu, X., Zhou, X., 2025. CRISPRoffT: comprehensive database of CRISPR/Cas off-targets. Nucleic Acids Res. 53 (D1), D914. https://doi.org/10.1093/nar/gkae1025.

Wang, W., Tian, B., Pan, Q., Chen, Y., He, F., Bai, G., Akhunova, A., Trick, H.N., Akhunov, E., 2021. Expanding the range of editable targets in the wheat genome using the variants of the Cas12a and Cas9 nucleases. Plant Biotechnol. J. 19 (12), 2428–2441. https://doi.org/10.1111/pbi.13669.

Wessels, H.H., Mendez-Mancilla, A., Guo, X., Legut, M., Daniloski, Z., Sanjana, N.E., 2020. Massively parallel Cas13 screens reveal principles for guide RNA design. Nat. Biotechnol. 38, 722–727. https://doi.org/10.1038/s41587-020-0456-9.

Wittmann, B.J., Alexanian, T., Bartling, C., Beal, J., Clore, A., Diggans, J., Flyangolts, K., Gemler, B.T., Mitchell, T., Murphy, S.T., Wheeler, N.E., Horvitz, E., 2024. Toward AI-resilient screening of nucleic acid synthesis orders: process, results, and recommendations. bioRxiv. https://doi.org/10.1101/2024.12.02.626439.

Wójcik-Gront, E., Zieniuk, B., Pawełkowicz, M., 2024. Harnessing AI-powered genomic research for sustainable crop improvement. Agriculture 14 (12). https://doi.org/10.3390/agriculture14122299.

Wright, D.A., Li, T., Yang, B., Spalding, M.H., 2014. TALEN-mediated genome editing: prospects and perspectives. Biochem. J. 462 (1), 15–24. https://doi.org/10.1042/BJ20140295.

Xiang, X., Corsi, G.I., Anthon, C., Qu, K., Pan, X., Liang, X., Han, P., Dong, Z., Liu, L., Zhong, J., et al., 2021. Enhancing CRISPR-Cas9 gRNA efficiency prediction by data integration and deep learning. Nat. Commun. 12, 3238. https://doi.org/10.1038/s41467-021-23576-0.

Xie, X., Ma, X., Zhu, Q., Zeng, D., Li, G., Liu, Y.G., 2017. CRISPR-GE: a convenient software toolkit for CRISPR-based genome editing. Mol. Plant 10 (9), 1246–1249. https://doi.org/10.1016/j.molp.2017.06.004.

Xu, Z., Kuang, Y., Ren, B., Yan, D., Yan, F., Spetz, C., Sun, W., Wang, G., Zhou, X., Zhou, H., 2021. SpRY greatly expands the genome editing scope in rice with highly flexible PAM recognition. Genome Biol. 22 (1). https://doi.org/10.1186/s13059-020-02231-9.

Yang, F., Wang, W., Wang, F., Fang, Y., Tang, D., Huang, J., Lu, H., Yao, J., 2022. scBERT as a large-scale pretrained deep language model for cell type annotation of single-cell RNA-seq data. Nat. Mach. Intell. 4 (10), 852–866. https://doi.org/10.1038/s42256-022-00534-z.

Zaman, W., Khalil, A.A.K., Amin, A., 2025. TALEN-interceded genome editing in plants: unveiling new frontiers in secondary metabolite improvement and genetic diversity. Plants 14 (19). https://doi.org/10.3390/plants14193024.

Zhao, H., Sheng, Y., Zhang, T., Zhou, S., Zhu, Y., Qian, F., Liu, M., Xu, W., Zhang, D., Hu, J., 2024. The CRISPR-Cas13a Gemini System for noncontiguous target RNA activation. Nat. Commun. 15, 2901. https://doi.org/10.1038/s41467-024-47281-w.

Zhang, T.Q., Chen, Y., Liu, Y., Lin, W.H., Wang, J.W., 2021. Single-cell transcriptome atlas and chromatin accessibility landscape reveal differentiation trajectories in the rice root. Nat. Commun. 12 (1). https://doi.org/10.1038/s41467-021-22352-4.

Zhang, G., Xie, H., Dai, X., 2024. DeepIndel: an interpretable deep learning approach for predicting CRISPR/Cas9-Mediated editing outcomes. Int. J. Mol. Sci. 25 (20). https://doi.org/10.3390/ijms252010928.

Zhang, G., Dai, Z., Dai, X., 2020. C-RNNCrispr: Prediction of CRISPR/Cas9 sgRNA activity using convolutional and recurrent neural networks. Comput. Struct. Biotechnol. J. 12 (18), 344–354. https://doi.org/10.1016/j.csbj.2020.01.013 eCollection 2020.

Zhang, T., Wang, Z., Liu, Q., Zhao, D., 2025. Genetic improvement of rice grain size using the CRISPR/Cas9 system. Rice 18 (1). https://doi.org/10.1186/s12284-025-00758-8.

Zhu, H., Liang, C., 2019. CRISPR-DT: designing gRNAs for the CRISPR-Cpf1 system with improved target efficiency and specificity. Bioinformatics 35, 2783–2789. https://doi.org/10.1093/bioinformatics/bty1061.

Zhu, Y.H., Liu, Z., Ji, Z., Yu, D.J., 2024. ULDNA: integrating unsupervised multi-source language models with LSTM-attention network for high-accuracy protein-DNA binding site prediction. Brief Bioinform. 25 (2), bbae040. https://doi.org/10.1093/bib/bbae040.

CHAPTER

26

Future landscape of CRISPR and AI in plant gene editing

Oseias R. Feitosa-Junior

Department of Plant Pathology, Microbiology and Biotechnology, North Dakota State University, Fargo, ND, United States

26.1 Introduction

Gene editing has become one of the primary drivers of modern plant science, enabling precise and targeted changes to crop genomes that traditional breeding cannot achieve with comparable speed or resolution. By enabling clean knockouts, allele replacements, and targeted insertions, the CRISPR[1] technology accelerates the development of traits critical for resilient agriculture—disease resistance, abiotic stress tolerance, improved nutritional profiles, and even the *de novo* domestication of wild relatives. After a decade of widespread implementation, it is also clear that CRISPR's promise is limited not by its original concept but also by a set of species- and process-specific bottlenecks: transformation and regeneration, delivery routes, multiplex allele editing, and precise gene targeting. These challenges set the terms for what the next wave of research must address in practice (Kumar et al., 2021; Cardi et al., 2023; Jung et al., 2024).

Two parallel waves of progress drive recent enthusiasm in the field. The first is molecular: reagent engineering and plant-tuned expression strategies have steadily improved on-target performance *in planta*. Engineered nucleases—from PAM[2]-flexible systems to enzymes adapted for low-temperature crops—now broaden editable loci and maximize targeting success. Alongside these, intron-containing, codon-optimized constructs and single-transcript guide RNA (sgRNA) architectures (tRNA- and ribozyme-based) enhance guide maturation and stability. The toolkit has also expanded into organelle genome editing, transgene-free workflows that avoid stable DNA integration, Virus-Induced Gene Editing (VIGE)to bypasses tissue culture, and genotype-specific strategies for hard-to-edit commercial lines.

[1] CRISPR: Clustered Regularly Interspaced Short Palindromic Repeats

[2] PAM: Protospacer Adjacent Motif

Intelligent Gene Editing in Plants
https://doi.org/10.1016/B978-0-443-44079-3.00007-3

Pioneering work across the field illustrates how aligning enzyme properties, construct architecture, and delivery routes with plant cellular biology produces substantial agronomic and technical gains (Schindele and Puchta, 2020; Son and Park, 2022; Schindele et al., 2023; Puchta and Houben, 2024).

The second wave is computational: machine learning (ML) and related artificial intelligence (AI) tools now substantially diminish the experimental search space. Deep learning models trained on high-throughput screens (including `DeepSpCas9`, `DeepCRISPR`, and their successors) predict guide RNA (gRNA) activity and off-target tendency significantly more accurately than heuristic rules, allowing for the prioritization of guides and the avoidance of likely dead ends. Network-based ML approaches can select candidate genes and transcription factors for complex traits, such as drought tolerance or disease resistance, thereby focusing wet-lab efforts on the most promising loci. Together, these tools reduce wasted cycles and increase the throughput at which conceptual targets lead to phenotyped lines validation (Chuai et al., 2018; Kim et al., 2019; Konstantakos et al., 2022).

Synergy, however, is not a replacement. Challenges, such as polyploid genomes, the plant cell wall, genotype-dependent tissue culture responses, and the cellular repair machinery's bias toward nonhomologous end joining (NHEJ), create limitations that neither enzyme engineering nor computation can fully eliminate on their own. Crops such as wheat, potato, and many brassicas often require near-simultaneous modification of multiple homologous alleles to produce stable, consistently inherited traits; while multiplex CRISPR makes this feasible in principle, it demands design, delivery, and genotyping complexity (Li et al., 2023; Schaart, van de Wiel and Smulders, 2021; Cardi et al., 2023).

AI's limitations are equally contextual. Modern systems are narrow: they perform consistently on well-defined tasks such as image classification, sequence-level prediction, and experimental scheduling; however, they lack the integrative, contextual judgment and normative reasoning that humans provide. The distinction between narrow AI and artificial general intelligence (AGI) matters for planning: narrow AI delivers immediate, measurable gains in guide design, off-target filtering, high-throughput phenotyping, and experiment optimization, whereas AGI—a speculative capability to reason across domains and contexts—would, if ever realized, permit autonomously creative science and context-sensitive decision. Consequentially, human expertise remains essential for hypothesis formation, ecological risk assessment, and ethical judgment, even though it is evident how narrow AI accelerates routine tasks (Nishant et al., 2024; Goertzel, 2014; Hao et al., 2024; Lennox, 2024).

Ethical perspectives critically refine these technical advancements. Observers caution that algorithmic competence must not obscure the socio-political implications of data ownership and model control. For plant gene editing, this means that model recommendations require transparent attribution and uncertainty estimates; training datasets for phenotyping or guide efficacy must be curated to ensure crop and environment representativeness; and decision placement should be within a pluralistic governance framework that considers socioeconomic, ecological, and regulatory contexts. Utimately, AI must be treated as an accelerant that still requires careful human stewardship (Macnaghten and Habets, 2020; Harfouche et al., 2024; Lennox, 2024).

AI–CRISPR synergy unfolds across the full editing pipeline. In pre-edit stages, ML and network models narrow candidate gene lists and suggest high-quality guides. During execution, computational tools inform construct architecture, plan experiments, and, when coupled with laboratory automation, reduce human error and increase throughput. In post-edit, computer projection and sensor networks enable high-throughput phenotyping, while integrative ML models map genotype to phenotype relationships, predict pleiotropy, and prioritize lines for field trials. At every stage, models generate hypotheses and priorities, laboratories test them, and careful phenotyping and multi-site field evaluation confirm agronomic utility and biosafety (Kim et al., 2019; Konstantakos et al., 2022).

This chapter describes how narrow AI can enhance CRISPR plant editing, highlighting advances in molecular and delivery techniques that make edits more practical in plant cells. It emphasizes the technical, ecological, and social limitations that still require human judgment, replication, and governance. It is shown how predictive algorithms, laboratory automation, and improved reagents form an integrated pipeline that moves edits to the field, and it is identified where research and policy must concentrate to translate technical possibilities into durable agricultural benefits (Son and Park, 2022; Cardi et al., 2023; Puchta and Houben, 2024). In the next sections, we examine gRNA and construct design (Section 26.2), ML-based outcome prediction (Section 26.3), AI-driven process optimization and automation (Section 26.4), ethical and regulatory frameworks (Section 26.5), and emerging technical and infrastructural trends (Section 26.6), culminating in a short synthesis of research, capacity-building, and governance priorities (Section 26.7).

26.2 AI and machine learning for CRISPR gRNA design in plants

The success of CRISPR-based editing in plants depends on selecting a gRNA that will cut the chosen locus efficiently in each crop, tissue, and condition, while avoiding undesirable off-target effects. Over the past decade, artificial intelligence and ML approaches have been transforming this process from largely exploratory prediction to AI-assisted candidate selection, greatly improving the likelihood that a chosen guide will perform well and reduce wasted laboratory effort. In plants, the risks are particularly high: many target traits require editing multiple alleles in polyploid genomes, working with cultivars that are difficult to transform, where efficiency and predictability are critical for both research and development costs and subsequent regulatory or field testing. Following, it will be shown how ML is applied to the design of gRNAs, and increasingly, entire editing constructs, for plants. It also discusses leading algorithmic approaches and available tools, examining evidence from crop studies, and addressing key limitations that still challenge plant experts.

26.2.1 What the models learn and relevance for plants

In principle, gRNA-efficacy models convert features of a candidate target into a numerical score (or probability distribution) that estimates on-target activity and, optionally, off-target risk. Features range from simple sequence descriptors (nucleotide identity at specific positions, GC content,[3] and predicted secondary structure) to biophysical terms (predicted gRNA–DNA hybrid free energy), as well as genomic context (chromatin accessibility, DNA methylation, and local nucleosome positioning), and organism-specific sequence variants. Models trained on large, consistent experimental datasets learn which combinations of features correlate with high editing frequency and can generalize to new targets when training data are sufficient and representative (Chuai et al., 2018; Kim et al., 2019; Xiang et al., 2021; Sun et al., 2024; Yuan et al., 2025).

For plant researchers, the implication is that models trained primarily on mammalian cell lines will capture many universal determinants of gRNA performance (such as PAM-proximal nucleotide biases or gRNA folding penalties) but may miss plant-specific influences. These include the transcriptional state of the target tissue, the performance of nucleases adapted for plant cellular conditions (as low-temperature-active or PAM-expanded systems), cultivar-specific sequence polymorphisms, and RNA-processing effects unique to plant single-transcript gRNA cassettes. Other factors — such as organelle genome accessibility, genotype-dependent transformation efficiency, or VIGE — can also shape editing outcomes. Consequently, the most effective pipelines combine generalist predictors with plant-tailored datasets and empirical validation (Malik et al., 2021; Son and Park, 2022; Abbasi et al., 2025).

26.2.2 Main algorithmic families and representative tools

Three broad classes of methods dominate gRNA prediction tooling: (i) *feature-based* or *classical ML methods* that encode hand-crafted sequence and genomic features and feed them to algorithms such as random forests, gradient-boosted trees, or logistic regression, which are robust in low-data settings, interpretable, and fast to run. Early scoring algorithms and many ensemble tools are in this category, and they remain useful when users want transparent feature contributions and rapid screening (Haeussler et al., 2016); (ii) *Deep learning approaches*, including Convolutional Neural Network (CNN), Recurrent Neural Network (RNN), or hybrid architectures, learn predictive features directly from raw sequences and auxiliary inputs. Models such as `DeepCRISPR` (Chuai et al., 2018) and `DeepSpCas9` (Kim et al., 2019) were important landmarks, demonstrating superior performance over earlier heuristics when large training sets were available. `CRISPRon` (Xiang et al., 2021) exemplifies a modern deep model trained on tens of thousands of gRNAs and provides both a web server and an API. (iii) *Hybrid* and *ensemble approaches* combine models, for example, a thermodynamics-informed feature model plus a CNN, to improve robustness. Ensembles also allow pragmatic trade-offs between sensitivity and interpretability. Newer multi-view networks, such as `CRISPR-M`, for off-target prediction (Sun et al., 2024), explicitly model mismatches and indels at off-target sites. Table 26.1 shows representative and widely used tools.

[3] GC content: the percentage of guanine (G) and cytosine (C) bases in a DNA or RNA sequence

TABLE 26.1 Summary of representative gDNA designing tools.

Tool	Method/ Architecture	Main inputs	Plant-readiness/notes	Reference
CRISPOR	Ensemble of scoring algorithms	Sequence + genome	Widely used, supports many genomes, transparent	Haeussler et al. (2016)
DeepCRISPR	Deep CNN + hybrid	Sequence, Off-target scan	Generalist; retrainable for plants	Chuai et al. (2018)
DeepSpCas9	Deep CNN	Sequence	High generalization; human cell data, but useful	Kim et al. (2019)
CRISPRon	Deep learning, large training set	Sequence + surrogate assay data	State-of-the-art on-target predictor; webserver	Xiang et al. (2021)
CRISPR-M	Multi-view deep network	Sequence alignments, mismatches	Improved off-target modeling, including indels	Sun et al. (2024)

26.2.3 Evidence from plants: what works, what is still provisional

A recurring pattern in the literature is that machine-learning models trained on large and high-quality datasets tend to generalize well within the data domain but often lose accuracy when applied to different organisms, tissues, or experimental protocols. Several practical points, supported by plant-focused studies and reviews, are worth emphasizing. Chromatin state and gene expression matter in plants as much as in animals: comparative work between endogenous editing rates and surrogate assays shows that local chromatin accessibility strongly influences cutting efficiency (Jensen et al., 2017; Xiang et al., 2021). In plants, where many important targets are regulated developmentally or in a tissue-specific manner, incorporating plant epigenomic maps such as ATAC-seq, DNase-seq, and histone mark profiles can improve guide ranking for active promoters or expressed genes (Mattiello et al., 2022; Candela-Ferre et al., 2024). Equally important are reagent and delivery choices: nuclease variants and temperature sensitivity constrain performance *in planta*, and temperature-tolerant Cas12a constructs (ttLbCas12a, intron-containing designs) demonstrably increase editing efficiency in *Arabidopsis thaliana* and cereals grown at lower temperatures (Schindele and Puchta, 2020; Schindele et al., 2023; Lawrenson et al., 2024). Recent reviews furthermore stress four complementary technical directions that affect predictability and applicability in crops: organellar genome editing, which requires different editors and delivery strategies; transgene-free workflows, VIGE-mediated delivery, and the optimization of protocols for recalcitrant elite inbred lines (Son and Park, 2022). Polyploidy and allelic variation add another layer of complexity in crops: a guide scoring highly for one homeolog may mismatch others or cut with unequal efficiencies, so effective pipelines must evaluate guides against all relevant allele sequences and cultivar polymorphisms—successful reports of mutating all six wheat alleles underscore both feasibility and design complexity (Wang et al., 2014; Zhu et al., 2020; Li et al., 2022). Finally, the scarcity of plant-specific labeled data remains limiting; transfer-learning strategies that pretrain on large mammalian datasets and fine-tune on smaller plant sets have improved performance *in planta*, but ultimately, practical field relevance requires coupling molecular predictions to agronomic phenotypes and network-based gene prioritization so that high predicted indel rates translate into durable trait gains, as will be further discussed in Section 26.3 (Shi et al., 2017; Konstantakos et al., 2022; Abbasi et al., 2025).

26.2.4 Practical workflow for a plant lab

A rational workflow for designing guides in plant projects typically begins with assembling the exact target sequences for the cultivar(s) of interest—including promoter and splice variants—and using that cultivar sequence as the reference for all off-target searches. The process usually proceeds with one or more established predictors (for example, `CRISPRon`, `DeepCRISPR`, `CRISPOR`), ideally, genome-wide off-target scans should be restricted to the cultivar assembly (Haeussler et al., 2016; Chuai et al., 2018; Xiang et al., 2021). Where available, chromatin or accessibility maps for the tissue of interest (ATAC-seq, DNase-seq, histone marks) can be overlaid, since local accessibility strongly modulates cutting efficiency and may reprioritize guides that otherwise score similarly on sequence alone (Jensen et al., 2017; Xiang et al., 2021). Guide selection is generally most robust when candidates perform consistently across multiple models, are conserved across homeologs for intended multiplexing, and avoid regulatory or splice motifs unless these are the intended targets. Experimental validation is commonly structured in tiers, beginning with rapid surrogate or biochemical assays (such as GUIDE-seq or CIRCLE-seq for specificity) and progressing to a limited number of stable transformants before large-scale field trials (Tsai et al., 2015; Cameron et al., 2017).

AI tools can accelerate prediction, chromatin integration, and ranking, but accurate cultivar sequence and annotation remain a human-led bottleneck. For this reason, empirical validation under crop growth conditions remains essential, since models trained on other genotypes or surrogate systems often show reduced performance when applied to different cultivars or tissues (Chuai et al., 2018; Konstantakos et al., 2022). Fig. 26.1 illustrates the recommended end-to-end workflow (from *cultivar input* to *Design–Build–Test–Learn* or DBTL and *human checkpoints*) that labs can adopt (*outcome modeling* is expanded in Section 26.3).

26.2.5 Limitations and failure modes

Models for gRNA design are powerful but still face structural limitations in real-world applications, as they often fail to account for complex biological variables. A primary failure mode is training-data bias and domain shift: predictors trained on mammalian or synthetic libraries may misestimate activity when plant chromatin, repair pathways, or growth temperature differ from the training domain (Eraslan et al., 2019). Allelic polymorphism and pangenome mismatches create false negatives in off-target lists if searches are run against a single reference rather than cultivar-aware assemblies. Equally important is the phenotype gap: high predicted cutting or indel rates do not guarantee durable, agronomically relevant trait gains because of pleiotropy and network compensation (Grinberg et al., 2020). Practical

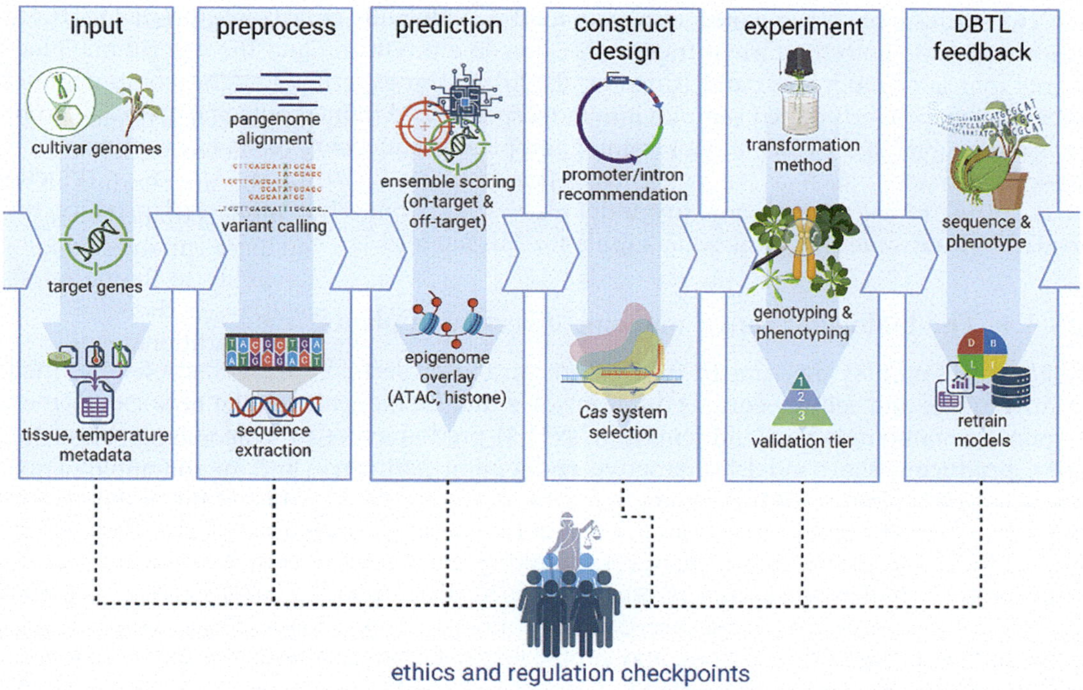

FIGURE 26.1 **AI-guided gRNA design pipeline for plants.** *Input*: Cultivar genome, gene targets, and tissue and temperature metadata. *Preprocessing*: Pangenome alignment and sequence extraction (±100 bp context). *Prediction*: ensemble scoring (on-target, off-target) with epigenome overlays. *Construct Design*: Promoter and intron recommendation, multiplex array, and *Cas* variant selection. *Experimental Plan*: Transformation, genotyping and phenotyping, validation tier, and *DBTL feedback:* Sequence and phenotype, retrain models. Human checkpoints for ethics and regulation are shown.

pipelines therefore require cultivar-specific sequence data, modest local validation sets, and integration with outcome-prediction and trait-prioritization modules (see 26.3 and 26.7).

26.3 Predicting gene editing outcomes with machine learning

The potential of using ML to predict CRISPR repair outcomes lies in an apparently simple empirical fact: the distribution of small insertions and deletions (indels) produced after a Cas-induced double-strand break is not random—it is strongly shaped by the local DNA sequence and by conserved mechanistic features of end processing—and therefore predictable. Large-scale profiling in mammalian systems demonstrated that, across hundreds to thousands of targets, the mix of deletions versus insertions, the favored deletion lengths (often using short microhomologies), and the surprisingly consistent occurrence of templated single-base insertions can be learned from sequence context alone. Work such as the one by Shen and colleagues (2018) (`inDelphi`) and the large `FORECasT` profiling effort (Allen et al., 2019) established that supervised models trained on sequence windows around the cut site can predict detailed outcome spectra with with remarkable precision. These studies provide both the formal framing for outcome prediction and the algorithmic blueprints that later models expand (Chen et al., 2019). The experimental and computational pipeline that converts edited samples into indel spectra and trained outcome models is summarized in. Fig. 26.2 Before model training, amplicon sequencing reads are routinely processed with tools such as `CrispRVariants` (Lindsay et al., 2016) or `CRISPResso` (Pinello et al., 2016) to call alleles, quantify indel classes and frameshift rates, and produce the locus-level outcome spectra used as inputs for ML models.

26.3.1 The biology of repair outcome diversity in plants

Plant cells employ the same broad DNA-repair toolset seen in other eukaryotes, but their relative usage and cellular context differ in ways that fundamentally alter repair outcomes. Canonical nonhomologous end joining (C-NHEJ) predominates in most somatic plant tissues, producing short indels; alternative end-joining pathways such as microhomology-mediated end joining (MMEJ) or single-strand annealing (SSA) generate deletion patterns that often exploit nearby short homologies; and homology-directed repair (HDR) is generally inefficient in differentiated plant cells, limiting the desired precise templated repair in many transformation workflows (Wada et al., 2020; Van Vu et al., 2021). Studies in plants show the same basic templated-insertion and microhomology biases documented in animals, but with organism-, tissue-, and protocol-specific modulations: the exact frequency of 1-bp insertions, the length spectrum of deletions, and the prevalence of larger deletions can shift with species, tissue type, regeneration protocol, and growth temperature (Lemos et al., 2018; Chen et al., 2019). For instance, a study using a common bean hairy-root system revealed that 1-bp insertions were primarily templated from the upstream nucleotide. The study also found that while indels were consistent for each guide, they varied among different guides. This resembles observations made in mammalian systems, but also highlights the unique experimental variance in plants (de Koning et al., 2023). These findings suggest that predicting gene editing outcomes in plants can be approached as a supervised learning problem over a defined sequence window, such as ±30 bp to ±60 bp around the cut site. Additionally, the results emphasize the importance of using plant-specific training data and context metadata to improve the accuracy of these predictions (Bortesi et al., 2016).

26.3.2 Early predictive models: inDelphi and FORECasT in plants

`inDelphi` and `FORECasT` represent two early and influential lines of work. These models were trained in large experimental libraries produced by pipelines like the one shown in Fig. 26.2. To model precise outcomes, `inDelphi` combines modules specialized for microhomology-mediated deletions with nearest-neighbor components for non-MMEJ indels. Conversely, `FORECasT` (and related large-scale logistic models) demonstrated that simpler, well-regularized statistical approaches trained on very large synthetic libraries could accurately recapitulate common indel classes (Shen et al., 2018; Allen et al., 2019). Those tools were trained on mammalian datasets, but when applied to plant assays, many of their high-level predictions prove informative. Plant validation studies (for example, hairy-root

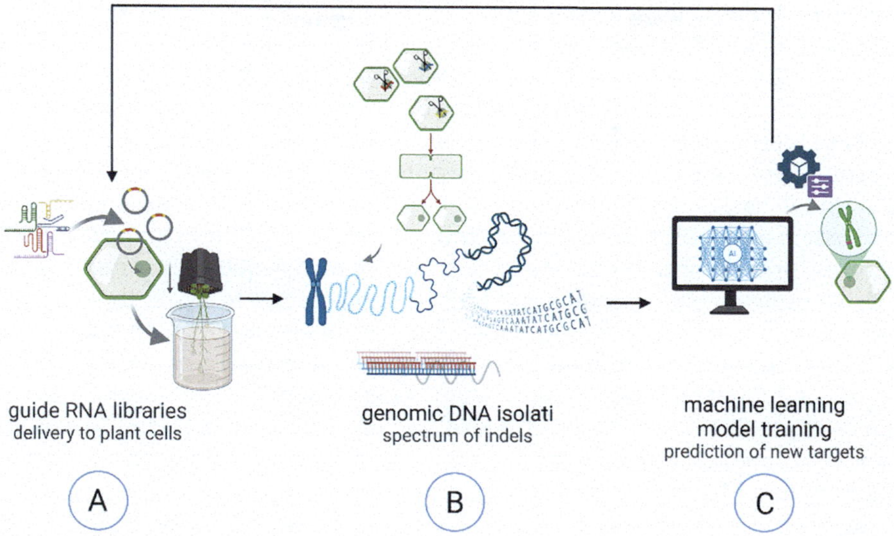

FIGURE 26.2 Workflow for generating and modeling CRISPR editing outcomes. (A) Deliver a gRNA library to plant cells (transformation, bombardment, VIGE, etc.). (B) Allow editing and growth, extract DNA, sequence target loci, and quantify indel spectra (tools include CrispRVariants and CRISPResso). (C) Train ML models on sequence windows plus metadata (tissue, nuclease, temperature) to predict outcome probabilities for new targets. Originally demonstrated in mammalian systems, this pipeline is increasingly adapted to plant assays (Bortesi et al., 2016; Chen et al., 2019; de Koning et al., 2023).

or protoplast screens) show that these models frequently capture framing-indel probabilities and dominant indel categories, making them useful first-pass predictors for plant projects (Chen et al., 2019; de Koning et al., 2023). Nevertheless, as highlighted in the previous section, systematic deviations are common: plant outcomes shift subtly with nuclease variant, growth temperature, tissue regeneration dynamics, and cultivar sequence differences, which limit raw cross-domain generalization without fine-tuning (Schindele and Puchta, 2020; Son and Park, 2022; Schindele et al., 2023; Lawrenson et al., 2024).

26.3.3 Modern deep-learning and attention models

The last several years have seen rapid progression from simple statistical predictors to expressive deep-learning architectures. Logistic-regression-style frameworks, such as `Lindel`, extended the number of modeled indel classes and encoded many binary sequence features. Deep-learning approaches—including convolutional hybrids and attention-based recurrent networks — learned distributed sequence representations and used attention weights to highlight the bases most informative for a given prediction (Li et al., 2022). `Apindel` is a representative attention-based recurrent model. It embeds DNA sequences using pre-trained k-mer vectors (GloVe embeddings) and processes them with Bi-LSTM (Bidirectional Long Short-Term Memory) plus a self-attention layer. Attention maps from `Apindel` frequently recover biologically meaningful signals — for example, highlighting the upstream base that templates 1-bp insertions — reassuring domain experts that the model is capturing true mechanistic biases rather than overfitting noise (Li et al., 2022). Applied to pooled datasets, Apindel and related models show strong performance on human benchmarks and, where tested, perform reasonably well with plant assay outcomes when supplemented with modest plant datasets (Xiang et al., 2021; Konstantakos et al., 2022).

These architectures offer two immediate advantages for plant research: (1) they can model hundreds of indel classes, not just summary statistics, increasing the granularity of predictions that breeders and molecular biologists might need; and (2) they can prioritize based on output features, providing interpretable clues about which sequence positions or microhomologies drive undesirable outcomes, guiding design changes to avoid them.

Other modern approaches leverage ensemble and deep-learning methods. `SPROUT` is a gradient-boosted trees model trained on ~1600 human T-cell targets. While not designed for plant-specific data, its tree-based approach is flexible and has been applied to other systems (Leenay et al., 2019). `CROTON` used neural architecture search (NAS) to automatically design a convolutional neural network for the task, achieving state-of-the-art performance with minimal manual tuning (Li et al., 2021; Abbasi et al., 2025).

Taken together, these advances demonstrate the power of modern ML methods in capturing the complexity of DNA repair. Models now routinely incorporate hundreds of indel classes and thousands of features, and deep networks (RNNs, CNNs, Transformers) can learn from large pooled datasets. The use of attention or tree ensembles often provides some interpretability. For plants, the most important lesson is that tools like `Lindel` or `Apindel`—even if trained on animal data—can already guide expectations. For example, using `Lindel/inDelphi` on bean data correctly anticipated that most 1-bp insertions would match the upstream base and that most deletions used short microhomologies, suggesting some conservation of repair bias across species (de Koning et al., 2023).

26.3.4 Transformers, hybrid networks, and emergent generalization

Transformer-style models and *Bidirectional Encoder Representations from Transformers* (or BERT)-derived approaches (as `DeepIndel`) have recently been adapted to indel prediction. Their self-attention layers natively capture both local sequence motifs and longer-range dependencies (for example, small repeated tracts that predispose to larger deletions), and they exhibit robust generalization on held-out human datasets (Zhang et al., 2024). Hybrid models that combine convolutional motif detectors with recurrent or transformer modules (for on-target efficiency and outcome prediction) use transfer learning to pretrain on large, heterogeneous datasets and then fine-tune on the smaller, context-specific plant datasets that are becoming available. These architectures are well-suited to the multi-task problems that plant editing presents (predicting efficiency, off-target risk, and full indel spectra simultaneously), but their successful application in crops depends on assembling curated plant training sets spanning tissues, genotypes, and nuclease variants (Luo et al., 2024).

Another hybrid approach focuses on sgRNA activity (on-target efficiency) rather than repair profiles. `CrnnCrispr` combines convolutional and recurrent branches to learn local sequence motifs and positional dependencies, outputting a single activity score. In tests on nine public human cell-line datasets, it outperformed prior tools, especially when fine-tuned on small datasets. The model uses transfer learning—pretraining on large benchmarks and fine-tuning on new data—a strategy that could be adapted for plants by pretraining on synthetic data and refining with a small set of real plant targets. This has not yet been applied to plant editing, although the architecture is generic. In the meantime, cross-application of human-trained models or transfer-learning approaches remains the most practical path until larger plant-specific datasets become available (Zhu et al., 2024).

26.3.5 Context-specific prediction

A central lesson from both human and plant genome editing is that context matters. Models pretrained in very large mammalian or synthetic libraries capture powerful general patterns, but their performance improves markedly when fine-tuned on a modest number of in-domain examples. A major challenge is that a model trained on 1 cell type or species may not generalize to another, such as chromatin state, DNA repair machinery levels, or cell-cycle profiles, which can differ substantially between, for example, human cell lines and maize protoplasts. This has driven the adoption of transfer learning approaches that adapt a pretrained model to a new context. `DeepCRISTL` (Deep Transfer Learning) (Elkayam et al., 2024) illustrates this in the on-target efficiency space: it first learns general gRNA efficiency patterns from large high-throughput datasets such as `DeepHF` and `CRISPRon`, then fully fine-tunes all network parameters on much smaller functional or endogenous datasets specific to a given cell type (Chen et al., 2020; Xiang et al., 2021). Elkayam and colleagues (2024) showed that fine-tuning a CRISPRon-based model on real experimental datasets consistently improved performance over the base model, underscoring the value of context-specific training. Fine-tuning requires dozens to hundreds of plant-specific targets produced by the sequencing and profiling pipeline in Fig. 26.2.

The same strategy applies to plant outcome prediction: a pretrained indel model can be adapted to a given crop or tissue by retraining on a small panel—dozens to low hundreds—of empirically measured targets collected under the transformation and growth conditions used by that lab. Such full-model fine-tuning allows the network to learn plant-specific biases, for example, if certain nucleotides are tolerated in plant NHEJ. To date, most plant CRISPR ML models have been one-size-fits-all, but as more plant data emerge, transfer learning will become essential. Moreover, context extends beyond species or cell type: even within a plant species, genotype or developmental stage can influence editing outcomes. Even within a plant species, genotype or developmental stage can influence editing outcomes. As said in 26.2.6, early studies suggest that highly expressed genes or open chromatin in plant cells tend to be edited more efficiently, indicating that incorporating such features could further refine outcome predictions (Lu et al., 2019; Pei et al., 2021).

26.3.6 Integrative and ensemble approaches

Because no single algorithm dominates all metrics, ensemble and stacking strategies improve robustness: combining logistic or ridge models (good for interpretable baseline patterns), tree ensembles (fast, feature-sensitive), and deep networks (powerful pattern discovery) can mitigate worst-case errors and adapt more gracefully to domain shifts. For plants, a pragmatic pipeline is therefore to apply a consensus of predictors (choosing those that allow cultivar-level off-target scanning and that can accept epigenomic inputs), then fine-tune a selected deep model using a small set of local validation targets, and finally use attention/feature analyses to inspect unusual or high-risk predictions. This approach balances the strengths of mature predictors (`inDelphi`, `FORECasT`, `Lindel`) with the flexibility of modern deep models and the domain specificity required by plant applications (Chen et al., 2019; He et al., 2021; Kick and Washburn, 2023).

26.4 Optimizing the gene editing process

AI and automation are converging to transform plant gene editing from a series of manual, disconnected steps into streamlined, data-driven pipelines. By linking computational design, robotic execution, and real-time data capture, these systems can accelerate every stage—from construct assembly and transformation to phenotyping and validation—while improving reproducibility and scale. Automation platforms now handle high-throughput molecular workflows, tissue culture, and transformation with minimal human intervention. Closed-loop DBTL cycles couple machine learning predictions directly to robotic experimentation, enabling rapid iteration toward desired traits. As these capabilities extend from the lab to smart greenhouses and field phenotyping, the path from concept to validated, deployable plant lines is becoming shorter, more precise, and more traceable than ever before (Farooq et al., 2024; Annese et al., 2025; Zhang et al., 2025).

26.4.1 Automation in plant gene editing workflows

Robotic automation provides the precision, scale, and consistency required to translate computational designs into validated edited plants. Standard liquid-handling platforms and automated workstations routinely process hundreds of samples in parallel across core molecular protocols—including DNA assembly, PCR setup, library preparation, and sequencing, thereby mitigating systemic bottlenecks. Automation now extends to tissue culture and transformation, with automated culture stations and bioreactors performing sterile passaging, media exchanges, and transplanting with minimal human intervention. *Temporary Immersion Bioreactors* (TIBs) such as RITA, SETIS, Ebb-and-Flow, and MATIS, and even other custom systems, have been adapted for large-scale micropropagation; in some crops (banana among them), TIBs produce substantially more shoots per explant than conventional gel culture while reducing labor input. These systems are coupled to inline sensors and pumps to automate nutrient delivery and dissolved-gas control (Alvard et al., 1993; Uma et al., 2021; Bethge et al., 2023).

New platforms now automate whole transformation workflows. Semiautomated *Agrobacterium* protocols enable the automation of cocultivation, washing, and selection in multiwell formats, allowing concurrent testing of dozens of plasmids. Similar automation applies to biolistic systems via programmatic delivery and collection cycles, as well as to protoplast transfection using robotic plating and cell recovery. Robotic imaging and phenotyping are maturing in parallel: gantry-mounted cameras, LIDAR (an acronym of ***LI****ght* ***D****etection* ***And***

Ranging), and hyperspectral or fluorescence sensors can repeatedly scan culture vessels and benches, producing rich time-series data on growth, chlorosis, fluorescence reporters, and morphology. Noninvasive platforms designed for in vitro plantlets permit continuous monitoring of shoot area and canopy metrics without opening vessels, thereby preserving sterility and generating dense phenotyping datasets for ML analysis (Atefi et al., 2021; Bethge et al., 2023; Annese et al., 2025).

After transformation, automation accelerates screening. Robotic pickers and manipulators isolate seedlings or explants flagged by fluorescence or growth signatures; automated DNA extraction and PCR systems prepare genotyping libraries at scale; capillary electrophoresis instruments (like Fragment Analyzer or TapeStation) and automated sequencers rapidly read edited loci. Together, these modules compress the interval from construct design to validated, genotyped plant, enabling many more experimental cycles per unit time and reducing human handling error (Ursache et al., 2021; Zaleski-Cox et al., 2023; Guertler et al., 2023).

Integration and traceability are critical enablers. *Laboratory Information Management Systems* (LIMS) and middleware standards (for example, SiLA) coordinate devices from multiple vendors and preserve sample origin across the pipeline, linking a gRNA design to its construct, transformation batch, imaging time points, and sequencing results. Cloud-connected automation models allow researchers to submit experimental designs even remotely; scripts then schedule and execute work on robotic stations, which return structured data for centralized analysis. This "cloud lab" concept, previously demonstrated in microbial and biochemical biofoundries, is extensible to plant bioengineering when appropriate containment and handling modules are available (Craig et al., 2017; Bromig et al., 2022; Arias and Taylor, 2024; Petzold and Mukhopadhyay, 2025).

26.4.2 Closed-loop AI–automation workflows

In plant gene editing, closed-loop AI–automation workflows couple computational design with automated execution in iterative DBTL cycles (Fig. 26.3), creating a self-optimizing pipeline spanning from initial *in silico* design to verified *in planta* integration. In such a pipeline, an ML model proposes a set of candidate edits (gRNAs, multiplex arrays, or regulatory variants), robots assemble DNA and perform transformation, automated assays collect molecular and phenotypic readouts, and ML ingests results to update predictions and propose the next experimental round. By iterating this loop at scale, the system self-optimizes toward desired molecular or phenotypic endpoints far faster than manual approaches (HamediRad et al., 2019; Liao et al., 2022).

A concrete example is the FAST-PB (Fast Automated Synthetic Biology for Plants and Bioproducts) concept: using a biofoundry, researchers assemble dozens of constructs in parallel, robotically transforming cells (using either protoplasts or calli), and apply high-content chemical profiling (for example, single-cell or bulk MALDI-MS). ML then identifies constructs that produce favorable metabolic changes, and the best candidates are moved rapidly into regeneration and validation. Such integrated pipelines have reported substantial improvements in target metabolites in weeks rather than months, illustrating how automated DBTL can compress R&D (or Research and Development) cycles for metabolic and trait engineering (Dong et al., 2025).

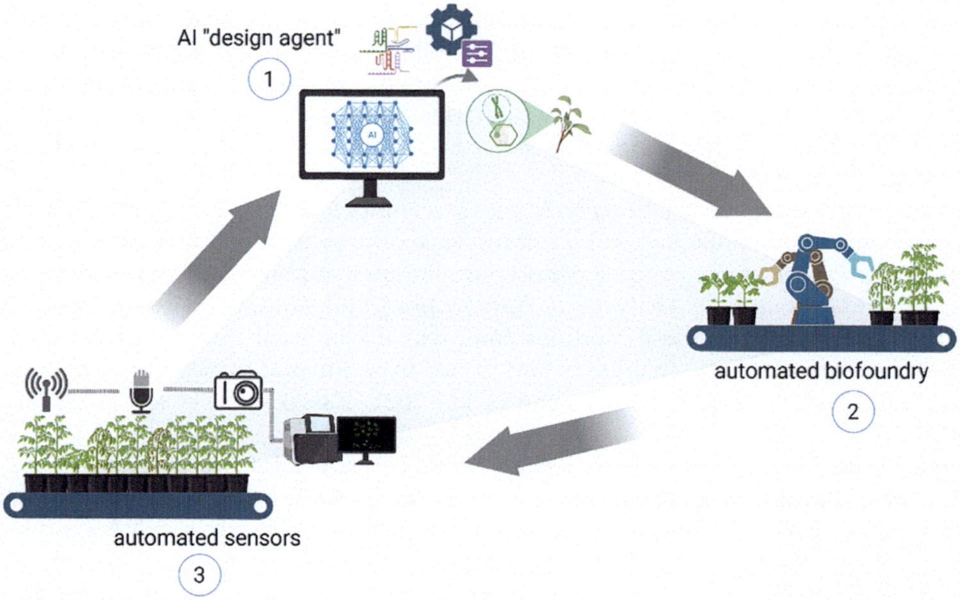

FIGURE 26.3 AI–automation pipeline for plant gene editing. In this envisioned system, an (1) AI "design agent" generates candidate edits (gRNAs, gene circuits, or promoter variants) using predictive models (as protein language models, epigenome predictors). (2) these designs are sent to an automated biofoundry, where robotic modules perform DNA assembly, plant transformation (protoplast electroporation or *Agrobacterium*), and growth of edited plants under controlled conditions. (3) automated sensors (such as imagers, sequencers, and mass spectrometers) collect phenotypic and molecular data from the plants. This data is fed back to the AI, which learns from the results and updates its models for the next design cycle. The closed loop repeats, efficiently converging on optimal edits.

Automated high-throughput CRISPR screens are another impactful use case: robots can perform genome-scale knockouts, CRISPRa or CRISPRi libraries across thousands of samples, while automated imaging and sensor arrays phenotype plants or cell cultures under controlled stressors (drought, salinity, pathogen challenge) (Mansoor et al., 2025; Bendixen et al., 2023). Hits identified by ML classifiers feed into secondary validation rounds that narrow candidates for breeding programs. Beyond screens, DBTL pipelines can support automated haploid induction, gene stacking, and multi-allelic editing strategies—tasks that are particularly valuable in polyploid crops where coordinated editing across homeologs is essential (Gaillochet et al., 2021).

Codesign of plants and machines is an emerging, but speculative frontier. Rather than merely adapting machines to plants, researchers may edit plant traits (architecture, flowering cues, reproductive structures) to facilitate mechanized handling, pollination, or harvesting. Early examples exist where breeding and engineering have altered floral traits to ease mechanical pollination; extending this idea to CRISPR-driven, machine-friendly genotypes suggests a future in which plant form and automated production systems are developed iteratively to maximize throughput and reduce labor (Nasti and Voytas, 2021; Kumbhakar et al., 2025).

26.4.3 Applications and impact in plant science

The combined AI-automation paradigm accelerates discovery and deployment across multiple domains. In crop improvement, automated CRISPR pipelines enable rapid functional validation and stacking of resistance, yield, or quality traits. In metabolic engineering, biofoundries iteratively combine enzyme variants and pathway architectures to identify high-producing constructs *in planta*. Synthetic biology benefits from the ability to build and test complex gene circuits and sensors at scale, with AI predicting circuit behavior and robots assembling constructs. In basic research, automated DBTL supports the generation of large mutant collections, enabling systems-level interrogation of genetic networks; when paired with single-cell transcriptomics and automated isolation, these platforms could lead to faster mechanistic discovery (Carbonell et al., 2019; Zhang et al., 2025).

In agriculture and breeding, automation increasingly extends into the field: autonomous drones and ground robots equipped with multispectral imaging and targeted application systems can monitor plant performance, deliver precise treatments, or even apply localized editing reagents in controlled trials. Smart greenhouses integrate inline sensors and AI controllers to maintain optimal conditions for regeneration and testing of edited lines. These applications not only shorten the path from lab bench to the field, but also enable more rigorous, high-resolution phenotyping that improves the predictive power of models that intend to improve genotype features, translating them to desirable phenotype traits (Roth et al., 2025; Singh et al., 2025).

26.5 Ethical and regulatory considerations in the age of AI-assisted gene editing

As we have learned so far through this chapter, AI seems to be accelerating the scale and precision of plant genome editing: models predict guide efficacy and off-target risk, network tools prioritize candidate loci for complex traits, and generative approaches are beginning to propose novel editors or construct designs. This combination increases both the practical benefits and the governance challenges. The social and legal dimensions—such as who decides which edits are positioned, under what evidentiary standards, and who bears long-term ecological and socio-economic responsibility—acquire new urgency as AI reduces experimental uncertainty and shortens product cycles, leaving regulatory oversight and public deliberation at risk of being behind technological capability. The balance societies strike between innovation and precaution will determine whether AI-assisted editing delivers broadly shared agricultural resilience or concentrates benefits and risks in just a few hands (Sparrow et al., 2021; Vindman et al., 2024).

26.5.1 Emerging ethical and societal challenges

Three areas of ethical concern are most immediate for plant applications. First, equity and control: advanced AI models, biofoundries, and both computing and storage infrastructure they require are expensive. Without deliberate policy, capability will concentrate in well-funded firms and institutes, raising the risk that trait priorities reflect commercial rather than public-good goals. The consequences include restricted access to improved varieties for smallholders and an erosion of seed sovereignty where intellectual property (IP) regimes are strict. These distributional risks have been repeatedly stressed in public-interest analyses of biotechnology governance (Harfouche et al., 2024).

Second, ecological stewardship: even precisely targeted edits, as briefly mentioned in previous sections, can create unwanted outcomes; broadly, the consequences might be a cascading of ecological effects—altered pest dynamics, unintended gene flow to wild relatives, or impacts on mutualists and soil microbiomes. More radical interventions, such as population control strategies or gene drives, magnify those concerns because they are designed to spread traits through populations and ecosystems; national and international bodies have already recommended special governance and precaution for such applications. Long-term monitoring, rigorous ecological risk assessment, and staged field testing are therefore ethical prerequisites for any release program (Kuzma, 2020).

Third, transparency and algorithmic accountability: AI methods range from simple, interpretable models to large and—most probably, for a large amount of our society—obscure neural networks. When an AI proposes a multiplex editing strategy that targets dozens of loci, regulators and stakeholders must be able to understand (at least at a practical level) why those targets were selected and what uncertainties remain. Explainability is not merely a technical detail but a governance requirement: it enables reproducibility, supports risk assessments, and underpins public trust (McDermid et al., 2021). Equally important is data representativeness: models trained predominantly on temperate-region cultivars or laboratory genotypes may embed biases that reduce performance (and safety) when applied in marginal or diverse agroecologies (Ayanoğlu et al., 2020).

26.5.2 Regulatory considerations

Current regulatory approaches for plant edits diverge substantially across jurisdictions, and AI complicates this fragmentation. The European Court of Justice ruled in 2018 that organisms obtained by mutagenesis fall within the European Union (EU) genetically modified organism (GMO) Directive, creating a process-based regulatory regime that subjects many edits to strict oversight. By contrast, other countries (for example, Argentina, the United States, Brazil, and Japan) have adopted product-based or case-by-case approaches that often exempt edits without foreign DNA from onerous GMO regulation; the USDA's[4] updated movement rule (SECURE[5]) is a notable example of this pragmatic, risk-based posture. These divergent frameworks create both regulatory friction for developers and opportunities for "jurisdictional arbitrage", where research and deployment concentrate in permissive environments. Reconciling scientific understanding, public values, and trade implications, therefore, requires international dialogue and proportionate, risk-based approaches (Sprink et al., 2016; Eriksson et al., 2019; Tachikawa and Matsuo, 2024).

AI adds two discrete regulatory complications. First, it changes the locus of design: an edit proposed by an algorithm may have been derived from aggregated, proprietary datasets or from models trained in other countries, complicating questions of origin and liability. Second,

[4] USDA: United States Department of Agriculture

[5] SECURE: Sustainable, Ecological, Consistent, Uniform, Responsible, Efficient

AI-driven designs can produce many plausible alternatives rapidly; regulators must therefore judge not only individual edits but development pipelines and decision rules (for example, whether an automated pipeline has safeguards to avoid high-risk multiplex strategies). To address these, regulatory systems should consider requirements for documentation of training data, uncertainty quantification in model outputs, and verification steps (as, for example, an independent laboratory confirmation) before approvals are granted. International fora (FAO, OECD, and WHO)[6] and regional networks are beginning to discuss such harmonization, but concrete standards are still embryonic (Larson et al., 2021; Judge et al., 2025).

26.5.3 Toward responsible governance: practical principles and tools

A pragmatic governance agenda for AI-assisted plant editing rests on four mutually reinforcing elements. First, transparency and traceability: developers should document datasets, model versions, and decision thresholds used to propose edits; registries of field trials and standardized reporting protocols improve public oversight and post-release monitoring (Hurlbut et al., 2018; Schiemann et al., 2020). Second, proportionality and risk stratification: low-risk edits (single-nucleotide changes mimicking natural variants) can be expedited under a proportionate framework, whereas high-consequence edits (gene drives, wide-host-range modifications) require staged testing, independent ecological assessment, and explicit societal consent (Macnaghten and Habets, 2020; Hoffman, 2021; González-Ortega et al., 2025). Third, capacity-building and equitable access: public investments (regional biofoundries, open datasets, federated model training initiatives) can democratize tool access and ensure trait agendas address food security and climate adaptation rather than narrow commercial interests (Farooq et al., 2024). Fourth, adaptive regulation and stakeholder engagement: regulatory "sandboxes" and iterative oversight mechanisms allow real-world data to inform policy while preserving safety; inclusive engagement with farmers and civil society is essential to legitimate decisions about which traits or interventions are acceptable. These elements echo long-standing prescriptions from both biosafety bodies and recent AI governance discussions (Undheim, 2024).

Philosophical and legal voices remind us that powerful tools carry amplified responsibility. As illustrated in Fig. 26.4, AI-assisted genome editing promises substantial agronomic and nutritional gains while introducing ecological, ethical, and legal challenges that require a proportionate and adaptive regulatory response. Contemporary commentators—from those calling attention to moral stewardship and institutional duty to those warning about the concentration of technological power—converge on the need for enforceable obligations: documented stewardship plans, independent monitoring, and legal clarity on liability and compensation (Jasanoff and Hurlbut, 2018; Lennox, 2024). For plant editing, this means integrating environmental safeguards into approval processes, requiring postmarket ecological surveillance, and ensuring farmers' rights and seed sovereignty are protected through appropriate licensing models. The challenge is not technical impossibility but collective will: designing laws and having institutions that make responsible innovation the default (Idris et al., 2023).

[6] FAO: Food and Agriculture Organization of the United Nations; OECD: Organization for Economic Cooperation and Development; and WHO: World Health Organization

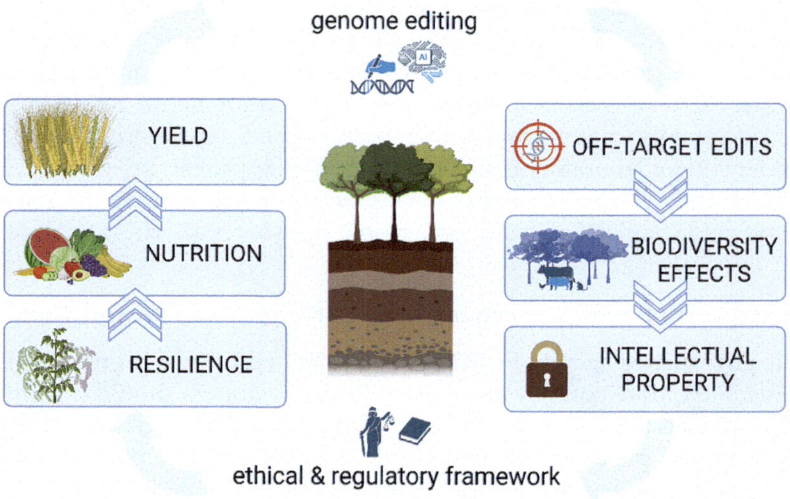

FIGURE 26.4 Plant crop genome editing conceptual diagram. This figure illustrates how AI-driven genome editing in plants can lead to improved crop traits (yield, nutrition, resilience) as well as potential concerns (off-target edits, biodiversity impacts, intellectual property). A strong ethical and regulatory framework (bottom) is needed to manage these impacts.

26.6 Future trends and emerging applications

The next decade will see tools from computing, robotics, and molecular biology merge into integrated pipelines that accelerate the journey from *in silico* design to validated, field-ready varieties. Two currents drive this convergence. One is the diversification of genome-editing modalities, double-strand break (or DSB)-free systems such as base and prime editors, engineered nucleases with expanded PAM compatibility or adaptation to challenging growth conditions, and improved delivery routes are widening the sequence space amenable to editing and easing biological constraints that have historically limited trait development in many crops. Some of the most prominent emerging editing strategies and challenges were mentioned already in Section 26.2 (Son and Park, 2022). The other current is the maturation of AI and automation into practical instruments for target discovery, guide and construct design, process optimization, high-throughput phenotyping, and closed-loop experimentation. Together, these developments shorten the DBTL cycle and increase the throughput and precision of plant genetic engineering (Wang and Doudna, 2023; Dong et al., 2025; Gilbertson et al., 2025; Saini et al., 2025).

26.6.1 Advanced AI and digital agriculture

AI is becoming deeply embedded in farming. Next-generation AI models will move beyond pattern recognition into causal reasoning and multimodal understanding, allowing the rise of new models, which should, in turn, be trained on diverse datasets to power agriculture. At the discovery stage, ML models identify candidate causal variants and regulatory nodes by integrating genomic, transcriptomic, and trait data, prioritizing edits likely to generate large trait effects in a given environment. In later stages, predictive models guide construct design and integrate phenotyping data from advanced sensor arrays, with automation enabling faster iteration—themes explored in detail in Sections 26.4 and 26.6.4 (Jiang et al., 2025; Nautiyal et al., 2025).

Digital twins and farm-scale modeling will make this pipeline context-aware. Virtual replicas of greenhouses or field plots that fuse weather, soil, remote-sensing, and genotype-specific growth models enable researchers and breeders to simulate how an edited allele will perform across seasons and management regimes before committing to long trials; recent reviews of digital twins in agriculture document the rapidly growing literature and prototype systems. Such simulation layers are especially valuable where climatic variability and genotype versus environment interactions dominate phenotype expression, because they focus experimental effort on edits with robust, translatable benefit (Peladarinos et al., 2023; Purcell and Neubauer, 2023; Amin et al., 2025).

26.6.2 Biotechnology and gene editing in crops

Targeted genome edits are already delivering measurable agronomic gains. For example, the modification of *ARGOS8* in maize improved yields under flowering stress in multi-site field trials (Shi et al., 2017), showing that precision edits can translate directly into deployable traits. As editing technologies advance, the scope of feasible changes is widening: base editors enable single-nucleotide alterations that replicate beneficial natural alleles, while prime editors allow short insertions or replacements without relying on the rare HDR pathway in somatic plant tissue. These approaches reduce dependence on complex donor templates and prolonged tissue-culture steps, easing a major bottleneck for many non-model crops (Molla et al., 2021; Li et al., 2023; Tian et al., 2025).

Future innovations in plant genome editing will focus on mimicking natural evolutionary processes by programming large structural variations such as insertions, duplications, deletions, inversions, and translocations, as well as controlling recombination and transposable elements to reshape genomes rapidly in the lab. Advances also include transgene-free methods like viral delivery of compact enzymes and RNA-guided systems via grafting, alongside enzyme enhancements with base and prime editors for precise modifications without DBS. In chromosome engineering, CRISPR/Cas enables targeted rearrangements to break genetic linkages or revert inversions, while *minichromosomes* serve as vectors for stacking complex traits, supported by centromere optimizations and novel recombinases for large DNA integrations. These developments promise enhanced agronomic traits, such as improved yield and stress resistance, by accelerating breeding and facilitating genetic isolation for biosafety (Son and Park, 2022; Kocsisova and Coneva, 2023; Puchta and Houben, 2024; Uranga et al., 2024; Gilbertson et al., 2025).

26.6.3 Integration of AI and genomic technologies

The convergence of AI and genomics is reshaping how targets are identified and validated. ML models trained in integrated genomic, transcriptomic, and phenotypic datasets can identify variants most likely to drive desirable traits, guiding editing priorities. Predictive frameworks already refine CRISPR design by linking gRNA sequence and context features to efficiency and specificity, while protein-structure models—exemplified by AlphaFold-class systems—assist in tailoring nucleases to plant cellular environments. Integration extends beyond target discovery: post-edit, AI-driven gene–trait association analyses confirm which modifications deliver the intended performance under field conditions. Multimodal models that fuse omics, microbiome, and environmental data can reveal intervention points that conventional breeding might overlook (Chia et al., 2025; Li et al., 2025).

One emerging paradigm is the "AI-assisted breeder": automated phenotyping platforms in greenhouses and field plots feed high-resolution data into ML models, which then prioritize edits or crosses. Coupled with biofoundry pipelines, this enables the rapid identification of beneficial alleles. Genomic selection methods, strengthened by deep-learning predictions of complex polygenic traits, become even more powerful when integrated with targeted editing, allowing breeders to combine precision and scale in ways inaccessible through classical crossing alone (Singh et al., 2025).

26.6.4 Robotics and automation in agriculture and breeding

Automation and robotics are migrating from specialized labs to production settings. Automated tissue-culture bioreactors, robotic handlers for explants, and imaging systems enable higher throughput and better reproducibility in regeneration and screening, directly addressing long-standing bottlenecks in many crops. Field robotics, from scouting drones to small autonomous ground vehicles, will not only monitor large experiments but also enable targeted, localized management of experimental plots—a capacity that both improves phenotyping reliability and reduces the cost of field testing edited lines (Spagnuolo et al., 2025). Codevelopment of plant traits and farm machinery—editing for architectures that ease mechanized handling—creates additional value by lowering downstream costs. These robotics platforms, when linked to AI-driven decision engines and remote-sensing networks, allow continuous, high-resolution evaluation of edited germplasm across environments, accelerating selection and deployment cycles (Atefi et al., 2021; Sheikh et al., 2024).

26.6.5 Sustainability, climate, and resource considerations

Sustainability and resource trade-offs must be explicit design constraints. Large AI models and their associated data centers have nontrivial energy and water footprints; the ML community has quantified training and inference costs and emphasized the importance of transparency and efficiency. Likewise, data center water use for cooling has become an operational and community issue in regions where agricultural water and public supply overlap (Natarajan, 2025; Asghar et al., 2025). To avoid counterproductive outcomes, AI-assisted plant engineering programs must plan deployments and computational infrastructure with energy and water constraints in mind, favoring energy-efficient models, green hosting, and edge computing when appropriate (Charmika et al., 2025).

Gene-edited crops themselves can improve sustainability: edits that reduce fertilizer need, increase water-use efficiency, or enhance carbon capture can lower agriculture's footprint. However, the manufacturing of robotics, reagents, and computing infrastructure entails material and energy costs that must be balanced against agronomic gains; a systems perspective — integrating renewable power, circular material flows, and local resource planning — is essential (Wilson, 2020; Pearson et al., 2022; "Genome editing and sustainable agriculture", 2024).

26.7 Conclusion

Plant gene editing has evolved into an engineering discipline at the intersection of molecular biology, computation, and systems design. The combination of advanced CRISPR toolkits — from base and prime editors to plant-optimized nucleases — and mature AI methods for guide selection, outcome prediction, and network-based gene prioritization is shortening development cycles and expanding the range of accessible targets. Laboratory innovation and computational advances now reinforce each other, translating edits into applicable traits.

Yet constraints limit this progress. Polyploidy and genetic complexity demand allele-wise design and multiplex strategies; delivery and regeneration remain bottlenecks for many crops; and ML models risk domain shift without large, plant-centric datasets. Operational and societal factors — infrastructure costs, uneven access, regulatory fragmentation, and public trust — will shape whether these capabilities benefit broadly.

A realistic agenda arises: build FAIR (Wilkinson et al., 2016) — i.e., *findable*, *accessible*, *interoperable*, and *reusable* — plant-focused datasets; advance delivery and regeneration science; adopt privacy-preserving model training; design computing and lab systems with sustainability in mind; and harmonize regulation around trait and ecological risk. If technical, infrastructural, and governance tracks advance together, the outcome will be faster cultivar development, robust AI-assisted breeding, and wider access to cutting-edge tools. Neglecting openness and sustainability risks might deepen existing biases. The capacity to edit plant genomes is here; directing it toward the common good is the defining task ahead.

Acknowledgments

The author would like to express gratitude to Drs. Aftab Ahmad, for the invitation to write this book chapter; Natalie do Valle Capelli, for providing inputs and for fruitful discussions; and Paulo Zaini, for insights into the figures.

Declaration

The author used generative artificial intelligence tools to assist with language refinement during the development of this chapter. All generated material was critically reviewed and validated by the author. The final content reflects the author's own scholarly work, who assumes full responsibility for the final content, interpretations, and conclusions presented.

References

Abbasi, A.F., Nabeel Asim, M., Dengel, A., 2025. Transitioning from wet lab to artificial intelligence: a systematic review of AI predictors in CRISPR. J. Transl. Med. 23 (1). https://doi.org/10.1186/s12967-024-06013-w.

Allen, F., Crepaldi, L., Alsinet, C., Strong, A.J., Kleshchevnikov, V., De Angeli, P., Páleníková, P., Khodak, A., Kiselev, V., Kosicki, M., Bassett, A.R., Harding, H., Galanty, Y., Muñoz-Martínez, F., Metzakopian, E., Jackson, S.P., Parts, L., 2019. Predicting the mutations generated by repair of Cas9-induced double-strand breaks. Nat. Biotechnol. 37 (1), 64–82. https://doi.org/10.1038/nbt.4317.

Alvard, D., Cote, F., Teisson, C., 1993. Comparison of methods of liquid medium culture for banana micropropagation - effects of temporary immersion of explants. Plant Cell Tissue Organ Cult. 32 (1), 55–60. https://doi.org/10.1007/BF00040116.

Amin, A., Zaman, W., Park, S.J., 2025. Harnessing multi-omics and predictive modeling for climate-resilient crop breeding: from genomes to fields. Genes 16 (7), 809. https://doi.org/10.3390/genes16070809.

Annese, D., Romani, F., Grandellis, C., Ives, L., Frangedakis, E., Buson, F.X., Molloy, J.C., Haseloff, J., 2025. Semi-automated workflow for high-throughput Agrobacterium-mediated plant transformation. Plant J. 122 (1). https://doi.org/10.1111/tpj.70118.

Arias, D.S., Taylor, R.E., 2024. Scientific discovery at the press of a button: navigating emerging cloud laboratory technology. Adv. Mater. Technol. 9 (16). https://doi.org/10.1002/admt.202400084.

Asghar, D.M., Usman, H., Mazhar Kharal, M., Mehboob, D.M., 2025. AI's cross border energy and water footprints: state duties, No harm thresholds, and Paris agreement compliance frameworks. Social Science Review Archives 3 (3), 398–408. https://doi.org/10.70670/sra.v3i3.875.

Atefi, A., Ge, Y., Pitla, S., Schnable, J., 2021. Robotic technologies for high-throughput plant phenotyping: contemporary reviews and future perspectives. Front. Plant Sci. 12. https://doi.org/10.3389/fpls.2021.611940.

Ayanoğlu, F.B., Elçİn, A.E., Elçİn, Y.M., 2020. Bioethical issues in genome editing by CRISPR-Cas9 technology. Turkish J. Biol. 44 (2), 110–120. https://doi.org/10.3906/biy-1912-52.

Bendixen, L., Jensen, T.I., Bak, R.O., 2023. CRISPR-Cas-mediated transcriptional modulation: the therapeutic promises of CRISPRa and CRISPRi. Mol. Ther. 31 (7), 1920–1937. https://doi.org/10.1016/j.ymthe.2023.03.024.

Bethge, H., Winkelmann, T., Lüdeke, P., Rath, T., 2023. Low-cost and automated phenotyping system "Phenomenon" for multi-sensor in situ monitoring in plant in vitro culture. Plant Methods 19 (1). https://doi.org/10.1186/s13007-023-01018-w.

Bortesi, L., Zhu, C., Zischewski, J., Perez, L., Bassié, L., Nadi, R., Forni, G., Lade, S.B., Soto, E., Jin, X., Medina, V., Villorbina, G., Muñoz, P., Farré, G., Fischer, R., Twyman, R.M., Capell, T., Christou, P., Schillberg, S., 2016. Patterns of CRISPR/Cas9 activity in plants, animals and microbes. Plant Biotechnol. J. 14 (12), 2203–2216. https://doi.org/10.1111/pbi.12634.

Bromig, L., Leiter, D., Mardale, A.-V., von den Eichen, N., Bieringer, E., Weuster-Botz, D., 2022. The SiLA 2 Manager for rapid device integration and workflow automation. SoftwareX 17, 100991. https://doi.org/10.1016/j.softx.2022.100991.

Cameron, P., Fuller, C.K., Donohoue, P.D., Jones, B.N., Thompson, M.S., Carter, M.M., Gradia, S., Vidal, B., Garner, E., Slorach, E.M., Lau, E., Banh, L.M., Lied, A.M., Edwards, L.S., Settle, A.H., Capurso, D., Llaca, V., Deschamps, S., Cigan, M., Young, J.K., May, A.P., 2017. Mapping the genomic landscape of CRISPR-Cas9 cleavage. Nat. Methods 14 (6), 600–606. https://doi.org/10.1038/nmeth.4284.

Candela-Ferre, J., Diego-Martin, B., Pérez-Alemany, J., Gallego-Bartolomé, J., 2024. Mind the gap: epigenetic regulation of chromatin accessibility in plants. Plant Physiol. 194 (4), 1998–2016. https://doi.org/10.1093/plphys/kiae024.

Carbonell, P., Radivojevic, T., Martín, H.G., 2019. Opportunities at the intersection of synthetic biology, machine learning, and automation. ACS Synth. Biol. 8 (7), 1474–1477. https://doi.org/10.1021/acssynbio.8b00540.

Cardi, T., Murovec, J., Bakhsh, A., Boniecka, J., Bruegmann, T., Bull, S.E., Eeckhaut, T., Fladung, M., Galovic, V., Linkiewicz, A., Lukan, T., Mafra, I., Michalski, K., Kavas, M., Nicolia, A., Nowakowska, J., Sági, L., Sarmiento, C., Yıldırım, K., Zlatković, M., Hensel, G., Van Laere, K., 2023. CRISPR/Cas-mediated plant genome editing: outstanding challenges a decade after implementation. Trends Plant Sci. 28 (10), 1144–1165. https://doi.org/10.1016/j.tplants.2023.05.012.

Charmika, R., Navya, R., Vijay, C., Madhusudhan, G., Bhavana, L., Maheshkumar, M., 2025. AI-driven automation for green buildings and sustainable agriculture: enhancing efficiency, scalability, and resource management.

International Research Journal of Innovations in Engineering and Technology 09 (Special Issue), 97–103. https://doi.org/10.47001/irjiet/2025.inspire16.

Chen, W., McKenna, A., Schreiber, J., Haeussler, M., Yin, Y., Agarwal, V., Noble, W.S., Shendure, J., 2019. Massively parallel profiling and predictive modeling of the outcomes of CRISPR/Cas9-mediated double-strand break repair. Nucleic Acids Res. 47 (15), 7989–8003. https://doi.org/10.1093/nar/gkz487.

Chen, Y., Zhang, L., Wang, H., Weinan, E., 2020. Ground state energy functional with Hartree–Fock efficiency and chemical accuracy. J. Phys. Chem. 124 (35), 7155–7165. https://doi.org/10.1021/acs.jpca.0c03886.

Chia, B.S., Seah, Y.F.S., Wang, B., Shen, K., Srivastava, D., Chew, W.L., 2025. Engineering a new generation of gene editors: integrating synthetic biology and AI innovations. ACS Synth. Biol. 14 (3), 636–647. https://doi.org/10.1021/acssynbio.4c00686.

Chuai, G., Ma, H., Yan, J., Chen, M., Hong, N., Xue, D., Zhou, C., Zhu, C., Chen, K., Duan, B., Gu, F., Qu, S., Huang, D., Jia, W., Qi, L., 2018. DeepCRISPR: optimized CRISPR guide RNA design by deep learning. Genome Biol. 19 (1). https://doi.org/10.1186/s13059-018-1459-4.

Craig, T., Holland, R., D'Amore, R., Johnson, J.R., McCue, H.V., West, A., Zulkower, V., Tekotte, H., Cai, Y., Swan, D., Davey, R.P., Hertz-Fowler, C., Hall, A., Caddick, M., 2017. Leaf LIMS: a flexible laboratory information management system with a synthetic biology focus. ACS Synth. Biol. 6 (12), 2273–2280. https://doi.org/10.1021/acssynbio.7b00212.

Dong, J., Croslow, S.W., Lane, S.T., Castro, D.C., Blanford, J., Zhou, S., Park, K., Burgess, S., Root, M., Cahoon, E.B., Shanklin, J., Sweedler, J.V., Zhao, H., Hudson, M.E., 2025. Enhancing lipid production in plant cells through automated high-throughput genome engineering and phenotyping. Plant Cell 37 (2). https://doi.org/10.1093/plcell/koaf026.

Elkayam, S., Tziony, I., Orenstein, Y., Valencia, A., 2024. DeepCRISTL: deep transfer learning to predict CRISPR/Cas9 on-target editing efficiency in specific cellular contexts. Bioinformatics 40 (8). https://doi.org/10.1093/bioinformatics/btae481.

Eraslan, G., Avsec, Ž., Gagneur, J., Theis, F.J., 2019. Deep learning: new computational modelling techniques for genomics. Nat. Rev. Genet. 20 (7), 389–403. https://doi.org/10.1038/s41576-019-0122-6.

Eriksson, D., Kershen, D., Nepomuceno, A., Pogson, B.J., Prieto, H., Purnhagen, K., Smyth, S., Wesseler, J., Whelan, A., 2019. A comparison of the EU regulatory approach to directed mutagenesis with that of other jurisdictions, consequences for international trade and potential steps forward. New Phytol. 222 (4), 1673–1684. https://doi.org/10.1111/nph.15627.

Farooq, M.A., Gao, S., Hassan, M.A., Huang, Z., Rasheed, A., Hearne, S., Prasanna, B., Li, X., Li, H., 2024. Artificial intelligence in plant breeding. Trends Genet. 40 (10), 891–908. https://doi.org/10.1016/j.tig.2024.07.001.

Gaillochet, C., Develtere, W., Jacobs, T.B., 2021. CRISPR screens in plants: approaches, guidelines, and future prospects. Plant Cell 33 (4), 794–813. https://doi.org/10.1093/plcell/koab099.

Genome editing and sustainable agriculture, 2024. Nat. Sustain. 7 (4), 369–370. https://doi.org/10.1038/s41893-024-01343-5.

Gilbertson, L., Puchta, H., Slotkin, R.K., 2025. The future of genome editing in plants. Nat. Plants 11 (4), 680–685. https://doi.org/10.1038/s41477-025-01956-4.

Goertzel, B., 2014. Artificial general intelligence: concept, state of the art, and future prospects. Journal of Artificial General Intelligence 5 (1), 1–48. https://doi.org/10.2478/jagi-2014-0001.

González-Ortega, E., Piñeyro-Nelson, A., Martínez-Debat, C., 2025. On techno-fixes in the era of crises: genomic interventions in food crops. Agroecology and Sustainable Food Systems 49 (4), 563–586. https://doi.org/10.1080/21683565.2024.2429620.

Grinberg, N.F., Orhobor, O.I., King, R.D., 2020. An evaluation of machine-learning for predicting phenotype: studies in yeast, rice, and wheat. Mach. Learn. 109 (2), 251–277. https://doi.org/10.1007/s10994-019-05848-5.

Guertler, P., Pallarz, S., Belter, A., Eckermann, K.N., 2023. Lutz Grohmann, Detection of commercialized plant products derived from new genomic techniques (NGT) - practical examples and current perspectives. Food Control 152, 109869. https://doi.org/10.1016/j.foodcont.2023.109869.

Haeussler, M., Schönig, K., Eckert, H., Eschstruth, A., Mianné, J., Renaud, J.B., Schneider-Maunoury, S., Shkumatava, A., Teboul, L., Kent, J., Joly, J.S., Concordet, J.P., 2016. Evaluation of off-target and on-target scoring algorithms and integration into the guide RNA selection tool CRISPOR. Genome Biol. 17 (1). https://doi.org/10.1186/s13059-016-1012-2.

HamediRad, M., Chao, R., Weisberg, S., Lian, J., Sinha, S., Zhao, H., 2019. Towards a fully automated algorithm driven platform for biosystems design. Nat. Commun. 10 (1). https://doi.org/10.1038/s41467-019-13189-z.

Hao, Q., Xu, F., Li, Y., Evans, J., 2024. AI expands scientists' impact but contracts science's focus. arXiv. https://doi.org/10.48550/arXiv.2412.07727.

Harfouche, A.L., Petousi, V., Jung, W., 2024. AI ethics on the road to responsible AI plant science and societal welfare. Trends Plant Sci. 29 (2), 104–107. https://doi.org/10.1016/j.tplants.2023.12.016.

He, W., Wang, H., Wei, Y., Jiang, Z., Tang, Y., Chen, Y., Xu, H., Martelli, P.L., 2021. GuidePro: a multi-source ensemble predictor for prioritizing sgRNAs in CRISPR/Cas9 protein knockouts. Bioinformatics 37 (1), 134–136. https://doi.org/10.1093/bioinformatics/btaa1068.

Hoffman, N.E., 2021. Revisions to USDA biotechnology regulations: the SECURE rule. Proc. Natl. Acad. Sci. U.S.A. 118 (22). https://doi.org/10.1073/pnas.2004841118.

Hurlbut, J.B., Jasanoff, S., Saha, K., Ahmed, A., Appiah, A., Bartholet, E., Baylis, F., Bennett, G., Church, G., Cohen, I. G., Daley, G., Finneran, K., Hurlbut, W., Jaenisch, R., Lwoff, L., Kimes, J.P., Mills, P., Moses, J., Park, B.S., Parens, E., Salzman, R., Saxena, A., Simmet, H., Simoncelli, T., Snead, O.C., Rajan, K.S., Truog, R.D., Williams, P., Woopen, C., 2018. Building capacity for a global genome editing observatory: conceptual challenges. Trends Biotechnol. 36 (7), 639–641. https://doi.org/10.1016/j.tibtech.2018.04.009.

Idris, S.H., Mat Jalaluddin, N.S., Chang, L.W., 2023. Ethical and legal implications of gene editing in plant breeding: a systematic literature review. J. Zhejiang Univ. - Sci. B 24 (12), 1093–1105. https://doi.org/10.1631/jzus.B2200601.

Jasanoff, S., Hurlbut, J.B., 2018. A global observatory for gene editing. Nature 555 (7697), 435–437. https://doi.org/10.1038/d41586-018-03270-w.

Jensen, K.T., Fløe, L., Petersen, T.S., Huang, J., Xu, F., Bolund, L., Luo, Y., Lin, L., 2017. Chromatin accessibility and guide sequence secondary structure affect CRISPR-Cas9 gene editing efficiency. FEBS (Fed. Eur. Biochem. Soc.) Lett. 591 (13), 1892–1901. https://doi.org/10.1002/1873-3468.12707.

Jiang, L., Xu, B., Husnain, N., Qi, W., 2025. Overview of agricultural machinery automation technology for sustainable agriculture. Agronomy 15 (6), 1471. https://doi.org/10.3390/agronomy15061471.

Judge, B., Nitzberg, M., Russell, S., 2025. When code isn't law: rethinking regulation for artificial intelligence. Pol. Soc. 44 (1), 85–97. https://doi.org/10.1093/polsoc/puae020.

Jung, W.J., Park, S.J., Cha, S., Kim, K., 2024. Factors affecting the cleavage efficiency of the CRISPR-Cas9 system. Anim. Cell Syst. 28 (1), 75–83. https://doi.org/10.1080/19768354.2024.2322054.

Kick, D.R., Washburn, J.D., 2023. Ensemble of best linear unbiased predictor, machine learning and deep learning models predict maize yield better than each model alone. In Silico Plants 5 (2). https://doi.org/10.1093/insilicoplants/diad015.

Kim, H.K., Kim, Y., Lee, S., Min, S., Bae, J.Y., Choi, J.W., Park, J., Jung, D., Yoon, S., Kim, H.H., 2019. SpCas9 activity prediction by DeepSpCas9, a deep learning–based model with high generalization performance. Sci. Adv. 5 (11). https://doi.org/10.1126/sciadv.aax9249.

Kocsisova, Z., Coneva, V., 2023. Strategies for delivery of CRISPR/Cas-mediated genome editing to obtain edited plants directly without transgene integration. Frontiers in Genome Editing 5. https://doi.org/10.3389/fgeed.2023.1209586.

de Koning, R., Daryanavard, H., Garmyn, J., Kiekens, R., Toili, M.E.M., Angenon, G., 2023. Fine-tuning CRISPR/Cas9 gene editing in common bean (Phaseolus vulgaris L.) using a hairy root transformation system and in silico prediction models. Front. Plant Sci. 14. https://doi.org/10.3389/fpls.2023.1233418.

Konstantakos, V., Nentidis, A., Krithara, A., Paliouras, G., 2022. CRISPR–Cas9 gRNA efficiency prediction: an overview of predictive tools and the role of deep learning. Nucleic Acids Res. 50 (7), 3616–3637. https://doi.org/10.1093/nar/gkac192.

Kumar, H.E., Nekrasov, V., Rymarquis, L.A., 2021. CRISPR-cas in Agriculture: Opportunities and Challenges. Frontiers Research Topics. https://doi.org/10.3389/978-2-88966-815-1.

Kumbhakar, R., Mondal, M., Thakro, V., Tripathi, S., Parida, S.K., 2025. Shaping the future: unravelling regulators modulating plant architecture for next-generation crops. Plant Sci. 358, 112534. https://doi.org/10.1016/j.plantsci.2025.112534.

Kuzma, J., 2020. Engineered Gene Drives: Ecological, Environmental, and Societal Concerns. Springer Science and Business Media LLC, pp. 371–399. https://doi.org/10.1007/978-3-030-53183-6_17.

Larson, D.B., Harvey, H., Rubin, D.L., Irani, N., Tse, J.R., Langlotz, C.P., 2021. Regulatory frameworks for development and evaluation of artificial intelligence–based diagnostic imaging algorithms: summary and recommendations. J. Am. Coll. Radiol. 18 (3), 413–424. https://doi.org/10.1016/j.jacr.2020.09.060.

Lawrenson, T., Clarke, M., Kirby, R., Forner, M., Steuernagel, B., Brown, J.K.M., Harwood, W., 2024. An optimised CRISPR Cas9 and Cas12a mutagenesis toolkit for barley and wheat. Plant Methods 20 (1). https://doi.org/10.1186/s13007-024-01234-y.

Leenay, R.T., Aghazadeh, A., Hiatt, J., Tse, D., Roth, T.L., Apathy, R., Shifrut, E., Hultquist, J.F., Krogan, N., Wu, Z., Cirolia, G., Canaj, H., Leonetti, M.D., Marson, A., May, A.P., Zou, J., 2019. Large dataset enables prediction of repair after CRISPR–Cas9 editing in primary T cells. Nat. Biotechnol. 37 (9), 1034–1037. https://doi.org/10.1038/s41587-019-0203-2.

Lemos, B.R., Kaplan, A.C., Bae, J.E., Ferrazzoli, A.E., Kuo, J., Anand, R.P., Waterman, D.P., Haber, J.E., 2018. CRISPR/Cas9 cleavages in budding yeast reveal templated insertions and strand-specific insertion/deletion profiles. Proc. Natl. Acad. Sci. U.S.A. 115 (9), E2010. https://doi.org/10.1073/pnas.1716855115.

Lennox, J.C., 2024. 2084 and the AI Revolution: How Artificial Intelligence Informs Our Future. Updated and expanded edition.

Li, J., Zhang, C., He, Y., Li, S., Yan, L., Li, Y., Zhu, Z., Xia, L., 2023. Plant base editing and prime editing: the current status and future perspectives. J. Integr. Plant Biol. 65 (2), 444–467. https://doi.org/10.1111/jipb.13425.

Li, S., Lin, D., Zhang, Y., Deng, M., Chen, Y., Lv, B., Li, B., Lei, Y., Wang, Y., Zhao, L., Liang, Y., Liu, J., Chen, K., Liu, Z., Xiao, J., Qiu, J.L., Gao, C., 2022. Genome-edited powdery mildew resistance in wheat without growth penalties. Nature 602 (7897), 455–460. https://doi.org/10.1038/s41586-022-04395-9.

Li, V.R., Zhang, Z., Troyanskaya, O.G., 2021. CROTON: an automated and variant-aware deep learning framework for predicting CRISPR/Cas9 editing outcomes. Bioinformatics 37, I342. https://doi.org/10.1093/bioinformatics/btab268.

Li, Z., Khan, W.U., Bai, G., Dong, C., Wang, J., Zhang, Y., Wang, C., Zhang, H., Wang, W., Luo, M., Chen, F., 2025. From code to life: the AI-driven revolution in genome editing. Adv. Sci. 12 (30). https://doi.org/10.1002/advs.202417029.

Li, C., Shu, Y., Hu, Q., 2023. Editorial: genome editing technology in polyploid crops. Front. Plant Sci. 14. https://doi.org/10.3389/fpls.2023.1340455.

Liao, X., Ma, H., Tang, Y.J., 2022. Artificial intelligence: a solution to involution of design–build–test–learn cycle. Curr. Opin. Biotechnol. 75, 102712. https://doi.org/10.1016/j.copbio.2022.102712.

Lindsay, H., Burger, A., Biyong, B., Felker, A., Hess, C., Zaugg, J., Chiavacci, E., Anders, C., Jinek, M., Mosimann, C., Robinson, M.D., 2016. CrispRVariants charts the mutation spectrum of genome engineering experiments. Nat. Biotechnol. 34 (7), 701–702. https://doi.org/10.1038/nbt.3628.

Lu, Z., Marand, A.P., Ricci, W.A., Ethridge, C.L., Zhang, X., Schmitz, R.J., 2019. The prevalence, evolution and chromatin signatures of plant regulatory elements. Nat. Plants 5 (12), 1250–1259. https://doi.org/10.1038/s41477-019-0548-z.

Luo, Y., Chen, Y., Xie, H.Z., Zhu, W., Zhang, G., 2024. Interpretable CRISPR/Cas9 off-target activities with mismatches and indels prediction using BERT. Comput. Biol. Med. 169, 107932. https://doi.org/10.1016/j.compbiomed.2024.107932.

Macnaghten, P., Habets, M.G.J.L., 2020. Breaking the impasse: towards a forward-looking governance framework for gene editing with plants. Plants People Planet 2 (4), 353–365. https://doi.org/10.1002/ppp3.10107.

Malik, A., Gul, A., Munir, F., Amir, R., Alipour, H., Babar, M.M., Bakhtiar, S.M., Paracha, R.Z., Khalid, Z., Hayat, M. Q., 2021. Evaluating the cleavage efficacy of CRISPR-Cas9 sgRNAs targeting ineffective regions of Arabidopsis thaliana genome. PeerJ 9. https://doi.org/10.7717/peerj.11409.

Mansoor, S., Karunathilake, E.M.B.M., Tuan, T.T., Chung, Y.S., 2025. Genomics, phenomics, and machine learning in transforming plant research: advancements and challenges. Hortic. Plant J. 11 (2), 486–503. https://doi.org/10.1016/j.hpj.2023.09.005.

Mattiello, L., Rütgers, M., Sua-Rojas, M.F., Tavares, R., Soares, J.S., Begcy, K., Menossi, M., 2022. Molecular and computational strategies to increase the efficiency of CRISPR-based techniques. Front. Plant Sci. 13. https://doi.org/10.3389/fpls.2022.868027.

McDermid, J.A., Jia, Y., Porter, Z., Habli, I., 2021. Artificial intelligence explainability: the technical and ethical dimensions. Philos. Trans. R. Soc. A Math. Phys. Eng. Sci. 379 (2207), 20200363. https://doi.org/10.1098/rsta.2020.0363.

Molla, K.A., Sretenovic, S., Bansal, K.C., Qi, Y., 2021. Precise plant genome editing using base editors and prime editors. Nat. Plants 7 (9), 1166–1187. https://doi.org/10.1038/s41477-021-00991-1.

Nasti, R.A., Voytas, D.F., 2021. Attaining the promise of plant gene editing at scale. Proc. Natl. Acad. Sci. U.S.A. 118 (22). https://doi.org/10.1073/pnas.2004846117.

Natarajan, A., 2025. The hydro-digital paradox: water scarcity in the age of artificial intelligence. Journal of Computer Science and Technology Studies 7 (7), 500–505. https://doi.org/10.32996/jcsts.2025.7.7.55.

Nautiyal, M., Joshi, S., Hussain, I., Rawat, H., Joshi, A., Saini, A., Kapoor, R., Verma, H., Nautiyal, A., Chikara, A., Ahmad, W., Kumar, S., 2025. Revolutionizing agriculture: a comprehensive review on artificial intelligence applications in enhancing properties of agricultural produce. Food Chem. X 29, 102748. https://doi.org/10.1016/j.fochx.2025.102748.

Nishant, R., Schneckenberg, D., Ravishankar, 2024. The formal rationality of artificial intelligence-based algorithms and the problem of bias. J. Inf. Technol. 39 (1), 19–40. https://doi.org/10.1177/02683962231176842.

Pearson, S., Carolina Camacho-Villa, T., Valluru, R., Gaju, O., Rai, M.C., Gould, I., Brewer, S., Sklar, E., 2022. Robotics and autonomous systems for net zero agriculture. Current Robotics Reports 3 (2), 57–64. https://doi.org/10.1007/s43154-022-00077-6.

Pei, L., Li, G., Lindsey, K., Zhang, X., Wang, M., 2021. Plant 3D genomics: the exploration and application of chromatin organization. New Phytol. 230 (5), 1772–1786. https://doi.org/10.1111/nph.17262.

Peladarinos, N., Piromalis, D., Cheimaras, V., Tserepas, E., Adrian Munteanu, R., Papageorgas, P., 2023. Enhancing smart agriculture by implementing digital twins: a comprehensive review. Sensors 23 (16), 7128. https://doi.org/10.3390/s23167128.

Petzold, C.J., Mukhopadhyay, A., 2025. From bench to biofactory: high-throughput technologies and automated workflows to accelerate biomanufacturing. Curr. Opin. Biotechnol. 94, 103320. https://doi.org/10.1016/j.copbio.2025.103320.

Pinello, L., Canver, M.C., Hoban, M.D., Orkin, S.H., Kohn, D.B., Bauer, D.E., Yuan, G.C., 2016. Analyzing CRISPR genome-editing experiments with CRISPResso. Nat. Biotechnol. 34 (7), 695–697. https://doi.org/10.1038/nbt.3583.

Puchta, H., Houben, A., 2024. Plant chromosome engineering – past, present and future. New Phytol. 241 (2), 541–552. https://doi.org/10.1111/nph.19414.

Purcell, W., Neubauer, T., 2023. Digital twins in agriculture: a state-of-the-art review. Smart Agric. Technol. 3, 100094. https://doi.org/10.1016/j.atech.2022.100094.

Roth, L., Marzougui, A., Walter, A., 2025. A review of the journey of field crop phenotyping: from trait stamp collections and fancy robots to phenomics-informed crop performance predictions. J. Plant Physiol. 311, 154542. https://doi.org/10.1016/j.jplph.2025.154542.

Saini, H., Johar, V., Mondal, R., Vyas, S., Bisht, V., 2025. Integration of artificial intelligence with genome editing: a new era for molecular plant breeding. Physiol. Mol. Plant Pathol. 140, 102952. https://doi.org/10.1016/j.pmpp.2025.102952.

Schaart, J.G., van de Wiel, C.C.M., Smulders, M.J.M., 2021. Genome editing of polyploid crops: prospects, achievements and bottlenecks. Transgenic Res. 30 (4), 337–351. https://doi.org/10.1007/s11248-021-00251-0.

Schiemann, J., Robienski, J., Schleissing, S., Spök, A., Sprink, T., Wilhelm, R.A., 2020. Editorial: plant genome editing – policies and governance. Front. Plant Sci. 11. https://doi.org/10.3389/fpls.2020.00284.

Schindele, P., Merker, L., Schreiber, T., Prange, A., Tissier, A., Puchta, H., 2023. Enhancing gene editing and gene targeting efficiencies in Arabidopsis thaliana by using an intron-containing version of tt Lb Cas12a. Plant Biotechnol. J. 21 (3), 457–459. https://doi.org/10.1111/pbi.13964.

Schindele, P., Puchta, H., 2020. Engineering CRISPR/Lb Cas12a for highly efficient, temperature-tolerant plant gene editing. Plant Biotechnol. J. 18 (5), 1118–1120. https://doi.org/10.1111/pbi.13275.

Sheikh, M., Iqra, F., Ambreen, H., Pravin, K.A., Ikra, M., Chung, Y.S., 2024. Integrating artificial intelligence and high-throughput phenotyping for crop improvement. J. Integr. Agric. 23 (6), 1787–1802. https://doi.org/10.1016/j.jia.2023.10.019.

Shen, M.W., Arbab, M., Hsu, J.Y., Worstell, D., Culbertson, S.J., Krabbe, O., Cassa, C.A., Liu, D.R., Gifford, D.K., Sherwood, R.I., 2018. Predictable and precise template-free CRISPR editing of pathogenic variants. Nature 563 (7733), 646–651. https://doi.org/10.1038/s41586-018-0686-x.

Shi, J., Gao, H., Wang, H., Lafitte, H.R., Archibald, R.L., Yang, M., Hakimi, S.M., Mo, H., Habben, J.E., 2017. ARGOS8 variants generated by CRISPR-Cas9 improve maize grain yield under field drought stress conditions. Plant Biotechnol. J. 15 (2), 207–216. https://doi.org/10.1111/pbi.12603.

Singh, H., Himanshi, Kaundal, M., 2025. AI-driven crop breeding-revolutionizing agriculture with smart technologies: a review. Journal of Advances in Biology & Biotechnology 28 (7), 1469–1486. https://doi.org/10.9734/jabb/2025/v28i72664.

Son, S., Park, S.R., 2022. Challenges facing CRISPR/Cas9-Based genome editing in plants. Front. Plant Sci. 13. https://doi.org/10.3389/fpls.2022.902413.

Spagnuolo, M., Todde, G., Caria, M., Furnitto, N., Schillaci, G., Failla, S., 2025. Agricultural robotics: a technical review addressing challenges in sustainable crop production. Robotics 14 (2), 9. https://doi.org/10.3390/robotics14020009.

Sparrow, R., Howard, M., Degeling, C., 2021. Managing the risks of artificial intelligence in agriculture. NJAS: Impact in Agricultural and Life Sciences 93 (1), 172–196. https://doi.org/10.1080/27685241.2021.2008777.

Sprink, T., Eriksson, D., Schiemann, J., Hartung, F., 2016. Regulatory hurdles for genome editing: process- vs. product-based approaches in different regulatory contexts. Plant Cell Rep. 35 (7), 1493–1506. https://doi.org/10.1007/s00299-016-1990-2.

Sun, J., Guo, J., Liu, J., Sinha, S., 2024. CRISPR-M: predicting sgRNA off-target effect using a multi-view deep learning network. PLoS Comput. Biol. 20 (3), e1011972. https://doi.org/10.1371/journal.pcbi.1011972.

Tachikawa, M., Matsuo, M., 2024. Global regulatory trends of genome editing technology in agriculture and food. Breed. Sci. 74 (1), 3–10. https://doi.org/10.1270/jsbbs.23046.

Tian, S., Yao, L., Zhang, Y., Rao, X., Zhu, H., 2025. Prime editing for crop improvement: a systematic review of optimization strategies and advanced applications. Genes 16 (8), 965. https://doi.org/10.3390/genes16080965.

Tsai, S.Q., Zheng, Z., Nguyen, N.T., Liebers, M., Topkar, V.V., Thapar, V., Wyvekens, N., Khayter, C., Iafrate, A.J., Le, L.P., Aryee, M.J., Joung, J.K., 2015. GUIDE-seq enables genome-wide profiling of off-target cleavage by CRISPR-Cas nucleases. Nat. Biotechnol. 33 (2), 187–198. https://doi.org/10.1038/nbt.3117.

Uma, S., Karthic, R., Kalpana, S., Backiyarani, S., Saraswathi, M.S., 2021. A novel temporary immersion bioreactor system for large scale multiplication of banana (Rasthali AAB—Silk). Sci. Rep. 11 (1). https://doi.org/10.1038/s41598-021-99923-4.

Undheim, T.A., 2024. The whack-a-mole governance challenge for AI-enabled synthetic biology: literature review and emerging frameworks. Front. Bioeng. Biotechnol. 12. https://doi.org/10.3389/fbioe.2024.1359768.

Uranga, M., Martín-Hernández, A.M., De Storme, N., Pasin, F., 2024. CRISPR–Cas systems and applications for crop bioengineering. Front. Bioeng. Biotechnol. 12. https://doi.org/10.3389/fbioe.2024.1483857.

Ursache, R., Fujita, S., Dénervaud Tendon, V., Geldner, N., 2021. Combined fluorescent seed selection and multiplex CRISPR/Cas9 assembly for fast generation of multiple Arabidopsis mutants. Plant Methods 17 (1). https://doi.org/10.1186/s13007-021-00811-9.

Van Vu, T., Doan, D.T.H., Kim, J., Sung, Y.W., Tran, M.T., Jong Song, Y., Das, S., Kim, J.-Y., 2021. CRISPR/Cas-based precision genome editing via microhomology-mediated end joining. Plant Biotechnol. J. 19 (2), 230–239. https://doi.org/10.1111/pbi.13490.

Vindman, C., Trump, B., Cummings, C., Smith, M., Titus, A.J., Oye, K., Prado, V., Turmus, E., Linkov, I., 2024. The convergence of AI and synthetic biology: the looming deluge. arXiv. https://doi.org/10.48550/arXiv.2404.18973.

Wada, N., Ueta, R., Osakabe, Y., Osakabe, K., 2020. Precision genome editing in plants: state-of-the-art in CRISPR/Cas9-based genome engineering. BMC Plant Biol. 20 (1). https://doi.org/10.1186/s12870-020-02385-5.

Wang, J.Y., Doudna, J.A., 2023. CRISPR technology: a decade of genome editing is only the beginning. Science 379 (6629). https://doi.org/10.1126/science.add8643.

Wang, Y., Cheng, X., Shan, Q., Zhang, Y., Liu, J., Gao, C., Qiu, J.L., 2014. Simultaneous editing of three homoeoalleles in hexaploid bread wheat confers heritable resistance to powdery mildew. Nat. Biotechnol. 32 (9), 947–951. https://doi.org/10.1038/nbt.2969.

Wilkinson, M.D., Dumontier, M., Aalbersberg, I.J., Appleton, G., Axton, M., Baak, A., Blomberg, N., Boiten, J.W., da Silva Santos, L.B., Bourne, P.E., Bouwman, J., Brookes, A.J., Clark, T., Crosas, M., Dillo, I., Dumon, O., Edmunds, S., Evelo, C.T., Finkers, R., Gonzalez-Beltran, A., Gray, A.J.G., Groth, P., Goble, C., Grethe, J.S., Heringa, J., t Hoen, P.A.C., Hooft, R., Kuhn, T., Kok, R., Kok, J., Lusher, S.J., Martone, M.E., Mons, A., Packer, A.L., Persson, B., Rocca-Serra, P., Roos, M., van Schaik, R., Sansone, S.A., Schultes, E., Sengstag, T.,

Slater, T., Strawn, G., Swertz, M.A., Thompson, M., Van Der Lei, J., Van Mulligen, E., Velterop, J., Waagmeester, A., Wittenburg, P., Wolstencroft, K., Zhao, J., Mons, B., 2016. Comment: the FAIR Guiding Principles for scientific data management and stewardship. Sci. Data 3. https://doi.org/10.1038/sdata.2016.18.

Wilson, A.K., 2020. Will Gene-Edited and Other GM Crops Fail Sustainable Food Systems? Rethinking Food and Agriculture: New Ways Forward. Elsevier, United States, pp. 247–284. https://doi.org/10.1016/B978-0-12-816410-5.00013-X.

Xiang, X., Corsi, G.I., Anthon, C., Qu, K., Pan, X., Liang, X., Han, P., Dong, Z., Liu, L., Zhong, J., Ma, T., Wang, J., Zhang, X., Jiang, H., Xu, F., Liu, X., Xu, X., Wang, J., Yang, H., Bolund, L., Church, G.M., Lin, L., Gorodkin, J., Luo, Y., 2021. Enhancing CRISPR-Cas9 gRNA efficiency prediction by data integration and deep learning. Nat. Commun. 12 (1). https://doi.org/10.1038/s41467-021-23576-0.

Yuan, H., Song, C., Xu, H., Sun, Y., Anthon, C., Bolund, L., Lin, L., Benabdellah, K., Lee, C., Hou, Y., Gorodkin, J., Luo, Y., 2025. An overview and comparative analysis of CRISPR-SpCas9 gRNA activity prediction tools. CRISPR J. 8 (2), 89–104. https://doi.org/10.1089/crispr.2024.0058.

Zaleski-Cox, M., Miklas, P.N., Soler-Garzón, A., Hoyos-Villegas, V., 2023. Automating high-throughput screening for anthracnose resistance in common bean using allele specific PCR. Plant Methods 19 (1). https://doi.org/10.1186/s13007-023-01071-5.

Zhang, D., Xu, F., Wang, F., Le, L., Pu, L., 2025. Synthetic biology and artificial intelligence in crop improvement. Plant Commun. 6 (2), 101220. https://doi.org/10.1016/j.xplc.2024.101220.

Zhang, G., Xie, H., Dai, X., 2024. DeepIndel: an interpretable deep learning approach for predicting CRISPR/Cas9-Mediated editing outcomes. Int. J. Mol. Sci. 25 (20), 10928. https://doi.org/10.3390/ijms252010928.

Zhu, H., Li, C., Gao, C., 2020. Applications of CRISPR–Cas in agriculture and plant biotechnology. Nat. Rev. Mol. Cell Biol. 21 (11), 661–677. https://doi.org/10.1038/s41580-020-00288-9.

Zhu, W., Xie, H., Chen, Y., Zhang, G., 2024. CrnnCrispr: an interpretable deep learning method for CRISPR/Cas9 sgRNA on-target activity prediction. Int. J. Mol. Sci. 25 (8), 4429. https://doi.org/10.3390/ijms25084429.

Index

'*Note:* Page numbers followed by "f" indicate figures and "t" indicate tables.'

A

B

C

D

E

H

T

U

Printed and bound by CPI Group (UK) Ltd, Croydon, CR0 4YY
25/09/2026
02222749-0018